Materials Engineering and Sciences Division 2017

Core Programming Area at the 2017 AIChE Annual Meeting

Minneapolis, Minnesota, USA
29 October – 3 November 2017

Volume 1 of 2

ISBN: 978-1-5108-5797-1

Printed from e-media with permission by:

Curran Associates, Inc.
57 Morehouse Lane
Red Hook, NY 12571

Some format issues inherent in the e-media version may also appear in this print version.

Printed by Curran Associates, Inc. (2018)

AIChE
120 Wall Street, FL 23
New York, NY 10005-4020

Phone: (800) 242-4363
Fax: (203) 775-5177

www.aiche.org

Additional copies of this publication are available from:

Curran Associates, Inc.
57 Morehouse Lane
Red Hook, NY 12571 USA
Phone: 845-758-0400
Fax: 845-758-2633
Email: curran@proceedings.com
Web: www.proceedings.com

TABLE OF CONTENTS

VOLUME 1

(9a) New Milestones and Challenges in High-Throughput Computation of Elastic Properties on the Materials Project 1
Joseph H. Montoya, Kristin Persson

(9b) Breaking Badly: A Comprehensive Assessment of Computational Methods for Predicting Tensile Strengths in Bulk Solids 2
Bryan M. Wong

(9c) Machine Learning the Thermochemistry of All Inorganic Crystalline Solids 3
Christopher J. Bartel, Ann M. Deml, Samantha L. Millican, John R. Rumptz, William Tumas, Alan W. Weimer, Stephan Lany, Vladan Stevanovic, Charles B. Musgrave, Aaron M. Holder

(9d) Designing Dopant Patterns in Indium-Doped Perovskite Oxygen Carriers 4
Christopher L. Hanselman, Dominic Alfonso, Jonathan W. Lekse, De Nyago Tafen, Christopher Matranga, David C. Miller, Chrysanthos E. Gounaris

(9f) In silico Structural Analyses of Borosilicate, Aluminosilicate, and Gallosilicate Zeolites Using Effective Tetrahedral Descriptors 5
Koki Muraoka, Tatsuya Okubo, Watcharop Chaikittisilp

(9g) Computational Identification of Hetero-Interpenetrated Porous Materials 6
Jihan Kim, Ohmin Kwon, Sanghoon Park

(16a) Lipid-like Materials for RNA Delivery: A How-to Guide for Hacking Gene Expression 7
Kathryn A. Whitehead

(16b) Triple Delivery Nanoscale Device for siRNA, Vismodegib and Gemcitabine Co-Delivery to Treat Pancreatic Cancer 8
Metin Uz, Satyanarayana Rachagani, Surinder Batra, Surya K. Mallapragada

(16c) Influence of Dextran and Surface Charge on Nanoparticle-Mediated siRNA Delivery 9
Daniel Vocelle, Olivia Chesniak, Mitch Smith, Christina Chan, S. Patrick Walton

(16d) Peg-Poly(beta-amino ester) Delivery Systems for Periodic shRNA 10
Connie Wu, Jiahe Li, Wade Wang, Paula T. Hammond

(16e) Structurally Programmed Assembly of Ribonucleoproteins for Superior mRNA Delivery 11
Jiahe Li, Wade Wang, Yanpu He, Yingzhong Li, Darrell J. Irvine, Paula T. Hammond

(16f) Cationic Peptide Amphiphile Micelles(PAMs) As Nucleic Acid Based Adjuvants Carriers for the Improvement of Subunit Vaccine Efficiency 12
Rui Zhang, Josiah Smith, Jake Kramer, Logan Morton, Brittany Allen, Caitlin Leeper, Xiaolei Li, Fabio Gallazzi, Tommi White, Bret Ulery

(31a) Understanding Peptide Assembly with Coarse-Grained Models Designed By Information Theory (Invited Talk) 13
M. Scott Shell

(31b) Modeling the Oxidative Consumption of Curcumin from Controlled Release Poly(beta amino ester) Microparticles in the Presence of a Free Radical Generating System 14
Carolyn T. Jordan, J. Zach Hilt, Thomas D. Dziubla

(31c) The Fusion and Spreading of Liposome with Different Sizes: Molecular Dynamics Simulation with Dry-Martini Force Field 15
Yanfei Lu, Lu Diannan

(31d) Design of Membrane-Embedded Amphiphilic Nanoparticles from Multiscale Simulations 16
Reid Van Lehn

(31e) Improvement on Self-Assembly of Virus-like Particles By the Introduction of Electrostatic Attraction 17
Lin Zhang, Xiaocui Guo, Yan Sun

(31f) Contemporary Modeling and Analysis of Human Blood Rheology 18
Matthew Armstrong, Tyler Helton, Evan Ousley, Michael Deegan

(31g) How Confinement and Hydrophobicity Affect CO2 Diffusion from the Bulk Phase to the Active Site of Human Carbonic Anhydrase II: A Study Based on Coarse-Grained Molecular Dynamics Simulation and the Markov-State Model 38
Gong Chen, Diannan Lu, Jianzhong Wu, Zheng Liu

(34a) Printing Fully Stretchable Thin-Film-Transistor Array 39
Jia Liu, Francisco M. Lopez, Jiechen Wang, Nathan G.-J. Wang, Jinyoung Oh, Sihong Wang, Zhenan Bao

(34b) Conjugated Macrocycles for Ultra-Sensitive Organic Photodetectors 40
Yu Zhong, Boyuan Zhang, Xiaoyang Zhu, Colin Nuckolls

(34c) Connecting Crystalline Domains: Blending Conjugated Polymers of Differing Molecular Weights for Enhanced Charge Transport 41
Michael McBride, Nils Persson, Danny Keane, Martha A. Grover, Elsa Reichmanis
(34d) Ultrafast Carrier Dynamics in Bimetallic Nanostructures-Enhanced Methylammonium Lead Bromide Perovskites 42
Rizia Bardhan
(34e) Aggregation Structure and Solvation of Solution-Phase PTB7 43
Daniel Reid, Nicholas Jackson, Juan J. De Pablo
(34f) Self-Aligned Strategy for Printed Electronics 44
Woo Jin Hyun, Lorraine F. Francis, C. Daniel Frisbie
(34g) Helical Perylene Diimide Ribbons for Molecular Electronics 45
Yu Zhong, Michael Steigerwald, Xiaoyang Zhu, Fay Ng, Colin Nuckolls
(36a) Engineering Pathways to New Functional Polyether Materials 46
Nathaniel A. Lynd
(36b) Topology Control of Bottlebrush Polymers 47
Damien Guironnet, Dylan Walsh
(36c) A New, Facile Approach to Epoxide Polymerization 48
Robert C. Ferrier Jr., Jennifer Imbrogno, Christina G. Rodriguez, Malgorzata Chwatko, Nathaniel A. Lynd
(36d) Dynamic Chemistry Leading to Full Property Recovery Associated with Crosslink Density in Reprocessed Polymer Networks and Network Composites: Network Synthesis By Step-Growth Reactions and By Nitroxide-Mediated Polymerization 49
John M. Torkelson, Xi Chen, Kailong Jin, Lingqiao Li
(36e) Hybrid Chain-Growth/Step-Growth Mechanism Observed in Heterofunctional Thiol-Ene Polymerizations 50
Dillon Love, Benjamin D. Fairbanks, Kang-Min Kim, Brady Worrell, Weixian Xi, J. Taylor Goodrich, Charles B. Musgrave, Mark Stoykovich, Christopher N. Bowman
(36f) High Pressure Ethylene Polymerization with a Post-Metallocene Bis-Phenyl Phenoxy Catalyst 51
Sean Ewart, Karjala Tom
(36g) A Simulation-Based Derivative-Free Optimization Framework Using the Kinetic Monte Carlo Method for Controlling Polymer Molecular Weight and Sequence Distribution Synthesized Via Free Radical Polymerization 52
Hanyu Gao, Andreas Waechter, Ivan Konstantinov, Steven G. Arturo, Linda J. Broadbelt
(36h) Experimental and Macroscopic Mechanistic Modeling Studies of the Methyl Acrylate Self-Initiation Reaction 53
Hossein Riazi, Ahmad Arabi Shamsabadi, Michael Grady, Andrew M. Rappe, Masoud Soroush
(36i) A Mathematical Model Based on Artificial Neural Network for Ethylene/Norbornene Copolymerization Catalyzed By 2-(tetramethylcyclopentadienyl)-4,6-Di-Tert-Butylphenoxytitanium Dichloride 54
Nikhil Prakash
(167a) Revealing Governing Mechanism in Directed Self-Assembly of sub 10 nm Particles with Single Particle Resolution 55
Zhen Luo, Shafigh Mehraeen
(167b) Uniform Thinning of Cu-Fe-Ni-Co Nanowires and Kinetic Monte Carlo Simulation 56
Xiaohua Geng, Elizabeth Podlaha
(167c) Experimental Assessment of Nucleation Theory at the Molecular Level 57
Matthew A. Gebbie, Nicholas A . Melosh
(167d) Low-Voltage Electrophoretic Deposition of All-Inorganic CZTS Nanocrystals for Fabrication of Thin Films 58
Andrew D. Dillon, Mohammad Mehdi Taheri, Shawn Mengel, Subham Dastidar, Jason B. Baxter, Aaron T. Fafarman
(167e) Preferential Binding of Polyvinylpyrrolidone (PVP) Is Not Responsible for Shape Control in Ag Nanoparticle Synthesis 59
Zhifeng Chen, Ji Woong Chang, Choumini Balasanthiran, Robert M. Rioux
(167f) Scalable Synthesis of Epitaxial, Oxidation-Proof Au@Ag Core-Shell Nanowires for Electronic and Photonic Applications 60
Ruoxue Yan, Yangzhi Zhu, Sanggon Kim, Peter Byrley
(167g) Synthesis of Bimetallic Alloy Nanoparticles through the Visible-Light Mediated Reduction of a Bimetallic Oxide Precursor: Case Study of Ag-Pt Nanoparticle Synthesis 61
Umar Aslam, Suljo Linic
(167h) Dynamic Control of Gold Nanoparticle-Conjugated DNA Origami Templates 62
Abhilasha Dehankar, Joshua Johnson, Carlos E. Castro, Jessica O. Winter

(167i) 'One-Pot' Multi-Scale Templating of Interdigitated Bi-Modal Porous Carbon Supercapacitors 63
Zheng Tian, Megha Sharma, Mark A. Snyder
(167j) Combined Experimental and Theoretical Study of Hexagonal Boron Nitride Crystal Growth 64
Song Liu, Bin Liu, James H. Edgar
(30a) Design of Aminopolymer Structure to Enhance Performance and Stability of CO2 Sorbents: Poly(propylenimine) Vs. Poly(ethylenimine) 65
Simon H. Pang, Ryan P. Lively, Christopher W. Jones
(30b) Development of 3D-Printed Aminosilica Monoliths for CO2 capture 66
Harshul Thakkar, Stephen Eastman, Fateme Rezaei
(30c) Engineering Supported Amine Adsorbents for CO2 Capture Applications 67
Zelong Xie, Christopher Cogswell, Sunho Choi
(30d) Impacts of Aminopolymer-Support Interaction on CO2 Sorption Performance Probed By Neutron Scattering Techniques 68
Adam Holewinski, Miles Sakwa-Novak, Matthew Potter, Nathan Ellebracht, Gernot Rother, Christopher W. Jones
(30e) Hierarchically-Structured Porous Materials for Enhanced Greenhouse Gas Capture 69
Tae-Hyun Bae
(30f) Natural Gas Adsorption in SSZ-13: Equilibrium and Dynamic Properties 70
Joshua A. Thompson
(30g) Zeolite on Demand: Design and Synthesis of Zeolites with Controlled Crystal Morphology and Location of Substituting Tetrahedral Atoms with the Aid of Theoretical Calculations 71
Watcharop Chaikittisilp, Koki Muraoka, Sye Hoe Keoh, Tatsuya Okubo
(30h) Tuning Solid Acids for the Target Catalytic Reactions 72
Jun Huang
(51a) Mechanical Properties of Organic Semiconductors for Mechanically Stable and Intrinsically Stretchable Solar Cells 73
Darren Lipomi
(51b) Digital Light Synthesis to Drive Additive Manufacturing 74
Joseph M Desimone
(51c) Symbiotic Pairing of Near-UV Solar Cells with Electrochromic Windows for Visible Light and Heat Management in Architectural Applications 75
Y. L. Lynn Loo
(51d) Taking Measure of Modern Polymer Synthesis 76
Kathryn L. Beers
(55a) Rational Design of Polyelectrolyte Complexes for Nucleic Acid Delivery 77
Jeffrey Vieregg, Matthew V. Tirrell
(55b) Microfabricated Immune-Isolating Devices for Long Term Cell Based Therapies 78
Suman Bose, Robert Langer, Daniel G. Anderson
(55c) Globular Protein Vesicles: Engineering Vesicle Size and Membrane Structure through a Tunable Molecular Packing Parameter 79
Yeongseon Jang, Julie A. Champion
(55d) Biomaterial Scaffolds for Scalable Differentiation and Transplantation of Hpsc-Derived Cells for Cell Replacement Therapy in the Central Nervous System 80
Maroof M. Adil, David V. Schaffer
(55f) Electrochemical Activation for DNA Attachment to Surfaces 81
Ariel Furst, Matthew Francis
(55g) Organizing Biochemical Reactions with Protein Droplets 82
Huaiying Zhang
(59a) Electron Energy Loss Spectroscopy for Optoelectronics and Thermal Dynamics at Nanocomposite Interfaces 83
D. Keith Roper, Jeremy Dunklin, Gregory T. Forcherio, Keith Berry, Carter Bodinger, Tyler Howard
(59b) On the Rice Husk Ash Admixing with Cement: Preparation, Characterization and Analysis 84
Nikhil Prakash
(59c) Tunable Magnetoresistance of Conductive Polymer Nanocomposites 85
Jiang Guo, Alexandra Galaska, Suying Wei, Brian J. Edwards, Bamin Khomami, Zhanhu Guo
(59d) Polydopamine Stabilized Fluorescent Nanozinc Oixde Reinforced Epoxy Nanocomposites Towards UV Shielding 86
Chaobo Liang, Ping Song, Hongbo Gu, Junwei Gu
(59e) Comparing the Toughening Effects of Modified Graphene Oxide and Core-Shell Rubber on Polyester Resins and Glass Fiber-Reinforced Polyester Composites 87
Kunwei Liu, Siyao He, Yuqiang Qian, Qi An, Andreas Stein, Christopher W. Macosko
(59f) Preparation and Characterizations of Barium Ferrite/Epoxy Nanocomposites 88
Hongyuan Zhang, Hongbo Gu

(59g) Effect of Compatibilizer on Strength and Toughness of Glass Fiber Mat Reinforced Polypropylene Composites 89
Chunyin Shen, Yadong Wu, Haiqing Wan, Junyan Wang, Gance Dai
(59h) A Facile Approach to Fabrication and Characterization of an Eco-Friendly Zein-Laponite Nanocomposite with Improved Mechanical, Thermal, Barrier and Surface Property 90
Tahrima B. Rouf
(78a) Solid Dispersions of Electroactive Materials for Energy Storage Applications 91
Gary M. Koenig Jr., Zhaoxiang Qi, Devanshi Gupta
(78b) Rational Design of the Cathode Materials in the Lithium-Sulfur Batteries 92
Tong Mou, Bin Wang
(78c) 3D Carbon Materials for Electric Double-Layer Capacitors with Ultrahigh Areal Capacitance 93
Liang Chang, Yun Hang Hu
(78d) 3D High-Surface-Area and Mesoporous Graphene Sheet-like Carbon for Supercapacitors 94
Haiyang Shen, Min Wei, Gang Wu
(78e) Stability Predictions for Dimethoxybenzene Based Catholyte Materials 95
Benjamin Silcox, Rajeev Assary, Jing Jing Zhang, Siu On Tung, Ilya Shkrob, Lu Zhang, Levi T. Thompson
(78f) Solvate Ionic Liquid-Based Gel Electrolytes Containing Functionalized Polymer-Based Networks for Use in Lithium Metal Battery Applications 96
Anthony D'Angelo, Matthew J. Panzer
(78g) Studies on Complex Electrolytes for Magnesium Batteries 97
Laura Merrill, Hunter Ford, Jennifer Schaefer
(78h) First-Principles Study of Temperature Dependence of Energy Gaps in Gas Sensor Materials 98
Yuning Wu, Yuhua Duan, Paul R. Ohodnicki, Benajmin T. Chorpening
(78i) Supercritical Fluid-Based Synthesis of Antimony Electrode Materials 99
Grant A. Williamson, Elena P. Pandres, Vincent C. Holmberg
(96a) Solution Combustion Synthesis of Porous CeO2 Nanopowders: Reaction Mechanism and Physical Properties 100
Wooram Kang, Derya Oncel Ozgur, Arvind Varma
(96b) Combining Pre- and Post-Nucleation Trajectories for the Design of Hierarchical FAU/EMT Materials from Organic-Free Sols 101
Dina Gaber, Safa Gaber, Issam Ismail, Saeed Alhassan, Maryam Khaleel
(96c) Synthesis of Sn-MFI Zeolite with Use of Mechanochemical Reaction 102
Kiyoshi Kanie, Moe Sakaguchi, Fumiya Muto, Masafumi Nakaya, Toshiyuki Yokoi, Atsushi Muramatsu
(96d) Crystallization of One-Dimensional Zeolites By Nonclassical Pathways 103
Rui Li, James Sutjianto, Aseem Chawla, Jeffrey D. Rimer
(96e) Expanding the Scope of Fluoride-Free Pure Silica Zeolite Syntheses 104
Vivek Vattipalli, Wei Fan
(96f) Atomic Resolution Imaging of MEL Intergrowth in 2-Dimensional MFI Nanosheets 105
Prashant Kumar, Han Zhang, Neel Rangnekar, Michael Tsapatsis, K. Andre Mkhoyan
(96g) Synthesis of Single-Unit-Cell Hierarchical Zeolites with Tunable Mesoporosity By Controlling Intergrowth Frequency 106
Dandan Xu, Anatoliy Kuznetsov, Prashant Kumar, Maryam Khaleel, Saeed Alhassan, Michael Tsapatsis
(96h) A Full Understanding of Microporous Vanadosilicate AM-6: The Crystal Quality and Structure of AM-6 107
Rumeysa Tekin, Juliusz Warzywoda, Albert Sacco Jr.
(98a) Rapid Wood Fractionation < 80° C for Sustainable and Economic Biorefinery 149
J. Y. Zhu
(98b) Insights into Biomass Recalcitrance 150
Arthur J. Ragauskas
(98c) Low Temperature and High Efficiency Biomass Fuel Cell and Bio-Hydrogen Production 151
Yulin Deng
(98d) Circular Economy: A Path Towards Innovation and Commercialization of Biocomposites for Sustainable Manufacturing 152
Amar K. Mohanty
(79a) Tuning the Molecular Design of Catalytic Materials to Increase Activity and Selectivity for Fine Chemical Production 153
Nicholas Brunelli, Aamena Parulkar, Nitish Deshpande, Mariah Whitaker, Rutuja Joshi
(79b) Deactivation of Zeolite Catalysts during Hydrodeoxygenation of Aromatic Oxygenates 154
Guo Shiou Foo, Mariana V. Rodrigues, Qandeel Almas, Chukwuemeka Okolie, Matthew M. Yung, Carsten Sievers
(79c) Improving Methanol-to-Olefins Conversion Performance of CHA Materials By Seeding the Hydrocarbon Pool 155
Praveen Bollini, Aditya Bhan

(79d) Tuning of Higher Alcohol Selectivity and Productivity in CO Hydrogenation Reactions over K/MoS2 Catalysts Supported on Mesoporous Activated Carbon and Mixed Mgal Oxide 156
Micaela Taborga Claure, Song-Hai Chai, Sheng Dai, Faisal M. Alamgir, Pradeep K. Agrawal, Christopher W. Jones
(79e) Kinetic and Spectroscopic Investigations of Alcohol Conversions over Metal Oxide Catalysts 157
Shuai Tan, Yongqiang Cheng, Luke L. Daemen, Ho Nyung Lee, Benjamin Doughty, Daniel Lutterman
(79f) Catalytic Upgrading of Fast Pyrolysis Bio-Oil for Renewable Hydrocarbon Production 158
Mariefel V. Olarte, Huamin Wang, Daniel Santosa, John G. Frye, Suh-Jane Lee, Jae-Soon Choi, Pimphan Aye Meyer, Susanne Jones, Corinne Drennan, Alan H Zacher
(79g) Chemical Reaction Engineering Principles of Continuous Flow Photoredox Catalysis 159
Eric G. Moschetta, Kaid Harper, Steve Richter, Steven J. Wittenberger
(79h) Factors Affecting Catalytic Performance in the Presence of Non-Thermal Plasmas 160
Jongsik Kim, Prateek Mehta, Patrick Barboun, William F. Schneider, David Go, Jason C. Hicks
(79i) Accelerating Innovation in Advanced Manufacturing 161
Michael McKittrick
(123a) Active Polymer Materials for Flexible Electronics: Molecular Design and Processing for Efficient Macroscale Charge Transport Pathways 162
Nils Persson, Michael McBride, Ping-Hsun Chu, Martha A. Grover, Elsa Reichmanis
(123b) Promoting Adhesion between Immiscible Polymers 163
Christopher W. Macosko
(123c) High Resolution Lithography Via Block Copolymers and Self-Assembling Surface Neutral Layers 164
Peter Trefonas III, Jong Keun Park, Mingqi Li, Janet Wu, Emad Aqad, Dan Millward, Valeriy Ginzburg, Phil Hustad
(123d) Field and Confinement-Directed Self-Assembly of Soft Mesophases to Create Useful Materials 165
Chinedum O. Osuji
(126a) Nanostructured Interfaces for Enhanced Biologic Transport and Immunomodulation 166
Tejal Desai
(126b) Nano- and Microfabricated Hydrogels for Regenerative Engineering 167
Ali Khademhosseini
(126c) Overcoming Obstacles to Brain Repair Using Biomaterials 168
Tatiana Segura
(152a) Digital Alchemy for Assembly Engineeringc 169
Sharon C. Glotzer
(152b) Improved Algebraic, Numerical, and Graphical Representations in Fluid Mechanics 170
Stuart W. Churchill, James C. Hill
(152c) The Scaling of Turbulence Near the Wall and the Churchill Turbulent Flux Correlation: Insights with Lagrangian Simulations 171
Dimitrios V. Papavassiliou, Quoc T. Nguyen, Chiranth Srinivasan
(152d) Flow Boiling Using a Piranha Pin Fin Heat Sink 172
Cory Woodcock, Xiangfei Yu, Yoav Peles, Joel L. Plawsky
(152e) Transport Problems in the Spirit of Stuart Churchill for Teaching and Research at the University of Michigan 173
Ronald G. Larson, Claudio Vilas Boas Favero
(176a) Solve this! Fundamental Approach to Problem Solving in Industrial Processes I (Invited Talks) 174
Zdravko Stefanov, Paul Chauvel, Jr., Eldad Herceg, Dana A. Livingston
(177a) Tuning External Surface and Textural Properties of Unit-Cell Thick Pillared MFI and Pillared MWW Zeolites By Atomic Layer Deposition and Its Consequence on Catalytic Reactions 175
Dongxia Liu, Junyan Zhang
(177b) Mesoporous TiO2 to TiO2-Cellulose Composite and its Derivatives for Environmental Remediation 176
Jinju Zhang, Lei Li, Yanxiang Li, Lixia Cao, Chuanfang Yang
(177c) Reliable Fabrication and Surface Modification of Beta Zeolite Membrane for Pervaporation of n-Butanol/Water Mixtures 177
Yun Li, Honghong Xu, Tao Chen, Xiufeng Liu, Baoquan Zhang
(177d) Investigation of Factors That Induce Cristobalite Formation during Titanosilicate Synthesis and Their Potential Impact on Heterogeneous Catalysis 180
Ayomi S. Perera, Haiyue Yu, Jeremy Cockcroft, Panagiotis Trogadas, Marc-Olivier Coppens
(177e) Novel Methods to Synthesize ZSM-11 As an Efficient Catalyst for Methanol-to-Hydrocarbon Reactions 181
Yufeng Shen, Thuy T. Le, Jeffrey D. Rimer

(177f) Preparation of Novel Al-MFI/Fe-MFI Core-Shell Catalysts and Their Catalytic Application for CH4 Conversion 182
Toshiyuki Yokoi, Yoshihiko Kimura, Takaya Kimura, Yusuke Kunitake, Atsushi Muramatsu
(177g) An Optimized Procedure for Selective Removal of Efal (Extra Framework Aluminum) in Y-Zeolites 183
Balasubramanian Vaithilingam, Gnana Pragasam Singaravel, Abdul Majed Al Katheeri, Stephane M., Mikael Berthod
(774a) Multifunctional Polymer Nanocomposites 184
Zhanhu Guo, Jiang Guo, Alexandra Galaska, Huige Wei, Suying Wei, Bin Qiu, Dawei Jiang, Hongbo Gu, Jiahua Zhu
(774b) Transparent Copper-Silica Nanoparticle-Chitosan Nanocomposite Coatings with Long-Term Antibacterial Efficacy 185
Debirupa Mitra, Min Li, En-Tang Kang, Koon Gee Neoh
(774c) Bio-Inspired Design of Stimuli Responsive Materials Based on a Bilayer Structure 186
Songshan Zeng, Rui Li, Dianyun Zhang, Wenhan Huang, Zhaofeng Wang, Stephan Freire, Andrew Smith, Emily Huang, Helen Nguon, Xiaoyuan Yu, Luyi Sun
(774d) Biomimetic Nanocoatings with Exceptional Mechanical, Barrier, and Flame Retardant Properties from Large Scale One-Step Co-Assembly 187
Fuchuan Ding, Jingjing Liu, Songshan Zeng, Yan Xia, Kacie M. Wells, Mu-Ping Nieh, Luyi Sun
(774e) Bioinspired Composite Materials with Stimuli-Responsive Color Changing Ability 188
Golnaz Isapour, Marco Lattuada
(774f) Investigation of Carbon Nanotubes & Cellulose Nanocrystals Composite for Potential Use in Microelectromechanical Systems 189
Mingzhe Jiang, Christopher L. Kitchens, Robert Seney, Bayliss Charles
(774g) Few Layers MoSe2 Incorporated with Nitrogen Doped Graphene Sheet for High Performance Lithium Sulfur Batteries 190
Hoi Lun Wong, Zhengtang Luo
(774h) Dual-Responsive Plasmonic Behavior of Gold Nanorods@PANI Core/Shell Nanostructures for Real-Time Control 191
Ju-Won Jeon, Jing Zhou, Jeffrey Geldmeier, James Ponder, Mahmoud A. Mahmoud, Mostafa El-Sayed, John Reynolds, Vladimir V. Tsukruk
(118b) Microwave Induced Heating of Carbon Nanotubes Localized at 3D-Printed Thermoplastic Interfaces 192
Charles Sweeney, Mohammad Saed, Micah Green
(196a) Theoretical Study of the Reaction Kinetics of Organosiloxane Polycondensation 193
Mona Bavarian, Siamak Nejati
(196aa) High-Ï‡ Block Copolymers with High Etch Selectivity for Sub-10 Nm Patterning 194
Sung-Soo Kim, Walter W. Young, Luis E. Oquendo, Michael Maher, Sunshine X. Zhou, Yusuke Asano, Marc A. Hillmyer, C. Grant Willson, Christopher J. Ellison
(196ab) Fabrication and Structural Analysis of Nanofibers Made By Syndiotactic Polypropylene with Ethylene-Comonomer Units 195
Fuyuaki Endo, Claudio De Rosa, Atsushi Hotta
(196ad) Interaction Between Supercritial CO2+Cosolvent and Poly(vinyl acetate) 196
Dong-Dong Hu, Lei Bao, Ling Zhao, Tao Liu
(196ae) Modeling of Distributions of Polymer Properties Using Parallel Computing in Julia 198
Esteban Pintos, Mariano Asteasuain
(196af) Transition Metal-Based Nanocrystals Confined-Growth on Heteroatom-Doped Graphene Toward Hydrogen Catalysis 199
Minghao Zhuang, Zhengtang Luo
(196d) From Process to Product - Enhancing the Understanding of Î±-Olefin-Polymerizations 200
Kristina M. Pflug, Jonas Nowottny, Markus Busch
(196e) Amine Effects on Radial-Mediated Thiol-Ene Reactions 201
Dillon Love, Kang-Min Kim, Johnathan Goodrich, Benjamin D. Fairbanks, Mark Stoykovich, Charles B. Musgrave, Christopher N. Bowman
(196f) Influence of Phosphate Salts and Solution pH on Aqueous-Phase NVP Free-Radical Polymerization 202
Fernando T P Borges, Fouad Teymour
(196h) Synthesis and Characterization of Crosslinked Polymers from Cottonseed Oil 204
Rangana Wijayapala, Deonante Frazier, Bill B. Elmore, Charles Freeman, Santanu Kundu
(196i) Date Pits as Cost-Effective, Renewable, and Efficient Fillers for Polymers 205
Fares Alsewailem, Yazeed Binkhodor

(196j) Biaxial and Shear Deformation of Simulated Amorphous Cis-, Trans-1, 4-Polybutadiene Chains 206
Suvrajyoti Kar, Michael L. Greenfield
(196k) Decomposition Behavior of Laponite/PLGA-PÎ•G-PLGA Nanocomposite Hydrogels at Body Temperature 207
Midori Kitagawa, Tomoki Maeda, Atsushi Hotta
(196l) Synthesis of Thermoplastic Polydimethylsiloxane with L-Phenylalanine-Based Hydrogen-Bond Network and its Self-Healing Property 208
Shunsuke Tazawa, Atsushi Shimojima, Tomoki Maeda, Atsushi Hotta
(196n) Controlled Swelling Rate Elastomer for Packers 209
Rostyslav Dolog, Darryl Ventura, Valery N. Khabashesku, Qusai Darugar
(196o) Effect of the Cross-Linking Agent (sodium polyphosphate) on Performances of Nacs-Wsc Microcapsules 215
Qing-Xi Wu, Yi-Xin Guan, Jun-Jie Yuan, Shan-Jing Yao
(196p) Swelling Behaviors of Cr(III)-Modified Acrylamide-Based Superabsorbent Polymer Microsphere in Brines 216
Jingyang Pu, Jiaming Geng, Na Zhang, Baojun Bai
(196q) Synthesis of Chemical Protective Elastomeric Barrier Materials 225
James Ogilvie-Battersby, Alessandra Molinaro, Christopher Zoto, Quoc Truong, Nese Orbey
(196r) Polysulfide-Based Nanofiber Prepared Via Inverse Vulcanization and Electrospinning for Effective Mercury (II) Sequestration 226
Lawrence A. Limjuco, Grace M. Nisola, Khino J. Parohinog, Kris Nino G. Valdehuesa, Wook-Jin Chung
(196s) Mechanical Properties of Maleated Polymer/Graphene Oxide Composites using Graphene Oxide as a Multi-Functional Crosslinker 227
Szu-Ming Yang, Heonjoo Ha, Christopher J. Ellison
(196t) Development and Analysis of a Thin Film Nanocomposite Membrane: Resistance to Chlorine 228
Abdulmajeed Altalhi, Holly A. Stretz
(196u) Improving Gas Transport Properties of Mixed Matrix Membranes Via Interfacial Improvement 229
Ahmad Arabi Shamsabadi, Morteza Sadeghi, Mohammad Dinari, Mahsa Salehi, Masoud Soroush
(196v) Antimicrobial Polymers: Present State of the Art 230
Nikhil Prakash
(196w) Layer-By-Layer Coated Microneedle Arrays for Staged Multi-Agent Immune Attack on Melanoma 231
Yanpu He, Jiahe Li, Hongkun He, Celestine Hong, Maylin Funkenbusch, Sheryl Wang, Maya Berlinger, Darrell J. Irvine, Paula T. Hammond
(196x) Direct Observation of Remarkable Nanoparticle Evolution during Aqueous Dissolution of Polymer/Drug Particles 232
Ralm Ricarte, Marc A. Hillmyer, Timothy P. Lodge
(196y) Polymerized Ionic Liquid Pentablock Terpolymer for Lithium-Metal Batteries 233
Tzu-Ling Chen, Yossef A. Elabd
(196z) Effect of Electric Field on the Structure and Dynamics of Model Ionomer Melts 234
Janani Sampath, Lisa M. Hall
(197a) Influence of Molecular Design on the Self-Assembly of Single-Stranded DNA Amphiphiles 235
Thomas Gartner III, Huihui Kuang, Efrosini Kokkoli, Arthi Jayaraman
(197b) Intracellular Trafficking of Enzyme-Cleavable Peptide Amphiphiles 236
Handan Acar, James L. Labelle, Matthew V. Tirrell
(197c) Polylactide-Based Biodegradable Zwitterionic Polymers and Their Conjugates with Drugs for Biomedical Applications 237
Haotian Sun, Michael Yu Zarng Chang, Wei-I Cheng, Qing Wang, Alex Commisso, Meghan Capeling, Yun Wu, Chong Cheng
(197d) Overcoming Obstacles to Brain Repair Using Biomaterials 238
Tatiana Segura
(197e) Aggregation Kinetics in Biological Environments as a Determinant of Nanoparticle Behavior in the Brain 239
Chad D. Curtis, Mike McKenna, Elizabeth Nance
(197f) Layer-By-Layer Nanoparticles for Interleukin-12 Delivery 240
Antonio E. Barberio, Santiago Correa, Erik Dreaden, Talar Tokatlian, Mariane B. Melo, Darrell J. Irvine, Paula T. Hammond
(197g) Implantable Biomaterials Produced By Complexing Chitosan to Alginate or Pectin: Surface Properties, Hemocompatibility and Cytotoxicity 241
Fernanda C. Bombaldi De Souza, Renata F. Bombaldi De Souza, Angela Maria Moraes, Diego Mantovani

(197h) Development of Sustainable Therapeutic Dressings Consisting of Chitosan-Alginate Films Incorporating Arrabidaea Chica Verlot Extract 242
Ana Luiza Resende Pires, Cecilia Buzatto Westin, Ilza Maria De Oliveira Sousa, Mary Ann Foglio, Angela Maria Moraes

(197i) Comparison of Chitosan Particles Produced By Ionic Gelation and By Supercritical Assisted Atomization 250
Julia Natalia Oliveira Mazoni, Paulo De Tarso Vieira E Rosa

(197j) Exploiting a Novel Aqueous-Two Phase Microfluidic System for Cell Encapsulation in GAG+Chitosan Microcapsules 251
Amin Vossoughi Shahvari, Howard W. T. Matthew

(197k) Laser-Activated Nanocomposites for Tissue Repair 252
Russell Urie, Deepanjan Ghosh, Mitzi Thelakkaden, Tanner Flake, Jerry Crum, Chengchen Guo, Jeff Yarger, Kaushal Rege

(197l) New Strategy for the Fabrication of Annular Cylindrical Polysaccharide-Based Scaffolds 253
Angela Maria Moraes, Renata F. Bombaldi De Souza, Fernanda C. Bombaldi De Souza

(197m) Formulation of Peptide Antimicrobials for Treatment of Wound Infections 254
Ritu Goyal, Michael Holloway, Pooja Patel, David Devore, Charles Roth

(197o) Synthesis and Degradation of Biodegradable Copolymers 255
Eswar Arunkumar Kalaga, Timothy Brenza

(197r) Ultra-High Surface Area Activated Carbon from a Renewable Resource 256
Ashli Polanco, Dmytro Volkov, Quoc Truong, Carl Lawton, Nese Orbey

(197t) A Facile Novel Fluorocarbon Copolymer Solution Coating Process for Improving Platelet Compatibility of Titanium 257
Sophia Chao-Wei Huang, Chi-Hui Cheng, Yun Chiu, Yi-Ching Lin, Jui-Che Lin

(198a) Alignment of Quantum Dot Nanorod/Silica Hybrid Particles on Glass Substrate for Luminescent Solar Concentrator 258
Kiju Um, Young-Geon Song, Kangtaek Lee

(198b) Effect of Graphene Oxide on Formation of Zirconium Tungstate Nanoparticles 259
Young-Geon Song, Kiju Um, Kangtaek Lee

(198d) Rapid Microwave-Assisted Synthesis of Hybrid Zeolitic-Imidazolate Frameworks 260
Febrian Hillman, John Zimmerman, Seung-Min Paek, Mohamad Hamid, Woo Taik Lim, Hae-Kwon Jeong

(198e) Facile Synthesis of Cd-Substituted Zeolitic-Imidazolate Framework Cd-ZIF-8 and Mixed-Metal Cdzn-ZIF-8 261
Jingze Sun, Hae-Kwon Jeong, Woo Taik Lim, Liya Semenchenko

(198f) A Study of Asymmetry Cu-MOFs Electrode Prepared in situ and Its Biomimetic Catalysis 263
Zhipeng Li, Liwei Ren, Diannan Lu

(198g) Protected SiC Catalyst Support for Steam Methane Reforming Reaction 264
Naftali Opembe, Seungdoo Park, Sergio Ibanez, Doug Mitchell, Matthew Seabaugh, Scott Swartz

(198h) Computational Screening of High Temperature Materials for Environmental Barrier Thin Films 265
Amanda Hoskins, Aidan Coffey, Charles B. Musgrave, Alan W. Weimer

(198i) Phase Transformation Induced By Tertragnility Variation of Metal-Redox Synthesised NiMn Nanoalloys 266
Jian Shen, Xin Jin

(198j) Fabrication of Electrospun Mesoporous Silica Nanomaterials for Water Vapor Adsorption 268
Soyoung Kim, Heechul Choi

(198k) Applications of Mesoporous RuCo2O4 Thin Film for High Performance Supercapacitor 269
Do-Heyoung Kim, Nilesh R. Chodankar

(198l) TiO2 Thin Film Deposition By Electrospray 270
Yaqun Zhu, Jong Hyun Shim, Junghyun Cho, Paul R. Chiarot

(198m) Green Synthesis of Copper Oxide Nanoparticles Using a Simple Microwave-Assisted Method 271
Prasad P Pawar, Shishir V Kumar, Adarsh Bafana, Ashiqur Rahman, Si A. Dahoumane, Clayton S Jeffryes

(198n) Comprehensive Thermodynamic Modeling of Mixed-Solvent Electrolyte Systems: An Investigation on the Quaternary System of FeCl2-FeCl3-HCl-H2o 272
Sina Hassanjani Saravi, Chau-Chyun Chen

(198o) Controllable Manipulation of Continuous AFI Membranes with Distinctive Microstructures on Macroporous Alpha-Alumina Substrates 273
Hongfeng Dong, Xiufeng Liu, Huiming Zhu, Baoquan Zhang, Jian Li

(198p) Magnetic Core Shell Microspheres for Extraction of Rare Earth Elements from Geothermal Brine Solution 274
B. Peter McGrail, Jian Liu

(198r) Encapsulation of Dye in NH2-Uio-66 Metal-Organic Framework for Photosensitized Oxidation of Benzyl Alcohol 275
Xiyi Li, Qingqing Hou, Neng Liao, Jing Xiao
(199a) The Fabrication of Graphene/Polyaniline Blended Fiber for Conducting and Flexible Energy Storage Devices 276
Yafei Feng, Jiaxin Shen, Cunliang Ma, Yidong Liu, Yong Min
(199b) Semiconducting Heterostructures for Photocatalytic Reduction of Carbon Dioxide 277
Debtanu Maiti, Johnnie Cairns, J. N. Kuhn, Venkat R. Bhethanabotla
(199c) Photoswitchable Quantum Dots Probes for Superresolution Microscopy 278
Abhilasha Dehankar, Kil Ho Lee, Abhijeet Marar, Karine Thate, Carol Lynn Alpert, Peter Kner, Jessica O. Winter
(199d) Bottom-up Synthesis of Nanoelectronic Titania Composites 279
Yang Lu, Evan K. Wujcik, Arijit Bose
(199h) Phase Diagrams, Defect Models and Thermoelectric Properties: ÃŸ-Ag2se and CoSb3 280
Sinn-Wen Chen, Zi-Yang Huang, Yang-Yuan Chen
(199i) Theoretical Study of a High Performance Thermoelectric Material: Stanene 281
Pabitra Choudhury, Charles Griego
(199j) Revealing the Enigmatic Interfacial Layer of Core/Shell Quantum Dots 282
Ajay Singh, Jennifer Hollingsworth
(199k) Spatial Manipulation of Thermal Flux Profiles Using Nanostructure Boundaries 283
Abhinav Malhotra, Martin Maldovan
(200a) Magnetic Polymer Nano-composites for Giant Magnetoresistance and Electromagnetic Shielding 284
Jiang Guo, Alexandra Galaska, Brian J. Edwards, Bamin Khomami, Zhanhu Guo
(200b) Constructing Ternary Conductive Polymer Composites with Cocontinuous Polymer Blends and Interfacial Graphene Nanoplatelets 285
Yangming Kou, Lian Bai, Xiang Cheng, Christopher W. Macosko
(200c) The Influence of Interfacial Graphene on the Morphological, Electrical and Mechanical Properties of Co-Continuous Polymer Blends 286
Catherine Esnaashari, Lian Bai, Christopher W. Macosko, Xiang Cheng
(200d) The Fabrication and Application of Composite Graphene Oxide Films 287
Jiaxin Shen, Yafei Feng, Cunliang Ma, Yidong Liu, Yong Min
(200e) Mechanically Stable Thermally Crosslinked Poly(acrylic acid)/ Reduced Graphene Oxide Aerogels 288
Heonjoo Ha, Han Xiao, Kadhiravan Shanmuganathan, Christopher J. Ellison
(200f) Ionophore-Decorated Magnetic Graphene Oxide As a Composite Adsorbent Material for Heavy Metal Ion Sequestration 289
Khino J. Parohinog, Grace M. Nisola, Wook-Jin Chung
(200g) Synthesis of Thermoresponsive Polymer/Fe3O4 Nanoparticle Composite and Its Application 290
Risako Sakai, Junichi Ida, Tatsushi Matsuyama
(200h) High-Performance, Ambient Phase Change Thermal Diodes for Energy Applications 291
Anton Cottrill, Song Wang, Albert Tianxiang Liu, Yuichiro Kunai, Michael Strano
(200i) A Study on the Preparation and Properties of Polymer Composites Using Amino Functionalized Microcrystalline Cellulose (MCC) As a Filler Material 292
Kiryong Ha, Hanna Kim, Yeokyung Yang, Kiseob Hwang, Kwang-Hee Lim
(200j) Electrically Conductive Films Made of Polythiophene and Fibrillated Wood Particles 293
Islam Hafez, Han-Seung Yang, Mehdi Tajvidi, Nicholas Seaton, William T. Y. Tze
(200l) Effect of Different Carbon Additives on Structure of Magnesium Composites for Hydrogen Storage 294
Yeboah Martin Luther
(200n) Low Loading of Grafted Thermoplastic Polystyrene Strengthened and Toughened Transparent Epoxy Composites 295
Chao Ma, Hongbo Gu
(200o) Preparation of Modified Graphene Oxide-Containing Styrene Masterbatches for Thermosetting Resin Composite 296
Siyao He, Yuqiang Qian, Kunwei Liu, Chris Macosko, Andreas Stein
(200p) Smart Windows Enabled By Buckling Instabilities in Periodic Composite Films 297
Peng Jiang, Zhuxiao Gu
(200q) Magnetic Polymer Nanocomposites for Electromagnetic Interference Shielding 298
Jiang Guo, Alexandra Galaska, Suying Wei, Brian J. Edwards, Bamin Khomami, Zhanhu Guo
(262a) Autoperforation of 2D Materials for Generating Two Terminal Memresistive Janus Particles 299
Albert Tianxiang Liu, Pengwei Liu, Michael Strano

(262b) Driven Morphological Evolution of Strained Thin Film Surfaces and Two-Dimensional Materials: Morphological Stability and Pattern Formation 300
Lin Du, Dimitrios Maroudas
(262c) Optimization of THz-Time Domain Spectroscopy Reflectivity for a 1THz Energy-Harvesting Metamaterial 301
Shendu Yang, Zachary Thacker, Evan Allison, Patrick J. Pinhero
(262d) Modeling Exciton Dynamics and Low-Frequency Vibrations in Quantum Dot Assemblies 302
Elizabeth M. Y. Lee, Adam P. Willard, William A. Tisdale
(262e) Extraordinarily Slow Electron-Hole Recombination in Perovskite Phase Cesium Lead Iodide 303
Subham Dastidar, Siming Li, Jason B. Baxter, Aaron T. Fafarman
(262f) Elucidating the Impact of Alcohol Post-Processing in High Performance Roll-to-Roll Printed Organic Photovoltaics 304
Kevin L. Gu, Xiaodan Gu, Hongping Yan, Zhenan Bao
(262g) Metamaterial and Rectenna Design and Testing for the Conversion of Blackbody Radiation to Electricity Using 5 THz Devices 305
Evan Allison, Zachary Thacker, Shendu Yang, Patrick J. Pinhero
(262h) Reusable Chromogenic Sensors Enabled By Novel Multi-Stimuli-Responsive Shape Memory Polymers 306
Sin-Yen Leo, Peng Jiang
(265a) Allyl-Sulfide Modified Hydrogels with Switchable Properties As Dynamic Cellular Niches 307
Kristi S. Anseth, Tobin E. Brown, Joseph Grim, Ian Marozas
(265b) Non-Linear Rheology and Fracture in Alginate Hydrogels 308
Seyed Meysam Hashemnejad, Rangana Wijayapala, Santanu Kundu
(265c) Analyzing the Effects of Time and Crosslinker Ratio on the Mechanical Properties of Biodegradable Zein Super Gels 309
Hazal Turasan, Jozef Kokini
(265d) Self-Assembly, Structure and Rheology of Polyelectrolyte Complex Hydrogels 310
Samanvaya Srivastava, Adam Levi, Matthew V. Tirrell
(265e) Evolution of Mechanics in Î±-Helix Peptide Bioconjugated Linear- and Star-Block Peg 311
Sean C. O'Neill, Raymond Tu
(265f) Modelling of the Degradation of Poly(ethylene glycol)-Co-(lactic acid)-Dimethacrylate Hydrogels 312
Marco Lattuada, Giuseppe Storti, Vincent Diederich
(265g) Tough, Rapidly Swelling Thermoplastic Elastomer Hydrogels for Hemorrhage Control 313
Erich Bain, Tyler R. Long, Frederick L. Beyer, Randy A. Mrozek, Joseph L. Lenhart
(265h) Synthesis and Characterization of Thermally Responsive N-Isopropylacrylamide Hydrogels Copolymerized with Novel Hydrophobic Polyphenolic Crosslinkers 314
Shuo Tang, Thomas Dziubla, J. Zach Hilt
(265i) Novel Biocompatible Thermo-Responsive Poly(N-vinyl Caprolactam)/Clay Nanocomposite Hydrogels with Macro-Porous Structure and High Mechanical Property 315
Zhuang Liu, Kun Shi, Xiao-Jie Ju, Wei Wang, Rui Xie, Liang-Yin Chu
(267a) Multiplicity of Morphologies in Poly (L-lactide) Bioresorbable Vascular Scaffolds 316
Artemis Ailianou, Karthik Ramachandran, Mary Beth Kossuth, James Paul Oberhauser, Julia A. Kornfield
(267b) Tissue Response and Integration in Biomaterial Implants Derived from Morphologically Unique Emulsion Gels 317
Todd Thorson, Ali Mohraz, Elliot Botvinick
(426h) The Unique Mechanism of Covalently Adaptable Hydrogel Degradation Characterized with Passive Microrheology 318
Francisco Escobar, Kristi S. Anseth, Kelly M. Schultz
(267d) Development of Gelatin and Graphene Based Conduits Using 3D Printing Strategies for the Transdifferentiation of Mesenchymal Stem Cells into Schwann Cell-like Phenotypes through Electrical Stimuli 319
Metin Uz, Maxsam Donta, Donald S. Sakaguchi, Surya K. Mallapragada
(267e) Effect of Molecular Weight and Degree of Functionality on Degradation, Biocompatibility and Two-Photon Polymerization of Acrylated Poly(caprolactone) 320
Brian J. Green, Jessica Thompson, Kristan S. Worthington, Budd A. Tucker, C. Allan Guymon
(267f) An Injectable and Anisotropic Hydrogel with Biomimetic Structures for Directed Cell and Nerve Growth 321
Jonas C. Rose, David B. Gehlen, Esther Jaekel, Jens Kohler, Khosrow Rahimi, Martin Moller, Laura De Laporte
(267g) Engineering Extracellular Matrix Mimetic Materials By Green Electrospinning of Collagen 322
Jorge Almodovar, David Castilla

(267h) Biomimetic Scaffolds for in vitro Bone Marrow Tissue Engineering 323
Yongkuk Park, Ryan Carpenter, Jungwoo Lee
(303a) Tapered Block Copolymers: Tuning Self-Assembly and Properties By Manipulating Monomer Segment Distributions 324
Thomas H. Epps
(303b) Photoplasiticity in Crosslinked Liquid Crystalline Networks: A Route to Reconfigurable Shape-Changing Materials 325
Matthew K. McBride, Matthew Hendrikx, Danqing Liu, Brady Worrell, Dick J. Broer, Christopher N. Bowman
(303c) Synthesis of Novel Nanostructured Copolymers with Alternating Linear Polymer and Dendrimer Blocks 326
Haotian Sun, Alex Commisso, Chong Cheng
(303d) Hierarchical Structures of PDMS-PU Copolymer and Particles for Hydrophobic Coatings 327
Marius Rutkevicius, Mackenzie Geiger, Tahira Pirzada, Saad A. Khan
(303e) Charged Polymer Conformations in Polyelectrolyte Complexes 328
Amanda B. Marciel, Samanvaya Srivastava, Matthew V. Tirrell
(303f) Solubility Parameters, Water Activity Coefficients and Proton Mobility of Sulfonated Poly (styrene-isobutylene-styrene), Sulfonated Poly (ether ether ketone), and Sulfonated Poly (2-ethoxyethyl methacrylate) Membranes 329
Maritza Perez Perez, David Suleiman
(303g) Single-Molecule Super-Resolution Microscopy in Nanostructured Polymer Thin Films 330
Muzhou Wang, James M. Marr, Jeffrey W. Gilman, J. Alexander Liddle
(303h) A Non-Equilibrium Molecular Dynamics (NEMD) Simulation of the Crosslinked Polyamide Membranein Water Desalination 331
Md Symon Jahan Sajib, Tao Wei
(306a) Linear and Nonlinear Rheology Predictions of Entangled Polymers in Complex Flows from First Principles 338
Jay D. Schieber
(306b) Melt Behavior-Chain Architecture-Polymer Composition Correlations in High Density Polyethylene 339
Rohan Hule, Derek W. Thurman, Antonios Doufas
(306c) Shear-Induced Conformational Changes of Flexible and Semi-Rigid Engineering Thermoplastics and Their Influence on Crystallization 340
Behzad Nazari, Jiho Seo, Ralph H. Colby, Alicyn M. Rhoades, Richard P. Schaake
(306d) Island-in-the-Sea Meltblown Nanofiber Nonwovens with Diverse Surface Properties for Filtration Applications 341
Iman Soltani, Satish Kumar, Frank S. Bates, Christopher W. Macosko
(306e) Enthalpic and Entropic Competitions in Solvent-Free Polymer-Grafted Nanoparticles 342
Snehashis Choudhury, Lynden A. Archer
(306f) Designing Material Dynamics in Polyelectrolyte Complexes 343
Yalin Liu, Brian Momani, Matthew Labbe, H. Henning Winter, Sarah L. Perry
(306g) Extensional Relaxation Times of Dilute and Semi-Dilute Polymer Solutions 344
Jelena Dinic, Leidy N. Jimenez, Madeleine Biagioli, Vivek Sharma
(306h) Nonequilibrium Molecular Dynamics Simulations of Entangled Polymer Solutions Undergoing Planar Elongational Flows 345
Mohammad Hadi Nafar Sefiddashti, Brian J. Edwards, Bamin Khomami
(306i) Ultra-High Performance Polymers Meet Ionic Liquids 346
Jason E. Bara, Kathryn O'Harra, Grayson P. Dennis, Marlow M. Durbin, Max Mittenthal, Enrique M. Jackson
(334a) Paper-Based Cell Culture Platforms for Personalized Medicine 347
Gulden Camci-Unal
(334b) Growth Factor Delivery from Silk-Extracellular Matrix Composite Sponges for Modulating Congenital Heart Defect Repair 348
Whitney L. Stoppel, Elizabeth C. Bender, Luke R. Perreault, Jonathan M. Grasman, Andrea Papait, David L. Kaplan, Lauren D. Black III
(334c) Oligodendrocyte Precursor Cell Maturation in a 3D Hydrogel System through the Incorporation of Drug Delivery Nanoparticles or Topographical Cues 349
Lauren Russell, Meghan Pinezich, Kyle Lampe
(334d) Resveratrol Releasing Scaffolds Protect Mice Against Diet Induced Obesity and Glucose Intolerance 350
Michael Hendley, Prakasam Annamalai, Michael Gower
(334f) Effects of Short-Term Magnetic Stimulation on MSCs Encapsulated in an Injectable, Magneto-Responsive Hydrogel 351
Adedokun Adedoyin, Adam Ekenseair

(334g) The Incorporation of Retinoic Acid-like Peptoids Onto an Artificial Extracellular Matrix for Increased Differentiation of Human Embryonic Stem Cells into Neural Cells 352
Jesse Roberts, German Perez, Shannon L. Servoss
(334h) 3D Culture of Trabecular Meshwork Cells 353
Matthew Osmond
(354b) Steps Toward Bio-Sourced Packaging: Effect of Composition and Processing on Properties of Co-Polyesters of Polyethylene Terephthalate (PET) with 2,5-Furan Dicarboxylic Acid 354
Anup Joshi, Maria Coleman
(354c) Synthesis and Solutionphasecharacterization of Hydroxylated Sulfonated Oligothioetheramides 355
Joseph Brown, Christopher A. Alabi
(354d) Predicting Stable and Metastable Frank-Kasper Phases in Block Polymers Using Self-Consistent Field Theory 356
Akash Arora, Kyungtae Kim, Morgan W. Schulze, Ronald M. Lewis III, Frank S. Bates, Kevin D. Dorfman
(354e) Spatiotemporal Evolution of Structure in Layer-By-Layer Assembled Thin Films Composed of Oppositely Charged Polyelectrolytes 357
Ali Salehi, Ronald G. Larson
(354f) High Performance Roll-to-Roll Printed PTB7-Th/PCBM Organic Solar Cells 358
Kevin L. Gu, Xiaodan Gu, Hongping Yan, Zhenan Bao
(354g) Pinch-Off Dynamics, Dripping-Onto-Substrate (DoS) Rheometry and Printability of Polymeric Complex Fluids 359
Jelena Dinic, Leidy N. Jimenez, Madeleine Biagioli, Vivek Sharma
(354h) Ultra-Fragile 'Granular Materials' Designed Via a Genetic Algorithm 360
Venkatesh Meenakshisundaram, Jui-Hsiang Hung, David S. Simmons
(354i) Functionalizing Surfaces with Zwitterionic Polymers to Control Cell Adhesion and Direct Neurite Growth 361
Braden Leigh, Elise Cheng, Corinne Andresen, Marlan Hansen, C. Allan Guymon
(354j) Understanding Facilitated Transport of Hydrogen in Polybenzimidazole Containing Palladium Nanoparticles Using an Integrated Experimental and Modeling Approach 364
Lingxiang Zhu, Deqiang Yin, Shailesh Konda, Mark T. Swihart, Haiqing Lin
(364a) Engineering Surfaces and Interfaces in Polymer Films with Bottlebrush Polymer Additives 365
Gila Stein, Rafael Verduzco
(364b) Effects of Extensional Flow and Nanoparticle Stabilization on Immiscible Polymer Blend Morphology 366
Matthew S. Thompson, Sushant Agarwal, Xueyan Song, Rakesh K. Gupta
(364c) Non-Isocyanate Polyurethane Thermoplastic Elastomer: Amide-Based Chain Extender Yields Enhanced Nanophase Separation and Properties in Polyhydroxyurethane 367
Goliath Beniah, David Fortman, William Heath, William Dichtel, John M. Torkelson
(364d) Molecular Simulations Study of Solvophobicity Effects on Assembled Structure in Solutions of Amphiphilic Block Copolymers and Nanoparticles 368
Daniel J. Beltran-Villegas, Arthi Jayaraman
(364e) Control Nano/Microstructure Using Photopolymerization-Induced Phase Separation (PhIPS) 369
Erion Hasa, Julie L. P. Jessop, Jeffrey W. Stansbury, C. Allan Guymon
(364f) Spontaneous Self-Assembly and Micellization of Random Copolymers in Organic Solvents 370
Ayse Asatekin
(364g) Synthesis and Antibacterial Study of Star-Shaped Poly[2-(dimethylamino)Ethyl Methacrylate]-Based Copolymers with an Inorganic Core 371
Hou Zheng, Yuji Pu
(364h) Understanding Molecular Exchange Kinetics in Polyelectrolyte Complex Micelles 372
Hao Wu, Jeffrey Ting, Samanvaya Srivastava, Matthew V. Tirrell
(364i) Molecular Simulation for the Prediction of Plasticizer Efficiency and Stability in a Polymer Matrix 373
Dongyang Li, Kushal Panchal, Li Xi
(375a) Copper-Silver Core-Shell Nanoparticles for Conductive Ink 374
Xiaofeng Dai
(375b) Electric Current-Induced Nanoscale Surface Roughness Reduction in Conducting Thin Films 375
Lin Du, Dimitrios Maroudas
(375c) Unusual Electronic Properties of Template-Directed Ï€-Conjugated Porphyrin and Phosphorene Nanotubes 376
Bryan M. Wong

(375d) Complex Pattern Formation from Current-Driven Dynamics of Single-Layer Epitaxial Islands on Crystalline Conducting Substrates 377
Ashish Kumar, Dwaipayan Dasgupta, Dimitrios Maroudas
(375e) The Infrared and Raman Spectra of Pure-Silica and Aluminosilicate Sodalite 378
Caio Peixoto, Amir M. Mofrad, Jack Blumeyer, Liu Jinrui, Karl D. Hammond, Heather K. Hunt
(375f) The Effect of Solvent Selection on the Optical Trapping, Manipulation, and Patterning of Nanomaterials 379
Matthew Crane, Elena P. Pandres, Patrick Whitham, E. James Davis, Daniel Gamelin, Vincent C. Holmberg, Peter Pauzauskie
(375g) Synthesis and Characterization of Electrochemically-Grown Zinc Oxide Nanowires for Use in Rectenna-Based Heat Harvesters 380
Adrian Haley, Shendu Yang, Patrick J. Pinhero
(381a) The Thiol-Thioester Exchange in Network and Linear Polymers 381
Christopher N. Bowman, Brady Worrell, Matthew K. McBride, Gayla Berg, Chen Wang
(381b) Nacre-Inspired Composite Gels for Biomedical Applications 382
Ayomi S. Perera, Richard Jackson, Mark Miodownik, Marc-Olivier Coppens
(381d) High Viscosity Polymer Gels Derived from Block Copolymer Nanocomposites 383
Sri Harsha Kalluru, Eric W. Cochran
(381e) Quantifying Topology, Gelation and Elasticity of Polymer Networks 384
Rui Wang, Bradley D. Olsen
(381f) Self-Assembly and Mechanical Properties of Di-Fmoc-L-Lysine Containing Molecular Gels 385
Seyed Meysam Hashemnejad, Md Masrul Huda, Neeraj Rai, Santanu Kundu
(381g) Thermoresponsive Sol-Gel Transitions of PÎ•G-Based Nanocomposite Hydrogels Controlled By Molecular Weights of Block Copolymers and Solute Concentrations 386
Tomoki Maeda, Midori Kitagawa, Keishi Tanimoto, Makoto Miyazaki, Koji Nagahama, Atsushi Hotta
(381h) Evaluating Reprocessability of Polymer Networks: Flory-Stockmayer Analysis 387
Lingqiao Li, Xi Chen, Kailong Jin, John M. Torkelson
(381i) Molecularly-Templated Reaction for Forming Poly(dimethyl siloxane)/Graphene Oxide Composite Elastomers 388
Heonjoo Ha, Kiryong Ha, Christopher J. Ellison
(411a) Graduate Student Award Session: A Two-Step Method for Transferring Single Wall Carbon Nanotubes Onto a Hydrogel Substrate 389
Mozhdeh Imaninezhad, Irma Kuljanishvili, Silviya Petrova Zustiak
(411b) Graduate Student Award Session: Understanding How Lipid Nanoparticle Structure Affects Immune Response 390
Lisa Kasiewicz, Sushant Kumar, Rahul Purwar, Kathryn A. Whitehead
(411c) Graduate Student Award Session: Photo- Induced Pinocytosis in Synthetic Liposomes 391
Danielle Konetski, Dawei Zhang, Christopher N. Bowman
(411d) Graduate Student Award Session: Sugar-Coating the Answers to Virus Binding: Glycocalyx-Mimetic Interfaces 392
Ramya Kumar, Domenic Kratzer, Kenneth Cheng, Irina Kopyeva, Joerg Lahann
(411e) Graduate Student Award Session: The Combined Effect of Matrix Microenvironment and Hypoxia on the Activity of Glioblastoma Stem Cells 393
Jee-Wei Emily Chen, Jann N. Sarkaria, Brendan A. Harley
(411f) Graduate Student Award Session: Pegylated Poly(beta-amino ester) Delivery Systems for Periodic shRNA 394
Connie Wu, Wade Wang, Paula T. Hammond, Jiahe Li
(411g) Graduate Student Award Session: Biodegradable Nano-Film Coated Self-Floating Hollow Glass Microspheres for Rapid Cell Isolation and Recovery 395
Ziye Dong, Caroline Ahrens, Dan Yu, Zhenya Ding, Hyuntaek Lim, Wei Li
(411h) Graduate Student Award Session: Enhancing Therapeutic Efficacy of Self-Assembling Prodrugs with Supramolecular Chemistry 396
Hao Su, Yuzhu Wang, Feihu Wang, Honggang Cui
(413a) Size-Selective Ionically Crosslinked Polymer Multilayer Films for Light Gas Separation 397
Jaime C. Grunlan, Benjamin Wilhite
(413b) Length Effects on Polyelectrolyte Complexation: How 'Poly' Must a Polyelectrolyte be? 398
Jeffrey Vieregg, Michael Lueckheide, Matthew V. Tirrell
(413c) Interaction and Dynamics of Polyelectrolytes in Polyzwitterionic Complexes 399
Y. Elaine Zhu, Benxin Jing, Kehua Lin
(413d) Tuning Complex Coacervation Using Sequence-Defined Polyelectrolytes: A Molecular Understanding 400
Tyler Lytle, Li-Wei Chang, Jason Madinya, Sarah L. Perry, Charles E. Sing

(413e) Photodirected Assembly and Self-Rupture of Polyelectrolyte-Based Soft Materials 401
Udaka K. De Silva, Amanda C. Bryant-Friedrich, Yakov Lapitsky
(413f) Coarse-Grained Model for Polyelectrolyte Complexation 402
Marat Andreev, Samanvaya Srivastava, Lu Li, Matthew V. Tirrell, Jack F. Douglas, Juan J. De Pablo
(413g) Synthesis and Solution-Phase Characterization of Hydroxylated Sulfonated Oligothioetheramides 403
Joseph Brown, Christopher A. Alabi
(413h) Effect of Nanoparticle on the Structure and Dynamics of Model Peaa Ionomers from Molecular Dynamics Simulations 404
Janani Sampath, Lisa M. Hall
(413i) Zwitterionic Copolymers As Novel Supporting Scaffolds for Ionic Liquid-Based Gel Electrolytes 405
Luis Rebollar, Fatin Lind, Matthew J. Panzer
(426a) Photoreversible Stiffness Modulation of Protein-Polymer Hydrogels 406
Luman Liu, Jared A. Shadish, Cole A. Deforest
(426b) Preparation and Characterization of Polypeptide Hydrogels as Synthetic Extracellular Matrices for Cellular Scaffolds 407
Hongkun He, Alex Wang, Marianna Sofman, Linda Griffith, Paula Hammond
(426c) Reinforced Hydrogel Fibers for Cell Encapsulation and Organ Printing 408
Suman Bose, Daniel G. Anderson, Robert Langer
(426f) Magnetically Templated Hydrogels for Peripheral Nerve Injury Repair 409
Ishita Singh, Christopher Lacko, Christine Schmidt, Carlos Rinaldi
(426g) A Two-Step Method for Transferring Single Wall Carbon Nanotubes Onto a Hydrogel Substrate 410
Mozhdeh Imaninezhad, Irma Kuljanishvili, Silviya Petrova Zustiak
(267c) Surface Tension-Assisted Additive Manufacturing of Multicomponent Biomaterials 411
Mark W. Tibbitt, Heloise Ragelle, Michael Cima, Robert Langer
(439a) Chiral Metamaterial Platform with Tunable Near and Far Field Chiroptical Response 412
Pavlos Pachidis, Vivian E. Ferry
(439b) Lateral Growth of Two-Dimensional 1H-WSe2/1T'-WTe2 Heterostructures 413
Mengqiang Zhao, Carl H. Naylor, Zhaoli Gao, William M. Parkin, A. T. Charlie Johnson
(439c) Nanostructured Optoelectronics Using Interfacially-Driven Assembly 414
Matthew G. Panthani
(439d) Tuning the Bandgap of Graphene Nanoribbons through Defect-Interaction-Driven Edge Patterning 415
Dimitrios Maroudas, Lin Du, Andre R. Muniz
(439e) Wrinkled MoS2 Field-Effect Transistors 416
Shikai Deng, Vikas Berry
(439f) Nanoantenna Enhanced Wavelength Mixing in Monolayer Transition Metal Dichalcogenide 417
D. Keith Roper, Gregory T. Forcherio, Mourad Benamara, Luigi Bonacina
(439g) Kinetics of Nanoring Formation from Quantum Dots in Epitaxial Thin Films 418
Lin Du, Dimitrios Maroudas
(441a) Design, Synthesis, and Characterization of Elastomeric and Mechanoresponsive Polymer Matrix Composites 419
Matthew D. Green
(441b) Molecular Simulation of Macromolecular Transport through Nanoporous Membranes 420
Noelia Almodovar Arbelo, Bryan W. Boudouris, David S. Corti
(441c) Unconventional Nanoscopic Shape Memory Effects Exhibited By Novel Multi-Stimuli-Responsive Shape Memory Polymers 421
Sin-Yen Leo, Peng Jiang
(441d) Direct Observation of Remarkable Nanostructure Evolution during Aqueous Dissolution of Polymer/Drug Blends 422
Ralm Ricarte, Marc A. Hillmyer, Timothy P. Lodge
(441e) Effect of Graphene and Graphene Oxide on the LCST of PNIPAM 423
Carter Berry, Sanket A. Deshmukh
(441f) Solvent Effects on the Structure and Thermodynamics of Polymer Blends with Varying Architectures 424
Thomas Gartner III, Arthi Jayaraman
(441g) Molecular Weight Dependence of the Intrinsic Size Effect on Tg in AAO Template-Supported Polymer Nanorods: A DSC Study 425
Tong Wei, Shadid Askar, Anthony Tan, John M. Torkelson

(441h) Versatile Cholesterol-Functionalized Block Copolymers in Aqueous Dispersions 426
Kenneth Mineart, Shrinivas Venkataraman, Yi Yan Yang, James L. Hedrick, Vivek M. Prabhu
(441i) Polymerization Thermodynamics Under Nanoconfinement 427
Haoyu Zhao, Qian Tian, Sindee L. Simon
(475a) Molecular Understanding, Design and Development of Ultra Low Fouling Zwitterionic Materials 428
Shaoyi Jiang
(475b) A Scalable Approach to Produce a Diversity of Structured Polymer Colloids 429
Rodney D. Priestley
(475c) Layered Morphologies in Precise Acid-Containing Polyethylenes: Hierarchical Structures and the Implications on Properties 430
Karen I. Winey
(475d) Examination of Thin Film Growth Using Supersonic Molecular Beams and in situ Real Time X-Ray Synchrotron Radiation: From Organic Small Molecule Semiconductors to Transition Metal Dichalcogenides 431
James R. Engstrom
(475e) Polymer Electrolytes for Electrochemical Energy Devices 432
Yushan Yan
(525c) TAT-Functionalized Liposomes for the Treatment of Meningitis 433
Caterina Bartomeu Garcia, Di Shi, Thomas J. Webster
(525d) Cartilage Penetrating Nanoparticles to Provide Sustained Delivery of Disease Modifying Drugs in Post-Traumatic Osteoarthritis 434
Brett C Geiger, Sheryl Wang, Alan Grodzinsky, Paula T. Hammond
(525e) Nanoparticle Tracking Analysis of Polymer Particle Aggregation in Blood Plasma 435
Kathleen McEnnis, Stephanie Christau, Sean McLoughlin, Joerg Lahann
(525f) High Throughput Screening of Biodegradable Nanogels with Tunable Size and Swelling for Intracellular Drug Delivery 436
David S. Spencer, Bryan C. Luu, David W. Beckman, Nicholas A. Peppas
(525g) Polyanhydride Nanoparticles Encapsulating Rifampicin Suppress Growth of Pathogenic Acanthamoeba in vitro 437
Nathan Peroutka-Bigus, Adam Mullis, Balaji Narasimhan, Bryan H. Bellaire
(525h) Design of Dual Encapsulated Biodegradable Nanoparticles for Cancer Treatment 438
Amber C. Jerke, Timothy Brenza
(526a) Understanding How Lipid Nanoparticle Structure Affects Immune Response 457
Lisa Kasiewicz, Sushant Kumar, Rahul Purwar, Kathryn A. Whitehead

VOLUME 2

(526b) Multi-Stage Drug Delivery System for Enhanced Payload Delivery to Lymph Node Cells 458
Alex Schudel, Cody Higginson, Mai Kwan Yau, M. G. Finn, Susan N. Thomas
(526c) Encapsulation and Thermal Stability of Immunological Biologics Using Complex Coacervation 459
Whitney C. Blocher, Rebecca Hershman, Sarah L. Perry
(526d) Evaluation of Biocompatibility of Novel Block Copolymer Gels/Micelles As Potential Vaccine Adjuvants 460
Justin Adams, Sujata Senapati, Michael J. Wannemuehler, Balaji Narasimhan, Surya Mallapragada
(526e) Intranasal Nanovaccine Provides Protection Against Homologous and Heterologous Influenza Virus 461
Zeb Zacharias, Kathleen Ross, Balaji Narasimhan, Thomas Waldschmidt, Kevin Legge
(526f) Design of a Combination Nanovaccine to Induce Rapid and Long Term Protective Immunity Against Bacillus Anthracis 462
Sean Kelly, Ross Darling, Nathan Peroutka-Bigus, Thomas Dubensky, Bryan H. Bellaire, Michael J. Wannemuehler, Balaji Narasimhan
(526g) Comprehensive Vaccine Design for Commensal Disease Progression 463
Charles H. Jones, Guojian Zhang, Roozbeh Nayerhoda, Marie Beitelshees, Andrew Hill, Yi Li, Bruce A. Davidson, Paul Knight III, Blaine Pfeifer
(526h) Oncofetal Antigen Peptide Nanoclusters for Cancer Vaccines 464
Alexandra Tsoras, Julie A. Champion
(536a) S Doped TiO2 for Photocatalytic Oxidation of CO in Visible Region Synthesized By Novel One Step Liquid Flame Spray Pyrolysis (LFSP): Kinetics and Mechanism 473
Siva Nagi Reddy Inturi, Panagiotis Smirniotis

(536b) Copper Hexacyanoferrate Hydrogel Electrodes for Electrochemically-Mediated Cation Separations 474
Kai-Jher Tan, Xiao Su, Johannes Elbert, T. Alan Hatton
(536c) Emulsion Templated Polymers As Structured Supports for MOF Adsorbents 475
Jacob I. Deneff, Krista S. Walton
(536d) Development of Novel Crosslinked Polymers for the Capture of Environmental Pollutants 476
Rishabh Shah, Thomas Dziubla, J. Zach Hilt
(536e) Grafted Polystyrene Controlled Formation of Magnetic Carbon Nanocomposites for Environmental Remediation 477
Hongbo Gu
(536f) Constructing Fenton-like Reaction over g-C3N4/NH2-MIL-88B(Fe) Photocatalyst to Degrade Organic Contamination in Aqueous Solution 478
Xiyi Li, Yunhong Pi, Zhong Li, Jing Xiao
(536g) Synthesis of Thermoresponsive Composite and Application for Water Treatment 488
Junichi Ida, Masanori Ochi, Ai Ishikawa, Atsushi Matsumoto, Tatsushi Matsuyama, Hideo Yamamoto
(536h) Integrated Adsorption and Visible-Light-Driven Photocatalytic Degradation of MB over Oxidized C60/NH2-MIL68(In) 489
Yunhong Pi, Xiyi Li, Zhong Li, Jing Xiao
(536i) In-Situ Studies on Anion Electroadsorption Mechanisms 504
Kelsey Hatzell, Marta Hatzell, Marm Dixit
(538a) Engineering Materials and Processes for Flexible Electronics 505
Antonio Facchetti
(538b) Spectroelectrochemistry of Conjugated Radical Polymers 506
Jodie L. Lutkenhaus
(538c) Conjugated Copolymers Based on Purines: Linking Monomer Design and Macromolecular Properties 507
Michael Kilbey, Graham Collier, Lauren Brown, Evan Boone, Brian Long
(538e) Pinch-Off Dynamics and Printability of Semi-Flexible and Rigid Rod Polymer Solutions 508
Leidy N. Jimenez, Jelena Dinic, Vivek Sharma
(538f) Surface-Directed Multi-Scale Assembly of Conjugated Polymers 509
Ying Diao, Erfan Mohammadi, Ge Qu
(538g) Interpenetrating Networks for Flexible Bulk Heterojunction Opvs 510
Jorge Mok, Zhiqi Hu, Rafael Verduzco
(538h) Development of an Anisotropic Coarse-Grained Conjugated Polymer Model for Optoelectronic Applications 511
Alec S. Bowen, Nicholas Jackson, Daniel Reid, Yamil J. Colon, Juan J. De Pablo
(538i) Automated Quantitative Analysis of Oriented Morphologies and Inter-Grain Connectivity in Conjugated Polymers 512
Nils Persson, Michael McBride, Martha A. Grover, Elsa Reichmanis
(545a) Spray Coating Chitin & Cellulose Nanomaterials for Enhancement of Barrier Properties 513
Chinmay C. Satam, Jerel Jallorina, J. Carson Meredith
(545b) Development of Low-Concentration Alkaline Treatment Method to Produce Thermally Stable Cellulose Nanofibrils (CNFs) 514
Hansol Lee
(545c) Tailored and Integrated Production of Carboxylated and Lignin Containing Cellulose Nanocrystals and Nanofibrils for Composite Applications 515
J. Y. Zhu
(545d) Nano Carbon Structures from Cellulosic Biomass for Use As Functional Materials 516
Ping Wang
(576a) Quantitative Three-Dimensional Morphological Characterization of Block Copolymer Films Enabled By Directed Self-Assembly 517
Paul F. Nealey
(576b) Examination of Line Edge Roughness of Directed Self-Assembled Block Copolymers: A Coarse-Grained Molecular Dynamics Study 518
Shubham Pinge, Durairaj Baskaran, Munirathna Padmanaban, Yong Lak Joo
(576c) Atomistic Simulations of Ion Transport in PsbP2VP Functionalized Lamellae 519
Weiwei Chu, Yamil J. Colon, Juan J. De Pablo
(576d) Performance of the Novel Ultra-Thin Proton Exchange Membrane Based on Poly(arylene ether sulfone)s for Pemfcs 520
Yang Zhao, Xue Li, Xiaofeng Xie
(576e) The Consequence of Morphology on Conductivity for Bolaamphiphiles 521
Mayank Misra, Christian Nowak, Yangyang Sun, Fernando Escobedo

(576f) Mechanical and Structural Analyses of Toughened Syndiotactic Polypropylene Gels: Effects of Gel-Preparation Temperature 522
Fuyuaki Endo, Ryusuke Okoshi, Naruki Kurokawa, Tomoki Maeda, Atsushi Hotta
(576g) Chemically Cross-Linked Poly(2-Hydroxyethyl Methacrylate)-Supported Deep Eutectic Solvent Gel Electrolytes 523
Huan Qin, Matthew J. Panzer
(576h) Ultrasonic Synthesis of Temperature Responsive Copolymer and Its Characterization 524
Masaki Kubo, Tomoyuki Koshimura, Takao Tsukada
(576i) Atomic-Level Comparisons of LCST Transition in Thermo-Sensitive Polymers 525
Yaxin An, Karteek K. Bejagam, Samrendra Singh, Sanket A. Deshmukh
(591a) ATRP-Grown Protein-Polymer Conjugates Selectively Enhance Transepithelial Protein Transport 526
Chad Cummings, Katherine Fein, Hironobu Murata, Rebecca Ball, Alan Russell, Kathryn A. Whitehead
(591b) Aptamer Micelles Targeting Cancer Cells Expressing the Chemokine Fractalkine 527
Michael A. Harris, Timothy R. Pearce, Thomas Pengo, Huihui Kuang, Colleen L. Forster, Efrosini Kokkoli
(591c) Non-Charged Cell-Penetrating Oligothioetheramides 528
Ngoc Phan, Christopher A. Alabi
(591d) Targeting, Delivery, and Immobilization of Therapeutic Factors with Native Free Radicals 529
Christopher J. Lowe, Keana Mirmajlesi, David I. Shreiber
(591e) Antimicrobial Peptide Amphiphile (AMPA) Medical Product Coatings for the Prevention of Nosocomial Infections 530
Josiah Smith, Alexis Dadelahi, Julie Nguyen, Fabio Gallazzi, John Dodam, Jeffrey Adamovicz, Roger De La Torre, Bret Ulery
(591f) Mitigating the Bioactivity Loss of Polymer- Insulin Conjugate 531
Zhiqiang Cao, Yang Lu
(591g) Enhancing Therapeutic Efficacy of Self-Assembling Prodrugs with Supramolecular Chemistry 532
Hao Su, Yuzhu Wang, Feihu Wang, Honggang Cui
(591h) Hydrogen Sulfide Donor Micelles: Synthesis, Characterization and Therapeutic Potential 533
Urara Hasegawa, Andre Van Der Vlies, Jerry J. Y. Chen, Tomoka Takatani-Nakase, Ikuhiko Nakase
(592a) Pegylation of Model Drug Carriers Enhances Uptake By Primary Human Neutrophils 534
William Kelley, Catherine A Fromen, Omolola Eniola-Adefeso
(592b) 'Smart' Nanoparticles for Immunotherapeutic Targeting of the Sting Pathway 535
Daniel Shae, Denise Buenrostro, Alyssa Merkel, Sema Sevimli, Julie A. Sterling, John Wilson
(592c) Biomaterial Nanoparticles Redistribute Therapeutic Antibodies to Lymph Node-Resident Cells to Enhance Cancer Immunotherapy Via Checkpoint Inhibition 536
David Francis, Alex Schudel, Nathan A. Rohner, Susan N. Thomas
(592d) Engineering Nanoparticle Artificial Antigen Presenting Cells Based on T Cell Biology Improves T Cell Enrichment and Activation for Cancer Immunotherapy 537
John Hickey, Fernando Vicente, Hai-Quan Mao, Jonathan Schneck
(592e) Biomaterial Scaffolds for Combined Focal Ablation and Immunotherapy to Target Disseminated Cancer 538
Francisco Pelaez, Stephen O'Flanagan, Qi Shao, Brandon Burbach, Tiffany Lam, John C. Bischof, Yoji Shimizu, Samira M. Azarin
(592f) Modulating the Immune Environment within Adipose Tissue with Polymer Scaffolds 539
Kendall Murphy, Michael Gower
(592g) Multi-Factor Microparticle Formulation for Local Induction of Regulatory Lymphocytes and Treatment of Periodontal Disease 540
Ashlee Greene, Sayuri Yoshizawa, Michelle Ratay, Charles Sfeir, Steven R. Little
(592h) Multivalent Soluble Antigen Arrays Target Antigen Presenting B Cells and Dampen Antigen-Specific Signaling to Promote Therapeutic Efficacy in Multiple Sclerosis 541
Brittany Hartwell
(593a) Novel Bio-Based Polyesters and Polycarbonates Derived from Xylochemicals 542
Joseph F. Stanzione III, Silvio Curia, Joseph Mauck, Alexander W. Bassett, John J. La Scala
(593b) Light Weight Biocomposite from Toughened Polyolefin and Biocarbon 555
Ehsan Behazin, Manju Misra, Amar K. Mohanty
(593c) Secondary Fermentation of Corn Ethanol Co-Products for Improved Amino Acid Qualities 556
Tanner Barnharst, Yanmei Zhang, Jingyu Wang, Bo Hu
(593d) Extraction High-Value Chemicals from Ethanol Co-Products: A Feasibility Assay on Phytate Extraction with Life Cycle and Techno Economic Assessment 557
Cristiano Reis, Aravindan Rajendran, Douglas Tiffany, Bo Hu
(604a) Highly Stable Perovskite Solar Cells Fabricated Using Aerosol-Based Technique 560
Shalinee Kavadiya, Pratim Biswas

(604b) Engineering Perovskite Solar Cell Interfaces to Realize > 1000 Hr, Unencapsulated Ambient Stability 561
Jeffrey A. Christians, Philip Schulz, Jonathan Tinkham, Tracy H. Schloemer, Bertrand Tremolet De Villers, Steve Harvey, Alan Sellinger, Joseph J. Berry, Joseph M. Luther
(604c) A Thermodynamic Basis for Engineering Enhanced Stability of the Perovskite Phase of Cesium Lead Iodide 562
Subham Dastidar, Aaron T. Fafarman
(604d) Synthesis of Nanostructured Inorganic Perovskites for Solar Cell Applications 563
Atefe Hadi, Rainie D. Nelson, Iver J. Cleveland, Jeremy M. Jacoby, Matthew G. Panthani
(604e) Discovery of Near-Infrared-Active Colloidal Multinary Lead Halide Perovskite Nanocrystals Using a Microfluidic Platform 564
Ioannis Lignos, Viktoriia Morad, Richard Maceiczyk, Loredana Protesescu, Maksym V. Kovalenko, Andrew J. Demello
(604f) Improved Charge Collection in Highly Efficient CsPbBrI2 Solar Cells with Light-Induced Dealloying 565
Joshua Choi
(604g) Synthesis and Characterization of Bi-Based Perovskite Semiconductors for Photovoltaic Application 566
Umar H. Hamdeh, Rainie D. Nelson, Bradley J Ryan, Ujjal Bhattacharjee, Jacob W. Petrich, Matthew G. Panthani
(617a) In-Situ Synthesis of Intergrown Uio-66 Membranes with Controlled Orientation 567
Bohan Shan, Bin Mu
(617b) Closed Packed, Oriented MOF Thin Films through Solution Shearing 568
Arian Ghorbanpour, Luke Huelsenbeck, Gaurav Giri
(617c) Mechanistic Insights into Low Temperature Ceramic Thin Film Growth and Crystallization Using Microwave Radiation 569
B. Reeja Jayan, Nathan Nakamura
(617d) Gold Nanoparticle Monolayers for Surface-Enhanced Raman Spectroscopy of Lithium Batteries 570
Daniel T. Hallinan Jr., Guang Yang, Jagjit Nanda, Boya Wang, Gang Chen
(617e) Understanding the Chemistry of Thiol-Amine Solutions: Versatile Solvents for Solution-Processed Thin Film Photovoltaics 571
Caleb Miskin, Priya Murria, Laurance Cain, Robert W. Boyne, Evan C. Wegener, Jeffrey T. Miller, Hilkka Kenttamaa, Rakesh Agrawal
(617f) In situ Grazing Incidence Small Angle X-Ray Scattering (GISAXS) Study of the Formation of Multilayered Ordered Mesoporous Titania Films 572
M. Arif Khan, Syed Z. Islam, Suraj Nagpure, Barbara L. Knutson, Stephen E. Rankin
(617g) 2-D Wulff Construction of FeOx Islands Grown on Pt(111) for Use in Catalysis 573
Joseph Kubal, Jeffrey Greeley
(621a) Molecular Simulation of Crystallization of Chain Molecules from the Melt 574
Gregory C. Rutledge
(621b) Anomalous Crystallization Behavior of Ring Polymers 575
Kiran S. Iyer, Murugappan Muthukumar
(621c) Thin Film Crystallization of Cyclic Polymers and Their Linear Analogues 576
Julie Albert, Giovanni Kelly, Scott Grayson, Farihah Haque
(621d) Synthesis, Characterization, and Structural Evolution of Designer Block Polyelectrolyte Complexes 577
Jeffrey Ting, Hao Wu, Abraham Herzog-Arbeitman, Samanvaya Srivastava, Matthew V. Tirrell
(621f) Solid and Liquid Core Polyelectrolyte Complex Micelles 578
Lorraine Leon
(621i) Uniqure Crystallization Behavior of Isotactic Polypropylene in the Presence of L-Isoleucine and Its Inhibition and Promotion Mechanism of Nucleation 579
Shicheng Zhao, Xiaoshan Peng
(621h) Scalable Nanocomposites Synthesis Via Electrospray-Mediated Electroemulsification and Flash Nanoprecipitation 580
Kil Ho Lee, Barbara E. Wyslouzil, Jessica O. Winter
(622a) Engineering Vapor-Deposited Polymers for Energy Conversion and Storage 581
Kenneth Lau
(622b) Charge Transfer Mechanisms in Organic Radical Polymer Batteries 582
Shaoyang Wang, Fei Li, Jodie L. Lutkenhaus
(622c) Free Volume Enhanced Anion Exchange Membranes from Triptycene Poly(Arylene Ether Sulfone) Copolymers 583
Yoonseob Kim, Timothy Swager

(622d) Surfactant-Polymer System Optimization in Heterogeneous Model with Mobility Control 584
Nai Cao, Pingchuan Dong, Brian McPherson, Xiaoxiao Liu
(622e) Breaking the Compensation Effect within the Vogel-Tammann-Fulcher Equation for Polymer-Based Electrolytes 590
Kyle M. Diederichsen, Hilda G. Buss, Bryan D. McCloskey
(622f) Carbon Derived from Polymerized Ionic Liquids 591
Rui Sun, Kelly M. Meek, Yossef A. Elabd
(622g) Tailoring Surface Functionalization of Silica Nanoparticles in Nafion Nanocomposites for Improved Ion Selectively in Vanadium Redox Flow Batteries 592
Allison Jansto, Eric M. Davis
(622h) Thermally Cross-Linked Poly(acrylic acid) / Reduced-Graphene Oxide Aerogels As a Replacement for Metal-Foil Current Collectors in Lithium Ion Batteries 593
Han Xiao, Joshua Pender, Mackenzie Meece-Rayle, Pedro De Souza, Kyle Klavetter, Heonjoo Ha, Jie Lin, Adam Heller, Christopher J. Ellison, C. Buddie Mullins
(622i) Graphene-Oxide/Polybenzimidazole Nanocomposite Membrane for High Temperature Fuel Cell Application 605
Shobha Mantripragada, Md. Tashfin Zayed Hossain, Khondker Sultana, Shamsuddin Ilias, Jianzhong Lou
(640a) Autoperforation of 2D Materials for Generating Two Terminal Memresistor Janus Particles with Nonvolatile Memory 606
Pingwei Liu, Albert Tianxiang Liu, Daichi Kozawa, Juyao Dong, Max Saccone, Volodymyr Koman, Song Wang, Minhao Wong, Michael Strano
(640b) Synthesis and Properties of Polymer/Graphene Oxide (GO) Thermosets with Multifunctional GO As a Crosslinker 607
Heonjoo Ha, Jaesung Park, Kiryong Ha, Benny D. Freeman, Christopher J. Ellison
(640c) Polymer Silica Composite Nanofibers Via Sol Gel Electrospinning 608
Tahira Pirzada, Sara A. Arvidson, Carl D. Saquing, S. Sakhawat Shah, Saad A. Khan
(640d) Graphene Oxide/Lipid Composite Material Towards a Multifunctional Drug Delivery Vehicle 609
Mohammad Shahadat Hussain Sarkar, Md. Alamin Miraz, Ashiqur Rahman, Yang Lu, Vu Phan, Clayton S Jeffryes, Evan K. Wujcik
(640e) Localizing Graphene at the Interface of HDPE/PLA Polymer Blends 610
Sung Cik Mun, Min Jae Kim, Liangliang Gu, Monica Cobos, Christopher W. Macosko
(640f) Cocontinuous Ternary Polymer Nanocomposites with Interfacial Graphene Nanoplatelets 618
Lian Bai, Radhika Sharma, Catherine Esnaashari, Christopher W. Macosko, Xiang Cheng
(640g) Investigating Catalytic Properties of ReSe2 Edge and Basal Plane for Hydrogen Evolution Reaction 619
Zhenjing Liu, Zhengtang Luo
(647a) Peg Hydrogels with Tunable Biodegradation Rate for Sustained Delivery of Platelet-Rich Plasma for Treatment of Osteoarthritis 626
Era Jain, Saahil Sheth, Nobuaki Chinzei, Natasha Case, Linda Sandell, Scott A. Sell, Muhammad Rai, Silviya Petrova Zustiak
(647b) Control over the Temporal Profile and Sequence of Anticancer Therapeutics from Magnetically Responsive Hydrogels 627
Tania Emi, Tanner Barnes, Emma Orton, Anne Reisch, Zachary Silveira, Miranda Mitchell, Celia Dunn, Anita E. Tolouei, Stephen Kennedy
(647c) Electrospun Gelatin Nanofibers As Carrier for Controlled and Sustained Release of a Hydrophobic Drug 628
Anindita Laha, Chandra Sharma, Saptarshi Majumdar
(647d) Engineering of Degradable Biopolymer Films Loaded with Imiquimod for Controlled Released in a Mucosal Environment 629
Lucas Garcia Camargo, Gabriela Souza Rezende, Stephany Di Carla Santos, Michelle Franz Montan Braga Leite, Renata Nishida Goto, Andreia Machado Leopoldino, Angela Maria Moraes
(647e) Microparticles for the Delivery of Anti-Diabetic Drugs to the Adipose Tissue 630
Christopher Isely, Michael Hendley, Kendall Murphy, Prakasam Annamalai, Michael Gower
(647f) Template Assisted Micro-Patterned Electrospun Nanofibrous Mat As a Potential Carrier for Controlled Drug Release 631
Manohar Kakunuri, Mudrika Khandelwal, Chandra S. Sharma, Stephen Eichhorn
(647g) PÎ•G-Based Nanocomposite Hydrogels: Thermoresponsive Sol-Gel Transitions and Decomposition Rates Regulated By the LÎ'/GA Ratio of PLGA-PÎ•G-PLGA 632
Midori Kitagawa, Tomoki Maeda, Atsushi Hotta
(648b) Electrospun Collagen Scaffold for Peripheric Nerve Regeneration 633
Carol Rivera Martinez, Janet Mendez, Jorge Almodovar
(648c) 3D Printing of Nerve Guidance Channels for Peripheral Nerve Repair 634
Wei Wu

(648d) Combined Physical and Biochemical Cues Direct the Growth of Inner Ear Neurites 635
Braden Leigh, Kristy Troung, Reid Bartholomew, Marlan Hansen, C. Allan Guymon
(648e) Harnessing Multi-Functional Microbial Cells for Designing Sweat-Responsive Bio-Hybrid Wearables 636
Wen Wang, Lining Yao, Chin-Yi Cheng, Teng Zhang, Hiroshi Atsumi, Luda Wang, Guanyun Wang, Oksana Anilionyte, Helene Steiner, Jifei Ou, Kang Zhou, Chris Wawrousek, Katherine Petrecca, Angela M. Belcher, Rohit Karnik, Daniel I. C. Wang, Xuanhe Zhao, Hiroshi Ishii
(648f) Development of Silk Protein Conjugates for Mucoadhesive Applications 637
Danielle L. Heichel, Kelly A. Burke
(648g) Laser-Activated Sealants for Skin Tissue Repair 638
Russell Urie, Deepanjan Ghosh, Mitzi Thelakkaden, Chengchen Guo, Jeff Yarger, Jacquelyn Kilbourne, Kaushal Rege
(648h) Layer-By-Layer Assembled Thin Film Biomaterials As Porous Biomolecular Delivery Systems 639
Adeline Gand, Mathilde Hindie, Michel Boissiere, Emmanuel Pauthe, Paul R. Van Tassel
(680a) Molecular Simulations of the Influence of Interfaces on the Dynamics of Polymers 640
Robert A Riggleman
(680b) Princess and the Pea Behavior in Polyelectrolyte Multilayers: Influence of the First Layer on Polyelectrolyte Multilayer Assembly and Properties 641
Xuejian Lyu, Amy M. Peterson
(680c) Instability of Liquid Crystal Thin Film on Topographically Patterned Substrates 642
Palash Dhara, Rabibrata Mukherjee
(680d) Electroless Deposition of Copper on Quaternized Chitosan Coatings for Antibacterial Application 643
Debirupa Mitra, En-Tang Kang, Ramanathan Kollengode, Matthew Cove, Koon Gee Neoh
(680e) Patterning Buckles in Polymer/Metal Thin Films By Laser Irradiation 644
Kunal Mondal, Ying Liu, Michael D. Dickey, Jan Genzer
(680f) High Performance Conducting Polymer Coatings for Corrosion Protection 645
Xinyu Zhang, Amit Nautiyal, Jonathan Cook
(680g) Enhanced Wetting Stability of Initiated Chemical Vapor Deposited (iCVD) Polydivinylbenzene Thin Films By Thermal Annealing 646
Junjie Zhao, Minghui Wang, Karen K. Gleason
(680h) Nanoscale Characterization of Water Penetration through Plasma Polymerized Coatings and Water at the Coating/Substrate Interface 647
Yang Zhou, Ali Dhinojwala, Mark Foster
(680i) Surface Topology and Modulus Effects on Adhesion in Novel Polyorganosilozane-Based Coatings 648
Ethan D. Smith, Stephen M. Martin, Vince Baranauskas
(687a) Templated Synthesis of Polymer - Gold Nanocomposites with Pluronic Gels 649
Srikanth Nayak, Surya Mallapragada, Wenjie Wang, David Vaknin
(687c) Predicting Surface Area in Green Synthesis of Sol-Gel Materials 650
Brian K. Peterson, Mobae Afeworki, David C. Calabro, Quanchang Li, Simon C. Weston
(687d) Leveraging Nanoparticle Template Assembly and Interfacial Phenomenon for Multi-Scale Control over Polyimide-Derived Carbon Molecular Sieve Films 651
Megha Sharma, Mark A. Snyder
(687f) Toward Rational Design of Hierarchical Beta Zeolites Via Cost-Effective Approaches 652
Ke Zhang, Sergio Fernandez, Michele L. Ostraat
(687g) Nanoscale Control of Homoepitaxial Growth on a Two-Dimensional Zeolite 653
Meera Shete, Manjesh Kumar, Donghun Kim, Neel Rangnekar, Dandan Xu, Berna Topuz, Kumar Varoon Agrawal, Evguenia Karapetrova, Benjamin Stottrup, Shaeel Al-Thabaiti, Sulaiman N. Basahel, Narasimharao Katabathini, Jeffrey Rimer, Michael Tsapatsis
(689a) Thermodynamics of Charging in Weak Polyelectrolytes 654
Jonathan K. Whitmer
(689b) Thermodynamics and Kinetics of Ordered, Stronly Segregated Diblock Copolymers 655
Ronald M. Lewis III
(689c) Effect of pH on the Interaction between Poly(vinyl alcohol) and Ice 656
Aaron A. Burkey, Nathaniel A. Lynd
(689d) Self Consistent Field Theory Study of Multivalent Cation Effect on Semiflexible End-Grafted Random Polyelectrolytes 657
Merina Jahan, Mark J. Uline
(689e) The Sequence-Dependence of the Persistence Length of DNA 658
Hui-Min Chuang, Jeffrey G. Reifenberger, Han Cao, Kevin D. Dorfman

(689f) Anomalous Hydrodynamic Radius of Polyethylene Glycol Molecules in Mixed Solvents Containing a Hydrotope 659
Xiong Zheng, Mikhail A. Anisimov, Jan V. Sengers, Mao-Gang He
(689g) Energy Saturation: An Alternative Mechanism for the Glass Transition 669
Isaac C. Sanchez, Sean O'Keefe
(689h) Molecular Dynamics Simulation of Polymerized Stockmayer Fluids: Effects of Chain Length and Connectivity on Saturated Dipoles Near Ions 670
Issei Nakamura, Lijun Liu
(689i) Probing Polymer Blend Phase Diagrams Via Oligomer Molecular Simulations 675
Qile Chen, Shuyi Xie, Timothy P. Lodge, J. Ilja Siepmann
(696a) Silica-Peg Gel Encapsulation Platform for Isolation of Dormant Cancer Cells 676
Julian Preciado, Hak Rae Lee, Emil Lou, Alptekin Aksan, Samira M. Azarin
(696b) Biodegradable Nano-Film Coated Self-Floating Hollow Glass Microspheres for Rapid Cell Isolation and Recovery 677
Ziye Dong, Caroline Ahrens, Dan Yu, Zhenya Ding, Hyuntaek Lim, Wei Li
(696c) Design of Electrohydrodynamic Sprayed Polyethylene Glycol Hydrogel Microspheres for Cell Encapsulation 678
Anisa Qayyum, Era Jain, Grant Kolar, Scott A. Sell, Silviya Petrova Zustiak
(696d) Quantum Dot-Based Biomarkers of Neuroinflammation in the Developing Brain 679
Mengying Zhang, Binh Dang, Kate Hildahl, Brittany Bishop, Reyn Aoki, Nicole Thompson, Vincent C. Holmberg, Elizabeth Nance
(696e) The Culprit of Gout: Triggering Factors and Formation of Monosodium Urate Monohydrate 680
Tzu-Hsuan Chen, Meng-Hsiu Chih, Hung-Lin Lee, Tu Lee
(696g) Fabrication of Polyethylene Glycol-Based Templated Macroporous Hydrogels for Cell Encapsulation 688
Mozhdeh Imaninezhad, Grant Kolar, Silviya Petrova Zustiak
(709a) Multicomponent Transport Models for Non-Electroneutral Solid Electrolytes 689
Charles W. Monroe
(709b) Diffusion and Sorption Phenomena of Organic Vapor Penetrants in Unmodified and Ethylenediamine Vapor Phase Cross-Linked Matrimid Thin Films 690
John P. Stanford, Peter Pfromm, Mary Rezac
(709c) Segmental Dynamics and Water Transport in Nafion-SiO2 Hybrid Membranes 691
Apoorv Balwani, Antonio Faraone, Eric M. Davis
(709d) Designing Core/Shell Metal Organic Framework/Polymer Films As Scalable Barrier Layers for Enhanced Protection on Photovoltaics 692
Fen Qiu, Zhuonan Song, Jeffrey Urban
(709e) PEO-Based Semi-Interpenetrating Polymer Networks (S-IPNs) for CO2-Selective Membranes 693
Gregory Kline, Qinnan Zhang, Jennifer Weidman, Ruilan Guo
(709f) An Experimental and Triple-Mode Sorption Modeling of Sorption and Diffusion in Polymers 694
Hom Sharma, Stephen Harley, Yunwei Sun, Elizabeth Glascoe
(709g) In situ Monitoring of Emergent Transport in Polymer Membranes 695
Daniel J. Miller, Breanna Dobyns, Bryan S. Beckingham
(709h) The Role of Chlorine Substituent in Gas Transport Properties of Polychlorotrifluoroethene (PCTFE) 696
Milad Yavari, Yoshi Okamoto, Haiqing Lin
(709i) Thermodynamic Model for Predicting Swelling of Poly(N-isopropyl acrylamide) Hydrogels in Solvent Mixtures 697
Sheik Tanveer, Fazle Hussain, Chau-Chyun Chen
(721a) Nanostructure-Driven Fatigue Resistance and Dynamic Recovery in Thermoplastic Elastomer Hydrogel Networks 708
Travis S. Bailey
(721b) Interfacial Crystallization of Polyolefins: An Improved Outlook for Polymer Blends 711
Alex M. Jordan, Kyungtae Kim, Frank S. Bates, Shaffiq Jaffer, Olivier Lhost, Christopher W. Macosko
(721c) Ultra-Stable Amorphous Teflon: Extreme Fictive Temperature Reduction As a Means to Probe Sub-Tg Dynamics 712
Gregory B. McKenna, Heedong Yoon, Yung P. Koh, Sindee L. Simon
(721d) Influence of Strucure and Dynamics in Matrix-Free Polymer-Grafted Nanocomposite Memebranes 713
Eileen Buenning, Sanat K. Kumar, Christopher J. Durning, Connor Bilchak, Brian C. Benicewicz, Dimitris Vlassopoulos

(721e) Role of Interfacial Adhesion in Rate-Dependent Deformation and Failure of Model Ring Opening Metathesis Polymer (ROMP) Filled Composites 714
Erich Bain, Daniel B. Knorr Jr., Joseph L. Lenhart
(721f) Mechanically Ductile and Stiff, Triazole-Based Glassy Photopolymer Network 715
Han Byul Song, Austin Baranek, Christopher N. Bowman
(721g) Controlled Topology Toughning Epoxy Via Incorporation of Partially Reacted Substructures 716
Jian Gao
(721h) Thermoplastic Polydimethylsiloxane Realized By Hydrogen-Bond Networks through L-Phenylalanine Terminals 717
Shunsuke Tazawa, Atsushi Shimojima, Tomoki Maeda, Atsushi Hotta
(721i) In vitro and silico Characterization of Grafted Hydrophobic Brush Membranes 718
John J. Keating, Mirco Sorci, Angelo Setaro, Patrick T. Underhill, Georges Belfort
(725a) Direct Synthesis and Morphology Control of Metal Organic Framework Nanosheets 719
Feng Xue, Prashant Kumar, Wenqian Xu, Michael Tsapatsis
(725b) Mixed-Linker MOF Synthesis and Remediation of Acid-Gas Degradation Using Linker-Exchange Techniques 720
Krishna Chandran Jayachandrababu, Souryadeep Bhattacharyya, David Sholl, Sankar Nair
(725c) Rapid Microwave-Assisted Synthesis of Hybrid Zeolitic-Imidazolate Frameworks with Mixed Metals and Mixed Linkers 721
Febrian Hillman, John Zimmerman, Seung-Min Paek, Mohamad Hamid, Woo Taik Lim, Hae-Kwon Jeong
(725d) Micro-, Meso-, and Macro-Scale Defects in Porous Organic Cages 722
Guanghui Zhu, Christopher W. Jones, Ryan P. Lively
(725e) Computational Studies of the Enhanced Acidity of Defect MOF 808: The Effect of Activation Process 723
Carolina Ardila-Suarez, Saul Perez Beltran, Gustavo Ramirez-Caballero, Perla B. Balbuena
(725f) Crystallization Process Development of Metal-Organic Frameworks By Linking Secondary Building Units, Lattice Nucleation and Luminescence: The Insight into Reproducibility 724
Tu Lee, Yun Hsuan Chang, Hung-Lin Lee
(725g) Fast Mechano-Chemical Synthesis N-Doped Uio-66 with Dopamine to Enhance Chlorobenzene Competitive Adsorption Under Humid Air 725
Zhenxia Zhao, Peng Hu, Zhongxing Zhao
(725h) High Yield Synthesis of ZIF-8 Nanoparticles Using Stoichiometric Reactants in a Jet-Mixing Reactor 726
Aamena Parulkar, Pinaki Ranadive, Nicholas Brunelli
(726a) Polymer Semiflexibility Induces Non-Universal Phase Transitions in Block Copolymers 727
Shifan Mao, Quinn Macpherson, Andrew J Spakowitz
(726b) A Strongly Coarse-Grained, Charge-Fluctuating Model for Polyelectrolytes 728
Nicholas Jackson, Marcel Langenberg, Marcus Muller, Juan J. De Pablo
(726c) Development of New Coarse-Grained Water Models Using Particle Swarm Optimization 729
Karteek K. Bejagam, Samrendra Singh, Yaxin An, Carter Berry, Sanket A. Deshmukh
(726e) Unified Polymer Erosion Model 730
Joel Coffel, Eric Nuxoll
(726f) Development of a Fused-Sphere SAFT-y Mie Force Field for Polymers and Application to Poly(vinyl butyral) Adsorption to Silica 731
Christopher Walker, Erik E. Santiso, Jan Genzer
(726g) Effects of Coarse-Graining on Simulations of Mechanical Properties of Polymers 732
Ting Ge, Mark Robbins
(726h) New Computational Methods for Rapid Simulation of Hydrodynamic Interactions in Polymer Solutions 739
James Swan, Andrew Fiore
(726i) Coarse-Grained Molecular Dynamics Simulations of PNIPAM Grafted Graphene Systems in an Aqueous Environment 740
Carter Berry, Karteek K. Bejagam, Sanket A. Deshmukh
(735a) Gel Assemblies of Colloidal Nanocrystals 741
Camila Saez Cabezas, Beth A. Lindquist, Ryan B. Jadrich, Thomas M. Truskett, Delia J. Milliron
(735b) Examining the Optical Effects of Chiral Carboxylic Acids Bound to the Surface of CdSe Nanoparticles 742
Mayank Puri, Vivian E. Ferry
(735c) Radial Elemental Distribution Analysis of Spherical Core/Shell Nanocrystals with STEM/EDX 743
Jacob Held, Katharine I. Hunter, Uwe R. Kortshagen, K. Andre Mkhoyan

(735d) Directional Carrier Transfer in Strongly Coupled Binary Nanocrystal Superlattice Films Formed By Assembly and in situ Ligand Exchange at a Liquid-Air Interface 744
Yaoting Wu, Siming Li, Natalie Gogotsi, Tianshuo Zhao, Blaise Fleury, Cherie R. Kagan, Christopher B. Murray, Jason B. Baxter

(735e) The Effect of CdS Shell Thickness on the Complex Index of Refraction of CdSe/CdS Core/Shell Nanocrystal Films 745
Dana Dement, Mayank Puri, Vivian E. Ferry

(735f) Engineering the Light-Matter Interactions of Ultrasmall CdSe Quantum Dots Via Modification of Surface Species for Light Harvesting Applications 746
Megan Webster, Kristi Pepa, Kevin Dominguez, Vinod Menon, Gustavo Lopez, Marco J. Castaldi, Ilona Kretzschmar

(740a) Computational Prediction and Evolutionary Design of Polymer Glass-Formation Behavior 747
David S. Simmons, Jui-Hsiang Hung, Venkatesh Meenakshisundaram, Tarak Kumar Patra

(740b) Domain Spacing and Phase Behavior of Salt-Doped Block Copolymers from Fluids Density Functional Theory 748
Jonathan R. Brown, Lisa M. Hall

(740c) Atomistic Simulations of Lamellae-Forming PS-b-P2VP 749
Yamil J. Colon, Weiwei Chu, Juan J. De Pablo

(740d) Rapid Conformational Fluctuations in a Model of Methylcellulose 750
Xiaolan Li, Frank S. Bates, Kevin D. Dorfman

(740e) Multi-Phase Coarse-Grained Models of Rod-like Polymers from Iterative Boltzmann Inversion 751
Christian Nowak, Fernando Escobedo

(740f) Coarse-Grained Model of Exciton Dynamics on Long-Chain Conjugated Polymer System 752
Elizabeth M. Y. Lee, William A. Tisdale, Adam P. Willard

(740g) Molecular Simulation of Thermoplastic Polyurethanes Under Large Mechanical Deformation 753
Shuze Zhu, Gregory C. Rutledge

(740h) Harnessing Virtual High-Throughput Screening and Machine Learning for the Discovery of Novel High Refractive Index Polymers 754
Johannes Hachmann

(740i) Development of Reaction Ensemble Monte Carlo (REMC) Algorithms to Study the Kinetics of Polymerization 755
Inderdip Shere, Ateeque Malani

(741a) Controlled Liquid-Liquid Phase Separation of Recombinant Oleosin 756
Ellen H. Reed, Daniel A. Hammer

(741c) Evaluation of the Procedure to Obtain Chitosan Based Gels with Potential Use As Bone Adhesive on Clinical Settings 757
Jairo A Jimenez, Juanita Echeverri, Paula A Sarmiento, J. German Vargas, Juan Carlos Briceno Triana, Camila Castro, Juan Pablo Casas, Felipe Salcedo

(741d) Deep Space Drug Shielding 760
Hannah Kim, Manosi Roy, Sutapa Barua

(741e) Synthesis of Designer Lipids Using "Click" Chemistries 761
Danielle Konetski, Dawei Zhang, Austin Baranek, Tao Gong, Brady Worrell, Christopher N. Bowman

(741f) Loading and Mobility of RNA in Porous Silica Nanoparticles for Delivery to Insects 762
Shanshan Zhou, Emily Nadeau, M. Arif Khan, Bruce Webb, Stephen E. Rankin, Barbara L. Knutson

(741g) Engineering and Functionalizing Protein-Based Materials 763
Sarah Bondos

(742a) Synthesis and Applications of Bio-Inspired Oligotea Peptidomimetics 764
Christopher A. Alabi

(742b) A Polyester Based Photoluminescent Hydrogel As Extracellular Matrix Mimics 765
Xiaoyang Xu

(742c) Alginate Nanoparticle Platform for Controlled Release of Biotherapeutics 766
Julia Vela Ramirez, Daniela Barrios Santos, Nicholas A. Peppas

(742f) Complex Bone Regeneration Via Controlled Release of Simple Signaling Molecules 767
Soheila Aliakbarighavimi, Brittany Allen, Jessica Stromsdorfer, Jake Kramer, Ram Rao Tata, Andrew Greenwald, Bret Ulery

(742g) Biomimetic Growth of a Pathologic Biomineral in Hydrogels 768
Gopichand Mallam, Marina Tsianou

(757a) The Influence of Intrinsic Framework Flexibility on Adsorption in Nanoporous Materials 769
Matthew Witman, Sanliang Ling, Sudi Jawahery, Peter Boyd, Maciej Haranczyk, Ben Slater, Berend Smit

(757b) Towards a Generalized Understanding of Acid Gas Interactions with ZIF Materials 770
Souryadeep Bhattacharyya, Rebecca Han, David S. Sholl, Sankar Nair

(757c) Heat-Treatment of Defective Uio-66 from Modulated Synthesis:Adsorption and Stability Studies 771
Yang Jiao, Yang Liu, Krista S. Walton, David S. Sholl
(757d) Design of Stratified Hybrid Metal Organic Frameworks for Chemical Detection and Destruction 772
Jonathan Ruffley, Tianyi Luo, Isabella Goodenough, Melissandre Richard, Eric Borguet, Nathaniel L. Rosi, J. Karl Johnson
(757e) Understanding Structure, Metal Distribution, and Water Adsorption in Mixed-Metal MOF-74 773
Joshua D. Howe, Cody R. Morelock, Yang Jiao, Karena W. Chapman, Krista S. Walton, David S. Sholl
(757f) Extraction of Rare Earth Elements from Geothermal Brine Solution Using Magneric Core Shell Microspheres 774
Praveen K. Thallapally, Nune Satish, Jian Liu, B. Peter McGrail
(757g) Cycloaddition Reaction of Epoxides with CO2 in a Copper Metal-Organic Framework: A Density-Functional Theory Study 775
Xu Li, Jianwen Jiang
(758a) Nanostructured Materials for Separations Based on Reactive Block Polymers 776
Marc A. Hillmyer
(758b) Block Copolymer Derived Nanostructured Surfaces: Templating Confined Surface Reactions 777
Katherine P. Barteau, Katharine W. Oleske, Ulrich Wiesner, Lara A. Estroff
(758c) Nanostructure Formation on Collapse of Polyelectrolyte Brushes 778
Blair Kathryn Brettmann
(758d) iCVD Deposition and Integration of Poly-(1H,1H,2H,2H-Perfluorodecylacrylate) (PPFDA) Under High Loading of TiO2 Nanoparticles 779
Zhengtao Chen, Kenneth Lau
(758e) Development of Resin-Containing Polymer Particles for Thermoset Powder Coatings 780
Guozhen Yang, Mengfei Huang, John Klier, Jessica D. Schiffman
(758f) Roll-to-Roll Micromolding of UV Curable Thiol-Ene Based Coatings 781
Yuyang Du, Alon McCormick, Lorraine F. Francis
(758g) Elastic Networks for Shape-Memory Contact Printing 782
Mitchell Anthamatten
(758h) Dynamics of Electric Double Layer Formation and Dissipation in Polyethylene Oxide:LiClO4 on Graphene Transistors 783
Susan Fullerton, Hua-Min Li, Ke Xu, Buchanan Bourdon, Hao Lu, Yu-Chuan Lin, Joshua Robinson, Alan Seabaugh
(758i) Redox-Active Organometallic Polymers for Small Molecule Separations 784
Xiao Su, Timothy Jamison, T. Alan Hatton
(765a) Disorder, Nonequilibrium Transport, and the Origin of Deep Traps in Quantum Dot Solids 785
William A. Tisdale
(765b) On the Molecular Origin of Intra-Gap Emission from CuInSe2-XSx Quantum Dots 786
Addis Fuhr, Nikolay Makarov, Hunter McDaniel, Hyeong Jin Yun, Hongbo Li, Anastassia Alexandrova, Phillippe Sautet, Victor I. Klimov
(765c) Accelerating Emission Dynamics in Perovskites Plasmonic Nanolasers 787
Sui Yang, Wei Bao, Xiaoze Liu, Xiang Zhang
(765d) Synthesis and Characterization of Plasmon-Resonant Hollow Gold Nanoshells 788
Joesph A. Zasadzinski, Jeongeun Shin
(765e) Solution-Based Synthesis of Cesium Bismuth Halide Perovskite Nanocrystals for Optoelectronic Applications 789
Rainie D. Nelson, Atefe Hadi, Alex Verburg, Matthew G. Panthani
(765f) Integrated Logic Gate Devices Fabricated Using Non-Toxic CuInSe2 Quantum Dots 790
Hyeong Jin Yun, Jaehoon Lim, Jeongkyun Roh, Darren Chi Jin Neo, Matt Law, Jeffrey M. Pietryga, Victor I. Klimov
(765g) Synthesis and Surface Functionalization of Group IV Quantum Dots 791
Yujie Wang, Michael Zembrzuski, Matthew G. Panthani
(766a) Understanding Rate Dependent Mechanical Properties of Supramolecular Hydrogels through Real Time SAXS Measurements during Stretching 792
Bryan D. Vogt
(766b) Microstructure and Mechano-Electrical Property of Ultra-Stretchable Iono-Elastomer Via Small Angle Scattering and Rheology 793
Ru Chen, Carlos R. Lopez-Barron, Norman J. Wagner
(766c) Toughening Isotactic Polypropylene with Block Copolymer Micelles 796
Jun Xu, Micah J. Howard, Frank S. Bates
(766d) The Importance of Crystalline Structure on the Tensile Properties of Uhmwpe 797
Nicolas J. Alvarez, Christopher Henry, Giuseppe Palmese

(766e) Controlling Polymerization Induced Phase Separation (PIPS) Using the Nonlinear Optical Properties of Light 798
Ian Hosein, Saeid Biria
(766f) Engineering Polymer Structure and Dispersity to Access Designer Materials with Exquisite Properties 799
Jimmy Lawrence, Eisuke Goto, Bernd Oschmann, Dongsub Kim, Jing Ming Ren, Craig J. Hawker
(766g) Thermal Bridges for Phonon Transport through Short Polymer Chains in a Hydrogen Bonded Polymer Composite 800
Nitin Mehra, Liwen Mu, Jiahua Zhu
(766h) The Effect of Pendant Alkyl Chain Length in Epoxy/Amine Thermosets Materiel Properties and Water Barrier Properties 801
John Vergara, Joshua Sadler, John La Scala, Santosh Kumar Yadav, Giuseppe Palmese
(766i) Novel Chemistries for the Replacement of Methylenedianiline in Composites 802
Joseph F. Stanzione III, Owen M. Stecca, Alexander W. Bassett, Jayson D. Cosgrove, Kevin M. Schmalbach, Joshua M. Sadler, John J. La Scala
(118a) Improvement of Thermal Conductivity for POSS Functionalized BN Fillers/Polyphenylene Sulfide Composites 810
Xutong Yang, Lin Tang, Yongqiang Guo, Junwei Gu
(118i) Structural Composites with Multiple Functionalities 811
Kenan Song
(118c) Fabrication of High-Dispersed Strawberry-like P(St-co-MAA)/SiO2 Composite Microspheres and Their Applications in Mortar 812
Guanzhi Cheng, Huajian Li, Xinguo Zheng, Yongjian Xie, Jing Liu
(118d) Hydrogen Bonded Thermal Highways Driving Heat Conduction in Polymer and Polymer Blends Films 813
Nitin Mehra, Jiahua Zhu
(118e) Low-Temperature Synthesis of Mn-Based Mixed Metal Oxides with Novel Fluffy Structures As Efficient Catalysts for Fto Reaction 814
Yi-Fan Han, Bo Meng
(118f) Structural Control of Polybenzoxazine/Epoxy Composites with Dual Crosslinking Network for Corrosion Protection 815
Changlu Zhou, Zhong Xin
(118g) Multifunctional Epoxy Conductive Nanocomposites 816
Xiaojiang Xu, Hongbo Gu
(118h) Nano-Structured Ceramic ALD Coatings to Stabilize SiC Against Reaction in High Temperature Steam 817
Amanda Hoskins, Aidan Coffey, Charles B. Musgrave, Alan W. Weimer
(118j) Multi-Scale Metrology for Visualization and Characterization of Interphase Failure 818
Richard Sheridan, Jeremiah Woodcock, Jeffrey W. Gilman, Gale Holmes, Catherine Brinson, Vamshi Gudapati, Dave Hartman, Amol Vaidya
(769a) Strategic Assemblies of Wood-Derived Building Blocks for the Sustainable Redesign of BPA-Based Polymers 819
Joseph F. Stanzione III, John J. La Scala
(769b) Preparation and Characterization of Cardanol Based Vinyl Ester Resins As Cross-Linker Units 832
Emre Kinaci
(769c) Ionic Liquids: Green Solvents for Dry Native Cellulose and Chitosan 833
Behzad Nazari, Nyalaliska Utomo, Sujyot Mony, Hyeonji Oh, Indira Saifuddin, Ralph H. Colby
(769d) Environmentally Friendly Flame Retardants Based on Adhesive Catecholamine 834
Hanim Kim, Joon Hee Cho, Kadhiravan Shanmuganathan, Amanda Jones, Sergei Nazarenko, Christopher J. Ellison
(769e) Soybean Oil Based Thermoset Materials with High Biorenewable Content 835
Sung-Soo Kim, Dustin Janes, Kadhiravan Shanmuganathan, Daniel Y. Chou, Christopher J. Ellison
(769g) Self Assembly of Coil-Hyperbranched Poly(styrene-block-acrylated epoxidized soybean oil) Block Copolymers 836
Fang-Yi Lin, Austin D. Hohmann, Nacu Hernandez, Eric W. Cochran
(769h) Physicochemical Properties of Polylactide/Delta-Valerolactone/Organosolv Lignin Atactic Terpolymers 837
Stephanie Harris, Ulrike Tschirner, Adam Gillespie, Madeleine Seeger
(769i) Oil Field Chemicals from Macromolecular Renewable Resources: Date Pit As a Case Study for Drilling Fluid Additive 838
Jimoh K. Adewole, Musa O. Najimu

(770a) Designing Synthetic Extracellular Matrices for the Creation of Controlled Culture Systems in the Study of Disease 859
April M. Kloxin
(770b) The Influence of Matrix Stiffness on the Behavior of Brain Metastatic Breast Cancer Cells 860
Akshay Narkhede, Shreyas Rao
(770c) The Combined Effect of Matrix Microenvironment and Hypoxia on the Activity of Glioblastoma Stemcells 861
Jee-Wei Emily Chen, Jann N. Sarkaria, Brendan A. Harley
(770d) Glioblastoma/Astrocyte Co-Culture on Polyelectrolyte Multilayer Films: A Template for Studying the Role of Astrocytes in Glioblastoma Progression 862
Kimberly M Stanke, Christina Wilson, Erin Eickman, Oleh Khalimonchuk, Srivatsan Kidambi
(770e) Three-Dimensional Tissues Using Human Pluripotent Stem Cell Spheroids As Biofabrication Building Blocks 863
Qiang Li, Haishuang Lin, Yuguo Lei
(770f) Dynamic Culture of Trabecular Meshwork Cells in 3D Biomimetic Scaffolds 880
Matthew Osmond, Mina Pantcheva, Melissa Krebs
(770g) Cellular Hitchhiking on Microparticles to Alleviate Skin Injury 891
Daniel Smith, Chase Herman, Sutapa Barua
(771a) Sugar-Coating the Answers to Virus Binding: Glycocalyx-Mimetic Interfaces 892
Ramya Kumar, Domenic Kratzer, Kenneth Cheng, Irina Kopyeva, Joerg Lahann
(771b) Glucose-Derived Cationic Block Poly(beta-peptides) Reverse Intrinsic Antibiotic Resistance in Gram-Negative Pathogens 893
Zhangyong Si, Hui Wen Lim, Damien Keogh, Moon Tay Yue Feng, Jo Thy Lachumy Subramanion, Yahua Chen, Guillermo C. Bazan, Everett Peter Greenberg, Yunn-Hwen Gan, Kevin Pethe, Mary B. Chan-Park
(771c) Mechanism Study of Selective Killing of Cationic Peptidopolysaccharide Nanoparticles with in vitro and in vivo Efficacy Against Multi-Drug Resistant Bacteria 894
Mary Chan, Hou Zheng, Yogesh Vikhe
(771d) Tether Supported Biomembranes with Phase Composition and Orientation Control for Biomaterial Ligand Displays and Membrane Protein Assays 895
William Houlihan, Yueming Li, Lane Gilchrist
(771e) Understanding the Interaction of the Polysaccharide, Chitosan, with a Novel Cuticular Protein, CPR 27, Using Quartz Crystal Microbalance with Dissipation 896
Aishik Chakraborty, Ellie K. Onstott, Neal T. Dittmer, Michael R. Kanost, Stevin H. Gehrke, Prajnaparamita Dhar
(771f) Probing Phase Transitions in Dynamic Biopolymer Complexation 897
Amanda B. Marciel, Handan Acar, Matthew V. Tirrell
(771g) Photo-Induced Pinocytosis in Synthetic Liposomes 898
Danielle Konetski, Dawei Zhang, Christopher N. Bowman
(771h) Protein Protection and Purification through Virus-like Particle Encapsidation: A Rapid 2-Step Cell-Free Protein Synthesis Approach 899
Bradley C. Bundy, Seung Ook Yang
(775a) Measurement and Modeling of Carrier Dynamics in Photovoltaic Cztse 900
Siming Li, Michael L. Lloyd, Brian E. McCandless, Jason B. Baxter
(775b) Impact of the Active Layer Morphology on Bimolecular Recombination Dynamics in Organic Solar Cells 901
Veaceslav Coropceanu, Jean-Luc Bredas, Shafigh Mehraeen
(775c) Atomistic Origin of the Concentration Dependence of Si Dopant Mobility in III-V Semiconductor Alloys 902
Mardochee Reveil, Paulette Clancy
(775d) Characterizing Defects in Photovoltaic Semiconductors with Optical Spectroscopy 903
Charles J. Hages, Thomas Unold
(775e) CIGS Nanocrystal Solar Cells on Plastics and Paper 904
Vikas Reddy Voggu, Sam Morehead, Brian A. Korgel, Taylor B. Harvey
(775f) Solution Processed Thin Film Photovoltaics Using Amine-Thiol Chemistry 905
Swapnil Dattatray Deshmukh, Xin Zhao, Ruihong Zhang, Caleb Miskin, David Rokke, Carol Handwerker, Rakesh Agrawal
(775g) Photonic Mirrors for Enhanced Optical Transport in Luminescent Solar Concentrators 906
Ryan Connell, Mayank Puri, Vivian E. Ferry
(775h) Dopant Mediated Assembly of Nanorods into Atomically Coupled 2D Sheets in Solution 907
Ajay Singh, Delia J. Milliron
(777a) Tailored Polymers for Structural Application in Large-Scale Extrusion Systems 908
Blake Marshall

(777b) Modeling Flow Phenomena in Fused Filament Fabrication Geometry 909
Eric L. Gilmer, Darren Miller, Jacob Fallon, Camden Chatham, Callie Zawaski, Allison M. Pekkanen, Timothy E. Long, Christopher B. Williams, Michael J. Bortner
(777c) Influence of Processing on Additively Manufactured Mechanically Adaptive Cellulose Nanocrystal Polymer Composites 910
Jacob Fallon, Michael J. Bortner, Earl J. Foster, Cara Herwig, Ben Kolb
(777d) 3D Printed Shape Memory Objects Based on Olefin Ionomer of Zinc-Neutralized Poly(ethylene-co-methacrylic acid) 911
Bryan D. Vogt
(777e) Three-Dimensional Printing By Multiphase Silicone/Water Capillary Inks 912
Sangchul Roh, Dishit Parekh, Bhuvnesh Bharti, Simeon Stoyanov, Orlin D. Velev
(777f) 3D Printing of Polymer-Bonded Magnets Using a Combination of Extrusion Direct Write and Stereolithography Methods 913
Alan Shen, Anson Ma, Sameh Dardona
(777g) PDMS-Based Ink Development for 3D Printing Applications 914
Kwan-Soo Lee, Joseph H. Dumont, Andrew M. Schmalzer, Chihoon Park, Andrea Labouriau
(777h) 3D Printed Active Microfluidic Elements for Portable Bioanalysis Assays 915
Duanduan Han, Victor M. Ugaz
(777i) 3D Printed Polymer-Based Bio-Inspired Neural Systems 916
Blake Johnson
Author Index

Some abstracts in this book refer to images that were not made available by the publisher

499890 New Milestones and Challenges in High-Throughput Computation of Elastic Properties on the Materials Project

Abstract:

Since its inception in 2010, The Materials Project has sought to use best-practices from information technology to make the process of computational simulation of solid compounds more efficient, well-documented, and shareable. To this effect, a high-throughput workflow for determining the second-order elastic tensor from density functional theory calculations was developed in 2014 and initiated with the publication of around 1000 elastic tensors, making it the largest online repository of computational elastic properties of materials. In this work, we report on the progress of our production workflow, refined from the original version, and the expansion of our database to over 4,500 elastic tensors. In addition, we report on the development of computational tools which may be used to compute properties derived from the elastic tensor, including sound velocity, debye temperature, and thermal conductivity. We report on the expansion of our infrastructure to include functionality that may be used to determine third and even fourth-order elastic constants from stress-strain data, and similarly on derived properties from these values like the gruneisen parameter and thermal expansion coefficients. Lastly, we provide an example of how elastic properties might be targeted in high-throughput searches from work in which we explore and identify new materials with negative poisson's ratios, or auxetic materials.

489251 Breaking Badly: A Comprehensive Assessment of Computational Methods for Predicting Tensile Strengths in Bulk Solids

Abstract:

Dispersion interactions play a crucial role in noncovalently-bound molecular systems, and recent studies have shown that dispersion effects are also critical for accurately describing covalently-bound solids. While most studies on bulk solids have solely focused on equilibrium properties (lattice constants, bulk moduli, and cohesive energies), there has been little work on assessing the importance of dispersion effects for solid-state properties far from equilibrium. In this work, we present a detailed analysis of both equilibrium and highly non-equilibrium properties (tensile strengths leading to fracture) of various palladium-hydride systems using representative DFT methods within the LDA, GGA, DFT-D2, DFT-D3, and nonlocal vdw-DFT families. Among the various DFT methods, we surprisingly find that the empirically-constructed DFT-D2 functional gives extremely anomalous and qualitatively-incorrect results for tensile strengths in palladium-hydride bulk solids.[1] We present a detailed analysis of these effects and discuss the ramifications of using these methods for predicting solid-state properties far from equilibrium. Most importantly, we suggest caution in using DFT-D2 (or other coarse-grained parameterizations obtained from DFT-D2) for computing material properties under large stress/strain loads or for evaluating solid-state properties under extreme structural conditions.

[1]N. V. Ilawe, J. A. Zimmerman, and B. M. Wong, *â€œBreaking Badly: DFT-D2 Gives Sizeable Errors for Tensile Strengths in Palladium-Hydride Solids.â€* Journal of Chemical Theory and Computation, **11**, 5426 (2015).

496458 Machine Learning the Thermochemistry of All Inorganic Crystalline Solids

Abstract:

High-throughput, theory-driven computational materials design and discovery has become an essential tool in materials science and engineering. The data-driven and cost-effective framework for accelerated discovery introduced by the Materials Genome Initiative (MGI) has transformed the scale and rate of materials development by exploiting the predictive ability of quantum chemical computational methods. In turn, the drive to rapidly predict, screen, and optimize materials using first-principles calculations has led to large datasets and the construction of open-source databases that are populated primarily by low-cost density functional theory (DFT) total energy calculations (inherently performed at 0 K). Although these datasets are an invaluable resource, their predictive ability at finite temperatures is limited and current methods for evaluating the missing temperature dependencies are computationally prohibitive, limiting the success of MGI approaches for existing and emerging high temperature applications. In this report, we show that compound Gibbs energies of formation can be predicted with accuracy comparable to the quasiharmonic approximation at finite temperatures of up to at least 1800 K, and at the same computational expense as a DFT total energy calculation (e.g., using PBE+U). Implementing our approach on the entirety of open materials databases (e.g., MaterialsProject) reveals interesting trends with respect to the stability (and metastability) of all inorganic crystalline solids. Specifically, we quantify the fraction of compounds in the Inorganic Crystal Structure Database (ICSD) predicted to be thermodynamically stable, quantify the magnitude of metastability for thermodynamically unstable structures, and assess chemistry-specific trends in these predictions.

496724 Designing Dopant Patterns in Indium-Doped Perovskite Oxygen Carriers

Abstract:

Doped perovskites are a highly versatile class of materials that exhibit tunable properties due to the choice and extent of doping in the lattice. In particular, some perovskites have found use as solid oxygen carriers for chemical looping applications [1,2]. These materials have promising properties such as rapid oxidation and reduction, tunable reduction temperature, and stability over a range of conditions [3]. As an example system, we study a $BaFeO_3$ perovskite doped with In for the purpose of rapid, tunable oxygen release in an oxygen carrier system. Since there are combinatorially many ways that dopant can arrange itself in a periodic lattice, it is generally impossible to rigorously to identify particular patterns of dopant that exhibit high performance for a target application. Fortunately, this combinatorial problem can be formulated and solved as a mathematical optimization problem.

The first step of translating the material design problem into a mathematical optimization model is the modelling of the material properties. To predict the reducibility of the doped perovskite, spin-polarized density functional theory (DFT+U) calculations were performed to identify the oxygen excess energy, the energy required to extract a particular oxygen from the lattice [4,5]. However, DFT calculations are computationally intensive and not tractable to be embedded in a mathematical optimization approach. To circumvent this, we propose to use a two-level approach predicated on a simplified structure-function relationship. First, we used DFT to generate a relevant training set of high-quality excess energy data. We then employed standard model identification tools to regress an algebraic model relating oxygen excess energy to some simple geometric descriptors of dopant placement.

The procedure of generating a simplified structure-function relationship was carried out by enumerating the possible conformations of dopant within a small neighborhood of a given lattice oxygen atom. These conformations were then screened to determine which motifs were actually unique with respect to rotational symmetry of the lattice. In this way, we were able to identify 47 conformations to evaluate via DFT that span the range of possible dopant loadings and placements. Geometric descriptors were tabulated using a combination of simple neighbor dopant-counting descriptors as well as descriptors to count the absolute value of the difference in dopant levels across a shared oxygen atom. This set of data points was then provided to ALAMO [6], a model identification tool, to determine the best algebraic model to represent the observed variance in oxygen excess energy with respect to our identified geometric descriptors.

Using the developed structure-function relationship, we were able to predict highly functional dopant patterns for a variety of target applications. To accomplish this, we formulated a mathematical optimization model with the structure-function relationship embedded in the algebraic constraints. Our approach is generic, enabling a variety of choices for objective function and constraints on the feasible design space.

The optimization model was used to generate a slate of designs that are optimal under various dopant loading levels and target values of oxygen excess energy. Our results highlight some distributions of In site reactivity that tend to be expressed when minimizing the overall oxygen excess energy of the lattice. These summary statistics on expressed site functionality can help elucidate the trends observed experimentally and will serve to guide the development of new materials. While the precise placement of dopant in the perovskite lattice is not possible with current experimental methods, these designs serve as a theoretical targets that can motivate further developments in synthesis methods.

References

[1] Adanez J., Abad A., Garcia-Labiano F., Gayan P., Diego L. F. De, "Progress in chemical-looping combustion and reforming technologies," *Progress in Energy and Combustion Science*, 38(2):215–282, 2012.

[2] Rydén M., Lyngfelt A., Mattisson T., Chen D., Holmen A., Bjørgum E., "Novel oxygen-carrier materials for chemical-looping combustion and chemical-looping reforming; LaxSr1-xFeyCo1-yO3-d perovskites and mixed-metal oxides of NiO, Fe2O3 and Mn3O4," *International Journal of Greenhouse Gas Control*, 2(1):21–36, 2008.

[3] Lekse J. W., Natesakhawat S., Alfonso D., Matranga C., "An experimental and computational investigation of the oxygen storage properties of BaLnFe2O5+d and BaLnCo2O5+d (Ln = La, Y) perovskites," *Journal of Materials Chemistry A*, 2(7):2397–2404, 2014.

[4] Anisimov V., Aryasetiawan F., Lichtenstein A. I., "First-principles calculations of the electronic structure and spectra of strongly correlated systems: the LDA+U method," *Journal of Physics: Condensed Matter*, 9:767–808, 1997.

[5] Kresse G., Furthmüller J., "Efficiency of ab initio total energy calculations for metals and semiconductors using a plane wave basis set," *Computational Materials Science*, 6(1):15, 1996.

[6] Cozad A., Sahinidis N. V., Miller D. C., "Learning Surrogate Models for Simulation-Based Optimization," *AIChE Journal*, 60(6):2211–2227, 2014.

494592 *In silico* Structural Analyses of Borosilicate, Aluminosilicate, and Gallosilicate Zeolites Using Effective Tetrahedral Descriptors

Abstract:

Zeolites, a class of crystalline microporous silicate-based materials, have been widely used as adsorbents and catalysts due to their uniform pore sizes, catalytic activities, and hydrothermal stability. Crystal structures of zeolites are built from the TO_4 tetrahedra (where T is Si, Al, B, Ga, etc.), resulting in more than 230 unique framework structures. The structure and the chemical composition of zeolites are two important factors for design of zeolite. Based on the empirical knowledge of zeolite synthesis, these two factors are known to be dependent on each other; however, their actual relationship has remained unclear due to the unclarified crystallization mechanism of zeolites.

We recently rationalized this structure-composition relationship by performing *in silico* analysis of aluminosilicate zeolites with almost all known zeolite structures and different Si/Al molar ratios [1]. The resulting energy-density landscape of modelled zeolites revealed that the relation between the relative framework energies versus the chemical compositions varies in accordance with the framework topologies. Extension of this approach to other trivalent atoms such as B and Ga is, however, not straightforward due to a lack of reasonable way to offset the energies of zeolites with different trivalent atoms.

Here, we analyze borosilicate, aluminosilicate and gallosilicate zeolites by utilizing a purely structural descriptor developed by Zimmerman et al., namely a tetrahedral ordering parameter, which can evaluate the distortion of TO_4 tetrahedra effectively [2]. Analysis of the ordering parameter of structurally optimized zeolites with uniform background charge reveals that B, Al, and Ga distort TO_4 tetrahedra themselves as well as the neighboring SiO_4 tetrahedra. Furthermore, this general behavior is inherent in the substituting trivalent atoms. The distortion caused by Al is relatively small because Al–O bond length (~1.72 Å) is relatively similar to Si–O (~1.61 Å). Compared to Al, introduction of B in silicate zeolite framework causes larger overall structural distortion that seems to be mainly taken over by the distortion of surrounding SiO_4 due to the lack of flexibility of BO_4. Borosilicate zeolites with relatively low Si content (e.g., Si/B < 3) are structurally unfavorable judging from their very low ordering parameter, while aluminosilicates and gallosilicates give reasonably high values at the same composition. The structural distortion induced by Ga seems to be mainly undertaken by GaO_4 itself, thereby minimizing the distortion of surrounding SiO_4.

Such different characteristics observed in trivalent atoms seemingly control their structure-directing functions to crystallization of zeolites and explain the empirical knowledge of zeolite synthesis such as;
- Aluminosilicate zeolites with high Al content favor the structures with even-numbered ring.
- Borosilicate zeolites tend to be synthesized as relatively high Si/B.
- In several cases, introduction of B and Ga in the synthetic system allows the crystallization of unconventional structures that cannot be synthesized as aluminosilicates and pure silica.

[1] K. Muraoka, W. Chaikittisilp, and T. Okubo, J. Am. Chem. Soc., 2016, 138 (19), 6184–6193.
[2] N. E. R. Zimmermann, B. Vorselaars, D. Quigley, and B. Peters, J. Am. Chem. Soc., 2015, 137 (41), 13352–13361.

495210 Computational Identification of Hetero-Interpenetrated Porous Materials

Abstract:

We have developed a computational algorithm that can identify whether two different porous materials can hetero-interpenetrate with one another. Initially, two prototypical MOFs (i.e. IRMOF-1 and IRMOF-8) were selected as target materials and tested for hetero-interpenetration against thousands of experimentally synthesized MOFs for matching candidates. As a result, PCN-68 and PCN-610 emerged as suitable candidates respectively in our screening studies. Subsequently, Grand Canonical Monte Carlo (GCMC) simulations were conducted on these hetero-interpenetrated materials to obtain the methane and hydrogen adsorption properties. Our computational results showed that there were enhancements in gravimetric hydrogen uptakes at 77 K for relatively narrow pressure range of less than around 1.5 (4.0) atm for IRMOF-1/PCN-68 (IRMOF-8/PCN-610) and enhancements in volumetric hydrogen uptakes for relatively wider pressure range of up to 18.8 (57.5) atm for IRMOF-1/PCN-68 (IRMOF-8/PCN-610) compared to the parent MOFs. The computational studies here suggest a simple way of identifying new porous materials that can have synergetic properties compared to the two parent materials.

488388 Lipid-like Materials for RNA Delivery: A How-to Guide for Hacking Gene Expression

Abstract:

Despite the promise of RNA therapeutics, progress towards the clinic has been slowed by the difficulty of delivering RNA drugs, such as short interfering RNA (siRNA) and messenger RNA (mRNA), into cellular targets within the body. RNA therapeutics are large (10^4 â€“ 10^6 g/mol) and negatively charged; they do not have favorable biodistribution properties in vivo nor an ability to cross the cellular membrane of target cells. In response to these challenges, our lab has created and tested large libraries of biodegradable lipid-like materials called â€˜lipidoidsâ€™ using high-throughput methodologies. Lipidoids are able to efficiently manipulate gene expression in a variety of biological systems, including hepatocytes, white blood cells and tumors. Furthermore, structure-function analysis has revealed material design criteria that reliably predict in vivo delivery efficacy without the need for any biological testing. Current efforts focused on extending lipidoid delivery capabilities to mRNA will be discussed.

497014 Triple Delivery Nanoscale Device for siRNA, Vismodegib and Gemcitabine Co-Delivery to Treat Pancreatic Cancer

Abstract:

Pancreatic cancer (PC) has an extremely poor prognosis with a five-year survival rate of less than 7%.[1] PC is characterized by extensive desmoplasia, which limits the delivery and efficacy of chemotherapy. Sonic hedgehog (Shh)-signaling plays an important role in the tumor growth through desmoplasia, metastasis and angiogenesis.[2] Shh also regulates epithelial to mesenchymal transition (EMT), is overexpressed in PC stem cells and contributes to resistance and PC progression. [3] In addition, previous studies from our lab have shown that MUC4 is aberrantly overexpressed in human PC and has a role in the progression and metastasis of PC cells.[4] Currently applied standard therapy for PC, involving surgical cytoreduction followed by systemic gemcitabine (GEM) based chemotherapy, has not provided desired outcome yet. As an alternative, small interfering RNA (siRNA) based therapy has been proposed for the PC treatment. However, targeted co-delivery of siRNA along with GEM and Vismodegib (VIS), an SHH antagonist, is challenging due to varying physicochemical properties, drug release rates, and pharmacokinetics. There is currently no ideal delivery system available to provide transport across vascular endothelial barriers, diffusion through extracellular matrix, and that can facilitate cellular entry and endosomal escape, while maintaining siRNA stability and providing site-specific release and activity.

In this study, we proposed the use of a temperature and pH responsive, cationic, amphiphilic and biocompatible pentablock (PB) copolymer based triple delivery nanoscale device (TDND) for *in vitro* and *in vivo* combined delivery of siRNA, VIS and GEM for the treatment of PC. This novel TDND facilitated cellular uptake due to amphiphilic character, endosomal escape due to protonatable tertiary amine groups in the cationic end blocks, and provided efficient simultaneous siRNA, VIS and GEM release in the cytoplasm through temperature responsive micellization. We hypothesized that targeting MUC4 expression through siRNA treatment and Shh signaling through VIS treatment via the TDNDs results in suppression of MUC4 expression, down regulation of SHH signaling pathway, inhibition of desmoplasia, pancreatic stellate cells and cancer stem cells. This may lead to an improved therapeutic outcome of GEM in PC through improving its perfusion in the tumor.

The PB copolymer consisting of temperature responsive PluronicF127 (poly(ethyleneoxide)-block-poly(propyleneoxide)-block poly(ethyleneoxide) (PEO-PPO-PEO)) middle block and pH responsive cationic (poly(2-diethylaminoethyl methacrylate)) (PDEAEM) end blocks was synthesized through atom transfer radical polymerization (ATRP). The TDND was prepared through the electrostatic complexation of PB copolymers with siRNA and subsequent self-assembly with VIS and GEM encapsulated PluronicF127 layers at different ratios. The TDND was characterized to evaluate size, zeta potential, serum and RNase stability. The simultaneous siRNA, VIS and GEM release test was conducted. *In vitro* toxicity, transfection efficiency, cellular uptake and endosomal escape capability of TDNDs were evaluated against Capan-1 PC line.

Our results indicated that at N/P ratios (N/P: molar ratios of nitrogens (N) in the pentablock copolymer to phosphates (P) in siRNA) above 7.5, complete complexation occurs which provides siRNA protection and stability while preventing the premature release. The stability of the TDND system was also tested in the presence of serum and RNase enzymes for 72h. It was observed that the TDNDs with N/P ratio of 12.5, 25 and 50 showed good siRNA protection and stability against serum and RNase enzyme. The designed TDND system with N/P ratio of 50 also provided efficient transfection and endosomal escape in Capan-1 human pancreatic adenocarcinoma cell line. This TDND system showed significantly higher transfection (~2.5 fold higher) than commercially available agent, Lipofectamine. The tumor cell killing efficiency of the TDND in Capan-1 pancreatic carcinoma cell line was also tested. TDND with N/P ratio of 50 (siRNA, GEM and VIS co-loaded) showed a synergistic effect in killing the cancer cells (80% of the cells were killed) compared to the control cases. We also investigated the simultaneous release of siRNA, GEM and VIS. It was observed that the thermoresponsive gelling TDND systems acted like a drug depot and were able to provide simultaneous release of siRNA, VIS and GEM and reached a plateau within 72h. Almost 90% of the siRNA and GEM were released within 5 days, while ~33% of VIS was released. The in vivo therapeutic efficacy of the TDNDs were also tested and showed significant reduction in tumor growth.

Overall, the TDND developed in this study was capable of overcoming the current hurdles of combination siRNA/drug delivery and showed significant activity in PC cell lines in vitro and in vivo for translational medicine and future clinical applications. This strategy provided an enhanced understanding of the mechanism of action underlying the synergistic activity of combined delivery.

1. Siegel, R. L.; Miller, K. D.; Jemal, A., Cancer statistics, 2016. *CA: a cancer journal for clinicians* **2016,** *66* (1), 7-30,

2. Bailey, J. M.; Swanson, B. J.; Hamada, T.; Eggers, J. P.; Singh, P. K.; Caffery, T.; Ouellette, M. M.; Hollingsworth, M. A., Sonic hedgehog promotes desmoplasia in pancreatic cancer. *Clinical cancer research : an official journal of the American Association for Cancer Research* **2008,** *14* (19), 5995-6004, PMID:PMC2782957.

3. Lee, C. J.; Li, C.; Simeone, D. M., Human Pancreatic Cancer Stem Cells: Implications for How We Treat Pancreatic Cancer. *Translational Oncology* **2008,** *1* (1), 14-18,

4. Rachagani, S.; Torres, M. P.; Kumar, S.; Haridas, D.; Baine, M.; Macha, M. A.; Kaur, S.; Ponnusamy, M. P.; Dey, P.; Seshacharyulu, P.; Johansson, S. L.; Jain, M.; Wagner, K. U.; Batra, S. K., Mucin (Muc) expression during pancreatic cancer progression in spontaneous mouse model: potential implications for diagnosis and therapy. *Journal of hematology & oncology* **2012,** *5*, 68, PMID:PMC3511181.

499306 Influence of Dextran and Surface Charge on Nanoparticle-Mediated siRNA Delivery

Abstract:

Understanding the relationship between delivery vehicle characteristics and the endocytosis and intracellular trafficking of short interfering RNA (siRNA) remains a critical bottle neck in the development of siRNA therapeutics. As a result, it is difficult to design a delivery vehicle for a novel target or specific cell type or to pinpoint the characteristics contributing to the inefficiency, toxicity, or immunogenicity of a vehicle.

Our aim is to explore the mechanism by which the chemical and physical characteristics of a nanoparticle influence siRNA delivery. As a model, we have chosen to use silica nanoparticles (sNPs) due to their straightforward and flexible synthesis. Our functional readouts are: silencing efficacy, accumulation of siRNA in the cells, intracellular location and degree of siRNA-sNP complex dissociation, and intracellular location and degree of siRNA strand separation. We are investigating how vehicle size, shape, charge density, and surface functionalization can be varied to synthesize optimal vehicles for delivery to a variety of cell types.

The presentation will discuss our current finding on the role of sNPs characteristics (dextran functionalization and zeta potential) on siRNA delivery. Our work explores the preferential mechanism of endocytosis (clathrin, caveolin, ARF6, GRAF1, flotillin, or macropinocytosis) across multiple cell types (HeLa, H1299, HEK293, and HepG2). Additionally, we will discuss findings from our confocal microscopy-based intracellular trafficking assay, which independently tracks the vehicle and siRNA strands across endosomal vesicles (early, late, slow/fast recycling, and lysosomal), the Endoplasmic Reticulum, and the Golgi apparatus. Our results to date demonstrate that multiple vehicle characteristics affect siRNA delivery and a critical role of intracellular trafficking in maximizing siRNA potency.

499886 Peg-Poly(beta-amino ester) Delivery Systems for Periodic shRNA

Abstract:

RNA interference (RNAi) provides a versatile therapeutic strategy via silencing of specific genes implicated in cancer and other diseases. However, clinical translation of RNAi for cancer therapeutics remains unrealized, due to challenges in delivery of small interfering RNA (siRNA) to tumors. The low valency and high rigidity of siRNA often requires high excesses of cationic delivery materials to condense into stable nanoparticles, leading to dose-limiting toxicities. To address this challenge, we adopt an RNAi platform based on periodic short hairpin RNAs (p-shRNAs). Consisting of siRNA sequences linked together, these polymeric RNAi molecules are generated by the repeated action of an RNA polymerase around a small circular DNA template. Through template design and selective enzymatic digestion, we can design p-shRNA structures that are efficiently processed inside cells into siRNAs and induce significant gene silencing.

With its much higher valency and flexibility compared to siRNA, p-shRNA requires far less polycationic material to be condensed, and forms more stable complexes. Poly(beta-amino ester)s (PBAEs) in particular have shown promise as gene delivery vehicles, due to their biodegradability and design versatility. To develop an optimal PBAE structure for p-shRNA delivery, we used factorial design to synthesize a library of PBAE variants of a base structure, poly-1. Screening of this library showed that p-shRNA silencing efficiency increases with increasing alkyl side chain percentage and decreasing molecular weight. Our designed poly-1 structures are able to fully condense p-shRNA into sub-100 nm complexes with high silencing efficiency, at much lower polymer-to-RNA ratios than those typically required for PBAE gene delivery. To enhance the colloidal stability of the complexes in physiological conditions, we added a poly(ethylene glycol) (PEG)-poly-1 copolymer containing alkyl side chains to the core complex. The resulting complexes, assembled via electrostatic and hydrophobic interactions, possess an outer PEG layer that can provide stability in the bloodstream and thereby increase tumor accumulation. We further introduced active tumor targeting by conjugating folate to the PEG terminus. Expanding our library to screen combinations of the core poly-1 and PEG-poly-1 structures revealed an optimal blend of hydrophobicities and molecular weights in the two polymer components. Thus, through nucleic acid engineering and rational carrier design, we have successfully developed a stable, potent RNAi delivery system that can trigger significant gene silencing at low doses, and potentially enable higher therapeutic efficacy *in vivo*.

500610 Structurally Programmed Assembly of Ribonucleoproteins for Superior mRNA Delivery

Abstract:

Messenger RNA (mRNA) represents a promising class of nucleic acid-based therapeutics. While numerous nanocarriers have been developed for mRNA delivery, the inherent labile nature of mRNA results in a very low transfection efficiency and poor expression of desired protein. Here we preassemble the mRNA translation initiation structure through an inherent molecular recognition between 7-methyl guanosine (m^7G) capped mRNA and eukaryotic initiation factor 4E (eIF4E) protein to form ribonucleoproteins (RNPs), thereby mimicking the first step of protein synthesis inside cells. Subsequent electrostatic stabilization of RNPs with structurally tunable cationic carriers leads to nano-sized complexes (nanoplexes), which elicit unprecedented levels of mRNA transfection in different cell types by enhancing intracellular mRNA stability and protein synthesis. By investigating a family of synthetic polypeptides bearing different side group arrangements of cationic charge, we find that the molecular structure modulates the nano-scale distance between the mRNA strand and the eIF4E protein inside the nanoplex, which directly impacts the enhancement of mRNA transfection. To demonstrate the biomedical potential of this approach, we first use this approach to introduce mRNA/eIF4E nanoplexes to murine dendritic cells, resulting in increased activation of cytotoxic CD8 T cells *ex vivo*. More importantly, eIF4E enhances gene expression in lungs following a systemic delivery of luciferase mRNA/eIF4E in mice. Collectively, this bio-inspired molecular assembly method could lead to a novel paradigm of gene delivery.

499772 Cationic Peptide Amphiphile Micelles(PAMs) As Nucleic Acid Based Adjuvants Carriers for the Improvement of Subunit Vaccine Efficiency

Abstract:

Molecular adjuvants have been widely used in recent vaccine development due to their excellent ability to stimulate Toll Like Receptors (TLRs) and activate Antigen Presenting Cells (APCs). Nucleic acid based adjuvants such as double stranded RNA polyinsosinic:polycytidylic acid (poly(I:C)) and single strand DNA CpG Oligodexynucleotides (CpG ODNs) are molecular adjuvants that work by mimicking bacterial nucleic acids to stimulate TLRs (TLR-3 and TLR-9, respectively) located in APC endosomes. However, because of their large molecular weight and negative charged backbone, nucleic acid based adjuvants require an effective delivery vehicle to reach their desired site of action.

Peptide amphilphiles (PAs) are a class of diblock biomaterials comprised of hydrophilic peptide(s) conjugated with hydrophobic lipid(s) which self-assemble into peptide amphiphile micelles (PAMs) in water. PAMs have been recently studied for subunit vaccine applications and have shown considerable promise in both prophylactic pathogenic vaccine and therapeutic cancer vaccine studies. For this study, instead of making traditional diblock PAs, we added a third block, oligo(lysine), on the other side of the hydrophobic lipid. When these novel triblock PAs self-assemble, this cationic oligo(lysine) block is exposed yielding cationic PAMs (Cat-PAMs). Due to their significant positive surface charge, Cat-PAMs can readily complex with negatively charged nucleic acid based adjuvants due to electrostatic interactions. Cellular experiments utilizing these micelles show they can effectively deliver nucleic based adjuvants to the endosome of APCs which induces stronger immune responses *in vitro* and *in vivo* than conventional vaccine formulations.

490383 Understanding Peptide Assembly with Coarse-Grained Models Designed By Information Theory (Invited Talk)

Abstract:

Peptide self-assembly is a major feature of both protein aggregation and emerging routes to synthetic nanostructured biomaterials. However, this inherently multiscale process remains challenging to model – and to predict as a function of amino acid sequence – due to the large associated length and time scales. Here, we discuss a novel modeling coarse-grained modeling approach to such problems based on information theory. A quantity called the relative entropy measures the information lost upon coarse graining and hence the (inverse) fitness of a coarse-grained model. Minimization of the relative entropy provides a universal principle for coarse-graining, and a way to "automatically" discover and design coarse models of many systems. This approach enables us to develop simple but surprisingly accurate coarse peptide models, and permits large-scale simulations of several kinds of self-assembly, including amyloid-like oligomerization and aggregation, and interactions of tethered peptides on surfaces.

489833 Modeling the Oxidative Consumption of Curcumin from Controlled Release Poly(beta amino ester) Microparticles in the Presence of a Free Radical Generating System

Abstract:

The balance between the basal level concentrations of both reactive oxygen species and antioxidants is an essential component to the regulation of a physiologically healthy cellular environment. Under a variety of pathophysiological conditions (e.g., radiation injury, inflammation, acute lung injury), cells are no longer able to maintain a balance between antioxidant and radical formation, which results in free radical overproduction and oxidative stress. Curcumin, a potent antioxidant, derived from the Indian spice, turmeric, has been a promising candidate for the control of inflammation and serves as a free radical scavenger. However, up to now, no double-blinded, placebo controlled clinical trial of curcumin has shown a positive benefit. The proper design of a controlled release system can enhance curcuminâ€™s bioavailability and eliminate failures seen in the delivery of curcumin thus far. The incorporation of curcumin into the backbone of a hydrolytically degradable cross-linked hydrogel that utilizes poly(beta amino ester) chemistry can provide a tunable protective network to release antioxidants at a controlled rate. The kinetics of this controlled release delivery system in the presence of oxidative environments has not yet been studied and can be used to investigate the consumption rates of active curcumin and degradation products to provide efficient therapy as a therapeutic.

Based on experimental data obtained, we developed a model that describes the concentration of curcumin as a free molecule in the presence of a general free radical generator, 2,2â€™-azobis(2-amidinopropane) dihydrochloride, over time. The set of mathematical expressions were solved simultaneously while the unknown consumption rate parameter of curcumin was minimized over the experimental data by sum of squared errors. These parameters were then used to find consumption rate parameters of additional degradation products found within the solution from a fully degraded microparticle system. From these results, we can quantify theoretical levels of free radicals present in an environment depending on the rate of release of curcumin provided. Modeling our drug delivery system will allow for improved translation and success overall in the delivery of curcumin as a therapeutic agent.

494846 The Fusion and Spreading of Liposome with Different Sizes: Molecular Dynamics Simulation with Dry-Martini Force Field

Abstract:

Liposome fusion is an essential phenomenon for cellular communication [1-2] and spreading of lipid vesicle is a potential way to build supported lipid bilayer membranes, which are of fundamental importance for many nano-devices [3-4]. Although great efforts have been applied to study the fusion and spreading of liposomes using both atomistic and coarse-grained simulation models [5-7], how the size of liposomes affects fusion and spreading process is still unknown due to the limited computing power. In this study, we conducted molecular dynamics simulations of the fusion and spreading of liposomes with different sizes based on dry-martini force field [8], which is recently developed for lipid membrane simulations. The effect of properties of liposome, including composition and size, and supported materials (charge, hydrophobicity) were investigated via molecular dynamics simulations. It is shown that a ringed structure composed of two identical sizes of liposomes occurs during the fusion of liposomes, and then the ranged structure transforms into the dumbbell-shaped structure. The mechanism and transient states were also investigated in term of the distribution of lipids, morphology analysis and energy analysis. It is also shown that the points with large curvature prefer to be the starting fusion sites. The simulation results reveal the important regulatory role of liposome size on fusion and spreading.

[1] Bonifacino, Juan S., and Benjamin S. Glick. "The mechanisms of vesicle budding and fusion." cell 116.2 (2004): 153-166.

[2] Martens, Sascha, and Harvey T. McMahon. "Mechanisms of membrane fusion: disparate players and common principles." Nature Reviews Molecular Cell Biology 9.7 (2008): 543-556.

[3] Jass, Jana, Torbjörn Tjärnhage, and Gertrud Puu. "From liposomes to supported, planar bilayer structures on hydrophilic and hydrophobic surfaces: an atomic force microscopy study." Biophysical journal 79.6 (2000): 3153-3163.

[4] Puu, Gertrud, and Inga Gustafson. "Planar lipid bilayers on solid supports from liposomesâ€"factors of importance for kinetics and stability." Biochimica et Biophysica Acta (BBA)-Biomembranes 1327.2 (1997): 149-161.

[5] Shillcock, Julian C., and Reinhard Lipowsky. "The computational route from bilayer membranes to vesicle fusion." Journal of Physics: Condensed Matter 18.28 (2006): S1191.

[6] Kawamoto, Shuhei, Michael L. Klein, and Wataru Shinoda. "Coarse-grained molecular dynamics study of membrane fusion: Curvature effects on free energy barriers along the stalk mechanism." The Journal of chemical physics 143.24 (2015): 243112.

[7] Kong, Xian, et al. "Spreading of a unilamellar liposome on charged substrates: a coarse-grained molecular simulation." Langmuir 32.15 (2016): 3785-3793.

[8] Arnarez, Cleìment, et al. "Dry Martini, a coarse-grained force field for lipid membrane simulations with implicit solvent." Journal of chemical theory and computation 11.1 (2014): 260-275.

494895 Design of Membrane-Embedded Amphiphilic Nanoparticles from Multiscale Simulations

Abstract:

Functionalized, monolayer-protected nanoparticles (NPs) are a versatile materials platform because the protecting surface monolayer can be engineered to tailor interactions with the surrounding environment. NPs are of particular interest in biomedical applications because they can be designed to mimic the physicochemical properties of typical biological macromolecules (i.e., soluble proteins). However, structure-function relationships correlating the nanoscale properties of engineered nanomaterials to their biological interactions remain unclear and can be difficult to determine experimentally. Here, we demonstrate that molecular modeling across multiple length scales can aid in the interpretation of experiments and guide the design of novel nanomaterial functionalities. We focus on understanding the ability of a class of amphiphilic NPs to enter cells via a non-endocytic, non-disruptive mechanism that is of broad interest for applications in drug or gene delivery. This behavior is surprising because bypassing the cell membrane requires the NPs to translocate charged groups across the hydrophobic core of the lipid bilayer. We use atomistic molecular dynamics simulations to gain molecular insight into these experimental observations and predict outcomes in model membrane systems. Our results show that the amphiphilic NPs insert into the bilayer to obtain configurations resembling membrane-embedded proteins due to favorable interactions between the NP surface and the hydrophobic bilayer core. We further develop an implicit solvent modeling approach that facilitates the high-throughput determination of how varying monolayer compositions affect the thermodynamics of bilayer insertion. Finally, we leverage this mechanistic understanding to provide design guidelines for monolayer compositions optimized for non-endocytic cellular uptake. This framework illustrates the power of multiscale modeling to understand and predict behavior at the nano-bio interface.

494275 Improvement on Self-Assembly of Virus-like Particles By the Introduction of Electrostatic Attraction

Abstract:

Virus-like particles (VLPs) are ideal nanoscale carrier beneficial from the biological compatibility and unique nanostructures. However, the application of VLPs is hindered because of the difficult control on self-assembly and loading position. In the present study, introduction of electrostatic attraction was proposed to assist the self-assembly of murinepolyomavirus (MPV) VLP through the construction of two mutants of subunit, capsomere (Cap). One was introduced with positive charge (m+Cap) while the other was introduced with negative charge (m-Cap). Monte Carlo simulations on a simplified model were performed to examine the charge distribution and the free energy change of VLP when mixing m+Cap and m-Cap within different molar ratio. The lowest free energy of mVLP was observed in the equimolecular mixture of m+Cap and m-Cap. Then the self-assembly of equimolecular mixture of m+Cap and m-Cap was examined experimentally. The introduction of electrostatic interaction in VLPs was confirmed effective for the improvement on self-assembly. It would facilitate the VLP-based drug delivery and development of biomaterials.

CONTEMPORARY MODELING AND ANALYSIS OF STEADY STATE AND TRANSIENT HUMAN BLOOD RHEOLOGY

29OCT17

Michael Clarion, Michael Deegan, Tyler Helton, Jesse Hudgins, Evan Ousley, Austin Breed, Matthew J. Armstrong[1]
[1]*Department of Chemistry and Life Science, United States Military Academy, West Point, NY 10996*

Collaborators: **Jeffrey Horner, Antony Beris, Norman Wager**[2]
[2]*Department of Chemical and Biomolecular Engineering, University of Delaware, Newark, DE 19716*

OUTLINE

- Introduction to blood and rheology...
- State of steady state blood rheology modeling
- Transient flows; not all models created equal
- Series of Physical Phenomena "fingerprinting"
- Future work
- Questions

MOTIVATION & OBJECTIVES

Motivation:

- How do we relate blood physiology to the rheology
- Use rheology to make predictions…long term project… "reverse problem"

- Objectives:
  - Probe/model both steady state and transient behavior
  - Improve existing thixotropic models
  - *Investigate diseased blood (diabetes, high cholesterol, etc.)*
  - *Visualize transient flow in terms of elastic and viscous comp.*

1. Cardiovascular Diseases (CVD's) World Health Organization ; http://www.who.int/mediacentre/factsheets/fs317/en/

BLOOD MAKEUP & COMPLEX RHEOLOGY

Rheology of a TEVP material:

- Shear Thinning
- Nonzero Yield Stress
- Viscoelasticity/plasticity
- Thixotropy

1. Components of Blood and Their Importance, American Society of Hematology, https://www.youtube.com/watch?v=R-sKZWqsUpw Fedosov and Gompper, *Proceedings of the 6th NIC Symposium*, (2012).
2. Ewoldt and McKinley, Rheol. Acta (2017). 56:195-210; 3. Horner et al. (2016) SoR Conference.

STEADY STATE BLOOD MODELING

-Not where the action is

- Not as relevant

- All steady state models work relatively well

- Only a few can model transient flow conditions without modifications...(ie. structure parameter/evolving viscosity term)

STEADY STATE MODELS (# PAR.)

$$\sigma_{tot} = \sigma_v(\dot{\gamma})$$

- Phenomenological
- Solve for viscosity as a function of shear rate

Carreau-Yasuda (5)

$$\frac{\eta - \eta_\infty}{\eta_0 - \eta_\infty} = \left[1 + (\lambda\dot{\gamma})^a\right]^{(n-1)/a}$$

Modified Cross (5)

$$\frac{\eta - \eta_\infty}{\eta_0 - \eta_\infty} = \frac{1}{\left(1 + (\lambda\dot{\gamma})^m\right)^a}$$

$$\sigma_{tot} = \sigma_e + \sigma_v(\dot{\gamma})$$

- Add a yield stress term
- Phenomenological

Herschel-Bulkley (3)

$$\sigma = \sigma_{y0} + K\dot{\gamma}^n$$

Casson (2)

$$\sigma^{1/2} = \sigma_{y0}{}^{1/2} + (\eta_\infty \dot{\gamma})^{1/2}$$

$$\sigma_{tot} = \sigma_e(\xi, \dot{\gamma}) + \sigma_v(\xi, \dot{\gamma})$$

-Separate elastic/plastic
-Add evolving thixo.
-Phenomenological

Apostolidis et al. (5)
(next slide)

$$\boldsymbol{\sigma} + \frac{\eta}{G}\dot{\boldsymbol{\sigma}} = \eta\dot{\gamma}$$

-Uses evolving fluidity term

BMP (4)

$$\boldsymbol{\sigma} + \frac{1}{G_0\varphi}\frac{d\boldsymbol{\sigma}}{dt} = \frac{\dot{\gamma}}{\varphi}$$

$$\frac{d\varphi}{dt} = \frac{(\varphi_0 - \varphi)}{\lambda} + k(\varphi_\infty - \varphi)\sigma\dot{\gamma}$$

http://homepages.engineering.auckland.ac.nz/~pkel015/SolidMechanicsBooks/Part_I/BookSM_Part_I/10_Viscoelasticity/10_Viscoelasticity_03_Rheological.pdf

5

APOSTOLIDIS ET AL. (AAB)

Algebraic Equations

$$\gamma = \gamma_e + \gamma_p \quad \leftrightarrow \quad \dot{\gamma} = \dot{\gamma}_e + \dot{\gamma}_p$$

$$\dot{\gamma}_p = \begin{cases} \dfrac{\dot{\gamma}}{\left(2 - \dfrac{|\gamma_e|}{\gamma_{max}}\right)} & \dot{\gamma} \geq 0 \\ \dfrac{\dot{\gamma}}{\left(2 + \dfrac{|\gamma_e|}{\gamma_{max}}\right)} & \dot{\gamma} < 0 \end{cases}$$

$$\gamma_{max} = \min\left(\frac{\gamma_0}{\xi^2}, \gamma_\infty\right)$$

ODEs

$$\frac{d\gamma_e}{dt} = \dot{\gamma}_p - \frac{\gamma_e}{\gamma_{max}}|\dot{\gamma}_p|$$

$$\frac{d\xi}{dt} = \frac{\sigma_{y0}}{\eta_\infty} k_\xi \left[(1-\xi) + \xi\sqrt{\frac{4\eta_\infty \overline{\dot{\gamma}_p}}{\sigma_{y0}}}\right]$$

$$\frac{dG}{dt} = -k_G \xi (G - \xi G_0)$$

Constitutive Equations

$$\sigma = G\gamma_e + \eta_\infty \dot{\gamma}_p$$

$$\sigma = G\gamma_e + \xi K_{ST} \dot{\gamma}_p^n + \eta_\infty \dot{\gamma}_p$$

Enhanced AAB

Correlations

$$\eta_\infty = n_p(1 + 2.0703 \times HCT + 3.7222 \times HCT^2)\exp\left\{-7.0276\left(1 - \frac{T_0}{T}\right)\right\}; \quad \sigma_{y0} = ((HCT - (0.3126 \times c_f^2 - .468 \times c_f + 0.1764))^2) \times (0.5084(c_f + 0.4517)^2)$$

1. Apostolidis et al. (2015) JoR; 2. Apostolidis and Beris (2014) JoR.

STEADY STATE DATA

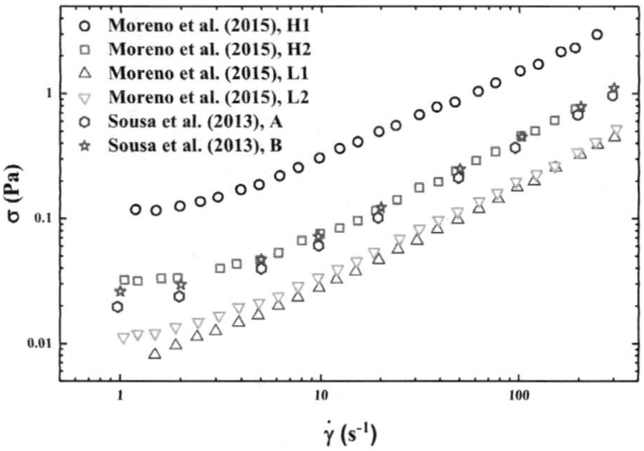

1. Sousa et al. (2013) Biorheology; 2. Moreno et al. (2015) Rheol. Acta

Microstructure Characterization
(model thixotropic material)

Parameter determination

Constitutive Models:

SSTM
MDTM (Development)
UAM
BMPM

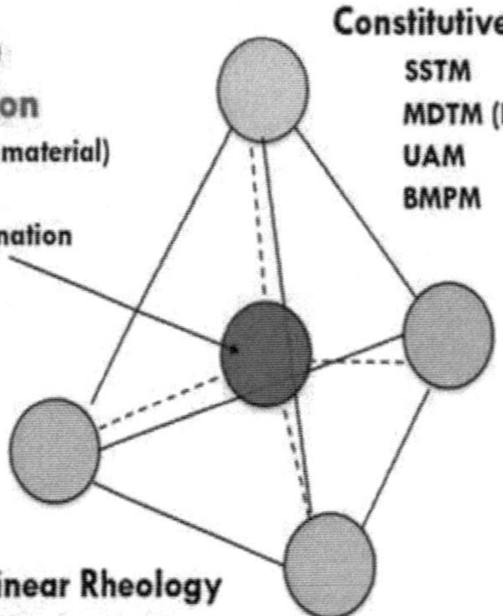

SANS/ Rheo. SANS/ SALS

Traditional Rheology:

Steady state flow curve
Stress relaxation
Start up
SAOS
Static yield stress

Non-Linear Rheology & (LAOS) Analysis:

Transients — Flow reversal
LAOS — *1-Dir. LAOS*
FFT/ Chebychev/ *SPP*

1. Armstrong et al. (2016) JoR; 2. Armstrong (2015) UDEL.

PARALLEL TEMPERING

Key features

- User-friendly, robust
- **4** adjustable algorithmic parameters
- Dynamic/static data
- Constraints
- Estimate error bar
- **"Level playing field"**

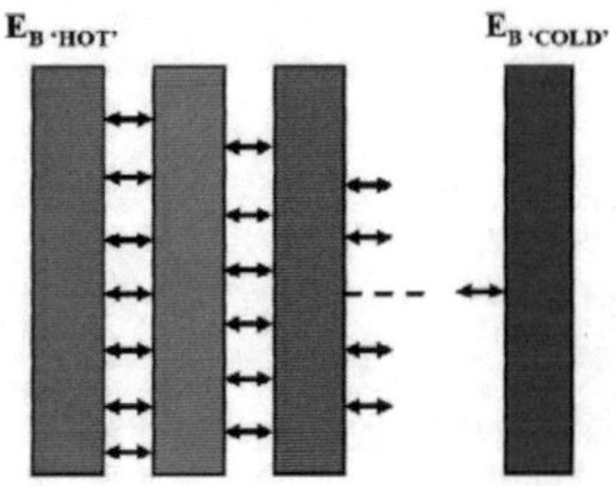

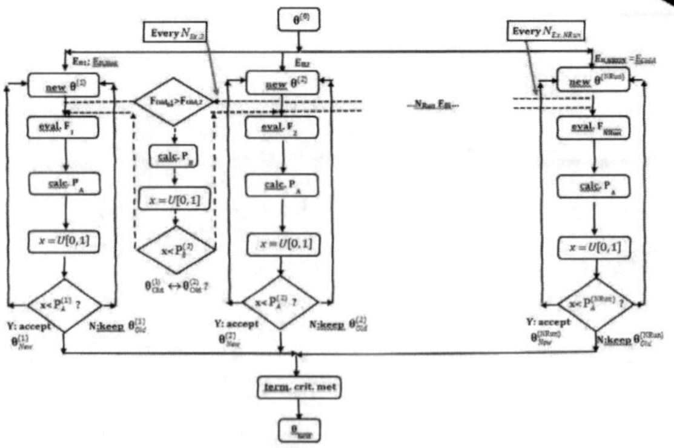

Preliminaries:

1. Define problem: identify ODE/DAE system; static vs. dynamic data
2. Choose continuum functional approximation
3. Choose coefficients for F_{obj} (based on zero model predictions)
4. Define statistical parameters μ, σ based on window of plausible parameter values
5. Perform test runs of model to determine h
6. Choose basic (if other than default) algorithmic parameters: $\frac{F_{n,Hot}}{F_{n,Cold}}$, N_{Iter}, N_{sm}, $Error_{Max}$
7. Perform test Monte Carlo runs to determine: $N_{ex,Hot}$, $N_{ex,Cold}$

STEADY STATE RESULTS (1/3)

- Fit all models to steady state data from literature using recently published stochastic, global optimization algorithm
- Compared models using cost function and Akaike Information Criteria
- AIC assigns a penalty for more parameters
- Simplest models out performed all models for steady state...but no simple models can handle transient flows in a meaningful way...

$$F_{ss} = \sum_{i}^{N}\left(\sqrt{\left(\frac{(y_i - f_i)}{y_i}\right)^2}\right)\Big/ N \qquad RSS = \sum_{i}^{N}\left(\sqrt{\left(\frac{(y_i - f_i)}{y_i}\right)^2}\right) \qquad AIC = 2k + 2\ln(RSS)$$

1. Armstrong et al. (2016) AIChE J.; 2. Akaike (1974) IEEE J.

STEADY STATE RESULTS (2/3)

Fitting Results

Par.	High Cholesterol		Low Cholesterol		Sousa		Tomaiuolo	Ave. Par.	Std. Par.	Par.	Std. Par.
	H1	H2	L1	L2	A	B	B				
σ_{y0}	0.0192	0.0026	0.0001	0.0001	0.0034	0.0030	0.0028	0.0046	0.00613	80	0.01658
η_∞	0.0100	0.0035	0.0013	0.0016	0.0033	0.0030	0.0065	0.0038	0.00269	13	0.01304
c_f	0.1471	0.1537	0.2484	0.2947	0.2312	0.2712	0.8770	0.2259	0.05186	41	0.04304
HCT	1.1418	0.4877	0.0156	0.1277	0.4611	0.4071	0.4959	0.4481	0.33307		
G	4.8334	5.1232	5.0226	4.8366	4.8169	5.0246	5.0148	4.9532	0.11283		
γ_0	0.0074	0.0010	0.0001	0.0002	0.0012	0.0009	0.0008	0.0017	0.00237		
γ_{MAX}	1	1	1	1	1	1	1				
ERROR	0.0126151	0.026727	0.00399525	0.0222902	0.0271794	0.0213317	0.00857	0.01753			
	TA Instrument AR-G2; controlled stress; double concentric cylinder; 5mL				MCR301, Anton Parr; 50mm plate-plate		MCR301; double gap				
RSS	0.3027621	0.6147208	0.09588601	0.5795463	0.2446145	0.1919856	0.20568	0.31931			
n (data points)	24	23	24	26	9	9	24				
k (parameters)	6	6	6	6	6	6	6				
AIC	9.6103839	11.026826	7.31080956	10.908981	9.1838561	8.6993297	8.8371326	9.36819			
AICc	16.610384	18.493492	14.3108096	17.131203	121.18386	120.69933	15.837133	46.3237			

STEADY STATE RESULTS (3/3)

Average Values

Model	Equations	Parameters	Cost Func.	RSS	AIC	AICc
Casson	$\sigma^{1/2} = \sigma_{y0}^{1/2} + (\eta_\infty \dot{\gamma})^{1/2}$	$2: \sigma_{y0}, \eta_\infty$	0.01058	0.2007	0.6269	2.8489
Herschel-Bulkley	$\sigma = \sigma_{y0} + K\dot{\gamma}^n$	$3: \sigma_{y0}, K, n$	0.00607	0.1171	1.5208	5.8697
Carreau-Yasuda	$\eta = (\eta_0 - \eta_\infty)\left(1 + (\lambda\dot{\gamma})^a\right)^{(n-1)/a} + \eta_\infty$ $\sigma = \eta\dot{\gamma}$	$5: \eta_0, \eta_\infty, \lambda, a, n$	0.01069	0.1905	6.5211	22.0203
Modified Cross	$\eta = (\eta_0 - \eta_\infty)\frac{1}{(1 + (\lambda\dot{\gamma})^m)^a} + \eta_\infty$ $\sigma = \eta\dot{\gamma}$	$5: \eta_0, \eta_\infty, \lambda, a, m$	0.01074	0.1913	6.5342	22.0334
BMP	$\varphi = 1/2\left[-(k\lambda\dot{\gamma}^2 - \varphi_0) + ((k\lambda\dot{\gamma}^2 - \varphi_0)^2 + 4k\lambda\dot{\gamma}\varphi_\infty)^{1/2}\right]$ $\sigma = \dot{\gamma}/\varphi$	$4: \varphi_0, \varphi_\infty, \lambda, k$	0.04169	0.7635	7.3986	15.4742
Apostolidis et al.	$\gamma = \gamma_e + \gamma_p \leftrightarrow \dot{\gamma} = \dot{\gamma}_e + \dot{\gamma}_p$ $\quad \dot{\gamma}_e = \dot{\gamma}_p - \frac{\gamma_e}{\gamma_{max}}\lvert\dot{\gamma}_p\rvert$ $\dot{\xi} = k_\lambda\left(-\xi\sqrt{\frac{4\mu\dot{\gamma}_p}{\sigma_{y0}}} + (1-\xi)\right)$ $\quad \gamma_{max} = \min\left(\frac{\gamma_0}{\lambda^2}, \gamma_{max}\right)$ $\dot{G} = k_G\xi(G_0 - \xi G)$ $\quad \sigma = G\gamma_e + \eta_\infty\dot{\gamma}_p$	$6: \gamma_\infty, \gamma_0, \sigma_{y0}, HCT, c_f, k_G, k_\xi$	0.0175	0.3193	9.3682	46.3237

TRANSIENT FLOW BLOOD MODELING

- More interesting…need a model that can "evolve" the viscosity term as function of shear rate, viscosity, fluidity or structure…

- Our best steady state models cannot predict all aspects of these flows with accuracy

1. Bureau et al. (1980) Biorheology; 2. Bird, Armstrong Hassager (1987) Dynamics of Polymeric Liquids; 3. Merger et al. (2016) Applied Rheology.

BUREAU ET AL. SAWTOOTH SHEAR RATE FUNCTION

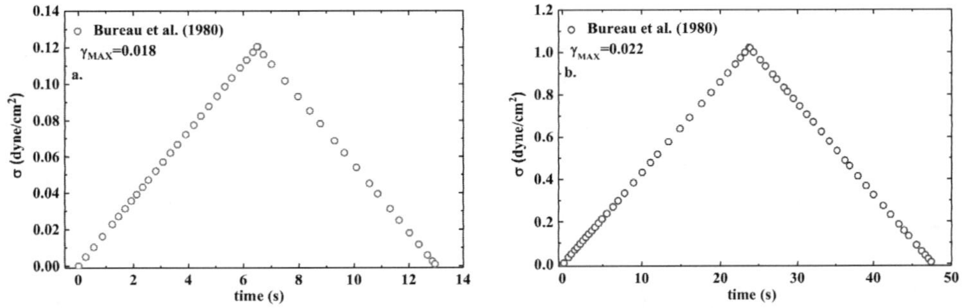

$$\dot{\gamma} = \gamma_{MAX} t \qquad \text{for } t \le t_{max}/2$$
$$\dot{\gamma} = \gamma_{MAX}(t_{max} - t) \quad \text{for } t > t_{max}/2$$

1. Bureau et al. (1980) Biorheology.

TRANSIENT MODELS

1. Combining White-Metzner and Carreau-Yasuda

$$\frac{\eta - \eta_\infty}{\eta_0 - \eta_\infty} = \left[1 + (\lambda\dot{\gamma})^a\right]^{(n-1)/a} \longleftrightarrow \boldsymbol{\sigma} + \frac{\eta}{G}\dot{\boldsymbol{\sigma}} = \eta\dot{\boldsymbol{\gamma}}$$

2. Stickel et al. "transient Herschel-Bulkley"

$$\frac{1}{G}\dot{\sigma} + \max\left(0, \frac{|\sigma| - \sigma_{y0}}{k|\sigma|^n}\right)^{1/n} \sigma = \dot{\gamma}$$

1. Bureau et al. (1980) Biorheology; 2. Bird, Armstrong Hassager (1987) Dynamics of Polymeric Liquids; 3. Merger et al. (2016) Applied Rheology; 4. Stickel et al. (2013) JoR.

NORMAL BLOOD TRANSIENT DATA

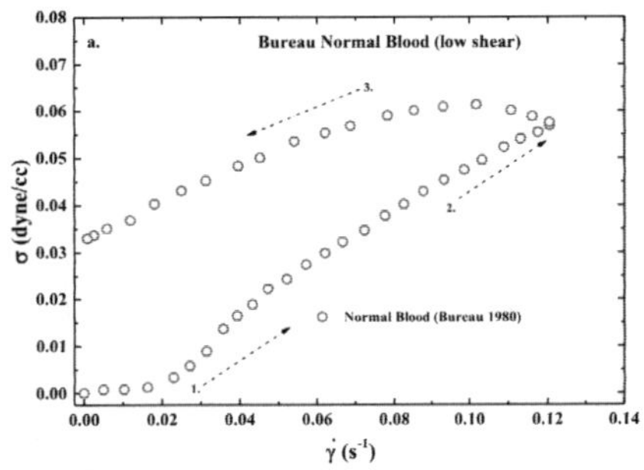

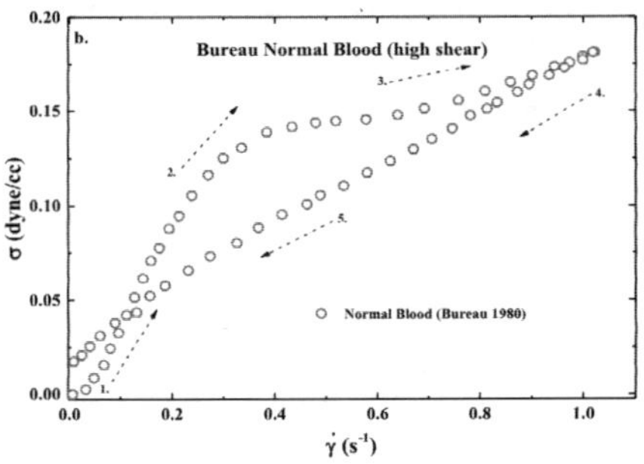

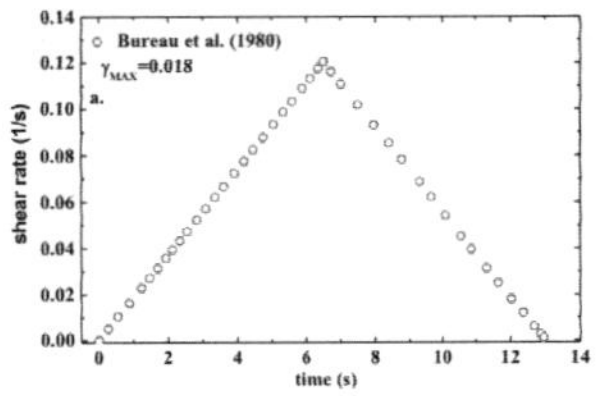

$$\dot{\gamma} = \gamma_{MAX} t \quad \text{for } t \leq t_{max}/2$$
$$\dot{\gamma} = \gamma_{MAX}(t_{max} - t) \quad \text{for } t > t_{max}/2$$

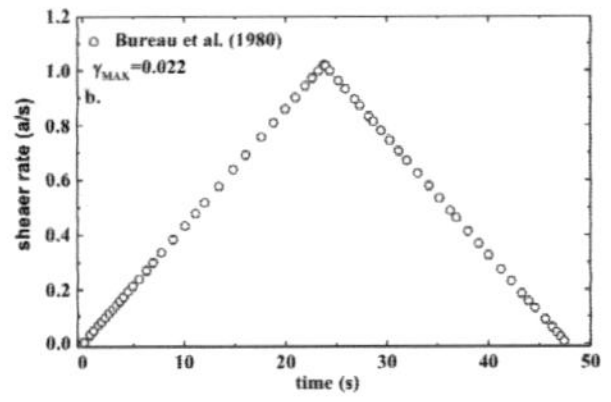

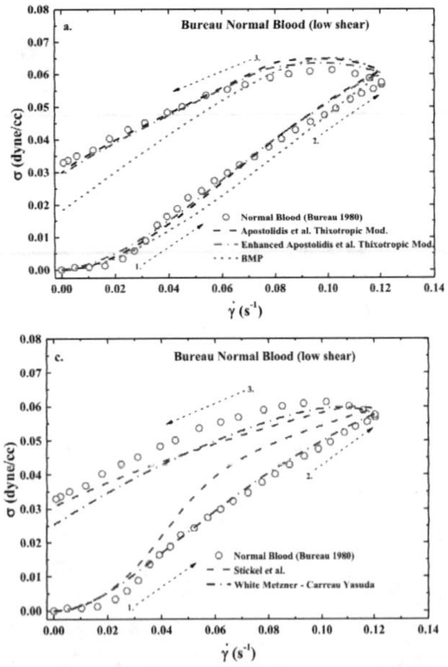

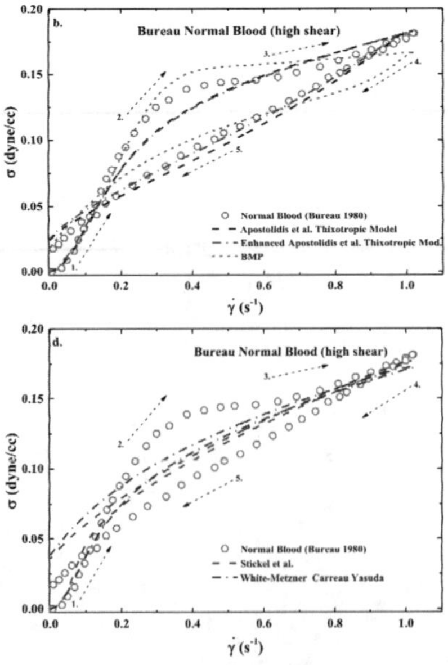

1. Bureau et al. (1980) Biorheology; 2. Armstrong et al. (2016) JoR; 3. Apostolidis et al. (2015) JoR.

STRUCTURE EVOLUTION

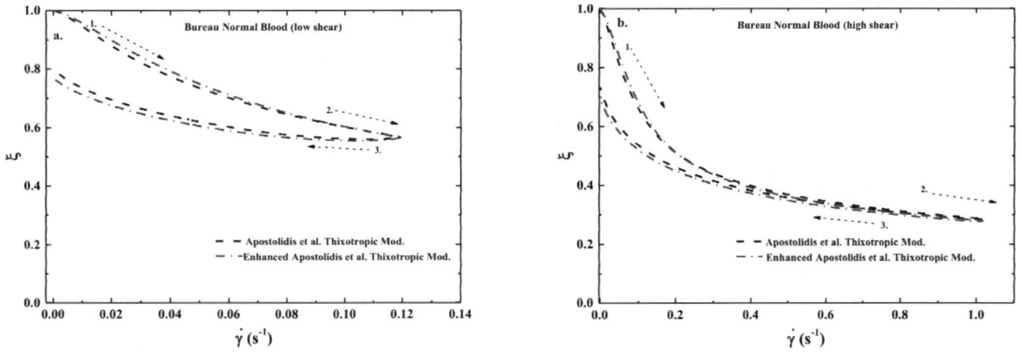

1. Bureau et al. (1980) Biorheology; 2. Armstrong et al. (2016) JoR; 3. Apostolidis et al. (2015) JoR.

BUREAU ET AL. BLOOD FITTING RESULTS

Enhanced Apostolidis et al. Thixotropic Blood Model						
Notes: 8 Par. ξ[0,1], (TCHC) Thixotropic model, 3 ODEs		Blood type (Bureau et al. 1980)				
Par.	Description	Normal	Diabetic	Umbilical	Anemic	Units
c_f	Fibrinogen concentration	0.186	0.100	0.124	0.100	g/dL
k_ξ	Time constant evolution of λ	1.316	2.350	2.865	2.622	s^{-1}
k_G	Time constant evolution of G	0.541	1.217	1.695	1.252	s^{-1}
$\tau_{y,0}$	Yield stress	0.025	0.016	0.018	0.016	dyne/cc
η_∞	Viscosity	0.043	0.043	0.043	0.043	dyne s/cc
G	Elastic Modulus	0.648	0.399	0.460	0.400	dyne/cc
K_{ST}	Structural contribution to viscosity	0.061	1.306	0.293	0.888	dyne s/cc
n	Power law of structural viscosity	0.843	1.191	1.262	1.144	-
	Results of parameter fitting:					Average:
	Cost Function:	2.115E-03	5.638E-03	1.900E-03	2.232E-03	2.97E-03
	AIC:	3.683	5.644	3.469	5.191	4.497

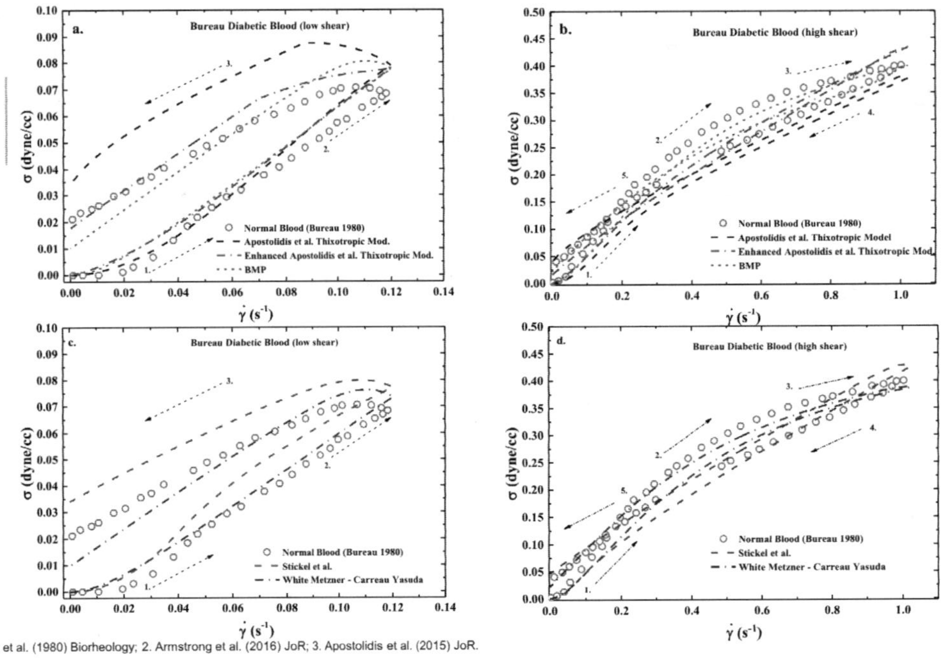

1. Bureau et al. (1980) Biorheology; 2. Armstrong et al. (2016) JoR; 3. Apostolidis et al. (2015) JoR.

TRANSIENT FITTING RESULTS

Cost Func:					
Model	Normal	Diabetic	Umblical	Anemic	AVE.
1. EAAB	2.115E-03	5.638E-03	1.900E-03	2.232E-03	2.97E-03
2. BMP	4.21E-03	4.99E-03	3.93E-03	2.23E-03	3.84E-03
*3. WM-CY	4.868E-03	5.639E-03	4.086E-03	4.843E-03	4.86E-03
4. AAB	2.29E-03	1.09E-02	2.58E-03	7.00E-03	5.70E-03
*5. DHB	6.537E-03	1.119E-02	4.141E-03	8.657E-03	7.63E-03

AIC:					
Model	Normal	Diabetic	Umblical	Anemic	AVE.
1. BMP	2.411	1.398	0.924	-0.210	1.131
2. AAB	-0.159	2.966	0.078	2.078	1.241
*3. DHB	1.939	3.014	1.026	2.501	2.120
4. EAAB	3.683	5.644	3.469	5.191	4.497
*5. WM-CY	5.350	5.644	4.999	5.339	5.333

***Dynamic HB and White-Metzner with Carreau-Yasuda expression for evolving viscosity thrown out due to fact that neither can replicate characteristic curve of the hysteresis loop**

SPP-"SAWTOOTH"

$$A = [\gamma(t)\ \ \dot{\gamma}(t)\ \ \sigma(t)]$$

(*Data vectors)

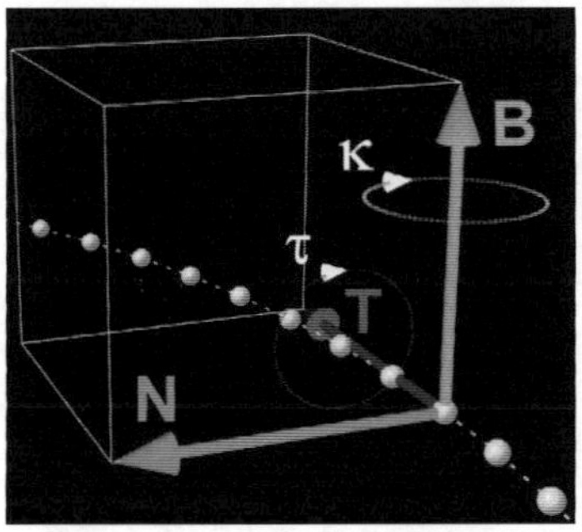

$$\mathbf{T} = \frac{\dot{\mathbf{A}}}{\|\dot{\mathbf{A}}\|} \qquad \mathbf{N} = \frac{\dot{\mathbf{T}}}{\|\dot{\mathbf{T}}\|} \qquad \mathbf{B} = \mathbf{T} \times \mathbf{N}$$

$$\mathbf{G}_t^{'}(\mathbf{t}) = -\frac{\mathbf{B}_{\gamma(t)}}{\mathbf{B}_{\sigma(t)}} \qquad \eta_t^{'}(\mathbf{t}) = -\frac{\mathbf{B}_{\dot{\gamma}(t)}}{\mathbf{B}_{\sigma(t)}} \qquad \boldsymbol{\sigma}_{\mathbf{off}}(\mathbf{t}) = \frac{\mathbf{1}}{\boldsymbol{\omega}^2}\ddot{\boldsymbol{\sigma}}(\mathbf{t})$$

$$\sigma(t) = G_t^{'}\gamma(t) + \eta_t^{'}\dot{\gamma}(t) + \sigma_{off}(t)$$

1. Rogers (2017) Rheol. Acta

ANALYZING SAWTOOTH

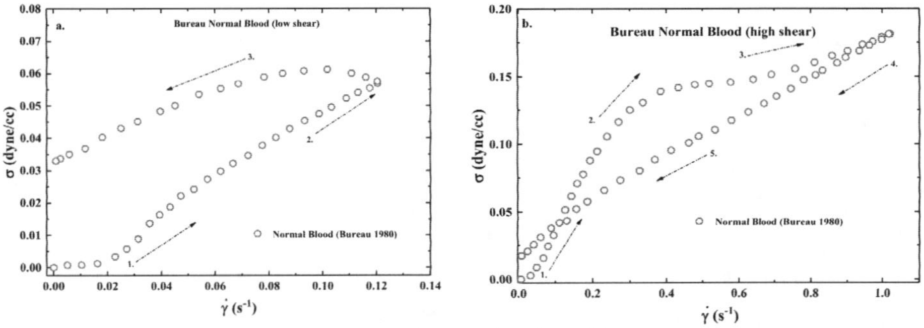

$$\dot{\gamma} = \gamma_{MAX} t \qquad \text{for } t \le t_{max}/2$$
$$\dot{\gamma} = \gamma_{MAX}(t_{max} - t) \quad \text{for } t > t_{max}/2$$

$$\sigma(t) = G_t^{'}\gamma(t) + \eta_t^{'}\dot{\gamma}(t) + \sigma_{off}(t) \qquad G_t^{'}(t) = -\frac{B_{\gamma(t)}(t)}{B_{\sigma(t)}(t)} \qquad \eta_t^{'}(t) = -\frac{B_{\dot{\gamma}(t)}(t)}{B_{\sigma(t)}(t)}$$

1. Bureau et al. (1980) Biorheology.

SAWTOOTH FINGERPRINTS

$\mathbf{A}_E = [\gamma\ \dot{\gamma}\ \mathbf{G}_t^{'}]$ $\mathbf{A}_V = [\gamma\ \dot{\gamma}\ \eta_t^{'}]$

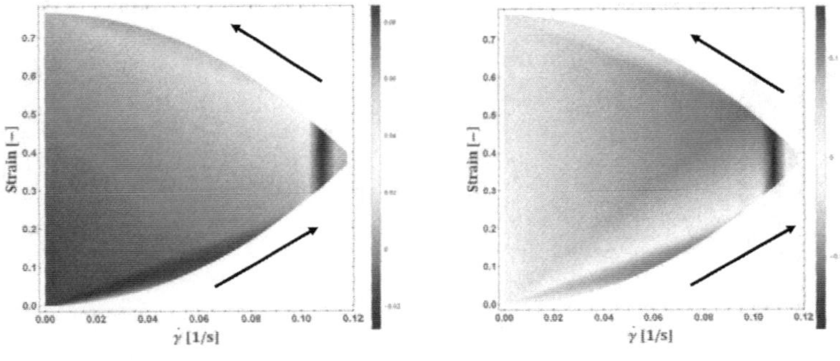

a. b.

Fig. 20a Elastic projection; **b** Viscous projection using Series of physical phenomena moduli fingerprinting normal blood Bureau et al (1980)

1. Bureau et al. (1980) Biorheology; Helton (2017) USMA.

SPP SAWTOOTH FINGERPRINTS

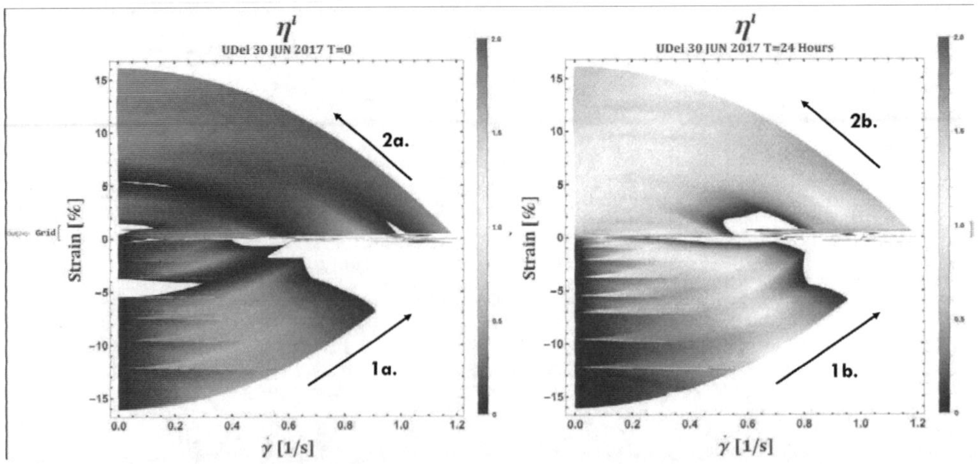

Blood sample age: 0 hours **Blood sample age: 24 hours**

1. Bureau et al. (1980) Biorheology; Helton (2017) USMA; Horner et al. (2017) UDEL.

Bureau Normal Blood; low shear rate

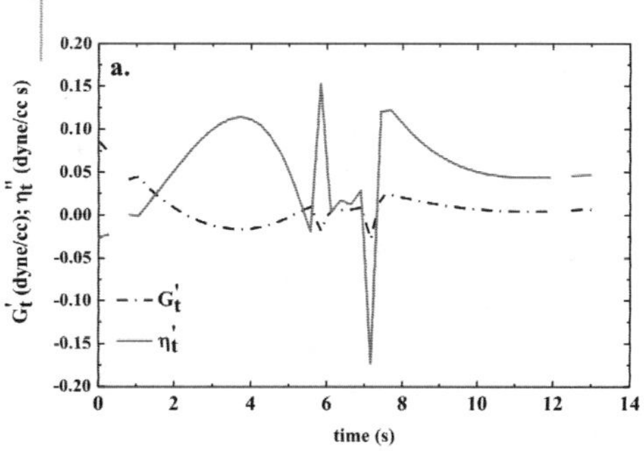

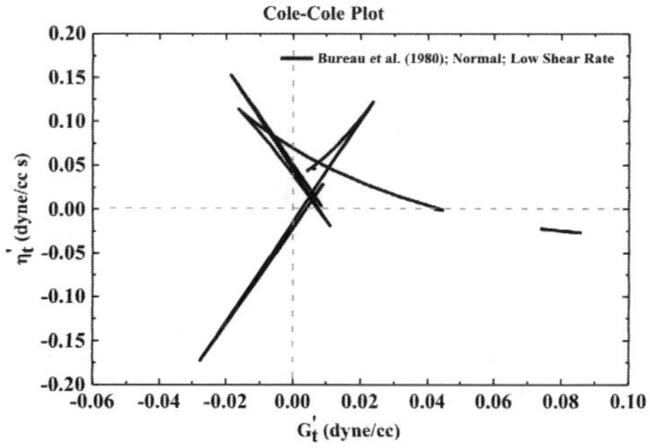

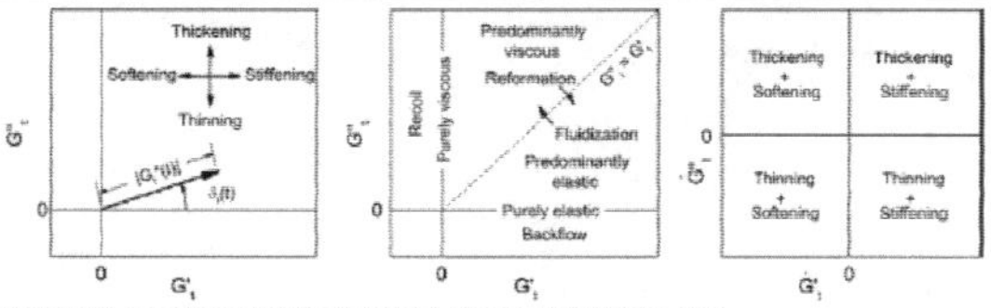

1. Sousa et al. (2013) Biorheology; 2. Armstrong et al. (2016) JoR; 3. Apostolidis et al. (2015) JoR; 4. Rogers (2017) Rheol. Acta.

SPP-"LAOS"

$$A = [\gamma(t) \;\; \dot{\gamma}(t)/\omega \;\; \sigma(t)]$$

(*Data vectors)

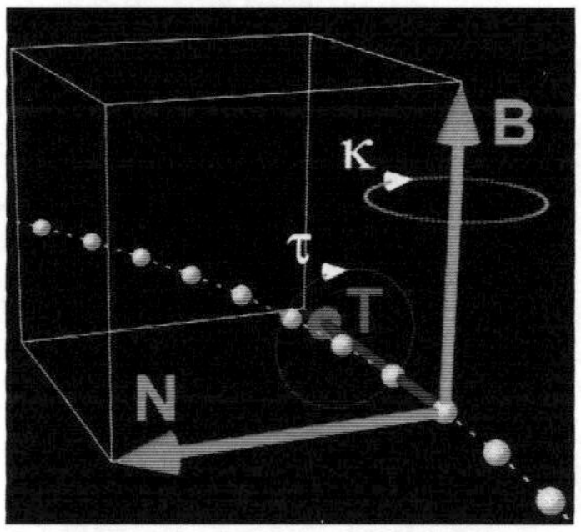

$$\mathbf{T} = \frac{\dot{\mathbf{A}}}{\|\dot{\mathbf{A}}\|} \qquad \mathbf{N} = \frac{\dot{\mathbf{T}}}{\|\dot{\mathbf{T}}\|} \qquad \mathbf{B} = \mathbf{T} \times \mathbf{N}$$

$$\mathbf{G}_t'(\mathbf{t}) = -\frac{\mathbf{B}_{\gamma(t)}}{\mathbf{B}_{\sigma(t)}} \qquad G_t''(\mathbf{t}) = -\frac{\mathbf{B}_{\dot{\gamma}(t)}}{\mathbf{B}_{\sigma(t)}} \qquad \boldsymbol{\sigma}_{\mathbf{off}}(\mathbf{t}) = \frac{1}{\boldsymbol{\omega}^2}\ddot{\boldsymbol{\sigma}}(\mathbf{t})$$

$$\sigma(t) = G_t'\gamma(t) + G_t''\dot{\gamma}(t)/\omega + \sigma_{off}(t)$$

$$\eta_t'(t) = \frac{G_t''(t)}{\omega}$$

1. Rogers (2017) Rheol. Acta

LAOS REFRESHER

$\gamma_0 = 1.0; \omega = 1\,(\mathrm{rad/s})$

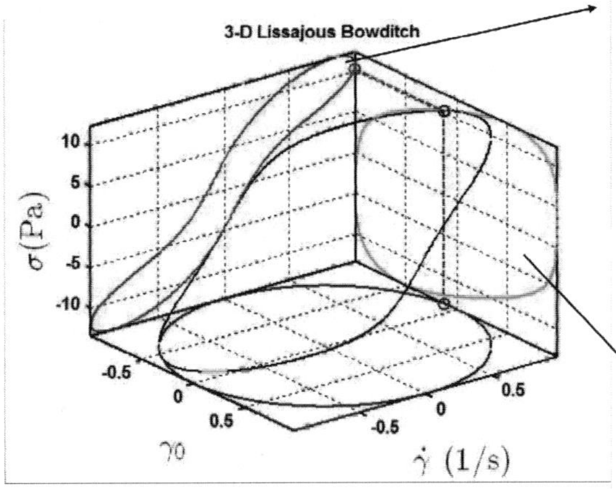

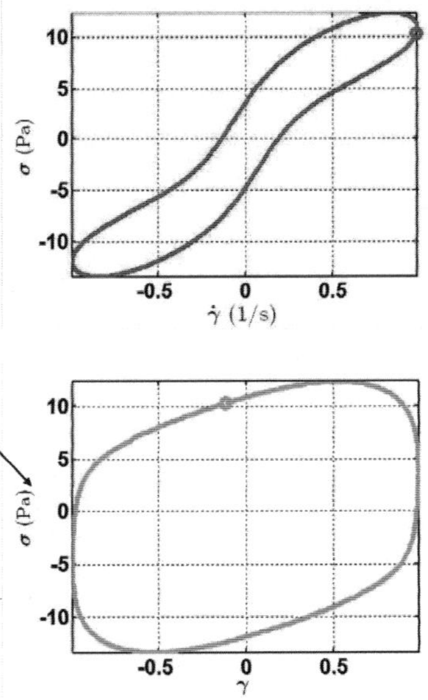

$\gamma = \gamma_0 \sin(\omega t)$ $\qquad \dot{\gamma} = \dot{\gamma}_0 \cos(\omega t)$

LAOS FINGERPRINTS

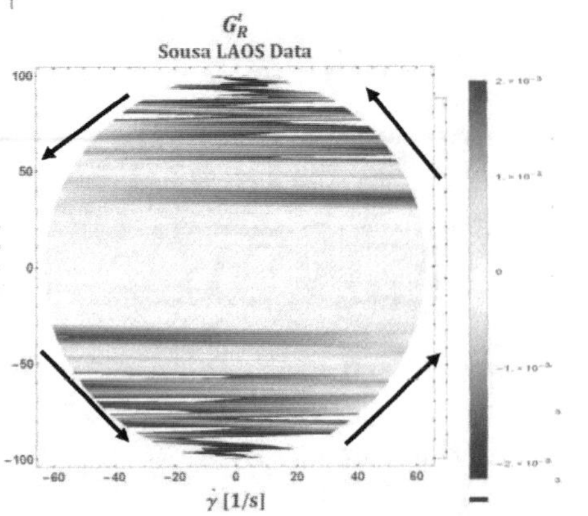

$\gamma_0 = 100; \omega = 0.631\,(\mathrm{rad/s})$

$\gamma(t) = \gamma_0 \sin(\omega t)$ $\qquad \dot{\gamma}(t) = \gamma_0 \omega \cos(\omega t)$

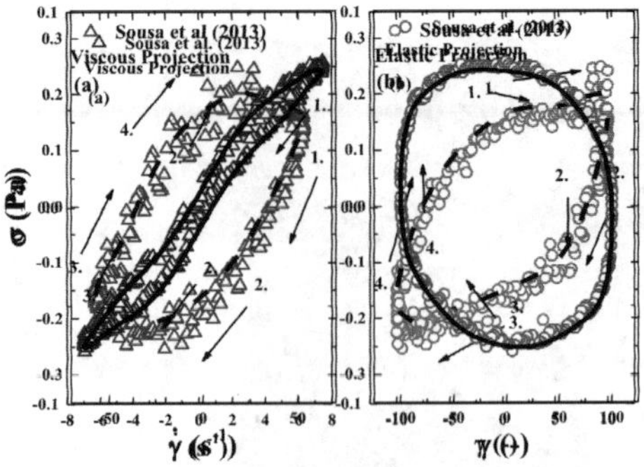

-Viscous projection Sousa et al. (2013) LAOS data; b. Elastic projection Sousa et al. (2013) LAOS data. (Sousa et al. 2013; Armstrong et al. 2016)

1. Sousa et al. (2013) Biorheology; 2. Armstrong et al. (2016) JoR; 3. Apostolidis et al. (2015) JoR; 4. Rogers (2017) Rheol. Acta.

COLE-COLE PLOT

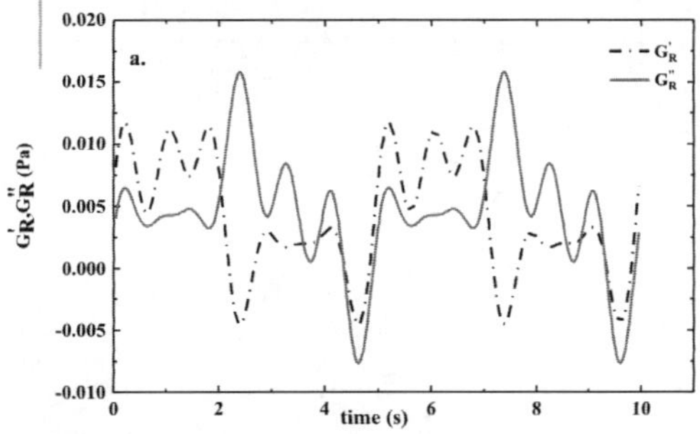

$$\gamma_0 = 10; \omega = 0.631(\text{rad/s})$$

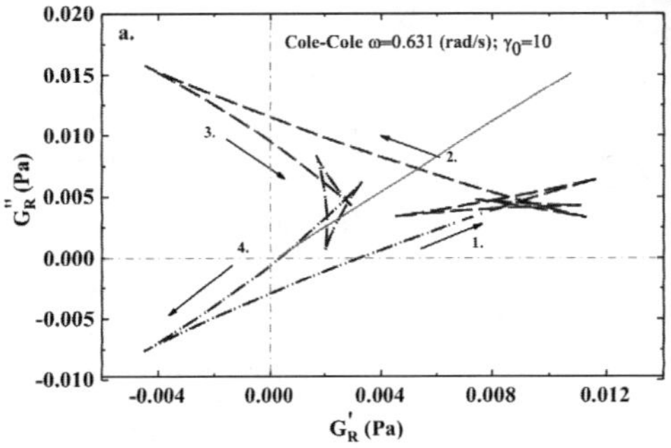

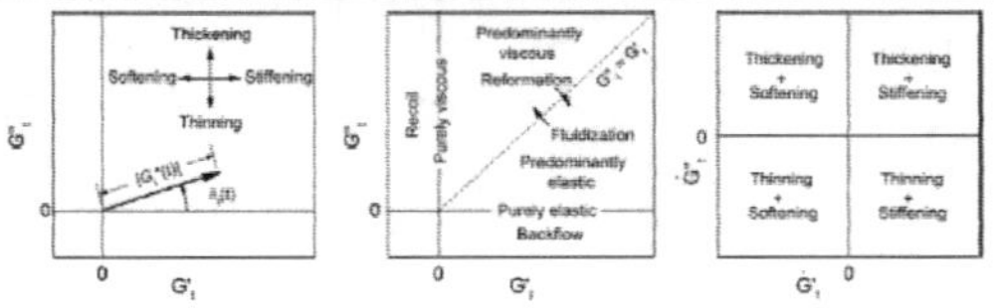

1. Rogers (2017) Rheol. Acta

- Evolution of the elastic vs. viscous properties
- Correlate to elastic/solid softening or stiffening
- Viscosity shear thickening or shear thinning

FUTURE WORK

- Blood rheological signatures as a function of temperature
- How does blood sugar affect the rheology?
- How does dehydration affect blood rheology
- Continued model development
- IRB/ lab testing....HDL/LDL/Triglycerides/sugar and hematocrit
- Step up/down fingerprints and mapping
- SPP-Structural

THANK YOU!

Collaborators at University of Delaware:

Jeff Horner, Dr. Norm Wagner, Dr. Antony Beris

Collaborators at USMA:

Dr. Ken Wickiser, Jaime Reilly, Jesse Trump,

LTC Tim Hill

Cadets:

CDT Michael Deegan, CDT Michael Clarion, CDT Tyler Helton, CDT Austin Breed, CDT Jesse Hudgins, CDT Evan Ousley

Dr. Simon Rogers

QUESTIONS

SPP SAWTOOTH FINGERPRINTS

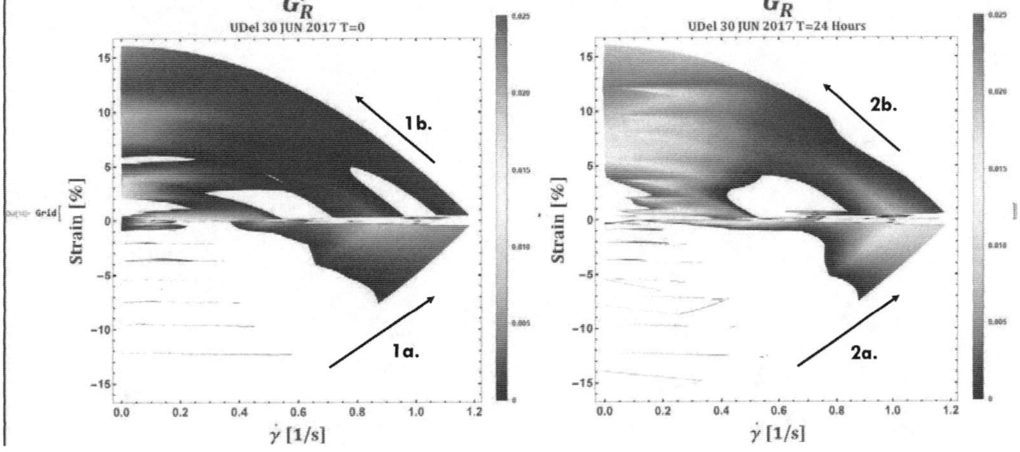

Blood sample age: 0 hours **Blood sample age: 24 hours**

FFT

$$\sigma(t) = \gamma_0 \sum_{n=1,odd}^{M} \left[G_n^{'}(\omega, \gamma_0)\sin(n\omega t) + G_n^{''}(\omega, \gamma_0)\cos(n\omega t) \right]$$

- fft(σ); Matlab (**imaginary part - elastic**; **real part - viscous**); alternance
- [G'; G"]=A\b... (for linear region only)
- Linear viscoelasticity; 1st harmonic response (ie. G' and G")
- 3rd harmonic signals departure from linearity
- Odd harmonics only: material response unchanged if coordinate system is reversed

[1] M. Wilhelm, Macromol. Mater. Eng. 287(2) (2002) 83-105. [2] A.J. Giacomin, J.M.Dealy, Techniques of Rheological Measurement. Chapman and Hall.

6

STRESS DECOMP.

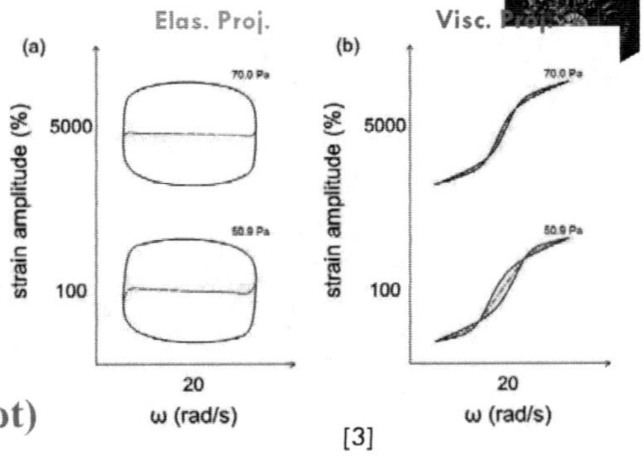

[3]

$$\sigma = \sigma' + \sigma''$$

$$\sigma' = \frac{\sigma(\gamma,\dot{\gamma}) + \sigma(\gamma,-\dot{\gamma})}{2} = \gamma_0 \sum_{n,odd}^{M} G_n'(\omega,\gamma_0)\sin(n\omega t)$$

$$\sigma'' = \frac{\sigma(\gamma,\dot{\gamma}) + \sigma(-\gamma,\dot{\gamma})}{2} = \gamma_0 \sum_{n=1,odd}^{M} G_n''(\omega,\gamma_0)\cos(n\omega t)$$

- Elastic contribution to stress has odd symmetry with respect to strain
  Odd $f(x): \sin(x) = -\sin(-x)$
  [1,2] [1,2]
- Viscous contribution to stress has even symmetry with respect to strain rate
  Even $f(x): \cos(x) = \cos(-x)$

[1] K.S. Cho, K. Hyun, K.H. Ahn, S.J. Lee. J. Rheol. 49(3) (2005) 747-758. [2] R.H. Ewoldt, A.E. Hasoi, G.H. McKinley. J. Rheol. 52(6) (2008) 1427-1458.
[3] J.M. Kim. Effect of the range of attraction on rheology. Mirostructure, thermodynamics of thermoreversible gel. PhD Thesis, UDEL (2013).

7

CHEBYCHEV

$$\sigma = \sigma' + \sigma'' \qquad x = \frac{\gamma}{\gamma_0} \qquad y = \frac{\dot{\gamma}}{\dot{\gamma}_0}$$

$$\sigma' = \gamma_0 \sum_{n,odd}^{M} e_n(\omega,\gamma_0)T_n(x) \qquad e_n = G_n'(-1)^{(n-1)/2} \quad n: odd$$

">0 strain stiffening"
"<0 strain hardening"

$$\sigma'' = \dot{\gamma}_0 \sum_{n=1,odd}^{M} v_n(\omega,\gamma_0)T_n(y) \qquad v_n = \frac{G_n''}{\omega} \quad n: odd$$

">0 shear thickening"
"<0 shear thinning"

***Apply only to the 3rd harmonic (when 3rd harmonic dominates)**

T_n ...nth order Chebychev polynomials of first kind

[1] K.S. Cho, K. Hyun, K.H. Ahn, S.J. Lee. J. Rheol. 49(3) (2005) 747-758. [2] R.H. Ewoldt, A.E. Hasoi, G.H. McKinley. J. Rheol. 52(6) (2008) 1427-1458.
[3] J.M. Kim. Effect of the range of attraction on rheology. Mirostructure, thermodynamics of thermoreversible gel. PhD Thesis, UDEL (2013).

8

LOW DIMENSIONAL INTRINSIC MATERIAL FUNCTIONS ...

$$\sigma(t;\gamma_0,\omega) = \sum_j \sum_n^j \gamma_0^j \left[G'^{(j)}_{jn}(\gamma_0,\omega)\sin(\omega t) + G''^{(j)}_{jn}(\gamma_0,\omega)\cos(\omega t) \right]$$

$$\updownarrow$$

$$\sigma(t;\gamma_0,\omega) = \gamma_0 \left\{ G'(\omega)\sin(\omega t) + G''(\omega)\cos(\omega t) \right\}$$
$$+ \gamma_0^3 \left\{ [e_1](\omega)\sin(\omega t) + \omega[v_1](\omega)\cos(\omega t) - [e_3](\omega)\sin(3\omega t) + \omega[v_3](\omega)\cos(3\omega t) \right\}$$
$$+ \mathrm{H.O.T.}$$

[1] R.H. Ewoldt, J. Rheol., 57(1) (2012) 177-195. [2] R.H. Ewoldt, N.A. Bharadwaj, Rheol. Acta, 57(2013) 210-219.

9

491337 How Confinement and Hydrophobicity Affect CO_2 Diffusion from the Bulk Phase to the Active Site of Human Carbonic Anhydrase II: A Study Based on Coarse-Grained Molecular Dynamics Simulation and the Markov-State Model

Abstract:

While nanoconfinement may enhance protein stability and reusability, its impact on the apparent activity of encapsulated enzymes in uncertain due to the complicated interactions among enzyme, substrate and product under confinement. To gain a knowledge for the design, synthesis and application of immobilized enzymes in nanopores, here we applied coarse-grained molecular dynamics simulation (CGMD) and the Markov-state model (MSM) to study CO2 diffusion from the bulk phase to the active site of human carbonic anhydrase II (hCA-II), an enzyme promising for intensified CO2 sequestration. Special efforts were directed at examining the effects of confinement and surface hydrophobicity on hCA-II conformation, orientation and interaction with CO2 molecules. The simulation results show that a hydrophobic pore gives an enhanced local CO2 density near the inner surface, but hinders the diffusion of CO2 from inner surface to active site of hCA-II. A hydrophilic pore hinders the adsorption of CO2 molecules due to the repulsion by the hydrophilic environment. However once CO2 enters the pore the repulsive interaction with the confinement wall would push CO2 toward the encapsulated CA, favouring the transport of CA to the active sites. MSM time-evolved probability analysis shows weak hydrophilic confinement gives the highest active-site-occupied probability. Above results offer a kinetic view with molecular insight and are useful for the design and utilization of encapsulated CA for CO2 sequestration.

495660 Printing Fully Stretchable Thin-Film-Transistor Array

Abstract:

Development of high-performance electronics that can intimately and seamlessly integrate with human bodies and biological systems at the cellular level are the key for the next generation consumable wearable electronics and biomedical devices, which requires device to possess biosystem-compatible mechanical and chemical properties. Recently, intrinsically stretchable polymeric electronic materials including semiconductors, conductors and dielectrics are intensively reported and considered as ideal candidates for such applications given their unique electrical and mechanical properties. However, the porous structures and poor chemical resistance prevent intrinsically stretchable polymeric materials to be integrated into functional electronics (e.g. thin-film transistor array) through the conventional fabrication process. Here, we present the first fully stretchable thin-film transistors array through the inkjet printing of intrinsically stretchable semiconductors. To enable this solution-processed assemble by printing, we invented a new perfluorinated dielectric material that is 1) extremely tolerate to most organic solvent for the fabrication, 2) directly photopatternable with sub 10 μm feature size by conventional lithographer and a chemically orthogonal development process and 3) highly stretchable with > 200% stretchability. In addition, we have developed an ink containing conjugated semiconducting polymer with siloxane-azide crosslink for the inkjet-printing patterning of semiconductor. Using perfluorinated dielectric material as substrate, dielectric layer and passivation, we can achieve the direct, reliable inkjet-printed pattern of intrinsically stretchable semiconductor into the transistor array with record-high active materials coverage, device yield and device density, and ultra-high stability in chemical and biological solutions. The electrical and mechanical characterizations show that the transistor array kept its original performance without significant degradation under 100% uniaxial strains and through multiple cycles. As the initial demonstration, the stretchable transistor array has been used as the backplane for fabrication of fully stretchable light-emitting diode (LED) display. We believe this unique dielectric materials and fabrication process can be applied to virtually any stretchable materials and impact on the low-cost fabrication of fully stretchable electronics for wearable and biomedical device in industry.

501615 Conjugated Macrocycles for Ultra-Sensitive Organic Photodetectors

Abstract:

Organic photodetectors (OPDs) are attractive for their high optical absorption coefficient, broad wavelength tunability, and compatibility with light-weight and flexible devices. One critical parameter that limits OPD's performance is the high dark current, because it determines the noise current level and sensitivity of an OPD. I will introduceÂ a new molecular design that enables ultra-sensitive organic photodetectors. We use a rigid, conjugated macrocycle as the electron acceptor in devices to obtain high photocurrent and low dark current. TheÂ macrocycle consists of redox-active diphenyl perylenediimides that are wrapped into a tetrameric structure.Â We make a direct comparison between the devices made with the macrocyclic acceptor and an acyclic control molecule; we find that the superior performance of the macrocycle originates from its rigid, conjugated, and cyclic structure. The macrocycleâ€™s rigid structure reduces the number of charged defects originating from deformed sp^2 carbons and covalent defects from photo/thermo-activation. With this molecular design we are able to suppress dark current density while retaining high responsivity in an ultra-sensitive non-fullerene OPD. Importantly, we achieve a detectivity of ~10^{14}Jones at near zero bias voltage. This is without the need for extra carrier blocking layers commonly employed in fullerene-based devices. Our devices are comparable to the best fullerene-based photodetectors, and the sensitivity at low working voltages (< 0.1 V) is a record for non-fullerene OPDs.

References:

[1] J. Am. Chem. Soc. 2016, 138 (39), 12861-12867.

[2] J. Am. Chem. Soc. 2016, 138 (50), 16426-16431.

499851 Connecting Crystalline Domains: Blending Conjugated Polymers of Differing Molecular Weights for Enhanced Charge Transport

Abstract:

Solution processability has brought significant attention to organic electronics as low cost alternatives to traditional inorganic electronics. Ordered self-assembly of semiconducting polymers in solution enhances charge transport in processed organic thin film transistors due to percolated crystalline domains on the mesoscale. However, methods to induce so-called â€œtie chainsâ€ between crystalline nanofibers are lacking. Here, we propose blending high (95 kDa) and low (37 kDa) molecular weight mixtures of poly(3-hexythiophene) (P3HT) with wide polydispersity to elucidate the relationship between tie chain formation and two-dimensional charge transport. Ultraviolet-irradiation and solution aging was utilized to induce aggregation of mixtures of high and low molecular weight P3HT. Solutions were subsequently blade-coated on bottom gate, bottom contact transistors to record field effect transistor charge mobility. Using atomic force microscopy (AFM) image analysis, it was found that increasing the percentage of high molecular P3HT resulted in increased fiber density and improved orientation of fibers in the direction of blade coating. However, optimal charge mobility occurred at a 50% mixture of high and low molecular weight P3HT, indicating a tradeoff between tie chain formation and aggregate phase separation. The development of shoulder peaks in solution UV-vis spectra and increased film UV-vis dichroic ratio values provide further support of this tradeoff. These further suggest the important of tie chains for enhanced charge transport in polymeric semiconductors but also provide new guidance for conjugated polymer synthesis.

486163 Ultrafast Carrier Dynamics in Bimetallic Nanostructures-Enhanced Methylammonium Lead Bromide Perovskites

Abstract:

An unprecedented rise in efficiency has been observed in methylammonium lead trihalide ($MAPbX_3$, X=I, Cl, Br) perovskite solar cells (PSCs) reaching >20% recently. While $MAPbI_3$ perovskites have been the primary focus in PSCs, the poor stability and rapid degradation in humidity has remained a major obstacle in $MAPbI_3$. $MAPbBr_3$ is a promising alternative to MAPbI3 with a large 2.2 eV bandgap which gives rise to a high open circuit voltage, long exciton diffusion length (>1.2 Î¼m) enabling good charge transport in devices, and higher moisture stability due to its stable cubic phase and low ionic mobility relative to the pseudocubic $MAPbI_3$. However, a relatively large exciton binding energy (76 meV) and poor light absorption beyond its band edge at 550 nm has limited the efficiencies for $MAPbBr_3$ solar cells.Â In this work, we examine the impact of hybrid bimetallic Au/Ag core/shell plasmonic nanostructures on the carrier dynamics of methylammonium lead tribromide ($MAPbBr_3$) mesoporous PSCs.[1] Plasmon-enhanced PSCs incorporated with Au/Ag nanostructures demonstrated improved light harvesting and increased power conversion efficiency by 26% relative to reference devices. Two complementary spectral techniques, transient absorption spectroscopy (TAS) and time-resolved photoluminescence (trPL), were employed to gain a mechanistic understanding of plasmonic enhancement processes. TAS revealed a decrease in photobleach formation time which suggests that the nanostructures improve hot carrier thermalization to an equilibrium distribution, relieving hot phonon bottleneck in $MAPbBr_3$ perovskites. TAS also showed a decrease in carrier decay lifetimes indicating nanostructures enhance photoinduced carrier generation and promote efficient electron injection into TiO_2 prior to bulk recombination. Further, nanostructure-incorporated perovskite films demonstrated quenching in steady-state PL and decreases in trPL carrier lifetimes, providing further evidence of improved carrier injection in plasmon-enhanced mesoporous PSCs.

1. H. F. Zarick, A. Boulesbaa, A. A. Puretzky, E. M. Talbert, Z. Debra, N. Soetan, D. B. Geohegan, and R. Bardhan*, ***Nanoscale***, 2017, 9, 1475-1483

500192 Aggregation Structure and Solvation of Â Solution-Phase PTB7

Abstract:

Organic photovoltaics are a low-cost, flexible alternative to traditional silicon-based photovoltaics. The complex kinetically trapped structure of the active layer in organic photovoltaics plays a determining role in the device efficiency. Arguably, the inability to rationally control the mesoscale morphology of the active layer is the limiting step in the successful proliferation of organic photovoltaics. A rigorous understanding of solvation properties of donor and acceptor materials is essential in order to control device morphology. Empirical quantities such as the Hansen solubility parameters have been proposed as a means to quantify solubility, however they have not been widely applied to organic electronics. In this work, we use atomistic molecular simulation coupled with free energy sampling techniques to quantify and understand solubility and structure in organic photovoltaics. We calculate free energies of dissociation for PTB7 oligomers in several widely-used solvents. We compare the resulting solubilities to both experiment and Hansen solubility parameters which we determine in simulation. We suggest that in-silico Hansen solubility parameters can be used for high-throughput screening of solvents in organic electronics due to their close agreement with free energy surfaces of solvation. We also examine the complex array of structures assumed by conjugated organic photovoltaic polymers in a variety of solvents and examine their scaling behaviors.

490393 Self-Aligned Strategy for Printed Electronics

Abstract:

Printed electronics is an emerging field for additive manufacturing of electronic devices with low cost and minimal material waste for a wide range of application areas including displays, distributed sensing, smart packaging, and energy management. Moreover, its compatibility with roll-to-roll production formats and flexible substrates is desirable for high-throughput manufacturing of flexible electronics. Despite the promise, however, the roll-to-roll production of printed electronics is quite challenging due to web movement hindering accurate ink registration and high-fidelity printing. To address this challenge, we have developed a self-aligned printing strategy compatible with roll-to-roll manufacturing by utilizing capillary action of liquid inks on microstructured surfaces. In the process that we term Self-aligned Capillarity-Assisted Lithography for Electronics (SCALE), microstructured ink reservoirs and capillary channels are imprinted on plastic substrates and filled by inkjet printing of functional materials into the reservoirs. The liquid inks move under capillary flow into the adjoining channels, allowing high-resolution patterning and self-alignment of functional materials to build electronic devices with generous printing tolerance. In this talk, I will show fabrication of key building blocks (*e.g.,* transistor, resistor, and capacitor) for electronic circuits using the SCALE process on plastics, suggesting a reliable, effective, and versatile platform to realize high-throughput manufacturing of printed and flexible electronics.

501097 Helical Perylene Diimide Ribbons for Molecular Electronics

Abstract:

In this presentation, I introduce a series of helical perylene diimide (PDI) ribbons as efficient electron acceptors. The ribbons are constructed by fusing perylene diimide units together with a two-carbon bridge. In the first part, we show that these helical ribbons are highly efficient non-fullerene acceptors for organic solar cells. By varying lengths of the ribbons and controlling film morphology in bulkjunction solar cells, we achieved a record power conversion efficiency of ~8.5 %. Femtosecond transient absorption spectroscopy reveals both electron and hole transfer processes at the donor acceptor interfaces. Atomic force microscopy reveals a mesh-like network of acceptors with pores that are tens of nanometers in diameter for efficient exciton separation and charge transport.

In the second part, we demenstrate that these ribbons represent a new molecular design that yields ultra-narrowband organic photodetectors. We fabricate charge collection narrowing photodetectors based on four helical ribbons that differ in the wavelength of their response. The photodetectors made from these materials have narrow spectral response with full-width at half maxima of < 20 nm. The devices shown here are superior by approximately a factor of 5 to those from traditional organic materials due to the narrowness of their response. Moreover, the active layers for the helical ribbon-based photodetectors are solution cast but have performance that is comparable to the state-of-the-art narrowband photodetectors made from methylammonium lead trihalide perovskite single crystals.

References:

[1]J. Am. Chem. Soc. 2014, 136 (22), 8122-8130.

[2]J. Am. Chem. Soc. 2014, 136 (43), 15215-21.

[3]Nat. Commun. 2015, 6, 8242.

[4]J. Am. Chem. Soc. 2017, DOI: 10.1021/jacs.6b13093.

[5]J. Am. Chem. Soc. 2017, DOI: 10.1021/jacs.6b13089.

492986 Engineering Pathways to New Functional Polyether Materials

Abstract:

Poly(ether)s derived from epoxides represent a versatile class of functional polymeric materials. Epoxide ring-strain depends little on monomer structure. This fact enables the potential for true compositional control of structure-property relationships in a macromolecular platform toolkit. Unfortunately, epoxides do not polymerize to high molecular weights and low dispersities by any method in common usage not requiring sophisticated synthetic skill and equipment. As a first step toward the democratization of poly(ether) chemistry, our lab has developed a class of stable, crystalline mono-μ-oxo-dialuminum (MOD) initiators that offer the control of anionic ring opening polymerization, but the tolerance to chemical functionality and access to high molecular weight of catalytic approaches. The MODs can be synthesized in a single step with purification afforded by direct crystallization from the reaction medium. The MOD-initiated living polymerizations create heterobifunctional materials at controlled molecular weights and with low dispersities. In this presentation, the historical context and inspiration, synthesis, mechanism, and finally the unique versatility of MOD chemistry will be presented.

488050 Topology Control of Bottlebrush Polymers

Abstract:

With the development of controlled polymerizations, polymer chemists have enabled the synthesis of more and more complex materials. Bottlebrush polymers are branched polymers with high branching density all originating from a linear backbone and are an example of a material with complex architecture that is accessible via controlled polymerizations. The rigidity of bottlebrush polymers provides a unique way to generate spatially self-organizing materials and have gained much interest in the soft materials community. Thus far bottlebrush polymers have been limited to linear cylindrical shape. Here we present a methodology to construct non-cylindrical bottlebrushes by using a combination of controlled polymerizations to generate a gradient of arm lengths. We implement two orthogonal chemistries, one for arm synthesis and another for backbone synthesis, to generate conical bottlebrushes as a proof of principle. Detailed kinetic studies provide support for the control and specificity of our approach. This methodology results in a material with unprecedented topology and opens a new avenue for more complex materials.

493833 A New, Facile Approach to Epoxide Polymerization

Abstract:

A new class of initiator for the living polymerization of epoxides is presented, which are characterized as mono-Âµ-oxo-dialuminum (MOD) species. MOD initiators are synthesized in high yield in a single step using as-received commercial reagents with purification being accomplished by direct crystallization from the reaction medium. We studied the characteristics of the new initiators by the polymerization and copolymerization of various functional epoxide monomers such as allyl glycidyl ether, propargyl glycidyl ether, and epichlorohydrin.Â *In-situ* FTIR and NMR spectroscopy demonstrated that the polymerization was living, and produced materials with defined end-group and backbone functionality. Molecular weight of the resultant polymers could be precisely controlled (5â€“75 kg/mol) with relatively low dispersity (1.04â€“1.40). We also studied the effect of initiator structure and were able to yield a four-fold improvement in polymerization rate over traditional anionic polymerization, as well as gain mechanistic insight. From this, we propose a mechanism that accounts for the changes in polymerization rate with MOD structure. This new synthetic tool does not require special equipment, or an inordinate amount of synthetic skill in order to create new, functional polyether-based materials from inexpensive and readily available precursors.

497170 Dynamic Chemistry Leading to Full Property Recovery Associated with Crosslink Density in Reprocessed Polymer Networks and Network Composites: Network Synthesis By Step-Growth Reactions and By Nitroxide-Mediated Polymerization

Abstract:

Conventional polymer networks are unable to be reprocessed in the melt state and recycled into high-value products because of permanent covalent crosslinks. Reprocessable networks, or covalent adaptable networks, are crosslinked polymers that contain sufficient reversible covalent bonds for network reconfiguration under proper conditions. Such networks contain dynamic linkages that can dissociate or exchange with others at the reprocessing condition. We focus on two reprocessable polymer networks that do not require pre- or post-polymerization functionalization but instead involve simple one-step chemistry with reaction of monomer and/or polymer. In one case, we show how novel use of nitroxide-mediated polymerization (NMP) involving reactions of a multifunctional radical initiator and a polymerizable monomer incorporating a stable nitroxide radical with polybutadiene and styrene lead to models for crosslinked tire rubber. (See Adv. Mater. 2016, 28, 6746.) The strongly temperature-dependent reversible capping/uncapping step in NMP provides crosslink reversibility. Based on the rubbery plateau modulus from DMA and tensile properties, these networks can be reprocessed multiple times in the melt state with the product exhibiting full recovery of properties related to crosslink density. Robust recovery is also observed in polymer composite networks made by NMP with filler such as carbon black. This one-step NMP method also yields polymer network structures that are possible with only a limited few other reaction methods or have yet to be obtained by any other method. In the second case, a step-growth polymer network has been synthesized and also exhibits full property retention after multiple reprocessing steps. In this case, the type of polymer that we employed has been described in literature as a vitrimer. We show that this particular network polymer is not a true vitrimer because both reversible and exchange reactions are present at reprocessing conditions. It is important to appreciate that reversible reactions may accompany exchange reactions in some vitirimers, because the extent of reversible reactions will alter the required reprocessing time. Also, if reversible reaction products are volatile and lost during reprocessing, it will be impossible for the reprocessed material to achieve full property recovery.

498026 Hybrid Chain-Growth/Step-Growth Mechanism Observed in Heterofunctional Thiol-Ene Polymerizations

Abstract:

Traditional methods for polymer synthesis usually afford minimal control over the primary sequence and solid-phase synthesis of sequence-controlled, biomimetic polymers is limited by scalability. Utilizing thiol-X chemistries, first generating sequence-controlled oligomers with the thiolate anion mediated thiol-Michael reaction and then polymerizing with the orthogonal thiyl radical mediated thiol-ene reaction, we have developed a method for producing sequence-ordered periodic copolymers that are capable of diverse side-group customization, with our primary focus being on attaching nucleobases and creating DNA-mimics. In this talk, I will focus on the characteristics of heterofunctional thiol-ene polymerizations in the context of synthesizing sequence-ordered polymers, particularly, the effects of functional group chemistries on the reaction kinetics and the monomer structure on molecular weight (MW) evolution.

Despite the popularity of thiol-ene coupling (TEC) chemistry in organic synthesis, in general, and polymer chemistry, particularly, heterofunctional thiol-ene monomers (i.e. low MW monomers containing both a thiol and an alkene functional group) have been relatively unexplored, likely due to the inherent instability of thiol and ene functional group chemistries when present in the same formulation. Initially, we investigated the polymerizability and MW evolution of heterofunctional monomers containing various side-group functionalities (e.g. alkyl chains, phenyl groups, nucleobases) and the effect of varying the number of atoms separating the thiol and ene functional groups. Surprisingly, while monitoring MW *vs* conversion, we observed that high MW species emerged at low conversions (< 20%) when the alkene and thiol are separated by 5 to 7 atoms, which is characteristic of a chain-growth mechanism rather than the expected step-growth mechanism associated with TEC reactions. Utilizing kinetics and MW profiles obtained by real-time Fourier transform (rt-FTIR) spectroscopy and gel permeation chromatography (GPC), respectively, it was determined that, due to the close proximity of the two functional groups, upon generation of the 2° C-radical by thiyl radical propagation into the alkene group, intramolecular chain-transfer to the thiol dominates over intermolecular chain-transfer resulting in the radical remaining on the same polymer chain and an apparent chain-growth/step-growth hybrid mechanism. Although, the potential for high MW polymers in comparison to the MWs achievable by traditional step-growth is desirable, the ease by which these monomers cyclize results in significant reductions in polymer yield (up to 50% monomer cyclization) leading us to develop strategies to eliminate cyclization while maintaining the chain-growth evolution. Incorporation of the monofunctional thiol, thiophenol, into the reaction mixture at an equivalency as low as 1%, facilitates this goal as the low bond dissociation energy of thiophenol's S—H bond, compared to the monomer contained thiol, ensures that, upon initiation, the radical population is partitioned exclusively into the monofunctional thiol additive, thus preventing cyclization. Further, thiol-ene reactions may be initiated without initiator by the direct photocleavage of the S—H bond with ~254 nm light, but due to the extension of the thiophenol UV absorption spectrum up to 330 nm, the thiophenol S—H bond may be selectively cleaved in the presence of alkyl thiols so that thiophenol can be reacted exclusively in the presence of alkyl thiols, which greatly increases the ability to incorporate thiol-ene chemistries for performing one-pot, sequential reactions.

484576 High Pressure Ethylene Polymerization with a Post-Metallocene Bis-Phenyl Phenoxy Catalyst

Abstract:

The use of post-metallocene bis-phenylphenoxy catalysts to polymerize ethylene under high ethylene pressures (>25,000 psi) results in some remarkable catalytic properties. The high ethylene pressure produces molar ethylene concentrations in the reactor as much as 40 times higher than in typical low pressure ethylene polymerizations. This high ethylene concentration results in high catalyst efficiency at high temperatures and low reactor residence time, between 180 and 240 °C the catalyst efficiency surprisingly increases with increasing temperature, allowing for use of these catalysts at temperatures much higher than can be utilized in the low pressure processes. It has further been demonstrated that under these conditions increasing hydrogen levels up to 0.5 mol% does not significantly affect the polymer molecular weight; however, polymer molecular weight control can be realized with varying reactor temperature. The polymer produced is shown to be high density polyethylene made from a single site catalyst and not free radical initiated low density polymer.

499717 A Simulation-Based Derivative-Free Optimization Framework Using the Kinetic Monte Carlo Method for Controlling Polymer Molecular Weight and Sequence Distribution Synthesized Via Free Radical Polymerization

Abstract:

The diversity of the potential arrangements of multiple monomers along the length of polymer chains and their impact on polymer properties spark interests in the design of polymer sequence characteristics for particular applications[1,2]. Kinetic Monte Carlo (KMC) is a technique that can track the explicit arrangement of monomers in the polymer chains, and recently there have been developments that significantly improve the calculation speed of KMC simulations[3,4]. Yet it is still difficult to integrate with conventional gradient-based optimization algorithms that are typically invoked to design polymer properties, because KMC does not supply derivative information directly as a stochastic simulation algorithm.

In this work, we developed an optimization framework based on a derivative-free method that incorporates KMC simulations to efficiently find synthesis conditions for property targets, including molecular weight and sequence. The derivative-free optimization algorithm is named Bounded Optimization BY Quadratic Approximation (BOBYQA) developed by Powell[5]. This algorithm finds the minimum of a differentiable "black box" function calculated from the output of a KMC simulation, which we have adopted to return properties of a polymer for given synthesis conditions.

The performance of our model is demonstrated on a copolymerization system of butyl acrylate (BA) and methyl methacrylate (MMA) in a batch reactor. We show by two case studies that the model can accurately and efficiently control molecular weight and sequence targets by changing reaction conditions such as temperature, monomer composition and overall conversion. The algorithm can also provide insight on designing reactivity ratios to better meet the property targets. In addition, the explicit sequence information generated from KMC allows for detailed analysis and control of the distribution of sequence characteristics among different chains.

Reference

1. Iedema PD, Hoefsloot HC. Synthesis of Branched Polymer Architectures from Molecular Weight and Branching Distributions for Radical Polymerisation with Long-Chain Branching, Accounting for Topology-Controlled Random Scission. *Macromolecular Theory and Simulations.* 2001;10(9):855-869.

2. Van Steenberge PHM, D'hooge DR, Wang Y, et al. Linear Gradient Quality of ATRP Copolymers. *Macromolecules.* 2012;45(21):8519-8531.

3. Gao H, Oakley LH, Konstantinov IA, Arturo SG, Broadbelt LJ. Acceleration of Kinetic Monte Carlo Method for the Simulation of Free Radical Copolymerization through Scaling. *Industrial & Engineering Chemistry Research.* 2015;54(48):11975-11985.

4. Tripathi AK, Sundberg DC. A Hybrid Algorithm for Accurate and Efficient Monte Carlo Simulations of Free-Radical Polymerization Reactions. *Macromolecular Theory and Simulations.* 2015;24(1):52-64.

5. Powell MJ. The BOBYQA Algorithm for Bound Constrained Optimization without Derivatives. *Cambridge Report 2009/06, University of Cambridge, Cambridge.* 2009.

500266 Experimental and Macroscopic Mechanistic Modeling Studies of the Methyl Acrylate Self-Initiation Reaction

Abstract:

Acrylic polymers are widely used as adhesives and functional additives in the coating and paint industries. Environmental regulations that have imposed upper bounds on the level of volatile organic compounds, have limited the allowable level of solvents in acrylic coatings and paints [1]. High-temperature (120-180 Â°C) polymerization has been employed to produce resins with low molecular weights and low solvent concentrations [2]. The low molecular weight of polymer chains compensates for the low solvent concentration, which ensures an adequately-low viscosity of the products and thus their brushability and sprayability. However, at these temperatures secondary reactions, such as monomer self-initiation, Î²-scission, inter/intra-molecular chain transfer, and backbiting, influence the polymerization strongly. The self-initiation reaction of acrylates has several positive attributes [3, 4]. As its presence decreases the need for conventional thermal initiators, operation costs decrease (thermal initiators are typically the most expensive components of paints and coatings), and it also lowers the level of unreacted thermal initiators in the final product, thus improving the quality of the products.

In this paper, an experimental study on the self-initiation reaction of methyl acrylate (MA) in free-radical polymerization is presented. To the best of our knowledge, this is the first reported experimental study of self-initiation of MA. MA conversion and polymer molecular weight measurements from free-radical MA homo-polymerization initiated by only the monomer are presented at different high temperatures. A macroscopic mechanistic model that is based on (i) already-known reaction mechanisms that contribute to the polymerization and (ii) a theoretically-identified MA self-initiation mechanism and a second-order reaction rate equation [5, 6], is presented. Using the macroscopic model, the frequency factor and activation energy of the MA self-initiation reaction are estimated from the measurements. The frequency factor and activation energy estimates are compared with predictions obtained via our purely-theoretical quantum chemical calculations (electronic-level modeling) [5, 6].

References

1. Economic impact and regulatory flexibility analyses of the final architectural coatings VOC rule. United States Environmental Protection Agency, Office of Air Quality Planning and Standards, EPA-452/R-98-002, 1998.
2. Grady, M.C., Simonsick, W.J., Hutchinson, R.A., Studies of higher temperature polymerization of n-butyl methacrylate and n-butyl acrylate. Macromolecular Symposia, 2002, 182, 149-168.
3. Arabi Shamsabadi, A., Moghadam, N., Corcoran, P., Srinivasan, S., Grady, M.C., Rappe, A.M., Soroush, M., Study of n-Butyl Acrylate Self-Initiation Reaction Experimentally and via Macroscopic Mechanistic Modeling. Processes, 2016, 4(2), 15.
4. Srinivasan, S., Lee, M. W., Grady, M. C., Soroush, M., A. M. Rappe, A.M., Computational Study of Self-Initiation Mechanism in Thermal Polymerization of Ethyl and n-Butyl Acrylates. J. Physical Chemistry A, 2010, 114(30), 7975-7983.
5. Srinivasan, S., Kalfas, G., Petkovska, V. I., Bruni, C., Grady, M. C., Soroush, M. Experimental Study of the Spontaneous Thermal Homopolymerization of Methyl and n-Butyl Acrylate. J. Appl. Polym. Sci. 2010, 118 (4), 1898-1909.
6. Liu, S., Srinivasan, S., Tao, J., Grady, M. C., Soroush, M., A. M. Rappe, A.M., Modeling Spin-Forbidden Monomer Self-initiation Reactions in Free-Radical Polymerization of Acrylates and Methacrylates. J. Physical Chem. A, 2014, 118(40), 9310-9318.

---

* To whom all correspondence should be addressed.

490294 A Mathematical Model Based on Artificial Neural Network for Ethylene/Norbornene Copolymerization Catalyzed By 2-(tetramethylcyclopentadienyl)-4,6-Di-Tert-Butylphenoxytitanium Dichloride

Abstract:

A Mathematical Model based on Artificial Neural Network for ethylene/norbornene copolymerization catalyzed by 2-(tetramethylcyclopentadienyl)-4,6-di-tert-butylphenoxytitanium dichloride

ABSTRACT

Ethylene/norbornene (E-N) copolymer is one of the most promising industrial thermoplastic polymers with high glass transition temperature (T_g), excellent moisture barrier properties, chemical resistance, and optical clarity. It is an attractive choice as a novel substrate material for high density data storage devices, packaging, and optical/biomedical applications.[1]

Titanium based metallocene catalysts with a pendant nitrogen donor or pendant oxygen donor on the cyclopentadienyl ring, are found to be good at catalyzing ethylene-norbornene co-polymerization with noteworthy catalytic activity and efficient norbornene incorporation for the ethylene/norbornene copolymerization.[2]

The use of neural networks (NNs) has become progressively popular for applications where the mechanistic description of the interrelation of dependent and independent variables is either obscure or very complex.[3] The neural network consists of processing neurons and information flow channels between the neurons, usually called 'interconnects'. Each processing neuron calculates the weighted sum of all interconnected signals from the previous layer plus a bias term and then generates an output through its activation transfer function. Training is done by assigning random weights to each neuron, evaluating the output of the network and calculating the error between the output of the network and the known results by means of an error or objective function. If the error is large, the weights are adjusted and the process goes back to evaluate the output of the network. This cycle is repeated till the error is small or a stop criterion is satisfied.[4]

In this work, an artificial neural network (ANN) as a supervised back-propagation model with different architecures is implimented to investigate the effect of reaction variables for ethylene/norbornene copolymerization catalyzed by 2-(tetramethylcyclopentadienyl)-4,6-di-tert-butylphenoxytitanium dichloride. The experimental data were taken from Jianguo *et al.*[5] The effect of temperature, monomer to co-monomer concentration ratio and cocatalyst to catalyst concentration ratio on catalyst activity, yields, molecular weights & distribution and glass transition temperature of the copolymer has been studied using neural network approach in MATLAB® 7.10.0 (R2015a) software.[6]

References

[1] Kaminsky W., Bark A. and Arndt M., *Makromol Chem Macromol Symp,* **47:83**, (1991).

[2] Mcknight A.L., Waymouth R.M., *Macromolecules*, **32:2816** (1999).

[3] Fernandes F. A. N. and Lona L. M.F., *Brazilian Journal of Chemical Engineering*, **22: 03**, p. 401-418 (2005).

[4] Haykin, S., *Neural Networks: A Comprehensive Foundation*, Prentice Hall, New York (1998).

[5] Jianguo N., Chunsheng L., Yuetao Z., Zaiqun L. and, Ying M., *Polymer*, **49**, p. 211-216. (2008)

[6] MATLAB version 7.10.0 (R2015a), computer software, Natick, Massachusetts: The MathWorks Inc., 2015.

503052 Revealing Governing Mechanism in Directed Self-Assembly of sub 10 nm Particles with Single Particle Resolution

Abstract:

There has been a great deal of interest in nanomanufacturing using directed self-assembly of nanoparticles into ordered arrays. Such arrays have potential applications in various emerging fields such as nanobiotechnology, nanoelectronics, nanosensors, and the like. In all these applications, conventional pattering techniques have already reached their limits in resolution. However, strategically placing sub 10 nm particles as active agents on textured substrates could potentially make devices with enhanced functionalities. Using dissipative particle dynamics simulations, we reveal the governing mechanism leading to nanoparticles self-assembly. Particularly, we show that sub 10 nm particles confinement before directed self-assembly is necessary to achieve a robust mechanism. Utilizing this computational tool, we are able to predict positioning of sub 10 nm particles with arbitrary shape onto desired templates and surface textures. Our method, simulation results, and computational tool will offer an innovative approach and fundamental understanding of the dynamics of directed self-assembly of sub 10 nm particles with single particle positioning.

494529 Uniform Thinning of Cu-Fe-Ni-Co Nanowires and Kinetic Monte Carlo Simulation

Abstract:

A novel procedure for thinning Cu-rich alloy nanowires is presented and described by a kinetic Monte Carlo simulation. Template-assisted electrodeposited Fe-Ni-Co nanowires were generated by pulse current deposition, followed by an acid, copper ion treatment to facilitate a displacement and simultaneous corrosion reaction within the nanowire. The resulting Cu-Fe-Ni-Co alloy nanowire exhibited interesting etching behavior, where the wire was uniformly thinned to dimension ten times thinner than the original wire diameter. Simulations suggest that the thinning behavior is a consequence of the alloy structure and that pure copper nanowires would break before they reach the same thinned diameter. A reconstruction of the local porous structure during etching was one of the primary reasons that etching of Cu-Fe-Ni-Co nanowires leads to a thinning effect without breakage.

500828 Experimental Assessment of Nucleation Theory at the Molecular Level

Abstract:

Nucleation is a core scientific principle, describing the formation of new phases and materials. Nevertheless, the applicability of classical nucleation at the molecular level remains open to debate. Currently, the direct observation of molecular proto-nuclei is an extreme challenge, leaving experimental studies of nucleation energetics restricted to colloids, proteins, and macroscopic systems. Here, we surpass this challenge by using atomically-defined sub- and post-critical nuclei to investigate the energetics of diamond nucleation during plasma-enhanced chemical vapor deposition and observe critical nucleation from 26 carbon clusters, confirming a key implication of classical theory. However, the nucleation barrier is two orders of magnitude lower than prior estimates from bulk properties. Rationalizing this discrepancy resolves long outstanding questions surrounding the growth of several industrially important materials and provides direct experimental verification of classical nucleation at the molecular level.

500287 Low-Voltage Electrophoretic Deposition of All-Inorganic CZTS Nanocrystals for Fabrication of Thin Films

Abstract:

Colloidal nanocrystals (NCs) have been proposed as the building blocks of prospective next-generation solar cells, due in-part to significantly reduced fabrication costs through their solution processability. However, fabrication of conductive, well-adhered films via low-waste, high-throughput methods has not been realized. Electrophoretic deposition (EPD) is one scalable method proposed for fabricating solution-processed absorbing layers for solar cells. In this study, we deposit films through low-voltage EPD of all-inorganic semiconductor NCs. In order to probe the transport processes and electrochemical reactions that occur during EPD, light scattering, absorbance spectroscopy, and electric measurements were used with special attention to understand the temporally separate current-driven processes that occur during an EPD run. Through choice of NC-ligand-solvent combinations and reactor design, improvements in the morphology and electronic properties of deposited films are realized. These insights are applied to EPD in both batch and continuous flow reactors.

499101 Preferential Binding of Polyvinylpyrrolidone (PVP) Is Not Responsible for Shape Control in Ag Nanoparticle Synthesis

Abstract:

The role of polyvinylpyrrolidone (PVP) in the shape control synthesis of Ag nanoparticles (NPs) has been attributed to its preferential binding to Ag (100) facets compared to Ag (111) facets. Theoretical studies have validated this hypothesis, however, only at the limit of molecular or small oligomeric analogs of PVP in the absence of solvent and modifiers (i.e., Cl^-) present during synthesis. We experimentally measured the surface density of PVP on different shaped Ag NPs in order to build equilibrium adsorption isotherms. Based on a Langmuir isotherm, a weak (5 M^{-1}) equilibrium adsorption constant for PVP adsorbed to Ag nanocubes was determined. For PVP molecular weights > 10K, constant PVP monomer surface densities were determined, implying a similar equilibrium adsorption constant. The equilibrium adsorption constant for Ag octahedron decreased to half the value of the Ag cubes, confirming the existing hypothesis that PVP selectively adsorbs on the cubic (100) surface over the (111) surface. The difference in apparent affinity is significantly less than the difference determined by first-principle simulations. The final Ag nanocube product contains a surface layer of adsorbed chlorine due to the inclusion of hydrochloric acid in a typical synthesis. However, the exact role of Cl^- has on shape control has not been fully elucidated. We combined various ex-situ characterization techniques such as scanning electron microscopy (SEM), x-ray diffraction (XRD) and x-ray photoemission spectroscopy (XPS) to probe the growth mechanisms of Ag nanocubes. The initially formed nanostructures during the early stages of synthesis were AgCl nanocubes which were subsequently etched by incoming Ag salt flux to form metallic Ag nanocubes containing a nanometer thick chloride layer. Variation of Cl^- concentration alone at constant PVP concentration led to modification of the nanoparticle shape. As the concentration of Cl^- decreased, truncated cubes, cuboctahedron and truncated octahedron formed, which confirms the importance of Cl^- for Ag NPs shape control. PVP serves only to stabilize Ag nanocubes in solution.

488665 Scalable Synthesis of Epitaxial, Oxidation-Proof Au@Ag Core-Shell Nanowires for Electronic and Photonic Applications

Abstract:

Ag nanowires (AgNWs) hold great promise for applications such as transparent and flexible displays, solar cells, chemical/biological sensors, photonic circuits and scanning probe microscopies, but their susceptibility to damage from oxidation significantly limit their device life-time and commercialization potential. To address this challenge, we developed a single-step, solution-phase, etch-free deposition technique to grow an ultra-thin, epitaxial, smooth layer of Au coating on the AgNW surface, which acts as a reliable and economical anti-oxidation barrier to provide long-term device stability. The Ag@Au core-shell NWs are stable in air and in physiological buffer solution for months. Transparent flexible electrodes fabricated from Ag@Au core-shell NWs shows steady performance in air at 80°C and 100% humidity for at least 3 months, while those fabricated with bare AgNWs died out within 2 weeks. The ultra-thin Au coating did not introduce significant Au fluorescence in the SERS spectrum, making them feasible for SERS and plasmonic sensing applications. It was also demonstrated that the ultra-thin Au coating does not have adverse effects on the propagation and coupling of surface plasmon polaritons (SPPs) in AgNW waveguides, and the device performance remains stable for weeks, a significant improvement against uncoated AgNW. AgNW-AFM probes are low-cost alternatives of high-aspect-ratio, high-resolution AFM probes, but the major drawback is their short shelf-life (typically a couple of weeks). It was demonstrated that the Ag@Au core-shell NWs functions similarly to bare-AgNW, with a much longer shelf-life for at least 4 weeks in the air.

501225 Synthesis of Bimetallic Alloy Nanoparticles through the Visible-Light Mediated Reduction of a Bimetallic Oxide Precursor: Case Study of Ag-Pt Nanoparticle Synthesis

Abstract:

Bimetallic nanoparticles are known to exhibit chemical and physical properties unique from their individual components. However, the synthesis of well-mixed alloy nanoparticles is often limited to highly miscible metals. This limitation has previously been overcome by using extreme synthesis conditions such as gamma radiation or high temperature reduction to kinetically trap the atoms of each metal in a single nanostructure.[1,2] In this presentation, we demonstrate a more facile synthesis approach for creating well-mixed bimetallic alloy nanoparticles of immiscible metals. We synthesize Ag-Pt nanoparticles through the visible light-induced reduction of a hollow bimetallic oxide precursor composed of Ag and Pt oxide. We show that Pt atoms are initially concentrated near the surface of the hollow precursor. The low-temperature reduction of the oxide causes the hollow precursor to become unstable and collapse into well-mixed alloy nanoparticles. We discuss the physical processes that lead to the light-mediated reduction of the oxide precursor and the wider applicability of synthesizing alloy nanoparticles composed of immiscible metals through the use of bimetallic precursors.[3]

1. Zhang, Z., Nenoff, T. M., Huang, J. Y., Berry, D. T. & Provencio, P. P. Room Temperature Synthesis of Thermally Immiscible Ag–Ni Nanoalloys. *J. Phys. Chem. C* **113,** 1155–1159 (2009).

2. Holewinski, A., Idrobo, J.-C. & Linic, S. High-performance Ag–Co alloy catalysts for electrochemical oxygen reduction. *Nat. Chem.* **6,** 828–834 (2014).

3. Aslam, U. & Linic, S. Kinetic Trapping of Immiscible Metal Atoms into Bimetallic Nanoparticles through Plasmonic Visible Light-Mediated Reduction of a Bimetallic Oxide Precursor: Case Study of Ag–Pt Nanoparticle Synthesis. *Chem. Mater.* **28,** 8289–8295 (2016).

500446 Dynamic Control of Gold Nanoparticle-Conjugated DNA Origami Templates

Abstract:

Fabrication of nano-sized electronic and photonic devices through lithographic techniques is becoming increasingly challenging and energy intensive. Additionally, these methods are primarily confined to 2D surfaces and often do not include dynamic components that can reconfigure in response to external stimuli. Bio-inspired assembly through molecular recognition, such as base pairing interactions between deoxyribonucleic acid (DNA) strands, can be exploited to create dynamic, as well as robust, 3-D platforms for the organization of nanoparticles (NPs). DNA origami platforms can be used to organize NPs in precise orientations and spacings in 2D and 3D, yielding tailored plasmonic interactions. In some cases, these templates can be dynamic, adopting new conformations in response to external stimulus and altering these interactions. However, using the gold standard strand invasion approach that relies on de/hybridization kinetics, the inherent actuation time scale for these structures ranges from minutes to hours. Heating can increase response kinetics; however, use of external energy input is limited by bulk heat transfer. Therefore, localized energy input is desired as a means to potentially provide faster dynamics. Properties of NPs, such as surface plasmon resonance, can be explored as a localized energy source. Current research is focused on a fundamental understanding of the impact of NPs on dynamics and conformations of DNA origami structures actuated using the energy responsive NPs.

Our present study employs DNA origami hinges with single stranded DNA overhangs on the hinge arms as a model system. The sequence of these overhangs is complementary to that of single stranded (ss)-DNA on modified gold NP (AuNP) surfaces. A comparative study was conducted between the conformations of a closed hinge with varying locations of ss-DNA on the hinge arm in the presence and absence of AuNPs and analyzed by Transmission Electron microscopy (TEM) to determine the free energy landscape. Additionally, actuation kinetics of two different actuation pathways, strand displacement and bulk heating, were compared. Kinetics was studied in the bulk phase using fluorescence spectroscopy to analyze a Forster Resonance Energy Transfer (FRET) reporter system that indicated both NP binding and hinge closure. These results provided a comprehensive understanding of mechanical energy stored in NP-DNA origami composites. Preliminary experiments were performed for a third actuation scheme, localized actuation of composites using laser excitation to plasmonically heat AuNPs. Future experiments are investigating kinetics of structures undergoing local, plasmonic heating utilizing an in-house two source fluorometer instrument with separate excitation and modulation sources. This research could lead to higher order NP-DNA origami composites, that can store and release energy in response to light, with potential applications in nanophotonics.

501470 'One-Pot' Multi-Scale Templating of Interdigitated Bi-Modal Porous Carbon Supercapacitors

Abstract:

Porous carbons show promise as electrodes for electrochemical double layer capacitors (EDLCs) or supercapacitors. Related efforts have focused on simultaneously maximizing accessible surface area while tuning molecular structure and charge transport. This work highlights the 'one-pot' nanocasting of carbons bearing a unique interdigitated bimodal three-dimensionally ordered mesopore (ib3DOm) topology constructed of thin (ca. 5 nm) but robust pore walls. This structure is achieved by the co-assembly of size-tunable (ca. 10-50 nm) silica nanoparticles with carbon molecular precursors followed by polymerization, pyrolysis and template sacrifice. Whereas we previously identified a narrow parametric envelope for achieving ib3DOm structures in film form [1], we will show that ib3DOm powder formation can be achieved over an expanded range of carbon precursor concentrations and silica nanoparticle solids contents. This expanded pseudo phase space establishes the versatility of this facile powder templating approach, and offers additional flexibility for exploiting interface-mediated synthesis-structure-function relations for tuning carbon microstructure [2]. Specifically, we will demonstrate how both carbon precursor concentration and silica nanoparticle solids content can be used to tailor the carbon allotrope distribution, namely sp^2 carbon content, within the carbon walls. This facile templating strategy enables tuning of surface areas to as high as 1330 m^2/g, pore volumes to as high as 5.8 cm^3/g, and sp^2 hybridized carbon content as high as 80%, with remarkable performance as electrodes for supercapacitors with specific capacitance of 275 F/g in H_2SO_4 electrolyte and 284 F/g in KOH electrolyte at 1 A/g, and retention as high as 90%.

References

[1] Z. Tian, M.A. Snyder, *Langmuir*, 30 (2014) 12411-12420.

[2] M. A. Snyder, *MRS Bulletin*, 41(9) (2016) 683-688.

496710 Combined Experimental and Theoretical Study of Hexagonal Boron Nitride Crystal Growth

Abstract:

A COMBINED EXPERIMENTAL AND THEORETICAL STUDY OF HIGH QUALITY HEXAGONAL BORON NITRIDE CRYSTAL GROWTH

Song Liu, Bin Liu, and J. H. Edgar

Department of Chemical Engineering, Kansas State University, Manhattan, Kansas 66506

Hexagonal boron nitride (hBN) is a layered, 2D material structurally analogous to graphite. Traditionally, it has been employed as a refractory ceramic, but recently new device applications of hBN have been envisioned, which will utilize its unique optical, phononic, nuclear, and electronic properties, with examples including deep ultraviolet light emitters (i.e., light emitting diodes, and laser diodes), neutron detectors, nanophotonics, and quantum computers. For these applications, precise control of the thickness, area, crystal perfection, crystallographic orientation, and purity of the hBN crystals is essential.

Here we discuss the growth of high purity, low defect density hBN crystals at atmospheric pressure by cooling a molten Ni-Cr flux saturated with boron and nitrogen. Specifically, through optimizing the process conditions including the dwell temperature (for molten solution preparation), and the cooling rate, single crystals up to 4 mm^2 in area and up to 500 μm thick were grown, some of the largest ever produced.

Reactive molecular dynamics (rMD) simulations, employing a ReaxFF force field established by our previous work on CVD hBN growth on Ni surfaces, have been designed and carried out to understand the molecular structures of dissolved N and B species in molten Ni-Cr solution, and the formation of hBN upon cooling. Furthermore, in order to reveal the role of Cr species in the Ni-Cr solution, the synthesis was simulated by density functional theory (DFT). In particular, the energetics and kinetics of elemental B and N on Ni-Cr alloy suggest that Cr increases the N solubility in Ni, which is growth rate limiting.

487711 Design of Aminopolymer Structure to Enhance Performance and Stability of CO_2 Sorbents: Poly(propylenimine) Vs. Poly(ethylenimine)

Abstract:

Research efforts in CO_2 capture technologies have dramatically increased due to interest in reducing the effect of anthropogenic emissions on atmospheric CO_2 concentrations. Much of this research has focused on capture from large point sources such as flue gas. Direct capture of CO_2 from ambient air is proposed as a complementary technology that has the advantage of not needing a flue gas feed stream and thus is more versatile in its deployment. However, at an ultradilute concentration of 400 ppm, the driving force for adsorption is low, necessitating the development of adsorbents that may be different from those used in flue gas capture. Additionally, these adsorbents must be stable under oxidizing conditions due to the presence of oxygen in the feed stream and the potential for oxygen to contact the adsorbent material during high temperature regeneration.

In this contribution, we report the finding that poly(propylenimine) (PPI) based structures have significantly improved sorption properties compared to the structurally-similar and well-studied poly(ethylenimine) (PEI). Specifically, we find that PPI-based sorbents are more effective at CO_2 capture, resulting in higher amine efficiencies, and perhaps more importantly, have improved stability under high temperature oxidizing conditions. Our experiments suggest that PPI-based CO_2 sorbents may allow for longer sorbent working lifetimes due to an increased tolerance to sorbent regeneration conditions and will have larger adsorption swing capacities, both of which will drive down the cost associated with CO_2 capture.

487844 Development of 3D-Printed Aminosilica Monoliths for CO_2 capture

Abstract:

Amine-based materials have represented themselves as a promising class of CO_2 adsorbents; however, their large-scale implementation requires their formulation into suitable structures. In this study, we report formulation of aminosilica adsorbents into monolithic structures through 3D printing technique. In particular, 3D-printed monoliths were fabricated using pre-synthesized silica-supported tetraethylenepentamine (TEPA) and poly(ethylenimine) (PEI) adsorbents using three different approaches. In addition, a 3D-printed bare silica monolith was prepared and post-functionalized with 3-aminopropyltrimethoxysilane (APS). Characterization of the obtained monoliths indicated that aminosilica materials retained their characteristics after being extruded into 3D-printed configurations. Adsorptive performance of amine-based structured adsorbents was also investigated in CO_2 capture. Our results indicated that aminosilica materials retain their structural, physical, and chemical properties in the monoliths. In addition, the aminosilica monoliths exhibited adsorptive characteristics comparable to their corresponding powders. This work highlights the importance of adsorbent materials formulations into practical contactors such as monoliths, as the scalable technology platform, that could facilitate rapid deployment of adsorption-based CO_2capture processes on commercial scales.

Reference

H. Thakkar et al. *ACS. Mater. Interfaces,* 2017, 9 (8), 7489-7498.

498196 Engineering Supported Amine Adsorbents for CO_2 Capture Applications

Abstract:

Engineering Supported Amine Adsorbents for CO_2 Capture Applications

Z. Xie, C. F. Cogswell and S. Choi

Department of Chemical Engineering, Northeastern University, Boston MA, 02115

Given current models showing a clear link between global climate change and rising atmospheric carbon dioxide levels, and the likelihood of our continued use of fossil fuel energy sources in the next decades, it has become one of the major goals of researchers to obtain effective methods for the capture of carbon dioxide.[1] While the use of liquid amine scrubbers in industrial settings is the current state of the art, many groups have focused on the use of supported amine solid sorbents because of their clear advantages including high capture capacities, low heats required for sorbent regeneration, and significantly less costly operation compared to volatile amine scrubbers.[2]

Supported amine research up to this point has mostly focused on the amine molecules, including the effect of amine incorporation mechanism and type of amines integrated within the hybrids. For instance, the way amine is added to the porous supports has been studied extensively, which led to the creation of different types of solid amines such as class 1, 2, and 3 hybrid adsorbents.[1,2] Major research interests in this research direction have been placed to the comparative studies on these different classes of supported amine sorbents in perspective of their adsorption capacity, adsorption kinetics, and sorbent regenerability that are affected by different amine-support interactions.

More recently, there have been growing research interests on the role of the support materials.[3] Considering the fact that the supported amine adsorbents are composed of both amine molecules and porous supports, this research direction is expected to enhance the performance of the current supported amine adsorbents, while improving our current understanding about the structure-property-relations of the supported amine materials even further.[4, 5] Major research interests in this research direction includes the effect of the pore structures of the support, the effect of the heteroatoms integrated in the support, etc. onto the CO_2 adsorption characteristics of the hybrids.

Through this presentation in honor of Prof. Christopher Jones for receiving the Andreas Acrivos Award for Professional Progress in Chemical Engineering, we will discuss the continuous research efforts we have been made on the supported amines, along with recent research results obtained in this research field.

References:

1. S. Choi, J. H. Drese and C. W. Jones, *ChemSusChem*, 2009, 2, 796–854.

2. D. Andirova, Y. Lei, X. D. Zhao, S. Choi, *ChemSusChem*, 2015, 8 (20), 3405-3409

3. D. Andirova, C. F. Cogswell, Y. Lei, S. Choi, *Micropor. Mesopor. Mat.*, 2016, 219, 276–305.

4. C. F. Cogswell, S. Choi et al., *Langmuir,* 2015, 31, 4534–4541.

5. C. F. Cogswell, S. Choi et al., *J. Mater. Chem. A*, 2017, DOI: 10.1039/C7TA01616F

497787 Impacts of Aminopolymer-Support Interaction on CO_2 Sorption Performance Probed By Neutron Scattering Techniques

Abstract:

Composites of poly(ethyleneimine) (PEI) and mesoporous silica are highly effective, reversible sorbents for CO_2, both from flue gas and in direct air-capture applications. The morphology of the PEI phase and its dynamic behavior (e.g. self-diffusion) within the silica can strongly impact both the kinetics and overall carbon capture efficiency of these sorbents. Here, we directly probe the shape of the supported polymer phase as a function of fill fraction using small-angle neutron scattering (SANS) and further investigate the mobility of PEI within this phase using quasi-elastic neutron scattering (QENS). Corroborated by textural characterization from physisorption analysis, the SANS data indicate that PEI forms a conformal coating on silica pore walls at low fill fraction, while at higher fillings the polymer aggregates and forms plugs that span the pore diameter. QENS additionally indicates that interaction with the silica surface dramatically decreases center-of-mass diffusion of PEI relative to the bulk phase and, to a lesser extent, also slows side-chain mobility. Hydrophobic functionalization of the silica leads to an increase of the PEI mobility above that in native silanol-terminated silica, and is also found to yield higher CO_2 capture efficiency (amine efficiency) as well as improved sorption kinetics. Thus, this work supports the existence of a link between the affinity of the support for PEI and the accessibility of active sorbent functional groups.

495560 Hierarchically-Structured Porous Materials for Enhanced Greenhouse Gas Capture

Abstract:

Owing to global concern on climate change, the technologies for capturing greenhouse gases such as CO_2, SF_6 and chlorofluorocarbons from anthropogenic sources have attracted a great amount of interests. In this work, a series of hierarchical LTA zeolites possessing various mesopore spaces that are decorated with alkylamines was designed and synthesized for an enhanced CO_2 adsorption. The highest CO_2 uptake capacity was observed when (3-aminopropyl)trimethoxysilane (APTMS) was grafted onto the hierarchical LTA zeolite having the largest mesopores. Owing to the contributions of both alkylamine groups grafted on the mesopore surfaces and active sites in zeolites LTA, the amount of CO_2 uptake on such material was observed to be much higher than those for conventional aminosilicas such SBA-15 and MCM-41. Furthermore, the adsorbent showed good CO_2 uptake capacity and recyclability under a dynamic flow condition.

Meanwhile, HKUST-1, an inexpensive metal-organic framework possessing open metal sites, has a great potential for capture and recovery of SF_6. In this work, the structural property of HKUST-1 was modified to yield a hierarchically-structured HKUST-1 nanocrystal exhibiting a superior performance with higher SF_6 uptake, better SF_6/N_2 selectivity, faster SF_6 adsorpton kinetics and lower energy penalty for regeneration than bulk HKUST-1 crystal as well as conventional zeolite and porous carbon adsorbents. Higher surface area and the presence of mesoporosity to facilitate the transport of SF_6 to active sites residing in microporous spaces were found to be key factors contributing to such enhancement. An outstanding potential utility of our HKUST-1 crystal in industrial applications was also validated with an idealized vacuum swing adsorption model.

492987 Natural Gas Adsorption in SSZ-13: Equilibrium and Dynamic Properties

Abstract:

In developing new adsorption separation processes, it is necessary to study both the equilibrium and dynamic adsorption properties of potential materials. In particular, experimental determination of isotherms and dynamic breakthrough properties aid in the development of modeling new adsorption systems toward process development. Here, the equilibrium adsorption properties of a small-pore zeolite, SSZ-13, is studied for its natural gas separation potential. In particular, using volumetric and gravimetric adsorption techniques, the equilibrium properties of CO_2, CH_4, C_2H_6, and H_2O are determined, including adsorption isotherms and heats of adsorption. In addition, using a dynamic breakthrough adsorption apparatus, high pressure breakthrough experiments demonstrate the mixed gas separation performance of SSZ-13 in mixtures containing CO_2, CH_4, C_2H_6 and H_2S. Simulation of these breakthrough experiments shows that ideal adsorbed solution theory adequately describes the mixed gas adsorption modeling for this zeolite, and other mixed gas adsorption models fail to account for roll-up effects of C_2H_6 observed during experiments. In gas mixtures containing both CO_2 and H_2S, there is mutual transformation of these gases to COS and H_2O, likely driven by the affinity of H_2O in the zeolite. Comparison with a benchmark zeolite, Na-13X, reveals superior performance for SSZ-13 due to a lower affinity toward H_2O, resulting in less conversion of gases and no early breakthrough of sulfur species.

491095 Zeolite on Demand: Design and Synthesis of Zeolites with Controlled Crystal Morphology and Location of Substituting Tetrahedral Atoms with the Aid of Theoretical Calculations

Abstract:

Zeolites, a series of microporous crystalline aluminosilicates and other metallosilicates, have found several applications across a wide range of industries, for example, as catalysts that can selectively crack heavy components in crude oil into smaller, useful chemicals. Synthesis of zeolites has been carried out by a "black-box" sequence of sophisticated hydrothermal reactions; therefore, "trial-and-error" is a routine practice for zeolite synthesis. Over decades, designed and targeted synthesis of useful zeolites with unique pore geometries and morphologies remains an elusive dream in zeolite synthesis, which, if succeed, would drive the communities toward on-demand applications needing more precisely tuned architectures and properties.

Among the 232 known zeolite frameworks, the vast majority of them require an organic structure-directing agent (OSDA) for their synthesis. OSDAs are organic compounds having size and shape that is commensurate with zeolite cavities, thereby providing the framework stabilization and facilitating the formation of zeolites. However, the structure direction of OSDAs at the molecular level has not been fully understood yet. In addition to OSDA, the substituting tetrahedral atoms (e.g., B, Al, and Ga) in the zeolite frameworks affect the formation of zeolites. Moreover, these substitutes are one of the key factors determining the physicochemical properties of zeolites.

Here, our attempts to understand the interactions between OSDAs and zeolite frameworks and the intrinsic influences of the contents and locations of the substituting atoms towards engineering of the physicochemical properties and morphologies of zeolites are presented [1–3].

[1] K. Muraoka, W. Chaikittisilp, T. Okubo, *J. Am. Chem. Soc.* **2016**, *138*, 6184–6193.

[2] S. H. Keoh, W. Chaikittisilp, K. Muraoka, R. R. Mukti, A. Shimojima, P. Kumar, M. Tsapatsis, T. Okubo, *Chem. Mater.* **2016**, *28*, 8997–9007.

[3] S. H. Keoh, W. Chaikittisilp, A. Endo, A. Shimojima, T. Okubo, *Bull. Chem. Soc. Jpn.* **2017**, *90*, 586–594.

492356 Tuning Solid Acids for the Target Catalytic Reactions

Abstract:

Solid acids facilitate cleaner and much easier reactions and thus have replaced toxic, corrosive, and unrecyclable liquid mineral acids in many catalytic applications, the most prominent being the cracking and refining of a billion tons of crude oil into useful chemical components. Brønsted and Lewis acid sites on solid acids have different functions and initiate catalytic reactions via proton transfer or hydride abstraction. By ion-exchange of parent zeolites with multivalent metal cations in solution, the type, strength, and concentration of acid sites could be tuned on zeolite via various dehydration temperatures. Amorphous silica/alumina (SA) is another popular solid acid that normally provide moderate Brøsted acidity, albeit weaker than that of zeolites. Via the homogeneous composition from precursors and formation of Al-O-Si bonds, the acidity of SAs can be tuned from moderate to zeolitic acid strength depending on the aluminum content. For the activation of C-H bonds of hydrocarbons, in-situ MAS NMR investigations indicated that Brønsted acid sites prefer to protonate aromatic rings of alkyl benzene with low activation energies of about 40 kJ/mol. Lewis acid sites are selectively activating the alkyl group by hydride abstraction with high activation energies of about 200 kJ/mol. Therefore, we tuned acidic properties of zeolites to switch on/off the petrochemical processes for target reaction intermediates, which normally have very short life time. To active the carbonyl groups for the fine-chemical synthesis or bio-oil upgrading, Brønsted and Lewis acid sites always contribute to the different reaction pathways and final products. Lots of efforts had been done to keep one type acid site only in the reaction to avoid competition and enhance the acid strength using expansive precursors to promote the conversion. However, we developed a cooperative catalysis system between Brønsted and Lewis acid sites for commercial catalysts, two acid centers can work together to promote desired reactions.

485054 Mechanical Properties of Organic Semiconductors for Mechanically Stable and Intrinsically Stretchable Solar Cells

Abstract:

This presentation describes my group's efforts to understand the molecular and microstructural basis for the mechanical properties of organic semiconductors for organic photovoltaic (OPV) devices. Our work is motivated by two goals. The first goal is to mitigate mechanical forms of degradation of printed modules during roll-to-roll fabrication, installation, and environmental forces—i.e., wind, rain, snow, and thermal expansion and contraction. Mechanical stability is a prerequisite for inexpensive processing on flexible substrates: to encapsulate devices in glass is to surrender this advantage. The second goal is to enable the next generation of ultra-flexible and stretchable solar cells for collapsible, portable, and wearable applications, and as low-cost sources of energy—"solar tarps"—for disaster relief and for the developing world. It may seem that organic semiconductors, due to their carbon framework, are already sufficiently compliant for these applications. We have found, however, that the mechanical properties (stiffness and brittleness) occupy a wide range of values, and can be difficult to predict from molecular structure alone. We are developing an experimental and theoretical framework for how one can combine favorable charge-transport properties and mechanical compliance in organic semiconductor films. In particular, we have explored the roles of the backbone, alkyl side chain, microstructural order, the glass transition, molecular packing with fullerenes, plasticizing effects of additives, extent of separation of [60]PCBM and [70]PCBM, structural randomness in low-bandgap polymers, and reinforcement by encapsulation, on the mechanical compliance. We are exploring the applicability of semi-empirical "back-of-the-envelope" models, along with multi-scale molecular dynamics simulations, with the ultimate goal of designing electroactive organic materials whose mechanical properties can be dialed-in. We have used the insights we have developed to demonstrate several new applications for OPV that demand extreme compliance, including biaxial stretching and conformal bonding of whole devices to hemispheres, and devices with ultrathin encapsulation mounted on human skin that survive significant cyclic mechanical deformation in the outdoor environment.

496979 Digital Light Synthesis to Drive Additive Manufacturing

Abstract:

Digital Light Synthesis to Drive Additive Manufacturing

Joseph M. DeSimone

Carbon, Inc

Redwood City, California

and

Department of Chemistry

University of North Carolina at Chapel Hill

and

Department of Chemical and Biomolecular Engineering

North Carolina State University

Despite the increasing popularity of 3D printing, also known as additive manufacturing (AM), the technique has not developed beyond the realm of rapid prototyping. This confinement of the field can be attributed to the inherent flaws of layer-by-layer printing, and in particular, anisotropic mechanical properties that depend on print direction, visible by the stair-casing surface finish effect. Indeed â€œ3D printingâ€ is a misnomer: it is actually 2D printing over and over again. This lecture will describe a new advance in additive manufacturing that is rapid, continuous and no longer layer-by-layer that promises to advance industry beyond basic prototyping to 3D manufacturing. The new Continuous Liquid Interface Production technology (CLIP) harnesses light and oxygen to continuously grow objects from a pool of resin instead of printing them layer-by-layer. CLIP capitalizes on the fundamental principle of oxygen inhibited photopolymerization to generate a continual liquid-interface of uncured resin between the growing part and the exposure window. This interface eliminates the necessity of an iterative layer-by-layer process allowing for continuous production. CLIP technology raises the state-of-the-art in additive manufacturing in three ways:

- GAME-CHANGING SPEED: 25-100 times faster than conventional 3D printing
- COMMERCIAL QUALITY: produces objects with consistent mechanical properties
- MATERIAL CHOICE: enables a broad range of polymeric materials

Moreover, continuous production enables advantages including the fabrication of large overhangs without the use of supports, reduction of the stair-casing effect without compromising print time, and isotropic mechanical properties. Combined, these advantages result in multiple indicators of layerless and monolithic fabrication using CLIP technology.

503212 Symbiotic Pairing of Near-UV Solar Cells with Electrochromic Windows for Visible Light and Heat Management in Architectural Applications

Abstract:

Energy use in residential and commercial buildings comprises about 40% of energy demand today and 30% of energy-related carbon emissions in the United States, with heating accounting for nearly twice the energy required for cooling and lighting combined. Increasing building energy efficiency will shave peak demands for electricity. In my talk, I will highlight our development of solar-powered electrochromic windows that can be integrated into windows to reduce electricity consumption.

The solar-powered electrochromic window comprises a polyelectrochromic conducting polymer that is optically transparent in its reduced state and dark blue in its oxidized state. Integration with a semitransparent organic solar cell provides the necessary power to switch between its transparent and colored states. Unique to this approach is our use of a single-junction organic solar cell. By designing and using materials that absorb exclusively in the ultra-violet and near-visible as the photoactive layers in our solar cells, the resulting devices exhibit open-circuit voltages that are unprecedented for single-junction organic devices (> 1.4 V). As such, we are able to drive switching of the electrochromic window without the need to construct solar cells with complex tandem architectures. Importantly, the active layers of our devices are pinhole- and defect-free. Coupled with inherently low resistive power losses, the photocurrents are scalable with the footprint of solar cells while our devices retain high fill factors. Whereas device optimization of typical solar cells to increase photovoltage has almost always come at the expense of decreasing photocurrents, the scalability of photocurrents in our devices has enabled us to break this paradigm. We can tune device photovoltage per application needs through judicious selection of the donor/acceptor materials pair whose optoelectronic properties can be tuned via molecular design, and access the necessary photocurrents by fabricating arbitrarily large devices.

501034 Taking Measure of Modern Polymer Synthesis

Abstract:

The last two decades have seen an explosion in the variety of molecular structures that can be prepared via advanced polymerization methods. This talk will survey the state of the art in polymer synthesis, and present some of the major unsolved challenges associated with confirming structure and quantifying purity of the highly sophisticated molecules anticipated to enable the next half century of soft materials innovation. In particular, the use of adsorption based separation methods hold tremendous promise, but are hindered by difficult method development, which has severely curtailed wide-spread adoption. This is an area where improved measurements and modeling of basic tethered polymer chain-surface interactions may enable tremendous insight and facilitate broader use of this method.

498809 Rational Design of Polyelectrolyte Complexes for Nucleic Acid Delivery

Abstract:

Therapeutic nucleic acids (genetic DNA, mRNA, siRNA, miRNA) hold immense potential for treating 'undruggable' disease targets and are ideally suited to personalized medicine, but their utility has been severely hampered by difficulties in delivering them to their intracellular sites of action. Present clinical efforts are dominated by lipid nanoparticle formulations, viral vectors, and ligand-conjugated, chemically-modified nucleic acids, all of which have significant limitations in biodistribution, toxicity, and off-target effects. Nanoparticle vectors assembled by polyelectrolyte complexation with hydrophilic polycations offer an attractive alternative. In particular, cationic peptides readily complex nucleic acids, with entropy gain from counterion release driving phase separation into liquid coacervates or solid precipitates. If a neutral hydrophilic polymer (e.g. PEG) is grafted to either polyelectrolyte, phase separation occurs on the microscale, producing micelles with the polyelectrolyte complex in the core and the neutral polymer in the corona. This protects the nucleic acid from nucleolytic degradation, reduces immunogenicity, and allows incorporation of targeting moieties on the outside of the corona block. A number of promising results have been obtained with this method in vitro and in animal models, but we currently lack the ability to predict polyelectrolyte complex and micelle properties from the characteristics of the constituent polymers, an essential requirement for translating laboratory successes to clinical application. We recently showed that the phase of nucleic acid complexes is controlled by the hybridization state of the nucleic acid, with single-stranded nucleic acids forming liquid coacervates and double-stranded nucleic acids forming solid precipitates. I will discuss these results, the underlying polymer physics, and the effect of polyelectrolyte length and structure on the characteristics of both bulk complexes and micelles. Developing a set of design rules for constructing micelles of designed shape and size should enable polyelectrolyte complexes to realize their potential as an improved delivery technology for therapeutic nucleic acids.

500315 Microfabricated Immune-Isolating Devices for Long Term Cell Based Therapies

Abstract:

Microfabricated immune-isolating devices for long term cell based therapies

Suman Bose, Robert Langer and Daniel Anderson

Koch Institute of Integrative Cancer Research,

Massachusetts Institute of Technology, Cambridge MA 02139.

Introduction: Cell based therapies has the potential to treat a variety of chronic disease including type 1 diabetes, anemia, liver failure and Parkinson disease [1]. Transplanting engineered or stem cell derived cells that secrete therapeutic factors for long periods of time have been an enduring goal. Implanted cells are often immunogenic, and in absence of immune-suppression are rapidly rejected by the host immune system. Cell encapsulation provides a safer alternative to immunosuppression for implanting foreign cells in vivo[2,3]. In this work we develop a microfabricated cell-encapsulating device having optimized geometry and pore-size, which protects the xenograft from immune cells while providing an optimal environment for their survival and growth. Through screening various biocompatible molecules, we develop a unique surface coating that results in dramatic reduction in foreign body reaction and fibrotic over growth on these devices. These devices could maintain engineered human cell lines secreting erythropoietin (EPO), in C57BL/6 mice for over 60 days. Rat islets implanted in these biocompatible devices restored normoglycaemia in STZ-diabetic mice for over 30 days.

Methods: Microfabricated PDMS chips were bonded to polycarbonate track etched membranes (PCTE) using a silanization-based chemical bonding method to make the final device. Surface modification was performed using surface initiated ATRP to create polymers from zwitterionic monomers[4] and small molecules[5]. For animal experiments, 500,000 293T cells engineered to secrete mouse erythropoietin (EPO) was encapsulated and implanted in 6 wk old C57BL6 mice and serum EPO levels and hematocrit was measured. Rat islets isolated from Sprague Dawley rats were encapsulated in our devices and implanted in STZ-induced C57BL/6 mice.

Results: The device contains chambers containing the graft cells that are protected from the host tissue by a thin (~10um) membrane having cylindrical pores of defined sizes (Fig 1a,b). Implanting devices with different pore sizes ranging from 1um Ð 100nm demonstrated that pore sizes < 600nm effectively prevents infiltration of all immune cells.

Grafting devices with a novel small molecule[5] showed remarkable reduction in collagen deposition over a period of 4 wk compared to untreated membranes (Fig 1c). Implanting EPO secreting human cells in C57BL/6 mice resulted in gradual increase in serum EPO and hematocrit over a period of 60 days while uncoated devices failed after 3 weeks (Fig 1d,e). In another model, we implanted 500 I.E of rat islets in STZ-induced diabetic mice and found that the biocompatible devices reversed diabetes for more than 30 days and significantly outperformed untreated devices (Fig 1f).

Conclusions: We have developed a biocompatible macro-device with optimized features that can maintain xenografts in immunocompetent mice, for extended periods of time. This work opens new opportunities to develop novel cell-based therapies for a number of chronic diseases; and forms the basis for experiments in non-human primate and eventually clinical trials.

References:

1.Allison, S. J. Nat. Rev. Nephrol. 6, 1Ð1 (2010).

2. Nyitray, C. E. et al. ACS Nano 9, 5675Ð5682 (2015).

3. Hunt, N. C. & Grover, L. M. Biotechnol. Lett. 32, 733Ð742 (2010).

4. Zhang, L. et.al. Nat Biotech, (2013).

5. Vegas, A. et al. Nat Biotech, (2016).

496748 Globular Protein Vesicles: Engineering Vesicle Size and Membrane Structure through a Tunable Molecular Packing Parameter

Abstract:

Vesicles containing globular, folded proteins have enormous potential for bio-functions different from traditional liposomes or polymersomes. We have designed two different types of recombinant fusion proteins, a globular domain (mCherry, enhanced green fluorescent protein (EGFP), or single chain antibody fragment (ScFv)) fused with a glutamic acid-rich leucine zipper (ZE) and an arginine-rich basic leucine zipper (ZR) fused with a thermo-responsive elastin-like polypeptide (ELP), to create globular protein vesicles through temperature-induced self-assembly. From systematic investigation of the protein-assembly process, this work suggests a simple practical strategy to engineer globular protein vesicles with desired size and membrane structure. The fusion proteins of globule-ZE and ZR-ELP form complexes via high affinity binding of the zippers, and transition through dynamic coacervate phases to stable hollow vesicles upon warming. The main parameters to decide protein vesicle size are the solution temperature and the protein concentration, which correlate with the thermal driving force behind the vesicular assembly. The thermal driving force dictates molecular packing parameter of the globule-zipper-ELP complex, which controls both of membrane curvature and structure. Unlike other vesicles, the tunable packing parameter enables formation of either single-layered or double-layered vesicles, as confirmed by localized encapsulation of hydrophobic dyes. These results provide critical information to create engineered globular protein vesicles for a number of bio-applications such as drug delivery, artificial cells, and micro-reactors.

494972 Biomaterial Scaffolds for Scalable Differentiation and Transplantation of Hpsc-Derived Cells for Cell Replacement Therapy in the Central Nervous System

Abstract:

Three major challenges facing effective cell replacement therapy in the central nervous system (CNS)- scalable differentiation, graft survival, and graft integration- may be addressed by optimally engineered biomaterials. Neurodegenerative disorders and traumatic injuries to the CNS, including Parkinsonâ€™s disease (PD), Huntingtonâ€™s disease (HD) and Spinal Cord Injury (SCI), affect more than 5M people in the US alone, cause a significant socio-economic burden, and currently lack effective long-term treatments or cures. Cell replacement therapy (CRT), using cells derived from human pluripotent stem cells (hPSCs), is a promising strategy to treat neurodegeneration and CNS trauma. Typically, a large number of cells are required for effective CRT in the clinic; however, current standards of generating cells on conventional 2D platforms lack scalability and reproducibility. Additionally, effective CRT is often hindered by poor post-transplantation survival of the relevant cell type and poor graft dispersion and integration with the host tissue. These challenges need to be addressed in order to facilitate clinical translation of CRT.

Towards this end, we rapidly and reproducibly generated large numbers of hPSC-derived cells in a fully-defined, scalable, thermoreversible poly-N-isopropyl acrylamide-co-poly-ethylene glycol (PNIPAAm-PEG)-based 3D biomaterial. Specifically, we generated ~2-5 fold higher yields (relative to cells differentiated on conventional 2D surfaces) of midbrain dopaminergic (mDA) neurons, medium spiny neurons (MSNs) and oligodendrocyte progenitor cells (OPCs), which are the relevant cell types for CRT in PD, HD and SCI, respectively. Interestingly, cells derived in 3D demonstrated potential for rapid functional maturation, which may facilitate downstream applications such as drug screening and disease modeling. Research into the mechanistic effects of the 3D biomaterial on stem cell biology, using bulk and single-cell RNAseq, is ongoing.

Next, to address the problem of low post-transplantation cell survival, we developed a biodegradable, bio-functionalized hyaluronic acid (HA)-based transplantation scaffold, with tunable stiffness and capable of rapid gelation. For these subsequent studies, we focused on improving CRT for PD. *In vitro*, HA hydrogels functionalized with RGD and heparin enhanced the neurite extension of mDA neurons. Following striatal transplantation of hydrogel-encapsulated hPSC-derived mDA neurons in rats, we obtained a 5-fold increase in cell survival, relative to unencapsulated cells transplanted as a suspension.

Subsequently, to address the challenge of poor graft dispersion and integration, we first used an *in vitro* screen to identify biomolecules that can disperse hPSC-derived mDA neurons. We then validated that these factors dispersed hPSC-derived mDA neurons encapsulated within HA-based hydrogels; furthermore, we discovered that stiffer hydrogels increased dispersion. Next, we developed a controlled release platform by physically incorporating dispersion factors into HA hydrogels. Controlled release of factors from HA-based hydrogels mediated effective dispersion of co-encapsulated hPSC-derived mDA neurons for at least two weeks *in vitro*. Finally, mDA neurons co-transplanted with dispersion factors within a protective HA hydrogel, optimized for controlled factor release and cell dispersion, rapidly alleviated disease symptoms in PD model rats 4 weeks after implantation. Importantly, treatment benefits correlated with increased mDA neuron survival, dispersion, and integration, with no graft overgrowth.

Overall, we developed a method for rapid generation of large numbers of hPSC-derived cells in a scalable biomaterial platform, for potential applications in cell replacement therapy, drug screening, and disease modeling. Additionally, we show that an optimally engineered cell-instructive transplantation platform holds promise for enhancing CRT in PD, and potentially in a range of other degenerative diseases or trauma.

495190 Electrochemical Activation for DNA Attachment to Surfaces

Abstract:

The ability to selectively attach biopolymers, including DNA and peptides, to surfaces is imperative for biological research. Of great interest is their specific placement on gold surfaces. While thiol self-assembly has been the standard for DNA assembly, it provides little control over the resulting thin film. Chemical coupling strategies to pre-formed monolayers enable more control but can require electrochemically active catalysts or reagents. These methods also make the amount of biopolymer deposited and its deposition location difficult to control. We have developed a reagentless coupling method to electrochemically attach biopolymers to surfaces using the specific oxidation of a catechol to a particular electrode. With our method, commercially-available DNA is rapidly attached to surfaces, yielding complete coupling within minutes. Furthermore, we have full control over the amount of biopolymer on the surface, which can be tailored to the specific application of the device. Using this attachment method, we have formed whole cell-based thin films. Additionally, we have optimized the platform for the specific detection of the endocrine disruptor bisphenol A (BPA), a problematic additive to plastics, using a DNA aptamer on the surface.

489017 Organizing Biochemical Reactions with Protein Droplets

Abstract:

Eukaryotic cells organize complex biochemistry by segregating reactions into functional compartments called organelles. Some of them, such as the nucleus and mitochondria, are bounded by membranes. But many others, like nucleoli, stress granules, P bodies, Cajal bodies, and signaling complexes, do not have membranes. Recent studies show that those membrane-free organelles are liquid droplets condensed from the cytoplasm, following the principles of liquid-liquid demixing. Long before the discovery that cells use liquid-liquid demixing to create membrane-free organelles, polymer-rich droplets were observed in lyophilic colloids and were termed as coacervate. Coacervates have been used as models for protocells, as soft materials, as reactors to increase reaction rates, and as delivery vehicles for proteins and small molecules. Protein-based droplets offer high biocompatibility and easy manipulation, are more advantageous than conventional coacervates and therefore have the potential to be used in many bioprocesses. Drawing on my experiences in soft matter and cell biology, I have been employing multidisciplinary tools including genetic manipulation, live cell imaging, biochemical reconstitution, optogenetic tools, microrheology, and fluorescence correlation spectroscopy to understand how intramolecular interactions drive liquid-liquid demixing, control droplet properties and determine specificity in cellular functions, which will continue to be one of my research directions. In parallel, I will take advantage of my training in biochemical engineering to design protein droplet-based bioprocesses including bioreactors in vitro and synthetic organelles in cells to increase reaction rates and product yields of pharmaceutically important molecules.

499288 Electron Energy Loss Spectroscopy for Optoelectronics and Thermal Dynamics at Nanocomposite Interfaces

Abstract:

Electron energy loss spectroscopy for optoelectronics and thermal dynamics at nanocomposite interfaces.

D. Keith Roper[a,b*], Jeremy R. Dunklin[d], Gregory T. Forcherio[c], Keith R. Berry[b], Carter Bodinger[b], Tyler Howard[e]

[a] *Microelectronics and Photonics Graduate Program, University of Arkansas, Fayetteville, AR 72701*

[b] *Department of Chemical Engineering, University of Arkansas, Fayetteville, AR 72701*

[c]*Army Research Laboratory, Adelphi, MD 20783*

[d]*National Renewable Energy Laboratory, Golden, CO 80401*

[e]*Department of Physics and Engineering Physics, Southeast Missouri State University, Cape Girardeau, Missouri 63701*

Optoelectronically active nanocomposite interfaces with programmable thermal dynamics could improve medical diagnostics, therapeutics and sustainable energy recovery. Nanoantenna at ceramic and polymer interfaces have been shown to enhance evaporative and pervaporative phase change. But challenges in simulating, fabricating and characterizing nanoantenna at solid-fluid interfaces has constrained their implementation in functional devices and systems.

This work introduced electron energy loss spectroscopy (EELS) coordinated with coupled dipole approximation (CDA), finite element modeling (FEM), microthermometry and scanning electron microscopy (SEM) to develop structure-property relations for optoelectronic and optothermal effects of nanoantenna (NA) in nanocomposites and at interfaces. NA were integrated with transition metal dichalcogenides, polymer thin films and ceramic surfaces by three methods (i) drop-casting, (ii) solvent dispersion and (iii) electroless plating. Electron energy loss spectroscopy (EELS) was used to predict and characterize modes, damping and electromagnetic near fields of NA at nanocomposite interfaces (Fig. 1). Assembled NA were simulated using CDA to compare optoelectronic damping with Mie theory, Maxwell Garnett effective medium theory (EMT), and measured spectra (Fig. 2). Optothermal dissipation simulated by FEM using SEM was compared with microthermometry to distinguish contributions from excitation, radiation, conduction and convection (Fig. 3). Integrating results from these simulation tools and novel integrated characterization approach supported relating effects of geometry and composition for both NA and nanocomposite on equilibrium as well as dynamic optoelectronic and optothermal properties of nanocomposites [3].

The structure-property relations obtained were validated over ultraviolet, visible, and near-infrared wavelengths and over greater than 10-fold changes in relative values of internal and external thermal time constants. Polymer thin films and ceramics exhibited consistent interfacial thermal dynamics in both liquid and gas media. Enhanced optical extinction and enhanced interfacial energy transport observed in super-wavelength scale nanocomposites with concentrated NA inclusions was attributable to optical diffraction.

Overall, coordination of EELS, CDA, FEM, and SEM with transmission UV-vis spectroscopy and microthermometry supported a comprehensive approach to simulate, fabricate and characterize nanocomposites and interfaces decorated with optoelectronically active nanoantenna. Structure-function relations that appear valid across a broad spectrum of material properties and operating conditions could guide design of nanoantenna-containing nanocomposites for implementation in a variety of uses.

[1] *G.T. Forcherio et al., Adv. Opt. Mat. (2016) 4(8) 1288* ***[2]*** *T.V. Howard et al., submitted.* ***[3]*** *D.K. Roper et al., in preparation.*

495770 On the Rice Husk Ash Admixing with Cement: Preparation, Characterization and Analysis

Abstract:

On the Rice Husk Ash Admixing with Cement: Preparation, Characterization and Analysis

ABSTRACT

Rice husk is the outermost part of rice paddy, it covers about 20 to 25 % of the rice weight. Rice husk can be converted into rice husk ash (RHA) by burning process. The chemical composition of rice husk is found to vary from one sample to another due to the differences in the type of paddy, crop year, climate and geographical conditions.[1]

Cement is a binder, a substance used in construction that sets, hardens and adheres to other materials, binding them together.[2] Cements are classified as calcium silicate and calcium aluminate cement. Calcium silicate cement is further classified into Portland and Slag, while calcium aluminate is classified into High alumina and Pozzolona cement.[3]

Portland cement is by far the most common type of cement in widespread use around the world. This cement is made by calcining limestone ($CaCO_3$) with other materials (such as clay) to 1450 °C in a kiln. Calcium oxide, so formed is then chemically fluxed with the other materials to form calcium silicates and other cementitious compounds. The resulting clinker is ground with a small amount of gypsum into a powder to make 'ordinary Portland cement', the most commonly used type of cement (often referred to as OPC).

Rice husk ash (RHA) fulfils the physical characteristics and chemical composition of mineral admixtures. RHA is composed mostly of silica in amorphous form (85-90%) and it has a highly micro-porous structure which makes it suitable to replace cement.

In the present work, rice husk ash is prepared and characterized. A varied percentage of RHA is used as an admixture with ordinary Portland cement in concrete to formulate samples. Further, compressive and flexural strength parameters of concrete are determined by Vicat test, float test, compression test and initial & final setting times. It has been found that rice husk ash substitution for ordinary Portland cement up to 30% is acceptable. On increase in RHA percentage, compressive strength decreases rapidly.

References

[1] Chandrasekhar S, Satyanarayan KG, Pramada PN and Raghavan P. Review processing, properties and applications of reactive silica from rice husk-an overview. *Journal of Materials Science (Norwell)*, 2003; 38(15):3159 - 3168.

[2] https://en.wikipedia.org on 11 April 2017.

[3] Jackson, N. and Dhir, R.K., *Civil Engineering Materials*, 4th Edition, Macmillan ELBS, Hong Kong, 1991; pp 144 –160.

503011 Tunable Magnetoresistance of Conductive Polymer Nanocomposites

Abstract:

Magnetoresistance (MR) describes a resistance change upon applying an external magnetic field. Over the past few decades, interest in the MR effect has grown rapidly due to its many applications in magnetic recording systems, especially computer memory and storage systems.[1] With unique physical and chemical property, conductive polymer nanocomposites show great potential for MR applications. In this work, the core-shell structural conductive polymer nanocomposites were synthesized by a surface initiated polymerization method. The chemical structure, crystallinity, morphology, and thermal stability of the polymer nanocomposites were reported. The nanofiller type (organic or inorganic, magnetic or nomagnetic, conductive to insulator), size, morphology (1, 2 and 3 D), and loading effects on the MR behavior of the conductive polymer nanocomposites were studied. The temperature dependent resistance of the magnetite nanocomposites indicated a 3-D variable range hopping (VRH) electrical conduction mechanism. Both positive and negative MR were observed in the polymer nanocomposites. The positive MR was analyzed by the wave function shrinkage model; the negative MR was analyzed by the orbital magnetoconductivity theory.[2]

Reference:

1. H. Gu, J. Guo, H. Wei, S. Guo, J. Liu, Y. Huang, M. A. Khan, X. Wang, D. P. Young, S. Wei and Z. Guo, *Advanced Materials*, 2015, **27**, 6277-6282.

2. J. Guo, H. Gu, H. Wei, Q. Zhang, N. Haldolaarachchige, Y. Li, D. P. Young, S. Wei and Z. Guo, *The Journal of Physical Chemistry C*, 2013, **117**, 10191-10202.

Acknowledgementsï¼š

This project is ï¬nancially supported by the American Chemical Society Petroleum Research Fund (ACS PRF# 53930-ND6).

488718 Polydopamine Stabilized Fluorescent Nanozinc Oixde Reinforced Epoxy Nanocomposites Towards UV Shielding

Abstract:

Epoxy based nanocomposites are widely used in the applications including aerospace, automobile, electromagnetic interference (EMI) shielding, and flexible electronics. Zinc oxide (ZnO) nanoparticles, as an inorganic ultraviolet (UV) absorber, have attracted increasing interests in the applications of UV lasers, light-emitting diodes (LED), and field-emission devices. The incorporation of ZnO nanoparticles into epoxy matrix can be employed in the standard LED technology which can significantly improve the UV light resistance *via* UV light shielding. However, the nanosized ZnO particles are tend to agglomerate since the crystal inorganic ZnO nanoparticles possess the organophobic surface, which would cause the severely phase separation and further damage the mechanical property of epoxy.In this work, the nano polydopamine (PDA) as the coulping agent for ZnO nanoparticles and epoxy have been prepared. The interaction between PDA and epoxy resin is explored. The enhanced tensile strength of the cured epoxy nanocomposites filled with the PDA functionalized ZnO nanoparticles(t-ZnO) is obtained. The results showed thatthe enhanced mechanical properties show an increased interfacial interaction between t-ZnO and epoxy matrix after surface modification by the nano PDA. The introduction of PDA layer has no obvious effect on the fluorescent property of ZnO.

References:

1. Gu, H.; Ma, C.; Gu, J.; Guo, J.; Yan, X.; Huang, J.; Zhang, Q.; Guo, Z. *J. Mater. Chem. C* **2016**, *4*, 5890-5906.

2. Gu, H.; Rapole, S.; Sharma, J.; Huang, Y.; Cao, D.; Colorado, H. A.; Luo, Z.; Haldolaarachchige, N.; Young, D. P.; Wei, S.; Guo, Z. *RSC Adv.* **2012**, *2*, 11007-11018.

3. Gu, H.; Tadakamalla, S.; Huang, Y.; Colorado, H. A.; Luo, Z.; Haldolaarachchige, N.; Young, D. P.; Wei, S.; Guo, Z. *ACS Appl. Mater. Inter.* **2012**, *4*, 5613-5624.

4. Gu, J.; Liang, C.; Zhao, X.; Gan, B.; Qiu, H.; Guo, Y.; Yang, X.; Zhang, Q.; Wang, D.-Y. *Compos. Sci. Technol.* **2017**, *139*, 83-89.

496220 Comparing the Toughening Effects of Modified Graphene Oxide and Core-Shell Rubber on Polyester Resins and Glass Fiber-Reinforced Polyester Composites

Abstract:

This paper investigates the translation of fracture toughness of polyester resins to the mode I interlaminar fracture toughness of polyester/glass fiber composites. An unsaturated polyester resin (UPR) and a vinyl ester resin (VER) were toughened by two very distinctive methods: (1) incorporating a very low loading (0.04 wt%) of alkyl or vinyl modified graphene oxide (mGO); or (2) adding 5 wt% of core-shell rubber (CSR). Glass fiber reinforced composites (GFRPs) were fabricated using the toughened resins as the matrix and various types of fiber mat as reinforcements. Results from compact tension tests revealed that 0.04 wt% loading of mGO increased the fracture toughness (G_{IC_Resin}) of UPR and VER by as much as 49% and 31%, respectively, without lowering the modulus and glass transition temperature (T_g). On the other hand, CSR is a more effective toughener, though at much higher loading. CSR enhanced the G_{IC_Resin} of UPR and VER by 375% and 600%, respectively, with a slight decrease in flexural properties and T_g. Flexural properties and mode I interlaminar fracture toughness (G_{IC_Comp}) of the GFRPs were evaluated. The incorporation of mGO and CSR did not alter the flexural properties of the GFRPs, regardless of resin type and fiber architecture. CSR was not able to increase the G_{IC_Comp} of UPR laminates; however, CSR increased the G_{IC_Comp} of VER laminates by 33%. The observed toughening effects do not translate directly from resin to GFRPs (*i.e.* a 600% increase in G_{IC_Resin} of VER by 5 wt% of CSR led to only a 33% enhancement in G_{IC_Comp}). Scanning electron microscopy of the fractured specimens revealed a weak fiber-matrix bonding for all GFRP specimens. During the interlaminar fracture toughness test, the cracks preferentially propagated through the relatively weak fiber-matrix interface; therefore, the matrix toughening effect provided by mGO or CSR was significantly diminished in the polyester/glass fiber composites.

494282 Preparation and Characterizations of Barium Ferrite/Epoxy Nanocomposites

Abstract:

Owing to the excellent physical, chemical properties and good processability, epoxy resin has been widely used in the adhesives, coatings, and aerospace. Recently, the materials miniaturization and light-weight demands the high performance for multifunctional epoxy resin.

With the rapid development of nano science and technology, nanocomposites have gradually occupied certain market share. Compared to the traditional materials, nanocomposite materials have unique physical and chemical properties. In this paper, barium ferrite nanoparticles were synthesized by citric acid sol-gel method. The polyaniline (PANI) was introduced via surface initiated polymerization method to improve the dispersion quality of nanoparticles within epoxy resin matrix. The X-ray diffraction analysis (XRD), FT-IR analysis data showed that PANI was successfully encapsulated on the surface of barium ferrite with the weight percentage of 12%. Finally, the epoxy nanocomposites with different loadings of nanoparticles were prepared. The results showed that the tensile strength of epoxy nanocomposites was obviously increased as the nanoparticle loading of 10 wt%. Relative to the pure epoxy resin, the tensile strength of unmodified barium ferrite/epoxy nanocomposite was increased by 2.8%, and for modified barium ferrite/epoxy nanocomposite increased by 12.2%.

Based on the experiments and data analysis, itâ€™s found that the PANI was favorable to the barium ferrite nanoparticles dispersion within epoxy matrix and provided epoxy with the enhanced mechanical property. In addition, these epoxy nanocomposite exhibited the good magnetic, dielectric properties and improved thermal stability.

References

[1] H. Cao, X. Wang, H. Gu, J. Liu, L. Luan, W. Liu, Y. Wang and Z. Guo, Carbon coated manganese monoxide octahedron negative-electrode for lithium-ion batteries with enhanced performance. *RSC Advances*, 2015, **5**, 34566-34571.

[2] R. Skomski, P. Manchanda, P. Kumar, B. Balamurugan, A. Kashyap and D. J. Sellmyer, Predicting the future of permanent-magnet materials, *IEEE Transactions on Magnetics*, 2013, **49**, 3215-3220.

[3] H. Gu, H. Lou, J. Tian, S. Liu and Y. Tang, Reproducible Magnetic Carbon Nanocomposites Derived from Polystyrene with Superior Tetrabromobisphenol A Adsorption Performance, *Journal of Materials Chemistry A*, 2016.

[4] H. Gu, Y. Huang, X. Zhang, Q. Wang, J. Zhu, L. Shao, N, Magnetoresistive polyaniline-magnetite nanocomposites with negative dielectrical properties, *Polymer*, 2012, **53**, 801-809.

[5] M. Verma, A. P. Singh, P. Sambyal, B. P. Singh, S. K. Dhawan and V. Choudhary, Barium ferrite decorated reduced graphene oxide nanocomposite for effective electromagnetic interference shielding, *Physical Chemistry Chemical Physics*, 2015, **17**, 1610-1618.

[6] H. Gu, S. Rapole, J. Sharma, Y. Huang, D. Cao, H. A. Colorado, Z. Luo, N, Magnetic polyaniline nanocomposites toward toxic hexavalent chromium removal, *RSC Advances*, 2012, **2**, 11007-11018.

[7] H. Gu, S. Tadakamalla, Y. Huang, H. A. Colorado, Z. Luo, N, Polyaniline stabilized magnetite nanoparticle reinforced epoxy nanocomposites, *ACS applied materials & interfaces*, 2012, **4**,5613-5624.

[8] H. Gu, C. Ma, J. Gu, J. Guo, X. Yan, J. Huang, Q. Zhang and Z. Guo, J. Mater. Chem. C, 2016, 4, 5890â€“5906.

490479 Effect of Compatibilizer on Strength and Toughness of Glass Fiber Mat Reinforced Polypropylene Composites

Abstract:

High performance glass fiber mat reinforced polypropylene composites were prepared by double steel press melting impregnation method. The effect of maleic anhydride grafting polypropylene (MPP) as compatibilizer on the mechanical properties of composites was investigated. It was found that both the tensile and flexural strength were efficiently improved with the certain amount of MPP contained in the matrix. On the other hand, the impact strength decreased slightly then sharply and then kept invariability with the increasing of MPP content in the matrix. It was interesting and salutary to note that there was an appropriate MPP content to balance the strength and toughness of materials. Based on the composites without MPP, the tensile and flexural strength were increased about 38.6% and 60.2% respectively with different MPP fraction, the former the 7% and the latter the 5%, meanwhile the impact strength decreased about 43.0%. However, the changing trend of mechanical properties versus the MPP content revealed that there was some concession on strength for toughness. This compromise was reached near the MPP content about 1%. In this case, though the tensile and flexural strength were increased only about 26.4% (96MPa) and 25.9% (146MPa), the impact strength kept 94.5% (108.95KJ/m2) compared with the composites without MPP.

The dramatically improved mechanical properties could be attributed to the simultaneously appropriate enhanced interfacial adhesion and homogenous micro crystals in both matrix and interfacial area under the condition of suitable MPP content. For clarity, The DSC cooling and second heating scan curves of the composites were determined. It was showed that the increasing of MPP content led to lower crystallization temperature Tc than that of composites with MPP loading of 1%, which was almost the same as without MPP ones, indicating reduced nucleation. This was verified by the in-situ polarized optical micrographs for composites samples during isothermal crystallization at 140°C. The detailed isothermal crystallization illustrated that nucleation ability was not decreased until the introduced MPP loading exceeding 1%, and this was consistent with lower Tc. Also, it was claimed that, with increasing MPP content, the diameter of crystal spherulite increased, typically more MPP anchored on the surface of glass fiber led to the formation of transcrystalline, which play a vital role on the improved strength but the deteriorated toughness because of the excessive interfacial adhesion and larger spherulite. Excitingly, in the condition of MPP content about 1% in matrix, the appropriate interfacial microstructure and the favourable nucleation yielding lower diameter crystal and higher crystallinity balanced the strength and toughness of composites.

A schematic mechanism of this strength and toughness with lower MPP content is depicted, in which the appropriate interfacial microstructure combined with the favourable micro crystal structure in matrix, and the longer glass fiber pulled absorbing more energy when failure occurred were considered as key factors. These were confirmed by polarized optical micrographs, also verified with the SEM and optical graphs of fracture surface morphology.

498639 A Facile Approach to Fabrication and Characterization of an Eco-Friendly Zein-Laponite Nanocomposite with Improved Mechanical, Thermal, Barrier and Surface Property

Abstract:

A Facile Approach to Fabrication and Characterization of an Eco-Friendly Zein-Laponite Nanocomposite with Improved Mechanical, Thermal, Barrier and Surface Property

Tahrima B. Rouf, Jozef L. Kokini
Department of Food Science, Purdue University, West Lafayette, IN-47907

Email: trouf@purdue.edu

Zein, a major by-product of ethanol production from maize, is a bio-renewable resource, with unique amphiphilic and film forming properties [1]. It has been extensively investigated as one of the excellent candidates for eco-friendly polymers for several applications [2–4]. It has the ability to change its conformation and physical properties from flexible films to providing tough, glossy, and hydrophobic materials [5, 6]. Since it was first commercialized [7], zein was the material of interest in several industries including adhesives, coatings, printings, fibers, and packaging films and plastics [5]. However, the brittleness of the standalone zein film requires it to be functionalized with different additives like plasticizers and filler agents [3, 8, 9]. The objective of this study was to functionalize zein, by using a, synthetic layered silicate nanoclay called laponite and characterize the resulting product, for providing an understanding of the effect of laponite nanofiller in the mechanical, thermal, barrier and surface properties of the biopolymer.

Investigated zein biopolymer was functionalized with 1%, 3%, 5 wt% and 10% laponite nanoclay using solvent casting method, where 70% ethanol was used as a solvent and glycerol as a plasticizer. The dispersion of layered laponite nanoparticles in zein matrix was examined using Transmission Electron microscopy (TEM), which showed that up to 5% addition of laponite, some intercalation and exfoliation of the nanofiller was visible, but at 10% loading some aggregates started to form. The surface properties of the zein-laponite nanocomposite films were examined using atomic force microscopy (AFM) and water contact angle measurement (WCA). AFM results showed that zein-laponite nanocomposites had reduced surface roughness compared to control zein films, which could be attributed to the uniform dispersion of laponite throughout zein matrix. The WCA measurements showed that addition of laponite did not significantly change the hydrophobic properties of zein. Chemistry of bond formation in zein-laponite nanocomposite was investigated using, Fourier transform infrared spectroscopy (FTIR). FTIR spectra of pristine laponite shows characteristic peaks at 3649 cm^{-1} (attributed to the -OH stretch of the lattice hydroxyl groups), 3450 cm^{-1} (attributed to OH stretching from free H_20), 1636cm^{-1} (OH bending) and 1011 cm^{-1} (attributed to Si-O stretching)[10]. A comparison of the FTIR spectra of pristine laponite and 5 different zein-laponite nanocomposites show a new peak at 1047cm^{-1} for Z-1% Lap, Z-3% Lap, Z-5% Lap and Z-10% is visible, which is very similar to characteristic laponite peak at 1011 cm^{-1} (Si-O). This peak formation and gradual increase in its intensity, with gradual increase in filler concentration indicated that laponite has been distributed uniformly throughout zein matrix. The mechanical properties had been investigated using texture analyzer, which showed significant improvement in the young's modulus and tensile strength, with increasing filler concentration. Similarly, water vapor permeability tests have shown decrease in permeability with increasing laponite concentration. This means successful filler dispersion throughout the matrix, created a tortuous path which obstructed the escape of water vapor molecules and increased the barrier property. However, Dynamic Scanning Calorimetry (DSC) showed that laponite did not have any effect on the glass transition temperature of zein.

In summary, this research comprises a detailed examination of the effect of nanofillers on the properties of zein biopolymers and aims at providing a complete picture of the properties of nanocomposites, with an ultimate goal to find new application for agro based food grade materials which are environmentally friendly, low-cost, renewable and abundantly available in nature.

References

1. Gezer PG, Brodsky S, Hsiao A, et al (2015) Modification of the hydrophilic/hydrophobic characteristic of zein film surfaces by contact with oxygen plasma treated PDMS and oleic acid content. Colloids and Surfaces B: Biointerfaces 135:433–440. doi: 10.1016/j.colsurfb.2015.07.006

2. Lai H-M, Padua GW (1997) Properties and Microstructure of Plasticized Zein Films. Cereal Chemistry Journal 74:771–775. doi: 10.1094/CCHEM.1997.74.6.771

3. Luecha J, Sozer N, Kokini JL (2010) Synthesis and properties of corn zein/montmorillonite nanocomposite films. J Mater Sci 45:3529–3537. doi: 10.1007/s10853-010-4395-6

4. Luecha J, Hsiao A, Brodsky S, et al (2011) Green microfluidic devices made of corn proteins. Lab on a Chip 11:3419–3425.

5. Shukla R, Cheryan M (2001) Zein: the industrial protein from corn. Industrial crops and products 13:171–192.

6. Lawton JW (2002) Zein: A History of Processing and Use. Cereal Chemistry Journal 79:1–18. doi: 10.1094/CCHEM.2002.79.1.1

7. Anderson TJ, Ilankovan P, Lamsal BP (2012) Two fraction extraction of α-zein from DDGS and its characterization. Industrial Crops and Products 37:466–472. doi: 10.1016/j.indcrop.2011.07.022

8. Manisara P, Parichat M (2012) Cellulose Microfibril from Banana Peels as a Nanoreinforcing Fillers for Zein Films. Open Journal of Polymer Chemistry. doi: 10.4236/ojpchem.2012.22007

9. Ozcalik O, Tihminlioglu F (2013) Barrier properties of corn zein nanocomposite coated polypropylene films for food packaging applications. Journal of Food Engineering 114:505–513. doi: 10.1016/j.jfoodeng.2012.09.005

10. Reza Mahdavinia G, Ettehadi S, Amini M, Sabzi M (2015) Synthesis and characterization of hydroxypropyl methylcellulose- g -poly(acrylamide)/LAPONITE® RD nanocomposites as novel magnetic- and pH-sensitive carriers for controlled drug release. RSC Advances 5:44516–44523. doi: 10.1039/C5RA03731J

485309 Solid Dispersions of Electroactive Materials for Energy Storage Applications

Abstract:

Flow batteries are a relatively less common battery system that provides the opportunity to decouple the power output and energy storage components within the cell. Such systems are well-suited to stationary energy applications, where the capability to independently scale power and energy provides operational advantages. These systems have primarily been restricted to stationary applications where the footprint of the battery is not restricted, due to the relatively low energy density for the commercial flow batteries. Established flow battery systems have been demonstrated at very large scales; however, one of their limitations for moving into applications that require higher energy density is that the energy stored per volume of electrolyte is limited by the solubility of the dissolved redox species within the electrolyte. Above the solubility limit, inactive precipitates form that do not contribute to electrochemical capacity but provide operational complications such as increased electrolyte viscosity.

Recently, our research group and others have explored flow battery systems where the electrolyte intentionally starts with solid material dispersed in the liquid phase, and the solid particles in the dispersion are the electroactive material contributing to the electrochemical capacity. Starting with solid electroactive materials in principle increases the energy density of the flow batteries substantially, but new challenges arise with regards to the rheological properties of the electrolyte and the power density in these systems. This talk will discuss the different strategies that have been proposed to take advantage of loading solid particles in electroactive flow battery systems. In addition, the use of similar flowing dispersions to characterize electroactive materials will be described.

495565 Rational Design of the Cathode Materials in the Lithium-Sulfur Batteries

Abstract:

Rational design of the sulfur electrode is essential to solve several practical problems in lithiumâ€"sulfur batteries. Great progress has been made by using carbon-based nanostructures for physically and chemically confining soluble polysulfide, however, the underlying mechanism remains unclear. Here using density functional theory calculations, we show that the interfacial interaction between Li_2S_8 â€" a typical lithium polysulfide formed at the beginning of discharge â€" and the oxygen/nitrogen modified graphitic surface and edges. Our results indicate that N and O impurities play a significant role in chemically stabilizing lithium polysulfide, which are otherwise soluble, onto the carbon surface. In addition, we find that O and N functional groups can work synergistically by forming the N-C-O complex, which can enhance the interfacial binding. Moreover, we show that, by manipulating the wettability of the cathode surface using oxygen-containing hydrophilic species, adsorption of the polysulfide and electronic conductivity can be controlled to improve the performance.

Acknowledgements: The calculations were performed at NERSC and OU Supercomputing Center for Education and Research (OSCER). We thank support from the U.S. DOE through a DOE/EPSCOR grant (DESC0004600).

494890 3D Carbon Materials for Electric Double-Layer Capacitors with Ultrahigh Areal Capacitance

Abstract:

Electric double-layer capacitors (EDLCs) store energy by charge accumulation on the electrode/electrolyte interface. The characteristic of electrode materials can significantly affect the EDLC performance. 3D carbon materials can provide large surface area and high electrical conductivity. It reveals that symmetrical EDLCs with 3D carbon electrodes exhibited gravimetric capacitance of 138.4 F/g at current density of 1 A/g. Furthermore, a critical issue that the enhancement of mass loading usually sacrifices the gravimetric capacitance was solved by 3D carbon materials, an ultrahigh areal capacitance up to 1.14 F/cm^2 can be obtained. The one-step synthesis procedure and scalable electrode make 3D carbon materials a promising candidate in large-scale application for capacitors.

497284 3D High-Surface-Area and Mesoporous Graphene Sheet-like Carbon for Supercapacitors

Abstract:

Current supercapacitors suffer from low energy density mainly due to the high degree of microporosity and insufficient hydrophilicity of their carbon electrodes. Development of a supercapacitor capable of simultaneously storing as much energy as a battery, along with providing sufficient power and long cycle stability would be valued for energy storage applications and innovations. Differing from commonly studied reduced graphene oxides, in this work we identified an inexpensive heteroatom polymer (polyaniline-PANI) as a carbon/nitrogen precursor, and applied a controlled thermal treatment at elevated temperature to convert PANI into 3D high-surface-area graphene-sheet-like carbon materials. During the carbonization process, various transition metals including Fe, Co, and Ni were added, which play critical roles in both catalyzing the graphitization and serving as pore forming agents. Factors including post-treatments, heating temperatures, and types of metal were found crucial for achieving enhanced capacitance performance on resulting carbon materials. Using $FeCl_3$ as precursor along with optimal heating temperature 1000 °C and mixed acid treatment ($HCl+HNO_3$), the highest Brunauer–Emmett–Teller (BET) surface area of 1645 m^2g^{-1} was achieved on the mesopore dominant graphene-sheet-like carbon materials. The unique morphologies featured with high-surface areas, dominant mesopores, proper nitrogen doping, and 3D graphene-like structures correspond to remarkably enhanced electrochemical specific capacitance up to 478 Fg^{-1} in 1.0 M KOH at a scan rate of 5 mV s^{-1}. Furthermore, in a real two-electrode system of a symmetric supercapacitor, a specific capacitance of 235 Fg^{-1} using Nafion binder is obtained under a current density of 1 Ag^{-1} by galvanostatic charge–discharge tests in 6.0 M KOH. Long-term cycle stability up to 5000 cycles by using PVDF binder in electrode was systematically evaluated as a function of types of metals and current densities.

500447 Stability Predictions for Dimethoxybenzene Based Catholyte Materials

Abstract:

Nonaqueous redox flow batteries show promise for use as large-scale energy storage devices, but further work must be done to develop stable electroactive species. Here we describe a computational method for the systematic molecular design of dimethoxybenzene based catholyte materials. In particular key predictors of electrochemical stability for materials with different functional groups have been developed. The approach provides a computational stability prediction and high throughput screening method for future synthesis and electrochemical testing. Dimethoxybenzene compounds were synthesized and tested according to the predictions, which show promise for functional group variation within the molecular family. The application of a robust molecular screening method will decrease the time needed to identify promising molecules for further testing and aid design of novel electroactive materials.

497731 Solvate Ionic Liquid-Based Gel Electrolytes Containing Functionalized Polymer-Based Networks for Use in Lithium Metal Battery Applications

Abstract:

Lithium metal batteries (LMBs) offer a promising advancement in battery technology by replacement of the graphite anode in Li-ion batteries with a more energy dense Li metal electrode. The use of a Li metal anode allows for the production of a battery with an increased energy density estimated to be about 1200 Wh/L (400 Wh/kg). LMBs require the use of an electrolyte that exhibits high electrochemical stability (> 4.2 V vs. Li/Li^+), high ionic conductivity (> 0.1 mS/cm), a Li-ion transference number of ~0.2-0.5, and a high elastic modulus (~10-1,000 MPa) to prevent dendrite formation at the Li metal anode-electrolyte interface. Ionic liquids (ILs) offer the potential to replace volatile aqueous- and organic-based electrolytes due to their moderate room temperature ionic conductivity (~1 mS/cm), negligible vapor pressure, and high electrochemical stability. Solvate ionic liquids (SILs) are an emergent subclass of IL electrolytes that offer the advantages of traditional ILs while incorporating an inorganic cation (or anion) species, which is coordinated by a neutral ligand to form a complex ion. For example, lithium cation-containing SILs have previously been demonstrated by complexing the Li^+ cation of an appropriate lithium salt with oligoether ligands (such as tetraglyme) to form a nonvolatile liquid electrolyte that is suitable for lithium-based electrochemical energy storage devices.

While SIL electrolytes hold great promise for many future applications, incorporating these liquids into solid-state gel electrolytes is a compelling direction of investigation due to the leakproof nature and robust flexibility of gels that can also potentially prevent dendrite formation for LMB applications. Recently, we fabricated mechanically robust solvate ionogel electrolytes via in situ free-radical polymerization/cross-linking of poly(ethylene glycol) diacrylate (PEGDA) to immobilize the solvate ionic liquid [Li(G4)][TFSI], which consists of an equimolar mixture of lithium bis(trifluoromethanesulfonyl)imide (Li[TFSI]) and tetraglyme (G4).[1] Solvate ionogels with varying PEGDA content have been prepared, and their electrochemical, ion transport, and mechanical properties have been measured. The solvate ionogels displayed enhanced lithium ion diffusivity values, which increased from 6.0 x 10^{-12} m^2/s to 10.0 x 10^{-12} m^2/s as the PEGDA content was increased from the minimum gelation point (8 vol.%) up to 21 vol.% polymer; the corresponding Li^+ transport numbers increased from 0.510 to 0.575, respectively. Meanwhile, room temperature ionic conductivity measurements reveal high conductivities, up to 0.89 mS/cm at 8 vol.% PEGDA and 0.39 mS/cm at 21 vol.% PEGDA. Compression testing verifies a widely tunable elastic modulus of these solvate ionogels, varying from 2 to 640 kPa. In this work, chemical modification of the polymer network has been investigated in order to increase the compressive elastic modulus up to ~10 MPa while maintaining a practical room temperature ionic conductivity of 0.43 mS/cm. Tensile testing, AC impedance spectroscopy, linear sweep voltammetry, DC polarization, and charge-discharge testing using a commercial cathode are performed in order to determine the most promising solvate ionogel electrolyte candidates for LMB applications.

[1]Dâ€™Angelo, A. J.; Panzer, M. J. *J. Phys. Chem. B* **2017**, *121*, 890-895.

496989 Studies on Complex Electrolytes for Magnesium Batteries

Abstract:

Magnesium-ion batteries are under investigation as a possible alternative platform to lithium-ion. Magnesium metal serves as a viable anode for post lithium-ion batteries due to its high volumetric charge capacity and widespread abundance. The electrolyte is identified as a limiting factor in magnesium batteries, as many analogous electrolytes to lithium-ion batteries can passivate the magnesium metal surface preventing rechargeability. Both electrolyte salt(s) and solvent(s) must be chosen to ensure chemical compatibility while facilitating desired electrochemical activity. Here we report our recent efforts to understand transport and electrochemical properties of complex liquid and polymer magnesium electrolytes. One area of interest is the effect of the anion electronic structure and solvent on the performance of magnesium bis(disilazide) based liquid electrolytes. This includes the electrochemical stability window and facileness of magnesium metal deposition and stripping. A secondary area of interest is facilitating transport in oligomeric and polymeric magnesium electrolytes with enhanced thermal stability.

498314 First-Principles Study of Temperature Dependence of Energy Gaps in Gas Sensor Materials

Abstract:

To better understand the function of optical gas sensor at high temperature (~1000 K), we use density functional theory to study the temperature dependence of the energy gaps in the sensor materials, including TiO_2 and $SrTiO_3$, due to the electron-phonon interaction and thermal expansion. The electron-phonon interaction is simulated using both Allen-Heine-Cardona (AHC) theory and finite displacement method, while thermal expansion study is based on the quasiharmonic approximation. Our simulation agrees with experiments. Our calculation also reveal that the thermal expansion contribution is as important as electron-phonon interaction in TiO_2.

497582 Supercritical Fluid-Based Synthesis of Antimony Electrode Materials

Abstract:

Antimony is one of the most promising high-rate-capability Na- and Li-ion battery electrode materials, demonstrating extraordinarily high rates of lithiation and sodiation, as well as a large theoretical capacity of 660 mAh/g. Antimony is one of the highest capacity single-element negative electrode materials for Na-ion batteries, with a theoretical capacity lower than only phosphorus and tin. In addition, antimony nanocrystals have exhibited charge/discharge rates as fast as 20C without a significant loss in capacity. Here, we report the supercritical fluid-based synthesis of nanostructured antimony electrode materials. By controlling reaction conditions, precursor concentration, and ligand choice, the synthesis can be tailored to produce an array of nanostructured antimony electrode materials with a variety of morphologies including platelets, truncated octahedra, and dendritic microstructures. This morphological selectivity can be used to evaluate the effects of structural anisotropy on the lithiation/delithation (and sodiation/desodiation) processes, and lead to structural optimization of high-rate-capability antimony-based electrodes.

493260 Solution Combustion Synthesis of Porous CeO_2 Nanopowders: Reaction Mechanism and Physical Properties

Abstract:

Cerium (IV) oxide (CeO_2) is an important material owing to its wide applications in polishing, catalysis, fuel cells and oxygen sensors. For all these cases, controlled synthesis of porous CeO_2 nanopowders is important since its redox and transport characteristics depend on the physical properties such as particle size, surface area, pore size and pore size distribution. In this work, CeO_2 nanopowders were prepared by solution combustion synthesis (SCS), varying the SCS parameters in terms of precursor oxidizer (cerium nitrate hexahydrate and cerium ammonium nitrate), fuel (glycine and hydrous hydrazine) and fuel-to-oxidizer ratio (0.5â€“3). These parameters influence the combustion behavior and, in turn, the properties of the CeO_2 powders. Based on the experimental results, the combustion reaction mechanism, and the correlation between the SCS parameters and the resulting powder morphology and properties is discussed.

493896 Combining Pre- and Post-Nucleation Trajectories for the Design of Hierarchical FAU/EMT Materials from Organic-Free Sols

Abstract:

Efforts to overcome diffusional limitation in microporous zeolites have been directed towards the design of hierarchical structures in said zeolites. Hierarchical zeolites contain highly interconnected networks of zeolitic micropores combined with meso- and/or macropores. Interest in these materials stems from the higher reaction rates,[1,2] improved selectivity,[3] resistance to deactivation,[4] and novel adsorption behavior[5] that they exhibit in comparison to the typical zeolites that only have micropores. Among the synthesis approaches, repetitive branching by rotational intergrowth[2,6Ð9] holds promise for industrial implementation due to its simplicity and lower cost as it is a one-step synthesis which uses simple structure-directing agents or additives compared to hard and dual-soft templating approaches.

As an important component of fluid catalytic cracking (FCC) catalysts, Faujasite is one of the most widely used zeolites. It consists of two polymorphs: FAU (cubic) and EMT (hexagonal), which share the same periodic building unit (PBU) known as the Faujasite sheet. The intergrowth of FAU and EMT in the same material results in the development of hierarchical FAU/EMT materials grown in a house-of-card assembly of nanosheets. Such materials have been reported using either organosilane surfactants[7,9] or Lithium or zinc salts.[8] However, both FAU and EMT can nucleate from inorganic sols containing only sodium ions, hence, the development of a house-of-card assembly of Faujasite sheets from sodium aluminosilicate mixtures is in principle possible thus avoiding extra costs of additional additives.

Earlier findings by Khaleel et.al[10] demonstrate the use of pre- and post-nucleation trajectories in the synthesis of high FAU content Faujasite nanocrystals. The aim of the present work is to gain further insight and extend this principle into designing house-of-card assembly of Faujasite nanosheets from organic-free sodium aluminosilicate sols. The effect of combining different synthesis trajectories was investigated and materials were characterized by high resolution transmission electron microscopy and X-ray diffraction. Such materials are type X (Si/Al < 1.5) FAU/EMT intergrowths that reduce the diffusion path length perpendicular to the sheets (ca. 200 nm) and also provide extra meso/macro-porosity between the sheets enhancing diffusion. Preliminary results indicate that different hierarchal zeolite materials can be designed by proper adjustment of the synthesis conditions. Findings demonstrate that it is possible to combine the effects of pre-and post-nucleation sol composition to steer crystal size and crystal structure, respectively. The initial sol preserves the original trajectory towards large crystals after composition change. Despite this memory effect, the sol at this stage is still agnostic towards FAU or EMT formation, the relative content of which is dominantly determined by the final composition. This expands the spectrum of hierarchical materials that can be designed. After optimizing the hierarchal zeolite structure, the performance of the prepared materials can be tested in

catalytic applications such as n-butane cracking

References

(1) van Donk, S.; Broersma, A.; Gijzeman, O. L. .; van Bokhoven, J. .; Bitter, J. .; de Jong, K. . *J. Catal.* **2001**, *204* (2), 272Ð280.

(2) Zhang, X.; Liu, D.; Xu, D.; Asahina, S.; Cychosz, K. A.; Agrawal, K. V.; Wahedi, Y. Al; Bhan, A.; Hashimi, S. Al; Terasaki, O.; Thommes, M.; Tsapatsis, M. *Science (80-.).* **2012**, *336* (6089), 1684Ð1687.

(3) Christensen, C. H.; Johannsen, K.; Schmidt, I.; Christensen, C. H. *J. Am. Chem. Soc.* **2003**, *125* (44), 13370Ð13371.

(4) Choi, M.; Na, K.; Kim, J.; Sakamoto, Y.; Terasaki, O.; Ryoo, R. *Nature* **2009**, *461* (7261), 246Ð249.

(5) Xu, D.; Swindlehurst, G. R.; Wu, H.; Olson, D. H.; Zhang, X.; Tsapatsis, M. *Adv. Funct. Mater.* **2014**, *24* (2), 201Ð208.

(6) Chaikittisilp, W.; Suzuki, Y.; Mukti, R. R.; Suzuki, T.; Sugita, K.; Itabashi, K.; Shimojima, A.; Okubo, T. *Angew. Chemie Int. Ed.* **2013**, *52* (12), 3355Ð3359.

(7) Inayat, A.; Knoke, I.; Spiecker, E.; Schwieger, W. *Angew. Chemie Int. Ed.* **2012**, *51* (8), 1962Ð1965.

(8) Inayat, A.; Schneider, C.; Schwieger, W. *Chem. Commun.* **2015**, *51* (2), 279Ð281.

(9) Khaleel, M.; Wagner, A. J.; Mkhoyan, K. A.; Tsapatsis, M. *Angew. Chemie - Int. Ed.* **2014**, *53* (36), 9456Ð9461.

(10) Khaleel, M.; Xu, W.; Lesch, D. A.; Tsapatsis, M. *Chem. Mater.* **2016**, *28* (12), 4204Ð4213.

495853 Synthesis of Sn-MFI Zeolite with Use of Mechanochemical Reaction

Abstract:

Introduction

Since the introduction of Sn(IV) into zeolite framework in MFI-type silicate leads to introduction of Lewis acidic property, it can be utilized as a catalyst on biomass conversion. However, the higher content of Sn into the framework seems too difficult to be obtained by the conventional hydrothermal synthesis method. The objective of the present study is to synthesize the MFI-type Sn-substitute silicate with higher Sn content. To introduce Sn into zeolite framework with higher Sn concentration,[1)] it is considerably important to create Sn-O-Si bond in the initial step of Sn-silicate particle formation. Hence, we have applied the mechanochemical treatment[2)] in the synthesis process of titano-silicate, reported in our laboratory, to prepare Sn-silicate precursors with atomic-level mixing of Sn and Si, and to synthesize Sn-MFI particles by the successive hydrothermal procedure. The remarkable increase in Sn content in the particles is attempted with the control in the synthesis conditions carefully.

Experimental

SiO_2 and amorphous SnO_2 was charged in Si_3N_4 vessels with Si_3N_4 balls. The mechanochemical reaction was carried out with 600 rpm to obtain SiO_2-SnO_2 precursors with their well mixing in atomic-level. Next, as-prepared precursors, SiOw (Caplex), H_2O, a 2.0 mol/dm^{-3} HCl aqueous solution, and a 2.0 mol/dm^{-3} tetrapropylammonium hydroxide (TPAOH) aqueous solution was put in Teflon-lined pressureâ€resistant vessels. The molar ration of the concentration was 1 SiO_2: 0.02 SnO_2: 30 H_2O: 0.5 TPAOH: 0-0.4 HCl. After the mixed solution was agitated well for 48 hours at room temperature, then, it was aged at 160 Â°C for 120 hours. Precipitates were recovered from concentrates after centrifugation of aged sample and then treated thermally at 540 Â°C for 12 hours so as to obtain Sn-MFI particles.

Results and Discussion

UV spectroscopic measurements of mechanochemically treated samples for 0-24 hours and Sn-MFI particles synthesized with HCl/Si = 0-0.3 revealed that the absorption peak of 260-300 nm corresponding to hexagonally coordinated Sn species was disappeared by the mechanochemical reaction for 24 hours. On the other hand, an absorbance at ca 210 nm corresponding to tetragonally coordinated Sn was increased. It can be explained by the introduction of Sn into SiO_2 framework in atomic level, leading to change in fine structure surrounding Sn.

Sn-MFI particles obtained by use of SiO_2-SnO_2 precursor formed in mechanochemical treatment for 24 hours was well characterized. Their XRD patterns showed MFI-type structure for HCl/Si = 0-0.3, but amorphous for HCl/Si = 0.4. The UV measurements revealed that tetragonally coordinated Sn species were confirmed for HCl/Si = 0-0.3

samples, so that Sn was undoubtedly located in the Sn-O-Si framework. The Sn content in Sn-MFI particles were determined by ICP-AES measurements. The Sn content was increased from 1.45 to 2.83 wt% with the increase in the ratio of HCl/Si from 0 to 0.3. Namely, the Sn content can be controlled by the HCl concentration. The particle shape was changed to sphere with increase in the ratio of HCl/Si from 0 to 0.3. The results suggested that mechanochemical pre-treatment is a promising procedure for the dense-doping of transition metals into zeolite frameworks.

1) N. K. Mal et al., *Microporous Mater.*, **12**, 331 (1997).

2) K. Yamamoto et al., *Micropor. Mesopor. Mater.*, **101**, 90 (2007).

497866 Crystallization of One-Dimensional Zeolites By Nonclassical Pathways

Abstract:

Zeolites are microporous materials that are heavily utilized in commercial applications such as catalysis, separations, ion exchange, and adsorption, among others. Their utility is based largely on their unique characteristics that include excellent stability, well-defined networks of pores, and tunable physicochemical properties (i.e., shape, size, and composition). Despite significant effort to understand the formation of zeolites,[1,2] their pathways of growth are not well understood owing to complex nonclassical mechanisms involving the assembly of precursors. For example, syntheses commonly lead to the formation of *worm-like particles* (WLPs), which are bulk amorphous precursors that serve as putative growth units during crystallization[3]. Evidence of crystallization by particle attachment (or CPA) is mounting for a range of materials that include biominerals[4], metal oxides[5], and zeolites[6]. Here, we will discuss the pathways of formation for two one-dimensional zeolites (LTL and TON types). We will discuss the results of standard synthesis protocols that incorporate alternative sources, crystal seeds and/or zeolite growth modifiers (ZGMs), which are molecules that alter the anisotropic growth rate(s) of zeolite crystallization. The results presented here focus on fundamental aspects of zeolite nucleation and growth, as well as describing how we can capitalize on this knowledge to rationally design materials with superior performance than those derived from conventional synthesis methods.

(1) Lupulescu, A. I.; Kumar, M.; Rimer, J. D.; *JACS.* 135 **(2013)** 6608-6617

(2) Choi, M.; Na, K.; Kim, J.; Sakamoto, Y.; Terasaki, O., et al.; *Nature.* 461 **(2009)** 246-249

(3) Kumar, M.; Li, R.; Rimer, J. D.; *Chem. Mater.* 28 **(2016)** 1714-1727

(4) Politi, Y.; Arad, T.; Klein, E.; Weiner, S.; Addadi, L.; *Science.* 306 **(2004)** 1161-1164

(5) Penn, R. L.; Banfield, J. F.; *Science.* 281 **(1998)** 969-971

(6) Lupulescu, A. I.; Rimer, J. D.; *Science.* 344 **(2014)** 729-732

497906 Expanding the Scope of Fluoride-Free Pure Silica Zeolite Syntheses

Abstract:

Zeolites are crystalline microporous materials, with wide applications as solid catalysts, adsorbents and ion exchangers. Pure silica zeolites show high hydrothermal stability, hydrophobicity and dielectric properties, making them ideally suited in applications such as gas separation membranes, low-*k* dielectrics and drug delivery. However, of the 232 zeolite frameworks recognized by the International Zeolite Association, less than 50 have been synthesized in pure silica form. Most pure silica zeolites are synthesized in the presence of hydrogen fluoride, which hinders their commercial applications because of cost and safety-related issues. Only a few zeolite structures can be synthesized in their pure silica form without the use of hydrogen fluoride.

Dry gel conversion (DGC) is one of the techniques that has been used in the past to synthesize pure silica forms of BEA, MFI and TON frameworks. The technique basically involves drying the alkaline synthesis solution to a dry solid, followed by steam-assisted crystallization (SAC). In the case of BEA framework, DGC is the only known method to synthesize the pure silica form without the use of hydrogen fluoride. In this study, we report the use of the DGC technique to synthesize pure silica CHA framework zeolite without the use of hydrogen fluoride for the first time. ^{29}Si MAS NMR studies were used to understand the synthesis mechanism involved and make improvements. In addition, we propose a new metric for analyzing pure silica zeolites and that the uniqueness of the DGC technique allows it to be used as a general technique for the fluoride-free synthesis of a number of new pure silica zeolites.

499137 Atomic Resolution Imaging of MEL Intergrowth in 2-Dimensional MFI Nanosheets

Abstract:

Structural aspects such as grain boundaries, defects and surface termination of two-dimensional (2D) zeolites as compared to other 2D materials like graphene, boron nitride, transition metal dichalcogenides, black phosphorus have not been fully explored.[1] Even though recent advancements in aberration-corrected transmission electron microscopy (TEM) have enabled routine atomic resolution imaging of several other 2D materials, 2D zeolites remain an exception primarily due to their lower structural stability under an electron beam.[2] In contrast to indirect methods of characterization like XRD, EXAFS, STM, we use aberration corrected annular dark field scanning transmission electron microscopy (ADF-STEM) to directly image the atomic arrangement in MFI nanosheet.[3,4] ADF-STEM involves scanning a focused ~ 0.8 Å size, aberration-corrected electron probe across the nanosheet. The scattered electrons transmitted through the nanosheet are collected by an annular detector to form an image. This direct method of imaging reveals that domains of MEL are intergrown with MFI framework at the atomic scale. The pattern of intergrowth is identified using a template matching algorithm over the TEM images. By comparing electron diffraction (ED) patterns with simulated ED patterns we estimate the width of MFI domains and the fraction of MEL contained within these nanosheets. We will discuss in detail, the damage rate of MFI nanosheets under electron beam, the strategies for overcoming the electron beam damage, the digital image processing algorithm employed to identify the pattern of intergrowth and the structure factor calculations for simulating reciprocal lattices for patterned MFI/MEL intergrown nanosheets.

References

[1] O.V. Yazyev et al, Nature Nanotechnology **9**, (2014), 755-767.

[2] O. Ugurlu *et al*, Phys. Rev. B **83** (2011), p. 113408.

[3] M. Choi *et al*, Nature **461**, (2009), 246-249.

[4] K. Varoon *et al*, Science **334,** (2011), 72-75.

499287 Synthesis of Single-Unit-Cell Hierarchical Zeolites with Tunable Mesoporosity By Controlling Intergrowth Frequency

Abstract:

The introduction of mesoporosity in microporous zeolites is advantageous for certain catalytic applications by providing improved accessibility to active sites through reduction of diffusion limitations. A single-unit-cell hierarchical (meso/microporous) zeolite, known as self-pillared-pentasil zeolite (SPP) (1, 2) can be made by direct one-step synthesis based on a crystal growth phenomenon known as twinning. The mesoporosity between the single-unit-cell thick branches of SPP is created through repetitive 90° rotational intergrowths of MFI nanosheets. In principle, the mesopore size can be controlled by the frequency of these rotational intergrowths. Here, we report such a controllable synthesis of SPP without using any surfactants or hard templates. A systematic shift of mesopore size distribution is observed by argon adsorption isotherms with variation of synthesis conditions. This more open structure of SPP allowed improved imaging by electron microscopy to elucidate the structure of the rotational intergrowth. Implications for catalysis and adsorption applications will be discussed.

References

1. Zhang, X. et al. Synthesis of Self-Pillared Zeolite Nanosheets by Repetitive Branching. Science 336, 1684 (2012).

2. Xu, D. et al. On the Synthesis and Adsorption Properties of Single-Unit-Cell Hierarchical Zeolites Made by Rotational Intergrowths.Â Adv. Funct. Mater. 24, 201â€"208 (2014).

The Crystal Quality and Structure of AM6

Rumeysa Tekin, Juliusz Warzywoda, Albert Sacco Jr.*

Materials Characterization Center, Whitacre College of Engineering, Texas Tech University, Lubbock, TX 79409, USA

Abstract

Microporous vanadosilicate AM-6, isostructural with titanosilicate ETS-10, is a novel promising material in the traditional zeolite applications such as catalysis, photocatalysis, gas separation, sorption and the monatomic –O-V-O-V-O chains have shown potential for use in quantum wire application. High purity vanadosilicate AM-6 products with different crystal quality (disorder, or average length of the V-O-V chains in the AM-6 framework) and sizes in the range of ~50 nm to ~6 μm were hydrothermally synthesized at 448-503 K using tetramethylammonium (TMA+) ions. The combined SEM, EDX, XRD, Raman, TGA, XPS, and nitrogen adsorption isotherm data showed that by treating these as-synthesized materials in gaseous ammonia at 673 K it is possible to completely remove TMA+ ions from the AM-6 pores without adversely affecting product crystallinity and the micropore structure of AM-6, and with only minimal degradation of crystal quality of the resulting products. However, the success in a complete removal of TMA+ ions without affecting other crystalochemical characteristics of the product depended on its initial crystal quality. TMA+ ions could not be completely removed from the lowest crystal quality product (FWHM=20.3 cm-1) without completely decomposing the V-O-V chains. The intermediate crystal quality product (FWHM=15.0 cm-1) could be successfully ammonia-treated only at 673 K to remove TMA+ ions, but higher temperatures resulted in a complete decomposition of the V-O-V chains. The highest crystal quality product (FWHM=12.7 cm-1) retained the integrity of V-O-V chains as well as the long range order and the micropore structure when ammonia-treated up to 748 K.

Keywords: AM-6; Crystal quality; Gaseous ammonia treatment; Hydrothermal synthesis; Nanocrystals; Vanadosilicates

1. Introduction

Inorganic microporous framework materials such as zeolites and zeolite-type materials are of great interest in a large number of industrial applications, and are used widely as molecular sieves, ion-exchangers, adsorbents, catalyst supports, and shape-selective catalysts [1-3]. AM-6 is a zeolite-type microporous vanadosilicate composed of corner-sharing VO_6 octahedra and corner-sharing SiO_4 tetrahedra [4]. The framework of AM-6 contains semiconductor V-O-V chains made of corner-sharing VO_6 octahedra, and these chains run in two orthogonal directions and are effectively isolated from one another by a silicate matrix composed of corner-sharing SiO_4 tetrahedra [5, 6]. AM-6 has a three-dimensional large-pore channel system with 12-membered ring apertures (7.6 Å × 4.9 Å), and displays a considerable degree of disorder resulting from the intergrowth of two ordered polymorphs [4, 7]. The disorder results in the interruption of the V-O-V chains in the AM-6 framework. Experimental and theoretical studies of the electronic and geometric structure of vanadium in AM-6 suggest that V^{4+} is located in the bulk of the chains, whereas V^{5+} preferentially occupies sites at the end of chains both in crystal defects within the framework and on the external crystal surface [5, 8].

Rocha et al. [4] were the first to synthesize vanadosilicate AM-6, but they used the isostructural titanosilicate ETS-10 [9] crystals as seeds in their synthesis. Synthesis of AM-6 through the use of ETS-10 seeds affects the product purity, and may compromise applications and complicate characterization of the product due to the presence of titanium atoms and ETS-10. It is therefore of interest and importance to synthesize seed-free AM-6 product. ETS-10-free AM-6 products were first synthesized by Ismail et al. [10] using tetramethylammonium (TMA^+) ions as an organic structure-directing agent. TMA^+ ions are located in the pores of AM-6, and, due to their large size (diameter ~6 Å [11]) relative to the pore size of AM-6 (~8 Å [12]), largely block the access to the pores [7, 10]. TMA^+ ions were removed from the pores of zeolite α by calcination for 17 h at 773 K in air [13]. AM-6 is not stable in air at such a temperature [4, 14]. Thus, Ismail et al. [10] removed TMA^+ ions from the pores of AM-6 via treatment of crystals for 50 min at 673 K in flowing (5 cm^3 s^{-1}) gaseous ammonia-helium mixture (2 vol% ammonia, 98 vol% helium). This procedure was a modified method developed by Kresnawahjuesa et al. [13] to completely remove TMA^+ ions from the pores of zeolite α at 673 K while minimizing damage to the zeolite structure. The nitrogen sorption, TGA, and XRD analyses of the ammonia-treated AM-6 showed that the porosity was fully developed, and the AM-6 structure was not adversely affected by the ammonia treatment at 673 K for 50 min [10]. Datta and Yoon [7] compared the quality of AM-6 crystals. This was accomplished by comparing the full-width-at-half-maximum (FWHM) of the 870 cm^{-1} Raman band of AM-6. This band is attributed to the V-O stretching vibration of the relatively undistorted VO_6 octahedra [4], and its FWHM is used as a measure of the quality or disorder of AM-6 crystals, i.e., the average V-O-V chain length in the AM-6 framework [7]. The high quality AM-6 crystals with a large average chain length have a narrow (i.e., small FWHM) 870 cm^{-1} Raman band, and the increasing FWHM is taken to mean the decreasing quality (increasing disorder) of crystals having shorter V-O-V chains on average [7]. The absence, short, or highly variable average length of V-O-V chains in the AM-6 framework can be expected to affect and/or cause the poor/variable performance of AM-6 in the potential applications of AM-6 crystals that require the ordered arrays of quantum wires, e.g., in optoelectronics. However, the effect of AM-6 crystal quality and structure on the (photo)catalytic applications is unknown. This is because

AM-6 crystals with the shorter V-O-V chains will have more V^{5+} atoms in their framework [8], and the visible light activity of AM-6 [14] was linked to the presence of V^{5+} along the V-O-V chains in the framework [5, 6]. Also, modeling the electronic structure of AM-6 suggests that the band gap energy can be optimized by changing the relative concentrations of V^{4+} and V^{5+} in the framework [6]. In addition, catalytic testing over AM-6 suggested that both V^{4+} and V^{5+} played a role in the reactivity and selectivity of glycerol oxidehydration to acrylic acid [15]. Thus, it is of interest to investigate the effect of gaseous ammonia treatment on the quality and structure of AM-6 crystallized in the presence of TMA^+ ions.

2. Experimental

2.1. Synthesis of AM-6

The synthesis parameters used in the crystallization of all AM-6 products are shown in Table 1. TMA^+-free product (AM-6 (75 nm)-AS) was synthesized using the modified method of Mani et al. [16]. TMA^+-free product (AM-6 (75 nm)-AS) was synthesized using the modified method of Mani et al. [16]. In a typical synthesis, the reaction mixture was obtained by mixing two precursor solutions: Solution A required 25.33 g sodium silicate (26.5% SiO_2, 10.6% Na_2O, Sigma-Aldrich), 2.13 g sodium chloride (Sigma-Aldrich), 1.45 g potassium chloride (Fisher), 0.73 g sodium fluoride (Fisher), 0.25 g sodium hydroxide (Fisher), and 7.5 g deionized water (resistivity > 18 MΩ cm). Solution B, prepared by dissolving 2.83 g vanadium (IV) oxide sulfate hydrate (97%, Aldrich) in 7.5 g deionized water, was added to the solution A. The resulting gel was hand-shaken for 15 min. After measuring pH, the gel was transferred to 10 mL Teflon-lined stainless steel autoclaves. The autoclaves were then sealed and placed in an oven at 448 K. After the crystallization was complete, the final products were purple in color, and were filtered, repeatedly washed with deionized water, and dried for 24 h at 333-343 K.

TMA^+-containing AM-6 products (TMA-AM-6 (20 nm)-AS, TMA-AM-6 (1 μm)-AS, and TMA-AM-6 (2 μm)-AS) were synthesized using the modified protocols of Ismail et al. [10, 17]. In a typical synthesis of TMA-AM-6 (1 μm)-AS, the reaction mixture was obtained by mixing two precursor solutions: Solution A required 22.44 g sodium silicate (26.5% SiO_2, 10.6% Na_2O, Sigma-Aldrich), 0.89 g sodium chloride (Sigma-Aldrich), 3.95 g potassium chloride (Fisher), 16.73 g tetramethylammonium (TMA) hydroxide (25wt.% solution in H_2O, Sigma-Aldrich), and 24.96 g deionized water (resistivity > 18 MΩ cm). Solution B, prepared by dissolving 5.16 g vanadium (IV) oxide sulfate hydrate (97%, Aldrich) in 20 g deionized water, was added to the solution A. The resulting gel was hand-shaken for 5 min. After measuring pH, the gel was transferred to 10 mL Teflon-lined stainless steel autoclaves. The autoclaves were then sealed and placed in an oven at 503 K. After the crystallization was complete, the final products were purple in color, and were filtered, repeatedly washed with deionized water, and dried for 24 h at 333-343 K.

TMA^+-containing AM-6 products (TMA-AM-6 (20 nm)-AS, TMA-AM-6 (1 μm)-AS, and TMA-AM-6 (2 μm)-AS) were synthesized using the modified protocols of Ismail et al. [10, 17].

The naming convention for the as-synthesized samples is as follows: TMA^+-free sample is denoted AM-6 (75 nm)-AS, where 75 nm stands for the mode of the particle size distribution of this product, and AS stands for as-synthesized. TMA^+-containing samples are denoted TMA-AM-6 (20 nm)-AS, TMA-AM-6 (1 μm)-AS, and TMA-AM-6 (2 μm)-AS, where TMA stands for the TMA^+ ions, 20 nm, 1 μm, and 2 μm stand for the mode of the particle size distribution of the respective product, and AS stands for as-synthesized.

2.2. Treatment of AM-6 with gaseous ammonia

Treatment of AM-6 with gaseous ammonia was carried out for 2 h at 573-773 K in a tubular stainless steel reactor (tube dimensions: 22 mm inner diameter × 350 mm length) using 2 vol.% mixture of ammonia in nitrogen (Airgas) flowing at 5 cm^3 s^{-1}. The heating rate to reach the reaction temperature was 15 K min^{-1}. AM-6 sample (0.75 g) to be treated was spread to a thickness of ~1 mm in a 70 mm × 14 mm × 10 mm (L × W × H) alumina combustion boat (CoorsTech). The boat was placed in the center of the stainless steel tube, and the temperature in the reaction zone was monitored by a thermocouple placed over the boat. The temperature variation of the flowing gas mixture over the length of the boat was ± 4 K from the set-point value.

The example of naming convention for the ammonia-treated samples is as follows: AM-6 (75 nm)-AS sample treated in ammonia at 673 K is denoted AM-6 (75 nm)-NH_3 (673 K), and TMA-AM-6 (20 nm)-AS sample treated in ammonia at 673 K is denoted TMA-AM-6 (20 nm)-NH_3 (673 K)

2.3. Characterization

X-ray powder diffraction (XRD, Bruker D5005 X-ray diffractometer, Cu Kα radiation, 40 kV, 30mA) was used to determine the purity, crystallinity (structural damage), and average crystallite size of the AM-6 products. The average crystallite size was calculated from the Scherrer equation [18], and the Williamson-Hall plot [19] using the full-width-at-half-maximum (FWHM) of the selected XRD peaks. The Williamson-Hall plot was also used to determine the average inhomogeneous strain [19]. The selected peaks were fitted using the Modified Lorentz function. The instrumental broadening was determined using lanthanum hexaboride (LaB_6) powder NIST standard reference material (SRM) 660a. The morphology and size of AM-6 particles were imaged using field emission scanning electron microscope (FE-SEM, Hitachi S-4700 FE-SEM, accelerating voltage 2 kV, beam current 10 μA, working distance 3 mm) in the secondary electron imaging mode. The FE-SEM images were used to determine the particle size distributions of TMA-AM-6 (20 nm)-AS and AM-6 (75 nm)-AS products. The particle size distributions of TMA-AM-6 (1 μm)-AS and TMA-AM-6 (2 μm)-AS products were determined using an API Aerosizer LD equipped with an API Aero-Disperser dry powder dispersion system (TSI, Inc., Particle Instruments/Amherst). The average density of AM-6 crystals used for these measurements was 1750 kg m^{-3} [10]. Energy dispersive X-ray spectroscopy (EDX, Phoenix EDAX X-ray analyzer attached to the Hitachi S-4700 FE-SEM, accelerating voltage 20 kV, beam current 10 μA) was conducted to estimate the Si/V ratios in the as-synthesized and ammonia-treated AM-6 crystals. Thermal gravimetric analysis (TGA) was carried out using a Mettler Toledo TGA/SDTA851^e module at a heating rate of 5 K min^{-1} under air flow (50 mL min^{-1}) to determine the effectiveness of TMA^+ ions removal from the pores of the TMA^+-containing AM-6 samples, and the water content in the TMA^+-free AM-6 samples. Prior to the TGA measurements, the samples were equilibrated at room temperature for more than 72 h in a 75% RH constant-humidity chamber. The nitrogen sorption measurements were conducted at 77 K using a Quantachrome Autosorb iQ (ASiQ) micropore analyzer using 0.03 g of sample for each measurement. Samples were outgassed at 423 K for 8 h under high vacuum. The multi-point specific surface area (S_{BET}) was determined from the adsorption isotherm using the Brunauer-Emmet-Teller (BET) method. The external specific surface area (S_{EXT}) and micropore volume (V_{MP}), were determined from the slope and intercept, respectively, of the de Boer t-plot. The specific surface area of micropores (S_{MICRO}) was determined as a difference between S_{BET} and S_{EXT}. Raman spectroscopy was used to determine the disorder (quality) of AM-6 samples. The Raman spectra were recorded on a Bruker Optics Senterra dispersive Raman microscope spectrometer with a spectral resolution of ~3-5 cm^{-1}, using 532 nm laser excitation. X-ray photoelectron spectroscopy (XPS) was used to confirm the extent of TMA^+ ions removal from the AM-6 pores. The XPS measurements were performed on a Physical Electronics PHI 5000 VersaProbe spectrometer (base pressure in the analysis chamber less than 1×10^{-7} Pa) using monochromatic Al Kα (hν = 1486.6 eV) X-ray source (25 W, 15 kV, 100 μm analysis spot size). The survey scans in the 0-1400 eV binding energy (BE) range were collected with a pass energy of 187.85 eV and a step of 0.8 eV. For the narrow energy scans the pass energy was 23.5 eV with a step of 0.1 eV. To correct for sample charging, the BE of the spectra was referenced to the adventitious carbon C 1s BE at 284.8 eV [20].

3. Results and discussion

The XRD patterns of all the as-synthesized AM-6 products are shown in Fig. 1, and indicate pure AM-6 [4, 7, 10, 14, 16, 17]. The reflections of TMA-AM-6 (20 nm)-AS and TMA-AM-6 (1 μm)-AS (Fig. 1a, b) showed a considerable line broadening, which was more pronounced than that observed for the reflections of TMA-AM-6 (2 μm)-AS and AM-6 (75 nm)-AS (Fig. 1c, d). Preliminary calculations using the Williamson-Hall plot, i.e., assuming that XRD line broadening is due to both the small particle size broadening and the inhomogeneous strain broadening [19], showed low values of strain (less than 0.3%) (Table 2). Thus, the observed line broadening was attributed primarily to the very small crystallites (average size smaller than 100 nm, Table 2) [18] present in all the as-synthesized AM-6 products. The average crystallite size values, calculated by applying the Scherrer equation [18] to the selected AM-6 reflections, are shown in Table 2. These results matched well the average crystallite size values obtained from the Williamson-Hall plot (Table 2), and revealed that the XRD-derived average crystallite size for TMA-AM-6 (20 nm)-AS < TMA-AM-6 (1 μm)-AS < AM-6 (75 nm)-AS ≈ TMA-AM-6 (2 μm)-AS. This is consistent with a more extensive XRD line broadening for the two former products, as illustrated in Fig. 1. For comparison, the modes of particle size distributions for the as-synthesized AM-6 products determined from the SEM (Fig. 5a) and Aerosizer (Fig. 5b) measurements are also included in Table 2. The Scherrer equation- and Williamson-Hall plot-calculated average crystallite sizes for TMA-AM-6 (20 nm)-AS and AM-6 (75 nm)-AS were similar to the modes of particle size distributions determined from the SEM observations for these products (Table 2). This suggests that the SEM-observed particles in the AM-6 (75 nm)-AS and TMA-AM-6 (20 nm)-AS products (Fig. 6a, b) are individual crystallites, and these crystallites are nanoparticles. In contrast, the modes of particle size distributions for the TMA-AM-6 (1 μm)-AS and TMA-AM-6 (2 μm)-AS products, determined from the Aerosizer measurements (Fig. 5b), and the particle sizes imaged by SEM in these products (Fig. 6c, e) were in the micrometer range; and were much larger than the average crystallite sizes calculated from the Scherrer equation (nanometer range) (Table 2). However, a closer inspection of SEM images revealed multiple nano-sized square-shaped structures present on the square faces of AM-6 particles in these products (Fig. 6d, f). These structures could be the coherently scattering domains/individual crystallites that are probed by XRD [18]. Thus, the micrometer-sized particles observed by SEM in the TMA-AM-6 (1 μm)-AS and TMA-AM-6 (2 μm)-AS products (Fig. 6c, e) are composed of multiple smaller nano-sized crystallites.

All the ammonia-treated AM-6 samples were analyzed by XRD. These XRD patterns matched the XRD patterns of the corresponding starting as-synthesized AM-6 products in terms of peak positions, intensities, and widths. The FWHM of AM-6 reflections in the ammonia-treated samples did not change within the measurement error from that measured for the reflections in the as-synthesized products (Table 3). This suggests that the ammonia treatment did not change the crystallinity/long range order, produce structural damage, or introduce additional inhomogeneous strain, as measured by XRD. A closer inspection of the XRD patterns of the TMA-AM-6 (20 nm)-NH_3 (673 K) and TMA-AM-6 (1 μm)-NH_3 (673 K) products revealed that the ammonia treatment of these products at 673 K produced small amounts of quartz.

The size and morphology of AM-6 particles in the as-synthesized products is shown in Fig. 6. In all cases, AM-6 particles exhibited truncated square bi-pyramidal habit. In addition, the SEM observations revealed various irregularities and faulting on particle surfaces for TMA-AM-6 (1 μm)-AS and TMA-AM-6 (2 μm)-AS (Fig. 6c-f). Particles in the AM-6 (75 nm)-AS product were in the form of well-defined, elongated truncated bi-pyramids, which typically did not exceed ~100 nm in the length of a square basal plane edge (Fig. 5a, Fig. 6a). Particles in the TMA-AM-6 (20 nm)-AS product appeared in the form of well-defined, nearly equiaxed truncated bi-pyramids, which typically did not exceed ~50 nm in the length of a square basal plane edge (Fig. 5a, Fig. 6b). Particle size distributions for the TMA-AM-6 (1 μm)-AS and TMA-AM-6 (2 μm)-AS products showed the modes of ~1 μm and ~2 μm, respectively (Fig. 5b). Particles in these products had their axial lengths considerably smaller than their basal widths; and therefore, exhibited a "plate-like" morphology (Fig. 6c, e). The SEM observations of these products showed multiple nano-sized square-shaped island-like structures on the square faces of individual particles (Fig. 6d, f). No differences in the particle morphology and size were observed by SEM after the as-synthesized products were treated with gaseous ammonia.

Chemical analysis of AM-6 crystals via EDX showed the identical within the error of measurement Si/V ratios in the as-synthesized and the corresponding ammonia-treated samples with an average value close to 5 for the AM-6 (75 nm), TMA-AM-6 (1 μm), and TMA-AM-6 (2 μm) products; and an average value close to 6 for the TMA-AM-6 (20 nm) product (Table 4). These values are in good agreement with the theoretical Si/V = 5 for AM-6 [7, 10, 14].

The Raman spectra of all the as-synthesized AM-6 products are illustrated in Fig. 7A. All these products showed the characteristic intense asymmetric Raman band at ~868 cm^{-1}, associated with the longitudinal V-O stretching vibration of the relatively undistorted VO_6 octahedra [4]. The FWHM of this band has been used as a measure of the quality or disorder/defects in AM-6 crystals, i.e., the presence and average length of the V-O-V chains in the AM-6 framework [7]. In addition, all the as-synthesized AM-6 products exhibited weaker Raman bands at ~640 cm^{-1}, ~940 cm^{-1}, and ~1040 cm^{-1}. This is consistent with the literature data [4, 7, 14, 15, 21]. Typically, the bands at ~940 cm^{-1} and ~1040 cm^{-1} are assigned to the short terminal V=O bonds in defect sites within the framework and on the external crystal surface [4, 15, 21]. The band at ~640 cm^{-1} is also thought to be associated with vibrations of VO_6 octahedra [16]. All the as-synthesized TMA-AM-6 products showed an additional weak band at ~750 cm^{-1} due to TMA^+ ions [7].

The Raman spectra of as-synthesized AM-6 products (Fig. 7A) revealed the FWHM variation of the ~868 cm^{-1} band (Table 5), and showed that FWHM of this band for TMA-AM-6 (2 μm)-AS < TMA-AM-6 (1 μm)-AS < AM-6 (75 nm)-AS < TMA-AM-6 (20 nm)-AS. This implies the opposite trend in the crystal quality or the average V-O-V chain length [7] for these products. The Raman spectra for all the as-synthesized AM-6 products recorded after the ammonia treatment at different temperatures are shown in Fig. 7B-E. Table 5 shows the effect of ammonia treatment temperature on FWHM of the ~868 cm^{-1} Raman band for these treated AM-6 products. The results revealed that the increasing ammonia treatment temperature up to 773 K resulted in the FWHM increase of the ~868 cm^{-1} Raman band, i.e., the increase of disorder/decomposition of the V-O-V chains for all products. However, the extent of the ~868 cm^{-1} Raman band broadening, and the ammonia treatment temperature that resulted in a disappearance of this band, i.e., the stability of the V-O-V chains towards the ammonia treatment, depended on the initial crystal quality (initial FWHM of the ~868 cm^{-1} Raman band). Thus, the TMA-AM-6 (20 nm)-AS product with the largest FWHM (i.e., the shortest chains on average, the poorest initial crystal quality) effectively lost the ~868 cm^{-1} Raman band (i.e., the complete decomposition of the V-O-V chains) at the lowest temperature (648 K). The V-O-V chains in the TMA-AM-6 (1 μm)-AS and TMA-AM-6 (2 μm)-AS products with the gradually better initial crystal quality were present in the crystal framework after the ammonia treatment at 648 K, and the complete chain decomposition temperature in these products increased to 698 K and 773 K, respectively. The results presented here clearly demonstrate that the stability of a TMA^+-containing AM-6 product towards the ammonia treatment strongly depends on its crystal quality; the better the crystal quality, the higher the ammonia treatment temperature that can be used to obtain a sample without a substantial decomposition of the V-O-V chains. The TMA^+-free product (AM-6 (75 nm)-AS) was also treated in gaseous ammonia (Fig. 7E, Table 5) to investigate whether the ammonia treatment temperature itself or the combination of the treatment temperature and the reaction between ammonia and TMA^+ ions (and the subsequent diffusion of the reaction products out of the micropores) caused the V-O-V chain decomposition. As shown in Fig. 7E and Table 5, the V-O-V chains in the TMA^+-free product started to decompose at the highest treatment temperature (723 K) among all the samples investigated. This has occurred despite the AM-6 (75 nm)-AS product having a poorer crystal quality than the TMA-AM-6 (1 μm)-AS and TMA-AM-6 (2 μm)-AS products, which had completely or significantly decomposed chains, respectively, at 723 K. This suggests that the reaction between ammonia and TMA^+ ions is an important contributor to the chain decomposition. The Raman spectra for the AM-6 (75 nm)-NH_3 (723 K) and TMA-AM-6 (20 nm)-NH_3 (623 K) samples clearly showed that the broadening of the ~868 cm^{-1} band, resulting from the increasing decomposition of the V-O-V chains, was accompanied by a simultaneous increase of the relative intensity of the ~1040 cm^{-1} band, and the development of a broad shoulder on the high frequency side of the ~868 cm^{-1} band. This confirms the assignment of the ~1040 cm^{-1} band to the short terminal V=O bonds in defect sites, and suggests that the broad shoulder on the high frequency side of the ~868 cm^{-1} band may be related to the vibration of V^{5+}-O bonds. The Raman spectra illustrated in Fig. 7B-D also showed that the weak band due to TMA^+ ions at ~750 cm^{-1} disappeared for the TMA^+-containing samples that were ammonia-treated at 673 K or higher temperatures. This suggests that the ammonia treatment at 673 K or higher temperatures was sufficient to remove all TMA^+ ions from the AM-6 pores. This is consistent with the TGA, XPS, and nitrogen adsorption results (see below).

The TGA curves of all the as-synthesized AM-6 products before and after the ammonia treatment at different temperatures are shown in Fig. 8. All samples showed a weight loss from 298 K to ~700-750 K, which was followed by a small weight gain, likely corresponding to the oxidation of V^{4+} to V^{5+} [7, 10]. All the as-synthesized TMA-AM-6 products showed similar two-step weight loss TGA profiles (Fig. 8a-c) between 298 K and ~750 K. The first step, from 298 K to 548 K, is attributed to the loss of adsorbed and structural water [16]. The second step, from 548 K to ~750 K, is attributed to the thermal decomposition/burning off the TMA^+ ions/carbonaceous deposits located in the AM-6 pores. The TMA^+-free (i.e., AM-6 (75 nm)-AS) product showed only one, water-related weight loss step between 298 K and ~700 K (Fig. 8d), which was followed by the weight gain due to the oxidation of V^{4+} to V^{5+}. This is consistent with the literature data [7, 16]. The weight loss data in the 298-548 K range, attributed to the water loss, are presented in Table 6. These data show that the as-synthesized TMA-AM-6 products contained less water than the as-synthesized TMA^+-free AM-6 product. This is consistent with the presence of large, pore-blocking TMA^+ ions in the pore system of the as-synthesized TMA-AM-6 products. The effect of ammonia treatment temperature on the TGA curves of all the as-synthesized TMA-AM-6 products is illustrated in Fig. 8a-c and Table 6. As expected, the increasing treatment temperature produced the increasing water weight loss in the 298-548 K range in the resulting TMA-AM-6 products (Table 6). This is due to the progressively more extensive removal of the TMA^+ ions from the TMA-AM-6 pores. However, the ammonia treatment temperatures lower than 673 K were typically not sufficient to completely remove TMA^+ ions from the TMA-AM-6 pores. This is illustrated by the second weight loss step, attributed to the thermal decomposition of TMA^+ ions, in the TGA profiles of the TMA-AM-6 samples ammonia-treated at temperatures lower than 673 K (Fig. 8a-c). It should be noted that the TMA-AM-6 (20 nm)-AS sample, which was ammonia-treated at 598-673 K, showed the larger water loss than the AM-6 (75 nm)-AS sample, and the TMA-AM-6 (1 μm) and TMA-AM-6 (2 μm) samples treated at 673 K and higher temperatures (Table 6). This implies that the water content in these TMA-AM-6 (20 nm) samples, three of which still contain the pore-blocking TMA^+ ions (Fig. 8a), was higher than in the TMA^+-free AM-6 samples. This is likely related to the very small particle size of the TMA-AM-6 (20 nm)-AS sample. Aggregation of the nano-sized particles in this sample results in the quasi-mesoporous structure (see the nitrogen adsorption isotherm for TMA-AM-6 (20 nm)-NH_3 (598 K), and the discussion of the nitrogen adsorption isotherms below), which could lead to the additional physisorption of water for this sample; and thus, higher water loss measured by TGA.

The nitrogen adsorption isotherms of various AM-6 samples are shown in Fig. 9. The shape of these isotherms is formed by a combination of Type I isotherm (microporous structures) and Type H3-like isotherm (mesoporous structures). The Type H3 hysteresis loop, which does not show limiting adsorption at high relative pressures, is associated with capillary condensation of nitrogen in the interparticle voids of aggregates (i.e., loosely coherent particle assemblies) [22]. The hysteresis loops were observed at relative pressures 0.8-1.0, and were more pronounced for the samples composed of nanoparticles (Fig. 9c, d) than for the samples comprised of microparticles (Fig. 9a, b). This further suggests that the mesoporous structures implied by the hysteresis loops in Fig. 9 are due to the aggregation of AM-6 nanoparticles and not because of the mesopores present within the particles. The mesopore-free AM-6 microparticles reported in Ref. 7 have the same, hysteresis loop-free shape of nitrogen adsorption isotherm as the TMA-AM-6 (2 μm)-NH_3 (673 K) sample shown in Fig. 9a. The identical trend to that shown in Fig. 9 for a relative contribution of the nitrogen volume adsorbed in the interparticle voids of aggregates to the total nitrogen volume adsorbed for the AM-6 samples with the variously sized particles is illustrated in Ref. 23 for different isostructural ETS-10 products with the variously sized particles.

The specific surface areas and micropore volumes of all the as-synthesized AM-6 samples before and after the ammonia treatment at different temperatures, obtained from the nitrogen adsorption isotherms, are shown in Table 7. The BET specific surface area (S_{BET}) and micropore volume (V_{MP}) of the AM-6 (75 nm)-AS sample match the literature data for AM-6 synthesized in the absence of TMA^+ ions [7, 12, 16]. The TMA-AM-6 (20 nm)-AS, TMA-AM-6 (1 μm)-AS, and TMA-AM-6 (2 μm)-AS samples had much lower S_{BET}, specific surface area of micropores (S_{MICRO}), and V_{MP} than the AM-6 (75 nm)-AS sample. This indicates that the pores in these samples are largely blocked by TMA^+ ions. This is consistent with the size of TMA^+ ions (diameter of ~0.6 nm [11]) being comparable to the size of AM-6 pores (~8 Å [12]). The ammonia treatment of the TMA^+-containing products at the increasing temperatures resulted in a partial to a complete TMA^+ ion removal from the pores, depending on the treatment temperature. This was concluded from comparing the S_{MICRO} and V_{MP} results for the ammonia-treated TMA-AM-6 samples and the AM-6 (75 nm)-AS sample (Table 7). Temperatures lower than 648-673 K were typically not sufficient to completely remove TMA^+ ions. This is consistent with the TGA results (Table 6, Fig. 8). However, at 673 K and higher treatment temperatures the S_{MICRO} and V_{MP} results for the ammonia-treated samples matched the corresponding results for the AM-6 (75 nm)-AS sample and the literature data [7, 12, 16]. This suggests that the micropore system of these ammonia-treated AM-6 samples was free of TMA^+ ions, and the AM-6 structure was not damaged. The latter conclusion is consistent with the XRD data for the ammonia-treated samples (SI 2, SI 3). However, very high treatment temperatures (e.g., 748 K) appeared to introduce additional microporosity (Table 7, sample TMA-AM-6 (2 μm)-NH_3 (748 K)).

As expected, the survey XPS spectrum of the AM-6 (75 nm)-AS sample did not reveal the N 1s peak, but the survey spectra of all the as-synthesized TMA-AM-6 products showed the N 1s peak due to the presence of TMA^+ in their pores (Fig. 10). This is consistent with the Raman, TGA, and nitrogen adsorption data. The survey spectrum of the TMA-AM-6 (20 nm)-NH_3 (598 K) sample showed the N 1s peak, unlike the survey spectra of the TMA-AM-6 (1 μm)-NH_3 (673 K) and TMA-AM-6 (2 μm)-NH_3 (673 K) samples, which did not show the N 1s peak. This again was consistent with the Raman, TGA, and nitrogen adsorption data, which implied the presence of TMA^+ ions in the former sample and the absence of TMA^+ ions in the two latter samples.

Typically, the samples that have been exposed to ambient air have a detectable amount of adventitious carbon contamination [20]. The C 1s spectrum for adventitious carbon contamination typically has C–C (BE = 284.8 eV), C–O–C (BE ≈ 286 eV), and O–C=O (BE ≈ 288.5 eV) components [20]. The high-resolution C 1s spectra of all the as-synthesized AM-6 samples and the TMA-AM-6 samples ammonia-treated at different temperatures were deconvoluted, and the results are shown in Fig. 11 and Table 8. Fig. 11a shows only adventitious carbon contamination for the AM-6 (75 nm)-AS sample, as expected. However, all the as-synthesized TMA-AM-6 products had a fourth peak at a BE ≈ 287 eV, in addition to the three adventitious carbon contamination components in their deconvoluted C 1s spectra (Fig. 11b-d, Table 8). This is consistent with the presence of TMA^+ ions in these samples, because the BE for C bonded with N is in the 285.2-288.4 eV range [24, 25]. The TMA-AM-6 (20 nm)-NH_3 (598 K) sample also showed an additional C 1s spectrum component at a BE ≈ 287 eV (Fig. 11e, Table 8), in agreement with the Raman, TGA, nitrogen adsorption, and survey N 1s XPS spectrum data for this sample. In contrast, the deconvoluted C 1s spectra of the TMA-AM-6 (1 μm)-NH_3 (673 K) and TMA-AM-6 (2 μm)-NH_3 (673 K) samples (Fig. 11f, g and Table 8) did not show the presence of the C–N bonds. This again is consistent with the Raman, TGA, nitrogen adsorption, and survey N 1s XPS spectra data for these samples.

4. Conclusions

The high purity AM-6 products with the average particle sizes between ~20 nm and ~2 μm were synthesized hydrothermally at 448-503 K using TMA^+ ions as a structure-directing agent. These products had different crystal quality (average length of the V-O-V chains in the AM-6 framework), as determined from the FWHM of the ~868 cm^{-1} Raman band. The TGA and nitrogen adsorption isotherm data confirmed that TMA^+ ions are present in the pores of as-synthesized products, and partially block these pores. Gaseous ammonia treatment at elevated temperatures was used to remove TMA^+ ions from the pores. The success in a complete removal of TMA^+ ions from the AM-6 products without substantially affecting their crystal quality depended on the initial crystal quality of the product and the treatment temperature. For the lowest initial crystal quality product (TMA-AM-6 (20 nm)-AS, FWHM=20.3 cm^{-1}), the highest ammonia treatment temperature (623 K), which did not completely decompose the V-O-V chains, was not high enough to completely remove TMA^+ ions from the pores. For the higher quality crystals in the TMA-AM-6 (1 μm)-AS and TMA-AM-6 (2 μm)-AS products, the lowest ammonia treatment temperature, necessary for the complete removal of TMA^+ ions (673 K), did not adversely affect the crystallinity, and the micropore structure of AM-6 with only minimal degradation of crystal quality of the resulting products. For both of these products the increase of ammonia treatment temperature above 673 K resulted in a concomitant decrease of crystal quality. The highest initial quality crystals in the TMA-AM-6 (2 μm)-AS product (FWHM=12.7 cm^{-1}) retained the integrity of V-O-V chains as well as the long range order and the micropore structure upon ammonia treatment at 748 K. Ammonia treatment of the intermediate initial quality crystals in the TMA-AM-6 (1 μm)-AS product (FWHM=15.0 cm^{-1}) at temperatures higher than 673 K resulted in a complete degradation of the V-O-V chains. The results of this investigation will allow to study the effects of both the external surface and the micropore system of AM-6 crystals in the photocatalytic applications by using the same products but having the pores blocked by TMA^+ ions or with the pores open by the gaseous ammonia treatment.

References

[1] D.W. Breck, Zeolite Molecular Sieves: Structure, Chemistry, and Use, Robert E. Krieger Publishing Company, Malabar, FL, 1984.

[2] J. Rocha, V. Felix, M.W. Anderson, Microporous Titanosilicates and Other Novel Zeolite-Type Solids, Rev. Port. Quím. 3 (1996) 35.

[3] J. Rocha, M.W. Anderson, Microporous Titanosilicates and other Novel Mixed Octahedral-Tetrahedral Framework Oxides, Eur. J. Inorg. Chem. 5 (2000) 801.

[4] J. Rocha, P. Brandão, Z. Lin, M.W. Anderson, V. Alfredsson, O. Terasaki, The First Large-Pore Vanadosilicate Framework Containing Hexacoordinated Vanadium, Angew. Chem. Int. Ed. Engl. 36 (1997) 100.

[5] A.M. Shough, D.J. Doren, M. Nash, R.F. Lobo, Effects of Vanadium Substitution on the Structure and Photocatalytic Behavior of ET-10, J. Phys. Chem. C 111 (2007) 1776.

[6] A.M. Shough, R.F. Lobo, D.J. Doren, A Visible Light Photocatalyst: Effects of Vanadium Substitution on ETS-10, Phys. Chem. Chem. Phys. 9 (2007) 5096.

[7] S.J. Datta, K.B. Yoon, Synthesis of Ideal AM-6 and Elucidation of V^{4+}-to-O Charge Transfer in Vanadate Quantum Wires, Angew. Chem. Int. Ed. 49 (2010) 4971.

[8] K. Ooms, T. Polenova, A.M. Shough, D.J. Doren, M.J. Nash, R.F. Lobo, Identification of Mixed Valence Vanadium in ETS-10 Using Electron Paramagnetic Resonance, ^{51}V Solid-State Nuclear Magnetic Resonance, and Density Functional Theory Studies, J. Phys. Chem. C 113 (2009) 10477.

[9] S.M. Kuznicki, Large-pored Crystalline Titanium Molecular Sieve Zeolites, US Patent 4,853,202, 1989.

[10] M.N. Ismail, N.D. Fraiman, D.M. Callahan Jr., G. Gursoy, E. Viveiros, O. Ozkanat, Z. Ji, R.J. Willey, J. Warzywoda, A. Sacco Jr., First Unseeded Hydrothermal Synthesis of Microporous Vanadosilicate AM-6, Micropor. Mesopor. Mater. 120 (2009) 454.

[11] E.W. McCleskey, W. Almers, The Ca Channel in Skeletal Muscle is a Large Pore, Proc. Natl. Acad. Sci. USA 82 (1985) 7149.

[12] F. Mani, J.A. Sawada, S.M. Kuznicki, Comparative Adsorption Study of EVS-10 and ETS-10, Micropor. Mesopor. Mater. 204 (2015) 43.

[13] O. Kresnawahjuesa, D.H. Olson, R.J. Gorte, G.H. Kühl, Removal of Tetramethylammonium Cations from Zeolites, Micropor. Mesopor. Mater. 51 (2002) 175.

[14] M.J. Nash, S. Rykov, R.F. Lobo, D.J. Doren, I. Wachs, Photocatalytic Activity of Vanadium-Substituted ETS-10, J. Phys. Chem. C 111 (2007) 7029.

[15] A.S. Paula, L.G. Possato, D.R. Ratero, J. Contro, K. Keinan-Adamsky, R.R Soares, G. Goobes, L. Martins, One-Step Oxidehydration of Glycerol to Acrylic Acid Using ETS-10-like Vanadosilicates, Micropor. Mesopor. Mater. 232 (2016) 151.

[16] F. Mani, L. Wu, S.M. Kuznicki, A Simplified Method to Synthesize Pure Vanadium Silicate Analogue of ETS-10, Micropor. Mesopor. Mater. 177 (2013) 91.

[17] M.N. Ismail, U.K. Ibe, T. Chernenko, M. Diem, J. Warzywoda, A. Sacco Jr., Synthesis and Characterization of Vanadosilicate AM-6 with Transition Metal Ions Isomorphously Substituted in the Framework, Micropor. Mesopor. Mater. 145 (2011) 118.

[18] B.D. Cullity, Elements of X-Ray Diffraction, second ed., Addison-Wesley, Reading, MA, 1978, p. 284.

[19] A.K. Zak, W.H.A. Majid, M.E. Abrishami, R. Yousefi, X-Ray Analysis of ZnO Nanoparticles by Williamson-Hall and Size-Strain Plot Methods, Solid State Sci. 13 (2011) 251.

[20] http://xpssimplified.com/elements/carbon.php. Thermo Fisher Scientific Inc. (2013).

[21] F. Mani, J.A. Sawada, S.M. Kuznicki, Spontaneous Formation of Silver Nanoparticles on the Vanadium Silicate EVS-10, Micropor. Mesopor. Mater. 224 (2016) 208.

[22] K.S.W. Sing, D.H. Everett, R.A.W. Haul, L. Moscou, R.A. Pierotti, J. Rouquérol, T. Siemieniewska, Reporting Physisorption Data for Gas/Solid Systems with Special Reference to the Determination of Surface Area and Porosity, Pure & Appl. Chem. 57 (1985) 603.

[23] C. Casado, Z. Amghouz, J.R. García, K. Boulahya, J.M. González-Calbet, C. Téllez, J. Coronas, Synthesis and Characterization of Microporous Titanosilicate ETS-10 Obtained with Different Ti Sources, Mater. Res. Bull. 44 (2009) 1225.

[24] J.F. Moulder, W.F. Stickle, P.E. Sobol, K.D. Bomben, in: J. Chastain, R.C. King Jr. (Eds.), Handbook of X-ray Photoelectron Spectroscopy, ULVAC-PHI, Inc., Chigasaki, Japan, 1995, p. 41.

[25] http://cnx.org/contents/fdddbf98-39ce-4edc-934e-dd1f7477b52b@2. A. Bratt, A.R. Barron, OpenStax CNX (2011).

Table 1
AM-6 synthesis parameters.

Sample	Molar composition of synthesis mixture	Synthesis temperature (K)	Synthesis time (days)	Product
AM-6 (75 nm)-AS	$1.44Na_2O : 0.19K_2O : 2.2SiO_2 : 0.11V_2O_5 : 35H_2O : 0.34F$	448	7	Pure AM-6
TMA-AM-6 (20 nm)-AS	$1.44Na_2O : 0.19K_2O : 2.2SiO_2 : 0.11V_2O_5 : 35H_2O : 0.34F : 0.44TMABr$	448	7	Pure AM-6
TMA-AM-6 (1 μm)-AS	$2.25Na_2O : 1.3K_2O : 4.85SiO_2 : 0.5V_2O_5 : 2.25TMAOH : 209H_2O$	503	2	Pure AM-6
TMA-AM-6 (2 μm)-AS	$5.6Na_2O : 1.3K_2O : 4.0SiO_2 : 0.5V_2O_5 : 3.0TMABr : 321.6H_2O$	503	2	Pure AM-6

Table 2
Average particle size of different AM-6 products.

Sample	Aerosizer measurements/ SEM observations	Scherrer equation[c]	Williamson-Hall plot	
TMA-AM-6 (20 nm)-AS	20 nm[a]	36 nm	21 nm	-0.0028[d]
AM-6 (75 nm)-AS	75 nm[a]	73 nm	66 nm	-0.0002[d]
TMA-AM-6 (1 μm)-AS	1 μm[b]	62 nm	30 nm	-0.0024[d]
TMA-AM-6 (2 μm)-AS	2 μm[b]	87 nm	77 nm	-0.0005[d]

[a] Mode of the particle size distribution obtained from the SEM observations
[b] Mode of the particle size distribution obtained from the Aerosizer measurements
[c] Average values from 105, 202, 204, and 208 reflections
[d] Average inhomogeneous strain

Table 3
The full-width-at-half-maximum (FWHM) of selected AM-6 reflections before and after ammonia treatment of different AM-6 products.

Sample	hkl	FWHM (°)	Sample	hkl	FWHM (°)
TMA-AM-6 (20 nm)-AS	105	0.3262±0.0089	TMA-AM-6 (20 nm)-NH_3 (598 K)	105	0.3357±0.0127
	202	0.1952±0.0026		202	0.2000±0.0034
	204	0.2185±0.0160		204	0.2252±0.0102
	208	0.2249±0.0291		208	0.2125±0.0065
AM-6 (75 nm)-AS	105	0.1203±0.0022	AM-6 (75 nm)-NH_3 (673 K)	105	0.1202±0.0023
	202	0.1312±0.0019		202	0.1314±0.0021
	204	0.1225±0.0054		204	0.1202±0.0032
	208	0.1223±0.0034		208	0.1232±0.0033
TMA-AM-6 (1 μm)-AS	105	0.2009±0.0114	TMA-AM-6 (1 μm)-NH_3 (673 K)	105	0.2104±0.0103
	202	0.1397±0.0022		202	0.1402±0.0022
	204	0.1277±0.0050		204	0.1257±0.0068
	208	0.1294±0.0072		208	0.1328±0.0132
TMA-AM-6 (2 μm)-AS	105	0.1132±0.0036	TMA-AM-6 (2 μm)-NH_3 (673 K)	105	0.1132±0.0031
	202	0.1062±0.0028		202	0.1059±0.0015
	204	0.1041±0.0054		204	0.1037±0.0033
	208	0.1107±0.0031		208	0.1108±0.0033

Table 4
EDX analysis results of different as-synthesized and ammonia-treated AM-6 samples.

Sample	Ammonia treatment temperature (K)	Si/V
AM-6 (75 nm)-AS	As-synthesized	5.33±1.43
AM-6 (75 nm)-NH_3 (673 K)	673	5.66±0.85
TMA-AM-6 (20 nm)-AS	As-synthesized	6.17±0.72
TMA-AM-6 (20 nm)-NH_3 (598 K)	598	6.31±0.51
TMA-AM-6 (1 μm)-AS	As-synthesized	5.55±0.15
TMA-AM-6 (1 μm)-NH_3 (673 K)	673	5.39±0.43
TMA-AM-6 (2 μm)-AS	As-synthesized	5.69±0.35
TMA-AM-6 (2 μm)-NH_3 (673 K)	673	5.57±0.81
TMA-AM-6 (2 μm)-NH_3 (748 K)	748	5.64±0.53

Table 5
The full-width-at-half-maximum (FWHM, in cm^{-1}) of the ~868 cm^{-1} Raman band for different as-synthesized and ammonia-treated AM-6 samples.

Ammonia treatment temperature (K)	TMA-AM-6 (20 nm)	TMA-AM-6 (1 μm)	TMA-AM-6 (2 μm)	AM-6 (75 nm)
As-synthesized	20.3±0.7	15.0±0.2	12.7±0.3	16.9±0.1
573	19.6±0.7	N/A[a]	N/A[a]	N/A[a]
598	19.7±0.7	N/A[a]	N/A[a]	N/A[a]
623	Broad band	15.5±0.5	N/A[a]	N/A[a]
648	Very broad band	17.1±0.9	15.0±0.4	N/A[a]
673	Very broad band	22.0±0.9	15.7±0.6	16.8±0.5
698	N/A[a]	No band	15.5±0.5	16.7±0.6
723	N/A[a]	N/A[a]	16.2±0.8	Broad band
748	N/A[a]	N/A[a]	18.9±0.8	N/A[a]
773	N/A[a]	N/A[a]	Very broad band	N/A[a]

[a] Not measured

Table 6
Results of the TGA analyses of different as-synthesized and ammonia-treated AM-6 samples.

Sample	Ammonia treatment temperature (K)	Weight loss (%, 298-548 K)
TMA-AM-6 (20 nm)-AS	As-synthesized	10.82
TMA-AM-6 (20 nm)-NH_3 (598 K)	598	13.36
TMA-AM-6 (20 nm)-NH_3 (623 K)	623	14.41
TMA-AM-6 (20 nm)-NH_3 (648 K)	648	15.17
TMA-AM-6 (20 nm)-NH_3 (673 K)	673	14.06
TMA-AM-6 (1 μm)-AS	As-synthesized	7.86
TMA-AM-6 (1 μm)-NH_3 (598 K)	598	10.11
TMA-AM-6 (1 μm)-NH_3 (623 K)	623	11.75
TMA-AM-6 (1 μm)-NH_3 (648 K)	648	12.11
TMA-AM-6 (1 μm)-NH_3 (673 K)	673	12.51
TMA-AM-6 (2 μm)-AS	As-synthesized	8.42
TMA-AM-6 (2 μm)-NH_3 (598 K)	598	10.01
TMA-AM-6 (2 μm)-NH_3 (648 K)	648	12.54
TMA-AM-6 (2 μm)-NH_3 (673 K)	673	12.58
TMA-AM-6 (2 μm)-NH_3 (748 K)	748	12.48
AM-6 (75 nm)-AS	As-synthesized	13.70

Table 7
Specific surface areas and micropore volumes of different as-synthesized and ammonia-treated AM-6 samples obtained from the nitrogen adsorption isotherms.

Sample	Ammonia treatment temperature (K)	S_{BET}[a] (m^2/g)	S_{EXT}[b] (m^2/g)	S_{MICRO}[c] (m^2/g)	V_{MP}[d] (cc/g)
AM-6 (75 nm)-AS	As-synthesized	410	53	357	0.139
AM-6 (75 nm)-NH_3 (673 K)	673	411	37	374	0.145
TMA-AM-6 (20 nm)-AS	As-synthesized	232	92	140	0.058
TMA-AM-6 (20 nm)-NH_3 (598 K)	598	320	103	217	0.089
TMA-AM-6 (20 nm)-NH_3 (673 K)	673	423	83	340	0.135
TMA-AM-6 (1 μm)-AS	As-synthesized	182	28	154	0.059
TMA-AM-6 (1 μm)-NH_3 (598 K)	598	278	29	249	0.094
TMA-AM-6 (1 μm)-NH_3 (648 K)	648	349	13	336	0.129
TMA-AM-6 (1 μm)-NH_3 (673 K)	673	376	24	352	0.134
TMA-AM-6 (2 μm)-AS	As-synthesized	218	31	187	0.072
TMA-AM-6 (2 μm)-NH_3 (598 K)	598	280	4	276	0.105
TMA-AM-6 (2 μm)-NH_3 (648 K)	648	384	33	351	0.135
TMA-AM-6 (2 μm)-NH_3 (673 K)	673	386	17	369	0.140
TMA-AM-6 (2 μm)-NH_3 (698 K)	698	386	10	376	0.142
TMA-AM-6 (2 μm)-NH_3 (748 K)	748	409	14	395	0.150

[a] BET specific surface area
[b] External specific surface area
[c] Specific surface area of micropores
[d] Micropore volume

Table 8
Mean peak positions of the bands deconvoluted from C 1s spectra of different as-synthesized and ammonia-treated AM-6 samples.

Sample	**Ammonia treatment temperature (K)**	**C-C (eV)**	**C-O-C (eV)**	**C-N (eV)**	**O-C=O (eV)**
AM-6 (75 nm)-AS	As-synthesized	284.8	286.21	---	289.34
TMA-AM-6 (20 nm)-AS	As-synthesized	284.8	286.11	287.06	288.66
TMA-AM-6 (20 nm)-NH_3 (598 K)	598	284.8	286.53	287.11	289.06
TMA-AM-6 (1 μm)-AS	As-synthesized	284.8	286.11	286.96	289.02
TMA-AM-6 (1 μm)-NH_3 (673 K)	673	284.8	285.89	---	288.87
TMA-AM-6 (2 μm)-AS	As-synthesized	284.8	286.55	286.95	289.33
TMA-AM-6 (2 μm)-NH_3 (673 K)	673	284.8	286.17	---	289.12

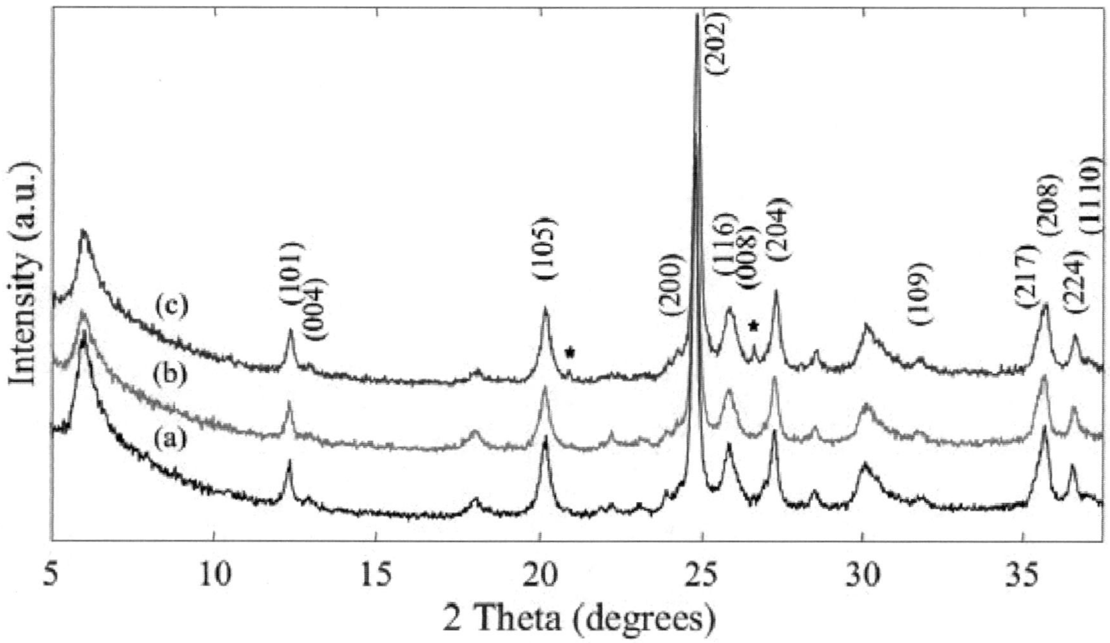

Fig. 1. XRD patterns of AM-6 products: (a) TMA-AM-6 (20 nm)-AS; (b) TMA-AM-6 (20 nm)-NH_3 (598 K); (c) TMA-AM-6 (20 nm)-NH_3 (673 K). Asterisks refer to the quartz peaks in the TMA-AM-6 (20 nm)-NH_3 (673 K) product.

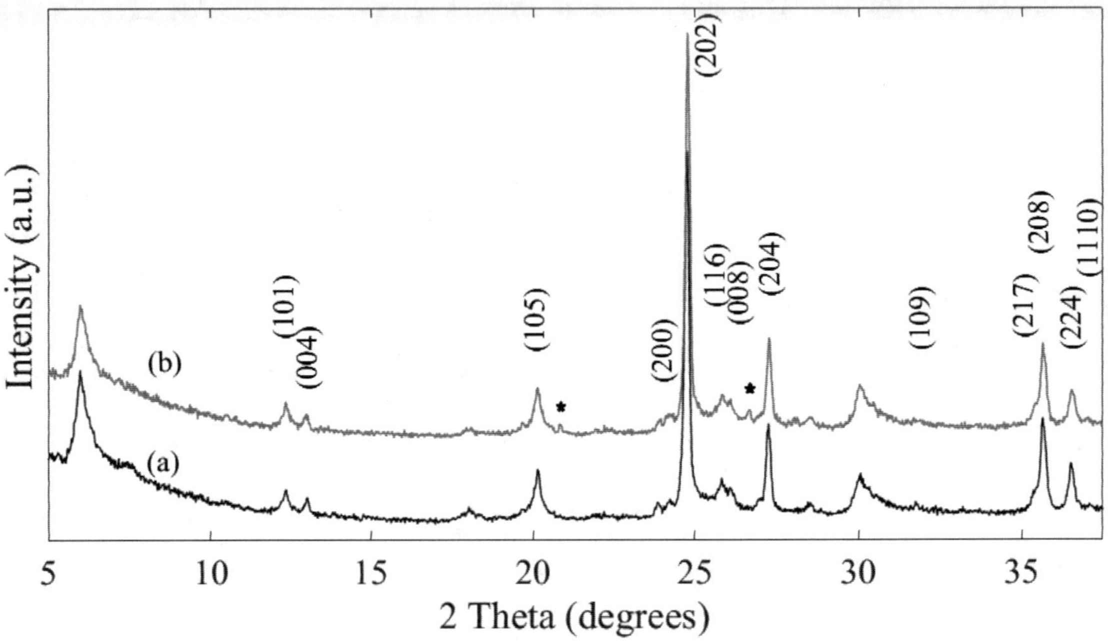

Fig. 2. XRD patterns of AM-6 products: (a) TMA-AM-6 (1 μm)-AS; (b) TMA-AM-6 (1 μm)-NH_3 (673 K). Asterisks refer to the quartz peaks in the TMA-AM-6 (1 μm)-NH_3 (673 K) product.

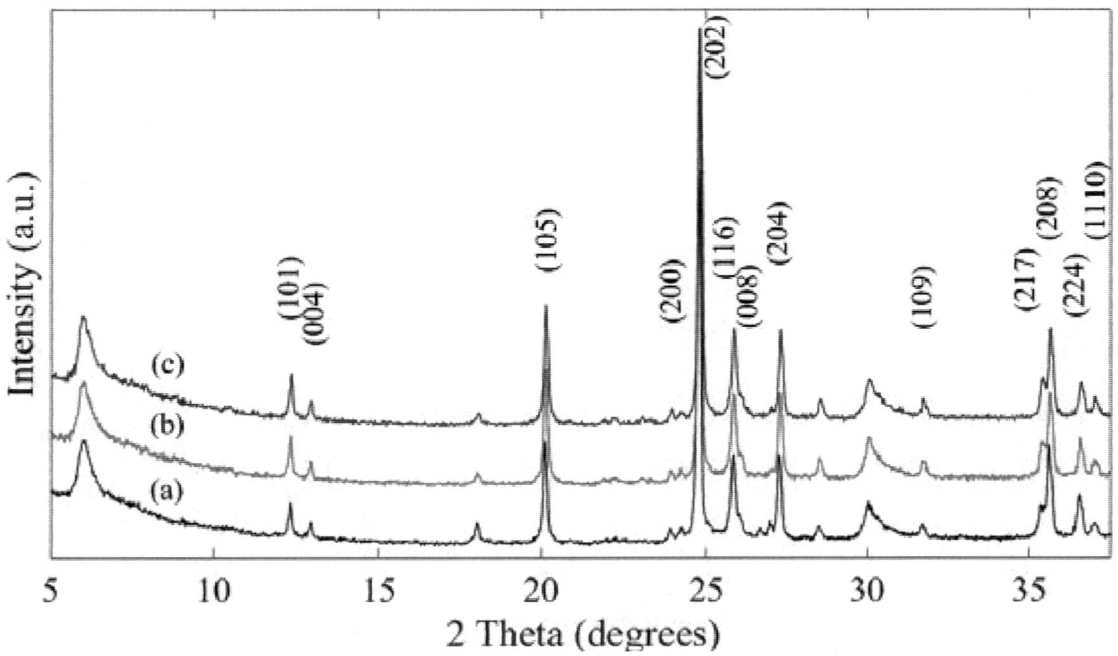

Fig. 3. XRD patterns of AM-6 products: (a) TMA-AM-6 (2 μm)-AS; (b) TMA-AM-6 (2 μm)-NH_3 (673 K); (c) TMA-AM-6 (2 μm)-NH_3 (748 K).

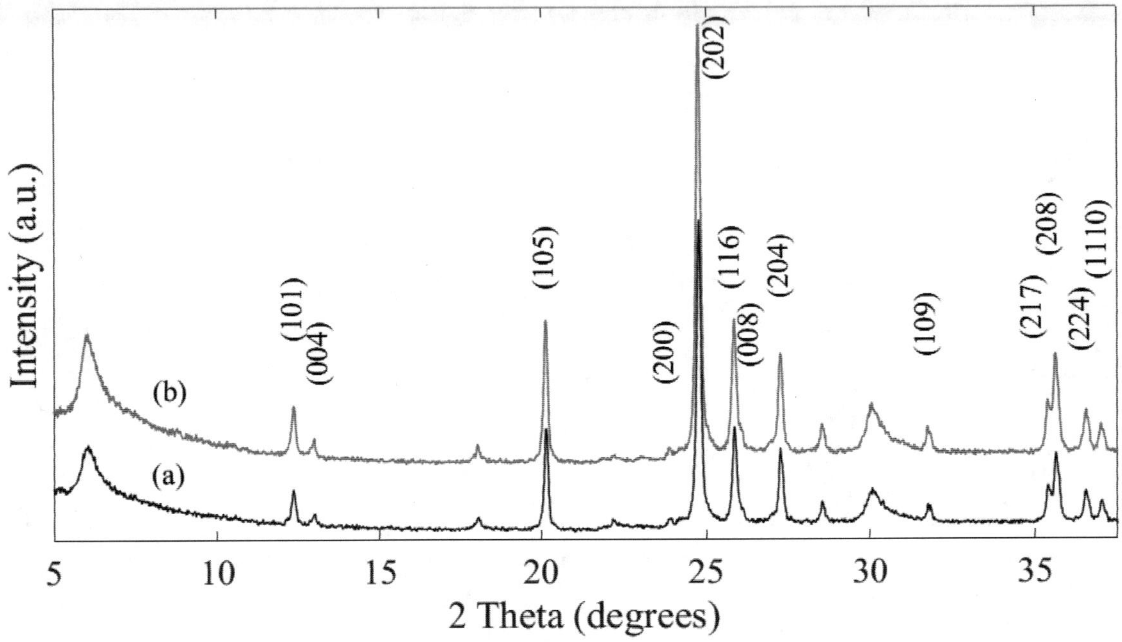

Fig. 4. XRD patterns of AM-6 products: a) AM-6 (75 nm)-AS; b) AM-6 (75 nm)-NH_3 (673 K).

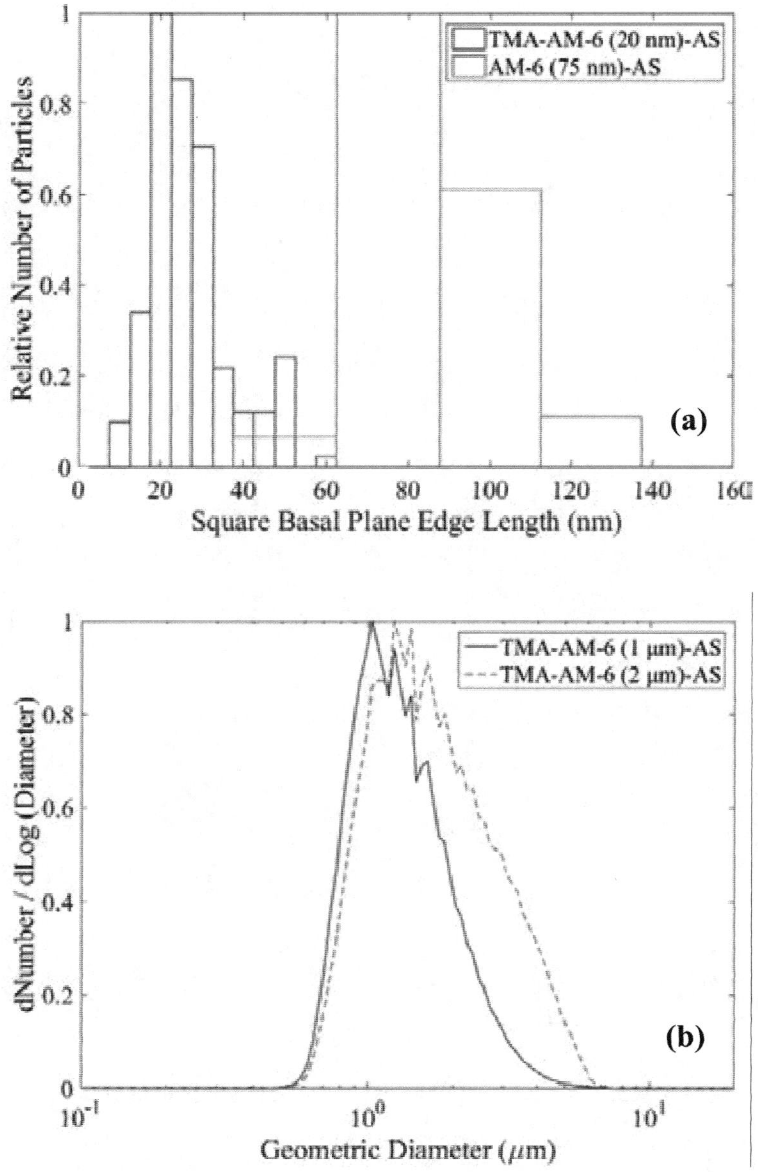

Fig. 5. Particle size distributions of AM-6 products: (a) TMA-AM-6 (20 nm)-AS and AM-6 (75 nm)-AS obtained from the SEM observations; (b) TMA-AM-6 (1 μm)-AS and TMA-AM-6 (2 μm)-AS obtained from the Aerosizer measurements. For TMA-AM-6 (20 nm)-AS c/a ≈ 1, and for AM-6 (75 nm)-AS c/a ≈ 1-2; where c is the axial length of the bi-pyramidal crystal and a is the square basal plane edge length of the bi-pyramidal crystal.

Fig. 6. FE- SEM images of AM-6 products: (a) AM-6 (75 nm)-AS; (b) TMA-AM-6 (20 nm)-AS; (c, d) TMA-AM-6 (1 μm)-AS; (e, f) TMA-AM-6 (2 μm)-AS.

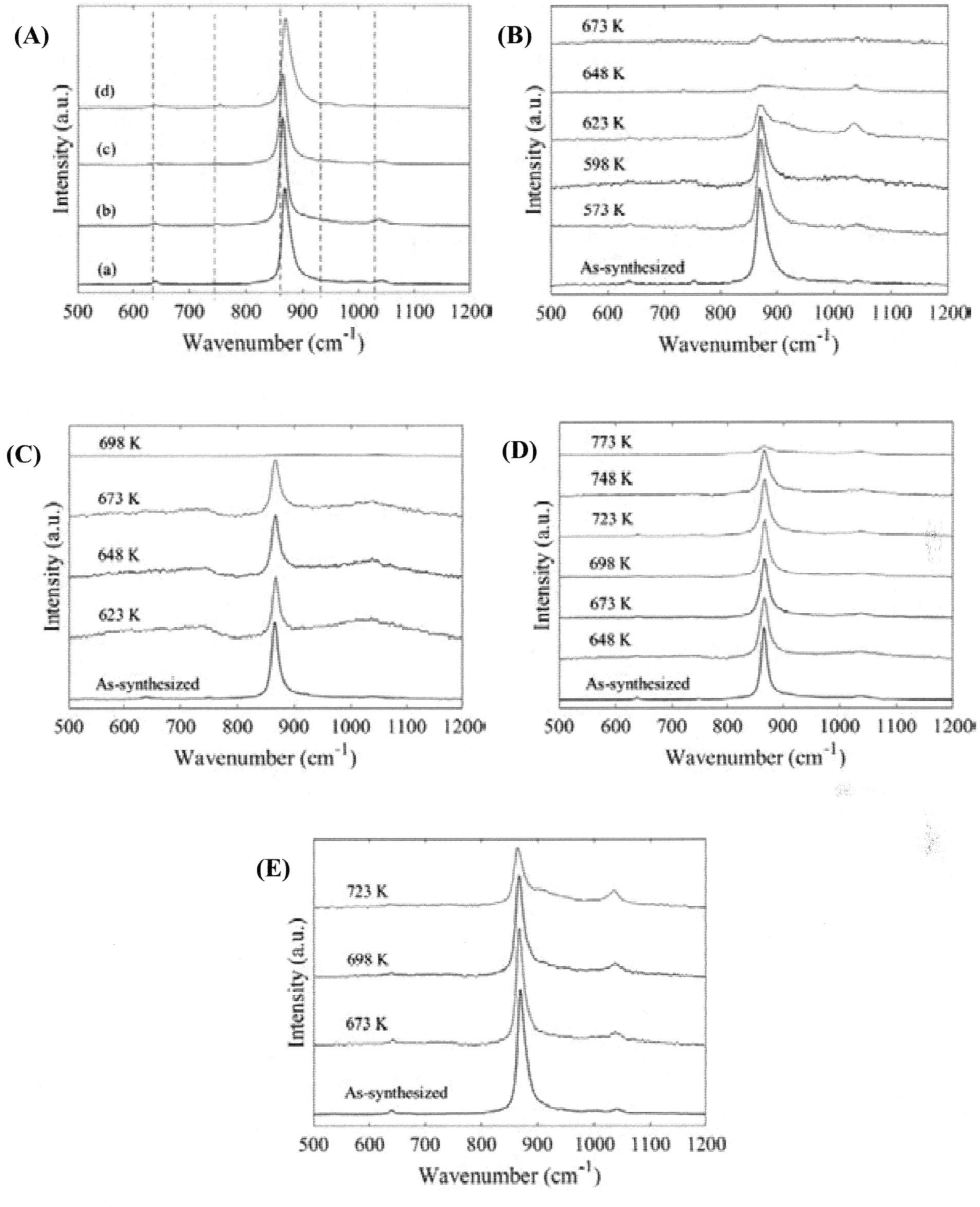

Fig. 7. (A) Raman spectra of the AM-6 samples: (a) AM-6 (75 nm)-AS; (b) TMA-AM-6 (2 μm)-AS; (c) TMA-AM-6 (1 μm)-AS; (d) TMA-AM-6 (20 nm)-AS. (B) Raman spectra of different

TMA-AM-6 (20 nm) samples before and after the ammonia treatment at different temperatures. (C) Raman spectra of different TMA-AM-6 (1 μm) samples before and after the ammonia treatment at different temperatures. (D) Raman spectra of different TMA-AM-6 (2 μm) samples before and after the ammonia treatment at different temperatures. (E) Raman spectra of different TMA-AM-6 (75 nm) samples before and after the ammonia treatment at different temperatures.

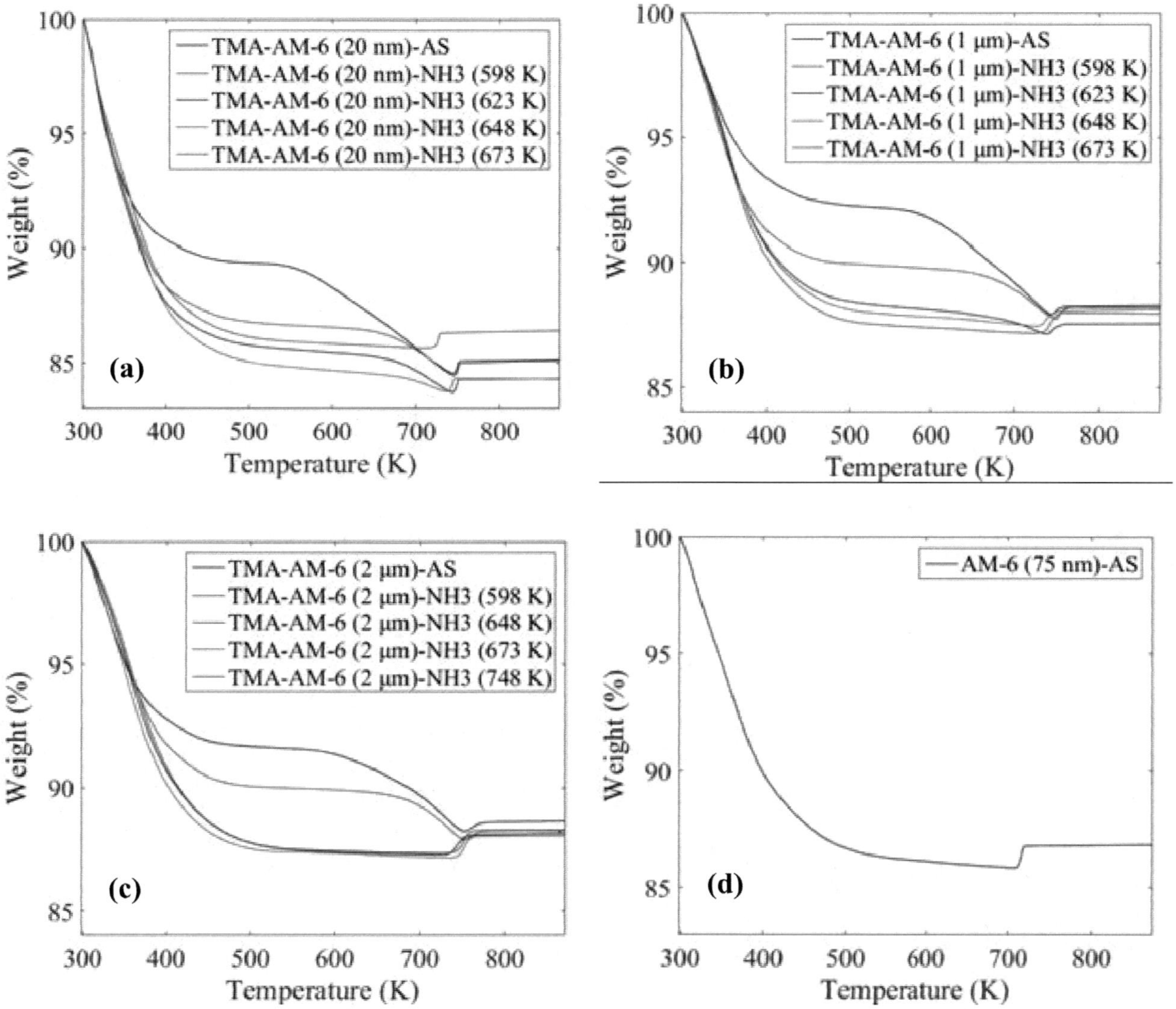

Fig. 8. TGA curves of AM-6 samples: (a) TMA-AM-6 (20 nm) as-synthesized and ammonia-treated at different temperatures; (b) TMA-AM-6 (1 μm) as-synthesized and ammonia-treated at different temperatures; (c) TMA-AM-6 (2 μm) as-synthesized and ammonia-treated at different temperatures; (d) AM-6 (75 nm)-AS.

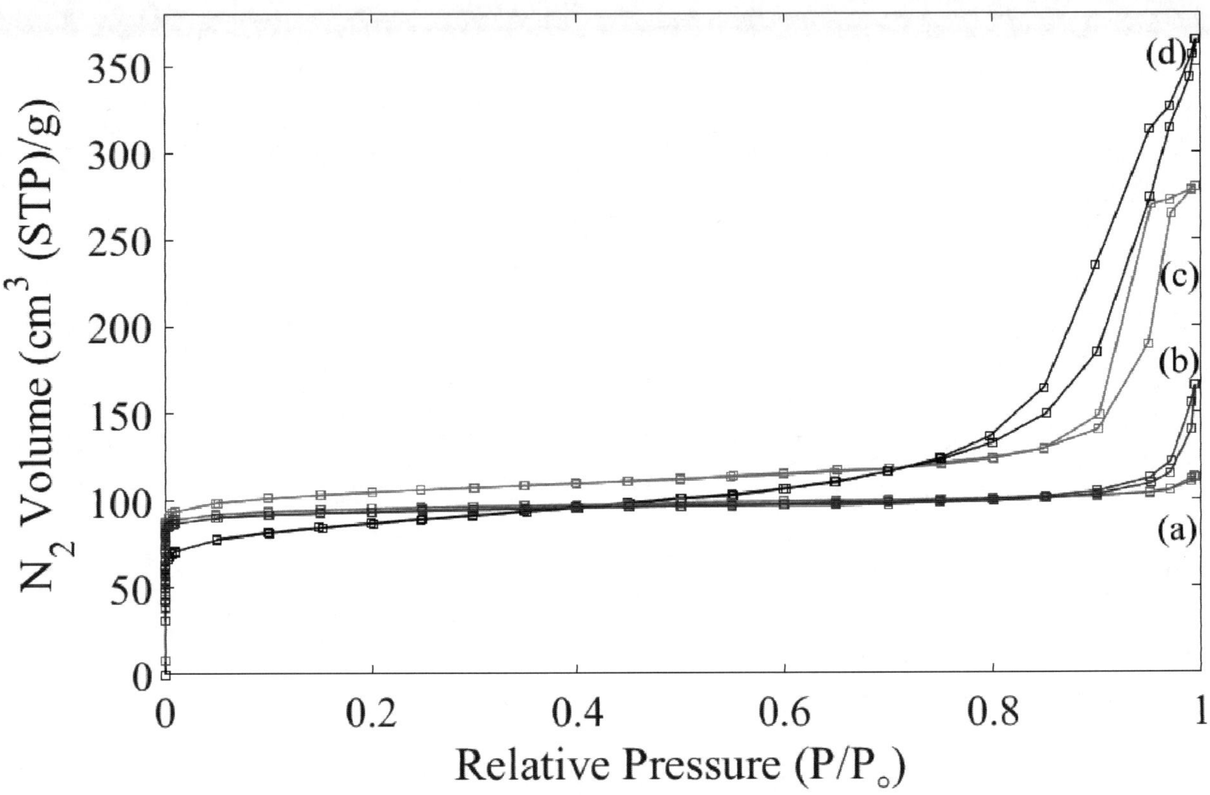

Fig. 9. N_2 adsorption isotherms of AM-6 samples: (a) TMA-AM-6 (2 μm)-NH_3 (673 K); (b) TMA-AM-6 (1 μm)-NH_3 (673 K); (c) AM-6 (75 nm)-AS; (d) TMA-AM-6 (20 nm)-NH_3 (598 K).

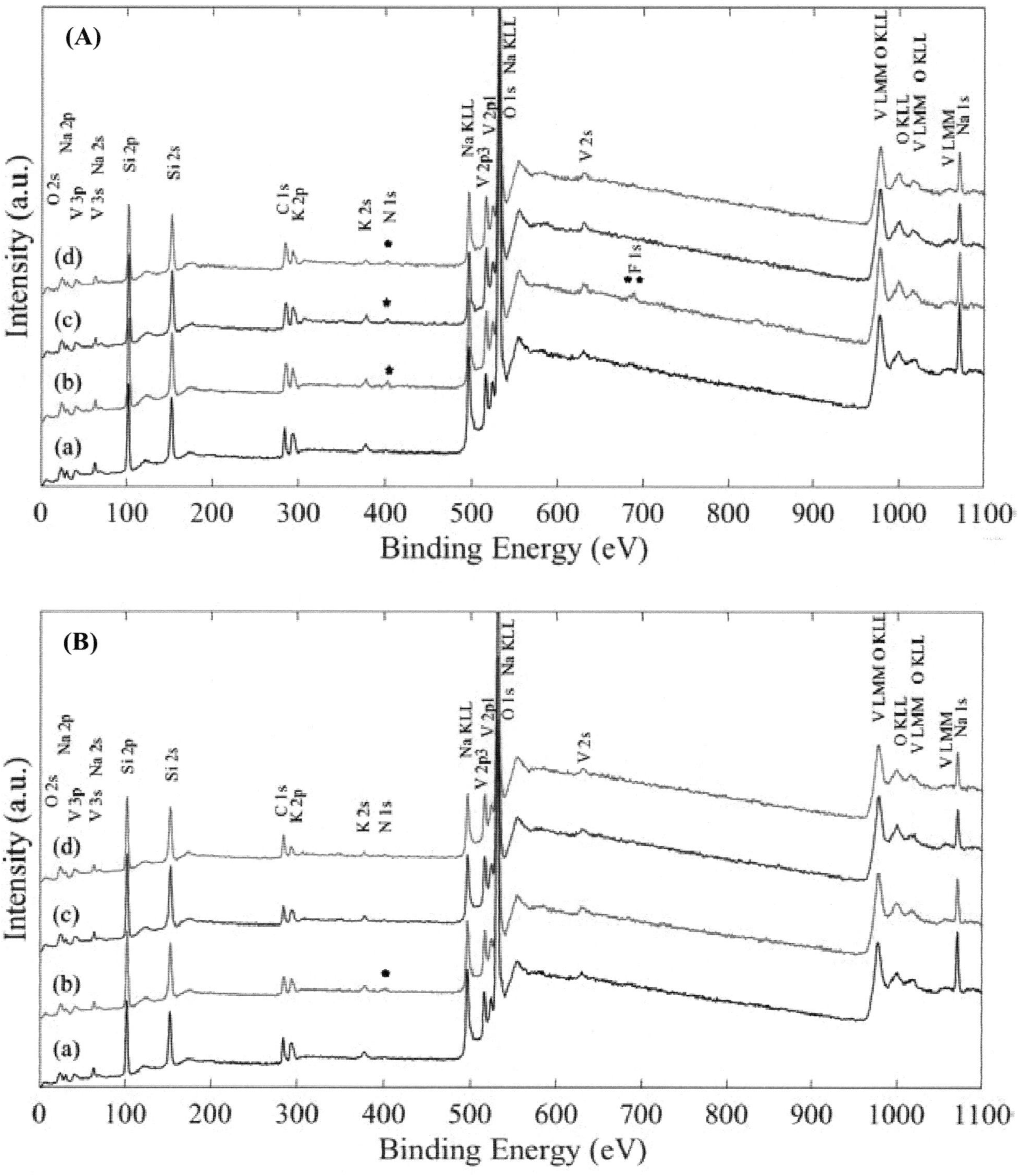

Fig. 10. XPS spectra of: (A) as-synthesized samples: (a) AM-6 (75 nm)-AS; (b) TMA-AM-6 (20 nm)-AS; (c) TMA-AM-6 (1 μm)-AS; (d) TMA-AM-6 (2 μm)-AS. XPS spectra of: (B) ammonia-

treated and as-synthesized samples: (a) AM-6 (75 nm)-AS; (b) TMA-AM-6 (20 nm)-NH_3 (598 K); (c) TMA-AM-6 (1 μm)-NH_3 (673 K); (d) TMA-AM-6 (2 μm)-NH_3 (673 K).

Asterisk refers to nitrogen. Double asterisks refer to fluorine.

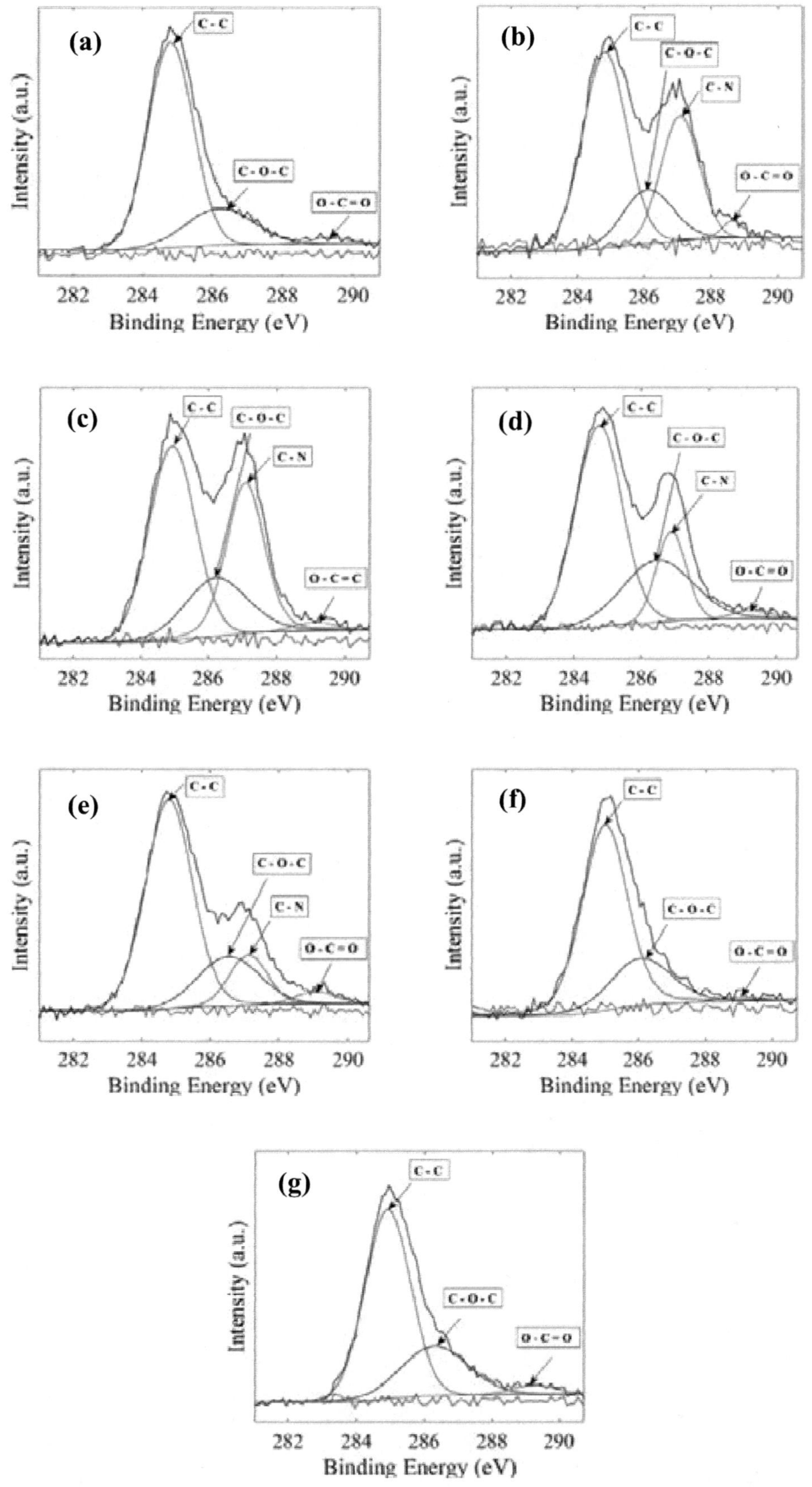

(a)
C - C
C - O - C
O - C = O
(b)
C - C
C - O - C
C - N
O - C = O
(c)
C - C
C - O - C
C - N
O - C = C
(d)
C - C
C - O - C
C - N
O - C = O
(e)
C - C
C - O - C
C - N
O - C = O
(f)
C - C
C - O - C
O - C = O
(g)
C - C
C - O - C
O - C = O
Intensity (a.u.)
282
284
286
288
290
Binding Energy (eV)

Fig. 11. Curve deconvolution of C 1s spectra for: (a) AM-6 (75 nm)-AS; (b) TMA-AM-6 (20 nm)-AS; (c) TMA-AM-6 (1 μm)-AS; (d) TMA-AM-6 (2 μm)-AS; (e) TMA-AM-6 (20 nm)-NH_3 (598 K); (f) TMA-AM-6 (1 μm)-NH_3 (673 K); (g) TMA-AM-6 (2 μm)-NH_3 (673 K).

491918 Rapid Wood Fractionation ≤ 80°C for Sustainable and Economic Biorefinery

Abstract:

Here we report a rapid and low temperature (≤ 80°C) wood fractionation process for producing wood pulp, cellulose and lignin nanomaterials, chemicals, and sugar/biofuels. Approximately 90% of wood lignin can be dissolved at 80°C for 20 min. Equivalent delignification using alkaline pulping can be achieved only at 170°C for 2 h. The novel process fractionated wood into two fractions: (1) a primarily carbohydrate-rich water-insoluble solid fraction that can be used for sustainable production of fibers such as dissolving pulp, lignocellulosic nanomaterials, and/or sugars through subsequent enzymatic hydrolysis, and (2) a spent acid liquor stream containing mainly dissolved lignin that can be easily precipitated as lignin nanoparticles simply by diluting the spent acid liquor. The dissolved hemicellulose sugars can be used to produce furfural after lignin precipitation. The chemical used has a low water solubility, which facilitates efficient recovery simply using commercially proven crystallization technology by cooling concentrated spent acid solution to ambient temperatures to achieve environmental sustainability.

502673 Insights into Biomass Recalcitrance

Abstract:

The utilization of biomass for biofuels and bio-derived chemicals and materials based on the biological platform is contingent on efficient separation, deconstruction and conversion. Todayâ€™s cellulosic ethanol plants, especially those based on the cellulase deconstruction, have demonstrated that the conversion of plant polysaccharides to ethanol can be accomplished on an industrial scale but current process challenges are hindering wide-spread adaption. One of the key issues is the recalcitrance of biomass and the need to utilize aggressive pretreatment technologies. Recent advances in our understanding of the plant cell wall structure and our ability to reduce biomass recalcitrance by altering these constituents by controlling genetic features of the plant and innovations in pretreatment chemistries provides innovative pathways to facilitate the biorefining of plant polysaccharides to biofuels and yield a lignin rich process stream that can be valorized to chemicals and/or materials. This seminar review our contributions in this field and highlight future opportunities.

492489 Low Temperature and High Efficiency Biomass Fuel Cell and Bio-Hydrogen Production

Abstract:

Title: Low temperature and high efficiency biomass fuel cell and bio-hydrogen production

Name: Yulin Deng

School of Chemical & Biomolecular Engineering, Georgia Institute of Technology, Atlanta, GA 30302-0620, yulin.deng@rbi.gatech.edu

A novel fuel cell which can directly use native polymeric biomasses, such as starch, cellulose, lignin, and even switchgrass and wood powders will be discussed. This fuel cell combines some features of solar cells, fuel cells, and redox flow batteries. Specifically, an oxidation/reduction ion pair, such as Fe^{2+}/Fe^{3+}, or polyoxomatelate, are used as catalyst to oxidize biomass by either absorbing solar light or heat energy. The power density of the solar-induced hybrid fuel cell powered with raw biomass could reach ~50 mW/cm^2. Unlike most cell technologies that are sensitive to impurities, the cell reported in this study is inert to most organic and inorganic contaminants present in the fuels. The fuel cell is completely noble metal free. The similar fuel cell system can also be converted to electrolysis cell for hydrogen production at low temperature using native biomass directly.

504077 Circular Economy: A Path Towards Innovation and Commercialization of Biocomposites for Sustainable Manufacturing

Abstract:

"Circular economy" is an emerging concept that targets waste minimizations through a closed loop economy system, unlike the traditional linear model ("make and dispose" type economy), and therefore supports sustainable development. The underlying principle of this emerging economy is "Nothing is waste; waste is a co-product waiting for another opportunity". In this system, the byproducts and co-products of one industry can find commercially viable applications in another industry, with appropriate design and engineering. There is a significant potential for utilization of biobased coproducts from agro and forest industries, food and beverage industry as well as biofuel and biomass-based pyrolysis industries. Our focus is on biocomposite applications that utilize these materials as reinforcing and filler materials for plastics, including biobased and biodegradable plastics. The developed eco-friendly biocomposites are suitable for use in consumer products, light-weight auto-parts and 100% compostable packaging.

Acknowledgements: Agriculture and Agri-Food Canada and Competitive Green Technologies (CGtech) through AgriInnovation Program (Project #052882 and 051910); the Ontario Ministry of Agriculture, Food, and Rural Affairs - University of Guelph Bioeconomy Industrial uses Research Program (Project # 200425); Natural Sciences and Engineering Research Council, Canada for the Discovery Grants (Project #401111); and the Ontario Research Fund, Research Excellence Program; Round-7 (ORF-RE07) from the Ontario Ministry of Research and Innovation (MRI), currently known as the Ontario Ministry of Research, Innovation and Science (MRIS) (Project # 052665 and #052644).

486459 Tuning the Molecular Design of Catalytic Materials to Increase Activity and Selectivity for Fine Chemical Production

Abstract:

Catalyst design can impact catalytic activity and selectivity. For heterogeneous catalytic materials, design is challenging since common synthetic methods tend to produce a distribution of catalytic sites. Through utilizing organic synthetic techniques, we demonstrate the ability to finely tune the catalytic properties of materials to increase catalytic activity. Specifically, we demonstrate that through controlling the alkyl linker length and the loading of the organosilane that we can tune molecular interactions to increase catalytic activity and selectivity for the isomerization of glucose to fructose. We extend this approach to creating well-defined catalytic sites in zeolites, which enables fine-tuning of the spacing between catalytic sites in the framework. These modifications improve catalytic activity relative to isolated catalytic sites.

497000 Deactivation of Zeolite Catalysts during Hydrodeoxygenation of Aromatic Oxygenates

Abstract:

Pyrolysis and liquefaction are promising processes for converting biomass into liquid bio-oils. However, upgrading of these bio-oils is typically needed to obtain properties that are compatible with processes for the production of chemicals and fungible gasoline additives. Hydrodeoxygenation (HDO) is an attractive approach for achieving this goal by replacing oxygen containing functional groups with hydrogen. In principle, water is formed as the only by-product. The process provides oils with reduced reactivity and corrosiveness and increases the energy density of the oil. Noble metal containing zeolite catalysts show considerable HDO activity, but deactivation remains an issue. In most studies, coking is assumed to be the primary cause of deactivation.

Here, we show that additional deactivation paths need to be considered. Bifunctional aromatics, like catechol and guaiacol, can strongly adsorb on Lewis acid sites and form rigid surface species, so-called roadblocks, that prevent access to active sites [1]. Roadblocks can drastically reduce the activity of zeolite catalysts even when only a fraction of the pore space and external surface area are occupied. In addition, dealumination processes during HDO will be discussed. The same deactivation phenomena occur during the conversion of pyrolysis vapors from pine indicating that model compound studies allow for accurate prediction of deactivation phenomena of zeolites in HDO reactions [2]. Based on these results, strategies for improving the design of HDO catalysts and processes will be discussed.

References:

[1] Foo, G. S.; Rogers, A. K.; Yung, M. M.; Sievers, C. ACS Catalysis 2016, 6, 1292â€“1307.

[2] Yung, M.M., Foo, G.S., Sievers, C. Catal. Today, accepted.

501532 Improving Methanol-to-Olefins Conversion Performance of CHA Materials By Seeding the Hydrocarbon Pool

Abstract:

Methanol-to-olefins conversion on zeotype materials proceeds via a mechanism described by a dual-cycle schematic comprising olefin and aromatic methylation/cracking cycles, with methylbenzenes formed during an induction period acting as co-catalytic scaffolds in ethene and propene formation. Decreasing catalyst lifetime with increasing methanol pressure at equivalent methanol weight hourly space velocities and increasing catalyst lifetime obtained when using dimethylether as the feed instead of methanol evidence formaldehyde, formed in hydride transfer steps, as a key intermediate effecting polycylic aromatic hydrocarbon formation leading to catalyst deactivation. Reactions unproductive to C-C bond formation predominate at low turnover numbers, characterized by a low concentration of active chain carriers and high effluent methane selectivities. We reasoned that introduction of active hydrocarbon chain carriers into an otherwise empty pore *before* the first turnover would lead simultaneously to a reduction in the length of the induction period, as well as a decrease in carbon loss due to the introduction, at low turnover numbers, of reaction pathways competing with methanol dehydrogenation to form formaldehyde.

We report experiments in which seeding the CHA zeotypic materials with acetaldehyde (0.1 kPa CH_3CHO, 673K, 0.02-0.5 C/H^+) reduces the initial rate of carbon loss by introducing reaction pathways that compete with methanol dehydrogenative transfer events. Increasing seed loadings and inlet methanol pressures lead to lower cumulative carbon loss at early turnovers, consistent with lower formaldehyde formation rates being responsible for the observed reduction in carbon loss. The introduced seed also acts as a co-catalyst, thereby reducing the length of the induction period, leading to higher turnover numbers with increasing seed loadings at equivalent times-on-stream. These studies exemplify a strategy for mitigating the deleterious effects of methanol hydride abstraction events that effect MTO deactivation.

493067 Tuning of Higher Alcohol Selectivity and Productivity in CO Hydrogenation Reactions over K/MoS_2 Catalysts Supported on Mesoporous Activated Carbon and Mixed Mgal Oxide

Abstract:

Tuning of Higher Alcohol Selectivity and Productivity in CO Hydrogenation Reactions Over K/MoS_2 Catalysts Supported on Mesoporous Activated Carbon and Mixed MgAl Oxide

Micaela Taborga Claure[1*], Song-Hai Chai[2], Sheng Dai[2], Faisal M. Alamgir[1], Pradeep K. Agrawal[1], and Christopher W. Jones[1]

1. School of Chemical & Biomolecular Engineering, Georgia Institute of Technology, Atlanta, GA, 30332 USA2.

2. Chemical Science Division, Oak Ridge National Laboratory, Oak Ridge, TN 37831 USA

The use of supports as a means of enhancing both activity and selectivity of MoS_2 has been investigated over the past four decades. Our previous work showed that carbon supports yield high methanol and ethanol productivities.[1] Mesoporous activated carbon, obtained via collaboration with ORNL, when used as a support, is confirmed to show greater alcohol productivity than commercial activated carbon. Hydrotalcite-derived mixed MgAl oxide (MMO) supports yield high $C_{3+}OH$ selectivity influenced by Mo-MMO interaction, but generally yield low productivity. It has been shown that the Mo:MMO ratio greatly affects selectivity while catalyst preparation methods do not.[2] In this work, we seek to evaluate the hypotheses that molybdenum supported on a family of MMO and carbon mixed-supports may allow tuning of higher alcohol productivity (due to carbon support) and selectivity (due to MMO support).

When Mo is first impregnated on the carbon support and the resulting material is physically ground with MMO, the selectivity trends lie in between that of the carbon and MMO single-support catalysts. Although not as selective toward higher alcohols compared to the single-support MMO catalyst, this novel mixed-support catalyst gives an increased selectivity toward higher alcohols compared to the single-support carbon catalyst, and a significant increase in higher alcohol productivity over both single-support catalysts, a result of Moâ€™s ability to migrate from the carbon to the MMO during sulfidation and reaction. In contrast, when Mo is first impregnated on MMO and the resulting material is ground with carbon, the catalyst has selectivity and productivity trends similar to the MMO single-support catalysts, indicating that strong Mo-MMO interactions limit the ability of Mo to migrate to the carbon support.

Detail characterization via STEM and XAS was used to probe the effect of the number of MoS_2 stacked layers on

reaction performance. The combined data shed light into the role of MoS_2 stacking on the selectivity towards higher alcohols and total hydrocarbons.

References

1. Morrill, M. R.; Thao, N. T.; Agrawal, P. K.; Jones, C. W.; Davis, R. J.; Shou, H.; Barton, D. G.; Ferrari, D., Mixed MgAl Oxide Supported Potassium Promoted Molybdenum Sulfide as a Selective Catalyst for Higher Alcohol Synthesis from Syngas. *Catalysis Letters* **2012,** *142* (7), 875-881.

2. Morrill, M. R.; Thao, N. T.; Shou, H.; Davis, R. J.; Barton, D. G.; Ferrari, D.; Agrawal, P. K.; Jones, C. W., Origins of Unusual Alcohol Selectivities over Mixed MgAl Oxide-Supported K/MoS2 Catalysts for Higher Alcohol Synthesis from Syngas. *ACS Catalysis* **2013,** *3* (7), 1665-1675.

*Current: ExxonMobil Research and Engineering, 1545 U.S. 22, Annandale, NJ, 08801

494697 Kinetic and Spectroscopic Investigations of Alcohol Conversions over Metal Oxide Catalysts

Abstract:

The next advancement in the field of catalysis is to move from *catalyst discovery* to *catalyst design*. For this to be realized, we must meet one of the grand challenges in science identified in the 2007 Basic Research Needs Workshop report on Catalysis for Energy, which is to â€œ*Understand reaction mechanisms and dynamics of catalyst transformations.*â€ One cannot draw reaction mechanism without quantitative and qualitative characteristic of the chemical composition/intermediate as well as the orientation at the catalyst surface. Therefore, monitoring these interactions *in situ* is necessary to understand different reaction steps and what transient intermediates/structures are present during the reaction.

In situ and *operando* vibrational spectroscopies have proven to be valuable tools to probe the chemical transformation as they occur in real time during heterogeneous catalysis. Here, we show how sum-frequency generation (SFG) spectroscopy can be used to probe surface-specific interactions that lead to a better understanding of reaction pathways of isopropyl alcohol (IPA) on the (100) surface of two metal oxides, cerium oxide and strontium titanate. Inelastic neutron scattering (INS), a complimentary vibrational technique, which offers unparalleled sensitivity towards monitoring hydrogenous materials on a variety of surfaces, was used to characterize and compare the reactivity of both synthesized and commercially-available samples. We use these two *in-situ* vibrational techniques to spectroscopically characterize catalyst samples during the reaction process. In addition, we correlate these spectroscopic observations with reaction kinetic studies to gain insights of the reaction mechanisms that govern the activity and selectivity.

501871 Catalytic Upgrading of Fast Pyrolysis Bio-Oil for Renewable Hydrocarbon Production

Abstract:

Hydrocarbons from the catalytic upgrading of biomass are a viable source of renewable liquid transportation fuel. Their production has been an important field of research for improving the environment, energy security, and energy productivity. Catalytic hydroprocessing of pyrolysis oil is one of the most promising thermochemical conversion routes for biomass. However, due to thermal instability, the treatment of pyrolysis oil requires a multi-step catalytic upgrading to allow long-term operation. Additionally, the presence of significantly larger amounts of water in pyrolysis oil compared to typical crude oil presents a challenging environment to typical commercial petroleum hydrotreating catalysts. Here, we report an overview of the research progress done at the Pacific Northwest National Laboratory for the fast pyrolysis-catalytic upgrading route in terms of catalyst and process development. Consideration for the techno-economic analysis of the technology and its impact will also be discussed.

494727 Chemical Reaction Engineering Principles of Continuous Flow Photoredox Catalysis

Abstract:

Chemical Reaction Engineering Principles of Continuous Flow Photoredox Catalysis Eric G. Moschetta,* Kaid Harper,* Steven M. Richter, Steve J. Wittenberger AbbVie, Inc. Process R&D, 1401 Sheridan Road North Chicago, IL 60064 *co-corresponding authors: eric.moschetta@abbvie.com, kaid.harper@abbvie.com Scaling up complex organic syntheses via flow chemistry is becoming common in the pharmaceutical industry. Careful reaction engineering of continuous flow chemistries allows for increased throughputs of desired chemical products and clever solutions to common problems observed in large scale batch reactors, such as improved heat transfer in highly exothermic reactions. One area of recent interest in organic synthesis is photoredox catalysis, reactions that involve species being excited to high energy states by visible light to undergo single electron transfer events that lead to the desired product(s). These reactions frequently use ruthenium and iridium polypyridyl complexes as the catalysts because they have long-lived excited states and high redox potentials, allowing access to many reaction pathways.[1] However, these precious metal catalysts would incur a large expense if such reactions are to be used at scale. As an alternative, various organic photocatalysts, such as Eosin Y and acridinium-based species, also possess high redox potentials and can facilitate other single electron transfer events that their transition metal counterparts cannot.[2] The main disadvantage of these organic photocatalysts is their shorter excited state lifetimes. Organic photocatalysts have excited state lifetimes on the order of nanoseconds while transition metal photocatalysts often have observed excited state lifetimes on the order of microseconds. While synthetic organic chemists have made, and continue to make, significant advances in the synthetic utility of transition metal and organic photocatalysts alike, the universal challenge in applying photoredox catalysis in the pharmaceutical industry is scaling up such reactions for kilogram-level production. Photoredox catalysis, and photochemistry in general, is not a part of the traditional chemical reaction engineering paradigm. For example, chemical engineers have many tools and solutions available for reactions where heat and mass transfer limitations, mixing limitations, multi-phase chemistry, and chemoselectivity all present problems in scale up of continuous flow reactors. Over the last decade, the enormous increase in the interest and study of photoredox catalysis has led to a broad and diverse array of chemical transformations. As such, a platform for scalable photoredox chemistry should be as versatile as possible to accommodate the many types of photocatalysts, reactants, and chemistries that are already developed. Additionally, such reactors should include high intensity light sources to supply the reactor with high photon fluxes to allow operation under the intrinsic kinetics of the reaction. Most importantly, the ideal reactor should operate in continuous flow to maximize production. This study implements a new continuous flow reactor for several photoredox chemistries and compares the observed rates of reactions to those obtained from other known reactor configurations, in both batch and flow. The discussion emphasizes how and why this reactor configuration outperforms the other reactors using fundamental principles of chemical reaction engineering and physical chemistry. The discussion concludes with a demonstration of the scale-up potential of the continuous flow reactor and how this new reactor configuration may be used across the entire discipline of photoredox catalysis in organic synthesis. **References**

[1] C. K. Prier, D. A. Rankic, D. W. C. MacMillan, *Chem. Rev.* **2013**, *113*, 5322-5363.

[2] N. A. Romero, D. A. Nicewicz, *Chem. Rev.* **2016**, *116*, 10075-10166. *This abstract presentation was sponsored by AbbVie. AbbVie contributed to the design, research, and interpretation of data, writing, reviewing, and approving the publication. All authors are AbbVie employees.*

496561 Factors Affecting Catalytic Performance in the Presence of Non-Thermal Plasmas

Abstract:

Plasmas have been used to enhance the selectivity, activity, and product yields in a variety of chemical reactions. Additionally, plasmas can be generated in the presence of a catalyst to assist in the dissociation or activation of molecules to promote subsequent reactions. A highly sought-after goal of plasma-assisted catalysis is the ability to control the reaction selectivity by tuning the plasma environment or plasma-catalyst interaction. We have been particularly interested in the use of non-thermal plasmas in the catalytic production of syn-gas from methane/carbon dioxide and the production of ammonia from nitrogen/hydrogen. Specifically, this presentation will focus on 1) the reaction regime where catalyst-plasma interactions are observed for the two reactions highlighted above, 2) the various catalyst-plasma interactions observed, and 3) the role of the plasma in enhancing reaction efficiency. For the plasma-assisted dry reforming of methane, we have performed detailed kinetic experiments to determine the catalytic consequences when the partial pressures of the reactants, flow rates, reaction temperatures, and plasma power are varied.Â From a variety of kinetic and control experiments, an energy barrier was obtained from a relationship between the reaction rate constant and input power (â‰¤ ~20 kJ/mol), which was lower than the ~70 kJ/mol obtained for thermal catalysis on Ni/Al_2O_3. Recent results for the ammonia synthesis reaction in the presence of non-thermal plasmas will also be discussed.

508501 Accelerating Innovation in Advanced Manufacturing

Abstract:

In the U.S. Department of Energy, the Advanced Manufacturing Office (AMO) has the mission to catalyze research, development and adoption of energy-related advanced manufacturing technologies and practices to drive U.S. economic competitiveness and energy productivity. AMO brings together manufacturers, institutes of higher education, national laboratories, and state and local governments in public-private R&D consortia to accelerate technology innovation. One such R&D Consortia is the Critical Materials Institute (CMI), established in 2013 and led by Ames Laboratory. CMI is a multidisciplinary effort to develop solutions across the materials lifecycle of materials essential to clean energy technologies and manufacturing, as well as reduce the impact of supply chain disruptions associated with these valuable resources. CMI is addressing challenges in critical materials, including mineral processing, manufacture, substitution, efficient use, and end-of-life recycling; integrating scientific research, engineering innovation, manufacturing and process improvements; and developing a holistic solution to the materials challenges facing the nation.

499001 Active Polymer Materials for Flexible Electronics: Molecular Design and Processing for Efficient Macroscale Charge Transport Pathways

Abstract:

Printed, flexible electronics have potential as low cost alternatives for devices in industries ranging from energy to health care to security. The successful introduction of these devices however, relies on the design and development of sustainable, robust and reliable materials and processes. Studies have shown that not only does device performance depend critically on semiconductor alignment at many length scales, materialsâ€™ mesostructure can be manipulated in solution prior to device fabrication. Recently, it has been demonstrated that even with no perturbations, polymer semiconductors self-assemble in solution over time. Observations surrounding the behavior of these materials suggest that requisite macroscopic long-range order required for high performance devices may be achieved through process optimization which utilizes knowledge associated with materials structure-process-property relationships. This presentation will explore how we can derive critical insights into the structure-process-property space using well-coupled modeling and experimental approaches, and how those can impact the design and development of all-printed, flexible electronic devices.

494336 Promoting Adhesion between Immiscible Polymers

Abstract:

Most plastic products consist of two or more polymers. Dramatic property improvements of packaging films, coextruded profiles, engineering plastics, and soft feel over-molded items are achieved via combining immiscible polymers. Controlling adhesion between these polymers is critical to optimizing their performance. We show how interfacial entanglements control the adhesion of bare interfaces. For amorphous polymers adhesion developed by thermal or solvent welding can be predicted from the entanglement molecular weight of each polymer and their interaction parameter, Ï‡. Shear flow at high stresses disentangles the interface, reducing adhesion. Reactions between immiscible chains can form bonds across interfaces, greatly increasing adhesion. In fact in coextrusion this is the main approach used to â€œstickâ€ polymers together, often via a thin, third component tie layer. We have found that extensional flow, such as occurs in compounding mixers and extrusion dies, can greatly accelerate these reactions (AIChE Journal, 2013). We have also found that small levels of catalyst can further accelerate reactions, especially if the catalyst preferentially wets the interface (ACS Macro Letters, 2015). This creates opportunities to use less reactive functional groups. Addition of premade block copolymers in principle can bridge interfaces but they often localize in micelles. We have recent success in promoting adhesion with small amounts of tetrablock copolymers (Science, 2017). We will also discuss current research on the role of crystallization at interfaces on polymer-polymer adhesion.

489706 High Resolution Lithography Via Block Copolymers and Self-Assembling Surface Neutral Layers

Abstract:

Novel materials which enable electronics manufacture helps to drive the innovation cycles underlying the Information Age. Commercial materials for fine line lithography are becoming increasingly challenging in design as the required features size shrink further, down to 10 nm size and beyond. As the pattern sizes required become smaller than the capability of existing imaging tools, block copolymers off a distinct route towards even finer feature generation through pattern frequency multiplication. We will discuss the motivation, technical approaches and results from two distinct materials approaches to block copolymer directed self-assembly (DSA): 1) Plasma etch resistant block copolymers for practical graphoepitaxy for 16 nm and smaller patterning[1,2], and 2) Formation of vertical chemoepitaxy patterns through the use of high cn polymeric surface active additives[3] that segregate to the top of the BCP film during casting and annealing, and balances the surface tensions at the top of the thin film.

1. **Impact of materials selection on graphoepitaxial directed self-assembly for line-space patterning**, Dung Quach, Valeriy V Ginzburg, Mingqi Li, Janet Wu, Shih-wei Chang, Peter Trefonas, Phillip D Hustad, Dan B Millward, Gurpreet S Lugani, Scott L Light, *Proc. SPIE Advanced Lithography*, 94230N-94230N-9 (2015).
2. **Graphoepitaxial and chemoepitaxial methods for creating line-space patterns at 33nm pitch: comparison to a HVM process,** Dan B Millward, Gurpreet S Lugani, Scott L Light, Ardavan Niroomand, Phillip D Hustad, Peter Trefonas, Dung Quach, Valeriy V Ginzburg, *Proc. SPIE Advanced Lithography*, 942304-12 (2015).
3. **Orientation control in thin films of a high-Ï‡ block copolymer with a surface active embedded neutral layer**, Jieqian Zhang, Michael B Clark, Chunyi Wu, Mingqi Li, Peter Trefonas III, Phillip D Hustad, *Nano Letters*, **16**, 728-735 (2015).

500184 Field and Confinement-Directed Self-Assembly of Soft Mesophases to Create Useful Materials

Abstract:

Self-assembly of block copolymers and small molecule mesogens gives rise to a rich phase behavior as a function of temperature, composition, and other relevant parameters. The ability to precisely control their chemical functionality combined with the tunable characteristic length scales (~1-100 nm) of their self-assembled mesophases position these systems as a versatile and attractive class of materials for compelling applications ranging from membranes for size and chemo-selective transport, to lithography. As a result there is intense interest in elucidating the physical processes relevant for directing self-assembly in these materials, with a goal of exploiting such fundamental understanding in scalable processes to create useful materials or devices. This presentation highlights recent advances in scalable directed self-assembly of soft nanostructured materials and emerging methods for generating highly ordered and heterostructured systems. In particular, we focus on the use of magnetic fields and physical confinement as scalable routes for creating useful nanostructured materials from block copolymers and polymerizable small molecule mesophases. A variety of new developments are discussed including alignment using low intensity magnetic fields, the use of orthogonal stimuli to create biaxial order, and the generation of alignment patterns by field screening using magnetic nanoparticles. Emerging applications in aligned nanofiltration membranes and sub-10 nm patterning are presented.

503528 Nanostructured Interfaces for Enhanced Biologic Transport and Immunomodulation

Abstract:

Drug delivery across epithelial barriers (oral, transdermal, mucosal) remains the preferred route for drug administration. However, therapeutic macromolecular drugs currently under development suffer from poor oral bioavailability, and consequently many of these macromolecules are delivered by injection. A variety of delivery paradigms have been developed, including chemical permeation enhancers, physical disruptors, and mucoadhesive materials, to enable more effective delivery of therapeutic macromolecules across epithelium but clinical utility has been limited thus far. Nanofabrication technology may offer potential advantages over conventional drug delivery strategies by enhancing molecular transport and local cellular response. The use of surface nanostructures, coupled to bulk materials, may permit enhanced transport of drugs, particularly protein therapeutics. In this talk, I will discuss the effect of nanostructured surfaces on the modulation of tight junction permeability and transport of key therapeutic molecules in vitro and in vivo. I will also discuss how nanostructures can be used modulate the immune microenvironment, presenting distinct biophysical cues to cells. The effect of geometry and the development of materials that can ultimately enhance therapeutic delivery is important for a broad range of diseases.

497694 Nano- and Microfabricated Hydrogels for Regenerative Engineering

Abstract:

Engineered materials that integrate advances in polymer chemistry, nanotechnology, and biological sciences have the potential to create powerful medical therapies. Our group aims to engineer tissue regenerative therapies using water-containing polymer networks, called hydrogels, that can regulate cell behavior. Specifically, we have developed photocrosslinkable hybrid hydrogels that combine natural biomolecules with nanoparticles to regulate the chemical, biological, mechanical and electrical properties of gels. These functional scaffolds induce the differentiation of stem cells to desired cell types and direct the formation of vascularized heart or bone tissues. Since tissue function is highly dependent on architecture, we have also used microfabrication methods, such as microfluidics, photolithography, bioprinting, and molding, to regulate the architecture of these materials. We have employed these strategies to generate miniaturized tissues. To create tissue complexity, we have also developed directed assembly techniques to compile small tissue modules into larger constructs. It is anticipated that such approaches will lead to the development of next-generation regenerative therapeutics and biomedical devices.

503544 Overcoming Obstacles to Brain Repair Using Biomaterials

Abstract:

Research in my laboratory over the past decade has focused on the generation of hydrogel biomaterials to support the formation of a reparative niche within diseased or injured sites that can block or prevent inhibitory signals from dominating the repair process, while providing pro-repair signals that can guide new tissue formation. The goal of our approach is to use engineered materials to "unlock" the regenerative capacity of damaged or diseased tissue to promote repair. The premise of our approach is that all tissues in the body have the capacity to repair through local stem or progenitor cells, but that due to unfavorable environmental conditions during the normal healing process they are not able to do so. In this talk, I will describe our approach to overcome physical and chemical barriers to brain repair after stroke using biomaterials.

503670 Digital Alchemy for Assembly Engineeringc

Abstract:

Back in the day, Stuart Churchill was a pioneer in the use of computers for modeling chemical and thermal processes. Since that time, the role of computing in chemical engineering has become indispensable. In this talk, we present new computational methods we call "digital alchemy" that allow for predictive modeling of thermodynamic assembly, and permit the design of building blocks optimized for assembly into target structures with desired properties. In this approach, standard thermodynamic ensembles are generalized to include alchemical variables and their thermodynamic conjugates, allowing for rapid, real-time computational investigation of large, complex design spaces.

503671 Improved Algebraic, Numerical, and Graphical Representations in Fluid Mechanics

Abstract:

The concept of equivalents is applied to fully developed flows through straight channels, and two applications are introduced: the equivalents that produce equal rates of flow through channels of different cross-section, and the diminishment that occurs between the regimes of laminar and turbulent flow. The validity of the eddy diffusivity and the invalidity of the mixing-length as measures of the shear stress due to turbulence have perhaps never before been demonstrated so clearly. The possible choices of primary variables and the consequences are also examined. The objective of these explorations is an improved understanding of the basic fluid mechanics by teachers, students, and industrial practitioners.

503672 The Scaling of Turbulence Near the Wall and the Churchill Turbulent Flux Correlation: Insights with Lagrangian Simulations

Abstract:

The evaluation of mean velocity and mean temperature in wall-bounded turbulent flows has been based on the "classical" theory of turbulence and the law of the wall. The viscous wall parameters, namely the friction velocity, the friction temperature, and the kinematic viscosity of the fluid have been used to scale not only the mean values of velocity, but also other mean turbulent quantities close to the wall, including the Reynolds stresses and the turbulent velocity and temperature fluctuations. Stuart Churchill, as he questioned the validity of the Reynolds analogy between momentum and heat or mass transfer in his works [1], he realized that extending the scaling of momentum to the scaling of scalar quantities in turbulence was not a sound approach. He proposed a scaling of turbulence that depends on the turbulent momentum flux and the turbulent heat flux. In his AIChE Institute Lecture (in Miami, 1998), Churchill presented a summary of his work on this topic, and closed form solutions for the mean velocity and temperature [2]. The turbulent Prandtl number, the most significant parameter for the modeling of turbulent transport not only in theory but also in modern CFD commercial software packages, was also interpreted physically based on these ideas [3]. We have used direct numerical simulation results and results from Lagrangian simulations [4] for the transport of passive scalars in turbulent channel flows to explore Churchill's ideas on turbulent flow scaling and to develop an updated theory for the scaling of heat transport in turbulent flows [5-7]. His passion to remove empiricisms and approximations from the development of predictive correlations in turbulent transport has been an inspiration. In this talk, we will revisit the Churchill scaling in view of our simulation results, and will discuss the counterintuitive finding that wall turbulence can be used to separate rather than mix substances under special circumstances [8].

1. Churchill, S. W., 1996, "A Critique of Predictive and Correlative Models for Turbulent Flow and Convection," Ind. Eng. Chem. Res., 35, pp. 3122–3140.
2. Churchill, S.W., 2000, "Progress in the Thermal Sciences: AICHE Institute Lecture," AIChE Journal, 46(9), pp. 1704-1722.
3. Churchill, S. W., 2002, "A reinterpretation of the Turbulent Prandtl Number," Ind. Eng. Chem. Res., 41, pp. 6393–6401.
4. Papavassiliou, D.V, 2002, "Turbulent transport from continuous sources at the wall of a channel," Int. J. Heat Mass Transfer, 45(17), pp. 3571-3583.
5. Le, P.M., and Papavassiliou, D.V., 2008, "On the scaling of heat transfer using thermal flux gradients for fully developed turbulent channel and Couette flows," Int. Commun. Heat Mass Transfer, 35(4), pp. 404-412.
6. Srinivasan, C., and Papavassiliou, D. V., 2011, "Prediction of Turbulent Prandtl Number in Wall Flows With Lagrangian Simulations," Ind. Eng. Chem. Res., 50(15), pp. 8881–8891.
7. Srinivasan, C., and Papavassiliou, D.V., 2013, "Heat transfer scaling for wall bounded turbulent flows," Applied Mechanics Review 65(3), Art. 031002.

8. Nguyen, Q., Srinivasan, C., and Papavassiliou, D.V., 2015, "Flow induced separation in wall turbulence", Phys Rev E, 91, Art. 033019.

503674 Flow Boiling Using a Piranha Pin Fin Heat Sink

Abstract:

Flow Boiling using a Piranha Pin Fin Heat Sink

Corey Woodcock, Xiangfei Yu, Yoav Peles[1] and Joel Plawsky

Rensselaer Polytechnic Institute; [1]University of Central Florida

The *Piranha Pin Fin* (*PPF*) microdevice is a heat exchanger engineered to dissipate extreme thermal loads using either dielectric fluids such as HFE7000 or water. Fabrication of the device is CMOS compatible and the first generations of the device have been fashioned from silicon substrates. The PPF operates on the principle that heat transfer would be more efficient if heated fluids could be removed nearly as fast as they are created. This process would keep the driving force as high as possible. Thus the heat exchanger is built as two layers; a primary substrate that dissipates the bulk of the heat and shunts the heated fluids to a booster section that continues to heat the fluid and extends the total surface area of the device. Three generations of devices were fabricated. The first two generations were single-layer systems designed to help understand the operation of the primary layer and optimize the shape and location of the pin fins. The third generation was the first, two-layer system to be fabricated and tested.

Figure 1 Representative PPF array

Correlations for the heat transfer coefficient, h_{tp}, based on the Boiling Number and the Weber Number in the PPF design was developed and validated with Gen II. The discrepancy of heat transfer coefficient between predicted and measured values was within 20%. The ratio of two-phase to single-phase pressure drop was correlated with the Boiling Number (Bo) and the Jakob Number (Ja). The discrepancy of the two-phase multiplier between prediction and experimental measurements were within 12%.

The third generation *PPF* heat sinks were the first series to provide both a primary substrate populated with *PPF*s, as well as a booster substrate populated with solid micro pin fins. The substrates are micro-aligned and fusion-bonded to provide a full three dimensional (3D) architecture with bulk silicon properties. Experimental results from the Gen III *PPF* microdevices provide a proof of concept of the true *PPF* heat sink concept with fluid venting into multiple substrates. At low flowrates (G = 670 [kg/s/m^2]) the Gen III *PPF* heat sink has been shown to dissipate heat loads approaching 1.1 [kW/cm^2] at an exit quality of 95%, while at higher flowrates (G = 3,300 [kg/s/m^2]) the thermal limit of the heat sink remains unknown due to exceeding the available power supply at >1.4 [kW/cm^2].

503673 Transport Problems in the Spirit of Stuart Churchill for Teaching and Research at the University of Michigan

Abstract:

Stuart Churchillâ€™s solutions to important heat transport problems have greatly influence both pedagogy and research. The correlation of Churchill and Chu for the natural convection heat transfer to or from a vertical plate, over an astonishing range of Rayleigh numbers is based partly on a similarity solution of a type common in heat, mass, and momentum transfer for problems in chemical engineering. One such problem is asphaltene or wax transport and deposition onto the wall of an oil pipeline, which can be solved using similar methods, as will be discussed. In addition, such problems, and the wealth of their applications make for excellent motivations and examples in the teaching of graduate or undergraduate transport, as will be discussed in the talk.

509054 Solve this! Fundamental Approach to Problem Solving in Industrial Processes I (Invited Talks)

Abstract:

Solving complex plant problems is an everyday exercise that affects all projects and budgets, and it is very rarely introduced to students at any level. In fact, by the time young engineers are few years into working in industry, most have lost the engineering skills that they developed in school. We hope to encourage future engineers to retain those skills and learn the importance of plant problem-solving.

502428 Tuning External Surface and Textural Properties of Unit-Cell Thick Pillared MFI and Pillared MWW Zeolites By Atomic Layer Deposition and Its Consequence on Catalytic Reactions

Abstract:

The pillared MWW (PMWW or MCM-36) and pillared MFI (PMFI) zeolites are 2-dimensional (2D) catalytically active materials made by pillaring of layered MCM-22(P) and multilamellar MFI precursors, respectively. The single- or near single-unit-cell thick 2D microporous layers in PMWW and PMFI expose comparable external surface areas to those in micropores, which become important or dominant contributor to the catalytic properties. Although the acidity and catalytic activities of PMWW and PMFI have been studied, modification of their external surfaces and its implications on catalytic reactions are not available. In this talk, we will present the tailoring of external surfaces of PMWW and PMFI zeolites by atomic layer deposition (ALD) of silicon (ALD-Si), titanium (ALD-Ti) and aluminum (ALD-Al), respectively. The textural, acidic and catalytic properties of the ALD modified pillared zeolites were investigated using a variety of characterization methods. The effects of ALD-Al, ALD-Ti and ALD-Si modifications on the pillared MFI and pillared MWW zeolites were comparatively quantified and analyzed.

486105 Mesoporous TiO_2 to TiO_2-Cellulose Composite and its Derivatives for Environmental Remediation

Abstract:

Mesoporous TiO2 nanoparticles and TiO2-cellulose composite fibers were synthesized via facile hydrolysis of TiOSO4 under microwave irradiation. The composite fibers were further converted to carbon fibers. These materials were endowed with high surface area and unique surface morphology that resulted in high photocatalytic activity for the decomposition of organics and the photocatalytic reduction of high-valence toxic metal ions. They are also efficient nanosorbents for the removal of micro-organic pollutants and heavy metal ions. The synthetic route and the structure of these materials were systematically studied, and the practical use of them was also illustrated by dynamic adsorption experiments.

490249 Reliable Fabrication and Surface Modification of Beta Zeolite Membrane for Pervaporation of n-Butanol/Water Mixtures

Abstract:

The depletion of crude oil resources is the main reason for motivating the different waves of interests in renewable energies including light-voltage, wind energy, and biofuels [1]. Compared with bioethanol, biobutanol has several advantages such as higher energy density, lower volatility, higher boiling point (117.7 °C) and flash point (29 °C), plus amenability to the formation of blends with gasoline in any proportion [2]. Currently, biobutanol is mainly produced through the fermentation of renewable biomass. In the fermentation process, the severe inhibition and toxicity of butanol usually lead to a significant decrease in its yield [3]. To remove butanol from the fermentation system, several separation techniques such as distillation, adsorption, liquid-liquid extraction (LLE), gas stripping and pervaporation have been employed [1]. The pervaporation technique with low energy requirement, easy operability and harmless to microorganisms is a promising technology for efficient separation of butanol/water mixtures [4].

The pervaporation membranes have attracted broad attention recently. Zeolite beta adopts a truncated bi-pyramidal shape with three-dimensional 12-ring interconnected channel system with pore diameters of 0.71 * 0.73 nm and 0.56 * 0.56 nm [5]. Zeolite beta membranes have the application potential in both separation and catalysis. At present, zeolite beta membranes are usually synthesized using TEAOH as the structure directing agent in either a basic medium (without fluoride) or a near-neutral medium (with fluoride). In the basic medium, the synthesis gel is in a liquid state, which can be fully adsorbed into the seed layer of zeolite beta crystals. Nevertheless, the crystallinity and intergrowth between vicinal crystals in zeolite beta membranes synthesized as such are not satisfactory [6]. In contrast to the alkaline medium, zeolite beta membranes prepared in a near-neutral gel containing fluoride often exhibits a truncated bi-pyramidal morphology with high crystallinity. The fluoride ion rich gels possess extremely high viscosity and low fluidity, therefore, the zeolite beta membranes formed in the near-neutral medium are generally non-uniform, the adhesion strength of which on their supports are pretty low.

In the present work, we propose a two-step method to fabricate zeolite beta membranes on porous alpha-alumina substrates, in which the secondary growth procedure is performed in the basic medium without fluoride followed by that in the near-neutral medium with fluoride. As shown in Figure 1(b), an intergrown and dense crystal layer was formed with the thickness of ca. 1 µm on the alpha-alumina supported zeolite beta crystal layer (Figure 1(a)) after the first step. Then, the thin layer of zeolite beta crystals was put into the near-neutral medium of synthesis gels. After the second step, the formed zeolite beta membranes ensured cross-linking with typical truncated bi-pyramidal morphology shown in Figure 1(c). The total thickness of the membrane was only 2.5~3 µm. As illustrated in Figure 2, the BEA topology was well-defined reflections from the characteristic XRD diffraction peaks. The zeolite beta membrane exhibited three reflections corresponding to (*101*), (*106*), and (*302*), indicating that the synthesized membrane was preferentially (*h0l*)-oriented. The EPMA elemental analysis of the membrane was given in Figure 3. The concentration of aluminum was gradually decreased from the substrate surface to the outer surface of the zeolite beta membrane, which demonstrated that there does not exist a transition layer inside the membrane. The pervaporation fluxes of several pure components through zeolite beta membranes were checked at 298 K. The separation of TIPB/ethanol mixtures by pervaporation was also used to evaluate the membrane quality. It could be concluded that the synthesized zeolite beta membranes by our two-step method are dense and uniform, especially the preferentially (*h0l*)-oriented membranes were well intergrown with little defects.

Figure 1 SEM images of the seed layer of beta crystals (a), zeolite beta membranes synthesized on the seed layer under the alkaline synthesis solution without fluoride (b, M1), and zeolite beta membranes synthesized on M1 under the near-neutral synthesis solution with fluoride (c, M2). Inset: cross-sectional view of the corresponding zeolite beta seed layer or membranes.

Figure 2 (a) XRD patterns of the substrate, seed layer and zeolite beta membranes. (A) alpha-alumina, (B) the seed layer, (C) zeolite beta membrane synthesized on the seed layer under the alkaline synthesis solution without fluoride (M1), and (D) zeolite beta membrane synthesized on M1 under the near-neutral synthesis solution with fluoride (M2). (b) EPMA elemental analysis of a (*h0l*)-oriented zeolite beta membrane. The detection limits for Al, Si, and F are about 0.001 wt%. EPMA elemental profile plotted in mass percentage versus distance with SEM image of the sample. The dashed line illustrates the EPMA scanning path.

Although zeolite beta membranes with little defects can be synthesized by the two-step method, their surface hydrophilicity/hydrophobicity is not satisfactory, and need to be improved to enhance the pervaporation performance. Mussel-inspired surface chemistry has attracted great interest due to its simplicity, versatility and wide applicability [7]. Herein, a facile method was proposed to functionalize the surface of zeolite beta membranes. A one-step procedure was developed to improve the surface hydrophilicity of zeolite beta membranes via co-deposition of polydopamine (PDA) and polyethyleneimine (PEI). Moreover, the PDA-functionalized surface of zeolite beta membranes could become hydrophobic after octadecylamine grafting *via* Schiff base reaction. The change in surface hydrophilic-hydrophobic properties was checked by contact angle measurements. As shown in Figure 3, the contact angle of synthesized zeolite beta membranes was ca. 80° (Figure 3(a)), which was awkward to be preferentially selective to either butanol or water. After hydrophilic modification, the average contact angle was decreased to ca. 40° (Figure 3(b)), leading to an increased hydrophilic surface. While the average contact angle would be increased to ca. 105° (Figure 3(c)) after hydrophobic modification, indicating that the surface of zeolite beta membranes became hydrophobic. The pervaporation performance of zeolite beta membranes before and after surface modification was demonstrated in Figure 4. The significant improvements in pervaporation of butanol/water mixtures were achieved by using the functionalized beta zeolite membranes. With respect to dilute butanol aqueous solution (5 wt% water, 70 °C), the separation factor of the zeolite beta membrane after hydrophilic modification was increased from ca. 19 to over 420 while keeping the total flux at ca. 1.5 kg/ (m^2 h). The pervaporation performance of surface hydrophobic modified zeolite beta membranes was measured for dilute n-butanol aqueous solution (5 wt% n-butanol, 70 °C). The membrane after surface modification exhibited a total flux of ca. 1.3 kg/ (m^2 h) with the separation factor of ca. 100, which increased from ca. 5. These results demonstrated that the surface property of zeolite beta membrane influenced the pervaporation performance when the membrane defects was little. Furthermore, the method we proposed endows the zeolite beta membrane with excellent surface properties, high separation factor and low flux reduction. Compared with current literature reports, the pervaporation performance of our zeolite beta membranes in this report was significantly enhanced for dilute and concentrated butanol aqueous solutions.

Figure 3 Static water contact angles of zeolite membranes. (a) without modification, (b) after hydrophilic modification, (c) after hydrophobic modification.

Figure 4 The enhancements in the pervaporation performance of beta zeolite membranes after surface modification. (a) after hydrophilic modification, (b) after hydrophobic modification.

Acknowledgements

This research was supported by the National Natural Science Foundation of China (Grant No. 21136008).

References

[1] Abdehagh N, Tezel F. H., Thibault J. Separation techniques in butanol production: challenges and developments. Biomass and Bioenergy. 60 (2014) 222-246.

[2] Liu G. P., Wei W, Wu H, Dong X. L., Jiang M, Jin W. Q. Pervaporation performance of PDMS/ceramic composite membrane in acetone butanol ethanol (ABE) fermentation¨CPV coupled process. Journal of Membrane Science. 373(1) (2011) 121-129.

[3] Fan H. W., Wang N. X., Ji S. L., Yan H, Zhang G. J. Nanodisperse ZIF-8/PDMS hybrid membranes for biobutanol permselective pervaporation. Journal of Materials Chemistry A. 2(48) (2014) 20947-20957.

[4] Li Y. K., Shen J, Guan K. C., Liu G. P., Zhou H. L., Jin W. Q. PEBA/ceramic hollow fiber composite membrane for high-efficiency recovery of bio-butanol via pervaporation. Journal of Membrane Science. 510 (2016) 338-347.

[5] Pham T. C. T., Kim H. S., Yoon K. B. Growth of uniformly oriented silica MFI and BEA zeolite films on substrates. Science. 334(6062) (2011) 1533-1538.

[6] Jakob A, Valtchev V, Soulard M, et al. Syntheses of zeolite beta films in fluoride media and investigation of their sorption properties. Langmuir. 25(6) (2009) 3549-3555.

[7] Lee H, Dellatore S. M., Miller W. M., Messersmith P. B. Mussel-inspired surface chemistry for multifunctional coatings. Science. 318(5849) (2007) 426-430.

490905 Investigation of Factors That Induce Cristobalite Formation during Titanosilicate Synthesis and Their Potential Impact on Heterogeneous Catalysis

Abstract:

Titanosilicates are industrially relevant zeolite derivatives with applications in heterogeneous catalysis, ion-exchange, and membranes in petrochemical, pharmaceutical, food and water purification fields.[1] They are potent and selective catalysts for alkene epoxidation, ammoximation and aromatic diol production, therefore, synthesis of new types of titanosilicates, targeting specialized industrially-relevant reactions, is an ever expanding field of research. However, their synthesis is challenging, due to requirement of specialized conditions and extreme sensitivity to changes in the synthesis environment, as well as occurrence of structural heterogeneities in the form of crystalline phases during synthesis. This, in turn, results in low catalytic performance. The physical structure of titanosilicates can be customized *via* temperature dependent phase transformations. Understanding the physicochemical consequences of such transformations is a key factor in the effective synthesis of novel and useful titanosilicates. The catalytic activity of titanosilicates is brought about by tetrahedrally coordinated, isolated Ti^{4+} sites, connected to the silica matrix *via* Ti-O-Si linkages. However, formation of Ti-O-Ti oligomers and TiO_2 crystalline phases is unavoidable during synthesis.[2] Formation of such phases is highly undesirable, as they are inactive in oxidative catalysis. Formation of crystalline silica phases such as quartz and cristobalite has also been reported. However, it is not completely understood whether or not these have any effect on the catalytic activity of titanosilicates.

A novel mesoporous titanosilicate was developed and optimized *via* the Doehlert matrix statistical approach,[3] as an effective and selective catalyst for the epoxidation of cyclohexene. During the synthesis optimization, it was revealed that, although the material is mostly amorphous, the most catalytically active samples contained small amounts of crystals. The crystalline phase was identified to be predominantly β-cristobalite, the lesser known, high-temperature crystalline form of silica, (the other being quartz), which can, however, exist in a metastable state at temperatures lower than 1470 °C. In order to understand the effect of this cristobalite phase in heterogeneous catalysis, different samples of the titanosilicate were synthesized and calcined at various temperatures ranging from 650 °C to 950 °C, to gradually increase the content of cristobalite. Control experiments were conducted with pure silica, at the same conditions. It was found that cristobalite formation is induced by Ti^{4+}, without any addition of alkali metals, which had not been previously reported in the literature. Catalytic experiments conducted with titanosilicate samples calcined at different temperatures and the corresponding pure silica samples show that the presence of a cristobalite phase correlates with catalytic activity of the titanosilicate. The materials were extensively characterized using XRD, XPS, BET, TEM and SEM analysis to help interpret the catalytic experiments. The effect of cristobalite on catalysis, as to whether it induces an impact or is merely a spectator species is discussed. Investigation of such structure-function relationships are of critical importance in understanding the factors that affect catalytic activity of titanosilicates, as well as provide insights into the importance of controlling calcination conditions in industrial synthesis.

1. A. S. Perera and M.-O. Coppens, in *Catalysis, Volume 28*, ed. J. J. Spivey, RSC Publishing, 2016, vol. 28, ch. 05, pp. 119-143

2. M. Baca, W. J. Li, P. Du, G. Mul, J. A. Moulijn and M.-O. Coppens, *Catalysis Letters*, 2006, **109**, 207-210.

3. D. H. Doehlert, *Applied Statistics*, 1970, **19**, 231–239.

497264 Novel Methods to Synthesize ZSM-11 As an Efficient Catalyst for Methanol-to-Hydrocarbon Reactions

Abstract:

ZSM-11 (MEL) is an emerging catalyst for a wide variety of applications. The MEL framework is similar to ZSM-5 (MFI), which is one of the most widely used zeolite catalysts in industry. Recent studies have shown that ZSM-11 performs better as a catalyst in several different reactions compared to ZSM-5 [1-3], possibly due to improved intra-crystalline diffusion within the less-tortuous straight channels of the MEL topology. However, there are relatively few systematic studies of ZSM-11 synthesis, specifically with respect to tailored physicochemical properties such as size, shape, and composition (i.e., Si/Al ratio, or SAR). Moreover, MEL is conventionally synthesized using tetrabutylammonium (TBA) as an organic structure-directing agent (OSDA), which is costly and yields ZSM-11 crystals with ZSM-5 impurity [4]. Here, we will present methods to prepare ZSM-11 catalysts as well as the results of testing for methanol-to-hydrocarbon (MTH) reactions. Through the judicious selection of synthesis parameters, we successfully developed a facile protocol to prepare ZSM-11 catalysts with sub-micron crystal size, high MEL phase purity, low OSDA usage, and tunable SAR. The ZSM-11 catalyst samples synthesized via these novel methods showed significantly improved catalyst lifetime in MTH reactions compared to those synthesized by conventional methods.

[1] Kustova, M. Y.; Rasmussen, S. B.; Kustov, A. L.; Christensen, C. H., *Applied Catalysis B-Environmental* **2006**, *67* (1-2), 60-67.

[2] Zhang, L.; Liu, H.; Li, X.; Xie, S.; Wang, Y.; Xin, W.; Liu, S.; Xu, L., *Fuel Processing Technology* **2010**, *91* (5), 449-455.

[3] Bleken, F.; Skistad, W.; Barbera, K.; Kustova, M.; Bordiga, S.; Beato, P.; Lillerud, K. P.; Svelle, S.; Olsbye, U., *Physical Chemistry Chemical Physics* **2011**, *13* (7), 2539-2549.

[4] Piccione, P. M.; Davis, M. E., *Microporous and Mesoporous Materials* **2001**, 49 (1-3), 163-169.

498539 Preparation of Novel Al-MFI/Fe-MFI Core-Shell Catalysts and Their Catalytic Application for CH_4 Conversion

Abstract:

Methane is a highly abundant and inexpensive source of fuel and chemicals. The development of novel technologies that can convert methane easily into chemicals has strongly been desired. However, its kinetic inertness and low reactivity limit the industrial utilization. We are tackling the development of novel catalysts for direct conversion of methane into methanol followed by lower olefins.

The **MFI**-type ferrisilicate zeolite (Fe-MFI) has exhibited a catalytic activity for the production of CH_3OH from CH_4. On the other hand, it is well-known that the **MFI**-type aluminosilicate zeolite (Al-MFI) is an excellent catalyst for the methanol to olefins (MFI) reaction. Based on these facts, we have focused on the development of a novel core-shell catalyst consisting of Fe-MFI and Al-MFI for direct conversion of CH_4 into CH_3OH followed by lower olefins.

First, we prepared the composite catalyst of Al-MFI and Fe-MFI as core and shell parts, respectively, as follows. Al-MFI (Si/Al=30) was added into the mother gel for Fe-MFI (Si/Fe = 50 in gel) with the weight ratio of Al-MFI / Fe-MFI of 1, and thus prepared mixture was hydrothermally treated at 443 K for 5 days. The product was designated as "Fe-MFI/Al-MFI". The "Al-MFI/Fe-MFI" was also prepared by using Fe-MFI (Si/Al=228). Both products had a pure **MFI**-type structure. The FE-SEM images of the Al-MFI/Fe-MFI core-shell catalyst indicated that the Fe-MFI 100-200 nm in size was covered with the Al-MFI nanocrystal 20-30 nm in size.

The catalytic reaction was performed in a fixed bed reactor. The flow rates of the reactants were $CH_4/O_2/Ar$ = 16/4/5 (SCCM). The catalyst amount was 100 mg. The reaction temperature was varied ranging from 300 to 600 °C, and the reaction time at each temperature was 1 min. The products including CO and CO_2 were analyzed by GC-TCD, and other hydrocarbon products were analyzed by GC-FID.

At the reaction temperature at 300 °C, the Fe-MFI/Al-MFI core-shell catalyst did not show any activity for CH_4 conversion. However, the Al-MFI/Fe-MFI core-shell catalyst gave the CH_4 conversion of 0.1 %, and it slightly produced propene, which would be formed from CH_3OH. The CH_4 conversion was increased along with the reaction temperature. At the reaction temperature of 500-600 °C, the Fe-MFI /Al-MFI core-shell catalyst exhibited a higher conversion than the Al-MFI/Fe-MFI core-shell catalyst, and it gave CO, CO_2, ethane and ethane as products. Note that in addition to these products, dimethyl ether (DME) was significantly produced over the Al-MFI/Fe-MFI core-shell catalyst with the CH_4 conversion of 5.5 %, while main products were still CO and CO_2 with their selectivities of almost 99 %.

It is considered that, in the Al-MFI/Fe-MFI core-shell catalyst, CH_4 was activated at Fe-MFI as core part, forming CH_3OH, which would be immediately converted into DME followed by lower olefins via the MTO recation. On the other hand, in the Fe-MFI/Al-MFI core-shell catalyst, CH_4 was activated at Fe-MFI as shell part, and successive oxidation would easily occur, resulting in the higher conversion. The impacts of the ratio of Al-MFI and Fe-MFI parts, and their components on the catalytic activity are currently under investigation.

501951 An Optimized Procedure for Selective Removal of Efal (Extra Framework Aluminum) in Y-Zeolites

Abstract:

Zeolites with low aluminum content (high Si/Al ratio) are more favorable for catalytic applications.Â Zeolite-Y is the prominent member of zeolite family cannot be directly synthesized with Si/Al ratios over ~2.5. Hence, post-synthetic modification such as dealumination is found to be most feasible way to increase the Si/Al ratio in the framework.[1-4] In general, dealumination refers to the removal of aluminum from zeolitic framework by chemical reactions. Most often, the â€˜Alâ€™ removed from zeolitic framework will fall into non-framework (extra framework). Due to EFAl species the degree of crystallinity to be reduced in zeolites, which leads to complicated issues on catalytic applications. To avoid this, selective removal of extra framework aluminum (EFAl) is important for various catalytic applications. In this study, an optimized procedure is developed to remove only EFAl species from commercial Zeolite-Y. Various concentration of water glass was added to commercial zeolites and treated at various temperatures (25 to 100 °C) and duration (0-6 hrs). An optimized procedure is being developed to remove just EFAl from zeolite materials. All the samples were well characterized by XRD, N_2adsorption-desorption measurements, ICP-AES, TEM and NMR.

Keywords: Dealumination, extra framework, Si/Al ratio, zeolites

References

1. P. J. Kooyman, P. van der Waal, and H. van Bekkum, Zeolites **1997**,18, 50-53.
2. R. Giudici, H.W. Kouwenhoven and R. Prins. Applied Catalysis A: Gen. **2000**, 203, 101-110.
3. C. S. Triantafillidis, A. G. Vlessidis and L. Nalbandian, N. P. Evmiridis, Micropor. Mesopor. Mater. **2001**, 47, 369-388.
4. E.T.C. Vogt and B. M. Weckhuysen, Chem.Soc.Rev. **2015**, 44, 7342-7370.

486797 Multifunctional Polymer Nanocomposites

Abstract:

With more awareness of energy conversion/storage and saving, different strategies have been developed to utilize the sustainable and renewable energy. Introducing nanoscale fillers can make inert polymer matrix possess unique properties to satisfy certain functions. For example, alumina nanoparticles have strengthened the weak thermosetting polymers [1]. A combined mixture of carbon nanofibers and magnetite nanoparticles have made the inert epoxy sensitive to magnetic field for sensing applications [2]. Introducing silica nanoparticles into conductive polymers such as polyaniline has enhanced the giant magnetoresistance (GMR) behaviors [3]. The nanoparticles can enhance the stability of the energy storage in the conductive polymer systems [4,5]. The introduced nanoparticles endow the transparent polymer electromagnetic interference (EMI) shielding function while reduce the density significantly [6-10]. With the desired miniaturization, the materials combining different functionalities have become importantly interesting [11].

In this talk, methodologies to prepare nanocomposites and their effects on the produced nanocomposites will be discussed. A variety of advanced polymer nanocomposites will be introduced. Unique properties including mechanical, electrical, optical, magnetoresistance etc. and their applications for environmental remediation, energy storage/saving, fire retardancy, electromagnetic interference (EMI) shielding, and electronic devices, etc. will be presented.

References

[1] Surface Functionalized Alumina Nanoparticle Filled Polymeric Nanocomposite with Enhanced Mechanical Properties, *Journal of Materials Chemistry*, 16, 2800-2808 (2006). Link to publisher; pdf file

[2] Strengthened Magnetoresistive Epoxy Nanocomposite Papers Derived from Synergistic Nanomagnetite-Carbon Nanofiber Nanohybrids, *Advanced Materials*, 27(40), 6277-6282, (2015) Link to publisher; pdf file

[3] Giant Magnetoresistive Phosphoric Acid Doped Polyaniline-Silica Nanocomposites, *Journal of Physical Chemistry C*, 117(12), 6426-6436 (2013) pdf

[4] Silica Doped Nano-polyaniline with Endured Electrochemical Energy Storage and the Magnetic Field Effects, *Journal of Physical Chemistry C*, 117(25), 13000–13010 (2013) pdf

[5] Electrochromic Polyaniline/Graphite Oxide Nanocomposites with Endured Electrochemical Energy Storage, *Polymer*, 54(7), 1820-1831 (2013) pdf

[6] Morphological Regulation Improved Electrical Conductivity and Electromagnetic Interference Shielding in Poly(L-lactide)/Poly(e-caprolactone)/Carbon Nanotubes Nanocomposites via Constructing Stereocomplex Crystallites in Poly(L-lactide) Phase, *Journal of Materials Chemistry C*, 5, 2807-2817 (2017), Link to publisher; pdf file

[7] Flexible, conductive, porous, fibrillar polymer–gold nanocomposites with enhanced electromagnetic interference shielding and mechanical properties, *Journal of Materials Chemistry C*, **5**, 1095 - 1105 (2017) Link to publisher; pdf file

[8] Electromagnetic Field Absorbing Polypropylene Nanocomposites with Tuned Permittivity and Permeability by Nanoiron and Carbon Nanotubes, *Journal of Physical Chemistry C*, 118(42) 24784-24796 (2014) Link to publisher; pdf file

[9] Electromagnetic Field Shielding Polyurethane Nanocomposites Reinforced with Core-Shell Fe-Silica Nanoparticles, *Journal of Physical Chemistry C*, 115, 15304-15310 (2011) pdf

[10] Magnetic and Electromagnetic Evaluation of the Magnetic Nanoparticle Filled Polyurethane Nanocomposites, *Journal of Applied Physics*, 10, 09M511 (2007). Link to publisher; pdf file

[11] An Overview of Multifunctional Epoxy Nanocomposites, *Journal of Materials Chemistry C*, **4**, 5890-5906 (2016) Link to publisher; pdf file

487620 Transparent Copper-Silica Nanoparticle-Chitosan Nanocomposite Coatings with Long-Term Antibacterial Efficacy

Abstract:

Bacteria-contaminated inanimate surfaces within hospitals/clinics result in transmission of pathogens via direct or indirect contact, leading to increased risk of healthcare-associated infections (HAIs). Although the use of antibacterial coatings is a potential way of reducing the bacterial burden, many surfaces like instrument panels and monitors necessitate the coatings to be transparent while being highly antibacterial. In this work, a layer of acrylated quaternized chitosan (AQCS) was covalently immobilized on commercially available transparent polyvinyl fluoride (PVF) films by UV-induced grafting. Silica nanoparticles (SiO_2 NPs) were grown *in situ* over the AQCS-coated PVF and followed with adsorption of copper ions on the modified substrates, where the NPs acted as â€œstorageâ€ sites for the copper ions. The coated PVF films were transparent and SiO_2 NP-containing coatings were found to possess higher visible light transmission as compared to AQCS-coated PVF. The copper content of the coatings increased from ~2 $\mu g/cm^2$ to ~16 $\mu g/cm^2$ with increase in the SiO_2 NP deposition time from 1 h to 4 h, when the copper adsorption time was fixed at 6 h. The copper-containing coatings reduced viable bacterial count by ~99 % and 100 %, when incubated with a bacteria-loaded droplet for 60 min and 120 min respectively. The SiO_2 NP-containing coatings after being wiped 100 times with a deionized water-wetted cloth still reduced bacterial count in bacteria-loaded droplets by ~96 %, after 60 min. The stability of these coatings were further improved with the deposition of another layer of quaternized chitosan (QCS) on the SiO_2 NPs, prior to copper adsorption. These coatings with an additional QCS layer showed no significant change in antibacterial efficacy even after 100 wipes, demonstrating its stability. Additionally, these coatings showed no significant toxicity to mouse fibroblast cells. Thus, the transparent coatings developed in this work hold great potential for application as long-term antibacterial coating for combating HAIs.

498328 Bio-Inspired Design of Stimuli Responsive Materials Based on a Bilayer Structure

Abstract:

A bilayer structure composed of polyvinyl alcohol (PVA) or PVA composite thin film atop thick polydimethylsiloxane (PDMS) substrate was prepared. The bilayer structure shows dynamic strain-responsive optical properties. The transition between a transparent state to an opaque state can be easily achieved by uni-axially stretching and releasing the device. Also, a series of derivative mechanochromisms with capabilities of switch â€œon/offâ€ fluorescence, change fluorescent color, reveal/hide information upon mechanical stimuli are prepared. Mechanical or/and moisture responsive wrinkling surface can also be achieved in a similar manner. By altering the chemical structure of the PVA top layer, three different moisture responsive dynamic wrinkling surfaces can be prepared. Selectively inducing chemical modification on the PVA surface can create moisture responsive patterned wrinkles, which can lead to applications in encryption, anti-counterfeit tab, and water indicator for circuits.

498267 Biomimetic Nanocoatings with Exceptional Mechanical, Barrier, and Flame Retardant Properties from Large Scale One-Step Co-Assembly

Abstract:

Large-scale biomimetic organic/inorganic hybrid nanocoatings with a nacre-like microstructure were prepared via a facile co-assembly process. Different from conventional polymer nanocomposites, such nanocoatings contain a high concentration of nanosheets, which can be well aligned along the substrate surface. Moreover, the nanosheets and polymer matrix can be chemically co-crosslinked. As a result, the nanocoatings exhibit exceptional mechanical properties (high stiffness and strength), barrier properties (to both oxygen and water vapor), and flame retardancy, but meanwhile they are highly transparent (maintaining more than 85% of their original transmittance to visible light). The nanocoatings can be applied to various substrates and regular or irregular surfaces (e.g., films as well as foams). Because of their excellent performance and high versatility, such nanocoatings are expected to find widespread application.

500774 Bioinspired Composite Materials with Stimuli-Responsive Color Changing Ability

Abstract:

The ability of changing colors is one of the most exciting features of advanced living organisms. Octopuses, chameleons, squids, all possess cells that can alter their colors. In some cases, the color changing mechanism is due to the ability of these cells to displace pigments over different areas, as in the case of chromatophores in the octopus. In other cases, the change in color is due to a change in the spacing of colloidal crystals, as in the chameleon iridophores. Inspired by this ability, we present here composite materials, consisting of a polymer matrix (PMMA or fluorinated acylates) containing water droplets. These droplets, with a size of tens of micrometers, contains different responsive particles. When magnetic colloids are incorporated, the application of magnetic field gradients will lead to a displacement of the colloids, similar to the displacement of pigments in chromatophores.

When instead temperature-responsive PNIPAm microgels are incorporated into the droplets, at sufficiently high volume fraction they will form colloidal crystals. These crystals create iridescent colors because of the length of the spacing between particles interfering with light. Because the size of the microgels are temperature responsive, the colloidal crystals can be destroyed at a temperature higher than the lower critical solubility temperature, leading to a change in color. Additional stimuli-responsive particles can be used to induce different stimuli-responsive color-changing mechanisms.

The proposed approach is thought to be a novel design for stimuli-responsive composite materials with color-changing ability.

499331 Investigation of Carbon Nanotubes & Cellulose Nanocrystals Composite for Potential Use in Microelectromechanical Systems

Abstract:

Investigation of Carbon Nanotubes & Cellulose Nanocrystals composite for Potential Use in Microelectromechanical Systems

Mingzhe Jiang, Christopher L. Kitchens*

Department of Chemical and Biomolecular Engineering, Clemson University

The use of cellulose nanocrystals (CNCs) in high performance coatings is attractive for micro-scale structures or device fabrication due to the anisotropic geometry, however CNC are insulating materials. Carbon nanotubes (CNTs) are also rod-shaped nanomaterials that display high mechanical strength and electrical conductivity. The hydrophobic regions of surface-modified CNCs can interact with hydrophobic CNT, and thus aid in association between CNTs and CNCs. CNCs' long-range electrostatic repulsion also plays a role in forming a stable CNT/CNC water dispersion. The sonication-driven dispersion and dispersing capacity of multi-wall carbon nanotubes (MWCNTs) and single wall carbon nanotube (SWCNTs) assisted by CNCs in aqueous solution has been measrued by UV–vis spectroscopy and atomic force microscopy. Conductive and anisotropic hybrid films (5 ~ 20 μm) are casted on piranha acid treated, glass microscope slides by blade coating with concentrated (above 12 wt%) CNT/CNC aqueous dispersion. The films were then examined for orientation of both CNTs and CNCs by polarized light and Wide Angle Xray Scattering. It was observed that CNCs with different concentration provide different dispersing capacity for Multi-Wall Carbon Nanotubes in aqueous solution and that the use of blade coating gives thin films an anisotropic conductive property, comparing to non-aligned film (naturally dried mixing solution on petri-dish). CNCs with different charge density have also been introduced as alternatives for improving the final hybrid film property.

494444 Few Layers $MoSe_2$ Incorporated with Nitrogen Doped Graphene Sheet for High Performance Lithium Sulfur Batteries

Abstract:

Rechargeable lithium sulfur (Li-S) batteries, which utilize sulfur as cathode materials, have been examined as one of the most promising energy storage device, owing to their high energy density of 2600 Wh/kg and theoretical capacity of 1672 mAh/g. Apart from that, the cathode has the advantages of sulfur abundance, low cost and environmental friendliness. However, the capacity fading in the battery performance attributed to the dissolution of long-chain lithium polysulfides (Li_2S_x, x = 4 - 8) causing damaging â€œshuttle effectâ€ becomes the main obstacle in Li-S batteries system. Introducing anchoring materials, transition mental dichalcogenides (TMDCs) to the cathode design has been confirmed as an efficient way to trap polysulfide species. Here, we present a novel material, a two-dimensional composite, molybdenum diselenide on nitrogen doped reduced graphene oxide sheet ($MoSe_2$/N-rGO), which is synthesized by CVD method to provide a strong chemisorption between the polysufides and $MoSe_2$. We found remarkable improvement in the capacity retention comparing to N-rGO electrode (without $MoSe_2$). It is observed that smaller percentage use of $MoSe_2$ could give a better battery performance due to its unpleasant electrical insulating properties. An optimal amount of $MoSe_2$ in the composite is discovered to be around 10% which provides a reference for future studies.

500707 Dual-Responsive Plasmonic Behavior of Gold Nanorods@PANI Core/Shell Nanostructures for Real-Time Control

Abstract:

Plasmonic nanoparticles have attracted great attention due to their unique localized surface plasmon resonance (LSPR) that can be utilized in various applications including solar cells, photocatalysis, and sensors. For target applications, specific LSPR properties can be achieved by controlling plasmonic particleâ€™s shape, composition, and size. However, for a given structure, it is generally challenging to control their plasmonic properties in a real time. In the present research, we demonstrate dual-responsive gold nanorod (AuNR)/polyaniline (PANI) core-shell nanostructures where the LSPR of AuNR cores can reversibly be controlled by changing electrochemical potential as well as environmental pH.Â The largest LSPR shift of the AuNR cores is achieved when environmental pH is changed. A significant plasmonic shift of 107 nm is obtained with only 8 nm thick PANI shell due to theÂ large refractive index change of PANI shells and the high LSPR sensitivity of AuNRs. Furthermore, the AuNR@PANI core/shell showed very stable and reversible LSPR dual-responsive properties during the cycling test without significant degradation.

495783 Microwave Induced Heating of Carbon Nanotubes Localized at 3D-Printed Thermoplastic Interfaces

Abstract:

A key obstacle facing 3D-printed plastic parts in engineering applications is the weak weld between successive filament traces, which often leads to delamination and mechanical failure. Here we report a novel concept for welding 3D-printed thermoplastic interfaces using microwave irradiation to induce intense heating of carbon nanotubes (CNTs) coated on filament interfaces. The microwave heating of the CNT-polymer composites are a function of both frequency and the CNT percolation threshold, as shown through in situ infrared imaging and simulation. We apply ink-coating (and/or coaxial extrusion) to deposit a dispersed CNT film at the interface of the filament; we then print the coated filaments such that CNTs are localized between each trace. Microwave irradiation induces polymer mobility at the interface and substantially improves the weld fracture strength. These results open up new design spaces for additive manufacturing and also yield new insight into the coupling between CNT network percolation and RF field heating response.

501272 Theoretical Study of the Reaction Kinetics of Organosiloxane Polycondensation

Abstract:

Silicone-based polymers are commonly used in coating industry because of the interesting characteristics of organosilane molecules in forming of covalent bonding between inorganic and organic compounds. They are widely used for encapsulation and packaging of electronic

components, in medicine, and for soft lithography[1-3]. The reaction mechanism involves an acid catalyzed hydrolysis followed by an SN2 type substitution. The polymerization reaction was modeled by including the two mechanisms of growth and recombination. At the early stage of the reaction, the growth mechanism is dominant as the concentrations of dimers and trimers are high. Several parameters change the reaction dynamics significantly, including the concentration of catalyst, evaporation rate and the temperature. It was found that all these factors have a direct effect on the molecular weight of polysiloxane and increase the polymer molecular weight as well as the polydispersity index. To achieve the desired molecular weight and physical properties, the tight control of the reaction is highly important.

References:

1) Schmid, H., and B. Michel. "Siloxane polymers for high-resolution, high-accuracy soft lithography." *Macromolecules* 33.8 (2000): 3042-3049.

2) Wong, C. P. "Thermal-mechanical enhanced high-performance silicone gels and elastomeric encapsulants in microelectronic packaging." *IEEE Transactions on Components, Packaging, and Manufacturing Technology: Part A* 18.2 (1995): 270-273.

3) Okoroanyanwu, Uzodinma. "Chemistry and Lithography." SPIE, 2010.

490815 High-Ï‡ Block Copolymers with High Etch Selectivity for Sub-10 Nm Patterning

Abstract:

As the demand for advanced device performance, feature density, and processing speed increases, the modern semiconductor industry is now relying on 193 nm immersion lithography, often augmented with multiple patterning techniques, to reduce the critical dimension of nanopatterns in integrated circuits. However, continued technological progress is challenged by high implementation costs and the intrinsic limitations of conventional top-down photolithography. Thus, alternative bottom-up lithographic techniques utilizing molecular self-assembly, including block copolymer (BCP) thin film lithography, have garnered substantial attention. BCPs can spontaneously assemble into nanometer scale periodic morphologies in thin films, which can later be used as an etch mask for transferring the pattern into underlying materials. Recently, a great deal of effort has been focused on the self-assembly of Si-containing blocks paired with more polar adjoining blocks to drive a large interaction parameter (Ï‡). When paired with more polar blocks, the added Si promotes microphase separation at low molecular weights into sub-10 nm domains. Incorporation of Si also imparts high etch selectivity to ensure effective pattern transfer into other materials.

In this work, poly(pentamethyldisilyl styrene)-*block*-poly(ethyl glycolide) (PDSS-PEGL) BCPs were used to fabricate perpendicular lamellar nanostructures with sub-10 nm linewidth. The top and the substrate interfacial interactions for the PDSS-PEGL thin films were precisely tuned to control the orientation of nanostructures across the entire substrate. The film was then treated with reactive ion etching protocols to remove the PEGL block and produce perpendicular line-space patterns with feature sizes of about 6 nm over a large area. The PDSS-PEGL BCP system is a promising candidate for sub-10 nm patterning processes in future lithographic applications.

487350 Fabrication and Structural Analysis of Nanofibers Made By Syndiotactic Polypropylene with Ethylene-Comonomer Units

Abstract:

Syndiotactic polypropylene with a small amount of ethylene comonomer (sPPEt) possesses an excellent elasticity [1]. In this study, the sPPEts with different ethylene comonomer compositions were synthesized and the fabrication of sPPEt nanofibers was attempted using the electrospinning (ES) method. sPPEt copolymers with different ethylene compositions were successfully fabricated into nanofiber membranes by electrospinning when using a mixture of methyl-cyclohexane/DMF/acetone (8:1:1) as an electrospinning solvent, resulting in thin fiber diameters: 730 nm for the sPPEt copolymer with the ethylene composition of 2 mol%, 324 nm for that with the ethylene composition of 10 mol%, and 467 nm for that with the ethylene composition of 15 mol%. However, the sPPEt copolymer with the ethylene composition of 25 mol% could not form fibers. The morphology of the electrospun sPPEt nanofibers was analyzed by the scanning electron microscopy (SEM). The electrospun nanofibers using sPPEt with the lower ethylene composition (2 mol%) had rough surface, whereas the sPPEt nanofibers with relatively higher ethylene compositions of 10 mol% and 15 mol% had smooth surfaces. The smooth surface of the sPPEt nanofibers with larger ethylene compositions could be due to the suppression of the gelation of the sPPEt solutions. The electropun sPPEt nanofibers were also analyzed by the Fourier-transfer infrared spectroscopy (FTIR) and the X-ray diffraction (XRD) measurements. From the FTIR analysis, it was revealed that the ratio of the trans-planar/helical conformations (T/R ratio) of the electrospun sPPEt nanofibers was higher than that of the simple sPPEt cast films regardless of the ethylene compositions in sPPEt. [1] Claudio De Rosa, Finizia Auriemma, Mechanical properties of syndiotactic propylene-ethylene copolymers, Macromolecules, Vol. 39, 249-256 (2006).

491239 Interaction Between Supercritial CO_2+Cosolvent and Poly(vinyl acetate)

Abstract:

Interaction between Supercritial CO_2+Cosolvent and Poly(vinyl acetate)

Dong-dong Hu, Lei Bao, Ling Zhao, Tao Liu*

(Shanghai Key Laboratory of Multiphase Materials Chemical Engineering, East China University of Science and Technology, Shanghai 200237, China)

hudd@ecust.edu.cn; liutao@ecust.edu.cn;

Phone: (+86) 21 64253470; Fax: (+86) 21 64253528

Supercritical carbon dioxide ($scCO_2$) is of interest as a potential alternative to conventional organic solvents. Unfortunately, the low solvating power of $scCO_2$ restricts its compatibility with polymers. Poly(vinyl acetate) (PVAc) was reported as one of the most CO_2-philic hydrocarbon homopolymers, which was a significant step forward. But even so, the solubility of PVAc is very low, making them unsuitable to apply on a large scale.

The cosolvents (ethanol, acetone, n-heptane, denoted by ET, AC and C7) were introduced to form CO_2+cosolvent systems for improving the solvent-solvent and solvent-solute interactions and enhancing the the compatibility with PVAc. Comparing the interaction of CO_2 with acetone, and n-heptane, it could be seen that ethanol and CO_2 provided stronger interaction energy from the ab initio calculation due to the shortest distance between the C atom of CO_2 and the O atom of ethanol (Fig. 1). The molecular dynamic modeling showed that the interaction between ethanol and CO_2 was strongest due to the bigger solubility parameter of ethanol (Fig. 2) and the obvious hydrogen bond bewteen ethanol and CO_2 (Fig. 3), which increased significantly the solubility parameter of solvent and improved the dissolution of polymers in CO_2. The dissolution behavior measurement (Fig. 4) confirmed that the addition of cosolvent could enhance the compatibility between CO_2 and PVAc to a certain extent, and the solubility of PVAc in $scCO_2$+cosolvent increased with the mole fraction of cosolvent, which has a good agreement with the simulated results of the solvent-solvent, solvent-solute interactions and the solubility parameter analysis.

Fig. 1 Most stable mode in ET/CO_2, AC/CO_2 and C7/CO_2systems.

(a) ET/CO_2/CO_2; (b) AC/CO_2; (c) C7/CO_2

Fig. 2 Solubility parameter of CO_2-ethanol binary systems with different compositions

(a) ET/CO_2/CO_2; (b) AC/CO_2; (c) C7/CO_2

Fig. 3 Comparison of rational correlation functions between CO_2-ethanol and CO_2-acetone systems.

Fig.4 Cloud point pressures of PVAc in CO_2+solvent at different temperatures

Acknowledgement:

The authors are grateful to the National Key Research and Development Program of China(2016YFB0302201) and the National Science Foundation of China (21376087) and Plan for Shanghai excellent technical leaders (14xd425500) for support.

References:

(1) DeSimone, J. M. *Science* 2002, *297*, 799-803.

(2) Beckman, E. J. *Chem. Commun.* 2004, 1885-1888.

(3) Lee, H.; Terry, E.; Zong, M.; Arrowsmith, N.; Perrier, S.; Thurecht, K. J.; Howdle, S. M. *J. Am. Chem. Soc.* 2008, *130*, 12242-12243.

(4) Cummings, S.; Xing, D.; Enick, R.; Rogers, S.; Heenan, R.; Grillo, I.; Eastoe, J. *Soft Matter* 2012, *8*, 7044.

(5) Girard, E.; Tassaing, T.; Camy, S.; Condoret, J. S.; Marty, J. D.; Destarac, M. *J. Am. Chem. Soc.* 2012, *134*, 11920-11923.

500758 Modeling of Distributions of Polymer Properties Using Parallel Computing in Julia

Abstract:

The polymer chain microstructure (molecular weight distribution, copolymer composition distribution, branching distribution, etc.) has a strong influence on the processing and end-use properties of the material. The operation of polymer processes aiming at achieving tight control of the molecular architecture of the polymer chains is very important for polymer manufacturers. Therefore, a significant effort has been devoted to the development of detailed predictive models of polymer processes. The calculation of the distribution of a single property, in most cases the MWD, has been extensively studied. On the other hand, a proper characterization of polymer samples sometimes requires simultaneous information on distributions of several properties. However, the development of mathematical models able to predict multivariate distributions leads to highly complex problems for which seldom solution approaches have been developed. In particular, large systems of differential-algebraic equations (DAE) are usually encountered. Therefore, model efficiency in terms of CPU time and memory requirements becomes a very significant issue. Besides, models to be used for optimization and on-line control should require short running times. In this context, parallel computing offers a great possibility for the development of fast models of polymer processes. In addition, the selection of a suitable programming language becomes an important decision, especially when working with problems that require high computation times and efficient parallel implementations.

Julia is an open-source, modern high-level dynamic programming language designed to achieve high-performance in technical applications, numerical analysis and scientific computing. The language was designed and developed to provide a good performance, approaching that of languages like C, while using a high-level programming style. In this way, it can also be effective for general-purpose programming. In order to achieve this goal, the language combines several modern programming language features, such as "multiple dispatch," "Just-In-time compilation," and a "Low-level Virtual Machine (LLVM) compiler infrastructure" among others. The language also has native parallel programming support. The code parallelization is quite straightforward compared to other languages where parallelization is not always native and the use of another application or libraries is needed. Julia may be seen as an open source, high-performance alternative to dynamic languages like MATLAB, R and Python, as well as a high-productivity alternative to static languages like C, C++ and FORTRAN. On top of providing excellent performance, the resulting code is easier to work with, modify and maintain.

The probability generating function (pgf) technique is a powerful method for modeling distributions of polymer properties.[1,2] The pgf technique is based on solving the pgf transform equations of the population balances describing a polymer process and the subsequent inversion of the resulting pgf transform of the distribution. A key step of this technique is the inversion of the pgf transform. In previous works, we developed this technique for modeling univariate distributions,[1,2] in particular the MWD, as well as bivariate distributions,[3-5] such as the copolymer composition distribution-molecular weight distribution (CCD-MWD), or the branching distribution-molecular weight distribution (BD-MWD).

For the prediction of univariate distributions, this work analyzes the performance of two variants of the pgf technique, each of them using a different inversion method. The first selected inversion method is the Papoulis (PAP) method.[2] It regards the pgf as a real valued function. In this method, the number of model equations is proportional to the number of points of the MWD that are computed. The second inversion method, called IFG,[1] regards the pgf as a complex valued function. This increases the number of model equations because the real and complex parts of the pgf transform are treated as two different variables. Besides, the number of model equations is proportional to the maximum significant chain length of the system, but it does not depend on the number of computed points of the MWD. For the prediction of bivariate distributions, three inversion methods are available: the 2D IFG, the 2D Papoulis and the 2D Pap-IFG methods.[3-5] The structure of the DAE of the pgf models make them suitable for parallel computation. The performance of the pgf technique with each of these methods using parallel computation in Julia is compared.

The pgf method was tested using different case studies of polymer systems. These were chosen to include different ranges of the independent variables of the distributions. It was found that the parallel implementation allowed reducing significantly the CPU time, in comparison to the conventional, sequential implementation. In the case of the univariate distributions, the two pgf inversion methods showed very good accuracy, but the IFG method was slightly superior. Regarding running time, this method was faster for systems of low molecular weight; on the other hand, Papoulis method was faster for high molecular weight systems. For the bivariate distributions, the 2D IFG method showed very good accuracy in all cases, the 2D Pap-IFG method yielded good accuracy, and the 2D Papoulis method showed the lowest level of accuracy. In terms of CPU time, the 2D IFG resulted the best option for systems with small to intermediate values of the distributed variables ($O(10^2)$), while the 2D Papoulis method was faster for large values of the distributed variables; the 2D Pap-IFG lied in between. The software Julia allowed a successful and efficient implementation of the pgf models using parallel computing.

References

1. Asteasuain, M. Ph.D. Thesis, Dpto. Ing. Qca., UNS, 2003.

2. Sarmoria, C.; Asteasuain, M.; Brandolin, A. CJChE 2012, 90, 263.

3. Asteasuain, M.; Brandolin, A. Macromol. Theory Simul. 2010, 19, 342.

4. Brandolin, A.; Asteasuain, M. Macromol. Theory Simul. 2013, 22, 273.

5. Brandolin, A.; Assini Balbueno, A.; Asteasuain, M. Comput. Chem. Eng. 2016, 94, 272.

494487 Transition Metal-Based Nanocrystals Confined-Growth on Heteroatom-Doped Graphene Toward Hydrogen Catalysis

Abstract:

With the combination of excellent electrical conductivity, high mechanical strength and surface area, graphene has been utilized as unique template for growth of nanomaterials, and further broadly applied in various research areas, such as environmental treatment and energy storage. Here, we will discuss and develop a new method of surface functionalization on graphene by UV-assisted *in situ* polymerization. In order to fully utilize the specific properties of both graphene and functional polymers, we will further explore the potential application by introducing graphene oxide and pristine graphene foam as templates, polyacrylamide (PAM) and gelatin with acryl groups (GelMA) as polymer scaffolds, to fabricate novel micro/nanostructure graphene-based composites, in which, not only the graphene can guide the polymerization as a template, but also the surface grafted polymer chains will turn the hydrophobic graphene surface to be highly hydrophilic, and simultaneously increase the integral mechanical strength and surface area with more active sites for storage of functional grains (like Transition Metal Phosphides and Chalcogenides).

494940 From Process to Product - Enhancing the Understanding of α-Olefin-Polymerizations

Abstract:

A fundamental task within polymer chemistry is getting a deeper insight of polymerization processes, especially those of industrial relevance. Only by understanding the fundamentals, such as the reaction networks, thermodynamics and the structure-property relationship, simulation-based process optimization and product design will be possible. This gains more and more relevance, as with increasing production scale on single reactors experimental based optimization becomes less of an option for costs and safety reasons.

In this context, the high-pressure polymerizations of α-olefins are a striking example for many reasons. Performing high-pressure experiments – both in mini-plants and industrially – is extremely expensive, thus the need for reliable models with a predictive ability is extraordinarily high. At the same time, the polymer has a very complex and inhomogeneous microstructure, consisting of randomly introduced short- and long-chain branches. Process conditions determine this microstructure, which has an enormous impact on processability and product properties. Thus, it is of high interest to understand the connection between process, microstructure and product properties.

We will introduce a three-step modeling approach, which is able to predict – to a certain extent – polymer properties, characterized by rheology, from process conditions. Numerical modeling of the kinetics on deterministic base is the first step; in a second step a hybrid Monte-Carlo algorithm is describing the microstructure and finally in the last step the entangled polymer dynamics of the branched material can be calculated via the branch-on-branch algorithm [1].

This modelling approach opens up a wide research area: the fundamentals can be investigated more thoroughly and globally and at the same time reliable simulation-based product design becomes more realistic.

Another important aspect within this context is investigating the thermo-physical properties of the system, as this enables for a proper thermodynamic evaluation of the process. However, in the high-pressure and high-temperature area reliable data on properties such as the heat capacity is very rare. In particular, excess effects due to mixtures of polymer and solvents have not been investigated. Thus, we apply transitiometry – a high-temperature, high-pressure calorimeter of high sensitivity – to investigate heat capacities at these extreme conditions.

By this, we do can significantly enhance the understanding of α-olefin-polymerizations and make another significant step towards simulation-based process optimization as well as process design.

[1] D. Read et al., *Science.* **2011**, 333(6051), 1871.

496928 Amine Effects on Radial-Mediated Thiol-Ene Reactions

Abstract:

Thiyl radical chemistry is of exceptional interest in biology and organic chemistry owing to the ubiquity of thiol containing moieties found in nature and the availability of a diverse range of alkyl and aryl thiols with similar reactivities. S-centered radicals undergo efficient, essentially barrierless compared to alkyl and alkoxy radicals, anti-Markovnikov addition into unsaturated X=C(R)R'bonds. Of particular interest is the thiyl radical reaction with alkenes, radical-mediated thiol-ene coupling (TEC), which is one of the so-called "click" reactions and has become a staple in polymerization and conjugation chemistry. TEC reactions may be performed under ambient conditions and without solvent, exhibiting rapid, often quantitative conversions, and an insensitivity to water and oxygen, and offers the ability for spatio-temporal control. Robustness of the thiol-ene reaction is amplified when used in concert with thiolate anion-mediated reactions with electron deficient alkenes (thiol-Michael coupling), isocyanates, thioesters, or epoxides to generate linear and network polymers with increasing compositional and topological complexity as these reactions may be performed orthogonally to thiol-ene reactive alkenes. In our present efforts, our group's focus is on fabricating polymers and networks with unique properties through the synergistic incorporation of thiyl radical and thiolate chemistries, including: sequence-ordered, DNA-mimetic polymers, covalent adaptable networks, and low dispersity oligomers and block copolymers.

Despite being well regarded as insensitive to most chemical functionalities present in organic and biological systems, we observed significant retardation in TEC reaction and reduced conversions (<50%) when performed in the presence of amines, which were incorporated into the formulation either as the polymer side-group functionality (e.g. nucleobases), buffering system, or catalyst for thiolate reactions in dual-cure networks. Interested in the mechanism of this phenomenon, we investigated the impact of amines, and ammonium and thiolate ions on the reaction kinetics of TEC reactions using real-time Fourier transform infrared (rt-FTIR) spectroscopy in order to expand the guidelines for developing thiol-ene systems and to further understand the nature of S-centered radicals in the context of general synthetic chemistry. We determined that amines indirectly retard radical-mediated thiol-ene reactions through the generation of the thiolate anion *in situ*, which sequesters the catalytic reactivity of the thiyl radical through a radical—lone pair, two-sulfur three-electron (2S-3e) interaction, effectively partitioning the radical population into an intermediate disulfide radical anion (DRA) species. The degree of retardation increases proportionally with the ratio of thiyl to carbon radicals present under steady-state reaction conditions, the concentration of thiolate anion, and depends highly on the structure of the thiolate anion and the resulting disulfide radical anion.

Solution phase, density functional theory (DFT) calculations were employed to validate the mechanism by estimating, and comparing to experimental results, the retardation activities of various thiolate anions on the TEC kinetics. Computation revealed that all DRA structures possible in our systems are metastable with respect to the thiol-ene product and that the electron withdrawing character of the parent thiol substituent is the main factor in determining the DRA's stability. Additionally, kinetic barriers to DRA formation were determined using parameters from Saveant

theory on electron-transfer (ET) reactions, where the DRA formation was treated as a step-wise, associative ET reaction. Interestingly, the relatively high viscosity and low polarity/dielectric character of the reaction medium led to the solvent reorganization energy to be the more contributive parameter to the kinetic barrier in the formation than the theory previously assumed; therefore, thiolate anions with a more rigid structure and lower internal reorganization requirements displayed minimal barriers to DRA formation, which is in agreement with experimental data.

The phenomena of amine/thiolate induced retardation discussed in this paper has important implications for the design of thiol-ene systems when the reaction is performed under alkaline conditions in aqueous environments or in the presence of basic moieties in polymer/organic synthesis. Although computations and experiments showed that the DRAs are, in general, thermodynamically favorable to form, near quantitative conversions may be obtained by choice of the reacting thiol and/or alkene substrate, i.e. using a thiol with low acidity or an alkene, like norbornene or vinyl ether, which have rapid curing kinetics and are not rate-limited by the thiyl radical propagation step, like allyl ether and vinyl siloxane. Further, the importance of the solvent reorganization energy on the rate of ET highlights that conventional ET theories may not adequately describe ET reactions in solvent environments that polymer chemists frequently encounter (i.e. organic, weakly polar, and viscous).

501163 Influence of Phosphate Salts and Solution pH on Aqueous-Phase NVP Free-Radical Polymerization

Abstract:

Background and Purpose:

Poly(N-vinylamides), such as N-vinyl pyrrolidone (NVP), have been extensively used as monomers in the synthesis of crosslinked hydrogels. NVP has been systematically used as an accelerator and co-monomer during free-radical polymerization of poly(ethylene) glycol diacrylate (PEGDA) hydrogels for biomedical applications. Our research has focused on the design and fabrication of monophosphate (Pi) and hexametaphosphate (PPi) loaded hydrogel nanoparticles copolymerized with PEGDA and NVP using inverse phase miniemulsion polymerization for applications in drug delivery. (1,2) Specifically, Pi- and PPi- loaded nanoparticles are utilized for sustained delivery of phosphates to the intestinal tract in an effort to suppress bacterial transition to virulence. (1) Even though aqueous-phase polymerization of NVP has been previously investigated, little is known regarding the effect of salts on the free-radical polymerization kinetics of NVP. NVP monomer and its radicals are believed to stabilize in aqueous solution by aggregating water molecules. *Santanakrishnan et al* (3) showed that the NVP propagation constant is dependent on monomer concentration in solution. The more dilute the monomer solution the higher the polymerization rate due to more favorable monomer/solvent(water) interactions as compared to monomer/monomer interactions. The addition of salts, including phosphates, to the monomer solution may also influence NVP polymerization kinetics as salts have a similar tendency to aggregate water molecules and to hydrate in aqueous solution. This phenomenon could cause a decrease in the number of water molecules available for interaction with NVP molecules resulting in increased monomer/monomer interactions and decreases in the propagation rate constant and the rate of polymerization. In theory, it would be possible to calculate an "effective" concentration of NVP that would match the concentration-dependent rate constant proposed by *Santanakrishnan et al* (3) to describe the system with the addition of salts.

Additionally, the pH of the precursor is also known to affect the rate of polymerization in these systems. NVP monomers have been previously shown to readily react under alkaline conditions. Furthermore, solution pH should have a significant impact on initiator dissociation kinetics. Many aqueous-phase free-radical initiators present a strong dependence on solution pH, being able to operate under a limited pH range. Thus, in selecting an initiator to polymerize NVP it is important to consider the pH effect on the dissociation of the specific initiator.

The goal of this study is to investigate the influence of two different phosphate salts (Pi and PPi), as well as variations in pH, on the kinetics of free-radical aqueous-phase NVP polymerization. These studies will provide insight to optimize the nanoparticle production process for targeted drug delivery of phosphates to the gut.

Methods:

In this study, potassium monophosphate and sodium hexametaphosphate were used. Potassium persulfate (KPS) and 2,2'-Azobis(2-methylpropionamidine) dihydrochloride (V-50) were used as free-radical initiators. Varying amounts of HCl and NaOH solutions were used to adjust the solution pH. The influence in the rate of polymerization by both the salts and the pH was measured over time by following conversion of NVP to poly-NVP.

Fixed amounts of salt and NVP were added to DI water in separate test tubes. After mixing and solubilization of the salt, each type of initiator was separately added to the solution. Samples were bubbled with nitrogen to prevent oxygen inhibition before the polymerization reaction proceeded in a water bath at constant 60ºC. At pre-determined times, the tubes were removed from the water bath, methyl hydroquinone was added to stop the polymerization reaction and the tubes stored at -80ºC until the samples were completely frozen. Samples were then lyophilized to remove water and unreacted NVP. The final mass of poly-NVP was then obtained and monomer conversion calculated as a function of time.

Results:

Initial experiments were conducted using the initiator KPS and revealed interesting interactions at both acidic and alkaline pH. The experiments showed no formation of poly-NVP. This was attributed to the fact that KPS tends to make the solution very acidic, along with the fact that NVP does not react under acidic conditions. With that in mind, pH adjustments were made to bring the pH back to neutral after the addition of KPS. Even with these adjustments no polymerization of NVP was observed. This was most likely attributed to the fact that KPS does not dissociate under neutral to basic conditions. We therefore identified a different initiator, V-50 as being ideal for aqueous phase polymerization of NVP.

Preliminary results for the aqueous NVP polymerization initiated by V-50 were obtained over a range of pH (3-13). These experiments indicated a clear correlation between the solution pH and the rate of reaction. Our data indicate that there is an optimum pH for the aqueous polymerization of NVP with the V-50 initiator. As in the case of KPS, the V-50 dissociation rate was also dependent on pH, with increases in pH decreasing the initiator dissociation constant and subsequently affecting the overall polymerization reaction rate.

Results on the kinetics of the monomer solution in the presence of the monophosphate salt indicate a slight decrease in reaction rate (on average 9.6% slower as compared to pure NVP kinetics). This effect was more pronounced initially being about 26% slower in the first 15 minutes of reaction. In the case of NVP kinetics in the presence of sodium hexametaphosphate (PPi) a similar but accentuated trend is observed. In the presence of PPi, there is an even greater decrease in reaction rate (~ 15.3% compared to the kinetics of the pure NVP) with ~ 47% slower kinetics in the first 15 minutes. These preliminary results confirm our hypothesis that the addition of salts influences the kinetics of the aqueous-phase NVP polymerization.

The polymerization model reported by *Santanakrishnan et al* was used as a reference for comparison. The model was shown to fit our experimental data and used to obtain the initiator efficiency.

Additional studies are being conducted to determine an "effective" NVP concentration after the addition of the salts and the optimum pH for NVP homopolymerization and to extend these findings to synthesize a stable emulsion using V-50 as the initiator to create Pi and PPi loaded nanoparticles.

References:

(1) Yin, Y. *et al*, "De Novo Synthesis and Functional Analysis of Polyphosphate-Loaded Poly(Ethylene) Glycol Hydrogel Nanoparticles Targeting Pyocyanin and Pyoverdin Production in Pseudomonas aeruginosa as a Model Intestinal Pathogen", Annals of Biomedical Engineering, **45** (4), pp. 1058-1068 (Apr. 2017)

(2) Vadlamudi, S. *et al.*, "Inverse miniemulsion polymerization of phosphate-loaded hydrogel nanoparticles for sepsis prevention", Unpublished master dissertation, Illinois Institute of Technology, Chicago, Illinois (2014)

(3) Santanakrishnan, S., *et al*, "Kinetics and Modeling of Batch and Semibatch Aqueous-Phase NVP Free-Radical Polymerization", Macromol. React. Eng., **4**, pp. 499–509 (2010)

497911 Synthesis and Characterization of Crosslinked Polymers from Cottonseed Oil

Abstract:

Traditional plastics are usually obtained from petroleum feedstocks. The petroleum-based plastics are not readily degradable and can stay in landfills for long time without any significant deterioration. In this study, a crosslinked polymer was synthesized from cottonseed oil. Unsaturated fatty acids, the major components of cottonseed oils, were initially epoxidized. Average epoxides content was estimated to be about 5.5% with yield of 87%, as determined by ^{1}H-NMR technique. A network polymer was then formed by crosslinking the epoxidized oil with maleic anhydrate. Mechanical properties of these polymers were altered by varying the amount of maleic anhydrate and the modulus increased with increasing maleic anhydride content. Polymerization method exhibited yield in the range of 68-80%. Our results provide a new synthetic strategy to obtain polymers from cottonseed oil.

503467 Date Pits as Cost-Effective, Renewable, and Efficient Fillers for Polymers

Abstract:

In this research, Date pits, by product with less value of the palm trees (Phoenix dactylifera L.), were successfully used as efficient and eco-friendly fillers and reinforcement for polymeric materials. A polymeric used to host date pits granules was general purpose polystyrene (PS). PS pellets were melt compounded with proper amount of date pits granules by extrusion. morphological and mechanical properties of the prepared composites were investigated to explore effect of date pit particle size on the properties of the composites. It was found that mechanical properties represented by Izod impact strength may be significantly enhanced by controlling date pit particle size within the range of 0.25-1.00 mm.

500001 Biaxial and Shear Deformation of Simulated Amorphous Cis-, Trans-1, 4-Polybutadiene Chains

Abstract:

Rolling resistance in rubber tires leads to losses in vehicle fuel economy. We hypothesize the rolling resistance on the macroscale connects directly to deformation-induced changes in elastomer chain conformations on the microscale through changes in elastic free energy. To study this phenomenon, random chain conformations were generated using Floryâ€™s Rotational Isomeric State (RIS) Approach, which weights particular torsional states along the backbone of polymer chains. Polybutadiene chains were generated under unperturbed conditions where chain size and shape characteristics were studied. Ensembles of unperturbed chains were then subjected to three separate deformation conditions: uniaxial along the x-direction, biaxial, and shear. For comparative analysis, Gaussian chains were generated and RIS chains were analyzed against them. The end-to-end distance vectors of the chains were computed before and after deformation, and the probability distribution of these vectors led to computations of elastic free energy. Changes in elastic free energy were studied for chain vectors under deformation at a single temperature with varying number of repeat units and for a single number of repeat units at varying temperatures. Change in elastic free energy of a system increased with increasing deformation. At a particular deformation, RIS chains showed higher elastic free energy changes than Gaussian chains. These changes in elastic free energy were then quantified to compute deformation force and stress acting on ensembles of chains. Good agreement was observed between deformation forces computed numerically and analytically for Gaussian chains. Under all conditions, RIS chains showed higher deformation force and stress as compared to Gaussian chains, and cis chains showed higher deformation force and stress than trans chains. For chains at a single temperature and varying number of repeat units, deformation force and stress decreased with increasing number of repeat units. For chains at a single repeat unit with varying temperature, the deformation force and stress decreased with increasing temperature. This approach allows for including specific chemistry effects on chain conformation and deformation by analyzing the stress-strain behavior of chains that surround filler particles.

500422 Decomposition Behavior of Laponite/PLGA-PÎ•G-PLGA Nanocomposite Hydrogels at Body Temperature

Abstract:

PLGA-PEG-PLGA (poly ($_{D,L}$-lactide-co-glycolide)-b-poly (ethylene glycol)-b-poly ($_{D,L}$-lactide-co-glycolide)) is one of the promising block-copolymers exhibiting interesting thermoresponsive sol-gel transitions and biodegradability. We have developed thermoresponsive hydrogels using PLGA-PEG-PLGA possessing a high PEG/PLGA ratio with laponite clay nanoparticles (laponite/PLGA-PEG-PLGA nanocomposites) for the drug delivery system (DDS) [1, 2]. The laponite/PLGA-PEG-PLGA nanocomposites presented a thermoresponsive sol-gel transition at the total solute concentration of 3.75 wt%, while the PLGA-PEG-PLGA single-body solution presented a sol-gel transition at the solution concentration of 15 wt% or higher at body temperature (37°C). To adjust the degradation rate for the control of the drug-release rate in the DDS, the decomposition behavior should be studied comprehensively. In this work, the decomposition behavior including the decomposition rate was investigated for the laponite/PLGA-PEG-PLGA nanocomposite hydrogels. The PLGA-PEG-PLGA triblock copolymers of the same molecular weight (600-1000-600) but with different LA/GA ratios (1.1, 4.0, and 8.8) were used for the experiments. The decomposition behavior was examined by analyzing the mass loss of the nanocomposites and by analyzing the composition of the remaining gels by NMR. The mass losses of the laponite/PLGA-PEG-PLGA nanocomposite gels with the LA/GA ratio of 4.0 and the LA/GA ratio of 8.8 were ~30% after 10 days for both samples, while the mass loss of the laponite/PLGA-PEG-PLGA nanocomposite gels with the LA/GA ratio of 1.1 was ~45% after 10 days. From the analyses by the NMR spectra, it was found that there was an increase in the LA content, a decrease in the PEG content, with no remarkable change in the GA content, regardless of the LA/GA ratios. Thus it was considered that more hydrophilic PEG blocks were preferentially diffused out of the nanocomposite gel during the decomposition, and that LA and GA remained in the nanocomposite hydrogels. If we consider the composition change of the nanocomposite hydrogels analyzed by NMR as the decomposition rate, the decomposition rate of the nanocomposite hydrogels with LA/GA=1.1 was higher than those with LA/GA=4.0 and LA/GA=8.8.

[1] Naho Oyama, Hiromasa Minami, Daichi Kawano, Makoto Miyazaki, Tomoki Maeda, Kazunori Toma, Atsushi Hotta, Koji Nagahama, A nanocomposite approach to develop biodegradable thermogels exhibiting excellent cell-compatibility for injectable cell delivery, Biomaterials Science, Vol. 2 (8), 1057-1062 (2014).

[2] Makoto Miyazaki, Tomoki Maeda, Kenji Hirashima, Naruki Kurokawa, Koji Nagahama, Atsushi Hotta, PEG-based nanocomposite hydrogel: Thermoresponsive sol-gel transition controlled by PLGA-PEG-PLGA molecular weight and solute concentration, Polymer, Vol. 115, 246-254 (2017).

501164 Synthesis of Thermoplastic Polydimethylsiloxane with L-Phenylalanine-Based Hydrogen-Bond Network and its Self-Healing Property

Abstract:

Polydimethylsiloxane (PDMS) with L-phenylalanine was found to possess thermoplastic property and self-healing property due to the crosslinkings through hydrogen bonding. The PDMS with L-phenylalanine was synthesized in two steps. The amino-acid terminated PDMS was first synthesized by adding L-phenylalanine to the end of amino-propyl terminated PDMS, followed by the condensation polymerization of the amino-acid terminated PDMS. In order to investigate the mechanical and the self-healing properties of PDMS with L-phenylalanine, the tensile testing and the dynamic mechanical analysis (DMA) were conducted. It was found by the tensile testing that the Young's modulus of the synthesized PDMS increased as the amount of L-phenylalanine increased. Especially, the Young's modulus of the thermoplastic PDMS containing L-phenylalanine was at most about 4 times higher than that of the thermoplastic PDMS without L-phenylalanine. From the DMA analyses, it was found that the synthesized PDMS with L-phenylalanine presented a relatively high storage modulus (~100 kPa) at around room temperature, which gradually decreased down to ~20 kPa at higher temperature of 150°C. The melting temperature of thermoplastic PDMS with L-phenylalanine was also confirmed to be ~120°C. Moreover, in order to investigate the self-healing property, the tensile testing was conducted with the self-healed sample as well as the original sample. The self-healed sample was prepared by cutting it into two pieces and by contacting the pieces at room temperature and leaving them for 24 h for the self-healing. Considering the Young's moduli of the self-healed sample and the original sample, the PDMS was found to possess the self-healing property.. The Young's modulus of the self-healed PDMS could maintain at least 70% of the Young's modulus of the original sample. In detail, the Young's modulus of the self-healed samples of PDMS without L-phenylalanine and that of the PDMS with L-phenylalanine were found to be 139 kPa and 856 kPa, respectively, while the Young's moduli of the original samples of PDMS without L-phenylalanine and the PDMS containing L-phenylalanine were 190 kPa and 790 kPa, respectively. (n=3)

CONTROLLED SWELLING RATE ELASTOMER FOR PACKERS

Rostyslav Dolog, Darryl Ventura, Qusai Darugar, and Valery Khabashesku, Baker Hughes, a GE Company, Houston, TX

Abstract

Packers are downhole tools, common to the oil and gas industry, that isolate wellbore zones during the completion of the well. This isolation enables a controlled production of hydrocarbons and safe well-treatment operations. Packers comprise an elastomeric packing element that expands upon tool actuation, filling the gap between the tool and the annulus (or casing) and isolating different zones. Reactive packing elements expand downhole due to the swelling in downhole fluids and do not require a setting force to deform and energize the elastomer.

Elastomers for reactive packing elements are designed to swell in water-based brines or hydrocarbons. Depending on the material design, the swelling rate of these elements is determined by environmental parameters such as temperature and fluids. This research focused on improving the control over the rate of elastomer swelling in downhole fluids. Slowing or delaying swelling of the packing element at the initial stages of deployment can provide additional time for a packer to reach its destination downhole, eliminating risks due to unforeseen delays in tool deployment and leading to improved reliability.

Introduction

Swellable elastomer compounds are used in reactive packer (RE) applications. These packers use two general types of swellable materials:

- Oil-swellable compounds, usually based on EPDM (ethylene propylene diene monomer rubber) elastomer.
- Water-swellable compounds, usually based on nitrile elastomers. These compounds comprise a super-absorbent polymer (SAP) component that serves as an active additive, enabling the elastomer to swell in water brines (nitrile elastomers are inherently resistant to water swelling).

RE packers are deployed downhole and do not require a setting force or tool to energize them. They swell in downhole fluids (oil or water, as designed), isolating the borehole zones. A challenge exist with controlling their rate of swelling, which currently depends on the temperature and type of downhole fluid deployed. There is a risk of premature RE packer swelling (and thus, setting) in long wellbore sections, before it reaches its destination, resulting in lengthy and expensive repairs of the completion setup. Consequently, slowing or delaying the swelling of the RE packer at the initial stages of deployment can provide enough time for a packer to reach its destination downhole.

Control of swelling in oil-swellable compounds

We proposed a solution to slow down the swelling of the oil-swellable compound in oil. We suggested to functionalize the surface of the rubber product with a polar group to make it repulsive towards oil and thus delay the swelling. Because only the outer layer will be oil repulsive, the matrix can swell at a lower, accelerating rate. When the elastomer begins to expand, the overall ratio of the polar groups to non-polar groups on the surface of the product decreases. Consequently, the material can expand to the designed limit, but the expansion process will be slower on the early stage of deployment, providing enough time for the tool operator to relocate packing element where needed.

Two types of functionalization reactions were conducted on 1-in. x 0.5-in. cured EPDM samples ("buttons"): sulfonation and fluorination.

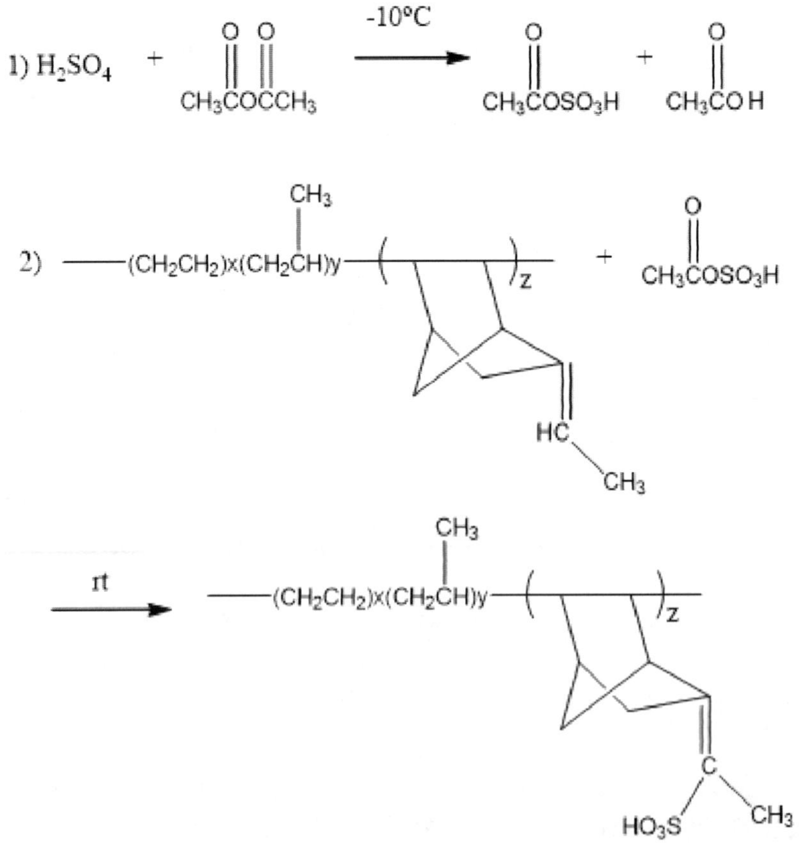

Figure 1. Reaction scheme for (1) formation of acetyl sulfate and (2) sulfonation of EPDM.[1]

[1] F. Barroso-Bujans et al./ Acta Materialia 56(2008) 4780-4788

As a result of fluorination, fluorine either substituted hydrogen atoms or attached to a double carbon-carbon bonds of the elastomer. Sulfonation followed the procedure shown in Figure 1. The buttons were sulfonated for 6 hours at room temperature.

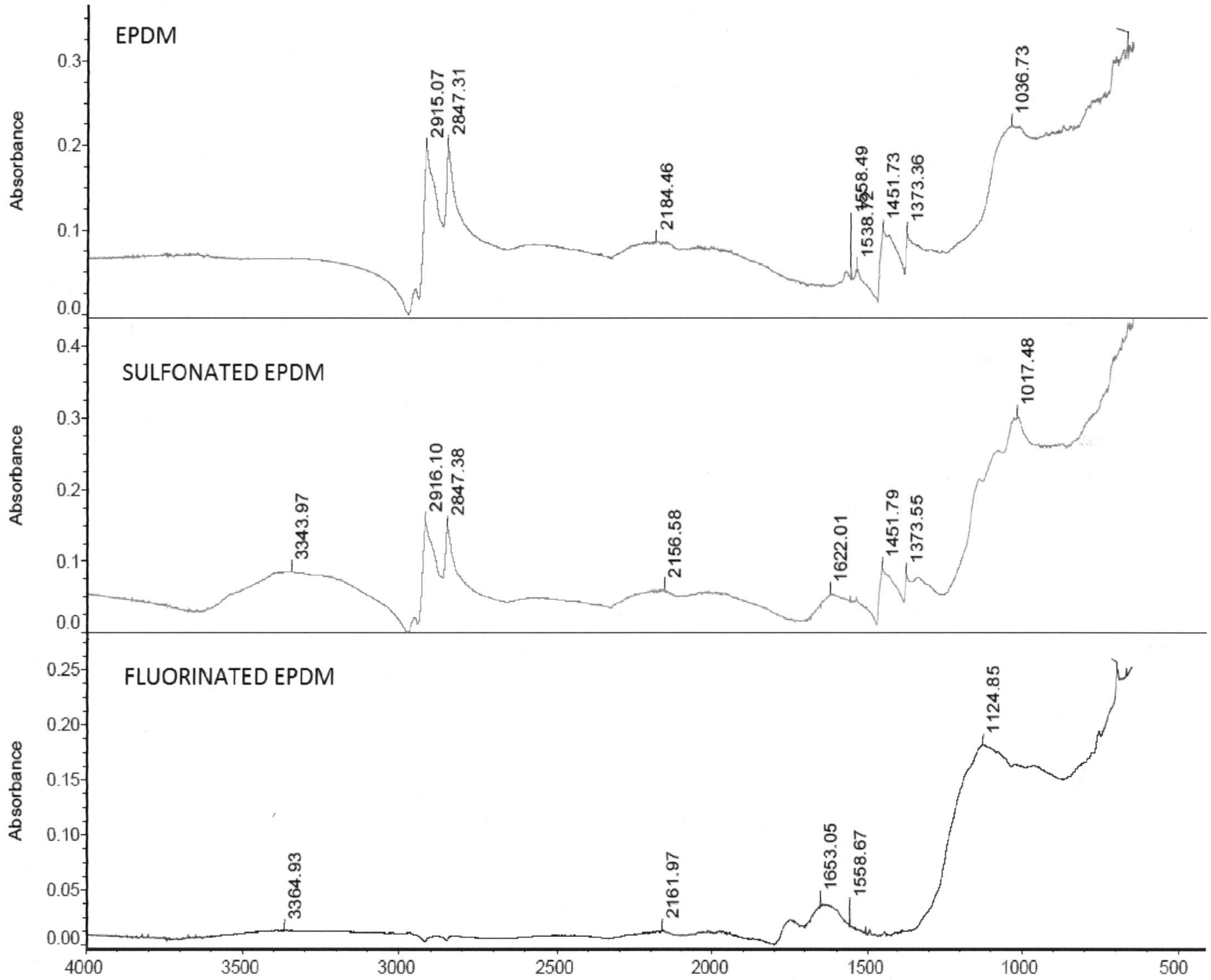

Figure 2. FTIR spectra of EPDM and functionalized EPDM

The efficiency of the functionalization was assessed via Fourier-transform infrared spectroscopy (FTIR) and energy-dispersive X-ray spectroscopy (EDS). EDS demonstrated that sulfonation and fluorination resulted in approximately 1% and 30% EPDM surface functionalization, respectively, as shown in Table 1. FTIR results are provided in Figure 2.

Furthermore, the FTIR study also clearly demonstrated the efficiency of surface fluorination, when comparing the IR spectra of neat EPDM with the IR spectra of the fluorinated EPDM: C-H stretching vibrations from –CH2- at ~2915 cm^{-1} and ~2847 cm^{-1} disappeared and a large broad peak appeared, corresponding to C-F bond stretching at ~1125 cm^{-1}.

Table 1. Summary of Elemental Analysis via EDS

EPDM			EPDM-SO3H			EPDM-F		
Element	Wt.%	Atomic%	Element	Wt.%	Atomic%	Element	Wt.%	Atomic%
C (K)	67	71.4	C (K)	88.8	91.8	C (K)	51.4	63.3
N (K)	18.3	16.7	O (K)	10	7.8	O (K)	4.5	4.2
O (K)	14.8	11.9	S (K)	1.1	0.4	F (K)	40.8	31.8
S (K)	0	0				Zn (L)	3.4	0.8

The results of the sulfonation study in FTIR were less conclusive. Often, sulfonic groups can be determined by the water absorbed due to $-SO_3H$ groups as well as by peaks that correspond directly to the $-SO_3H$ group. For example, the broad band of O-H stretching that corresponds to hydrogen bonding at ~3200-3600 cm^{-1}, which we observe at ~3344 cm^{-1} (Sulfonated EPDM), is an indication that water bonded to the surface after sulfonation. The peak at ~1622 cm^{-1} may also be attributed to H-O-H deformation mode of physically adsorbed water. However, the peak due to stretching of O=S=O bonds, which usually appears in the 1030-1060 cm^{-1} range, was not observed, possibly due to low sulfonation level and overlapping with other peaks.

A relatively small decrease in the swelling rate demonstrates the proof of concept, and this approach can be further developed and improved by increasing polarity of the surface. It is important to note that a decrease in the rate of swelling should not affect total final swelling capability of the packing element compound, and after ~2-7 days material must swell to the designed volume.

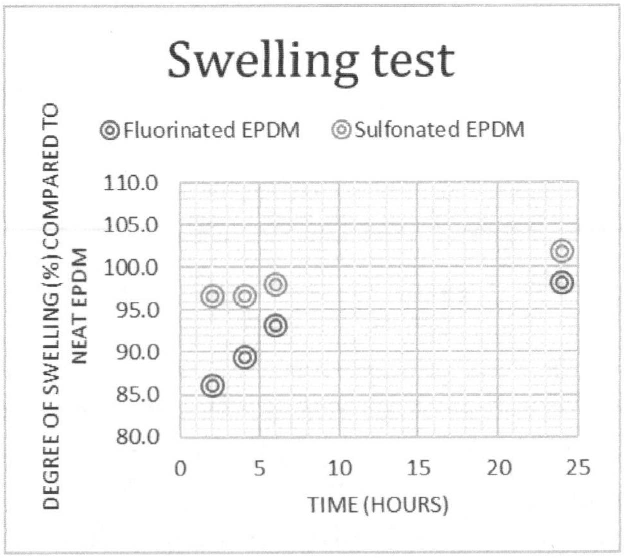

Figure 3. Swelling of surface functionalized buttons as a degree (%) of swelling of a neat button at the select point of time.

Unfunctionalized EPDM buttons, surface-sulfonated EPDM buttons, and surface-fluorinated EPDM buttons were tested for swelling performance. Week-long swelling tests were conducted at 200°F in LVT 200 oil to simulate downhole production fluid. The volume of the buttons was measured using a densimeter after 2, 4, 6, 24, 48, and 96 hours of swelling, and the results are shown in Figure 3. From this figure it is shown that functionalized buttons swelled 5-15% slower than neat buttons. The swelling rate can be slowed further by increasing the degree of functionalization (sulfonation). A more significant decay in swelling rate is desired.

Control of swelling in water-swellable compounds

The swelling performance of water-swellable compounds greatly depends on the salinity of the brine and the salt type that is used to produce the brine, e.g., their swelling ability in the brines based on salts of multivalent cations is several times lower than in KCl or NaCl brines.[2] Swelling also decreases with the increasing salt concentration. This effect can be explained by the dependence of the swelling of super-adsorbent polymers, used in swellable rubbers as a swellable component, on the pH, the ionic strength and the presence of certain cations in the solution. The issue becomes increasingly important as industry is shifting towards high-temperature/high-pressure (HTHP) applications that require heavier brines typically based on halides of bivalent cations.

We suggest resolving the issue by incorporating an additive to the rubber that can precipitate the salt from the solution that penetrated the rubber. To mitigate the effect of multivalent cations (e.g., Ca^{2+} or Zn^{2+}), one can chose a soluble inorganic salt as an additive, such as its cation and anion can form insoluble compounds with the cation of the brine, anion of the brine, or both. In addition, organic acid and/or salt could also be used to precipitate the salt from the brine inside the swellable rubber compound or act as a chelating agent.

Various compounds could be used for the purposes described above. To evaluate this hypothesis, Sodium Citrate was used as an additive to the water-swellable compound. This salt is inexpensive, commercially available, and thermally stable at temperature up to 350°F,

In this example, a modified swellable compound comprised a mixture of base compound (80 wt.%) and sodium citrate (20 wt.%). Both formulations, modified compound and a reference base compound, were molded and cured at 300°F for 75 minutes. The swelling experiment was conducted at 200°F in 10% $CaCl_2$ brine. The volume of the rubber buttons was measured daily. After four days, the modified compound demonstrated more than a double volume increase compared to the neat compound, essentially demonstrating the proof of concept. This increase was achieved due to the ability of sodium citrate to bind calcium cations by chelating and/or an ion-exchange mechanism. The results are shown in Figure 4. This concept can be expanded to other chelating agents or precipitating agents that can bind calcium or other multivalent cations present in brines.

[2] Water swelling rubber compound for use in reactive packers and other downhole tools, US8181708 B2, J.R. Korte et.al.

Furthermore, compression tests on rubber buttons demonstrated that sodium citrate provided significant reinforcement to the compound in the pre-swollen state, see Table 2.

Table 2. Compression Test Results

Sample	Compression Modulus (PSI)				
	C_{10}	C_{20}	C_{30}	C_{50}	C_{70}
Base compound	72	153	234	651	2929
Modified compound	317	538	815	2708	5725

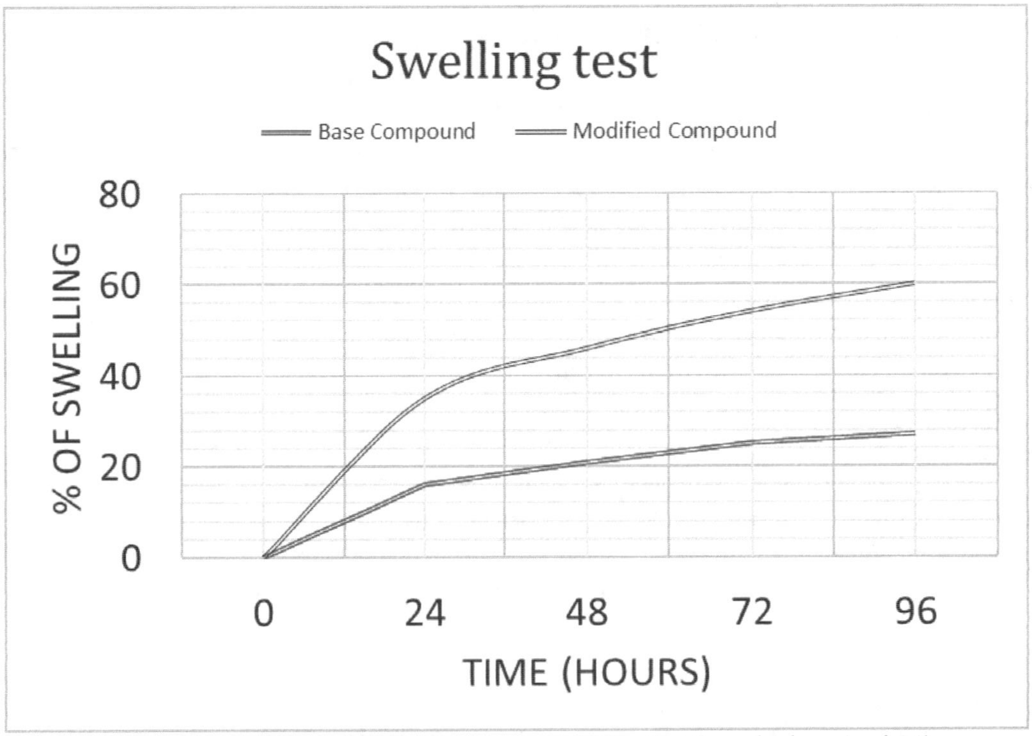

Figure 4. Swelling of compounds in 10% $CaCl_2$ solution (vol.%)

Conclusions

Swellable elastomer technology is actively implemented in reactive packing elements that are widely used across the industry and are common for open case completions. Oil-swellable elastomers and water-swellable elastomers are based on different technologies: the former uses the affinity of cross-linked elastomers to oil, while the latter introduces an active component, prone to swelling in water, to the inert water-resistant elastomers. This research focused on development of methods to control swelling of elastomers in different fluids to improve technology versatility and expand their application.

491058 Effect of the Cross-Linking Agent (sodium polyphosphate) on Performances of Nacs-Wsc Microcapsules

Abstract:

Effect of the cross-linking agent (sodium polyphosphate) on performances of NaCS-WSC microcapsules

Qing-Xi Wu[1,2], Yi-Xin Guan[1], Jun-Jie Yuan[1], and Shan-Jing Yao[1*]

([1]College of Chemical and Biological Engineering, Zhejiang University, Hangzhou 310027, China; [2]School of Life Sciences, Anhui University, Hefei 230601, China)

Microcarriers prepared with polyelectrolyte complexes (PECs) using natural polysaccharides as raw materials have attracted great attentions due to their mild process conditions, favorable properties and vital applications in biomedical and bioengineering fields. Among which, chitosan-related polysaccharides such as water soluble chitosan (WSC) and cellulose-related substances such as sodium cellulose sulfate (NaCS) are representatives of challenging biomaterials. However, the poor mechanical property of the PECs becomes the obstacle encountered in applications.

In the previous work, we found that small cross-linking agent like sodium polyphosphate (PPS) had evident effects on microcapsule¡¯s morphology. Thus, it¡¯s important to study the PEC formation process under conditions of using cross-linking agent. In this work, the effect of PPS on performances of NaCS-WSC microcapsules was investigated in detail, while NaCS-Chitosan (CS) microcapsules were chosen as the control. Compared with the microcapsules that formed without cross-linking agent, PPS as the cross-linking agent could penetrate through the PEC film and form tighter interior structures. The NaCS-WSC-PPS microcapsules had lower swelling ratios which may be ascribed to their tighter interior microstructures compared with NaCS-WSC microcapsules. PPS also had important effect on the total mass of PECs produced; and the decline of zeta potential of NaCS-WSC-PPS microcapsules was lower than that of NaCS-WSC microcapsules, indicating PPS participated in the interactions and played an important role in the microcapsules¡¯ formation process. Similar results were found in the controls. Due to the unique physicochemical properties of WSC and CS, the NaCS-WSC microcapsules and NaCS-CS microcapsules that were prepared with or without PPS had distinct microstructures. As is known, CS can only be dissolved in water under acidic solutions, while WSC can be directly dissolved and protonated in acid-free aqueous providing ¨CNH_3^+ groups quickly, which resulted in the microstructure¡¯s difference.

The NaCS-WSC microcapsules prepared with the PECs could be good candidates as enzyme-triggered controlled release drug carriers targeted for colon. Therefore, it may be a strategy to improve the mechanical properties and performances of the PEC microcapsules using PPS as the cross-linking agent.

Keywords: Polyelectrolyte complexes; Cross-linking agent; Microcarriers

[1]

Supported by the National Natural Science Foundation of China and Natural Science Foundation of Anhui Province (CN) (1708085QC64)

Tel: +86-571-87951982, E-mail: yaosj@zju.edu.cn

Swelling Behaviors of Cr^{3+}-modified Acrylamide-Based Superabsorbent Polymer Microsphere in Brines

Jingyang Pu, Jiaming Geng, Na Zhang and Baojun Bai
Missouri University of Science and Technology, Rolla, MO, US.

Abstract

Ion-specificity effect between cations existed in aqueous system and hydrophilic polymer gel is a typical limitation and challenge for hydrophilic polymer utilization during tertiary oil recovery. This study presents a novel cation-insensitive microsphere system (M-SAP) which is developed with controllable initial size and stable swelling ratio under varying salinity solvents. Polymerization behaviors were depended on varying concentration of each compounds and the result shows that the microsphere formed via 6.36mM N, N'-methylene bisacrylamide initiated by 0.859mM ammonium persulfate shows most stable structures and possess ideal sphericity (sample #2). Instead of directly crosslinking the polymer as other references, the utilization of sealed chromium cation (Cr^{3+}) as a shield of M-SAP to prevent ion-specificity effect from out-coming cation is reported. The micro-structure of M-SAPs were discussed which helped analyze Cr^{3+} preserving and shielding mechanism. Cr^{3+} preserving and shielding mechanism can be interpreted as that Cr^{3+} helps eliminate the electrostatic repulsions on carboxylate group from out-coming cations.

Introduction

Hydrophilic polymer gels, also named as superabsorbent polymer (SAP), have generated extensive commercial and academic interest in recent years. In existed hydrophilic polymer gel systems, the gels swell, collapse and even show volumetric phase transition in response to changes of surrounding environments, e.g. temperature[1,2], solvent composition[3,4], ion valence and concentration[5] and pH[6] of the immersing solution.

Among them, ion-specificity behaviors which observed in aqueous system for polymer gel are an open problem. In some studies the effect of cations to polymer gel swelling are interpreted by testing the swelling behavior of the polymer gel in varying solvents. It is found that different extent reduction of final swollen size[7,8] was detected for charged cations[9]. For example, calcium cation (Ca^{2+}) has a marginally greater effect per mole than the inorganic monovalent cations. while the effectiveness of the trivalent cation (e.g. aluminum (Al)) is appreciably bigger, These free cations decrease the specific levorotation[10] more effectively than the inorganic monovalent cation, with the molar effectiveness increasing linearly. Because of facile synthesis process, low price and its tunable property positioned, ion-specific polymer or gel as a promising material for various applications, ranging from poisonous ion remove[11] to fuel cell[12]. Moreover, ion-specific polymer system represents a large family of materials where the metal/polymer type or ratio and spatial organization can be tuned during polymerization, which in turn alter their phase transition, rheology and thermodynamic properties.

Gel treatments have been used as chemical conformance improvement methods to increase sweeping efficiency of mature reservoirs and control high water cut. Ion-specificity effect between cation and polyacrylamide-based gel is a typical limitation and challenge for hydrophilic polymer utilization during tertiary oil recovery. Preformed microspheres (a kind of SAP gel) have been successfully and economically applied to control water production and increase oil production[13-15]. The success degree of this microsphere treatment depends heavily on the stability of microsphere initial size and swelling. However, most studies interpreted ion-specificity to these microspheres swelling in the aqueous environment, but merely mentioned possibility for inhibiting this phenomenon. Although the influence of metal cations in solvent is disclosed in some studies[16,17], to our knowledge, there is no comprehensive fabrication and evaluation study to overcome cations effect on the polyacrylic chain and microsphere swelling behavior.

Some cations has been utilized as crosslinker for bridging polymer chains during polymerization reaction, e.g., of xanthan[18], chitosan[19] and polyacrylamide[20]. The positive result shown in these studies provides a possibility of pre-loading cations into SAP. Herein, we report a facile and simple one-step procedure to prepare chromium acetate mixed microsphere (M-SAP) with very stable swelling ratio compared to a reference microsphere. However, the purpose of adding these cations is to reduce ion-specificity effect from solvent instead of crosslinking the polymers. Our results show that most chromium ions (Cr^{3+}) are sealed in the microsphere and functions indeed as the shield by studying the swollen size distribution of the microsphere in varying salinities. The mechanism of shielding was probed. Furthermore, we carefully studied the effect of crosslinker concentration (N, N'-methylene bisacrylamide), initiator (ammonium persulfate) and stirring rate during suspension polymerization.

Experimental

Materials:

Acrylamide (AM, purity above >98%), chromium acetate ($(CH_3COO)_7(OH)_2$, [Cr^{3+}] 24%) and n-Decane (purity above 99%, a continuous phase) were all purchased from Alfa Aesar (USA). N, N'-methylene bisacrylamide (MBAA, purity above 99%) and Span80 (viscosity among 1200~2000 mPa·s) were obtained from Sigma Aldrich (USA). Ammonium persulfate (APS, purity above 98%) and acetone (ACS grade) were supplied by Fisher Scientific (USA). All reagents were used without any further purification. Deionized water (DIW) was used for the preparation of all aqueous solutions.

Polymerization:

Reagent compositions were listed in Table 1. Suspension polymerization of M-SAP was carried out in a three-neck flask. The flask was equipped with a mechanical stirrer, a reflux condenser, a thermometer and a nitrogen inlet, and was added dropwise to a constant temperature water bath. In suspension polymerization, n-Decane was used as continuous phase with Span-80 added (nonionic polymeric surfactant). A synthesis example (Sample #1) for M-SAP is supplied: AM (20 g, 0.281M), MBA (0.1 g, 0.649 mM) and chromium acetate (0.02 g, 0.053 mM) were dissolved in 20mL of distilled water as water phase. Air was excluded from the aqueous solution by introducing nitrogen for at least 20 minutes. After mixing with APS (0.026 g, 0.114mM), AM aqueous solution was then immediately poured into the 50 mL of n-Decane containing Span 80 (2 g, 4.667 mM). Air was flushed from the reactor by introducing nitrogen until the entire process was completed. Polymerization was allowed to proceed under 50°C with stirring rate of 500 RPM for 3h, after which stirring was stopped and the precipitate was washed by acetone (100 g, 1.722 M). The product was dried in a vacuum oven at 50°C for 24 hours to give as a white powder. Other components didn't mentioned in Table 2 were kept constant (20mL DIW, 20mL n-Decane and 2mL Span80, 500rpm).

Table 1. Contents of M-SAP samples

Sample NO.	[AM] (M)	[MBAA] (mM)	[Cr] (mM)	[APS] (mM)
#1	2.76	3.18	0.521	0.859
#2	2.76	6.36	0.651	0.859
#3	2.76	6.36	0.781	0.859

Initial Size Determination:

The morphology and initial size distribution of M-SAP samples were evaluated by a scanning electron microscope (SEM) (Hitachi S4700, Tokyo, Japan. Dry M-SAP which was used to calculate the initial M-SAP diameters time. All samples were sputter-coated with gold for 120 seconds before observation.

Swelling Equilibrium Measurement in Varying Salinities:

Size distribution and mean size average of swollen SAPG were measured using a laser scattering microsphere size distribution analyzer Microtrac 3000 (LSPSDA) at 23°C. Samples with varying Cr^{3+} concentration were used to study the effect of Cr^{3+} on equilibrium swelling M-SAP or equilibrium swelled (ES) M-SAP size. The measurements were performed in a cation concentration controlled aqueous environment. sodium chloride (Na^{+}) solution, calcium chloride (Ca^{2+}) solution, aluminum chloride (Al^{3+}) solution with different cation concentration (0.01 M, 0.1 M and 0.5 M) were prepared and separated into two parts (solution 1 and solution 2). Microspheres were measured and put into the different brine solution (solution 1) for fully swelling for 24 hours. Solution 2 was drained by instrument's cell as the aqueous environment. During the measurement, a certain amount of microsphere solutions (solution 1) were added into the cell after each drainage stage. The environmental temperature has been controlled under static conditions at 23°C.

Results and Discussion

ES behavior in varying salinity solutions:

Swelling is one of the most important properties of SAP. For regular polyacrylamide-based microsphere, cation type and intensity affect the swelling. Addition of charges by Cr^{3+} in the M-SAP is expected to maintain the swelling ratio of SAP since the new osmatic pressure balance approached. Detailed studies of the M-SAP equilibrium swelling in solvent were interpreted by investigating M-SAP equilibrium swelling sizes in varying salinities and compared these result with N-SAP performances under the same situations. Initial size 11.4 μm (±0.3 μm) M-SAP sample #2 were selected for this comparison experiment.

After put into solvent for 24 hours, swollen M-SAP size peaks distributions and average size distributions as a function of cation concentration for different types of solvent were shown in Figure 1. It can be observed that for M-SAP swelled in Na^{+}, Ca^{2+} or Al^{3+} solvent, no peaks left shit phenomenon appeared with increasing specific cation concentration of saline solutions which mean M-SAPs' hydrodynamic size distributions didn't reduce (Figure 1a-1c). When swelling of M-SAP were performed in these solvent, no significant change was observed, which proves that in the specific controlled range of cation concentration (0.01 M-0.5 M) M-SAP equilibrium swelling is independent to solvent types, different from the SAP mentioned in introduction whose ES sizes were depended on the salinity of solution[15]. Average size distribution of M-SAP indicated that M-SAP swelling followed polymer inhomogeneity swelling theory as shown in Figure 1d. When applied to a solvent, the tie-line resulting from the Maxwell construction represents a region of coexistence of phases, which means the swelling transition region could represent a region of coexistence of phases. It's conceivable that when a microsphere is partially swollen, a core-shell structure will be formed[21]. Inhomogeneity has been identified in the structure of the microsphere of various compositions in other studies[22]. Co-existence of phases allows the pressure to remain constant as the volume changes in a short. This theory can be used to explain the slightly increasing of hydrodynamic sizes of microsphere measured in 0.1% saline solvent or higher. After swelling in the solvent, inner free water cation concentration of M-SAP was higher than original saline solutions. Once put them back to the same solvent, referring to osmotic pressure change, the hydrodynamic size of microsphere was larger.

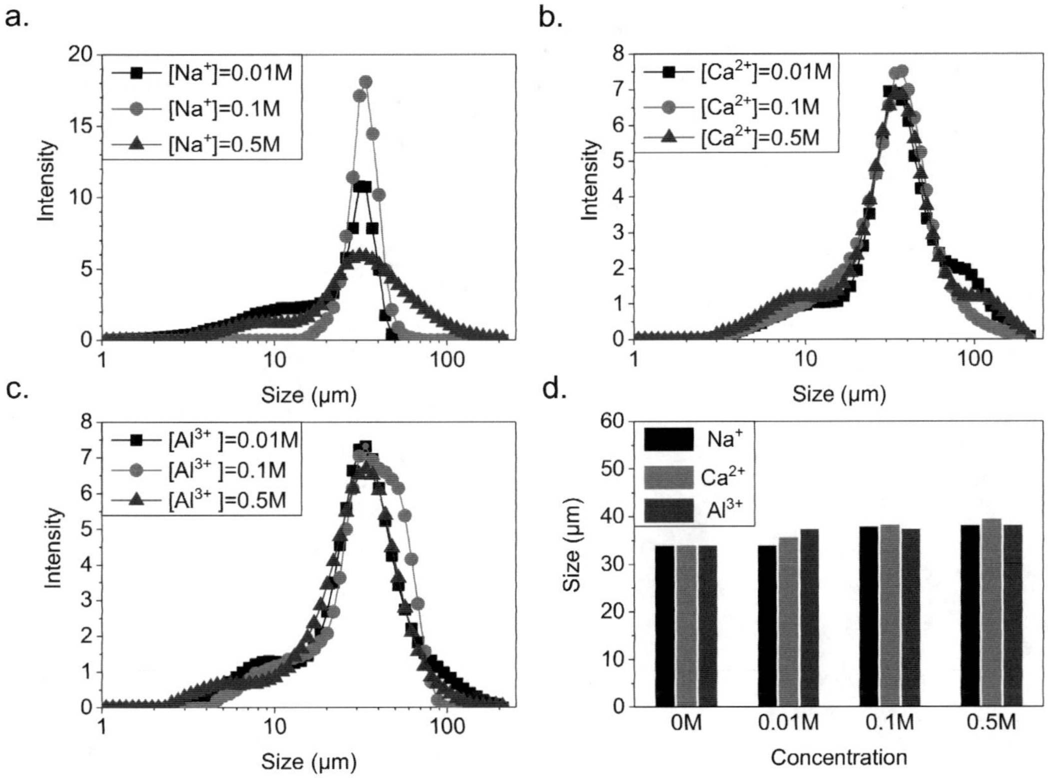

Figure 1. Effects of cation type and concentration of solvent on M-SAP equilirbium swollen size distribution and mean size at 23°C: (a)Na^+, (b)Ca^{2+} and (c)Al^{3+} with cation concentration: 0.01 M, 0.1 M and 0.5 M and (d)average sizes of M-SAP under different cation concentrations.

Elucidation of M-SAP Cation Resistance:

From an absorbent point of view, the main question is the role of Cr^{3+} in the system. First, we investigated whether the Cr^{3+} was consumed bonding onto the polymer chains. Gelation of chromium-polyacrylamide systems was prevented by organic chromium cations and excess organic ligands at low temperatures and was delayed at elevated temperatures[23]. The use of excess ligands to control the rate of gelation in bulk samples was claimed and explained in these studies[24-26]. Generally, addition of acetate to a polyacrylamide/ Cr^{3+} based gelling system would be expected to limit the amount of Cr^{3+} uptake as well as the crosslinking reaction. As a result, even after crosslinker MBAA running out, carboxylate groups won't immediately alter free state Cr^{3+} for continually crosslinking.

Samples with different concentration of Cr^{3+} (Sample #1-3) were investigated to verify the independence of Cr^{3+} to M-SAP polymer crosslinking during polymerization (Figure 2). Comparing the

M-SAP with varying Cr^{3+} concentrations by the resulting values of size distribution, it was obvious that all three samples performed the similar initial average size (13.3 μm ± 0.3 μm) and ES average size (54.1 μm ± 5.5 μm) respectively. This result proved that Cr^{3+} had no influence on both initial and ES M-SAP sizes, which mainly reflected crosslinking density. Different from MBAA, the consumed Cr^{3+} amount on bonding or crosslinking polymer were ignorable.

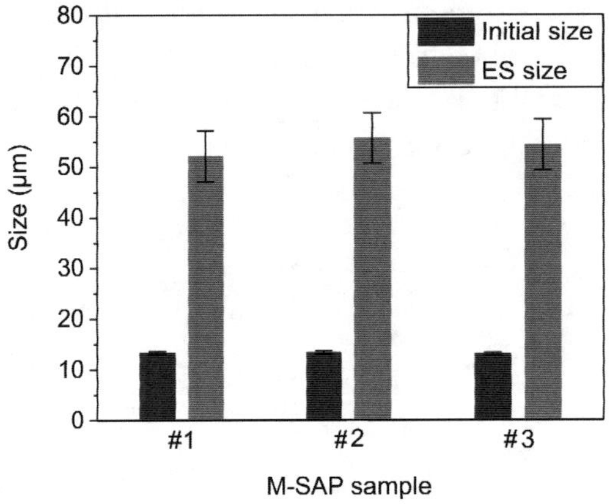

Figure 2. Effect of Cr^{3+} concentration to M-SAP microsphere initial and ES average sizes measured by SEM and LSPSDA respectively.

Another important question is how did free-state Cr^{3+} been saved into the M-SAP. To determine the location where free-state Cr^{3+} can stay in the M-SAP, inner fluid environment was studied. In fact, previous reports regarding the three states of water existed within SAP to explain the ion stability[27,28]. One example of previous discrete singular convolution (DSC) study shows that three states of water are existed with in SAP: free water, freezable bound water, and unfrozen water. Ideally, freezable bound water and unfrozen water were found near polymer chains like "sealed" environment in which no fluxible liquid can be found or exchanged with other areas of SAP. SEM measurement was introduced to prove the three states of water structures and the result was shown in Figure 3. After polymerization and purity stage, M-SAP exhibits stratified structure as Figure 3a shown, with numerous micro-folds on each layer. Equilibrium swollen M-SAP was prepared followed by freeze dry process for 24 hours before SEM examination. The result dedicated that densely sponge network structures were obtained as shown in Figure 3b. On the platy wall of M-SAP no micro-folds were found where micro-folds were saturated by freezable bound water and unfrozen water. To study liquid existed in this area, 240 hours long-time freeze-dry was used and the result was shown as Figure 3c. Interestingly, platy-wall folds were obtained with small crystallites homogenously distributed on the surface of each layer. One may infer from these SEM observations that after short-time frozen dry, freezable bound water and unfrozen water were still remained on the surface of polymer platy wall. Additional long-time drove the rest fluid out with leaving the pre-dissolute crystals stay there. Since no other metals were added into the system, free state Cr^{3+} were supposed existed in this "sealed" environment.

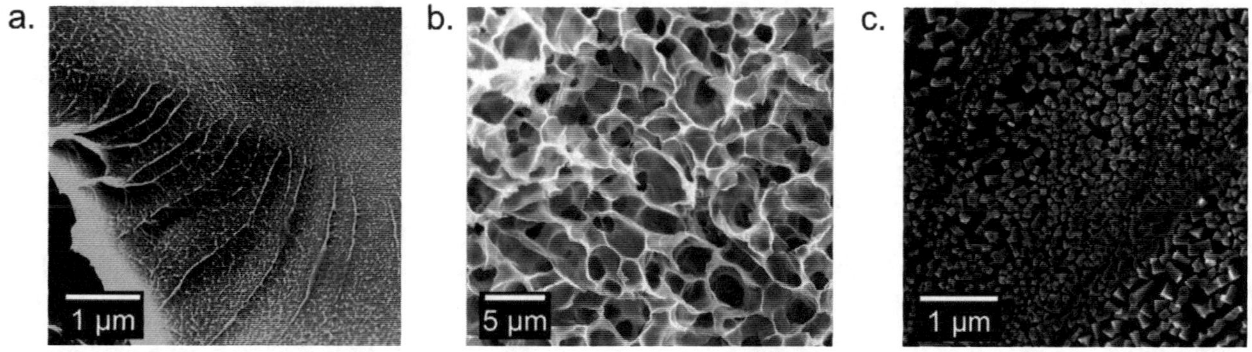

Figure 3. SEM micrographs for M-SAP #2: (a)surface of initial SPAG; (b)network structure of equilibrium swollen M-SAP with DIW; (c)layer-by-layer strucure of M-SAP with Cr^{3+} crystals inserted.

Since M-SAP's network structure bear a resemblance to multiple layer membranes[29], surface of each layer was considered as one independent osmotic pressure system. Therefore, one may infer that Gouy-Chapman-Stern model[30] can also be used to interpret the ionic environment in the vicinity of a charged surface. Charges adhere close to the electrode surface in the stern layer and decreases exponentially with distance from the surface[31]. Previous development in the theory of Stern-layer transport was used to study the effect of surface conduction on the low-frequency dielectric response properties of colloidal suspensions. Simply stated, if cations are put into the system, it will compress the stern layer and inhibit the effect of electrostatic repulsion (Figure 4). When the cation changed the apparent electric charges along the main polymer chains, brine neutralization effect was eliminated. The polymer chain of M-SAP contains negative charge with carboxylate group. Positive charged ions will neutralize the negative charge of carboxylate group by electrostatic effect[32]. Each positive charge ion will attract multiple polymer chains resulting in neutralizing the electrostatic repulsion between each polymer chains respectively, which as a result reduce the hydrodynamic size of the microspheres. Namely, On top of the associated crosslinking agent, Cr^{3+} entailed a role of shield against cations in solvent.

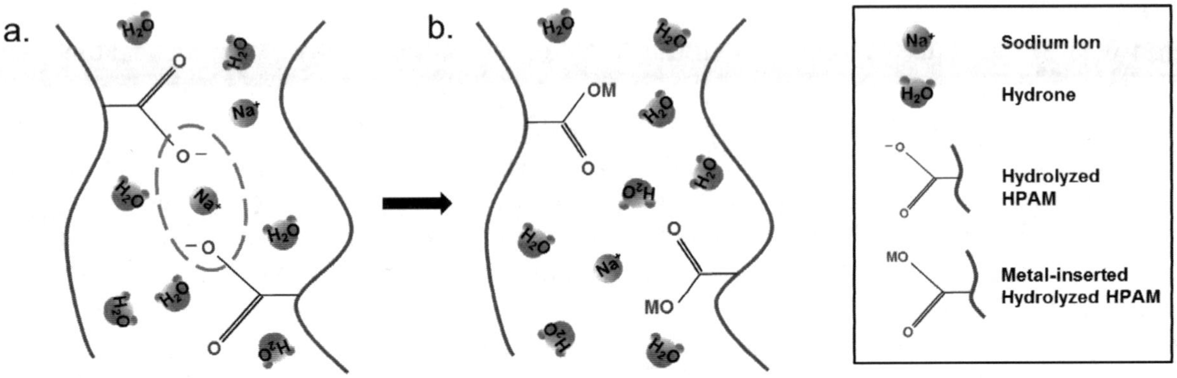

Figure 4. Schematic representation of differences between general hydrogel polymer (a) and metal-ion introduced hydrogel polymer influenced by environmental Na^+ (b). polymer with pre-inserted metal ions don't affected by environmental Na^+.

Conclusion

In the present study, we investigated a stable swelling ratio microsphere (M-SAP) with a controllable initial size via one-step suspension polymerization. For reducing errors during measurement, M-SAP with size around 10μm was selected as candidate for following swelling ratio measurement which showed best sphericity and uniform network structure. The micro-structure of M-SAPs were studied which helped optimize the M-SAP initial size and analyze Cr^{3+} preserving mechanism.

Acknowledgments

The authors would like to express their grateful acknowledge to the financial support from DOE under the contract of DE-FE0024558.

Reference

1. Hirotsu S, Hirokawa Y, Tanaka T. Volume-phase transitions of ionized N-isopropylacrylamide gels. *The Journal of Chemical Physics.* 1987;87(2):1392-1395.
2. Otake K, Inomata H, Konno M, Saito S. Thermal analysis of the volume phase transition with N-isopropylacrylamide gels. *Macromolecules.* 1990;23(1):283-289.
3. Muta H, Miwa M, Satoh M. Ion-specific swelling of hydrophilic polymer gels. *Polymer.* 2001;42(14):6313-6316.
4. Philippova O, Pieper T, Sitnikova N, Starodoubtsev S, Khokhlov A, Kilian H. Conformational transitions in polyelectrolyte networks in binary solvents: microheterogeneities in the collapsed state. *Macromolecules.* 1995;28(11):3925-3929.
5. Suzuki M, Hirasa O. An approach to artificial muscle using polymer gels formed by micro-phase separation. *responsive gels: volume transitions II.* 1993:241-261.
6. Kabiri K, Zohuriaan-Mehr M, Mirzadeh H, Kheirabadi M. Solvent-, ion-and pH-specific swelling of poly (2-acrylamido-2-methylpropane sulfonic acid) superabsorbing gels. *Journal of polymer research.* 2010;17(2):203-212.

7. Chen J, Park K. Synthesis and characterization of superporous hydrogel composites. *Journal of Controlled Release.* 2000;65(1):73-82.
8. Ricka J, Tanaka T. Swelling of ionic gels: quantitative performance of the Donnan theory. *Macromolecules.* 1984;17(12):2916-2921.
9. Collins KD. Charge density-dependent strength of hydration and biological structure. *Biophysical journal.* 1997;72(1):65-76.
10. Schleich T, Von Hippel PH. Specific ion effects on the solution conformation of poly-L-proline. *Biopolymers.* 1969;7(6):861-877.
11. Freeman DH. The gels for liquid chromatography: Oxford University Press; 1973.
12. Xu J, Antonietti M, Shalom M. Moving Graphitic Carbon Nitride from Electrocatalysis and Photocatalysis to a Potential Electrode Material for Photoelectric Devices. *Chemistry–An Asian Journal.* 2016;11(18):2499-2512.
13. Bai B, Zhang H. Preformed-particle-gel transport through open fractures and its effect on water flow. *SPE Journal.* 2011;16(02):388-400.
14. Chauveteau G, Omari A, Tabary R, Renard M, Veerapen J, Rose J. New size-controlled microgels for oil production. Paper presented at: SPE international symposium on oilfield chemistry2001.
15. Lei G, Li L, Nasr-El-Din HA. New gel aggregates to improve sweep efficiency during waterflooding. *SPE Reservoir Evaluation & Engineering.* 2011;14(01):120-128.
16. Sydansk R. A new conformance-improvement-treatment chromium (III) gel technology. Paper presented at: SPE enhanced oil recovery symposium1988.
17. Amiri A, Øye G, Sjöblom J. Influence of pH, high salinity and particle concentration on stability and rheological properties of aqueous suspensions of fumed silica. *Colloids and Surfaces A: Physicochemical and Engineering Aspects.* 2009;349(1):43-54.
18. Nolte H, John S, Smidsrød O, Stokke BT. Gelation of xanthan with trivalent metal ions. *Carbohydrate Polymers.* 1992;18(4):243-251.
19. Maruca R, Suder BJ, Wightman J. Interaction of heavy metals with chitin and chitosan. III. Chromium. *Journal of Applied Polymer Science.* 1982;27(12):4827-4837.
20. Sydansk RD. Acrylamide-polymer/chromium (III)-carboxylate gels for near wellbore matrix treatments. *SPE Advanced Technology Series.* 1993;1(01):146-152.
21. Stavland A, Ersdal T, Kvanvik B, Lohne A, Lund T, Vikane O. Evaluation of xanthan-Cr (III) gels for deep emplacement: retention of Cr (III) in north sea sandstone reservoirs. Paper presented at: Seventh European Symposium on Improved Oil Recovery, Moscow, Russia,(27-29 October 1993)1993.
22. Guorong S, Zhihai C. A new polymerization method and kinetics for acrylamide: Aqueous two-phase polymerization. *Journal of Applied Polymer Science.* 2009;111(3):1409-1416.
23. Burrafato G, Lockhart TP. A New, Mild Chemical Method For The Degelation Of Cr (+ 3)-Polyacrylamide Gels. 1989.
24. Lockhart TP, Burrafato G. Gellable aqueous composition and its use in enhanced petroleum: Google Patents; 1992.
25. Sydansk RD. Process for reducing permeability in a subterranean hydrocarbon-bearing formation utilizing a gelation solution having a controlled gelation rate: Google Patents; 1995.
26. Natarajan D, McCool C, Green D, Willhite G. Control of in-situ gelation time for HPAAM-chromium acetate systems. Paper presented at: SPE/DOE improved oil recovery symposium1998.
27. Save M, Schappacher M, Soum A. Controlled Ring-Opening Polymerization of Lactones and Lactides Initiated by Lanthanum Isopropoxide, 1. General Aspects and Kinetics. *Macromolecular chemistry and physics.* 2002;203(5-6):889-899.

28. Kizilay MY, Okay O. Effect of hydrolysis on spatial inhomogeneity in poly (acrylamide) gels of various crosslink densities. *Polymer.* 2003;44(18):5239-5250.
29. Enick RM, Olsen DK, Ammer JR, Schuller W. Mobility and Conformance Control for CO2 EOR via Thickeners, Foams, and Gels--A Literature Review of 40 Years of Research and Pilot Tests. Paper presented at: SPE improved oil recovery symposium2012.
30. Olajire AA. Review of ASP EOR (alkaline surfactant polymer enhanced oil recovery) technology in the petroleum industry: Prospects and challenges. *Energy.* 2014;77:963-982.
31. Levitt D, Pope GA. Selection and screening of polymers for enhanced-oil recovery. Paper presented at: SPE Symposium on Improved Oil Recovery2008.
32. Gall J. Subterranean formation permeability correction: Google Patents; 1973.

489701 Synthesis of Chemical Protective Elastomeric Barrier Materials

Abstract:

Microporous polyurethane coatings with chemical barrier properties are being developed for potential use in industrial chemical protective clothing. Ether-rich microporous polyurethane coatings/membranes are known to offer excellent moisture vapor transport properties; however, toxic chemicals can easily permeate their structures. In this effort, to protect from toxic chemicals, super activated carbon is added to the coating. Efforts to produce this unique coating will be discussed, where moisture vapor permeability is achieved by introducing microsized pores throughout the membrane using a water soluble pore former. After curing, the pore former is removed by washing with water. The membrane's moisture vapor permeability and barrier properties to selected toxic chemicals are tested using a vapor absorptive capacity apparatus and a chemical permeation setup. The structure and the morphology of the membrane are determined using electron microscopy and a nitrogen sorption apparatus, and will be discussed.

PAO #: U17-172

496273 Polysulfide-Based Nanofiber Prepared Via Inverse Vulcanization and Electrospinning for Effective Mercury (II) Sequestration

Abstract:

Recent progress on inverse vulcanization has opened various opportunities to develop advanced materials with useful electrochemical and optical properties which have been investigated as Li-S batteries, for IR imaging technology, and as self-healing materials. Herein, the first extensive study on the preparation of an Hg^{2+} adsorbent based on inverse vulcanized polysulfide (PS) is presented.

Sulfur-rich poly(sulfur-random-acrylate) (poly(S-r-Ac)) was successfully prepared via direct addition of Î²-carboxyethyl acrylate to molten sulfur as confirmed via FTIR and ^{1}H and ^{13}C NMR analyses. The viscoelastic nature (i.e. 0< phase angle (Î´) <90Â°) and thermal properties of (poly(S-r-Ac) can be directly manipulated by adjusting molar ratio of components (i.e. S:CEA molar ratio). That is, increasing sulfur content decreased the viscosity, and T_g, and % weight remaining in simultaneous thermal analysis profiles of the PS. This trend agrees with the hypothesis that with greater amount of sulfur, sulfur rank in poly(S-r-Ac) is longer which was also confirmed by higher S-S XPS peak area of PS in S:CEA = 2:1 molar ratio. Given that S-S (226 kJ mol^{-1}) bond energy is weaker than other intramolecular bonds present such as C-S (272 kJ mol^{-1}), C-C (346 kJ mol^{-1}), C-O (358 kJ mol^{-1}), and C=O (799 kJ mol^{-1}), it is expected that poly(S-r-Ac) with greater amount of sulfur (i.e. S:CEA=2:1) degrades at relatively lower temperature and has lower T_g than those with lesser amount of sulfur component.

Poly(S-r-Ac) is soluble to most common solvents which include dimethyl formamide (DMF), enabling it to form homogenous solution with polyacrylonitrile (PAN). Polysides blended in polyacrylonitrile (PAN) were prepared as dope solutions in DMF and were electrospun into nanofibers (NFs). The NFs with different PS loadings were characterized and were evaluated as an adsorbent for Hg^{2+} sequestration. Batch adsorption experiments showed that the PS (S:CEA=2:1) follows a monolayer Langmuir-type adsorption isotherm with a maximum adsorption capacity of 612 mg g^{-1}. Kinetics study showed that the PS efficiently sequestered Hg^{2+} attaining adsorption equilibrium within 1.5 hrs. Lastly, prepared PS exhibited selectivity towards Hg^{2+} having highest concentration factor (CF=1.79 L g^{-1}) and distribution coefficient (K_d =1.95 x 10^6 mL g^{-1}) despite having the least concentration amidst other competing ions. Little affinity towards Cr^{2+}, Zn^{2+}, Pb^{2+}, and Cu^{2+} and no affinity towards Cd^{2+} and Ni^{2+} were observed. Reusability test exhibited the ability of the prepared PS NF to be recycled using 6 M HCl as an eluent to sequester adsorbed Hg^{2+} in the NF.

This work demonstrates new insight for the preparation of polysulfide-based adsorbents from inverse vulcanization. It also presents new approach for easy, cheap, and effective preparation of adsorbents for Hg^{2+} sequestration from contaminated water or wastewater. This research was supported by NRF funded by the Ministry of Science, ICT and future Planning (2017R1A2B2002109) and the Ministry of Education (2009-0093816).

490003 Mechanical Properties of Maleated Polymer/Graphene Oxide Composites using Graphene Oxide as a Multi-Functional Crosslinker

Abstract:

Many previous studies have shown that mechanical properties of a polymer, such as tensile strength and modulus, can be improved by adding fillers at the cost of reducing the ductility of the polymer. The vast majority of composites that have been examined have been physically mixed or compounded. In this research, we propose a method to utilize graphene oxide (GO) as a multi-functional crosslinker where covalent bonds may be formed with maleic anhydride grafted styrene-ethylene-butylene-styrene triblock copolymer (SEBS-g-MA). This study illustrates reaction can occur even at room temperature to form a mechanically robust elastomer. Owing to the hydrophobicity of the composite accompanied by its excellent mechanical properties, this composite material could be used practically as an anti-corrosion coating.

491449 Development and Analysis of a Thin Film Nanocomposite Membrane: Resistance to Chlorine

Abstract:

The degradation of polyamide nanofiltration media by chlorine decreases the efficiency and life time an expensive membrane which can be used in conjunction with reverse osmosis as one stage in the purification of produced water. Adding Montmorillonite nanoparticles may increase the chlorine-resistance of thin film composite polyamide membrane. However, understanding the degradation and the protection requires quantitative analysis of a membrane which is microns thin and grown in-situ on a polysulfone support. Thermogravimetric analysis can be used as an indirect method to detect the degradation state of the polyamide. Early controls have been analyzed using nonisothermal degradation models to determine kinetic parameters for both control polyamide composed of m-phenylenediamine and trimesoyl chloride and the nanocomposite. Future work will involve using TGA kinetic parameters as a correlative measure of the effects of chlorination on the membrane.

Improving Gas Transport Properties of Mixed Matrix Membranes via Interfacial Improvement

Ahmad Arabi Shamsabadi, Morteza Sadeghi, Mohammad Dinari, Mahsa Salehi, and Masoud Soroush

Drexel UNIVERSITY

17AIChE Annual Meeting, Minneapolis, MN

Motivation

With the increasing average global atmospheric CO_2 level (>403 ppm), the membrane technology has attracted significant attention. It is less inexpensive, has a simpler process, and consumes less energy, compared to other separation and purification technologies. Polymers are widely used for membrane fabrication. Currently polymer membranes suffer from two major challenges: (1) The trade-off relationship between the permeability and selectivity, and (2) Membrane instability in the form of physical aging and plasticization. The fabrication of membranes made from polymers embedded with nanoparticles can lead to mixed matrix membranes (MMMs) with high gas permeability and excellent sieving ability. However, the agglomeration of nanoparticles in the polymer matrix, the formation of defects in the membranes, and the incompatibility between the polymer and nanoparticles are still challenges in the fabrication of nanocomposite membranes. In designing mixed-matrix membranes, improvement of interfacial interactions between the particles and polymer results in high-performance defect-free membranes for gas separation. In this work, we investigate how the surface chemistry of nanoparticles affect the interfacial interactions and gas transport of MMMs.

Mixed Matrix Membranes

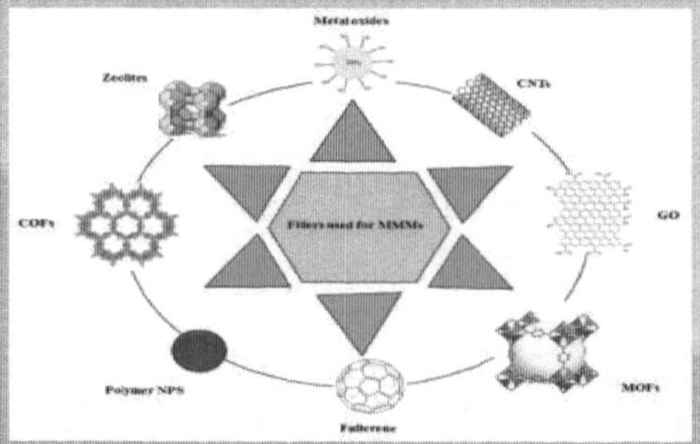

Figure 1. Different fillers used for MMMs fabrication.

- MMMs are commonly made of a dispersed nanomaterial phase and a continuous polymer matrix, hoping to combine the high permeability and sieving ability of the advanced nanomaterial.
- In designing nanocomposite membranes to surpass the Robeson upper bound, filler and polymer compatibility, particle size and configuration, and uniform particle dispersion are important factors.
- In case of no interfacial voids, which deteriorate the membranes sieving ability, the inclusion of nanomaterials can improve membrane performance by increasing selectivity, permeability, or both.

Membranes Fabrication and Characterization

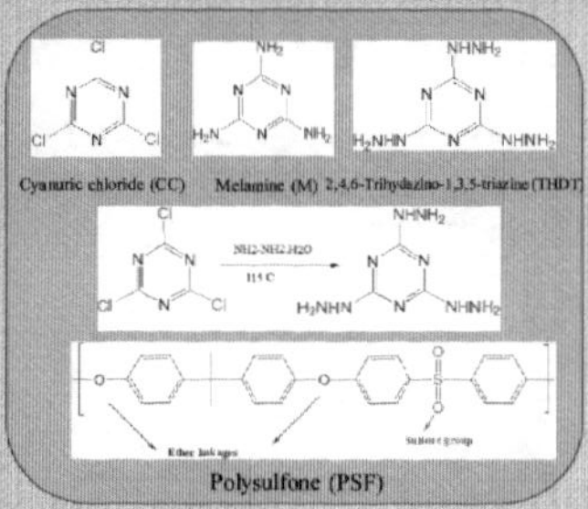

Figure 2. Nanoparticles and polymer used in this study.

- Membranes were fabricated via solvent evaporation.
- PSF is commercially available, it has high thermal and mechanical resistance, and it contains ether and sulfone groups as acceptors for hydrogen bonding.
- No donor hydrogen bonding group is in the PSF structure.
- Compared to the pure PSF, no considerable change in FTIR of the MMMs was observed.
- XRD tests exhibited no change in amorphousness of PSF.
- XRD tests showed higher d-spacing for MMMs compared to the pure PES, which shows incorporating all particles created more pathways for gas transport.

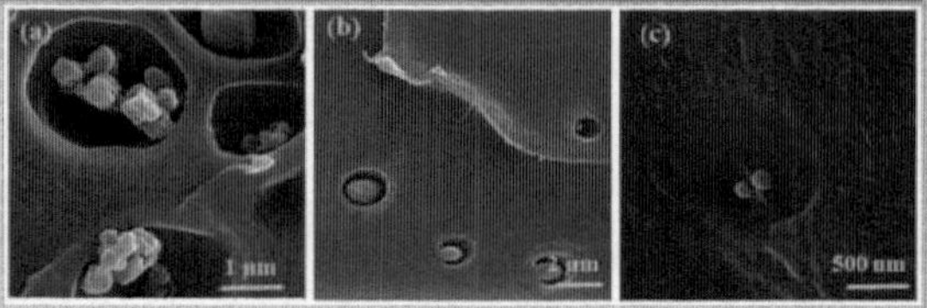

Figure 3. Cross-sectional SEMs of (a) PSF-CC-2%, (b) PSF-M-2%, and (c) PSF-THDT-2% membranes.

The agglomeration of CC nanoparticles and poor PSF/CC interactions resulted in defects in the membrane, melamine has no agglomeration but interfacial voids can be observed in the PSF-M membranes. Good THDT-PSF interactions led to defect-free nanocomposite membranes.

Membranes Gas Transport Properties

Table 1. Permeability and selectivity of MMMs.

Sample	Permeability (barrer)				Selectivity		
	CO_2	CH_4	O_2	N_2	CO_2/CH_4	CO_2/N_2	O_2/N_2
PSF	**2.82**	**0.12**	**0.64**	**0.14**	**23.86**	**20.02**	**4.56**
PSF-CC-0.5	3.12	0.27	0.77	0.29	11.49	10.59	2.62
PSF-CC-1	4.18	0.42	0.99	0.43	10.01	9.7	2.29
PSF-CC-2	4.51	0.6	1.2	0.62	7.57	7.23	1.27
PSF-M-0.5	2.86	0.16	0.65	0.17	17.7	16.51	3.73
PSF-M-1	3.03	0.17	0.68	0.2	17.48	15.31	3.45
PSF-M-2	3.54	0.23	0.75	0.25	15.37	13.97	2.97
PSF-THDT-0.5	3.3	0.18	0.67	0.2	17.93	16.57	3.38
PSF-THDT-1	3.75	0.17	0.66	0.19	22.65	20.02	3.53
PSF-THDT-2	3.92	0.09	0.54	0.12	41.6	31.72	1.27

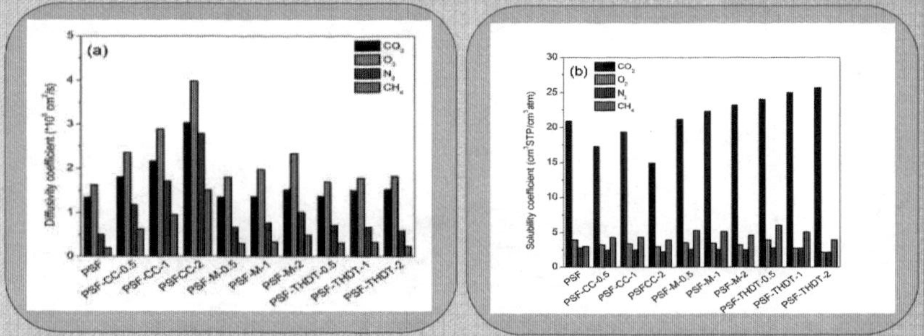

Figure 4. (a) Diffusivity and (b) solubility coefficients of the MMMs.

- Significant selectivity reduction was observed for the PSF-CC membranes due to poor PSF/CC interactions.
- No hydrogen bonds were formed between PSF and CC particles.
- The permeability increase with a low selectivity drop demonstrates better interaction of melamine with PSF than with cyanuric chloride.
- 19% CO_2 permeability, 132% CO_2/CH_4 selectivity, and 91% CO_2/N_2 selectivity enhancement were observed for the PSF-THDT-2% membrane.
- Compared to other MMMs, the performance of the PSF-THDT-2% membrane is closer to the 2008 CO_2/N_2 and CO_2/CH_4 Robeson upper bounds.

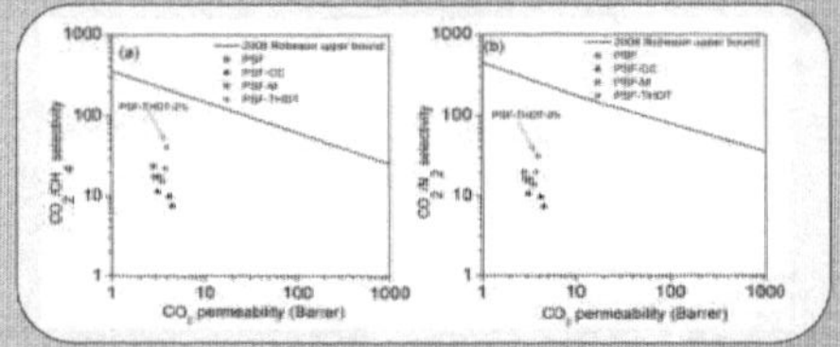

Figure 6. The PSF-THDT-2% membrane performance compared to the 2008 Robeson upper bound: (a) CO_2/CH_4, and (b) CO_2/N_2.

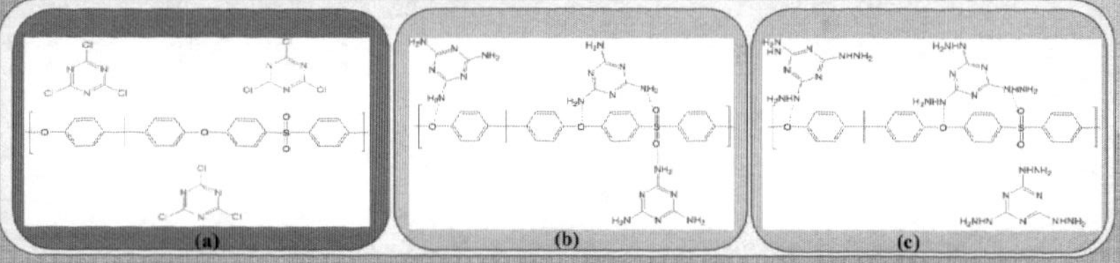

Figure 5: (a) Possible interactions between cyanuric chloride and PSF, (b) possible interactions between melamine and PSF, and (c) possible interactions between THDT and PSF.

Conclusions

- Formation of detrimental voids leads to significant selectivity reduction in PSF-CC membranes.
- PSF has stronger interactions with melamine nanoparticles than with CC nanoparticles.
- Strong interfacial interactions between PSF and THDT result in considerable CO_2/CH_4 and CO_2/N_2 selectivity improvements.
- This work shows the importance of interfacial interactions in fabricating highly-selective MMMs.

501747 Antimicrobial Polymers: Present State of the Art

Abstract:

Antimicrobial Polymers: Present State of the Art.

ABSTRACT

Antimicrobial polymers represent a very assuring class of therapeutics with unique characteristics for treating microbial infections.[1] These polymers usually exhibit long-term activity and limited residual toxicity, are chemically stable, non-volatile and do not permeate through the skin.[2] Antimicrobial polymers were discovered in 1965 and since then, have attracted considerable attention in academic as well as industrial research.

Antimicrobial polymers are classified as bound or leaching antimicrobials depending on the polymer material. These are divided into three types: polymeric biocides, biocidal polymers, and biocide-releasing polymers.[3] These antimicrobial polymers find major applications in the medical, food packaging, and textile industries.

This work is an attempt to portray the present state of the art and the developments in the domain of antimicrobial polymers. Antimicrobial mechanisms exhibiting passive and active action and substrate material types containing bound and leaching antimicrobials are discussed. This work also deals with canvassing the applications of these antimicrobial polymers in the medical, food, and textile industries.

References

[1] Santos, M. R. E.; Fonseca, A. C.; Mendonça, P. V.; Branco, R.; Serra, A. C.; Morais, P. V. and Coelho, J. F. J. Recent Developments in Antimicrobial Polymers: A Review. *Materials*, 2016; 9, pp. 599-631.

[2] Xue, Y.; Xiao, H. and Zhang, Y. Antimicrobial polymeric materials with quaternary ammonium and phosphonium salts. *Int. J. Mol. Sci.*, 2015; 16, pp. 3626–3655.

[3] Siedenbiedel, F. and Tiller, J.C. Antimicrobial polymers in solution and on surfaces: Overview and functional principles. *Polymers*, 2012; 4, pp. 46–71.

499825 Layer-By-Layer Coated Microneedle Arrays for Staged Multi-Agent Immune Attack on Melanoma

Abstract:

This work aims to develop a new, facile, and rapidly applied melanoma treatment for post-surgery melanoma patients to prolong or preserve their lives. Micro-scale needle arrays will be used as the drug delivery platform, for a highly controlled transcutaneous delivery of cancer vaccines. Layer-by-layer polyelectrolyte assembly will be used to generate the drug releasing thin film coatings on microneedles.

A microneedle platform for effective gene delivery is designed, with an effective release polymer that assists LbL film construction and implantation upon application in the skin. The DNA and/or RNA therapeutic cargos will be engineered with functional proteins and polymer scaffolds to enhance expression. This application seeks to achieve a staged release of melanoma cancer vaccines in order to activate a robust immune response; through direct insertion of microneedles into the dermis at or near the melanoma tumor site, a combination of tumor antigen present at the site and encapsulated nucleic acid encoding for tumor antigens can be released, with co-delivery of adjuvants that upregulate dendritic cell antigen uptake ability and antibodies that block tumor immunosuppression.

492456 Direct Observation of Remarkable Nanoparticle Evolution during Aqueous Dissolution of Polymer/Drug Particles

Abstract:

Amorphous solid dispersions (SDs), metastable blends of an active pharmaceutical ingredient and a polymer excipient, can significantly enhance the aqueous solubility of hydrophobic drugs, but the mechanism is not well understood. To resolve this ambiguity, we explored the dissolution of hydroxypropyl methylcellulose acetate succinate (HPMCAS) SDs in phosphate buffered saline (PBS). Light scattering measurements of HPMCAS solutions showed the polymer forms a mixture of ~10 and ~100 nm sized structures (assigned to linear and covalently coupled polymer chains, respectively) in both PBS and tetrahydrofuran. The measurements also demonstrated that PBS is a poor solvent for HPMCAS at and below 37 Â°C, suggesting the solvent induces the polymer to associate with itself or other hydrophobic species in solution. *In vitro* dissolution of HPMCAS SDs â€“ containing either phenytoin or probucol as the encapsulated hydrophobic drug â€“ in PBS was studied using a combination of cryogenic transmission electron microscopy (both bright-field imaging and electron diffraction) and small-angle X-ray scattering. These tools revealed that HPMCAS and drug form <100 nm amorphous nanoparticles in solution. A direct correlation between the SD dissolution profiles and nanostructure evolution was uncovered for both drugs, implying that a significant portion of the drug measured in the supernatant during the dissolution assay is retained in the form of nanoparticles. Drug identity, loading, and targeted concentration influenced nanoparticle size, shape, and lifespan. These discoveries confirm that persistent nanostructures are important for enhancing drug supersaturation during SD dissolution.

492734 Polymerized Ionic Liquid Pentablock Terpolymer for Lithium-Metal Batteries

Abstract:

Lithium-ion batteries are currently the leading commercial technology for portable electronic devices and electric vehicles. However, new solid polymer electrolytes (SPEs) that comprehensively possess the preferred properties of high ion conductivity, high electrochemical stability, robust mechanical properties, flexibility, and good film forming properties, are required to improve the overall capacitance and lifetime of batteries and overcome the limitations of current liquid electrolytes (flammability, low electrochemical stability, and short battery lifetime). Compared with liquid electrolytes, SPEs still suffer from poor room temperature ionic conductivity and have been limited from commercial applications. Polymerized ionic liquid block copolymers (PILBCPs) are new type of SPEs combining the benefits of both ionic liquids (*e.g.*, high ionic conductivity and electrochemical stability) and block copolymers (*e.g.*, self-assembled nanostructured morphologies), which make them the promising materials for electrochemical devices. In this study, a polymerized ionic liquid (PIL) ABCBA pentablock terpolymer (PILPTP) containing tert-butylstyrene as the A outer blocks, ethylene-propylene as the B blocks, and a pyrrolidinium functionalized styrene PIL as the inner C midblock was investigated for batteries. The electrochemical properties of this PILPTP were characterized by linear sweep voltammetry, cyclic voltammetry, and impedance spectroscopy. Preliminary lithium-metal batteries performance with this PILPTP as the electrolyte and separator were also investigated.

490770 Effect of Electric Field on the Structure and Dynamics of Model Ionomer Melts

Abstract:

Ionomers are polymers containing a small amount of charged groups attached to a relatively non-polar backbone, used in packaging and other applications. We consider a melt of ionomers and their counterions with no solvent, in which the aggregation of ions significantly impacts overall material properties. A prior coarse-grained model has been successful in capturing the low wavevector peak observed in x-ray scattering known as the ionomer peak, as well accurately predicting other structural and dynamic properties of these materials. This model represents poly(ethylene-co-acrylic) acid neutralized completely by sodium salt, using neutral monomer beads, charged beads that represent COO^- , and charged unbound beads that represent Na^+. Commercially available ionomers however are only partially neutralized, with some COOH (carboxyl groups) along the backbone. These carboxyl groups play an important role in aggregate morphology and influence the overall properties of the polymer. Thus, in order to better model typical experimental systems that are partially neutralized, we use additional "sticker" groups that represent COOH. These stickers are similar to uncharged monomers but with adjusted short-range interaction strengths with each other and with ionic groups. Although sticker-sticker interactions are radially symmetric, aggregate morphologies obtained using the sticker-based model are in good agreement with prior atomistic simulation results. By including stickers, we can now more accurately represent Surlyn®, which is widely used commercially. To this modified model, we apply a static electric field. Aggregate morphology under the field was analyzed and compared with zero field morphology. Local and long range order of the aggregates was also studied. All analyses were performed on different chain architectures and neutralization levels.

490821 Influence of Molecular Design on the Self-Assembly of Single-Stranded DNA Amphiphiles

Abstract:

Amphiphilic molecules composed of a single-stranded DNA (ssDNA) headgroup, a hydrophobic spacer, and a dialkyl hydrophobic tail have been demonstrated to self-assemble into exotic nanostructures in aqueous solution including twisted nanotapes, helical nanotapes, and nanotubes [Pearce & Kokkoli, *Soft Matter,* 2015, **11**, 109-117]. These assemblies have utility in a wide range of applications such as templates for nanofabrication, biosensing, gene delivery and drug delivery. In recent experiments, Kokkoli and coworkers showed that controlling the building blocks of the ssDNA-amphiphile can alter the size, shape, and relative abundance of the above assembled structures. In this poster we will show coarse-grained molecular dynamics simulations that explain why these changes in the amphiphile molecular design result in a specific assembled structure observed experimentally. We will present our coarse-grained model of the amphiphilic molecule in explicit solvent, and show how specific changes in the building blocks of the amphiphiles affect the shape of the assembled structure, area per amphiphile molecule, thickness of the assembled hydrophobic and hydrophilic regions, and surface tension of bilayers formed by these amphiphiles.

493313 Intracellular Trafficking of Enzyme-Cleavable Peptide Amphiphiles

Abstract:

Peptides are promising therapeutics to promote apoptosis by modulating protein-protein interactions (PPIs), offering superior specificity and fewer off-target effects than existing drugs. Peptide amphiphile technology that combines a peptide with a hydrophobic tail efficiently delivers peptide payloads into cells, obstacles remain for viable clinical translation. For example, PAs frequently accumulate within endosomal compartments thus limiting the ability of delivered peptides to target proteins within the cytosol, thereby lowering their therapeutic potency. Furthermore, the intracellular trafficking and ultimate fates of the hydrophobic tails and peptides are largely unknown.

Recently, we showed enhanced intracellular controlled release and accumulation of the peptides by introducing endosome-specific enzyme cleavable linkers between the hydrophobic tail and peptide of the PA building blocks. In addition to enhanced accumulation, the delivery platform that we showed offers the ability to track the individual therapeutic and hydrophobic components in real time using fluorescent labels to better understand and manipulate intracellular drug trafficking. Thus, this platform is a superior nanomedicine tool for enhanced intracellular delivery and tracking of therapeutics.

In this study, we studied the trafficking of the peptides after endosomal enzyme cleavage. We synthesized the peptide amphiphiles, as described in our previous study. In this follow-up study, we added endosomal escape groups to the peptides. Different organelle targeting peptides were selected and their co-localization was studied with different organelle fluorescent labels.

501313 Polylactide-Based Biodegradable Zwitterionic Polymers and Their Conjugates with Drugs for Biomedical Applications

Abstract:

Incorporating functionalities with biodegradable macromolecular architectures, functional polylactides (PLAs) have important applications in polymer-based therapeutics. In particular, the integration of functional PLAs with hydrophilic poly(ethylene glycol) (PEG) chains and other biomedical-relevant structures has generated a variety of novel biomaterials with significant application potentials in therapeutic delivery. However, reduced bioactivity and immunogenicity of PEGylated therapeutics severely restrict their biomedical efficacy. Recently zwitterionic polymers (ZPs) have attracted significant interests as alternatives to PEG due to their special properties. Several studies on integrating non-degradable ZPs with degradable polymeric structures have been reported, but the non-biodegradable ZPs still would cause side effects and systemic toxicity resulted from polymer accumulation. Therefore, it is critically important to develop biodegradable ZPs for modification of therapeutics. Herein we report fully biodegradable ZPs and the corresponding polymer-drug conjugates (PDCs). PLA-based main-chain was designed for the ZPs and the PDCs because of its biodegradability. Carboxybetaine and sulfobetaine were selected as the zwitterions because they are superhydrophilic. Clinic anticancer drugs, such as paclitaxel (PTX), were chosen as the drug moieties of the PDCs. The ZPs and PDCs were synthesized by thiol-ene reactions of thiol-functionalized zwitterions and/or drugs with allyl-functionalized PLA. These materials showed enzymatic degradability and well-suppressed nonspecific interaction with biomolecules, and the PDCs further exhibited acid-sensitive sustained drug release behavior. Moreover, cytotoxicity and cellular uptake studies illustrated high biocompatibility of the ZPs, as well as remarkable anticancer effectiveness and ready cellular internalization of the PDCs. Overall, this work indicates the promising application potentials of biodegradable ZP-based biomaterials.

503543 Overcoming Obstacles to Brain Repair Using Biomaterials

Abstract:

Research in my laboratory over the past decade has focused on the generation of hydrogel biomaterials to support the formation of a reparative niche within diseased or injured sites that can block or prevent inhibitory signals from dominating the repair process, while providing pro-repair signals that can guide new tissue formation. The goal of our approach is to use engineered materials to "unlock" the regenerative capacity of damaged or diseased tissue to promote repair. The premise of our approach is that all tissues in the body have the capacity to repair through local stem or progenitor cells, but that due to unfavorable environmental conditions during the normal healing process they are not able to do so. In this talk, I will describe our approach to overcome physical and chemical barriers to brain repair after stroke using biomaterials.

493357 Aggregation Kinetics in Biological Environments as a Determinant of Nanoparticle Behavior in the Brain

Abstract:

Aims: Develop an assay to assess colloidal stability of nanotherapeutic platforms in biologically relevant environments. Investigate the impact of key physiological variables on nanoparticle diffusion within the brain microenvironment.

Methods: Use dynamic light scattering and confocal microscopy to measure aggregation dynamics of nanoparticles in biologically complex milieu. Use high-speed multi-particle tracking and fluorescence microscopy to measure diffusion coefficients of nanoparticles in an agarose gel model of the brain.

Results: The presence of divalent cations, including Mg^{2+} and Ca^{2+} can cause aggregation of electrocratically stabilized particles. Even particles coated with PEG, a common agent used to provide both steric stabilization and stealth properties, experience aggregation under certain conditions. Additional factors that can vary in biological environments, including temperature, CO_2 partial pressure, and pH, have a significant impact on nanoparticle aggregation kinetics. Diffusion studies in agarose gels demonstrate that PEGylated and non-PEGylated polystrene nanoparticles have similar diffusive behavior at short time scales in the absence of proteins and divalent cations. However, when divalent cations are introduced, PEGylated particles maintain similar diffusion coefficients, while non-PEGylated nanoparticles exhibit significantly hindered behavior.

Conclusions: Reduced diffusive capabilities of non-PEGylated paticles has been largely attributed to interactions with proteins. This study demonstrates that an additional mechanism, decreased colloidal stability due to interactions with divalent cations, also plays a role. Even in protein-poor environments, such as cerebrospinal fluid, nanoparticles can become hindered and lose their ability to provide improved delivery to target sites. Using the assays demonstrated in this poster, engineers can design nanoparticle platforms and predict their behavior in the brain and other organs before performing *in vivo* and *ex vivo* studies.

499783 Layer-By-Layer Nanoparticles for Interleukin-12 Delivery

Abstract:

Title: Layer-by-Layer Nanoparticles for Interleukin-12 delivery in Ovarian Cancer

Antonio Barberio, Santiago Correa, Erik C. Dreaden, Mariane B Melo, Darrell J. Irvine, Paula T. Hammond

1 Department of Chemical Engineering, MIT

2 Koch Institute for Integrated Cancer Research, MIT

3 Department of Biological Engineering, MIT

Abstract:

Immunotherapy has become a popular target for cancer treatment as it allows for a more diverse application of treatment. Rather than target treatment towards the many highly specific types of cancer, immunotherapies aim to trigger the immune system to fight the malignancy, making them applicable across many tissues and oncogenic mutations. One option for this type of immune treatment is the delivery of a general immune stimulator such as a cytokine. One such cytokine is interleukin-12 (IL-12). Systemic IL-12 delivery was attempted in the past but led to intolerable off target toxicities at effective doses. This work focuses on delivering a single-chain version of IL-12 in a layer-by-layer nanoparticle to reduce the off-target toxicity issue, using principles of electrostatic assembly to create a protective coating that can also provide targeting and stealth properties for systemic delivery. The particle will incorporate layer-by-layer technology to tailor its delivery properties to maximize the efficacy of IL-12 in the tumor and reduce its off target effects. Layer by layer nanoparticles are created by using alternating material intermolecular interactions such as electrostatic charge to build a multi-component coating on a nanoparticle core. The addition of layered materials allows for optimized targeting, delivery, and combinations of different payloads. The delivery system is designed to deliver IL-12 both from the core of the particle and from the layers. IL-12 is loaded into the core of the particle by encapsulation into a negatively charged liposome. The cytokine is loaded into the layers of the particle via interactions with the layering materials, including serum proteins and carrier polyelectrolyte backbones. One of the main challenges of this work is achieving effective delivery of a cytokine that must achieve activity by binding to an external membrane receptor in a nanoparticle that may be internalized. To address this issue materials that trigger exocytosis or rapidly degrade in the tumor micro environment to release the drug in the extracellular space are tested in the layers. In addition to optimizing the materials for drug release, different terminal layer materials are tested for optimal targeting of the particle. The particle must effectively target the tumor, but it must also deliver the cytokine to T and natural killer cells that express the Il-12 receptor for the treatment to be effective. This work attempts to produce a clinically relevant form of systemic IL-12 therapy and make future combination therapies with IL-12 a possibility to obtain in a systemically deliverable package.

499006 Implantable Biomaterials Produced By Complexing Chitosan to Alginate or Pectin: Surface Properties, Hemocompatibility and Cytotoxicity

Abstract:

Scaffolds are artificial extracellular matrices, capable of supporting cell growth and three-dimensional tissue formation. Along with cells and growth factors, scaffolds are one of the three key components for tissue engineering. The chemical composition and structure of the scaffolds, together with their surface properties, may affect their biocompatibility, limiting their range of clinical applications. Interfacial free energy, surface hydrophilicity, presence of functional groups, density of charges and surface topography are among the characteristics of the scaffolds that affect their biological performance. Hemocompatibility and cytotoxicity are also important aspects of biocompatibility that need to be evaluated prior to scaffold implantation. In this work, different aspects of the biocompatibility of scaffolds constituted of nature-derived polymers were evaluated. In this sense, surface properties as roughness and hydrophilicity, as well as thrombogenicity and *in vitro* cytotoxicity were assessed for chitosan-alginate (Ch-A) and chitosan-pectin (Ch-P) scaffolds produced in the presence or not of the porogenic agent KolliphorÂ® P188 (K). Chitosan is a linear polysaccharide obtained by deacetylation of chitin, a polymer vastly found in the exoskeleton of crustaceans, consisting of units of N-acetyl-D-glucosamine and D-glucosamine. At low pH values, chitosan can behave as a polycation and interact with negatively charged polymers, forming polyelectrolyte complexes. Alginate is an anionic linear polysaccharide extracted from brown algae, and its structure is composed of two types of repeating subunits, Î²-(1,4)-D-mannuronate and Î±-(1,4)-L-guluronate. Pectin is an essentially linear anionic polysaccharide extracted from citrus fruits consisting mainly of D-galacturonic acid units joined by Î±-(1-4) glycosidic linkages. The use of alginate in combination with other compounds, such as natural and synthetic polymers, is widespread in the literature for the production of scaffolds for tissue engineering applications. Pectin, on the other hand, is less explored for this use. Combining chitosan with alginate, or pectin can provide matrices with improved properties when compared to those constituted of the isolated polymers, e.g. mechanical properties, pH stability and fluid uptake. Micro-scale roughness was observed for all formulations, with porous samples presenting more irregular surfaces. The surfaces of alginate-containing samples are more hydrophilic than those of the pectin-containing samples and the addition of K increases the hydrophilicity in both cases. For the hemocompatibility assessment, the free hemoglobin test was performed, in which red blood cells not entrapped in a thrombus formed in the sample surface are hemolyzed and free hemoglobin molecules are colorimetrically measured. The scaffolds presented between 68% and 95% of free hemoglobin when compared to the control Teflon®, commonly used in implantable devices. Ch-A formulations were found to be less thrombogenic than Ch-P, what can be attributed to their surface wettability, as the adsorption of blood proteins would be less intense over their more hydrophilic surface. *In vitro* direct cytotoxicity of scaffolds to fibroblasts varied from 15% and 30%. No trend for the influence of the addition of K to the scaffolds was observed on cytotoxicity. Considering the studied properties of the materials, all formulations presented satisfactory performance and each of them may be directed to particular applications in tissue engineering. Complementary studies on other aspects of the biomaterials biocompatibility should be performed to assure the feasibility of the use of these devices as implantable materials.

The authors are grateful for the finantial support of this work by the Coordination for the Improvement of Higher Education Personnel (CAPES) and the São Paulo Research Foundation (FAPESP), both from Brazil, and from the Natural Sciences and Engineering Research Council (NSERC) of Canada.

DEVELOPMENT OF SUSTAINABLE THERAPEUTIC DRESSINGS CONSISTING OF CHITOSAN-ALGINATE FILMS INCORPORATING *ARRABIDAEA CHICA* VERLOT EXTRACT

Pires, A. L. R.[1]; Westin, C. B.[1]; Sousa, I. M. O.[2]; Foglio, M. A.[2]; Moraes, A. M.[1]

[1]School of Chemical Engineering and University of Campinas (UNICAMP), Campinas, SP, Brazil

[2]School of Pharmaceutical Sciences – University of Campinas (UNICAMP), Campinas, SP, Brazil

Abstract

The incorporation of natural bioactive compounds in wound dressings produced from biopolymers can have a significant impact on the prevention and therapy of infections and inflammatory processes. The complexation of the polysaccharides chitosan (C) and alginate (A) is already relevant in the production of dressings for skin lesions, since both compounds are capable of interfering positively in the healing process. Such natural polymers interact by electrostatic attraction and form thin, homogeneous, transparent and water-insoluble biocompatible films. The addition of the extract of the plant *Arrabidaea chica* Verlot (AcV) to CA films is of particular interest, given that it may exert a synergistic effect in the treatment of lesions due to the anti-inflammatory, anti-hemorrhagic, anti-allergic, antioxidant and antiseptic properties of the plant extract. In this sense, the purpose of the present work was to analyze the effect of incorporating different proportions of the standardized extract of *Arrabidaea chica* Verlot into dense CA films on the characteristics of the resulting biomaterial. The visual aspect, the morphology, the thickness, the uptake of phosphate-buffered saline, the cytotoxicity to fibroblasts (Vero cells) as well as the extract incorporation efficiency and release kinetics were evaluated. The results attained showed that the presence of the AcV extract turned the CA film red, moderately affecting its transparency. The analysis of the surface and cross-sectional morphology of the films by scanning electron microscopy (SEM) indicated that the plant extract was effectively dispersed in the structure of the material, also being noticed a slight increase in the thickness of the material by the incorporation of the AcV extract. Incorporation efficiency values above 80% were observed, and control of *in vitro* release kinetics was noticed for at least 4 hours. In addition, mild stimulation of fibroblast proliferation was observed. In conclusion, films produced by the complexation of chitosan and alginate incorporating the extract of *Arrabidaea chica* Verlot have great applicability potential in the treatment of skin lesions, representing an attractive alternative for the therapy of lesions using dressings obtained from sustainable feedstocks.

Introduction

The natural polymers chitosan (C) and alginate (A) have been widely used in the production of wound dressings, since they can positively assist in the processes of healing and repairing of damaged tissues. Moreover, they are biocompatible and biodegradable.[1,2,3]

At appropriate pH conditions, chitosan and alginate may behave as polymers of opposing charges (positive and negative, respectively), forming, when associated, a polyelectrolyte complex (PEC) that may be shaped as gels, fibers, dense and porous membranes.[4] In addition, the PEC of CA presents greater stability to pH variation, as well as improved swelling when compared to chitosan alone and appropriate control when releasing compounds incorporated in their structure.[5]

The incorporation of natural bioactive compounds into biodegradable polymer matrices can have a significant impact on the prevention and therapy of infections and inflammatory processes occurring on the skin if adequate drug release control takes place, increasing the effectiveness of the treatment.[6] In the present work, the incorporation of the extract of the plant named *Arrabidaea chica* Verlot (AcV) into CA dense membranes was explored.

The genus *Arrabidaea* belongs to the Bignoniaceae family and is generally found in tropical America, from southern Mexico to Brazil, being very common in the Amazon Forest.[7] The extract of the leaves of AcV has been frequently used by the Indians due to its insoluble pigments, which are dark red to brown in colour. The extract has been used directly on the skin as a dye, insect repellent and also for the treatment of diseases, while the leaves are used to cleanse chronic wounds, treat fungal infections and herpes.[8,9]

The medicinal properties of the AcV plant extract are already reported[10,11], including its efficiency in promoting fibroblast growth and collagen synthesis, as well as its antioxidant action and the considerable reduction its use causes in the size of the dermal lesions in rats.[11] Medina et al.[10] also noted significant contraction of the injured area after topical application of nanoparticles loaded with the AcV extract. In addition, the authors observed that low concentrations of the compound were sufficient to maintain cell viability in vitro and that higher concentrations resulted in improvement of cell proliferation.

Thus, due to its widespread use by native populations and the corroboration of its healing effects, this species is present in the list of the Brazilian National Health System (SUS) as a medicinal plant of interest for popular use for the treatment of skin lesions.[12]

Based on the data mentioned above, in the present work we proposed to analyze, through physical-chemical and biological characterization, the potential of dense membranes produced by the complexation of chitosan and alginate and containing a standardized extract of *Arrabidaea chica* Verlot as a dressing skin lesions.

Materials and Methods

Materials

For the production of the membranes, chitosan (product code C3646, lot 109K0043V) and sodium alginate (product code A2033, lot 058K0126) from Sigma-Aldrich were used. The standardized extract of *Arrabidaea chica* Verlot was kindly donated by Prof. Dr. Mary Ann Foglio of the School of Pharmaceutical Sciences of the University of Campinas (Campinas, SP, Brazil). All other reagents were at least analytical grade and the water used was distilled and deionized in the Milli-Q® Millipore system.

Preparation of chitosan-alginate membranes

The preparation of dense chitosan-alginate membranes containing the standardized extract of A. *chica* Verlot was performed according to the procedures proposed by Rodrigues et al.[13] Bueno and Moraes[14] and adapted by Pires e Moraes.[15]

To obtain the membranes, 180 mL of 1% (m/v) chitosan solution in 2% acetic acid (m/v) were mixed with 360 mL of 0.5% (w/v) alginate solution at 25 ° C and 500 rpm. Following, the suspension was further mixed at an stirring rate of 1000 rpm for 10 minutes. Thereafter, aqueous 2M NaOH solution was added to adjust the pH to 7.0, and then 7.2 mL of 2% $CaCl_2$ aqueous solution were added for the primary crosslinking of the alginate carboxyl groups not associated with chitosan amines. Subsequently, aliquots of the AcV extract dissolved in ethanol were added, under mixing at 1000 rpm, aiming to achieve extract to polysaccharide mass ratios of 33.3 to 133.3 mg/g.

The obtained material was then deaerated in a vacuum pump for 120 minutes, poured into four 15 cm diameter polystyrene Petri dishes and dried for 20 hours at 37 ° C.

After drying, each membrane was immersed for 30 minutes in 150 mL of 2% $CaCl_2$ aqueous solution (for a secondary carboxyl crosslinking step), and finally, after being copiously washed with deionized water, the membranes were and dried at room temperature under light protection.

When required, the membranes were sterilized by exposure to ethylene oxide.

Visual aspect and morphology of the membranes

The aspect of the membranes was recorded using a Nikon camera (model COOLPIX S3300). The surface and cross-section morphology of the membranes was analyzed using a scanning electron microscope (Leo 440i model, Leica) after coating the samples with gold.

Thickness and absorption of phosphate buffered solution by the membranes

The mean thickness of the membranes was determined with an IP54 digital micrometer (Mitutoyo, model MDC-25S). The degree of swelling of the dried membranes in phosphate buffered saline (PBS) was analyzed by gravimetry[14].

Toxicity to Vero cells

The indirect cytotoxicity of the formulation considered the most promising to Vero cells was evaluated by the MTT reduction method (3-(4,5-dimethylthiazol-2-yl)-2,5-diphenyltetrazolium bromide). Extracts of the membranes were obtained by exposing them to supplemented RPMI culture medium for 48, at the concentration of 0.05 g of dry material per milliliter of medium.

Incorporation efficiency and analysis of the "Arrabidae chica" Verlot release from the membranes

The evaluation of the extract incorporation efficiency and the analysis of its release from the membranes were performed based on the methodologies used by Bierhalz et al.[16]

Extracts from the membrane samples obtained during the secondary cross-linking procedure and the water washing step were analyzed by spectrophotometry at 470 nm to indirectly determine the mass of extract retained in the membranes. The extract release was evaluated by exposing previously weighed test specimens (2 cm x 2 cm) to a mixture consisting of 8 mL of PBS and 2 mL of ethanol. Periodically, 1 mL of the solution in which the samples were immersed was withdrawn for absorbance analysis at 470 nm and then returned to the vial.

Results and Discussion

Membrane aspect and morphology

Figure 1 shows the visual and morphological aspects of the chitosan-alginate membranes produced in the presence of the standardized extract of *Arrabidaea chica* Verlot in the proportions of 33.3, 66.7 and 133.3 mg of extract per gram of polysaccharide. The samples were respectively denominated as $CA_{33.3}$, $CA_{66.7}$ and $CA_{133.3}$.

The extract of AcV contains pigments of the category of anthocyanins, named carajurin and carajurone. These insoluble pigments are dark red to brown, and these characteristics can be clearly observed in the membranes containing the extract. It is further visually noted that by increasing the concentration of the compound, the reddish color gradually intensifies. Although the transparency of the materials is affected by the increase of the concentration of the bioactive compound, it is still possible to consider that all the devices are sufficiently transparent to allow direct inspection of the wound bed without the need of removing the dressing.

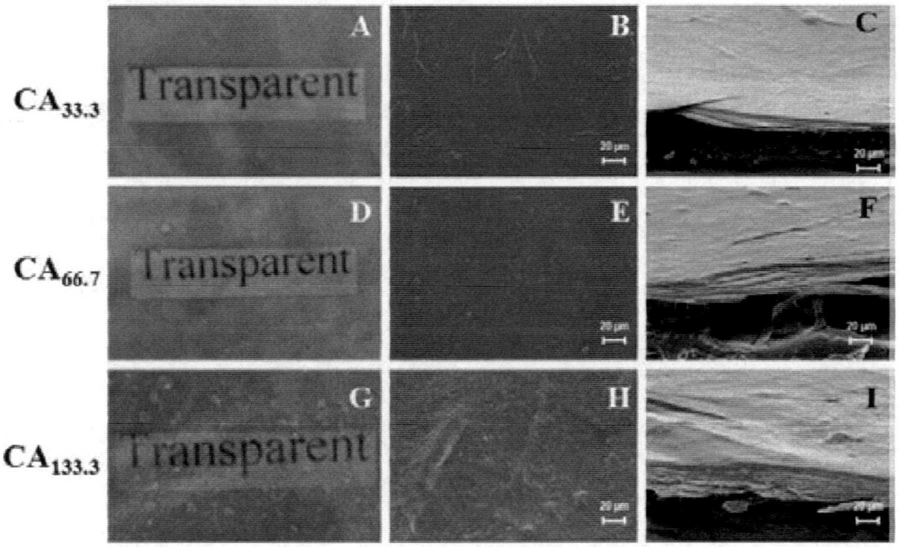

Figure 1 - Visual and morphological aspect of the membranes of CA produced in the presence of AcV extract at the proportions of 33.3, 66.7 and 133.3 mg of extract per gram of polysaccharide.

The surface morphologies and cross-section of the samples show that the addition of AcV extract did not significantly changed the structure of the membranes, which remained homogeneous and had the typical lamellar structure observed for the CA dense matrix free of the extract.[13,14,15] Moreover, the extract can be considered as adequately dispersed in the polymer matrix.

Thickness an absorption of PBS by the membranes

The values obtained for thickness and absorption of PBS by the membranes are shown in Figure 2. Thicknesses varied from 0.06 to 0.08 mm; the formulation $CA_{133.3}$ was significantly thicker than the others, with an expressive increase of 74% when compared to a membrane free of AcV extract. Thus, despite the previous dissolution of the powder plant extract in ethanol, its presence exerted strong influence in this property.

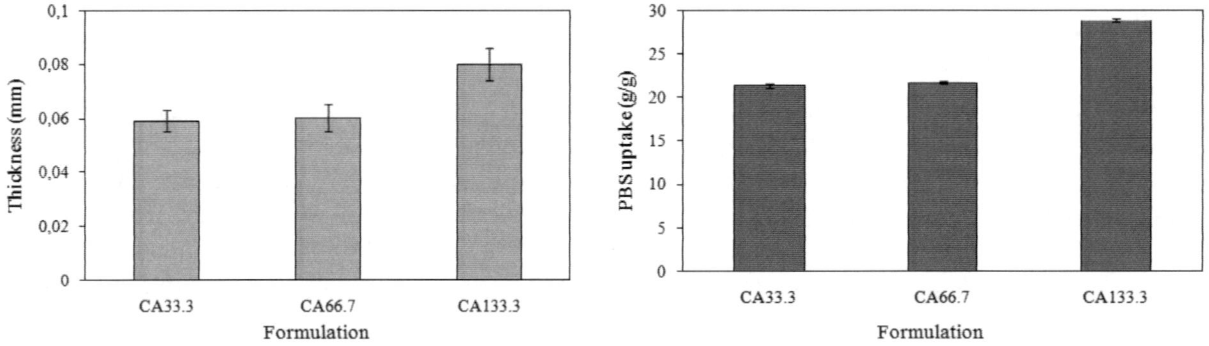

Figure 2 – Thickness and absorption of CA membranes containing different proportions of the AcV extract.

The values of PBS absorption showed the same behavior observed for thickness. The greatest difference was found for the $CA_{133.3}$ formulation. Generally, the amount of fluid absorbed by membranes prepared with the same mass of polymers is directly related to the thicknesses of the specimen, since the greater the thickness of the membrane, the more easily the liquid penetrates the matrix, due to the greater distance of the lamellae.

Efficiency of incorporation, AcV extract release kinetics and cytotoxicity

Figure 3 shows the efficiency of incorporation of the extract added to the mixture of polysaccharides in different proportions, evaluated considering the total anthocyanins as reference compounds in the extract, due to their high light absorption at 470 nm.

In general, the extract incorporation efficiency values were high, varying from about 80 to 88%. The largest value was noted for the $CA_{133.3}$ formulation. Considering that the plant extract is rich in tannins, which are phenolic compounds with the ability to bind strongly to other molecules, it can be assumed that there was an interaction, possibly by hydrogen bonds, between the tannins from the extract and the polysaccharides of the membrane. Such behavior may have contributed to the high efficiency values of incorporation of the AcV extract in the CA matrices.

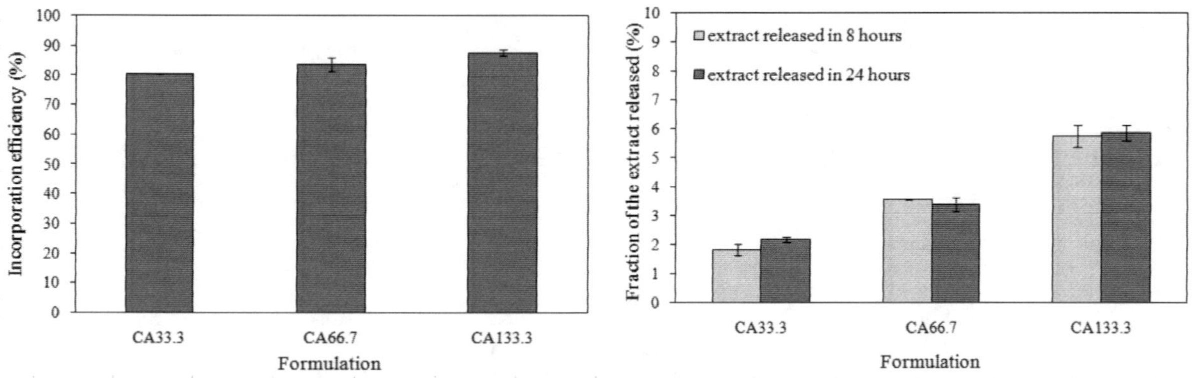

Figure 3 – Efficiency of incorporation and fraction of extract released from the CA membranes containing different proportions of the AcV extract.

Figure 3 also shows the results of the percentage of extract released after 8 and 48 h of exposure to PBS. A tendency of increase in the fraction of extract released with the increase of the initial concentration of the AcV extract added to the membranes was observed. The formulation prepared with the ratio of 133.3 mg of extract per gram of polysaccharides showed the highest values, reaching a release of about 6% of the extract, equivalent to 126 μg/mL. For the formulations prepared with the ratios of 33.3 and 66.7 mg/g, equilibrium was reached already in the first hours of exposure to PBS (4 hours), while for the membrane produced with 133.3 mg of extract per gram of polysaccharides, equilibrium was attained only after 8 hours of exposure.

The $CA_{133.3}$ formulation seems to be the most attractive among those tested due to its high efficiency of incorporation and release of the extract *in vitro*. However, this observation must also be analyzed considering the desirable concentration of active agent released in the lesion, meaning the concentration that results in the closure of the lesion without being irritant or toxic to the patient.

Thus, the cytotoxicity of the eluate collected by exposing the formulation considered as the most promising to supplemented RPMI culture medium was analyzed on Vero cells (fibroblasts) after 24 h by MTT analysis, taking into account that such formulation releases about 126 μg/mL, as mentioned before. The result attained indicated that this formulation not only does not induce deleterious effects on the cells, but also seems to stimulate cell growth.

Conclusion

The incorporation of different proportions of the standardized extract of *Arrabidaea chica* Verlot to CA membranes by direct addition to the polysaccharide mixture turned the biomaterials red and discretely reduced their opacity. The micrometric structure of the material showed no significant variations due to the incorporation of the extract, but the thickness of the membranes tended to increase, as well as their ability to absorb PBS. Efficiencies of incorporation of the extract in the range of 80 to 88% were observed, and the formulation prepared with the proportion of 133.3 mg of AcV per gram of polysaccharides was selected as the most promising because it showed both a high incorporation efficiency and a greater amount of extract released (126 μg/ml). The maximum amount released from the extract did not show cytotoxicity to Vero cells, indicating that CA membranes containing AcV are potentially safe for the proposed purpose, having high applicability as dressings for skin lesions.

Acknowledgements: The authors acknowledge the support from CAPES-Brazil, CNPq-Brazil and FAPESP-Brazil.

References

1. Li, X.; Xie, H.; Lin, J.; Xie, W.; Ma, X. (2009), "Characterization and biodegradation of chitosan-alginate polyelectrolyte complexes," *Polymer Degradation and Stability*, v.94, pp.1-6.
2. Costache, M.C.; Qu, H.; Ducheyne, P.; Devore, D.I. (2010), "Polymer-xerogel composites for controlled release wound dressings," *Biomaterials*, v.31, pp.6336-6343.
3. Dantas, M.D.M.; Cavalcante, D.R.R.; Araújo, F.E.N.; Barreto, S.R.; Aciole, G.T.S.; Pinheiro, A.L.B.; Ribeiro, M.A.G.; Lima-Verde, I.B.; Melo, C.M.; Cardoso, J.C.; Albuquerque Júnior, R.L.C. (2011), "Improvement of dermal burn healing by combining sodium alginate/chitosan-based films and low level laser therapy," *Journal of Photochemistry and Photobiology B: Biology*, v.105, pp.51-59.
4. Khan, F.; Ahmad, S.R. (2013), "Polysaccharides and their derivates for versatile tissue engineering application," *Macromolecular Bioscience*, v.13, pp.395-421.
5. Verma, D.; Desai, M. S.; Kulkarni, N.; Langrana, N. (2011), "Characterization of surface charge and mechanical properties of chitosan/alginate based biomaterials," *Materials Science and Engineering C*, v.31, pp.1741-1747.
6. Dash, M.; Chiellini, F.; Ottenbrite, R.M.; Chiellini, E. (2011), "Chitosan-A versatile semi-synthetic polymer in biomedical application," *Progress in Polymer Science*, v.36, pp.981-1014.
7. Sá, J.C.; Almeida-Souza, F.; Mondêgo-Oliveira, R.; Oliveira, I.S.S.; Lamarck, L.; Magalhães, I.F.B.; Ataídes-Lima, A.F.; Ferreira, H.S.; Abreu-Silva, A.L. (2016), "Leishmanicidal, cytotoxicity and wound healing potential of *Arrabidaea chica* Verlot," *BMC Complementary and Alternative Medicine*, v.16, pp.1-11.
8. Barbosa, W.L.R.; Pinto, L.N.; Quignard, E.; Vieira, J.M.S.; Silva Jr., J.O.C.; Albuquerque, S. (2008), "*Arrabidaea chica* (HBK) Verlot: phytochemical approach, antifungal and trypanocidal activities," *Brazilian Journal of Pharmacognosy*, v.18, pp.544-548.
9. Mafioleti, L.; Silva Junior, I.F.; Colodel, E.M.; Flack, A.; Martins, D.T.O. (2013), "Evaluation of the toxicity and antimicrobial activity of hydroethanolic extract of Arrabidaea chica (Humb. & Bonpl.) B. Verl," *Journal of Ethnopharmacology*, v.150, pp.576-582.
10. Medina, L.S.; Gómez, G.A.; Ortega, A.G.; Sousa, M.O.; Queiroz, N.C.A.; Zago, P.M.W.; Jorge, M.P.; Monteiro, K.M.; Carvalho, J.E.; Román, J.S.; Foglio, M.A.(2015), "Chitosan–

tripolyphosphate nanoparticles as *Arrabidaea chica* standardized extract carrier: synthesis, characterization, biocompatibility, and antiulcerogenic activity," *International Journal of Nanomedicine*, v.10, pp.3897-3909.

11. Jorge, M.P.; Madjarof, C.; Ruiz, A.L.T.G.; Fernandes, A.T.; Rodrigues, R.A.F.; Sousa, I.M.O.; Foglio, M.A.; Carvalho, J.E. (2008), "Evaluation of wound healing properties of *Arrabidaea chica* Verlot extract," *Journal of Ethnopharmacology*, v.118, p.361-366.
12. BRASIL. Ministério da Saúde. RENISUS – Relação de Plantas Medicinais de Interesse ao SUS. Brasília: Ministério da Saúde, 2009. Disponível em: < http://bvsms.saude.gov.br/bvs/sus/pdf/marco/ms_relacao_plantas_medicinais_sus_0603.pdf> [Acessado em: 12/09/2017].
13. Rodrigues, A.P.; Sanchez, E.M.S.; Costa, A.C.; Moraes, A.M. (2008), "The influence of preparation condition on the characteristics of chitosan-alginate dressings for skin lesions," *Journal of Applied Polymer Science*, v.109, pp.2703-2710.
14. Bueno, C.Z.; Moraes, A.M. (2011), "Development of porous lamellar chitosan-alginate membranes: Effect of different surfactants on biomaterial properties," *Journal of Applied Polymer Science*, v.122, pp.624-631.
15. Pires, A.L.R.; Moraes, A.M. (2015), "Improvement of the mechanical properties of chitosan-alginate wound dressings containing silver through the addition of a biocompatible silicone rubber," *Journal of Applied Polymer Science*, v.132, pp.41686-41695.
16. Bierhalz, A.C.K.; Lopes, S.A.; Pires, A.L.R.; Moraes, A.M. (2016), "Development of polysaccharide-based membranes incorporating the bioactive compound aloin," *International Journal of Polymeric Materials and Polymeric Biomaterials*, v.66, pp.193-202.

500970 Comparison of Chitosan Particles Produced By Ionic Gelation and By Supercritical Assisted Atomization

Abstract:

Chitosan is a biodegradable, biocompatible, biological polymer with low toxicity, antimicrobial activity and mucoadhesive properties. Such characteristics make it interesting for use as a controlled release system for a wide variety of drugs. The production of pure chitosan particles or incorporating active compounds can be performed by different methods. In the conventional techniques, for example emulsion, there are some disadvantage such as the difficulty of controlling the particles size and the use of organic solvents, requiring post-processing steps of the particles. Due to these limitations, new techniques such as supercritical technologies, which have the advantage of obtaining particles with greater purity, the possibility of processing thermosensitive molecules and the exemption or drastic reduction in the amount of organic solvents used in production. In this work, a comparison of the particles obtained by ionic gelation methods with the use of TPP obtained (sodium tripolyphosphate) and SAA (Supercritical assisted atomization) technique is carried out. The particles were prepared by ionic gelation, according to the method established by Calvo et al. (1997) where acid solutions of chitosan (0.5-2.0 mg/ml) (low molar mass, deacetylation grade of 93.1%) are mixed with TPP solution by HPLC pump. Stirring was maintained at 11000 rpm (Ultraturrax). The preparation of the particles by SAA was based on the techniques of Reverchon (2006) and Wu et al. (2015). This process is based on the use of a saturator (filled with glass beads) with a temperature ranging from 40-90°C, where the solution of ethanol
(50% v/v), chitosan (0.5-2.0 mg/ml) in acid medium contacts a supercritical CO_2 stream for a period of time to ensure saturation. This mixture is atomized by a capillary tube and sprayed on the precipitator (CO_2 velocity 4.0-9.6 cm^3/min with CO_2 / solution ration of 1.8) together with nitrogen (at high temperature), which has an auxiliary purpose in the evaporation of the used solvent. The characterization of the chitosan particles produced by ionic gelling and SAA were carried out by Dynamic Light Scattering (DLS) analysis and Scanning Electron Microscope (SEM). The particles had zeta potential at least +30 mV, preventing aggregation of the particles. Particles of smaller sizes were obtained by the ionic gelation process at low concentrations of chitosan, the increase of the particle size with the decrease of the zeta potential was also observed. In SAA, the smallest particles were obtained in lower concentrations of chitosan, where it was observed the predominance of the nucleation process in detriment to the growth process.

References:

1. Wu, H.; Lee, H.; Chen, H.; Chien, L. Precipitation kinetics and biological properties of chitosan microparticles produced using supercritical assisted atomization. *Chemical engineering Research and Design*, 104, 615â€"625, 2015.
2. Reverchon, E., Adami, R., Caputo, G. Supercritical assisted atomization: Performance comparison between laboratory and pilot scale. *Journal of Supercritical Fluids*, 37, 298â€"306, 2006.
3. Calvo, P., RemuÃ±Ã¡n-Lopez, C., Vila-Jato, J.L., Alonso, M.J. Novel hydrophilic chitosan-polyethylene oxide

nanoparticles as protein carriers. *International Applied Polymers Science*, 63, 125-132, 1997.

500522 Exploiting a Novel Aqueous-Two Phase Microfluidic System for Cell Encapsulation in GAG+Chitosan Microcapsules

Abstract:

â€œExploiting a novel aqueous-two phase microfluidic system for cell encapsulation in GAG+chitosan microcapsulesâ€

Improving the efficiency of cell encapsulation techniques is a necessary step for the wider application of capsule-based modular tissue engineering. Control over the capsules size and distribution which are the two most important parameters in cell encapsulation can be greatly facilitated using microfluidic techniques. In this study, a new microfluidic system using aqueous two phase system (ATPS) was devised for making fine and uniform capsules for cell encapsulation. ATPS systems have traditionally been used for protein separation due to the moderate conditions they make for the cells. The mild condition is vital for higher cell viability and functionality which is a great advantage compared to the traditional water-oil microfluidics. Dextran and polyethylene glycol (PEG) is the most widely used ATPS used due to their smooth environment for protein separation. In the present work, GAG solution, consisting of 1.5wt% carboxymethylcellulose (CMC), 4 wt% chondroitin sulfate (CSA) and 10-20 wt% dextran forms the disperse fluid in the microfluidic system which phase separated within a 5-10 wt% polyethylene glycol (PEG) leading to the formation of droplets inside the microfluidic. This method allows a core fluid (GAG) to be surrounded by sheath stream flowing (PEG). The surface tension generated between these two immiscible fluids lead to the formation of the droplets. The droplets then go into a stirred 0.6 wt% chitosan solution to generate a polyelectrolyte complex shell around the capsule core. In order to enhance the uniformity and decrease the size of the GAG droplets, an optimized PDMS microfluidic system was utilized. The challenging part in making droplets in the microfluidic system using ATPS systems is the ultra-low interfacial tension of these systems often less than 10^{-4}N/m. This characteristic exerts extra considerations to be taken while making droplets in the ATPS microfluidics. Hence, ultra-low flowrates for both GAG and PEG solutions in the range of 1-100 mlit/min was used. The effect of four influencing parameters including PEG concentration, dextran concentration, PEG flow rate and GAG flow rate on the size and uniformity of the microcapsules were also investigated. Most of the encapsulated cells were in the range of 100-300 Î¼m in diameter which is ideal for encapsulation of cells including hepatocytes and mesenchymal stem cells. It was observed that by increasing the PEG flowrate in fixed GAG flowrates, the capsules diameter would increase and vice versa. This method gives a stable control over the shape, size and thickness of the capsules formed which finally will boost the diffusion of nutrients and oxygen to the cell and secretion of wastes. Reliability, reproducibility and accuracy of the technique leaves it amongst the best novel encapsulation techniques.

499386 Laser-Activated Nanocomposites for Tissue Repair

Abstract:

Sutures and staples are ubiquitous in surgery, but present their own complications and cause local trauma to tissue during application. Glues, adhesives and sealants have been investigated as means for approximating wounds, but many suffer from toxicity and poor strength. Photothermal tissue sealing and repair, facilitated by near-infrared laser-responsive nanoglues, has emerged as a potential alternative to existing tissue apposition methods. Here, we demonstrate the photothermal sealing of lacerted skin and other ruptured tissues using laser activation of novel nanoglues in which, gold nanorods (GNRs) are embedded throughout a polypeptide matrix. Critical factors including laser power density, exposure time, and laser type (continuous wave or pulsed wave) were investigated in order to determine optimal treatment parameters that led to the highest recovery in sealing strength. Laser-activated sealing of full thickness lacerations in skin of live mice resulted recovery of tissue mechanical properties and integrity that were comparable to that seen with commercial sutures. Local inflammation caused by photothermal sealing approach was minimal and was compared to that seen with commercial sutures. Furthermore, laser-activated sealing was also employed to reinforce suture lines in skin, as a combination approach that can provide significant benefit for rapid sealing and healing of skin lacerations. Our results demonstrate that polypeptide-GNR nanoglues may have applications in photothermal sealing and repair of ruptured tissues and may emerge as attractive alternatives to sutures.

496206 New Strategy for the Fabrication of Annular Cylindrical Polysaccharide-Based Scaffolds

Abstract:

Different types of organs and tissues show annular cylindrical geometry *in vivo*, such as ureters, trachea, blood vessels and heart valves. In the case of diseases related to atherosclerosis, direct replacement of damaged blood vessels through surgical intervention is already frequently performed. However, most of the materials currently available for this type of therapy are efficient only when used to substitute blood vessels with diameters greater than six millimeters, as complications may occur in smaller diameter devices, leading to vessel reocclusion. Recent advances in the development of vascular substitutes are consequences of tissue engineering approaches, focusing on the culture of vascular cells on biodegradable scaffolds. Once the seeded cells have deposited an extracellular matrix and the original scaffold is biodegraded, the tissue resembles and behaves as native tissue. Since the scaffold will act as a structural cast for neotissue formation, devices with annular cylindrical geometry are a prerequisite for vascular tissue engineering. Different approaches have been used for the fabrication of tubular scaffolds, including methods based on thermally induced phase-separation, particulate-leaching, dipping in polymeric solution, 3D printing, electrospinning, solvent casting/ freeze-drying and others. In this work, an alternative technique for the fabrication of tubular-shaped scaffolds was used, based on the casting of polymeric solutions by controlled crosslinking mediated by a semipermeable cast. A chitosan-alginate polymeric mixture was used for the production of the scaffold and $CaCl_2$ was used for alginate crosslinking. The crosslinking was performed in two steps, as a single-step can lead to incomplete, heterogeneous reticulation and impairment of the scaffold mechanical properties. $CaCl_2$ concentration and temperature for the secondary crosslinking step were varied. The surfactant Kolliphor® P188 was added to the polymeric mixture to improve the scaffolds porosity by foam formation. The use of a Ca^{+2} ions/polymers ratio of 12.3 g/g resulted in semi-solid tubular structures with *c.a.* 5 mm of internal diameter and 1.5 mm of wall thickness when the reticulation was performed at 4°C. The presence of the surfactant resulted in highly porous structures with both macro and micropores.

The authors are grateful for the financial support of this work by the Coordination for the Improvement of Higher Education Personnel (CAPES) and by the São Paulo Research Foundation (FAPESP), both from Brazil.

498105 Formulation of Peptide Antimicrobials for Treatment of Wound Infections

Abstract:

Cationic antimicrobial peptides (CAMPs) have shown great potential for the treatment of infections involving gram-negative bacteria. However, they are hindered by their short biological life due to hydrolytic degradation in human plasma by proteolytic enzymes. Furthermore, they are not particularly effective once biofilm formation occurs. In order to overcome these challenges, we are developing polyelectrolyte copolymers for the nanoencapsulation of CAMPs. These complexes are formulated to extend the biological lifetime of CAMPs, to provide controlled release from the delivery complex, and to disrupt bacterial biofilms thereby augmenting the activity of the antimicrobial peptide.

Poly(alkylacrylic acid) polymers were prepared via free radical polymerization under N_2 using azobisisobutyronitrile as the initiator. Next, grafted copolymers were prepared by attaching polyetheramine chains to the poly(alkylacrylaic acid) backbones through carbodiimide coupling. All polymers were characterized by NMR and GPC and tested for their ability to form sub-micron particles encapsulating CAMPs that include well established CAMPs such as KSL-W and polymyxin B as well as novel cyclic lipopeptide provided by a collaborator. In contrast to our previous work using these polymers to coat lipoplexes for oligonucleotide delivery, here graft copolymers with poly(methacrylic acid) backbones proved more effective than those built on poly(propylacrylic acid) in encapsulation of peptides and forming stable nano-sized complexes. These complexes are to protect KSL-W, from degradation in human plasma for up to 24 hours while maintaining the free peptideâ€™s activity against the gram-positive bacteria, S. aureus. By varying the graft percentage in the range of 0-10%, we are able to tune the release rate of polymyxin B from 1-10 days. Furthermore, the nanocomplexes were active against *Klebsiella pnemoniae* biofilms, with additive activity resulting from the polymer and the peptide in the nanocomplex.

Together, these results indicate that our graft copolymer formulation is effective at protecting, releasing and potentiating the activity of CAMPs. Future work will explore the formulation of these nanocomplexes within gels for topical administration.

500490 Synthesis and Degradation of Biodegradable Copolymers

Abstract:

Abstract:

Over the last few decades, synthetic biodegradable polymers captured the attention of drug delivery scientists with their prominent features-biocompatibility and *in vivo* degradation into benign materials. Different classes of synthetic biodegradable polymers (polyesters, polyanhydrides and polyethers) have different properties (degradation method and rate, hydrophobicity, and thermal stability) which can be tailored for specific applications. The aim of this project is the optimization of polymer synthesis procedures to tailor biodegradation rate and thermal stability of innovative biodegradable polymers.

The optimization was performed by synthesizing homopolymers of poly(lactic acid) and poly(sebacic acid) by melt polycondensation. The operating parameters investigated were reaction time, temperature, pressure and the effect of acetylation of monomers on degree of polymerization. The resulting polymers were characterized by H-NMR for purity and to determine the average molecular weights of the homopolymers synthesized. Other properties such as Morphology, thermal properties, molecular structure will be studied using Scanning Electron Microscopy (SEM), Dynamic Scanning Calorimetry (DSC) and Fourier transform infrared spectroscopy (FTIR). Lactic acid and sebacic acid were copolymerized to obtain random copolymers. In this study, we are also studying the effect of acetylation on degree of polymerization for lactic acid and sebacic acid. Moreover, we are looking at different conjugating options for coupling blocks of homopolymers of lactic acid and sebacic acid.

To study the influence of monomer composition on biodegradation we are developing a COMSOL Multiphysics model which considers: 1) time dependent variables - number average molecular weight of long chains, number of chain scissions per unit volume, total number of hydrolysable units of all the long chain per unit volume, average degree of polymerization of all the short chains, total number of hydrolysable units of all the short chains formed; and 2) kinematic parameters - diffusion coefficient of short chains in a degrading polymer, porosity of the polymer caused by loss of short chains.

Future work will look at expanding design space of biodegradation rate and thermal stability by introducing additional monomers in the synthesis process.

485235 Ultra-High Surface Area Activated Carbon from a Renewable Resource

Abstract:

High surface area activated carbons are widely utilized in a variety of industrial applications such as purification or separation of liquids and gases, elimination of toxic substances, reaction catalysis, and solvent or natural gas recovery/storage. Traditionally, activated carbons are prepared through gradual gasification of char with either steam, CO_2, or a combination of both, resulting in highly porous structures, and ultimately significant mass reduction (normally 80-90% weight loss) of the starting material. Therefore, agricultural by-products, such as cellulose, are considered as very important precursors to synthesizing activated carbons because they are renewable and available at low-cost. The aim of the present study is to apply lignocellulosic aerogels prepared with ionic liquid solutions as precursors for ultra-high surface area activated carbon. The lignocellulosic aerogels are pyrolyzed and subsequently acid-etched in a tubular furnace under inert atmosphere to form the ultra-high surface area nanostructure. The samples are characterized at each state using scanning electron microscopy (SEM) and nitrogen sorption to study the morphology and the pore structure. BET analysis of the dehydrated aerogels show initial surface areas up to 440 m^2/g with pore diameters of 47nm. The effects of synthesis conditions and time-temperature profile during pyrolysis on the structure of the resulting activated carbon will be studied and the results will be compared with commercial activated carbons derived from coal. PAO #: U17-173

494388 A Facile Novel Fluorocarbon Copolymer Solution Coating Process for Improving Platelet Compatibility of Titanium

Abstract:

Titanium (Ti) and its alloys possess mechanical properties that are desirable in many biomedical applications compared to other metals. Furthermore, the native metal oxide layer that prevents further oxidation are also known to be biocompatible. However, clinical findings have shown that titanium and its alloys are prone to adverse bioreactions such as platelet adhesion and activation which could lead to thrombogenic complications. It has been found that surfaces modified with fluorocarbons could reduce the degree of both platelet adhesion and activation. Nevertheless, direct fluorocarbon deposition onto titanium substrates would require significant technical efforts. Instead, this research utilized a facile coating process with novel copolymers containing 2,2,2-trifluoroethyl methacrylate (TFEMA) and vinylphosphonic acid (VPA) to modify the titanium surface, giving the surface lower surface energy and higher hydrophobicity, significantly reducing the thrombus formation while exhibiting good cytocompatibility. The anchorage group, phosphonic acid provided by VPA, can be covalently bound to the oxide surface of titanium metal. Via free radical polymerization, VPA and TFEMA formed copolymers with different hydrophobicity were then used to modify titanium substrates, on which a series of surface characterization, *in vitro* platelet adhesion tests, and cytotoxicity assays were performed. Nuclear magnetic resonance (NMR) and X-ray photoelectron spectroscopy (XPS) confirmed the synthesis of the copolymers and the modification of Ti substrates. The platelet adhesion tests showed significantly reduced amount of adherent platelets on certain copolymer-modified Ti substrates with low degrees of activation. The *in vitro* cytotoxicity assays further highlighted that the modifications conducted on Ti does not induce cytotoxicity.

490099 Alignment of Quantum Dot Nanorod/Silica Hybrid Particles on Glass Substrate for Luminescent Solar Concentrator

Abstract:

Luminescent solar concentrator (LSC) is a solar energy harvest device which converts light energy into electrical energy. It mainly consists of luminescent materials embedded in polymer matrix and photovoltaic cells at the edge. The luminescent materials can absorb incident light and deliver it to the edge as emitted light at a certain wavelength. Due to the broad absorption and narrow emission, quantum dots (QDs) have been widely used as luminescent materials in LSC. Among various types of QDs, quantum dot nanorods (QDNRs) have recently drawn considerable interest. Unlike spherical QDs, QDNRs emit the polarized light along the long axis. Therefore, if QDNRs are vertically aligned into the polymer matrix in LSC, the light energy can be efficiently delivered to the edge photovoltaic cells and conversion efficiency is enhanced. To align the QDNRs on glass substrate, we prepare QDNR/silica hybrid particles by sol-gel reaction and vertically align hybrid particles on glass substrate. We use SEM and TEM techniques to confirm the alignment and characterized the optical properties by UV and PL spectrometer. We believe that vertical align method using QDNRs/silica hybrid particles have potential in luminescent solar concentrator.

490122 Effect of Graphene Oxide on Formation of Zirconium Tungstate Nanoparticles

Abstract:

Development of novel inorganic nanofillers is important in the processing of semiconductors. For instance, incorporation of inorganic nanoparticles in polymer composites can alleviate thermal stresses by reducing a difference in the thermal expansion coefficients between polymer and metal or ceramic materials. In particular, zirconium tungstate (ZrW_2O_8) has attractive physical properties for a nanofiller because it has a large isotropic negative thermal expansion coefficient in the temperature range of 0.3 K to 1050 K. Although several methods for synthesis of ZrW_2O_8 particles have recently been reported, most of the proposed protocols provide rod-shaped particles in micrometer range. To achieve desired particle properties, however, control on the size and shape of these particles is required. In this study, we investigate roles of graphene oxide on the formation of ZrW_2O_8 particles. First, we prepare $ZrW_2O_7(OH)_2$ that is a precursor of ZrW_2O_8 by a hydrothermal method. Next, $ZrW_2O_7(OH)_2$ is calcined at 900 K, which transforms $ZrW_2O_7(OH)_2$ to ZrW_2O_8. We investigate effects of graphene oxide and other reaction parameters on the size and shape of ZrW_2O_8 particles. We expect this study to help control the size and morphology of ZrW_2O_8 particles for semiconductor applications.

492105 Rapid Microwave-Assisted Synthesis of Hybrid Zeolitic-Imidazolate Frameworks

Abstract:

Separating a crude component into a more valuable purer product is a major part of many industrial chemical plants. Typically these processes utilize distillation, which account to 10 – 15 % of world's energy consumption.[1] One alternative solution that can reduce energy consumption up to 90 % relative to distillation is through membrane separation using porous solid material such as zeolites and metal organic frameworks (MOFs).[2] Zeolitic-imidazolate frameworks (ZIFs),[3] a subclass of MOFs, in particular have attracted many attentions due to their chemical/thermal stabilities, their ultra-microporosities, and high surface area when compared to other MOFs material.[3] ZIFs possess zeolite-like topologies in which the tetrahedral Si or Al and the bridging O in zeolite structure are replaced by transition metals (such as Zn,[3] Co,[4] and Cd[5]) and imidazolate-derived ligands. A common drawback for membrane gas separation is the limited availability of aperture pore size and functionalities of the porous materials. Studies have shown that through mixing metals[6,7] and ligands,[8,9] one can continuously tune the MOFs/ZIFs (termed hybrid ZIFs) properties to match with the characteristic of specific gas mixture. However, the common synthesis of hybrid ZIFs has generally been through slow diffusion techniques,[4,10] conventional hydrothermal,[7,11] and solvothermal methods,[9,12,13] requiring several hours up to days of synthesis time. In addition, the yields of those synthesis protocols are often low, an economical issue for hybrid ZIFs to be used in a large industrial scale.

Herein we plan to present a new microwave-assisted synthetic strategy to rapidly prepare hybrid zeolitic-imidazolate frameworks (ZIFs): ZIFs with mixed metal centers and/or mixed linkers.[14] The microwave-based method significantly shortens synthesis time, produces higher yield, substantially reduces the amount of ligand, and eliminates the use of deprotonating agents. X-ray diffraction pattern reveals a mixed metal CoZn-ZIF-8 (i.e., ZIF-8 with both Co and Zn centers) maintains the sodalite (SOD) zeolitic topology from ZIF-8 parent. Elemental mapping using Energy-dispersive X-ray spectroscopy (EDS) and electronic/geometric information obtained from X-ray absorption spectroscopy (XAS) confirm the uniform distribution of tetrahedral Co and Zn metal centers within the same framework of the mixed-metal ZIF. The metal to nitrogen (M-N) stretching frequencies on IR band were observed to be systematically blue-shifted as the Co/Zn ratio in the mixed metal ZIF increases. Furthermore, for the first time, a hybrid ZIF with both mixed metal centers (Co and Zn) and mixed linkers (2-methylimidazolate and benzimidazolate) was prepared through one-step microwave synthesis. Finally, a mixed metal CoZn-ZIF-8 with the Co/Zn ratio of ~ 1 was grown as membranes on porous α-Al2O3 supports, showing higher propylene/ propane separation factor (~120) when compared to pure Zn-ZIF-8 membranes (~63) prepared with similar method.

References:

1. *Materials for Separation Technologies: Energy and Emission Reduction Opportunities*; Technical Report: Oak Ridge, TN **2005**.

2. Sholl, D. S.; Lively, R. P. Seven Chemical Separations to Change the World. *Nature* **2016**, *532*, 435-437.

3. Park, K. S.; Ni, Z.; Cote, A. P.; Choi, J. Y.; Huang, R.; Uribe-Romo, F. J.; Chae, H. K.; O'Keeffe, M.; Yaghi, O. M. Exceptional chemical and thermal stability of zeolitic imidazolate frameworks. *Proc. Natl. Acad. Sci. U. S. A.* **2006**, *103*, 10186-10191.

4. Banerjee, R.; Phan, A.; Wang, B.; Knobler, C.; Furukawa, H.; O'Keeffe, M.; Yaghi, O. M. High-Throughput Synthesis of Zeolitic Imidazolate Frameworks and Application to CO2 Capture. *Science* **2008**, *319*, 939-943.

5. Karagiaridi, O.; Bury, W.; Sarjeant, A. A.; Stern, C. L.; Farha, O. K.; Hupp, J. T. Synthesis and characterization of isostructural cadmium zeolitic imidazolate frameworks via solvent-assisted linker exchange. *Chem. Sci.* **2012**, *3*, 3256-3260.

6. Wang, C.; Yang, F.; Sheng, L.; Yu, J.; Yao, k.; Zhang, L.; Pan, Y. Zinc-Substituted ZIF-67 Nanocrystals and Polycrystalline Membranes for Propylene/Propane Separation. *Chem. Comm.* **2016**, *52*, 12578-12581.

7. Kaur, G.; Rai, R. K.; Tyagi, D.; Yao, X.; Li, P.-Z.; Yang, X.-C.; Zhao, Y.; Xu, Q.; Singh, S. K. Room-temperature synthesis of bimetallic Co-Zn based zeolitic imidazolate frameworks in water for enhanced CO2 and H2 uptakes. *J. Mater. Chem. A* **2016**, *4*, 14932-14938.

8. Eum, K.; Jayachandrababu, K. C.; Rashidi, F.; Zhang, K.; Leisen, J.; Graham, S.; Lively, R. P.; Chance, R. R.; Sholl, D. S.; Jones, C. W.; Nair, S. Highly Tunable Molecular Sieving and Adsorption Properties of Mixed-Linker Zeolitic Imidazolate Frameworks. *J. Am. Chem. Soc.* **2015**, *137*, 4191-4197.

9. Thompson, J. A.; Blad, C. R.; Brunelli, N. A.; Lydon, M. E.; Lively, R. P.; Jones, C. W.; Nair, S. Hybrid Zeolitic Imidazolate Frameworks: Controlling Framework Porosity and Functionality by Mixed-Linker Synthesis. *Chem. Mater.* **2012**, *24*, 1930-1936.

10. Phan, A.; Doonan, C. J.; Uribe-Romo, F. J.; Knobler, C. B.; O'Keeffe, M.; Yaghi, O. M. Synthesis, Structure, and Carbon Dioxide Capture Properties of Zeolitic Imidazolate Frameworks. *Acc. Chem. Res.* **2010**, *43*, 58-67.

11. Roshan, R. K.; Babu, R.; Jeong, H.-M.; Hwang, G.-Y.; Jeong, G.-S.; Kim, M.-I.; Kim, D.; Park, D.-W. A solid solution zeolitic imidazolate framework as a room temperature efficient catalyst for the chemical fixation of CO2. *Green Chem.* **2016**, *18*, 6349-6356.

12. Wu, T.; Bu, X.; Zhang, J.; Feng, P. New Zeolitic Imidazolate Frameworks: From Unprecedented Assembly of Cubic Clusters to Ordered Cooperative Organization of Complementary Ligands. *Chem. Mater.* **2008**, *20*, 7377-7382.

13. Rashidi, F.; Blad, C. R.; Jones, C. W.; Nair, S. Synthesis, characterization, and tunable adsorption and diffusion properties of hybrid ZIF-7-90 frameworks. *AIChE J.* **2016**, *62*, 525-537.

14. Hillman, F.; Zimmerman, J. M.; Paek, S.-M.; Hamid, M. R. A.; Lim, W. T.; Jeong, H.-K. Rapid microwave-assisted synthesis of hybrid zeolitic-imidazolate frameworks with mixed metals and mixed linkers. *J. Mater. Chem. A* **2017**, *5*, 6090-6099.

501141 Facile Synthesis of Cd-Substituted Zeolitic-Imidazolate Framework Cd-ZIF-8 and Mixed-Metal Cdzn-ZIF-8

Abstract:

Facile Synthesis of Cd-substituted Zeolitic-imidazolate Framework Cd-ZIF-8 and Mixed-metal CdZn-ZIF-8

Jingze Sun,[1] Liya Semenchenko,[2] Woo Taik Lim,[3] and Hae-Kwon Jeong*[1, 4]

[1]Artie McFerrin Department of Chemical Engineering and [4]Department of Materials Science and Engineering, Texas A&M University, College Station, TX 77843-3122

[2]Department of Chemical Engineering, University of Florida,

Gainesville, FL 32611

[3]Department of Applied Chemistry, College of Natural Science, Andong National University, 388 Song-chon Dong, Kyungpook 760-749, Korea

* Corresponding author: hjeong7@tamu.edu

Zeolitic-imidazolate frameworks (ZIFs),[1] are of great practical interest due to their stable chemical nature and ultra-microporosities. ZIFs are composed of tetrahedrally coordinated transition-metals (i.e., Zn, Co, Cd,) and bridged by imidazolate-derived linkers. Among many ZIFs, ZIF-8,[1] constituted of zinc and 2-methylimidazole (hereafter mIm) forming sodalite (SOD) topology, has been most extensively studied for its use in gas separation applications[2, 3] mainly due to its robust synthesis protocols, thermal/chemical stability, and well-defined aperture size. ZIF-8 has shown to be extremely effective for the kinetic separation of propylene from propane.[2, 4]

As with any other crystalline molecular sieves such as zeolites, the available aperture sizes of ZIFs are fixed and discrete, thereby limiting their gas separation applications. For example, while high-quality ZIF-8 membranes show impressive propylene/propane selectivity (as high as ~ 200),[5] their ethylene/ethane separation performances are negligible.[6]

Various strategies have been applied for changing the effective aperture size of ZIF-8 systematically, including varying organic linkers, varying metal nodes, thermal treatment for removing methyl groups,[7] and mixed linker approaches.[8] Jeong and his co-workers[9] have demonstrated mixed metal approach to be effective by introducing Co centers along with Zn centers in ZIF-8 (i.e., mixed metal CoZn-ZIF-8). However, mixed metal approach is limited by the number of metal elements that can be incorporated into ZIF-8 framework.

Cadmium can be added into the toolbox for mixed metal approaches to systemically vary the effective apertures of ZIF-8 and its iso-structures. There is, however, only one synthesis recipe[10]reported for Cd-ZIF-8 (formally as CdIF-1). Unfortunately, the reported recipe is not compatible with those recipes for Zn- and Co-ZIF-8 (formally named ZIF-67), thereby limiting its use in mixed metal approach.

Here, we would like to present a facile solvothermal recipe for phase-pure Cd-ZIF-8 in methanol, thereby compatible with the recipes of Zn- and Co-ZIF-8. The presence of an organic base, tetraethylamine (TEA), was found critical not only for the formation of Cd-ZIF-8 but also for the inhibition of phase transformation from Cd-ZIF-8 into other phases. The single crystal X-ray structure of Cd-ZIF-8 was solved performed and compared with those of Zn-ZIF-8 and Co-ZIF-8 at the same condition. Finally, for the first time, mixed-metal CdZn-ZIF-8 crystals with a various Cd/Zn ratio were solvothermally synthesized, demonstrating the opportunity for varying the effective aperture sizes of ZIF-8 iso-structures.

References and Notes:

(1) Park, K. S.; Ni, Z.; Cote, A. P.; Choi, J. Y.; Huang, R. D.; Uribe-Romo, F. J.; Chae, H. K.; O'Keeffe, M.; Yaghi, O. M., Exceptional chemical and thermal stability of zeolitic imidazolateframeworks. *PNAS* **2006,** 103, (27), 10186-10191.

(2) Kwon, H. T.; Jeong, H. K., Highly propylene-selective supported zeolite-imidazolate framework (ZIF-8) membranes synthesized by rapid microwave-assisted seeding and secondary growth. *Chem. Commun.* **2013,** 49, (37), 3854-6.

(3) Zhang, K.; Lively, R. P.; Zhang, C.; Koros, W. J.; Chance, R. R., Investigating the Intrinsic Ethanol/Water Separation Capability of ZIF-8: An Adsorption and Diffusion Study. *J. Phys. Chem. C* **2013,** 117, (14), 7214-7225.

(4) Pan, Y. C.; Li, T.; Lestari, G.; Lai, Z. P., Effective separation of propylene/propane binary mixtures by ZIF-8 membranes. *J. Membr. Sci.* **2012,** 390, 93-98.

(5) Lee, M. J.; Kwon, H. T.; Jeong, H. K., Defect-dependent stability of highly propylene-selective zeolitic-imidazolate framework ZIF-8 membranes. *J. Membr. Sci.* **2017,** 529, 105-113.

(6) Bux, H.; Chmelik, C.; Krishna, R.; Caro, J., Ethene/ethane separation by the MOF membrane ZIF-8: molecular correlation of permeation, adsorption, diffusion. *J. Membr. Sci.* **2011,** 369, (1), 284-289.

(7) Gadipelli, S.; Travis, W.; Zhou, W.; Guo, Z., A thermally derived and optimized structure from ZIF-8 with giant enhancement in CO 2 uptake Energ. Environ. Sci.

2014, 7, (7), 2232-2238.

(8) Liao, Y. T.; Dutta, S.; Chien, C. H.; Hu, C. C.; Shieh, F. K.; Lin, C. H.; Wu, K. C. W., Synthesis of Mixed-Ligand Zeolitic Imidazolate Framework (ZIF-8-90) for CO2 Adsorption. *J. Inorg. Organomet. Polym. Mater.* **2015,** 25, (2), 251-258.

(9) Hillman, F.; Zimmerman, J. M.; Paek, S.-M.; Hamid, M. R. A.; Lim, W. T.; Jeong, H.-K., Rapid microwave-assisted synthesis of hybrid zeolitic_imidazolate frameworks with mixed metals and mixed linkers. *J. Mater. Chem. A* **2017,** 5, (13), 6090-6099.

(10) Tian, Y. Q.; Yao, S. Y.; Gu, D.; Cui, K. H.; Guo, D. W.; Zhang, G.; Chen, Z. X.; Zhao, D. Y., Cadmium imidazolate frameworks with polymorphism, high thermal stability, and a large surface area. *Chemistry* **2010,** 16, (4), 1137-41.

495679 A Study of Asymmetry Cu-MOFs Electrode Prepared *in situ* and Its Biomimetic Catalysis

Abstract:

A study of asymmetry Cu-MOFs electrode prepared *in situ* and its biomimetic catalysis

Zhipeng Li, Liwei Ren, Diannan Lu*

Department of Chemical Engineering, Tsinghua University, Beijing 100084, China

To whom it corresponds: ludiannan@tsinghua.edu.cn

Metal–organic frameworks (MOFs) are compounds consisting of metal center or clusters coordinate to organic ligands. MOFs have been widely used in almost essential fields like luminescence, catalysis, storage or separation of gases [1-3]. Traditional MOFs with symmetric structure are synthesized using hydrothermal method [4]. In this work, we *in situ* synthesized the asymmetric MOFs, Cu-bipy-BTC, on the surface of gold electrode directly. The structure of Cu-bipy-BTC MOFs on the electrode surface was characterized using FT-IR, XRD, SEM and EDS. It is shown that the newly *in situ* synthesized Cu-bipy-BTC MOFs have different active metal centers, which are located in different hydrophobic environment. More importantly, the MOFs exhibit flexible structure, which responses to the electric field applied on it. This flexible structure is very similar to that of active site of laccase [5], giving better catalytic activity. This MOFs electrode enables the catalytic oxidation of catechol at + 0.4 V(vs Ag/AgCl). The amperometric responses are linear with concentrations of catechol ranged from 10 to 250 μM and 250 to 1000 μM with the sensitivity of 1.6224 $\mu A \cdot cm^{-2} \cdot \mu M^{-1}$ and 0.2595 $\mu A \cdot cm^{-2} \cdot \mu M^{-1}$, respectively. Compared with other catechol electrochemical sensor, this asymmetric MOFs-based electrode has advantages of higher sensitive and easy-made, which has potential application in the detection of phenols.

Reference:

with Expanding Pore Size and Tunable Luminescent Properties[J]. Chemistry of Materials, 2015, 27(15): 5332-5339.

[2] Schlesinger M, Schulze S, Hietschold M, et al. Evaluation of synthetic methods for microporous metal–organic frameworks exemplified by the competitive formation of $[Cu_2(btc)_3(H_2O)_3]$and$[Cu_2(btc)(OH)(H_2O)]$[J]. Microporous and Mesoporous Materials, 2010, 132(1): 121-127.

[3] Ren J, Musyoka N M, Langmi H W, et al. Hydrogen storage in Zr-fumarate MOF[J]. International Journal of Hydrogen Energy, 2015, 40(33): 10542-10546.

[4] Mao J, Yang L, Yu P, et al. Electrocatalytic four-electron reduction of oxygen with Copper (II)-based metal-organic frameworks[J]. Electrochemistry Communications, 2012, 19: 29-31.

[5] Song L F, Jiang C H, Jiao C L, et al. Two new metal– organic frameworks with mixed ligands of carboxylate and bipyridine: synthesis, crystal structure, and sensing for methanol[J]. Crystal Growth & Design, 2010, 10(12): 5020-5023.

498414 Protected SiC Catalyst Support for Steam Methane Reforming Reaction

Abstract:

There is an ever-growing need for hydrogen for use in petroleum refining, and in the production of methanol and ammonia production. Steam methane reforming (SMR) is an important reaction due to its importance in hydrogen production. SMR is a highly endothermic reaction (ΔH_R = 206 kJ/mol). This implies that heat must be delivered to surfaces onto which SMR occurs efficiently. Thus, high thermal conductivity catalysts and catalyst supports will enable reactors to operate with lower energy input which will lead to lowering the operating costs. The spillover effect is that hydrogen can be produced at lower cost and this leads to cost savings and higher profitability for hydrogen manufacturers. Current catalyst designs don't offer the advantage of efficient heat conductivities.

Nexceris has designed an innovative catalyst, trademarked *HeatPath*^{TM}. This technology was developed after a series of modeling studies suggested that the impact of increasing the efficiency of heat delivery on hydrogen production was significantly larger than the impact of increasing catalyst activity. Being an equilibrium-controlled reaction, conventional catalysts deliver the needed activity in SMR and so there is less value in designing a higher activity catalyst. However, in terms of lowering the cost of hydrogen production, energy input is a major variable that ought to be controlled. *HeatPath*^{TM} has enhanced heat transfer, which leads to increased methane reforming at relatively lower temperatures. Here, we will review the development pathway, as well as characterization results of *HeatPath*^{TM}, and will discuss why this technology offers the best opportunity for lowering the cost of hydrogen production.

492745 Computational Screening of High Temperature Materials for Environmental Barrier Thin Films

Abstract:

Environmental barrier coatings are an important technology for the commercialization of many high temperature refractory materials in oxidative environments. These films improve the lifetime of corrosion sensitive materials for application such as gas turbine engines, space vehicles, and soar-chemical applications. We present a computational approach to compare candidate coating materials and screen for ideal chemical properties based on the desired application. We have targeted a variety of high temperature coating materials to model chosen based on thermal properties, structural characteristics, and stability in oxidative environments. We have grown ALD films of a selection of materials predicted to be promising based on the computational results, and measured the oxidation resistance with thermogravimetric analysis. We present that these coatings improve the oxidation resistance of SiC by up to 64% at 1000°C. We expect to utilize this computational screening method to develop more advanced stabilizing coatings for SiC and the myriad of other applications that require high-temperature oxidation resistant materials.

494562 Phase Transformation Induced By Tertragnility Variation of Metal-Redox Synthesised NiMn Nanoalloys

Abstract:

Phase Transformation Induced by Tertragnility Variation of Metal-redox Synthesised NiMn Nanoalloys

Jian Shen[1,2*], Jin Xin[2*]

1 Water Environment Technology Research Institute, MCC, Nanjing 210019, China

2 State Key Laboratory for Heavy Oil Processing, China University of Petroleum, Qingdao 266580, China

Introduction

Advanced permanent magnets have emerged as a critical component for modern society from power electronics and data storage to green energy-related generators and electric motors. Manganese (Mn), as a rare-earth-free metal, attracts wide interest because of the largest magnetic moment among all transition metals.Such large unpaired electrons and high magnetic moment in Mn-based alloys are believed to be responsible for their complex structural and magnetic phases. One of the long range chemically ordered MnNi alloys is found in the antiferromagnetic equiatomic $Mn_{50}Ni_{50}$ (MnNi ¦È phase) compound, which adopts a face centered tetragonal (fct) L1 0 AuCu structure. Generally, the MnNi ordered structure tends to phase transform from a body centered cubic (bcc) structure after annealing at 973 K. At present, most studies in tetragonal MnNi growth are reliant on the energy-consuming processing (ball milling or arcmelting). The challenge of MnNi nanosynthesis arises from the simultaneous reducing and alloying between manganese and nickel due to their different reduction potentials and rates.

Iron cobalt (FeCo), as a ¡°soft¡± magnetic materials, has a body-centered cubic (bcc) nanostructure with small magnetocrystal-line anisotropy energy (MAE), limiting their application in energy-critical technologies requiring high energy product. Though computational efforts have predicted a giant uniaxial MAE in tetragonal FeCo nanostructures, it is still a challenging task to grow metastable tetragonal FeCo nanostructures. Here we adopt a unique synthesis utilizing the metal-red[1]ox method for MnNi synthesis[1]. Moreover, we report a facile epitaxial metal-redox growth method to deposit the FeCo shell onto NiMn core (Figure 1) to enable the core/shell nanostructures, where we select the lattice-matched NiMn and FeCo phases as a prototypical example to illustrate the effect of tetragonal-effects-induced phase transformation on the magnetic performance of FeCo shell.

Materials and Methods

Manganese carbonyl, nickel(II) acetylacetonate, iron pentacarbonyl, cobalt carbonyl, 1-octadecene, oleic acid, oleylamine, hexane, chloroform, all the chemicals and solvent were purchased from Sigma-Aldrich, and used without further purification. All synthesis procedures were operated in the Schlenk-line and the final samples were stored in the glove-box. TEM and VSM was adopted to characterize structural and magnetic properties, respectively.

Figure 1. The schematic structural evolution of an FeCo shell induced by the tetragonality change of the MnNi phase under phase transformation.

Results and Discussion

To improve its MAE and utilize tetragonality change of MnNi phase, the FeCo shell is epitaxially grown on the MnNi nanocrystals. Figures 2(a)-(f) show the structural and elemental mapping images of MnNi-FeCo core-shell nanostructures. As shown in figure 2(g), the optimum coercivity of FeCo reaches 714Oe with an average shell thickness of 2.0 nm after thermal annealing at 673 K.

Figure 2. (a)–(f) TEM image, HRTEM image, STEM image and elemental mapping images of MnNi–FeCo nanoalloys. (g) The shell thickness (the stoichiometry of FeCo shell at 45:55, blue) and stoichiometry of FeCo (the average FeCo shell thickness of 2.0 nm, red) dependent coercivity of MnNi@FeCo core@shell nanoalloys.

The annealing temperature dictates the tetragonality of MnNi ¦È phase, which directly controls the tetragonal distortion in the FeCo shell. As shown in Figure 3(a)-(b), the results matches the concepts on temperature-dependent tetragonality of the MnNi ¦È phase. In order to verify phase transformation induced by tertragnility variation, we calculated the Bain strain mismatch of the (111)-(110) structure relationship between MnNi core and FeCo shell[2]. The calculated mismatches between L10-MnNi/bct-FeCo and L10-MnNi/bcc-FeCo were 3.7% and 6.9%, respectively, which confirms the stable interfacial structure in (111) L10-MnNi core and (110) bct-FeCo shell.. Therefore, the tetragonality change of the L10 MnNi phase triggers tetragonal distortion in the FeCo shell phase for the controlled magnetic coercivity. Ultimately, MnNi-FeCo nanocrystals with high coercivity and saturation magnetization were obtained (940 Oe and 154 $emug^{-1}$).

Figure 3. (a) The magnetic hysteresis (M–H) loops of MnNi-FeCo nanocrystals with an average shell thickness of 2.0 nm and thestoichiometry of FeCo 45:55, annealed at temperatures from 573 K to 873 K. (b) The annealing temperature-dependent coercivity of MnNi-FeCo nanocrystals with an average shell thickness of 2.0 nm and the stoichiometry of FeCo 45:55. (c)–(g) HRTEM image, STEM image and elemental mapping images of MnNi–FeCo nanoalloys with an average shell thickness of 2.0 nm and the stoichiometry of FeCo 45:55 after annealing at 673 K.

Significance

Phase transformation induced by tetragonality change for tunable MAE, which open up a new strategy to grow metastable tetragonal nanostructures with high MAE.

Reference

1£®Alec, K., Shen, J., Gong, M., Cui, J., Ren, S. Chemistry of Materials 27 (2015) 4677-4681.

2£®Shen, J., Dai, Q., Ren, S. Nanotechnology 27 (2016) 10LT01.

---

* Corresponding Author: shenj1983@126.com (J.S.); james_jinxin@hotmail.com (X.J.)

495324 Fabrication of Electrospun Mesoporous Silica Nanomaterials for Water Vapor Adsorption

Abstract:

To enhance the efficiency of thermally driven adsorption/desorption cycles for desalination or chillers, the development of novel nanoporous adsorbents which have high adsorption capacity as well as the efficiency of heat transfer across the adsorbents and heat exchangers has long been considered. Electrospinning process is a simple and versatile technique to produce inorganic materials into nanostructures combined with sol-gel method. Indeed, nanostructured mesoporous fibers are under investigation of the porous and hollow structure formation and thermal stability for applications. Here, we describe a systematic approach for the fabrication of electrospun mesoporous silica materials for water vapor adsorption by considering that structural parameters and nonwoven architecture. Surfactant-templated mesoporous silica materials were prepared via electrospinning process by adjusting the parameters: applied voltage, feeding rate, and etc. The physico-chemical and mechanical features have been studied in more detailed. The water vapor adsorption capacities of the prepared mesoporous silica fibers were evaluated for assorted temperature conditions. The results showed that the water vapor adsorption characteristics were dominated by the crystalline phase and the morphology of the mesoporous silica fibers. Hence, electrospun mesoporous silica materials are promising candidates for the adsorption desalination and the cooling applications.

496777 Applications of Mesoporous $RuCo_2O_4$ Thin Film for High Performance Supercapacitor

Abstract:

Applications of mesoporous $RuCo_2O_4$ thin film for high performance supercapacitor

*Nilesh R. Chodankar, Do-Heyoung Kim**

**kdhh@chonnam.ac.kr*

School of Chemical Engineering, Chonnam National University, Gwangju 500-757, South Korea
Supercapacitors have high power capability, excellent cycling lifetime and better efficiency; however, to enable their use at wider level it is essential to step-up their energy storing capacity. The design of nanostructured electrode material with higher capacitance, rate capability and excellent cycling stability has emerged as prime theme of the supercapacitor research. In present work, we are reporting the facial electrodeposition of binary metal oxide ($RuCo_2O_4$) on the low cost stainless steel mesh substrate. The nanostructured $RuCo_2O_4$ electrode contain the large number of mesoporous clusters, leaving an ample of free space for easy electrochemical reactions. Strikingly, this unique nanostructure provides an excellent electrochemical performance in terms of high energy storing capacitance (more than 1000 F/g) with excellent rate proficiency and long-lasting operating cycling stability (around 10000 cycles). More importantly, the electrochemical investigation is carried out to confirm the origin of capacitive behavior from the dual surface-controlled and diffusion-controlled process. Furthermore, a complete flexible solid-state supercapacitor device was assembled with $RuCo_2O_4$ electrodes which produce remarkable values of energy density, which is far superior as compared to the carbon and unitary metal oxide based symmetric supercapacitors.

497312 TiO_2 Thin Film Deposition By Electrospray

Abstract:

Electrospray atomization has unique capabilities for deploying colloidal suspensions to create nanoparticle films and coatings using high electric potentials. In this research, electrospray was used to deposit TiO_2 prepared from $TiCl_4$ via a low-temperature solution process. The thickness of the TiO_2 thin film was modulated by varying the electrospray time. The growth mechanism of the TiO_2 film was investigated using scanning electron microscopy. The crystallinity of the film before and after annealing was measured to study the effect of heat treatment on the deposit structure. To evaluate its function, a TiO_2 film was deposited on to a transparent conductive oxide to investigate its application for dye sensitized solar cells.

499948 Green Synthesis of Copper Oxide Nanoparticles Using a Simple Microwave-Assisted Method

Abstract:

In this study, a facile microwave-assisted synthesis method for colloidal copper oxide nanoparticles (CuNPs) has been designed using an environmentally benign approach with glucose and starch as the reducing and capping agents. The electrical, catalytic, sensing and surface properties of copper based nanomaterials find increasing applications in conductive films, supercapacitors, nanofluids, catalysis and as a potential antimicrobial agent. Due to their low cost and unique properties copper based nanoparticles are good candidates for the afore mentioned applications. Stability is of crucial importance for synthesis of colloidal nanoparticles. Without proper protection, the CuNPs tend to aggregate severely. Stable CuNPs were obtained using starch as a capping agent resulting into small-sized nanoparticles preventing their aggregation. The colloidal CuNPs produced were characterized using X- ray diffraction (XRD), scanning electron microscopy (SEM), UV- visible spectrophotometry and inductively coupled plasma-atomic emission spectrometry (ICP- AES); whereÂ the crystalline structure and size of the CuNPs was confirmed by XRD peaks, SEM analysis provided the morphology of prepared nanoparticles, Surface Plasmon Resonance (SPR) of the synthesized CuNPs obtained by UVâ€“Vis spectroscopy showed the reduction of copper ions and reaction conversion. Also, to study the reaction yield, unreacted copper was detected by ICP-AES. We further looked at potential electronic applications of the nanoparticles which were studied using cyclic voltammetry (CV). The effects of temperature, pressure, time, power requirement and molar concentrations of reactants were studied to fix the conditions of the reaction system. We have explored a possible mechanism for the reaction, along with an energy balance on the system, thereby identifying a scope for development of standardized technology using microwave and biodegradable sugars for synthesis of oxide, bimetallic and alloy nanoparticles. Prospective applications in catalysis and bio-sensing will be explored.

Keywords: Copper oxide nanoparticles, microwave, green method

501701 Comprehensive Thermodynamic Modeling of Mixed-Solvent Electrolyte Systems: An Investigation on the Quaternary System of $FeCl_2$-$FeCl_3$-HCl-H_2o

Abstract:

To treat and remove phosphorus from wastewater, the primary method has been through iron salts, such as ferrous and ferric chloride [1, 2] which are also present in leaching of chalcopyrite and etching of nickel-lead frame [3, 4]. Moreover, iron species exist in the electronic waste (e-waste) associated with computer and cell phones which requires treatment before disposal [5]. In this context, a rigorous, accurate, and widely applicable thermodynamic model is developed to describe the vapor-liquid equilibria (VLE) and solid-liquid equilibria (SLE) behaviors of $FeCl_2$ and $FeCl_3$ in water, as well as in the binary mixed-solvents of HCl-H_2O. The model is based on eNRTL equation, with a set of chemical reactions depicting complete dissociation of iron (II) and (III) chloride before saturation, and salt precipitation equilibria with hydrated salt crystals. Furthermore, the mixed-solvent electrolyte systems of $FeCl_2$-HCl-H_2O, and $FeCl_3$-HCl-H_2O were thoroughly investigated by directly inheriting molecule-molecule, and molecule-ion interaction parameters of the binary system of HCl-H_2O from previously established model by present authors. By correlating experimental data including mean ionic activity coefficient, osmotic coefficient, enthalpy of formation, heat capacity, and solubility, the model predicts VLE and SLE behaviors, calorimetric properties, and speciation of ternary systems of $FeCl_2$-HCl-H_2O, $FeCl_3$-HCl-H_2O, and $FeCl_2$-$FeCl_3$-H_2O over the entire concentration range and temperatures up to 400 K successfully. Hence, the model should present the most reliable predictions for the quaternary system of $FeCl_2$-$FeCl_3$-HCl-H_2O.

Keywords: Electrolyte NRTL model, Ferrous Chloride, Ferric Chloride

References:

1. P. A. Moore, and D. M. Miller, "Decreasing phosphorus solubility in poultry litter with aluminum, calcium, and iron amendments", *Journal of Environmental Quality*, vol. 23(2), pp. 325-330, 1994.
2. O. Gutierrez et al., "Iron salts dosage for sulfide control in sewers induces chemical phosphorus removal during wastewater treatment", *Water Research*, vol. 44 (11), pp. 3467-3475, 2010.
3. H. Majima et al., "The leaching of chalcopyrite in ferric chloride and ferric sulfate solutions", *Canadian Metallurgical Quarterly*, vol. 24 (4), pp. 283-291, 1985.
4. M. Lee and K. Lee, "Separation of iron and nickel from a spent $FeCl_3$ etching solution by solvent extraction", *Hydrometallurgy*, vol. 80 (3), pp. 163-169, 2005.
5. R. Widmer et al., "Global Perspectives on e-waste", *Environmental Impact Review*, vol. 25 (5), pp. 436-458,

   2005.

502032 Controllable Manipulation of Continuous AFI Membranes with Distinctive Microstructures on Macroporous Alpha-AluminaÂ Substrates

Abstract:

Zeolite and molecular sieve membranes with uniform and molecular-sized pores have excellent thermal, mechanical and chemical stabilities, and they have acquired potential applications as separation membranes and membrane catalysts. Besides, new application areas such as low-dielectric constant materials as well as anti-corrosion and anti-microbial coatings have been developed. The microstructure of zeolite and molecular sieve membranes including continuity, crystal orientation and thickness is of great significance in actual applications. However, it is still a challenging issue to manipulate the microstructure of zeolite membranes today.

The availability of continuous and well-defined AFI-type molecular sieve membranes (AFI membranes) would trigger their applications in innovative materials, microreactors and even microelectronic devices. Currently, the synthesis of AFI membranes has been focused on controlling the crystal orientation to achieve *c*-oriented channels perpendicular to the substrate surface, leading to the uniform residence time of involved molecules across diffusion channels. Up to present, there have been several routes for fabricating AFI arrays, films and membranes with a preferred orientation, including electric-field-driven assembly, geometry-confined growth, epitaxial growth, manipulation of synthesis conditions, microwave-enhanced growth, and induced growth at the substrate surface. It should be noted that the majority of AFI membranes reported to date were prepared on dense substrates such as silicon wafers or glass plates that would not be usable in separation and catalysis processes. Herein, we wish to manipulate the microstructure of AFI membranes in the in-situ crystallization. As such, the continuous and dense AFI membranes of single- and double-layered structure covered with either highly *c*-oriented hexagonal prisms or spherical agglomerates could be synthesized at will. There are not relevant reports yet.

A well-designed manipulation strategy is proposed to produce continuous AFI membranes with four different microstructures over porous alpha-alumina substrates, as shown in Figure 1. A double-layer and highly *c*-oriented AFI membrane of hexagonal prisms is obtained when a thin layer of medium molecular weight (MMW) chitosan is employed as the structure-directing matrix together with aluminum isopropoxide (AIP) as the Al source, it can be transformed to a single-layer and highly *c*-oriented AFI membrane of hexagonal prisms if the structure-directing matrix is replaced by a thin layer of low molecular weight (LMW) chitosan. When the Al source is changed to pseudo-boehmite, the single-layer AFI membrane is composed of highly ordered spherical agglomerates of small crystals instead of vertical hexagonal prisms; while the membrane will turn to the double-layer AFI membrane of highly-ordered crystal agglomerates if a thin layer of MMW chitosan is used once again and keeping pseudo-boehmite as the Al source.

Figure 1. Schematic diagram of manipulating microstructure of AFI membranes using a thin chitosan layer of different molecular weight and different Al sources.

The existence of a thin chitosan layer ensures the formation of the continuous and well-intergrown AFI membrane of either highly c-oriented hexagonal prisms or highly-ordered crystal agglomerates. The coverage, denseness and orientation of all synthesized AFI membranes can be further optimized by adjusting the water content in the precursor solution. It has been demonstrated that the manipulation method established in this study is rather reliable with pretty high reproducibility.

Acknowledgements

This work was supported by the National Natural Science Foundation of China under a Key-Program Grant (21136008).

487881 Magnetic Core Shell Microspheres for Extraction of Rare Earth Elements from Geothermal Brine Solution

Abstract:

Rare-earth elements (REEs) have significant commercial uses in the high-tech sector. However, difficulties associated with REE extraction in the conventional mining process has resulted in the search for REEs from alternative resources in a cost effective, environmentally-friendly manner. Among various sources, geothermal brine, used for generating geothermal energy was found to possess high concentration (~ppm level) of REEs in +3 ionic forms along with other dissolved metal ions. A system that can selectively trap the REEs using the existing setup of a geothermal power plant will lead to an additional revenue stream that will accelerate the development and deployment of geothermal plants in the United States and the rest of the world. Here, i present a magnetic core-shell approach that can effectively extract REEs in their ionic form from aqueous solution with up to 99.99% removal efficiency. The shell, composed of thermally and chemically stable metal-organic frameworks (MOFs) surrounds a Fe_3O_4 magnetic core. Majority of the Fe^{2+} species of the magnetite (Fe_3O_4) microsphere is oxidized to Fe^{3+} species (maghemite) during *in-situ* coating of the MOF materials as evident by magnetometry, Mössbauer spectroscopy, and X-ray absorption spectroscopy. The core-shell particles can be completely removed from the mixture under an applied magnetic field, offering a practical, near-term efficient REEs-removal process.

ENCAPSULATION OF DYE IN NH_2-UiO-66 METAL-ORGANIC FRAMEWORK FOR PHOTOSENSITIZED OXIDATION OF BENZYL ALCOHOL

Xiyi Li[1,2], Qingqing Hou[1,2], Yunhong Pi[1,2], Qibin Xia[1], Zhong Li[1], Jing Xiao[1,2]*
1. School of Chemistry and Chemical Engineering, South China University of Technology, Guangzhou 510640, China
2. Guangdong Provincial Key Laboratory of Atmospheric Environment and Pollution Control, South China University of Technology, Guangzhou 510640, China

INTRODUCTION

Dye-sensitization has become a relatively mature technology for visible-light harvesting for the photocatalysts[1,2]. However, it always suffers from desorption problem during the photocatalytic reaction which impedes it acts as a sustained and stable photocatalyst[3,4]. Our work makes good use of the characteristic during the self-assembly process of metal-organic framework, synthetizing the RhB@NH_2-UiO-66 to prevent the leaching of RhB without losing the crystalline lattice.

The introduction of RhB significantly enhanced the light-absorption properties, and in the visible-light-induced aerobic oxidation of benzyl alcohol, RhB@NH_2-UiO-66 exhibits 3.5 times higher conversion than the pristine NH_2-UiO-66 without losing the high selectivity.

SYNTHESIS & CHARACTERIZATION

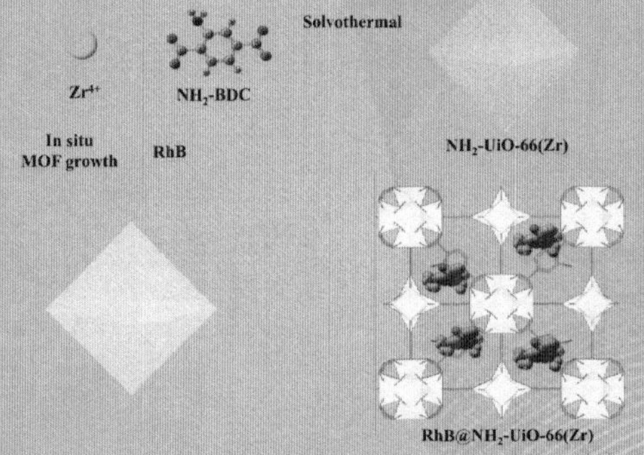

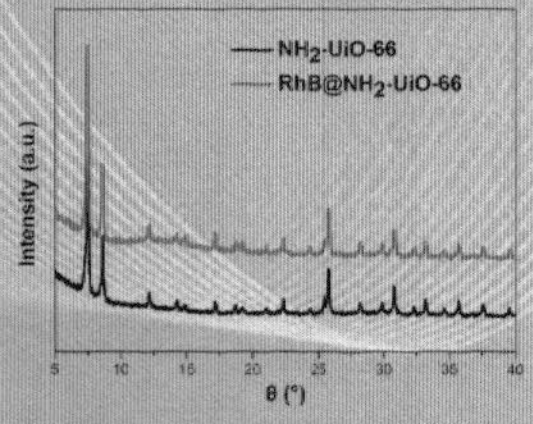

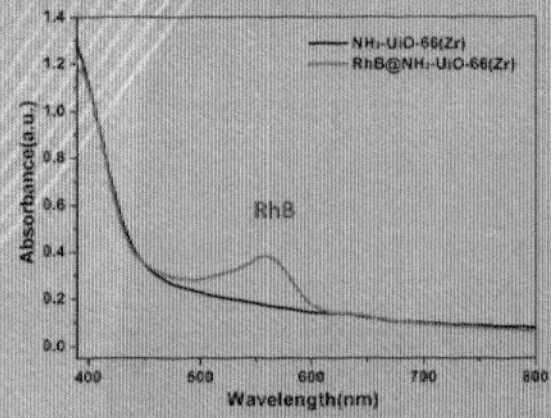

APPLICATION IN PHOTOCATALYSIS

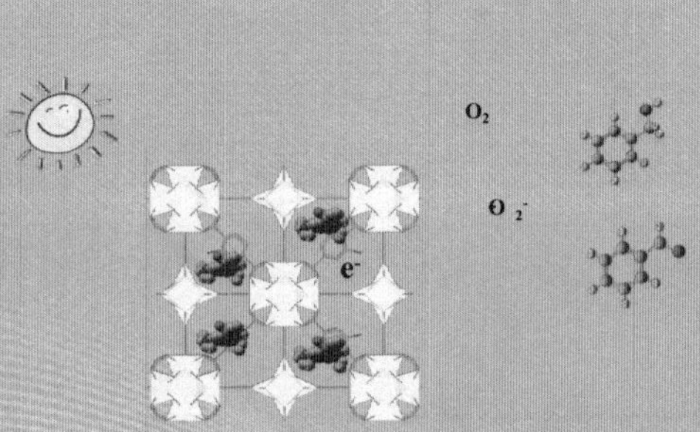

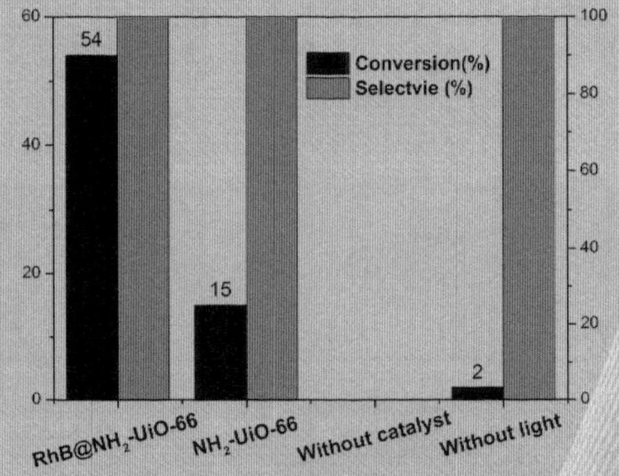

FUTURE WORK

The substrate to be expand

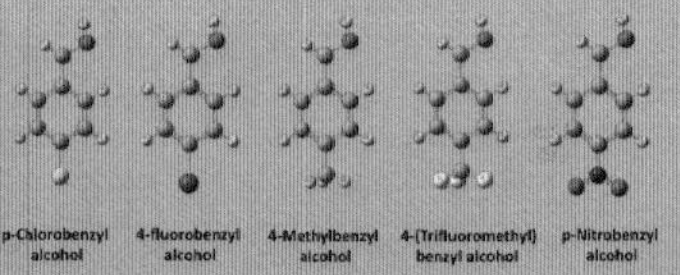

REFERENCES

[1] He, J.; Wang, J.-Q.; Chen, Y.-J.; Zhang, J.-P.; Duan, D.-L.; Wang, Y.; Yan, Z.-Y. *Chem. Commun.*, **2014**, *50, 7063–7066.*
[2] Xiong, Z.-G.; Zhang, L.-L.; Ma, J.-Z.; Zhao, X.-S. *Chem. Commun.*, **2010**, *46, 6099–6101.*
[3] Gu, Z.-Z.; Chen, L.-Y.; Duan, B.-H.; Luo, Q.; Liu, J.; Duan, C.-Y. *Chem. Commun.*, **2016**, *52, 116–119.*
[4] Shen, T.; Zhao, Z.-G.; Yu, Q.; XU, H.-J. *J. Photoch. Photobio. A* **1989**, *47, 203–212.*

ACKNOWLEDGMENT

The authors are grateful for financial support
- National Natural Science Foundation of China (21576093)
- Guangdong Natural Science Funds for Distinguished Young Scholar
- The Scientific Research Foundation for the Returned Overseas Chinese Scholars

489913 The Fabrication of Graphene/Polyaniline Blended Fiber for Conducting and Flexible Energy Storage Devices

Abstract:

Graphene fibers were fabricated through wet-spinning process using graphene oxide aqueous solution, which was used as starting materials to produce graphene/polyaniline blended fiber through in-situ polymerization of aniline on the surface of the them. Those blended fibers were applied on various conducting and flexible electronic devices for energy storage as well as wearable electronics.
The raw materials and corresponding process were analyzed through various analytic equipment, as FTIR, Ramon, TGA/DSC, XRD, SEM, TEM, XPS, and CV, etc. The preliminary results indicated that the blended fiber had core-shell structures, which presented superior performance of transporting electrons, and enhance the charge storage capabilities. Those materials show great potential on the applications of energy storage and wearable electronics.

498530 Semiconducting Heterostructures for Photocatalytic Reduction of Carbon Dioxide

Abstract:

The current environmental and energy scenario demands significant interest in renewable energy. With decreasing fossil fuel resources globally and increased concern over energy security, harvesting solar energy seems imperative. The environmental concerns are mainly due to the rising carbon dioxide levels in the atmosphere. Hence a rational solution to this problem is to convert waste CO_2 to hydrocarbons using solar energy. Amongst the different routes to achieve this purpose, solar photocatalytic conversion has garnered a lot of interest. Photocatalytic processes are associated with advantages of low temperature reactions. Moreover the prospect of utilizing the abundant solar energy to convert CO_2 to any useful hydrocarbon is appealing. A majority of the photocatalysts being used currently can only be used under UV light making it less useful for solar applications. Hereby, we study solid solutions and heterostructures of zinc oxide, aluminum nitride and gallium nitride, which present the possibility to harvest the solar irradiation. These materials can easily be tuned to be better solar photocatalysts. The key parameters necessary for a successful photocatalyst are optimum band gap and band edges of the material, its stability under reaction conditions and the effect of vacancy and interstitials on the material. We ensued a density functional theory (DFT) based investigation to study the effect of different heterojunctions with varying layer thickness towards the band gap and band edge tunability. We found consistent trends of band gap variation with material composition. The effect of anionic vacancies and strain on the materials was also studied. The effect of varying elemental composition at any heterojunction was studied via charge analysis. These materials thus present a versatile platform for material property engineering and present themselves as potential candidates for solar photocatalytic conversion of CO_2.

500268 Photoswitchable Quantum Dots Probes for Superresolution Microscopy

Abstract:

Superresolution microscopy has been developed as a tool to image native states of biological structures beyond the diffraction limit of light. Although electron microscopy techniques have surpassed this limit, they lack live imaging capability. Stochastic optical reconstruction microscopy (STORM) involves reconstruction of multiple images of localized fluorescent probes stochastically switching between light and dark states. Unlike many other superresolution imaging techniques, STORM permits discrete imaging at high densities. Unfortunately, most fluorescent probes designed for STORM provide limited fluorescence signals as a result of quenching, low brightness and contrast between dark and light states, poor stability, and short lifetimes. This research aims to develop a robust fluorescent probe to overcome prior limitations by using photoswitchable quantum dots (QDs).

QD probes were designed using the principles of Förster resonance energy transfer (FRET) by radiative transfer of the QD (donor) fluorescence to a light sensitive material (acceptor) to achieve a dark state. Previous studies have shown that gold nanoparticles (AuNPs) are excellent acceptors, especially for QDs. Additionally, FRET is strongly dependent on the distance between the donor and acceptor. Therefore, the probe consisted of QD-AuNP pairs formed using a DNA linker, whose length can be precisely controlled by the number of base pairs. The QD-AuNP pairs were therefore synthesized by conjugating the nanoparticle surfaces with complementary single stranded-deoxyribonucleic acid (ss-DNA). Reversible photoswitching was induced in these probes by utilizing photo-sensitive azobenzene modified DNA strands that can induce reversible DNA hybridization through cis-trans conversions induced by light. In this research, QD coatings were optimized to reduce the QD-AuNP separation distance and QD-AuNP spectral overlap were optimized. Both are crucial for efficient FRET. These probes are being analyzed using fluorescence spectroscopy by simultaneously modulating the photo-sensitive-DNA with a laser source to measure quenching efficiency. The optimized probes developed from this study will then be tested by comparative STORM imaging.

493002 Bottom-up Synthesis of Nanoelectronic Titania Composites

Abstract:

Materials science is becoming a more and more important influencer in electronics, as new synthesis methods and new materials are consistently coming to fruition. In particular, bottom-up synthesis schemes offer unique material options, various alignments, and micro- to nanoscale (long- and short-range) control over morphology. Surfactant-templated and electrospun schemes have been developed here for room temperature and timely synthesis options of titania nanocomposites. Particularly those used in nanoelectronics applications.

494258 Phase Diagrams, Defect Models and Thermoelectric Properties: ÃŸ-Ag_2se and $CoSb_3$

Abstract:

Thermoelectric materials have attracted tremendous research efforts owing to their potential contributions in enhancing energy usage efficiency and as renewable energy. ß-Ag_2Se and $CoSb_3$ are promising thermoelectric materials for room- temperature and mid-temperature applications. It has been reported that their thermoelectric propertied can be further enhanced with Ga-doping and In-doping, respectively. Phases and microstructures are critical to the thermoelectric properties. Phase diagrams provide fundamental phase equilibria information, and are important for the understanding and illustration of phases and microstructural evolution of materials. Ternary Co-Ga-Sb and Ag-In-Se alloys are prepared, and their equilibrium phases are determined. The Co-Ga-Sb and Ag-In-Se phase diagrams are determined based on the phase diagrams of their constitute binary systems and the experimental phase equilibria results of ternary alloys. The crystal structure of the ß-Ag_2Se phase is orthorhombic and that of the $CoSb_3$ phase is skutterudite. The solubilities of indium in ß-Ag_2Se and gallium in $CoSb_3$ are small, and are less than 0.5at.% and about 1at.%, respectively. Since the third element doping is important for the thermoelectric properties of both compounds, efforts are carried out to describe the small solubilities of indium in b-Ag_2Se and gallium in $CoSb_3$. Defect-type models are used for thermodynamic modeling. The formation energies of perfect ß-Ag_2Se and $CoSb_3$ phases are determined by the Calphad-type modeling. The defect types of the compounds are proposed, and the Gibbs energies of the compounds related with temperatures and concentrations are formulated. Phase diagrams are then calculated using the Calphad-type approach. Besides thermodynamic properties, thermoelectric properties can also be calculated based on the defect models. The calculated thermoelectric properties are compared with elemental measurements. Based on the results of both thermodynamic and thermoelectric calculations, it is confirmed that Ga atoms occupy both the void and Sb sites in $CoSb_{3.}$ The defect type of the binary ß-Ag_2Se is a cationic Frenkel type disorder, and indium atoms are assumed to be on both Ag and Se sites.

497414 Theoretical Study of a High Performance Thermoelectric Material: Stanene

Abstract:

Improving the efficiency of power generation while reducing the amount of energy waste is a crucial step towards sustainability. High performance thermoelectric materials with the capability of converting waste heat to electricity are a potential solution to move away from an energy crisis. In recent years, the research towards finding new candidates for such applications has gone to great lengths. Specifically, the reduction of low-dimensional materials to the nanoscale has produced fascinating results regarding the improvement of thermoelectric performance. In this study, we investigate the enhancement of the thermoelectric figure of merit for stanene between nanoribbon form and the bulk using first-principles calculations. The thermopower properties were calculated using combined first-principles calculations and Boltzmann transport theory within the relaxation time approximation calculated from deformation potential theory. We find that bulk stanene exhibits a thermoelectric figure of merit that is near half of unity at the optimum doping level. We propose that the thermoelectric capabilities of stanene are enhanced further for armchair and zigzag edged nanoribbon structures.

495762 Revealing the Enigmatic Interfacial Layer of Core/Shell Quantum Dots

Abstract:

Colloidal quantum dots (QDs) are attractive materials for lasers, displays and other light-emitting applications due to their narrow spectral emission bandwidth, size-tunable bandgap and high-photoluminescence quantum yield (PLQY). However, QDs undergo inevitable degradation of their unique optical properties over time due to their sensitive surface chemistry. To overcome these limitations, several approaches have been used, such as overcoating with an inorganic semiconductor shell of a wider band gap (core/shell hetrostructures), surface functionalization with organic and inorganic ligands, or polymer coating and composite formation. In particular, core/shell heterostructured QDs with thick shells [so called "giant" QDs (g-QDs)] have shown dramatically modified single-dot properties (non-blinking, non-photobleaching) and clear suppression of non-radiative recombination pathways (Auger recombination, photoexcited-carrier surface trapping). However, we have recently determined that the outstanding properties afforded by the g-QD's thick shell are sensitive to both the shell-structure quality (crystal defects) and the properties of the interfacial layer (sharp or smooth core/shell interface). In particular, the useful "lifetime" of the g-QD under conditions of elevated temperature and exposed to air or water is strongly dependent on these structural and interface properties. Significantly, we have further revealed that these aspects of the internal nanoscale structure can be precisely tailored by choice of synthetic parameters. Here, we will present our recent results on understanding this critical synthesis-structure-function nexus.

497079 Spatial Manipulation of Thermal Flux Profiles Using Nanostructure Boundaries

Abstract:

The precise manipulation of phonon (thermal carrier) transport properties is central to controlling thermal transport in semiconductor nanostructures. The control over thermal phonon transport to yield low thermal conductivities lies at the heart of creating efficient thermoelectrics with a high figure-of-merit (ZT). The physical understanding, prediction, and control of thermal phonon heat spectra and thermal conductivity accumulation functions has attracted significant attention in recent years. In this talk, we offer a unique insight into the nanoscale thermal transport phenomena by establishing the existence of thermal flux profiles nearly analogous to momentum flux profiles, and we utilize this understanding to advance the possibilities of manipulating heat by spatially modulating thermal transport in nanostructures. We show that phonon-boundary scattering influences the preferred thermal pathway used by energy during its flow in thermal transport nanostructures. The role of introducing boundaries with different surface conditions on resultant thermal flux is presented and methodologies to enhance these spatial modulations are discussed. This talk aims to advance the fundamental understanding on the nature of heat transport at nanoscale with potential applications in multiple research areas including thermoelectrics and optoelectronics.

503009 Magnetic Polymer Nano-composites for Giant Magnetoresistance and Electromagnetic Shielding

Abstract:

Due to the incomparable advantages such as light weight, optical, magnetic and enhanced mechanical properties, polymer nanocomposites have been widely studied by the researchers for different applications such as environmental remediation, energy (or information) storage and sensing.[1-4] Giant Magnetoresistance (GMR) phenomenon can be designed for information storage, our previous work shows that polymer nanocomposites show unique GMR behavior, indicating great potential applications for information storage.[5] Meanwhile, as the development of electrical devices, the electromagnetic wave radiation is a potential threat for human beingâ€™s health.[1] Thus, the materials with electromagnetic wave absorption are urgently needed. In this project, we found the negative GMR in the magnetite polypyrrole nanocomposites synthesized by the surface initiated polymerization method. The temperature dependent resistance of the magnetite nanocomposites indicates a 3-D variable range hopping (VRH) electrical conduction mechanism. The positive GMR is observed in the synthesized pure polypyyole at room temperature and analyzed by the wave function shrinkage model. The negative MR is obtained in the synthesized magnetic PNCs at room temperature and analyzed by the orbital magnetoconductivity theory. Meanwhile, the enhanced electromagnetic wave shielding is observed in its epoxy nanocomposites, which is due to the increased interface area, eddy current loss and increased anisotropic energy. Moreover, the significantly reduced flammability is observed and analyzed as well.

Reference:

1. J. Guo, H. Song, H. Liu, X. Liu, C. Luo, X. Zhang, J. Kong, Z. Guo, Y. Ren, T. Ding, M. A. Khan and D. P. Young, *Journal of Materials Chemistry C*, 2017, DOI: 10.1039/C7TC01502J.

2. J. Guo, L. Guan, H. Wei, M. A. Khan, X. Zhang, B. Li, Q. Wang, B. L. Weeks, D. P. Young, T. Shen, S. Wei and Z. Guo, *Journal of The Electrochemical Society*, 2016, **163**, H664-H671.

3. H. Gu, J. Guo, H. Wei, S. Guo, J. Liu, Y. Huang, M. A. Khan, X. Wang, D. P. Young, S. Wei and Z. Guo, *Advanced Materials*, 2015, **27**, 6277-6282.

4. B. Qiu, J. Guo, X. Zhang, D. Sun, H. Gu, Q. Wang, H. Wang, X. Wang, X. Zhang and B. L. Weeks, *ACS applied materials & interfaces*, 2014, **6**, 19816-19824.

5. J. Guo, H. Gu, H. Wei, Q. Zhang, N. Haldolaarachchige, Y. Li, D. P. Young, S. Wei and Z. Guo, *The Journal of Physical Chemistry C*, 2013, **117**, 10191-10202.

489088 Constructing Ternary Conductive Polymer Composites with Cocontinuous Polymer Blends and Interfacial Graphene Nanoplatelets

Abstract:

Conductive polymer composites have a wide range of industrial applications such as electrostatic discharge (ESD) protection and electromagnetic interference (EMI) shielding. Current technology requires high loadings of conductive fillers to achieve a percolated structure inside the polymer matrix, thereby sacrificing material processability. In this study, we construct a class of conductive polymer composites using polymer blends with lower critical solution temperatures and interfacial graphene nanoplatelets. When a polymer blend is raised above its critical temperature, it spontaneously phase separates into a cocontinuous structure via spinodal decomposition. By controlling the wetting property of graphene nanoplatelets, we selectively localize the graphene nanoplatelets at the interface of cocontinuous polymer blends, which gives rise to a percolated graphene structure with ultra-low percolation thresholds for electrical conductivity. Strong electric fields have also been applied during sample annealing, allowing us to test the influence of electric fields on the graphene localization and the alignment of graphene nanoplatelets in cocontinuous polymer composites.

489199 The Influence of Interfacial Graphene on the Morphological, Electrical and Mechanical Properties of Co-Continuous Polymer Blends

Abstract:

Interfacial localization of conductive fillers at the interfaces in co-continuous polymer blends is shown to be effective in suppressing the coarsening of co-continuous structure during annealing and increasing conductivity with a low electrical percolation threshold. In this study, two different conductive fillersâ€"graphene nanoplatelets (GnPs) and thermally reduced graphene oxide (r-GO)â€"were melt compounded with polyethylene terephthalate glycol-modified (PETG) and polystyrene (PS) blends to prepare ternary co-continuous polymer composites. The morphological, electrical and mechanical properties of the resulting composites were systemically investigated. Through transmission electron microscopy (TEM) micrographs, we showed that GnPs transferred from the premixed PETG phase to the thermodynamically favored PS phase during the melt compounding, whereas most r-GO trapped at the interface of the polymer blends. Such a difference in the distribution of conductive fillers leads to drastic differences in the morphological, electrical and mechanical properties of the ternary composites. Particularly, we found that the interfacial r-GO greatly improved the Youngâ€™s modulus and fracture toughness of the co-continuous polymer blends.

489903 The Fabrication and Application of Composite Graphene Oxide Films

Abstract:

Graphene film shows excellent performance on not only electric and thermal conductivity, but also mechanical strength, which brought it to be a hot research topic in the last few years. In our study, a free-standing graphene films were prepared from concentrated graphene oxide aqueous solution through solution casting processï¼Œwhich gives out excellent mechanical and electrical performance.

Firstly, the graphene oxide aqueous dispersion was prepared and evenly coated on the glass substrate. Then the solidification and drying process were applied to form graphene films, following with various ionic compounds treatments. In our study, we investigated various additives, which can form various chemical bindings to enhance the film surface morphology, mechanical strength, and electric performance, which presented a great potential for industrial application.

490406 Mechanically Stable Thermally Crosslinked Poly(acrylic acid)/ Reduced Graphene Oxide Aerogels

Abstract:

Porous carbonaceous materials, such as graphene-based aerogels (GAs), are very attractive because they possess a unique combination of properties including high surface area, excellent thermal and electrical conductivity, and low mass density (a few milligrams per cubic centimeter) compared to dense analogs. Such aerogels have a wide range of potential applications in energy storage devices, polymer composites, sensors, absorbents, catalysts, etc. However, one major drawback for GAs is that they are often very fragile, due to the weak Van der Waals interactions between adjacent two dimensional graphene sheets, resulting in a delicate structure. Numerous attempts to increase the mechanical properties of GAs have been made previously, such as introducing physically mixed additives like carbon nanotubes or incorporating short chemical crosslinks between graphene sheets. However, the previous additives or crosslinks used in GAs usually comprise a sizeable weight fraction of the final material, which often causes the collapse of the pore structure, or partially degrades some of the attractive properties of GAs.

We recently demonstrate a method [1] to enhance the mechanical properties of a specific type of GA composed of graphene oxide (GO) using a small amount of inexpensive, commercially available and thermally cross-linkable poly(acrylic acid) (PAA). Additionally, GO can be reduced to form reduced graphene oxide (rGO) in a single step using hydroiodic acid (HI) vapor. The thermally crosslinked graphene oxide aerogels (termed XPAA/rGO) exhibit dramatically improved mechanical properties, while all the other attractive features of GAs (high surface area, high porosity, high electrical conductivity, low density) are maintained. Finally, experiments revealing the performance of XPAA/rGO as superabsorbents, electrochemical energy storage electrodes and pressure sensors will also be discussed.

[1] Mechanically Stable Thermally Crosslinked Poly(Acrylic Acid)/Reduced Graphene Oxide Aerogels, H. Ha, K. Shanmuganathan, and C.J. Ellison^{+}, *ACS Applied Materials and Interfaces*, 7, 6220â€"6229 (Feb. 2015).

http://pubs.acs.org/doi/abs/10.1021/acsami.5b00407

496346 Ionophore-Decorated Magnetic Graphene Oxide As a Composite Adsorbent Material for Heavy Metal Ion Sequestration

Abstract:

Heavy Metal (HM) ion contamination of water resources from human activities represents a major global concern. These ions must be reduced to an acceptably low level according to increasingly stringent environmental regulations. Due to the hazard of the HM contamination of water, there has been a growing interest in the development of materials that are capable of removing low concentrations of HM ions from contaminated waters. Adsorbent materials with a strong affinity and high adsorption capacity for the targeted metal ions need to be developed for HM removal from aqueous sources. In addition, multi-functional adsorbent materials which can aid in its efficient recovery after use, is highly desirable.

A multi-functional adsorbent was successfully synthesized and used as an HM adsorbent. The composite material was decorated with crown ethers (CE) for HM ion-selective removal, magnetite (Fe_3O_4) for material recovery and recyclability, and graphene oxide (GO) as a two-dimensional, high-aspect ratio support material for the CE and magnetite. The GO was prepared by oxidizing graphite powder through a modified Hummers' method. The magnetite nanoparticles were immobilized on the surface of the GO via the solvothermal process (rGO-Fe_3O_4). Alkyne moieties were grafted on the rGO-Fe_3O_4 surface via diazonium chemistry using 4-ethynylaniline. The alkyne terminals in 4-ethynylaniline served as "clickable" CE-specific attachment sites. On the other hand, the CE (i.e. 1-Aza-18-Crwon-6) was modified in order to introduce an azide-functional group. The azide-CE was then "clicked" on the Alkyne-rGO-Fe_3O_4 at 60 °C via the 1,3 cycloaddition click chemistry reaction.

The synthesized composite adsorbent material (CE-rGO-Fe_3O_4) was characterized using Boehm Titration, Transmission Electron Microscopy (TEM), X-ray Diffraction Spectrometry (XRD), Thermogravimetric Analysis (TGA), Fourier-Transform Infrared Spectroscopy (FT-IR), Raman Spectroscopy, and X-ray Photoelectron Spectroscopy (XPS). Boehm Titration was used to quantity the amount of oxygenous groups on GO. The morphology of the adsorbent was examined under TEM, which revealed the presence of Fe_3O_4 NPs on the GO surface. The XRD and FTIR were used to confirm the presence of different functional groups on the GO support, which were consistent with the TGA results. Raman Spectroscopy also reflected the degree of functionalization of the GO-based materials by quantifying the changes in the D- and G-bands and their corresponding I_D/I_G ratios. XPS was utilized to monitor the changes in the chemical states of the elements present in the adsorbent material upon subsequent modification and functionalization. XPS results revealed the presence of the Fe_3O_4 in the adsorbent (CE-rGO-Fe_3O_4) as indicated by the Fe2p signals. On the other hand, the successful attachment of the CEs via click reaction was confirmed with the presence of the peak attributed to the triazole functional group found in the N1s spectra. The adsorbent material was systematically tested in order to determine the heavy metal adsorption capacity and selectivity, and finally the material recyclability performance.

This research was supported by NRF funded by the Ministry of Science, ICT and future Planning (2017R1A2B2002109 and 2015R1A2A1A15055407) and the Ministry of Education (2009-0093816).

493638 Synthesis of Thermoresponsive Polymer/Fe_3O_4 Nanoparticle Composite and Its Application

Abstract:

For heavy metal ion recovery from wastewater, various methods such as the usage of absorbents incorporating ion exchange groups or chelating groups have been developed. However, main disadvantage of the methods is generation of secondary wastes during recycling process. To solve the problem, we focused on poly(*N*-isopropylacrylamide) (PNIPAM) which was known as one of the thermoresponsive polymer. Since PNIPAM has lower critical solution temperature (LCST) around 32 ºC, if functional groups to capture heavy metal ions are copolymerized into the polymer chain, repeatable adsorption/desorption of heavy metal ions will be possible by swinging temperature and using the resulting conformation change. Therefore, in this study, thermoresponsive polymer/Fe3O4 nanoparticle composite was developed and applied for heavy metal ion recovery with the aim of achieving (1) repeatable adsorption/desorption property of heavy metal ion by only swinging temperature, and (2) reusability due to magnetic property. In the experiment, free radical copolymerization of *N*-isopropylacrylamide and acrylic acid (AA) was performed for the synthesis of straight-chained thermoresponsive copolymers. Then, an amine modified Fe3O4 nanoparticle was synthesized by the co-precipitation method. Finally, the copolymer was immobilized onto the amine modified Fe3O4 nanoparticle through amide bound formation. The results of X-ray diffractometer (XRD) analysis, Fourier transform infrared spectroscopy (FT-IR) measurement and thermogravimetric analysis (TGA) indicated that poly(NIPAM-*co*-AA) was immobilized on Fe3O4 nanoparticle and the maximum of amount of the copolymer was 0.185 g/g-Fe3O4. Cu(II) adsorption experiments exhibited that the saturated adsorption amounts were 0.16 mmol/g-adsorbent at 10 (below LCST) and 0.31 mmol/g-adsorbent at 50 (above LCST), respectively, for the sample which had maximum immobilized copolymer amount. In addition, repeated Cu(II) adsorption/desorption by swinging temperature was also possible. These results indicated that it was successful to synthesize reusable adsorbent which can recover heavy metal ions from wastewater without producing secondary waste during its Cu(II) adsorption/desorption process.

495207 High-Performance, Ambient Phase Change Thermal Diodes for Energy Applications

Abstract:

Thermal diodes, or devices that transport thermal energy asymmetrically, analogous to electrical diodes, hold promise for thermal energy harvesting and conservation, as well as for phononics or information processing. Thermal diodes enabled by phase change materials are ideal candidates for such applications due to their potential to efficiently operate at ambient conditions under low temperature biases (< 20 K). Theory and experiment for high-performance, ambient thermal diodes, enabled by junctions of phase change materials, is presented. It is shown that such diodes possess ideal temperature biases and material dimensions for optimal operation and design, and their performance greatly exceeds that of ambient thermal diodes operating under low temperature biases in the literature to date. In addition, the application of such thermal diodes for ambient energy harvesting and conservation applications is addressed both experimentally and theoretically.

A Study on the Preparation and Properties of Polymer Composites Using Amino Functionalized Microcrystalline Cellulose (MCC) As a Filler Material

KiRyong Ha[1], KiSeob Hwang[2], Kwang-Hee Lim[3], Hanna Kim[1] and Yeokyung Yang[1]
1. Keimyung University, South Korea
2. IT Convergence Materials R&D Group, South Korea
3. Daegu University, South Korea

Introduction

Biodegradable microcrystalline cellulose (MCC) can be used as fillers to manufacture polymer composites.[1] However, due to the hydrophilic nature of the MCC, significant challenge remains in order to disperse MCCs uniformly in hydrophobic polymer matrixes, such as LDPE or PU. Herein, we would like to prepare a surface modified MCC filler that can improve the compatibility of MCC with hydrophobic LDPE or PU[2] by using a silane coupling agent.[3]

In this study, 3-aminopropyltriethoxy silane (APS) coupling agent was used to modify the surface of MCC by silanization reaction. The degree of modification of MCC surface was confirmed by FT-IR, Elemental Analysis (EA) and X-ray photoelectron spectroscopy (XPS), and the mechanical properties of LDPE or PU (polytetrahydrofuran + MDI + 1,4-butanediol)/APS modified MCC composites were measured by UTM. The modulus of the polymer composites filled with APS modified MCC was increased substantially, due to the improved compatibility between filler and polymer matrix or a strong urea bond formation between PU and APS modified MCC fillers.

Experimental

Materials

1) Microcrystallinecellulose (MCC, Sigma-Aldrich)
2) (3-Aminopropyl)triethoxysilane (APS, Sigma-Aldrich)
3) Tetrahydrofuran (THF, Daejung)
4) Ethyl alcohol (EtOH, Duksan)
5) Low density polyethylene (LDPE, Lotte Chemical)
6) Polytetrahydrofuran (PTHF, Sigma-Aldrich)
7) (4,4'-Methylenbis)phenyl isocyanate (MDI, Sigma-Aldrich)
8) 1,4-butanediol (1,4-BD, Duksan)
9) Dimethylformamide (DMF, Daejung)

Analysis

1) FT-IR (Jasco®, FT-IR 620)
2) XPS (ThermoFisher Scientific, K-Alpha)
3) EA(Thermo Fisher Scientific, FLASH 1112)
4) UTM (QMESYS, QM100SE))
5) SEM (SEC, SNE3200M)

- Modification of MCC with APS

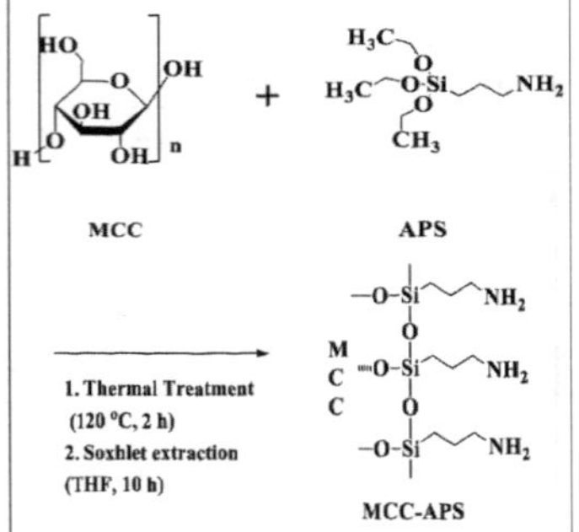

- Preparation of LDPE/MCC composites

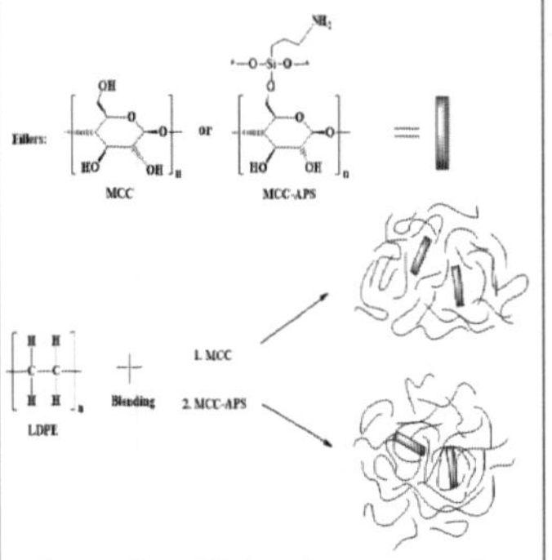

- Preparation of Polyurethane composite with MCC-APS.

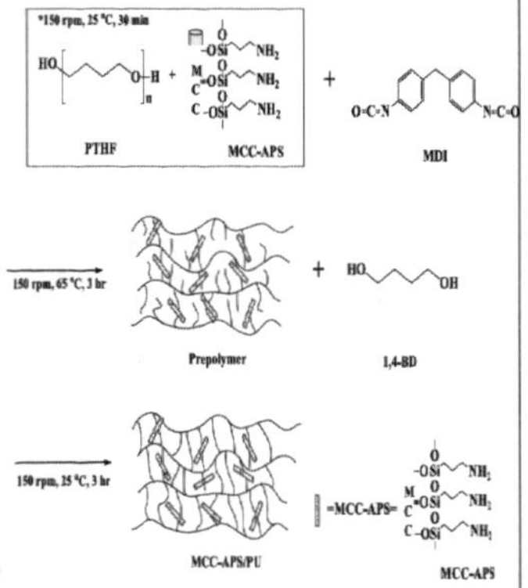

MCC modified APS (FT-IR , EA, XPS)

(1) FT-IR results

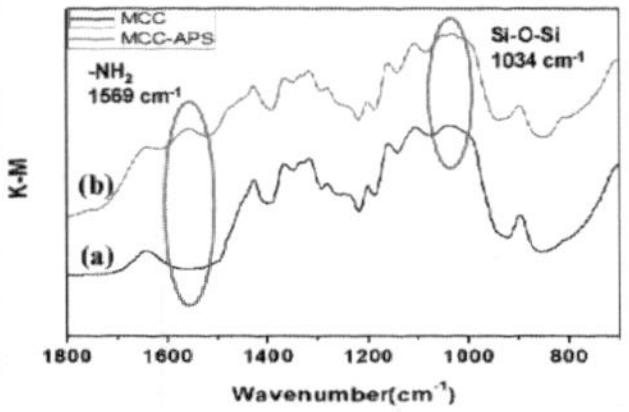

Fig. 1. DRIFT spectra of (a) pure MCC and (b) MCC-APS.

(2) EA results

Table 1. N/C ratio of MCC and MCC-APS

Fillers	N(%)	C(%)	N/C
Pure MCC	-	42.05	-
MCC-APS	0.89	41.54	0.02

(3) XPS results

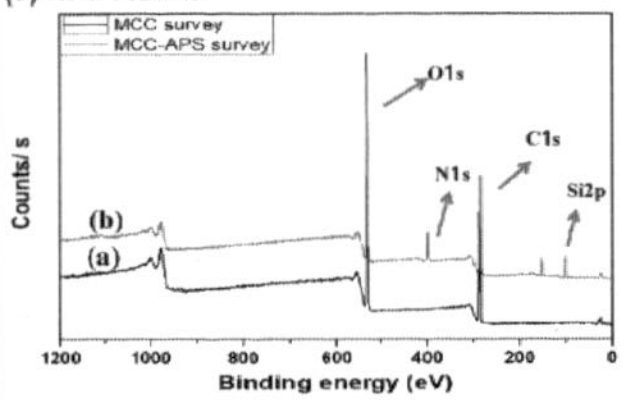

Fig. 2. XPS survey spectra of (a) MCC and (b) MCC-APS.

LDPE composite results

(1) UTM results

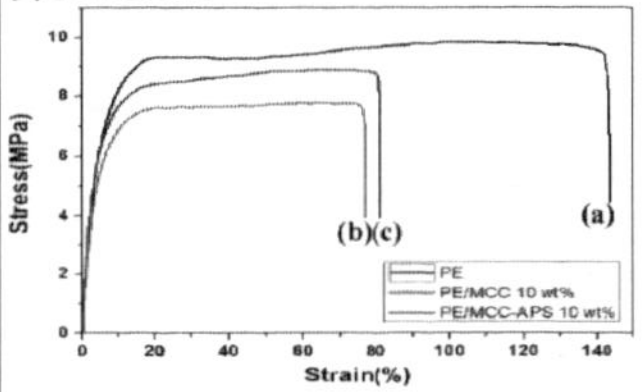

Fig. 3. Strain-Stress curves of (a) LDPE, (b) LDPE/MCC composite, (c) LDPE/MCC-APS composite.

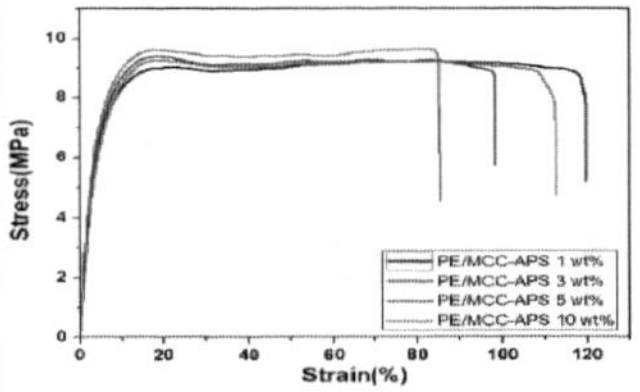

Fig. 4. Effects of filler(MCC-APS) amounts on the mechanical properties of LDPE composites.

(2) SEM images

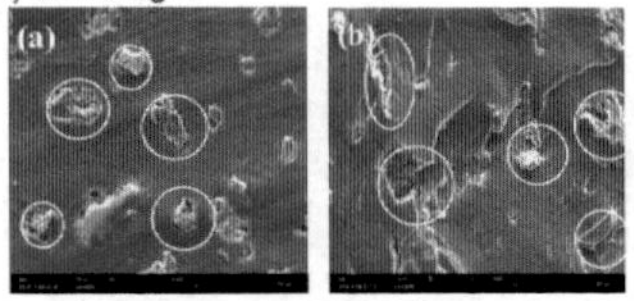

Fig. 5. SEM images of (a) LDPE/MCC composite, (b) LDPE/MCC-APS composite, and (c) LDPE/MCC-TPPI composite.

PU composite results

(1) UTM results

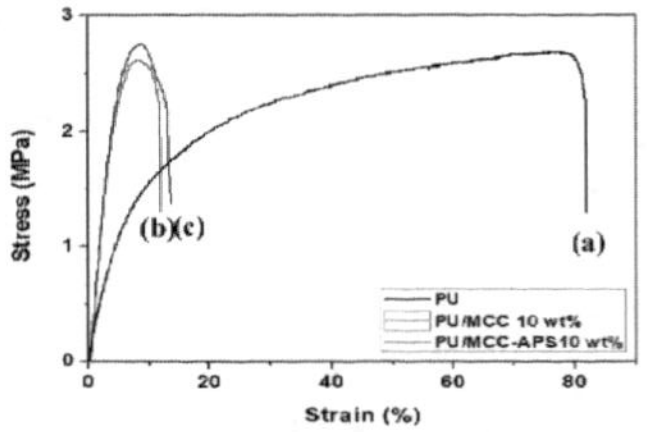

Fig. 6. Strain-Stress curves of (a) PU, (b) PU/MCC composite, and (c) PU/MCC-APS composite.

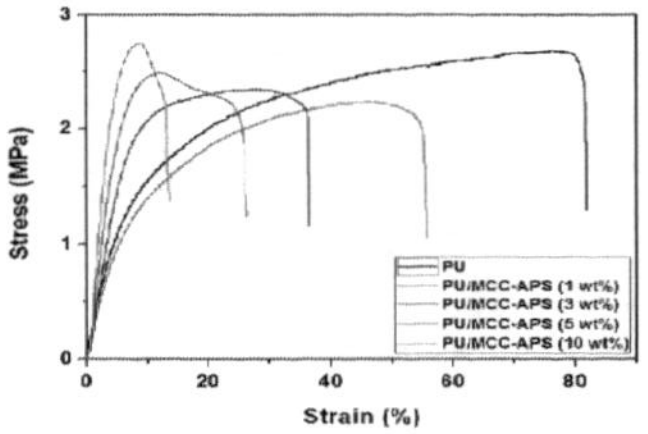

Fig. 7. Effects of filler(MCC-APS) amounts on the mechanical properties of Polyurethane composites.

(2) SEM images

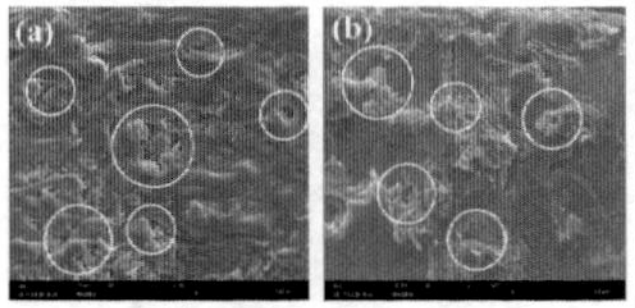

Fig. 8. SEM images of (a) PU/MCC composite and (b) PU/MCC-APS composite.

Conclusions

1) APS silanization reaction introduced $-NH_2$ groups on the MCC surface supported by FT-IR, EA and XPS analysis results.

2) We found increase in mechanical properties of MCC-APS composites due to increased compatibility between MCC-APS fillers and LDPE matrix and chemical bonding formation at interface of MCC-APS filler and PU matrix.

References

[1] Lee, S.; Ha, K., *Polymer (Korea)*, 2013, **37**, 777.
[2] Koga, H.; Kitaoka T.; Isogai, A., *Journal of Materials Chemistry*, 2011, **25**, 9356.
[3] Chen, T. -K.; Tien, Y. -I.; Wei, K. -H., *Polymer*, 2000, **41**, 1345.

500843 Electrically Conductive Films Made of Polythiophene and Fibrillated Wood Particles

Abstract:

The objective of this study was to investigate the formation and properties of composite films made of fibrillated wood microfibers deposited with poly(3,4-ethylenedioxythiophene)/poly(styrenesulfonate) (PEDOT:PSS). Hardwood particles were reduced in size and fibrillated using an ultra-fine friction grinder following an alkaline pretreatment step. Oxidative polymerization was performed on 3,4-ethylenedioxythiophene (EDOT) monomer in the presence of PSS and microfibers. The formation of PEDOT:PSS and its interaction with wood microfibers were verified through scanning electron microscopy and infrared spectroscopy of the resulted films. The composite films were found to exhibit a range of tensile properties and electrical resistivity depending on the ratio between EDOT monomer and microfibers in the formulation. An EDOT:microfibers ratio of 33 wt.% resulted in films having the best combination of sheet resistivity (340 Î©/sq.) and tensile strength (38 MPa). The selected films were also tested for solvent resistance and thermal stability to enable considerations of their use in different environments. Overall, this study showed that low-cost wood particles after simple treatments could be used as a green and environmentally friendly reinforcing material for PEDOT:PSS.

484530 Effect of Different Carbon Additives on Structure of Magnesium Composites for Hydrogen Storage

Abstract:

In this work the effect of different carbon materials: carbonized palm nut shells (PNC), graphite (G) and carbonized anthracite coal (CAC), on the structural changes and hydrogen storage properties of magnesium were investigated and compared. Samples of Mg containing 20wt. % additive were prepared by milling in a rotary ball mill and later hydrogenated using a PCTPro hydrogenation equipment. Structural and phase evolution were investigated using X-ray diffractometry (XRD).Milling magnesium with carbonized palm nut shells for 2 hours showed much structural change in magnesium compared with other carbon materials in this study.

The hydrogen storage capacity of the resulting nanocomposites were ranked in this manner 80Mg-20PNC>80Mg-20CAC>80Mg-20G.

488524 Low Loading of Grafted Thermoplastic Polystyrene Strengthened and Toughened Transparent Epoxy Composites

Abstract:

Transparent epoxy composites strengthened and toughened by thermoplastic polystyrene grafted with epichlorohydrin (*g*-PS) have been prepared at low loading levels. The polymer backbone of PS was manipulated by the epoxide and hydroxyl groups confirmed by Fourier transform infrared spectroscopy (FT-IR), thermogravimetric analysis (TGA), and X-ray photoelectron spectroscopy (XPS). Contact angle and differential scanning calorimetry (DSC) tests indicated that the grafting process could decrease the surface tension and increase the compatibility between PS and epoxy resins. The effects of *g*-PS loading and the grafting process on both the viscosity of liquid epoxy resin suspensions, as well as the physicochemical properties of cured epoxy composites have been systematically investigated. The cured *g*-PS/epoxy composites demonstrated an enhanced tensile strength (maximum of 97.4 MPa) compared to either cured pure epoxy (77.6 MPa) or PS/epoxy composites (79.1 MPa). The modulus of toughness for *g*-PS/epoxy composites reaches values of up to 355.9 MJ m^{-3}, which is respectively 176.6 and 141.1% higher than those for cured pure epoxy and PS/epoxy composites. The uniform *g*-PS distribution in the cured *g*-PS/epoxy composites was observed by scanning electron microscopy (SEM). The glass transition temperature (*Tg*) of cured *g*-PS/epoxy composites was shifted to a higher temperature (increased by 16.3) in the dynamic mechanical analysis (DMA) compared with that of cured pure epoxy (111.7). The strong interfacial interaction obtained between *g*-PS and the epoxy matrix was responsible for the enhanced mechanical and thermal mechanical properties. This work provides a new insight into the investigation of interaction and compatibility between thermoplastic and thermoset materials.

1. H. Gu, C. Ma, J. Gu, J. Guo, X. Yan, J. Huang, Q. Zhang and Z. Guo, J. Mater. Chem. C, 2016, **4**, 5890–5906.

2. H. Gu, J. Guo, H. Wei, M. A. Khan, X. Wang, D. P. Young and S. Wei, *Adv. Mater.*, 2015, **27**, 6277-6282.

3. H. Gu, H. Lou, J. Tian, S. Liu and Y. Tang, J. Mater. Chem. A, 2016, **4**, 10174–10185.

4. V. A. Agubra and H. V. Mahesh, *J. Polym. Sci., Part B: Polym. Phys.*, 2014, **52**, 1024-1029.

5. W. Tian, L. Liu, F. Meng, Y. Liu, Y. Li and F. Wang, *Corros. Sci.*, 2014, **86**, 81-92.

6. H. Gu, J. Guo, H. Wei, S. Guo, J. Liu, Y. Huang, M. A. Khan, X. Wang, D. P. Young and S. Wei, Adv. Mater., 2015, **27**, 6277–6282.

498477 Preparation of Modified Graphene Oxide-Containing Styrene Masterbatches for Thermosetting Resin Composite

Abstract:

Incorporating graphene-based nanomaterials into thermosetting resin matrices is one of the key challenges in preparing resin nanocomposites for industrial scale applications. Thermosets, unlike thermoplastics, cannot be processed by melt blending or solvent casting, so graphene nanosheets can only be added to a low-viscosity resin prepolymer, which does not provide high shear during mixing to break down graphene aggregates. The physical form of graphene nanomaterials can also significantly affect the dispersion and the final properties of the composites. Drying of graphene materials often results in significant stacking of nanosheets and makes them hard to redisperse. Freeze-drying has been utilized to mitigate sheet stacking during drying of graphene, but the lengthy treatment is not industrially friendly. In this research, a new process of dispersing chemically modified graphene oxide (mGO) was evaluated to address the above-mentioned issues. This process entails the preparation of an mGO/styrene masterbatch by solvent exchange and a simple mechanical mixing step to generate a homogeneous mGO/resin dispersion. For comparison, two other post-synthesis processes, namely oven drying and freeze-drying, were also studied. The dried mGO was blended into resin using the same mechanical mixing conditions. At low mGO loading levels of 0.02â€“0.08wt.%, the three processing routes produce resin composites that differ slightly in mechanical properties. Composites made with oven-dried mGO show the highest increase in fracture toughness and the most severe decrease in flexural strength, while the composites prepared from the mGO masterbatch exhibit less toughness increase but a better flexural strength retention. Freeze-dried mGO provides a toughening effect and a reduction in flexural strength with a magnitude in between that offered by oven-dried and masterbatch-derived mGO. However, significant differences in dispersion homogeneity can be observed by visible light microscopy (VLM) and scanning electron microscopy (SEM). The masterbatch process offers an economical way of producing high quality mGO dispersions in unsaturated polyester and vinyl ester resins.

501067 Smart Windows Enabled By Buckling Instabilities in Periodic Composite Films

Abstract:

Smart windows that can regulate the transmission of some (or all) wavelengths of light by changing between a translucent and a transparent state can greatly reduce the costs on heating, cooling, and daylighting, highly promising for next-generation net-zero energy buildings. Unfortunately, almost all state-of-the-art smart windows require expensive transparent conductors to operate and they suffer from relatively low light transmittance in the transparent states, impeding their overall costs and energy efficiencies. Here report a novel smart window technology that eliminates expensive transparent conductors in the final devices and enables high optical transmittance in the transparent state (with performance close to traditional window glass). Self-assembled colloidal crystals are assembled using a scalable electrostatics-assisted Langmuir-Blodgett technology, followed by embedding in elastic polymers to create bilayer elastomeric composites comprising rigid skin layers with well-controlled thicknesses (determined by the sizes of the colloids). Buckling instabilities induced by bending the bilayer composites lead to strong light scattering that significantly reduces light transmittance, rendering dynamic light regulation between highly transparent and translucent states. Optical modeling and mechanical finite element analysis have also been conducted to complement the systematic experimental studies. Ultimately, the experimental and theoretical results match reasonably well.

503010 Magnetic Polymer Nanocomposites for Electromagnetic Interference Shielding

Abstract:

Due to the development of electrical devices, the electromagnetic wave radiation is a potential threat to human beingâ€™s health.[1] Polymer nanocomposites show great potential for electromagnetic interference (EMI) shielding application. In this project, epoxy nanocomposites reinforced with polypyrrole functionalized nano-magnetitie (Fe_3O_4-PPy) showed significantly enhanced electromagnetic wave absorption performance and flame retardancy. The Fe_3O_4-PPy were prepared by a surface initiated polymerization method. The epoxy/(30.0wt%)Fe_3O_4-PPy nanocomposites possessed a minimum reflection loss (RL) value of -35.7 dB, much lower than either epoxy/(7.5wt%)PPy nanocomposites with a minimum RL value of -11.0 dB or epoxy/(30.0wt%)Fe_3O_4 with a minimum RL value of -17.8 dB at the same thickness (1.7 mm). Meanwhile, the bandwidth of epoxy/(30.0wt%)Fe_3O_4-PPy nanocomposites for RL<-10 dB and RL<-20 dB (which means about the 90% and 99% attenuation of the incident electromagnetic wave) was 4.0 and 0.8 GHz, respectively. The increased interface area, eddy current loss and anisotropic energy were essentially important to acquire higher reflection loss and broader absorption bandwidth for epoxy/(30.0wt%)Fe_3O_4-PPy nanocomposites. Moreover, the significantly reduced flammability was observed in the epoxy/(30.0 wt%)Fe_3O_4-PPy nanocomposites compared with pure epoxy. The total heat release of epoxy/(30.0 wt%)Fe_3O_4-PPy decreased from 25.5 kJ/g of pure epoxy to just 12.3 kJ/g. The mechanical property, magnetic property, electrical conductivity and particle dispersion quality of the epoxy nanocomposites were reported as well. These new nanocomposites with enhanced electromagnetic wave absorption property and flame retardancy possess great potential to be deployed for safer electromagnetic wave absorber in electronic industry to satisfy stringent industrial standards.[2]

Reference:

1. K. Zhang, H.-O. Yu, Y.-D. Shi, Y.-F. Chen, J.-B. Zeng, J. Guo, B. Wang, Z. Guo and M. Wang, *Journal of Materials Chemistry C*, 2017, **5**, 2807-2817.

2. J. Guo, H. Song, H. Liu, X. Liu, C. Luo, X. Zhang, J. Kong, Z. Guo, Y. Ren, T. Ding, M. A. Khan and D. P. Young, *Journal of Materials Chemistry C*, 2017, DOI: 10.1039/C7TC01502J.

Acknowledgementsï¼š

This project is ï¬nancially supported by the American Chemical Society Petroleum Research Fund (ACS PRF# 53930-ND6).

503763 Autoperforation of 2D Materials for Generating Two Terminal Memresistive Janus Particles

Abstract:

Due to its inherent stochasticity, brittle fracture is seldom used as a nanofabrication method. However, the imposition or templating of a specific strain field can guide fracture along a pre-set design. Herein, we show that this autoperforation provides a means of spontaneous assembly for 2D surfaces. Chemical vapor deposited mono- and bi-layer graphene, molybdenum disulfide (MoS2), or hexagonal boron nitride (hBN) autoperforate into circular envelopes when sandwiching a microprinted polymer spot, allowing lift-off and assembly into solution. The resulting composite microparticles have two independently addressable, external Janus faces that we show can function as an intraparticle array of parallel, two terminal electronic devices. As an example, we print a 0.9 wt% black phosphorous (BP) nanoflake in polystyrene (PS) latex ink into mono-layer graphene sandwich particles, resulting in micro-particles possessing non-volatile, 15 bit memory storage via a spatially addressable memresistor array throughout the particle interior. Such particles form the basis of free floating devices capable of collecting and storing digital information in their environment. The 2D envelopes demonstrate remarkable chemical and mechanical stability during 4 months of preservation in aqueous media. They also survive a highly acidic gastrointestinal environment, as well as aerosolization over 0.3 meters and re-collection. Autoperforation of 2D materials into such envelope structures open the door to precise compositing of particulate devices, extending nanoelectronics into previously inaccessible environments.

495629 Driven Morphological Evolution of Strained Thin Film Surfaces and Two-Dimensional Materials: Morphological Stability and Pattern Formation

Abstract:

Mass transport in solid materials driven by externally applied fields, such as stresses, electric fields, and temperature gradients, can cause morphological instabilities, leading to failure of materials used in electronic and optoelectronic devices. However, properly controlled applied fields can also stabilize planar surface morphology, reduce surface roughness, and drive the formation of intriguing nanoscale morphological features, providing a path toward precise nanopatterning for the development of electronic and photonic materials with optimal functionality. In this presentation, we report our recent work on driven morphological evolution of epitaxial thin film surfaces and two-dimensional materials.

Thin film surface roughness is responsible for various materials reliability problems in microelectronics and nanofabrication technologies, which requires the development of surface roughness reduction strategies. Toward this end, we have conducted a modeling and simulation study that established the electrical surface treatment of conducting thin films as a viable physical processing strategy for surface roughness reduction. We have developed a continuum model of surface morphological evolution that accounts for the residual stress in the film, surface diffusional anisotropy and film texture, the film's wetting of the layer that is deposited on, and surface electromigration. Supported by linear stability theory, self-consistent dynamical simulations based on the model have demonstrated that the action over several hours of a sufficiently strong and properly directed electric field on a conducting thin film can reduce its nanoscale surface roughness and lead to an atomically smooth planar film surface. The modeling predictions are in agreement with experimental measurements on copper thin films deposited on silicon nitride layers.

In addition, we have developed a theory for the experimentally observed formation of multiple quantum dots (QDs) in strained-layer heteroepitaxy based on surface morphological stability analysis of a coherently strained epitaxial thin film on a crystalline substrate. Using a fully nonlinear model of surface morphological evolution that accounts for a wetting potential contribution to the epitaxial film's free energy as well as surface diffusional anisotropy, we have demonstrated the formation of multiple QD patterns in self-consistent dynamical simulations of the evolution of the epitaxial film surface perturbed from its planar state. The simulation predictions are supported by weakly nonlinear analysis of the epitaxial film surface morphological stability. We have found that, in addition to the Stranski-Krastanow instability, long-wavelength perturbations from the planar film surface morphology can trigger a nonlinear instability, resulting in the splitting of a single QD into multiple QDs of smaller sizes, and predicted the critical wavelength of the film surface perturbation for the onset of the nonlinear tip-splitting instability. The theory provides a fundamental interpretation for the observations of "QD pairs" or "double QDs" and other multiple QDs reported in experimental studies of epitaxial growth of semiconductor strained layers, such as the multiple QD patterns on the surface of an epitaxial film grown on a patterned substrate. The theory also sets the stage for precise engineering of tunable-size nanoscale surface features in strained-layer heteroepitaxy by exploiting film surface nonlinear, pattern forming phenomena. We have also used this model for detailed simulation of the kinetics of nanoring formation from QDs in coherently strained epitaxial films upon their thermal annealing and to conduct a systematic exploration of the resulting nanoring structures upon variation of the processing conditions.

Moreover, we have conducted a systematic analysis of pore-edge interactions in graphene nanoribbons (GNRs) using first-principles density functional theory (DFT) calculations and molecular-statics computations based on reliable interatomic potentials, which has revealed strongly attractive interactions for nanopores in the vicinity of GNR edges. These attractive interactions provide the thermodynamic driving force for nanopore migration toward the GNR edge, leading to its coalescence with the GNR edge through a sequence of carbon ring reconstructions. We have studied nanopore dynamics near GNR edges in detail through molecular-dynamics (MD) simulations at high temperature. We have constructed the optimal kinetic pathways of the mechanisms mediating the coalescence of the nanopore and the GNR edge, as identified by the MD simulations, using climbing-image nudged elastic band calculations. The post-coalescence morphological evolution of an armchair GNR edge leads to the formation of a V-shaped edge pattern consisting of zigzag linear segments (facets). DFT calculations of the electronic band structure of such patterned GNRs show that the zigzag segments formed at the armchair edges can be used to tune the bandgap of the GNR. The bandgap of the patterned GNRs exhibits a linear dependence on the density of the zigzag edge atoms, which is controlled by the size and concentration of the pores introduced in the defect-engineered GNR.

498573 Optimization of THz-Time Domain Spectroscopy Reflectivity for a 1THz Energy-Harvesting Metamaterial

Abstract:

Metamaterials, a new synthetic set of materials with repeating patterns of metals and dielectrics, have the ability to manipulate electromagnetic waves. In this study, three-dimensional metamaterials are designed and fabricated to manipulate 0.1 - 1 THz radiation into a well-organized, functional source of plasmons that are rectified by rectenna into a direct current. The metamaterial consists of copper and a photosensitive polymer, SU-8, in an alternating micron-scale patterns fabricated using various techniques, including photolithography, wet and dry etching of metals and oxides, thermal evaporation of thin-layer metals, and electrodeposition of bulk copper.

In order to investigate the absorption of 1 THz radiation for metamaterial, a THz-Time Domain Spectroscopy (THz-TDS) is used to measure the reflectivity of the metamaterial. The geometries of the metamaterials, including the height and the spacing of SU-8 pillars, the thickness of copper, the flatness of the metamaterial surface, have an impact on the reflectivity. In this study, an optimization of the metamaterial geometry is achieved.

499186 Modeling Exciton Dynamics and Low-Frequency Vibrations in Quantum Dot Assemblies

Abstract:

Colloidal quantum dots (QDs) are a promising material platform for solution-processable optoelectronic devices such as solar cells, light-emitting diodes, thermoelectric modules, and flexible electronics. Central to the operation of these devices is the formation, transport, and conversion of energy carriers at the nanoscale, such as in the form of excitons and phonons, for transferring electronic and vibrational energies, respectively.

In the first part of the talk, I will present a computational study of exciton transport in QD solids comprised of polydisperse dots that gives rise to energetic disorder. To gain insight into the relationship between energetic disorder and the spatiotemporal dynamics of excitons, we performed kinetic Monte Carlo (KMC) simulations. We find that a model in which excitons hop between energetically (and spatially) disordered QDs with transition rates determined using Förster theory is capable of reproducing the experimental trends in both exciton displacement and transient energies. Moreover, we use KMC method to investigate the role of energetic disorder in seemingly subdiffusive behavior of excitons that lead to time-varying diffusivity, which we observe in time-resolved optical microscopy experiments.

In the second part of the talk, I will show how the characteristics of collective vibrational modes or low-frequency phonon modes depend on mechanical and physical properties of QDs, such as QD size, ligand identity, and core-shell structure. Here we propose an improvement over the traditional continuum elastic theories for describing confined acoustic phonons in colloidal systems by including the mass-loading effect of the ligands. Calculated $l=0$ phonon energies from our model agree very well with those probed by low-frequency Raman spectroscopy for various CdSe QD samples. Lastly, we demonstrate how our model can help infer elastic properties, such as sound velocities and elastic moduli, of the surrounding ligand materials that may be difficult or impossible to directly probe otherwise.

499643 Extraordinarily Slow Electron-Hole Recombination in Perovskite Phase Cesium Lead Iodide

Abstract:

The exceptional photovoltaic performance of hybrid organic – inorganic perovskites, exhibiting power conversion efficiencies beyond 22%, has inspired a surge in research over the last few years into the broader family of lead-halide perovskites. More recently, all-inorganic $CsPbX_3$ (where X is a halide) has emerged as a promising alternative to the hybrid perovskites due to its greater chemical stability. Herein, polycrystalline thin films of all-inorganic, perovskite-phase cesium lead iodide ($CsPbI_3$) have been investigated by pump-probe transient absorption (TA) and time-resolved terahertz spectroscopies (TRTS). By modeling the carrier recombination kinetics as a function of the initial photo-induced carrier concentration, the rate constants are determined, revealing an unusually slow bi-molecular rate constant (k_2) of 10^{-11} cm^3s^{-1} and a mono-molecular recombination life time as long as 160 ns. From the TRTS measurement, the carrier mobility (m) has been estimated to be >10 cm^2/Vs. The combination of these parameters leads to an estimated carrier diffusion length (L_D), a critical parameter in solar absorber, well over one micrometer in $CsPbI_3$. Remarkably, these excellent optoelectronic parameters are exhibited by a solution-deposited thin film, and are equivalent or even slightly superior to the examples of highly optimized hybrid perovskites in the literature, processed in the same manner. The surprising similarity between $CsPbI_3$ and its hybrid equivalent strongly implies that the organic cation does not provide any explicit benefit in photovoltaic performance. In conjunction with recent successes in improving the structural stability of the perovskite phase of $CsPbI_3$ these results argue for a greater role for $CsPbI_3$ in future photovoltaic technologies.

500078 Elucidating the Impact of Alcohol Post-Processing in High Performance Roll-to-Roll Printed Organic Photovoltaics

Abstract:

Despite having surpassed 10% power conversion efficiency (PCE), which has long been held as the threshold for commercial viability, high performance organic photovoltaics (OPVs) are still mostly constrained to lab-scale devices fabricated on rigid substrates by spin coating. In order to fully realize the often-touted advantages of being lightweight, flexible, and scalable, it is necessary to translate this high efficiency to roll-to-roll printing methods. Efforts to produce scalable printed OPVs trail significantly in efficiency, highlighting the need to better understand the processing-morphology-performance relationship in the context of linear printing processes. Here we investigate the promising OPV system PTB7-Th/PC71BM, which has demonstrated >10% PCE via spincoating but only exhibits ~1% PCE when continuously roll-to-roll printed. Of particular interest is the ubiquitous alcohol wash post-treatment applied to the dried active layer, which induces a significant improvement in device performance, and its crucial role for printed films. While it has been speculated that the primary utility of the alcohol post-treatment is to remove the additive 1,8-diiodooctane (DIO) in spin coated films, we find that the wash process itself dramatically impacts morphology in printed films regardless of the presence of DIO. Here we employ various x-ray characterization techniques to probe phase separation, crystallinity, and molecular orientation, as well as *in-situ* grazing-incidence wide-angle x-ray scattering (*in-situ* GIWAXS) to monitor morphological evolution during the alcohol post-treatment process. It is discovered that isopropanol induces significant polymer alignment and enhanced Ï€-Ï€ relative degree of crystallinity. Through the understanding gained in this study, we are able to achieve a roll-to-roll printed OPV with 5% PCE, which is to our knowledge the highest reported performance for a roll-to-roll printed single junction photoactive layer on a flexible substrate.

500481 Metamaterial and Rectenna Design and Testing for the Conversion of Blackbody Radiation to Electricity Using 5 THz Devices

Abstract:

Metamaterials, a new synthetic set of materials with properties that do not exist on their own in nature, typically consist of various composite materials such as metals and dielectrics. Their repeating geometric structure and material set provide the metamaterial with a negative index of refraction. For this class of metamaterials, the negative index of refraction induces their primary function: their ability to manipulate electromagnetic waves. In this study, metamaterials are designed and fabricated to manipulate incoherent blackbody radiation into a coherent, functional source of electromagnetic waves that are rectified to a DC voltage through a rectenna diode.

The metamaterial-rectenna unit is designed to convert heat into electricity using the optical properties of the blackbody radiation emitted from heat source. In general, as the temperature of the object increases, the wavelength and intensity of the blackbody radiation increase. The metamaterial-rectenna units fabricated in this study are designed to capture electromagnetic radiation with a frequency of 5 THz.

The constructed device uses a rectenna as the rectifying device to convert the alternating current (AC) induced from the electromagnetic waves emitted as blackbody radiation to a direct current (DC). The rectenna, containing a metal insulator-insulator metal (MIIM) tunneling diode, utilizes the ordered electromagnetic radiation from the metamaterial. The metamaterial consists of copper and a photosensitive polymer in an alternating pattern that focuses the electromagnetic radiation over onto the rectenna. The metal of choice, copper, is selected due to its high thermal and electrical conductivity. The negative photopolymer SU-8 is selected polymer substrate due to its well-characterized optical properties as well as use in photolithography. The metamaterial is fabricated using various techniques of photolithography, including wet and dry etching of metals and oxides, bilayer lift-off processes, thin-layer deposition of metals using evaporation, and electrodeposition.

In order to test the functionality of the metamaterial, the reflection spectrum of the metamaterial showing a resonance peak at 5 THz was obtained. The metamaterial-rectenna pair was tested using a thermal source while measuring the electrical response. Matlab and Comsol modeling was used to analyze, as well as, predict the response and characteristics of the devices. Using test pads that have been built into the structure and connected to the rectenna, the diodes are characterized using four-point probe IV measurements. The thermal measurements utilize two of these probes to detect any voltage that is generated by the rectenna-metamaterial device when exposed to a controlled heat source. In this study, further examination of the metamaterial-rectenna pair is performed with devices designed for 5 THz electromagnetic radiation.

501235 Reusable Chromogenic Sensors Enabled By Novel Multi-Stimuli-Responsive Shape Memory Polymers

Abstract:

Here we report novel chromogenic sensors that exhibit easily perceived color changes when exposed to different external stimuli, such as pressure, shear stress, ballistic impact, a large variety of vapors and liquids, heat, and acoustic wave. These multifunctional sensors are reusable, inexpensive, light weight, consuming no electrical power, and very small footprint, promising for a spectrum of applications ranging from user-friendly environmental monitoring to specifically sensing chemicals. This new technology is enabled by integrating scientific principles drawn from two disparate fields that do not typically intersect - the fast-growing photonic crystal and shape memory polymer (SMP) technologies. The active components of the SMPs are thin macroporous photonic crystal layers (only a few mm thick) which are fabricated by using self-assembled, 3-D highly ordered colloidal crystals as structural templates. This microscopic thin-film configuration renders orders of magnitude faster response speed than bulky SMP samples in traditional applications. In addition, by leveraging easily perceived color changes associated with the unconventional all-room-temperature shape memory cycles enabled by the recent discovery of a new series of multi-stimuli-responsive SMPs, sensitive and specific detection of an analyte in a multicomponent solution, such as ethanol in gasoline with a detection limit of 10 ppm, has been demonstrated.

503373 Allyl-Sulfide Modified Hydrogels with Switchable Properties As Dynamic Cellular Niches

Abstract:

allyl-sulfide modified HYDROGELS with switchable properties

as dynamic cellular niches

Tobin Brown, Joseph Grim, Ian Marozas, and Kristi Anseth

Department of Chemical and Biological Engineering and the BioFrontiers Institute

University of Colorado, Boulder, CO, USA 80303

Hydrogels possess a number of properties that render them useful as synthetic extracellular matrices for 3D cell culture and/or cell delivery systems for regenerative medicine; these properties include cytocompatibility, ease of functionalization, and physical properties similar to many soft tissues. Most synthetic hydrogel systems are highly tunable, both in terms of their mechanical and biochemical properties, and when combined with advanced manufacturing methods (e.g., photochemical reactions, 3D printing, microfluidics), experimenters can create hydrogels with gradients, patterned ligands or other hierarchical structures found in native tissues. While the ability to control and manipulate hydrogel properties across many size scales has been powerful, the native extracellular matrix (ECM) is a dynamic environment, particularly during development, wound healing, and disease, and biological and mechanical signals can change dramatically with time. The ability to dynamically tune hydrogel properties can be lost in many synthetic ECM mimics, especially if they are crosslinked and functionalized by irreversible covalent bonds.

In an effort to capture the inconstant nature of native cellular niches, we have investigated hydrogels containing allyl sulfide functionalities. In the presence of photogenerated radicals, allyl sulfides undergo reversible addition and fragmentation. When incorporated as a pendant group within a polymer network, this allows for reversible photopatterning, which we have demonstrated with thiol-containing peptides [1]. More recently, we have expanded this approach to achieve reversible protein immobilization in hydrogels (Figure 1a). Thiolated proteins were covalently tethered to hydrogels through a photoinitiated thiol-ene reaction with the alkene of the allyl sulfide. To release the protein, a subsequent thiol-ene reaction was performed using a small molecule thiol. Importantly, this process is cytocompatible and can be repeated numerous times as the allyl sulfide is regenerated with each thiol-ene reaction. Next, to achieve dynamic control over matrix mechanical properties, allyl sulfides were incorporated directly into the network crosslinks. During photoinitiated radical generation, crosslinks where exchanged with pendant thiols or soluble monofunctional thiols to achieve rapid stress relaxation or photodegradation, respectively. Figure 1b shows temporal regulation of the network viscoelasticity in the presence of photogenerated radicals. During light exposure, pendant thiyl radicals exchange with network allyl sulfides, resulting in deformation under an applied stress. Similarly, soluble monofunctional thiyl radicals can exchange with network allyl sulfides, resulting in rapid photodegradation [2]

(Figure 1c). Together, these materials constitute a class of hydrogels with dynamic chemical and mechanical properties, and this presentation include recent progress in using these materials as a dynamic culture platform for regulating the functional properties of human mesenchymal stem cells.

Figure 1: Hydrogels with dynamic biochemical and mechanical properties. a) A full-length protein is first tethered to the hydrogel in a stripe pattern. The protein is then released in defined regions in the presence of soluble thiol and photoinitiator LAP. b) A creep experiment demonstrates light-mediated changes in viscoelasticity. During UV exposure (shaded regions) the gel deforms in response to an applied stress, and the gel reverts to purely elastic behavior when the light is shuttered and when the photoinitiator is consumed. c) Photodegradation in the presence of soluble thiol. Increased concentration of free thiol leads to faster photodegradation.

References –

(1) N. R. Gandavaparu, M. A. Azagarsamy and K.S. Anseth, *Advanced Materials*, *26*, 2521-26 (2014).

(2) T.E. Brown, I.A. Marozas and K.S. Anseth, *Advanced Materials*, 10.1002/adma.201605001 (2017)

497404 Non-Linear Rheology and Fracture in Alginate Hydrogels

Abstract:

Polysaccharide networks are widely used in biomedical and food applications. In this presentation, results on alginate hydrogels, a model polysaccharide network, will be presented. Effect of two different types of crosslinking, ionic and covalent, on the mechanical properties will be reported. Calcium salts have been used to obtain ionically crosslinked gels. Diamines and amine functionalized carbon nanodots have been used to synthesizing covalently crosslinked gels. Alginate gels with carbon nanodots display photoluminescent behavior, when excited with an UV light of 365 nm. Small angle x-ray scattering was used to capture the structural differences in ionic and covalently crosslinked gels. Shear rheology data capture strain stiffening behavior for these gels. In addition, negative normal stress has been captured, which has not been reported earlier for any polysaccharide networks. Cavitation rheology technique was used to capture the failure behavior of these gels and the results indicate distinctly different behavior for the ionic and covalently crosslinked gels. The results presented here provide new understanding of the deformation behavior of alginate hydrogels.

498445 Analyzing the Effects of Time and Crosslinker Ratio on the Mechanical Properties of Biodegradable Zein Super Gels

Abstract:

Analyzing the Effects of Time and Crosslinker Ratio on the Mechanical Properties of Biodegradable Zein Super Gels

Hazal Turasan, Jozef Kokini, Department of Food Science, Purdue University, West Lafayette, IN

Zein is the main storage protein found in corn and comprises almost 80% of all the corn proteins. Due to the lack of essential amino acids and insolubility in water, zein is not a good choice as a nutrition source for humans. However, its unique properties, such as film forming capability, make zein one of the most commonly chosen plant-based materials for many applications including textile production, adhesive industry, creating various analytical platforms or tissue engineering. Zein also has the ability to undergo gelation in different solutions. Many researchers have explored the solubility of zein in different solutions[1]-[4]. Based on the need for another solvent, the solvents for zein are classified in three. The solvents that does not need a secondary solvent to dissolve zein are classified as the primary solvents, and those that require the addition of either water or another solvent were defined as secondary solvents. Acetic acid is under the category of primary solvents. Ethanol, on the other hand, is classified in the secondary solvent for zein since it requires the addition of water to dissolve zein completely [4]. It has been reported that gelling properties of zein in solution depends on the type of the solution zein dissolved in, such that zein has less tendency to form a gel in its primary solvents [4]. It has also been reported that the addition of crosslinker to zein solutions increase the gelation process [5]. In this study, the gelation process of zein protein in a primary solvent, acetic acid solutions, have been explored to produce a stiff biomacromolecule gel from a plant-based polymer. The zein solutions have been crosslinked with glutaraldehyde solution and the results have been compared to control solutions with no glutaraldehyde. Rheological measurements were conducted on zein solutions to observe the changes in the chemistry and to determine the time of gelation. In addition, compression tests have been conducted on the fully gelled zein solutions to understand the highest compression the gels can endure without breaking.

The zein solutions were prepared with 70% aqueous acetic acid solution. The concentration of zein was kept constant at 1:5 (g/ml) ratio. Oleic acid and monoglyceride was added as plasticizer and the emulsifier at a ratio of 1:1 (g/g zein) and 1:0.15 (g/g zein), respectively. Different crosslinking ratios (0%, 4%, 8%, and 12% (w/w of zein)) have been tested to understand the minimum effective concentration of glutaraldehyde for stiff gel formation. The solutions were sonicated to remove the gas bubbles for 30 seconds for every 10 ml solution and left in room temperature without stirring for 30 days. Starting from Day 0 amplitude sweep tests and frequency sweep tests were conducted in replicates. The results showed that all the solutions showed liquid like behavior in Day 0, such that the loss modulus (G") values were greater than the storage modulus (G') values for all the crosslinker ratios. Uncrosslinked zein solutions did not show any sign of crosslinking throughout the experiment, such that neither the loss modulus values nor the storage modulus values showed an increasing trend in 30 Days. In addition, the loss modulus values were always greater than the storage moduli values for 0% solutions the entire experiment. This result suggests that zein in acetic acid solution does not gel without crosslinker. For 4% crosslinked solutions the loss moduli were greater than the storage moduli for 28 Days, indicating no gelation. However, at Day 30, the G' values crossed the G" values, showing gelation onset for 4% crosslinked solution at Day 30. For 8% crosslinked zein solutions, the transition from sol to gel phase occurred at Day 12, shown by the crossover of G' and G" values. For 12% crosslinked zein solutions, the gelation started the earliest, at Day 9. The highest G' value reached in 12% crosslinked zein gels was 2000 Pa where it was 1100 Pa for 8% crosslinked zein gels. The mechanical compression tests showed that the gels formed from 4% were not stiff enough to resist compression; however, gels with 8% and 12% crosslinker ratio were stiff enough to take 70% compression to fully recover back to their original structure after the compression. At 80% strain, the gels collapsed irreversibly (Figure 1).

Figure 1. Compression test on zein gels. A) Before 70% strain, B) at 70% strain, C) Recovery after 70% strain, D) Before 80% strain, E) at 80% strain moment before failure and F) at 80% strain gel failure.

In summary, this study focuses on producing bio-based polymer super gels from zein-acetic acid solutions and understanding the rheological behavior of zein solutions during gelation. The results showed that with enough concentration of crosslinker, a super gel could be formed from corn protein. The uncrosslinked zein-acetic acid solutions did not show any gelation. 4% crosslinking was enough to start gelation, however, the formed gel was not stiff enough for compression test. Both 8% and 12% crosslinking caused gelation at day 12 and day 9 respectively resulting in stiff gels. The mechanical compression tests showed that both gels recovered from strains up to 70%. However, 80% compression caused gel failure.

[1] Y. Li, Q. Xia, K. Shi, and Q. Huang, "Scaling Behaviors of alpha-Zein in Acetic Acid Solutions," *J. Phys. Chem. B*, vol. 115, no. 32, pp. 9695–9702, Aug. 2011.

[2] R. H. Manley and C. D. Evans, "Binary Solvents for Zein," *Ind. Eng. Chem.*, vol. 35, no. 6, pp. 661–665, Jun. 1943.

[3] D. Fu, "Zein properties and alternative recovery methods," *ETD Collect. Univ. Neb. - Linc.*, pp. 1–188, Jan. 2000.

[4] J. W. Lawton, "Zein: A History of Processing and Use," *Cereal Chem. J.*, vol. 79, no. 1, pp. 1–18, Jan. 2002.

[5] D. J. Sessa, A. Mohamed, and J. A. Byars, "Chemistry and Physical Properties of Melt-Processed and Solution-Cross-Linked Corn Zein," *J. Agric. Food Chem.*, vol. 56, no. 16, pp. 7067–7075, Aug. 2008.

501494 Self-Assembly, Structure and Rheology of Polyelectrolyte Complex Hydrogels

Abstract:

Polyelectrolyte complexes (PEC) form when oppositely charged polyelectrolyte chains spontaneously associate and phase separate in aqueous media. Combining one or both polyelectrolyte chains with a neutral polymer restricts bulk phase separation of the PECs, and thus leads to self-assembled structures with PEC domains surrounded by neutral polymer coronae, forming micelles and hydrogels. The PEC domains in these assemblies can encapsulate therapeutics as well as genetic materials and thus have tremendous potential in drug delivery and tissue engineering applications.

We will present insights on the equilibrium structure and bulk properties of PEC hydrogel assemblages comprising oppositely charged block copolyelectrolytes. Detailed investigations of structure-property relationships of hydrogels comprising model block copolyelectrolytes with oppositely charged functionalized polyallyl glycidyl ethers as ionic and polyethylene glycol as neutral blocks will be discussed. Large-scale ordering of the nanoscale PEC domains is observed at high polymer loadings, characterized by a disorder-order transition followed by morphological transitions with increasing polymer loading. Complementary X-ray and neutron scattering investigations providing a comprehensive structural description of these materials will be presented, strongly supporting our hypothesis on the charged block playing a structure-defining role and the neutral block undertaking a structure-directing role. The effect of key parameters such as polymer block lengths and concentration on the relationship between the nanoscale PEC morphology and equilibrium material properties, elucidated via detailed structure characterization and rheology studies, will also be discussed.

498801 Evolution of Mechanics in α-Helix Peptide Bioconjugated Linear- and Star-Block Peg

Abstract:

We have designed a peptide conjugated poly-ethylene glycol (PEG) bioconjugate system that allows us to examine intra- and inter-molecular dynamics of gelation. We measure the kinetics of gelation for end-functionalized linear- and star-architectures, and we correlate the gelation behavior with the molecular structure and self-association. The 23-amino acid peptide sequence is known to form a coiled-coil structure as a function of the solution's electrolyte concentration, and the two topologies of the PEG are peptide end-functionalized to examine formation of supramolecular assemblies. Subsequently, microrheology is used to evaluate the dynamics of self-assembly and the gelation time-scales. This study shows that the dynamics of peptide folding and assembly for linear-PEG conjugated systems yield a percolated network, but the star-PEG conjugated systems yield discrete assemblies and remain viscous. The results suggest that the degree of intra- and inter-molecular folding defines the critical gel behavior of the supramolecular system.

501101 Modelling of the Degradation of Poly(ethylene glycol)-Co-(lactic acid)-Dimethacrylate Hydrogels

Abstract:

Due to their similarity with extracellular matrix, hydrogels are ideal substrates for cell growth. Hydrogels made of synthetic polymers are excellent alternatives to natural ones and offer the key advantage of precisely controllable degradation times. In this work, we have prepared hydrogels from modified poly(ethylene glycol) macromonomers, functionalized on both ends first with a few lactic acid units, and then with methacrylate groups. A library of hydrogels has been prepared using free-radical polymerization of the macromonomers, by changing both the macromonomer concentration and their type, *i.e.*, the number of lactic acid units contained in them. The degradation kinetics of these hydrogels, caused by the hydrolysis of the lactic acid units, have been carefully monitored in terms of swelling ratio, mass loss and Youngâ€™s modulus. A complete mathematical model, accounting for hydrogel degradation, swelling and reverse gelation has been developed and used to predict all the measured quantities until complete disappearance of the gels. The model has been shown to be capable of accurately predicting the time evolution of all the properties investigated experimentally. To the best of our knowledge, this is the first study where such a systematic comparison between model predictions and experimental data has been presented.

488905 Tough, Rapidly Swelling Thermoplastic Elastomer Hydrogels for Hemorrhage Control

Abstract:

Uncontrolled hemorrhage accounted for 80-90% of potentially survivable deaths of US soldiers over the past 15 years, a majority of which involved truncal or junctional wounds that could not be treated by tourniquet. Materials that rapidly absorb large amounts of water when inserted into a wound have shown promise to control hemorrhage by concentrating clotting factors and sealing the wound cavity. However, materials generally become weaker in proportion to swelling, causing first and second generation granular hemostatic agents to break apart in the wound, leaving debris that must be surgically removed from the body. We present here a tough, rapidly swelling polymer hydrogel based on commodity thermoplastic elastomers with potential application as a component in portable, durable, simple to apply, rapidly effective, and low cost hemostatic dressings. The swollen material has comparable toughness to current â€œtoughâ€ hydrogels, supporting intact removal from wounds, but can absorb up to 800 wt% water in just seconds, potentially facilitating rapid hemorrhage stabilization. Fast swelling and toughness are achieved via a continuous ionic soft phase supported by nanophase-separated hydrophobic glassy domains, as probed by transmission electron microscopy and neutron scattering. A material with optimized composition takes up 250 wt% water in < 1 minute under physiological conditions, and can be stretched up to double its length at 1 MPa stress without breaking while swollen. The composition based on simple free-radical polymerization is easily tunable, making this a low cost and versatile material platform.

499739 Synthesis and Characterization of Thermally Responsive N-Isopropylacrylamide Hydrogels Copolymerized with Novel Hydrophobic Polyphenolic Crosslinkers

Abstract:

Thermosensitive hydrogels were synthesized by copolymerizing N-isopropylacrylamide (NIPAAm) with novel hydrophobic crosslinkers, including curcumin multiacrylate (CMA), quercetin multiacrylate (QMA), and 4,4'-dihydroxybiphenyl multiacrylate (44BMA). Gels were synthesized through free radical polymerization. The crosslinking process was investigated under differential scanning calorimetry (DSC) to obtain the optimized reaction conditions. Hydrogel compositions were then characterized using solid state-NMR (ss-NMR), and the temperature dependent swelling behavior and lower critical solution temperature (LCST) were characterized using swelling studies and differential scanning calorimetry (DSC). Increasing the crosslinker content resulted in a significant decrease in the LCST and swelling ratio of hydrogels, which could be attributed to the increased hydrophobicity introduced by crosslinkers. All of the hydrogels have demonstrated temperature responsive swelling with the extent of swelling decreasing with increasing crosslinker content. The lower crosslinker content gels displayed sharper phase transitions, while the high crosslinker content gels had broader phase transitions.

495486 Novel Biocompatible Thermo-Responsive Poly(*N*-vinyl Caprolactam)/Clay Nanocomposite Hydrogels with Macro-Porous Structure and High Mechanical Property

Abstract:

Poly(*N*-vinyl caprolactam) (PVCL) hydrogels usually suffer from the imporous structure and poor mechanical property as well as the toxicity of the crosslinker, although PVCL itself is biocompatible. In this paper, novel biocompatible thermo-responsive poly(*N*-vinyl caprolactam)/clay nanocomposite (PVCL-Clay) hydrogels with macro-porous structure and high mechanical property are developed for the first time. The macro-porosity in the hydrogel is introduced by using Pickering emulsions as templates, which contain *N*-vinyl caprolactam (VCL) monomer as dispersed phase and clay nanosheets as stabilizers at the interface. The three-dimensional hydrogel network is crosslinked by the clay nanosheets, and thus the hydrogel exhibits high mechanical property originating from the nanocomposite structure. The prepared PVCL-Clay hydrogel possesses remarkable temperature-responsive characteristics with a volume phase transition temperature (VPTT) around 35 °C, and also provides a feasible platform for cell culture. With macro-porous structure and high mechanical property as well as flexible assembly property, the proposed biocompatible thermo-responsive PVCL-Clay nanocomposite hydrogels are ideal material candidates for biomedical, analytical and other applications such as entrapment of enzymes, cell culture, tissue engineering, and affinity and displacement chromatography.

497955 Multiplicity of Morphologies in Poly (L-lactide) Bioresorbable Vascular Scaffolds

Abstract:

Poly(L-lactide), PLLA, is the structural material of the first clinically approved bioresorbable vascular scaffold (BVS), a promising alternative to permanent metal stents for treatment of coronary heart disease. BVSs are transient implants that support the occluded artery for 6 months, and are completely resorbed in 2 years. Clinical trials of BVSs report restoration of arterial vasomotion and elimination of serious complications such as Late Stent Thrombosis. It is remarkable that a scaffold made from PLLA, known as a brittle polymer, does not fracture when crimped onto a balloon catheter or during deployment in the artery. We used x-ray microdiffraction to discover how PLLA acquired ductile character and found that the crimping process creates localized regions of extreme anisotropy; PLLA chains in the scaffold change orientation from the hoop direction to the radial direction on micron-scale distances. This multiplicity of morphologies in the crimped scaffold works in tandem to enable a low-stress response during deployment, which avoids fracture of the PLLA hoops and leaves them with the strength needed to support the artery. Thus, the transformations of the semicrystalline PLLA microstructure during crimping explain the unexpected strength and ductility of the current BVS and point the way to thinner resorbable scaffolds in the future.

491559 Tissue Response and Integration in Biomaterial Implants Derived from Morphologically Unique Emulsion Gels

Abstract:

The long-term success of biomaterial implants is dependent largely on the ability to mediate the host tissue foreign body response and allow adequate void for new tissue and vasculature formation. Modulating the material chemistry of implants has been shown to produce varying levels of inflammatory tissue response. Additionally, micro-topological features and overall 3-D morphology have been implicated as crucial factors in long-term implant success. In this talk, we will demonstrate how a new class of morphologically unique emulsion gels can be utilized for synthesizing biomaterial implants, and how their unique morphological attributes translate to foreign body response mitigation and host tissue integration. Bicontinuous interfacially jammed emulsion gels (bijels) are formed via colloidal silica self-assembly during spinodal decomposition in partially miscible binary liquid solutions. Particles carefully tuned to mutually wet the respective fluid phases adsorb at the fluid-fluid interface at the onset of temperature-driven phase separation. Phase separation is kinetically halted once the interface is fully occupied by particles, and a bijel is formed as the mixture undergoes a gelation transition. The resulting gel is comprised of co-continuous, fully percolating fluid micro-channels with consistent curvature and uniform, tunable channel diameter that are separated by a silica monolayer. The bijel offers a robust template for synthesizing uniquely structured biomaterials from various hydrogel precursors with a number of advantages for long-term implant applications. Firstly, micro-channel domains may be tuned to prevent cells from organizing in dense, encapsulating layers at the tissue-implant interface. Secondly, the consistent curvature of the penetrating micro-channels may promote a favorable interaction between immune cells and implant surfaces. Lastly, the fully penetrating nature of the micro-channels allows for full utilization of the void phase in generating new tissue and dense vascular networks desired in long-term implantable materials.

497844 The Unique Mechanism of Covalently Adaptable Hydrogel Degradation Characterized with Passive Microrheology

Abstract:

Covalently adaptable hydrogels (CAHs) mimic aspects of the native extracellular matrix enabling cells to shear, degrade and apply traction similar to in vivo motility. Understanding the degradation of this material is key to designing it for applications such as drug delivery and 3D culture platforms. In this work, we use multiple particle tracking microrheology (MPT) to measure the dynamic scaffold properties during CAH degradation due to a push out of equilibrium. The CAH is pushed out of equilibrium by changes in the incubation pH. MPT measures the movement of particles embedded in the CAH and the resulting mean-squared displacement describes the network state and connectivity. The CAH is formed by self-assembly of 8-arm polyethylene glycol (PEG)-aldehyde and 8-arm PEG-hydrazine to form covalently adaptable hydrazone bonds. CAHs are formed at both acidic and physiological pHs. Degradation at acidic pH is measured for both CAHs formed in acidic and physiological buffers. MPT measures that the scaffold undergoes at least two cycles of near complete degradation and gelation over several hours. The last degradation event results in irreversible degradation. An acidic CAH incubated in a physiological buffer degrades over a longer time period, approximately a week. During degradation, this scaffold undergoes bond breakage and reformation several times before irreversible degradation occurs. The extent of bond breakage and reformation is much smaller than measured at acidic conditions and, instead, the number of bonds oscillates over the period of several days, with bond hysteresis each cycle. Time-cure superposition (TCS) is used to determine the critical degradation time of each cycle, t_c, and the critical relaxation exponent, n. TCS is the superposition of viscoelastic functions at different extents of reaction. Using TCS, we determine that the value of n for degradation in acidic buffer is 0.48±0.11 and in physiological buffer 0.86±0.04. This indicates that during degradation at pH 4.3 the scaffold is a tightly cross-linked network that can store energy and during degradation at pH 7.4 the scaffold has a more open porous structure and dissipates energy. Additionally, the degradation is modeled using first and second order reaction kinetics. The unique rheological properties and mechanism of degradation of this CAH makes it an ideal candidate for biological applications such as sustained release of active molecules within the body.

492454 Development of Gelatin and Graphene Based Conduits Using 3D Printing Strategies for the Transdifferentiation of Mesenchymal Stem Cells into Schwann Cell-like Phenotypes through Electrical Stimuli

Abstract:

Transdifferentiated mesenchymal stem cells (MSCs) possessing Schwann Cell (SC) like phenotypes have recently been transplanted in different conduits for peripheral nerve regeneration to enable autologous transplants for patients with peripheral nerve damage. However, current difficulties in controlling the final fate of the implanted cell population and providing scalable differentiation protocols along with an ideal 3D scaffold completely mimicking the extracellular matrix limit the clinical use of these transdifferentiated MSCs. In addition, the current knowledge of the effect of mechanical and structural properties of 3D scaffolds on the differentiation behavior of MSCs is limited.[1]

In our previous study[2], we developed gelatin based 3D conduits possessing different mechanical properties and microstructures for the chemical transdifferentiation of MSCs into SC-like phenotypes. The comparison between conduits fabricated with nanofibrous, macroporous and ladder-like microstructures showed that the ladder-like conduits provided the most favorable environment for MSC transdifferentiation to SC-like phenotypes.Â Our results indicated that the 3D ladder-like and macroporous structures enhanced MSC attachment, proliferation and spreading, creating interconnected cellular networks. The 3D ladder-like conduit structure with complex modulus of ~0.4x10^6 Pa and pore size of ~150 nm provided the most favorable microenvironment for MSC differentiation leading to ~ 85% immunolabeling of all SC markers (Î±-S100, Î±-S100Î² and Î±-p75). The transdifferentiated MSCs in ladder-like 3D conduits also secreted significant amounts of neurotrophic growth factors (NGF: 2.5 pg/mL and GDNF: 0.7 pg/mL per cell). This work demonstrated the importance of controlling the 3D microstructure and mechanical properties to facilitate tissue engineering strategies involving stem cells that can serve as promising approaches for peripheral nerve regeneration.[2]

The procedures regularly applied for the transdifferentiation of MSCs to SC-like phenotypes requires the use of high cost chemicals or growth factors. Our recent work showed that graphene-based materials can serve as an effective alternative to transdifferentiate MSCs in to SC-like phenotypes using electrical stimuli. We observed significant enhancement in MSCs transdifferentiation (~85% of the MSCs differentiated into SCs-like phenotypes) and paracrine activity (differentiated MSCs secreted ~80 ng/mL NGF, 10 ng/mL BDNF and GDNF) via electrical stimuli as compared to the conventional chemical differentiation treatment (~75% of the MSCs differentiated into SCs like phenotypes and ~55 ng/mL of NGF secretion) using graphene based materials.[3]

Based on this background, in this study, gelatin and graphene based conduits possessing different microstructures and mechanical properties were fabricated through 3D printing. The effect of 3D conduit microstructure and mechanical properties along with the applied electrical stimuli on MSCs behavior and transdifferentiation into SC-like phenotypes were investigated. The results indicated that the application of electrical stimuli through the conductive graphene layers within the gelatin based 3D microstructure provided a profound effect on the MSCs to SC-like phenotype differentiation and paracrine activity. We observed significant SC marker staining and enhanced NGF secretion (5.2 pg/mL per cell) compared to our previous works. These results suggest that the electrical stimuli applied within the 3D gelatin matrix enables enhanced differentiation and paracrine activity compared to transdifferentiation procedures involving electrical stimuli applied in 2D substrates[3] and chemical stimuli applied in 3D gelatin scaffolds[2], leading to promising nerve regeneration strategies.

References:

1. Elizabeth J Sandquist, Metin Uz, Anup D Sharma, Bhavika B Patel, Surya K Mallapragada, Donald S Sakaguchi, Stem Cells, Bioengineering, and 3-D Scaffolds for Nervous System Repair and Regeneration, Neural Engineering **2016**, 25-81.

2. Uz, M.; Büyüköz, M.; Sharma, A. D.; Sakaguchi, D. S.; Altinkaya, S. A.; Mallapragada, S. K., Gelatin-based 3D conduits for transdifferentiation of mesenchymal stem cells into Schwann cell-like phenotypes. *Acta Biomaterialia* **2017.**

3. Das, S. R.*; Uz, M.*; Ding, S.; Lentner, M. T.; Hondred, J. A.; Cargill, A. A.; Sakaguchi, D. S.; Mallapragada, S.; Claussen, J. C., Electrical Differentiation of Mesenchymal Stem Cells into Schwann-Cell-Like Phenotypes Using Inkjet-Printed Graphene Circuits. *Advanced Healthcare Materials* **2017**, 6 (7). (* indicates equal contribution)

487211 Effect of Molecular Weight and Degree of Functionality on Degradation, Biocompatibility and Two-Photon Polymerization ofÂ AcrylatedÂ Poly(caprolactone)

Abstract:

AICHE Abstract 2017

Effect of molecular weight and degree of functionality on degradation, biocompatibility and two-photon polymerization of acrylated poly(caprolactone)

Brian J. Green,[1] Jessica R. Thompson,[2] Kristan S. Worthington,[2] C. Allan Guymon,[1] Budd A. Tucker[2]

[1]Department of Chemical and Biochemical Engineering, The University of Iowa, 4133 Seamans Center, Iowa City, IA 52242, USA

[2]Wynn Institute of Vision Research, Department of Ophthalmology and Visual Science, Carver College of Medicine, The University of Iowa, 4111 Medical Education and Research Facility, Iowa City, IA 52242, USA

Tissue scaffolds are materials that provide structure capable of supporting cell growth for tissue engineering applications. An ideal scaffold has properties that are application specific and must be tailored for individual cell and tissue types, making polymeric materials especially viable because of the broad range of chemistries and material properties available. Additionally, precise control of scaffold structure in three-dimensions is integral for the design of effective scaffolds. One area in which advanced tissue scaffolds are needed is reversing retinal degeneration associated with age-related and inherited eye diseases. These diseases cause the loss or death of photoreceptor cells, cells that the human body is incapable of regenerating. Stem cell therapy is a promising approach to replacing the degenerate cells, but scaffolds capable of directing the delivery and proliferation of stem cells in three-dimensions must be designed to effectively restore vision.

In this research, we functionalize poly(caprolactone) (PCL), a degradable polymer commonly used in biomedical applications, with acrylate groups to enable photopolymerization and two-photon lithography, a fabrication method that allows precise control by utilizing photopolymerization to create 3D structures with features as small as 100 nm. To fully understand the effects of PCL molecular weight and number of functional groups on each monomer, a number of properties were measured. Reaction kinetics and functional group conversion were used to determine suitability for 3D printing and confirmed by printing structures and measuring the polymerization threshold and resolution. Polymer degradation profile, mechanical properties, and biocompatibility were then used to determine the monomer from which large scaffold structures would be fabricated. It was found that lower molecular weight PCL monomers with a higher number of functional groups had the fastest reaction kinetics and subsequently the highest suitability for 3D printing scaffolds using two-photon lithography. Of the monomers best suited to 3D printing, the higher molecular weight monomer had more desirable mechanical properties, degradation rate, and similar biocompatibility and was therefore selected to fabricate large-scale retinal tissue scaffolds.

490349 An Injectable and Anisotropic Hydrogel with Biomimetic Structures for Directed Cell and Nerve Growth

Abstract:

Many biological tissues consist of complex, anisotropic, and hierarchical structures, such as nerve, bone, enamel, and heart. For these tissues, regenerative matrices have to mimic the biological architecture to guide cell organization during the healing process. However, up to now, most injectable materials, which allow a minimal-invasive application, are isotropic and therefore, lack the ability to template complex tissues with directionally organized functions and mechanical properties. Here, we demonstrate a new injectable composite material, which has the ability to direct cell growth. The composite hydrogel contains rod-shaped microgel objects with variable mechanical properties, geometry, and porosity, which are incorporated into a surrounding biocompatible hydrogel. The microgels are doped with a low amount of superparamagnetic iron oxide nanoparticles (SPIONs), which induce longitudinal alignment *in situ* within a magnetic field in the milli tesla range. Moreover, defined amounts of cell adhesive peptides can be covalently coupled to the microgel precursor to foster a cell-specific interaction and thus mimics the features of endogenous structural proteins. Both nonadhesive and biofunctionalized microgels are applied to create a low-invasive, injectable, anisotropic hydrogel ('Anisogel'). This study encloses the development and characterization of the Anisogel in regard to the cellular response of fibroblasts, primary nerve cells and dorsal root ganglions (DRGs).[1]

Rod-shaped microgels were fabricated with a mold-based soft lithography approach. The microgels consist of a UV-crosslinking star-shaped poly(EO-stat-PO)-acrylate (sPEG-A), which was supplemented with SPIONs. The harvested microgels can be tuned in regard to their size, aspect ratio, stiffness, and porosity, as well as modified in their interaction with cells by covalently attached cell adhesion peptides. For coupling of cell-adhesive peptides the Michael-type addition between thiol-containing biomolecules and the acrylate end group of the sPEG-A in basic conditions was utilized. Furthermore, doping microgels with SPIONs allows their orientation within less than 60 s in a magnetic field of ~100 mT. Interestingly, microgels that have an aspect ratio of 10 exhibit an ultrahigh magnetic response, which allows significant microgel alignment in only 1.9 mT, which corresponds to only 19 times the earth's magnetic field. In order to create a regenerative anisotropic hydrogel, microgels were aligned in a fibrin gel, which enzymatically polymerizes within 120 s, allowing prior microgel alignment. After fibrin gelation, the microgels maintained their position and orientation in the absence of a magnetic field. To investigate the material's functionality in regard to neural tissue, fibroblasts, primary neurons or DRGs were inserted into the Anisogel. When the microgels were randomly oriented, cells infiltrated the matrix less and showed isotropic morphologies. In the case of oriented microgels, extending cells were affected by the aligned physical barrier, resulting in linear cell outgrowth. Interestingly, very low amounts between 1 to 3 vol % of structural guidance elements were sufficient to align the cells. In some cases a single contact with an oriented microgel was enough to align the nerve cells over tens of micrometer distances, propelling the hypothesis that the cells decide to orient inside the Anisogel. The attachment of adhesion peptides to microgels led to a stronger interaction with the surrounding cells, while maintaining the induced cellular directionality. Therefore, the Anisogel with biointeractive microgels can superiorly mimic the ECM of complex tisssues.

By applying magnetoceptive, tailorable microgel rods in a fibrin hydrogel, a global material anisotropy was created after injection. The biomaterial is the first that can achieve highly controlled and ordered structures *in situ* and demonstrated to guide neurite growth in a linear manner. This feature could be groundbreaking as supporting therapeutic material for spinal cord repair.

1. Rose, J. C.; Cámara-Torres, M.; Rahimi, K.; Köhler, J.; Möller, M.; De Laporte, L., Nerve Cells Decide to Orient inside an Injectable Hydrogel with Minimal Structural Guidance. *Nano Lett.* **2017**.

496814 Engineering Extracellular Matrix Mimetic Materials By Green Electrospinning of Collagen

Abstract:

The fabrication of biomimetic scaffolds that replicate the extracellular matrix of tissues has been limited due to the use of toxic solvents in its preparation. Using the electrospinning technique, it is possible to mimic the morphology, nano-scale, geometry and orientation of the extracellular matrix of tissues in the human body. However, typically toxic solvents are used to prepare the polymeric solution used in the electrospinning process, which limits its potential use as a scaffold in biomedical applications. In our work, a collagen nanofibrous scaffold and a composite nanofibrous scaffold of collagen/hydroxyapatite were prepared by electrospinning using a mild solvent. Electron microscopy revealed that it was possible to fabricate an aligned or randomly oriented nanofibrous structure of collagen, with a range of diameters between from 36 to 130 nm. Moreover, the secondary structure of the collagen is preserved after electronspinning as confirmed by FTIR. Hydroxyapatite dispersed into a collagen/acetic acid/water solution was electrospun to yield composite nanofibers. These fibers supports mammalian cell adhesion and growth. Electron microscopy reveals nanofibers with uniform morphology, with an average diameter of 342 nm and a rough surface brought upon by the hydroxyapatite. Both X-ray and infrared spectroscopy confirm the presence of the hydroxyapatite embedded in the collagen fibers. The inclusion of hydroxyapatite does not alter the native collagen structure. Lastly, these composite nanofibers support pre-osteoblast adhesion. These results showed how â€œgreen electrospinningâ€ could be used to generate either pure collagen scaffolds or nanocomposite scaffolds with potential biomedical applications.

501070 Biomimetic Scaffolds for *in vitro* Bone Marrow Tissue Engineering

Abstract:

In vitro human bone marrow tissue replicas represent unique opportunity for studying bone marrow biology and realizing ex vivo expansion of hematopoietic stem cells. Major cellular and molecular components constituting bone marrow microenvironment have been identified, but recapitulating full-range bone marrow functions in vitro remains an elusive task. This is in part due to the lack of relevant culture platforms that reflect bone marrow-like extracellular milieu. From a mechanical aspect, marrow is the softest tissue that is formed inside of the hardest tissue in the body, bone. We hypothesized that extreme contrast of biophysical properties in a microscale porous geometry could be critical in delineating stromal cell function and subsequently supporting hematopoiesis. To test this hypothesis, we have developed a glass-bead and methylcellulose hybridized 3D porous hydrogel scaffold that reflects both soft and hard part of bone marrow in a controlled and analytical manner. The hybrid scaffold design enables to study the role of biophysical, biochemical, and structural cues in directing primary human bone marrow derived stromal cells behavior. Integration of soft and hard materials in a single 3D platform would be an effective strategy to create functional in vitro bone marrow tissue models. We further increased the model complexity by introducing a well-plate based 3D bioreactor and a hypoxic chamber that provide marrow-like mechanical and oxygen milieu. We envision that established ex vivo bone marrow tissue models can advance various aspects of bone marrow related studies serving as an enabling tool.

487496 Tapered Block Copolymers: Tuning Self-Assembly and Properties By Manipulating Monomer Segment Distributions

Abstract:

The self-assembly of block copolymers (BCPs) presents unique opportunities to design materials with attractive chemical and mechanical properties based on the ability of BCPs to form periodic structures with nanoscale domain spacings. One area of recent progress in our group focuses on the behavior of tapered BCPs in which the segment distribution at the interface between blocks is synthetically varied to tune morphology, domain density profiles, thermal transitions as well as mechanical and transport properties. Two application targets for these materials are lithium-ion conducting membranes for batteries and nanostructured thin films for nanotemplates and barrier membranes. In the first target area, we found that the taper volume fraction and composition allow us to manipulate the self-assembly of salt-doped BCPs in a well-defined manner that permits optimization of morphology and ion-content. Additionally, we found that the tapered interfaces influence the glass-transition behavior of the ion-conducting block leading to significant changes in lithium-ion transport (ion conductivity). In the second target area, we found the taper content alters the rate of self-assembly as well as the rate of island/hole formation (and ultimate island/hole size) upon thermal annealing. Additionally, using reflectivity techniques, we probed the domain density profiles as a function of taper composition and linked these profiles to changes in domain spacing and glass transition temperature. Overall, these studies show the versatility of tapering to provide a unique handle for simultaneously optimizing multiple materials properties.

495394 Photoplasiticity in Crosslinked Liquid Crystalline Networks: A Route to Reconfigurable Shape-Changing Materials

Abstract:

Reversible addition fragmentation chain transfer (RAFT) presents a powerful approach to control the bond connectivity, molecular orientation, and three-dimensional shape of crosslinked networks *postpolymerization* with spatial and temporal control. Here, we demonstrated its implementation in crosslinked liquid crystalline networks (LCNs) and used RAFTÕs light responsive bond exchange process coupled with a mechanical or thermal bias to permanently alter the alignment of the LCN. The radical mediated mechanism of RAFT bond exchange was initiated through photo-driven cleavage of radical photoinitiators, molecules that fragment into radical species upon excitation with the appropriate wavelength. Because of this light initiated process, the bond exchange process was decoupled from the thermotropic phase behavior of the LCN, unlike other covalent adaptable LCNÕs, allowing unprecedented control over the alignment and stress relaxation. The RAFT bond exchange process and thermoresponsive LCN behavior gives rise to a multitude of controllable variables that were tuned to alter birefringence, develop surface topographies, and program thermoreversible shape changes. We programmed thermoreversible optical images, which are written in the birefringence, by using temperature to drive disruption of pre-aligned LCNs coupled with varying light intensity and exposure time. The programmed LCN disruption occurred in a continuous manner allowing more complex grayscale images to stored in the birefringence. Extending this approach to surface coating, we found that mismatches in alignment, programmed with RAFT bond exchange, lead to protrusions and wrinkling that was used to develop ordered surface topography on an other flat surface. In freestanding RAFT-LCNs, complex and hierarchical shape change was programmed and erased repeatable demonstrating the power of light responsive processes to program shapes independent of the bias being applied to the LCN. By decoupling the thermal behavior and the bond exchange process in RAFT-LCNs, we have demonstrated broad applications and powerful control over thermoreversible shape changes in elastomeric LCNs.

497350 Synthesis of Novel Nanostructured Copolymers with Alternating Linear Polymer and Dendrimer Blocks

Abstract:

Linear polymers and dendrimers are essential classes of polymers that have been intensively investigated over the past decades. In particular, with monodispersity and well-defined nanoscopic morphologies, dendrimers are also important modules for nanoscale polymer syntheses. A variety of macromolecular architectures using dendrimers as building blocks have been studied. However, nanostructured copolymers with alternating linear polymer and dendrimer blocks so far have not been reported. Herein the synthesis of such a novel alternating copolymer composed of dendrimer and linear poly(ethylene glycol) (PEG) blocks via click chemistry is presented. Firstly, poly(thio-ether) dendrimers with two azide functional groups were synthesized via a robust and efficient divergent approach based on alternating thiol-ene and esterification reactions. Subsequently, the alkyne-azide click reactions of Î±,Ï‰-diacetylene PEGs with the functional dendrimers were performed to yield the linear-dendrimer alternating copolymers with high molecular weight. Both the final alternating copolymers and the polymeric precursors were systematically characterized. The self-assembly properties of the alternating copolymers were studied using microscopy techniques. Unusual assembled nanostructures, such as nanotubes, were observed. Overall, this work elucidated an innovative approach for the synthesis of a new type of dendrimer-based nanostructured polymer, and this approach may potentially be expanded and utilized to access a broad variety of novel nanostructured polymers.

500104 Hierarchical Structures of PDMS-PU Copolymer and Particles for Hydrophobic Coatings

Abstract:

Development of durable functional coatings is a key focus for many industries. Tighter environmental regulations on the use of fluorinated compounds open a new avenue for green approaches to synthesize high performance, superhydrophobic coatings. To prepare such coatings both chemical and physical properties must be addressed. By combining hydrophobic properties of polydimethylsiloxane (PDMS) with those of mechanically robust polyurethane (PU) we have designed a durable hydrophobic polymer. The individual monomers were first linked into a copolymer using isocyanate and hydroxyl groups followed by dispersing the polymer in water. We used dip coating and layer-by-layer deposition approaches to coat cotton fabrics with this PDMS-PU polymer dispersion.

We assessed the effects of monomer molecular weight, particle size and hydrophobic particle effect to fabric coatings. Variation in molecular weight of individual PDMS and PU monomer units had a direct effect on the contact angle of the polymer coating. Lower molecular weight of the hydrophilic PU increased the water contact angle by ca. 20 °, while higher molecular weight of PDMS raised the contact angle by ca. 10 °. Addition of as low as 0.1 wt% of hydrophobized silica particles into the formulation also increased water contact angle by 10 ° and further increase was observed with additional silica content. Antisolvent precipitation of the polymer produced particles with different sizes, which allowed us to achieve hierarchical structure on the coated substrates. Controlling water addition rate enabled us to obtain different particle sizes, ranging from 70 nm to ~3 μm. Narrow particle size distributions allowed to establish a link between the particle size and water contact angle. The versatility of this polymer allows deposition of environmentally friendly hydrophobic coatings on many different substrates for various industries.

489996 Charged Polymer Conformations in Polyelectrolyte Complexes

Abstract:

Polyelectrolyte complexes form when oppositely charged polymers are mixed together in aqueous media, resulting in a polymer-rich complex phase coexisting with a polymer-depleted supernatant phase. This process is driven by gains in enthalpy from electrostatic associations between oppositely charged polymers and entropy from the release of bound counter-ions. The complex phase exists as an equilibrium liquid coacervate or a glassy precipitate, depending on the strength of electrostatic interactions. These interactions are in turn mediated by acidity/basicity of the monomers and their distribution along the polymer backbone, as well as the ionic strength and pH of the solution. However, it remains unclear when and why one phase is favored over the other and what conformations individual polyelectrolytes adopt in either complex phase.

In this study, we examined the effects of polyelectrolyte chain length, as well as solution pH and ionic strength on complexation behavior and polyelectrolyte microstructure using a variety of spectroscopic and scattering techniques. Specifically, we used charged homopolymer polypeptides â€“ (poly)-lysine and (poly)-glutamic acid. This model system allows the chain length (20 â€“ 400-mer), side-chain functionality and chirality (L, D) to be tuned while keeping the backbone chemistry constant, thus enabling a systematic investigation of polyelectrolyte chain conformation in both the liquid coacervate and glassy precipitate phases at various ionic strengths. Overall, understanding the microstructure and the underlying forces that drive polyelectrolyte complexation will enable design of novel materials for applications ranging from drug delivery to coatings.

501177 Solubility Parameters, Water Activity Coefficients and Proton Mobility of Sulfonated Poly (styrene-isobutylene-styrene), Sulfonated Poly (ether ether ketone), and Sulfonated Poly (2-ethoxyethyl methacrylate) Membranes

Abstract:

Solubility Parameters, Water Activity Coefficients and Proton Mobility of Sulfonated Poly (styrene-isobutylene-styrene), Sulfonated Poly (ether ether ketone), and Sulfonated Poly (2-ethoxyethyl methacrylate) Membranes

Maritza Pérez-Pérez and David Suleiman

Chemical Engineering Department

University of Puerto Rico

Mayagüez, PR 00681-9000

Solubility parameter,s water activity coefficients and proton mobility were studied for various sulfonated block copolymer membranes used as proton exchanged membranes (PEM) for direct methanol fuel cell (DMFC) applications. Equilibrium swelling and absorption properties of Sulfonated Poly (styrene-isobutylene-styrene) (S-SIBS), Sulfonated Poly (ether ether ketone) (SPEEK), and Sulfonated Poly (2-ethoxyethyl methacrylate) (SPEEM) were measured using 14 different solvents. The differences in the microstructures were studied using Small Angle X-Ray Scattering (SAXS) and Atomic Force Microscopy (AFM). Two solubility parameters were obtained, one corresponding to the polymer backbone (20.31 $J^{1/2}$ $cm^{-3/2}$) and the other corresponding to the solubility parameter of the sulfonic acid domains (36.1 $J^{1/2}$ $cm^{-3/2}$). The effective water activity coefficients and proton mobility in these ionomers were determined varying the temperature and relative humidity in the polymers. The ionic concentration dependence of proton mobility is attributed to the differences in the microstructure of the polymer, because the different environment can influence the sulfonic, ether, and ester groups inside of the hydrated nano-channels. The membranes possessed nano-channels between 1 nm and 25 nm and water content values between 10 and 31. Because of that, at high sulfonation levels the polymer present higher tortuosity and in some cases higher proton mobility than the state-of-the-art Nafion.

498175 Single-Molecule Super-Resolution Microscopy in Nanostructured Polymer Thin Films

Abstract:

Super-resolution microscopy offers an exciting method to image nanostructures non-invasively, opening the possibility of direct *in situ* real-space observation at the nanoscale. In one form of these methods, sparse fractions of single fluorophores are switched into a bright state from a large population of dark molecules, and the locations of these molecules are determined by fitting their images to the expected point-spread function. Most previous studies have applied these techniques to important scientific problems in the biological community, but little work has explored their use in materials science, where important effects such as dye chemistry and fluorophore orientation introduce new demands.

We explore the capabilities of single-molecule super-resolution microscopy using lithographically patterned polymer thin films as model samples. Line and space patterns of varying pitch were written using electron-beam lithography, and the resulting super-resolution images were compared to the expected structures. Features as small as 20 nm half-pitch are clearly resolvable. From these results, we developed a theory that predicts the ultimate attainable resolution as a function of fluorophore localization precision, accuracy, and density. We then explore the effect of fluorophore orientation, which can alter the single-molecule point-spread function and reduce localization accuracy, but can also be used to report nanoscale chemical environment and dynamics in materials. This capability is demonstrated using polymer films that have been deformed by nanoimprint lithography, where mechanical stress is detected in areas as small as 15 nm through changes in the fluorophore orientation distribution.

A Non-Equilibrium Molecular Dynamics (NEMD) Simulation of the Crosslinked Polyamide Membrane in Water Desalination

Md Symon Jahan Sajib[1]
Tao Wei[1,2]

1. Department of chemical engineering, Lamar University
2. Department of chemical engineering, Howard University

Introduction

A concern about future access to fresh drinking water

- Industrial Development
- Unprecedented growth of Population

Different Membrane Technologies–

- Multi stage flash
- Ultra-filtration
- Reverse Osmosis
- Nano-Filtration

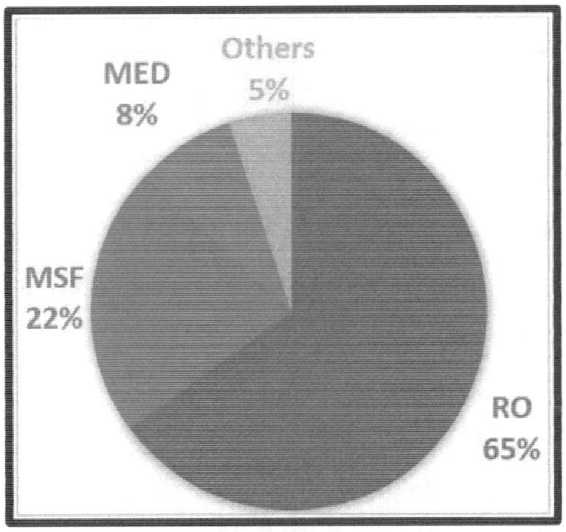

Figure : Total global desalination capacity by technology in 2013

❖Energy efficiency, eco-friendliness gives RO advantage over others.

Minxia DING, Molecular Simulations of Reverse Osmosis Membranes, 15th April,2015

Two main family of RO membrane

- ➢ Aromatic Polyamide (PA)
- ➢ Cellulose acetate (CA)

PA is more promising due to -

- ➢ Less microbiological attack
- ➢ Less compaction under high pressure
- ➢ Better chemical stability,
- ➢ larger fluxes
- ➢ ease of large scale fabrication.

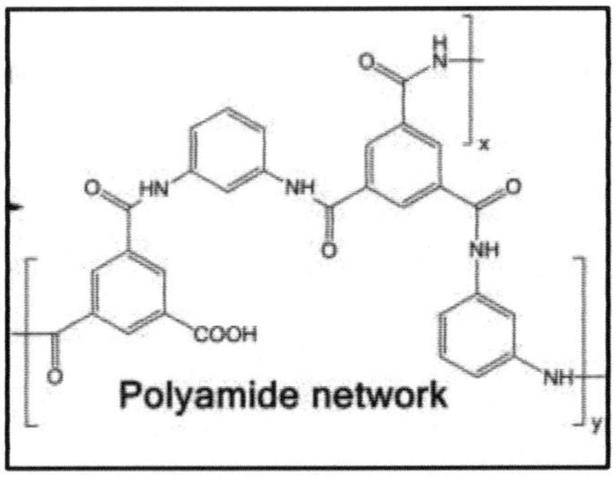

Method:

Membrane Buildup:

➢ Cross-linked based on distance criteria (<0.5nm).

➢ Final achieved crosslink degree : 68.8%

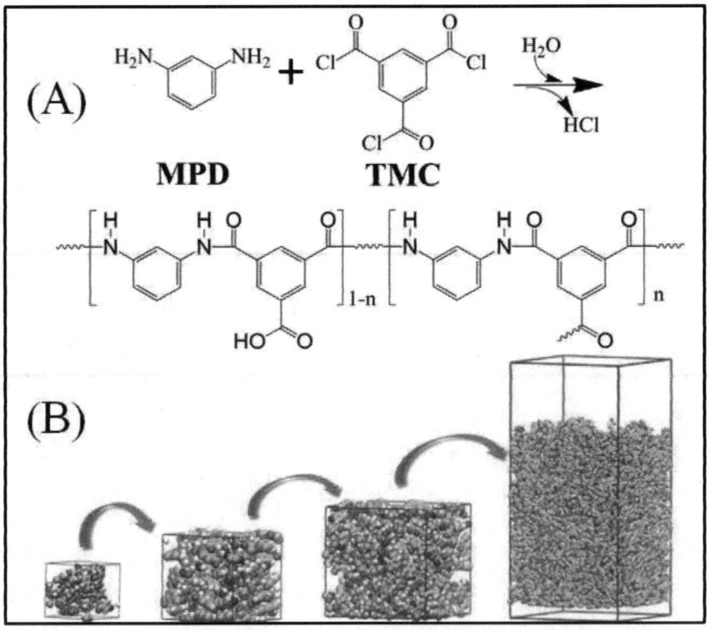

Figure 1. (A) Crosslinking reactions and definition of polymer degree, n ($0 \leq n \leq 1$; $n = 1$ for fully cross-linked polyamide and $n = 0$ for fully linear one); (B) Hierarchical crosslinking process.

Md Symon Jahan Sajib, Tao Wei et al , *J. Phys. Chem. B*, **2016**, *120* (39),10311–10318

Methods

EMD

NEMD

- GROMACS package (version 4.6.5)
- CHARMM General Force Field
- TIP3P water model
- 100MPa, 50 MPa & 5MPa used for NEMD by accelerating graphene wall.

$$a = F/m$$

$$P = F/A$$

Experimental Design of Polyamide Membrane

water permeability:
3.5×10^{-12} m Pa^{-1} s^{-1};
salt rejection:
over 99.6%.

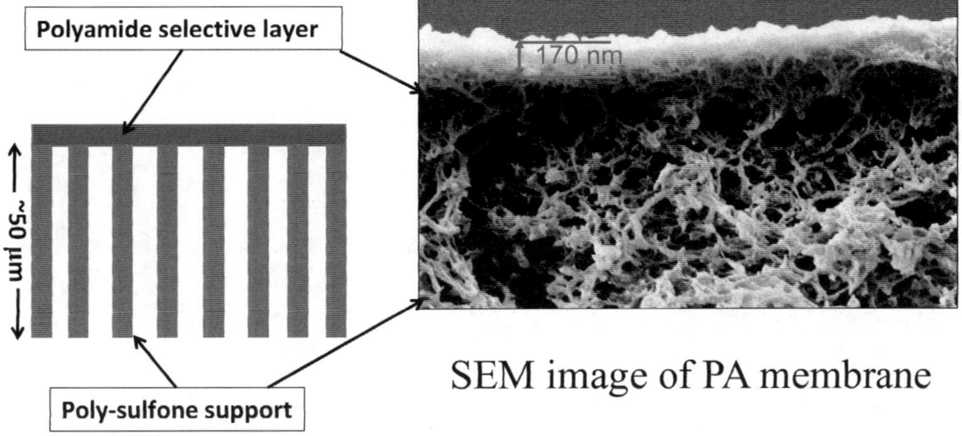

SEM image of PA membrane

Membrane Properties

Properties	Experimental [1,2]	This work [3]
Density, g/cm^3	1.38	1.34
Water content ,wt%	12-23	20
MPD:TMC	1.30	1.32
Crosslinking degree, %	60-80	68.6
Residual –COOH, wt%	4.1-7.1	6.2
O:N	1.2-1.7	1.27

➢ Our crosslinked membrane resembles closely with experimental results.

1.Zhang, X. et al, *J Membr. Sci.*, 2009, **331**, 143-151
2. Lee, J., et al., *J Membr Sci.*, 2013, **425–426**, 217-223
3. Md Symon Jahan Sajib, Tao Wei et al , *J. Phys. Chem. B*, **2016**, *120* (39),10311–10318

Water diffusion in Equilibrium Condition

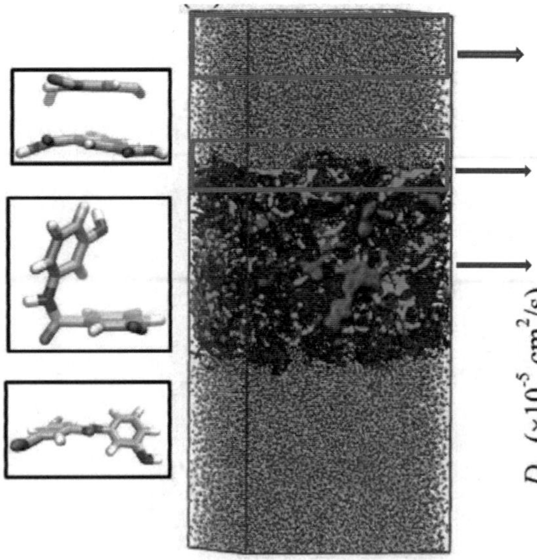

D_{water}: 5.0×10^{-5} cm^2/s

D_{inter}: 2.4×10^{-5} cm^2/s

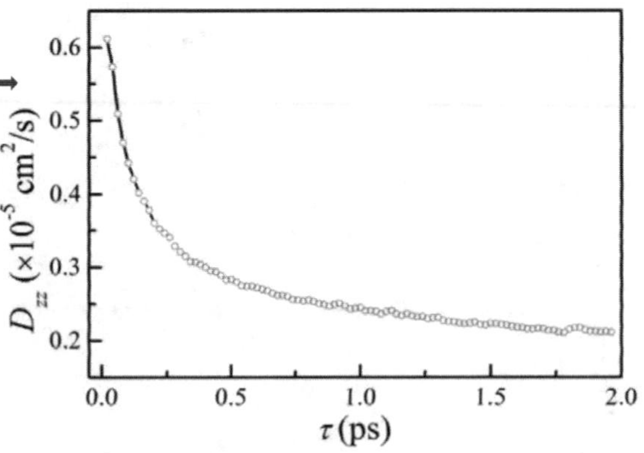

➢ Three structure component – Parallel, T-shape, Slip structure

➢ Heterogeneous water dynamics

Md Symon Jahan Sajib, Tao Wei et al , *J. Phys. Chem. B*, **2016**, *120* (39),10311–10318

5 MPa

Density Profile Analysis

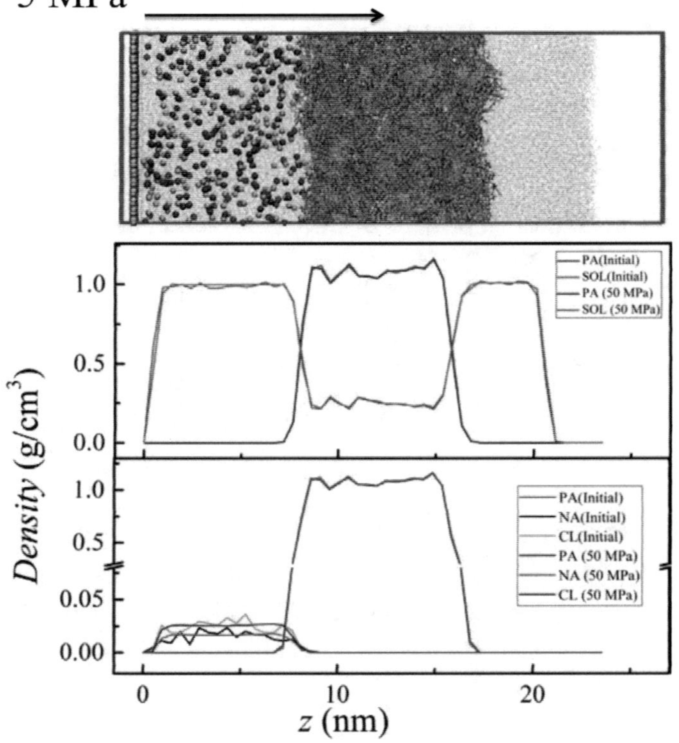

- Rejection of Ions
- Flux = 0.0003 nm/ns
- Thickness of PA is 10.56 nm
- No concentration gradient build up inside the membrane

50 MPa

Density Profile Analysis

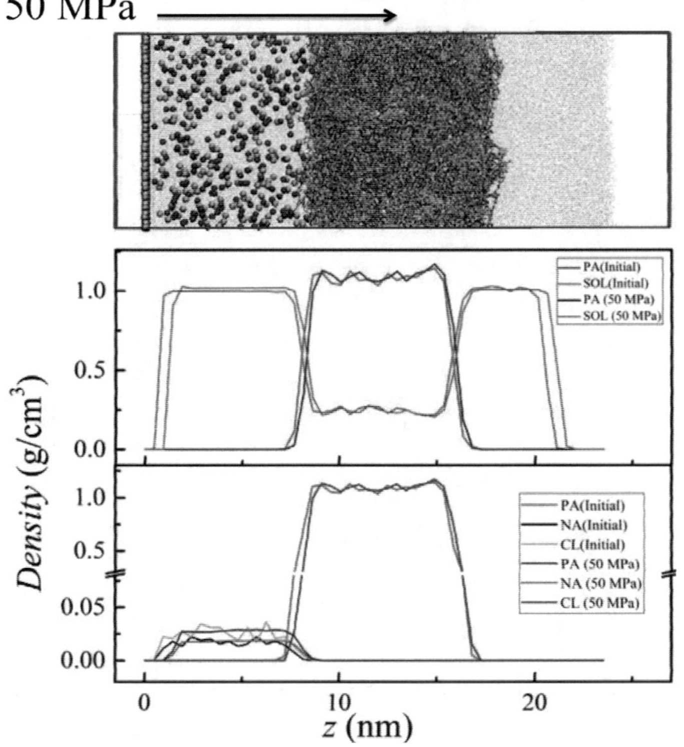

- Rejection of Ions
- Flux = 0. 002 nm/ns
- Thickness of PA is 10.08 nm and shifting of PA is evident.
- No concentration gradient build up inside the membrane

Density Profile Analysis

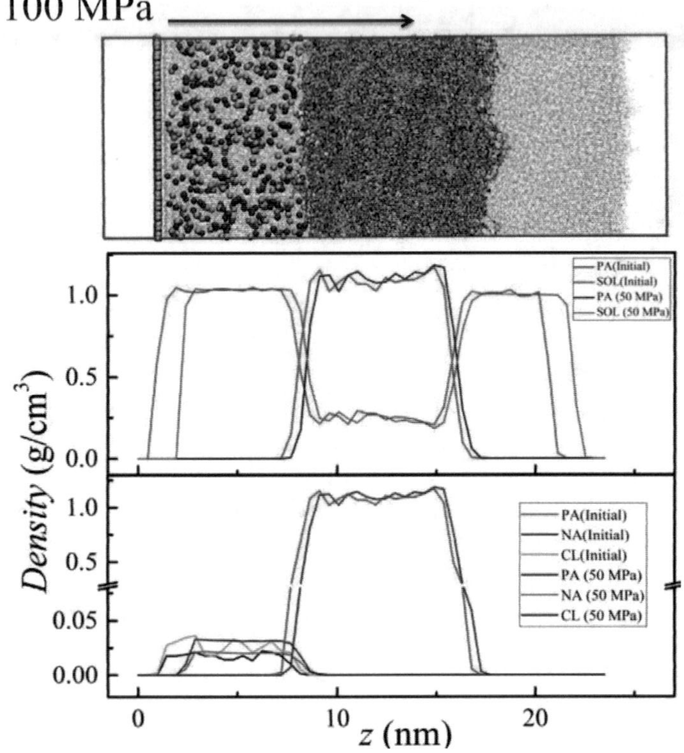

- ➢ Full Rejection of Ions
- ➢ Flux = 0.003 nm/ns
- ➢ Thickness of PA is 10.08 nm and larger shifting of PA is evident
- ➢ No concentration gradient build up inside the membrane

Pore Size Distribution Analysis

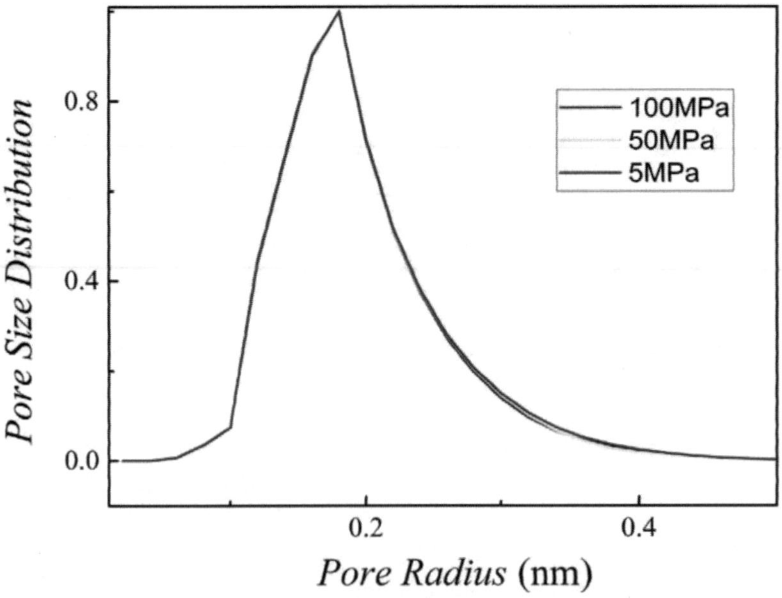

- ➢ No considerable change in pore size distribution in different pressure

Conclusion

- Our simulated crosslinked membrane resembles properties of real membrane.

- Heterogeneous diffusivity of water inside polymer membrane. Faster diffusion is observed through less crosslinked area.

- The increased pressure difference can facilitate higher water permeability inside the membrane, however the change of pore size distribution is negligible, which is responsible for water-salt separation.

Acknowledgement

TACC
TEXAS ADVANCED COMPUTING CENTER

XSEDE
Extreme Science and Engineering Discovery Environment

Any Question?

492732 Linear and Nonlinear Rheology Predictions of Entangled Polymers in Complex Flows from First Principles

Abstract:

We present a novel approach to modeling entangled polymer dynamics, which makes quantitative predictions without parameter adjustment. Nearly since its introduction by Doi and Edwards in 1978, the "tube model" has been the dominant approach to understanding relaxation and rheology in polymer melts. However, despite great efforts and countless additions, consistent quantitative description of data has remained elusive. Moreover, the model still lacks a microscopic basis. Instead, we adopt an alternative approach, mentioned by Doi, Edwards and de Gennes, but never before pursued quantitatively, called "slip-links".

We use a series of hypothesis-driven coarse-graining steps to create a hierarchy of integrated slip-link models. This procedure produces a mathematical model whose calculations are 3 million times faster than the most-detailed level of description, and 20 billion times faster than atomistic-level calculations. Predictions of experiment are quantitative.

As an example, we show that the theory predicts the formation of an unusually shaped vortex in a journal bearing flow, which might be examined experimentally. More important than computational speed up is the dramatic reduction in the number of dynamic variables necessary to describe the system, which suggests a deep understanding of the physics of entangled polymers, justifying the postulates made by Sam Edwards and Pierre-Gilles de Gennes more than 40 years ago.

496633 Melt Behavior-Chain Architecture-Polymer Composition Correlations in High Density Polyethylene

Abstract:

Realization of suitable melt behavior is an important need to be addressed as part of the final properties in HDPEs. Slight variation in factors such as polymer architecture can lead to considerable changes in the final melt behavior. Current work undertakes understanding these disparate mechanisms, establishing correlations between notable rheological aspects and their molecular origins, and establishes molecular design strategies to tune melt behavior.

This study highlights experimental HDPEs synthesized from dual, single-site, co-supported catalysts exhibiting enhanced melt properties such as shear thinning, melt strength and extensibility compared to a few commercial samples. Commonality amongst the resins with the best melt performance is bimodality, and broad molecular weight distribution with well-defined splits and spreads. Additionally, investigations on other relevant processing parameters such as melt elasticity and strain hardening were also carried out, corroborating the need for multiple, pertinent analyses to establish performance enhancement per application

To summarize, the effort highlights improvements in melt properties attributing them to appropriate tuning of polymer chain architecture and molecular weight distribution, thus determining ultimate applicability to selective applications. Ultimately, this may serve as a pathway for producing innovative products that deliver business success and growth.

499878 Shear-Induced Conformational Changes of Flexible and Semi-Rigid Engineering Thermoplastics and Their Influence on Crystallization

Abstract:

Brief intervals of shear flow at rates exceeding the reciprocal of the Rouse time of the longest chains create precursors that nucleate orders of magnitude more crystals and change the morphology from ~30 Î¼m spherulites to far smaller ~1 Î¼m crystallites. This flow-induced crystallization (FIC) at farily low shear rates builds with shearing time and eventually saturates. In contrast, at much higher stress levels that might occur in processing flows, a second morphology transition to shish-kebabs is observed when a critical shear stress (~0.14 MPa) is exceeded. These two transitions are studied in detail for isotactic polypropylenes and poly(ether ether ketone)s representing flexible and semi-rigid polymers, and for Nylon6,6 representing a flexible polymer with strong hydrogen bonding, to see which aspects of FIC are universal to all polymers and which aspects are polymer-specific. The fact that the precursors are quite stable allows the sheared samples to be removed from the rheometer and studied extensively with DSC and optical microscopy, wherein annealing at elevated temperatures allows the study of precursor stability.

498731 Island-in-the-Sea Meltblown Nanofiber Nonwovens with Diverse Surface Properties for Filtration Applications

Abstract:

Melt-blowing is the most usual technique to directly produce nonwovens from polymeric resins. The relatively low thickness of meltblown fibers, 1-2 µm, makes them superior to the competing melt-processing technique, spun-bonding, regarding separation performance. To further improve this feature of meltblowing technique, by decreasing the fibers diameters, the island-in-the-sea approach was investigated in this study. Therefore, initially, immiscible blends were made of polymers with different dissolution, rheological, and surface properties. The successive meltblowing process resulted in microfiber nonwovens with internal island-in-the-sea structures. Finally, upon dissolution of the polymeric matrix with appropriate solvents, nanofiber nonwovens remained. It was noticed that, while meltblown microfiber diameter decreases with reducing their blends components viscosities, thinner extracted nanofibers are processed with decreasing both relative viscosity and interfacial tension of blends. In addition, by using a water-soluble polymer, sulfonated polyethylene terephthalate, existing concerns about cost and environmental problems related to chemical solvents are alleviated. Despite a significant increase in the droplet size in the blend, upon increasing the minor phase fraction, the nanofiber diameter increases almost slightly. Thus, it can be conjectured that nanofiber diameter is mainly controlled by the fiber pinch-off, induced by quick stretching during the melt blowing process, rather than being affected by the initial blend droplet size. Nanofibers resulted from the blends with high relative shear viscosity, Î·$_{minor}$/Î·$_{matrix}$>3, showed a higher level of regularities, particularly when the polymer flow rate decreases. Moreover, with an increase in the dispersed phase content irregularities like bundling of nanofibers increases, which may be a result of their coalescence, while being extruded and meltblown. Also, in the case of fibers from polyvinylidene fluoride (PVDF), due to its lower surface energy, compared with polybutylene terephthalate (PBT), bundling decreases to result in its more significant attenuation of droplets into nanofibers upon meltblowing. In summary, nonwoven mats of hydrophilic PBT and hydrophobic PVDF nanofibers of 100-200 nanometer diameters were made, by applying island-in-the-sea approach through meltblowing process, to provide the opportunity of engineering them towards desired filtration applications.

498613 Enthalpic and Entropic Competitions in Solvent-Free Polymer-Grafted Nanoparticles

Abstract:

Polymer grafted nanoparticles, in the past, have drawn significant attention owing to their ability of forming uniformly distributed polymer-nanoparticle mixtures. One class of such grafted nanoparticles is the â€˜self-suspended hairy nanoparticlesâ€™, where polymer chains are densely grafted onto nanoparticle surface, such that they show good phase stability even in absence of a dispersing medium. Here, we report on the dynamics and structure of covalently grafted polyethylene glycol (PEG) onto silica nanoparticles, which show several interesting phenomena like temperature induced jamming, slow-down of polymer chain dynamics as well as stress overshoot during start-up of steady shear. The entropic attractions between the polymer chains owing to the particle curvature can explain many of these observed behaviors. Further, we show that the attraction between the polymer chains can be manipulated in mixtures of nanoparticles with different polymer chemistry to design materials with unusual characteristics. Specifically, we blend nanoparticles with PEG and PMMA corona chemistries, which are known to have negative Flory Huggins interaction parameter. These blends show similar soft glassy rheological behavior but with a significantly higher plateau modulus and yielding energy. SAXS measurements and DFT calculations further reveal that the grafted chains are highly stretched and interdigitated, owing to the favorable enthalpic attractions. As an extension to these ideas, we study the polymer-grafted nanoparticle blends with polymer chains (PEG and PI) having positive Flory-Huggins interaction parameter. Previous studies on block-copolymers have shown interesting micro-phase separation and self-assembly in contrast to polymer-blends, owing to their higher entropic penalty. Such nanostructures in the block-copolymers have led to their usage in various bulk applications including energy storage, water purification as well as thin-film coatings. In SiO_2-PEG/SiO_2-PI blends, the rheological experiments reveal large reduction of storage modulus and energy of dissipation compared to their single component counterparts. We explain this phenomenon in terms of competitions between enthalpic repulsions among incompatible corona chains and the entropic attractions due to the space filling requirement. More subtle effects of the enthalpic-entropic competition are well captured by quantifying the stretching and terminal relaxation dynamics of tethered polymer chains.

495173 Designing Material Dynamics in Polyelectrolyte Complexes

Abstract:

Polyelectrolyte complexes (PECs) are formed through the electrostatic interaction of oppositely charged polymers. Depending on the solution conditions and choice of polymers, polyelectrolyte complexation can result in the formation of both solid precipitates and/or a liquid-liquid phase separation known as complex coacervation. We have recently used linear viscoelasticity to explore the nature of this liquid-to-solid transition for the model system of poly(4-styrenesulfonic acid, sodium salt) (PSS) and poly(diallyldimethyl ammonium chloride) (PDADMAC) in the presence of potassium bromide (KBr). The use of time-salt superposition facilitated the unambiguous characterization of this transition as a salt-driven physical gelation. Here, we extend these efforts, performing a systematic investigation into the effect of polymer chemistry and molecular weight on the resulting material dynamics. The goal of this work is to establish a molecular-level understanding of the dynamic mechanical response of polyelectrolyte complex materials to facilitate their use across a wide range of applications.

501022 Extensional Relaxation Times of Dilute and Semi-Dilute Polymer Solutions

Abstract:

Stream-wise velocity-gradients associated with extensional flows arise in thinning liquid necks spontaneously formed during capillary-driven thinning and break-up of liquid sheets, jets or stretched liquid bridges. Though shear flows only weakly perturb the chain dimensions, extensional flows can strongly stretch and orient the chains, leading to coil-stretch transition and even full extension. Understanding, characterizing and controlling the extensional viscosity that quantifies the resistance to stream-wise gradients and extensional relaxation time of polymer solutions that determines the pinch-off time are of critical importance for varied applications that rely on drop formation and liquid transfer. Fortunately, visualization and analysis of capillary-thinning dynamics allows for quantitative comparison of transient as well as steady terminal extensional viscosity, extensional relaxation time, and pinch-off time for applications involving jetting, printing, coating, spraying, atomization and microfluidic-based drop/ particle/ emulsion formation. In this contribution, we employ the recently developed dripping-onto-substrate (DoS) rheometry protocols for visualizing and analyzing capillary-driven thinning and pinch-off dynamics of a columnar neck formed between a nozzle and a sessile drop. We examine the effect of increasing polymer concentration on pinch-off dynamics and extensional relaxation times of polymer solutions considered semi-dilute even in quiescent state as the concentration exceeds the critical overlap concentration computed using the criteria based on close packing of unperturbed coils. Despite the relative simplicity and overall price, DoS rheometry overcomes longstanding challenges associated with characterization of low viscosity fluids, including aqueous polymer solutions. We find that the magnitude and concentration-dependence displayed by extensional relaxation time and extensional viscosity are quite distinct from the values obtained in response to shear flow, as the progressive screening of excluded volume and hydrodynamic interactions depends on polymer concentration as well as on degree of stretching and conformational anisotropy.

501823 Nonequilibrium Molecular Dynamics Simulations of Entangled Polymer Solutions Undergoing Planar Elongational Flows

Abstract:

Experiments have shown that concentrated solutions of polymer chains in small molecule solvents exhibit different flow dynamics when subject to elongational flows from polymer melts and polymer solutions that use short chain oligomers as the solvent. This difference has been attributed to the monomeric friction reduction induced by the anisotropic environment around the polymer chains at high extension rates that occur in polymer melts and polymer solutions that which oligomeric solvents. Molecular simulation of these liquids offers a complementary perspective of individual molecule dynamics under flow.

In this work, we performed nonequilibrium molecular dynamics (NEMD) simulations of entangled linear polyethylene solutions subject to planar elongational flows. Through this study, we investigate how flow dynamics is affected by the structure of the solvent and its interaction with polymer molecules. To achieve this, we used benzene as a small molecule solvent and $C_{16}H_{32}$ as an oligomeric solvent in two different sets of simulations. Rheological, structural, and topological (entanglement network) properties of the systems are studied as functions of extension rate. Specifically, segmental orientation and stretch of polymer molecules, as well as solvent oligomers, are examined within a wide range of Weissenberg numbers. We will also discuss steady state in elongational flows from a molecular perspective.

499715 Ultra-High Performance Polymers Meet Ionic Liquids

Abstract:

High-performance (HP) and ultra-high-performance (UHP) polymers represent important and valuable classes of materials for use in demanding applications where thermal, mechanical and/or chemical stresses preclude the use of commodity and mid-range polymers. UHP/HP polymers also provide high strength-to-weight ratios that allow them to be used in place of metals and ceramics, which can reduce weight (and increase fuel economy) in automotive and aerospace applications, which will be important for greenhouse gas reductions and planned missions to Mars. Expanding the types of UHP/HP polymers that can be used in additive manufacturing (3-D printing) is also a critical need as material selection is limited and very few printers support UHP/HP polymers. Simultaneously, interest in polymers that are based upon, derived from and/or able to interface with ionic liquids (ILs) has significantly accelerated in recent years. However, almost all of these â€œpoly(IL)â€ materials are produced via free radical polymerizations yielding polyolefin backbones with pendant cations. Ionenes are an analogous class of ionic polymers wherein ammonium (or similar) cations reside in the backbone rather than as pendants. Yet, nearly all ionenes have been built with long alkyl chains linking cations and lack robust thermal, mechanical and chemical properties. Here, we take a step forward in expanding the chemistry, properties and applications of UHP/HP polymers by building ionic UHP/HP polymers and composites containing ionic liquids (ILs), performing associated characterizations of structure-property relationships and demonstrating the utility of these unprecedented materials for 3-D printing.

500882 Paper-Based Cell Culture Platforms for Personalized Medicine

Abstract:

Two-dimensional (2D) cell culture platforms do not recapitulate the events that occur within the native microenvironment of tissues. Current tissue engineering models utilize sophisticated instrumentation or expensive set-ups for fabrication of three-dimensional (3D) scaffolds, require extensive optimization procedures, and do not provide physiologically relevant size structures in centimeter-scale. The resulting constructs usually contain heterogeneously distributed cells and demonstrate mass transport limitations. It is, therefore, challenging to fabricate biocompatible constructs for personalized medicine. In this work, we developed paper-based cell culture platforms for a wide range of applications that involve mammalian cells, bacteria, fungi, and plants cells. These new cell culture platforms are flexible, tunable, simple, low-cost, and amenable to high-throughput sample preparation and analysis.

We used paper-based scaffolds as matrices to support cells and hydrogels in 3D. We have grown various types of cells (e.g. stem cells, primary cells from animals and humans, cells from patient biopsies, immune cells, fibroblasts, osteoblasts, epithelial cells, tumor cells, bacteria, fungi, or plant cells) in paper scaffolds. After the desired cell culture period, we monitored and characterized the behavior of cells through standard analytical assays such as cytotoxicity, metabolic activity, proliferation, DNA content, protein content, apoptosis, immunostaining, high-resolution imaging, and mechanical tests.

We fabricated and utilized paper-based scaffolds for different applications in personalized medicine. For example, we investigated migration of primary human tumor cells (isolated from patients) in a multilayered paper-based cell culture platform. This approach can be adapted to screen different doses of chemotherapeutics or radiation in a patient-specific fashion. We also used paper scaffolds to induce template-guided biomineralization in origami-inspired structures. This method can be used to fabricate constructs for patients who have irregular size and shape bone defects. In addition, we generated wax-printed patterns in paper scaffolds that are for high-throughput sample preparation and analysis of osteoblast cultures. We have shown that we can easily form and control gradients of oxygen, nutrients, and other biological molecules in paper-based cell culture platforms. We also used these systems to culture bacteria, fungi, and plant cells and developed *in vitro*disease models. Due to porous nature of paper, there is no mass transport limitations to the cells. Our results demonstrated that the paper scaffolds enable patterning from micron to cm scale, adapt modular configurations, can provide physiologically relevant tissue models, and cells can easily be recovered from these scaffolds for further analysis. Paper scaffolds can also be used for origami-inspired tissue engineering for a host of applications.

We recently developed new paper-based cell culture platforms to fabricate multicellular and compartmentalized tissue-mimetics for clinical applications. To overcome the major limitations of the traditional tissue models, we adapted a layer-by layer strategy to assemble tissue-like structures from low-cost and biocompatible paper-based materials. This approach offers unique opportunities from understanding fundamental biology to developing disease models for personalized medicine, and assembling different tissues for organ-on-paper configurations.

501570 Growth Factor Delivery from Silk-Extracellular Matrix Composite Sponges for Modulating Congenital Heart Defect Repair

Abstract:

Congenital heart defects are the leading cause of death in live-born infants and the most severe CHDs require surgery within the first few days of life.[1-2] While these surgical procedures extend the life of the patient, they do not provide a complete correction of the defect. For example, in the case of Tetralogy of Fallot, the standard repair utilizes a patch made of polytetrafluorethylene (PTFE) or polyethylene terephthalate (Dacron™) to widen the right ventricular outflow tract (RVOT). These synthetic patch materials do not mimic the native mechanical properties of the heart, do not degrade over reasonable time periods, and do not promote new tissue formation. In particular, the Dacron™ patches interfere with proper patient growth, impede electrical conduction in the heart, and are inert and non-contractile. Thus, these synthetic patch materials do not meet the needs of the organ during physiologic heart development. In addition, Dacron™ patches are prone to aneurysm, calcification, and infection.[3-5] Therefore, there exists a clinical need to develop a patch or construct that promotes the growth of the patient's own tissue to minimize the cost and risks associated with the re-operative procedures necessitated by the inadequacies of currently available synthetic materials. To address these issues, functional, cellularized silk-extracellular matrix (ECM) composite cardiac grafts[6]are under development.

In the work presented here, we aim to utilize growth factor (GF) loaded silk-ECM patches to promote functional restoration of the right ventricular heart wall while minimizing scarring at the injury site in a rodent model system of RVOT repair. We have optimized silk-ECM patches with tunable growth factor release, evaluating the role of heparin, ECM, and GF concentration in release kinetics following protease degradation *in vitro* and via subcutaneous implantation. *In vitro* and subcutaneous analyses of these GF delivering silk-ECM materials provide key data regarding the importance of timing, concentration, and GF specificity that we hypothesize to lead to improved repair, delivered cell viability/retention, vascular ingrowth, and native cell infiltration. Pilot studies are underway to evaluate the GF delivering silk-ECM cardiac grafts in a full thickness RVOT defect in a rodent model, where we aim to evaluate functional recovery and histological outcomes compared to a standard Dacron™ patch.

References

1.Stumper, O., Hypoplastic left heart syndrome. *Postgraduate Medical Journal* **2010,** *86*(1013), 183-188.

2.Jonsson, H.; Ivert, T.; Jonasson, R.; Holmgren, A.; Bjork, V. O., Work capacity and central hemodynamics thirteen to twenty-six years after repair of tetralogy of Fallot. *J Thorac Cardiovasc Surg* **1995,** *110*(2), 416-26.

3.Parks, W. J.; Ngo, T. D.; Plauth, W. H., Jr.; Bank, E. R.; Sheppard, S. K.; Pettigrew, R. I.; Williams, W. H., Incidence of aneurysm formation after Dacron patch aortoplasty repair for coarctation of the aorta: long-term results and assessment utilizing magnetic resonance angiography with three-dimensional surface rendering. *J Am Coll Cardiol* **1995,** *26*(1), 266-71.

4.Wainwright, J. M.; Hashizume, R.; Fujimoto, K. L.; Remlinger, N. T.; Pesyna, C.; Wagner, W. R.; Tobita, K.; Gilbert, T. W.; Badylak, S. F., Right ventricular outflow tract repair with a cardiac biologic scaffold. *Cells Tissues Organs* **2012,** *195*(1-2), 159-70.

5.Fujimoto, K. L.; Guan, J.; Oshima, H.; Sakai, T.; Wagner, W. R., In vivo evaluation of a porous, elastic, biodegradable patch for reconstructive cardiac procedures. *Ann Thorac Surg* **2007,** *83*(2), 648-54.

6.Stoppel, W. L.; Hu, D.; Domian, I. J.; Kaplan, D. L.; Black, L. D., 3rd, Anisotropic silk biomaterials containing cardiac extracellular matrix for cardiac tissue engineering. *Biomed Mater* **2015,** *10* (3), 034105.

499092 Oligodendrocyte Precursor Cell Maturation in a 3D Hydrogel System through the Incorporation of Drug Delivery Nanoparticles or Topographical Cues

Abstract:

In the central nervous system, oligodendrocyte precursor cells are capable of differentiating into oligodendrocytes responsible for generating the myelin sheath. This electrically insulating layer enables neuronal axons to send electrical signals rapidly and efficiently to surrounding neurons. Following diseases or injuries this myelin sheath is broken down as oligodendrocytes are damaged and oligodendrocyte precursor cells (OPCs) are often unable to differentiate into mature myelinating oligodendrocytes in the damaged tissue, leaving patients with long-term disabilities and pain. While many biomaterial systems enable neuronal maturation and axon extension, few systems have been investigated in terms of oligodendrocyte precursor cells and the biomaterial properties that are necessary for regenerating functional, myelinating oligodendrocytes.

Here, we investigate the influence of controlled drug delivery and topographical cues on oligodendrocyte precursor cell maturation in a 3D polyethylene glycol based hydrogel. OPCs were encapsulated in poly(ethylene glycol)-dimethacrylate gels made through photoinitiation with lithium phenyl(2,4,6-trimethylbenzoyl)phosphinate. In some gels, nanoparticles releasing platelet derived growth factor-alpha (PDGFR-Î±) were encapsulated along with cells. Poly(lactic acid-co glycolic acid) nanoparticles were made using the double emulsion, W/O/W method with varying lactic acid to glycolic acid (L:G) ratios and resulting release profiles. Nanoparticles had diameters ranging from 0.959 to 3.46 micrometers, measured via scanning electron microscopy. The fastest release system was the 50:50 (L:G) 0.59IV nanoparticle which released 90% of the load in the first 24 hours. Interestingly, cells responded to faster release rates by upregulating RNA for the mature oligodendrocyte marker, myelin basic protein. This was found with a four fold increase in myelin basic protein RNA, calculated through RT-PCR, in gels containing the fastest release rate nanoparticles compared to those without nanoparticles. In addition to nanoparticles, topographical cues were incorporated into some gels through the addition of electrospun fibers. On going work is investigating how the incorporation of polystyrene electrospun fibers affect the differentiation of oligodendrocyte precursor cells as well, however it is believed that topographical cues play an important role in oligodendrocyte differentiation and maturation. In 2D cells plated on fibers are observed extending processes towards the polystyrene fibers, much like oligodendrocytes would to axons in the central nervous system. These results taken together indicate the importance of biochemical and topographical cues on the differentiation of oligodendrocyte precursor cells in 3D biomaterial systems and their potential application towards nervous system regeneration.

500203 Resveratrol Releasing Scaffolds Protect Mice Against Diet Induced Obesity and Glucose Intolerance

Abstract:

Despite current treatments, the number of people diagnosed with type 2 diabetes continues to rise, demonstrating a need for new therapeutic strategies. Resveratrol, a natural polyphenol found in the skin of red grapes, has come into the spotlight as a promising anti-diabetic agent after success in many preclinical diabetes models; however, oral and intravenous administration have had mixed success in clinical trials, in part due to poor bioavailability and patient compliance. To address these issues we encapsulated resveratrol within biodegradable polymeric scaffolds for implant into the adipose tissue, a common target of anti-diabetic drugs. We fabricated scaffolds by encapsulating resveratrol within poly(lactide-co-glycolide) (PLG) microparticles using a single emulsion technique. We then coated sodium chloride crystals with the microparticles and pressed the mixture into a cylindrical disk. Gas foaming followed my repeated washings in water produced an interconnected polymer matrix containing resveratrol (i.e., a scaffold). Scaffolds integrate with the epididymal adipose tissue of C57BL/6 mice within seven days after implant and histological analysis indicates the polymer matrix is in contact with adipocytes. Interestingly, resveratrol scaffolds reduce fat gain and improve glucose intolerance caused by a 60% fat diet as measured by dual-energy x-ray absorptiometry and an intraperitoneal glucose tolerance test (IPGTT), respectively. To address how scaffolds are working, we collected the liver of the mice fourteen days after scaffold implant into the adipose tissue (twenty-one days after beginning the 60% fat diet). Western blot analysis indicated that protein kinase B (PKB or Akt) was elevated in the liver of mice receiving resveratrol scaffold implants compared to controls, suggesting enhanced insulin sensitivity in the liver, which might account for the improved IPGTT performance. Currently, we are investigating gene expression at the implant site to address how scaffolds implanted into the adipose tissue are protecting mice from high fat diets. Collectively, we expect this work to impact our understanding of how biomaterial implant modulates adipose tissue function and aid in the development of new treatments for diabetes, the most common metabolic disorder in the United States.

499080 Effects of Short-Term Magnetic Stimulation on MSCs Encapsulated in an Injectable, Magneto-Responsive Hydrogel

Abstract:

Osteoarthritis (OA) is a degenerative joint disease that occurs when articular cartilage begins to naturally degrade and causes severe disability in the lower extremities of the body. Every year, 3.1 million surgeries are done to treat the damage caused by OA[1], and the cost of non-steroidal anti-inflammatory drugs (NSAIDs) exceeds $4 billion annually[2]. The regeneration of damaged articular cartilage has been a main area of interest in the field of tissue engineering due to its inability to naturally regenerate. This lack of functional regeneration is primarily attributed to the innate avascular and heterogeneous tissue structure. Stimuli-responsive biomaterials, capable of responding to changes in the local environment, have emerged as promising scaffolds for applications in guiding the repair of a cartilage defect[3]. In particular, hydrogels that are susceptible to variations in external magnetic fields could be used to spatiotemporally deliver the localized mechanical forces needed to promote and maintain growth of healthy cartilage[4].

An injectable, magneto-responsive hydrogel was developed by chemically incorporating functional paramagnetic iron (III) oxide (Fe_3O_4) nanoparticles into a dual-gelling pNiPAAm-based hydrogel with degradable PAMAM-based crosslinking macromers. Rheological results showed that the nanoparticles played an active role as chemical crosslinkers[EA1]Â . Furthermore, the nanocomposite hydrogelâ€™s physical and magnetic properties were determined utilizing differential scanning calorimetry, scanning electron microscopy (SEM), superconducting quantum interference device (SQUID) magnetometry, and tangential force measurements. Finally, the effects of short-term magnetic stimulation generated by an external magnetic field on encapsulated mesenchymal stem cells were evaluated through biochemical assays and gene expression. Together, these results demonstrated that this magneto-responsive hydrogel system holds promise as a non-invasive means to externally stimulate cell activity and promote regeneration *in situ*.

1. *Arthritis Brochure. American Academy of Orthopaedic Surgeons*. Available from: www.aaos.org.

2. Van Manen, M.D., J. Nace, and M.A. Mont, *Management of primary knee osteoarthritis and indications for total knee arthroplasty for general practitioners.* J Am Osteopath Assoc, 2012. **112**(11): p. 709-15.

3. Li, Y., et al., *Magnetic hydrogels and their potential biomedical applications.* Advanced Functional Materials, 2013. **23**(6): p. 660-672.

4. Pankhurst, Q.A., et al., *Applications of magnetic nanoparticles in biomedicine.* Journal of physics D: Applied physics, 2003. **36**(13): p. R167.

[EA1]

500679 The Incorporation of Retinoic Acid-like Peptoids Onto an Artificial Extracellular Matrix for Increased Differentiation of Human Embryonic Stem Cells into Neural Cells

Abstract:

Neural cells differentiated from human embryonic stem cells (hESCs) provide a compelling component for tissue engineering and regenerative medicine. The extracellular matrix (ECM) plays a major role in the differentiation and proliferation of neural cells. The hESCs are extremely sensitive to properties of the matrix in that even small changes to the chemical composition, physical makeup, or topography of the ECM can drastically effect neural cell differentiation. Promising studies have focused on the development of nano-sized scaffolds through which neural differentiation enhancers can be adapted. Classically, the vitamin A pathway has been explored as a promoter of neural cell differentiation and proliferation. Retinoic acid, a metabolite of vitamin A, can be incorporated into the ECM, and results have shown an increased number of neurons differentiated. We have proposed the formation of a distinct artificial extracellular matrix coated with retinoic acid-like peptoid microspheres to promote the differentiation of hESCs into neurons and astrocytes.

492546 3D Culture of Trabecular Meshwork Cells

Abstract:

Motivation:

Glaucoma is a disease in which damage to the optic nerve leads to progressive, irreversible vision loss. The primary risk factor for glaucoma is intraocular pressure (IOP), which is associated with resistance to aqueous humor outflow. The tissue closely associated with IOP is the trabecular meshwork (TM). To understand the relation between the TM and its effect on IOP it has been proposed to create a biomimetic scaffold that exhibits the same behavior as the native TM. To mimic the tissue structure of the trabecular meshwork, tissue scaffolds have been constructed using the same materials found in the native TM. The scaffold architecture is tuned to promote TM cells to exhibit the same morphology as the native TM.

Problem Statement:

Uniaxially aligned collagen scaffolds containing chrondoitin sulfate have been used to reconstruct the extracellular matrix found in native TM. TM cells proliferate on these scaffolds in static culture and become restricted after one week. The pore sizes of the collagen scaffolds restrict the proliferation of TM cells to the surface. The effect of different glycosaminoglycans (GAGs) in the trabecular meshwork ECM is not understood. It has been shown that TM cells perfused with media with at least 1 μL/min show significantly more cell growth for a longer period of culture. We are now investigating dynamic cell culture to simulate perfusion conditions found in the native TM tissue, and investigating the effects on cell morphology, gene expression and proliferation. Pressure change across the perfusion chamber is due to cellular activity resulting in a decrease in porosity.

Approach:

Scaffolds were made from collagen type I and chondroitin sulfate, using freeze casting to produce uniaxially aligned pores. Trabecular meshwork (TM) cells were isolated from porcine eyes.

TM cells were grown on collagen scaffolds inside a perfusion reactor to measure the effect of dynamic culture on cell proliferation. A perfusion reactor was built so that scaffolds could be placed inside and cell media perfused through it in the axial direction. Flow rates varying from 10 μL/min to 100 μL/min were used to measure diffusion and convection controlled growth. A pressure transducer was used to quantify the pressure change across the perfusion reactor. By measuring pressure change, we can measure changes in porosity, and obtain information about cell interactions with the scaffold over time, as well as screen the effects of potential therapeutics for glaucoma.

The cells were also grown on scaffolds with varying pore sizes. This increase in pore size from ~10 μm to ~50 μm allows cells to proliferation deeper into the scaffold.

TM cells were grown on collagen scaffolds containing chondroitin sulfate (CS), and hyaluronic acid (HA) and a

mixture of CS and HA to measure the effect that each GAG had on the growth and proliferation.

Results:

TM cells cultured on collagen-GAG scaffolds in the perfusion reactor were found to proliferate more quickly compared to those cultured on collagen-GAG scaffolds in static culture. Additionally, the cells migrated deeper into the scaffolds under perfusion compared to those in static culture. Once the cells were confluent, they exhibited morphology more closely resembling native TM. The pressure change of the perfusion reactor increased as the growth of the TM cells decreased the porosity of the scaffold.

TM cells cultured on collagen scaffolds with large pore sizes had increased growth compared to those of the small pore sizes. The larger pore sizes allowed for proliferation deeper in to the interior of the scaffold.

TM cells cultured on the collagen-CS, collagen-HA, and collagen-CS-HA all had similar rates of proliferation showing that the effect of different GAG components have a small effect on the proliferation of cells. (Here I am not sure what these different GAGs will do but I hypothesize they will have a greater effect on gene expression than on proliferation alone)

Conclusions:

The perfusion reactor is a powerful tool to be used in the study of the Trabecular meshwork. As shown above, TM cells grow much better under perfuse conditions. The use of the perfusion reactor replicates the perfuse conditions of the native TM tissue. This setup can be used to model both healthy and glaucomatous tissue by measuring pressure change in real time over a longer period of time. Potential therapeutics for glaucoma can be investigated *in vitro* before use in clinical studies. Further studies will show the benefit of tissue culture under perfuse conditions.

The microarchitecture of tissue scaffolds has been shown to have a large effect on cell growth and behavior. As shown above cells have a much higher growth rate on larger pores that on smaller pores.

The composition of tissue scaffolds can have a large impact on how cells interact and differentiate, however in the TM the presence of different GAG molecules has shown to have little effect on the overall proliferation of TM cells.

503229 Steps Toward Bio-Sourced Packaging: Effect of Composition and Processing on Properties of Co-Polyesters of Polyethylene Terephthalate (PET) with 2,5-Furan Dicarboxylic Acid

Abstract:

Polyethlylene terephthalate (PET) is widely used for packaging applications due to its unique thermal, mechanical, and barrier properties, along with excellet optical clarity. It is synthesized from petroleum based monomers, terephthalic acid (TPA) and ethylene glycol (EG), by melt polymerization. In recent years, there has been increased academic and industrial interest in developing bio-sourced monomers for synthesis of PET and PET analogues. Upsurge in the field of biomass conversion has made economic production of furanics from non-food feedstocks feasible and opened a pathway for a host of novel polymer systems.

491892 Synthesis and Solutionphasecharacterization of Hydroxylated Sulfonated Oligothioetheramides

Abstract:

Nature has long demonstrated the importance of chemical sequence to develop structure and tune physical interactions. By balancing electrostatic charge, length, hydrophobicity, and entropy, naturally advanced structures (e.g. stem loops, alpha helices) create sensitive stimuli-response, allostery, and signaling. Thus, understanding the strength and balance of these physical forces can lead toward the development of structured macromolecules for biological function. We have recently described the rapid and efficient assembly of sequence-defined oligothioetheramides (oligoTEAs), featuring a tunable abiotic thioether backbone and diverse pendant groups. Structural characterization can now facilitate sequence-structure-function relationships by observing the oligomer chain dynamics and conformation.

In this work, we have systematically characterized linear sulfonated oligoTEAs. To characterize structure, we have developed methodology to describe flexible macromolecules with variable temperature diffusion ordered spectroscopy (VT-DOSY) and electron paramagnetic resonance (EPR) to constrain the Stokes-Einstein-Sutherland (SES) equation. With this characterization, we have examined the contributions of synthetic length (2-12mer), permanently positive and negative charged pendant groups, backbone hydroxylation and hydrophobicity, as well as the use of structured monomers on the resulting oligomer structure. We confirm the strength of entropy and hydrophobicity to create polymer â€œcollapseâ€ as a function of synthetic length and composition, described by increased flexibility and decreased aspect ratio (shape). Our results also suggest that hydrophobicity and entropy can overwhelm intramolecular electrostatic repulsion. While low resolution, this solution-phase characterization appears robust in its ability to describe flexible structures that can jeopardize other techniques, especially at this length-scale. As such, we have begun to apply this structural characterization to develop sequence-structure-function relationships with flexible, but biologically functional materials.

492976 Predicting Stable and Metastable Frank-Kasper Phases in Block Polymers Using Self-Consistent Field Theory

Abstract:

Frank-Kasper (FK) phases are topologically closed-packed structures containing particles with disparate shapes and sizes. Hence, these phases are typically found in multicomponent systems such as intermetallic alloys and surfactant solutions. Recently, one member of the FK-phases family, the FK Ïƒ phase, was observed in sphere-forming diblock copolymers, which are single-component materials. Inspired by this experimental discovery, self-consistent field theory (SCFT) calculations were performed to understand the origins of such complex structures in block polymers. We developed a physically informed and robust approach to perform successful SCFT calculations of complex structures. Using this approach, we found that many such structures exist on the free energy surface of diblock copolymers. Importantly, the calculations show that various FK phases have nearly degenerate free energies, suggesting that different structures can be observed experimentally by altering the nucleation pathway on multidimensional free energy surface. We will demonstrate the significance of these calculations in the context of our recent discovery of two new FK phases in poly(isoprene-b-lactide) diblock copolymers.

488664 Spatiotemporal Evolution of Structure in Layer-By-Layer Assembled Thin Films Composed of Oppositely Charged Polyelectrolytes

Abstract:

Despite the numerous applications of polyelectrolyte (PE) multilayer films and coatings, a framework to model the dynamics of this process quantitatively is still lacking. We here deploy a stress-diffusion multi-fluid framework to integrate transient unidirectional migration of strongly-dissociating PE chains in the presence of gradients of electrochemical potential as well as of elastic stress with previously reported quasi-equilibrium descriptions of PE Layer-by-Layer (LbL) assembly. The significant role of chemical specificity of PE functional groups is accounted for by explicitly tracking the formation of the mean-field density of ionic pairs between the two segment types. Ionic pairs alter the local diffusion driving force, slow down mutual diffusion of chains and control the relaxation of elastic stress generated by the ingress of solvent. The film structure and the interface between PE multilayer and external solution is tracked continuously by numerically solving a set of coupled and nonlinear equations, namely the Poisson equation, Maxwell constitutive equation, solvent osmotic pressure equilibrium, flux equations derived from multi-fluid model as well as material conservation equations. Additionally, the Landau-Ginzburg non-local free energy density functional employed here allows for the structure of diffuse interfaces in the system to be resolved dynamically. The proposed model could drastically accelerate engineering of the structure and properties of multilayered coatings.

488326 High Performance Roll-to-Roll Printed PTB7-Th/PCBM Organic Solar Cells

Abstract:

Despite having surpassed 10% power conversion efficiency (PCE), widely held as the threshold for commercial viability, high performance organic photovoltaics (OPVs) are still mostly constrained to lab-scale devices fabricated by spin coating. Efforts to produce scalable printed OPVs trail significantly in efficiency, highlighting the need to better understand the processing-morphology-performance relationship in the context of linear printing processes. Here we investigate the promising OPV system PTB7-Th/PC71BM, which has demonstrated >10% PCE via spincoating but only exhibits ~1% PCE when roll-to-roll printed due to differences in drying dynamics. Of particular interest is the ubiquitous alcohol wash post-treatment applied to the dried active layer, which induces a significant improvement in device performance, and its crucial role for printed films. While it has been speculated that the primary utility of the alcohol post-treatment is to remove the additive 1,8-diiodooctane (DIO) residue in the dried film, we find here that the wash process itself dramatically impacts morphology in printed films regardless of the presence of DIO. Here we employ various x-ray characterization techniques to probe phase separation, crystallinity, and molecular orientation, as well as *in-situ* grazing-incidence x-ray diffraction (*in-situ* GIXD) to monitor morphological evolution during the isopropanol post-treatment process. It is discovered that isopropanol induces significant donor polymer alignment and enhanced Ï€-Ï€ degree of crystallinity. Through the understanding gained in this study, we are able to achieve a roll-to-roll printed OPV with 5% PCE, which is to our knowledge the highest reported performance for a roll-to-roll printed single junction photoactive layer on a flexible substrate.

500488 Pinch-Off Dynamics, Dripping-Onto-Substrate (DoS) Rheometry and Printability of Polymeric Complex Fluids

Abstract:

Liquid transfer and drop formation/deposition processes involve complex free-surface flows including the formation of columnar necks that undergo spontaneous capillary-driven instability, thinning and pinch-off. For simple (Newtonian and inelastic) fluids, a complex interplay of capillary, inertial and viscous stresses determines the nonlinear dynamics underlying finite-time singularity as well as self-similar capillary thinning and pinch-off dynamics. In rheologically complex fluids, extra elastic stresses as well as non-Newtonian shear and extensional viscosities dramatically alter the nonlinear dynamics. Stream-wise velocity gradients that arise within the thinning columnar neck create an extensional flow field, and many complex fluids exhibit a much larger resistance to elongational flows than Newtonian fluids with similar shear viscosity. Characterization of pinch-off dynamics and the response to both shear and extensional flows that influence drop formation/ deposition in microfluidic and printing applications requires bespoke instrumentation not available, or easily replicated, in most laboratories. Here we show that dripping-onto-substrate (DoS) rheometry protocols that involve visualization and analysis of capillary-driven thinning and pinch-off dynamics of a columnar neck formed between a nozzle and a sessile drop can be used for measuring shear viscosity, power law index, extensional viscosity, relaxation time and the most relevant processing timescale for printing. We showcase the versatility of DoS rheometry by characterizing and contrasting the pinch-off dynamics of a wide spectrum of simple and complex fluids: water, printing inks, semi-dilute polymer solutions, yield stress fluids, food materials and cosmetics. We show that DoS rheometry enables characterization of low viscosity printing inks and polymer solutions that are beyond the measurable range of commercially-available capillary break-up extensional rheometer (CaBER). We show that for high viscosity fluids, DoS rheometry can be implemented relatively inexpensively using an off-the-shelf digital camera, and for many complex fluids, similar power law scaling exponent describes both neck thinning dynamics and the shear thinning response. Using a particular example of aqueous polymer solutions, we show the measurement of both the extensional relaxation time and extensional viscosity of weakly elastic, polymeric complex fluids with low shear viscosity Î· < 20 mPaÂ· s and relatively short relaxation time, Î» < 1 ms. Lastly, we will elucidate how polymer composition, flexibility and molecular weight determine the kinetics of capillary-driven thinning and pinch-off in our experiments.

496548 Ultra-Fragile "Granular Materials" Designed Via a Genetic Algorithm

Abstract:

Ultra-Fragile "Granular Materials" Designed via a Genetic Algorithm

Venkatesh Meenakshisundaram, Jui-Hsiang Hung, and David S. Simmons

Understanding glass formation behavior in soft matter materials such as polymers and colloids has held the interest of researchers for several decades and still remains a grand challenge. A key attribute to complexity of glass formation behavior is the local structure of these materials. While there are extensive studies using binary Lennard-Jones binary mixtures and Kremer-Grest model polymers to elucidate the role of local structures on glass formation, a conclusive answer is yet to be determined. In this work, we employ molecular dynamics simulations within the framework of a genetic algorithm to optimize shapes of coarse models, similar to granular materials, for largest fragility. Fragility is a measure of deviation from Arrhenius behavior at the glass transition temperature. We used four different systems in this study where granular materials were composed of 2, 3, 4, and 6 spherical beads. In general, the study revealed that the larger fragility can be achieved by increasing the number of beads within a granular material. The shapes determined for different systems in this study are non-intuitive and can provide further insights into role of local structures in glass formation. Shapes determined from this work can also be used as blueprint to develop macromers that can be polymerized to design new materials.

This work is made possible by the generous funding of W. M. Keck foundation.

489721 Functionalizing Surfaces with Zwitterionic Polymers to Control Cell Adhesion and Direct Neurite Growth

Abstract:

The most successful neural prosthesis to date is the cochlear implant (CI), which restores hearing to patients with profound hearing loss. To perceive sound in a CI, a wearable external device processes and transmits encoded wavelength information to a subcutaneous receptor-stimulation unit. The internal component decodes the information and transmits the signal, in the form of electrical current, to an electrode array positioned in the scala tympani of the cochlea. This electrical signal effectively bypasses the dysfunctional residual hair cells of the cochlea and directly stimulates the spiral ganglion neurons (SGNs), the primary neural receptors in the ear. Tonal stimulation is achieved through positioning electrode signaling along the length of the array, thereby mimicking the tonotopic organization of the cochlea. While this hearing restoration for profoundly deaf individuals is quite a remarkable feat, CI patients have difficulty distinguishing complex tonal information such as voice inflection. This limited resolution is due, in large part, to the distance between the CI electrode array and target neurons. Electrical signals must be strong enough to reach target neurons positioned hundreds to thousands of micrometers from electrode arrays, which causes significant current spread and consequent stimulation of a broad range of SGNs. Thus, even devices with advanced electrical design suffer from low resolution information. Additionally, dense fibrous tissue forms on the surface of implanted electrodes which is common to all implantable devices. This dense tissue attenuates applied currents and increases stimulation thresholds, which further decreases the functional outcomes of CIs. Nonspecific protein adsorption to implant surfaces may trigger the foreign body response: the inflammatory response evoked by the presence of a foreign object in host tissue. Macrophages recognize these nonspecifically-protein labeled implants as foreign, attach to the surface, and attempt to ingest them. Because the implant is large and not biodegradable, these macrophages fuse together and form foreign body giant cells, which secrete inflammatory cytokines and induce fibroblast adhesion. This process permanently isolates the device and renders the implant nonfunctional or less effective.

Eliminating nonspecific protein adsorption to device surfaces could serve as one approach to mitigate the biological response to a material. Recently, zwitterionic materials such as carboxybetaine methacrylate (CBMA) and sulfobetaine methacrylate (SBMA) have been developed as robust ultra-low fouling biomaterials. Zwitterions are inherently net neutral molecules containing an equal number of positively and negatively charged atoms. When used as monomers, the resulting highly polar zwitterionic polymers maintain high wettability, while their zero-charge nature prevents attraction of negatively- or positively-charged molecules to material surfaces. The high energy of hydration is theorized to give zwitterionic materials their antifouling characteristics.

Surface properties are a key component when engineering low fouling materials because the surface comes into direct contact with host fluid, blood protein, and other immunogenic factors. These biomolecules and their nonspecific adsorption to implant surfaces play an integral role in fibrous tissue formation. A common approach to achieve nonfouling properties involves introducing functional groups onto the surface, which then act as anchors for polymer growth. In this work, glass was used as a proof of concept substrate because it can be easily modified with a variety of functional groups using silane coupling chemistry. Glass samples were first exposed to oxygen plasma to generate hydroxyl surface groups followed by reaction with a silane-containing coupling agent incorporating a reactive methacrylate group. Zwitterionic methacrylates were then covalently attached to glass surface methacrylates by illuminating a monomer- and photoinitiator-containing solution in contact with the substrate. UV-light was used to induce the polymerization reaction and covalently build a grafted zwitterionic polymer from the substrate. For our experiments, SBMA and CBMA were used as zwitterionic monomers to alter the surface properties of functionalized glass substrates. The concentration of zwitterionic monomer in solution was varied to change surface coverage and probe the physicochemical properties of the antifouling polymer-grafted surface.

Grafting of the methacrylate silane coupling agent and zwitterionic SBMA and CBMA polymers was characterized through X-ray photoelectron spectroscopy (XPS) analysis, which provides elemental composition of the first 5-10 nm of a surface. XPS measurements showed an increase in carbon content upon addition of the grafting agent. Grafting of the SBMA and CBMA to the glass surface was characterized through nitrogen content. Both SBMA and CBMA showed appearance of the characteristic nitrogen peak in the spectrum. SBMA also contained sulfur further corroborating the grafting of SBMA. To understand how the elemental composition of the surface changes as a function of SBMA and CBMA solution concentration, percent nitrogen was quantified. At low SBMA and CBMA solution concentrations, the atomic nitrogen content is less than 1%, indicating low levels of grafting and surface coverage with the polymer. As the zwitterion solution concentration increases, the atomic composition increases and approaches the theoretical maximum of a pure zwitterionic polymer, calculated using atomic percent. These observations imply that the surface coverage of the SBMA and CBMA polymers is strongly dependent on the concentration of monomer in the solution. Further, for both the SBMA- and CBMA-coated samples, the percent silicon decreased with increasing solution concentration of monomer and showed very little, if any, signal at higher concentrations. These observations are consistent with increases in thickness with higher monomer solution concentrations. On the basis of the XPS data collected, it is estimated that the thickness of the polymer layers measured using all monomer solution concentrations is less than the penetration depth of XPS data collection. XPS only examines the atomic composition the first 5-10 nm of the surface, and thus it is possible that changes (a few nanometers) in film thickness occur giving rise to changes in percent nitrogen.

Surface grafting of a polymer or other molecule is accompanied by a corresponding change in surface properties including surface energy that can be evaluated using water contact angle measurements. The silanized glass was relatively hydrophobic with a contact angle of over 57 degrees as would be expected given the nonpolar nature of the coupling agent used to functionalize the glass substrates. Conversely, zwitterionic polymers are hydrophilic, due to the presence of charged atoms along the polymer side chains and are expected to decrease contact angles. The contact angles of the SBMA- and CBMA-coated samples decreased sharply with increasing wt% monomer in solution. At higher concentrations, the contact angles remained unchanged at approximately 15 degrees. The marked decrease in contact angle for zwitterion-coated surfaces indicates increases in hydrophilicity and further corroborates that the surface was successfully functionalized with the zwitterionic polymer. With large changes in surface properties, it is reasonable to believe that the zwitterionic polymer will affect adsorption of molecules including proteins. As mentioned previously, the first step in foreign body response is nonspecific protein adsorption, such as would occur when a surface comes into contact with serum during implantation. Hydrophilic, net neutral surfaces such as zwitterionic polymers have been shown to effectively prevent the nonspecific adsorption of protein. Fibrinogen adsorption was quantified using immunofluorescence. Adsorption of fibrinogen to SBMA- and CBMA-coated surfaces correlates to water contact angles with sharp decreases up to solution concentrations of 1 wt%. These observations suggest that, similar to the contact angle, a maximum in graft density is reached at low solution concentrations. The protein adsorption data, in combination with the XPS data, suggest that only a very thin film (a few nanometers) of the zwitterionic polymer brushes is necessary in order to achieve a significant decrease in protein adsorption and water contact angle.

Fibroblasts, the primary cell type that contributes to fibrosis on implanted devices, were seeded onto SBMA- and CBMA-coated substrates. An almost 50% decrease in cell density was observed on SBMA-coated surfaces with a remarkable 99% reduction in cell density for CBMA-coated samples. Additionally, astrocytes and Schwann cells both exhibited similar repulsion by SBMA- and CBMA-coated surfaces. To determine the ability of zwitterionic-functionalized glass to spatially control and direct cell growth, micropatterned substrates were fabricated with alternating zwitterion-coated and uncoated bands. Due to the inherent spatial control afforded by photopolymerization, areas of functionalization can be precisely defined through use of a photomask. Fibroblasts, astrocytes, and Schwann cells all elongated along uncoated regions avoiding interactions with zwitterion-coated substrates. Quantification of cell density in uncoated compared to zwitterion-coated regions revealed significant decreases in cell density for all cell types with CBMA exhibiting lower cell density than SBMA. Zwitterionic micropatterns were also generated to be used in directing the growth of Schwann cells and SGN neurites. Neurite extensions grew directly between zwitterion-coated bands avoiding interaction with the functionalized regions. While SGN neurites extend along SBMA stripes, CBMA patterns significantly improve alignment compared to SBMA patterns. Therefore, CBMA-coated surfaces demonstrated a superior ability to resist cell adhesion and to direct cell growth. The described photopolymerizable surfaces enabled investigation of cell-material interactions that could enable engineering of surfaces that resist fibrotic tissue formation. Additionally, generated micropatterns provide potential in directing regeneration of neurites for neural prosthetic development.

493321 Understanding Facilitated Transport of Hydrogen in Polybenzimidazole Containing Palladium Nanoparticles Using an Integrated Experimental and Modeling Approach

Abstract:

Membrane technology is an energy-efficient and low-cost approach for pre-combustion CO_2 capture and H_2 purification in the integrated gasification combined cycle (IGCC) processes. Conventional membranes are based on rigid polymers with strong size sieving ability, such as poly[2,2'-(m-phenylene)-5,5'-bisbenzimidazole] (PBI) that provides high H_2/CO_2 diffusion selectivity. In this study, we demonstrate enhanced H_2 sorption and diffusion in PBI films with embedded palladium (Pd) nanoparticles, which have strong affinity towards H_2. Pd nanoparticles with uniform diameters of 6 - 8 nm are prepared via a hot-injection colloidal synthesis. The loading of Pd nanoparticles in PBI increases H_2 sorption by almost 1,000 times, and at high Pd loadings, the Pd nanoparticles may form fast channels allowing the H_2 molecules to jump from one particle to another and thus increasing the effective H_2 diffusivity. For example, adding 70 wt.% Pd in PBI increases H_2 permeability from 25 to 70 Barrers, and H_2/CO_2 selectivity from 13 to 29 at 150 °C. Such performance is above the Robeson's upper bound for H_2/CO_2 separation, demonstrating the potential of these new materials for industrial H_2/CO_2 separation. The gas transport in these PBI-Pd nanocomposites is being modeled using computational fluid dynamics (CFD) to elucidate the mechanisms for the facilitated H_2 transport. This presentation will also discuss the visualized microstructure of the nanocomposites and provide a unified view of the H_2 and CO_2 transport in the nanocomposites using an integrated experimental and simulation approach.

502569 Engineering Surfaces and Interfaces in Polymer Films with Bottlebrush Polymer Additives

Abstract:

I will discuss the design of bottlebrush polymer additives for thin polymer films. Bottlebrush polymers are architecturally similar to comb polymers, but with a higher density of side chains (1 per backbone monomer). The highly branched architecture leads to unusual physics in polymer blends, such as limited miscibility with long linear polymers and a very strong attraction to surfaces and interfaces. First, we examined bottlebrush copolymer additives for thin films of immiscible polymers. These ternary blends can form two or three phases depending on the bottlebrush architecture. In three-phase systems, the bottlebrushes rapidly accumulate at the polymer/polymer interface and encapsulate the minority domains. This encapsulation layer traps the microstructure and suppresses coarsening under high temperature annealing. Second, we designed surface-active bottlebrush additives for commodity thermoplastics. In these systems, a balance between bulk miscibility and bottlebrush surface attraction controls the segregation profiles through the film thickness. We identify conditions where bottlebrush additives spontaneously accumulate at the free surface and substrate, offering a simple route to generate brush-like coatings with unique interfacial properties.

498908 Effects of Extensional Flow and Nanoparticle Stabilization on Immiscible Polymer Blend Morphology

Abstract:

Polymer blending facilitates the combination of the attractive attributes of two or more polymers while compensating for the unfavorable ones. Most polymers are thermodynamically incompatible with one another, and their blending yields a two-phase microstructure. This morphology generally determines the mechanical and rheological properties of the blend system which then determine its applications. Morphology development typically involves deformation of the dispersed phase followed by drop breakup. However, drop coalescence competes with this process, and ultimately a balance must be reached between these two competing processes. Extensional flow fields are known to promote drop breakup and are especially important for blends with high viscosity ratios, that is for blends where the viscosity of the dispersed phase is at least about 3.8 times greater than that of the matrix phase. Coalescence may be attenuated by compatibilizers that modify the interface between the polymer phases. Nanoparticles with tuned surface chemistry may also be used as compatibilizers. A combination of extensional flow and nanoparticle stabilization should, therefore, result in a fine, stable morphology.

To begin the investigation toward the effects of extensional flow blending with and without the incorporation of nanoparticles, preliminary results were obtained using two different polymer blend systems: polycarbonate (PC)/styrene acrylonitrile (SAN) and polystyrene (PS)/linear low-density polyethylene (LLDPE). However, the majority of the presented results involve blends of high-density polyethylene (HDPE) dispersed in PS. With this blend system, with the material grades selected, the viscosity ratio exceeded 3.8 over the entire domain of deformation rates anticipated in the processing used. Coarse blends of various compositions were formulated using shear flow in an internal mixer or in a twin-screw extruder. These blends were subjected to extensional flow in converging dies of different geometries and where more than one stretching episode was possible; the temperature, total strain, and flow rate were varied, among other factors, in a systematic manner. Experiments were repeated in the presence of various grades of fumed nanosilica of different sizes and surface treatments, which imparted different surface tension and relative surface polarity (hydrophilic versus hydrophobic) for the nanoparticles. The mixing sequence was varied including premixing the nanosilica into the thermodynamically non-preferred polymer phase.

Scanning electron microscopy (SEM) was used to determine the size and size distribution of the dispersed polymer phase. The material was typically sectioned in the flow direction, but sectioning in the direction perpendicular to flow and etching, or selectively dissolving, one phase or the other was also investigated. The primary effect of extensional flow blending was to reduce the volume-average diameter of the dispersed polymer phase, especially with increasing strains and flow rates, or strain rates, which is directly dependent on both. Finding suitable conditions for the nanoparticles to selectively localize at the HDPE/PS interface was challenging, but relatively small amounts of nanoparticles dispersed in the PS matrix decreased the volume-average diameter of HDPE drops. When the nanosilica was preloaded into the HDPE dispersed phase, very coarse initial blends were produced which then exhibited dramatic decreases in phase size with extensional flow. These and other results are properly organized and presented.

494040 Non-Isocyanate Polyurethane Thermoplastic Elastomer: Amide-Based Chain Extender Yields Enhanced Nanophase Separation and Properties in Polyhydroxyurethane

Abstract:

Polyurethane (PU) represents an important class of commodity polymers with a broad range of uses such as foams, elastomers, coatings, films and adhesives. It is produced from a step-growth reaction between isocyanates and alcohols.[1] Isocyanates are facing increasing regulatory scrutiny across the world, e.g., from the United States Environmental Protection Agency and the European Union REACH regulation, necessitating the search for alternative route to PU or PU-like material without employing isocyanates.

Cyclic carbonate aminolysis is a very promising chemistry to produce polyhydroxyurethane (PHU), a PU-like material with an additional hydroxyl group adjacent to every urethane linkage.[2,3] Significant fractions of commercial PU are segmented type of copolymers, but most PHU studies to-date often focus on single-phase thermoplastics or crosslinked thermosets.[1-3] There are currently very limited studies on segmented, nanophase-separated PHUs and how to control and tune their properties.[4-7] Very recent studies by Torkelson and coworkers on segmented PHUs demonstrate that the hydroxyl groups and soft-segment choice play a critical role on the development of nanophase separation and elastomeric character in polyether-based PHUs.[4,5] Polyethylene oxide (PEO)-based PHUs are single-phase materials due to significant phase mixing whereas polytetramethylene oxide (PTMO)-based PHUs exhibit nanophase separation with broad interphase having a wide range of local compositions.[4] Furthermore, the thermal and mechanical properties of segmented PHUs still needs improvement to match those of conventional thermoplastic polyurethane (TPU) elastomers.[4-9] Achieving competitive properties in PHU elastomer is essential to produce replacement materials for isocyanate-based TPU elastomer.

Here, we report our approach in designing non-isocyanate polyurethane (NIPU) thermoplastic elastomers with properties competitive to conventional TPU elastomer by incorporating amide-based, diamine diamide (DDA) chain extender. Above a certain hard-segment concentration, nanophase separation can be achieved in PEO-based PHUs, a system previously known to yield only single phase PHUs, by using DDA chain extender. In the PTMO-based system, we obtained PHU elastomers with wide, relatively temperature-independent, rubbery plateau regions with flow temperature up to 200 Â°C. These PTMO-based PHUs exhibit tunable mechanical properties with Youngâ€™s modulus ranging from 6.6 to 43.2 MPa, tensile strength ranging from 2.4 to 6.7 MPa, and elongation at break of ~300%.Â They also possess elastomeric recovery similar to that of conventional TPUs. The amide linkages in DDA chain extender lead to stronger hydrogen bonding in the hard segment, thereby overcoming the competing effect of intersegmental hydrogen bonding from the hydroxyl groups. The regular structure of amide-based segments in the DDA chain extender helps to drive hard-segment crystallization, thereby resulting in significantly enhanced nanophase separation and thermal properties over PHUs made without DDA chain extender. Incorporation of DDA chain extender is a powerful strategy to produce PHU thermoplastic elastomer with competitive properties rivaling those of conventional TPUs.

References

1) Engels, H.-W.; Pirkl, H.-G.; Albers, R.; Albach, R. W.; Krause, J.; Hoffmann, A.; Casselmann, H.; Dormish, J. *Angew. Chem. Int. Ed.* **2013**, *52*, 9422.

2) Guan, J.; Song, Y.; Lin, Y.; Yin, X.; Zuo, M.; Zhao, Y.; Tao, X.; Zheng, Q. *Ind. Eng. Chem. Res.* **2011**, *50*, 6517.

3) Blattmann, H.; Fleischer, M.; Bähr, M.; Mülhaupt, R. *Macromol. Rapid Commun.* **2014**, *35*, 1238.

4) Leitsch, E. K.; Beniah, G.; Liu, K.; Lan, T.; Heath, W. H.; Scheidt, K. A.; Torkelson, J. M. *ACS Macro Lett.* **2016**, *5*, 424.

5) Beniah, G.; Liu, K.; Heath, W. H.; Miller, M. D.; Scheidt, K. A.; Torkelson, J. M. *Eur. Polym. J.* **2016**, 84, 770.

6) Beniah, G.; Uno, B. E.; Lan, T.; Jeon, J.; Heath, W. H.; Scheidt, K. A.; Torkelson, J. M. *Polymer* **2017**, 110, 218.

7) Beniah, G.; Heath, W. H.; Jeon, J.; Torkelson, J. M. *J. Appl. Polym. Sci.* **2017**, 134, 44942.

8) Krijgsman, J.; Husken, D.; Gaymans, R. J. *Polymer* **2003**, 44, 7043.

9) Klinedinst, D. B.; Yilgor, I.; Yilgor, E.; Zhang, M.; Wilkes, G. L. *Polymer* **2012**, *53*, 5358.

10) Qi, H. J.; Boyce, M. C. *Mech. Mater.* **2005**, *37*, 817.

11) Buckley, C. P.; Prisacariu, C.; Martin, C. *Polymer.* **2010**, *51*, 3213.

489866 Molecular Simulations Study of Solvophobicity Effects on Assembled Structure in Solutions of Amphiphilic Block Copolymers and Nanoparticles

Abstract:

Hybrid materials composed of inorganic nanoparticles (NPs) and amphiphilic block copolymers (BCPs) combine technologically desirable characteristics of both inorganic nanoparticles and the rich phase behavior of block copolymers, and have found uses in biological or non-biological cargo delivery, imaging contrast agents, pollutant capture, chemical sensing, and separation/purification applications. In recent years, there has been growing interest in mixing solvents and utilizing the effective solvophobicity of the BCP block(s) to tailor the assembled structure of the NP-BCP hybrid materials, namely the size and shape, composition, and spatial arrangement of the components. In this talk, we will present a comprehensive coarse-grained molecular dynamics (MD) simulations study explaining the impact of varying solvophobicity on assembly of amphiphilic BCP and NP as a function of BCP compositions and sequence and NP affinity to either or both block(s) of BCP. We quantify the effective solvophobicity needed for assembly/micellization, and then as a function of varying solvophobicity quantify and visualize the shape and size of assembled structures with and without NPs, amount of NP uptake and the spatial arrangement of the NPs in the assembled NP-BCP structure.

487210 Control Nano/Microstructure Using Photopolymerization-Induced Phase Separation (PhIPS)

Abstract:

Abstract ID: 487210

Photopolymerization-induced phase separation (PhIPS) is one of the techniques to combine the thermo-mechanical and optical properties of different constituents in a single polymer. Controlling phase separation, polymer networks could be generated with enhanced physical and mechanical characteristics, including increases in material toughness, reduction of polymerization-induced shrinkage stress, and enhanced abrasion resistance compared to equivalent homogeneous polymer networks. In this study the effect of composition and polymerization rate for photocurable crosslinked systems for the creation of phase-separated polymers will be examined. Changing the composition, thermodynamic instability may allow the polymer system to become incompatible and phase separate. The polymerization kinetics are also important as cross-linked network structures may be trapped before complete phase separation occurs, affecting the morphology and domain size of polymers. A mixture composed of butyl acrylate (BA) and di-functional oxetane (DOX) was investigated at different monomer compositions and light intensities. The formation of different polymer domains was observed using atomic force microscopy and dynamic mechanical analysis. The photopolymerization kinetics for different ratios of DOX/BA were characterized using real-time Fourier transform infrared spectroscopy. BA conversions between 95-100% were reached within 2 minutes, independent of light intensity. On the other hand, DOX shows very different conversion behavior compared to those of BA. The final DOX conversions were only between 20 and 30% which can be attributed to a low propagation rate constant. Higher light intensities not only increased DOX conversion to as high as 50%, but also changed the final polymer morphology by decreasing the domain size and swapping the co-continuous phase. At low light intensity, BA soft domain was the co-continuous phase, with DOX hard domain being dispersed in the network. The mechanical properties of these polymers were very weak. However, increasing to higher light intensity, the co-continuous phase changed to DOX harder domain and the size of each domain was decreased. These changes in the morphology affected the thermo-mechanical properties of the polymers such as showing more segregated tanÎ´ peaks, higher modulus and stress-strain values. Since light intensity affected both conversion and morphology, it was important to investigate which of these two parameters had the greatest impact on the thermomechanical properties. To obtain better control of our system, DOX conversion was brought to the same value (~70%) using UV and post thermal curing. Our investigation indicated that there were not significant changes in the polymer morphology and mechanical properties. Consequently, it is reasonable to believe that the physical properties are dependent on the initial obtained morphology from the photocuring of the samples and not to higher DOX conversions. These results demonstrate the ability to control the physical properties of phase separated polymers by changing the morphology through different light intensities.

497089 Spontaneous Self-Assembly and Micellization of Random Copolymers in Organic Solvents

Abstract:

Spontaneous self-assembly of random and statistical copolymers in solution is unusual due to the structural irregularity of the copolymer chain and close proximity of short incompatible segments. This study describes the first observation of supramolecular structures such as micelles and vesicles formed by a random copolymer in organic solvents. Upon dissolution in methanol or tetrahydrofuran, the random copolymer poly(trifluoroethyl methacrylate-random-methacrylic acid) forms small spherical micelles, worm-like assemblies and large vesicles spontaneously. Self-assembly is driven by the high incompatibility between the fluorinated and acidic, partially ionized repeat units. Micelle size can also be altered by the addition of metal ions, which interact with the carboxylic acid groups on methacrylic acid through complexation or Coulombic forces. These findings demonstrate an easy, single-step approach to creating nano-scale structures with tunable size and morphologies in organic solvents from easily synthesized random copolymers, with potential applications in coatings, selective membranes, catalysis, and drug delivery.

494586 Synthesis and Antibacterial Study of Star-Shaped Poly[2-(dimethylamino)Ethyl Methacrylate]-Based Copolymers with an Inorganic Core

Abstract:

Star polymers with poly[2-(dimethylamino)ethyl methacrylate] as the arms and POSS as the core (POSS-*g*-PDMA) were synthesized by atom transfer radical polymerization (ATRP). The effect of molecular weight on the antibacterial activity was studied and lower molecular weight POSS-g-PDMA has better bactericidal activity as measured by the minimum inhibition concentration. POSS-*g*-PDMA was further modified by various techniques to increase hydrophilicity in attempting to improve their antifouling activity without compromising bactericidal activity. POSS-g-PDMA was quaternized to different degrees and the antibacterial activities of the obtained quaternary polymers were studied; the antibacterial activity decreased as the degree of quaternization increased. Finally, cationic-zwitterionic polymers with both random and block structures, where PDMA and poly(sulfobetaine) were cationic and zwitterionic blocks respectively, were synthesized. The random cationic-zwitterionic polymers showed poor antibacterial activity while the block polymers retained the antibacterial activity of the pristine POSS-*g*-PDMA. The block copolymers of POSS-g-(PDMA-*b*-polysulfobetaine) showed enhanced antifouling property and serum stability as seen by their nanoparticle size stability in the presence of serum and reduced red blood cell aggregation. The antibacterial kinetics showed that *E. coli* can be killed within 30 min by both random and block polymers. Finally, block polymers showed low toxicity to zebrafish embryo and could be potentially used in aquaculture antibacterial applications.

494914 Understanding Molecular Exchange Kinetics in Polyelectrolyte Complex Micelles

Abstract:

Polyelectrolyte complex (PEC) micelles form when oppositely charged block polyelectrolytes are mixed together in aqueous media. The polyelectrolyte blocks electrostatically associate and phase separate, leading to a dense, polymer-rich PEC core stabilized by a neutral block corona. Considerable theoretical and experimental effort has been focused on the structure of PEC micelles. However, much less is known about the stability and kinetics of the PEC micelles, particularly the mechanism governing the fundamental molecular exchange kinetics between the micelles.

In this work, we developed a PEC micelle platform that is formed by a diblock copolyelectrolyte, poly(ethylene oxide) (PEO) block poly(vinylbenzyl trimethylammonium chloride) (PEO-b-PVBTMA), and a homopolyelectrolyte poly(styrene sulfonate) (PSS), with specifically tailored block lengths and change densities. The structural stability and molecular kinetics in this system were studied using microscopy, dynamic light scattering, small-angle X-ray scattering (SAXS) and time-resolved small-angle neutron scattering (TR-SANS), as a function of various parameters, including core-forming block length, charge density, and salt concentration.

Understanding these processes is pivotal for understanding the long-term stability of the PEC micelles in complex environments, and has important implications for the various intended applications of the PEC micelles, for example the payload release profile of the drugs being carried by the micelles.

499462 Molecular Simulation for the Prediction of Plasticizer Efficiency and Stability in a Polymer Matrix

Abstract:

Despite the widespread use of plasticizers in the polymer industry, the mechanism of plasticization remains poorly understood. Industrial selection of plasticizers largely relies on semi-empirical rules and even for successful candidates, it is often difficult to answer why they outperform other molecules in terms of effectively reducing the glass transition temperature and elastic modulus of the material. In addition to plasticization efficiency, plasticizer migration is an issue that is causing increasing concern. Plasticizers that are not chemically bonded to the host polymer often migrate out of the matrix, either evaporating into the atmosphere or being absorbed into any liquids or other polymers in contact with the plastic. The loss of plasticizers over time not only causes the deterioration of material properties, it is also an environmental and safety concern. In particular, it is recently found that phthalates, the most consumed family of plasticizers on the market, are hazardous to human health. We use molecular simulation at both the full-atom and coarse-grained levels to understand the plasticization mechanism. From this knowledge, a number of molecular descriptors will be identified for a semi-quantitative prediction of the efficiency of any plasticizer based on its chemical structure. In addition, predictions will also be made on the thermodynamic compatibility between the plasticizer and its host polymer, as well as the diffusion coefficient of the plasticizer in the latter; these will help us estimate the tendency and speed of plasticizer migration. The end goal is to develop a computational protocol for screening plasticizers and finding high-performance and stable alternatives to phthalates.

486141 Copper-Silver Core-Shell Nanoparticles for Conductive Ink

Abstract:

As the printed electronics technology rapidly developed, there is an increasing need for lowing the cost of electronic conductive ink. Copper nanoparticles can be utilized as a low-cost replacement for silver nanoparticles which are currently used in the commercial conductive ink. However, the main obstacle for using copper nanoparticles is their spontaneous oxidation at ambient conditions. In this work, we prepared the uniform Cu-Ag core-shell nanoparticles with a particle size of 12.6nm using a novel two-step process, which consists of the thermal decomposition of Cu complex and the galvanic displacement of Cu with Ag. HR-TEM and EDS confirm the core-shell structure and particle size. XRD analyses confirm that the Cu-Ag nanoparticles have considerably higher oxidation stability than Cu nanoparticles. The prepared Cu-Ag core-shell nanoparticles is used to formulate the conductive ink, and then to form the conductive pattern on the substrate. The results showed that the resistivity of Cu-Ag line on polyimide film obtained from the nano-Cu-Ag ink was low to 15Î¼Î©â€¢cm. To verify the continuity of conductive pattern, we demonstrated the lighting of LED when conductive track was connected to 3V battery.

488859 Electric Current-Induced Nanoscale Surface Roughness Reduction in Conducting Thin Films

Abstract:

Thin film surface roughness is responsible for various materials reliability problems in nanoelectronics and nanofabrication technologies, which requires the development of surface roughness reduction strategies. For example, in present and next-generation electronic technologies, with feature scales on the order of 10 nm, there is a great need for nanoscale surface roughness reduction, since even atomic-scale roughness can lead to significant reduction of the electrical and thermal conductivity of copper thin films.

Toward this end, in this presentation, we report a modeling and simulation study that has established the electrical surface treatment of conducting thin films as a viable physical processing strategy for surface roughness reduction.Â We have developed a continuum model of surface morphological evolution that accounts for the residual stress in the deposited conductor film, surface diffusional anisotropy and film texture, the filmâ€™s wetting of the layer that is deposited on, and surface electromigration. Supported by linear stability theory, self-consistent dynamical simulations based on the model demonstrate that the action over several hours of a sufficiently strong and properly directed electric field on a conducting thin film can reduce its nanoscale surface roughness and lead to a smooth planar film surface. The modeling predictions are in good agreement with experimental measurements on copper thin films deposited on silicon nitride layers. Moreover, through systematic linear stability analyses and dynamic simulation protocols, we have examined in detail the effects of film texture and applied electric field direction and optimized the electrical surface treatment by minimizing the electric current density required for film surface smoothening. We have found that the critical electric field strength requirement for surface roughness reduction on {110}, {100}, and {111} copper film surfaces exhibits a very strong dependence on the applied electric field direction, i.e., the electric field alignment with respect to the principal residual stress directions in the film and the fast surface diffusion direction.

489274 Unusual Electronic Properties of Template-Directed π-Conjugated Porphyrin and Phosphorene Nanotubes

Abstract:

Using large-scale DFT calculations (up to 1476 atoms and 18 432 orbitals), we present the first detailed analysis on the unusual electronic properties of recently synthesized porphyrin[1] and phosphorene nanotubes.[2] We surprisingly observe extremely large oscillations in the bandgap of these nanostructures as a function of size, in contradiction to typical quantum confinement effects (i.e., the bandgap increases with size in several of these nanotubes). In particular, we find that these intriguing electronic oscillations arise from a size-dependent alternation of aromatic/nonaromatic characteristics in these porphyrin nanotubes. Our analyses of band structures and orbital diagrams indicate that the electronic transitions in these nanostructures are direct-bandgap, optically active "bright" states that can be readily observed in photoelectron spectroscopic experiments. Most importantly due to their unusual bandgap oscillations, we find that both type I and type II donor–acceptor p–n heterojunctions are possible in these template-directed, "bottom-up synthesized" porphyrin nanotubes—a unique property that is not present in conventional carbon nanotubes.

[1]S. I. Allec, N. V. Ilawe, and B. M. Wong, *"Unusual Bandgap Oscillations in Template-Directed π-Conjugated Porphyrin Nanotubes."* Journal of Physical Chemistry Letters, **7**, 2362 (2016)

[2]S. I. Allec and B. M. Wong, *"Inconsistencies in the Electronic Properties of Phosphorene Nanotubes: New Insights from Large-Scale DFT Calculations."* Journal of Physical Chemistry Letters, **7**, 4340 (2016)

490738 Complex Pattern Formation from Current-Driven Dynamics of Single-Layer Epitaxial Islands on Crystalline Conducting Substrates

Abstract:

The ability to assemble nanostructures using externally applied macroscopic forcing is an emerging paradigm in the field of nanofabrication. Macroscopic forcing provided in the form of an applied electric field can be used to drive atomic transport in a cluster of atoms through edge electromigration. A very interesting problem in this regard is of the current-driven dynamical response of single-layer adatom and vacancy clusters, *i.e.*, islands and voids of single-layer thickness/depth, on conducting substrate surfaces.

In this presentation, we discuss the development and validation of a fully nonlinear model for the current-driven morphological evolution of single-layer epitaxial islands on elastic substrates of face-centered cubic (FCC) crystals in the regime where diffusional atomic transport is limited to the island edge. The model also accounts for edge diffusional anisotropy. We focus on the driven dynamics of large isolated rounded islands on crystalline substrates under the action of an electric field oriented along the fast edge diffusion direction. First, we analyze the edge morphological stability of these islands using linear stability theory. The theory predicts that, under the action of an external field, the edge of islands larger than a critical size becomes unstable. Direct dynamical simulations based on the model of the evolution of islands with larger-than-critical sizes shows that edge morphological instabilities lead to necking and breakup of the parent islands into a population of daughter islands, which evolves to form complex island patterns. For large islands on {110} and {100} FCC substrates, we show that multiple necking instabilities generate island patterns that may include not-simply-connected void-containing islands. This pattern formation is mediated by sequences of island breakup and coalescence events with the islands of the pattern distributed symmetrically with respect to the electric field direction; the corresponding equilibrium nanopatterns, obtained by switching the electric field off, are intriguing and may include single-layer nanorings with very appealing properties for technological applications. We analyze the dependence of the formed patterns on the original island size and the duration of application of the external field and characterize the evolution of the generated number of daughter islands as well as their average size and uniformity. We show that the evolution of the average island size follows a universal scaling relation and that the evolution of the total edge length of the formed island pattern follows Kolmogorov-Johnson-Mehl-Avrami kinetics. Our study makes a strong case for the use of electric fields, as precisely controlled macroscopic forces, toward surface patterning involving complex nano-scale features.

496250 The Infrared and Raman Spectra of Pure-Silica and Aluminosilicate Sodalite

Abstract:

Porous materials may find a wide range of applications in the optical and photonics fields in the near future, including filtration and separation systems capable of providing real-time information about the efficacy of such processes through opto-electronic transduction mechanisms, entrapment of optically active molecules (e.g., lasing polymers and fluorescent dyes), and sensor coatings that possess size- and shape-selectivity. Crystalline, porous materials, such as zeolites, exhibit such selectivity through their molecular-sieving abilities. Additionally, zeolites have other highly-desirable properties for commercial opto-electronic applications, including high surface area, high pore volume, excellent thermal and chemical stability, and innate long-range order. However, current challenges in this area include the difficulty of scaling up zeolite synthesis from laboratory to industrial scale, particularly for zeolites synthesized as thin films and coatings; the small number of successful designs that take full advantage of the properties of zeolites, such as pore size and shape; and, most importantly, the lack of specific knowledge about the optical properties (refractive indices, vibrational modes, absorbance and emission bands, etc.) of zeolitic materials. We have begun a systematic effort to characterize and catalog such properties. Understanding the intricate relationships between synthesis parameters and final optical properties will provide system designers with the ability to fine-tune zeolite synthesis to suit particular applications. Moreover, unusual and heretofore unknown material properties may come to light during this process. For example, we have recently demonstrated that some zeolites are capable of producing photoluminescence if synthesized under the correct conditions [1]. As part of an ongoing, systematic effort to understand and predict the fundamental interactions between infrared light and zeolitic structures, the goal of this study is to examine the complex interrelationships among composition, structure, crystallization parameters, and the vibrational spectra of pure-silica and aluminosilicate sodalite (zeolite framework code **SOD**).

We report the synthesis and optical characterization of these materials with varying crystallization times and temperatures, and varying aging times. A software-aided design of experiments was applied to assist in the statistical analysis of the samples, and to determine the most important synthetic factors that impact their infrared and Raman spectra. These experimentally-determined spectra will be assisted by Density Functional Theory (DFT) calculations to interpret the vibrational spectra of a zeolite framework with any composition. Together, computational models and experimental data will assist the design of optoelectronic systems utilizing zeolites and their special optical properties.

References:
1. Planells, A., et al., *Investigation of the photoluminescence of microporous silicalite-1 (MFI) films.* Microporous and Mesoporous Materials, 2016. **220**: p. 73-80.

500389 The Effect of Solvent Selection on the Optical Trapping, Manipulation, and Patterning of Nanomaterials

Abstract:

The optical trap's ability to manipulate nanoparticles in three dimensions using the radiation pressure of light and to measure piconewton forces, has enabled the investigation of a wide range of effects, including the laser-cooling of solids, the Casimir effect, and protein folding. In addition, the contactless control of a particle's location makes the optical trap a premier technique to select individual particles from a dispersion and to pattern nanoparticles. However, there are no reports of optical trapping in non-aqueous liquid media, which limits the available library of nanoparticles for patterning.

Here, we demonstrate the optical trapping and patterning of nanowires and microspheres in a range of non-aqueous solvents for the first time. To characterize the trapping conditions, we employ the temperature-dependent photoluminescence of ratiometric Mn^{2+}:ZnCdSe nanocrystals to measure the local temperature of trapped nanowires, observing temperatures >100°C above ambient conditions. We find that the choice of solvent influences the trap's local properties, which drastically affects the optical trapping and patterning efficiency. Finally, we discuss the initial optimization of these parameters and model the heat transport, including convection, around the trapped particle.

500663 Synthesis and Characterization of Electrochemically-Grown Zinc Oxide Nanowires for Use in Rectenna-Based Heat Harvesters

Abstract:

Synthesis and Characterization of Electrochemically-grown Zinc Oxide Nanowires for Use in Rectenna-based Heat Harvesters

Haley, Adrian Lucas; Yang, Shendu; Pinhero, P.J.

Department of Chemical Engineering, University of Missouri, Columbia, MO 65211

Spintronic devices offer a possibility of enhanced functionality, higher speed, and reduced power consumption compared with conventional electronic materials. Zinc oxide has been proven to show noticeable spin properties after spin polarization which can be useful in fields such as information storage, photovoltaics, and thermoelectrics.

In this study, ZnO nanostructures are electrodeposited on a silver coated n-type (100) silicon wafer in a metallic salt solution. Scanning electron microscopy (SEM) and transmission electron microscopy (TEM) are used to study the effect of experimental parameters of concentration, growing voltage, and growing time on the morphology of ZnO nanowires. Energy-dispersive X-ray spectroscopy (EDS) is used to characterize the purity of ZnO nanostructures. This is followed by the fabrication of a metal-insulator-metal (MIM) layered sandwich, which can be potentially used as an electron-tunneling diode device. The current-voltage properties of MIM devices are characterized using conductive atomic force microscopy and four-point probe measurements.

499997 The Thiol-Thioester Exchange in Network and Linear Polymers

Abstract:

Dynamic bonds are covalent linkages which, if given specific stimuli, can form an equilibrium between exchanging reactants and their products. Recent work by our lab has developed dynamic functional groups which can be seemlessly incorporated into network or linear polymers and harnesses the supurb reactivity and chemoselectivity of thiyl radicals and thiolate anions. These dynamic functionalities act as crosslinks in network polymers or in the backbone of linear polymers and are easily adapted to various stimuli including light, heat, or chemical. Here we show recent developments towards the formation of dynamic liquid crystalline networks and thioester-based networks both responsive towards light in order to develop shape shifting and remoldable photopolymers. Additionally, we demonstrate a sequence shifting, dynamic thioester-linked DNA-like polymer by implementing an covalently exchangeable thioester backbone. We believe that the broad implementation of this chemistry demonstrates the breadth of this approach towards the formation of smart, responsive polymers.

490568 Nacre-Inspired Composite Gels for Biomedical Applications

Abstract:

Nacre (or mother of pearl) is found in the inner shell layer of certain molluscs & has a complex hierarchical structure made of hard ($CaCO_3$) and soft (proteins) components. Layers of hexagonal plates of aragonite, which is a crystalline form of $CaCO_3$ are held together by a matrix of proteins, in a manner that resembles a "brick and mortar" structure. Incorporation of the soft protein in between the hard aragonite plates imparts superior mechanical strength to nacre, which is three to nine time higher than that of pure aragonite. This ingenious combination of materials result in remarkable properties, such as, resistance against fracture, high pressure absorption, prevention of crack propagation, and being incredibly light weight.

Inspired by these features, we have developed robust, flexible materials with self-adaptive and self-healing properties. Polymer hydrogels were incorporated with functional nanoparticles to produce composites with unique physicochemical and mechanical properties. 3D printing was utilized as an effective technique to combine the nanoparticles with polymer gels, with greater dimension control. Moulds with hexagonal wells were printed and filled with a "hard" component comprising of the nanoparticles in the polymer gel. The compressibility of these composites was customized using a crosslinking agent. These hexagonal templated materials were then arranged into layered structures, which were held together by a "soft", flexible polymer matrix. The physical and mechanical properties of these composites could be greatly varied by changing the polymer concentration of the soft matrix. Due to 3D printing, a wide range of dimensions could be achieved for making such composites. Techniques were also developed that allowed 3D printing of the nanoparticles directly onto polymer gels, according to desired patterns. These novel composites were extensively characterized (TEM, SEM/EDX, FTIR, tensile strength, compressibility, and other techniques to be discussed in the presentation) and displayed enhanced capabilities in mechanical strength, self-adaptation to environmental stresses and in self-healing ability. These features, together with biocompatibility, allow for a myriad of biomedical applications.

500070 High Viscosity Polymer Gels Derived from Block Copolymer Nanocomposites

Abstract:

Here we report the synthesis and characterization of high viscosity nanocomposite polymer gels (NC gels) formed via highly exfoliated block copolymer nanocomposites dissolved in an organic solvent along with an ionic salt. Poly(ethylidene norbornene-b-cyclopentene) nanocomposites synthesized via surface initiated Ring Opening Metathesis Polymerization (ROMP) using functionalized montmorillonite (MMT) clay were used to form NC gels. These nanocomposites, when dissolved in tetrahydrofuran solvent along with an ionic salt (for example NaCl) resulted in very high viscosity NC gels. NC gels formed with nanocomposite loading as low as 0.75 wt% showed measured viscosities in the range of 100 Pa.s at room temperature. Morphological characterization of three-dimensional network of NC gels was conducted via TEM. Mechanical properties of NC gels formed with different nanocomposite and ionic salt loadings were characterized via rheological measurements.

502904 Quantifying Topology, Gelation and Elasticity of Polymer Networks

Abstract:

Despite the ubiquity of applications, much of our fundamental knowledge about polymer networks is based on an assumption of ideal end-linked structure. However, polymer networks invariably possess topological defects: loops of different orders which have profound effects on network properties. Here, we develop a kinetic graph theory which demonstrates that all different orders of cyclic topologies are a universal function of a single dimensionless parameter characterizing the conditions for network formation. The theory is in excellent agreement with both experimental measurements of hydrogel loop fractions and Monte Carlo simulations without any fitting parameters. The one-to-one correspondence between the network topology and primary loop fraction demonstrates that the entire network topology is characterized by measurement of just primary loops; different cyclic defects cannot vary independently, in contrast to the intuition that the densities of all topological species are freely adjustable. Using kinetic Monte Carlo simulation, we are able to predict the gel point for real polymer networks which is in good agreement with experimental results with no fitting parameters. The critical exponents increases with the loop defects, in contrast to the prediction from both the classical Flory-Stockmayer theory and percolation theory. Furthermore, to bridge the network topology and gel elasticity, we develop a real elastic network theory (RENT) that systematically accounts for the impacts of cyclic defects. We demonstrate that small loops have vital effect on the modulus; whereas this negative impact decreases rapidly as the loop order increases, especially for networks with higher junction functionalities. RENT provides predictions that are highly consistent with experimental observations of polymer network elasticity, providing a quantitative theory of elasticity that is based on molecular details of polymer networks.

497616 Self-Assembly and Mechanical Properties of Di-Fmoc-L-Lysine Containing Molecular Gels

Abstract:

Molecular gels formed by self-assembly of low molecular weight gelators are of significant interest because of their potential applications in drug delivery, tissue engineering, materials for sensors, etc. Here, we present gelation of a low molecular weight gelator, di-Fmoc-L-lysine, in various water-organic solvent mixtures through solvent-trigger approach. Di-Fmoc-L-lysine has two aromatic moieties and multiple hydrogen bond donors and acceptors making it an interesting material to study. FTIR technique was used to capture change in chemical fingerprints during the gelation process. AFM and SEM results indicate fibrous structure in these gel with the evidence of branching. Association of gelator molecules during the early stage of gelation leading to such structures has also been predicted by molecular dynamics simulation. Shear and cavitation rheology were used to investigate the mechanical properties of these gels. The gels display strain-softening behavior at moderate strain values. Cavitation rheology technique captures the critical pressure for fracture. Critical energy release rate (G_c) was estimated from the critical pressure and the magnitude of G_c was found to be an order of magnitude lower than that reported for the polymer gels indicating different failure mechanism in these gels.

500056 Thermoresponsive Sol-Gel Transitions of PÎ•G-Based Nanocomposite Hydrogels Controlled By Molecular Weights of Block Copolymers and Solute Concentrations

Abstract:

PEG-based amphiphilic copolymers such as PLGA-PEG-PLGA (PLGA: poly($_{D,L}$-lactide-co-glycolide) and PEG-PPO-PEG (PPO: poly(propylene oxide)) are promising candidates for biomedical applications due to the thermoresponsive sol-gel transition properties and the biocompatibility originating from the PEG block in the PEG-based amphiphilic copolymers. For the maximum utilization of the highly biocompatible PEG block, the hydrophilic PEG block should be relatively longer than the hydrophobic block. In other words, the ratio of PEG to the hydrophobic block (the PEG/hydrophobic-block ratio) should be higher. However, it is known that PEG-based copolymers with a high PEG/hydrophobic-block ratio lacked the thermoresponsive sol-gel transition properties. In fact, it was found that the aqueous solution of PLGA-PEG-PLGA with a high PEG/PLGA ratio of above 0.5 would not exhibit thermoresponsive sol-gel transitions. In this work, PLGA-PEG-PLGA with higher PEG/PLGA ratios were synthesized and laponite clay nanoparticles were added to the PLGA-PEG-PLGA solutions to realize the thermoresponsive sol-gel transition in the temperature range between 25 and 37°C [1]. We comprehensively studied the sol-gel transition behavior by varying the molecular weights of PLGA-PEG-PLGA, the PEG/PLGA block ratios of PLGA-PEG-PLGA, and the solute concentrations [2]. The sol-gel transition behavior was studied by the test-tube inversion method and the dynamic mechanical analysis (DMA). It was found that PLGA-PEG-PLGA with a high PEG/PLGA ratio of 3.0 could exhibit the sol-gel transition by adding 1.0 wt% of laponite. It was also found that the sol-gel transition temperature and the rheological behavior were effectively regulated by varying the molecular weight of PLGA-PEG-PLGA from 2.0 k to 6.0 k, and the laponite concentration from 0.75 wt% to 1.5 wt%. Moreover, in order to discuss the thermoresponsive sol-gel transition, PEG-PLGA diblock copolymers were also synthesized and the sol-gel transition behavior was studied.

[1] Naho Oyama, Hiromasa Minami, Daichi Kawano, Makoto Miyazaki, Tomoki Maeda, Kazunori Toma, Atsushi Hotta, Koji Nagahama, A nanocomposite approach to develop biodegradable thermogels exhibiting excellent cell-compatibility for injectable cell delivery, Biomaterials Science, Vol.2 (8), 1057-1062 (2014).

[2] Makoto Miyazaki, Tomoki Maeda, Kenji Hirashima, Naruki Kurokawa, Koji Nagahama, Atsushi Hotta, PEG-based nanocomposite hydrogel: Thermoresponsive sol-gel transition controlled by PLGA-PEG-PLGA molecular weight and solute concentration, Polymer, Vol.115, 246-254 (2017).

493738 Evaluating Reprocessability of Polymer Networks: Flory-Stockmayer Analysis

Abstract:

Reprocessable polymer networks can undergo structure rearrangement through dynamic chemistries under proper conditions, making them promising candidates for recyclable crosslinked materials, e.g. tires. See, for example, â€œRecyclable Crosslinked Polymer Networks via One-Step Controlled Radical Polymerizationâ€, Adv. Mater. 2016, 28, 6746-6750. This research field has been focusing on various chemistries. However, there has been lacking of a fundamental theory explaining the relationship between the abundancy of dynamic linkages and polymer network reprocessability. Based on the classical Flory-Stockmayer analysis on network gelation, we developed a theory for reprocessable polymer networks to quantitatively predict the critical condition of being reprocessable. Our analysis indicates that it is unnecessary for all linkages between crosslinks to be dynamic to make the resulting network reprocessable. As long as there is no percolated permanent network in the system, the material can fully rearrange and thus be reprocessable. To experimentally validate our theory, we also used a thiol-epoxy network model system with various dynamic linkage compositions. The stress relaxation behavior and reprocessing of resulting materials support our theoretical prediction: only 50 % of linkages between crosslinks need to be dynamic for a tri-arm network to be reprocessable. Therefore, this analysis provides the first theoretical criterion for designing and evaluating reprocessability of polymer networks that contain both dynamic and permanent linkages between crosslinks.

489392 Molecularly-Templated Reaction for Forming Poly(dimethyl siloxane)/Graphene Oxide Composite Elastomers

Abstract:

Graphene oxide (GO) is a carbonaceous material consisting of various oxidative functional groups, such as carboxylic acids, hydroxyls, and epoxides. Owing to the inherent oxidative functionality and high surface area of GO, our group has recently investigated [1] many methods to crosslink polymers in the presence of a small amount of GO, using it as a multifunctional crosslinker. Among many possible chemical routes proposed in the past few years, research implementing polymers with secondary amide groups to form highly elastic composites is practically important for many applications. A short list of important materials containing secondary amide groups include DNA, amino acids, nylons, *etc*. However, a major challenge emerges from the fact that secondary amide groups are less reactive than other functional groups, for example, primary amines, due to their stable resonance structure. Therefore, introducing a catalytic functional group that can enhance the chemical reaction may introduce a new platform to synthesize highly crosslinked, robust polymer composites. Such catalytic functional groups could simply assist in co-locating reactive functional groups to assist chemical reaction. Herein, this catalytic function is served by pyrene groups that form Ï€-Ï€ interactions with the basal plane of GO.

This study [2] explores the molecularly-templated reaction of pyrene-terminated telechelic poly(dimethyl siloxane) (PDMS) with GO. The polymer was terminated with pyrene ends connected to the main polymer chain through a secondary amide linkage. These pyrene ends form dynamic Ï€-Ï€ interactions with GO surfaces, which reduce intermolecular spacing between amide groups near epoxides on the surface of GO, thereby templating the reaction. The key focus of this research was to generate a new method to crosslink polymers containing secondary amide groups with GO *via* an epoxy ring-opening reaction, and investigate the critical role of the pyrene end group participating as a catalyst during the crosslinking reaction. This chemical crosslinking reaction was confirmed and supported by infrared spectroscopy, oscillatory shear rheology, gel content, swelling ratio, and mechanical property measurements. We have also tuned the concentration of the secondary amide groups by modulating the molecular weight of the polymer main chain, and were able to prepare elastomers that are highly crosslinked (*e.g.*, up to 96 wt % non-dissolving gel) but highly extensible (*e.g.*, extensional strains of more than 200%) as well. For comparison, methoxy-terminated telechelic PDMS with the same secondary amide linkages and molecular weight was synthesized. As expected, this control material did not exhibit appreciable crosslinking with GO, illustrating the importance of pyrene end groups. Considering the crosslinked nature of the composites and generality of the proposed reaction, we believe this molecular-templating capability could introduce a wide range of cost-effective engineered composite elastomers beyond PDMS.

[1] Gas Permeation and Selectivity of Poly(dimethylsiloxane)/Graphene Oxide Composite Elastomer Membranes, H. Ha, J. Park, S. Ando, C.B. Kim, K. Nagai, B.D. Freeman and C.J. Ellison, *Journal of Membrane Science*, 518, 131-140 (Jun. 2016). http://dx.doi.org/10.1016/j.memsci.2016.06.028

[2] Molecularly-Templated Reaction for Forming Poly(dimethyl siloxane)-Graphene Oxide Composite Elastomers, H. Ha, K. Ha, and C.J. Ellison, *under review*.

489869 Graduate Student Award Session: A Two-Step Method for Transferring Single Wall Carbon Nanotubes Onto a Hydrogel Substrate

Abstract:

In recent years, significant progress has been made in elucidating the unique properties of carbon nanotube (CNTs)-hydrogel nanocomposites for biomedical applications. Due to the extremely hydrophobic nature of CNTs and the hydrophilic nature of the hydrogels, there are impediments to developing nanocomposites, including attaining a homogeneous dispersion of CNTs in the polymer solutions or alignment and patterning of CNTs in the hydrogels. Several approaches have been reported on patterning CNTs on various polymeric substrates. However, these methods include high temperature, high vacuum or the use of a sacrificial layer and hence, are incompatible with hydrogel substrates as they can lead to irreversible collapse in hydrogel structure. In this study, we designed a novel method to transfer CNTs onto the hydrogels with a single-step stamping technique. First, dense and aligned single wall carbon nanotubes (SWCNTs) were grown on single-crystal quartz substrate. Subsequently, polyethylene glycol (PEG), poly acrylamide (PA), and polyvinyl alcohol (PVA) hydrogel solutions were deposited on the CNTs grown on the quartz substrate. Upon hydrogel gelation, the CNTs were transferred from the quartz onto the hydrogel. The successful transfer of CNTs was confirmed by scanning electron microscopy (SEM). The transferring efficiency was found to be ~99%. Furthermore, results showed efficient cell attachment on the CNT-hydrogel composites that indirectly confirmed the successful transfer. CNT-hydrogel nanocomposites, such as the ones presented here, can be applied for various biomedical applications such as tissue engineering, drug delivery, and biomedical devices.

489751 Graduate Student Award Session: Understanding How Lipid Nanoparticle Structure Affects Immune Response

Abstract:

Lipid-based nanoparticles have been widely demonstrated as efficacious drug delivery systems, particularly for applications in nucleic acid delivery. Despite their extensive development, however, there is limited understanding of the effect that these delivery vehicles have on the immune system. In fact, a major reason why therapeutics fail in clinical trials is due to unforeseen toxicity and/or immunogenicity. In particular, the immune system can potentially mount an antibody response to the delivery system, trigger the release of liver enzymes, or prompt an inflammatory cytokine response. Here, we describe a study of the interactions between lipidoid nanoparticles, a potent class of lipid-like RNA delivery vehicles, and the components of the innate and humoral immune system. A small combinatorial lipidoid library of 20 lipidoids was synthesized and tested for immunostimulatory effects in cell culture and animal models. In vitro analysis of RAW264.7 mouse macrophages after nanoparticle exposure indicates statistically significant rises in the gene expression of several cytokines, including tumor necrosis factor Î± (TNFÎ±), for select lipid nanoparticles. Increased cytokine expression persisted for over 24 hours and depended largely on the tail structure of the lipidoid. Furthermore, the effect of lipid nanoparticles on peripheral blood mononuclear cells (PBMCs) isolated from healthy human donors was assessed via ELISA screens of TNFÎ±, IL-1Î², IFN-Î³, IL-6, and IL-2. Antibody and cytokine responses were also measured following single and/or multiple intravenous injection(s) of lipid nanoparticles into C57BL/6 mice. In all cases, both the tail structure and amine structure of the lipid nanoparticle were found to be important in provoking an immune response. These structure-immune response relationships will inform the design of future drug delivery carriers and potentially mitigate failures encountered in clinical stages.

501042 Graduate Student Award Session: Photo- Induced Pinocytosis in Synthetic Liposomes

Abstract:

An essential step in the development of artificial cells is making these cells capable of taking up resources for sustained transcription and translation. Previous work on incorporation of appropriate machinery into the membrane to make this possible has proven to be complex and complicated. Here, we present the introduction of photocleavable units into the phospholipid structure through which photo-induced pinocytosis behavior is achieved. Upon exposure to 365nm or 405nm light, cleavage products build up, leading non-spherical vesicles to engulf external fluid during their transition to a more spherical shape. This process would allow future artificial cellular systems to uptake resources that range from dNTPs and energy to whole proteins from the external environment.

503555 Graduate Student Award Session: Sugar-Coating the Answers to Virus Binding: Glycocalyx-Mimetic Interfaces

Abstract:

The glycocalyx is a membranous structure composed of glycoproteins and carbohydrate residues. This sugary forest that sheaths our cells is responsible for protecting us from pathogen invasions and forms a crucial component of our cellsâ€™ immune machinery. Virus-glycocalyx interactions are poorly understood and a detailed study of these interactions is hindered by the tendency of the native glycocalyx to collapse duringÂ in vitroÂ studies. We have developed a bio-inspired model system for the glycocalyx to identify and understand physicochemical parameters that shape its interactions with viral pathogens. Like the native glycocalyx, our tunable surface engages in both specific and non-specific modes of viral binding. We have developed a library of carbohydrate-based polymer brushes, including mannose, glucose and galactose, which recapitulate the chemical complexity of the glycocalyx. Though our carbohydrate-functionalized surfaces are inert to non-specific protein adsorption, they recognized lectin. Spatially controlled immobilization of brushes helped us achieve precisely patterned presentation of lectins and viruses. In addition to specific affinity based interactions, we have also engineered non-specific electrostatically mediated interactions into our surface to elucidate the role of surface charge on viral binding. Our model surface is composed of positively charged affinity sites and polymerization initiation sites for the growth of polymer brushes bearing carbohydrate residues. By varying the ratio of affinity sites to initiation sites, we were able to observe changes in adsorption profiles of model proteins, influenza viruses and adenoviruses. Quartz crystal microbalance, scanning electron microscopy and fluorescence microscopy were employed to visualize these interactions and develop design rules for virus-resistant coatings.

492095 Graduate Student Award Session: The Combined Effect of Matrix Microenvironment and Hypoxia on the Activity of Glioblastoma Stem Cells

Abstract:

Introduction Glioblastoma (GBM) is the most common and lethal form of brain cancer. GBM is known to diffusely infiltrate throughout the brain tissue, which contributes to poor patient prognosis. The complex and heterogeneous cell population, structural and nutrient gradients along the tumor margin, complicates the study of GBM. Despite the standard-of-care treatment, which includes surgical debulking tumor mass as much as safely for the patient normal brain function followed by chemo- and/or radiotherapy, the short 14.6 months median survival rate, low long term (5+ year < 5%) survival rate and the fast and high recurrence rate (~6.9 months after standard treatment) suggesting the little-to-no effect of current therapy. Currently, there are limited tools approaches to rigorously assess the relationship between tissue microenvironment and the growth, invasion, and therapeutic outcomes. Several subsets of cells within the heterogenous GBM tumor niche has been considered to contribute to the poor prognosis. GBM stem cells (GSCs) also known as GBM-initiating cells are thought to be therapeutic resistant and hence are important for curing and avoiding recurrence of GBM. Tumor-associated-macrophages/microglias are immune cells recruited into tumor sites and the polarization (proinflammatory M1 and anti- inflammatory M2) is recently thought to be important to tumor growth and related to GBM invasive phenotype. This study focuses on developing an engineered hydrogel platform to mimic the brain tumor environment to investigate how matrix composition, metabolic environment and drug treatments alter the cellular heterogeneity of patient-derived GBM specimens.

Materials and Methods Hydrogels are generated from methacrylamide-functionalized gelatin (Gel-MA) and methacrylated hyaluronic acid (HAMA). Patient-derived xenograft (PDX) GBM cells (GBM6, GBM10, GBM12, GBM15, GBM26, GBM39) were obtained from our collaborator (Dr. Jann Sarkaria, Mayo Clinic). These specimens maintain patient-specific morphological and molecular characteristics and span a range of invasive and therapy resitant phenotypes. Hydrogels were maintained for 4 wt% and made for with or without 15 w/w% matrix-immobilized HA and photo-initiator (PI) was adjusted (-HAMA: 0.1%, +HAMA:0.02%) for maintaining elastic modulus. Pre-polymerized solutions were then mixed with cell suspension homogenously and polymerized using UV lamp (7.1 mW/mm^2, 30 sec). Hydrogels were then put into culture for either normoxia (20% O_2) or hypoxia (1% O_2) and culture media were made with different treatments: +Dimethyl sulfoxide (0.1 vol%, control), +Erlotinib (Erl, receptor tyrosine kinase inhibitor, 10 Î¼M) or Temozolomide (TMZ, standard oral chemotherapy drug, 30 Î¼M). Therapeutic outcomes were determined by overall metabolic reactivity MTT assay (ThermoFisher). The heterogeneity of the GBM cell populations were determined using flow cytometer before each experiment and after degradation of hydrogels at timepoints throughout culture. Hydrogels were obtained after culture and put into collagenase type IV (100 unit/ 500uL) and hyaluronidase (100 unit/ 500uL) containing PBS for 30 minutes for fully degraded hydrogel. Four cell populations were stained after degradation: GSCs (CD133), TAMs (CD68), astrocytes (GFAP) and endothelial cells (ECs, CD31).

Results and Discussion Hyrogel matrix microenvironment significantly influences the kinetics of glioblastoma stem cells within the PDX specimens. Interestingly, the subfraction of GSCs within the invasive GBM6 PDX popualtion showed a significant increase under hypoxia (1% O_2) culture for up to 7 days compare to groups cultured in normoxia (20% O_2) condition. Specimens cultured in the presence of matrix-bound HA showed a significant higher fraction of GSC$^+$ cells, suggesting that GBM GSC% is influenced by both matrix composition as well as the metabolic environment. To study the complex heterogenous cell population between different PDX cells, a screening experiment of GSC%(CD133$^+$) and TAM%(CD68$^+$) was performed on 4 PDX lines: GBM6, GBM10, GBM12 and GBM39, each having different invasive potential (lo:0; 7:hi), Erlotinib response (0:lo, 100:hi) and molecular subtype (M: mesenchymal, C:classical). These speimens showed a wide range of initial and final cell compositions, suggesting profiling the heterogeneity of PDX specimens in biomimietic hydrogels may yield new, therapeutically actionable data. Ongoing work is looking at expansion of cell markers into four interesting cell groups (GSCs, TAMs, astrocytes and ECs) as well as in the presence of two therapeutic molecules (erlotinib, Temozolomide) to cell how the heterogeneous cell population changes as a response to microenvironmental cues and therapies.

Conclusion This study provides a rigorous way to study cell population, matrix influence and therapeutic outcome in a defined and controlled biomaterial in vitro platform. Data showed a huge heterogeneity existing in GBM and future work will examine more.

495430 Graduate Student Award Session: Pegylated Poly(beta-amino ester) Delivery Systems for Periodic shRNA

Abstract:

RNA interference (RNAi) provides a versatile therapeutic strategy via silencing of specific genes implicated in cancer and other diseases. However, clinical translation of RNAi for cancer therapeutics remains unrealized, due to challenges in delivery of small interfering RNA (siRNA) to tumors. The low valency and high rigidity of siRNA often requires high excesses of cationic delivery materials to condense into stable nanoparticles, leading to dose-limiting toxicities. To address this challenge, we adopt an RNAi platform based on periodic short hairpin RNAs (p-shRNAs). Consisting of siRNA sequences linked together, these polymeric RNAi molecules are generated by the repeated action of an RNA polymerase around a small circular DNA template. Through template design and selective enzymatic digestion, we can design p-shRNA structures that are efficiently processed inside cells into siRNAs and induce significant gene silencing.

With its much higher valency and flexibility compared to siRNA, p-shRNA requires far less polycationic material to be condensed, and forms more stable complexes. Poly(beta-amino ester)s (PBAEs) in particular have shown promise as gene delivery vehicles, due to their biodegradability and design versatility. To develop an optimal PBAE structure for p-shRNA delivery, we used factorial design to synthesize a library of PBAE variants of a base structure, poly-1. Screening of this library showed that p-shRNA silencing efficiency increases with increasing alkyl side chain percentage and decreasing molecular weight. Our designed poly-1 structures are able to fully condense p-shRNA into sub-100 nm complexes with high silencing efficiency, at much lower polymer-to-RNA ratios than those typically required for PBAE gene delivery. To enhance the colloidal stability of the complexes in physiological conditions, we added a poly(ethylene glycol) (PEG)-poly-1 copolymer containing alkyl side chains to the core complex. The resulting complexes, assembled via electrostatic and hydrophobic interactions, possess an outer PEG layer that can provide stability in the bloodstream and thereby increase tumor accumulation. We further introduced active tumor targeting by conjugating folate to the PEG terminus. Expanding our library to screen combinations of the core poly-1 and PEG-poly-1 structures revealed an optimal blend of hydrophobicities and molecular weights in the two polymer components. Thus, through nucleic acid engineering and rational carrier design, we have successfully developed a stable, potent RNAi delivery system that can trigger significant gene silencing at low doses, and potentially enable higher therapeutic efficacy *in vivo*.

497646 Graduate Student Award Session: Biodegradable Nano-Film Coated Self-Floating Hollow Glass Microspheres for Rapid Cell Isolation and Recovery

Abstract:

The ability to accurately and efficiently isolate specific cell populations from heterogeneous mixtures has enabled researchers to advance the fields of experimental cell biology and translational medicine. Established cell isolation and purification techniques such as fluorescence-activated cell sorting, microfluidic devices and immunomagnetic micro/nano-particles are either expensive and non-portable, involving sophisticated microfabrication and excessive lab equipment or having negative impact on cell viability and function, which limit their practical usage in resource-limited settings. We present a simple yet effective method for cell isolation and recovery that requires neither specialized lab equipment nor any form of power source. Specifically, self-floating hollow glass microspheres were coated with an enzymatically degradable Nano-film, and conjugated with antibodies to allow both fast capture and release of subpopulations of cells from a cell mixture. Targeted cells were captured by the microspheres and allowed to float to the top of the hosting liquid thereby isolating targeted cells. An antifouling polymer brush layer was grafted onto the Nano-film to minimize non-specific cell adhesion. Using the EpCAM expressing cancer cell line, PC-3, in blood as a model system, we have demonstrated the isolation and recovery of cancer cells without compromising cell viability or proliferative potential. The whole process takes less than one hour. To support the rational extension of this platform technology we introduce extensive characterization of the film fabrication and application. Our approach is expected to overcome practical hurdles and provide viable targeted cells for downstream analyses in resource-limited settings.

489479 Graduate Student Award Session: Enhancing Therapeutic Efficacy of Self-Assembling Prodrugs with Supramolecular Chemistry

Abstract:

Chemical modification of small molecule hydrophobic drugs is a clinically proven strategy to devise prodrugs with enhanced treatment efficacy. While this prodrug strategy improves the parent drug's water solubility and pharmacokinetic profile, it typically compromises the drug's potency against cancer cells due to the retarded drug release rate and reduced cellular uptake efficiency. We report here the supramolecular design of self-assembling prodrugs with much improved water solubility while maintaining high potency against cancer cells. The direct use of anticancer drugs to create their own nanostructures allows for fine-tuning and optimization of the drug loading and their physicochemical properties in a precise manner through molecular engineering of the underlying prodrug building units. We found that our rationally designed self-assembing prodrugs are capable of associating into one-dimensional nanostructures of various physiochemical properties in aqueous solution with a width of approximately 10 nm and lengths on the scale of several micrometers. We further evaluated the stability of resulting filaments of different surface chemistries in both cell medium and human serum, and their gelation ability in physiological condition. More importantly, our results suggest that this platform of prodrugs exhibit much greater efficacy against various cancer cells relative to the clinically used prodrug both *in vitro* and *in vivo*. We believe these findings open a new avenue for rational design of supramolecular prodrugs for more effective cancer treatment.

501633 Size-Selective Ionically Crosslinked Polymer Multilayer Films for Light Gas Separation

Abstract:

The production of inexpensive, high purity hydrogen remains a critical challenge to improving the sustainability of fossil fuels and for realizing renewable and clean energy sources capable of displacing fossil fuels. Polymeric gas separation membranes combining low production costs with robust mechanical properties have found broad industrial application for O_2 and N_2 enrichment of air, upgrading of natural gas and hydrogen recovery from ammonia. In the absence of structure (i.e. in amorphous glassy or rubbery films), increasing the free volume of size-selective polymer membranes improves the permeability at the cost of reducing the permselectivity. This translates to an upper bound in separation performance achievable by dense, homogeneous and amorphous polymeric films. We recently demonstrated that layer-by-layer (LbL) deposition provides a robust, low-cost means to introduce size-selective functionality via manipulation of polymer processing conditions. Extension of the LbL assembly technique to gas purification membranes promises the ability to tune overall gas permeability and permselectivity through manipulation of film nanostructure by varying deposition conditions. The alternate deposition of branched polyethylenimine (PEI) and poly (acrylic acid) (PAA) produces a "scrambled salt" structure, resulting in a highly interpenetrating network of high density which exhibits size selective gas separation in excess of Robeson's 'upper boundary,' which we attribute to the structure imparted by a high degree of 'ionic-crosslinking.' Specifically, this ionically crosslinked assembly exhibits H2/N2 and H2/CO2 selectivities in excess of 1000:1 and 100:1, respectively, which are superior to the properties of any organic, inorganic or mixed-matrix membrane outside of dense metallic (e.g. Pd) or electroceramic (e.g., BaYxCe1-xO3) films reported in the open literature. This unique thin film overcomes the drawbacks of common polymeric membranes (i.e., low selectivity and poor mechanical properties), making it a significant advance in polymeric membranes for gas separation.

501546 Length Effects on Polyelectrolyte Complexation: How 'Poly' Must a Polyelectrolyte be?

Abstract:

The term 'polyelectrolyte complexation' describes the associative phase separation that occurs when oppositely charged polymers are mixed in aqueous solution, resulting in formation of liquid coacervates or solid precipitates. Measurements indicate that the free energy gain of polyelectrolyte complexation is primarily due to increase in entropy upon mixing, and this is usually described in terms of counterion release: by directly neutralizing each other, the polyelectrolytes release their screening counterions into solution. Consistent with this picture, complexation is found to be inhibited by high concentrations of salt. Similar physics drives the condensation of genomic DNA into compact structures in vitro and in vivo: low valence cations are displaced by binding of multivalent ions and cationic proteins. Theoretical treatments of both problems also typically assume the polyelectrolytes are quite long. Recent measurements aimed at understanding polyelectrolyte complexes as delivery vehicles for therapeutic oligonucleotides and as protocells in early life scenarios challenge this understanding, however: coacervation is observed for mixtures of short oligomers and even singly-charged nucleotide monophosphates. These molecules are not polyelectrolytes by conventional description, raising the question of what drives complexation when the translational entropy of all the constituent molecules is similar. In order to address this question, we measured the complexation properties of oligonucleotides and mononucleotides by short peptides and polyamines. I will present results of these studies, including phase diagrams and thermodynamic measurements that compare the effect of counterion release to other molecular interactions such as base stacking, in order to define the length scale at which our conventional understanding of polyelectrolyte physics begins to operate.

501207 Interaction and Dynamics of Polyelectrolytes in Polyzwitterionic Complexes

Abstract:

Oppositely charged polyelectrolytes can associate in aqueous solution form distinct condensed phase such as liquid-like coacervates and solid/gel-like complexes. While a great deal is known about their phase behaviors, the interaction and structural dynamics of polyelectrolyte complexes remain unclear. Use zwitterionic polymer, poly (sulfobetaine methacrylate) (PSBMA), we compare its interaction and complex formation with polycation, poly(2-vinylpyridine) (P2VP) and polyanion, poly(styrene sulfonate) (PSS) in aqueous solution. We observe the formation of biphasic PSBMA-P2VP coacervates at low salt, in sharp contrast to the formation of PSBMA-PSS gel at the same solution condition, given that PSBMA is net negatively charged in the salted solution. To examine the interplay between polyelectrolyte interaction and structural dynamic of polyelectrolyte complexes, we investigate the self-diffusion of single P2VP and PSS in PSBMA-based complexes by fluorescence correlation spectroscopy and fluorescence recovery after photobleaching. Fickian diffusion behavior is observed with PSS in PSBMA-PSS gel. However, fast and slow diffusion modes are observed in the dense PSBMA-P2VP coacervate after the removal of supernatant, suggesting inhomogeous structure of dense coacervates. Fickian diffusion of P2VP in PSBMA-P2VP coacervates could be recovered at high salt. A simple model is proposed to describe the impact of competing electrostatic and entropic forces on the dynamic heterogeneity in macroscopically homogenous polyzwitterionic complexes.

492585 Tuning Complex Coacervation Using Sequence-Defined Polyelectrolytes: A Molecular Understanding

Abstract:

Oppositely-charged polyelectrolytes can undergo an associative phase separation to form a polymer-dense 'complex coacervate' phase. This electrostatically-driven phase separation is used in a broad array of applications, ranging from food science to self-assembled materials. A fundamental understanding of these materials is necessary for materials design, however a comprehensive physical picture of complex coacervation is still emerging. Our work has demonstrated that the phase behavior of complex coacervate-forming systems is very sensitive to molecular detail, such as short-range interactions and charge spacing. We use computer simulation to demonstrate that this connection between monomer-level structure and coacervation extends to charge sequence. This is reinforced by experimental data, which links the 'blockiness' of a coacervate sequence to phase behavior in qualitative and near-quantitative agreement with simulations. We demonstrate that the physical origin of this sequence-dependent behavior is due to non-uniform counterion condensation effects that can be captured in simple theoretical models. This provides a route to realizing materials that utilize sequence-defined polymers to dictate bulk material properties.

494752 Photodirected AssemblyÂ and Self-Rupture of Polyelectrolyte-Based Soft Materials

Abstract:

Polyelectrolyte complexes (PECs) form spontaneously through the association of oppositely charged polymers and attract widespread interest as materials for an array of existing and emerging applications. Their shapes, however, are typically limited to rather simplistic (spherical, cylindrical or planar) structural motifs. Here, to extend PEC formation to a broader range of custom-designed structures, we present a simple photolithographic approach to preparing PECs with customized shapes. In this approach, a polyelectrolyte is mixed with an oppositely charged monomer and a photoinitiator. Upon photoirradiation, the monomer undergoes site-specific polymerization and, by locally generating an associating polyanion/polycation mixture, forms solid PECs with shapes that match the photoirradiation pattern. Like other PECs, the resulting custom-shaped complexes are stimulus-responsive and can be either swollen or dissolved on demand by varying the pH and ionic strength of their environments. Further, by generating gradients in irradiation intensity within the regions of photoirradiation, PECs with differential swelling and, thus, stimulus-responsive actuation properties can be prepared.

Besides demonstrating the preparation and properties of PECs formed through photodirected assembly, we will describe how polyelectrolyte gels with self-rupturing properties (i.e., the ability to break into fragments after a predetermined delay) can be prepared by synthesizing them to have highly non-uniform crosslink densities. This self-rupture is caused by highly-differential swelling, which generates stresses strong enough to overcome covalent bonds. The time required for these polyelectrolyte gels to rupture can be tubed by varying their: (1) crosslink content/uniformity; (2) size and internal morphology; and (3) ambient pH and ionic strength. In addition to their ability to self-destruct, these self-rupturing gels also can be prepared to contain liquid-filled payload compartments, which enable them to potentially serve as timed/delayed release devices.

499985 Coarse-Grained Model for Polyelectrolyte Complexation

Abstract:

Under appropriate circumstances, a mixture of oppositely charged polyelectrolytes can separate into two distinct liquid phases: a polymer-rich complex coacervate phase and a polymer-deficient supernatant phase.
The coacervation process and the structural properties of the resulting phases can be strongly influenced by variations in pH, the type and concentration of added salt, temperature, and ionic strength.
We have developed a coarse-grained molecular model to describe and predict polyelectrolyte complexation.

For fully charged homopolymers, we present three key findings. First, we show that the model is capable of capturing the dependence of polymer concentration, molecular weight, and salt concentration on coacervation. The results of simulations compare favorably with available experimental data. Second, we show that the concentration of salt in the coacervate complexes is generally different than in the supernatant phase. This observation also supports several recent experimental reports. Third, our model allows us to calculate the dynamic properties, including moduli, of the coacervate. The shapes of the predicted dynamic moduli also compare well with experimental measurements.

Building on our results for homopolymers, we next apply the model to mixtures of complexating polymers consisting of short charged blocks and long uncharged blocks.
For diblock and triblock copolymers, our model reproduces the mean and distribution of micelle aggregation numbers observed in experiments. Notably, for triblocks we predict the formation of low-polymer concentration gels whose structure is confirmed in neutron scattering experimental data. By relying on extensive calculations, we are able to identify several key features of these gels, as well as several distinguishing characteristics that do not arise in conventional amphiphilic gels.

491620 Synthesis and Solution-Phase Characterization of Hydroxylated Sulfonated Oligothioetheramides

Abstract:

Nature has long demonstrated the importance of chemical sequence to develop structure and tune physical interactions. By balancing electrostatic charge, length, hydrophobicity, and entropy, many naturally advanced structures (e.g. stem loops, alpha helices) create sensitive stimuli-response, allostery, and signaling. Thus, understanding the strength and balance of these physical forces can lead toward the development of structured macromolecules for biological function. We have recently described the rapid and efficient assembly of sequence-defined oligothioetheramides (oligoTEAs), featuring a tunable abiotic thioether backbone and diverse pendant groups. Structural characterization can now facilitate sequence-structure-function relationships by observing the oligomer chain dynamics and conformation.

In this work, we have systematically characterized linear sulfonated oligoTEAs. To characterize structure, we have developed methodology to describe flexible macromolecules with variable temperature diffusion ordered spectroscopy (VT-DOSY) and electron paramagnetic resonance (EPR) to constrain the Stokes-Einstein-Sutherland (SES) equation. With this characterization, we have examined the contributions of synthetic length (2-12mer), permanently positive and negative charged pendant groups, backbone hydroxylation and hydrophobicity, as well as the use of structured monomers on the resulting oligomer structure. We confirm the strength of entropy and hydrophobicity to create polymer â€œcollapseâ€ as a function of synthetic length and composition, described by increased flexibility and decreased aspect ratio (shape). Our results also suggest that hydrophobicity and entropy can overwhelm intramolecular electrostatic repulsion. While low resolution, this solution-phase characterization appears robust in its ability to describe flexible structures that can jeopardize other techniques, especially at this length-scale. As such, we have begun to apply this structural characterization to develop sequence-structure-function relationships with flexible, but biologically functional materials.

490755 Effect of Nanoparticle on the Structure and Dynamics of Model Peaa Ionomers from Molecular Dynamics Simulations

Abstract:

Ionomers are polymers that contain a small fraction of ionic groups covalently bound to a non-polar backbone. These bound ions along with free counterions tend to aggregate strongly, holding the polymer chains together as temporary crosslinks. This gives rise to interesting properties different from the parent, uncharged polymer. Ionomers are used in dental fixtures, fuel cell membranes and packaging, among other applications. We consider a dense melt of ionomers and unbound counterions, with no solvent and above the glass transition temperature. Prior work established a useful coarse-grained model for acetic acid based ionomers, but considered only fully neutralized materials. To allow clearer mapping to typical experimental systems, we developed a new model that includes unneutralized acid groups with adjusted Lennard-Jones interaction strengths with each other and with ionic groups. Using this modified model, we simulate a single spherical nanoparticle (whose diameter is ten times that of a monomer bead) surrounded by partially neutralized ionomers, and we consider two different nanoparticle â€“ monomer and nanoparticle â€“ ion interaction strengths. Detailed aggregate morphologies were calculated and compared with similar systems without nanoparticles. These analyses were done systematically for differing mole content of ions and different levels of neutralization. We find that aggregates are depleted near the nanoparticle surface, increasing the segmental dynamics near the surface relative to the bulk.

493900 Zwitterionic Copolymers As Novel Supporting Scaffolds for Ionic Liquid-Based Gel Electrolytes

Abstract:

Copolymers featuring zwitterionic pendant groups have been previously reported to offer several functional advantages for developing advanced materials, including anti-fouling membrane coatings and biocompatible, self-healing hydrogels. Recently, we reported the first demonstration of a class of sulfobetaine-type zwitterionic copolymer-supported ionic liquid gel (ionogel) composites, which offer excellent mechanical stiffness tunability while maintaining an excellent level of room temperature ionic conductivity.[1] Ionogel electrolytes are an emerging class of nonvolatile, nonflammable ion conductors that can enable the realization of safer electrochemical energy storage devices, as well as other applications. In this work, we describe the synthesis and characterization of a suite of novel zwitterionic copolymer-supported ionogels featuring scaffolds that are formed *in situ* within different hydrophobic ionic liquids via UV-initiated free radical copolymerization. Variation of both the ionic liquid and zwitterionic monomer chemical identities has facilitated the creation of polymer-supported ionogels that exhibit room temperature ionic conductivities as high as 12 mS/cm and compressive elastic modulus values that can be tuned between 0.9 kPa and 1.0 MPa. Experimental evidence of increased ion pair dissociation, as well as the formation of dipole-dipole physical cross-links that enhance gel stiffness due to the presence of zwitterionic groups, has been obtained. These findings serve as an early step towards understanding the largely unexplored realm of zwitterion/ionic liquid intermolecular interactions and the properties of zwitterionic copolymer-based ionogel composites.

[1]F. Lind, L. Rebollar, P. Bengani-Lutz, A. Asatekin, M.J. Panzer, *Chem. Mater.* 2016, *28*, 8480.

497693 Photoreversible Stiffness Modulation of Protein-Polymer Hydrogels

Abstract:

Polymer-based hydrogels have emerged as a powerful class of biomaterials that enable cells to be cultured in 3D within near-physiological, synthetic microenvironments. Recognizing that biology is neither homogenous nor static, we have developed unique strategies that permit reversible and user-defined modulation of gel properties in the presence of live cells. As our employed methods are photochemical in nature, hydrogel alteration can be performed at any point in time and 3D space (i.e., 4D) through directed light exposure. In this presentation, I will detail our recent efforts in the development of photoresponsive protein-polymer hydrogels whose local stiffness can be fully dynamically and reversibly, thereby capturing the biophysical changes that often accompany disease and healing. By recapitulating the dynamic heterogeneity of native tissue within well-defined polymer-based culture platforms, we believe that our efforts will be useful in probing 4D cell fate decisions and in the engineering of complex multicellular functional tissues.

496976 Preparation and Characterization of Polypeptide Hydrogels as Synthetic Extracellular Matrices for Cellular Scaffolds

Abstract:

The synthetic polypeptides, poly(γ-propargyl-L-glutamate) (PPLG), provide great potential for the development of novel hydrogels as extracellular matrix (ECM) for biomedical applications. Incorporating PPLG macromers into conventional step-growth PEG hydrogels results in a hybrid PEG-polypeptide hydrogel platform. This allows for highly clustered adhesion ligands on an α-helical scaffold via side chain modifications and a more diverse spectrum of swelling and mechanical properties than standard PEG gels.

PPLG was synthesized by ring-opening polymerization of the N-carboxy anhydride of γ-propargyl-L-glutamate and adopts a rigid α-helix conformation. The PPLG macromer was synthesized by grafting the "clickable" side chains of PPLG with a mixture of short ethylene glycol chains and norbornene groups. The PPLG hydrogels were prepared by step-growth UV-crosslinking using PPLG macromer and 4-arm polyethylene glycol (PEG) crosslinker bearing thiol end groups. Two control hydrogels were prepared by replacing PPLG macromers with 8-arm PEG crosslinker bearing norbornene end groups or poly(γ-propargyl-D,L-glutamate) (PPDLG) macromer with a flexible random coil conformation, which was a random copolymer produced from a 50:50 mixture of γ-propargyl-L-glutamate and γ-propargyl-D-glutamate monomers.

The physical and mechanical properties of these hydrogels, including swelling ratio, permeability, osmotic pressure, elastic modulus, and rheological properties were characterized by swelling test, fluorescence recovery after photo-bleaching (FRAP), osmotic swelling measurement, atomic force microscopy (AFM) indentation in fluid, and rheometry, respectively. The microstructure of the hydrogels were examined by cryogenic transmission electron microscopy (cryo-TEM) and small-angle X-ray scattering (SAXS).

The hydrogels were modified with PHSRN-K-RGD cell attachment motif for use as synthetic extracellular matrices. 2D cell culture studies were conducted with human induced pluripotent stem cell-derived endothelial cells (iPS-ECs), envisioning applications involving monolayers of endothelial cells. In addition, peptide-degradable hydrogels were prepared using linear or 4-arm PEG crosslinkers bearing matrix metalloproteinase (MMP)-degradable sequences with PEG, PPLG, or PPDLG macromers. 3D cell culture studies are currently underway on the peptide-degradable hydrogels for cell encapsulation and vascularization.

501769 Reinforced Hydrogel Fibers for Cell Encapsulation and Organ Printing

Abstract:

Introduction: Assembling cells into long fiber like structures have a great potential for building complex higher order 3D tissues and organoids. Cell laden fibers have been made using microfluidics by encapsulating cells in a continuous stream of hydrogels such as Ca-Alginate and gelatin. However such fibers are very delicate with poor mechanical properties and handling them is challenging. In this work we present a novel method to fabricate cell-laden alginate fibers reinforced with polymeric or metallic filament core. These reinforced fibers have superior mechanical strength combined with the advantaged of the hydrogel shell.

Materials and Methods: The fibers are made by custom made 3D printed nozzle that coaxially injects a steady stream of alginate over a spooling polymeric thread (which forms the core). The spooling system consisted of a 3D printed spool-drum rotated though a stepper motor driven at a precise predetermined rate by a microcontroller. The alginate flow rate together with the spooling rate determined the final dimension of the composite fiber. For encapsulating cells, cells were suspended in 1.4% alginate and introduced into the nozzle, while the alginate was crosslinked in a bath having 20mM Ba++ for 10 min. Cell loaded fibers were cultured in petridishes using normal medial and imaged after 3 weeks to observe cell growth.

Results: Reinforced fibers were made using braided nylon suture (diameter 150um) as the core and alginate shell. The fibers showed excellent mechanical properties with good adhesion of the hydrogel to the polymeric core. No peeling of the hydrogel was observed even upon bending of the fibers. SEM analysis of the fibers reveled the core-shell structure of the composite fibers with the polymeric core concentric to the shell. The thickness of the hydrogel shell could be precisely controlled by varying the flowrate or the spooling speed, and had excellent agreement with theoretical predictions. Hela cells laden reinforced fibers supported cell survival and growth for at least 3 weeks *in vitro*. Curiously, cells proliferated within the alginate shell forming linear 'string' like patterns travelling from within the core towards the surface having uncanny resemblance to the fluidic streamlines formed within the alginate just before crosslinking.

Conclusion: In this work we have demonstrated a new fabrication method for creating composite hydrogel fibers. These fibers would not only be useful in creating higher order cellular constructs but also as implantable systems for *in vivo* applications.

497996 Magnetically Templated Hydrogels for Peripheral Nerve Injury Repair

Abstract:

Peripheral nerve injuries (PNI) are a significant health problem and economic burden, costing $150 million dollars annually. There are various options currently on market for peripheral nerve injury (PNI) repair. Nerve wraps, conduits, and connects can be used for nerve regeneration for short distances between transected nerve stumps,[1] however, as the gap size increases, the only viable options for PNI repair are autografts and decellularized nerve allografts â€“ cadaveric tissue that has been stripped of cellular content using detergents. Nevertheless, autografts require multiple surgeries and can cause functional defects at the donor site if multiple, large grafts are required, and decellularized nerve tissue is limited by donor availability and can only be used in PNI gaps less than 5 cm. Thus, there is a need for engineered solutions for PNIs with gaps greater than 2 cm. The success of allografts and autografts have been attributed to the microtubular, channel-like structure of the natural peripheral nerve extracellular matrix (ECM) that is maintained in both cases.[2]Our hypothesis is that a hydrogel-based scaffold templated with porous microchannels could be viable for PNI repair of large (> 2 cm) nerve gaps by mimicking the tubular structure of the peripheral nerve ECM and guiding axon growth in a unidirectional fashion.

We introduce a novel method of patterning three-dimensional, anisotropic porosity: magnetic templating. This process involves suspending magnetic alginate microparticles (MAMs) in a hydrogel precursor solution, aligning the MAMs in the presence of a magnetic field, crosslinking the hydrogel, and finally degrading the MAMs, leaving behind a tubular porous architecture in the templated hydrogel. Our hydrogels are composed of collagen to encourage cell adhesion and glycidyl methacrylate hyaluronic acid (GMHA), which we have chosen due to the presence of hyaluronan in the natural peripheral nerve ECM, and due to its biodegradability, tunable mechanical properties, and capability for chemical modifications. MAMs are comprised of calcium-crosslinked alginate and encapsulate magnetic iron oxide nanoparticles that lend the MAMs their magnetic properties. Previous work in this project has been conducted using MAMs made via an emulsion method. While good preliminary results were obtained using emulsion MAMs (eMAMs), this platform is limited due to issues with poor control over MAM size and magnetic content and batch-to-batch variability. We have developed a protocol for fabricating MAMs using microfluidic droplet generation to produce MAMs with uniform size and magnetic content. This method was chosen for improved reproducibility and tunability of the ultimate microarchitecture of the magnetically templated hydrogels.

This microfluidic MAM (μMAM) fabrication platform has allowed us to better tune MAM composition and examine corresponding effects on alignment and degradation. Using nano-computed tomography (nanoCT) to create 3D re-constructions of magnetically templated hydrogels, utilizing the magnetic iron oxide nanoparticles as the contrast agents in the scan, we can measure MAM chain length, alignment, and areal density within the MAM-templated hydrogel. To examine channel morphology, we used confocal microscopy of templated hydrogels after MAM degradation and subsequent filling of the channels with dextran-FITC for visualization. An orthophenanthroline absorbance assay and electron paramagnetic resonance (EPR) were both used as iron quantification methods to track

clearance of iron oxide from the templated hydrogels. A ten-fold decrease in iron oxide content was measured after MAM dissolution. Using finite element analysis simulations in COMSOL to simulate arrays of permanent NdFeB magnets, we designed a magnetic templating setup that produces long, uniform magnetically-templated hydrogels (3 cm long currently, but longer fields are possible in theory) to demonstrate scalability for clinical translation.

Finally, we have conducted a preliminary *in vivo* study by inducing a 10 mm gap in a rat sciatic nerve to test three types of implants: a magnetically templated hydrogel, a non-templated hydrogel, and a fresh nerve isograft (which would represent the clinical treatment). For this pilot study, eMAMs were used to template the hydrogel implants. As this was a pilot study, a short four-week time point was chosen â€“ although this timespan was not long enough for complete nerve reconnection in a rat model, we still observed promising results. The magnetically templated hydrogel implants achieved substantial cell infiltration and re-modeling while the non-templated hydrogel implants did not. Using immunofluorescent staining for neurofilament, epifluorescence microscopy, and image analysis, the areal density of axons was measured in cross-sections taken throughout the implants. With n = 2, the study showed that after four weeks, the magnetically templated hydrogel had an areal density of axons comparable to that of the isograft near the proximal end of the implant. Our results show promise for a microstructured biomaterial that could aid in PNI repair and has other potential tissue engineering applications.

1. Kehoe S, Zhang XF, Boyd D. FDA approved guidance conduits and wraps for peripheral nerve injury: A review of materials and efficacy. *Injury*. 2012;43(5):553-572. doi:10.1016/j.injury.2010.12.030.

2. Spivey EC, Khaing ZZ, Shear JB, Schmidt CE. The fundamental role of subcellular topography in peripheral nerve repair therapies. *Biomaterials*. 2012;33(17):4264-4276. doi:10.1016/j.biomaterials.2012.02.043.

489859 A Two-Step Method for Transferring Single Wall Carbon Nanotubes Onto a Hydrogel Substrate

Abstract:

In recent years, significant progress has been made in elucidating the unique properties of carbon nanotube (CNTs)-hydrogel nanocomposites for biomedical applications. Due to the extremely hydrophobic nature of CNTs and the hydrophilic nature of the hydrogels, there are impediments to developing nanocomposites, including attaining a homogeneous dispersion of CNTs in the polymer solutions or alignment and patterning of CNTs in the hydrogels. Several approaches have been reported on patterning CNTs on various polymeric substrates. However, these methods include high temperature, high vacuum or the use of a sacrificial layer and hence, are incompatible with hydrogel substrates as they can lead to irreversible collapse in hydrogel structure. In this study, we designed a novel method to transfer CNTs onto the hydrogels with a single-step stamping technique. First, dense and aligned single wall carbon nanotubes (SWCNTs) were grown on single-crystal quartz substrate. Subsequently, polyethylene glycol (PEG), poly acrylamide (PA), and polyvinyl alcohol (PVA) hydrogel solutions were deposited on the CNTs grown on the quartz substrate. Upon hydrogel gelation, the CNTs were transferred from the quartz onto the hydrogel. The successful transfer of CNTs was confirmed by scanning electron microscopy (SEM). The transferring efficiency was found to be ~99%. Furthermore, results showed efficient cell attachment on the CNT-hydrogel composites that indirectly confirmed the successful transfer. CNT-hydrogel nanocomposites, such as the ones presented here, can be applied for various biomedical applications such as tissue engineering, drug delivery, and biomedical devices.

491575 Surface Tension-Assisted Additive Manufacturing of Multicomponent Biomaterials

Abstract:

Additive manufacturing, including 3D printing and 3D bio-printing, are enabling the production of geometrically-defined, patient specific biomaterials.[1] Despite the many recent advances in additive manufacturing technologies,[2] limitations remain to the speed of production, access to a broad range of materials, and easy integration of multiple materials and cells within a single device. Here, we present a facile method to fabricate multicomponent biomaterials with defined geometry incorporating both structural and mechanical support from polymeric constituents (*e.g.*, (meth)acrylates, polycaprolactone, poly(lactic-*co*-glycolic acid)) as well as cellular and tissue integration from hydrogels (*e.g.*, poly(ethylene glycol) based, collagen/elastin based, gelatin methacryloyl based). The method relies on surface tension-driven surface wetting, which allows for rapid coating of a reticulated frame with suspended films. We have developed a physical understanding of the coating process and demonstrate how it can be exploited to fabricate anisotropic biomaterials. In addition, the process can access a variety of applications on account of its scalability and versatility, with a special emphasis on tracheal replacements. We demonstrate the efficacy of this approach to fabricate mechanically supported and cell-laden biomaterials for *ex vivo* models of functional tissue as well as for human-scale replacement tissue for regenerative medicine applications. Overall, this presentation will present a facile process of multicomponent biomaterial fabrication and illustrate how this contributes to the emerging field of additive manufacturing in biotechnology and medicine.

References

1. Murphy SV *et al. Nat Biotech* **32** (2014) 773-785
2. Kang HW *et al. Nat Biotech* **34** (2016) 312-319

497149 Chiral Metamaterial Platform with Tunable Near and Far Field Chiroptical Response

Abstract:

The development of artificially structured electromagnetic materials, called metamaterials, has led to the realization of optical phenomena that cannot be obtained with natural materials. Chiral plasmonic metamaterials have been proposed as a promising platform for optoelectronic devices with exotic applications, such as negative index of refraction materials and superlenses that can break the diffraction limit. Chiral metamaterials have unit cells that lack mirror symmetry and inversion centers, and exhibit asymmetries in response to circularly polarized light that are orders of magnitude greater than the ones observed from naturally occurring chiral molecules. In the far-field, these asymmetries are manifested in the circular dichroism (CD) signal that quantifies the difference in absorption of left and right handed light. Despite the extensive literature on the far-field optical response of chiral metamaterials, our understanding of the chiral electromagnetic light-matter interactions remains limited. To develop design principles for chiral metamaterials, it is important to characterize and tune the optical chirality C of the electromagnetic fields in the vicinity of the nanostructures in the chiral unit cell, and elucidate the complex relationship between the near- and far-field optical response of chiral systems.

Here, we used Finite-Difference Time-Domain calculations to simulate the optical response of a stacked gold L-shape resonator system. In this system, the sign of the CD spectrum can be controllably changed through lateral shifts in the relative position of two gold L- resonators, which tunes the interaction strength. These small structural reconfigurations change the energetic ordering of the hybridized modes and alter the CD response of the system without changing its handedness. We examined the sensitivity of this system to structural perturbations, and demonstrated that the CD spectrum can abruptly reverse under small (less than 1 nm) reconfigurations of the L-resonators.

We have also calculated the chirality of the local electromagnetic fields and revealed superchiral hotspots with optical chirality 15 times larger than that of circular polarized light. More importantly, our simulations demonstrated that the optical chirality and the volume of the superchiral fields are much larger for the mode with enhanced extinction cross section. Lateral shifts of the upper resonator relative to the bottom one tuned the magnitude, volume, and handedness - given by the sign of optical chirality- of the local superchiral fields excited by right and left circularly polarized light.

The ability of our chiral L-shape assembly to abruptly change the far-field chiroptical response makes it ideal for experimentally fabricating a fast switchable chiral platform with dynamically tunable CD response. Meanwhile, the tunable chiral local electromagnetic fields could be utilized to examine how the change in the chiral interactions between the nanostructures translates into a change in the far field response of chiral systems.

498567 Lateral Growth of Two-Dimensional 1H-WSe_2/1T'-WTe_2 Heterostructures

Abstract:

Two-dimensional (2D) transition metal dichalcogenides (TMDs) such as MoS_2, $MoSe_2$, WS_2, and WSe_2, have attracted great attention due to their unique properties and atomic thickness. Until recently, great success was achieved in synthesizing 2D TMDs in 1T' phase, such as 1T'-$MoTe_2$ and WTe_2, which is particularly interesting due to they are predicted to be large-gap quantum spin Hall topological insulators [Qian et al. Science 2014, 346, 1344–1347]. Exploring the full potential of these 2D TMDs requires precise spatial modulation of their chemical composition and electronic properties to create well-defined heterostructures.

Here we report for the first time the growth of 1H-WSe_2/1T'-WTe_2 heterostructures by a two-step chemical vapor deposition method. The 1H-WSe2 was synthesized first, followed by an epitaxial growth of 1T'-WTe_2 on the edge. Monolayer thickness of the two flakes was confirmed by atomic force microscopy. Raman mapping studies demonstrated that the resulting heterostructures exhibited clear structure and phase modulation. Transmission electron microscopy revealed a single crystalline structure and a clear transition between the two different phases. Electrical transport studies demonstrated the formation of lateral p-type semiconductor and semi-metal heterojunctions by the combination of WSe_2 and WTe_2. This work provides an important advance in developing layered semiconductor-semimetal heterostructures and is an essential step towards next-generation nanoelectronic devices.

502208 Nanostructured Optoelectronics Using Interfacially-Driven Assembly

Abstract:

Nanostructured Optoelectronics have been of interest due to their remarkable properties, including size- shape, and surface-dependent absorption and emission. Our research group has developed interfacial chemistry approaches to synthesize novel nanostructured materials such as embedded nanocrystals and self-assembled layered structures. Using spectrscopic techniques, we probe the effects of surface chemistry and structure on the optoelectronic properties of these nanostructured solids.

This talk will focus on our recent synthetic efforts in two material systems: (1) Group IV quantum dots and (2) inorganic halide perovskites. Our results show that these materials have unique size-, composition-, and structure-dependent properties that can make them useful for a variety of optoelectronic applications.

488863 Tuning the Bandgap of Graphene Nanoribbons through Defect-Interaction-Driven Edge Patterning

Abstract:

Graphene nanoribbons (GNRs) with widths narrower than 10 nm have outstanding electronic, thermal, and mechanical properties and are considered as very promising two-dimensional material structures for both front-end and back-end technologies in future generations of high-performance and low-power-consumption electronics. The electronic band structure of the GNRs can be modified and their bandgap can be tuned by controlling the GNR width and the type, i.e., the orientation, of its edges.

In this presentation, we report a defect engineering strategy for tuning the bandgap of GNRs based on patterning of the GNR edges. The defects of interest are nanopores, or vacancy clusters, which can be introduced into GNRs in regular arrangements by ion etching or electron irradiation. We have conducted a systematic analysis of pore-edge interactions in GNRs using molecular-statics computations based on reliable interatomic potentials, which have revealed strongly attractive interactions for nanopores in the vicinity of GNR edges. These attractive interactions provide the thermodynamic driving force for nanopore migration toward the GNR edge, leading to its coalescence with the GNR edge through a sequence of carbon ring reconstructions. We have studied nanopore dynamics near GNR edges in detail through molecular-dynamics (MD) simulations at high temperature. We have constructed the optimal kinetic pathways of the mechanisms mediating the coalescence of the nanopore and the GNR edge, as identified by the MD simulations, using climbing-image nudged elastic band calculations. We have found that the post-coalescence morphological evolution of an armchair GNR edge leads to the formation of a V-shaped edge pattern consisting of zigzag linear segments (facets). First-principles calculations of the electronic band structure of such patterned GNRs based on density functional theory show that the zigzag segments formed at the armchair edges can be used to tune the bandgap of the GNR. We find a linear dependence of the bandgap of the patterned GNRs on the linear density of the zigzag edge atoms, which is controlled by the size and concentration of the pores introduced in the defect-engineered GNR.

500096 Wrinkled MoS2 Field-Effect Transistors

Abstract:

Manipulating the band structure through local-strain is an effective route for tailoring properties of 2D transition metal dichalcogenides (TMDs). Here, we show that wrinkles (wavelength = 100 nm - 10 Î¼m) can induce uniaxial strain in multilayer MoS_2 flakes (more than 20 layers), a resultant phonon softening (2 cm^{-1}/% strain for Raman E^1_{2g} mode), and reduction of the optical bandgap (40 meV/% strain). We also report on the influence of the geometric shapes (symmetric, asymmetric and folded) of wrinkles on the strain and optical band gap in the wrinkled MoS_2. In the field-effect transistor and optoelectronic device studies, a larger transport barrier (= 0.815 meV) is exhibited in the transverse direction of the MoS_2 wrinkle, a 1000-fold improvement of on/off ratio in dark, and a 10-fold photocurrent enhancement compared to the flat MoS_2 devices are also observed. This is attributed to the exciton funnel effect and built-in potential induced by band gap reduction and doping variation in wrinkled devices. The wrinkle induced tunability of bandgap and transport in thin film devices can potentially evolve flexible electronic and optoelectronic devices based on 2D TMDs.

500282 Nanoantenna Enhanced Wavelength Mixing in Monolayer Transition Metal Dichalcogenide

Abstract:

Nanoantenna enhanced wavelength mixing in monolayer transition metal dichalcogenide

D. Keith Roper[a,b*], Greg Forcherio[c], Mourad Benamara[d], Luigi Bonacina[e]

[a] *Microelectronics and Photonics Graduate Program, University of Arkansas, Fayetteville, AR 72701*

[b] *Department of Chemical Engineering, University of Arkansas, Fayetteville, AR 72701*

[c] *Army Research Laboratory, Adelphi, MD 20783*

[d] *Institute for Nanoscience and Engineering, University of Arkansas, Fayetteville, AR 72701*

[e] *GAP-Biophotonics, Université de Genève, Genève, CH 1211*

Monolayer (1L) transition metal dichalcogenide (TMD) crystals offer compelling new functionalities, e.g., gate tunability of p-n junction, valleytronics and anti-ambipolarity as well as unique optical, electronic and transport properties attributable to distinct atom-scale heterointerfaces. These features could enable atomically-thin flexible integrated circuits, field effect transistors and closed-loop resonators. However, present utility of 2D semiconductors is limited by difficulty in tuning their optoelectronic properties.

This work compared wavelength mixing at heterointerfaces of 1LTMD exhibiting broken inversion symmetry with and without an adjacent resonant nanoantenna (NA). Heterostructures of 1LTMD and NA were self-assembled via exfoliation and dropcasting. Nanometer- and femtosecond-resolved electron energy loss spectroscopy (EELS) was used to simulate and measure low-energy NA plasmon modes, damping and electric near fields at heterointerfaces. Discrete dipole approximation (DDA) was used to predict and induce bright, dark, and hybrid modes, damping, and electromagnetic near fields (Fig. 1). EELS allowed quantitative femtosecond-scale measurement of spectroscopic plasmon dephasing and nanometer-resolved mapping of electric fields on the NA and 2DNC-NA hybrids. A tunable laser scanning confocal microscope (LSCM) was used to measure wavelength mixing of NA, TMD and TMD-NA heterostructures (Fig. 2). Hyper Rayleigh Scattering (HRS) was used to measured second order nonlinear coefficient of NA-1LTMD. From HRS and LSCM data, values of second harmonic efficiency were compared for TMD and TMD-NA heterostructures. A computational scaffold supporting analysis spanning atom to nanometer scales was used to develop compact structure-function relations in order to rationalize measured values.

Coordination of DDA, EELS, LSCM and HRS supports comprehensive characterization of wavelength mixing in 1LTMD and 1LTMD-NA heterostructures at high spatiotemporal resolution.

[1] *G.T. Forcherio and D.K. Roper et al., in preparation.*

488845 Kinetics of Nanoring Formation from Quantum Dots in Epitaxial Thin Films

Abstract:

Solid material nanostructures in the form of nanorings exhibit many interesting properties, including tunable plasmon resonance and spin-polarized current switching, due to the unique nature of their electronic confinement. Such nanostructures have potential for numerous technological applications in optoelectronic and magnetic data storage devices. Recent experimental studies have reported the formation of gold and silver nanorings from the corresponding epitaxial nanoclusters or quantum dots upon thermal annealing. This presentation aims at a fundamental understanding of the kinetics of nanoring formation from quantum dots in coherently strained epitaxial films and a systematic exploration of the resulting nanoring structures upon variation of the processing conditions.

We have developed a three-dimensional kinetic model for the surface morphological evolution of coherently strained heteroepitaxial thin films, accounting for curvature-driven diffusion on the film surface, biaxial lattice misfit strain in the film, and the wetting potential between the film and the substrate. Self-consistent dynamical simulations show that the competition of the surface energy with the elastic strain energy lead through a Stranski-Krastanow instability to the formation of nanoclusters on the film surface, with the surface morphology gradually reaching a steady state under the action of the wetting potential. However, due to the difference in the thermal expansion coefficients between the film and substrate materials, the temperature increase during the heating of the film induces additional biaxial strain in the film due to thermal mismatch, triggering further morphological evolution of the nanoclusters upon thermal annealing and their transformation into nanorings. Our simulation results provide a fundamental comprehensive interpretation of the experimental reports in the literature.

Having validated our model, we explored further the morphological dynamics of nanoclusters in epitaxial layers under thermal annealing. We have found that the annealing temperature, the size of the nanocluster prior to thermal annealing, and the thickness of the wetting layer of the heteroepitaxial system have a strong effect on the morphological evolution of the nanoclusters during thermal annealing. Our dynamical simulations demonstrate that multiple concentric nanorings, structurally similar with the reported concentric nanorings fabricated by liquid droplet epitaxy, can form from sufficiently large nanoclusters at sufficiently high annealing temperatures. Finally, we provide a fundamental explanation of multiple concentric nanoring formation from quantum dots on epitaxial film surfaces based on weakly nonlinear stability analysis. Our modeling and simulation study sets the stage for designing experimental protocols toward precise control of complex nanoring structures on epitaxial film surfaces.

494292 Design, Synthesis, and Characterization of Elastomeric and Mechanoresponsive Polymer Matrix Composites

Abstract:

The development of multifunctional polymer matrix composites (PMCs) is highly desirable for high strength materials, armor, weapons systems, and machine parts. Research over the past couple of decades has focused on loading carbon nanotubes (CNTs) due to the synergistic increases in mechanical properties and conductivity.[1-5] However, two problems continue to persist: 1) insufficient dispersive forces lead to nanoparticle aggregation, and 2) the failure mechanism of high performance composites is still not well documented. Modifying the surface chemistry of nanoparticles has shown to improve the distribution throughout the matrix and, subsequently, enhance the tensile strength and/or conductivity of the matrix.[6] Furthermore, mechanoresponsive groups that produce light when activated by stress or strain have been incorporated to better understand the failure of composites, which enables spatial and temporal observation of molecular changes to the thermoset under a mechanical force.[7-11] This talk will focus on the synthesis and characterization of novel elastomeric PMCs containing surface functionalized CNTs and mechanoresponsive functionalities. The nanoparticle dispersion, PMC microstructure, composite modulus and thermomechanical properties, and the mechanoresponsive behavior will be discussed.

References:

1. Park, O.-K.; Hwang, J.-Y.; Goh, M.; Lee, J. H.; Ku, B.-C.; You, N.-H., Mechanically Strong and Multifunctional Polyimide Nanocomposites Using Amimophenyl Functionalized Graphene Nanosheets. *Macromolecules* **2013,** *46* (9), 3505-3511.

2. Jingjing Qiu and Chuck Zhang and Ben Wang and Richard, L., Carbon nanotube integrated multifunctional multiscale composites. *Nanotechnology* **2007,** *18* (27), 275708.

3. Moniruzzaman, M.; Winey, K. I., Polymer Nanocomposites Containing Carbon Nanotubes. *Macromolecules* **2006,** *39* (16), 5194-5205.

4. Mutiso, R. M.; Winey, K. I., Electrical properties of polymer nanocomposites containing rod-like nanofillers. *Progress in Polymer Science* **2015,** *40* (0), 63-84.

5. Winey, K. I.; Vaia, R. A., Polymer Nanocomposites. *MRS Bulletin* **2007,** *32* (04), 314-322.

6. Liu, Y.; Kumar, S., Polymer/Carbon Nanotube Nano Composite Fibersâ€"A Review. *ACS Applied Materials & Interfaces* **2014,** *6* (9), 6069-6087.

7. Zou, J.; Liu, Y.; Shan, B.; Chattopadhyay, A.; Dai, L. L., Early damage detection in epoxy matrix using

cyclobutane-based polymers. *Smart Materials and Structures* **2014,** *23* (9), 095038.

8. Black, A. L.; Lenhardt, J. M.; Craig, S. L., From molecular mechanochemistry to stress-responsive materials. *Journal of Materials Chemistry* **2011,** *21* (6), 1655-1663.

9. Davis, D. A.; Hamilton, A.; Yang, J.; Cremar, L. D.; Van Gough, D.; Potisek, S. L.; Ong, M. T.; Braun, P. V.; Martinez, T. J.; White, S. R.; Moore, J. S.; Sottos, N. R., Force-induced activation of covalent bonds in mechanoresponsive polymeric materials. *Nature* **2009,** *459* (7243), 68-72.

10. Weder, C., Mechanoresponsive Materials. *Journal of Materials Chemistry* **2011,** *21* (23), 8235-8236.

11. MakyÅ,a, K.; MÃ¼ller, C.; LÃ¶rcher, S.; Winkler, T.; Nussbaumer, M. G.; Eder, M.; Bruns, N., Fluorescent Protein Senses and Reports Mechanical Damage in Glass-Fiber-Reinforced Polymer Composites. *Advanced Materials* **2013,** *25* (19), 2701-2706.

493729 Molecular Simulation of Macromolecular Transport through Nanoporous Membranes

Abstract:

To design efficient nanoporous membranes a critical need exists to establish how pore size and pore chemistry affect the interfacial transport of macromolecules across membranes. A novel hybrid Monte Carlo and molecular dynamics simulation technique was implemented to elucidate the equilibrium behavior and transport properties of a model macromolecule as it navigated across a nanoporous polymer thin film (i.e., a nanofiltration membrane). A two-dimensional system, where a nano-pore separates two reservoirs on each side filled with water-like solvent particles was simulated. A steady-state flux of the solvent was established by imposing different chemical potentials within both reservoirs. The density and velocity profiles, viscosity, and flux of the solvent particles was measured in the nanopore system as a function of pore size. A non-constant density profile across the pore was found for small pore sizes, which results in a slip flow velocity profile. A linear polymer chain was added into the top reservoir to study structural rearrangements of the chain as it moved into the nano-confined spaces. The chosen model linear homopolymer was one that had interactions representative of poly(ethylene oxide) (PEO) when a PEO chain is dissolved in an aqueous environment. The structural rearrangements of the PEO chain as it passes through the nanopore under an imposed chemical potential gradient was quantified by measuring spatial configurations (i.e. end-to-end distance, radius of gyration, etc.) of the polymer as a function of solvent quality, polymer chain length, nanopore diameter, and PEO-nanopore wall interactions. These computational studies provide a more detailed picture of the underlying physical mechanisms that drive macromolecular transport through nanopores, and, in particular, how dimensionally-large macromolecules (i.e., with large radii of gyration) enter and move through dimensionally-small pores (i.e., small radii nanopores). The insights gained from these studies will aid in the development of more cost-effective water purification systems in separation technologies for a range of industrial applications.

496489 Unconventional Nanoscopic Shape Memory Effects Exhibited By Novel Multi-Stimuli-Responsive Shape Memory Polymers

Abstract:

Shape memory polymers (SMPs) are stimuli-responsive materials that can change their shapes upon application of an external stimulus (e.g., heat and light). They have been extensively exploited for a wide spectrum of technological applications ranging from smart biomedical devices to aerospace morphing structures. Unfortunately, most of the currently available SMPs are thermoresponsive and they suffer from heat-demanding programming and recovery processes. Additionally, the intriguing nanoscopic shape memory effects, which could significantly enhance the response speed and greatly expand the application aspects of SMPs, are largely unexplored. Here we report unconventional nanoscopic shape memory effects exhibited by a new series of multi-stimuli-responsive SMPs. When combined with a unique macroporous photonic crystal structure templated from self-assembled colloidal crystals, these smart SMPs enable unusual â€œcoldâ€ programming and instantaneous shape recovery at ambient conditions triggered by multiple untraditional stimuli, such as static pressure, lateral shear stress, a large variety of liquids and vapors, and ultrasonic waves. Importantly, striking chromogenic responses associated with a full all-room-temperature shape memory cycle could render a new and sensitive optical technology for investigating the intriguing nanoscopic deformation and recovery of SMPs.

490006 Direct Observation of Remarkable Nanostructure Evolution during Aqueous Dissolution of Polymer/Drug Blends

Abstract:

Amorphous solid dispersions (SDs), metastable blends of an active pharmaceutical agent and a polymer excipient, can significantly enhance the aqueous solubility of hydrophobic drugs, but the mechanism is not well understood. To address this gap in knowledge, we investigated the dissolution of hydroxypropyl methylcellulose acetate succinate (HPMCAS) SDs in phosphate buffered saline (PBS). Light scattering measurements of HPMCAS solutions showed the polymer forms a mixture of ~10 and ~100 nm sized structures (assigned to linear and covalently coupled polymer chains, respectively) in both PBS and tetrahydrofuran. The measurements also indicated that PBS is a poor solvent for HPMCAS at and below 37 Â°C, suggesting the solvent induces the polymer to associate with itself or other hydrophobic species in solution. *In vitro* dissolution of HPMCAS SDs â€“ containing either phenytoin or probucol as the hydrophobic drug â€“ in PBS was studied using a combination of cryogenic transmission electron microscopy (both imaging and electron diffraction) and small-angle X-ray scattering. These techniques revealed that HPMCAS and drug form <100 nm amorphous nanoparticles in solution. A direct correlation between the SD dissolution profiles and nanostructure evolution was discovered for both drugs, suggesting a significant portion of the drug measured in the supernatant during the dissolution assay is retained in the form of nanoparticles. Nanoparticle size, shape, and lifespan were influenced by drug identity, loading, and targeted concentration. These observations establish that persistent nanostructures are important for maintaining drug supersaturation during SD dissolution.

501031 Effect of Graphene and Graphene Oxide on the LCST of PNIPAM

Abstract:

Poly(N-isopropylacrylamide) (PNIPAM) is a thermo-sensitive polymer displaying a coil-to-globule transition above its lower critical solution temperature (LCST) of 305 K. As its LCST is close to the physiological temperature, PNIPAM is an attractive candidate for in-vivo applications such as drug delivery. Graphene nanostructures have also shown great potential in the area of biomedicine with applications ranging from drug delivery to tissue engineering. Additionally, complexes of graphene oxide and various aromatic drugs have yielded vastly improved water solubility allowing for increased efficacy. In this work, we have conducted all-atom molecular dynamics (MD) simulations of a PNIPAM chain consisting of 30 monomer units along with sheets of either graphene or graphene oxide in the presence of water. All simulations were carried out at 325 K and 275 K, which are above and below the LCST of PNIPAM, respectively. The impact of sheet width and its initial angle with respect to PNIPAM on the LCST of PNIPAM is investigated.

489811 Solvent Effects on the Structure and Thermodynamics of Polymer Blends with Varying Architectures

Abstract:

Advances in polymer synthesis have unlocked a widening library of nonlinear polymer architectures, such as cyclic, star, hyperbranched, and bottlebrush polymers. Blends of these architectures exhibit interesting and complex physics, as entropic effects related to the interplay between chain ends and branch points can be leveraged to tune blend structure and properties such as chain entanglements, segregation to surfaces, rheology, and others. Further tunability in structure and properties can be achieved through solvent processing of these materials, as the presence of solvent alters polymer blend thermodynamics by screening interactions between polymer chains and between polymer and surfaces. Herein, we will explore the thermodynamic implications of the solvent processing of polymer blends of varying architecture (linear, cyclic, star, and bottlebrush) using molecular dynamics simulation and polymer reference interaction site model (PRISM) theory. We will first present our coarse-grained model for polymer blends swollen by solvents of varying quality. We will then describe the application of the above computational techniques to calculate, both in real and Fourier space, the total intermolecular pair correlation functions, direct pair correlation functions, and effective interactions between polymer(s) and solvent(s). In particular, we will show how both entropic and enthalpic effects play a role in how solvents tune structure and thermodynamics in blends of polymer architectures.

491360 Molecular Weight Dependence of the Intrinsic Size Effect on T_g in AAO Template-Supported Polymer Nanorods: A DSC Study

Abstract:

Many studies have established a major effect of nanoscale confinement on the glass transition temperature (T_g) of polystyrene (PS), most commonly in thin films with one or two free surfaces. Here, we characterize smaller yet significant intrinsic size effects (in the absence of free surfaces or significant attractive polymer-substrate interactions) on the T_g and fragility of linear PS (see *J. Chem. Phys.* 2017, 146, 203323). Melt infiltration of various molecular weights (MWs) of PS into anodic aluminum oxide (AAO) templates is used to create nanorods supported on AAO with rod diameter (d) ranging from 24 to 210 nm. The T_g (both as $T_{g,onset}$ and fictive temperature) and fragility values are characterized by differential scanning calorimetry. No intrinsic size effect is observed for 30 kg/mol PS in template-supported nanorods with d = 24 nm. However, effects on T_g are present for PS nanorods with M_n and $M_w \geq \sim$ 175 kg/mol, with effects increasing in magnitude with increasing MW. For example, in 24-nm-diameter template-supported nanorods, $T_{g,rod} - T_{g,bulk}$ = -2.0 to -2.5 °C for PS with M_n = 175 kg/mol and M_w = 182 kg/mol, and $T_{g,rod} - T_{g,bulk} = \sim$ -8 °C for PS with M_n = 929 kg/mol and M_w = 1420 kg/mol. In general, reductions in T_g occur when $d \leq \sim 2R_g$, where R_g is the bulk polymer radius of gyration. Thus, intrinsic size effects are significant when rod diameter is smaller than the diameter ($2R_g$) associated with the spherical volume pervaded by coils in bulk. We hypothesize that the T_g reduction occurs when chain segment packing frustration is sufficiently perturbed by confinement in the nanorods. This explanation is supported by observed reductions in fragility with increasing extent of confinement. We also explain why these small intrinsic size effects do not contradict reports that the T_g-confinement effect in supported PS films with one free surface exhibits little or no MW dependence. This point can be understood by comparing template-supported nanorod and thin film T_g reductions at a similar characteristic length scale, t^*. For PS nanorods with d_m = 63 nm or t^* = 15.8 nm, as measured by DSC there is no reduction in T_g relative to bulk response when M_n and $M_w \leq$ ~500 kg/mol, and $T_{g,rod} - T_{g,bulk} = \sim$ -2 °C for PS with M_n = 929 kg/mol and M_w = 1420 kg/mol. As determined by ellipsometry, supported PS films with thickness of 16 nm (t^* = 16 nm) exhibit $T_{g,film} - T_{g,bulk} = \sim$ -19 °C (*Europhys. Letter.*1994, 27,59). Thus, at MWs of PS most commonly used in studies of T_g-confinement effects in the presence of a free surface, intrinsic size effects contribute nothing within error to the measured effect; at extremely high PS MW, the contribution of intrinsic size effects is only a very small percentage of the reported T_g reduction.

The characterization of intrinsic size effect on T_g is being extended to a crosslinked, supported PS nanorod system. Linear precursor polystyrene-co-vinylbenzoycyclobutene (VBCB) nanorods with M_n = 30 kg/mol

exhibit no intrinsic size effect on T_g, where $d > 2R_g$ even in template with 24-nm pore diameter. Upon crosslinking the VBCB units, the network structure exhibit effectively infinite MW. The comparison of T_g-confinement effects in linear and crosslinked PS nanorods is being studied and will be discussed.

494468 Versatile Cholesterol-Functionalized Block Copolymers in Aqueous Dispersions

Abstract:

Cholesterol plays an important role in many biological processes, namely regulating the mechanical and transport properties of biomembranes. Biotechnology research attempts to exploit the inherent benefits of cholesterol through its use, for example, in liposomes for drug delivery. The potential applicability of cholesterol can be further extended through its chemical incorporation into random and block copolymers. The controlled segregation of cholesteryl moieties into one or more portions (*i.e.* blocks) of a copolymer produces well-controlled nanostructures (micelles and polymersomes) through self-assembly. Here, we first present the self-assembly behavior for a set of novel cholesteryl block copolymers, poly[(ethylene glycol)-*b*-(cholesteryl 2-oxo-1,3,6-dioxazocane-6-carboxylate)] (PEG-*b*-P(8C-Chol)), which have relatively short P(8C-Chol) block lengths and therefore are directly dispersible in water. The results provide a relation between the number of cholesteryl groups per polymer chain ($X_{n,Chol}$) and the size and geometry of structures formed. Next, we combine the same PEG-*b*-P(8C-Chol) copolymers in a minority fraction (~5 mol%) with a common phospholipid in order to form surface-functionalized liposomes varying in insertion moieties per copolymer chain (*via* $X_{n,Chol}$). Again, the major implications relate the molecular structure to the self-assembly behavior. Both studies provide overarching design rules regarding cholesteryl-containing block copolymers and could have a major impact in the future synthesis of drug delivery devices and other naoncarriers that utilize cholesterol.

493312 Polymerization Thermodynamics Under Nanoconfinement

Abstract:

The behavior of materials confined at the nanoscale has been of considerable interest over the past several decades. Here, our focus is the influence of nanoconfinement on the thermodynamics of free radical n-alkyl methacrylate polymerizations. Using differential scanning calorimetry to follow the reaction, we observe that the ceiling temperature decreases when the methacrylate reactions are performed in nanopores and that the total entropy loss on going from monomer to polymer increases upon nanoconfinement (i.e., the entropy of propagation, Î"S_p, becomes a larger negative value). These confinement effects are observed to become weaker as temperature increases and as the n-alkyl group increases in size, from methyl to ethyl to butyl. The entropy loss on confining a chain is expected to scale with nanoconfinement size and with chain length, and the scaling relationships will be tested and compared to the literature. The results will also be discussed in the context of other nanoscale confinement effects, such as the T_g depression.

495621 Molecular Understanding, Design and Development of Ultra Low Fouling Zwitterionic Materials

Abstract:

An important challenge in many applications, ranging from medical devices to ship hulls, is the prevention of nonspecific protein adsorption on surfaces. To address this challenge, our goals are twofold. First, we strive to provide a fundamental understanding of nonfouling mechanisms at the molecular level using an integrated experimental and simulation approach. Second, we aim to develop biocompatible and environmentally benign ultra low fouling materials based on the molecular principles we have learned. As a result, we have demonstrated that zwitterionic and mixed charge materials and surfaces are highly resistant to nonspecific protein adsorption, cell adhesion and bacteria adhesion/biofilm formation from complex media. Both simulation and experimental results show that the strong hydration of zwitterionic materials is responsible for their excellent nonfouling properties (JACS, **127**, 14473, 2005. and Advanced Materials, **27**, 15, 2015). Recent results show that zwitterionic materials induce no capsule formation upon implantation (Nature Biotechnology, **31**, 553, 2013) and no immunological response in blood circulation (PNAS, **112**, 12046, 2015) and are able to preserve protein (Nature Chemistry, **4**, 59, 2012) and cell (Angewandte Chemie, **53**, 12729, 2014) bioactivity. At present, zwitterionic materials, as alternatives to poly(ethylene glycol) (PEG)-based materials, have been applied to a number of applications, including medical devices, drug delivery carriers, antimicrobial coatings, and marine coatings.

503560 A Scalable Approach to Produce a Diversity of Structured Polymer Colloids

Abstract:

In this talk, we demonstrate a strategy for the self-assembly of multi-faced nanocolloids through the precipitation-induced, rapid demixing of polymer phases within a confined solution volume. Using a two-faced Janus colloid assembled from two simple homopolymers as a model system, we demonstrate the ability of the method to provide simultaneous control over particle size, surface functionality, and compositional anisotropy as the assembly process is scaled. Furthermore, we illustrate that tuning the molecular weight of the homopolymers and increasing the number of polymer components in the system can facilitate the formation of multi-faced and protruded nanocolloids, respectively. Our results demonstrate that incompatible polymers with different properties can be self-assembled into nanocolloids with controllable surface topology by simultaneously reducing the timescale and solution volume over which they undergo self-assembly. Finally, we demonstrate the ability to tune geometry in polymer-based colloids as well produce hybrid polymer/metal colloidal systems by our approach

503566 Layered Morphologies in Precise Acid-Containing Polyethylenes: Hierarchical Structures and the Implications on Properties

Abstract:

Acid-containing polymers have specific interactions that produce both hierarchical nanoscale morphologies containing acid-rich aggregates and remarkable bulk properties. Untangling the correlations between the primary structure of such associating polymers and their morphologies and properties has long been a challenge in polymer physics, because most associating polymers have random sequences of polar and non-polar monomeric units. New synthetic methods can increasingly produce polymers with greater molecular precision that provide greater uniformity of and control over the hierarchical structure morphologies and even yield new morphologies. The talk will focus on precise polyethylenes with acid groups evenly spaced along linear chains and with spacer lengths of 21 carbons such that the polymers produce hierarchical structures with acid layers. Using a combination of atomistic molecular dynamics, X-ray scattering and FTIR, the chain conformations near the acid groups have been determined to be hairpins. Moreover, this acid-layer motif with hairpin chain conformations packs into a semi-crystalline morphology with the acid-layers perpendicular to the crystalline lamellae. The implications of this morphology will be explored for tensile deformation and proton transport.

503565 Examination of Thin Film Growth Using Supersonic Molecular Beams and *in situ* Real Time X-Ray Synchrotron Radiation: From Organic Small Molecule Semiconductors to Transition Metal Dichalcogenides

Abstract:

The study of thin film growth using real time techniques has always represented an important challenge to the experimentalist. Over the past several years our group has been employing supersonic molecular beam techniques and in situ real time X-ray synchrotron radiation to examine a variety of systems in two important emerging areas: organic small molecule semiconductors and transition metal dichalcogenides (TMDs). In a number of cases we have supplemented the information we obtain from X-ray scattering, with results from in situ real time X-ray fluorescence, and atomic force microscopy. In this presentation we will review some of our more intriguing findings. We will address issues such as: does the sequence of deposition matter concerning the growth of heterostructures involving two small molecule organic semiconductors? do lower growth rates and higher substrate temperatures always produce smoother thin films? and in cases where both a thin-film and bulk phase form, which grows first? The growth of thin films of TMDs has attracted a great deal of interest lately, yet in situ real time studies of these materials are few and far between. We will show that in situ X-ray synchrotron radiation is a powerful probe of the growth of these materials, and we will show for the first time direct, real-time observation of layer-by-layer growth of a TMD 2D semiconductor.

503564 Polymer Electrolytes for Electrochemical Energy Devices

Abstract:

Hydroxide exchange membranes (HEMs) can enable platinum-group-metal-free HEM fuel cells (HEMFCs) and HEM electrolyzers (HEMELs) that are potentially economically viable. These membranes are also useful in other electrochemical devices like solar hydrogen electrodialysis cells. The basic structure of a HEM is an organic cation tethered to a polymer backbone. We have systematically studied the cation, the tether and the backbone in the past decade, and developed HEMs with quaternary phosphonium,[1] sulfonium,[2] cobaltocenium,[3] and stabilized imidazolium[4] that have shown high chemical stability, hydroxide conductivity and reasonably low water uptake. In this presentation, I review our past work and present our most recent work on a new class of poly(aryl piperidinium) (PAP) HEMs. The PAP polymers are designed by combining cyclic quaternary ammonium cations and ether-bond-free backbones, which together provide unprecedented chemical stability, mechanical robustness and ionic conductivity. The rigid and hydrophobic aryl backbone enables achievement of both high ion exchange capacities and low swelling ratios, leading to high hydroxide conductivities and mechanical robustness. I will also discuss our work on the development of a HEM based on molecular sieving crystalline porous polymers (organic zeolites).

References:

1. S. Gu, R. Cai, T. Luo, Z. W. Chen, M. W. Sun, Y. Liu, G. H. He and Y. S. Yan, A Soluble and Highly Conductive Ionomer for High-Performance Hydroxide Exchange Membrane Fuel Cells, *Angew. Chem.-Int. Edit.*, 2009, **48**, 6499.

2. B. Z. Zhang, S. Gu, J. H. Wang, Y. Liu, A. M. Herring and Y. S. Yan, Tertiary sulfonium as a cationic functional group for hydroxide exchange membranes, *RSC Advances*, 2012, **2**, 12683.

3. S. Gu, J. Wang, R. B. Kaspar, Q. Fang, B. Zhang, E. Bryan Coughlin and Y. Yan, Permethyl Cobaltocenium (Cp*2Co+) as an Ultra-Stable Cation for Polymer Hydroxide-Exchange Membranes, *Sci Rep*, 2015, **5**, 11668.

4. J. H. Wang, S. Gu, R. B. Kaspar, B. Z. Zhang and Y. S. Yan, Stabilizing the Imidazolium Cation in Hydroxide-Exchange Membranes for Fuel Cells, *Chemsuschem*, 2013, **6**, 2079.

490039 TAT-Functionalized Liposomes for the Treatment of Meningitis

Abstract:

Brain inflammatory diseases such as bacterial meningitis have recently become a global concern in clinical care due to the recent emergence of antibiotic-resistant bacteria, making it increasingly difficult to treat these infections. Treatment is made further difficult once these bacteria form a biofilm, a complex polysaccharide matrix that both protects the bacteria from environmental stimuli and allows for the spread of antibiotic resistance within the bacterial community. Broad-spectrum antibiotics, which are commonly used to treat such infections, face difficulties penetrating these biofilms and the outer bacterial cell wall. More importantly, the overuse of broad-spectrum antibiotics in clinical care is prevalent and often ineffective, leading to increased drug resistance in bacteria and poorer patient care outcomes. Therefore, there is a growing need to develop an alternative means for treating bacterial meningitis that do not rely on the use of broad-spectrum antibiotics.

To address this need, we report the use of fusogenic liposomes to deliver targeted antibiotics at the site of infection, notably inside the bacterial cell using cell penetrating peptides (TAT 47-57). Liposomes have recently been demonstrated as effective drug carriers, though combining this approach with TAT peptides to deliver drugs inside bacteria as a way to combat bacterial meningitis which has yet to be fully explored and would be a novel means of treatment. To this end, we investigated the effect of TAT-functionalized liposomes to inhibit the growth of bacteria commonly associated with bacterial meningitis, including *Streptococcus pneumoniae*, methicillin-resistant *Staphylococcus aureus* (MRSA), and *Escherichia coli.*

Liposomes were prepared by lipid film rehydration, functionalized with TAT (47-57) using wet chemistry, and characterized using transmission electron microscopy to both determine spherical diameter (~100 nm) and verify successful liposome formation. Liposomes were loaded with one of three commonly used antibiotics to treat these bacterial infections, including vancomycin, methicillin, and ampicillin. The inhibitory effect on bacterial growth of the liposomes was determined using bacterial growth curves and determining bacteria colony forming units, and this system was compared against non-TAT-functionalized liposomes and against free antibiotic.

The results from this study demonstrated excellent growth inhibitory effects of TAT-functionalized liposomes loaded with methicillin for treating MRSA, with a reported minimum inhibitory concentration (MIC) of 1.7 μg/mL, well below the needed dosage for free antibiotic (5 μg/mL). A similar decrease in bacteria population was also demonstrated using TAT-functionalized liposomes loaded with vancomycin to treat *S. pneumoniae* (MIC of 1 μg/mL). The TAT-functionalized liposomes presented the highest antibacterial activity in all assays performed, followed by non-functionalized liposomes, and lastly free antibiotics. Cytotoxicity assays with endothelial cells and astrocytes demonstrated a notable increase in percent cell viability using functionalized and non-functionalized liposomes in comparison with free antibiotics. Results obtained with a methicillin concentration of 5 μg/mL showed only 20% cell viability for the free antibiotic, while for liposomes it was significantly higher at 90%.

All results obtained for TAT-functionalized liposomes, including their low MIC with respect to the use of free antibiotics plus their remarkable high percentage of cell viability with astrocytes and endothelial cells, provides evidence that this liposomal system can be a safe, alternative means for treating bacterial meningitis.

This work was supported by funds from Northeastern University.

500847 Cartilage Penetrating Nanoparticles to Provide Sustained Delivery of Disease Modifying Drugs in Post-Traumatic Osteoarthritis

Abstract:

Osteoarthritis is a debilitating disease of the joint which manifests as a loss of articular cartilage, causing serious pain and impeding mobility. There are a variety of causes of the disease – however, one particular etiology, post-traumatic osteoarthritis (PTOA), is of particular interest for clinical treatment because it has a biologically and temporally well-defined onset – a traumatic joint injury. As PTOA does not induce loss of cartilage until years after the traumatic injury, there is potential for immediate intervention with a disease modifying osteoarthritis drug to delay or even halt the progression of the disease. However, no such disease modifying drug currently exists and the current standard of care is merely palliative.

There are several classes of molecules that show promise *in vitro* and in animal studies, such as anti-inflammatory small molecules, cytokine receptor antagonists, anabolic growth factors, and inhibitors of matrix metalloproteinases. Unfortunately, they have not been useful as therapeutics in humans because even when injected directly into the joint, they are unable to penetrate and reside in cartilage for long enough to have disease modifying effect. The high clearance rate of molecules in the joint space and small pore size of cartilage tissue (<15 nm) pose considerable delivery challenges for any drug of interest.

However, the high sulfated proteoglycan content in cartilage provides an opportunity to harness electrostatic interactions between the negatively charged tissue and a positively charged nanocarrier. We hypothesized that such interactions, if strong enough, could bind the nanocarrier to the cartilage faster than it could be cleared from the joint. Yet the interactions must also be weak enough to allow for dissociation and diffusion throughout the tissue. Thus, tight control of positive charge and small (<15 nm) size were crucial design criteria for these nanocarriers.

We chose to conjugate cationic dendrimers to an anabolic growth factor, insulin like growth factor 1 (IGF1), to test this hypothesis. The terminal groups of dendrimers provide a high degree of positive charge, which can be tightly controlled by conjugation with different amounts of PEG. To find an optimal degree of positive charge for binding and penetration of cartilage, a small library of partially PEGylated dendrimers was constructed. This library was screened with cartilage binding and toxicity assays, and the top two formulations were selected for further study. One formulation was about 3 times as charged as the other.

The top two dendrimer – IGF1 formulations were tested for penetration of both *ex vivo* bovine cartilage biopsies and *in vivo* rat joint cartilage. Both formulations showed vastly superior binding and penetration of cartilage compared to IGF1 alone. As expected, the more charged formulation was more effective than the less charged formulation at binding cartilage, but took longer to diffuse through the depth of the tissue.

This was followed by a pharmacokinetic study in which IGF1 alone or one of the top two dendrimer-IGF1 formulations were injected into rat joints (n=8 per group) and monitored by an *in vivo* imaging system over 28 days. Half-lives were calculated for each formulation based on single phase exponential decay fits of the data. IGF-1 alone had a joint half-life of 0.41 days, the less charged dendrimer-IGF1 had a joint half-life of 1.08 days, and the more charged dendrimer-IGF1 had a joint half-life of 4.21 days, nearly 10 times that of unconjugated IGF1.

We then sought to determine if the sustained delivery and cartilage penetration capability of these IGF1 nanocarriers could rescue cartilage in a surgically induced rat model of osteoarthritis. Rat joints were destabilized by ACL transection and partial meniscectomy. 48 hours after surgery, dendrimer-IGF1 conjugates, IGF1 alone, or no treatment was injected. 4 weeks later, the rats were sacrificed and joint histology was scored for various parameters relevant to osteoarthritis.

498209 Nanoparticle Tracking Analysis of Polymer Particle Aggregation in Blood Plasma

Abstract:

Polymer particles, such as polylactic-co-glycolic acid (PLGA), are often studied as potential drug delivery vehicles, but further understanding of their behavior in the biological environment is necessary to achieve a successful drug delivery system. One major hurdle for drug delivery vehicles is evading clearance by the immune system before reaching the intended targets. If the particles aggregate in blood, then the aggregated particles will be removed by the immune system. The surface chemistry of a particle is critical to its aggregation behavior, yet measuring aggregation in a complex solution such as blood is difficult. Nanoparticle Tracking Analysis (NTA) is a powerful tool for analyzing nanoparticles in solution. By using fluorescently labeled particles, particles can be analyzed in complex solutions such as blood plasma. Analysis in blood plasma allows for quantification of heterogenous aggregation of the particles (i.e. particles with proteins) that wouldnâ€™t be captured by other methods.

In this work PLGA particles were made via electrohydrodynamic co-jetting. They were subsequently modified to add either poly(ethylene glycol) monolayers, poly(ethylene glycol) methacrylate (PEGMA) brushes, or poly(carboxybetaine methacrylate) (PCBMA) brushes to the particle surface. The particles were incubated in blood plasma and analyzed with NTA. The aggregation behavior of the particles was observed and quantified and compared to the heterogeneous aggregation behavior observed with unmodified PLGA particles.

500080 High Throughput Screening of Biodegradable Nanogels with Tunable Size and Swelling for Intracellular Drug Delivery

Abstract:

The design and synthesis of biodegradable, intelligently responsive polymeric nanomaterials in which size and swelling can be tuned without changing material composition is a distinct advantage of cross-linked nanomaterials synthesized by emulsion polymerization. Nanogels have enhanced stability to dilution and lower drug leakage compared to their self-assembled counterparts and emulsion polymerization offers the advantage of controlling nanogel size through surfactant choice and concentration. Furthermore, the use of controlled radical polymerization has the advantage of imparting non-statistical architectures through reaction from a macroinitiator or incorporation of functional initiators.

Recently, PEG graft monomers were shown to have poor conversion in organic solvents below their equilibrium monomer concentrations in atom transfer radical polymerization (ATRP) reactions. In this work, biodegradable cationic nanogels composed of 2-diethylaminoethyl methacrylate, methacrylate comonomers, and a biodegradable disulfide crosslinker (bis(2-methacryloyl)oxyethyl disulfide) were synthesized via activators regenerated by electron transfer (ARGET) ATRP in a high throughput format. First, nanogels were synthesized with or without PEG in an emulsion polymerization. Select nanogel formulations were then modified with PEG or PEG copolymers using copper-catalyzed alkyne-azide cycloaddition. Reaction parameters including surfactant hydrophilic-lipophilic balance, monomer hydrophobicity, crosslinking density, and monomer concentration were systematically investigated to synthesize nanoparticles with predictable size, swelling, and responsive behaviors. Following reaction parameter optimization, a library of nanogels containing a range of hydrophobic comonomers, PEG graft density, and architecture modifications were analyzed for their ability to load and release model chemotherapeutic agents and oligonucleotides. Leading candidates were evaluated in vitro for cytotoxic effects against a doxorubicin resistant cell line.

Nanogel composition was monitored as a function of time via purification and degradation of nanogels under reducing conditions followed by ^{1}H NMR. Molecular weight of degraded nanogels was determined by GPC to be < 40 kDa. Nanogel hydrodynamic size was determined by dynamic light scattering and swelling volume was calculated as the cube of the ratio of swollen to collapsed hydrodynamic diameters. Nanogel collapsed hydrodynamic diameter was modulated from 50 â€“ 150 nm with average polydispersity values < 0.100. Volume swelling ratios ranged from 1.5-9.1 in 1x PBS demonstrating the ease of tunability. Critical swelling pH onset ranged from pH 7.5 to 6.0 depending on the degree of hydrophobic modification. TEM was used to determine dry size, morphology and verify low polydispersity. Nanogel formulations were screened with respect to the loading of doxorubicin and siRNA. Loading and release profiles were related to materials properties to select a diverse set of candidate formulations for dose dependent cytotoxicity against a doxorubicin resistant cell line *in vitro*.

The use of controlled radical polymerization and click chemistry allowed for efficient synthesis of nanogels with either

statistical or non-statistical architectures. Moreover, this approach demonstrates the ability to independently tune the responsive properties of a library of biodegradable nanogels for intracellular drug delivery.

500588 Polyanhydride Nanoparticles Encapsulating Rifampicin Suppress Growth of Pathogenic *Acanthamoeba in vitro*

Abstract:

Acanthamoeba spp. infections represent a growing problem for developed countries. In immunosuppressed individuals, *Acanthamoeba* spp. can cause granulomatous amebic encephalitis (GAE), which is often fatal due to a prodromal period, limited diagnostic tools, and a lack of effective antimicrobial treatments. Additionally, *Acanthamoeba* can cause *Acanthamoeba* keratitis (AK), a painful corneal infection that can result in vision impairment and can require enucleation if left untreated. AK is associated with corneal trauma and improper cleaning of contact lens. Infections by amoeba demonstrate several challenges to antimicrobial treatment. Delivery of antimicrobial therapeutics to amoebae is limited by the blood-brain barrier in GAE, restricted internalization and rapid efflux of drugs by the amoeba. An effective antimicrobial therapy capable of killing both the metabolically-active trophozoite and cyst forms of *Acanthamoeba* remains elusive.

Free-living amoeba feed on organic matter including bacteria that are typically 0.5 â€“ 2 microns. Prior research demonstrated that through co-polymerization of hydrophobic and hydrophilic monomers, polyanhydride nanoparticles exhibit bacteria-like properties. By virtue of their hydrolytic degradation and surface erosion, they typically demonstrate a short burst followed by slower, sustained release of their encapsulated payload, which can be tuned by altering the composition of the polymer. Previous research has shown that polyanhydride nanoparticles can encapsulate and stabilize hydrophilic and hydrophobic payloads, enabling a wider selection of therapeutics for delivery. Given that amoebae acquire nutrients through similar phagocytic mechanisms, we hypothesize that polyanhydride nanoparticles could achieve high local concentrations of antimicrobial drugs at intracellular compartments within amoebae using a Trojan horse mechanism.

We report that rifampicin encapsulated in 20:80 1,8-bis(p-carboxyphenoxy)-3,6-dioxaoctane: 1,6-bis(p-carboxyphenoxy)hexane (CPTEG:CPH) nanoparticles shows a burst release over one day followed by a slow, near-zero-order release in aqueous *in vitro* experiments. We show that rifampicin-encapsulating nanoparticles significantly reduce the growth of two pathogenic *Acanthamoeba* strains at 72 hours. These results show that 20:80 CPTEG:CPH polyanhydride nanoparticles improve rifampicin efficacy in *Acanthamoeba*, which we hypothesize is a function of their controlled, local release to the amoebae. By improving the activity of extant antimicrobial therapies, we seek to improve the tool set available to combat amoebae infections.

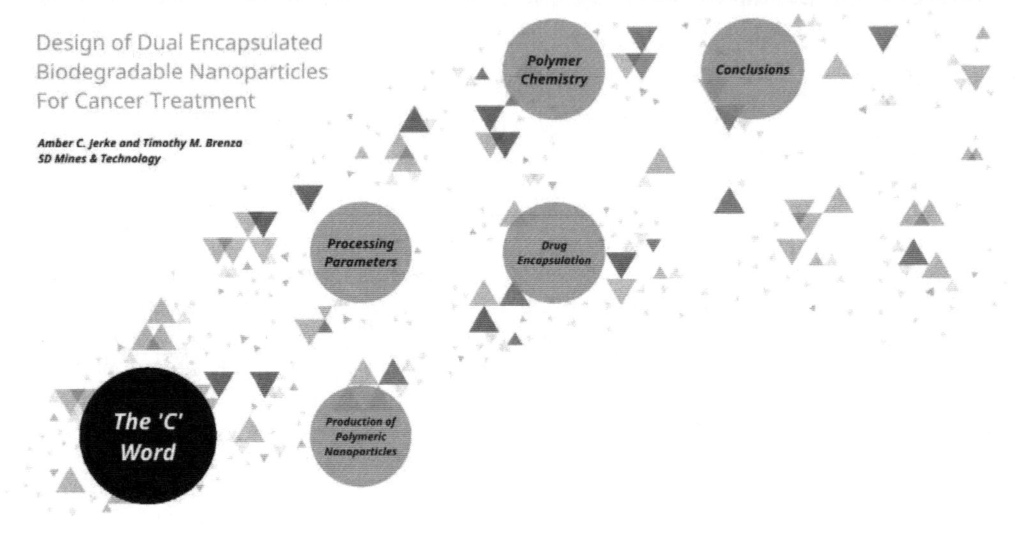
Design of Dual Encapsulated Biodegradable Nanoparticles For Cancer Treatment
Amber C. Jerke and Timothy M. Brenza
SD Mines & Technology
Polymer Chemistry
Conclusions
Processing Parameters
Drug Encapsulation
The 'C' Word
Production of Polymeric Nanoparticles

1.

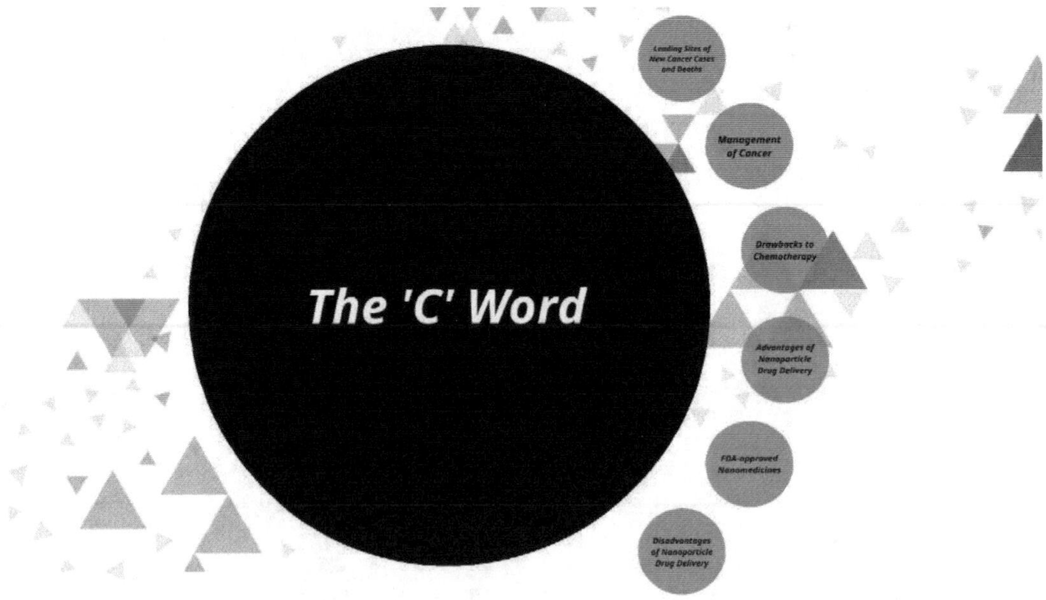
The 'C' Word
Leading Sites of New Cancer Cases and Deaths
Management of Cancer
Drawbacks to Chemotherapy
Advantages of Nanoparticle Drug Delivery
FDA-approved Nanomedicines
Disadvantages of Nanoparticle Drug Delivery

2.

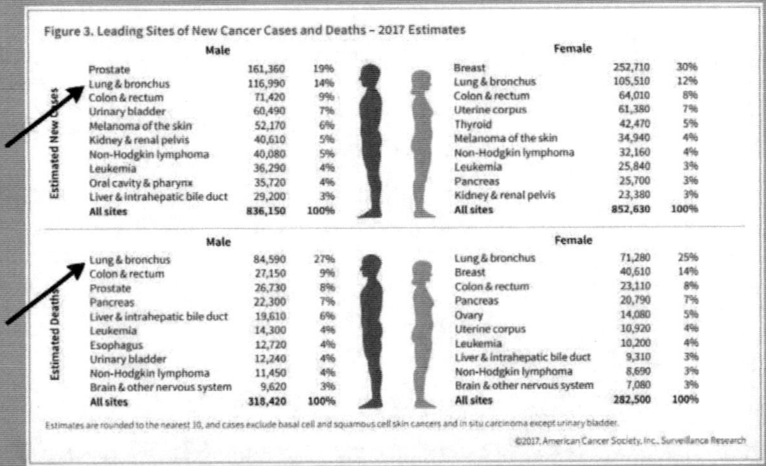

Figure 3. Leading Sites of New Cancer Cases and Deaths – 2017 Estimates

Estimated New Cases

Male			Female		
Prostate	161,360	19%	Breast	252,710	30%
Lung & bronchus	116,990	14%	Lung & bronchus	105,510	12%
Colon & rectum	71,420	9%	Colon & rectum	64,010	8%
Urinary bladder	60,490	7%	Uterine corpus	61,380	7%
Melanoma of the skin	52,170	6%	Thyroid	42,470	5%
Kidney & renal pelvis	40,610	5%	Melanoma of the skin	34,940	4%
Non-Hodgkin lymphoma	40,080	5%	Non-Hodgkin lymphoma	32,160	4%
Leukemia	36,290	4%	Leukemia	25,840	3%
Oral cavity & pharynx	35,720	4%	Pancreas	25,700	3%
Liver & intrahepatic bile duct	29,200	3%	Kidney & renal pelvis	23,380	3%
All sites	**836,150**	**100%**	**All sites**	**852,630**	**100%**

Estimated Deaths

Male			Female		
Lung & bronchus	84,590	27%	Lung & bronchus	71,280	25%
Colon & rectum	27,150	9%	Breast	40,610	14%
Prostate	26,730	8%	Colon & rectum	23,110	8%
Pancreas	22,300	7%	Pancreas	20,790	7%
Liver & intrahepatic bile duct	19,610	6%	Ovary	14,080	5%
Leukemia	14,300	4%	Uterine corpus	10,920	4%
Esophagus	12,720	4%	Leukemia	10,200	4%
Urinary bladder	12,240	4%	Liver & intrahepatic bile duct	9,310	3%
Non-Hodgkin lymphoma	11,450	4%	Non-Hodgkin lymphoma	8,690	3%
Brain & other nervous system	9,620	3%	Brain & other nervous system	7,080	3%
All sites	**318,420**	**100%**	**All sites**	**282,500**	**100%**

Estimates are rounded to the nearest 10, and cases exclude basal cell and squamous cell skin cancers and in situ carcinoma except urinary bladder.

Figure 1: Leading Sites of New Cancer Cases and Deaths - 2017 Estimates (from American Cancer Society. *Cancer Facts & Figures 2017*. Atlanta: American Cancer Society; 2017.)

Overall five-year survival rate of 19.5 percent

3.

Management of Cancer

- Treatment includes:
  - Surgery
  - Radiation therapy
  - Chemotherapy
  - ...and more!
- No impact
  - Relative survival +7.3 percent since 1975

4.

Drawbacks to Chemotherapy

- Poor aqueous solubility
- Require large "cocktail" drug dose
- Require administration by a trained professional

5.

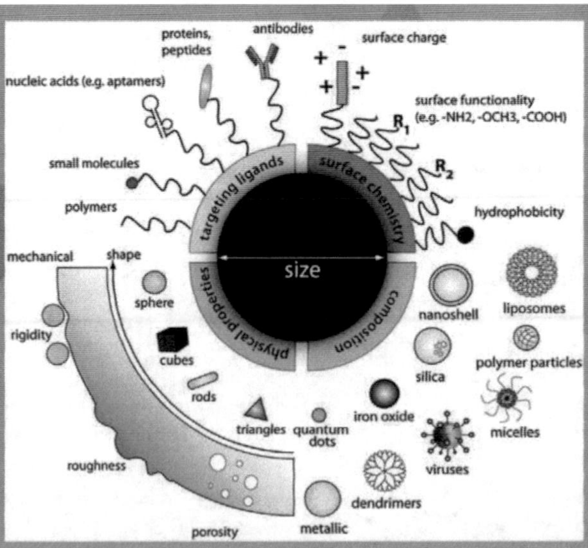

Figure 2: Flexible design allows for nanoparticles to be tailored to specific clinical applications (from Chou, L. Y., et al. (2011). "Strategies for the intracellular delivery of nanoparticles." Chem Soc Rev 40(1): 233-245.)

6.

FDA-approved Nanomedicines

- Over 50 nanomedicines approved for various clinical applications
- Approximately 8 approved for the therapeutic treatment of cancer
- Only Abraxane® (albumin-bound paclitaxel) is approved for use in non-small cell lung cancer (NSCLC)
- Must be consistent in biological behavior and pharmacological profile

Bobo, D., et al. (2016). "Nanoparticle-Based Medicines: A Review of FDA-Approved Materials and Clinical Trials to Date." Pharm Res 33(10): 2373-2387.

7.

Disadvantages of Nanoparticle Drug Delivery

- Produced in small, non-homogenous batches
- Difficulty in experimental reproducibility
- Difficulty in scale up for commercial production

8.

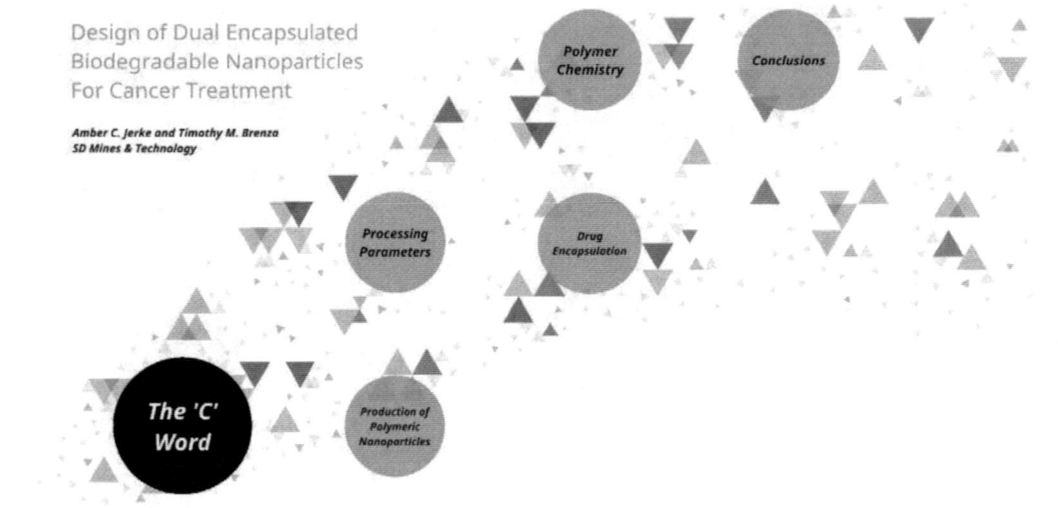

Design of Dual Encapsulated
Biodegradable Nanoparticles
For Cancer Treatment
Amber C. Jerke and Timothy M. Brenza
SD Mines & Technology
Polymer Chemistry
Conclusions
Processing Parameters
Drug Encapsulation
The 'C' Word
Production of Polymeric Nanoparticles

9.

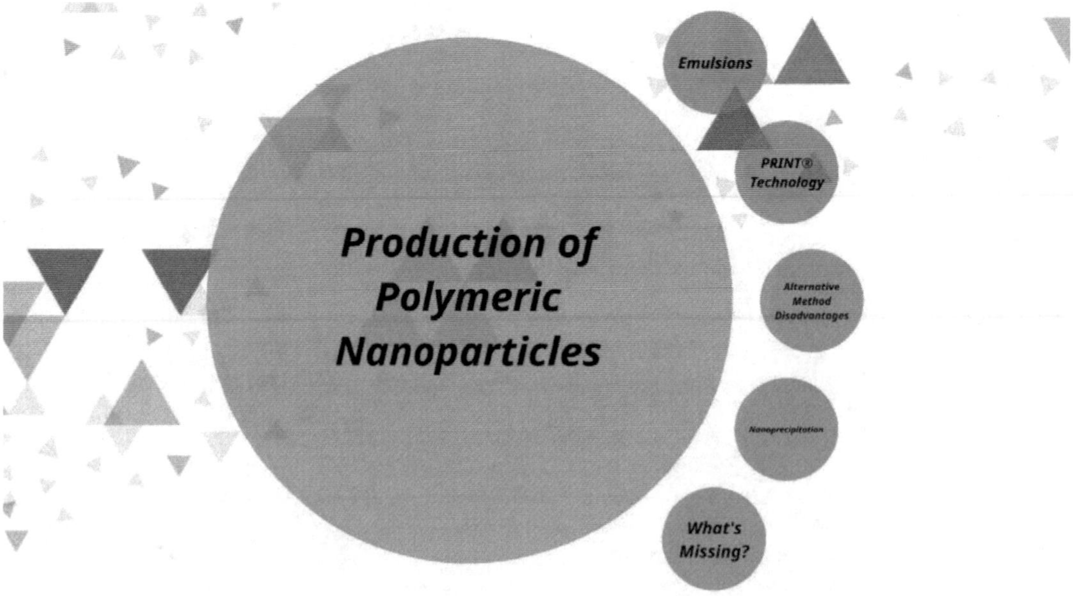

Production of Polymeric Nanoparticles
Emulsions
PRINT® Technology
Alternative Method Disadvantages
Nanoprecipitation
What's Missing?

10.

Emulsions

- Español et al. prepared encapsulated poly(lactic-*co*-glycolic acid) (PLGA) nanoparticles with a size under 150 nm
- Employed factorial experimental design to study effects of drug to polymer ratio, Pluronic® concentration, and sonication time
- Reached encapsulation efficiency of 67.7 percent for dexamethasone and 54.2 percent for diclofenac sodium

Español, L., et al. (2016). "Dual encapsulation of hydrophobic and hydrophilic drugs in PLGA nanoparticles by a single-step method: drug delivery and cytotoxicity assays." RSC Adv. 6(112): 111060-111069.

11.

PRINT® Technology

- Particle Replication in Non-wetting Templates
- Bowerman et al. prepared rod-shaped docetaxel encapsulated PLGA nanoparticles with a size of 215.2 nm and encapsulation effiency of 39.1 percent

Bowerman, C. J., et al. (2017). "Docetaxel-Loaded PLGA Nanoparticles Improve Efficacy in Taxane-Resistant Triple-Negative Breast Cancer." Nano Lett 17(1): 242-248.

12.

Alternative Method Disadvantages

- Successful in deliverying nanoparticles' payload to respective target
- Success in marketplace depends strongly on encapsulation efficiency and ease of scale up
- Emulsions require surfactants
  - Adds complexity, requires additional characterization

13.

Nanoprecipitation

- Can be operator independent, scalable, and used to encapsulate hydrophobic drugs
- Fonseca et al. studied effect of preparation method, ratio of organic to aqueous phase, and PLGA molecular weight on encapsulation efficiency
- Rietscher et al. showed semi-automated system affected by flow rate, injection rate, and PLGA concentration

Fonseca, C., S. Simoes, and R. Gaspar, Paclitaxel-loaded PLGA nanoparticles: preparation, physicochemical characterization and in vitro anti-tumoral activity. J Control Release, 2002. 83(2): p. 273-286.

Rietscher, R., et al., Semi-automated nanoprecipitation-system--an option for operator independent, scalable and size adjustable nanoparticle synthesis. Pharm Res, 2015. 32(6): p. 1859-63.

14.

Nanoprecipitation

- Johnson and Prud'homme used confined impinging jets (CIJ) mixer to prepare B-carotene and polystyrene-*block*-poly(ethylene oxide) composite nanoparticles
- Zhu also used CIJ mixer to prepare paclitaxel-loaded PLGA-*block*-PEG nanoparticles

Johnson, B.K. and R.K. Prud'homme, Flash NanoPrecipitation of Organic Actives and Block Copolymers using a Confined Impinging Jets Mixer. Australian Journal of Chemistry, 2003. 56(10): p. 1021.

Zhu, Z., Flash nanoprecipitation: prediction and enhancement of particle stability via drug structure. Mol Pharm, 2014. 11(3): p. 776-86.

15.

What's Missing?

- Little work has been done on the molecular level
  - Polymer chemistry
  - Drug class
- Process and molecular variables have not been combined to predict desired outcomes

16.

Design of Experiments

- Study processing parameters using polystyrene
  - Syringe flow rate
  - Solvent to anti-solvent ratio
  - Stir rate
- Study polymer-drug interactions using several dyes with varying degrees of hydrophobicity (e.g. DiD' oil, fluorescein, rhodamine B)
  - Drug class
  - Addition methods
- Study polymer chemistry using poly(lactic-*co*-glycolic acid) (PLGA)
  - Ratio of lactic acid to glycolic acid
  - Molecular weight

17.

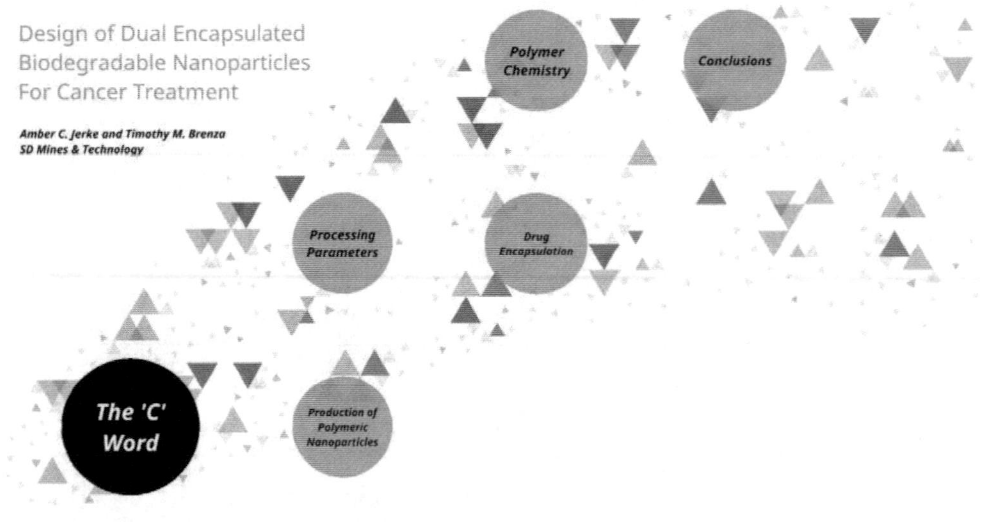

18.

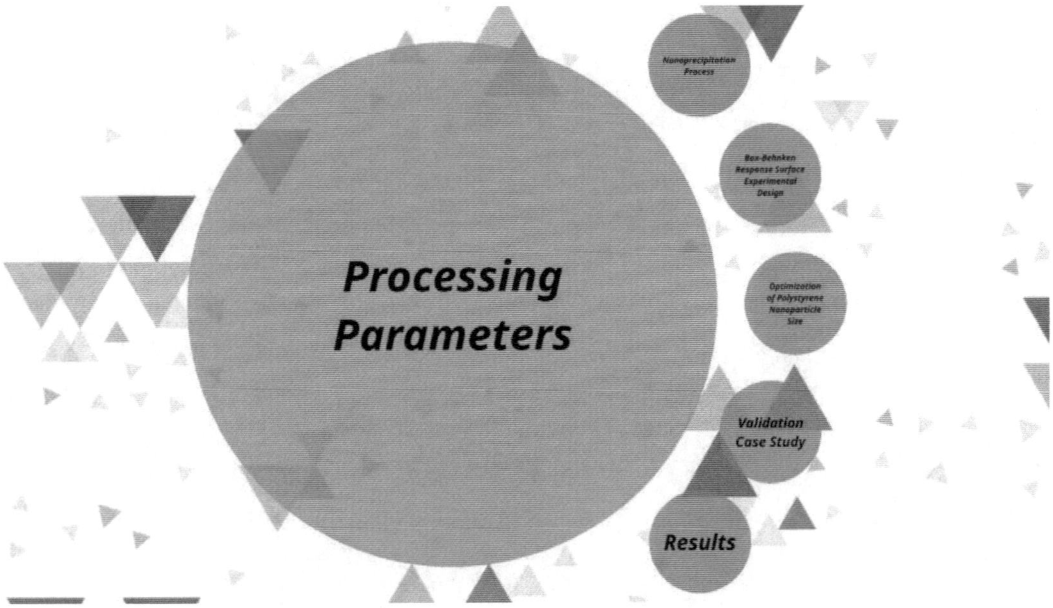
Processing Parameters
Nanoprecipitation Process
Box-Behnken Response Surface Experimental Design
Optimization of Polystyrene Nanoparticle Size
Validation Case Study
Results

19.

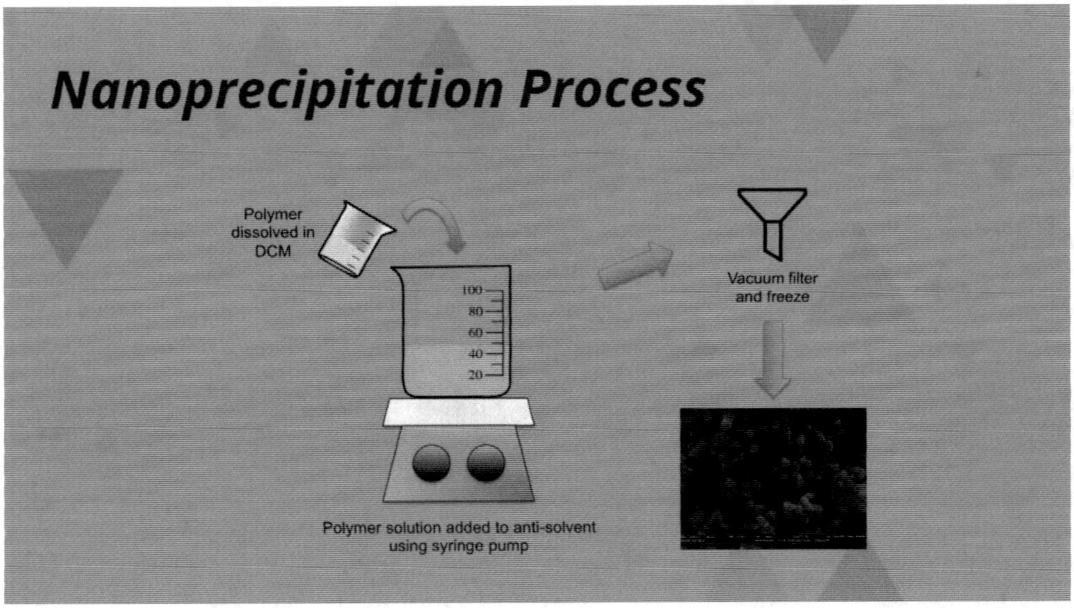
Nanoprecipitation Process
Polymer dissolved in DCM
100
80
60
40
20
Polymer solution added to anti-solvent using syringe pump
Vacuum filter and freeze

20.

Box-Behnken Response Surface Experimental Design

Table 1. Experimental Design

Run Order	Factor 1	Factor 2	Factor 3	Response 1	Response 2	Response 3
	A: Syringe Flow Rate	B: Anti-solvent/ Solvent Ratio	C: Stir Rate	Average Diameter*	Polydispersity*	Average Zeta Potential
	ml/min		rpm	nm		mV
1	0.1	120	60	256.66	0.09	-27.8
2	13.0	120	605	206.06	0.08	-28.5
3	26.0	120	60	146.10	0.07	-21.8
4	0.1	40	605	191.47	0.09	-24.7
5	0.1	120	1150	187.79	0.09	-24.8
6	26.0	120	1150	125.75	0.06	-26.0
7	26.0	200	605	116.46	0.09	-21.7
8	13.0	120	605	189.35	0.11	-25.8
9	13.0	40	1150	198.83	0.10	-25.8
10	13.0	120	605	182.60	0.11	-27.3
11	13.0	200	60	213.18	0.07	-28.2
12	13.0	200	1150	186.37	0.08	-26.3
13	0.1	200	605	193.55	0.13	-28.0
14	13.0	40	60	351.49	0.39	-38.8
15	26.0	40	605	283.56	0.08	-28.6

** Size and polydispersity determined from analysis of SEM pictographs using Image J software*

21.

Optimization of Polystyrene Nanoparticle Size

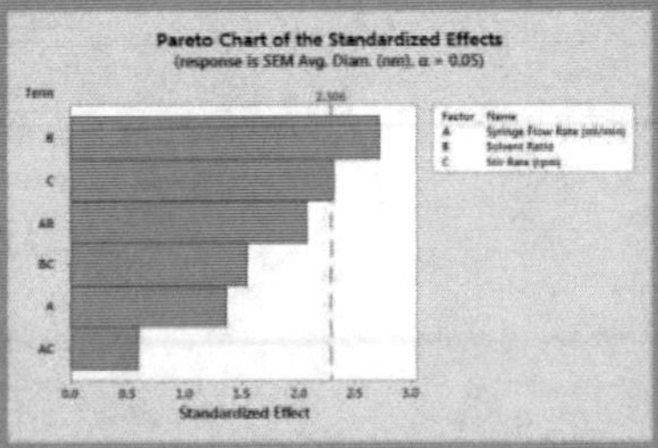

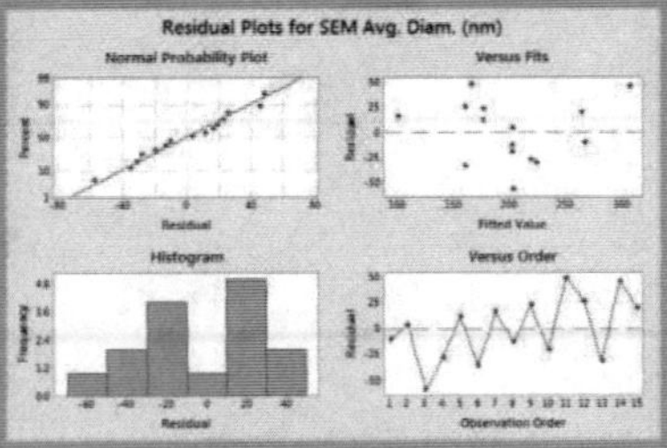

Predictive Equation:

Diameter = 319.8 + 2.35**A** - 0.397**B** - 0.1702**C** - 0.0408**AB** + 0.00169**AC** + 0.000722**BC**

Where:

A = syringe flow rate (ml/min)
B = solvent/anti-solvent ratio
C = stir rate (rpm)

22.

Validation Case Study

Desired Size	Syringe Flow Rate (ml/min)	Solvent Ratio	Stir Rate (rpm)
Small (100.3 nm)	26.0	193	60
Medium (200.0 nm)	26.0	100	650
Large (301.2 nm)	26.0	40	220

Results

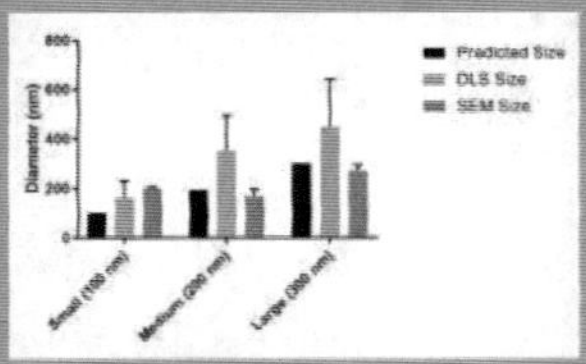

Figure 3: Presented is the mean particle size of n=3 batches with error bars representing the standard deviation.

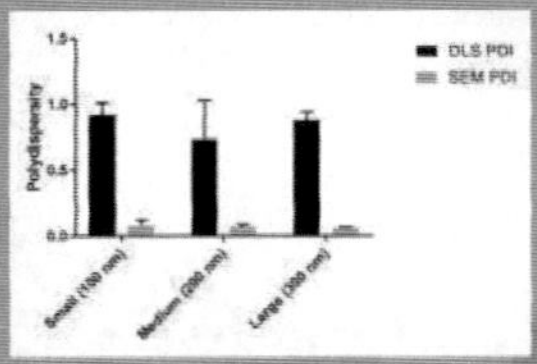

Figure 4: Presented is the mean polydispersity of n=3 batches.

Figure 5: SEM images of a) small, b) medium, and c) large polystyrene nanoparticles.

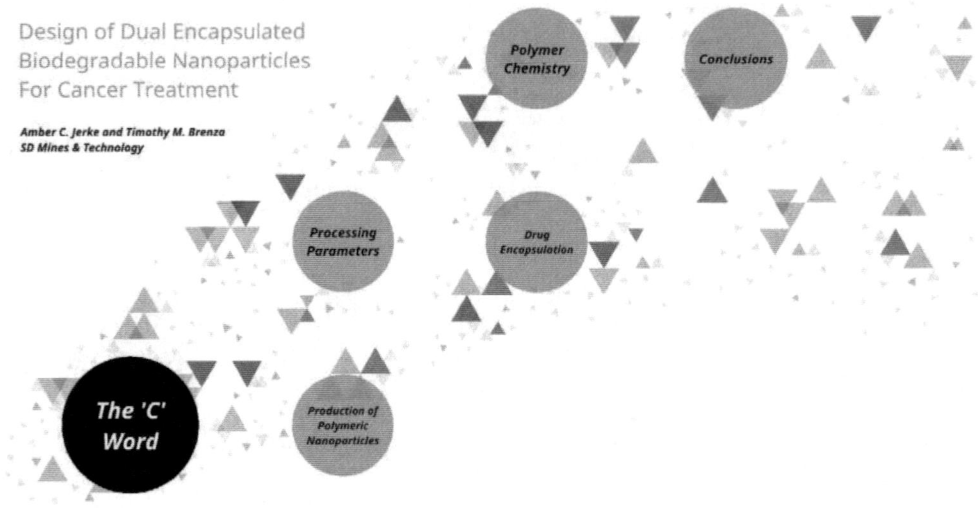
Design of Dual Encapsulated
Biodegradable Nanoparticles
For Cancer Treatment
Amber C. Jerke and Timothy M. Brenza
SD Mines & Technology
Polymer Chemistry
Conclusions
Processing Parameters
Drug Encapsulation
The 'C' Word
Production of Polymeric Nanoparticles

25.

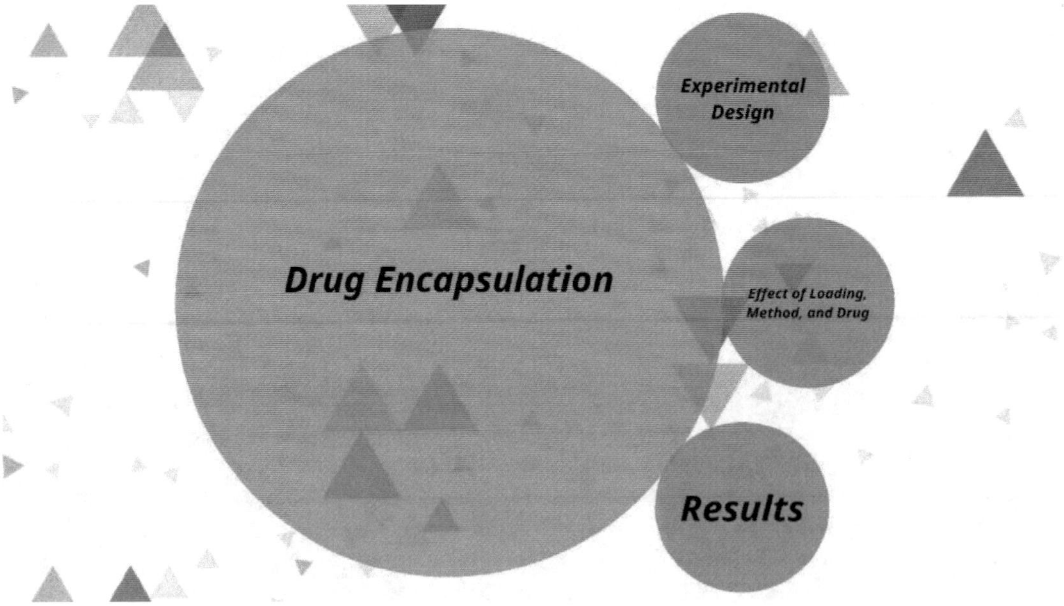
Drug Encapsulation
Experimental Design
Effect of Loading, Method, and Drug
Results

26.

Experimental Design

Std Order	Run Order	Drug Loading (%)	Loading Method	Drug
10	1	1%	B	DiD' Oil
12	2	1%	B	Rhodamine B
11	3	1%	B	FITC
2	4	10%	A	FITC
7	5	1%	A	DiD' Oil
3	6	10%	A	Rhodamine B
1	7	10%	A	DiD' Oil
9	8	1%	A	Rhodamine B
5	9	10%	B	FITC
8	10	1%	A	FITC
6	11	10%	B	Rhodamine B
4	12	10%	B	DiD' Oil

Effect of Loading, Method, and Drug

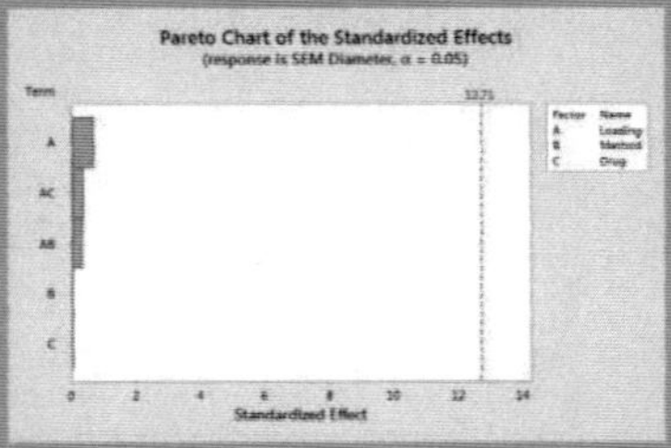

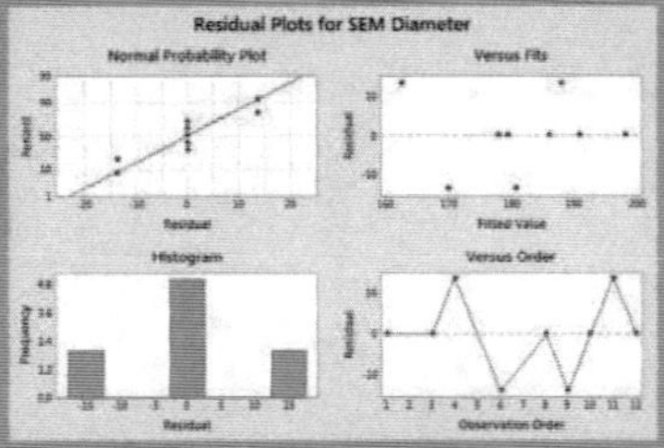

- Drug loading percent, loading method, and drug type did not significantly affect average particle diameter

Results

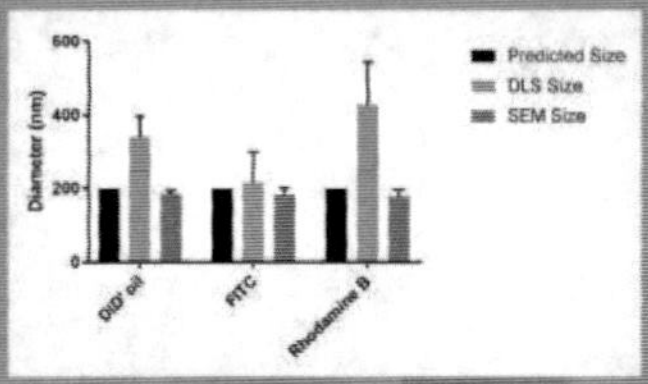

Figure 6: Presented is the mean particle size of n=2-4 batches with error bars representing the standard deviation.

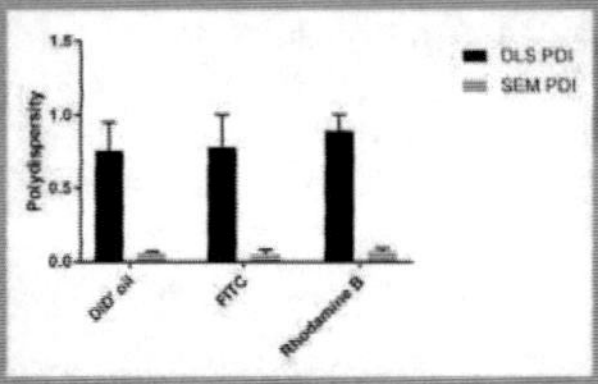

Figure 7: Presented is the mean polydispersity of n=2-4 batches.

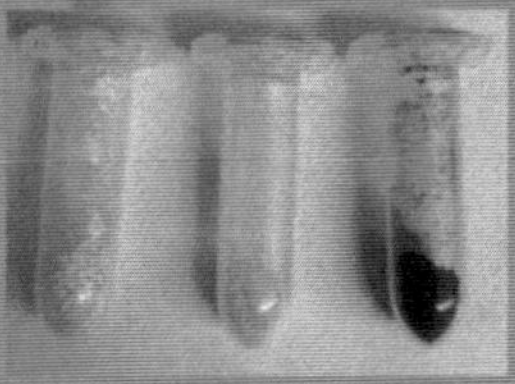

Figure 8: Polystyrene nanoparticles encapsulating DiD' oil (left), FITC (middle), and Rhodamine B (right)

29.

Design of Dual Encapsulated Biodegradable Nanoparticles For Cancer Treatment

Amber C. Jerke and Timothy M. Brenza
SD Mines & Technology

Polymer Chemistry

Conclusions

Processing Parameters

Drug Encapsulation

The 'C' Word

Production of Polymeric Nanoparticles

30.

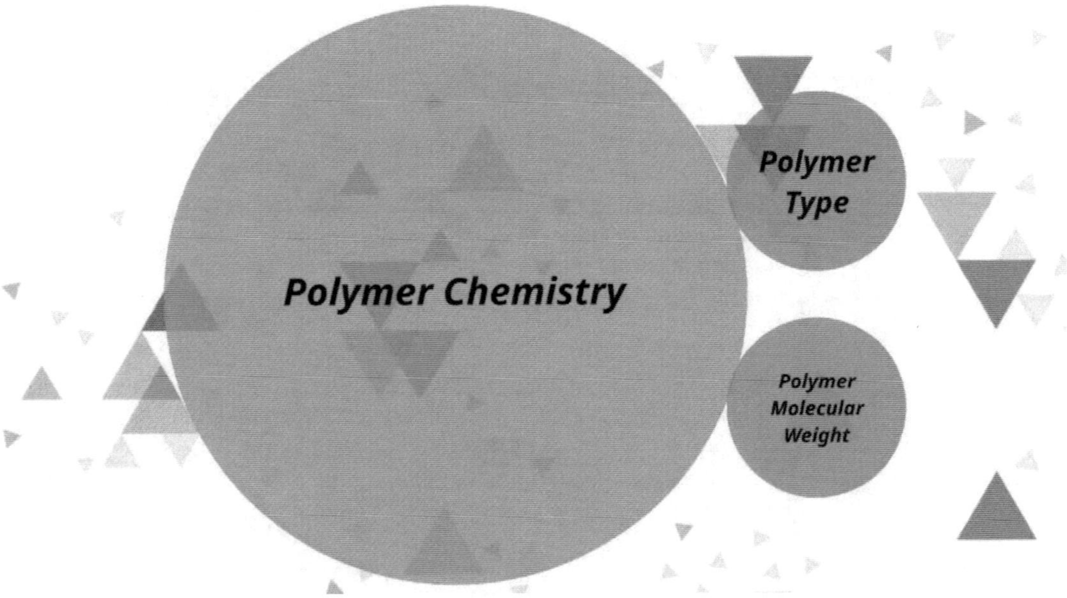

Polymer Type

Figure 9: RESOMER RG 502H (PLGA 50/50 with molecular weight 7,000-17,000)

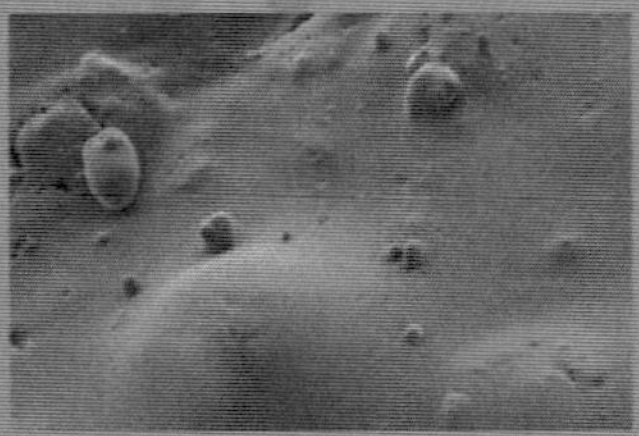

Figure 10: RESOMER RG 752H (PLGA 75/25 with molecular weight 4,000-15,000)

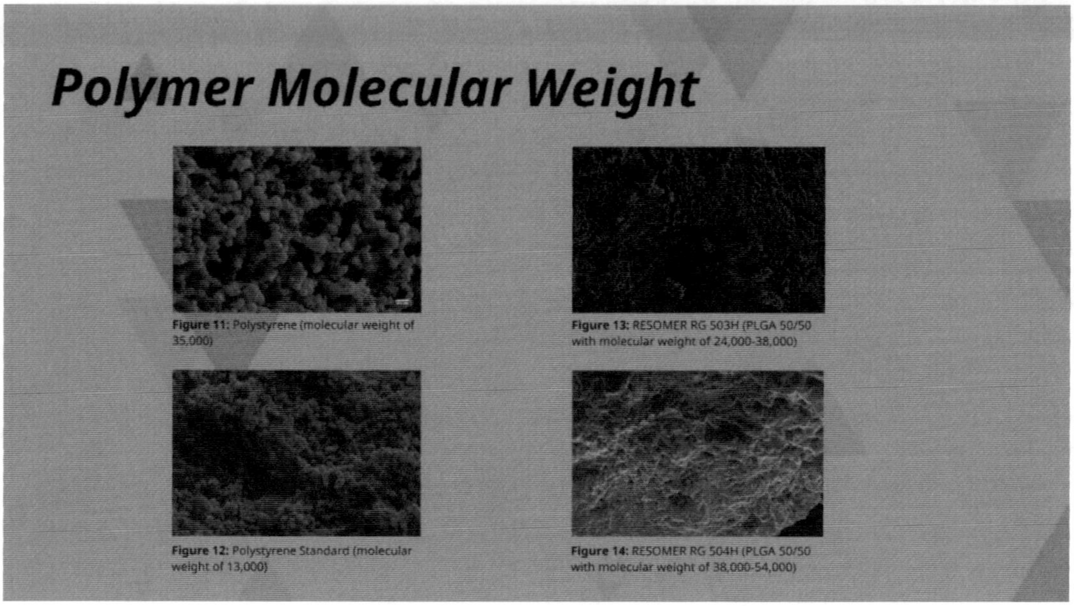
Polymer Molecular Weight
Figure 11: Polystyrene (molecular weight of 35,000)
Figure 13: RESOMER RG 503H (PLGA 50/50 with molecular weight of 24,000-38,000)
Figure 12: Polystyrene Standard (molecular weight of 13,000)
Figure 14: RESOMER RG 504H (PLGA 50/50 with molecular weight of 38,000-54,000)

33.

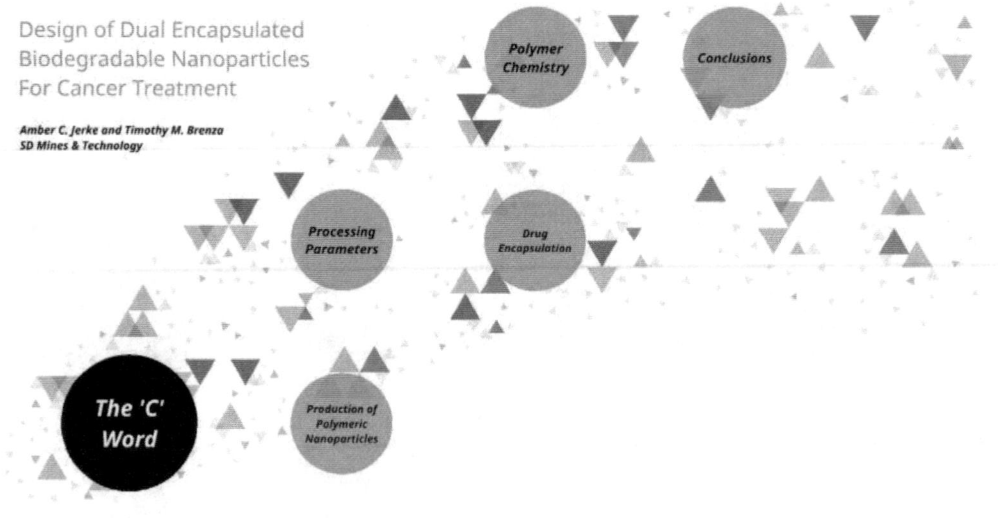
Design of Dual Encapsulated Biodegradable Nanoparticles For Cancer Treatment
Amber C. Jerke and Timothy M. Brenza
SD Mines & Technology
Polymer Chemistry
Conclusions
Processing Parameters
Drug Encapsulation
The 'C' Word
Production of Polymeric Nanoparticles

34.

Conclusions

- Developed a predictive model based on operating parameters for nanoparticle synthesis by controlled precipitation
- Both 200 and 300 nm particles were synthesized with low batch to batch variability in particle size and distribution
- Drug loading percent, loading method, and drug type do not have a significant effect on particle size
- Polymer type and molecular weight has an effect on particle formation that has yet to be determined

35.

Future Work

- Molecular modeling to understand interaction between solvent and polymer
- Optimize manufacturing for enhanced therapeutic encapsulation and nanoparticle yield
- Dual encapsulation of chemotherapeutic (paclitaxel) and antiangiogenesis agent (fumagillin)
- Investigate efficacy of chemotherapeutic loaded biodegradable nanoparticles for the treatment of lung cancer

36.

Acknowledgements

- Advisor and Lab Members
  - Dr. Timothy M. Brenza
  - Eswar Kalaga
  - Jordan Hoops
  - Brianna Jaward
- Dr. Edward Duke and EMES SEM Facility
- Funding: Start-up funding from SDSM&T Chemical & Biological Engineering Department

37.

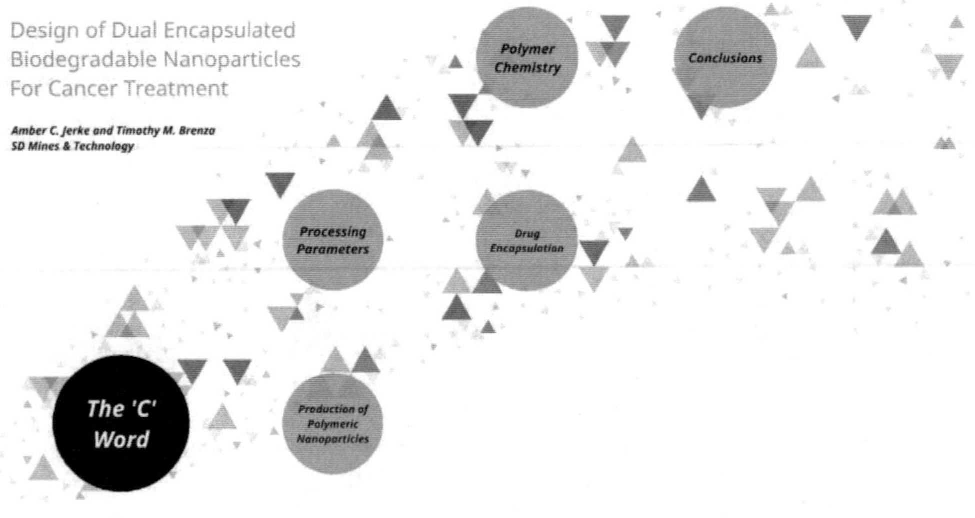

38.

489737 Understanding How Lipid Nanoparticle Structure Affects Immune Response

Abstract:

Lipid-based nanoparticles have been widely demonstrated as efficacious drug delivery systems, particularly for applications in nucleic acid delivery. Despite their extensive development, however, there is limited understanding of the effect that these delivery vehicles have on the immune system. In fact, a major reason why therapeutics fail in clinical trials is due to unforeseen toxicity and/or immunogenicity. In particular, the immune system can potentially mount an antibody response to the delivery system, trigger the release of liver enzymes, or prompt an inflammatory cytokine response. Here, we describe a study of the interactions between lipidoid nanoparticles, a potent class of lipid-like RNA delivery vehicles, and the components of the innate and humoral immune system. A small combinatorial lipidoid library of 20 lipidoids was synthesized and tested for immunostimulatory effects in cell culture and animal models. In vitro analysis of RAW264.7 mouse macrophages after nanoparticle exposure indicates statistically significant rises in the gene expression of several cytokines, including tumor necrosis factor Î± (TNFÎ±), for select lipid nanoparticles. Increased cytokine expression persisted for over 24 hours and depended largely on the tail structure of the lipidoid. Furthermore, the effect of lipid nanoparticles on peripheral blood mononuclear cells (PBMCs) isolated from healthy human donors was assessed via ELISA screens of TNFÎ±, IL-1Î², IFN-Î³, IL-6, and IL-2. Antibody and cytokine responses were also measured following single and/or multiple intravenous injection(s) of lipid nanoparticles into C57BL/6 mice. In all cases, both the tail structure and amine structure of the lipid nanoparticle were found to be important in provoking an immune response. These structure-immune response relationships will inform the design of future drug delivery carriers and potentially mitigate failures encountered in clinical stages.

AUTHOR INDEX

Acar, Handan 236, 897
Adamovicz, Jeffrey 530
Adams, Justin 460
Adedoyin, Adedokun 351
Adewole, Jimoh K. 838
Adil, Maroof M. 80
Afeworki, Mobae 650
Agarwal, Sushant 366
Agrawal, Kumar Varoon 653
Agrawal, Pradeep K. 156
Agrawal, Rakesh 571, 905
Ahrens, Caroline 395, 677
Ailianou, Artemis 316
Aksan, Alptekin 676
Al Katheeri, Abdul Majed 183
Alabi, Christopher A. 355, 403, 528, 764
Alamgir, Faisal M. 156
Albert, Julie 576
Alexandrova, Anastassia 786
Alfonso, Dominic 4
Alhassan, Saeed 101, 106
Aliakbarighavimi, Soheila 767
Allen, Brittany 12, 767
Allison, Evan 301, 305
Almas, Qandeel 154
Almodovar, Jorge 322, 633
Alpert, Carol Lynn 278
Alsewailem, Fares 205
Altalhi, Abdulmajeed 228
Al-Thabaiti, Shaeel 653
Alvarez, Nicolas J. 797
An, Qi 87
An, Yaxin 525, 729
Anderson, Daniel G. 78, 408
Andreev, Marat 402
Andresen, Corinne 361
Anilionyte, Oksana 636
Anisimov, Mikhail A. 659
Annamalai, Prakasam 350, 630
Anseth, Kristi S. 307, 318
Anthamatten, Mitchell 782
Aoki, Reyn 679
Aqad, Emad 164
Arbelo, Noelia Almodovar 420
Archer, Lynden A. 342
Ardila-Suarez, Carolina 723
Armstrong, Matthew 18
Arora, Akash 356
Arturo, Steven G. 52
Arvidson, Sara A. 608
Asano, Yusuke 194
Asatekin, Ayse 370
Askar, Shadid 425
Aslam, Umar 61
Assary, Rajeev 95
Asteasuain, Mariano 198
Atsumi, Hiroshi 636
Azarin, Samira M. 538, 676
Bae, Tae-Hyun 69
Bafana, Adarsh 271
Bai, Baojun 216
Bai, Lian 285, 286, 618
Bailey, Travis S. 708
Bain, Erich 313, 714
Balasanthiran, Choumini 59
Balbuena, Perla B. 723
Ball, Rebecca 526
Balwani, Apoorv 691
Bao, Lei 196
Bao, Wei 787
Bao, Zhenan 39, 304, 358
Bara, Jason E. 346
Baranauskas, Vince 648
Baranek, Austin 715, 761
Barberio, Antonio E. 240
Barboun, Patrick 160
Bardhan, Rizia 42
Barnes, Tanner 627
Barnharst, Tanner 556
Barteau, Katherine P. 777
Bartel, Christopher J. 3
Bartholomew, Reid 635
Barua, Sutapa 760, 891
Basahel, Sulaiman N. 653
Baskaran, Durairaj 518
Bassett, Alexander W. 542, 802
Bates, Frank S. 341, 356, 711, 750, 796
Batra, Surinder 8
Bavarian, Mona 193
Baxter, Jason B. 58, 303, 744, 900
Bazan, Guillermo C. 893
Beckingham, Bryan S. 695
Beckman, David W. 436
Beers, Kathryn L. 76
Behazin, Ehsan 555
Beitelshees, Marie 463
Bejagam, Karteek K. 525, 729, 740
Belcher, Angela M. 636
Belfort, Georges 718
Bellaire, Bryan H. 437, 462
Beltran, Saul Perez 723
Beltran-Villegas, Daniel J. 368
Benamara, Mourad 417
Bender, Elizabeth C. 348
Beniah, Goliath 367
Benicewicz, Brian C. 713
Berg, Gayla 381
Berlinger, Maya 231
Berry, Carter 423, 729, 740
Berry, Joseph J. 561

AUTHOR INDEX

Berry, Keith ... 83
Berry, Vikas ... 416
Berthod, Mikael ... 183
Beyer, Frederick L. ... 313
Bhan, Aditya ... 155
Bharti, Bhuvnesh ... 912
Bhattacharjee, Ujjal ... 566
Bhattacharyya, Souryadeep ... 720, 770
Bhethanabotla, Venkat R. ... 277
Biagioli, Madeleine ... 344, 359
Bilchak, Connor ... 713
Binkhodor, Yazeed ... 205
Biria, Saeid ... 798
Bischof, John C. ... 538
Bishop, Brittany ... 679
Biswas, Pratim ... 560
Black III, Lauren D. ... 348
Blocher, Whitney C. ... 459
Blumeyer, Jack ... 378
Bodinger, Carter ... 83
Boissiere, Michel ... 639
Bollini, Praveen ... 155
Bonacina, Luigi ... 417
Bondos, Sarah ... 763
Boone, Evan ... 507
Borges, Fernando T P ... 202
Borguet, Eric ... 772
Bortner, Michael J. ... 909, 910
Bose, Arijit ... 279
Bose, Suman ... 78, 408
Botvinick, Elliot ... 317
Boudouris, Bryan W. ... 420
Bourdon, Buchanan ... 783
Bowen, Alec S. ... 511
Bowman, Christopher N. ... 50, 201, 325, 381, 391, 715, 761, 898
Boyd, Peter ... 769
Boyne, Robert W. ... 571
Bredas, Jean-Luc ... 901
Brenza, Timothy ... 255, 438
Brettmann, Blair Kathryn ... 778
Brinson, Catherine ... 818
Broadbelt, Linda J. ... 52
Broer, Dick J. ... 325
Brown, Jonathan R. ... 748
Brown, Joseph ... 355, 403
Brown, Lauren ... 507
Brown, Tobin E. ... 307
Brunelli, Nicholas ... 153, 726
Bryant-Friedrich, Amanda C. ... 401
Buenning, Eileen ... 713
Buenrostro, Denise ... 535
Bundy, Bradley C. ... 899
Burbach, Brandon ... 538
Burke, Kelly A. ... 637
Burkey, Aaron A. ... 656
Busch, Markus ... 200
Buss, Hilda G. ... 590
Byrley, Peter ... 60
Cabezas, Camila Saez ... 741
Cain, Laurance ... 571
Cairns, Johnnie ... 277
Calabro, David C. ... 650
Camargo, Lucas Garcia ... 629
Camci-Unal, Gulden ... 347
Cao, Han ... 658
Cao, Lixia ... 176
Cao, Nai ... 584
Cao, Zhiqiang ... 531
Capeling, Meghan ... 237
Carpenter, Ryan ... 323
Casas, Juan Pablo ... 757
Case, Natasha ... 626
Castaldi, Marco J. ... 746
Castilla, David ... 322
Castro, Camila ... 757
Castro, Carlos E. ... 62
Chai, Song-Hai ... 156
Chaikittisilp, Watcharop ... 5, 71
Chakraborty, Aishik ... 896
Champion, Julie A. ... 79, 464
Chan, Christina ... 9
Chan, Mary ... 894
Chang, Ji Woong ... 59
Chang, Liang ... 93
Chang, Li-Wei ... 400
Chang, Michael Yu Zarng ... 237
Chang, Yun Hsuan ... 724
Chan-Park, Mary B. ... 893
Chapman, Karena W. ... 773
Charles, Bayliss ... 189
Chatham, Camden ... 909
Chauvel, Paul ... 174
Chawla, Aseem ... 103
Chen, Chau-Chyun ... 272, 697
Chen, Gang ... 570
Chen, Gong ... 38
Chen, Jee-Wei Emily ... 393, 861
Chen, Jerry J. Y. ... 533
Chen, Qile ... 675
Chen, Ru ... 793
Chen, Sinn-Wen ... 280
Chen, Tao ... 177
Chen, Tzu-Hsuan ... 680
Chen, Tzu-Ling ... 233
Chen, Xi ... 49, 387
Chen, Yahua ... 893
Chen, Yang-Yuan ... 280
Chen, Zhengtao ... 779
Chen, Zhifeng ... 59

AUTHOR INDEX

Cheng, Chi-Hui ... 257
Cheng, Chin-Yi ... 636
Cheng, Chong ... 237, 326
Cheng, Elise ... 361
Cheng, Guanzhi ... 812
Cheng, Kenneth ... 392, 892
Cheng, Wei-I ... 237
Cheng, Xiang ... 285, 286, 618
Cheng, Yongqiang ... 157
Chesniak, Olivia ... 9
Chiarot, Paul R. ... 270
Chih, Meng-Hsiu ... 680
Chinzei, Nobuaki ... 626
Chiu, Yun ... 257
Cho, Joon Hee ... 834
Cho, Junghyun ... 270
Chodankar, Nilesh R. ... 269
Choi, Heechul ... 268
Choi, Jae-Soon ... 158
Choi, Joshua ... 565
Choi, Sunho ... 67
Chorpening, Benajmin T. ... 98
Chou, Daniel Y. ... 835
Choudhury, Pabitra ... 281
Choudhury, Snehashis ... 342
Christau, Stephanie ... 435
Christians, Jeffrey A. ... 561
Chu, Liang-Yin ... 315
Chu, Ping-Hsun ... 162
Chu, Weiwei ... 519, 749
Chuang, Hui-Min ... 658
Chung, Wook-Jin ... 226, 289
Churchill, Stuart W. ... 170
Chwatko, Malgorzata ... 48
Cima, Michael ... 411
Clancy, Paulette ... 902
Claure, Micaela Taborga ... 156
Cleveland, Iver J. ... 563
Cobos, Monica ... 610
Cochran, Eric W. ... 383, 836
Cockcroft, Jeremy ... 180
Coffel, Joel ... 730
Coffey, Aidan ... 265, 817
Cogswell, Christopher ... 67
Colby, Ralph H. ... 340, 833
Coleman, Maria ... 354
Collier, Graham ... 507
Colon, Yamil J. ... 511, 519, 749
Commisso, Alex ... 237, 326
Connell, Ryan ... 906
Cook, Jonathan ... 645
Coppens, Marc-Olivier ... 180, 382
Coropceanu, Veaceslav ... 901
Correa, Santiago ... 240
Corti, David S. ... 420
Cosgrove, Jayson D. ... 802
Cottrill, Anton ... 291
Cove, Matthew ... 643
Crane, Matthew ... 379
Crum, Jerry ... 252
Cui, Honggang ... 396, 532
Cummings, Chad ... 526
Curia, Silvio ... 542
Curtis, Chad D. ... 239
Dadelahi, Alexis ... 530
Daemen, Luke L. ... 157
Dahoumane, Si A. ... 271
Dai, Gance ... 89
Dai, Sheng ... 156
Dai, Xiaofeng ... 374
Dang, Binh ... 679
D'Angelo, Anthony ... 96
Dardona, Sameh ... 913
Darling, Ross ... 462
Darugar, Qusai ... 209
Dasgupta, Dwaipayan ... 377
Dastidar, Subham ... 58, 303, 562
Davidson, Bruce A. ... 463
Davis, E. James ... 379
Davis, Eric M. ... 592, 691
De La Torre, Roger ... 530
De Laporte, Laura ... 321
De Pablo, Juan J. ... 43, 402, 511, 519, 728, 749
De Rosa, Claudio ... 195
De Silva, Udaka K. ... 401
De Souza, Fernanda C. Bombaldi ... 241, 253
De Souza, Pedro ... 593
De Souza, Renata F. Bombaldi ... 241, 253
De Villers, Bertrand Tremolet ... 561
Deegan, Michael ... 18
Deforest, Cole A. ... 406
Dehankar, Abhilasha ... 62, 278
Demello, Andrew J. ... 564
Dement, Dana ... 745
Deml, Ann M. ... 3
Deneff, Jacob I. ... 475
Deng, Shikai ... 416
Deng, Yulin ... 151
Dennis, Grayson P. ... 346
Desai, Tejal ... 166
Deshmukh, Sanket A. ... 423, 525, 729, 740
Deshmukh, Swapnil Dattatray ... 905
Deshpande, Nitish ... 153
Desimone, Joseph M ... 74
Devore, David ... 254
Dhar, Prajnaparamita ... 896
Dhara, Palash ... 642
Dhinojwala, Ali ... 647
Diannan, Lu ... 15
Diao, Ying ... 509

AUTHOR INDEX

Dichtel, William 367
Dickey, Michael D. 644
Diederich, Vincent 312
Diederichsen, Kyle M. 590
Dillon, Andrew D. 58
Dinari, Mohammad 229
Ding, Fuchuan 187
Ding, Zhenya 395, 677
Dinic, Jelena 344, 359, 508
Dittmer, Neal T. 896
Dixit, Marm 504
Dobyns, Breanna 695
Dodam, John 530
Dolog, Rostyslav 209
Dominguez, Kevin 746
Dong, Hongfeng 273
Dong, Juyao 606
Dong, Pingchuan 584
Dong, Ziye 395, 677
Donta, Maxsam 319
Dorfman, Kevin D. 356, 658, 750
Doufas, Antonios 339
Doughty, Benjamin 157
Douglas, Jack F. 402
Dreaden, Erik 240
Drennan, Corinne 158
Du, Lin 300, 375, 415, 418
Du, Yuyang 781
Duan, Yuhua 98
Dubensky, Thomas 462
Dumont, Joseph H. 914
Dunklin, Jeremy 83
Dunn, Celia 627
Durbin, Marlow M. 346
Durning, Christopher J. 713
Dziubla, Thomas 14, 314, 476
Eastman, Stephen 66
Echeverri, Juanita 757
Edgar, James H. 64
Edwards, Brian J. 85, 284, 298, 345
Eichhorn, Stephen 631
Eickman, Erin 862
Ekenseair, Adam 351
Elabd, Yossef A. 233, 591
Elbert, Johannes 474
Ellebracht, Nathan 68
Ellison, Christopher J. 194, 227, 288, 388, 593, 607, 834, 835
Elmore, Bill B. 204
El-Sayed, Mostafa 191
Emi, Tania 627
Endo, Fuyuaki 195, 522
Engstrom, James R. 431
Eniola-Adefeso, Omolola 534
Epps, Thomas H. 324
Escobar, Francisco 318
Escobedo, Fernando 521, 751
Esnaashari, Catherine 286, 618
Estroff, Lara A. 777
Ewart, Sean 51
Facchetti, Antonio 505
Fafarman, Aaron T. 58, 303, 562
Fairbanks, Benjamin D. 50, 201
Fallon, Jacob 909, 910
Fan, Wei 104
Faraone, Antonio 691
Favero, Claudio Vilas Boas 173
Fein, Katherine 526
Feng, Moon Tay Yue 893
Feng, Yafei 276, 287
Fernandez, Sergio 652
Ferrier Jr., Robert C. 48
Ferry, Vivian E. 412, 742, 745, 906
Finn, M. G. 458
Fiore, Andrew 739
Flake, Tanner 252
Fleury, Blaise 744
Foglio, Mary Ann 242
Foo, Guo Shiou 154
Forcherio, Gregory T. 83, 417
Ford, Hunter 97
Forster, Colleen L. 527
Fortman, David 367
Foster, Earl J. 910
Foster, Mark 647
Francis, David 536
Francis, Lorraine F. 44, 781
Francis, Matthew 81
Frazier, Deonante 204
Freeman, Benny D. 607
Freeman, Charles 204
Freire, Stephan 186
Frisbie, C. Daniel 44
Fromen, Catherine A 534
Frye, John G. 158
Fuhr, Addis 786
Fullerton, Susan 783
Funkenbusch, Maylin 231
Furst, Ariel 81
Gaber, Dina 101
Gaber, Safa 101
Galaska, Alexandra 85, 184, 284, 298
Gallazzi, Fabio 12, 530
Gamelin, Daniel 379
Gan, Yunn-Hwen 893
Gand, Adeline 639
Gao, Hanyu 52
Gao, Jian 716
Gao, Zhaoli 413
Garcia, Caterina Bartomeu 433

AUTHOR INDEX

Gartner III, Thomas 235, 424
Ge, Ting 732
Gebbie, Matthew A. 57
Gehlen, David B. 321
Gehrke, Stevin H. 896
Geiger, Brett C 434
Geiger, Mackenzie 327
Geldmeier, Jeffrey 191
Geng, Jiaming 216
Geng, Xiaohua 56
Genzer, Jan 644, 731
Ghorbanpour, Arian 568
Ghosh, Deepanjan 252, 638
Gilchrist, Lane 895
Gillespie, Adam 837
Gilman, Jeffrey W. 330, 818
Gilmer, Eric L. 909
Ginzburg, Valeriy 164
Giri, Gaurav 568
Glascoe, Elizabeth 694
Gleason, Karen K. 646
Glotzer, Sharon C. 169
Go, David 160
Gogotsi, Natalie 744
Gong, Tao 761
Goodenough, Isabella 772
Goodrich, J. Taylor 50
Goodrich, Johnathan 201
Goto, Eisuke 799
Goto, Renata Nishida 629
Gounaris, Chrysanthos E. 4
Gower, Michael 350, 539, 630
Goyal, Ritu 254
Grady, Michael 53
Grasman, Jonathan M. 348
Grayson, Scott 576
Greeley, Jeffrey 573
Green, Brian J. 320
Green, Matthew D. 419
Green, Micah 192
Greenberg, Everett Peter 893
Greene, Ashlee 540
Greenfield, Michael L. 206
Greenwald, Andrew 767
Griego, Charles 281
Griffith, Linda 407
Grim, Joseph 307
Grodzinsky, Alan 434
Grover, Martha A. 41, 162, 512
Grunlan, Jaime C. 397
Gu, Hongbo 86, 88, 184, 295, 477, 816
Gu, Junwei 86, 810
Gu, Kevin L. 304, 358
Gu, Liangliang 610
Gu, Xiaodan 304, 358
Gu, Zhuxiao 297
Guan, Yi-Xin 215
Gudapati, Vamshi 818
Guironnet, Damien 47
Guo, Chengchen 252, 638
Guo, Jiang 85, 184, 284, 298
Guo, Ruilan 693
Guo, Xiaocui 17
Guo, Yongqiang 810
Guo, Zhanhu 85, 184, 284, 298
Gupta, Devanshi 91
Gupta, Rakesh K. 366
Guymon, C. Allan 320, 361, 369, 635
Ha, Heonjoo 227, 288, 388, 593, 607
Ha, Kiryong 292, 388, 607
Hachmann, Johannes 754
Hadi, Atefe 563, 789
Hafez, Islam 293
Hages, Charles J. 903
Haley, Adrian 380
Hall, Lisa M. 234, 404, 748
Hallinan Jr., Daniel T. 570
Hamdeh, Umar H. 566
Hamid, Mohamad 260, 721
Hammer, Daniel A. 756
Hammond, Karl D. 378
Hammond, Paula 10, 11, 231, 240, 394, 407, 434
Han, Duanduan 915
Han, Rebecca 770
Han, Yi-Fan 814
Handwerker, Carol 905
Hanselman, Christopher L. 4
Hansen, Marlan 361, 635
Haque, Farihah 576
Haranczyk, Maciej 769
Harley, Brendan A. 393, 861
Harley, Stephen 694
Harper, Kaid 159
Harris, Michael A. 527
Harris, Stephanie 837
Hartman, Dave 818
Hartwell, Brittany 541
Harvey, Steve 561
Harvey, Taylor B. 904
Hasa, Erion 369
Hasegawa, Urara 533
Hashemnejad, Seyed Meysam 308, 385
Hatton, T. Alan 474, 784
Hatzell, Kelsey 504
Hatzell, Marta 504
Hawker, Craig J. 799
He, Hongkun 231, 407
He, Mao-Gang 659
He, Siyao 87, 296
He, Yanpu 11, 231

AUTHOR INDEX

Heath, William 367
Hedrick, James L. 426
Heichel, Danielle L. 637
Held, Jacob 743
Heller, Adam 593
Helton, Tyler 18
Hendley, Michael 350, 630
Hendrikx, Matthew 325
Henry, Christopher 797
Herceg, Eldad 174
Herman, Chase 891
Hernandez, Nacu 836
Hershman, Rebecca 459
Herwig, Cara 910
Herzog-Arbeitman, Abraham 577
Hickey, John 537
Hicks, Jason C. 160
Higginson, Cody 458
Hildahl, Kate 679
Hill, Andrew 463
Hill, James C. 170
Hillman, Febrian 260, 721
Hillmyer, Marc A. 194, 232, 422, 776
Hilt, J. Zach 14, 314, 476
Hindie, Mathilde 639
Hohmann, Austin D. 836
Holder, Aaron M. 3
Holewinski, Adam 68
Hollingsworth, Jennifer 282
Holloway, Michael 254
Holmberg, Vincent C. 99, 379, 679
Holmes, Gale 818
Hong, Celestine 231
Hosein, Ian 798
Hoskins, Amanda 265, 817
Hossain, Md. Tashfin Zayed 605
Hotta, Atsushi 195, 207, 208, 386, 522, 632, 717
Hou, Qingqing 275
Houlihan, William 895
Howard, Micah J. 796
Howard, Tyler 83
Howe, Joshua D. 773
Hu, Bo 556, 557
Hu, Dong-Dong 196
Hu, Peng 725
Hu, Yun Hang 93
Hu, Zhiqi 510
Huang, Emily 186
Huang, Jun 72
Huang, Mengfei 780
Huang, Sophia Chao-Wei 257
Huang, Wenhan 186
Huang, Zi-Yang 280
Huda, Md Masrul 385
Huelsenbeck, Luke 568
Hule, Rohan 339
Hung, Jui-Hsiang 360, 747
Hunt, Heather K. 378
Hunter, Katharine I. 743
Hussain, Fazle 697
Hustad, Phil 164
Hwang, Kiseob 292
Hyun, Woo Jin 44
Ibanez, Sergio 264
Ida, Junichi 290, 488
Ilias, Shamsuddin 605
Imaninezhad, Mozhdeh 389, 410, 688
Imbrogno, Jennifer 48
Inturi, Siva Nagi Reddy 473
Irvine, Darrell J. 11, 231, 240
Isapour, Golnaz 188
Isely, Christopher 630
Ishii, Hiroshi 636
Ishikawa, Ai 488
Islam, Syed Z. 572
Ismail, Issam 101
Iyer, Kiran S. 575
Jackson, Enrique M. 346
Jackson, Nicholas 43, 511, 728
Jackson, Richard 382
Jacoby, Jeremy M. 563
Jadrich, Ryan B. 741
Jaekel, Esther 321
Jaffer, Shaffiq 711
Jahan, Merina 657
Jain, Era 626, 678
Jallorina, Jerel 513
Jamison, Timothy 784
Janes, Dustin 835
Jang, Yeongseon 79
Jansto, Allison 592
Jawahery, Sudi 769
Jayachandrababu, Krishna Chandran 720
Jayan, B. Reeja 569
Jayaraman, Arthi 235, 368, 424
Jeffryes, Clayton S. 271, 609
Jeon, Ju-Won 191
Jeong, Hae-Kwon 260, 261, 721
Jerke, Amber C. 438
Jessop, Julie L. P. 369
Jiang, Dawei 184
Jiang, Jianwen 775
Jiang, Mingzhe 189
Jiang, Peng 297, 306, 421
Jiang, Shaoyi 428
Jiao, Yang 771, 773
Jimenez, Jairo A 757
Jimenez, Leidy N. 344, 359, 508
Jin, Kailong 49, 387
Jin, Xin 266

AUTHOR INDEX

Jing, Benxin ... 399
Jinrui, Liu ... 378
Johnson, A. T. Charlie ... 413
Johnson, Blake ... 916
Johnson, J. Karl ... 772
Johnson, Joshua ... 62
Jones, Amanda ... 834
Jones, Charles H. ... 463
Jones, Christopher W. ... 65, 68, 156, 722
Jones, Susanne ... 158
Joo, Yong Lak ... 518
Jordan, Alex M. ... 711
Jordan, Carolyn T. ... 14
Joshi, Anup ... 354
Joshi, Rutuja ... 153
Ju, Xiao-Jie ... 315
Kagan, Cherie R. ... 744
Kakunuri, Manohar ... 631
Kalaga, Eswar Arunkumar ... 255
Kalluru, Sri Harsha ... 383
Kang, En-Tang ... 185, 643
Kang, Wooram ... 100
Kanie, Kiyoshi ... 102
Kanost, Michael R. ... 896
Kaplan, David L. ... 348
Kar, Suvrajyoti ... 206
Karapetrova, Evguenia ... 653
Karnik, Rohit ... 636
Kasiewicz, Lisa ... 390, 457
Katabathini, Narasimharao ... 653
Kavadiya, Shalinee ... 560
Keane, Danny ... 41
Keating, John J. ... 718
Kelley, William ... 534
Kelly, Giovanni ... 576
Kelly, Sean ... 462
Kennedy, Stephen ... 627
Kenttamaa, Hilkka ... 571
Keogh, Damien ... 893
Keoh, Sye Hoe ... 71
Khabashesku, Valery N. ... 209
Khademhosseini, Ali ... 167
Khaleel, Maryam ... 101, 106
Khalimonchuk, Oleh ... 862
Khan, M. Arif ... 572, 762
Khan, Saad A. ... 327, 608
Khandelwal, Mudrika ... 631
Khomami, Bamin ... 85, 284, 298, 345
Kidambi, Srivatsan ... 862
Kilbey, Michael ... 507
Kilbourne, Jacquelyn ... 638
Kim, Do-Heyoung ... 269
Kim, Donghun ... 653
Kim, Dongsub ... 799
Kim, Hanim ... 834
Kim, Hanna ... 292
Kim, Hannah ... 760
Kim, Jihan ... 6
Kim, Jongsik ... 160
Kim, Kang-Min ... 50, 201
Kim, Kyungtae ... 356, 711
Kim, Min Jae ... 610
Kim, Sanggon ... 60
Kim, Soyoung ... 268
Kim, Sung-Soo ... 194, 835
Kim, Yoonseob ... 583
Kimura, Takaya ... 182
Kimura, Yoshihiko ... 182
Kinaci, Emre ... 832
Kitagawa, Midori ... 207, 386, 632
Kitchens, Christopher L. ... 189
Klavetter, Kyle ... 593
Klier, John ... 780
Klimov, Victor I. ... 786, 790
Kline, Gregory ... 693
Kloxin, April M. ... 859
Kner, Peter ... 278
Knight III, Paul ... 463
Knorr Jr., Daniel B. ... 714
Knutson, Barbara L. ... 572, 762
Koenig Jr., Gary M. ... 91
Koh, Yung P. ... 712
Kohler, Jens ... 321
Kokini, Jozef ... 309
Kokkoli, Efrosini ... 235, 527
Kolar, Grant ... 678, 688
Kolb, Ben ... 910
Kollengode, Ramanathan ... 643
Koman, Volodymyr ... 606
Konda, Shailesh ... 364
Konetski, Danielle ... 391, 761, 898
Konstantinov, Ivan ... 52
Kopyeva, Irina ... 392, 892
Korgel, Brian A. ... 904
Kornfield, Julia A. ... 316
Kortshagen, Uwe R. ... 743
Koshimura, Tomoyuki ... 524
Kossuth, Mary Beth ... 316
Kou, Yangming ... 285
Kovalenko, Maksym V. ... 564
Kozawa, Daichi ... 606
Kramer, Jake ... 12, 767
Kratzer, Domenic ... 392, 892
Krebs, Melissa ... 880
Kretzschmar, Ilona ... 746
Kuang, Huihui ... 235, 527
Kubal, Joseph ... 573
Kubo, Masaki ... 524
Kuhn, J. N. ... 277
Kuljanishvili, Irma ... 389, 410

AUTHOR INDEX

Kumar, Ashish 377
Kumar, Manjesh 653
Kumar, Prashant 105, 106, 719
Kumar, Ramya 392, 892
Kumar, Sanat K. 713
Kumar, Satish 341
Kumar, Shishir V 271
Kumar, Sushant 390, 457
Kunai, Yuichiro 291
Kundu, Santanu 204, 308, 385
Kunitake, Yusuke 182
Kurokawa, Naruki 522
Kuznetsov, Anatoliy 106
Kwon, Ohmin 6
La Scala, John J. 542, 801, 802, 819
Labbe, Matthew 343
Labelle, James L. 236
Labouriau, Andrea 914
Lacko, Christopher 409
Laha, Anindita 628
Lahann, Joerg 392, 435, 892
Lam, Tiffany 538
Lampe, Kyle 349
Langenberg, Marcel 728
Langer, Robert 78, 408, 411
Lany, Stephan 3
Lapitsky, Yakov 401
Larson, Ronald G. 173, 357
Lattuada, Marco 188, 312
Lau, Kenneth 581, 779
Law, Matt 790
Lawrence, Jimmy 799
Lawton, Carl 256
Le, Thuy T. 181
Lee, Elizabeth M. Y. 302, 752
Lee, Hak Rae 676
Lee, Hansol 514
Lee, Ho Nyung 157
Lee, Hung-Lin 680, 724
Lee, Jungwoo 323
Lee, Kangtaek 258, 259
Lee, Kil Ho 278, 580
Lee, Kwan-Soo 914
Lee, Suh-Jane 158
Lee, Tu 680, 724
Leeper, Caitlin 12
Legge, Kevin 461
Lei, Yuguo 863
Leigh, Braden 361, 635
Leite, Michelle Franz Montan Braga 629
Lekse, Jonathan W 4
Lenhart, Joseph L. 313, 714
Leo, Sin-Yen 306, 421
Leon, Lorraine 578
Leopoldino, Andreia Machado 629
Levi, Adam 310
Lewis III, Ronald M. 356, 655
Lhost, Olivier 711
Li, Dongyang 373
Li, Fei 582
Li, Hongbo 786
Li, Huajian 812
Li, Hua-Min 783
Li, Jiahe 10, 11, 231, 394
Li, Jian 273
Li, Lei 176
Li, Lingqiao 49, 387
Li, Lu 402
Li, Min 185
Li, Mingqi 164
Li, Qiang 863
Li, Quanchang 650
Li, Rui 103, 186
Li, Siming 303, 744, 900
Li, Wei 395, 677
Li, Xiaolan 750
Li, Xiaolei 12
Li, Xiyi 275, 478, 489
Li, Xu 775
Li, Xue 520
Li, Yanxiang 176
Li, Yi 463
Li, Yingzhong 11
Li, Yueming 895
Li, Yun 177
Li, Zhipeng 263
Li, Zhong 478, 489
Liang, Chaobo 86
Liao, Neng 275
Liddle, J. Alexander 330
Lignos, Ioannis 564
Lim, Hui Wen 893
Lim, Hyuntaek 395, 677
Lim, Jaehoon 790
Lim, Kwang-Hee 292
Lim, Woo Taik 260, 261, 721
Limjuco, Lawrence A. 226
Lin, Fang-Yi 836
Lin, Haiqing 364, 696
Lin, Haishuang 863
Lin, Jie 593
Lin, Jui-Che 257
Lin, Kehua 399
Lin, Yi-Ching 257
Lin, Yu-Chuan 783
Lind, Fatin 405
Lindquist, Beth A. 741
Ling, Sanliang 769
Linic, Suljo 61
Lipomi, Darren 73

AUTHOR INDEX

Little, Steven R. 540
Liu, Albert Tianxiang 291, 299, 606
Liu, Bin 64
Liu, Danqing 325
Liu, Dongxia 175
Liu, Jia 39
Liu, Jian 274, 774
Liu, Jing 812
Liu, Jingjing 187
Liu, Kunwei 87, 296
Liu, Lijun 670
Liu, Luman 406
Liu, Pengwei 299
Liu, Pingwei 606
Liu, Song 64
Liu, Tao 196
Liu, Xiaoxiao 584
Liu, Xiaoze 787
Liu, Xiufeng 177, 273
Liu, Yalin 343
Liu, Yang 771
Liu, Yidong 276, 287
Liu, Ying 644
Liu, Zheng 38
Liu, Zhenjing 619
Liu, Zhuang 315
Lively, Ryan P. 65, 722
Livingston, Dana A. 174
Lloyd, Michael L. 900
Lodge, Timothy P. 232, 422, 675
Long, Brian 507
Long, Timothy E. 909
Long, Tyler R. 313
Loo, Y. L. Lynn 75
Lopez, Francisco M. 39
Lopez, Gustavo 746
Lopez-Barron, Carlos R. 793
Lou, Emil 676
Lou, Jianzhong 605
Love, Dillon 50, 201
Lowe, Christopher J. 529
Lu, Diannan 38, 263
Lu, Hao 783
Lu, Yanfei 15
Lu, Yang 279, 531, 609
Lueckheide, Michael 398
Luo, Tianyi 772
Luo, Zhen 55
Luo, Zhengtang 190, 199, 619
Luther, Joseph M. 561
Luther, Yeboah Martin 294
Lutkenhaus, Jodie L. 506, 582
Lutterman, Daniel 157
Luu, Bryan C. 436
Lynd, Nathaniel A. 46, 48, 656
Lytle, Tyler 400
Lyu, Xuejian 641
Ma, Anson 913
Ma, Chao 295
Ma, Cunliang 276, 287
Maceiczyk, Richard 564
Macosko, Chris 296
Macosko, Christopher W. 87, 163, 285, 286, 341, 610, 618, 711
Macpherson, Quinn 727
Madinya, Jason 400
Maeda, Tomoki 207, 208, 386, 522, 632, 717
Maher, Michael 194
Mahmoud, Mahmoud A. 191
Maiti, Debtanu 277
Majumdar, Saptarshi 628
Makarov, Nikolay 786
Malani, Ateeque 755
Maldovan, Martin 283
Malhotra, Abhinav 283
Mallam, Gopichand 768
Mallapragada, Surya 460, 649
Mallapragada, Surya K. 8, 319
Mantovani, Diego 241
Mantripragada, Shobha 605
Mao, Hai-Quan 537
Mao, Shifan 727
Marar, Abhijeet 278
Marciel, Amanda B. 328, 897
Maroudas, Dimitrios 300, 375, 377, 415, 418
Marozas, Ian 307
Marr, James M. 330
Marshall, Blake 908
Martin, Stephen M. 648
Martinez, Carol Rivera 633
Matranga, Christopher 4
Matsumoto, Atsushi 488
Matsuyama, Tatsushi 290, 488
Matthew, Howard W. T. 251
Mauck, Joseph 542
Mazoni, Julia Natalia Oliveira 250
McBride, Matthew K. 325, 381
McBride, Michael 41, 162, 512
McCandless, Brian E. 900
McCloskey, Bryan D. 590
McCormick, Alon 781
McDaniel, Hunter 786
McEnnis, Kathleen 435
McGrail, B. Peter 274, 774
McKenna, Gregory B. 712
McKenna, Mike 239
McKittrick, Michael 161
McLoughlin, Sean 435
McPherson, Brian 584
Meece-Rayle, Mackenzie 593

AUTHOR INDEX

Meek, Kelly M. 591
Meenakshisundaram, Venkatesh 360, 747
Mehra, Nitin 800, 813
Mehraeen, Shafigh 55, 901
Mehta, Prateek 160
Melo, Mariane B. 240
Melosh, Nicholas A 57
Mendez, Janet 633
Meng, Bo 814
Mengel, Shawn 58
Menon, Vinod 746
Meredith, J. Carson 513
Merkel, Alyssa 535
Merrill, Laura 97
Meyer, Pimphan Aye 158
Miller, Daniel J. 695
Miller, Darren 909
Miller, David C. 4
Miller, Jeffrey T. 571
Millican, Samantha L. 3
Milliron, Delia J. 741, 907
Millward, Dan 164
Min, Yong 276, 287
Mineart, Kenneth 426
Miodownik, Mark 382
Miraz, Md. Alamin 609
Mirmajlesi, Keana 529
Miskin, Caleb 571, 905
Misra, Manju 555
Misra, Mayank 521
Mitchell, Doug 264
Mitchell, Miranda 627
Mitra, Debirupa 185, 643
Mittenthal, Max 346
Miyazaki, Makoto 386
Mkhoyan, K. Andre 105, 743
Mofrad, Amir M. 378
Mohammadi, Erfan 509
Mohanty, Amar K. 152, 555
Mohraz, Ali 317
Mok, Jorge 510
Molinaro, Alessandra 225
Moller, Martin 321
Momani, Brian 343
Mondal, Kunal 644
Monroe, Charles W. 689
Montoya, Joseph H. 1
Mony, Sujyot 833
Morad, Viktoriia 564
Moraes, Angela Maria 241, 242, 253, 629
Morehead, Sam 904
Morelock, Cody R. 773
Morton, Logan 12
Moschetta, Eric G. 159
Mou, Tong 92
Mrozek, Randy A. 313
Mu, Bin 567
Mu, Liwen 800
Mukherjee, Rabibrata 642
Muller, Marcus 728
Mullins, C. Buddie 593
Mullis, Adam 437
Mun, Sung Cik 610
Muniz, Andre R. 415
Muramatsu, Atsushi 102, 182
Muraoka, Koki 5, 71
Murata, Hironobu 526
Murphy, Kendall 539, 630
Murray, Christopher B. 744
Murria, Priya 571
Musgrave, Charles B. 3, 50, 201, 265, 817
Muthukumar, Murugappan 575
Muto, Fumiya 102
Nadeau, Emily 762
Nagahama, Koji 386
Nagpure, Suraj 572
Nair, Sankar 720, 770
Najimu, Musa O. 838
Nakamura, Issei 670
Nakamura, Nathan 569
Nakase, Ikuhiko 533
Nakaya, Masafumi 102
Nance, Elizabeth 239, 679
Nanda, Jagjit 570
Narasimhan, Balaji 437, 460, 461, 462
Narkhede, Akshay 860
Nautiyal, Amit 645
Nayak, Srikanth 649
Nayerhoda, Roozbeh 463
Naylor, Carl H. 413
Nazarenko, Sergei 834
Nazari, Behzad 340, 833
Nealey, Paul F. 517
Nejati, Siamak 193
Nelson, Rainie D. 563, 566, 789
Neo, Darren Chi Jin 790
Neoh, Koon Gee 185, 643
Ng, Fay 45
Nguon, Helen 186
Nguyen, Julie 530
Nguyen, Quoc T. 171
Nieh, Mu-Ping 187
Nisola, Grace M. 226, 289
Nowak, Christian 521, 751
Nowottny, Jonas 200
Nuckolls, Colin 40, 45
Nuxoll, Eric 730
Oberhauser, James Paul 316
Ochi, Masanori 488
O'Flanagan, Stephen 538

AUTHOR INDEX

Ogilvie-Battersby, James....225
Oh, Hyeonji....833
Oh, Jinyoung....39
O'Harra, Kathryn....346
Ohodnicki, Paul R....98
Okamoto, Yoshi....696
O'Keefe, Sean....669
Okolie, Chukwuemeka....154
Okoshi, Ryusuke....522
Okubo, Tatsuya....5, 71
Olarte, Mariefel V....158
Oleske, Katharine W....777
Olsen, Bradley D....384
O'Neill, Sean C....311
Onstott, Ellie K....896
Opembe, Naftali....264
Oquendo, Luis E....194
Orbey, Nese....225, 256
Orton, Emma....627
Oschmann, Bernd....799
Osmond, Matthew....353, 880
Ostraat, Michele L....652
Osuji, Chinedum O....165
Ou, Jifei....636
Ousley, Evan....18
Ozgur, Derya Oncel....100
Pachidis, Pavlos....412
Padmanaban, Munirathna....518
Paek, Seung-Min....260, 721
Palmese, Giuseppe....797, 801
Panchal, Kushal....373
Pandres, Elena P....99, 379
Pang, Simon H....65
Pantcheva, Mina....880
Panthani, Matthew G....414, 563, 566, 789, 791
Panzer, Matthew J....96, 405, 523
Papait, Andrea....348
Papavassiliou, Dimitrios V....171
Parekh, Dishit....912
Park, Chihoon....914
Park, Jaesung....607
Park, Jong Keun....164
Park, Sanghoon....6
Park, Seungdoo....264
Park, Yongkuk....323
Parkin, William M....413
Parohinog, Khino J....226, 289
Parulkar, Aamena....153, 726
Patel, Pooja....254
Patra, Tarak Kumar....747
Pauthe, Emmanuel....639
Pauzauskie, Peter....379
Pawar, Prasad P....271
Pearce, Timothy R....527
Peixoto, Caio....378
Pekkanen, Allison M....909
Pelaez, Francisco....538
Peles, Yoav....172
Pender, Joshua....593
Peng, Xiaoshan....579
Pengo, Thomas....527
Pepa, Kristi....746
Peppas, Nicholas A....436, 766
Perera, Ayomi S....180, 382
Perez, German....352
Perez, Maritza Perez....329
Peroutka-Bigus, Nathan....437, 462
Perreault, Luke R....348
Perry, Sarah L....343, 400, 459
Persson, Kristin....1
Persson, Nils....41, 162, 512
Peterson, Amy M....641
Peterson, Brian K....650
Pethe, Kevin....893
Petrecca, Katherine....636
Petrich, Jacob W....566
Pfeifer, Blaine....463
Pflug, Kristina M....200
Pfromm, Peter....690
Phan, Ngoc....528
Phan, Vu....609
Pi, Yunhong....478, 489
Pietryga, Jeffrey M....790
Pinezich, Meghan....349
Pinge, Shubham....518
Pinhero, Patrick J....301, 305, 380
Pintos, Esteban....198
Pires, Ana Luiza Resende....242
Pirzada, Tahira....327, 608
Plawsky, Joel L....172
Podlaha, Elizabeth....56
Polanco, Ashli....256
Ponder, James....191
Potter, Matthew....68
Prabhu, Vivek M....426
Prakash, Nikhil....54, 84, 230
Preciado, Julian....676
Priestley, Rodney D....429
Protesescu, Loredana....564
Pu, Jingyang....216
Pu, Yuji....371
Puri, Mayank....742, 745, 906
Purwar, Rahul....390, 457
Qayyum, Anisa....678
Qi, Zhaoxiang....91
Qian, Yuqiang....87, 296
Qin, Huan....523
Qiu, Bin....184
Qiu, Fen....692
Qu, Ge....509

AUTHOR INDEX

Rachagani, Satyanarayana....8
Ragauskas, Arthur J....150
Ragelle, Heloise....411
Rahimi, Khosrow....321
Rahman, Ashiqur....271, 609
Rai, Muhammad....626
Rai, Neeraj....385
Rajendran, Aravindan....557
Ramachandran, Karthik....316
Ramirez, Julia Vela....766
Ramirez-Caballero, Gustavo....723
Ranadive, Pinaki....726
Rangnekar, Neel....105, 653
Rankin, Stephen E....572, 762
Rao, Shreyas....860
Rappe, Andrew M....53
Ratay, Michelle....540
Rebollar, Luis....405
Reed, Ellen H....756
Rege, Kaushal....252, 638
Reichmanis, Elsa....41, 162, 512
Reid, Daniel....43, 511
Reifenberger, Jeffrey G....658
Reis, Cristiano....557
Reisch, Anne....627
Ren, Jing Ming....799
Ren, Liwei....263
Reveil, Mardochee....902
Reynolds, John....191
Rezac, Mary....690
Rezaei, Fateme....66
Rezende, Gabriela Souza....629
Rhoades, Alicyn M....340
Riazi, Hossein....53
Ricarte, Ralm....232, 422
Richard, Melissandre....772
Richter, Steve....159
Riggleman, Robert A....640
Rimer, Jeffrey....103, 181, 653
Rinaldi, Carlos....409
Rioux, Robert M....59
Robbins, Mark....732
Roberts, Jesse....352
Robinson, Joshua....783
Rodrigues, Mariana V....154
Rodriguez, Christina G....48
Roh, Jeongkyun....790
Roh, Sangchul....912
Rohner, Nathan A....536
Rokke, David....905
Roper, D. Keith....83, 417
Rosa, Paulo De Tarso Vieira E....250
Rose, Jonas C....321
Rosi, Nathaniel L....772
Ross, Kathleen....461
Roth, Charles....254
Rother, Gernot....68
Rouf, Tahrima B....90
Roy, Manosi....760
Ruffley, Jonathan....772
Rumptz, John R....3
Russell, Alan....526
Russell, Lauren....349
Rutkevicius, Marius....327
Rutledge, Gregory C....574, 753
Ryan, Bradley J....566
Sacco Jr., Albert....107
Saccone, Max....606
Sadeghi, Morteza....229
Sadler, Joshua....801, 802
Saed, Mohammad....192
Saifuddin, Indira....833
Sajib, Md Symon Jahan....331
Sakaguchi, Donald S....319
Sakaguchi, Moe....102
Sakai, Risako....290
Sakwa-Novak, Miles....68
Salcedo, Felipe....757
Salehi, Ali....357
Salehi, Mahsa....229
Sampath, Janani....234, 404
Sanchez, Isaac C....669
Sandell, Linda....626
Santiso, Erik E....731
Santos, Daniela Barrios....766
Santos, Stephany Di Carla....629
Santosa, Daniel....158
Saquing, Carl D....608
Saravi, Sina Hassanjani....272
Sarkar, Mohammad Shahadat Hussain....609
Sarkaria, Jann N....393, 861
Sarmiento, Paula A....757
Satam, Chinmay C....513
Satish, Nune....774
Sautet, Phillippe....786
Schaake, Richard P....340
Schaefer, Jennifer....97
Schaffer, David V....80
Schieber, Jay D....338
Schiffman, Jessica D....780
Schloemer, Tracy H....561
Schmalbach, Kevin M....802
Schmalzer, Andrew M....914
Schmidt, Christine....409
Schneck, Jonathan....537
Schneider, William F....160
Schudel, Alex....458, 536
Schultz, Kelly M....318
Schulz, Philip....561
Schulze, Morgan W....356

AUTHOR INDEX

Seabaugh, Alan....783
Seabaugh, Matthew....264
Seaton, Nicholas....293
Seeger, Madeleine....837
Sefiddashti, Mohammad Hadi Nafar....345
Segura, Tatiana....168, 238
Sell, Scott A....626, 678
Sellinger, Alan....561
Semenchenko, Liya....261
Senapati, Sujata....460
Seney, Robert....189
Sengers, Jan V....659
Seo, Jiho....340
Servoss, Shannon L....352
Setaro, Angelo....718
Sevimli, Sema....535
Sfeir, Charles....540
Shadish, Jared A....406
Shae, Daniel....535
Shah, Rishabh....476
Shah, S. Sakhawat....608
Shahvari, Amin Vossoughi....251
Shamsabadi, Ahmad Arabi....53, 229
Shan, Bohan....567
Shanmuganathan, Kadhiravan....288, 834, 835
Shao, Qi....538
Sharma, Chandra....628, 631
Sharma, Hom....694
Sharma, Megha....63, 651
Sharma, Radhika....618
Sharma, Vivek....344, 359, 508
Shell, M. Scott....13
Shen, Alan....913
Shen, Chunyin....89
Shen, Haiyang....94
Shen, Jian....266
Shen, Jiaxin....276, 287
Shen, Yufeng....181
Shere, Inderdip....755
Sheridan, Richard....818
Shete, Meera....653
Sheth, Saahil....626
Shi, Di....433
Shi, Kun....315
Shim, Jong Hyun....270
Shimizu, Yoji....538
Shimojima, Atsushi....208, 717
Shin, Jeongeun....788
Shkrob, Ilya....95
Sholl, David....720
Sholl, David S....770, 771, 773
Shreiber, David I....529
Si, Zhangyong....893
Siepmann, J. Ilja....675
Sievers, Carsten....154
Silcox, Benjamin....95
Silveira, Zachary....627
Simmons, David S....360, 747
Simon, Sindee L....427, 712
Sing, Charles E....400
Singaravel, Gnana Pragasam....183
Singh, Ajay....282, 907
Singh, Ishita....409
Singh, Samrendra....525, 729
Slater, Ben....769
Smirniotis, Panagiotis....473
Smit, Berend....769
Smith, Andrew....186
Smith, Daniel....891
Smith, Ethan D....648
Smith, Josiah....12, 530
Smith, Mitch....9
Snyder, Mark A....63, 651
Sofman, Marianna....407
Soltani, Iman....341
Song, Han Byul....715
Song, Kenan....811
Song, Ping....86
Song, Xueyan....366
Song, Young-Geon....258, 259
Song, Zhuonan....692
Sorci, Mirco....718
Soroush, Masoud....53, 229
Sousa, Ilza Maria De Oliveira....242
Spakowitz, Andrew J....727
Spencer, David S....436
Srinivasan, Samanvaya....171
Srivastava, Samanvaya....310, 328, 372, 402, 577
Stanford, John P....690
Stanke, Kimberly M....862
Stansbury, Jeffrey W....369
Stanzione III, Joseph F....542, 802, 819
Stecca, Owen M....802
Stefanov, Zdravko....174
Steigerwald, Michael....45
Stein, Andreas....87, 296
Stein, Gila....365
Steiner, Helene....636
Stephane, M....183
Sterling, Julie A....535
Stevanovic, Vladan....3
Stoppel, Whitney L....348
Storti, Giuseppe....312
Stottrup, Benjamin....653
Stoyanov, Simeon....912
Stoykovich, Mark....50, 201
Strano, Michael....291, 299, 606
Stretz, Holly A....228
Stromsdorfer, Jessica....767
Su, Hao....396, 532

AUTHOR INDEX

Su, Xiao....474, 784
Subramanion, Jo Thy Lachumy....893
Suleiman, David....329
Sultana, Khondker....605
Sun, Haotian....237, 326
Sun, Jingze....261
Sun, Luyi....186, 187
Sun, Rui....591
Sun, Yan....17
Sun, Yangyang....521
Sun, Yunwei....694
Sutjianto, James....103
Swager, Timothy....583
Swan, James....739
Swartz, Scott....264
Sweeney, Charles....192
Swihart, Mark T....364
Tafen, De Nyago....4
Taheri, Mohammad Mehdi....58
Tajvidi, Mehdi....293
Takatani-Nakase, Tomoka....533
Tan, Anthony....425
Tan, Kai-Jher....474
Tan, Shuai....157
Tang, Lin....810
Tang, Shuo....314
Tanimoto, Keishi....386
Tanveer, Sheik....697
Tata, Ram Rao....767
Tazawa, Shunsuke....208, 717
Tekin, Rumeysa....107
Teymour, Fouad....202
Thacker, Zachary....301, 305
Thakkar, Harshul....66
Thallapally, Praveen K....774
Thate, Karine....278
Thelakkaden, Mitzi....252, 638
Thomas, Susan N....458, 536
Thompson, Jessica....320
Thompson, Joshua A....70
Thompson, Levi T....95
Thompson, Matthew S....366
Thompson, Nicole....679
Thorson, Todd....317
Thurman, Derek W....339
Tian, Qian....427
Tian, Zheng....63
Tibbitt, Mark W....411
Tiffany, Douglas....557
Ting, Jeffrey....372, 577
Tinkham, Jonathan....561
Tirrell, Matthew V....77, 236, 310, 328, 372, 398, 402, 577, 897
Tisdale, William A....302, 752, 785
Tokatlian, Talar....240
Tolouei, Anita E....627
Tom, Karjala....51
Topuz, Berna....653
Torkelson, John M....49, 367, 387, 425
Trefonas III, Peter....164
Triana, Juan Carlos Briceno....757
Trogadas, Panagiotis....180
Troung, Kristy....635
Truong, Quoc....225, 256
Truskett, Thomas M....741
Tsapatsis, Michael....105, 106, 653, 719
Tschirner, Ulrike....837
Tsianou, Marina....768
Tsoras, Alexandra....464
Tsukada, Takao....524
Tsukruk, Vladimir V....191
Tu, Raymond....311
Tucker, Budd A....320
Tumas, William....3
Tung, Siu On....95
Turasan, Hazal....309
Tze, William T. Y....293
Ugaz, Victor M....915
Ulery, Bret....12, 530, 767
Uline, Mark J....657
Um, Kiju....258, 259
Underhill, Patrick T....718
Unold, Thomas....903
Urban, Jeffrey....692
Urie, Russell....252, 638
Utomo, Nyalaliska....833
Uz, Metin....8, 319
Vaidya, Amol....818
Vaithilingam, Balasubramanian....183
Vaknin, David....649
Valdehuesa, Kris Nino G....226
Van Der Vlies, Andre....533
Van Lehn, Reid....16
Van Tassel, Paul R....639
Vargas, J. German....757
Varma, Arvind....100
Vattipalli, Vivek....104
Velev, Orlin D....912
Venkataraman, Shrinivas....426
Ventura, Darryl....209
Verburg, Alex....789
Verduzco, Rafael....365, 510
Vergara, John....801
Vicente, Fernando....537
Vieregg, Jeffrey....77, 398
Vikhe, Yogesh....894
Vlassopoulos, Dimitris....713
Vocelle, Daniel....9
Voggu, Vikas Reddy....904
Vogt, Bryan D....792, 911

AUTHOR INDEX

Volkov, Dmytro 256
Waechter, Andreas 52
Wagner, Norman J. 793
Waldschmidt, Thomas 461
Walker, Christopher 731
Walsh, Dylan 47
Walton, Krista S. 475, 771, 773
Walton, S. Patrick 9
Wan, Haiqing 89
Wang, Alex 407
Wang, Bin 92
Wang, Boya 570
Wang, Chen 381
Wang, Daniel I. C. 636
Wang, Feihu 396, 532
Wang, Guanyun 636
Wang, Huamin 158
Wang, Jiechen 39
Wang, Jingyu 556
Wang, Junyan 89
Wang, Luda 636
Wang, Minghui 646
Wang, Muzhou 330
Wang, Nathan G.-J. 39
Wang, Ping 516
Wang, Qing 237
Wang, Rui 384
Wang, Shaoyang 582
Wang, Sheryl 231, 434
Wang, Sihong 39
Wang, Song 291, 606
Wang, Wade 10, 11, 394
Wang, Wei 315
Wang, Wen 636
Wang, Wenjie 649
Wang, Yujie 791
Wang, Yuzhu 396, 532
Wang, Zhaofeng 186
Wannemuehler, Michael J. 460, 462
Warzywoda, Juliusz 107
Wawrousek, Chris 636
Webb, Bruce 762
Webster, Megan 746
Webster, Thomas J. 433
Wegener, Evan C. 571
Wei, Huige 184
Wei, Min 94
Wei, Suying 85, 184, 298
Wei, Tao 331
Wei, Tong 425
Weidman, Jennifer 693
Weimer, Alan W. 3, 265, 817
Wells, Kacie M. 187
Westin, Cecilia Buzatto 242
Weston, Simon C. 650
Whitaker, Mariah 153
White, Tommi 12
Whitehead, Kathryn A. 7, 390, 457, 526
Whitham, Patrick 379
Whitmer, Jonathan K. 654
Wiesner, Ulrich 777
Wijayapala, Rangana 204, 308
Wilhite, Benjamin 397
Willard, Adam P. 302, 752
Williams, Christopher B. 909
Williamson, Grant A. 99
Willson, C. Grant 194
Wilson, Christina 862
Wilson, John 535
Winey, Karen I. 430
Winter, H. Henning 343
Winter, Jessica O. 62, 278, 580
Witman, Matthew 769
Wittenberger, Steven J 159
Wong, Bryan M. 2, 376
Wong, Hoi Lun 190
Wong, Minhao 606
Woodcock, Cory 172
Woodcock, Jeremiah 818
Worrell, Brady 50, 325, 381, 761
Worthington, Kristan S. 320
Wu, Connie 10, 394
Wu, Gang 94
Wu, Hao 372, 577
Wu, Janet 164
Wu, Jianzhong 38
Wu, Qing-Xi 215
Wu, Wei 634
Wu, Yadong 89
Wu, Yaoting 744
Wu, Yun 237
Wu, Yuning 98
Wujcik, Evan K. 279, 609
Wyslouzil, Barbara E. 580
Xi, Li 373
Xi, Weixian 50
Xia, Yan 187
Xiao, Han 288, 593
Xiao, Jing 275, 478, 489
Xie, Rui 315
Xie, Shuyi 675
Xie, Xiaofeng 520
Xie, Yongjian 812
Xie, Zelong 67
Xin, Zhong 815
Xu, Dandan 106, 653
Xu, Honghong 177
Xu, Jun 796
Xu, Ke 783
Xu, Wenqian 719

AUTHOR INDEX

Xu, Xiaojiang....816
Xu, Xiaoyang....765
Xue, Feng....719
Yadav, Santosh Kumar....801
Yamamoto, Hideo....488
Yan, Hongping....304, 358
Yan, Ruoxue....60
Yan, Yushan....432
Yang, Chuanfang....176
Yang, Guang....570
Yang, Guozhen....780
Yang, Han-Seung....293
Yang, Seung Ook....899
Yang, Shendu....301, 305, 380
Yang, Sui....787
Yang, Szu-Ming....227
Yang, Xutong....810
Yang, Yeokyung....292
Yang, Yi Yan....426
Yao, Lining....636
Yao, Shan-Jing....215
Yarger, Jeff....252, 638
Yau, Mai Kwan....458
Yavari, Milad....696
Yin, Deqiang....364
Yokoi, Toshiyuki....102, 182
Yoon, Heedong....712
Yoshizawa, Sayuri....540
Young, Walter W.....194
Yu, Dan....395, 677
Yu, Haiyue....180
Yu, Xiangfei....172
Yu, Xiaoyuan....186
Yuan, Jun-Jie....215
Yun, Hyeong Jin....786, 790
Yung, Matthew M.....154
Zacharias, Zeb....461
Zacher, Alan H....158
Zasadzinski, Joesph A.....788
Zawaski, Callie....909
Zembrzuski, Michael....791
Zeng, Songshan....186, 187
Zhang, Baoquan....177, 273
Zhang, Boyuan....40
Zhang, Dawei....391, 761, 898
Zhang, Dianyun....186
Zhang, Guojian....463
Zhang, Han....105
Zhang, Hongyuan....88
Zhang, Huaiying....82
Zhang, Jing Jing....95
Zhang, Jinju....176
Zhang, Junyan....175
Zhang, Ke....652
Zhang, Lin....17
Zhang, Lu....95
Zhang, Mengying....679
Zhang, Na....216
Zhang, Qinnan....693
Zhang, Rui....12
Zhang, Ruihong....905
Zhang, Teng....636
Zhang, Xiang....787
Zhang, Xinyu....645
Zhang, Yanmei....556
Zhao, Haoyu....427
Zhao, Junjie....646
Zhao, Ling....196
Zhao, Mengqiang....413
Zhao, Shicheng....579
Zhao, Tianshuo....744
Zhao, Xin....905
Zhao, Xuanhe....636
Zhao, Yang....520
Zhao, Zhenxia....725
Zhao, Zhongxing....725
Zheng, Hou....371, 894
Zheng, Xinguo....812
Zheng, Xiong....659
Zhong, Yu....40, 45
Zhou, Changlu....815
Zhou, Jing....191
Zhou, Kang....636
Zhou, Shanshan....762
Zhou, Sunshine X.....194
Zhou, Yang....647
Zhu, Guanghui....722
Zhu, Huiming....273
Zhu, J. Y.....149, 515
Zhu, Jiahua....184, 800, 813
Zhu, Lingxiang....364
Zhu, Shuze....753
Zhu, Xiaoyang....40, 45
Zhu, Y. Elaine....399
Zhu, Yangzhi....60
Zhu, Yaqun....270
Zhuang, Minghao....199
Zimmerman, John....260, 721
Zoto, Christopher....225
Zustiak, Silviya Petrova....389, 410, 626, 678, 688

AIChE
120 Wall Street, FL 23
New York, NY 10005-4020

ISBN 978-1-5108-5797-1

CURRAN ASSOCIATES INC.
proceedings.com

Materials Engineering and Sciences Division 2017

Core Programming Area at the 2017 AIChE Annual Meeting

Minneapolis, Minnesota, USA
29 October – 3 November 2017

Volume 2 of 2

AIChE
The Global Home of Chemical Engineers

Materials Engineering and Sciences Division 2017

Core Programming Area at the 2017 AIChE Annual Meeting

Minneapolis, Minnesota, USA
29 October – 3 November 2017

Volume 2 of 2

ISBN: 978-1-5108-5797-1

Printed from e-media with permission by:

Curran Associates, Inc.
57 Morehouse Lane
Red Hook, NY 12571

Some format issues inherent in the e-media version may also appear in this print version.

Printed by Curran Associates, Inc. (2018)

For permission requests, please contact AIChE
at the address below.

AIChE
120 Wall Street, FL 23
New York, NY 10005-4020

Phone: (800) 242-4363
Fax: (203) 775-5177

www.aiche.org

Additional copies of this publication are available from:

Curran Associates, Inc.
57 Morehouse Lane
Red Hook, NY 12571 USA
Phone: 845-758-0400
Fax: 845-758-2633
Email: curran@proceedings.com
Web: www.proceedings.com

TABLE OF CONTENTS

VOLUME 1

(9a) New Milestones and Challenges in High-Throughput Computation of Elastic Properties on the Materials Project 1
Joseph H. Montoya, Kristin Persson

(9b) Breaking Badly: A Comprehensive Assessment of Computational Methods for Predicting Tensile Strengths in Bulk Solids 2
Bryan M. Wong

(9c) Machine Learning the Thermochemistry of All Inorganic Crystalline Solids 3
Christopher J. Bartel, Ann M. Deml, Samantha L. Millican, John R. Rumptz, William Tumas, Alan W. Weimer, Stephan Lany, Vladan Stevanovic, Charles B. Musgrave, Aaron M. Holder

(9d) Designing Dopant Patterns in Indium-Doped Perovskite Oxygen Carriers 4
Christopher L. Hanselman, Dominic Alfonso, Jonathan W. Lekse, De Nyago Tafen, Christopher Matranga, David C. Miller, Chrysanthos E. Gounaris

(9f) In silico Structural Analyses of Borosilicate, Aluminosilicate, and Gallosilicate Zeolites Using Effective Tetrahedral Descriptors 5
Koki Muraoka, Tatsuya Okubo, Watcharop Chaikittisilp

(9g) Computational Identification of Hetero-Interpenetrated Porous Materials 6
Jihan Kim, Ohmin Kwon, Sanghoon Park

(16a) Lipid-like Materials for RNA Delivery: A How-to Guide for Hacking Gene Expression 7
Kathryn A. Whitehead

(16b) Triple Delivery Nanoscale Device for siRNA, Vismodegib and Gemcitabine Co-Delivery to Treat Pancreatic Cancer 8
Metin Uz, Satyanarayana Rachagani, Surinder Batra, Surya K. Mallapragada

(16c) Influence of Dextran and Surface Charge on Nanoparticle-Mediated siRNA Delivery 9
Daniel Vocelle, Olivia Chesniak, Mitch Smith, Christina Chan, S. Patrick Walton

(16d) Peg-Poly(beta-amino ester) Delivery Systems for Periodic shRNA 10
Connie Wu, Jiahe Li, Wade Wang, Paula T. Hammond

(16e) Structurally Programmed Assembly of Ribonucleoproteins for Superior mRNA Delivery 11
Jiahe Li, Wade Wang, Yanpu He, Yingzhong Li, Darrell J. Irvine, Paula T. Hammond

(16f) Cationic Peptide Amphiphile Micelles(PAMs) As Nucleic Acid Based Adjuvants Carriers for the Improvement of Subunit Vaccine Efficiency 12
Rui Zhang, Josiah Smith, Jake Kramer, Logan Morton, Brittany Allen, Caitlin Leeper, Xiaolei Li, Fabio Gallazzi, Tommi White, Bret Ulery

(31a) Understanding Peptide Assembly with Coarse-Grained Models Designed By Information Theory (Invited Talk) 13
M. Scott Shell

(31b) Modeling the Oxidative Consumption of Curcumin from Controlled Release Poly(beta amino ester) Microparticles in the Presence of a Free Radical Generating System 14
Carolyn T. Jordan, J. Zach Hilt, Thomas D. Dziubla

(31c) The Fusion and Spreading of Liposome with Different Sizes: Molecular Dynamics Simulation with Dry-Martini Force Field 15
Yanfei Lu, Lu Diannan

(31d) Design of Membrane-Embedded Amphiphilic Nanoparticles from Multiscale Simulations 16
Reid Van Lehn

(31e) Improvement on Self-Assembly of Virus-like Particles By the Introduction of Electrostatic Attraction 17
Lin Zhang, Xiaocui Guo, Yan Sun

(31f) Contemporary Modeling and Analysis of Human Blood Rheology 18
Matthew Armstrong, Tyler Helton, Evan Ousley, Michael Deegan

(31g) How Confinement and Hydrophobicity Affect CO2 Diffusion from the Bulk Phase to the Active Site of Human Carbonic Anhydrase II: A Study Based on Coarse-Grained Molecular Dynamics Simulation and the Markov-State Model 38
Gong Chen, Diannan Lu, Jianzhong Wu, Zheng Liu

(34a) Printing Fully Stretchable Thin-Film-Transistor Array 39
Jia Liu, Francisco M. Lopez, Jiechen Wang, Nathan G.-J. Wang, Jinyoung Oh, Sihong Wang, Zhenan Bao

(34b) Conjugated Macrocycles for Ultra-Sensitive Organic Photodetectors 40
Yu Zhong, Boyuan Zhang, Xiaoyang Zhu, Colin Nuckolls

(34c) Connecting Crystalline Domains: Blending Conjugated Polymers of Differing Molecular Weights for Enhanced Charge Transport 41
Michael McBride, Nils Persson, Danny Keane, Martha A. Grover, Elsa Reichmanis
(34d) Ultrafast Carrier Dynamics in Bimetallic Nanostructures-Enhanced Methylammonium Lead Bromide Perovskites 42
Rizia Bardhan
(34e) Aggregation Structure and Solvation of Solution-Phase PTB7 43
Daniel Reid, Nicholas Jackson, Juan J. De Pablo
(34f) Self-Aligned Strategy for Printed Electronics 44
Woo Jin Hyun, Lorraine F. Francis, C. Daniel Frisbie
(34g) Helical Perylene Diimide Ribbons for Molecular Electronics 45
Yu Zhong, Michael Steigerwald, Xiaoyang Zhu, Fay Ng, Colin Nuckolls
(36a) Engineering Pathways to New Functional Polyether Materials 46
Nathaniel A. Lynd
(36b) Topology Control of Bottlebrush Polymers 47
Damien Guironnet, Dylan Walsh
(36c) A New, Facile Approach to Epoxide Polymerization 48
Robert C. Ferrier Jr., Jennifer Imbrogno, Christina G. Rodriguez, Malgorzata Chwatko, Nathaniel A. Lynd
(36d) Dynamic Chemistry Leading to Full Property Recovery Associated with Crosslink Density in Reprocessed Polymer Networks and Network Composites: Network Synthesis By Step-Growth Reactions and By Nitroxide-Mediated Polymerization 49
John M. Torkelson, Xi Chen, Kailong Jin, Lingqiao Li
(36e) Hybrid Chain-Growth/Step-Growth Mechanism Observed in Heterofunctional Thiol-Ene Polymerizations 50
Dillon Love, Benjamin D. Fairbanks, Kang-Min Kim, Brady Worrell, Weixian Xi, J. Taylor Goodrich, Charles B. Musgrave, Mark Stoykovich, Christopher N. Bowman
(36f) High Pressure Ethylene Polymerization with a Post-Metallocene Bis-Phenyl Phenoxy Catalyst 51
Sean Ewart, Karjala Tom
(36g) A Simulation-Based Derivative-Free Optimization Framework Using the Kinetic Monte Carlo Method for Controlling Polymer Molecular Weight and Sequence Distribution Synthesized Via Free Radical Polymerization 52
Hanyu Gao, Andreas Waechter, Ivan Konstantinov, Steven G. Arturo, Linda J. Broadbelt
(36h) Experimental and Macroscopic Mechanistic Modeling Studies of the Methyl Acrylate Self-Initiation Reaction 53
Hossein Riazi, Ahmad Arabi Shamsabadi, Michael Grady, Andrew M. Rappe, Masoud Soroush
(36i) A Mathematical Model Based on Artificial Neural Network for Ethylene/Norbornene Copolymerization Catalyzed By 2-(tetramethylcyclopentadienyl)-4,6-Di-Tert-Butylphenoxytitanium Dichloride 54
Nikhil Prakash
(167a) Revealing Governing Mechanism in Directed Self-Assembly of sub 10 nm Particles with Single Particle Resolution 55
Zhen Luo, Shafigh Mehraeen
(167b) Uniform Thinning of Cu-Fe-Ni-Co Nanowires and Kinetic Monte Carlo Simulation 56
Xiaohua Geng, Elizabeth Podlaha
(167c) Experimental Assessment of Nucleation Theory at the Molecular Level 57
Matthew A. Gebbie, Nicholas A . Melosh
(167d) Low-Voltage Electrophoretic Deposition of All-Inorganic CZTS Nanocrystals for Fabrication of Thin Films 58
Andrew D. Dillon, Mohammad Mehdi Taheri, Shawn Mengel, Subham Dastidar, Jason B. Baxter, Aaron T. Fafarman
(167e) Preferential Binding of Polyvinylpyrrolidone (PVP) Is Not Responsible for Shape Control in Ag Nanoparticle Synthesis 59
Zhifeng Chen, Ji Woong Chang, Choumini Balasanthiran, Robert M. Rioux
(167f) Scalable Synthesis of Epitaxial, Oxidation-Proof Au@Ag Core-Shell Nanowires for Electronic and Photonic Applications 60
Ruoxue Yan, Yangzhi Zhu, Sanggon Kim, Peter Byrley
(167g) Synthesis of Bimetallic Alloy Nanoparticles through the Visible-Light Mediated Reduction of a Bimetallic Oxide Precursor: Case Study of Ag-Pt Nanoparticle Synthesis 61
Umar Aslam, Suljo Linic
(167h) Dynamic Control of Gold Nanoparticle-Conjugated DNA Origami Templates 62
Abhilasha Dehankar, Joshua Johnson, Carlos E. Castro, Jessica O. Winter

(167i) 'One-Pot' Multi-Scale Templating of Interdigitated Bi-Modal Porous Carbon Supercapacitors 63
Zheng Tian, Megha Sharma, Mark A. Snyder
(167j) Combined Experimental and Theoretical Study of Hexagonal Boron Nitride Crystal Growth 64
Song Liu, Bin Liu, James H. Edgar
(30a) Design of Aminopolymer Structure to Enhance Performance and Stability of CO2 Sorbents: Poly(propylenimine) Vs. Poly(ethylenimine) 65
Simon H. Pang, Ryan P. Lively, Christopher W. Jones
(30b) Development of 3D-Printed Aminosilica Monoliths for CO2 capture 66
Harshul Thakkar, Stephen Eastman, Fateme Rezaei
(30c) Engineering Supported Amine Adsorbents for CO2 Capture Applications 67
Zelong Xie, Christopher Cogswell, Sunho Choi
(30d) Impacts of Aminopolymer-Support Interaction on CO2 Sorption Performance Probed By Neutron Scattering Techniques 68
Adam Holewinski, Miles Sakwa-Novak, Matthew Potter, Nathan Ellebracht, Gernot Rother, Christopher W. Jones
(30e) Hierarchically-Structured Porous Materials for Enhanced Greenhouse Gas Capture 69
Tae-Hyun Bae
(30f) Natural Gas Adsorption in SSZ-13: Equilibrium and Dynamic Properties 70
Joshua A. Thompson
(30g) Zeolite on Demand: Design and Synthesis of Zeolites with Controlled Crystal Morphology and Location of Substituting Tetrahedral Atoms with the Aid of Theoretical Calculations 71
Watcharop Chaikittisilp, Koki Muraoka, Sye Hoe Keoh, Tatsuya Okubo
(30h) Tuning Solid Acids for the Target Catalytic Reactions 72
Jun Huang
(51a) Mechanical Properties of Organic Semiconductors for Mechanically Stable and Intrinsically Stretchable Solar Cells 73
Darren Lipomi
(51b) Digital Light Synthesis to Drive Additive Manufacturing 74
Joseph M Desimone
(51c) Symbiotic Pairing of Near-UV Solar Cells with Electrochromic Windows for Visible Light and Heat Management in Architectural Applications 75
Y. L. Lynn Loo
(51d) Taking Measure of Modern Polymer Synthesis 76
Kathryn L. Beers
(55a) Rational Design of Polyelectrolyte Complexes for Nucleic Acid Delivery 77
Jeffrey Vieregg, Matthew V. Tirrell
(55b) Microfabricated Immune-Isolating Devices for Long Term Cell Based Therapies 78
Suman Bose, Robert Langer, Daniel G. Anderson
(55c) Globular Protein Vesicles: Engineering Vesicle Size and Membrane Structure through a Tunable Molecular Packing Parameter 79
Yeongseon Jang, Julie A. Champion
(55d) Biomaterial Scaffolds for Scalable Differentiation and Transplantation of Hpsc-Derived Cells for Cell Replacement Therapy in the Central Nervous System 80
Maroof M. Adil, David V. Schaffer
(55f) Electrochemical Activation for DNA Attachment to Surfaces 81
Ariel Furst, Matthew Francis
(55g) Organizing Biochemical Reactions with Protein Droplets 82
Huaiying Zhang
(59a) Electron Energy Loss Spectroscopy for Optoelectronics and Thermal Dynamics at Nanocomposite Interfaces 83
D. Keith Roper, Jeremy Dunklin, Gregory T. Forcherio, Keith Berry, Carter Bodinger, Tyler Howard
(59b) On the Rice Husk Ash Admixing with Cement: Preparation, Characterization and Analysis 84
Nikhil Prakash
(59c) Tunable Magnetoresistance of Conductive Polymer Nanocomposites 85
Jiang Guo, Alexandra Galaska, Suying Wei, Brian J. Edwards, Bamin Khomami, Zhanhu Guo
(59d) Polydopamine Stabilized Fluorescent Nanozinc Oixde Reinforced Epoxy Nanocomposites Towards UV Shielding 86
Chaobo Liang, Ping Song, Hongbo Gu, Junwei Gu
(59e) Comparing the Toughening Effects of Modified Graphene Oxide and Core-Shell Rubber on Polyester Resins and Glass Fiber-Reinforced Polyester Composites 87
Kunwei Liu, Siyao He, Yuqiang Qian, Qi An, Andreas Stein, Christopher W. Macosko
(59f) Preparation and Characterizations of Barium Ferrite/Epoxy Nanocomposites 88
Hongyuan Zhang, Hongbo Gu

(59g) Effect of Compatibilizer on Strength and Toughness of Glass Fiber Mat Reinforced Polypropylene Composites 89
Chunyin Shen, Yadong Wu, Haiqing Wan, Junyan Wang, Gance Dai
(59h) A Facile Approach to Fabrication and Characterization of an Eco-Friendly Zein-Laponite Nanocomposite with Improved Mechanical, Thermal, Barrier and Surface Property 90
Tahrima B. Rouf
(78a) Solid Dispersions of Electroactive Materials for Energy Storage Applications 91
Gary M. Koenig Jr., Zhaoxiang Qi, Devanshi Gupta
(78b) Rational Design of the Cathode Materials in the Lithium-Sulfur Batteries 92
Tong Mou, Bin Wang
(78c) 3D Carbon Materials for Electric Double-Layer Capacitors with Ultrahigh Areal Capacitance 93
Liang Chang, Yun Hang Hu
(78d) 3D High-Surface-Area and Mesoporous Graphene Sheet-like Carbon for Supercapacitors 94
Haiyang Shen, Min Wei, Gang Wu
(78e) Stability Predictions for Dimethoxybenzene Based Catholyte Materials 95
Benjamin Silcox, Rajeev Assary, Jing Jing Zhang, Siu On Tung, Ilya Shkrob, Lu Zhang, Levi T. Thompson
(78f) Solvate Ionic Liquid-Based Gel Electrolytes Containing Functionalized Polymer-Based Networks for Use in Lithium Metal Battery Applications 96
Anthony D'Angelo, Matthew J. Panzer
(78g) Studies on Complex Electrolytes for Magnesium Batteries 97
Laura Merrill, Hunter Ford, Jennifer Schaefer
(78h) First-Principles Study of Temperature Dependence of Energy Gaps in Gas Sensor Materials 98
Yuning Wu, Yuhua Duan, Paul R. Ohodnicki, Benajmin T. Chorpening
(78i) Supercritical Fluid-Based Synthesis of Antimony Electrode Materials 99
Grant A. Williamson, Elena P. Pandres, Vincent C. Holmberg
(96a) Solution Combustion Synthesis of Porous CeO2 Nanopowders: Reaction Mechanism and Physical Properties 100
Wooram Kang, Derya Oncel Ozgur, Arvind Varma
(96b) Combining Pre- and Post-Nucleation Trajectories for the Design of Hierarchical FAU/EMT Materials from Organic-Free Sols 101
Dina Gaber, Safa Gaber, Issam Ismail, Saeed Alhassan, Maryam Khaleel
(96c) Synthesis of Sn-MFI Zeolite with Use of Mechanochemical Reaction 102
Kiyoshi Kanie, Moe Sakaguchi, Fumiya Muto, Masafumi Nakaya, Toshiyuki Yokoi, Atsushi Muramatsu
(96d) Crystallization of One-Dimensional Zeolites By Nonclassical Pathways 103
Rui Li, James Sutjianto, Aseem Chawla, Jeffrey D. Rimer
(96e) Expanding the Scope of Fluoride-Free Pure Silica Zeolite Syntheses 104
Vivek Vattipalli, Wei Fan
(96f) Atomic Resolution Imaging of MEL Intergrowth in 2-Dimensional MFI Nanosheets 105
Prashant Kumar, Han Zhang, Neel Rangnekar, Michael Tsapatsis, K. Andre Mkhoyan
(96g) Synthesis of Single-Unit-Cell Hierarchical Zeolites with Tunable Mesoporosity By Controlling Intergrowth Frequency 106
Dandan Xu, Anatoliy Kuznetsov, Prashant Kumar, Maryam Khaleel, Saeed Alhassan, Michael Tsapatsis
(96h) A Full Understanding of Microporous Vanadosilicate AM-6: The Crystal Quality and Structure of AM-6 107
Rumeysa Tekin, Juliusz Warzywoda, Albert Sacco Jr.
(98a) Rapid Wood Fractionation < 80° C for Sustainable and Economic Biorefinery 149
J. Y. Zhu
(98b) Insights into Biomass Recalcitrance 150
Arthur J. Ragauskas
(98c) Low Temperature and High Efficiency Biomass Fuel Cell and Bio-Hydrogen Production 151
Yulin Deng
(98d) Circular Economy: A Path Towards Innovation and Commercialization of Biocomposites for Sustainable Manufacturing 152
Amar K. Mohanty
(79a) Tuning the Molecular Design of Catalytic Materials to Increase Activity and Selectivity for Fine Chemical Production 153
Nicholas Brunelli, Aamena Parulkar, Nitish Deshpande, Mariah Whitaker, Rutuja Joshi
(79b) Deactivation of Zeolite Catalysts during Hydrodeoxygenation of Aromatic Oxygenates 154
Guo Shiou Foo, Mariana V. Rodrigues, Qandeel Almas, Chukwuemeka Okolie, Matthew M. Yung, Carsten Sievers
(79c) Improving Methanol-to-Olefins Conversion Performance of CHA Materials By Seeding the Hydrocarbon Pool 155
Praveen Bollini, Aditya Bhan

(79d) Tuning of Higher Alcohol Selectivity and Productivity in CO Hydrogenation Reactions over K/MoS2 Catalysts Supported on Mesoporous Activated Carbon and Mixed Mgal Oxide 156
Micaela Taborga Claure, Song-Hai Chai, Sheng Dai, Faisal M. Alamgir, Pradeep K. Agrawal, Christopher W. Jones
(79e) Kinetic and Spectroscopic Investigations of Alcohol Conversions over Metal Oxide Catalysts 157
Shuai Tan, Yongqiang Cheng, Luke L. Daemen, Ho Nyung Lee, Benjamin Doughty, Daniel Lutterman
(79f) Catalytic Upgrading of Fast Pyrolysis Bio-Oil for Renewable Hydrocarbon Production 158
Mariefel V. Olarte, Huamin Wang, Daniel Santosa, John G. Frye, Suh-Jane Lee, Jae-Soon Choi, Pimphan Aye Meyer, Susanne Jones, Corinne Drennan, Alan H Zacher
(79g) Chemical Reaction Engineering Principles of Continuous Flow Photoredox Catalysis 159
Eric G. Moschetta, Kaid Harper, Steve Richter, Steven J. Wittenberger
(79h) Factors Affecting Catalytic Performance in the Presence of Non-Thermal Plasmas 160
Jongsik Kim, Prateek Mehta, Patrick Barboun, William F. Schneider, David Go, Jason C. Hicks
(79i) Accelerating Innovation in Advanced Manufacturing 161
Michael McKittrick
(123a) Active Polymer Materials for Flexible Electronics: Molecular Design and Processing for Efficient Macroscale Charge Transport Pathways 162
Nils Persson, Michael McBride, Ping-Hsun Chu, Martha A. Grover, Elsa Reichmanis
(123b) Promoting Adhesion between Immiscible Polymers 163
Christopher W. Macosko
(123c) High Resolution Lithography Via Block Copolymers and Self-Assembling Surface Neutral Layers 164
Peter Trefonas III, Jong Keun Park, Mingqi Li, Janet Wu, Emad Aqad, Dan Millward, Valeriy Ginzburg, Phil Hustad
(123d) Field and Confinement-Directed Self-Assembly of Soft Mesophases to Create Useful Materials 165
Chinedum O. Osuji
(126a) Nanostructured Interfaces for Enhanced Biologic Transport and Immunomodulation 166
Tejal Desai
(126b) Nano- and Microfabricated Hydrogels for Regenerative Engineering 167
Ali Khademhosseini
(126c) Overcoming Obstacles to Brain Repair Using Biomaterials 168
Tatiana Segura
(152a) Digital Alchemy for Assembly Engineeringc 169
Sharon C. Glotzer
(152b) Improved Algebraic, Numerical, and Graphical Representations in Fluid Mechanics 170
Stuart W. Churchill, James C. Hill
(152c) The Scaling of Turbulence Near the Wall and the Churchill Turbulent Flux Correlation: Insights with Lagrangian Simulations 171
Dimitrios V. Papavassiliou, Quoc T. Nguyen, Chiranth Srinivasan
(152d) Flow Boiling Using a Piranha Pin Fin Heat Sink 172
Cory Woodcock, Xiangfei Yu, Yoav Peles, Joel L. Plawsky
(152e) Transport Problems in the Spirit of Stuart Churchill for Teaching and Research at the University of Michigan 173
Ronald G. Larson, Claudio Vilas Boas Favero
(176a) Solve this! Fundamental Approach to Problem Solving in Industrial Processes I (Invited Talks) 174
Zdravko Stefanov, Paul Chauvel, Jr., Eldad Herceg, Dana A. Livingston
(177a) Tuning External Surface and Textural Properties of Unit-Cell Thick Pillared MFI and Pillared MWW Zeolites By Atomic Layer Deposition and Its Consequence on Catalytic Reactions 175
Dongxia Liu, Junyan Zhang
(177b) Mesoporous TiO2 to TiO2-Cellulose Composite and its Derivatives for Environmental Remediation 176
Jinju Zhang, Lei Li, Yanxiang Li, Lixia Cao, Chuanfang Yang
(177c) Reliable Fabrication and Surface Modification of Beta Zeolite Membrane for Pervaporation of n-Butanol/Water Mixtures 177
Yun Li, Honghong Xu, Tao Chen, Xiufeng Liu, Baoquan Zhang
(177d) Investigation of Factors That Induce Cristobalite Formation during Titanosilicate Synthesis and Their Potential Impact on Heterogeneous Catalysis 180
Ayomi S. Perera, Haiyue Yu, Jeremy Cockcroft, Panagiotis Trogadas, Marc-Olivier Coppens
(177e) Novel Methods to Synthesize ZSM-11 As an Efficient Catalyst for Methanol-to-Hydrocarbon Reactions 181
Yufeng Shen, Thuy T. Le, Jeffrey D. Rimer

(177f) Preparation of Novel Al-MFI/Fe-MFI Core-Shell Catalysts and Their Catalytic Application for CH4 Conversion 182
Toshiyuki Yokoi, Yoshihiko Kimura, Takaya Kimura, Yusuke Kunitake, Atsushi Muramatsu
(177g) An Optimized Procedure for Selective Removal of Efal (Extra Framework Aluminum) in Y-Zeolites 183
Balasubramanian Vaithilingam, Gnana Pragasam Singaravel, Abdul Majed Al Katheeri, Stephane M., Mikael Berthod
(774a) Multifunctional Polymer Nanocomposites 184
Zhanhu Guo, Jiang Guo, Alexandra Galaska, Huige Wei, Suying Wei, Bin Qiu, Dawei Jiang, Hongbo Gu, Jiahua Zhu
(774b) Transparent Copper-Silica Nanoparticle-Chitosan Nanocomposite Coatings with Long-Term Antibacterial Efficacy 185
Debirupa Mitra, Min Li, En-Tang Kang, Koon Gee Neoh
(774c) Bio-Inspired Design of Stimuli Responsive Materials Based on a Bilayer Structure 186
Songshan Zeng, Rui Li, Dianyun Zhang, Wenhan Huang, Zhaofeng Wang, Stephan Freire, Andrew Smith, Emily Huang, Helen Nguon, Xiaoyuan Yu, Luyi Sun
(774d) Biomimetic Nanocoatings with Exceptional Mechanical, Barrier, and Flame Retardant Properties from Large Scale One-Step Co-Assembly 187
Fuchuan Ding, Jingjing Liu, Songshan Zeng, Yan Xia, Kacie M. Wells, Mu-Ping Nieh, Luyi Sun
(774e) Bioinspired Composite Materials with Stimuli-Responsive Color Changing Ability 188
Golnaz Isapour, Marco Lattuada
(774f) Investigation of Carbon Nanotubes & Cellulose Nanocrystals Composite for Potential Use in Microelectromechanical Systems 189
Mingzhe Jiang, Christopher L. Kitchens, Robert Seney, Bayliss Charles
(774g) Few Layers MoSe2 Incorporated with Nitrogen Doped Graphene Sheet for High Performance Lithium Sulfur Batteries 190
Hoi Lun Wong, Zhengtang Luo
(774h) Dual-Responsive Plasmonic Behavior of Gold Nanorods@PANI Core/Shell Nanostructures for Real-Time Control 191
Ju-Won Jeon, Jing Zhou, Jeffrey Geldmeier, James Ponder, Mahmoud A. Mahmoud, Mostafa El-Sayed, John Reynolds, Vladimir V. Tsukruk
(118b) Microwave Induced Heating of Carbon Nanotubes Localized at 3D-Printed Thermoplastic Interfaces 192
Charles Sweeney, Mohammad Saed, Micah Green
(196a) Theoretical Study of the Reaction Kinetics of Organosiloxane Polycondensation 193
Mona Bavarian, Siamak Nejati
(196aa) High-Ï‡ Block Copolymers with High Etch Selectivity for Sub-10 Nm Patterning 194
Sung-Soo Kim, Walter W. Young, Luis E. Oquendo, Michael Maher, Sunshine X. Zhou, Yusuke Asano, Marc A. Hillmyer, C. Grant Willson, Christopher J. Ellison
(196ab) Fabrication and Structural Analysis of Nanofibers Made By Syndiotactic Polypropylene with Ethylene-Comonomer Units 195
Fuyuaki Endo, Claudio De Rosa, Atsushi Hotta
(196ad) Interaction Between Supercritial CO2+Cosolvent and Poly(vinyl acetate) 196
Dong-Dong Hu, Lei Bao, Ling Zhao, Tao Liu
(196ae) Modeling of Distributions of Polymer Properties Using Parallel Computing in Julia 198
Esteban Pintos, Mariano Asteasuain
(196af) Transition Metal-Based Nanocrystals Confined-Growth on Heteroatom-Doped Graphene Toward Hydrogen Catalysis 199
Minghao Zhuang, Zhengtang Luo
(196d) From Process to Product - Enhancing the Understanding of Î±-Olefin-Polymerizations 200
Kristina M. Pflug, Jonas Nowottny, Markus Busch
(196e) Amine Effects on Radial-Mediated Thiol-Ene Reactions 201
Dillon Love, Kang-Min Kim, Johnathan Goodrich, Benjamin D. Fairbanks, Mark Stoykovich, Charles B. Musgrave, Christopher N. Bowman
(196f) Influence of Phosphate Salts and Solution pH on Aqueous-Phase NVP Free-Radical Polymerization 202
Fernando T P Borges, Fouad Teymour
(196h) Synthesis and Characterization of Crosslinked Polymers from Cottonseed Oil 204
Rangana Wijayapala, Deonante Frazier, Bill B. Elmore, Charles Freeman, Santanu Kundu
(196i) Date Pits as Cost-Effective, Renewable, and Efficient Fillers for Polymers 205
Fares Alsewailem, Yazeed Binkhodor

(196j) Biaxial and Shear Deformation of Simulated Amorphous Cis-, Trans-1, 4-Polybutadiene Chains 206
Suvrajyoti Kar, Michael L. Greenfield
(196k) Decomposition Behavior of Laponite/PLGA-PÎ•G-PLGA Nanocomposite Hydrogels at Body Temperature 207
Midori Kitagawa, Tomoki Maeda, Atsushi Hotta
(196l) Synthesis of Thermoplastic Polydimethylsiloxane with L-Phenylalanine-Based Hydrogen-Bond Network and its Self-Healing Property 208
Shunsuke Tazawa, Atsushi Shimojima, Tomoki Maeda, Atsushi Hotta
(196n) Controlled Swelling Rate Elastomer for Packers 209
Rostyslav Dolog, Darryl Ventura, Valery N. Khabashesku, Qusai Darugar
(196o) Effect of the Cross-Linking Agent (sodium polyphosphate) on Performances of Nacs-Wsc Microcapsules 215
Qing-Xi Wu, Yi-Xin Guan, Jun-Jie Yuan, Shan-Jing Yao
(196p) Swelling Behaviors of Cr(III)-Modified Acrylamide-Based Superabsorbent Polymer Microsphere in Brines 216
Jingyang Pu, Jiaming Geng, Na Zhang, Baojun Bai
(196q) Synthesis of Chemical Protective Elastomeric Barrier Materials 225
James Ogilvie-Battersby, Alessandra Molinaro, Christopher Zoto, Quoc Truong, Nese Orbey
(196r) Polysulfide-Based Nanofiber Prepared Via Inverse Vulcanization and Electrospinning for Effective Mercury (II) Sequestration 226
Lawrence A. Limjuco, Grace M. Nisola, Khino J. Parohinog, Kris Nino G. Valdehuesa, Wook-Jin Chung
(196s) Mechanical Properties of Maleated Polymer/Graphene Oxide Composites using Graphene Oxide as a Multi-Functional Crosslinker 227
Szu-Ming Yang, Heonjoo Ha, Christopher J. Ellison
(196t) Development and Analysis of a Thin Film Nanocomposite Membrane: Resistance to Chlorine 228
Abdulmajeed Altalhi, Holly A. Stretz
(196u) Improving Gas Transport Properties of Mixed Matrix Membranes Via Interfacial Improvement 229
Ahmad Arabi Shamsabadi, Morteza Sadeghi, Mohammad Dinari, Mahsa Salehi, Masoud Soroush
(196v) Antimicrobial Polymers: Present State of the Art 230
Nikhil Prakash
(196w) Layer-By-Layer Coated Microneedle Arrays for Staged Multi-Agent Immune Attack on Melanoma 231
Yanpu He, Jiahe Li, Hongkun He, Celestine Hong, Maylin Funkenbusch, Sheryl Wang, Maya Berlinger, Darrell J. Irvine, Paula T. Hammond
(196x) Direct Observation of Remarkable Nanoparticle Evolution during Aqueous Dissolution of Polymer/Drug Particles 232
Ralm Ricarte, Marc A. Hillmyer, Timothy P. Lodge
(196y) Polymerized Ionic Liquid Pentablock Terpolymer for Lithium-Metal Batteries 233
Tzu-Ling Chen, Yossef A. Elabd
(196z) Effect of Electric Field on the Structure and Dynamics of Model Ionomer Melts 234
Janani Sampath, Lisa M. Hall
(197a) Influence of Molecular Design on the Self-Assembly of Single-Stranded DNA Amphiphiles 235
Thomas Gartner III, Huihui Kuang, Efrosini Kokkoli, Arthi Jayaraman
(197b) Intracellular Trafficking of Enzyme-Cleavable Peptide Amphiphiles 236
Handan Acar, James L. Labelle, Matthew V. Tirrell
(197c) Polylactide-Based Biodegradable Zwitterionic Polymers and Their Conjugates with Drugs for Biomedical Applications 237
Haotian Sun, Michael Yu Zarng Chang, Wei-I Cheng, Qing Wang, Alex Commisso, Meghan Capeling, Yun Wu, Chong Cheng
(197d) Overcoming Obstacles to Brain Repair Using Biomaterials 238
Tatiana Segura
(197e) Aggregation Kinetics in Biological Environments as a Determinant of Nanoparticle Behavior in the Brain 239
Chad D. Curtis, Mike McKenna, Elizabeth Nance
(197f) Layer-By-Layer Nanoparticles for Interleukin-12 Delivery 240
Antonio E. Barberio, Santiago Correa, Erik Dreaden, Talar Tokatlian, Mariane B. Melo, Darrell J. Irvine, Paula T. Hammond
(197g) Implantable Biomaterials Produced By Complexing Chitosan to Alginate or Pectin: Surface Properties, Hemocompatibility and Cytotoxicity 241
Fernanda C. Bombaldi De Souza, Renata F. Bombaldi De Souza, Angela Maria Moraes, Diego Mantovani

(197h) Development of Sustainable Therapeutic Dressings Consisting of Chitosan-Alginate Films Incorporating Arrabidaea Chica Verlot Extract 242
Ana Luiza Resende Pires, Cecilia Buzatto Westin, Ilza Maria De Oliveira Sousa, Mary Ann Foglio, Angela Maria Moraes
(197i) Comparison of Chitosan Particles Produced By Ionic Gelation and By Supercritical Assisted Atomization 250
Julia Natalia Oliveira Mazoni, Paulo De Tarso Vieira E Rosa
(197j) Exploiting a Novel Aqueous-Two Phase Microfluidic System for Cell Encapsulation in GAG+Chitosan Microcapsules 251
Amin Vossoughi Shahvari, Howard W. T. Matthew
(197k) Laser-Activated Nanocomposites for Tissue Repair 252
Russell Urie, Deepanjan Ghosh, Mitzi Thelakkaden, Tanner Flake, Jerry Crum, Chengchen Guo, Jeff Yarger, Kaushal Rege
(197l) New Strategy for the Fabrication of Annular Cylindrical Polysaccharide-Based Scaffolds 253
Angela Maria Moraes, Renata F. Bombaldi De Souza, Fernanda C. Bombaldi De Souza
(197m) Formulation of Peptide Antimicrobials for Treatment of Wound Infections 254
Ritu Goyal, Michael Holloway, Pooja Patel, David Devore, Charles Roth
(197o) Synthesis and Degradation of Biodegradable Copolymers 255
Eswar Arunkumar Kalaga, Timothy Brenza
(197r) Ultra-High Surface Area Activated Carbon from a Renewable Resource 256
Ashli Polanco, Dmytro Volkov, Quoc Truong, Carl Lawton, Nese Orbey
(197t) A Facile Novel Fluorocarbon Copolymer Solution Coating Process for Improving Platelet Compatibility of Titanium 257
Sophia Chao-Wei Huang, Chi-Hui Cheng, Yun Chiu, Yi-Ching Lin, Jui-Che Lin
(198a) Alignment of Quantum Dot Nanorod/Silica Hybrid Particles on Glass Substrate for Luminescent Solar Concentrator 258
Kiju Um, Young-Geon Song, Kangtaek Lee
(198b) Effect of Graphene Oxide on Formation of Zirconium Tungstate Nanoparticles 259
Young-Geon Song, Kiju Um, Kangtaek Lee
(198d) Rapid Microwave-Assisted Synthesis of Hybrid Zeolitic-Imidazolate Frameworks 260
Febrian Hillman, John Zimmerman, Seung-Min Paek, Mohamad Hamid, Woo Taik Lim, Hae-Kwon Jeong
(198e) Facile Synthesis of Cd-Substituted Zeolitic-Imidazolate Framework Cd-ZIF-8 and Mixed-Metal Cdzn-ZIF-8 261
Jingze Sun, Hae-Kwon Jeong, Woo Taik Lim, Liya Semenchenko
(198f) A Study of Asymmetry Cu-MOFs Electrode Prepared in situ and Its Biomimetic Catalysis 263
Zhipeng Li, Liwei Ren, Diannan Lu
(198g) Protected SiC Catalyst Support for Steam Methane Reforming Reaction 264
Naftali Opembe, Seungdoo Park, Sergio Ibanez, Doug Mitchell, Matthew Seabaugh, Scott Swartz
(198h) Computational Screening of High Temperature Materials for Environmental Barrier Thin Films 265
Amanda Hoskins, Aidan Coffey, Charles B. Musgrave, Alan W. Weimer
(198i) Phase Transformation Induced By Tertragnility Variation of Metal-Redox Synthesised NiMn Nanoalloys 266
Jian Shen, Xin Jin
(198j) Fabrication of Electrospun Mesoporous Silica Nanomaterials for Water Vapor Adsorption 268
Soyoung Kim, Heechul Choi
(198k) Applications of Mesoporous RuCo2O4 Thin Film for High Performance Supercapacitor 269
Do-Heyoung Kim, Nilesh R. Chodankar
(198l) TiO2 Thin Film Deposition By Electrospray 270
Yaqun Zhu, Jong Hyun Shim, Junghyun Cho, Paul R. Chiarot
(198m) Green Synthesis of Copper Oxide Nanoparticles Using a Simple Microwave-Assisted Method 271
Prasad P Pawar, Shishir V Kumar, Adarsh Bafana, Ashiqur Rahman, Si A. Dahoumane, Clayton S Jeffryes
(198n) Comprehensive Thermodynamic Modeling of Mixed-Solvent Electrolyte Systems: An Investigation on the Quaternary System of FeCl2-FeCl3-HCl-H2o 272
Sina Hassanjani Saravi, Chau-Chyun Chen
(198o) Controllable Manipulation of Continuous AFI Membranes with Distinctive Microstructures on Macroporous Alpha-Alumina Substrates 273
Hongfeng Dong, Xiufeng Liu, Huiming Zhu, Baoquan Zhang, Jian Li
(198p) Magnetic Core Shell Microspheres for Extraction of Rare Earth Elements from Geothermal Brine Solution 274
B. Peter McGrail, Jian Liu

(198r) Encapsulation of Dye in NH2-Uio-66 Metal-Organic Framework for Photosensitized Oxidation of Benzyl Alcohol 275
Xiyi Li, Qingqing Hou, Neng Liao, Jing Xiao
(199a) The Fabrication of Graphene/Polyaniline Blended Fiber for Conducting and Flexible Energy Storage Devices 276
Yafei Feng, Jiaxin Shen, Cunliang Ma, Yidong Liu, Yong Min
(199b) Semiconducting Heterostructures for Photocatalytic Reduction of Carbon Dioxide 277
Debtanu Maiti, Johnnie Cairns, J. N. Kuhn, Venkat R. Bhethanabotla
(199c) Photoswitchable Quantum Dots Probes for Superresolution Microscopy 278
Abhilasha Dehankar, Kil Ho Lee, Abhijeet Marar, Karine Thate, Carol Lynn Alpert, Peter Kner, Jessica O. Winter
(199d) Bottom-up Synthesis of Nanoelectronic Titania Composites 279
Yang Lu, Evan K. Wujcik, Arijit Bose
(199h) Phase Diagrams, Defect Models and Thermoelectric Properties: ÃŸ-Ag2se and CoSb3 280
Sinn-Wen Chen, Zi-Yang Huang, Yang-Yuan Chen
(199i) Theoretical Study of a High Performance Thermoelectric Material: Stanene 281
Pabitra Choudhury, Charles Griego
(199j) Revealing the Enigmatic Interfacial Layer of Core/Shell Quantum Dots 282
Ajay Singh, Jennifer Hollingsworth
(199k) Spatial Manipulation of Thermal Flux Profiles Using Nanostructure Boundaries 283
Abhinav Malhotra, Martin Maldovan
(200a) Magnetic Polymer Nano-composites for Giant Magnetoresistance and Electromagnetic Shielding 284
Jiang Guo, Alexandra Galaska, Brian J. Edwards, Bamin Khomami, Zhanhu Guo
(200b) Constructing Ternary Conductive Polymer Composites with Cocontinuous Polymer Blends and Interfacial Graphene Nanoplatelets 285
Yangming Kou, Lian Bai, Xiang Cheng, Christopher W. Macosko
(200c) The Influence of Interfacial Graphene on the Morphological, Electrical and Mechanical Properties of Co-Continuous Polymer Blends 286
Catherine Esnaashari, Lian Bai, Christopher W. Macosko, Xiang Cheng
(200d) The Fabrication and Application of Composite Graphene Oxide Films 287
Jiaxin Shen, Yafei Feng, Cunliang Ma, Yidong Liu, Yong Min
(200e) Mechanically Stable Thermally Crosslinked Poly(acrylic acid)/ Reduced Graphene Oxide Aerogels 288
Heonjoo Ha, Han Xiao, Kadhiravan Shanmuganathan, Christopher J. Ellison
(200f) Ionophore-Decorated Magnetic Graphene Oxide As a Composite Adsorbent Material for Heavy Metal Ion Sequestration 289
Khino J. Parohinog, Grace M. Nisola, Wook-Jin Chung
(200g) Synthesis of Thermoresponsive Polymer/Fe3O4 Nanoparticle Composite and Its Application 290
Risako Sakai, Junichi Ida, Tatsushi Matsuyama
(200h) High-Performance, Ambient Phase Change Thermal Diodes for Energy Applications 291
Anton Cottrill, Song Wang, Albert Tianxiang Liu, Yuichiro Kunai, Michael Strano
(200i) A Study on the Preparation and Properties of Polymer Composites Using Amino Functionalized Microcrystalline Cellulose (MCC) As a Filler Material 292
Kiryong Ha, Hanna Kim, Yeokyung Yang, Kiseob Hwang, Kwang-Hee Lim
(200j) Electrically Conductive Films Made of Polythiophene and Fibrillated Wood Particles 293
Islam Hafez, Han-Seung Yang, Mehdi Tajvidi, Nicholas Seaton, William T. Y. Tze
(200l) Effect of Different Carbon Additives on Structure of Magnesium Composites for Hydrogen Storage 294
Yeboah Martin Luther
(200n) Low Loading of Grafted Thermoplastic Polystyrene Strengthened and Toughened Transparent Epoxy Composites 295
Chao Ma, Hongbo Gu
(200o) Preparation of Modified Graphene Oxide-Containing Styrene Masterbatches for Thermosetting Resin Composite 296
Siyao He, Yuqiang Qian, Kunwei Liu, Chris Macosko, Andreas Stein
(200p) Smart Windows Enabled By Buckling Instabilities in Periodic Composite Films 297
Peng Jiang, Zhuxiao Gu
(200q) Magnetic Polymer Nanocomposites for Electromagnetic Interference Shielding 298
Jiang Guo, Alexandra Galaska, Suying Wei, Brian J. Edwards, Bamin Khomami, Zhanhu Guo
(262a) Autoperforation of 2D Materials for Generating Two Terminal Memresistive Janus Particles 299
Albert Tianxiang Liu, Pengwei Liu, Michael Strano

(262b) Driven Morphological Evolution of Strained Thin Film Surfaces and Two-Dimensional Materials: Morphological Stability and Pattern Formation ... 300
Lin Du, Dimitrios Maroudas
(262c) Optimization of THz-Time Domain Spectroscopy Reflectivity for a 1THz Energy-Harvesting Metamaterial ... 301
Shendu Yang, Zachary Thacker, Evan Allison, Patrick J. Pinhero
(262d) Modeling Exciton Dynamics and Low-Frequency Vibrations in Quantum Dot Assemblies ... 302
Elizabeth M. Y. Lee, Adam P. Willard, William A. Tisdale
(262e) Extraordinarily Slow Electron-Hole Recombination in Perovskite Phase Cesium Lead Iodide ... 303
Subham Dastidar, Siming Li, Jason B. Baxter, Aaron T. Fafarman
(262f) Elucidating the Impact of Alcohol Post-Processing in High Performance Roll-to-Roll Printed Organic Photovoltaics ... 304
Kevin L. Gu, Xiaodan Gu, Hongping Yan, Zhenan Bao
(262g) Metamaterial and Rectenna Design and Testing for the Conversion of Blackbody Radiation to Electricity Using 5 THz Devices ... 305
Evan Allison, Zachary Thacker, Shendu Yang, Patrick J. Pinhero
(262h) Reusable Chromogenic Sensors Enabled By Novel Multi-Stimuli-Responsive Shape Memory Polymers ... 306
Sin-Yen Leo, Peng Jiang
(265a) Allyl-Sulfide Modified Hydrogels with Switchable Properties As Dynamic Cellular Niches ... 307
Kristi S. Anseth, Tobin E. Brown, Joseph Grim, Ian Marozas
(265b) Non-Linear Rheology and Fracture in Alginate Hydrogels ... 308
Seyed Meysam Hashemnejad, Rangana Wijayapala, Santanu Kundu
(265c) Analyzing the Effects of Time and Crosslinker Ratio on the Mechanical Properties of Biodegradable Zein Super Gels ... 309
Hazal Turasan, Jozef Kokini
(265d) Self-Assembly, Structure and Rheology of Polyelectrolyte Complex Hydrogels ... 310
Samanvaya Srivastava, Adam Levi, Matthew V. Tirrell
(265e) Evolution of Mechanics in Î±-Helix Peptide Bioconjugated Linear- and Star-Block Peg ... 311
Sean C. O'Neill, Raymond Tu
(265f) Modelling of the Degradation of Poly(ethylene glycol)-Co-(lactic acid)-Dimethacrylate Hydrogels ... 312
Marco Lattuada, Giuseppe Storti, Vincent Diederich
(265g) Tough, Rapidly Swelling Thermoplastic Elastomer Hydrogels for Hemorrhage Control ... 313
Erich Bain, Tyler R. Long, Frederick L. Beyer, Randy A. Mrozek, Joseph L. Lenhart
(265h) Synthesis and Characterization of Thermally Responsive N-Isopropylacrylamide Hydrogels Copolymerized with Novel Hydrophobic Polyphenolic Crosslinkers ... 314
Shuo Tang, Thomas Dziubla, J. Zach Hilt
(265i) Novel Biocompatible Thermo-Responsive Poly(N-vinyl Caprolactam)/Clay Nanocomposite Hydrogels with Macro-Porous Structure and High Mechanical Property ... 315
Zhuang Liu, Kun Shi, Xiao-Jie Ju, Wei Wang, Rui Xie, Liang-Yin Chu
(267a) Multiplicity of Morphologies in Poly (L-lactide) Bioresorbable Vascular Scaffolds ... 316
Artemis Ailianou, Karthik Ramachandran, Mary Beth Kossuth, James Paul Oberhauser, Julia A. Kornfield
(267b) Tissue Response and Integration in Biomaterial Implants Derived from Morphologically Unique Emulsion Gels ... 317
Todd Thorson, Ali Mohraz, Elliot Botvinick
(426h) The Unique Mechanism of Covalently Adaptable Hydrogel Degradation Characterized with Passive Microrheology ... 318
Francisco Escobar, Kristi S. Anseth, Kelly M. Schultz
(267d) Development of Gelatin and Graphene Based Conduits Using 3D Printing Strategies for the Transdifferentiation of Mesenchymal Stem Cells into Schwann Cell-like Phenotypes through Electrical Stimuli ... 319
Metin Uz, Maxsam Donta, Donald S. Sakaguchi, Surya K. Mallapragada
(267e) Effect of Molecular Weight and Degree of Functionality on Degradation, Biocompatibility and Two-Photon Polymerization of Acrylated Poly(caprolactone) ... 320
Brian J. Green, Jessica Thompson, Kristan S. Worthington, Budd A. Tucker, C. Allan Guymon
(267f) An Injectable and Anisotropic Hydrogel with Biomimetic Structures for Directed Cell and Nerve Growth ... 321
Jonas C. Rose, David B. Gehlen, Esther Jaekel, Jens Kohler, Khosrow Rahimi, Martin Moller, Laura De Laporte
(267g) Engineering Extracellular Matrix Mimetic Materials By Green Electrospinning of Collagen ... 322
Jorge Almodovar, David Castilla

(267h) Biomimetic Scaffolds for in vitro Bone Marrow Tissue Engineering 323
Yongkuk Park, Ryan Carpenter, Jungwoo Lee
(303a) Tapered Block Copolymers: Tuning Self-Assembly and Properties By Manipulating Monomer Segment Distributions 324
Thomas H. Epps
(303b) Photoplasiticity in Crosslinked Liquid Crystalline Networks: A Route to Reconfigurable Shape-Changing Materials 325
Matthew K. McBride, Matthew Hendrikx, Danqing Liu, Brady Worrell, Dick J. Broer, Christopher N. Bowman
(303c) Synthesis of Novel Nanostructured Copolymers with Alternating Linear Polymer and Dendrimer Blocks 326
Haotian Sun, Alex Commisso, Chong Cheng
(303d) Hierarchical Structures of PDMS-PU Copolymer and Particles for Hydrophobic Coatings 327
Marius Rutkevicius, Mackenzie Geiger, Tahira Pirzada, Saad A. Khan
(303e) Charged Polymer Conformations in Polyelectrolyte Complexes 328
Amanda B. Marciel, Samanvaya Srivastava, Matthew V. Tirrell
(303f) Solubility Parameters, Water Activity Coefficients and Proton Mobility of Sulfonated Poly (styrene-isobutylene-styrene), Sulfonated Poly (ether ether ketone), and Sulfonated Poly (2-ethoxyethyl methacrylate) Membranes 329
Maritza Perez Perez, David Suleiman
(303g) Single-Molecule Super-Resolution Microscopy in Nanostructured Polymer Thin Films 330
Muzhou Wang, James M. Marr, Jeffrey W. Gilman, J. Alexander Liddle
(303h) A Non-Equilibrium Molecular Dynamics (NEMD) Simulation of the Crosslinked Polyamide Membranein Water Desalination 331
Md Symon Jahan Sajib, Tao Wei
(306a) Linear and Nonlinear Rheology Predictions of Entangled Polymers in Complex Flows from First Principles 338
Jay D. Schieber
(306b) Melt Behavior-Chain Architecture-Polymer Composition Correlations in High Density Polyethylene 339
Rohan Hule, Derek W. Thurman, Antonios Doufas
(306c) Shear-Induced Conformational Changes of Flexible and Semi-Rigid Engineering Thermoplastics and Their Influence on Crystallization 340
Behzad Nazari, Jiho Seo, Ralph H. Colby, Alicyn M. Rhoades, Richard P. Schaake
(306d) Island-in-the-Sea Meltblown Nanofiber Nonwovens with Diverse Surface Properties for Filtration Applications 341
Iman Soltani, Satish Kumar, Frank S. Bates, Christopher W. Macosko
(306e) Enthalpic and Entropic Competitions in Solvent-Free Polymer-Grafted Nanoparticles 342
Snehashis Choudhury, Lynden A. Archer
(306f) Designing Material Dynamics in Polyelectrolyte Complexes 343
Yalin Liu, Brian Momani, Matthew Labbe, H. Henning Winter, Sarah L. Perry
(306g) Extensional Relaxation Times of Dilute and Semi-Dilute Polymer Solutions 344
Jelena Dinic, Leidy N. Jimenez, Madeleine Biagioli, Vivek Sharma
(306h) Nonequilibrium Molecular Dynamics Simulations of Entangled Polymer Solutions Undergoing Planar Elongational Flows 345
Mohammad Hadi Nafar Sefiddashti, Brian J. Edwards, Bamin Khomami
(306i) Ultra-High Performance Polymers Meet Ionic Liquids 346
Jason E. Bara, Kathryn O'Harra, Grayson P. Dennis, Marlow M. Durbin, Max Mittenthal, Enrique M. Jackson
(334a) Paper-Based Cell Culture Platforms for Personalized Medicine 347
Gulden Camci-Unal
(334b) Growth Factor Delivery from Silk-Extracellular Matrix Composite Sponges for Modulating Congenital Heart Defect Repair 348
Whitney L. Stoppel, Elizabeth C. Bender, Luke R. Perreault, Jonathan M. Grasman, Andrea Papait, David L. Kaplan, Lauren D. Black III
(334c) Oligodendrocyte Precursor Cell Maturation in a 3D Hydrogel System through the Incorporation of Drug Delivery Nanoparticles or Topographical Cues 349
Lauren Russell, Meghan Pinezich, Kyle Lampe
(334d) Resveratrol Releasing Scaffolds Protect Mice Against Diet Induced Obesity and Glucose Intolerance 350
Michael Hendley, Prakasam Annamalai, Michael Gower
(334f) Effects of Short-Term Magnetic Stimulation on MSCs Encapsulated in an Injectable, Magneto-Responsive Hydrogel 351
Adedokun Adedoyin, Adam Ekenseair

(334g) The Incorporation of Retinoic Acid-like Peptoids Onto an Artificial Extracellular Matrix for Increased Differentiation of Human Embryonic Stem Cells into Neural Cells 352
Jesse Roberts, German Perez, Shannon L. Servoss
(334h) 3D Culture of Trabecular Meshwork Cells 353
Matthew Osmond
(354b) Steps Toward Bio-Sourced Packaging: Effect of Composition and Processing on Properties of Co-Polyesters of Polyethylene Terephthalate (PET) with 2,5-Furan Dicarboxylic Acid 354
Anup Joshi, Maria Coleman
(354c) Synthesis and Solutionphasecharacterization of Hydroxylated Sulfonated Oligothioetheramides 355
Joseph Brown, Christopher A. Alabi
(354d) Predicting Stable and Metastable Frank-Kasper Phases in Block Polymers Using Self-Consistent Field Theory 356
Akash Arora, Kyungtae Kim, Morgan W. Schulze, Ronald M. Lewis III, Frank S. Bates, Kevin D. Dorfman
(354e) Spatiotemporal Evolution of Structure in Layer-By-Layer Assembled Thin Films Composed of Oppositely Charged Polyelectrolytes 357
Ali Salehi, Ronald G. Larson
(354f) High Performance Roll-to-Roll Printed PTB7-Th/PCBM Organic Solar Cells 358
Kevin L. Gu, Xiaodan Gu, Hongping Yan, Zhenan Bao
(354g) Pinch-Off Dynamics, Dripping-Onto-Substrate (DoS) Rheometry and Printability of Polymeric Complex Fluids 359
Jelena Dinic, Leidy N. Jimenez, Madeleine Biagioli, Vivek Sharma
(354h) Ultra-Fragile 'Granular Materials' Designed Via a Genetic Algorithm 360
Venkatesh Meenakshisundaram, Jui-Hsiang Hung, David S. Simmons
(354i) Functionalizing Surfaces with Zwitterionic Polymers to Control Cell Adhesion and Direct Neurite Growth 361
Braden Leigh, Elise Cheng, Corinne Andresen, Marlan Hansen, C. Allan Guymon
(354j) Understanding Facilitated Transport of Hydrogen in Polybenzimidazole Containing Palladium Nanoparticles Using an Integrated Experimental and Modeling Approach 364
Lingxiang Zhu, Deqiang Yin, Shailesh Konda, Mark T. Swihart, Haiqing Lin
(364a) Engineering Surfaces and Interfaces in Polymer Films with Bottlebrush Polymer Additives 365
Gila Stein, Rafael Verduzco
(364b) Effects of Extensional Flow and Nanoparticle Stabilization on Immiscible Polymer Blend Morphology 366
Matthew S. Thompson, Sushant Agarwal, Xueyan Song, Rakesh K. Gupta
(364c) Non-Isocyanate Polyurethane Thermoplastic Elastomer: Amide-Based Chain Extender Yields Enhanced Nanophase Separation and Properties in Polyhydroxyurethane 367
Goliath Beniah, David Fortman, William Heath, William Dichtel, John M. Torkelson
(364d) Molecular Simulations Study of Solvophobicity Effects on Assembled Structure in Solutions of Amphiphilic Block Copolymers and Nanoparticles 368
Daniel J. Beltran-Villegas, Arthi Jayaraman
(364e) Control Nano/Microstructure Using Photopolymerization-Induced Phase Separation (PhIPS) 369
Erion Hasa, Julie L. P. Jessop, Jeffrey W. Stansbury, C. Allan Guymon
(364f) Spontaneous Self-Assembly and Micellization of Random Copolymers in Organic Solvents 370
Ayse Asatekin
(364g) Synthesis and Antibacterial Study of Star-Shaped Poly[2-(dimethylamino)Ethyl Methacrylate]-Based Copolymers with an Inorganic Core 371
Hou Zheng, Yuji Pu
(364h) Understanding Molecular Exchange Kinetics in Polyelectrolyte Complex Micelles 372
Hao Wu, Jeffrey Ting, Samanvaya Srivastava, Matthew V. Tirrell
(364i) Molecular Simulation for the Prediction of Plasticizer Efficiency and Stability in a Polymer Matrix 373
Dongyang Li, Kushal Panchal, Li Xi
(375a) Copper-Silver Core-Shell Nanoparticles for Conductive Ink 374
Xiaofeng Dai
(375b) Electric Current-Induced Nanoscale Surface Roughness Reduction in Conducting Thin Films 375
Lin Du, Dimitrios Maroudas
(375c) Unusual Electronic Properties of Template-Directed Ï€-Conjugated Porphyrin and Phosphorene Nanotubes 376
Bryan M. Wong

(375d) Complex Pattern Formation from Current-Driven Dynamics of Single-Layer Epitaxial Islands on Crystalline Conducting Substrates 377
Ashish Kumar, Dwaipayan Dasgupta, Dimitrios Maroudas
(375e) The Infrared and Raman Spectra of Pure-Silica and Aluminosilicate Sodalite 378
Caio Peixoto, Amir M. Mofrad, Jack Blumeyer, Liu Jinrui, Karl D. Hammond, Heather K. Hunt
(375f) The Effect of Solvent Selection on the Optical Trapping, Manipulation, and Patterning of Nanomaterials 379
Matthew Crane, Elena P. Pandres, Patrick Whitham, E. James Davis, Daniel Gamelin, Vincent C. Holmberg, Peter Pauzauskie
(375g) Synthesis and Characterization of Electrochemically-Grown Zinc Oxide Nanowires for Use in Rectenna-Based Heat Harvesters 380
Adrian Haley, Shendu Yang, Patrick J. Pinhero
(381a) The Thiol-Thioester Exchange in Network and Linear Polymers 381
Christopher N. Bowman, Brady Worrell, Matthew K. McBride, Gayla Berg, Chen Wang
(381b) Nacre-Inspired Composite Gels for Biomedical Applications 382
Ayomi S. Perera, Richard Jackson, Mark Miodownik, Marc-Olivier Coppens
(381d) High Viscosity Polymer Gels Derived from Block Copolymer Nanocomposites 383
Sri Harsha Kalluru, Eric W. Cochran
(381e) Quantifying Topology, Gelation and Elasticity of Polymer Networks 384
Rui Wang, Bradley D. Olsen
(381f) Self-Assembly and Mechanical Properties of Di-Fmoc-L-Lysine Containing Molecular Gels 385
Seyed Meysam Hashemnejad, Md Masrul Huda, Neeraj Rai, Santanu Kundu
(381g) Thermoresponsive Sol-Gel Transitions of PÎ•G-Based Nanocomposite Hydrogels Controlled By Molecular Weights of Block Copolymers and Solute Concentrations 386
Tomoki Maeda, Midori Kitagawa, Keishi Tanimoto, Makoto Miyazaki, Koji Nagahama, Atsushi Hotta
(381h) Evaluating Reprocessability of Polymer Networks: Flory-Stockmayer Analysis 387
Lingqiao Li, Xi Chen, Kailong Jin, John M. Torkelson
(381i) Molecularly-Templated Reaction for Forming Poly(dimethyl siloxane)/Graphene Oxide Composite Elastomers 388
Heonjoo Ha, Kiryong Ha, Christopher J. Ellison
(411a) Graduate Student Award Session: A Two-Step Method for Transferring Single Wall Carbon Nanotubes Onto a Hydrogel Substrate 389
Mozhdeh Imaninezhad, Irma Kuljanishvili, Silviya Petrova Zustiak
(411b) Graduate Student Award Session: Understanding How Lipid Nanoparticle Structure Affects Immune Response 390
Lisa Kasiewicz, Sushant Kumar, Rahul Purwar, Kathryn A. Whitehead
(411c) Graduate Student Award Session: Photo- Induced Pinocytosis in Synthetic Liposomes 391
Danielle Konetski, Dawei Zhang, Christopher N. Bowman
(411d) Graduate Student Award Session: Sugar-Coating the Answers to Virus Binding: Glycocalyx-Mimetic Interfaces 392
Ramya Kumar, Domenic Kratzer, Kenneth Cheng, Irina Kopyeva, Joerg Lahann
(411e) Graduate Student Award Session: The Combined Effect of Matrix Microenvironment and Hypoxia on the Activity of Glioblastoma Stem Cells 393
Jee-Wei Emily Chen, Jann N. Sarkaria, Brendan A. Harley
(411f) Graduate Student Award Session: Pegylated Poly(beta-amino ester) Delivery Systems for Periodic shRNA 394
Connie Wu, Wade Wang, Paula T. Hammond, Jiahe Li
(411g) Graduate Student Award Session: Biodegradable Nano-Film Coated Self-Floating Hollow Glass Microspheres for Rapid Cell Isolation and Recovery 395
Ziye Dong, Caroline Ahrens, Dan Yu, Zhenya Ding, Hyuntaek Lim, Wei Li
(411h) Graduate Student Award Session: Enhancing Therapeutic Efficacy of Self-Assembling Prodrugs with Supramolecular Chemistry 396
Hao Su, Yuzhu Wang, Feihu Wang, Honggang Cui
(413a) Size-Selective Ionically Crosslinked Polymer Multilayer Films for Light Gas Separation 397
Jaime C. Grunlan, Benjamin Wilhite
(413b) Length Effects on Polyelectrolyte Complexation: How 'Poly' Must a Polyelectrolyte be? 398
Jeffrey Vieregg, Michael Lueckheide, Matthew V. Tirrell
(413c) Interaction and Dynamics of Polyelectrolytes in Polyzwitterionic Complexes 399
Y. Elaine Zhu, Benxin Jing, Kehua Lin
(413d) Tuning Complex Coacervation Using Sequence-Defined Polyelectrolytes: A Molecular Understanding 400
Tyler Lytle, Li-Wei Chang, Jason Madinya, Sarah L. Perry, Charles E. Sing

(413e) Photodirected Assembly and Self-Rupture of Polyelectrolyte-Based Soft Materials 401
Udaka K. De Silva, Amanda C. Bryant-Friedrich, Yakov Lapitsky
(413f) Coarse-Grained Model for Polyelectrolyte Complexation 402
Marat Andreev, Samanvaya Srivastava, Lu Li, Matthew V. Tirrell, Jack F. Douglas, Juan J. De Pablo
(413g) Synthesis and Solution-Phase Characterization of Hydroxylated Sulfonated Oligothioetheramides 403
Joseph Brown, Christopher A. Alabi
(413h) Effect of Nanoparticle on the Structure and Dynamics of Model Peaa Ionomers from Molecular Dynamics Simulations 404
Janani Sampath, Lisa M. Hall
(413i) Zwitterionic Copolymers As Novel Supporting Scaffolds for Ionic Liquid-Based Gel Electrolytes 405
Luis Rebollar, Fatin Lind, Matthew J. Panzer
(426a) Photoreversible Stiffness Modulation of Protein-Polymer Hydrogels 406
Luman Liu, Jared A. Shadish, Cole A. Deforest
(426b) Preparation and Characterization of Polypeptide Hydrogels as Synthetic Extracellular Matrices for Cellular Scaffolds 407
Hongkun He, Alex Wang, Marianna Sofman, Linda Griffith, Paula Hammond
(426c) Reinforced Hydrogel Fibers for Cell Encapsulation and Organ Printing 408
Suman Bose, Daniel G. Anderson, Robert Langer
(426f) Magnetically Templated Hydrogels for Peripheral Nerve Injury Repair 409
Ishita Singh, Christopher Lacko, Christine Schmidt, Carlos Rinaldi
(426g) A Two-Step Method for Transferring Single Wall Carbon Nanotubes Onto a Hydrogel Substrate 410
Mozhdeh Imaninezhad, Irma Kuljanishvili, Silviya Petrova Zustiak
(267c) Surface Tension-Assisted Additive Manufacturing of Multicomponent Biomaterials 411
Mark W. Tibbitt, Heloise Ragelle, Michael Cima, Robert Langer
(439a) Chiral Metamaterial Platform with Tunable Near and Far Field Chiroptical Response 412
Pavlos Pachidis, Vivian E. Ferry
(439b) Lateral Growth of Two-Dimensional 1H-WSe2/1T'-WTe2 Heterostructures 413
Mengqiang Zhao, Carl H. Naylor, Zhaoli Gao, William M. Parkin, A. T. Charlie Johnson
(439c) Nanostructured Optoelectronics Using Interfacially-Driven Assembly 414
Matthew G. Panthani
(439d) Tuning the Bandgap of Graphene Nanoribbons through Defect-Interaction-Driven Edge Patterning 415
Dimitrios Maroudas, Lin Du, Andre R. Muniz
(439e) Wrinkled MoS2 Field-Effect Transistors 416
Shikai Deng, Vikas Berry
(439f) Nanoantenna Enhanced Wavelength Mixing in Monolayer Transition Metal Dichalcogenide 417
D. Keith Roper, Gregory T. Forcherio, Mourad Benamara, Luigi Bonacina
(439g) Kinetics of Nanoring Formation from Quantum Dots in Epitaxial Thin Films 418
Lin Du, Dimitrios Maroudas
(441a) Design, Synthesis, and Characterization of Elastomeric and Mechanoresponsive Polymer Matrix Composites 419
Matthew D. Green
(441b) Molecular Simulation of Macromolecular Transport through Nanoporous Membranes 420
Noelia Almodovar Arbelo, Bryan W. Boudouris, David S. Corti
(441c) Unconventional Nanoscopic Shape Memory Effects Exhibited By Novel Multi-Stimuli-Responsive Shape Memory Polymers 421
Sin-Yen Leo, Peng Jiang
(441d) Direct Observation of Remarkable Nanostructure Evolution during Aqueous Dissolution of Polymer/Drug Blends 422
Ralm Ricarte, Marc A. Hillmyer, Timothy P. Lodge
(441e) Effect of Graphene and Graphene Oxide on the LCST of PNIPAM 423
Carter Berry, Sanket A. Deshmukh
(441f) Solvent Effects on the Structure and Thermodynamics of Polymer Blends with Varying Architectures 424
Thomas Gartner III, Arthi Jayaraman
(441g) Molecular Weight Dependence of the Intrinsic Size Effect on Tg in AAO Template-Supported Polymer Nanorods: A DSC Study 425
Tong Wei, Shadid Askar, Anthony Tan, John M. Torkelson

(441h) Versatile Cholesterol-Functionalized Block Copolymers in Aqueous Dispersions 426
Kenneth Mineart, Shrinivas Venkataraman, Yi Yan Yang, James L. Hedrick, Vivek M. Prabhu
(441i) Polymerization Thermodynamics Under Nanoconfinement 427
Haoyu Zhao, Qian Tian, Sindee L. Simon
(475a) Molecular Understanding, Design and Development of Ultra Low Fouling Zwitterionic Materials 428
Shaoyi Jiang
(475b) A Scalable Approach to Produce a Diversity of Structured Polymer Colloids 429
Rodney D. Priestley
(475c) Layered Morphologies in Precise Acid-Containing Polyethylenes: Hierarchical Structures and the Implications on Properties 430
Karen I. Winey
(475d) Examination of Thin Film Growth Using Supersonic Molecular Beams and in situ Real Time X-Ray Synchrotron Radiation: From Organic Small Molecule Semiconductors to Transition Metal Dichalcogenides 431
James R. Engstrom
(475e) Polymer Electrolytes for Electrochemical Energy Devices 432
Yushan Yan
(525c) TAT-Functionalized Liposomes for the Treatment of Meningitis 433
Caterina Bartomeu Garcia, Di Shi, Thomas J. Webster
(525d) Cartilage Penetrating Nanoparticles to Provide Sustained Delivery of Disease Modifying Drugs in Post-Traumatic Osteoarthritis 434
Brett C Geiger, Sheryl Wang, Alan Grodzinsky, Paula T. Hammond
(525e) Nanoparticle Tracking Analysis of Polymer Particle Aggregation in Blood Plasma 435
Kathleen McEnnis, Stephanie Christau, Sean McLoughlin, Joerg Lahann
(525f) High Throughput Screening of Biodegradable Nanogels with Tunable Size and Swelling for Intracellular Drug Delivery 436
David S. Spencer, Bryan C. Luu, David W. Beckman, Nicholas A. Peppas
(525g) Polyanhydride Nanoparticles Encapsulating Rifampicin Suppress Growth of Pathogenic Acanthamoeba in vitro 437
Nathan Peroutka-Bigus, Adam Mullis, Balaji Narasimhan, Bryan H. Bellaire
(525h) Design of Dual Encapsulated Biodegradable Nanoparticles for Cancer Treatment 438
Amber C. Jerke, Timothy Brenza
(526a) Understanding How Lipid Nanoparticle Structure Affects Immune Response 457
Lisa Kasiewicz, Sushant Kumar, Rahul Purwar, Kathryn A. Whitehead

VOLUME 2

(526b) Multi-Stage Drug Delivery System for Enhanced Payload Delivery to Lymph Node Cells 458
Alex Schudel, Cody Higginson, Mai Kwan Yau, M. G. Finn, Susan N. Thomas
(526c) Encapsulation and Thermal Stability of Immunological Biologics Using Complex Coacervation 459
Whitney C. Blocher, Rebecca Hershman, Sarah L. Perry
(526d) Evaluation of Biocompatibility of Novel Block Copolymer Gels/Micelles As Potential Vaccine Adjuvants 460
Justin Adams, Sujata Senapati, Michael J. Wannemuehler, Balaji Narasimhan, Surya Mallapragada
(526e) Intranasal Nanovaccine Provides Protection Against Homologous and Heterologous Influenza Virus 461
Zeb Zacharias, Kathleen Ross, Balaji Narasimhan, Thomas Waldschmidt, Kevin Legge
(526f) Design of a Combination Nanovaccine to Induce Rapid and Long Term Protective Immunity Against Bacillus Anthracis 462
Sean Kelly, Ross Darling, Nathan Peroutka-Bigus, Thomas Dubensky, Bryan H. Bellaire, Michael J. Wannemuehler, Balaji Narasimhan
(526g) Comprehensive Vaccine Design for Commensal Disease Progression 463
Charles H. Jones, Guojian Zhang, Roozbeh Nayerhoda, Marie Beitelshees, Andrew Hill, Yi Li, Bruce A. Davidson, Paul Knight III, Blaine Pfeifer
(526h) Oncofetal Antigen Peptide Nanoclusters for Cancer Vaccines 464
Alexandra Tsoras, Julie A. Champion
(536a) S Doped TiO2 for Photocatalytic Oxidation of CO in Visible Region Synthesized By Novel One Step Liquid Flame Spray Pyrolysis (LFSP): Kinetics and Mechanism 473
Siva Nagi Reddy Inturi, Panagiotis Smirniotis

(536b) Copper Hexacyanoferrate Hydrogel Electrodes for Electrochemically-Mediated Cation Separations 474
Kai-Jher Tan, Xiao Su, Johannes Elbert, T. Alan Hatton
(536c) Emulsion Templated Polymers As Structured Supports for MOF Adsorbents 475
Jacob I. Deneff, Krista S. Walton
(536d) Development of Novel Crosslinked Polymers for the Capture of Environmental Pollutants 476
Rishabh Shah, Thomas Dziubla, J. Zach Hilt
(536e) Grafted Polystyrene Controlled Formation of Magnetic Carbon Nanocomposites for Environmental Remediation 477
Hongbo Gu
(536f) Constructing Fenton-like Reaction over g-C3N4/NH2-MIL-88B(Fe) Photocatalyst to Degrade Organic Contamination in Aqueous Solution 478
Xiyi Li, Yunhong Pi, Zhong Li, Jing Xiao
(536g) Synthesis of Thermoresponsive Composite and Application for Water Treatment 488
Junichi Ida, Masanori Ochi, Ai Ishikawa, Atsushi Matsumoto, Tatsushi Matsuyama, Hideo Yamamoto
(536h) Integrated Adsorption and Visible-Light-Driven Photocatalytic Degradation of MB over Oxidized C60/NH2-MIL68(In) 489
Yunhong Pi, Xiyi Li, Zhong Li, Jing Xiao
(536i) In-Situ Studies on Anion Electroadsorption Mechanisms 504
Kelsey Hatzell, Marta Hatzell, Marm Dixit
(538a) Engineering Materials and Processes for Flexible Electronics 505
Antonio Facchetti
(538b) Spectroelectrochemistry of Conjugated Radical Polymers 506
Jodie L. Lutkenhaus
(538c) Conjugated Copolymers Based on Purines: Linking Monomer Design and Macromolecular Properties 507
Michael Kilbey, Graham Collier, Lauren Brown, Evan Boone, Brian Long
(538e) Pinch-Off Dynamics and Printability of Semi-Flexible and Rigid Rod Polymer Solutions 508
Leidy N. Jimenez, Jelena Dinic, Vivek Sharma
(538f) Surface-Directed Multi-Scale Assembly of Conjugated Polymers 509
Ying Diao, Erfan Mohammadi, Ge Qu
(538g) Interpenetrating Networks for Flexible Bulk Heterojunction Opvs 510
Jorge Mok, Zhiqi Hu, Rafael Verduzco
(538h) Development of an Anisotropic Coarse-Grained Conjugated Polymer Model for Optoelectronic Applications 511
Alec S. Bowen, Nicholas Jackson, Daniel Reid, Yamil J. Colon, Juan J. De Pablo
(538i) Automated Quantitative Analysis of Oriented Morphologies and Inter-Grain Connectivity in Conjugated Polymers 512
Nils Persson, Michael McBride, Martha A. Grover, Elsa Reichmanis
(545a) Spray Coating Chitin & Cellulose Nanomaterials for Enhancement of Barrier Properties 513
Chinmay C. Satam, Jerel Jallorina, J. Carson Meredith
(545b) Development of Low-Concentration Alkaline Treatment Method to Produce Thermally Stable Cellulose Nanofibrils (CNFs) 514
Hansol Lee
(545c) Tailored and Integrated Production of Carboxylated and Lignin Containing Cellulose Nanocrystals and Nanofibrils for Composite Applications 515
J. Y. Zhu
(545d) Nano Carbon Structures from Cellulosic Biomass for Use As Functional Materials 516
Ping Wang
(576a) Quantitative Three-Dimensional Morphological Characterization of Block Copolymer Films Enabled By Directed Self-Assembly 517
Paul F. Nealey
(576b) Examination of Line Edge Roughness of Directed Self-Assembled Block Copolymers: A Coarse-Grained Molecular Dynamics Study 518
Shubham Pinge, Durairaj Baskaran, Munirathna Padmanaban, Yong Lak Joo
(576c) Atomistic Simulations of Ion Transport in PsbP2VP Functionalized Lamellae 519
Weiwei Chu, Yamil J. Colon, Juan J. De Pablo
(576d) Performance of the Novel Ultra-Thin Proton Exchange Membrane Based on Poly(arylene ether sulfone)s for Pemfcs 520
Yang Zhao, Xue Li, Xiaofeng Xie
(576e) The Consequence of Morphology on Conductivity for Bolaamphiphiles 521
Mayank Misra, Christian Nowak, Yangyang Sun, Fernando Escobedo

(576f) Mechanical and Structural Analyses of Toughened Syndiotactic Polypropylene Gels: Effects of Gel-Preparation Temperature ... 522
Fuyuaki Endo, Ryusuke Okoshi, Naruki Kurokawa, Tomoki Maeda, Atsushi Hotta
(576g) Chemically Cross-Linked Poly(2-Hydroxyethyl Methacrylate)-Supported Deep Eutectic Solvent Gel Electrolytes ... 523
Huan Qin, Matthew J. Panzer
(576h) Ultrasonic Synthesis of Temperature Responsive Copolymer and Its Characterization ... 524
Masaki Kubo, Tomoyuki Koshimura, Takao Tsukada
(576i) Atomic-Level Comparisons of LCST Transition in Thermo-Sensitive Polymers ... 525
Yaxin An, Karteek K. Bejagam, Samrendra Singh, Sanket A. Deshmukh
(591a) ATRP-Grown Protein-Polymer Conjugates Selectively Enhance Transepithelial Protein Transport ... 526
Chad Cummings, Katherine Fein, Hironobu Murata, Rebecca Ball, Alan Russell, Kathryn A. Whitehead
(591b) Aptamer Micelles Targeting Cancer Cells Expressing the Chemokine Fractalkine ... 527
Michael A. Harris, Timothy R. Pearce, Thomas Pengo, Huihui Kuang, Colleen L. Forster, Efrosini Kokkoli
(591c) Non-Charged Cell-Penetrating Oligothioetheramides ... 528
Ngoc Phan, Christopher A. Alabi
(591d) Targeting, Delivery, and Immobilization of Therapeutic Factors with Native Free Radicals ... 529
Christopher J. Lowe, Keana Mirmajlesi, David I. Shreiber
(591e) Antimicrobial Peptide Amphiphile (AMPA) Medical Product Coatings for the Prevention of Nosocomial Infections ... 530
Josiah Smith, Alexis Dadelahi, Julie Nguyen, Fabio Gallazzi, John Dodam, Jeffrey Adamovicz, Roger De La Torre, Bret Ulery
(591f) Mitigating the Bioactivity Loss of Polymer- Insulin Conjugate ... 531
Zhiqiang Cao, Yang Lu
(591g) Enhancing Therapeutic Efficacy of Self-Assembling Prodrugs with Supramolecular Chemistry ... 532
Hao Su, Yuzhu Wang, Feihu Wang, Honggang Cui
(591h) Hydrogen Sulfide Donor Micelles: Synthesis, Characterization and Therapeutic Potential ... 533
Urara Hasegawa, Andre Van Der Vlies, Jerry J. Y. Chen, Tomoka Takatani-Nakase, Ikuhiko Nakase
(592a) Pegylation of Model Drug Carriers Enhances Uptake By Primary Human Neutrophils ... 534
William Kelley, Catherine A Fromen, Omolola Eniola-Adefeso
(592b) 'Smart' Nanoparticles for Immunotherapeutic Targeting of the Sting Pathway ... 535
Daniel Shae, Denise Buenrostro, Alyssa Merkel, Sema Sevimli, Julie A. Sterling, John Wilson
(592c) Biomaterial Nanoparticles Redistribute Therapeutic Antibodies to Lymph Node-Resident Cells to Enhance Cancer Immunotherapy Via Checkpoint Inhibition ... 536
David Francis, Alex Schudel, Nathan A. Rohner, Susan N. Thomas
(592d) Engineering Nanoparticle Artificial Antigen Presenting Cells Based on T Cell Biology Improves T Cell Enrichment and Activation for Cancer Immunotherapy ... 537
John Hickey, Fernando Vicente, Hai-Quan Mao, Jonathan Schneck
(592e) Biomaterial Scaffolds for Combined Focal Ablation and Immunotherapy to Target Disseminated Cancer ... 538
Francisco Pelaez, Stephen O'Flanagan, Qi Shao, Brandon Burbach, Tiffany Lam, John C. Bischof, Yoji Shimizu, Samira M. Azarin
(592f) Modulating the Immune Environment within Adipose Tissue with Polymer Scaffolds ... 539
Kendall Murphy, Michael Gower
(592g) Multi-Factor Microparticle Formulation for Local Induction of Regulatory Lymphocytes and Treatment of Periodontal Disease ... 540
Ashlee Greene, Sayuri Yoshizawa, Michelle Ratay, Charles Sfeir, Steven R. Little
(592h) Multivalent Soluble Antigen Arrays Target Antigen Presenting B Cells and Dampen Antigen-Specific Signaling to Promote Therapeutic Efficacy in Multiple Sclerosis ... 541
Brittany Hartwell
(593a) Novel Bio-Based Polyesters and Polycarbonates Derived from Xylochemicals ... 542
Joseph F. Stanzione III, Silvio Curia, Joseph Mauck, Alexander W. Bassett, John J. La Scala
(593b) Light Weight Biocomposite from Toughened Polyolefin and Biocarbon ... 555
Ehsan Behazin, Manju Misra, Amar K. Mohanty
(593c) Secondary Fermentation of Corn Ethanol Co-Products for Improved Amino Acid Qualities ... 556
Tanner Barnharst, Yanmei Zhang, Jingyu Wang, Bo Hu
(593d) Extraction High-Value Chemicals from Ethanol Co-Products: A Feasibility Assay on Phytate Extraction with Life Cycle and Techno Economic Assessment ... 557
Cristiano Reis, Aravindan Rajendran, Douglas Tiffany, Bo Hu
(604a) Highly Stable Perovskite Solar Cells Fabricated Using Aerosol-Based Technique ... 560
Shalinee Kavadiya, Pratim Biswas

(604b) Engineering Perovskite Solar Cell Interfaces to Realize > 1000 Hr, Unencapsulated Ambient Stability 561
Jeffrey A. Christians, Philip Schulz, Jonathan Tinkham, Tracy H. Schloemer, Bertrand Tremolet De Villers, Steve Harvey, Alan Sellinger, Joseph J. Berry, Joseph M. Luther
(604c) A Thermodynamic Basis for Engineering Enhanced Stability of the Perovskite Phase of Cesium Lead Iodide 562
Subham Dastidar, Aaron T. Fafarman
(604d) Synthesis of Nanostructured Inorganic Perovskites for Solar Cell Applications 563
Atefe Hadi, Rainie D. Nelson, Iver J. Cleveland, Jeremy M. Jacoby, Matthew G. Panthani
(604e) Discovery of Near-Infrared-Active Colloidal Multinary Lead Halide Perovskite Nanocrystals Using a Microfluidic Platform 564
Ioannis Lignos, Viktoriia Morad, Richard Maceiczyk, Loredana Protesescu, Maksym V. Kovalenko, Andrew J. Demello
(604f) Improved Charge Collection in Highly Efficient CsPbBrI2 Solar Cells with Light-Induced Dealloying 565
Joshua Choi
(604g) Synthesis and Characterization of Bi-Based Perovskite Semiconductors for Photovoltaic Application 566
Umar H. Hamdeh, Rainie D. Nelson, Bradley J Ryan, Ujjal Bhattacharjee, Jacob W. Petrich, Matthew G. Panthani
(617a) In-Situ Synthesis of Intergrown Uio-66 Membranes with Controlled Orientation 567
Bohan Shan, Bin Mu
(617b) Closed Packed, Oriented MOF Thin Films through Solution Shearing 568
Arian Ghorbanpour, Luke Huelsenbeck, Gaurav Giri
(617c) Mechanistic Insights into Low Temperature Ceramic Thin Film Growth and Crystallization Using Microwave Radiation 569
B. Reeja Jayan, Nathan Nakamura
(617d) Gold Nanoparticle Monolayers for Surface-Enhanced Raman Spectroscopy of Lithium Batteries 570
Daniel T. Hallinan Jr., Guang Yang, Jagjit Nanda, Boya Wang, Gang Chen
(617e) Understanding the Chemistry of Thiol-Amine Solutions: Versatile Solvents for Solution-Processed Thin Film Photovoltaics 571
Caleb Miskin, Priya Murria, Laurance Cain, Robert W. Boyne, Evan C. Wegener, Jeffrey T. Miller, Hilkka Kenttamaa, Rakesh Agrawal
(617f) In situ Grazing Incidence Small Angle X-Ray Scattering (GISAXS) Study of the Formation of Multilayered Ordered Mesoporous Titania Films 572
M. Arif Khan, Syed Z. Islam, Suraj Nagpure, Barbara L. Knutson, Stephen E. Rankin
(617g) 2-D Wulff Construction of FeOx Islands Grown on Pt(111) for Use in Catalysis 573
Joseph Kubal, Jeffrey Greeley
(621a) Molecular Simulation of Crystallization of Chain Molecules from the Melt 574
Gregory C. Rutledge
(621b) Anomalous Crystallization Behavior of Ring Polymers 575
Kiran S. Iyer, Murugappan Muthukumar
(621c) Thin Film Crystallization of Cyclic Polymers and Their Linear Analogues 576
Julie Albert, Giovanni Kelly, Scott Grayson, Farihah Haque
(621d) Synthesis, Characterization, and Structural Evolution of Designer Block Polyelectrolyte Complexes 577
Jeffrey Ting, Hao Wu, Abraham Herzog-Arbeitman, Samanvaya Srivastava, Matthew V. Tirrell
(621f) Solid and Liquid Core Polyelectrolyte Complex Micelles 578
Lorraine Leon
(621i) Uniqure Crystallization Behavior of Isotactic Polypropylene in the Presence of L-Isoleucine and Its Inhibition and Promotion Mechanism of Nucleation 579
Shicheng Zhao, Xiaoshan Peng
(621h) Scalable Nanocomposites Synthesis Via Electrospray-Mediated Electroemulsification and Flash Nanoprecipitation 580
Kil Ho Lee, Barbara E. Wyslouzil, Jessica O. Winter
(622a) Engineering Vapor-Deposited Polymers for Energy Conversion and Storage 581
Kenneth Lau
(622b) Charge Transfer Mechanisms in Organic Radical Polymer Batteries 582
Shaoyang Wang, Fei Li, Jodie L. Lutkenhaus
(622c) Free Volume Enhanced Anion Exchange Membranes from Triptycene Poly(Arylene Ether Sulfone) Copolymers 583
Yoonseob Kim, Timothy Swager

(622d) Surfactant-Polymer System Optimization in Heterogeneous Model with Mobility Control 584
Nai Cao, Pingchuan Dong, Brian McPherson, Xiaoxiao Liu
(622e) Breaking the Compensation Effect within the Vogel-Tammann-Fulcher Equation for Polymer-Based Electrolytes 590
Kyle M. Diederichsen, Hilda G. Buss, Bryan D. McCloskey
(622f) Carbon Derived from Polymerized Ionic Liquids 591
Rui Sun, Kelly M. Meek, Yossef A. Elabd
(622g) Tailoring Surface Functionalization of Silica Nanoparticles in Nafion Nanocomposites for Improved Ion Selectively in Vanadium Redox Flow Batteries 592
Allison Jansto, Eric M. Davis
(622h) Thermally Cross-Linked Poly(acrylic acid) / Reduced-Graphene Oxide Aerogels As a Replacement for Metal-Foil Current Collectors in Lithium Ion Batteries 593
Han Xiao, Joshua Pender, Mackenzie Meece-Rayle, Pedro De Souza, Kyle Klavetter, Heonjoo Ha, Jie Lin, Adam Heller, Christopher J. Ellison, C. Buddie Mullins
(622i) Graphene-Oxide/Polybenzimidazole Nanocomposite Membrane for High Temperature Fuel Cell Application 605
Shobha Mantripragada, Md. Tashfin Zayed Hossain, Khondker Sultana, Shamsuddin Ilias, Jianzhong Lou
(640a) Autoperforation of 2D Materials for Generating Two Terminal Memresistor Janus Particles with Nonvolatile Memory 606
Pingwei Liu, Albert Tianxiang Liu, Daichi Kozawa, Juyao Dong, Max Saccone, Volodymyr Koman, Song Wang, Minhao Wong, Michael Strano
(640b) Synthesis and Properties of Polymer/Graphene Oxide (GO) Thermosets with Multifunctional GO As a Crosslinker 607
Heonjoo Ha, Jaesung Park, Kiryong Ha, Benny D. Freeman, Christopher J. Ellison
(640c) Polymer Silica Composite Nanofibers Via Sol Gel Electrospinning 608
Tahira Pirzada, Sara A. Arvidson, Carl D. Saquing, S. Sakhawat Shah, Saad A. Khan
(640d) Graphene Oxide/Lipid Composite Material Towards a Multifunctional Drug Delivery Vehicle 609
Mohammad Shahadat Hussain Sarkar, Md. Alamin Miraz, Ashiqur Rahman, Yang Lu, Vu Phan, Clayton S Jeffryes, Evan K. Wujcik
(640e) Localizing Graphene at the Interface of HDPE/PLA Polymer Blends 610
Sung Cik Mun, Min Jae Kim, Liangliang Gu, Monica Cobos, Christopher W. Macosko
(640f) Cocontinuous Ternary Polymer Nanocomposites with Interfacial Graphene Nanoplatelets 618
Lian Bai, Radhika Sharma, Catherine Esnaashari, Christopher W. Macosko, Xiang Cheng
(640g) Investigating Catalytic Properties of ReSe2 Edge and Basal Plane for Hydrogen Evolution Reaction 619
Zhenjing Liu, Zhengtang Luo
(647a) Peg Hydrogels with Tunable Biodegradation Rate for Sustained Delivery of Platelet-Rich Plasma for Treatment of Osteoarthritis 626
Era Jain, Saahil Sheth, Nobuaki Chinzei, Natasha Case, Linda Sandell, Scott A. Sell, Muhammad Rai, Silviya Petrova Zustiak
(647b) Control over the Temporal Profile and Sequence of Anticancer Therapeutics from Magnetically Responsive Hydrogels 627
Tania Emi, Tanner Barnes, Emma Orton, Anne Reisch, Zachary Silveira, Miranda Mitchell, Celia Dunn, Anita E. Tolouei, Stephen Kennedy
(647c) Electrospun Gelatin Nanofibers As Carrier for Controlled and Sustained Release of a Hydrophobic Drug 628
Anindita Laha, Chandra Sharma, Saptarshi Majumdar
(647d) Engineering of Degradable Biopolymer Films Loaded with Imiquimod for Controlled Released in a Mucosal Environment 629
Lucas Garcia Camargo, Gabriela Souza Rezende, Stephany Di Carla Santos, Michelle Franz Montan Braga Leite, Renata Nishida Goto, Andreia Machado Leopoldino, Angela Maria Moraes
(647e) Microparticles for the Delivery of Anti-Diabetic Drugs to the Adipose Tissue 630
Christopher Isely, Michael Hendley, Kendall Murphy, Prakasam Annamalai, Michael Gower
(647f) Template Assisted Micro-Patterned Electrospun Nanofibrous Mat As a Potential Carrier for Controlled Drug Release 631
Manohar Kakunuri, Mudrika Khandelwal, Chandra S. Sharma, Stephen Eichhorn
(647g) PÎ•G-Based Nanocomposite Hydrogels: Thermoresponsive Sol-Gel Transitions and Decomposition Rates Regulated By the LÎ'/GA Ratio of PLGA-PÎ•G-PLGA 632
Midori Kitagawa, Tomoki Maeda, Atsushi Hotta
(648b) Electrospun Collagen Scaffold for Peripheric Nerve Regeneration 633
Carol Rivera Martinez, Janet Mendez, Jorge Almodovar
(648c) 3D Printing of Nerve Guidance Channels for Peripheral Nerve Repair 634
Wei Wu

(648d) Combined Physical and Biochemical Cues Direct the Growth of Inner Ear Neurites 635
Braden Leigh, Kristy Troung, Reid Bartholomew, Marlan Hansen, C. Allan Guymon
(648e) Harnessing Multi-Functional Microbial Cells for Designing Sweat-Responsive Bio-Hybrid Wearables 636
Wen Wang, Lining Yao, Chin-Yi Cheng, Teng Zhang, Hiroshi Atsumi, Luda Wang, Guanyun Wang, Oksana Anilionyte, Helene Steiner, Jifei Ou, Kang Zhou, Chris Wawrousek, Katherine Petrecca, Angela M. Belcher, Rohit Karnik, Daniel I. C. Wang, Xuanhe Zhao, Hiroshi Ishii
(648f) Development of Silk Protein Conjugates for Mucoadhesive Applications 637
Danielle L. Heichel, Kelly A. Burke
(648g) Laser-Activated Sealants for Skin Tissue Repair 638
Russell Urie, Deepanjan Ghosh, Mitzi Thelakkaden, Chengchen Guo, Jeff Yarger, Jacquelyn Kilbourne, Kaushal Rege
(648h) Layer-By-Layer Assembled Thin Film Biomaterials As Porous Biomolecular Delivery Systems 639
Adeline Gand, Mathilde Hindie, Michel Boissiere, Emmanuel Pauthe, Paul R. Van Tassel
(680a) Molecular Simulations of the Influence of Interfaces on the Dynamics of Polymers 640
Robert A Riggleman
(680b) Princess and the Pea Behavior in Polyelectrolyte Multilayers: Influence of the First Layer on Polyelectrolyte Multilayer Assembly and Properties 641
Xuejian Lyu, Amy M. Peterson
(680c) Instability of Liquid Crystal Thin Film on Topographically Patterned Substrates 642
Palash Dhara, Rabibrata Mukherjee
(680d) Electroless Deposition of Copper on Quaternized Chitosan Coatings for Antibacterial Application 643
Debirupa Mitra, En-Tang Kang, Ramanathan Kollengode, Matthew Cove, Koon Gee Neoh
(680e) Patterning Buckles in Polymer/Metal Thin Films By Laser Irradiation 644
Kunal Mondal, Ying Liu, Michael D. Dickey, Jan Genzer
(680f) High Performance Conducting Polymer Coatings for Corrosion Protection 645
Xinyu Zhang, Amit Nautiyal, Jonathan Cook
(680g) Enhanced Wetting Stability of Initiated Chemical Vapor Deposited (iCVD) Polydivinylbenzene Thin Films By Thermal Annealing 646
Junjie Zhao, Minghui Wang, Karen K. Gleason
(680h) Nanoscale Characterization of Water Penetration through Plasma Polymerized Coatings and Water at the Coating/Substrate Interface 647
Yang Zhou, Ali Dhinojwala, Mark Foster
(680i) Surface Topology and Modulus Effects on Adhesion in Novel Polyorganosilozane-Based Coatings 648
Ethan D. Smith, Stephen M. Martin, Vince Baranauskas
(687a) Templated Synthesis of Polymer - Gold Nanocomposites with Pluronic Gels 649
Srikanth Nayak, Surya Mallapragada, Wenjie Wang, David Vaknin
(687c) Predicting Surface Area in Green Synthesis of Sol-Gel Materials 650
Brian K. Peterson, Mobae Afeworki, David C. Calabro, Quanchang Li, Simon C. Weston
(687d) Leveraging Nanoparticle Template Assembly and Interfacial Phenomenon for Multi-Scale Control over Polyimide-Derived Carbon Molecular Sieve Films 651
Megha Sharma, Mark A. Snyder
(687f) Toward Rational Design of Hierarchical Beta Zeolites Via Cost-Effective Approaches 652
Ke Zhang, Sergio Fernandez, Michele L. Ostraat
(687g) Nanoscale Control of Homoepitaxial Growth on a Two-Dimensional Zeolite 653
Meera Shete, Manjesh Kumar, Donghun Kim, Neel Rangnekar, Dandan Xu, Berna Topuz, Kumar Varoon Agrawal, Evguenia Karapetrova, Benjamin Stottrup, Shaeel Al-Thabaiti, Sulaiman N. Basahel, Narasimharao Katabathini, Jeffrey Rimer, Michael Tsapatsis
(689a) Thermodynamics of Charging in Weak Polyelectrolytes 654
Jonathan K. Whitmer
(689b) Thermodynamics and Kinetics of Ordered, Stronly Segregated Diblock Copolymers 655
Ronald M. Lewis III
(689c) Effect of pH on the Interaction between Poly(vinyl alcohol) and Ice 656
Aaron A. Burkey, Nathaniel A. Lynd
(689d) Self Consistent Field Theory Study of Multivalent Cation Effect on Semiflexible End-Grafted Random Polyelectrolytes 657
Merina Jahan, Mark J. Uline
(689e) The Sequence-Dependence of the Persistence Length of DNA 658
Hui-Min Chuang, Jeffrey G. Reifenberger, Han Cao, Kevin D. Dorfman

(689f) Anomalous Hydrodynamic Radius of Polyethylene Glycol Molecules in Mixed Solvents Containing a Hydrotope ... 659
Xiong Zheng, Mikhail A. Anisimov, Jan V. Sengers, Mao-Gang He
(689g) Energy Saturation: An Alternative Mechanism for the Glass Transition ... 669
Isaac C. Sanchez, Sean O'Keefe
(689h) Molecular Dynamics Simulation of Polymerized Stockmayer Fluids: Effects of Chain Length and Connectivity on Saturated Dipoles Near Ions ... 670
Issei Nakamura, Lijun Liu
(689i) Probing Polymer Blend Phase Diagrams Via Oligomer Molecular Simulations ... 675
Qile Chen, Shuyi Xie, Timothy P. Lodge, J. Ilja Siepmann
(696a) Silica-Peg Gel Encapsulation Platform for Isolation of Dormant Cancer Cells ... 676
Julian Preciado, Hak Rae Lee, Emil Lou, Alptekin Aksan, Samira M. Azarin
(696b) Biodegradable Nano-Film Coated Self-Floating Hollow Glass Microspheres for Rapid Cell Isolation and Recovery ... 677
Ziye Dong, Caroline Ahrens, Dan Yu, Zhenya Ding, Hyuntaek Lim, Wei Li
(696c) Design of Electrohydrodynamic Sprayed Polyethylene Glycol Hydrogel Microspheres for Cell Encapsulation ... 678
Anisa Qayyum, Era Jain, Grant Kolar, Scott A. Sell, Silviya Petrova Zustiak
(696d) Quantum Dot-Based Biomarkers of Neuroinflammation in the Developing Brain ... 679
Mengying Zhang, Binh Dang, Kate Hildahl, Brittany Bishop, Reyn Aoki, Nicole Thompson, Vincent C. Holmberg, Elizabeth Nance
(696e) The Culprit of Gout: Triggering Factors and Formation of Monosodium Urate Monohydrate ... 680
Tzu-Hsuan Chen, Meng-Hsiu Chih, Hung-Lin Lee, Tu Lee
(696g) Fabrication of Polyethylene Glycol-Based Templated Macroporous Hydrogels for Cell Encapsulation ... 688
Mozhdeh Imaninezhad, Grant Kolar, Silviya Petrova Zustiak
(709a) Multicomponent Transport Models for Non-Electroneutral Solid Electrolytes ... 689
Charles W. Monroe
(709b) Diffusion and Sorption Phenomena of Organic Vapor Penetrants in Unmodified and Ethylenediamine Vapor Phase Cross-Linked Matrimid Thin Films ... 690
John P. Stanford, Peter Pfromm, Mary Rezac
(709c) Segmental Dynamics and Water Transport in Nafion-SiO2 Hybrid Membranes ... 691
Apoorv Balwani, Antonio Faraone, Eric M. Davis
(709d) Designing Core/Shell Metal Organic Framework/Polymer Films As Scalable Barrier Layers for Enhanced Protection on Photovoltaics ... 692
Fen Qiu, Zhuonan Song, Jeffrey Urban
(709e) PEO-Based Semi-Interpenetrating Polymer Networks (S-IPNs) for CO2-Selective Membranes ... 693
Gregory Kline, Qinnan Zhang, Jennifer Weidman, Ruilan Guo
(709f) An Experimental and Triple-Mode Sorption Modeling of Sorption and Diffusion in Polymers ... 694
Hom Sharma, Stephen Harley, Yunwei Sun, Elizabeth Glascoe
(709g) In situ Monitoring of Emergent Transport in Polymer Membranes ... 695
Daniel J. Miller, Breanna Dobyns, Bryan S. Beckingham
(709h) The Role of Chlorine Substituent in Gas Transport Properties of Polychlorotrifluoroethene (PCTFE) ... 696
Milad Yavari, Yoshi Okamoto, Haiqing Lin
(709i) Thermodynamic Model for Predicting Swelling of Poly(N-isopropyl acrylamide) Hydrogels in Solvent Mixtures ... 697
Sheik Tanveer, Fazle Hussain, Chau-Chyun Chen
(721a) Nanostructure-Driven Fatigue Resistance and Dynamic Recovery in Thermoplastic Elastomer Hydrogel Networks ... 708
Travis S. Bailey
(721b) Interfacial Crystallization of Polyolefins: An Improved Outlook for Polymer Blends ... 711
Alex M. Jordan, Kyungtae Kim, Frank S. Bates, Shaffiq Jaffer, Olivier Lhost, Christopher W. Macosko
(721c) Ultra-Stable Amorphous Teflon: Extreme Fictive Temperature Reduction As a Means to Probe Sub-Tg Dynamics ... 712
Gregory B. McKenna, Heedong Yoon, Yung P. Koh, Sindee L. Simon
(721d) Influence of Strucure and Dynamics in Matrix-Free Polymer-Grafted Nanocomposite Memebranes ... 713
Eileen Buenning, Sanat K. Kumar, Christopher J. Durning, Connor Bilchak, Brian C. Benicewicz, Dimitris Vlassopoulos

(721e) Role of Interfacial Adhesion in Rate-Dependent Deformation and Failure of Model Ring Opening Metathesis Polymer (ROMP) Filled Composites 714
Erich Bain, Daniel B. Knorr Jr., Joseph L. Lenhart
(721f) Mechanically Ductile and Stiff, Triazole-Based Glassy Photopolymer Network 715
Han Byul Song, Austin Baranek, Christopher N. Bowman
(721g) Controlled Topology Toughning Epoxy Via Incorporation of Partially Reacted Substructures 716
Jian Gao
(721h) Thermoplastic Polydimethylsiloxane Realized By Hydrogen-Bond Networks through L-Phenylalanine Terminals 717
Shunsuke Tazawa, Atsushi Shimojima, Tomoki Maeda, Atsushi Hotta
(721i) In vitro and silico Characterization of Grafted Hydrophobic Brush Membranes 718
John J. Keating, Mirco Sorci, Angelo Setaro, Patrick T. Underhill, Georges Belfort
(725a) Direct Synthesis and Morphology Control of Metal Organic Framework Nanosheets 719
Feng Xue, Prashant Kumar, Wenqian Xu, Michael Tsapatsis
(725b) Mixed-Linker MOF Synthesis and Remediation of Acid-Gas Degradation Using Linker-Exchange Techniques 720
Krishna Chandran Jayachandrababu, Souryadeep Bhattacharyya, David Sholl, Sankar Nair
(725c) Rapid Microwave-Assisted Synthesis of Hybrid Zeolitic-Imidazolate Frameworks with Mixed Metals and Mixed Linkers 721
Febrian Hillman, John Zimmerman, Seung-Min Paek, Mohamad Hamid, Woo Taik Lim, Hae-Kwon Jeong
(725d) Micro-, Meso-, and Macro-Scale Defects in Porous Organic Cages 722
Guanghui Zhu, Christopher W. Jones, Ryan P. Lively
(725e) Computational Studies of the Enhanced Acidity of Defect MOF 808: The Effect of Activation Process 723
Carolina Ardila-Suarez, Saul Perez Beltran, Gustavo Ramirez-Caballero, Perla B. Balbuena
(725f) Crystallization Process Development of Metal-Organic Frameworks By Linking Secondary Building Units, Lattice Nucleation and Luminescence: The Insight into Reproducibility 724
Tu Lee, Yun Hsuan Chang, Hung-Lin Lee
(725g) Fast Mechano-Chemical Synthesis N-Doped Uio-66 with Dopamine to Enhance Chlorobenzene Competitive Adsorption Under Humid Air 725
Zhenxia Zhao, Peng Hu, Zhongxing Zhao
(725h) High Yield Synthesis of ZIF-8 Nanoparticles Using Stoichiometric Reactants in a Jet-Mixing Reactor 726
Aamena Parulkar, Pinaki Ranadive, Nicholas Brunelli
(726a) Polymer Semiflexibility Induces Non-Universal Phase Transitions in Block Copolymers 727
Shifan Mao, Quinn Macpherson, Andrew J Spakowitz
(726b) A Strongly Coarse-Grained, Charge-Fluctuating Model for Polyelectrolytes 728
Nicholas Jackson, Marcel Langenberg, Marcus Muller, Juan J. De Pablo
(726c) Development of New Coarse-Grained Water Models Using Particle Swarm Optimization 729
Karteek K. Bejagam, Samrendra Singh, Yaxin An, Carter Berry, Sanket A. Deshmukh
(726e) Unified Polymer Erosion Model 730
Joel Coffel, Eric Nuxoll
(726f) Development of a Fused-Sphere SAFT-γ Mie Force Field for Polymers and Application to Poly(vinyl butyral) Adsorption to Silica 731
Christopher Walker, Erik E. Santiso, Jan Genzer
(726g) Effects of Coarse-Graining on Simulations of Mechanical Properties of Polymers 732
Ting Ge, Mark Robbins
(726h) New Computational Methods for Rapid Simulation of Hydrodynamic Interactions in Polymer Solutions 739
James Swan, Andrew Fiore
(726i) Coarse-Grained Molecular Dynamics Simulations of PNIPAM Grafted Graphene Systems in an Aqueous Environment 740
Carter Berry, Karteek K. Bejagam, Sanket A. Deshmukh
(735a) Gel Assemblies of Colloidal Nanocrystals 741
Camila Saez Cabezas, Beth A. Lindquist, Ryan B. Jadrich, Thomas M. Truskett, Delia J. Milliron
(735b) Examining the Optical Effects of Chiral Carboxylic Acids Bound to the Surface of CdSe Nanoparticles 742
Mayank Puri, Vivian E. Ferry
(735c) Radial Elemental Distribution Analysis of Spherical Core/Shell Nanocrystals with STEM/EDX 743
Jacob Held, Katharine I. Hunter, Uwe R. Kortshagen, K. Andre Mkhoyan

(735d) Directional Carrier Transfer in Strongly Coupled Binary Nanocrystal Superlattice Films Formed By Assembly and in situ Ligand Exchange at a Liquid-Air Interface 744
Yaoting Wu, Siming Li, Natalie Gogotsi, Tianshuo Zhao, Blaise Fleury, Cherie R. Kagan, Christopher B. Murray, Jason B. Baxter

(735e) The Effect of CdS Shell Thickness on the Complex Index of Refraction of CdSe/CdS Core/Shell Nanocrystal Films 745
Dana Dement, Mayank Puri, Vivian E. Ferry

(735f) Engineering the Light-Matter Interactions of Ultrasmall CdSe Quantum Dots Via Modification of Surface Species for Light Harvesting Applications 746
Megan Webster, Kristi Pepa, Kevin Dominguez, Vinod Menon, Gustavo Lopez, Marco J. Castaldi, Ilona Kretzschmar

(740a) Computational Prediction and Evolutionary Design of Polymer Glass-Formation Behavior 747
David S. Simmons, Jui-Hsiang Hung, Venkatesh Meenakshisundaram, Tarak Kumar Patra

(740b) Domain Spacing and Phase Behavior of Salt-Doped Block Copolymers from Fluids Density Functional Theory 748
Jonathan R. Brown, Lisa M. Hall

(740c) Atomistic Simulations of Lamellae-Forming PS-b-P2VP 749
Yamil J. Colon, Weiwei Chu, Juan J. De Pablo

(740d) Rapid Conformational Fluctuations in a Model of Methylcellulose 750
Xiaolan Li, Frank S. Bates, Kevin D. Dorfman

(740e) Multi-Phase Coarse-Grained Models of Rod-like Polymers from Iterative Boltzmann Inversion 751
Christian Nowak, Fernando Escobedo

(740f) Coarse-Grained Model of Exciton Dynamics on Long-Chain Conjugated Polymer System 752
Elizabeth M. Y. Lee, William A. Tisdale, Adam P. Willard

(740g) Molecular Simulation of Thermoplastic Polyurethanes Under Large Mechanical Deformation 753
Shuze Zhu, Gregory C. Rutledge

(740h) Harnessing Virtual High-Throughput Screening and Machine Learning for the Discovery of Novel High Refractive Index Polymers 754
Johannes Hachmann

(740i) Development of Reaction Ensemble Monte Carlo (REMC) Algorithms to Study the Kinetics of Polymerization 755
Inderdip Shere, Ateeque Malani

(741a) Controlled Liquid-Liquid Phase Separation of Recombinant Oleosin 756
Ellen H. Reed, Daniel A. Hammer

(741c) Evaluation of the Procedure to Obtain Chitosan Based Gels with Potential Use As Bone Adhesive on Clinical Settings 757
Jairo A Jimenez, Juanita Echeverri, Paula A Sarmiento, J. German Vargas, Juan Carlos Briceno Triana, Camila Castro, Juan Pablo Casas, Felipe Salcedo

(741d) Deep Space Drug Shielding 760
Hannah Kim, Manosi Roy, Sutapa Barua

(741e) Synthesis of Designer Lipids Using "Click" Chemistries 761
Danielle Konetski, Dawei Zhang, Austin Baranek, Tao Gong, Brady Worrell, Christopher N. Bowman

(741f) Loading and Mobility of RNA in Porous Silica Nanoparticles for Delivery to Insects 762
Shanshan Zhou, Emily Nadeau, M. Arif Khan, Bruce Webb, Stephen E. Rankin, Barbara L. Knutson

(741g) Engineering and Functionalizing Protein-Based Materials 763
Sarah Bondos

(742a) Synthesis and Applications of Bio-Inspired Oligotea Peptidomimetics 764
Christopher A. Alabi

(742b) A Polyester Based Photoluminescent Hydrogel As Extracellular Matrix Mimics 765
Xiaoyang Xu

(742c) Alginate Nanoparticle Platform for Controlled Release of Biotherapeutics 766
Julia Vela Ramirez, Daniela Barrios Santos, Nicholas A. Peppas

(742f) Complex Bone Regeneration Via Controlled Release of Simple Signaling Molecules 767
Soheila Aliakbarighavimi, Brittany Allen, Jessica Stromsdorfer, Jake Kramer, Ram Rao Tata, Andrew Greenwald, Bret Ulery

(742g) Biomimetic Growth of a Pathologic Biomineral in Hydrogels 768
Gopichand Mallam, Marina Tsianou

(757a) The Influence of Intrinsic Framework Flexibility on Adsorption in Nanoporous Materials 769
Matthew Witman, Sanliang Ling, Sudi Jawahery, Peter Boyd, Maciej Haranczyk, Ben Slater, Berend Smit

(757b) Towards a Generalized Understanding of Acid Gas Interactions with ZIF Materials 770
Souryadeep Bhattacharyya, Rebecca Han, David S. Sholl, Sankar Nair

(757c) Heat-Treatment of Defective Uio-66 from Modulated Synthesis:Adsorption and Stability Studies 771
Yang Jiao, Yang Liu, Krista S. Walton, David S. Sholl
(757d) Design of Stratified Hybrid Metal Organic Frameworks for Chemical Detection and Destruction 772
Jonathan Ruffley, Tianyi Luo, Isabella Goodenough, Melissandre Richard, Eric Borguet, Nathaniel L. Rosi, J. Karl Johnson
(757e) Understanding Structure, Metal Distribution, and Water Adsorption in Mixed-Metal MOF-74 773
Joshua D. Howe, Cody R. Morelock, Yang Jiao, Karena W. Chapman, Krista S. Walton, David S. Sholl
(757f) Extraction of Rare Earth Elements from Geothermal Brine Solution Using Magneric Core Shell Microspheres 774
Praveen K. Thallapally, Nune Satish, Jian Liu, B. Peter McGrail
(757g) Cycloaddition Reaction of Epoxides with CO2 in a Copper Metal-Organic Framework: A Density-Functional Theory Study 775
Xu Li, Jianwen Jiang
(758a) Nanostructured Materials for Separations Based on Reactive Block Polymers 776
Marc A. Hillmyer
(758b) Block Copolymer Derived Nanostructured Surfaces: Templating Confined Surface Reactions 777
Katherine P. Barteau, Katharine W. Oleske, Ulrich Wiesner, Lara A. Estroff
(758c) Nanostructure Formation on Collapse of Polyelectrolyte Brushes 778
Blair Kathryn Brettmann
(758d) iCVD Deposition and Integration of Poly-(1H,1H,2H,2H-Perfluorodecylacrylate) (PPFDA) Under High Loading of TiO2 Nanoparticles 779
Zhengtao Chen, Kenneth Lau
(758e) Development of Resin-Containing Polymer Particles for Thermoset Powder Coatings 780
Guozhen Yang, Mengfei Huang, John Klier, Jessica D. Schiffman
(758f) Roll-to-Roll Micromolding of UV Curable Thiol-Ene Based Coatings 781
Yuyang Du, Alon McCormick, Lorraine F. Francis
(758g) Elastic Networks for Shape-Memory Contact Printing 782
Mitchell Anthamatten
(758h) Dynamics of Electric Double Layer Formation and Dissipation in Polyethylene Oxide:LiClO4 on Graphene Transistors 783
Susan Fullerton, Hua-Min Li, Ke Xu, Buchanan Bourdon, Hao Lu, Yu-Chuan Lin, Joshua Robinson, Alan Seabaugh
(758i) Redox-Active Organometallic Polymers for Small Molecule Separations 784
Xiao Su, Timothy Jamison, T. Alan Hatton
(765a) Disorder, Nonequilibrium Transport, and the Origin of Deep Traps in Quantum Dot Solids 785
William A. Tisdale
(765b) On the Molecular Origin of Intra-Gap Emission from CuInSe2-XSx Quantum Dots 786
Addis Fuhr, Nikolay Makarov, Hunter McDaniel, Hyeong Jin Yun, Hongbo Li, Anastassia Alexandrova, Phillippe Sautet, Victor I. Klimov
(765c) Accelerating Emission Dynamics in Perovskites Plasmonic Nanolasers 787
Sui Yang, Wei Bao, Xiaoze Liu, Xiang Zhang
(765d) Synthesis and Characterization of Plasmon-Resonant Hollow Gold Nanoshells 788
Joesph A. Zasadzinski, Jeongeun Shin
(765e) Solution-Based Synthesis of Cesium Bismuth Halide Perovskite Nanocrystals for Optoelectronic Applications 789
Rainie D. Nelson, Atefe Hadi, Alex Verburg, Matthew G. Panthani
(765f) Integrated Logic Gate Devices Fabricated Using Non-Toxic CuInSe2 Quantum Dots 790
Hyeong Jin Yun, Jaehoon Lim, Jeongkyun Roh, Darren Chi Jin Neo, Matt Law, Jeffrey M. Pietryga, Victor I. Klimov
(765g) Synthesis and Surface Functionalization of Group IV Quantum Dots 791
Yujie Wang, Michael Zembrzuski, Matthew G. Panthani
(766a) Understanding Rate Dependent Mechanical Properties of Supramolecular Hydrogels through Real Time SAXS Measurements during Stretching 792
Bryan D. Vogt
(766b) Microstructure and Mechano-Electrical Property of Ultra-Stretchable Iono-Elastomer Via Small Angle Scattering and Rheology 793
Ru Chen, Carlos R. Lopez-Barron, Norman J. Wagner
(766c) Toughening Isotactic Polypropylene with Block Copolymer Micelles 796
Jun Xu, Micah J. Howard, Frank S. Bates
(766d) The Importance of Crystalline Structure on the Tensile Properties of Uhmwpe 797
Nicolas J. Alvarez, Christopher Henry, Giuseppe Palmese

(766e) Controlling Polymerization Induced Phase Separation (PIPS) Using the Nonlinear Optical Properties of Light 798
Ian Hosein, Saeid Biria
(766f) Engineering Polymer Structure and Dispersity to Access Designer Materials with Exquisite Properties 799
Jimmy Lawrence, Eisuke Goto, Bernd Oschmann, Dongsub Kim, Jing Ming Ren, Craig J. Hawker
(766g) Thermal Bridges for Phonon Transport through Short Polymer Chains in a Hydrogen Bonded Polymer Composite 800
Nitin Mehra, Liwen Mu, Jiahua Zhu
(766h) The Effect of Pendant Alkyl Chain Length in Epoxy/Amine Thermosets Materiel Properties and Water Barrier Properties 801
John Vergara, Joshua Sadler, John La Scala, Santosh Kumar Yadav, Giuseppe Palmese
(766i) Novel Chemistries for the Replacement of Methylenedianiline in Composites 802
Joseph F. Stanzione III, Owen M. Stecca, Alexander W. Bassett, Jayson D. Cosgrove, Kevin M. Schmalbach, Joshua M. Sadler, John J. La Scala
(118a) Improvement of Thermal Conductivity for POSS Functionalized BN Fillers/Polyphenylene Sulfide Composites 810
Xutong Yang, Lin Tang, Yongqiang Guo, Junwei Gu
(118i) Structural Composites with Multiple Functionalities 811
Kenan Song
(118c) Fabrication of High-Dispersed Strawberry-like P(St-co-MAA)/SiO2 Composite Microspheres and Their Applications in Mortar 812
Guanzhi Cheng, Huajian Li, Xinguo Zheng, Yongjian Xie, Jing Liu
(118d) Hydrogen Bonded Thermal Highways Driving Heat Conduction in Polymer and Polymer Blends Films 813
Nitin Mehra, Jiahua Zhu
(118e) Low-Temperature Synthesis of Mn-Based Mixed Metal Oxides with Novel Fluffy Structures As Efficient Catalysts for Fto Reaction 814
Yi-Fan Han, Bo Meng
(118f) Structural Control of Polybenzoxazine/Epoxy Composites with Dual Crosslinking Network for Corrosion Protection 815
Changlu Zhou, Zhong Xin
(118g) Multifunctional Epoxy Conductive Nanocomposites 816
Xiaojiang Xu, Hongbo Gu
(118h) Nano-Structured Ceramic ALD Coatings to Stabilize SiC Against Reaction in High Temperature Steam 817
Amanda Hoskins, Aidan Coffey, Charles B. Musgrave, Alan W. Weimer
(118j) Multi-Scale Metrology for Visualization and Characterization of Interphase Failure 818
Richard Sheridan, Jeremiah Woodcock, Jeffrey W. Gilman, Gale Holmes, Catherine Brinson, Vamshi Gudapati, Dave Hartman, Amol Vaidya
(769a) Strategic Assemblies of Wood-Derived Building Blocks for the Sustainable Redesign of BPA-Based Polymers 819
Joseph F. Stanzione III, John J. La Scala
(769b) Preparation and Characterization of Cardanol Based Vinyl Ester Resins As Cross-Linker Units 832
Emre Kinaci
(769c) Ionic Liquids: Green Solvents for Dry Native Cellulose and Chitosan 833
Behzad Nazari, Nyalaliska Utomo, Sujyot Mony, Hyeonji Oh, Indira Saifuddin, Ralph H. Colby
(769d) Environmentally Friendly Flame Retardants Based on Adhesive Catecholamine 834
Hanim Kim, Joon Hee Cho, Kadhiravan Shanmuganathan, Amanda Jones, Sergei Nazarenko, Christopher J. Ellison
(769e) Soybean Oil Based Thermoset Materials with High Biorenewable Content 835
Sung-Soo Kim, Dustin Janes, Kadhiravan Shanmuganathan, Daniel Y. Chou, Christopher J. Ellison
(769g) Self Assembly of Coil-Hyperbranched Poly(styrene-block-acrylated epoxidized soybean oil) Block Copolymers 836
Fang-Yi Lin, Austin D. Hohmann, Nacu Hernandez, Eric W. Cochran
(769h) Physicochemical Properties of Polylactide/Delta-Valerolactone/Organosolv Lignin Atactic Terpolymers 837
Stephanie Harris, Ulrike Tschirner, Adam Gillespie, Madeleine Seeger
(769i) Oil Field Chemicals from Macromolecular Renewable Resources: Date Pit As a Case Study for Drilling Fluid Additive 838
Jimoh K. Adewole, Musa O. Najimu

(770a) Designing Synthetic Extracellular Matrices for the Creation of Controlled Culture Systems in the Study of Disease 859
April M. Kloxin
(770b) The Influence of Matrix Stiffness on the Behavior of Brain Metastatic Breast Cancer Cells 860
Akshay Narkhede, Shreyas Rao
(770c) The Combined Effect of Matrix Microenvironment and Hypoxia on the Activity of Glioblastoma Stemcells 861
Jee-Wei Emily Chen, Jann N. Sarkaria, Brendan A. Harley
(770d) Glioblastoma/Astrocyte Co-Culture on Polyelectrolyte Multilayer Films: A Template for Studying the Role of Astrocytes in Glioblastoma Progression 862
Kimberly M Stanke, Christina Wilson, Erin Eickman, Oleh Khalimonchuk, Srivatsan Kidambi
(770e) Three-Dimensional Tissues Using Human Pluripotent Stem Cell Spheroids As Biofabrication Building Blocks 863
Qiang Li, Haishuang Lin, Yuguo Lei
(770f) Dynamic Culture of Trabecular Meshwork Cells in 3D Biomimetic Scaffolds 880
Matthew Osmond, Mina Pantcheva, Melissa Krebs
(770g) Cellular Hitchhiking on Microparticles to Alleviate Skin Injury 891
Daniel Smith, Chase Herman, Sutapa Barua
(771a) Sugar-Coating the Answers to Virus Binding: Glycocalyx-Mimetic Interfaces 892
Ramya Kumar, Domenic Kratzer, Kenneth Cheng, Irina Kopyeva, Joerg Lahann
(771b) Glucose-Derived Cationic Block Poly(beta-peptides) Reverse Intrinsic Antibiotic Resistance in Gram-Negative Pathogens 893
Zhangyong Si, Hui Wen Lim, Damien Keogh, Moon Tay Yue Feng, Jo Thy Lachumy Subramanion, Yahua Chen, Guillermo C. Bazan, Everett Peter Greenberg, Yunn-Hwen Gan, Kevin Pethe, Mary B. Chan-Park
(771c) Mechanism Study of Selective Killing of Cationic Peptidopolysaccharide Nanoparticles with in vitro and in vivo Efficacy Against Multi-Drug Resistant Bacteria 894
Mary Chan, Hou Zheng, Yogesh Vikhe
(771d) Tether Supported Biomembranes with Phase Composition and Orientation Control for Biomaterial Ligand Displays and Membrane Protein Assays 895
William Houlihan, Yueming Li, Lane Gilchrist
(771e) Understanding the Interaction of the Polysaccharide, Chitosan, with a Novel Cuticular Protein, CPR 27, Using Quartz Crystal Microbalance with Dissipation 896
Aishik Chakraborty, Ellie K. Onstott, Neal T. Dittmer, Michael R. Kanost, Stevin H. Gehrke, Prajnaparamita Dhar
(771f) Probing Phase Transitions in Dynamic Biopolymer Complexation 897
Amanda B. Marciel, Handan Acar, Matthew V. Tirrell
(771g) Photo-Induced Pinocytosis in Synthetic Liposomes 898
Danielle Konetski, Dawei Zhang, Christopher N. Bowman
(771h) Protein Protection and Purification through Virus-like Particle Encapsidation: A Rapid 2-Step Cell-Free Protein Synthesis Approach 899
Bradley C. Bundy, Seung Ook Yang
(775a) Measurement and Modeling of Carrier Dynamics in Photovoltaic Cztse 900
Siming Li, Michael L. Lloyd, Brian E. McCandless, Jason B. Baxter
(775b) Impact of the Active Layer Morphology on Bimolecular Recombination Dynamics in Organic Solar Cells 901
Veaceslav Coropceanu, Jean-Luc Bredas, Shafigh Mehraeen
(775c) Atomistic Origin of the Concentration Dependence of Si Dopant Mobility in III-V Semiconductor Alloys 902
Mardochee Reveil, Paulette Clancy
(775d) Characterizing Defects in Photovoltaic Semiconductors with Optical Spectroscopy 903
Charles J. Hages, Thomas Unold
(775e) CIGS Nanocrystal Solar Cells on Plastics and Paper 904
Vikas Reddy Voggu, Sam Morehead, Brian A. Korgel, Taylor B. Harvey
(775f) Solution Processed Thin Film Photovoltaics Using Amine-Thiol Chemistry 905
Swapnil Dattatray Deshmukh, Xin Zhao, Ruihong Zhang, Caleb Miskin, David Rokke, Carol Handwerker, Rakesh Agrawal
(775g) Photonic Mirrors for Enhanced Optical Transport in Luminescent Solar Concentrators 906
Ryan Connell, Mayank Puri, Vivian E. Ferry
(775h) Dopant Mediated Assembly of Nanorods into Atomically Coupled 2D Sheets in Solution 907
Ajay Singh, Delia J. Milliron
(777a) Tailored Polymers for Structural Application in Large-Scale Extrusion Systems 908
Blake Marshall

(777b) Modeling Flow Phenomena in Fused Filament Fabrication Geometry 909
Eric L. Gilmer, Darren Miller, Jacob Fallon, Camden Chatham, Callie Zawaski, Allison M. Pekkanen, Timothy E. Long, Christopher B. Williams, Michael J. Bortner

(777c) Influence of Processing on Additively Manufactured Mechanically Adaptive Cellulose Nanocrystal Polymer Composites 910
Jacob Fallon, Michael J. Bortner, Earl J. Foster, Cara Herwig, Ben Kolb

(777d) 3D Printed Shape Memory Objects Based on Olefin Ionomer of Zinc-Neutralized Poly(ethylene-co-methacrylic acid) 911
Bryan D. Vogt

(777e) Three-Dimensional Printing By Multiphase Silicone/Water Capillary Inks 912
Sangchul Roh, Dishit Parekh, Bhuvnesh Bharti, Simeon Stoyanov, Orlin D. Velev

(777f) 3D Printing of Polymer-Bonded Magnets Using a Combination of Extrusion Direct Write and Stereolithography Methods 913
Alan Shen, Anson Ma, Sameh Dardona

(777g) PDMS-Based Ink Development for 3D Printing Applications 914
Kwan-Soo Lee, Joseph H. Dumont, Andrew M. Schmalzer, Chihoon Park, Andrea Labouriau

(777h) 3D Printed Active Microfluidic Elements for Portable Bioanalysis Assays 915
Duanduan Han, Victor M. Ugaz

(777i) 3D Printed Polymer-Based Bio-Inspired Neural Systems 916
Blake Johnson

Author Index

Some abstracts in this book refer to images that were not made available by the publisher

502749 Multi-Stage Drug Delivery System for Enhanced Payload Delivery to Lymph Node Cells

Abstract:

Targeted drug delivery to lymph nodes (LN), tissues where organized lymphocyte accumulation in the body occurs, has the potential to improve the treatment of a variety of pathologies, such as B and T cell malignancies and immunodeficiency viral infections, as well as enhance the efficacy of vaccines and cancer immunotherapy. To date, size-based principles of lymphatic-mediated transport have emerged as an effective means to achieve facile LN delivery of formulated agents after injection in the periphery. However, within the LN, solutes and particulates are filtered based on size, with smaller molecules penetrating more deeply into LN follicles and the paracortex than large particulates by virtue of their increased diffusivity and uptake within LN conduits. Herein, we report the design and implementation of a multistage lymphatic delivery and payload release system combining 30 nm thiolated poly(propylene sulfide) nanoparticles that exhibit prodigious uptake into lymphatics after administration with the thiol-reactive oxanorbornadiene (OND) linkage chemistry that degrades in a pH- and polarity-insensitive manner at programmable half-lives ranging from hours to days to release small-molecule cargo. Post administration in the skin, nanoparticle tethering dramatically enhances model cargo accumulation within LN draining the site of injection that when released via the OND penetrates deeply throughout the collagenous LN interstitium due to its increased diffusivity relative to polymer nanoparticles. As a result, small molecule cargo is delivered to cell subtypes localized within the LN cortex and paracortex at high levels, a 10- to 1000-fold relative to free cargo alone or cargo tethered to the nanoparticles via an irreversible linkage. These data suggest that by utilizing an engineered multi-stage system to overcome barriers to both lymphatic and lymph node uptake, payload can be delivered to previously undruggable cell subpopulations resident within lymph nodes and has the potential to dramatically enhance drug bioactivity and resulting therapeutic effect.

495334 Encapsulation and Thermal Stability of Immunological Biologics Using Complex Coacervation

Abstract:

A major obstacle when working with biomolecules is their stability at varying temperatures. Relief programs often struggle to transport sufficient vaccines and other cold chain-dependent medical supplies due to the need to keep these materials carefully refrigerated. The cold chain, currently, accounts for 80% of the cost of vaccination and breaks in the cold chain resulting in the loss of nearly half of vaccines produced. Thus, by mitigating the need for the cold chain, vaccine campaigns and disaster relief efforts will no longer be confined by cold chain logistics.

Complex coacervation offers a novel means for the delivery of proteins and other bioactive materials. We hypothesize that control over the chemical and physical aspects of the molecular microenvironment can be used to modulate protein function and stability with the potential to mimic the biological milieu. In particular, we posit that polypeptide-based coacervates can be used to create a highly crowded environment where the chemistry of the polypeptides can be used to tune the stabilizing character of the encapsulating environment. Furthermore, encapsulation via coacervation is a fully aqueous process that is compatible with biologic materials and can achieve extremely high levels of protein encapsulation via preferential partitioning. We have investigated the encapsulation and resulting thermal stability of three model proteins, bovine serum albumin (BSA), hen egg white lysozyme (HEWL), and a novel monoclonal antibody (mAb), into poly(lysine) and poly(glutamate)-based complex coacervates. Encapsulation efficiency of these model systems, specifically lysozyme, has gone as high as 80% with partitioning coefficients above 200. Preliminary results also show that the encapsulated proteins retain their secondary structure. Additionally, data suggests that even at higher temperatures, the secondary structure remains the same. Polypeptide-based coacervates are a powerful platform technology in understanding the role of the chemical environment on the stability of a protein formulation. Ultimately, we look to harvest these materials to enable the development of highly stable therapeutics and biologics such as refrigeration-free vaccines.

497123 Evaluation of Biocompatibility of Novel Block Copolymer Gels/Micelles As Potential Vaccine Adjuvants

Abstract:

Evaluation of biocompatibility of novel block copolymer gels/micelles as potential vaccine adjuvants

Justin R. Adams[1], Sujata Senapati[1], Michael J. Wannemuehler[2], Balaji Narasimhan[1], Surya Mallapragada[1]

[1]Chemical & Biological Engineering; [2]Veterinary Microbiology and Preventive Medicine;

Iowa State University, Ames, IA

In addition to being delivery systems for antigens, an ideal adjuvant should be able to modulate an effective innate immune response and enhance the induction of long-lasting immunity. Adjuvant platforms may also allow for fewer doses required and less invasive routes of vaccine administration. Even after decades of research, presently only a few vaccine adjuvants, such as, alum and oil in water emulsion, are approved for safe human usage. However, vaccines adjuvanted with these compounds have long been associated with inflammation at the injection site. Therefore, it is essential to discard the practice of using off-the-shelf adjuvants for vaccines and move towards the design and implementation of novel vaccine adjuvants. Biocompatible polymers have been recently investigated as novel adjuvants and delivery systems for next generation vaccines. However, as research into new polymeric adjuvants progresses, the safety of these compounds is of great concern. There is a need to balance the immunomodulatory properties of vaccine adjuvants with the desire to improve patient compliance by avoiding adverse reactions such as pain, redness, and swelling (i.e., reactogenicity) at the site of administration.

Amphiphilic block copolymers have been used for applications in gene therapy and drug delivery *in vivo*. Our research group has previously synthesized cationic pentablock copolymers following a new method of catalysis for atom transfer radical polymerization (ATRP), which significantly reduces *in vitro* cellular toxicity of the polymer. These pentablock copolymers are composed of FDA approved Pluronic F127 and pendant amine groups of poly(2-diethylaminoethylmethacrylate) with reversible micelle and gel formation capabilities in response to external stimuli such as temperature and/or pH. In the micellar phase, these pentablock copolymer can act as a cargo hold for various protein payloads allowing for sustained release of the antigen.

In this work, we investigate the safety and biocompatibility of the pentablock copolymer gels/micelles in a murine model. This study included serum and urine analysis, assessment of inflammation at the injection site, proinflammatory cytokine analysis, and histopathological analysis. Mice were subcutaneously immunized with micelle/gel formulations and were compared to separate groups of mice treated with saline or incomplete FreundÕs adjuvant (IFA). For the urinalysis, little to no statistical difference in the quantity of creatinine in urine of mice administered with the pentablock copolymer formulations was observed when compared to the saline alone group. The

histopathological studies of the administration site tissue, kidney, and liver showed no adverse reactions. Using a live animal imager, local inflammation was assessed and was shown to be markedly less inflammatory post-administration of the block copolymer when compared to IFA.

Combined with previous studies demonstrating the induction of robust humoral immunity in mice, the results of this study highlight the safety and adjuvanticity of the pentablock copolymer micelle/gel vaccine delivery platform.

491488 Intranasal Nanovaccine Provides Protection Against Homologous and Heterologous Influenza Virus

Abstract:

Each year influenza A virus (IAV) causes significant illness and economic burden. While current inactivated vaccines may induce production of virus-neutralizing antibodies against antigenically matched strains, their protection is limited against heterologous infection. In addition, intranasally administered live-attenuated influenza vaccines, which were designed to promote cell-mediated immunity, are no longer recommended by the Centers for Disease Control and Prevention due to current poor vaccine efficacy. Therefore, next generation vaccines that can induce protective antibody- and T cell-mediated immunity hold the promise of increasing the breadth of protection against IAV.

Polyanhydride nanoparticles are a versatile platform suitable for the delivery of IAV vaccines that have been demonstrated to promote robust antibody production and T cell responses. In this work, the efficacy of an intranasal polyanhydride nanovaccine encapsulating hemagglutinin, nucleoprotein, and CpG was examined in mice. The nanovaccine provided robust protection during a homologous influenza challenge and induced the formation of germinal centers within the lung, draining lymph nodes, and nasal-associated lymphoid tissue post-vaccination. In addition, influenza-specific and hemagglutination inhibition antibodies were detected in the serum and bronchoalveolar lavage fluid indicating the induction of both local/mucosal and systemic immunity.

While IAV-specific antibodies as well as T cells are protective against homologous influenza viruses, studies suggest that T cells are critical for cross-protection against heterologous infection. Mice administered the polyanhydride nanovaccine demonstrated enhanced accumulation of both $CD4^+$ and $CD8^+$ influenza-specific T cells within the lungs. Additionally, influenza-specific tissue-resident memory T cells were observed through at least 45 days post-vaccination. Consistent with the induction of IAV-specific T cells after vaccination, we observed that protection against heterologous challenge was greatly enhanced in mice administered the polyanhydride nanovaccine (80% survival) compared to controls (0% survival). Together, these data demonstrate the ability of polyanhydride nanovaccines to induce robust local and systemic adaptive immunity as well as provide protection against homologous and heterologous influenza challenge.

501123 Design of a Combination Nanovaccine to Induce Rapid and Long Term Protective Immunity Against *Bacillus Anthracis*

Abstract:

Bacillus anthracis, the causative agent of anthrax, is a highly lethal Gram-positive bacteria due to its ability to produce lethal and edema toxins. In addition to its virulence, B. anthracis has the ability to form durable endospores, allowing it to survive in the environment for decades. Because of these attributes, this pathogen has a history of use as a biological warfare agent, as in the events following 9/11, which resulted in the deaths of five civilians and possible exposure to 30,000 others. Currently, there is an FDA-approved vaccine formulation, Anthrax Vaccine Absorbed (AVA), which consists of culture supernatants of predominantly protective antigen (PA), the receptor-binding component for both lethal toxin and edema toxin, adsorbed to aluminum hydroxide. Although efficacious, the vaccine regimen consists of three immunizations within the first six months to achieve protective immunity followed by boosters at 12 and 18 months and yearly boosters afterward. Therefore, there still remains a need for an anthrax vaccine formulation that can provide both rapid and long-lived protective immunity in the event of exposure of warfighters or civilians to B. anthracis, while also improving patient compliance through reducing the number of immunizations required.

Polyanhydride nanoparticle-based vaccines (i.e., nanovaccines) consisting of 1,6-bis(p-carboxyphenoxy) hexane (CPH) and 1,8-bis(p-carboxyphenoxy)-3,6-dioxaoctane (CPTEG) provide a solution to these challenges. These biodegradable compounds are safe, biocompatible, and have demonstrated the ability to maintain the structural stability and functional activity of encapsulated PA over the course of four months. The current study investigated the ability of polyanhydride nanovaccines containing PA and the cyclic dinucleotide, cyclic di-GMP, a known activator of the STING pathway, to elicit rapid and long-lived protective immunity against B. anthracis. In this study, three separate groups of female C57BL/6 mice were immunized with the PA nanovaccine and serum samples were collected and analyzed for antibody titer, avidity, and epitope specificity at specified time points post-immunnization. In addition, a lethal factor neutralization assay was performed to evaluate the functional antibody titer of the immunized mice. Mice were later challenged at 14, 42, or 70 days PI with a lethal dose of Ames strain B. anthracis and survival rates of the vaccinated mice were assessed.

500467 Comprehensive Vaccine Design for Commensal Disease Progression

Abstract:

Commensal organisms with the potential to cause disease pose a challenge in developing treatment options. Using the example featured in this study, pneumococcal disease begins with Streptococcus pneumoniae colonization followed by triggering events that prompt release of a virulent sub-population of bacteria. Current vaccines focus upon colonization prevention, which poses unintended consequences of serotype niche replacement. In the enclosed work, non-covalent co-localization of two classes of complementary antigens, one to prevent colonization of the most aggressive S. pneumoniae serotypes and another to restrict virulence transition, provides complete vaccine effectiveness in animal subjects and the most comprehensive coverage of disease reported to date. As a result, the proposed vaccine formulation offers universal pneumococcal disease prevention with the prospect of effectively managing a disease that afflicts 10s-100s of millions globally. The approach more generally puts forth a balanced prophylactic treatment strategy in response to complex commensal-host dynamics.

Oncofetal Antigen Peptide Nanoclusters for Cancer Vaccines

Alexandra N. Tsoras, Julie A. Champion.

School of Chemical & Biomolecular Engineering.
Georgia Institute of Technology

AIChE National Meeting, Minneapolis, MN.
November 1, 2017
Biomaterials for Immunological Applications:
Immune Activation and Vaccination

Georgia Institute of Technology

Outline of Today's Talk:

- Review Subunit Vaccines: Peptide vs. Protein
- Peptide Nanoclusters
- Synthesis/Characterization
- *In vitro* and *in vivo* results

1

Shift In Research Towards Subunit Vaccines

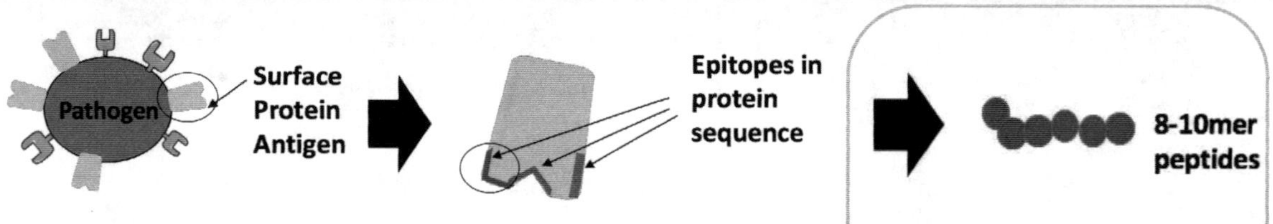

Attenuated/inactivated whole pathogen
- Widely used
- Immunogenic
- Reversion potential
- Off target responses

Surface protein antigens
- Safe
- Less but some immunogenicity
- Off target response potential
- Difficulty expressing

Peptide epitope antigens
- Correct target
- Synthesized
- Low immunogenicity
- Too small

Skwarczynski, M. and I. Toth (2011). Current Drug Delivery 8: 282-289 2

Oncofetal Antigen as a Blood Cancer Vaccine

- Oncofetal Antigen (OFA) is an immature laminin receptor protein
  - surface protein highly expressed on many blood cancer cells but not normal cells
  - Multiple identified epitopes to promote T-cell-mediated immunity
  - Low immunogenic response

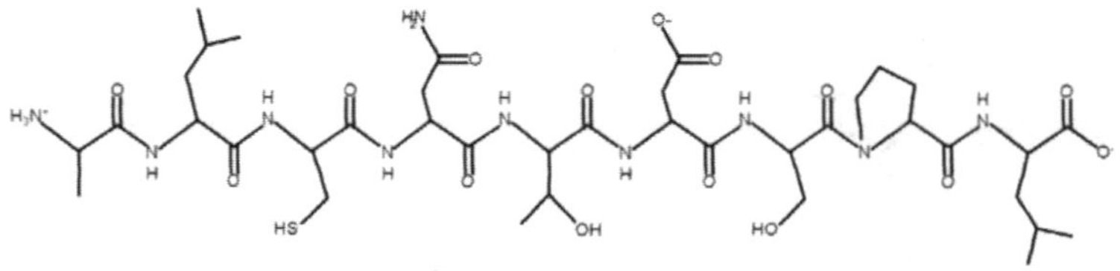

OFA 2: ALCNTDSPL

Siegel S, et al. *(2006) J Immunol* 176:6935-6944 3

Techniques for Improving Peptide Vaccine Immunogenicity

Adjuvants
- Induce innate immune responses
- Approved for commercial use
- Overstimulation
- High diffusion rates

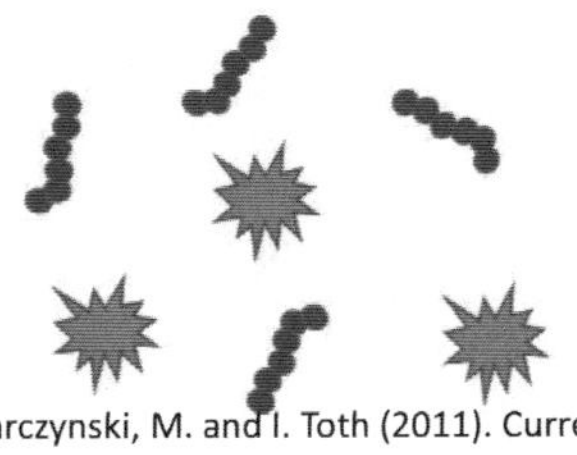

Nanoparticles or self-assembling materials
- Slower Diffusion
- Increased Stimulation
- Off target materials
- High carrier:target

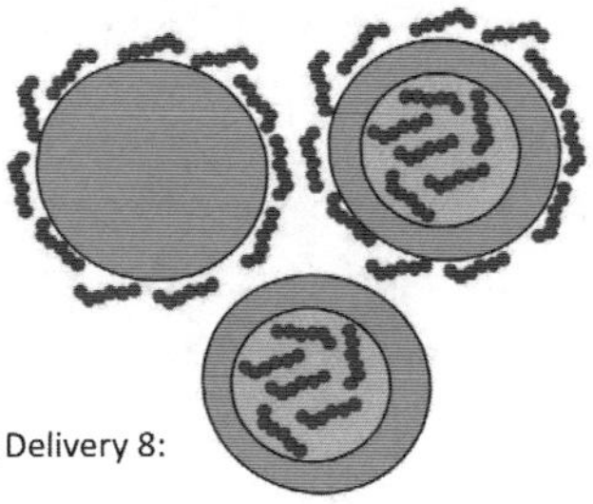

Peptide Nanoclusters (PNCs)
- Low carrier:target
- Antigen throughout

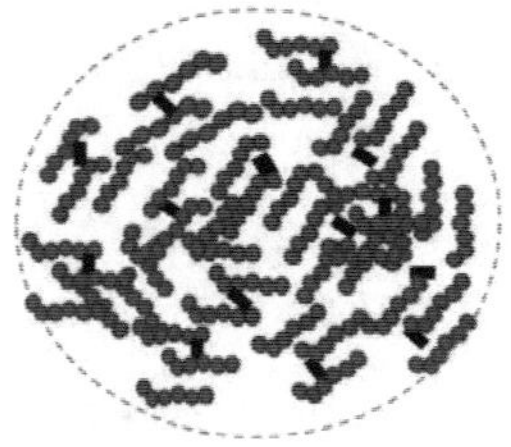

Skwarczynski, M. and I. Toth (2011). Current Drug Delivery 8: 282-289

4

Nanocluster Synthesis by Desolvation

Desolvation Steps/Parameters:

Pre-processing
1. Solvent
2. Desolvent
3. Ratio D:S
4. Crosslinker
- linker concentration
Post-processing

Target size: 20-200nm

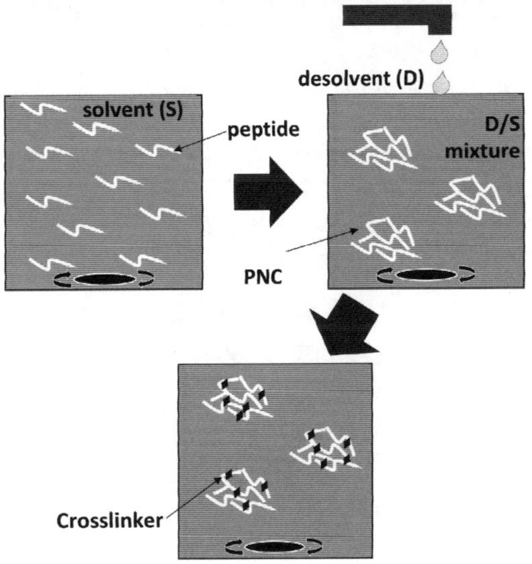

Weber C, Kreuter J, Langer K. (2000). Int. J. of Pharm. 194: 91-102.

5

Parameter Screening for OFA 2 PNC Optimization

Optimizing size and stability PNC with Dynamic Light Scattering (DLS)

Final formulation:
Monodisperse, significant number of PNC

CHES solvent:
Lack of PNC formation

Methanol desolvent:
Large aggregates

Acetone desolvent:
Polydisperse

Too much crosslinker:
Some Aggregation

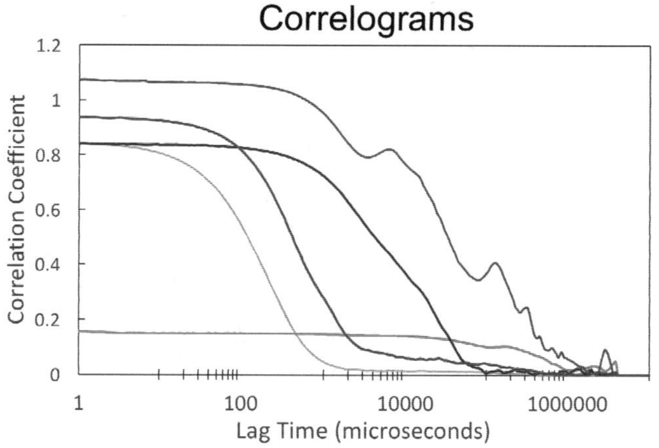

6

Final Protocol Yielded Acceptable Size and Stable PNC

- **Preprocessing:** add cysteine to C-terminal (OFA 2C)
- **Solvent:** hexofluoroisopropanol (HFIP)
- **Desolvent**: diethyl ether (DEE)
- **D:S Ratio**: 19:1
- **Crosslinker**: trimethylolpropane tris(3-mercaptopropionate) (TTM)

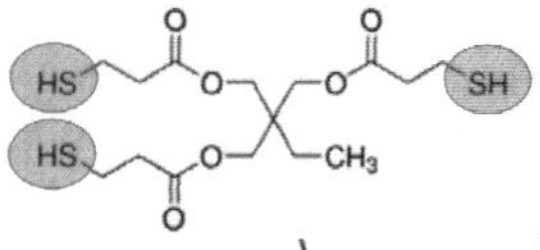

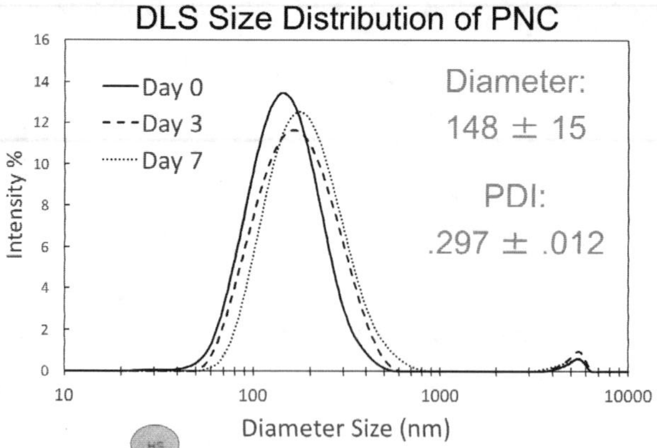

ALCNTDSPLC

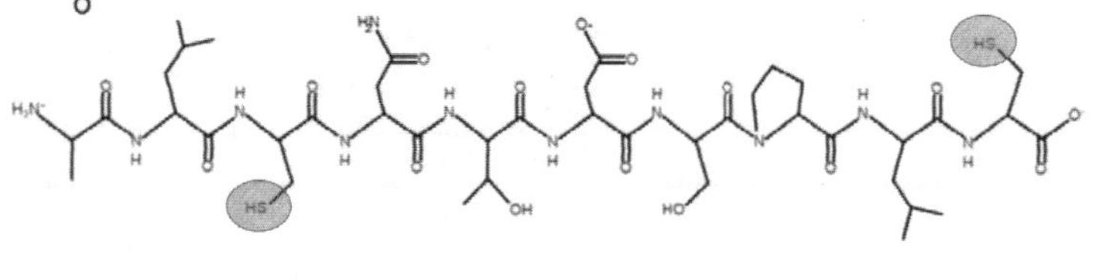

Yield: ~99.8% peptide desolvated

7

Atomic Force Microscopy Images Confirm Size of PNC

- PNC are sizes within desired range (20-200nm)

- DLS biased toward larger PNC because of intensity measurement

- Lyophilized PNC
  - Dehydration may collapse and shrink

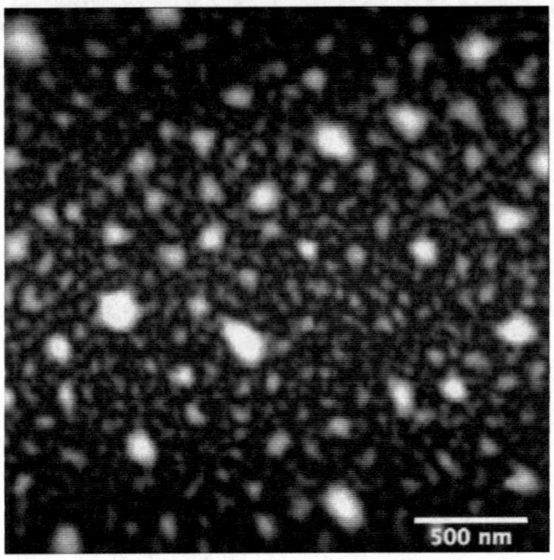

Slightly smaller sizes than DLS

8

Immature DC Internalization of PNC and Soluble Peptide is Comparable

10% FITC-tagged OFA 2 in PNC/soluble peptide

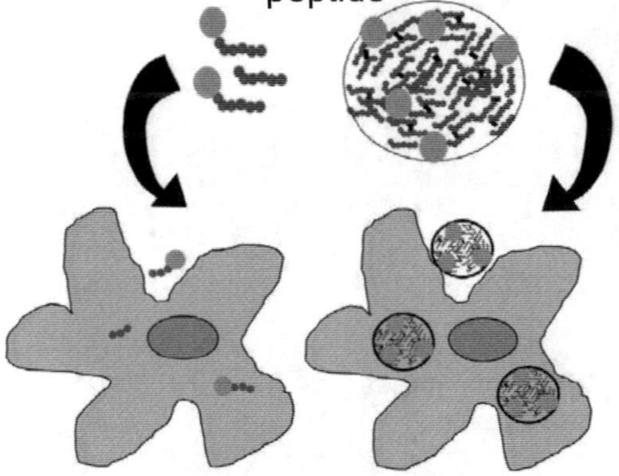

DC cell line: MUTZ-3

Expresses MHC I haplotype HLA-A2

DC Internalization of OFA 2C

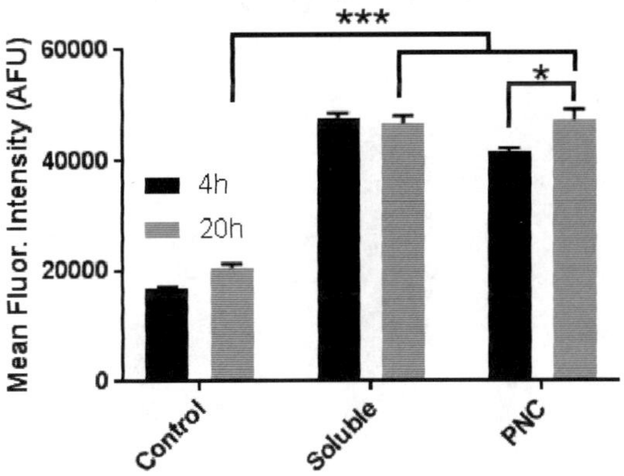

*p<.05,***p<.001

9

Presentation of OFA 2C on DC Surface Significant Only with PNC Administration

10% biotinylated OFA 2 in PNC/soluble peptide

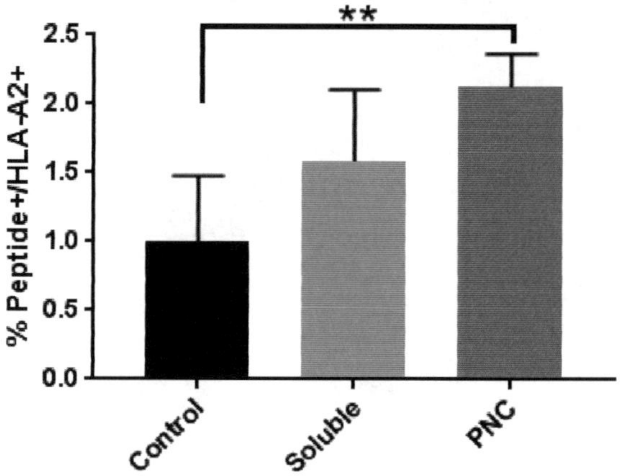

**p<.01

10

In vivo Injection Site Retention Increased with PNC

Intradermal Injection of 5% Near-infrared-tagged OFA 2C PNC or soluble peptide

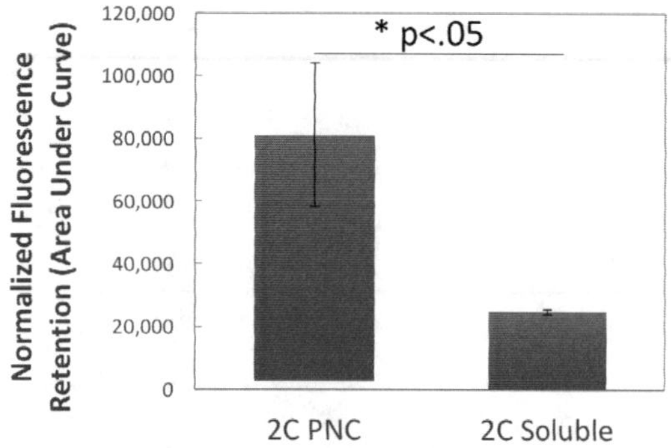

Significantly higher retention in region of interest over 54h

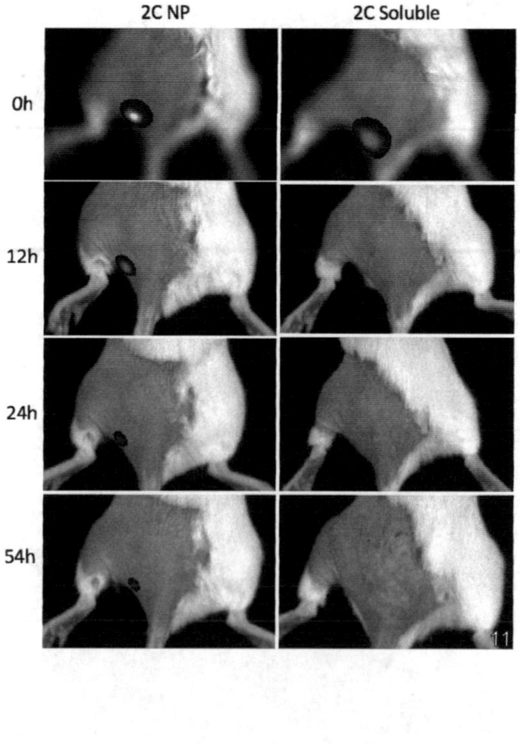

11

Conclusions

- Stable OFA 2C PNC synthesized in target size range 20-200nm using desolvation and optimizing 4 parameters
- PNC internalized similarly to soluble peptide *in vitro*, but presented on DC surface significantly more
- *In vivo* retention also increased with intradermal PNC injection

12

Future Work

- T-cell activation *in vitro* when co-cultured with PNC or soluble peptide-pulsed DC
- PNC from other peptide epitopes
  - Different solubilities
  - Different chemistries to crosslink
  - Different desired adaptive immune responses

13

Acknowledgements

- Dr. Julie Champion
- Champion Research Group

- University of Michigan School of Medicine – Varani Lab

Funding:
- Benovus Bio, Inc.
- MT Campagna

Georgia Tech | Chemical & Biomolecular Engineering

BENOVUSBIO
Innovative Immunotherapy

14

DC Presentation broken down by Fluorophore: *in vitro*

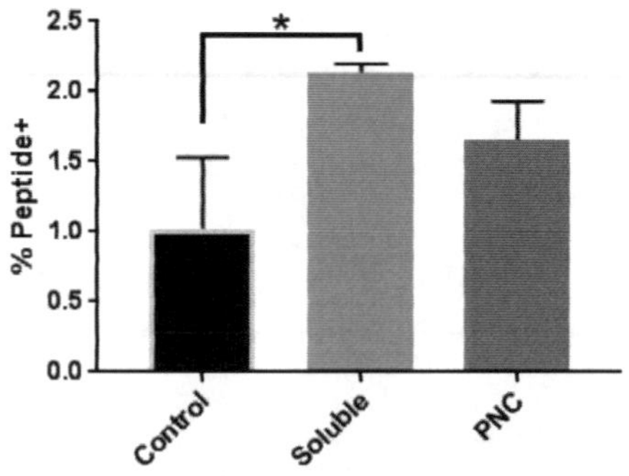

Technically more peptide on surface with soluble

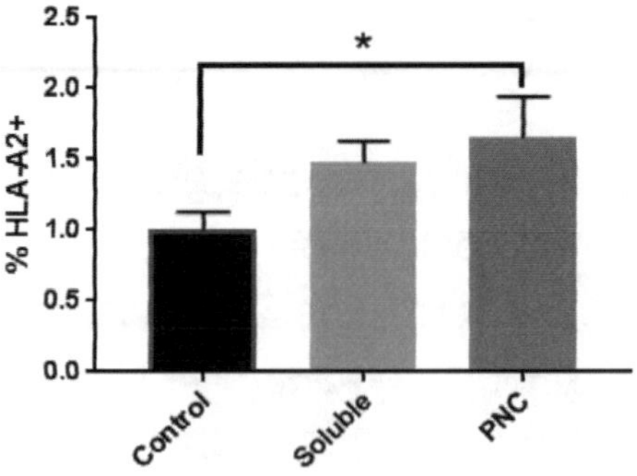

More HLA-A2 on surface with PNC

Additional Modifications Possible for *in vitro/ in vivo* Studies

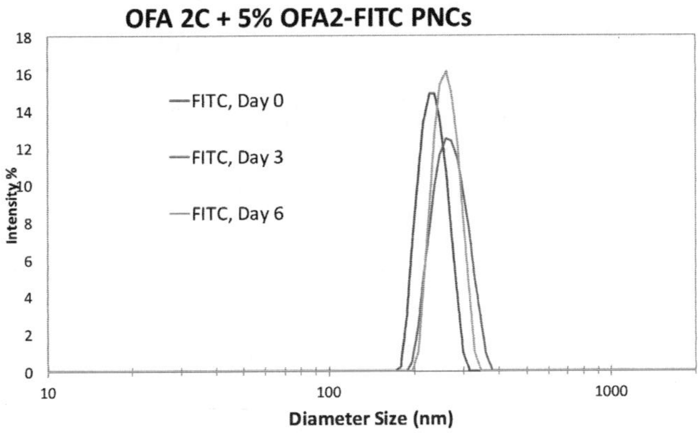

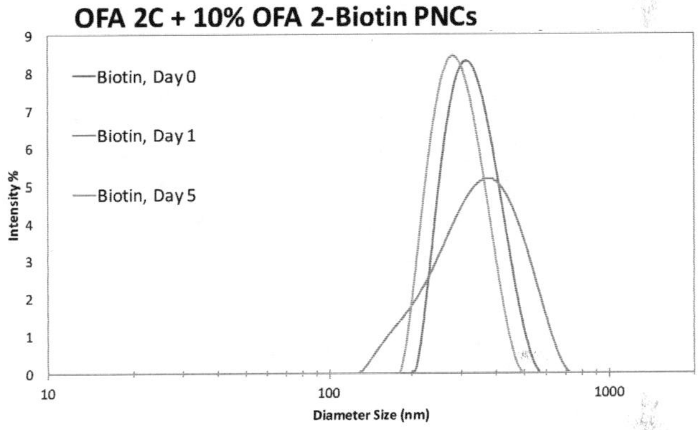

DLS: 260-280nm
Stable over 1 week

DLS: 290-350nm
Stable over 1 week

499395 S Doped TiO_2 for Photocatalytic Oxidation of CO in Visible Region Synthesized By Novel One Step Liquid Flame Spray Pyrolysis (LFSP): Kinetics and Mechanism

Abstract:

A novel single step Liquid Spray pyrolysis (LFSP) method is used to synthesize a series of S-TiO_2 catalyst of nanosized crystals with dimensional homogeneity and high monodispersity. A series of S-TiO_2 catalyst of nanosized crystals with high monodispersity and dimensional homogeneity are synthesized by using the novel single step Liquid Spray pyrolysis (LFSP) method. Different quantities of S was incorporated into TiO_2 catalyst, which is beneficial to the generation of the successive energy bands inside the TiO_2 band gap. Our XPS results indicate that the S doping concentration could reach to 8 wt% owing to the homogeneous doping in the mesoporous TiO_2 catalysts. The S doped TiO_2 catalyst exhibit such excellent characteristics as high specific area, relatively small particle size, pure anatase phase and excellent UV-vis absorption capacity in the range of 400-800 nm, which are all beneficial to the photocatalytic oxidation of CO under the visible light irradiation. The photo conversion of CO to CO_2 is quantitative in the presence of O_2. The photocatalytic CO oxidation rates and the amounts of S deposited on TiO_2 surface show a close correlation. The CO photooxidation is strongly inhibited in the absence of O_2 or in the presence of alternative oxidants such as N_2O and H_2O. The addition of water vapor little affects the CO photooxidation kinetics in the presence of O_2. Hydroxyl radicals do not seem to play a significant role in CO photooxidation on S-TiO2. The photocatalytic oxidation kinetics was investigated in detail as a function of S loading, coated TiO_2 mass, CO concentration, O_2 concentration, light intensity, additives, and humidity.

500131 Copper Hexacyanoferrate Hydrogel Electrodes for Electrochemically-Mediated Cation Separations

Abstract:

Copper Hexacyanoferrate Hydrogel Electrodes for Electrochemically-Mediated Cation Separations

Kai-Jher Tan, Xiao Su, Johannes Elbert, and T. Alan Hatton*

Department of Chemical Engineering, Massachusetts Institute of Technology

25 Ames Street, Cambridge, MA, U.S.A. 02139

**E-mail: tahatton@mit.edu*

Chemical separations are of crucial importance and are used ubiquitously in the large-scale production of commodity chemicals and valuable species. However, as the purification steps in industrial manufacturing are often highly material and energy-intensive, there is a clear urgency for pursuing green engineering to develop and implement clean energy-efficient separation processes to reduce the economic and environmental impacts of high global energy consumption.[1,2] In particular, separations from aqueous media are of great interest. The removal of trace contaminants from water has been highlighted recently as a critical area for improvement, with applications to heavy metal recovery and seawater desalination.[2]

Electrochemical processes are an attractive platform for conducting separations, as they do not require solvents that change solution chemistry or impart environmental effects downstream, as well as heat and pressure inputs like conventional techniques.[1] Furthermore, they can be designed to target certain chemical species, whereas at the molecular scale, processes like distillation or chromatography can be inefficient.[1,2] Previously, redox-active materials such as metallopolymers have been shown to perform highly selective faradaic separations[3], as well as intercalating-type materials. One such crystalline candidate is the mixed-valence transition metal hexacyanoferrate class of compounds, whose analogues have been used in ion exchangers, ion-sensing, electrochromic applications, and lately, as an energy storage material for batteries[4,5,6].

We report a new surfmer-based method to produce functionalized redox-active copper hexacyanoferrate (CuHCF) composite hydrogels for electrochemically separating ions in aqueous media. The developed technique is a straightforward, inexpensive, one-pot preparation process that can consistently produce self-contained redox-active electrodes, without the need for manual filtration or grinding steps. The resulting electrodes are uniform, stable in water, and capable of withstanding up to ~100 separation cycles without gel leaching.

CuHCF exhibits strong redox response as well as affinity for cesium cations, whose radioactive isotope, ^{137}Cs, is present in nuclear waste and is both an environmental and health concern.[7] CuHCF nanoparticles can be further applied to remove other metal cations with similar hydrated ionic radii.[4,6] We also investigate methods to further improve the operation of the hydrogel in an electrochemical system by using specific dual-functionalized design schemes to tune the counter electrode[8] for increasing adsorption and energy efficiency.

In conclusion, we present a novel technique to conveniently generate stable CuHCF composite hydrogels, which demonstrate robust redox activity and applicability in separating heavy metal ions from aqueous media for both water remediation and valuable cation recovery.

References:

(1) Anastas, P. T.; Zimmerman, J. B. *Environ. Sci. Technol.* **2003**, *37*, 94A.

(2) Sholl, D. S.; Lively, R. P. *Nature*, **2016**, *532*, 435.

(3) Su, X.; Kulik, H. J.; Jamison, T. F.; Hatton, T. A. *Adv. Funct. Mater.* **2016**, *26*, 3394.

(4) de Tacconi, N. R.; Rajeshwar, K.; Lezna, R. O. *Chem. Mater.* **2003**, *15*, 3046.

(5) Wessells, C. D.; Huggins, R. A.; Cui, Y. *Nat. Commun.* **2011**, *2*, 550.

(6) Wang, R. Y.; Shyam, B.; Stone, K. H.; Weker, J. N.; Pasta, M.; Lee, H.; Toney, M. F.; Cui, Y. *Adv. Energy Mater.* **2015**, *5*, 1401869.

(7) Chen, R.; Tanaka, H.; Kawamoto, T.; Asai, M.; Fukushima, C.; Kurihara, M.; Ishizaki, M.; Watanabe, M.; Arisaka, M.; Nankawa, T. *ACS Appl. Mater. Interfaces.* **2013**, *5*, 12984.

(8) Su, X.; Tan, K. J.; Elbert, J.; Rüttiger, C.; Gallei, M.; Jamison, T. F.; Hatton, T. A. *Energy Environ Sci.* **2017**, *Accepted.*

498147 Emulsion Templated Polymers As Structured Supports for MOF Adsorbents

Abstract:

Metal organic frameworks are well suited to a variety of applications, including adsorption, catalysis, and gas storage. However, their potential applications are limited while in pellet or powder form, and they can lack the mechanical stability required for some applications. High internal phase emulsion templated polymers (polyHIPEs) are permanently and highly macroporous polymer structures that have shown promise as catalyst and adsorbent supports. PolyHIPEs could provide a platform for the creation of large, three dimensional bodies containing these active materials while maintaining high permeability and accessibility to the internal portions of the body through large, interconnected pores.

We have successfully incorporated the MOF UiO-66-NH_2 into a 2-hydroxyethyl methacrylate polyHIPE crosslinked with N,Nâ€™-methylene bisacrylamide. The porous structure of the polyHIPE has been preserved despite the incorporation of MOF particles. While the lower than expected BET surface areas of the composites suggested that the MOF was obstructed by the polymer, CO_2 adsorption was not negatively affected at low pressures suggesting that the MOF is accessible to adsorbates even within the polymer. Dye uptake confirms this in the aqueous phase. The MOF-polyHIPE composites show promise as adsorbents for both gases and liquids, with the polymer enhancing the structural stability of the MOF and allowing the creation of three dimensional monoliths rather than powder or pellets.

497267 Development of Novel Crosslinked Polymers for the Capture of Environmental Pollutants

Abstract:

Development of Novel Crosslinked Polymers for the Capture of Environmental Pollutants

Rishabh Shah, Thomas D. Dziubla, J. Zach Hilt

Department of Chemical and Materials Engineering

Â University of Kentucky Lexington, KY 40506

Corresponding Author: zach.hilt@uky.edu

The ubiquitous nature of persistent organic pollutants (POP), such as polychlorinated biphenyls (PCBs), have gained great attention, and many efforts are being made to eliminate these contaminants from the environment. Capture of these pollutants using polymeric materials from the environment is one remediation strategy.Â In this research, aromatic monomers and crosslinkers based on polyphenolic compounds have been studied, since we expect these functionalities to have high affinity for PCBs due to Ï€-Ï€ stacking interactions.

The overall goal of this project was to develop novel crosslinked polymers and their composites that have high affinity towards PCBs and compare their binding properties. Novel monomers and crosslinkers were initially synthesized by acrylation of 4,4-dihydroxybiphenyl, 4-phenylphenol, 2-phenylphenol, and the acrylated forms were identified as 44BMA, 4PPMA, 2PPMA respectively. These monomers were characterized by various techniques, including differential scanning calorimetry (DSC) for their melting temperature and NMR to confirm their structure. Subsequently, crosslinked polymer films were synthesized using styrene, 4PPMA, and 2PPMA as monomers and crosslinked with different amounts of 44BMA. These films were then characterized using swelling studies in different solvents, fourier transform infrared spectroscopy (FTIR), and binding studies with PCBs, etc.

488537 Grafted Polystyrene Controlled Formation of Magnetic Carbon Nanocomposites for Environmental Remediation

Abstract:

The reduction of pollution emissions and remediation of polluted natural resources are common challenges faced by both developed and developing countries in the world. Hexavalent chromium (Cr(VI)) ions and tetrabromobisphenol A (TBBPA) are widespread in environment. The scientists and engineers in all countries have been seeking for efficient and economical methods to remove them from polluted water systems. In this work, the polystyrene controlled formation of different carbon structure nanocomposites have been prepared by calcination of the epoxide group-functionalized polystyrene as the carbon source under the different atmospheres. These carbon nanocomposites have been evaluated for the environmental remediation including the toxic Cr(VI) removal and TBBPA adsorption. Various factors influencing the Cr(VI) removal and TBBPA adsorption efficiency, such as treatment time, and initial pollutants concentration have been systematically studied. The removal kinetics and thermodynamics was determined. The results showed that these carbon nanocomposites showed a unique capability to remove pollutants from polluted water over a wide pH range. The Cr(VI) removal and TBBPA adsorption kinetics was followed the pseudo-first-order behaviour and pseudo-second-order behaviour, respectively. The calculated thermodynamic parameters suggested that the Cr(VI) removal and TBBPA adsorption process is spontaneous. These pollutants removal mechanism was also explored.

[1] H. Gu, H. Lou, J. Tian, S. Liu and Y. Tang, *Journal of Materials Chemistry A*, 2016, **4**, 10174-10185.

[2] B. Qiu, H. Gu, X. Yan, J. Guo, Y. Wang, D. Sun, Q. Wang, M. Khan, X. Zhang, B. L. Weeks, D. P. Young, Z. Guo and S. Wei, *J. Mater. Chem. A*, **2**, 2014, 17454-17462.

[3] F. Gao, H. Gu, H. Wang, X. Wang, B. Xiang and Z. Guo, *RSC Adv.*, **5**, 2015, 60208-60219.

[4] H. Gu,* H. Lou, D. Ling, B. Xiang, Z. Guo, *RSC Adv.*, **6**, 110134-110145, (2016)

[5] W. Jiang, Q. Cai, W. Xu, M. Yang, Y. Cai, D. D. Dionysiou and K. E. O'Shea, *Environ. Sci. Technol.*, 2014, **48**, 8078-8085.

華南理工大學

AIChE
The Global Home of Chemical Engineers

Constructing Fenton-like Reaction over g-C_3N_4/NH_2-MIL-88B(Fe) Photocatalyst to Degrade Organic Contamination in Aqueous Solution

Xiyi Li, Yunhong Pi, Zhong Li, Jing Xiao*

School of Chemistry and Chemical Engineering
South China University of Technology
November 1, 2017, AIChE, Minneapolis, America

Outline

1. Background
2. Research objective
3. Experimental and discussion
4. Conclusions

AIChE
The Global Home of Chemical Engineers

Background | Objective | Experiment | Conclusion

Industrial wastewater

3

Pollutants are widely present in wastewater from various industrial sources.

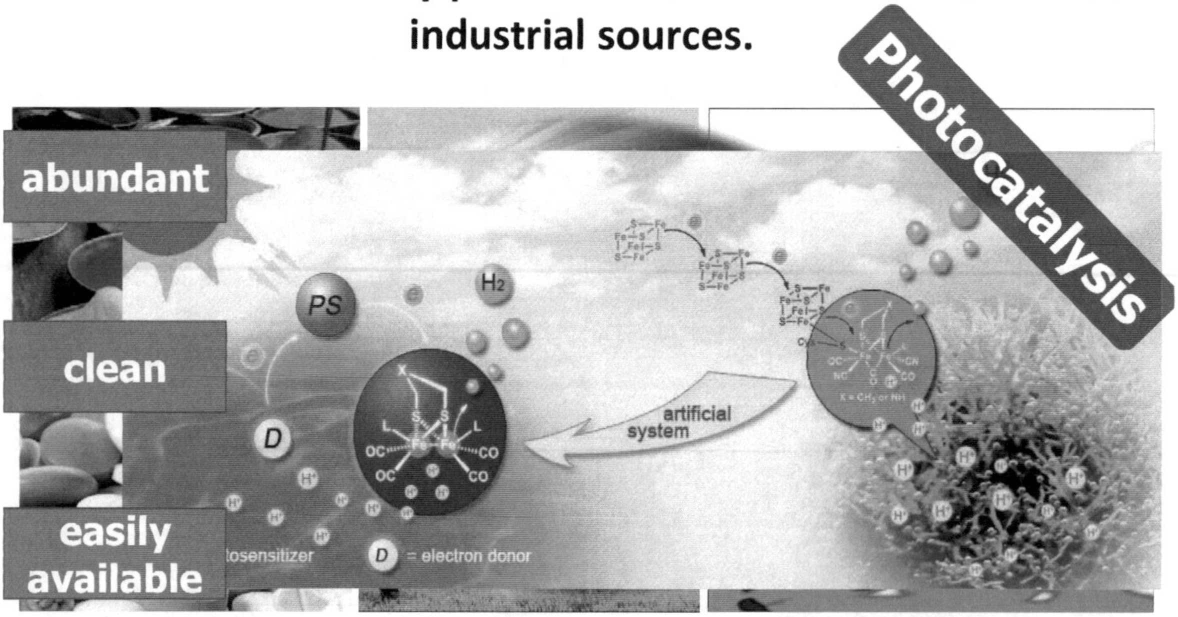

4

MOFs

◆ **As a new class of versatile porous solid materials, MOFs show promises to be tuned as effective photocatalysts via post-synthetic modification (PSM).**

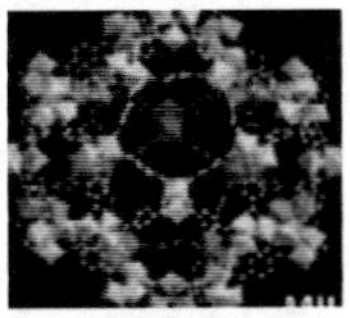

High specific surface area & Pore volume

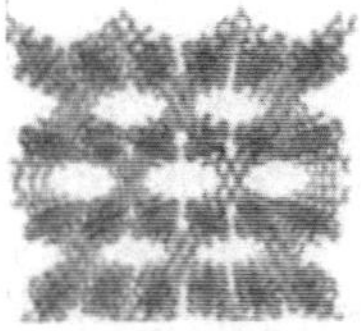

Tailorable pore structure & surface properties

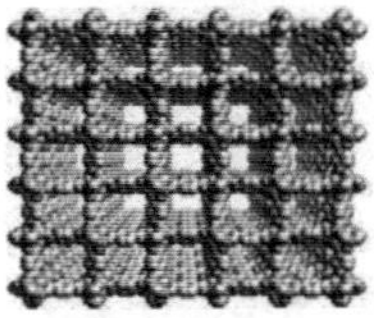

Uniform pore size distribution

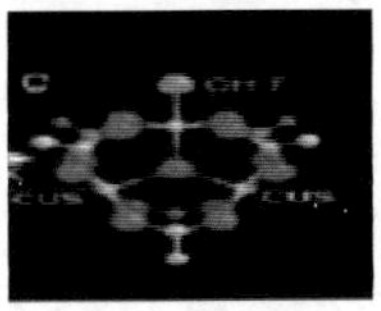

Diverse ways of coordination

5

$g-C_3N_4$

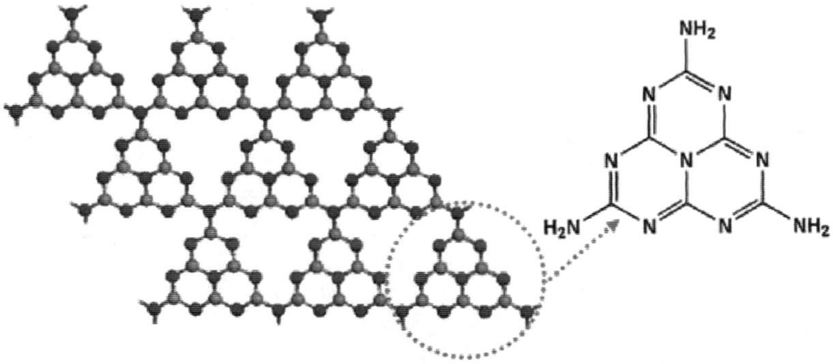

◆ Photosensitizer: **visible-light-active polymeric semiconductor (~2.7 eV)**

◆ Unique structure:

- **2D frameworks of tri-s-triazine connected via tertiary amines**
- **high stable thermal and chemical stability**

◆ Drawback:

- **The recombination of photo-induced e-/h+ pairs**

6

AIChE
The Global Home of Chemical Engineers

Background | Objective | Experiment | Conclusion

1 The integration of MOFs photocatalyst with g-C_3N_4 materials

2 Clarification of the structure and property

3 Investigation of the photocatalytic activity

4 Clarification of the mechanistic pathways

7

AIChE
The Global Home of Chemical Engineers

Background | Objective | Experiment | Conclusion

◆ **Synthesis process of g-C_3N_4/NH_2-MIL-88(Fe)**

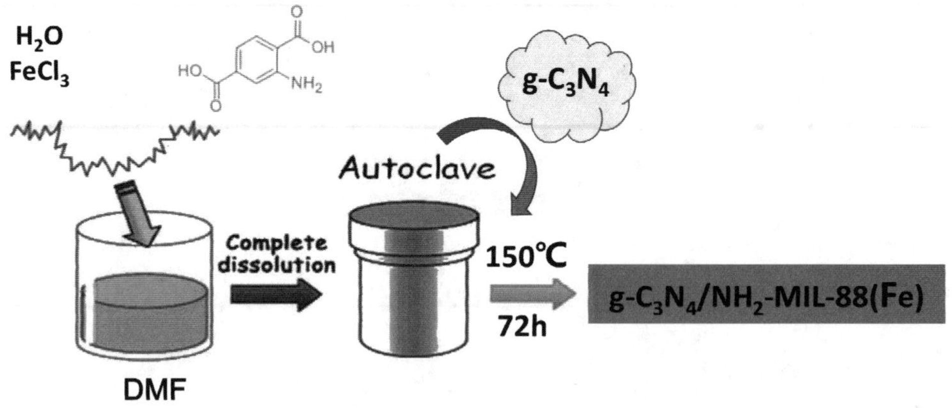

8

Characterizations of $g\text{-}C_3N_4/NH_2\text{-}MIL\text{-}88(Fe)$ (Ip-x)

◆ XRD & FT-IR

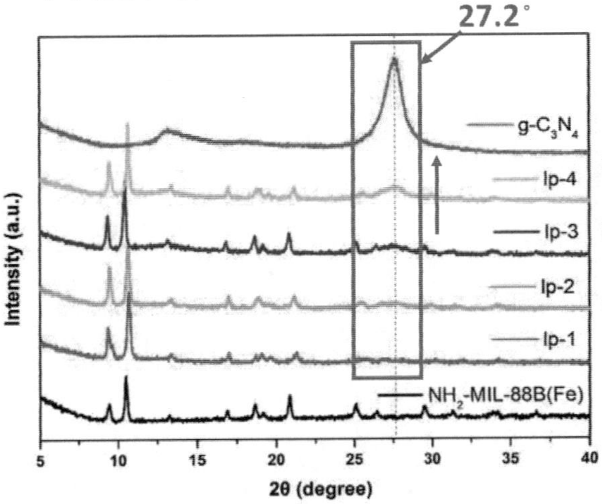

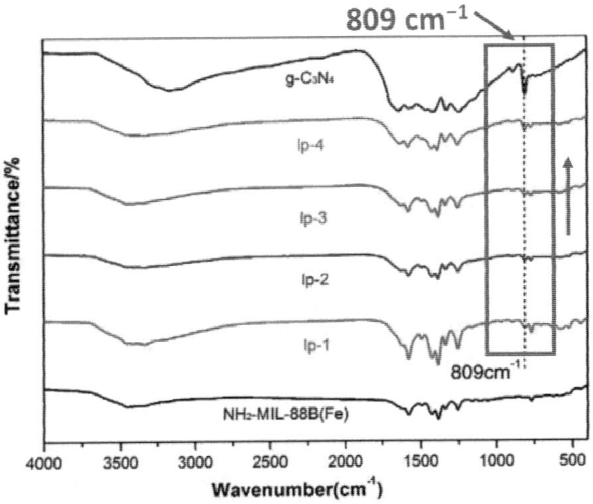

Crystal structure of NH_2-MIL-88B(Fe) remained intact after introducing $g\text{-}C_3N_4$

9

Characterizations of $g\text{-}C_3N_4/NH_2\text{-}MIL\text{-}88(Fe)$ (Ip-x)

◆ SEM & TEM

NH_2-MIL-88B(Fe)

$g\text{-}C_3N_4$

Ip-2

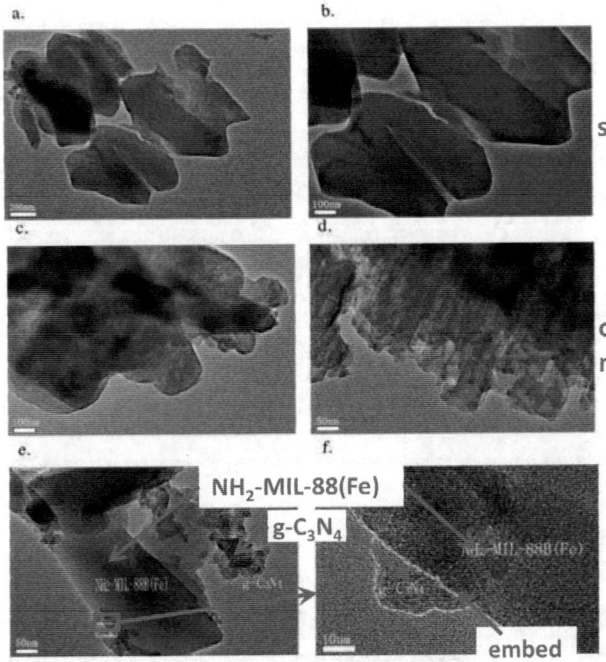

spindleshaped

overlapping lamellar nanosheets

successful incorporation of $g\text{-}C_3N_4$ into NH_2-MIL-88B(Fe)

10

Characterizations of g-C_3N_4/NH_2-MIL-88(Fe) (lp-x)

◆ XPS

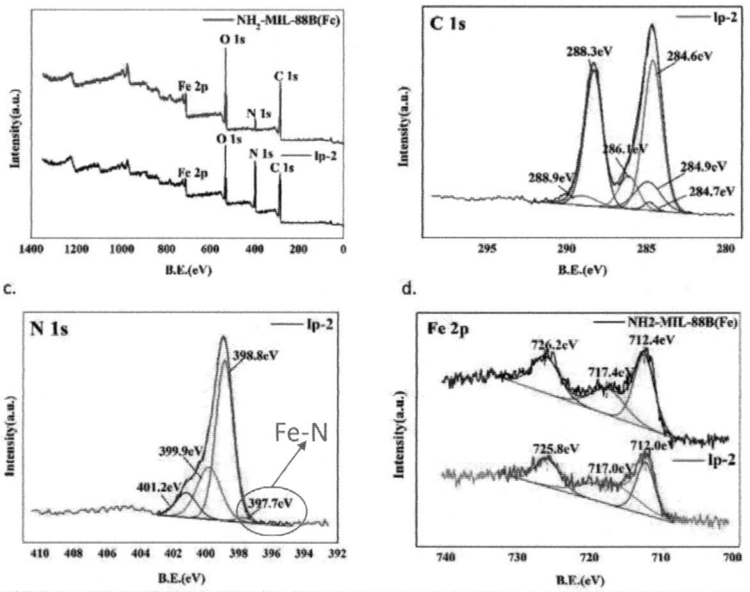

The XPS results further indicated the neighboring interaction between g-C_3N_4 and NH_2-MIL-88B(Fe) in the lp-2 composite, consistent with the TEM analyses.

11

Characterizations of g-C_3N_4/NH_2-MIL-88(Fe) (lp-x)

◆ UVDRS

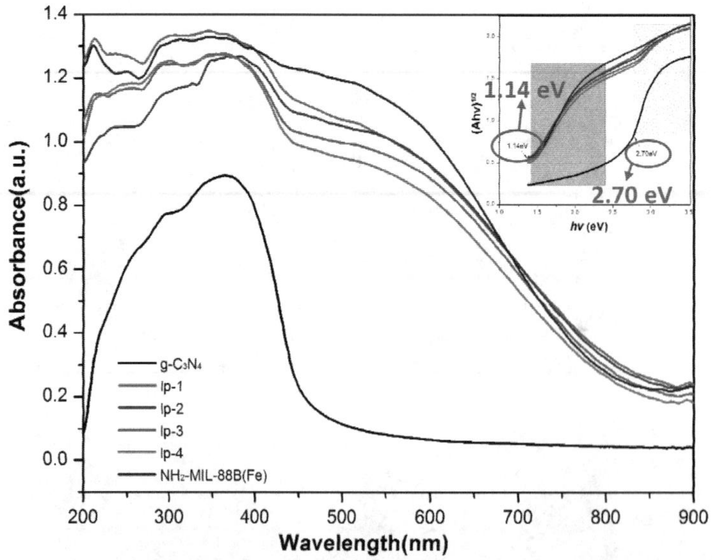

The Eg edges of lp-x composites are located at between 1.14 and 2.70 eV, suggesting the formation of heterojunction between NH_2-MIL-88B(Fe) and g-C_3N_4.

12

Photocatalytic activity of g-C_3N_4/NH_2-MIL-88(Fe) (Ip-x)

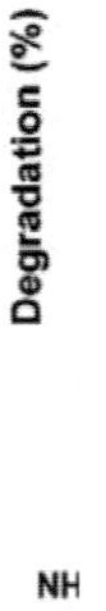

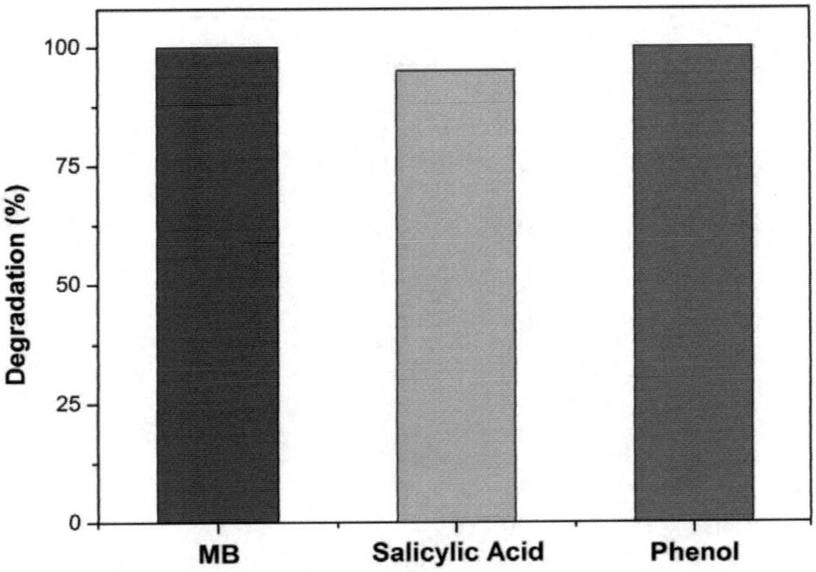

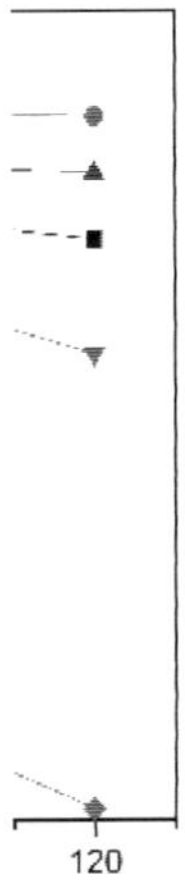

13

Performance of Ip-2

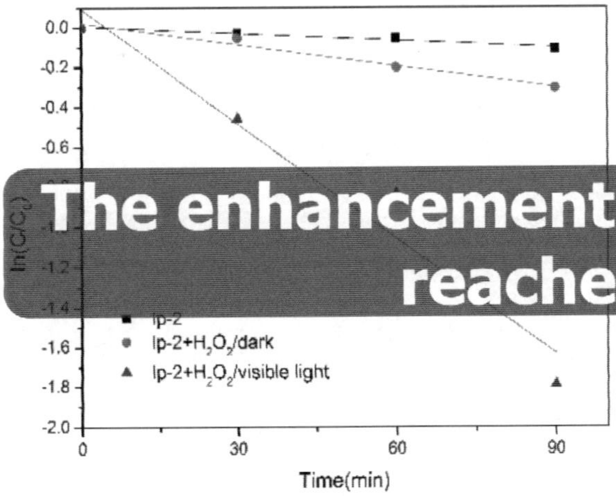

Pseudo-first-order kinetics

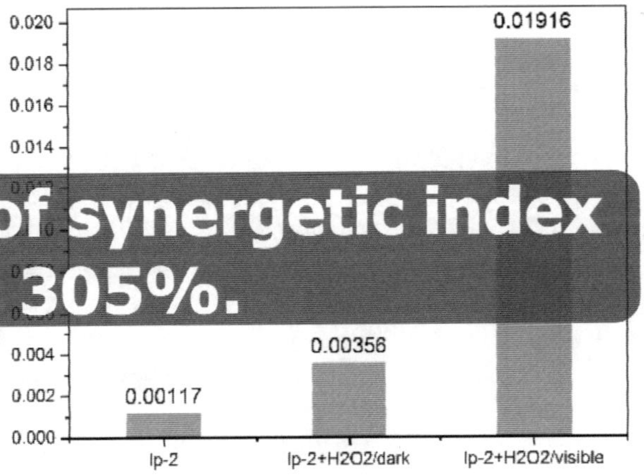

Reaction rate constants

The enhancement of synergetic index reaches 305%.

14

Reusability of Ip-2

◆ **Regeneration cycles**

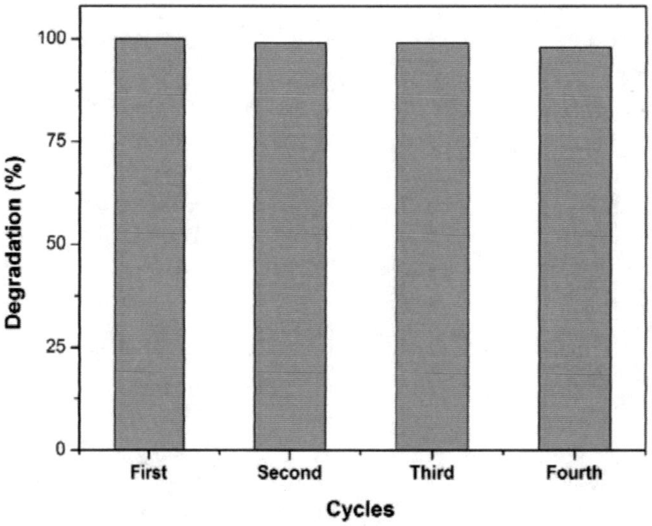

high reusability and stability

15

Characterization of mechanism on Ip-2

◆ **ESR spectra of DMPO-•OH adducts formed with irradiation time of visible light**

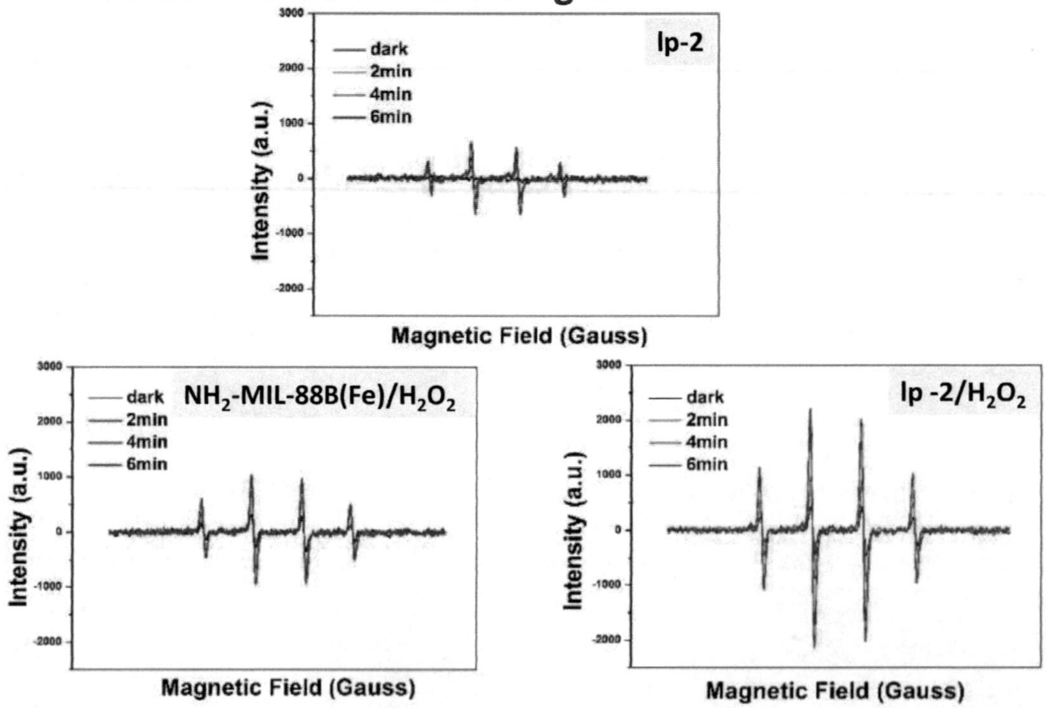

16

Two primary mechanisms on lp-2

◆ **Illustration of Fenton-like excitation of H_2O_2 for •OH generation under visible light.**

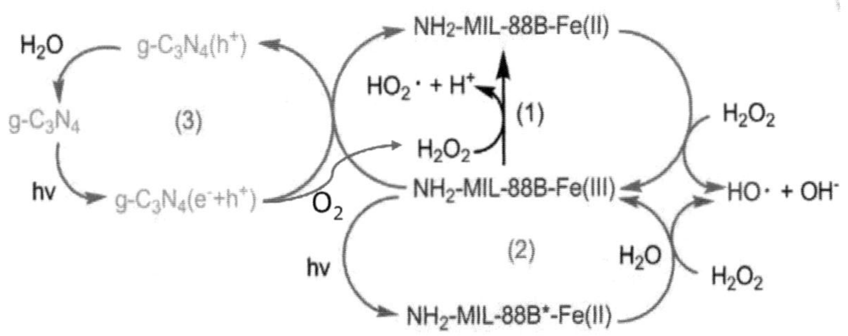

(i) Direct Fenton-like excitation of H_2O_2 (path (1));
(ii) photo-induced Fenton-like excitation of H_2O_2 (path (2) and (3)).

17

AIChE
The Global Home of Chemical Engineers

Background | Objective | Experiment | Conclusion

01 **Highly active $g\text{-}C_3N_4/NH_2$-MIL-88B(Fe) Fenton-like heterojunction photocatalysts were successfully fabricated for the effective photodegradation under visible light.**

02 **lp-2 can degrade MB completely in 120 min under visible light, much greater than the parent $g\text{-}C_3N_4$ and NH_2-MIL-88B(Fe), individually.The synergetic index in the lp-2/visible-light/H_2O_2 system reached as high as 305%.**

◆ The formation of the $g\text{-}C_3N_4/NH_2$-MIL-88B(Fe) heterojunction enhance dramatically the transfer of photo-induced electrons, and facilitate the excitation of H_2O_2 .

03 **This work demonstrates the first example of facilitating Fenton-like excitation of H_2O_2 via introduction of $g\text{-}C_3N_4$ to stable amine functionalized Fe-centered MOF for visible light-induced photodegradation.**

18

Acknowledgement

- **Group members at SCUT**

Graduate Students:

Prof. Li   Prof. Xiao   Xiyi Li   Yunhong Pi

- **Funding supports**
  - National Natural Science Foundation of China (21576093)
  - Guangdong Natural Science Funds for Distinguished Young Scholar
- **The organizing committee of AIChE meeting**

Thanks for your attention!

495674 Synthesis of Thermoresponsive Composite and Application for Water Treatment

Abstract:

Poly (*N*-isopropylacrylamide) (PNIPAM) is known as representative thermoresponsive polymer that exhibits lower critical solution temperature (LCST) around 32 °C. In our laboratory, we have been synthesizing thermoresponsive composite in which PNIPAM polymer or gel and inorganic nanoparticles were combined, and also have been tried to apply them for heavy metal ion recovery or organic compound decomposition in the solution. For heavy metal ion recovery, poly(PNIPAM-*co*-AA)/Fe_3O_4 composite adsorbents were synthesized under various conditions. The result of Cu(II) adsorption experiment showed that the adsorption capacity at above LCST (60 °C) was larger than that at below LCST (10 °C). Then, repeated Cu(II) adsorption/desorption operation was performed by swinging temperature. The result showed that the adsorption/desorption behavior was reversible and recovery of Cu(II) ion could be achieved by only swinging temperature. This would eliminate the production of secondary waste such as acid waste during recovering process of heavy meal ions. For organic compound decomposition, highly functionalized thermoresponsive composites gel beads in which two kinds of functional inorganic particles and thermoresponsive polymer work concertedly were prepared. In this study, poly(N-isopropylacrylamide) and calcium alginate were used as the thermoresponsive polymer and structure support polymer, respectively. TiO_2 and Fe_3O_4 were used as functional inorganic nanoparticles. The resulting composites exhibited thermoresponsive volume change and photocatalytic activity to decompose methyl orange. Localized heating of the thermoresponsive bead containing Fe_3O_4 was also achieved by applying an alternating current (AC) magnetic field on the bead. Due to the localized heating property, repeated shrinking-swelling movement (i.e., pumping movement) of the composite was achieved by applying an AC magnetic field intermittently. Finally, based on the experimental results, the effect of the promoted mass transfer of the substrate and product due to thermoresponsive pumping on the enhancement of the apparent photocatalytic activity was simulated. The results showed the effectiveness of thermoresponsive pumping in improving the apparent photocatalytic activity of TiO_2 nanoparticles embedded in the composite gel.

華南理工大學

AIChE
The Global Home of Chemical Engineers

Integrated adsorption and visible-light-driven photocatalysis over C_{60}/NH_2-MIL68(In) and MWCNT/NH_2-MIL68(In)

Yunhong Pi, Xiyi Li, Zhong Li, Jing Xiao*

School of Chemistry and Chemical Engineering
South China University of Technology

November 1, 2017, AIChE, Minneapolis, America

Outline

1. Background
2. Research objective
3. Experimental and discussion
4. Conclusions

AIChE
The Global Home of Chemical Engineers

Background | Objective | Experiment | Conclusion

Industrial wastewater

Clean water scarcity

3

Pollutants are widely present in wastewater from various industrial sources.

Organic dyes

Fertilizers

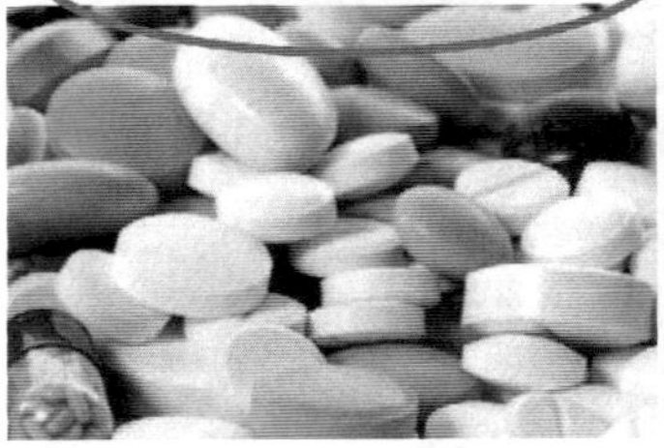

Pharmaceuticals

Pesticides

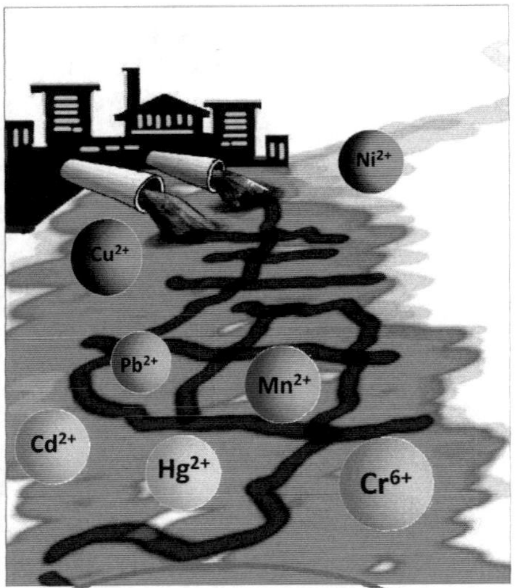

Heavy metal ions

Highly toxic and difficult to be removed.

4

MOFs

- As a new class of versatile porous solid materials, MOFs show promises to be tuned as effective photocatalysts via post-synthetic modification (PSM).

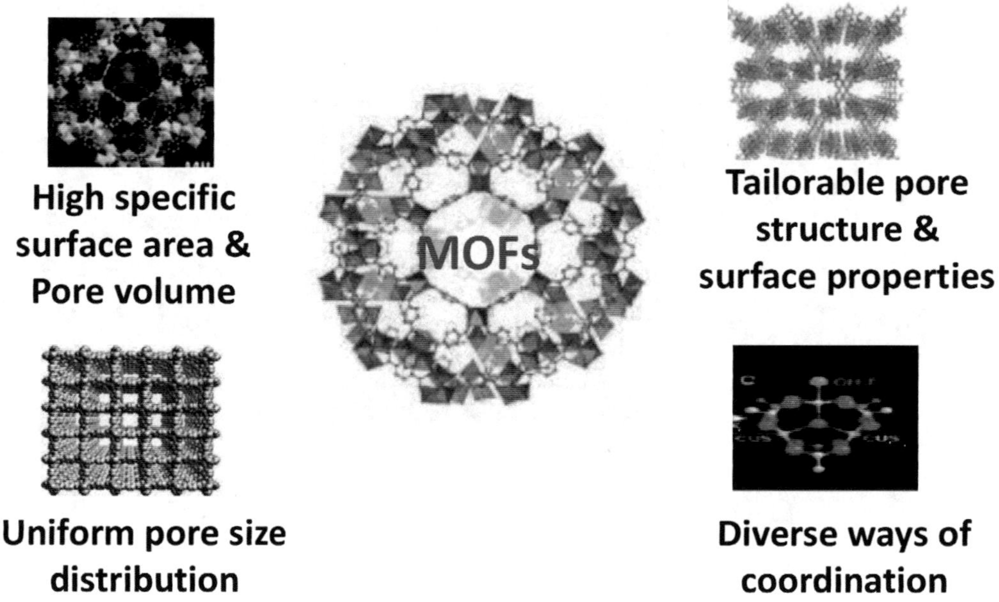

5

Removal of pollutants with MOFs

6

Carbon Materials

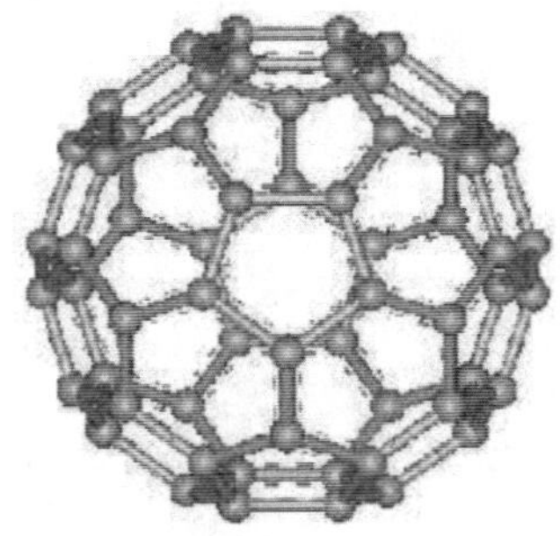

C_{60}

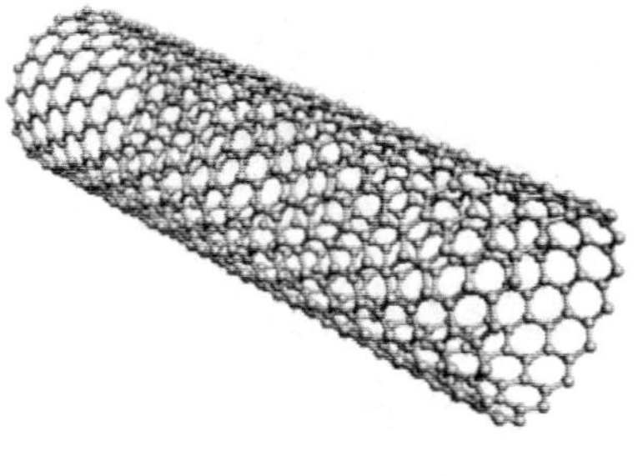

MWCNT

- **Delocalized conjugated structures: with a closed-shell configuration consisting of 30 bonding molecular orbitals with 60 π-electrons;**
- **Electron reservoir : trap photoexcited electrons from semiconductors, efficiently arouse rapid photoinduced charge separation.**

- **Electron acceptor: cause an efficient charge transfer and retard the charge recombination;**
- **Photosensitizer: expand the visible light absorption and thus enhance the light utilization efficiency;.**

7

MOF-base: MIL-68(In)-NH2

- ✓ **windows: 0.6 and 1.6 nm**
- ✓ **excellent stability**
- ✓ **High photosensitivity.**

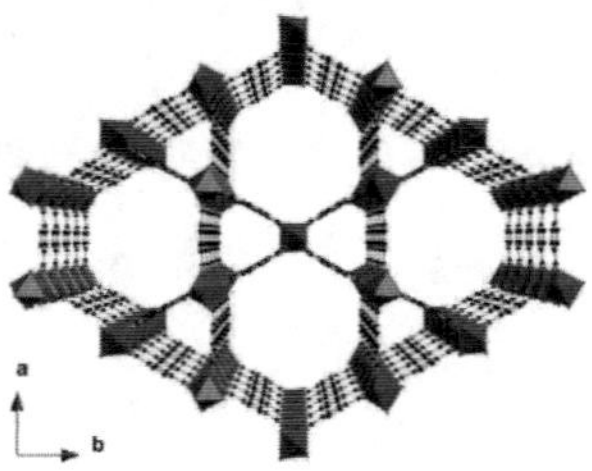

Hypothesis

C_{60}-NH_2-MIL-68(In)(LP-4)

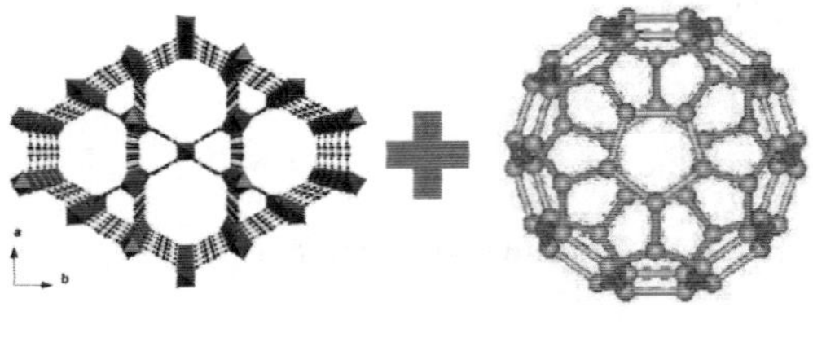

MWCNT-NH_2-MIL-68(In)(PL-1)

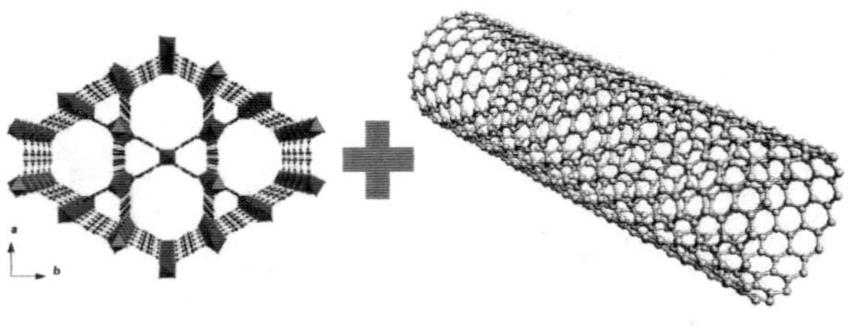

8

1 The integration of MOFs photocatalyst with carbon materials

2 Clarification of the structure and property

3 Investigation of the improved adsorption and photocatalytic activity

4 Clarification of the mechanistic pathways

9

◆ Integrated adsorption and visible-light-driven photocatalytic degradation of MB over oxidized C_{60}/NH_2-MIL68(In)

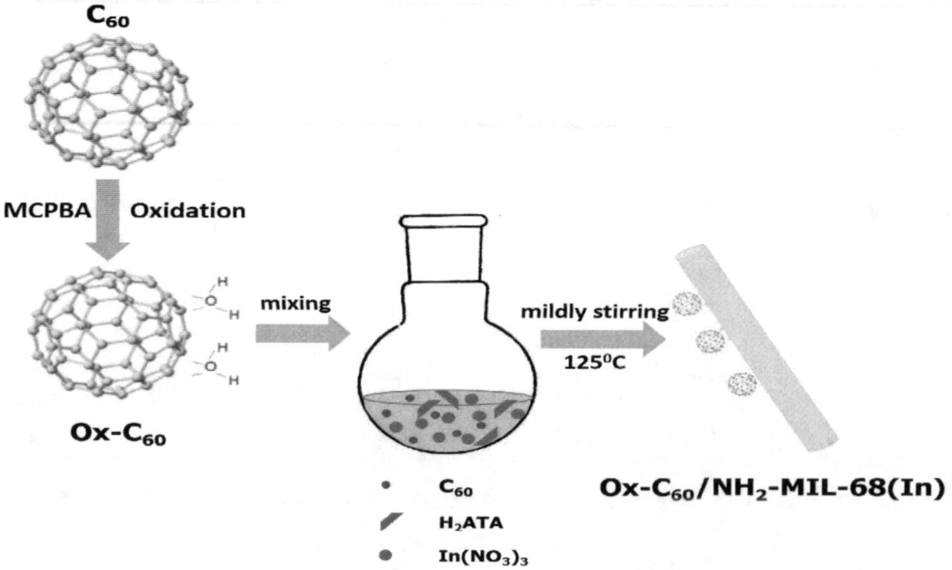

Synthesis process

10

Characterizations of C_{60}/NH_2-MIL68(In)

◆ FT-IR

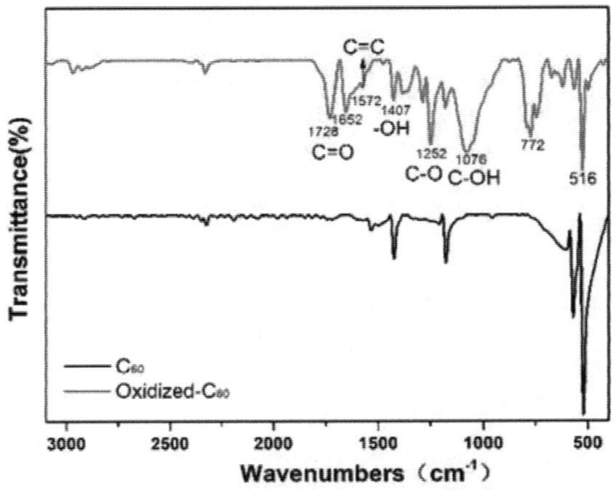

The success of C_{60} oxidation

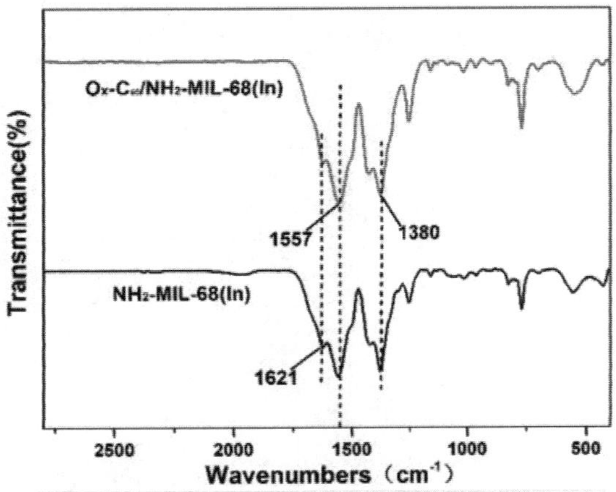

The introduction of Ox-C_{60} did not disturb the NH_2-MIL-68(In) framework structure.

11

Characterizations of C_{60}/NH_2-MIL68(In)

◆ XRD & UVDRS

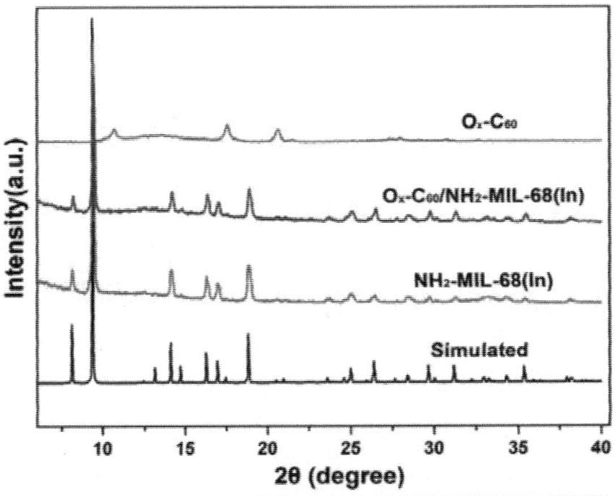

Crystal structure of NH_2-MIL-68(In) remained intact after introducing Ox-C_{60}.

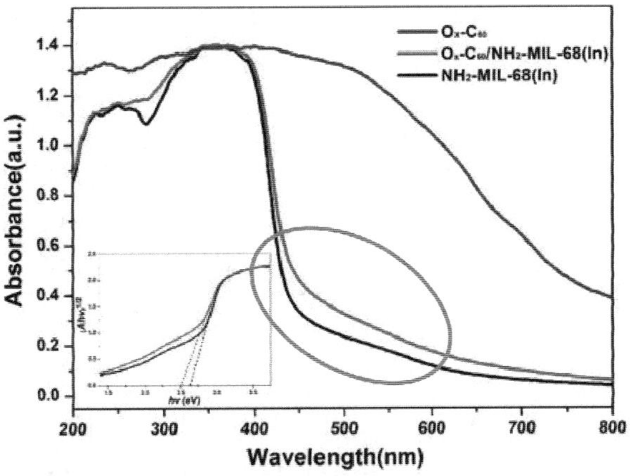

Light absorption intensity in the visible regioı increased after introducing Ox-C_{60} into MOF.

12

Characterizations of C_{60}/NH_2-MIL68(In)

◆ SEM & TEM

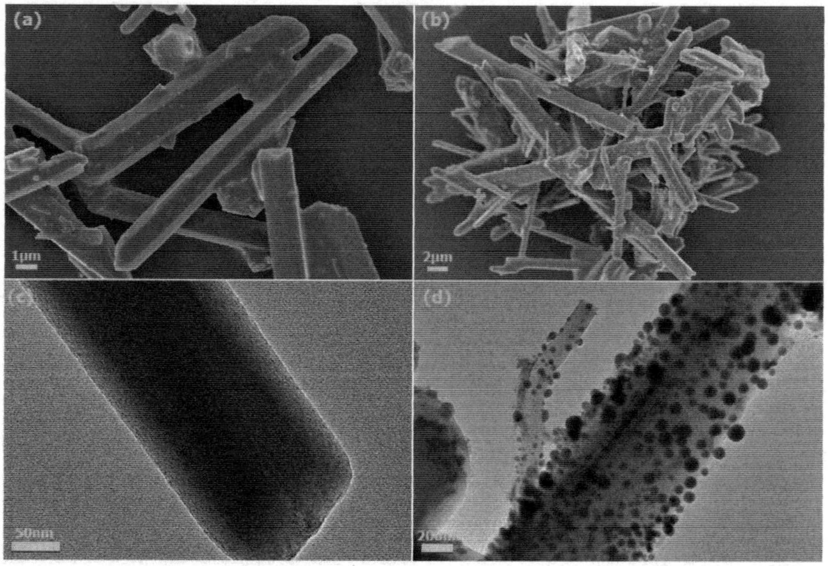

Clean needlelike structures of NH_2-MIL-68(In)

Rough surface with spherical C_{60} particles

13

Photocatalytic activity of C_{60}/NH_2-MIL68(In)

◆ Degradation of MB

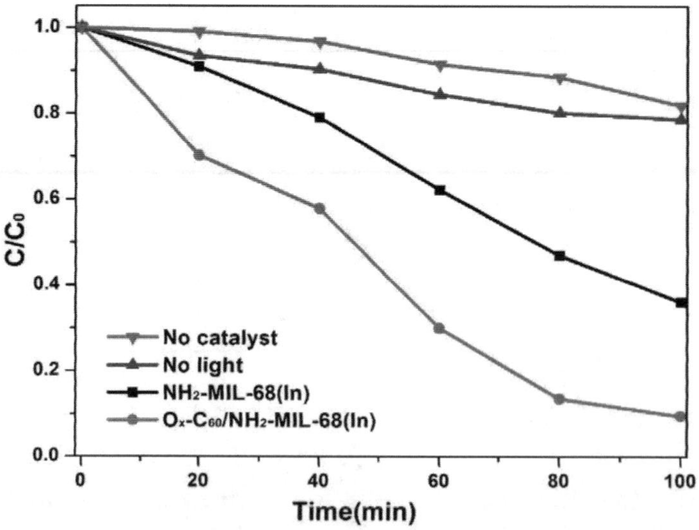

Degradation efficiency of MB under visible light

14

Possible mechanism of C_{60}/NH_2-MIL68(In)

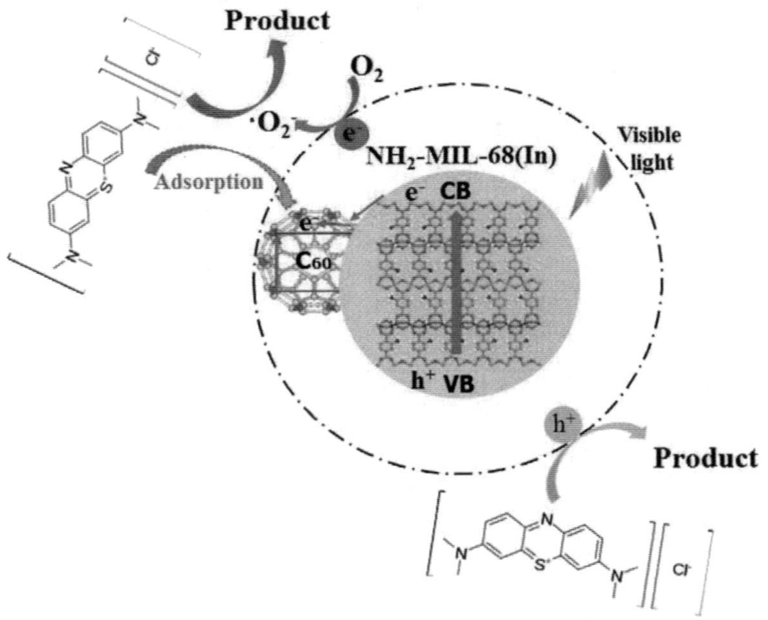

Mechanistic pathway of the degradation of MB on C_{60}/NH_2-MIL68(In) under visible light

15

AIChE
The Global Home of Chemical Engineers

Background | Objective | Experiment | Conclusion

◆ Formation of NH_2-MIL68(In) on MWCNTs backbone for enhanced photocatalytic reduction of Cr(VI)

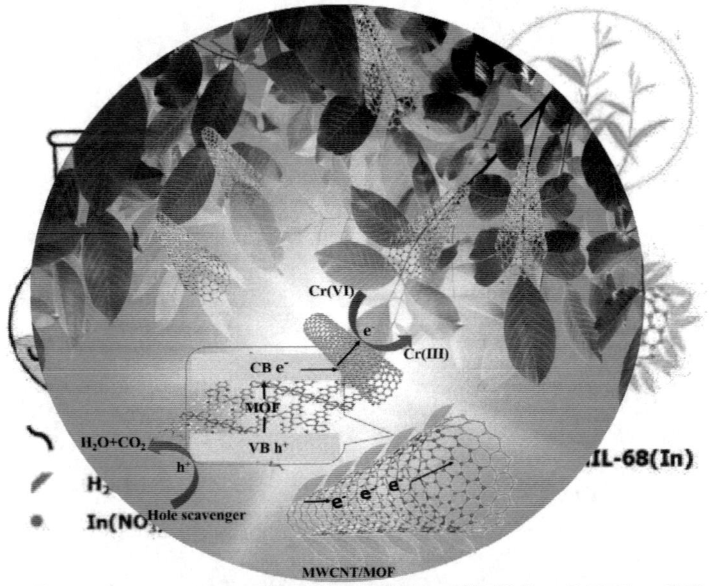

Synthesis process of MWCNT/NH_2-MIL-68(In)
Perspective

16

Characterizations of MWCNT/NH_2-MIL68(In)

◆ XRD & FT-IR

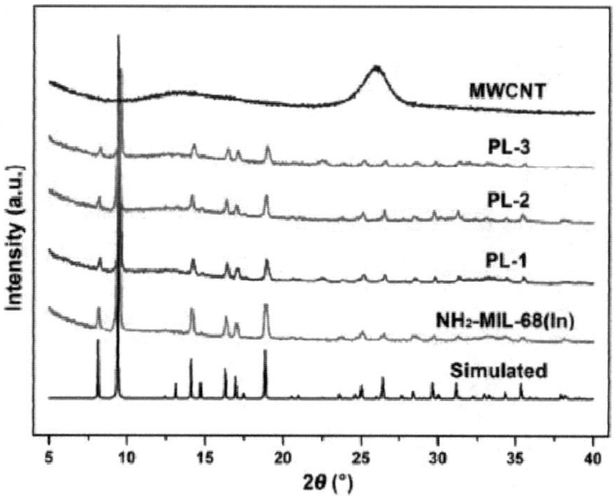

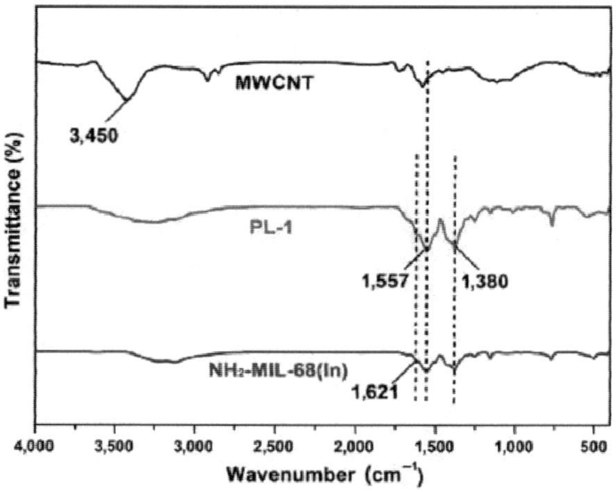

The structure of NH_2-MIL-68(In) remained intact after introducing MWCNT.

17

Characterizations of MWCNT/NH_2-MIL68(In)

◆ N_2 adsorption

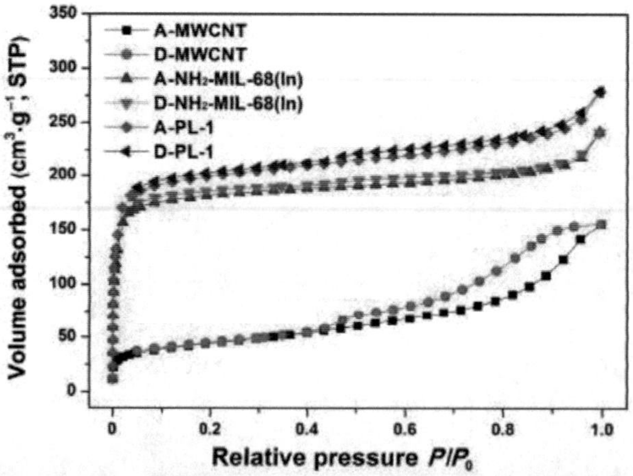

Samples	S_{BET} ($m^2\ g^{-1}$)	V_{BJH} ($cm^3\ g^{-1}$)	$V_{t\text{-}plot}$ ($cm^3\ g^{-1}$)
MWCNT	152.3	0.218	0.018
NH_2-MIL-68(In)	740.6	0.12	0.23
MWCNT /NH_2-MIL-68(In) (PL-1)	800.9	0.18	0.24

18

Characterizations of MWCNT/NH_2-MIL68(In)

◆ SEM & TEM

SEM images for MWCNT

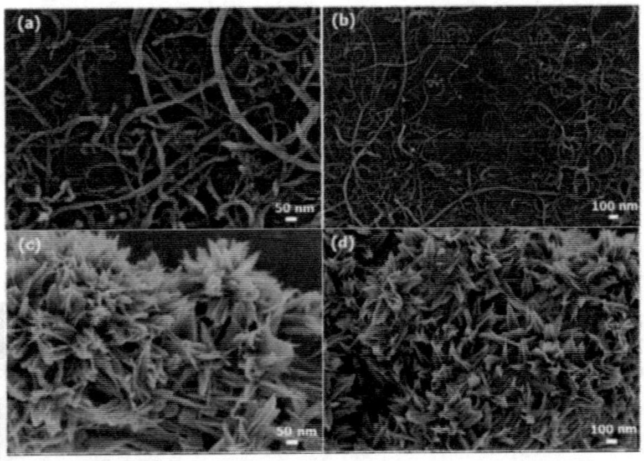

SEM images for NH_2-MIL-68(In)

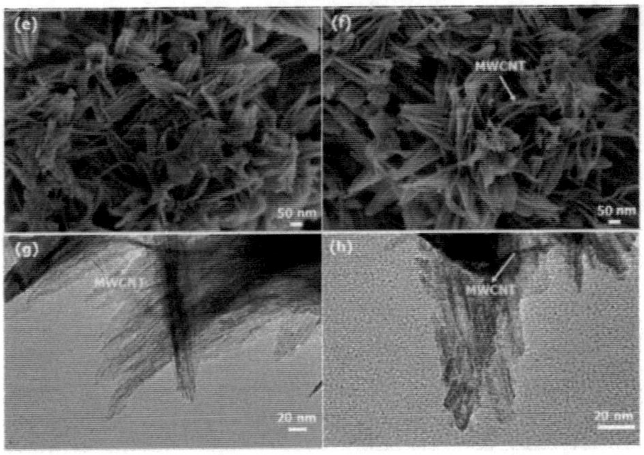

TEM images for MWCNT

TEM images for NH_2-MIL-68(In)

19

Characterizations of MWCNT/NH_2-MIL68(In)

◆ XPS

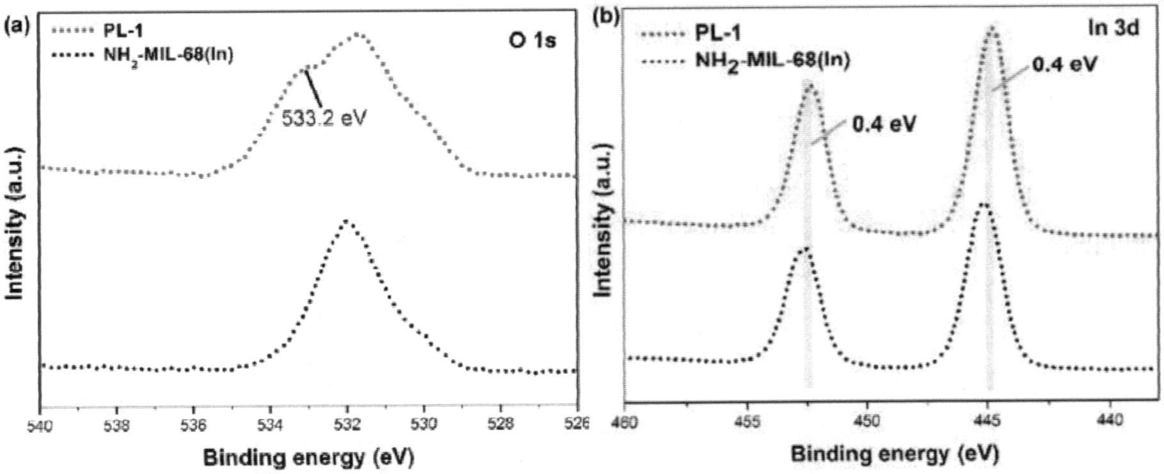

XPS of NH2-MIL-68(In) and PL-1, (a) O 1s and (b) In 3d.

20

Characterizations of MWCNT/NH_2-MIL68(In)

◆ **UVDRS**

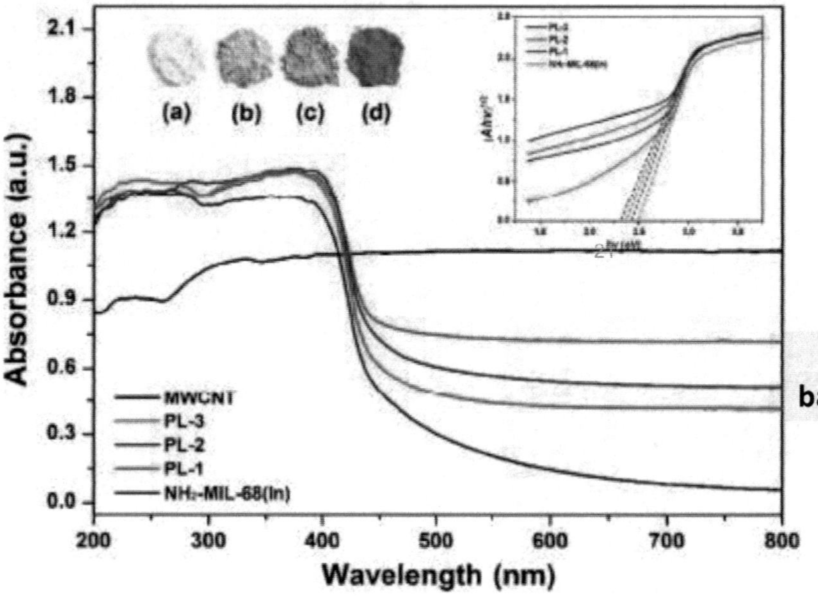

Red-shift band-gap narrowing

21

Performance of MWCNT/NH_2-MIL68(In)

◆ **Reduction efficiency of Cr(VI) over NH_2-MIL-68(In) and the MWCNT/NH_2-MIL-68(In) composites (PL-x; x = 1,2,3)**

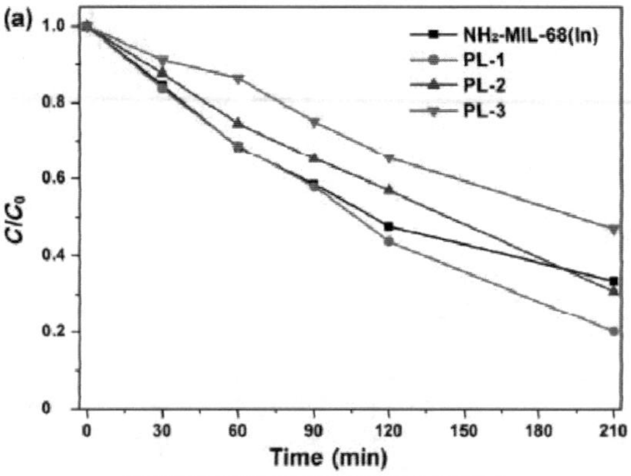

(a) Under visible light

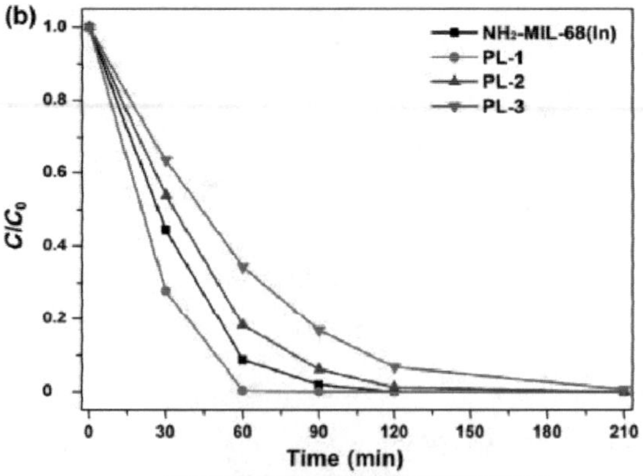

(b) Under UV-vis light

22

Performance of MWCNT/NH_2-MIL68(In)

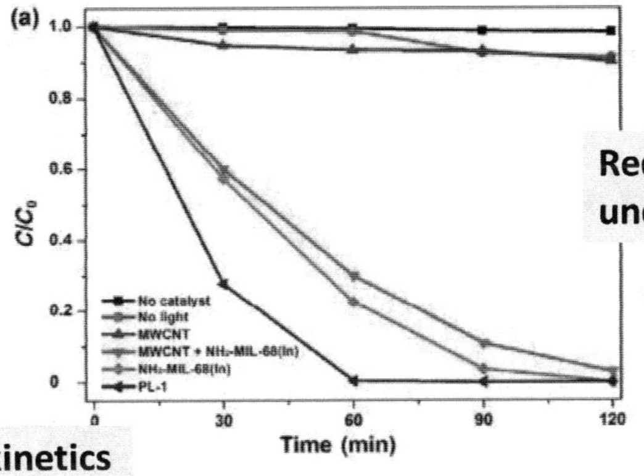

Reduction efficiency of Cr(VI) under different conditions.

Pseudo-first-order kinetics

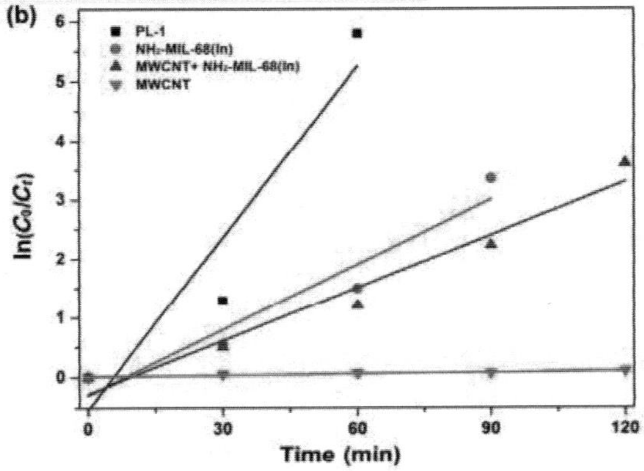

Reaction rate constants

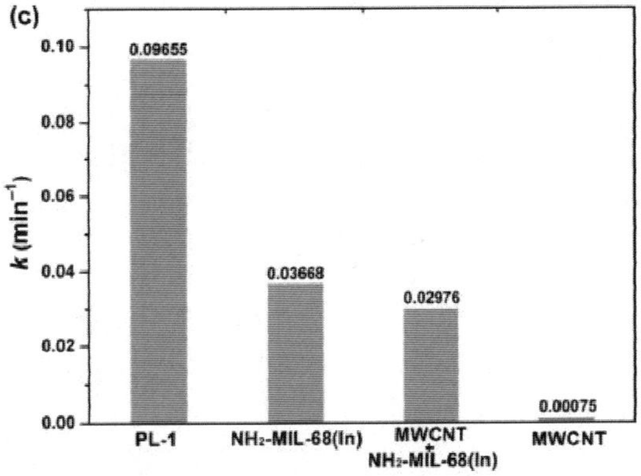

23

Reusability of MWCNT/NH_2-MIL68(In)

◆ Regeneration cycles & XRD & FT-IR

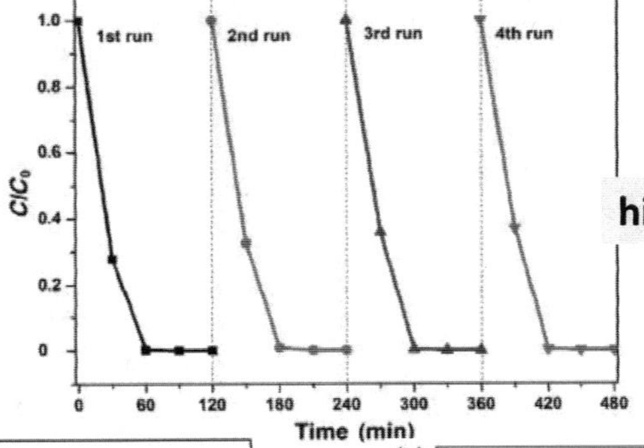

high reusability and stability

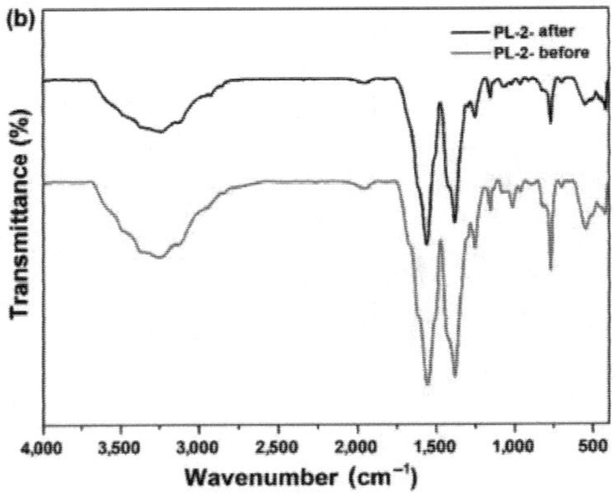

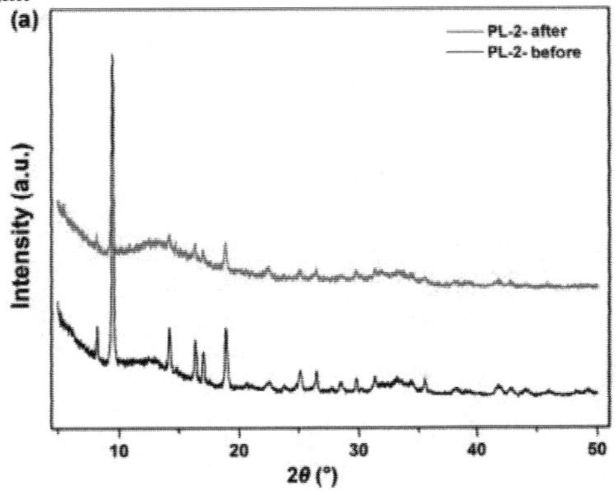

24

Characterization of mechanism on MWCNT/NH_2-MIL68(In)

Pore size distribution

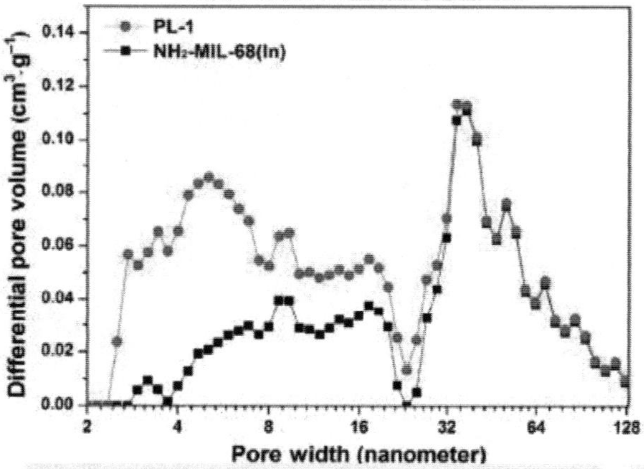

VB XPS

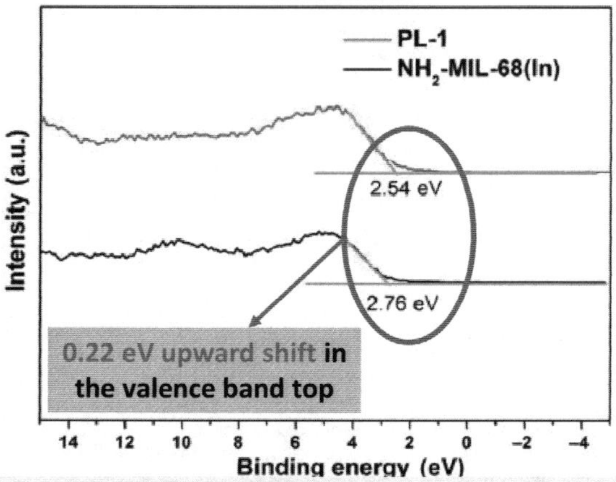

The incorporated MWCNT generates new mesopores to facilitate Cr(VI) diffusion.

The position of CB of PL-1 kept almost the same, high reduction potential and photoreduction capability was retained.

25

Characterization of mechanism on MWCNT/NH_2-MIL68(In)

PL spectra

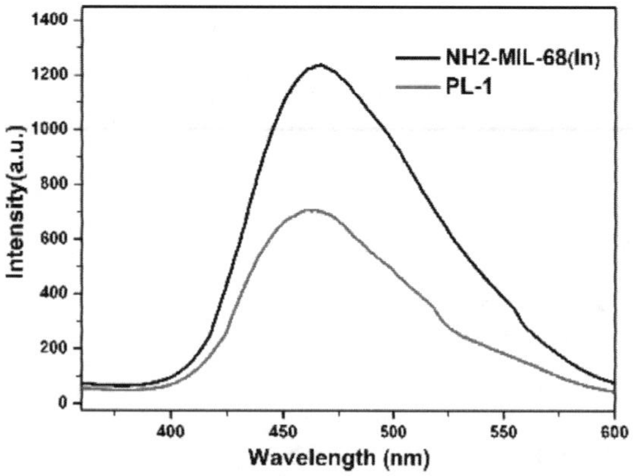

Photocurrent response

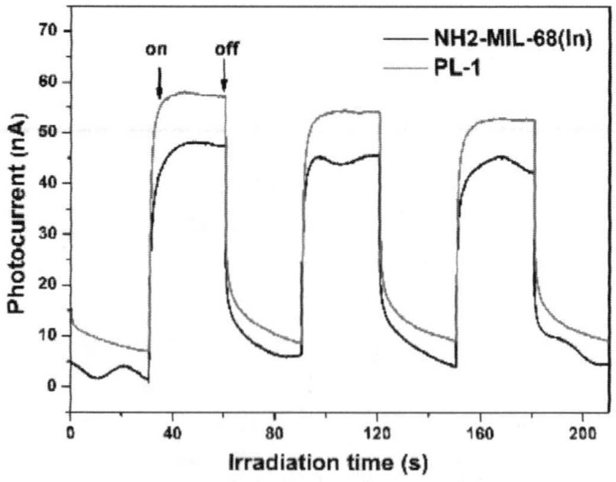

Charge transfer was facilitated by incorporation of MWCNT in the PL-1 composite, resulting in the decreased recombination of photo-induced e^-/h^+, leading to improved Cr(VI) photoreduction efficiency.

26

Possible mechanism on MWCNT/NH_2-MIL68(In)

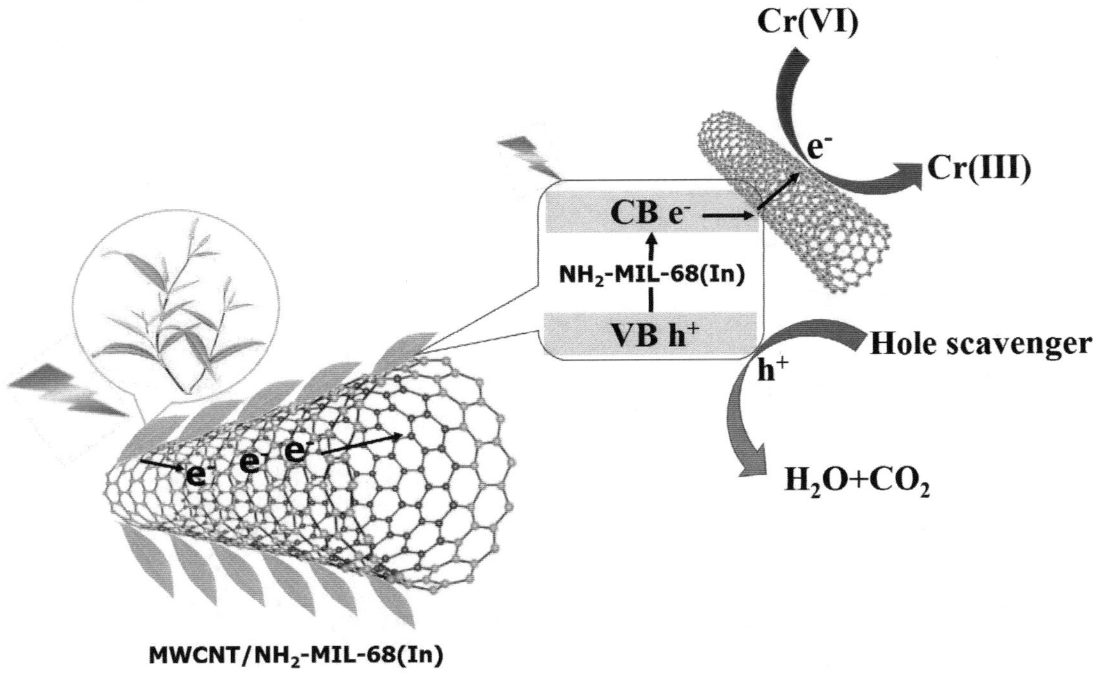

Schematic illustration of the reduction of Cr(VI) on PL-1 under irradiation

27

AIChE
The Global Home of Chemical Engineers

Background Objective Experiment **Conclusion**

01 Carbon materials were introduced into the NH_2-MIL-68(In) by a facile dynamic solvothermal method to enhance the photocatalytic performance.

02 PL-1 can reduce Cr(VI) completely within 60 min under UV-Vis irradiation, and the rate constant of the PL-1(0.096 min-1) nearly tripled that of the NH2-MIL-68(In) (0.036 min-1).

- Provides new mesopores to facilitate Cr(VI) diffusion.
- Serve as an excellent photosensitizer to enhance visible light absorption and without lowering the conduction band position.
- Not only accepts the photogenerated electrons from NH_2-MIL-68(In), but also provides a platform for them to migrate, ensuring efficient charge transfer and retarding e^-

03 The work provides the first example using MWCNT/MOF composite for photocatalytic reduction, and demonstrates the potential of designing MWCNT/MOF hybrid for photocatalysis.

Acknowledgement

- **Group members at SCUT**

Graduate Students:

Prof. Li **Prof. Xiao** **Yunhong Pi** **Xiyi Li**

- **Funding supports**
  - ◆ National Natural Science Foundation of China (21576093)
  - ◆ Guangdong Natural Science Funds for Distinguished Young Scholar
- **The organizing committee of AIChE meeting**

Thanks for your attention!

497106 In-Situ Studies on Anion Electroadsorption Mechanisms

Abstract:

Phosphorus is present in many agricultural waste streams, and is appearing in water in an increasing concentration. The increased presence in surface water can contribute to eutrophication and be harmful to ecosystems. Thus, it is desirable to treat these water sources to either remove or recover the phosphorus. *While there are many methods for separation, herein, we examine how materials can be engineered for electric field driven (electrochemical) recovery of phosphate anions from aqueous solutions[1].* This work will present a fundmental study using electrochemical atomic force microscopy to understand how electroadsorption processes are altered by varying phosphate anion charges (43−, 42−, 34−) [1,2,3]. The overarching goal of this work is to understand how phosphate anions electrostatically interact with porous materials exposed to an electrical bias in the formation of an electric double layer. Fundamental electrochemistry experiments will be presented on the effect of phosphate adsorption for energy storage in concentrated electrolytes as well as phosphate removal applications (dilute solutions). In-situ dilatometry experiments will be presented in order to quanitfy mechanical stresses due to the electroadsorption processes.

[1] Xie, Jie, et al. "Removal and recovery of phosphate from water by activated aluminum oxide and lanthanum oxide." *Powder Technology* 269 (2015): 351-357.

[2] Alliata, D., et al. "In situ AFM study of interlayer spacing during anion intercalation into HOPG in aqueous electrolyte." *Langmuir* 15.24 (1999): 8483-8489.

[3] Black, Jennifer M., et al. "Strain-Based In Situ Study of Anion and Cation Insertion into Porous Carbon Electrodes with Different Pore Sizes." *Advanced Energy Materials* 4.3 (2014).

503477 Engineering Materials and Processes for Flexible Electronics

Abstract:

Printed electronics is a new technology envisioning the fabrication of electronic devices using solution processable materials and printing methodologies. In this presentation I will describe the materials synthesis, development, and process engineering enabling the fabrication of unconventional electronic and opto-electronic devices, such as transistors, circuits and solar modules, all on flexible foils using several new materials. Materials development include new synthetic, green routes to semiconducting polymers. Examples of unconventional electronic materials include organic small molecular and polymeric semiconductors, metal chelates and complexes, and hybrid organic-inorganic metal oxides. Thus, we will show the fabrication of polymer based complementary circuits with excellent performance (field-effect mobilities ~ 1-6 cm^2/Vs), flexible amorphous oxides TFTs (field-effect mobilities ~ 5-40 cm^2/Vs) as well as polymeric donor-fullerene solar cells with efficiencies approaching 11% in inverted architectures.

503526 Spectroelectrochemistry of Conjugated Radical Polymers

Abstract:

Conjugated radical polymers are an interesting marriage of a conjugated backbone with radical pendant groups. This brings new opportunities for organic electronics, energy storage, and thermoelectrics. The challenge is to understand how the singly occupied molecular orbital (SOMO) of the radical interacts with the properties of the conjugated backbone, as this can strongly affect the resultant conductivity, mobility and redox activity. Here, a series of polythiophenes with pendant TEMPO (2,2,6,6-Tetramethylpiperidin-1-yl)oxyl) groups of varying alkyl spacer length are compared in the context of electrochemical energy storage. The conjugated radical polymers are examined as cathodes for nonaqueous batteries. Energy storage could potentially occur at both the radical group and the conjugated backbone. However, the overall capacity of the conjugated radical polymers is lower than expected, as is the conductivity. The explanation for this reveals an important design criteria for conjugated radical polymers in that the redox potential (or energy level) of the radical group and the conjugated backbone should be finely tuned so as to control electron exchange between the two moieties. This is further evidenced by in situ spectroelectrochemistry, in which the oxidation state of the polythiophene backbone is observed to change with time as electrons are transferred from one moiety to the other.

503527 Conjugated Copolymers Based on Purines: Linking Monomer Design and Macromolecular Properties

Abstract:

Purines seem to be compelling building blocks for conjugated polymers: They are susceptible to a wide range of site-specific chemical transformations that affect energy levels and emission wavelengths, and they possess the ability to self-assembly through non-bonded interactions. Motivated by these ideas, we describe the preparation of â €œpoly(purine)â€ copolymers in which purine-derived monomers are introduced directly into a polymer backbone via Stille cross-coupling polymerization with benzodithophene (BDT), as well as purine-based donor-acceptor small molecules. Variations in the nature of the BDT comonomer and substitution pattern of purinyl comonomer lead to changes in thermal and charge-transfer characteristics. Specifically, UV-Vis and fluorescence spectroscopies show that poly(purine)s exhibit intramolecular charge-transfer characteristics and display distinct changes in emission wavelengths based on purinyl comonomer design. These efforts are extended toward fully conjugated poly(purine)s through the development of small molecule donor/acceptor chromophores in which purines behave either as electron-donating or electron-accepting unit, which tunes the charge transfer behavior, excited state lifetime, and quantum yield of the building block. These demonstrations of main chain poly(purine)s and pi-conjugated, purine-based chromophores highlight the potential of purines as useful motifs in conjugated polymers for optoelectronic applications.

503541 Pinch-Off Dynamics and Printability of Semi-Flexible and Rigid Rod Polymer Solutions

Abstract:

Understanding and controlling drop formation and liquid transfer is crucial for designing printable photovoltaic and electronic devices using semi-flexible and rigid rod polymer solutions. Applications involving jetting, coating, spraying, atomization and various printing techniques (ink-jet, gravure printing, roll-to-roll) all involve formation of liquid necks that undergo capillary-driven thinning and pinch-off. Stream-wise velocity-gradients associated with extensional flows arise in thinning liquid necks, and as the response of polymer solutions to extensional flows influences both their processability, as well as final orientation and morphology. Characterizing and controlling the extensional viscosity that quantifies the resistance to stream-wise gradients and extensional relaxation time of polymer solutions that determines the pinch-off time are of critical importance for selecting and optimizing semi-flexible and rigid rod polymer solutions used in functional inks. However, due to the low viscosity and low elasticity of most printable inks, the quantitative analysis of pinch-off dynamics and extensional rheology response is beyond the measurement rage of commercially extensional rheometer called Capillary Break-up Extensional Rheometer (CaBER). In this contribution, we employ the recently developed dripping-onto-substrate (DoS) rheometry protocols for visualizing and analyzing capillary-driven thinning and pinch-off dynamics of a columnar neck formed between a nozzle and a sessile drop for typical functional inks including aqueous PSS-PEDOT solutions, aqueous NaPSS solutions and polythiophene solutions in organic solvents. We elucidate the influence of flexibility and extensibility on pinch-off dynamics and processability, emphasizing how printability can be quite distinct even if different polymer solutions exhibit a dynamically similar response in shear rheology characterization.

501384 Surface-Directed Multi-Scale Assembly of Conjugated Polymers

Abstract:

Organic electronic and photoelectronic materials that are light-weight, flexible and can be manufactured using energy-efficient and high-throughput methods. The solution printability at near ambient conditions enables deposition on flexible polymer substrates to create wearable, stretchable, imperceptible electronic devices for use in applications unimagined before. On the other hand, key challenges remain: how does molecular assembly proceed during solution printing and how to control the resulting thin film morphology? The significance of this challenge lies in the fact that charge transport in printed thin films is highly sensitive to their morphological parameters from molecular, mesoscopic to device scale. Addressing this challenge can open up new avenues for attaining high electronic performances, facilitating the much needed structure-property relationship studies in polymer-based electronic devices.

In this talk, I will present a new strategy we recently developed for controlling multi-scale assembly of conjugated polymers that are directly compatible with solution printing. Central to our method is the design of surfaces for overcoming the barrier to polymer nucleation, thereby directing the nucleation-triggered multiscale assembly process during printing. Using this method, we achieved high degree of global and local alignment over large area. In certain cases, we even observed small-molecule-like morphology for high molecular weight conjugated polymers, which has been rarely observed before. With high degree of control over thin film alignment and molecular packing, we correlate these morphological characteristics with anisotropic charge transport properties towards establishing structure-property relationships. By systematically tuning the degree of alignment, we show that the charge transport anisotropy can be switched to favor either transport along the polymer backbone, or along the pi-pi stacking direction.

503523 Interpenetrating Networks for Flexible Bulk Heterojunction Opvs

Abstract:

Bulk heterojunction organic photoltaic (OPV) devices can be fabricated fully though solution processing on a variety of substrates, but the active layer typically exhibits very poor flexibility and robustness in response to strains and deformation. Here, we present a general approach to fabricating flexible bulk heterojunction OPVs through the incorporation of an interpenetrating polymer network. We show that this provides P3HT/PCBM bulk heterojunction OPVs that are fully flexible and able to sustain stresses in excess of 50 %. Small molecular OPV blends are more robust, and we flexible OPV devices with power conversion efficiencies exceeding 5 % power conversion efficiencies.

500100 Development of an Anisotropic Coarse-Grained Conjugated Polymer Model for Optoelectronic Applications

Abstract:

Conjugated polymers have attracted attention for their uses in a variety of optoelectronic applications; their electronic properties can be altered to meet a variety of needs by fine-tuning both molecular structure and processing conditions. Developing models for simulating conjugated polymers that are fast yet contain the fine structural details necessary for examining the influence of morphology on electronic structure are crucial for understanding how to optimize conjugated polymer processing conditions. Specifically, conventional coarse-graining techniques fail to address the anisotropic interactions of the conjugated polymer backbone fundamental to both polymer pi-stacking and optoelectronic transport. In order to address this problem, we have developed a coarse-grained conjugated polymer model wherein monomers are represented as disks that interact through an anisotropic Gay-Berne potential and the backbone is modeled as a twistable wormlike chain. These two features allow this coarse-grained model to capture the anisotropic pi-stacking and semi-flexibility properties that are crucial to understanding these polymers, while substantially simplifying computations. We have parameterized the coarse-grained model from an atomistic model of a thiophene-containing conjugated polymer through free energy surface calculations and persistence length fitting. The anisotropic structural detail, including dihedral degrees of freedom, maintained by our coarse-grained model allows for the straightforward computation of x-ray scattering profiles and other relevant correlation functions, without the need for atomistic backmapping. Moreover, the form of our coarse-grained model allows for a simple mapping to an electronic tight-binding hamiltonian wherein the electronic structure of the valence bands may be evaluated. We perform structural and optoelectronic benchmarking of our coarse-grained model and compare these results to atomistic simulations and experimental data.

500244 Automated Quantitative Analysis of Oriented Morphologies and Inter-Grain Connectivity in Conjugated Polymers

Abstract:

A wide range of conjugated polymers form thin films with oriented morphologies, typically characterized with an imaging technique such as AFM. To provide quantitative metrics to describe and compare these morphologies, we developed an automated image analysis software package called GTFiber. GTFiber was originally designed for the high-throughput extraction and measurement of fibers from images, and has enabled the determination of quantitative process-structure-property relationships for poly(3-hexylthiophene)-based field effect transistors. These included both the effect of the spin- and blade-coating flow fields on fiber alignment during thin film deposition as well as the correlation between fiber alignment and field effect mobility. Data visualizations produced using GTFiber revealed striking nanofiber bundling behavior during solution processing. More recently, we have demonstrated GTFiber's generalizability to fibrillar systems in other materials, as well as the oriented morphologies observed in thin films of PNDI or strain-stretched thin films of conjugated polymer blends.

The extraction and vectorization of fibers from images also enabled us to simulate tie chain behavior on a length scale unattainable by molecular dynamics. By performing a Monte Carlo simulation of P3HT chain packing overlaid on the real, extracted fiber backbones, we were able to define a quantitative measure of grain interconnectivity: the tie-chain length density. Tie chain length density quantifies the expected number of tie chain connections to a given grain per unit lenth, and is sensitive to fiber packing, orientational order, and the assumed molecular weight distribution. While experimental observation of tie chains remains difficult, our simulation provides an upper bound on grain interconnectivity, while laying the quantitative groundwork necessary to describe this important feature of conjugated polymer thin films.

494252 Spray Coating Chitin & Cellulose Nanomaterials for Enhancement of Barrier Properties

Abstract:

Chitin is an abundantly occurring biopolymer like cellulose. The annual natural chitin production is estimated to be around 10^{10} - 10^{12} tons per year (Barikani et al., 2014) and is about the same order of magnitude as cellulose, whose annual bio-production by photosynthesis is estimated to be of the order of 10^{11}-10^{12} tons per year (Klemm et al.,2004). Similar to cellulose, chitin can also occur as chitin nanofibers (ChNFs) and when it is subjected to high shear treatment, nanofibers of 10-20 nm diameter and several microns in length are obtained (Wu et al., 2014; Ifuku & Saimoto, 2012).Â Wu et al. (2014) were able to produce chitin nanofibers in suspension from purified crab Î±-chitin, and when these suspensions were cast into films the resultant films had O_2 & CO_2 gas permeabilities of 0.006 & 0.018 barrer respectively. These good barrier properties as well as chitinâ€™s natural bacterial resistance make chitin an excellent candidate for sustainable barrier packaging applications. But packaging materials are generally multilayered, and a good method for incorporating these chitin films in continuous multilayer packaging processing needs to be developed. Spray coating is one of the most versatile methods of applying liquid onto surfaces. Due to its frequent use in painting, spray coating processes are widely encountered in everyday life (McKeen, 2016). One of the main features of a spray coating process is the contactless delivery of the material to be coated onto the surface. This implies that the actual delivery process is not influenced by the surface topography and thus uneven, or delicate surfaces can be coated. The delivered load of coated material is independent of the surface and the lack of a metering gap leads to decreased sensitivity to defects and especially helps in reduction of web breaks in roll to roll process (Czerwonatis, 2008). Beneventi, et al. (2014) were able to coat slurries of micro fibrillated cellulose (MFC) onto wet highly porous papers allowing complete retention of MFC on the surface of the substrate and improved barrier properties of the final product. In the present study we demonstrate the successful spray coating of chitin nanofiber & cellulose nanocrystal (CNC) suspensions onto polylactic acid films. The spray coating of these materials was found to lead to a reduction in the oxygen barrier properties of the final composite film by as much as 75 % of its original value depending on the loading and the material used (CNC, Chitin or both). Unlike the study by Beneventi, et al. the substrate in this case is non porous and the suspensions do not wet the surface. The resulting de-wetting problems were successfully solved by heating the substrate surface. This process could open up a path for the use of more sustainable polymers with otherwise excellent properties but are lacking in barrier properties. Addition of a ChNF or CNC layer could also lead to a reduction in the amount of the more expensive polymer in the multilayer (with the sole function of providing oxygen barrier) all the while increasing the sustainability.

500752 Development of Low-Concentration Alkaline Treatment Method to Produce Thermally Stable Cellulose Nanofibrils (CNFs)

Abstract:

The thermal stability of cellulose nanofibrils (CNFs) can be improved by converting cellulose crystal structure from cellulose I to cellulose II using alkaline pretreatment method. The conventional pretreatment method requires 17.5% or higher concentrations of sodium hydroxide and large volume of water for neutralization to produce thermally stable, but interdigitated or aggregated CNFs. The main objective of this study was to develop a low concentration alkaline pretreatment method to produce thermally stable and highly dispersible CNFs. Cellulose nanofibrils with improved thermal stability were successfully prepared by treating cellulose powder with 2wt.% NaOH solution at low temperature (below 0), followed by mechanical disintegration through a French cell press with only ten passes. The thermally stable CNFs had an average diameter of 90 nm with the mean onset and maximum thermal decomposition temperature of 305 and 343 respectively, compared with those of 283 and 310 respectively, for the CNFs prepared without the NaOH treatment. The thermally stable CNFs can create new opportunities for the development of CNFs based packaging composites and high-end electronic applications.

491943 Tailored and Integrated Production of Carboxylated and Lignin Containing Cellulose Nanocrystals and Nanofibrils for Composite Applications

Abstract:

Here we demonstrate di-carboxylic acid hydrolysis for the integrated production of lignin containing cellulose nanocrystals (LCNC) and nanofibrils (LCNF) using two unbleached mixed hardwood chemical pulps of lignin contents of 3.9 and 17.2%. Acid hydrolysis experiments used maleic acid solution of 60 wt% concentration at 120°C for 120 min under ambient pressure. Yields of LCNC were low of less than 6% under this set of condition. The higher lignin content samples produced LCNC with greater height (diameter) of 25 nm but similar length of approximately 230 nm to that from the lower lignin content fibers (height of 20 nm). Interestingly, the higher lignin content sample resulted in LCNF with smaller height (diameter) of 7 nm but longer length of > 1 μm, or greater aspect ratio than the LCNF from the lower lignin fibers of height 10 nm and length < 1 μm. Lignin protected cellulose from esterification which resulted in LCNC and LCNF that was less carboxylated compared to those lignin-free CNC and CNF and therefore had lower charges. However, lignin is more hydrophobic and thermally stable than carbohydrates therefore LCNC and LCNF are favorable for composite applications.

504076 Nano Carbon Structures from Cellulosic Biomass for Use As Functional Materials

Abstract:

We report herein the production of hierarchical nanocarbon materials from cellulose fibers. The carbonized cellulosic fibers with branching out CNTs (CC-CNT) offer interesting surface and electrochemical properties that are of interests for chemical processing and biosensing applications. In our tests for chemical adsorption tests, CC-CNT offered highly selective adsorption of hydrocarbons from gas mixtures, especially chemicals with C=C bond and aromatic structures. More interestingly, the chemicals absorbed and accumulated can be quickly released from the surface of the materials upon application of modest electrical current. Such an electrically switchable adsorption behavior may promise applications in a broad areas including air quality control, chemical sensing and fuel storage.

503374 Quantitative Three-Dimensional Morphological Characterization of Block Copolymer Films Enabled By Directed Self-Assembly

Abstract:

DSA of block copolymer films on chemically nanopatterned surfaces is an emerging technology that is well-positioned for commercialization in nanolithography and nanomanufacturing. DSA of (PS-b-PMMA) films on lithographically defined chemically nanopatterned surfaces is one focus of our activities in which the main research objectives revolve around understanding the fundamental thermodynamics and kinetics that governs assembly, and therefore patterning properties such as 3D structure, perfection, and processing latitude. A second focus is to use the physical and chemical principles that we have elucidated for DSA of PS-b-PMMA towards the development of block copolymer systems capable of self-assembling into the sub 10 nm regime and continuing to meet the stringent constraints of manufacturing. The research is enabled by the recent development of techniques to combine metrology tools (TEM tomography, GISAXS, RSoXS), theoretically informed course grained models, and evolutionary algorithms to quantitatively determine and predict the independent process and material parameters that result in different 3D structures of assembled domains.

502696 Examination of Line Edge Roughness of Directed Self-Assembled Block Copolymers: A Coarse-Grained Molecular Dynamics Study

Abstract:

Over the past decade, with the ever-increasing demand for miniaturization of magnetics storage devices, flash memory devices, semiconductor capacitors among others, directed self-assembly (DSA) of block copolymers (BCPs) has attracted the interest of both academia and industry to develop nanolithography to form defect free patterns with optimal control and tolerance on the lithographic features. DSA is not restricted by inherent diffraction based limitations of conventional lithography and has a much lower cost of ownership compared to the existing alternatives. However, the theoretical harmonic spatial arrangement and periodicity of the micro-phases are seldom perfectly achieved due to a variety of defects causing potential limitations in micro-electronics applications. In particular, the lamellae forming symmetric BCPs used to create line / space patterns show roughness features at the interface of the two blocks. This roughness, quantified as Line Edge Roughness (LER) is defined as the deviation of a feature edge in magnitude and frequency from the ideal desired shape. Morphologies with high LERs often inhibit the smooth functioning of transistors by affecting the electron flow through the spaced channels. It is thus extremely crucial to minimize the LER to harness the full potential of DSA techniques to develop the next-gen devices.

In this work, we have adopted a large-scale coarse-grained molecular dynamics (CGMD) framework to analyze the LER variation for the line / space patterns formed by symmetric polystyrene-*block*-polymethyl methacrylate (PS-*b*-PMMA). Three types of substrate models have been considered to direct the BCP. 1) Flat patterned substrate 2) Wedged patterned substrate and 3) Wedged substrate with a random PS-b-PMMA polymeric brush. For the first type of substrate, the enthalpic contribution of the chemoepitaxial substrate is the driving force to form the lamellae. The wedged patterned has an entropic contribution from the topography of the substrate in addition to the patterning. The third type of substrate employs random PS-*b*-PMMA co-polymer brushes to mimic the neutral layer of the chemoepitaxial substrate. The BCP morphology formed by each type of the substrate has been optimized by varying the polymer and surface properties as well as the annealing conditions. The morphology obtained from large scale CGMD simulations is further processed by etching the PMMA domains resulting in the line / space patterns. Line edge roughness analysis is then performed to quantify the high and low frequency noise of the features by plotting the power spectrum density curves. We demonstrate that the correlation factors from fitting the power spectrum gives values comparable to experiments by accounting for instrumentation / measurement noise. Lastly, the effect of additives for reduction in LER and defect annihilation is also presented.

495395 Atomistic Simulations of Ion Transport in Ps*b*P2VP Functionalized Lamellae

Abstract:

In microphase-separated electrolytes formed by block copolymers, ion mobility in bulk as well as at the interface is crucial to conductivity. Computational simulations provide the opportunity to understand the mechanism of ion transport. Coarse-grained models and field theory can capture the phase behavior of polymer electrolytes but miss detailed interactions between polymers and ions. On the other hand, computational limitations make it difficult to obtain equilibrated morphologies with atomistic details. Current atomistic simulations of polymer electrolytes mainly focus on ion transport in the bulk at small length scales. Little is known about ion behavior at interfaces between different polymer domains. To study ion transport of polymer electrolytes with nanostructure, we generate an equilibrated morphology with the theoretically-informed coarse-grained (TICG) model and then map atom-level resolution pertaining to the poly(styrene-block-2-vinylpyridine/n-methyl pyridinium, iodide) (PS-b-P2VP/NMP^+ I^-) system.

Molecular dynamics (MD) simulations were performed on functionalized atomistic PS-b-P2VP/ NMP^+ I^- â€"lamellae and disorderedâ€"and P2VP/NMP^+ I^- homopolymer systems. We study the functionalized PS-b-P2VP at various percentages of functionalization and with different water weight percentage, which changes the coordination environment of the ions. Finally, we investigate the density distribution and mobility of polymers, ions, and water molecules throughout the lamellae. Specifically, we find the residence time of ions in solvation sites at the interface to be shorter than those in the bulk.

495602 Performance of the Novel Ultra-Thin Proton Exchange Membrane Based on Poly(arylene ether sulfone)s for Pemfcs

Abstract:

Performance of the Novel Ultra-thin Proton Exchange Membrane Based on Poly(arylene ether sulfone)s for PEMFCs

Yang ZHAO, Xue LI, Xiaofeng XIE*

Institute of Nuclear and New Energy Technology, Tsinghua University, Beijing, 100084, China

xiexf@tsinghua.edu.cn

Several perceived problems of Nafion membranes including high cost, high fuel permeability and restricted properties under the more stringent operating conditions (at temperatures above 80°C and relative humidity (RH) conditions below 50%), have been gradually emerging in the developing process. In order to solve these problems, the novel ultra-thin proton exchange membrane based on poly(arylene ether sulfone)s is designed, prepared, and investigated. We use the atom transfer radical polymerization (ATRP) to prepare the membrane based on sulfonated poly(arylene ether sulfone) (SPAES), and the chemical structure of the membrane is shown in Scheme 1. This graft copolymer, of which architecture is comprised of a hydrophobic backbone that contain considerable flexible ionic side chains, has the potential to provide well-interconnected proton transporting pathways due to enhanced phase-separated morphology. The cells based on same hydration degree and different membranes were shown in Scheme 2. The diffusion coefficient of hydronium ions was used to forecast the proton conductivity of SPAES membranes, while the diffusion coefficient of water molecules was used to investigate the diffusion of water from cathode to anode.

Scheme 1. The chemical structure of the membrane.

Scheme 2. The cells based on same hydration degree and different membranes:

A) Nafion 211; B) SPAES.

This method could induce phase-separated morphology between hydrophilic ion-conducting channels and the

hydrophobic matrix into the novel ultra-thin membranes. The morphology of SPAES membrane by TEM and SAXS showed that a wormlike and interconnected hydrophilic network of small ionic clusters were observed, higher than that of Nafion, which had a cluster-network morphology by narrow ionic nanochannels. Measurement of physical properties showed that these novel ultra-thin membranes were available for PEMFCs applications. The water uptake, swelling ratio and fuel permeability were each improved by lengths of the side chain. Future studies will focus on improving the oxidative stability of these materials, further enhancing conductivity by improving IEC, and evaluating PEM performance in a single cell PEMFC test.

Keywords£º Proton Exchange Membrane£» Ultra-thin membrane£» Atom Transfer Radical Polymerization£» Phase Separated Morphology.

Acknowledgements: This work was supported by the National Natural Science Foundation of China (51573083) and Intergovernmental International Scientific and Technological Innovation Cooperation Key Projects (2016YFE0102700).

497727 The Consequence of Morphology on Conductivity for Bolaamphiphiles

Abstract:

Polyelectrolytes have gained popularity as potential candidates for rechargeable batteries, membrane fuel cells, and drug delivery carriers. Here we consider ionic-electronic conductors in rod-coil block copolymers and bolaamphiphiles as a possible next step in the development of these class of materials. These proposed materials decouple electrical conductivity, ionic conductivity, and mechanical properties through the use of distinct but complementary functionalities. Mixing ionically and electronically conducting materials takes advantage of ionic movement combined and electronic transport while the presence of a stiff polymer backbone provides mechanical stability. However, the combination of flexible and rigid moieties also leads to complex phase behavior which presents significant challenges to molecular modeling. Using a combination of mesoscopic and atomistic models, we investigate the effect of composition and placement of charged and non-charged functionalities on the conductivity of the materials. We thus elucidate how changes in morphology correlate with changes in conductive properties. The ultimate goal is to unveil basic principles for the development and optimization of these class of materials.

487346 Mechanical and Structural Analyses of Toughened Syndiotactic Polypropylene Gels: Effects of Gel-Preparation Temperature

Abstract:

Physical gels, with physical crosslinkings such as hydrogen bonds or polymer crystals, could be readily re-molded due to their reversible crosslinking points. However, from the viewpoint of mechanical properties, physical gels are generally inferior to chemical gels that have strong crosslinking points made of chemical bonds. Our recent studies revealed that the mechanically-tough syndiotactic polypropylene (sPP) gel (one of the physical gels with crosslinking points made of sPP crystals) could be obtained by quenching the sPP/decahydronaphthalene (decalin) solution with liquid nitrogen (LN). In this study, the effects of the gel-preparation temperature on the mechanical properties of the sPP/decalin gels were investigated and the structural analyses of the sPP gels were also conducted by the polarized optical microscopy (POM), the scanning electron microscopy (SEM), and the small-angle neutron scattering (SANS) measurements to discuss the relationship between the microstructures and the physical properties of the sPP gels. The Youngâ€™s modulus and the fracture stress of the 10 wt% sPP/decalin gel prepared by the LN quenching (Gel_LN) were 12 times and 96 times, respectively, higher than those of the 10 wt% sPP/decalin gel prepared at room temperature (Gel_RT). sPP spherulites with the diameter of ~15 Î¼m were observed by POM and by SEM micrographs for Gel_RT. For Gel_LN, however, no spherulites but much smaller sphere-like structures with the diameter of 100 nm were observed, assembling to form a flat plate. The SANS analyses showed that the structure in Gel_LN was smaller than that in Gel_RT. These results suggested that the quenching caused the formation of dense and smaller structures of the sPP gels, resulting in the relatively high toughness of the sPP gels.

493027 Chemically Cross-Linked Poly(2-Hydroxyethyl Methacrylate)-Supported Deep Eutectic Solvent Gel Electrolytes

Abstract:

Nonvolatile, ion-dense electrolytes, such as ionic liquids (ILs) and deep eutectic solvents (DESs), are attractive candidates for safe, high performance electrochemical/electrostatic energy storage devices. Creating solid-state, low volatility gel polymer electrolytes offers an appealing way to further leverage ILs/DESs for future electrical energy storage device applications due to their leak-proof nature and robust flexibility. In this work, we describe the first-time incorporation of an inexpensive and eco-friendly DES (a 1:2 molar ratio mixture of choline chloride:ethylene glycol) into a solid-state gel electrolyte by employing UV-initiated free radical copolymerization of 2-hydroxyethyl methacrylate (HEMA) and poly(ethylene glycol) diacrylate (PEGDA) *in situ* within the DES to create a chemically cross-linked polymer scaffold. A relatively small amount of PEGDA (1:30 molar ratio with HEMA) is included to serve as a chemical cross-linker in order to create a covalently-bonded polymer network within the DES. A high room temperature ionic conductivity of 5.7 mS/cm has been obtained for a DES gel containing a 13.2 vol.% copolymer scaffold. Tunable DES gel compressive elastic modulus values, which could be varied between 14 kPa and 1.0 MPa, were obtained by varying the total polymer content. Clear evidence of hydrogen bonding interactions between the poly(HEMA) scaffold and the DES anion was revealed through species self-diffusivity measurements, as well as by FTIR spectroscopy. Using a 22.9 vol.% polymer-supported DES gel as electrolyte/separator in a supercapacitor prototype with activated carbon fabric electrodes, a specific capacitance of 33.3 F/g and an energy density of 15.8 Wh/kg (based on the mass of two electrodes) were obtained during discharge at a current density of 10 mA/g. This study supports the notion that chemically cross-linked polymer-supported DES gel electrolytes are suitable for energy storage applications, and opens the door to future investigations of how polymer-DES interactions can be exploited in order to maximize the performance of polymer-supported DES gel composites.

495920 Ultrasonic Synthesis of Temperature Responsive Copolymer and Its Characterization

Abstract:

Poly(*N*-isopropylacrylamide) (PNIPAM) is a temperature-responsive polymer with a lower critical solution temperature (LCST). Poly (2-hydroxyethyl methacrylate) (PHEMA) has a high mechanical strength, so that a copolymer of Poly(NIPAM-*co*-HEMA) is expected to be applied as advanced functional polymeric materials in various fields, such as sensors, actuators, drug delivery system, and so on. The properties of the copolymer are dependent on the molecular weight distribution and copolymer composition.

Ultrasonic irradiation to liquid results in cavitation phenomena, i.e., formation and collapse of micro scale bubbles. Collapsing bubbles cause local high temperature and high pressure field, and extremely high shear flow in liquid. When ultrasound is irradiated to a solution containing monomer, polymerization proceeds by generating radicals owing to local high temperature, and the synthesized polymer decomposes due to high shear stress, simultaneously. Thus, the molecular weight and polydispersity of polymer is expected to be controlled by changing the condition of ultrasonic irradiation. Ultrasonic irradiation to monomer solution was applicable to produce various vinyl polymers.

In this research poly (*N*-isopropylacrylamide-co-2-hydroxyethyl methacrylate) was synthesized using ultrasonic irradiation in a mixed solvent of water and ethanol without chemical initiator. The effects of ultrasonic power intensity, and monomer molar ratio on time course of conversion to polymer, number average molecular weight, and polydispersity were investigated in order to synthesize the polymer with low molecular weight distribution (i.e., low polydispersity). The responsive temperature of synthesized polymer was also evaluated.

Higher ultrasonic power intensity resulted in higher conversion to polymer and lower molecular weight. This is because the higher amount of radical is generated at higher ultrasonic power intensity. The polydispersity was less than 1.5 for all the case. Thus, the molecular weight can be controlled by ultrasonic power intensity while maintaining low polydispersity. Higher NIPAAM ratio resulted in higher polymerization rate, faster polymer degradation rate, and lower polydispersity. PNIPAAM radical has lower reaction activity with ethanol than PHEMA radical, so that PNIPAAM polymerizes faster than PHEMA.

The transmittance of aqueous solution of copolymer decreased abruptly at a lower critical solution temperature (LCST) which depends on the monomer molar ratio. The lower the molar ratio of NIPAM, the lower the LCST because HEMA is more hydrophobic than NIPAM. The higher polymerization rate resulted in the higher LCST. Those results suggested that the responsive temperature of the copolymer can be controlled by changing the molar ratio of NIPAM and the ultrasound intensity.

495242 Atomic-Level Comparisons of LCST Transition in Thermo-Sensitive Polymers

Abstract:

Stimuli-sensitive polymers due to their ability to respond to changes in surroundings, have a wide range of applications in biotechnology and medicine. Here, we have conducted all-atom molecular dynamics (MD) simulations of three polymers with different lower critical solution temperatures (LCST): Poly(N-isopropylacrylamide) (PNIPAM) (LCST=305K), poly(dimethylaminoethyl methacrylate) (PDMAEMA) (LCST = 323K), and poly(*N,Nâ€™*-diethyl acrylamide) (PDEAAm) (LCST = 299- 308K). Charmm force field was employed to run the simulations. Different sets of charges of the atoms in the charm force field was employed to study their effects on the structure of the polymers. And the MD simulations were conducted below and above the LCST of respective polymers to investigate the origin of their LCSTs. The structural and dynamical properties of the polymers and proximal water were obtained: radius distribution of function (RDF) between water and polymer, radius of gyration (Rg) of polymers, residence time probability of water and free energy.

488391 ATRP-Grown Protein-Polymer Conjugates Selectively Enhance Transepithelial Protein Transport

Abstract:

Despite its patient-friendliness, the oral route is not yet a viable strategy for the delivery of biomacromolecular therapeutics. This is, in part, due to the large size of proteins, which greatly limits their absorption across the intestinal epithelium. Although chemical permeation enhancers can improve macromolecular transport, their positive impact is often accompanied by toxicity. One element potentially contributing to this toxicity is the lack of co-localization of the enhancer with the protein drug, which can result in non-specific permeation of the intestine as well as enhancer overdosing in some areas due to non-uniform distribution.

To circumvent these issues, this study describes a new way of increasing protein permeability via a polymer conjugation process that co-localizes permeation enhancer with the protein. Based on previous reports demonstrating the utility of 1-phenylpiperazine as an intestinal permeation enhancer, we synthesized protein-polymer conjugates with a phenylpiperazine-containing polymer using polymer-based protein engineering. A novel phenylpiperazine acrylamide monomer was synthesized and chain extended using atom transfer radical polymerization from the model protein bovine serum albumin (BSA). At non-cytotoxic doses, the protein-polymer conjugates induced a dose dependent reduction in the trans-epithelial electrical resistance of Caco-2 monolayers and an impressive ~30-fold increase in BSA permeability. Furthermore, this permeability increase was selective, as the permeability of the small molecule calcein co-incubated with the protein-polymer conjugate increased only 5-fold. Together, these data represent an important first step in the development of protein polymer conjugates that facilitate selective protein transport across membranes that are typically impermeable to macromolecules.

490707 Aptamer Micelles Targeting Cancer Cells Expressing the Chemokine Fractalkine

Abstract:

There is a growing need for cancer therapeutics to be delivered without off-target side effects; in order to do this, the cancer cells must express a unique target. We have previously developed a single stranded DNA (ssDNA) aptamer, FKN-S2, that binds to a novel extracellular chemokine, fractalkine, with high affinity and specificity. Unlike other chemokines, fractalkine has been shown to act as an adhesion molecule and is expressed on inflamed and cancerous cells thus making it a promising target for cancer therapeutics. We synthesized the FKN-S2 as an aptamer amphiphile and showed that self-assembles into 18 nm micelles. The FKN-S2 micelles decorated with polyethylene glycol (PEG) were delivered to mouse colon adenocarcinoma cells that expressed human fractalkine (MCA-38.FKN) as well as control healthy cells in vitro, and only the MCA-38.FKN cells were shown to internalize the micelles. The MCA-38.FKN cells were subcutaneously injected into nude female mice to create xenograft tumors for in vivo experiments. The biodistribution and uptake of FKN-S2 PEGylated micelles was compared to random ssDNA-amphiphile PEGylated micelles using 64Cu-labeling for in vivo micro positron emission tomography (Î¼PET)/computerized tomography (CT) imaging, and fluorescence labeling for ex vivo confocal microscopy imaging. Although the biodistribution of the two micelles was the same, ex vivo confocal microscopy showed that only the FKNS2 micelles were capable of internalizing into the tumor cells. This is the first demonstration of a ssDNA aptamer, or any other ligand being used to target fractalkine for in vitro and in vivo delivery of nanoparticles to fractalkine expressing cells, and shows that the FKN-S2 micelles may serve as a valuable tool for developing targeted therapies in the future.

493890 Non-Charged Cell-Penetrating Oligothioetheramides

Abstract:

The discovery of cell-penetrating peptides (CPPs) over two decades ago uncovered a novel method for transporting a variety of cargoes into cells.[1] Although promising, CPPs have several drawbacks that hinder their use for *in vivo* therapeutic applications such as rapid metabolic degradation by proteases and a propensity for mounting an immune response. Recently, the Alabi research group developed a unique approach for the rapid assembly of sequence-defined oligothioetheramides (oligoTEAs).[2] OligoTEAs have three distinct advantages over native peptides. First, sequence-defined oligomers are abiotic and not susceptible to protease degradation, thus increasing their bioavailability. Second, access to direct modification of the oligoTEA backbone enables direct control over their conformation, rigidity, and pendant group spacing to ultimately tune interactions between the binding motifs and the cell membrane. Last, the use of readily accessible synthetic monomers will allow for massive compositional diversity.

Studies conducted on several CPPs thus far indicates that a combination of cationic and hydrophobic residues are critical for translocation across cellular membrane.[3,4,5] Based on these and many other studies, we embarked on the assembly of an 8-residue oligoTEA library composed of a hydrophilic backbone (DL-1,4-dithiothreitol) and distinct combinations of two different pendant groups: a cationic (guanidinium) monomer and a hydrophobic (benzyl) monomer. Probing this initial library of 16 oligomeric structures for cellular uptake led to the conclusion that for this class of macromolecules, cationic residues are not a requirement for efficient uptake. In fact, they seem to impede uptake. Expanding our library with further studies reveals the unusual discovery of a non-charged cell penetrating oligoTEA that undergoes efficient cellular uptake with no cytotoxicity, and outperforms R9K, a well-known and widely used CPP. This talk will focus on the sequence-controlled assembly of this non-charged cell penetrating oligoTEA, its uptake efficiency across a variety of cell lines, and its mechanism of cell entry.

References: **1.** K.M. Stewart et al. *Org Biomol Chem* 6, 2242-2255 (2008) **2.** M. Porel et al. *J Am Chem Soc* 136, 13162-13165 (2014) **3.** D.J. Mitchell et al. *J Pept Res* 56, 318-325 (2000) **4.** N. Svensen et al. *Angew Chem Int Ed* 50, 6133-6136 (2011) **5.** C.B. Cooley et al. *J Am Chem Soc* 131, 16401â€"16403 (2009).

494483 Targeting, Delivery, and Immobilization of Therapeutic Factors with Native Free Radicals

Abstract:

Elevated concentrations of free radicals are characteristic of a wide variety of tissue injuries and disease states, such as inflammatory diseases, neurological disorders, ischemia, burn wounds, and traumatic brain injuries. Free radicals are extremely reactive and can cause significant damage to proteins, lipid membranes, and nucleic acids. Left unchecked, these free radicals can perpetuate inflammatory states, induce the continued generation of further free radicals, and overall serve as an obstacle to healing. Delivery of exogenous compounds, such as antioxidants, to scavenge or otherwise detoxify these free radicals has been frequently investigated, but typically antioxidants alone are insufficient in remediating complex injury environments. Drugs or growth factors that promote healing through other mechanisms in these conditions and have similarly shown promise, but are often cleared too rapidly from the injury site to have a prolonged effect. Although the heightened reactivity of free radicals is deleterious in vivo, it is of great benefit in polymer chemistry, propagating polymerization reactions, and initiating crosslinking reactions between specific functional groups. We hypothesized that free radicals generated through injury and disease can be leveraged to induce crosslinking of acrylated groups on polymers. The advantages of this reaction are two-fold: 1) sufficient crosslinking of acrylated polymers will reduce their diffusivity, thereby localizing them to the target tissue and providing a vehicle for delivery of therapeutic factors; and 2) the crosslinking reaction may reduce or sequester radicals, reducing their ability to damage the tissue. Acrylated PEGs reduced the concentration of a radical used in standard scavenging assays (DPPH) as well as with reactive oxygen species (ROS) and reactive nitrogen species (RNS) commonly found in vivo. Further, in cellular assays, acrylated PEGs provided a measure of protection to cultures from hydrogen peroxide injury. These ROS also induced acrylate-acrylate crosslinking. Together these results strongly support our hypothesis and indicate that acrylated PEGs can be used to target and sustain the presentation of therapeutic factors within injured or diseased tissues and also sequester the free radicals to protect the tissue from further damage. Current efforts aim to identify optimal sizes and configurations of the acrylated carrier polymers and to test these polymers in a tissue explant model. Ultimately, this proposed acrylated PEG based system represents a versatile drug delivery platform that targets regions of injured or diseased tissue and can be amended to carry nearly endless therapeutic payloads.

This work is supported by a pilot research grant (CBIR16PIL015) and a graduate fellowship (CBIR14FEL004) generously provided by the New Jersey Commission on Brain Injury Research (NJCBIR).

500711 Antimicrobial Peptide Amphiphile (AMPA) Medical Product Coatings for the Prevention of Nosocomial Infections

Abstract:

Nosocomial (hospital-acquired) infections are a significant issue in the clinical medicine today. One common origin of infection is through bacterial growth on equipment and materials in contact with patients. The most common nosocomial infection is ventilator-associated pneumonia (VAP), for which an endotracheal tube (ETT) provides a surface that facilitates bacteria colonization of deep lung tissue. Approximately 21% of patients in Intensive Care Units (ICUs) acquire VAP and around 10% of those will die from the infection. While treatment with antibiotics can sometimes help, many bacteria are becoming drug resistant and limit drug exposure by creating a protective biofilm.

Antimicrobial peptides show considerable promise as a novel, effective alternative to traditional antibiotics in combating VAP. Recent advances in peptide and peptide amphiphile therapeutics have resulted in products that have been successfully translated to the clinic. With inspiration from a collaborator and the literature, we have designed, synthesized, and evaluated a suite of novel antimicrobial peptide amphiphiles (AMPAs), including a cyclic AMPA, a linear biologically-homologous AMPA, and a polypeptide AMPA produced via N-carboxyanhydride ring-opening polymerization. We have designed an inexpensive and simple additive manufacturing method for coating medical-grade plastics with AMPAs to prevent bacterial growth, overcoming key challenges associated with the prolonged, localized delivery and stability of peptide amphiphiles. Using atomic force microscopy, we have shown successful coating of a thin AMPA films on ETTs. Submersion coating experiments have established our ability to apply this method of coating to several different AMPAs as well as with multiple kinds of medical-grade plastics. We have established that our suite of AMPA coatings are capable of reducing bacterial adhesion and biofilm growth through preliminary *in vitro* analysis against clinically-relevant bacteria but show minimal toxicity to mammalian cells. These preliminary *in vitro* results indicate that AMPAs outperform their peptide counterparts and, due to the stable nature of the AMPA film, prevent bacterial adhesion for significantly longer than peptide alone. These results provide strong evidence that AMPA-coated ETTs will be an effective method of preventing VAP clinically.

500802 Mitigating the Bioactivity Loss of Polymer- Insulin Conjugate

Abstract:

There is significant interest in pursuing polymerâ€"protein conjugation to obtain a better protein drug with improved pharmacokinetics. However, a generic problem with polymerâ€"protein conjugation is that the polymers drastically reduce the bioactivity of the modified protein. Take the widely known PEGylation strategy for example, conjugating polyethylene glycol (PEG) to protein through conventional covalent coupling chemistry resulting in polymer chains non-selectively attached on the protein surface groups (e.g., amines), and potential attachment near to the proteinâ€™s active site significantly inhibit the protein from performing designated biological function (e.g., amine group on Gly A1 of insulin). Alternatively, PEG can be used to modify the protein through site-specific conjugation technique, such as involving re-engineering the proteins. But the resulting conjugation product is still subject to bioactivity loss. In general, it appears that no matter which site of the protein the polymer is conjugated to, most resulting polymer-protein conjugates inevitably suffer from bioactivity loss. Here we present a potential strategy using zwitterionic polymers for protein conjugation to mitigate the bioactivity loss, and the development of zwitterionic polymer-insulin conjugate as a novel pharmaceutical agent.

489477 Enhancing Therapeutic Efficacy of Self-Assembling Prodrugs with Supramolecular Chemistry

Abstract:

Chemical modification of small molecule hydrophobic drugs is a clinically proven strategy to devise prodrugs with enhanced treatment efficacy. While this prodrug strategy improves the parent drug's water solubility and pharmacokinetic profile, it typically compromises the drug's potency against cancer cells due to the retarded drug release rate and reduced cellular uptake efficiency. We report here the supramolecular design of self-assembling prodrugs with much improved water solubility while maintaining high potency against cancer cells. The direct use of anticancer drugs to create their own nanostructures allows for fine-tuning and optimization of the drug loading and their physicochemical properties in a precise manner through molecular engineering of the underlying prodrug building units. We found that our rationally designed self-assembing prodrugs are capable of associating into one-dimensional nanostructures of various physiochemical properties in aqueous solution with a width of approximately 10 nm and lengths on the scale of several micrometers. We further evaluated the stability of resulting filaments of different surface chemistries in both cell medium and human serum, and their gelation ability in physiological condition. More importantly, our results suggest that this platform of prodrugs exhibit much greater efficacy against various cancer cells relative to the clinically used prodrug both *in vitro* and *in vivo*. We believe these findings open a new avenue for rational design of supramolecular prodrugs for more effective cancer treatment.

501892 Hydrogen Sulfide Donor Micelles: Synthesis, Characterization and Therapeutic Potential

Abstract:

Hydrogen Sulfide Donor Micelles: Synthesis, Characterization and Therapeutic Potential

Urara Hasegawa, André J. van der Vlies, Jerry J.Y. Chen, Tomoka Takatani-Nakase, Ikuhiko Nakase

Hydrogen sulfide (H_2S) is produced in the human body and serves as a gaseous signal-transmitter molecule that regulates inflammation, relaxes vascular smooth muscles, promotes angiogenesis and mediates neurotransmission. However, due to the difficulty in handling H_2S, its therapeutic use has not been well explored. Therefore, development of H_2S delivery technologies has attracted growing attention. One common approach is to use H_2S donor compounds such as anethole dithiolethione (ADT) derivatives which generate H_2S by the interaction with intracellular enzymes. However, this approach suffers from uncontrolled gas generation which limits their applications.

To address this issue, we developed nano-sized gas donors based on polymeric micelles (H_2S donor micelles, Figure 1) which enable sustained H_2S generation under physiological conditions. In this presentation, we report the design and synthesis of polymeric micelles containing anethole dithiolethione (ADT) moieties. The micelles were prepared from amphiphilic diblock copolymers consisting of a hydrophilic segment and a hydrophobic segment bearing ADT moieties. The micelles generated H_2S slowly and continuously for sustained periods of time compared to the small H_2S donor compound ADT. Since the disruption of the micelles by adding a surfactant increased the rate of gas generation, the micellar structure seems to be a key to successful control of gas generation. Furthermore, the micelles showed various biological effects including pro-angiogenic and anti-apoptotic effects in *in vitro* cell culture assays as well as the chicken chorioallantoic membrane (CAM) model.

References

1. U. Hasegawa, A. J. van der Vlies, *Bioconjugate Chem.*, 25 (2014) 1290-1300.

2. U. Hasegawa, A. J. van der Vlies, *Med. Chem.Comm.*, 6 (2015) 273-276.

500163 Pegylation of Model Drug Carriers Enhances Uptake By Primary Human Neutrophils

Abstract:

Introduction: The advent of targeted drug carriers has opened many new avenues for the delivery of therapeutics directly to the site of disease, reducing systemic side effects and enhancing the efficacy of therapeutic molecules. However, the packaging of therapeutics into particulate carriers for delivery comes with its own set of challenges and barriers. Among these, a great deal of research effort has focused on protecting carriers from clearance by phagocytes by altering carrier surface chemistry. Many groups have explored the use of polyethylene glycol (PEG) chain coatings to mitigate unwanted phagocytosis, as PEG is highly hydrophilic and is well-known for its anti-fouling properties. However, this strategy has seen limited success in the clinic, as carriers are often still rapidly cleared with little improvement in therapeutic benefit. Notably, very few papers have explored the effects of PEG on uptake by freshly obtained primary human phagocytes in physiological conditions, creating a disconnect between the prevailing literature and ultimate applications. In this work, we investigate the effect of PEGylation on uptake by primary human neutrophils *in vitro*, and compare these effects to several cell lines and other model phagocytic cells systems in evaluating the effects of surface chemistry on phagocytosis.

Materials and Methods: Human blood was obtained from healthy donors via venipuncture per a protocol approved by the University of Michigan Internal Review Board. Mouse blood was obtained via cardiac puncture per a protocol approved by the University of Michigan Institutional Animal Care and Use Committee (IACUC) and in accordance with the National Institute of Health guidelines for the care and use of laboratory animals. Carboxylated polystyrene spheres (PS-COOH) were conjugated with methoxy-terminated polyethylene glycol (PEG, 5k molecular weight, Fisher) using *N*-(3-Dimethylaminopropyl)-*N*â€²-ethylcarbodiimide hydrochloride (EDC, Sigma) chemistry. PEG chain density was quantified using rhodamine-labeled PEG and a standard curve created using a fluorescence plate reader. Particles of varying size and PEG densities were introduced to human blood, mouse blood, and other phagocytic cells (mouse bone marrow-derived macrophages, THP-1 monocytes, HL-60 derived neutrophils) for varying time periods (0-2 hours). Post-phagocytosis, cells were stained (for CD45, CD11b, Ly-6G, etc.) and fixed for flow cytometry.

Results and Discussion: Surprisingly, we found that primary human neutrophils preferentially phagocytose PEGylated carriers relative to their carboxylated counterparts (~40% particle positive neutrophils for PEGylated particles, compared to ~20% for carboxylated) (see Figure 1). This phenomenon was observed in nearly every donor used in the study (15/16), and was found to be consistent for individual donors. Additionally, this effect was consistently observed for 2m, 500 nm, and 200 nm particles with a wide variety of PEG conjugation densities. This phenomenon was not observed, however, in THP1 monocytes, mouse bone marrow-derived macrophages, or freshly-isolated mouse neutrophils; however, addition of human plasma to model phagocytic cells dramatically increased uptake of PEGylated particles, suggesting the involvement of factors in human plasma in elevated uptake. Further work to elucidate the precise factors responsible for this phenomenon is currently underway.

Conclusions: The addition of PEG chains to model drug carriers results in increased uptake by primary human neutrophils. This phenomenon is observed across a range of carrier sizes and PEG densities, but is not observed in typical model phagocytic cells. This result is seemingly linked to factors in human plasma resulting in preferential uptake of PEGylated carriers as compared to carboxylated counterparts. The surprising results have implications in the design of effective drug carriers, and possibly shed light as to why PEGylation has not proven to be the â€œmagic bulletâ€ in creating stealth drug carriers, and may lead to a more thorough understanding of the mechanisms of phagocytosis by neutrophils, which are thus far less studied than other phagocytes such as macrophages and monocytes.

501585 "Smart" Nanoparticles for Immunotherapeutic Targeting of the Sting Pathway

Abstract:

Immune checkpoint inhibitors are transforming cancer treatment, yet the majority of patients fail to achieve complete and durable responses. Resistance to checkpoint blockade is associated with non-immunogenic tumors that lack significant T cell infiltration and are instead characterized by a highly immunosuppressive tumor microenvironment (TME). To reprogram such "cold" TMEs to a more "hot" T cell-inflamed phenotype, we are developing pH-responsive, "smart" polymer nanoparticles (NPs) for efficient cytosolic delivery of cyclic dinucleotides (CDNs), including 2'3'-cGAMP (cyclic [G(2',5')pA(3,5')p]), the natural and endogenous high affinity ligand of the stimulator of interferon genes (STING) pathway. The STING pathway is the primary mechanism by which the innate immune system senses cancer cells to stimulate a localized inflammatory response that is essential to endogenous anti-tumor T cell adaptive immunity. This critical role of STING in cancer immune surveillance provides compelling rationale for the use of cGAMP and structurally related CDN STING agonists as therapeutics for reversing immunosuppression in the TME and generating robust anti-tumor T cell responses in the setting of non-immunogenic tumors. However, as an anionic compound, cGAMP does not readily cross cellular membranes, is poorly endocytosed, and, critically, enters the cytosol where STING is localized with very low efficiency. Moreover, CDNs are rapidly cleared from the body with modest delivery to tumors and/or lymph nodes. These major drug delivery barriers limit the therapeutic potential of cGAMP and other structurally related CDNs. To solve these challenges, we have designed polymer vesicles - polymersomes - that promote efficient endosomal escape of cGAMP to the cytosol. cGAMP can be efficiently loaded into NPs (38% encapsulation efficiency) with an average diameter of 100 nm, a neutral surface charge, and low cytotoxicity. NP delivery of cGAMP dramatically enhances STING activation in vitro relative to free cGAMP, which elicits only low levels of type-I interferon (IFN-I) production even at a 500-1000x higher dose *in vitro*. In established subcutaneous (SC) B16F10 melanoma tumors, intratumoral (IT) administration of nanoparticulate cGAMP formulations converted tumors into an inflamed phenotype, characterized by elevated IFN-I production, dendritic cell maturation, T-cell infiltration, and polarization of immunosuppressive M2 macrophages towards an anti-tumor M1 phenotype. These shifts in the TME corresponded with an inhibition of tumor growth for mice treated with STING-activating nanoparticles, with a subset of mice demonstrating complete tumor regression, long-term survival, and systemic antitumor immunity that protected against tumor re-challenge. Collectively, these data demonstrate that endosomolytic polymer nanoparticles significantly improve the therapeutic potency of cGAMP, and establish a pre-clinical foundation for the use of STING-activating nanoparticles as a strategy for increasing tumor immunogenicity and improving responses to immune checkpoint blockade.

502759 Biomaterial Nanoparticles Redistribute Therapeutic Antibodies to Lymph Node-Resident Cells to Enhance Cancer Immunotherapy Via Checkpoint Inhibition

Abstract:

Immunotherapy has emerged as the most promising new treatment approach for an array of advanced malignancies. In particular, monoclonal antibody (mAb) checkpoint inhibitor drugs are increasingly utilized as cancer immunotherapeutic agents due to their capacity to modulate immune cell signaling to enhance anti-tumor adaptive immunity. Although responses are often durable, a large majority of patients do not respond to such treatments. Furthermore, several treatment-associated toxicities remain to be minimized. Since lymph nodes (LNs) are tissues where adaptive immune responses are initiated and regulated, we hypothesized that targeted delivery to LNs has significant potential to improve the therapeutic effects of immunotherapeutic mAb drugs for cancer immunotherapy. To accomplish this goal, we leveraged ~30 nm nanoparticles (NPs) comprised of a poly(propylene sulfide) core surface stabilized with block co-polymer Pluronics that, by virtue of their formulation and ultrasmall size, exhibit greatly enhanced profiles of lymphatic uptake into lymphatic tissues and accumulation within LNs post administration. Since mAbs require their 3D conformation to recognize cognate antigen, a mild and aqueous phase-based conjugation strategy was employed to maximally preserve the functionality of the mAb Fab. We found NP conjugation did not negatively affect mAb binding capacity and significantly enhanced mAb accumulation within dLN 6-50-fold and lowered accumulation in systemic off-target tissues. Localized immunotherapy utilizing checkpoint inhibitor mAb conjugated to the NP resulted in slowed tumor growth relative to saline-treated controls and animals treated with free mAb administered systemically. Our results suggest lymphatic draining NPs can be used to improve the immunotherapeutic effects of checkpoint blockade therapy.

499125 Engineering Nanoparticle Artificial Antigen Presenting Cells Based on T Cell Biology Improves T Cell Enrichment and Activation for Cancer Immunotherapy

Abstract:

Engineering therapies that modulate the bodyâ€™s immune system is an attractive and effective approach to fighting diseases such as cancer. Antigen presenting cells, such as dendritic cells, are at the heart of the immune response, yet are often immunosuppressed in cancer which results in defective antigen-presentation or even immunosuppressive signaling to CD8+ killer T cells. To overcome this barrier in the cancer immunity cycle we have engineered a particle-based artificial antigen presenting cell (aAPC).

Our design of the aAPCs is centered on a reductionist approach to biomimicry. Originally, our aAPCs were micron-sized to mimic the size of cells; however, this poses a problem for in vivo therapeutics as there is a possibility for embolization. Therefore, our studies herein focus on quantifying the effect of aAPC size and designing an effective nano-sized aAPC based on our understanding of T cells. As a part of our design criteria, we aimed to engineer these aAPCs from iron oxide nanoparticles so that we could magnetically enrich antigen-specific T cells because of their very low frequencyâ€"1 in 10^5-10^6 T cells.

We found consistent with T cell biology that, particles larger than T cell receptor (TCR) clusters of around 300 nm in diameter were more effective at activating CD8+ T cells. These nano-aAPCs also had increased binding through multivalent interactions, which led to an increase in ability to magnetically enrich antigen-specific T cells. Enrichment was also enhanced by tuning particle concentration and only adding antigen-specific signaling molecules to the particles. Synthesizing our findings, we report much higher numbers of antigen-specific T cell generated for cancer therapy with our 300 nm aAPCs. Beyond immunotherapeutic applications, our quantitative and biologically inspired particles provide a case study to guide the design of other cell-modulating technologies.

500553 Biomaterial Scaffolds for Combined Focal Ablation and Immunotherapy to Target Disseminated Cancer

Abstract:

Metastasis continues to be the primary cause of mortality for patients suffering from cancer. Focal therapies such as hyperthermia, cryosurgery, and irreversible electroporation (IRE) have been successful at eliminating solid tumors, but often fail to mitigate local recurrence and have a very limited effect on disseminated tumor cells. With recent advances in immunotherapy, both solid tumor vaccine and immune checkpoint blockade have emerged as potential cancer therapies. Furthermore, ongoing efforts to combine focal therapy with immunotherapy to address both local and systemic cancer are beginning to show some promise. However, the mechanisms by which focal tumor ablation therapy and immunotherapy synergize are not known. Polymeric scaffolds have previously been shown to capture metastasizing cancer cells *in vivo* and are likely to provide a biomechanically consistent platform suitable for the clinical application of a variety of focal therapies *in vivo*. We sought to identify the optimal focal therapy for generating a robust $CD8^+$ cytotoxic T cell-mediated immune response when applied to tumor cells within the scaffold. We chose to evaluate this strategy using the B16 mouse melanoma model, taking advantage of transgenic mouse T cells which are specific for the melanoma antigen tyrosinase-related-protein-2 (TRP-2). Focal therapies were first studied *in vitro* to quantify differences in solubilized protein released from B16 cells upon treatment. BCA total protein assays performed on the released supernatant following heating, cryo-freezing, and IRE showed that cryo and IRE achieved similar levels of protein release as the full lysis control, while heat-treated samples yielded 29 ± 16% of the protein amount in the control. To assay immune presentation of the TRP-2 antigen, the soluble proteins released from heat, cryo and IRE lysates were incubated with bone marrow-derived dendritic cells, which were then combined with purified naive ($CD44^{low}$) TRP-2-specific T cells labeled with CellTrace Violet (CTV). Cryo and IRE derived B16 extracts promoted similar levels of T cell activation, as measured by upregulation of CD44 and dilution of CTV (proliferation) at high cell equivalents. Interestingly, IRE achieved significantly higher T cell activation at lower cell equivalents, suggesting that the mode of cell lysis can promote differential tumor antigen release or immune presentation. Applying the focal therapies to tissue-laden polycaprolactone scaffolds *in vivo* showed at least a 79% reduction in viability by WST-1 assay and visible cellular damage by histology following all three methods of ablation. Future *in vivo* studies will study the effects of focal treatment on scaffolds containing B16 cells in combination with local and systemic immunomodulation to generate a cytotoxic T cell response capable of targeting both local and systemic disease and protecting against secondary challenge.

499896 Modulating the Immune Environment within Adipose Tissue with Polymer Scaffolds

Abstract:

Biomaterial scaffolds are central to many tissue engineering strategies as they create a space for tissue growth and provide support for cell adhesion and migration. However, biomaterial implant results in an unavoidable inflammatory response, which can impair the intended interactions of the scaffold with the host tissue. We hypothesized that scaffold-based delivery of small molecules could modulate the number of leukocytes that infiltrate the implant site, reducing the resultant inflammation. To test this hypothesis, we encapsulated resveratrol, an anti-inflammatory polyphenol, within poly(lactide-co-glycolide) (PLG) microparticles using a single emulsion technique and fabricated scaffolds from the microparticles with gas foaming. Histological analysis indicates that scaffolds integrate with the epididymal adipose tissue within fourteen days and the polymer matrix is in contact with resident adipocytes and CD45+ immune cells. Flow cytometry reveals that immune cell populations within the implant site include macrophages, NK cells, T cells, B cells and monocytes. Interestingly, resveratrol delivery significantly decreases $CD11b^+$ monocyte infiltration relative to control scaffolds and this correlates with lower IL-6 expression within the epididymal tissue. Molecular targets for resveratrol include the NAD-dependent deacetylase SIRT1, which modulates cellular signaling in both adipocytes and immune cells. This presentation will focus on how resveratrol delivery modulates gene expression in adipocytes, immune cells and other stromal vascular cells following biomaterial implant. Furthermore, we will characterize resveratrol release from the scaffold, its residence time in the tissue, and how long anti-inflammatory effects are expected to last. Collectively, this work will provide an understanding of how PLG-based drug delivery targets various cell types within the adipose tissue and is expected to lay the foundation for future work focused on modulating adipose tissue function as a treatment for metabolic disease.

493120 Multi-Factor Microparticle Formulation for Local Induction of Regulatory Lymphocytes and Treatment of Periodontal Disease

Abstract:

According to the American Dental Association, periodontal disease is a critical oral health care issue affecting over 65 million adults in the United States alone. Characterized by an overburden of invasive bacteria, gum inflammation and plaque buildup, periodontitis can result in severe loss of gingival tissue attachment and bone resorption. Current treatment focuses on the removal of the oral bacterial load through procedures such as scaling and root planing often accompanied by antibiotic treatments. However, literature suggests periodontal disease is not merely a consequence of the overburden of invasive bacteria but a result of immune imbalance caused by the inflammatory response to the invasive bacteria. Therefore, we've sought to reverse the underlying immune imbalance through the local presence of regulatory lymphocytes (Tregs) that are known to restore immunological homeostasis. Previously, we showed that local recruitment of endogenous regulatory lymphocytes aids in reversing the progression of periodontitis. The local, controlled release of the chemokine CCL22, induced Tregs recruitment which re-established periodontal homeostasis. However, this approach relies on the availability of functional, natural, regulatory T cells, and it is becoming more apparent that many inflammatory diseases may result from dysfunctional or deficient Tregs. Therefore, we hypothesize that the local induction of regulatory lymphocytes may provide a more robust mechanism to change the ratio of regulatory T cells in the periodontium, and reverse disease conditions. To this end, we have fabricated controlled release microparticles containing Treg-inducing factors; TGF-ß (growth factor essential for Treg differentiation) , Rapamycin (mTorr inhibitor) and IL-2 (cytokine that promotes the differentiation of immature T cells into regulatory T cells) which have been shown to locally induce regulatory lymphocytes. Using a mouse model for periodontal disease, we have shown that the use of these controlled release microparticles significantly reduced disease outcomes. Hence, our multi-factor microparticle formulation to restore immunological balance through the induction of Tregs may be able to serve as a platform delivery system to treat other inflammatory diseases.

501023 Multivalent Soluble Antigen Arrays Target Antigen Presenting B Cells and Dampen Antigen-Specific Signaling to Promote Therapeutic Efficacy in Multiple Sclerosis

Abstract:

Multivalent soluble antigen arrays target antigen presenting B cells and dampen antigen-specific signaling to promote therapeutic efficacy in multiple sclerosis

Brittany L. Hartwell, Chad J. Pickens, Martin Leon, Laura Northrup, Francisco Martinez-Becerra, Cory Berkland

The University of Kansas

Many current therapies for autoimmune diseases such as multiple sclerosis (MS) act through nonspecific targeting of the immune response, rendering global immunosuppression, poor efficacy, and adverse side effects. To address a pressing need for safer and more effective therapies, we developed multivalent soluble antigen arrays (cSAgAs) to induce antigen-specific tolerance in MS. cSAgA$_{PLP:LABL}$ consists of a flexible hyaluronic acid (HA) linear polymer backbone click-conjugated with multiple copies of autoantigen (PLP) and adhesion inhibitor (LABL) peptides. We hypothesized that multivalent presentation of both peptides would enable antigen-specific targeting and enhanced binding with autoreactive antigen-presenting cells (APCs), thereby interrupting signaling between APCs and T cells to halt propagation of an autoimmune response. *In vivo* studies established that cSAgA$_{PLP:LABL}$ is therapeutically effective against experimental autoimmune encephalomyelitis (EAE), a murine model of relapsing-remitting MS. To elucidate the moleculeÕs therapeutic cellular mechanism, flow cytometry binding and calcium flux signaling assays were performed alongside live cell imaging to evaluate binding, specificity, and signaling modulation in different cell populations.

Studies performed *in vitro* in a model B cell system revealed that cSAgA$_{PLP:LABL}$ exhibited enhanced binding over polymer alone (cHA) and polymer conjugated with PLP only (cHA$_{PLP}$) or LABL only (cHA$_{LABL}$). Furthermore, cSAgA$_{PLP:LABL}$ exhibited high avidity antigen-specific binding, targeting the B cell receptor (BCR) to dampen BCR-mediated calcium signaling. These observations suggested the moleculeÕs action at the cell surface led to B cell anergy, a state of antigen unresponsiveness induced by continuous BCR binding and marked by reduced calcium flux signaling.

cSAgA$_{PLP:LABL}$ binding and signaling were then evaluated *ex vivo* in mixed primary splenocytes isolated from EAE and healthy mice to determine which immune cell subsets were targeted and whether this resulted in an anergic phenotype. Flow cytometry binding studies were performed in splenocytes antibody-labeled for B cells (CD19$^+$), T cells (CD3$^+$), dendritic cells (CD11c$^+$), and autoimmune-associated B cells (CD19$^+$CD11c$^+$, ABCs), a subclass of B cells that act as potent antigen presenting cells in the spleen during autoimmune disease. cSAgA$_{PLP:LABL}$ exhibited significantly higher binding in B cells, particularly ABCs, over other cell types (Figure 1AB). Additionally, B cell binding was significantly higher in EAE cells than healthy cells, suggesting disease-specific targeting that was not present in T cells (Figure 1CD). These results support our hypothesis that cSAgA$_{PLP:LABL}$ targets B cells and APCs over other cell types in a disease-specific manner.

In addition to reduced calcium flux signaling, another result of B cell anergy is down-regulation of costimulatory signals (i.e., CD80 and CD86) whose presentation is essential for autoreactive T cell activation. Thus, we investigated CD80/CD86 expression in splenocytes under two conditions: (1) harvested from EAE mice at peak of disease and treated *in vitro* during simultaneous PLP antigen challenge to evaluate the effect on costimulatory signaling, and (2) harvested from mice treated *in vivo* and re-challenged with PLP antigen *ex vivo* to evaluate lasting tolerance. First, *in vitro* treatment with cSAgA$_{PLP:LABL}$ caused significant down-regulation of CD86 in EAE splenocytes harvested from mice at peak of disease, causing B cells and ABCs to signal more like healthy cells rather than activated APCs (Figure 2AB). Furthermore, B cells and ABCs from mice treated with cSAgA$_{PLP:LABL}$ *in vivo* (Figure 2C) did not respond to *ex vivo* PLP rechallenge with a significant increase in CD80/CD86 expression, again resembling healthy cells (Figure 2DE). Splenocytes from the treated mice also exhibited elevated levels of regulatory cytokines IL-10 and IL-4. Combined, these results indicate that treatment with cSAgA$_{PLP:LABL}$ achieved efficacy by inducing PLP-specific anergy in B cells, resulting in reduced APC capacity and lasting tolerance to autoantigen. We conclude that cSAgA$_{PLP:LABL}$ offers a promising option for antigen-specific immunotherapy to repress autoimmune disease.

Figure 1. Maximum steady state (max. SS) binding in antibody-labeled splenocytes isolated from EAE and healthy mice determined by flow cytometry, comparing binding of fluorescently-labeled molecules in **(A)** CD19$^+$ B cells, **(B)** CD19$^+$CD11c$^+$ autoimmune-associated B cells (ABCs), **(C)** CD19$^+$ B cells, and **(D)** CD3$^+$ T cells. Splenocytes were isolated from mice at peak of disease and cultured for 72 hr prior to the experiment. Statistical significance was determined by ANOVA followed by TukeyÕs post hoc test with $p<0.05$ and n=3 (*$p<0.05$, **$p<0.01$, ***$p<0.001$, ****p<0.0001).

Figure 2. Costimulatory marker expression was evaluated in antibody-labeled EAE splenocytes following *in vitro* or *in vivo* treatment with cSAgA$_{PLP:LABL}$, compared to EAE and healthy controls. **(A/B)** Splenocytes isolated from EAE mice at peak of disease were treated with cSAgA$_{PLP:LABL}$ for 72 hr ***in vitro*** in the presence or absence of PLP. CD86 expression (anti-CD86 mean fluorescence intensity, MFI) was evaluated by flow cytometry in **(A)** CD19$^+$ B cells and **(B)** CD19$^+$CD11c$^+$B220$^+$ autoimmune-associated B cells (ABCs) to evaluate APC costimulatory signaling. **(C)** EAE mice were treated ***in vivo*** with cSAgA$_{PLP:LABL}$ versus PBS negative control on days 4, 7, and 10, where higher clinical score indicated greater disease severity (n=5). **(D/E)** Splenocytes isolated from mice on day 25 following ***in vivo*** treatment in (C) were rechallenged with/without PLP for 72 hr. CD86 expression was evaluated by flow cytometry in **(C)** CD19$^+$ B cells and **(D)** CD19$^+$CD11c$^+$B220$^+$ autoimmune-associated B cells (ABCs) to evaluate lasting tolerance to autoantigen. Statistical significance (compared to the negative PBS control) was determined by ANOVA followed by TukeyÕs post hoc test with $p<0.05$ and n=3 (*$p<0.05$, **$p<0.01$, ***$p<0.001$, ##/****p<0.0001).

Rowan University

Novel Bio-Based Polyesters & Polycarbonates from Xylochemicals

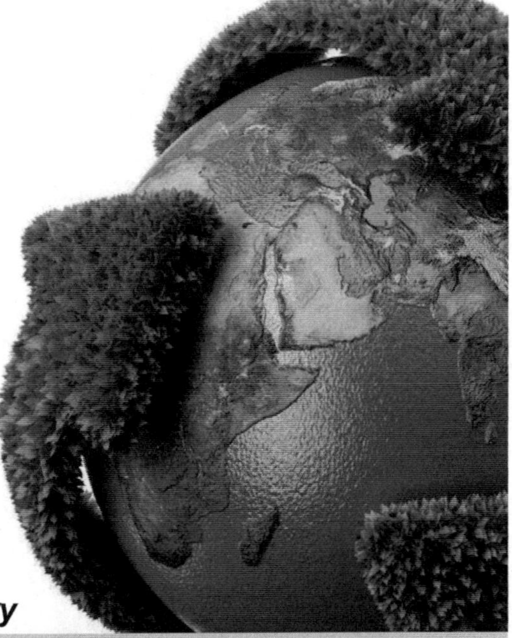

2017 AIChE Annual Meeting
Minneapolis, MN October 29 – November 3 2017
"Bioplastics, Biocomposites and Value-Added Uses of Biofuel Coproducts for Sustainable Manufacturing"

Joseph F. Stanzione, III, Ph.D.
Department of Chemical Engineering
Sustainable Materials Research Laboratory

 1

Plastics: The Facts

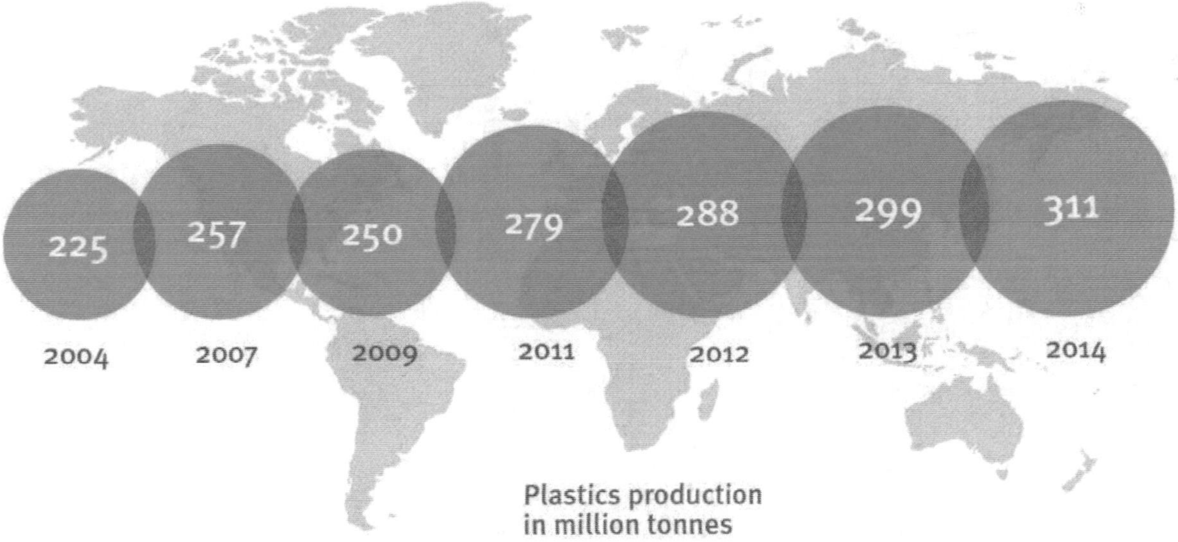

- **Plastic production is growing globally[1]**
- **Currently around 330 Mtonnes**

[1] PlasticsEurope, *Plastics - the Facts* (2015)

 2

Plastics: The Facts

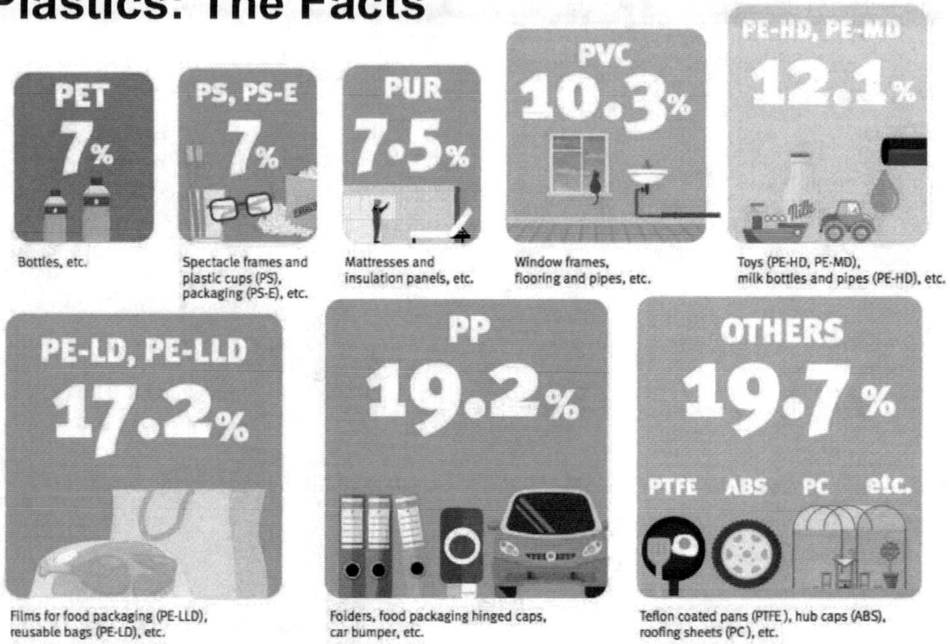

Less than 1.5% are renewably sourced[2]

[2] European Bioplastics, *Bioplastics facts and figures* (2017)

Plastics: The Future

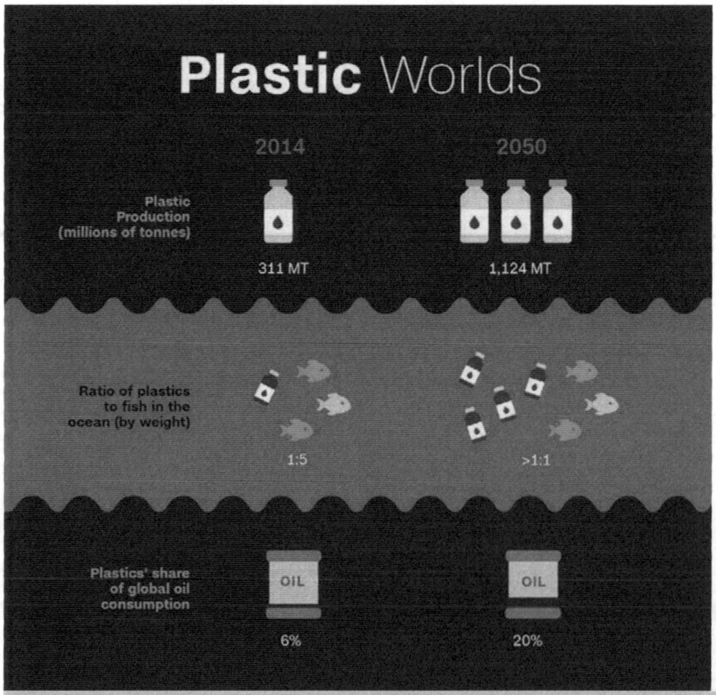

By 2050:

- Expected to triple
- More plastic than fish in the ocean
- 20% of global annual oil consumption

[3] World Economic Forum (2016)

Trying to Replace PET

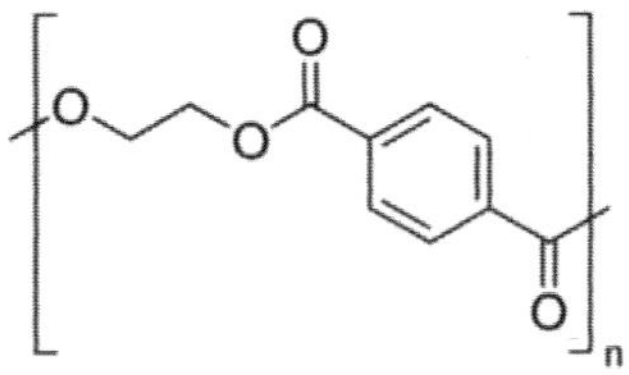

- **Very good mechanical properties**
- **T_g = (67-81) °C, $T_m \approx$ 267 °C**
- **Petroleum-derived (green routes in progress)**
- **Not easily degraded**

High Tg Bio-Based Aliphatic Polyesters from Bicyclic D-Mannitol

Cristina Lavilla, Abdelilah Alla, Antxon Martínez de Ilarduya, and Sebastián Muñoz-Guerra*

Departament d'Enginyeria Química, Universitat Politècnica de Catalunya, ETSEIB, Diagonal 647, 08028 Barcelona, Spain

- Sugar derived
- T_gs range from -40 to 70 °C

Renewable (semi)aromatic polyesters from symmetrical vanillin-based dimers†

Audrey Llevot,[a,b] Etienne Grau,[a,b] Stéphane Carlotti,[a,b] Stéphane Grelier[a,b] and Henri Cramail*[a,b]

- Lignin derived
- T_gs up to 140 °C
- Insoluble in any solvent

Bio-based alternatives in the synthesis of aliphatic–aromatic polyesters dedicated to biodegradable film applications

Nicolas Jacquel [c], René Saint-Loup [c], Jean-Pierre Pascault [a,b], Alain Rousseau [a,b], Françoise Fenouillot [a,b,*]

[a] Université de Lyon, F-69003, France
[b] INSA-Lyon, Ingénierie des Matériaux Polymères, IMP, UMR5223, F-69621 Villeurbanne, France
[c] Roquette Frères, Division Recherche, F-62080 Lestrem, France

- Sugar derived (isosorbide)
- Negative T_gs, high T_ms

etc…!

 5

Additional Concerns: Toxicity

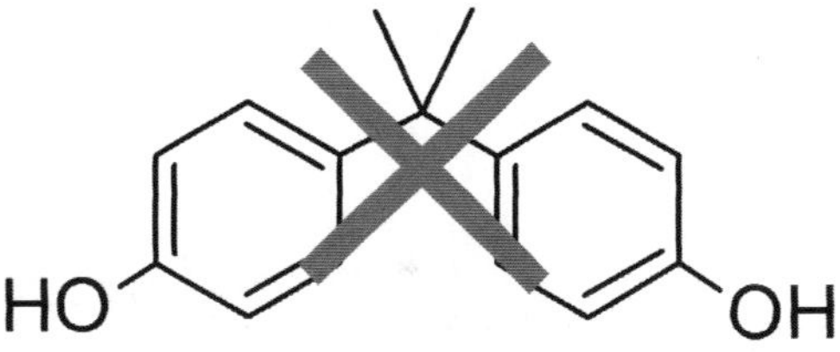

Bisphenol A (BPA)

 6

Bisguaiacol (BG): A Lignin-Derived Bisphenol

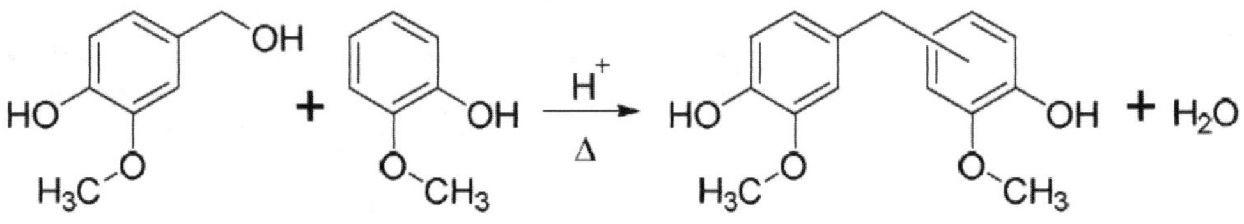

- **Similar structure to BPA[4]**
- **Fully renewable**
- **Reduced toxicity[5]**

- **Isomeric distribution**
  - **82%** ***p,p'***
  - **15%** ***m,p'***
  - **3%** ***o,p'***

[4] Hernandez *et al.*, *ACS Sustain Chem Eng* (2016)
[5] Hong *et al.*, *Int. J. Environ. Res. Publ. Health(* 2016)

 7

Q: Can we use BG to synthesize high-performance renewable thermoplastics?

 8

BG-based Polycarbonates

Joseph Mauck, M.S. CHE

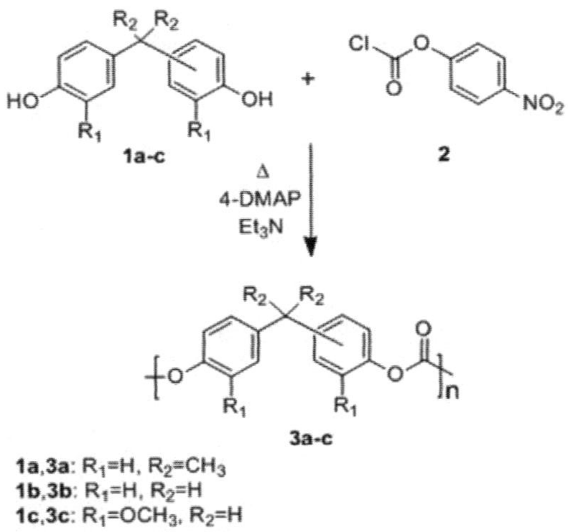

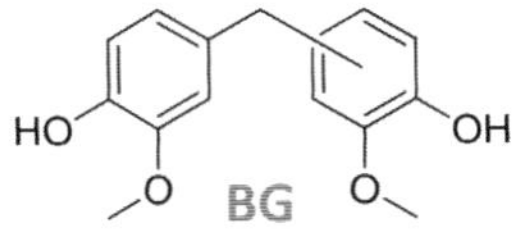

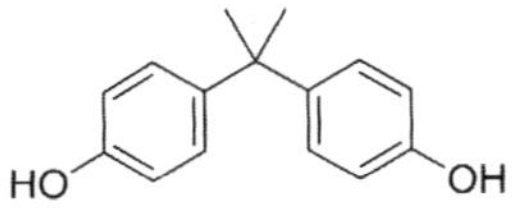

BPA

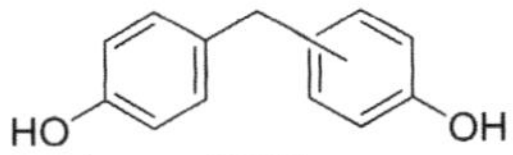

BPF

[6] J. Mauck *et al.*, *J Biobased Materials & Bioenergy* (2017)

BG-based Polycarbonates: Thermal Properties

Sample	*IDT* (°C)	$T_{50\%}$ (°C)	T_{max} (°C)	Char Content (%)	T_g (°C)
BG-PC	375	455	450	24.2	100
BPA-PC	440	543	545	23.2	150
BPF-PC	369	546	501	40.5	119

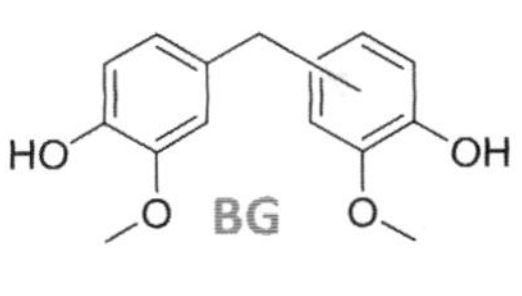

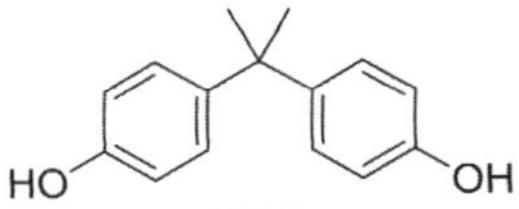

BPA

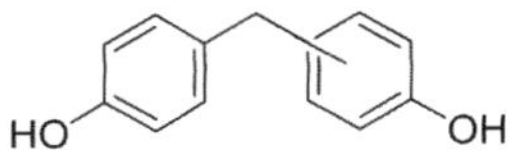

BPF

[6] J. Mauck *et al.*, *J Biobased Materials & Bioenergy* (2017)

Q: Can we use BG to synthesize polyesters?

Preparation of BG-based Polyesters

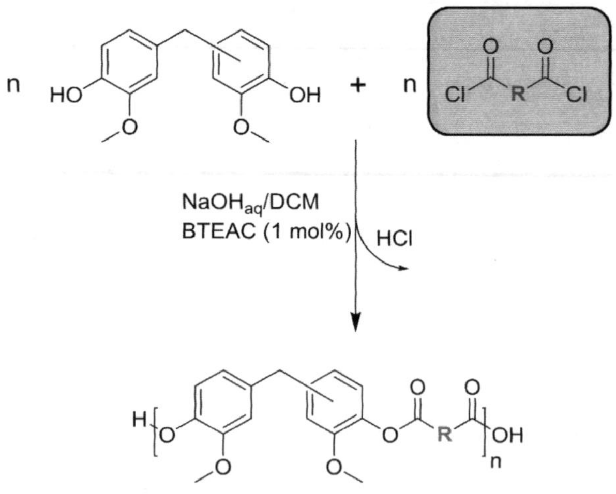

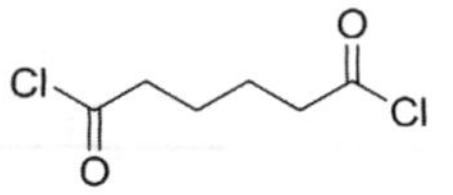

Adipoyl Chloride

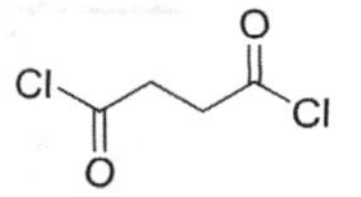

Succinyl Chloride

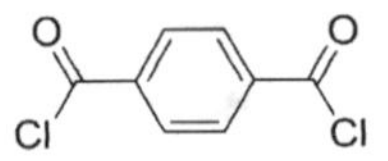

Terephthaloyl Chloride

[7] S. Curia *et al.*, *In preparation* (2017)

BG-based polyesters

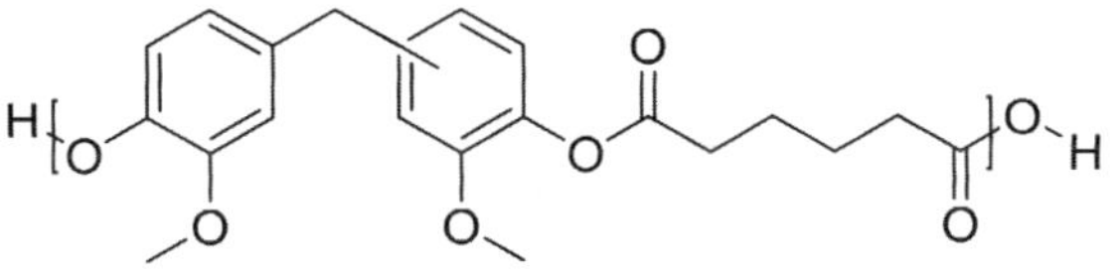

Poly(BG adip)
M_n=6.6 kg mol^{-1} (Đ=1.82)

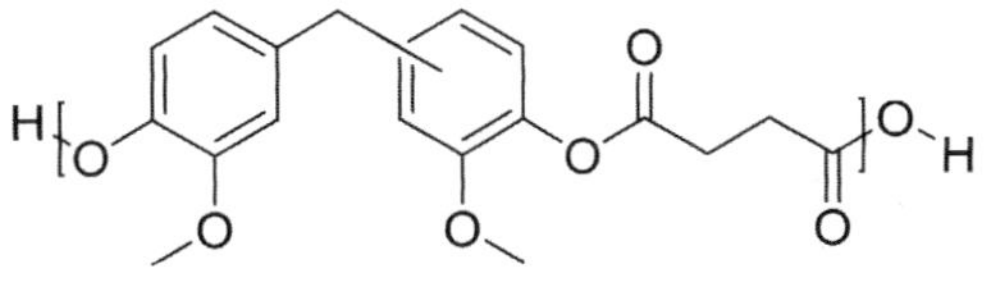

Poly(BG succ)
M_n=7.2 kg mol^{-1} (Đ=2.16)

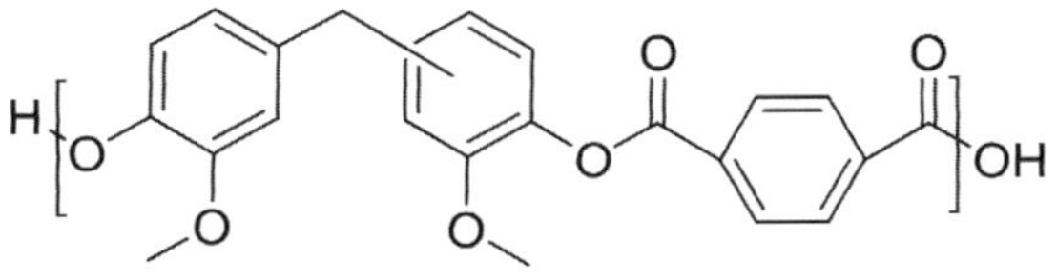

Poly(BG tere)
M_n=26.9 kg mol^{-1} (Đ=2.29)

[7] S. Curia *et al.*, *In preparation* (2017)

IR Analysis

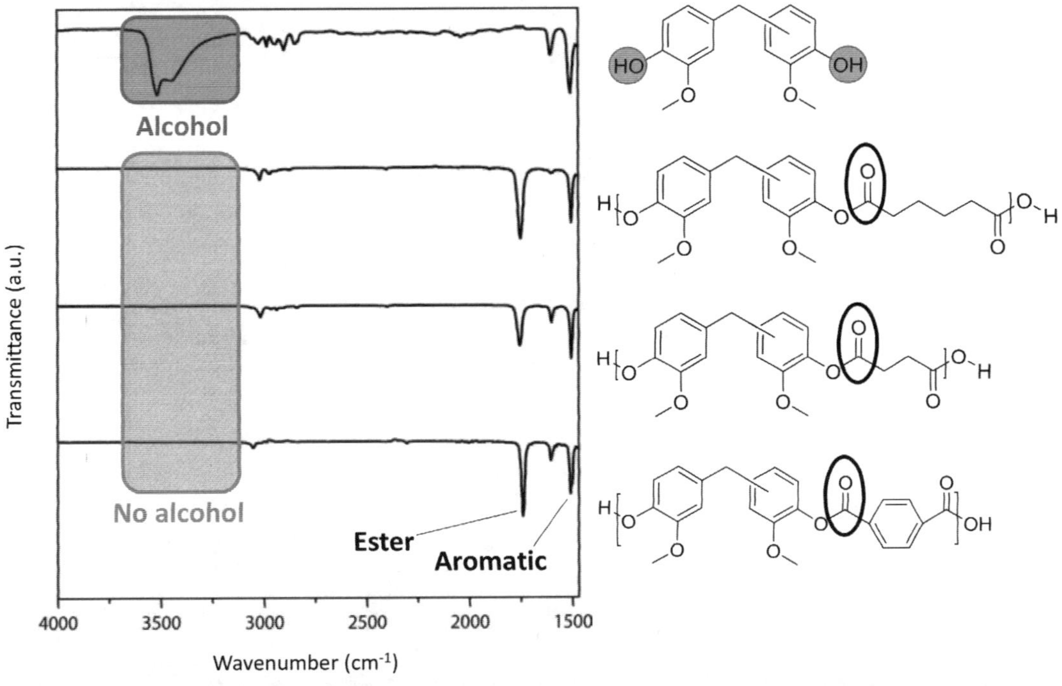

^{1}H-NMR Analysis (in $CDCl_3$)

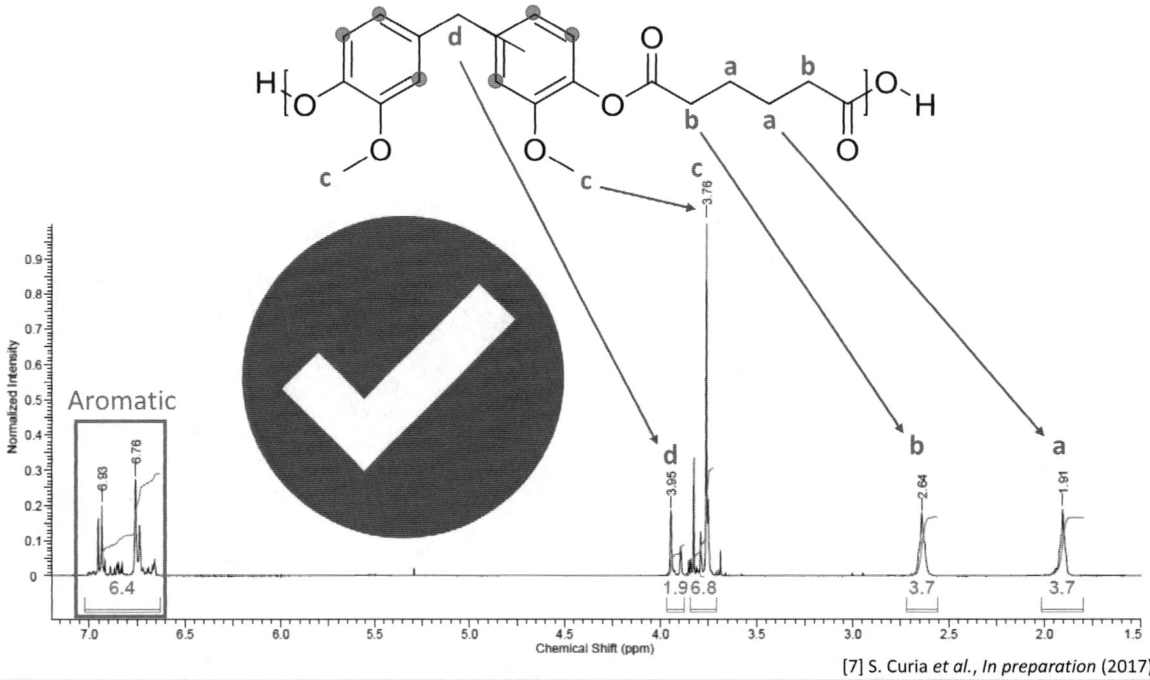

[7] S. Curia *et al.*, *In preparation* (2017)

BG-based Polyesters: DSC Analysis

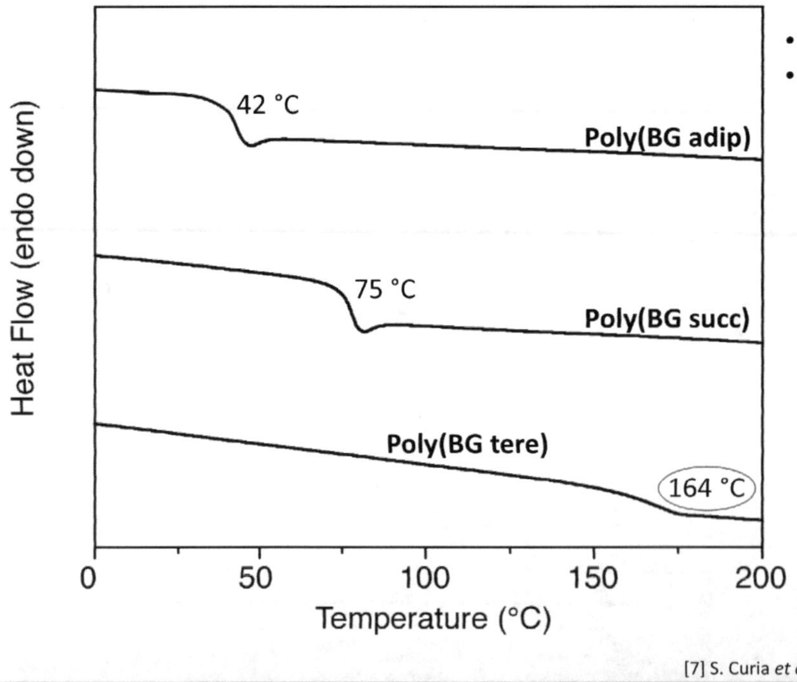

- **10 °C min^{-1}**
- **N_2 50 mL min^{-1}**

[7] S. Curia *et al.*, *In preparation* (2017)

BG-based Polyesters: Thermal Stability (N_2)

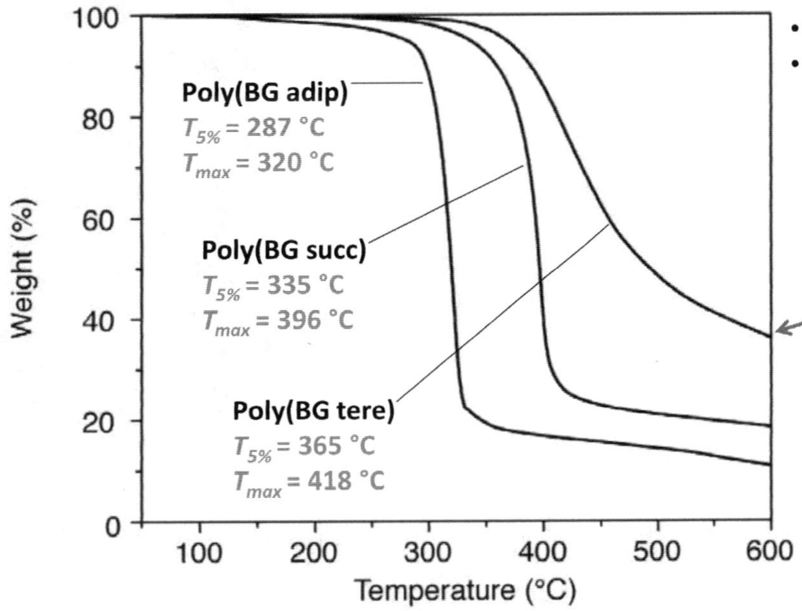

- **10 °C min^{-1}**
- **N_2 25 mL min^{-1}**

36% char @ 600 °C

[7] S. Curia *et al.*, *In preparation* (2017)

 17

BG-based Polyesters: Thermal Stability (Air)

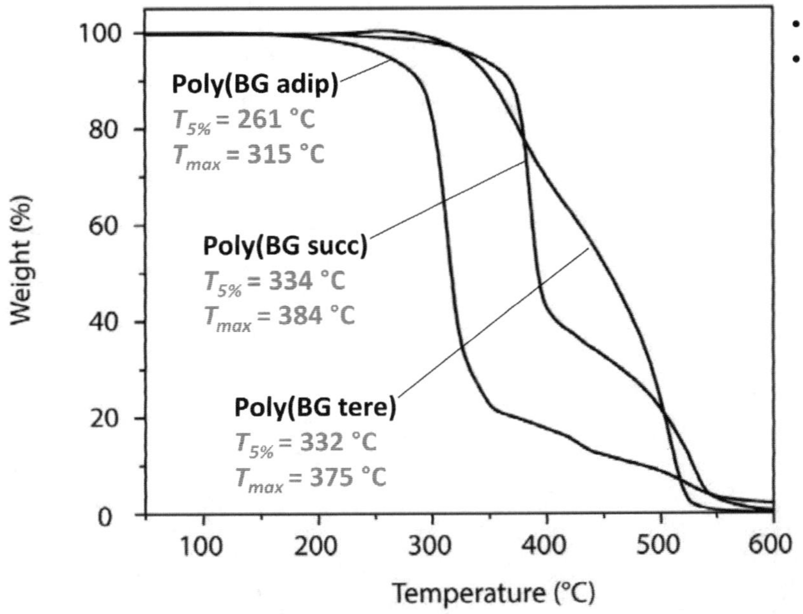

- **10 °C min^{-1}**
- **Air 25 mL min^{-1}**

T_{max} still above 300 °C

[7] S. Curia *et al.*, *In preparation* (2017)

 18

BG+FDCA: a Fully Renewable Polyarylate

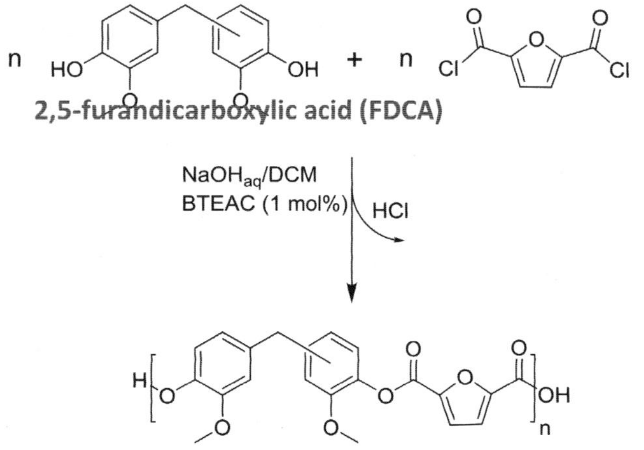

Poly(BG FDCA)
M_n=6.3 kg mol^{-1} (Đ=1.57)

[7] S. Curia *et al.*, *In preparation* (2017)

Poly(BG-FDCA): DSC

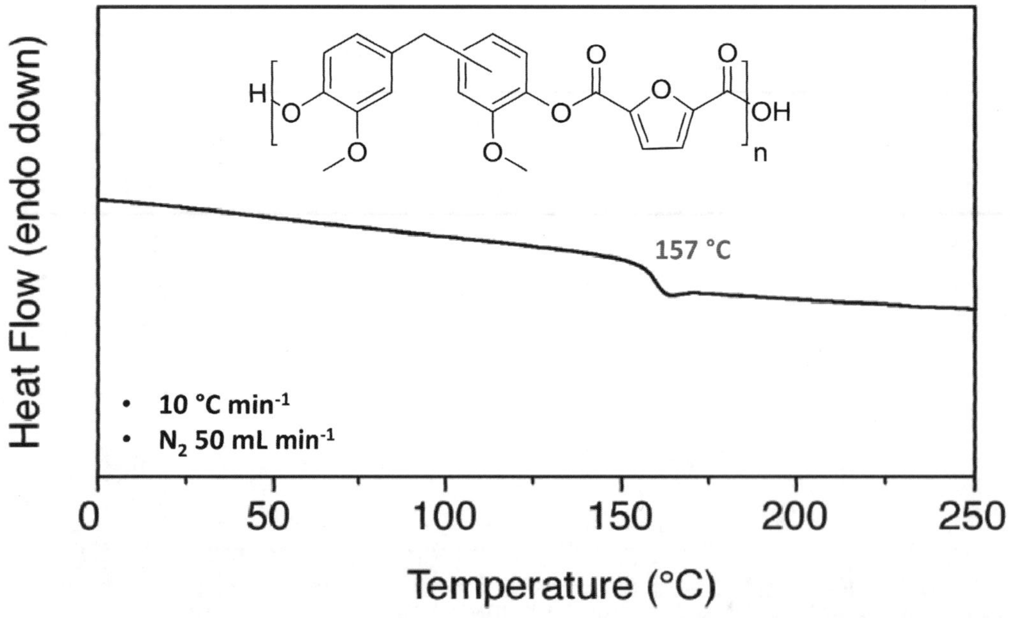

[7] S. Curia *et al.*, *In preparation* (2017)

Poly(BG-FDCA): TGA N_2 *vs* air

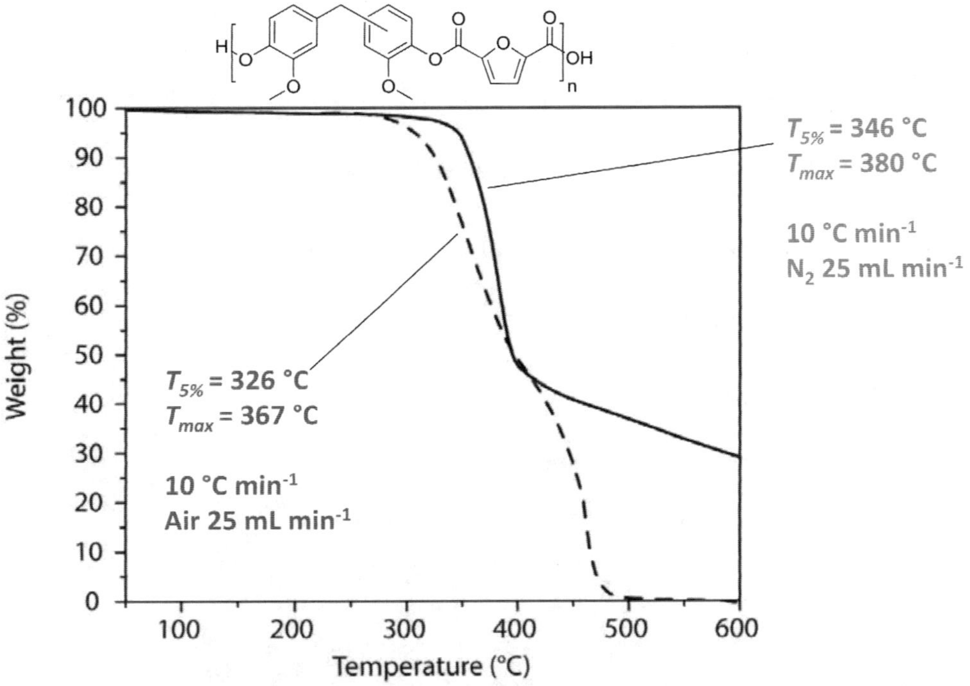

[7] S. Curia *et al.*, *In preparation* (2017)

Conclusions

- **BG** can be used to synthesise **useful polycarbonates** & **polyesters**
- **BG-based polyesters** show **high thermal resistance** and **tuneable T_g**
- **Poly(BG FDCA)** has a remarkably **high T_g (157 °C)**

Xylochemicals — High T_g Bio-Based Thermoplastics

Future Investigations

- Hydrolytic degradation & biodegradation tests
- End-cap with functional molecules
- Chain extension, melt polymerization (achieve higher MW)
- Gas permeability/diffusivity
- Mechanical tests

Xylochemicals High T_g Bio-Based Thermoplastics

 23

Acknowledgements

- Sustainable Materials Research Laboratory
- Joseph R. Mauck
- Silvio Curia, Ph.D.
- John J. La Scala, Ph.D. (ARL)
- Rowan University
  - ❖ Department of Chemical Engineering
  - ❖ Department of Chemistry & Biochemistry
- U.S. ARL for Financial Support via Cooperative Agreement W911NF-16-2-0225

Silvio Curia, Ph.D.

Rowan University

ARL U.S. ARMY RDECOM

South Jersey Technology Park at Rowan University

 24

Thank you!
Any questions?

Crater Lake National Park

Rocky Mountain National Park

503981 Light Weight Biocomposite from Toughened Polyolefin and Biocarbon

Abstract:

In converting biomass to bioenergy through thermochemical processes, a considerable amount of solid residue generates in each conversion cycle (up to 35 wt.%) which mainly consists of renewable carbon (biocarbon). However, the biocarbon is considered as a byproduct of the pyrolysis units. Unless a high-volume application exists for the biocarbon, the significant amount of this byproduct can challenge the long-term sustainability of pyrolysis units. In this project, biocarbon is intended to be used as filler in plastic industry targeting toughened polypropylene (PP) based automotive parts. In order to achieve a suitable balance of mechanical properties, biocomposites were compatibilized using a maleated polypropylene. Furthermore, the PP phase was nucleated using a Î² nucleating agent to increase the impact toughness without significant loss in the stiffness. Tensile properties and notched Izod impact strength were measured as response variable. Atomic force microscopy (AFM) was utilized to analyze the PP-biocarbon interface in presence and absence of the compatibilizer. The effect of the nucleating agent was investigated by differential scanning calorimetry (DSC) and correlated to the improvement of biocomposites impact toughness. The results suggest that the morphological changes caused by the compatibilizer and toughening effect of Î² crystals can generate high impact toughness in composites with 20 wt.% loaded biocarbon. Therefore, the biocarbon can be utilized in high loading levels similar to current petroleum-based composites with comparable mechanical properties. At the end, substituting conventional mineral filled composites with biocarbon-based biocomposites can provide the required mechanical properties with added benefits of lower density (â‰ˆ15%), 30% higher bio-based content and cheaper price. These benefits would translate into improved fuel economy of cars and reduction in greenhouse gas production. Considering the high-volume usage of plastic parts especially in automotive industry, this approach would make the biomass to bioenergy conversion more sustainable.

Keywords: Biocarbon, Toughened biocomposites, Polypropylene.

Acknowledgement: We gratefully acknowledge the financial support of the Ontario Ministry of Agriculture, Food and Rural Affairs (OMAFRA)/University of Guelph - Bioeconomy for Industrial Uses Research Program Theme Project # 030055; Ontario Research Fund, Research Excellence Program; Round-7 (ORF-RE07) from the Ontario Ministry of Research, Innovation and Science (MRIS) (Project #052644 and 052665), and the Natural Sciences and Engineering Research Council (NSERC), Canada Discovery Grant Project # 401111 to carry out this research work

497443 Secondary Fermentation of Corn Ethanol Co-Products for Improved Amino Acid Qualities

Abstract:

Secondary Fermentation of Corn Ethanol Co-Products for improved Amino Acid Qualities

In 2016 5.28 billion bushels of corn were used to produce about 14.79 billion gallons of ethanol in the United States. As a result, about 36 million tons of Dried Distillers Grains with Solubles (DDGS) were manufactured and fed to livestock (U.S. Dept. of Energy, 2017). DDGS are a common feed supplement in cattle rations as it is inexpensive and has positive feeding characteristics. One of the drawbacks of DDGS is that it is lacking in in key amino acids such as tryptophan, arginine, and lysine. Historically these amino acids have been supplemented by external addition of feed grade amino acids to rations. The research carried out attempts to fortify DDGS with higher amounts of tryptophan through secondary fermentation of Whole Stillage (WS) by bacteria. When cultured on WS the bacteria secretes extracellular tryptophan, which serves to close the amino acid gap in corn ethanol co-products. The bacteria used is Generally Regarded as Safe (GRAS) and has been used to produce feed ingredients historically. Because of the research, profitability of corn ethanol operations will be increased due to the improved feeding value of the co-products produced.

References

United State Department of Energy, â€œEthanol Fuel Basics,â€ Energy Efficiency & Renewable Energy http://www.afdc.energy.gov/fuels/ethanol_fuel_basics.html 03/16/2017

500583 Extraction High-Value Chemicals from Ethanol Co-Products: A Feasibility Assay on Phytate Extraction with Life Cycle and Techno Economic Assessment

Abstract:

Background

The well-established dry-grind process today accounts for also over 80% of the operating ethanol facilities in the United States, producing not only ethanol, but also distiller's grains. Distiller's grains, commonly commercialized as dried distiller's grains with solubles (DDGS) accounts for most of the heavy fraction from the ethanol distillation and rectification operations, and is processed through a complex downstream processing after the fermentation is unloaded. Despite being a nutritious animal feed, inclusion of DDGS as feed ration is still limited due to compositional properties, such as the high presence of non-starch polysaccharides, which are non-digestible to monogastric animals, the fatty acid profile, and the overdose of phosphorus, which is largely non-available to some animals. Phosphorus behavior in ethanol downstream processing is of particular interest, not only because it can lead to environmental issues, but also because it can potentially be an additional economic source to ethanol producers. Over 40% of the phosphorus present in ethanol co-products is present as phytate, the main form of organic phosphorus in cereals and grains, which is not completely hydrolyzed during the whole ethanol process. Phytate has a high market value, and today is a platform chemical utilized in a wide array of industries, and it is mostly exploited because of its high chelating potential (He et al. 2017). Phytate is also the raw material to produce *myo*-inositol, a cyclic polyol with a variety of uses by pharmaceutical and nutraceutical companies.

Phytate is found to be dissolved on the ethanol co-products(He et al. 2017). During the dry-grind downstream processing, thin stillage, i.e., the liquid-rich fraction derived from the centrifugation of whole stillage, accounts for the highest concentration of phytate among all the co-products. Recent research (He et al. 2017) has proposed an extraction method using ion exchange separation, which can efficiently separate phytate from thin stillage, with little to no effect on the residual nutritional profile of the final DDGS. Even though the concentration of phytate being relatively low on ethanol co-products, at around 1 g L^{-1} in thin stillage, due to the enormous volumetric production by dry-grind plants, the production leveraging costs in extracting phytate might become an industrial alternative.

The high volumes of DDGS produced by ethanol plants, alongside with its limited dietary inclusion, pose a challenge for higher revenues to ethanol plants. The second-most important co-product from dry-grind ethanol facilities, distiller's corn oil, represent today a mature product suitable for biodiesel production and other uses of low-grade vegetable oil(Reis et al. 2017). Removal of distiller's corn oil, on the other hand, has been criticized by animal nutritionists claiming a loss in energy value from DDGS. Despite still representing about 20% of the total revenue stream from dry-grind ethanol plants, DDGS are still ranking in low positions when compared to higher end animal feeds, such as soybean meal. The extraction of high-value chemicals through mature sound technologies may represent, thus, an alternative for higher economic security of ethanol plants. This current research describes a techno-economic analysis and a life-cycle assessment of a phytate extraction process on a large-scale dry-grind corn-to-ethanol plant in the United States, and its posterior *in-loco* conversion to *myo*-inositol.

Methods

The scope of this work analyzed the utilization of corn-to-ethanol thin stillage produced through the dry-grind process as raw material to extract phytate using an ion exchange process, followed by hydrolysis to generate *myo*-inositol. The process was modelled according to He et al. (2017) , and by additional empirical analysis, when required. Thin stillage was obtained from a dry-grind ethanol plant in the state of Iowa, USA. The process of converting thin stillage to a phytate brine solution is divided into the major steps: (1) filtration, (2) flow through ion exchange column, (3) desorption of phytate using brine solution, (4) reconditioning of column using sodium hydroxide, and (5) column washing using water. The conversion of phytate brine solution to calcium phytate is done by (6) addition of calcium chloride and sodium hydroxide, followed by product filtration, and production of *myo*-inositol is achieved by (7) hydrolysis of phytate brine solution, (8) precipitation of phosphate with calcium chloride and sodium hydroxide, and (9) evaporation and drying of *myo*-inositol.

The life cycle assessment of calcium phytate extracted from corn-to-ethanol thin stillage was estimated using GREET model. Calcium phytate extraction, and posterior conversion to *myo*-inositol were implemented on two stationary models on GREET – Dry mill ethanol production w/ corn oil extraction, and Dry mill ethanol production w/o corn oil extraction. Techno-economic assessment for ethanol plants was considered using the BuGal model, with added market and technical data for the modified phytate extraction and conversion units. Since the basis of the BuGal model for techno-economic analysis has been modified to analyze both conditions (with and without distiller's corn oil extraction), the two stationary models on GREET to describe corn-to-ethanol process were utilized.

Results and discussion

LCA results indicate that the proposed phytate extraction does not impact significantly most of the environmental impact factors of ethanol. As a gallon of ethanol was considered as functional unit, the results indicate that for a system with phytate extraction and oil extraction, CO_2 emissions will increase from 3.21 kg to 3.28 kg per gallon of ethanol, whilst for a system without distiller's corn oil extraction, 3.23 kg of CO_2 will be increased to 3.30 kg of CO_2 per gallon of ethanol produced. NOx, SOx, N2O, and other gases also increased with phytate extraction by a factor of about 1%. Analyses were carried out for techno economic assessment in 4 different cases, (1) with distiller's corn oil extraction and phytate extraction, (2) with distiller's corn oil extraction and without phytate extraction, (3) without oil extraction and without phytate extraction extraction, (4) without oil extraction and with phytate extraction, and were consistently all across different cases.

Unlike major recent upgrades to the dry-grind ethanol industry, such as distiller's corn oil removal, extraction of phytate and production of *myo*-inositol are related to products found in dilute concentrations. The low concentration of phytate in ethanol co-products requires low energy input on an ethanol basis, while still providing a high-value co-product. BuGal model calculates the operating expenses per gallon of denatured ethanol sold as USD 0.2673 for the control conditions, i.e., without phytate extraction, and USD 0.2686 for a system considering 20% of thin stillage being redirected to phytate extraction, due to increase in energy consumption for additional processing. Regarding chemical costs, phytate extraction and conversion to inositol likely increased the value from USD 0.1201 to USD 0.1226 per gallon of denatured ethanol, mainly due to the requirement of NaCl for phytate desorption from the ion exchange process, additional process water, and calcium chloride addition. These costs are balanced by the calcium phytate, *myo*-inositol, and calcium phosphate as fertilizer (derived from the phytate hydrolysis) sales, at prices estimated at USD 12, USD 10, and USD 0.20 per kg, respectively. BuGal estimates an average large-scale dry-grind operation, with factor of equity and debt being 0.7 and 0.3, respectively, with an interest rate charged on debt as 0.07, and the rate of return required by investors on the plant's equity as 0.12. Even though these numbers are variable, the economic considerations on such assumptions are beyond the scope of this work, and remain constant through all assumptions. Following similar numbers as Kwiatkowski et al. (2006), 2.795 gallons of ethanol are estimated to be produced from each bushel of corn, and 17.5 lb of

DDGS, and CO_2 are produced on a bushel basis. Denatured ethanol sales are estimated to be at USD 1.70 per gallon, an estimate within current market prices , and that phytate extraction would not affect the price of DDGS, thus, being marketed at a price of USD 161 per ton, also within current market estimates.

Conclusion

The removal of phytate from thin stillage may represent a novel process in the dry-grind ethanol industry. The removal of phytate via ion exchange does not increase significantly (around 1% of increase) the emissions involved on the production of one gallon of ethanol, and that it can be a source of profit if operated within a certain range of thin stillage processed within a plant. The results also presented herein describe the potential of converting the extracted phytate to myo-inositol. Myo-inositol has a more flexible market when compared to phytate, and could also represent another source of revenue for ethanol plants.

References

Kwiatkowski JR, McAloon AJ, Taylor F, Johnston DB (2006) Modeling the process and costs of fuel ethanol production by the corn dry-grind process Industrial crops and products 23:288-296

He Q, Rodrigues Reis, CE, Wang F, Hu B (2017) Phytate extraction from coproducts of the dry-grind corn ethanol process. RSC Advances. doi:10.1039/C6RA27409A

Reis CER, Rajendran A, Hu B (2017) New technologies in value addition to the thin stillage from corn-to-ethanol process Reviews in Environmental Science and Bio/Technology 16:175-206 doi:10.1007/s11157-017-9421-6

494042 Highly Stable Perovskite Solar Cells Fabricated Using Aerosol-Based Technique

Abstract:

Organometallic perovskite (based on $CH_3NH_3PbI_3$ and its analogues) have attracted substantial attention due to their low-cost synthesis and various applications. Perovskite-based solar cells have shown rapid increase in the efficiency over a short period of time. However, material's instability under ambient condition is a key challenge for their large-scale applications. Previous studies focused on improving the stability use additives or encapsulation methods and are performed in controlled environment, which restricts the large-scale production of perovskite solar cells. We present a simple, scalable and novel aerosol-based technique, electro-hydro atomization, to fabricate highly stable perovskite solar cells under ambient conditions (room temperature, humidity and atmosphere), without use of any additives or encapsulation. A two-step deposition method is used to form the perovskite layer, where PbI_2 is spin coated and CH_3NH_3I (MAI) is electro-sprayed on the PbI_2 layer. The electro-sprayed perovskite film was then annealed at 100 °C for 5 min. A detailed investigation of the effect of various electrospray parameters such as, deposition time, MAI concentration, substrate-to-nozzle distance, and flow rate (droplet size), on the performance of the solar cells is carried out.

The key features of the electrospray-based deposition are: 1) precisely controlled reaction between the two precursors, by gradually supplying MAI nanoparticles onto the PbI_2 layer and 2) intermediate phase formation during the electrospray of MAI nanoparticles. Both of these characteristics lead to the formation of ultra-smooth and moisture-resistant perovskite film. Smoothness of the film is examined by atomic force microscopy (root mean square roughness is 31 nm for electrosprayed perovskite film and 58.9 nm for spin coated film over a large area of 50 µm × 50 µm) and moisture-resistance is examined by contact angle measurement (66.09° for electrospray film and 33.01° for spin coated sample). The intermediate phase formation is characterized by X-ray diffraction (XRD) and change in the XRD pattern is observed during annealing. The cells fabricated using the electrospray are more efficient (up to 12%) and highly stable (up to 5.5 months) under ambient condition than the cells prepared using a conventional spin coating method. Additionally, the electrospray-deposited film exhibits self-healing behavior when exposed to moisture. Material stability is also investigated by monitoring the amount of PbI_2 generating from the perovskite degradation and is correlated to the device stability. Electrospray also provides precise control over the material required to form the perovskite layer, reducing material loss. Therefore, electrospray technique is promising for the large-scale fabrication of stable and efficient perovskite solar cells under ambient condition. Details of the electrospray-assisted system and highly stably perovskite formation using this technique will be presented.

Reference:

Kavadiya, S. et al., Advanced Energy Materials, 10.1002/aenm.201700210

501620 Engineering Perovskite Solar Cell Interfaces to Realize > 1000 Hr, Unencapsulated Ambient Stability

Abstract:

Because the power conversion efficiency of organic-inorganic halide perovskite solar cells has reached parity with commercial thin film photovoltaic absorbers, their future commercial prospects appear to hinge on stability. Much recent work has focused on the intrinsic stability of the $APbX_3$ perovskite absorber materials (A = $CH_3NH_3^+$, $CH(NH_2)_2^+$, Cs^+, Rb^+ and X = Cl^-, Br^-, I^-). Recent work has provided insight into the moisture instability,[1] thermal instability,[2] phase instability,[3] and phase segregation[4] of halide perovskite materials. To some extent, improved perovskite stability has translated to improved device stability, yet further gains are still necessary, motivating a systematic look at the interfaces of the device stack.

In this study, we systematically investigate the interfaces in the typical n-i-p perovskite solar cell, identify key degradation mechanisms, and systematically engineer these interfaces to improve operational stability. Replacing spiro-OMeTAD with a Li^+-free hole transport layer (HTL) material, EH44,[5] we can achieve power conversion efficiencies of approximately 18% in a standard TiO_2/perovskite/HTL/Au device stack. Unencapsulated devices utilizing EH44 show a factor of 4 better ambient operational stability compared to spiro-OMeTAD devices, after a similar initial burn-in of the devices. Analyzing the initial burn-in, Time of Flight Secondary Ion Mass Spectrosocpy (ToF-SIMS) on the device stack shows a redistribution of the perovskite active layer during the initial burn-in which is independent of the HTL. This redistribution, and the corresponding burn-in, is driven by interface effects and is significantly reduced when the TiO_2 electron transport layer (ETL) is replaced with SnO_2. When combined with MoO_x/Al electrodes,[6] this device stack allows for a ~3 order of magnitude increase in the operational stability of unencapsulated devices in ambient conditions. We observe only a 12% decrease in efficiency after 1000 hrs of continual ambient operation (6% degradation in best device). The ability of these devices to withstand the combined stresses of UV-light, oxygen, and moisture, demonstrates the importance of carefully designed interfaces for realizing true long-term stability.

References

(1) Christians, J. A.; Miranda Herrera, P. A.; Kamat, P. V. *J. Am. Chem. Soc.* **2015**, *137* (4), 1530.

(2) Nenon, D.; Christians, J. A.; Wheeler, L. M.; Blackburn, J.; Sanehira, E. M.; Dou, B.; Zhu, K.; Berry, J. J.; Luther, J. M. *Energy Environ. Sci.* **2016**, *9*, 2072.

(3) Li, Z.; Yang, M.; Park, J.; Wei, S.; Berry, J. J.; Zhu, K. *Chem. Mater.* **2016**, *28*, 284.

(4) Yoon, S. J.; Draguta, S.; Manser, J. S.; Sharia, O.; Schneider, W. F.; Kuno, M.; Kamat, P. V. *ACS Energy Lett.*

2016, *1* (1), 290.

(5) Leijtens, T.; Giovenzana, T.; Habisreutinger, S. N.; Tinkham, J. S.; Noel, N. K.; Kamino, B. A.; Sadoughi, G.; Sellinger, A.; Snaith, H. J. *Appl. Mater. Interfaces* **2016**, *8*, 5981.

(6) Sanehira, E. M.; Tremolet de Villers, B. J.; Schulz, P.; Reese, M. O.; Ferrere, S.; Zhu, K.; Lin, L. Y.; Berry, J. J.; Luther, J. M. *ACS Energy Lett.* **2016**, *1*, 38.

500570 A Thermodynamic Basis for Engineering Enhanced Stability of the Perovskite Phase of Cesium Lead Iodide

Abstract:

The perovskite phase of cesium lead iodide (α-$CsPbI_3$ or "black" phase), possesses excellent absorptivity, charge carrier mobility and recombination kinetics for photovoltaic applications. However, the stable phase at room temperature is a non-functional, "yellow" phase (δ-$CsPbI_3$). Using differential scanning calorimetry, it is shown herein that the desired black phase can be captured in a metastable state that is maintained in the absence of moisture, up to a temperature of 100°C. A reversible phase-change enthalpy of 14.2 (± 0.5) kJ/mol is measured, implying a large entropic stabilization of the black phase. In order to engineer enhanced stability of the perovskite phase, particularly toward atmospheric moisture, this entropic contribution to the stability is intentionally maximized through nanostructuring of the thin film. These results enhance the utility of this promising photovoltaic material and contribute more generally to our fundamental understanding of the effects of common processing steps on the larger family of lead-halide perovskites.

494782 Synthesis of Nanostructured Inorganic Perovskites for Solar Cell Applications

Abstract:

The fabrication of stable, efficient, and low-cost solar cells has been an ongoing challenge in the field of photovoltaic. Perovskite materials have been able to partly address these characteristics. In particular, lead-based hybrid organic-inorganic perovskites have demonstrated high power conversion efficiency (up to 22%). However, these materials suffer from poor stability in the presence of oxygen and humidity, and upon exposure to light and heat. To solve this, we are trying to synthesize layered 2D and quasi-2D structures of $CsPbI_3$ perovskites. We implemented bulky organic cations such as PEA^+ and nBA^+ in the structure of $CsPbI_3$ perovskites to stabilize their more efficient phase (cubic) and protect them from degradation by moisture and oxygen. Our absorbance data suggest the improved stability of thin-films prepared by such materials. Addressing the stability issues of perovskites without compromising their efficiency could be the first step toward their commercialization for solar cell applications.

501447 Discovery of Near-Infrared-Active Colloidal Multinary Lead Halide Perovskite Nanocrystals Using a Microfluidic Platform

Abstract:

Lead-based metal halide perovskites, $APbX_3$ structures, with organic-inorganic A-site cations (A = methylammonium (MA) $CH_3NH_3^+$, formamidinium (FA), $CH_3(NH_2)_2^+$, Cs, Rb) and an anion (X = Cl, Br, I) have opened exciting opportunities in the perovskite-based solar cell devices with unprecedented efficiencies which reach up to 22 %.[1] However, red and near infrared regions still present a challenge for perovskite materials: all three archetypical A-site monocationic perovskites – $CH_3NH_3PbI_3$, $FAPbI_3$ and $CsPbI_3$ – suffer from either chemical or thermodynamic instabilities. A promising avenue towards mitigation of these challenges lies in the formation of multinary compositions.[2] Recently, lead based perovskites with mixed-cations on the A site have shown great potential for the fabrication of thermally stable perovskite solar cells with high photoelectric conversion efficiencies.[3] In particular, the addition of Cs to FA-based perovskites generally influence the stability and optoelectronic properties of their pure perovskite counterparts.[4] The intrinsic challenge in the synthesis of mixed-cation and mixed- halide NCs is to identify the effect of various reaction parameters such as precursor ratios, ligand concentrations, temperature and reaction time on optical and structural characteristics. Recently, we demonstrated the use of microfluidic reactors for the rapid parametric screening and on-line characterization of the optical characteristics of cesium lead halide perovskite NCs while identifying the optimum reaction parameters for their synthesis.[5]

Herein, we demonstrate for the first time the formation of stable $Cs_xFA_{1-x}Pb(Br_{1-y}I_y)_3$ NCs with a concurrent band gap tuning between 690 and 775 nm while presenting the underlying reaction mechanisms using a fully automated droplet-based microfluidic platform. The synthesized Cs/FA perovskite compounds with narrow emission linewidths (45 – 65 nm) overcome the instability issues facing $CsPbX_3$ and $FAPbX_3$ NCs, exhibiting finely stable emission and absorption spectra over long term storage (several weeks). The combination of online photoluminescence and absorption measurements allows for identifying compositional and size effects on the optical properties of the synthesized NCs. Last, the reactions can be directly transferable in conventional laboratory reactors for the large-scale synthesis of $Cs_xFA_{1-x}Pb(Br_{1-y}I_y)_3$NCs positioning them as strong candidates for the fabrication of solar cells.

1. NREL Chart. http://www.nrel.gov/pv/assets/images/efficiency_chart.jpg (05.01.2017),
2. Li, Z.; Yang, M. J.; Park, J. S.; Wei, S. H.; Berry, J. J.; Zhu, K. Chem. Mater. 2016, 28, 284-292.
3. Wang, Z.; Shi, Z.; Li, T.; Chen, Y.; Huang, W. Angew. Chem Int. Ed. 2017, 56, 1190-1212.
4. McMeekin, D. P.; Sadoughi, G.; Rehman, W.; Eperon, G. E.; Saliba, M.; Horantner, M. T.; Haghighirad, A.; Sakai, N.;

   Korte, L.; Rech, B.; Johnston, M. B.; Herz, L. M.; Snaith, H. J. Science 2016, 351, 151-155.
5. Lignos, I.; Stavrakis, S.; Nedelcu, G.; Protesescu, L.; Demello, A. J.; Kovalenko, M. V. Nano Lett. 2016, 16, 1869-1877.

495011 Improved Charge Collection in Highly Efficient $CsPbBrI_2$ Solar Cells with Light-Induced Dealloying

Abstract:

The meteoritic rise of solar cell efficiencies based on low-cost solution processed metal halide perovskites (MHPs) is raising an exciting possibility of accelerating wide-spread usage of solar power. To realize this potential, the poor stability of MHP solar cells has been identified as a major challenge to be overcome. The majority of MHP solar cells to date have been based on hybrid organic-inorganic MHPs that undergo thermal degradation at temperatures which can be commonly reached at geographic locations. Moreover, interaction between the organic moieties and water has been identified as a predominant degradation pathway.

Recently, all-inorganic, Cs^+ ion-based lead halide perovskites have been proposed as a means to overcome the instability issues of organic cations in hybrid MHPs. In this talk, I will present our results that show the importance of halide dealloying in $CsPbBrI_2$ perovskite solar cells in solar cell performance. This dealloying occurs under steady illumination and is a reversible process, suggesting transitions between equilibrium states under dark and illuminated environments. Through decoupling charge collection in the electron- and hole-transporting layers, our photoluminescence, X-ray diffraction, and solar cell device characterization results suggest that the light-induced halide dealloying improves hole collection in solar cells, resulting in increased efficiency (9.2% ± 0.64 on average with a champion efficiency of 10.3%). Our results provide deeper insights on the impact of dealloying on perovskite solar cell performance and highlight the growing potential of all-inorganic perovskite solar cells.

495064 Synthesis and Characterization of Bi-Based Perovskite Semiconductors for Photovoltaic Application

Abstract:

Recent reports of efficient solar cells and other optoelectronic devices based on halide perovskite semiconductors has generated broad interest in these materials; however, the halide perovskites have major issues related to their stability and toxicity. For this reason, Pb-free halide perovskites are being explored as nontoxic, and more stable alternatives. Within this family of semiconductors, Bi-based halide perovskites offer the potential to have similar optoelectronic performance as Pb-halide perovskites due to their similar electronic structures. Our recent findings demonstrate that Bi-halide perovskites have superior thermal and environmental stability compared to methylammonium lead iodide. Proof-of-concept devices demonstrate the potential of Bi-based perovskites for photovoltaics and other optoelectronic applications. We demonstrate that the morphology of Bi-halide films can be improved by annealing in solvent vapor, resulting in increased current density in photovoltaic devices. The improved understanding in the chemistry and processing of Bi-based semiconductors is a step towards next-generation optoelectronic devices with the performance and stability necessary for widespread commercialization.

493590 In-Situ Synthesis of Intergrown Uio-66 Membranes with Controlled Orientation

Abstract:

Metal-organic frameworks (MOFs) have attracted enormous attention from researchers due to its tremendous chemical and structural diversities, which span a large variety of properties and pore sizes. In the synthesis of ZrMOF-based film/membrane, one of the most significant challenges is achieving full intergrowth between ZrMOF particles to form a continuous compact layer. In this presentation, we will discuss the cause of intergrowth between UiO-66 particles based on a bi-phase synthesis study and present UiO-66 membranes with controllable orientations synthesized in both free-standing and Î±-Al_2O_3 supported forms via employing proposed mechanism.

500190 Closed Packed, Oriented MOF Thin Films through Solution Shearing

Abstract:

Metal organic frameworks (MOFs) are a class of nanoporous crystalline materials that are composed of metal clusters connected through organic ligands. Their ultrahigh porosity and internal surface area, tunability of the organic and inorganic components, and possibility of the incorporation of various functional groups make them suitable candidates in a variety of applications such as gas storage and separation, catalysis, sensing, etc. Whereas the majority of MOFs have been produced and tested in the form of powder, the deposition of MOF crystals or their in-situ formation on an appropriately functionalized substrate to make thin films has gained significant attention. Such films could then be used as catalytic or separation membranes, sensors, electronic devices, etc. MOF film fabrication through the deposition of solvothermally synthesized crystals, however, frequently results in rough, randomly oriented, low density, or poorly adhered coatings of undesired size/morphology crystals, which could be inefficient in practical applications. On the other hand, some of the state-of-the-art techniques for the controlled deposition of MOFs on substrates such as the Layer-by-Layer (LbL) method typically suffer from slow kinetics. Solution shearing is a novel thin film formation and crystal deposition technique, in which the precursor solution or crystal suspension is sandwiched between a shearing blade and a substrate. Crystals form in-situ or deposit from the suspension over the appropriately-heated substrate while the blade moves at an adjusted speed. This method has already shown its potential in providing fine control on film density, thickness, and alignment in the field of organic electronics and can readily be employed for large-scale fabrications. Using solution shearing and characterization techniques such as grazing incidence X-ray diffraction (GIXD), we have managed to make highly aligned dense films of MOF crystals and investigated the crystalline characteristics and shearing parameters that control the quality of final products.

496121 Mechanistic Insights into Low Temperature Ceramic Thin Film Growth and Crystallization Using Microwave Radiation

Abstract:

This work combines experimental and computational methods to gain key mechanistic insights into the low temperature crystallization of ceramic (e.g., TiO_2) thin films by selective microwave-radiation (MWR) heating of solid materials within a growth solution. Preliminary results show that selective heating of a conducting layer (e.g., metal) by MWR assists in assembling thin films directly on various substrates (e.g., glass, metal, plastic) and significantly lowers the solution temperatures (150 oC) required for growing crystalline anatase TiO_2 thin films. In contrast, conventional sintering and hydrothermal processes require high post-deposition annealing temperatures (> 400 oC). This new processing marks a significant decrease in energy consumption, as microwave heating of the solution to 150 °C uses only 0.18% of the energy as heating a furnace to 450 °C, and demonstrates the advantages of MWR-assisted heating reported previously: reduced temperatures and reaction time, and improved product yield when compared with conventional synthesis routes.

499894 Gold Nanoparticle Monolayers for Surface-Enhanced Raman Spectroscopy of Lithium Batteries

Abstract:

Lithium battery safety and quality are largely limited by contamination and side reactions that lead to capacity fade or even hazardous runaway conditions. Electrochemical reactions are exceptionally sensitive to trace contamination that is difficult to detect with spectroscopic methods. Reliable performance of a battery relies on extremely thin passivation layers that form between electrodes and electrolyte. We have developed a substrate for surface-enhanced Raman spectroscopy (SERS) that enables detection at concentrations relevant for studying battery contamination and passivation layers. SERS relies on surface plasmon resonance to enhance Raman scattering by many orders of magnitude. It has traditionally suffered from lack of reproducibility due to lack of control of surface structure. We report a self-assembly technique to prepare gold nanoparticle monolayers with reproducible long-range hexagonal structure. Our approach allows for precise control of particle size and spacing, which are important for Raman enhancement. We report the performance of our substrates with model analytes and with components used in lithium batteries.

500298 Understanding the Chemistry of Thiol-Amine Solutions: Versatile Solvents for Solution-Processed Thin Film Photovoltaics

Abstract:

Solution processing of inorganic electronic materials is an attractive alternative to vacuum based methods due to its promise of high-throughput, roll-to-roll fabrication at low cost. Recently, significant work has been done to advance the so-called molecular precursor route to inorganic thin films. This route involves the dissolution of appropriate precursors to form a homogenous solution, followed by an appropriate coating method, and subsequent heat treatments to promote the growth of a polycrystalline film of the desired final phase. This method has been applied with great success for thin film PV such as $Cu(In,Ga)Se_2$,
$Cu_2ZnSn(S,Se)$
diverse applications.

An especially promising solvent system, a mixture of an amine and a thiol, has been identified. Recent work has demonstrated the incredible flexibility of the amine-thiol solvent system in dissolving a host of salts and precursors that are insoluble in either solvent by itself.[1,2] The system has found significant use in processing photovoltaic absorber layers, luminescent quantum dot films, and a variety of other thin-film materials, making it a very general solvent system for processing of inorganic thin films.[3–7]

Currently, little is known about the solution chemistry enabling this powerful solvent. In this study we present electrospray ionization mass spectrometry, X-ray absorption spectroscopy, and Raman spectroscopy results obtained for copper chloride precursors dissolved in thiol-amines to reveal the species formed in solution. In short, they were found to be copper thiolates, copper chlorides, copper thiolate-chlorides, and alkylammonium chlorides. We also show that the final oxidation state of the complexed Cu is +1, even when the initial state of the Cu is +2. We also present similar analysis on Se and Te solutions and elucidate the differences when chalcogens are dissolved in contrast to metal salts—as ultimately chalcogens and metal precursors are typically co-dissolved or combined in a single solution prior to deposition.

We also present SEM-EDS, XRD, and Raman spectroscopy studies of annealed films prepared from these solutions to understand the conditions under which precursor impurities (such as chlorides) are removed from the films making them suitable for use in high efficiency devices. We find that chlorides are difficult remove from the final films due to the species formed in solution and suggest future straegies to avoid this contamination. We also present results obtained by applying this film preparation method to CdTe,[7] CIGS,[6] and other metal chalcogenide photovoltaic devices.[4,5]

References

1 D. H. Webber and R. L. Brutchey, *J. Am. Chem. Soc.*, 2013, **135**, 15722–15725.

2 B. C. Walker and R. Agrawal, *Chem. Commun.*, 2014, **50**, 8331–8334.

3 Q. Tian, G. Wang, W. Zhao, Y. Chen, Y. Yang, L. Huang and D. Pan, *Chem. Mater.*, 2014, **26**, 3098–3103.

4 R. Zhang, S. Cho, D. G. Lim, X. Hu, E. A. Stach, C. A. Handwerker and R. Agrawal, *Chem. Commun.*, 2016, **52**, 5007–5010.

5 R. Zhang, S. M. Szczepaniak, N. J. Carter, C. A. Handwerker and R. Agrawal, *Chem. Mater.*, 2015, **27**, 2114–2120.

6 X. Zhao, M. Lu, M. J. Koeper and R. Agrawal, *J. Mater. Chem. A*, 2016, **4**, 7390–7397.

7 C. K. Miskin, A. Dubois-Camacho, M. O. Reese and R. Agrawal, *J. Mater. Chem. C*, 2016, **4**, 9167-9171.

500373 *In situ* Grazing Incidence Small Angle X-Ray Scattering (GISAXS) Study of the Formation of Multilayered Ordered Mesoporous Titania Films

Abstract:

Mesoporous titania (TiO_2) thin films have a unique combination of chemical, optical and electronic properties that make it a promising material in diverse fields including energy conversion and storage, environmental remediation, and biomaterials. All of these applications benefit from being able to produce accessible nanoscale channels due to their high surface area, well defined size and shape, and short diffusion paths for reactants and charge carriers. Mesoporous titania thin films with vertically oriented hexagonal cylindrical nanopores have been synthesized in our group by evaporation induced self-assembly, and we have shown that alignment of the pores orthogonal to the substrate can be achieved by modifying substrates with crosslinked layer of the Pluronic surfactant pore template (Koganti et al., 2006). A previous *in situ* grazing angle small angle x-ray scattering (GISAXS) study showed that aging at low temperature (4 °C) is important for the development of the orthogonal HCP (o-HCP) mesostructure in 60 nm thick TiO_2 films (Nagpure et al., 2015). However, a number of additional variables are important for eventual application of o-HCP TiO_2 films. First, for applications requiring high mass loading of TiO_2, the film thickness needs to be scaled up. We have found that this can be done using a layer-by-layer deposition technique, but the fidelity of epitaxial orthogonal alignment in the layers remains to be investigated. Second, humidity has been hypothesized to be an important parameter dictating orthogonal orientation by controlling the evaporation of solvent from the as-coated film. Well-ordered mesoporous structures have been obtained by aging at 4 °C and 94% RH but the limits and effects of RH are not known.

Here, *in situ* GISAXS is performed to determine if and how orthogonally oriented columnar templated layers form for multilayer Pluronic F127-templated TiO_2 films. GISAXS is performed on each layer of multilayer films up to 8 layers (having a total thickness of 1000 nm). Results show that films cast onto modified glass slides maintain vertically oriented mesophase order as multiple layers are deposited. On the other hand, horizontally oriented micelles on unmodified slides lose their preferred orientation as more layers are deposited, and by the third layer a powder pattern is obtained indicating a random mix or orientations in the film. The mesostructure growth kinetics of every layer are analyzed using the Avrami equation to determine the growth rate, order and direction. *In situ* GISAXS is carried out with varying RH (35% and 70% RH) for F127-templated TiO_2films cast onto modified glass. While about 70% relative humidity provides well-ordered, oriented micelles, uniform but randomly oriented micelles are found at 35% RH. These studies, together with complementary SEM imaging, confirm that high relative humidity is important to the formation of orthogonally oriented mesoporous structures, and that thick films with orthogonally oriented pores can be formed by epitaxial growth of o-HCP mesostructures on top of existing o-HCP composite films.

References:

Koganti, V.R., Dunphy, D., Gowrishankar, V., McGehee, M.D., Li, X., Wang, J., Rankin, S.E. 2006. Generalized Coating Route to Silica and Titania Films with Orthogonally Tilted Cylindrical Nanopore Arrays. *Nano Letters*, **6**(11), 2567-2570.

Nagpure, S., Das, S., Garlapalli, R.K., Strzalka, J., Rankin, S.E. 2015. In Situ GISAXS Investigation of Low-Temperature Aging in Oriented Surfactant-Mesostructured Titania Thin Films. *The Journal of Physical Chemistry C*, **119**(40), 22970-22984.

501090 2-D Wulff Construction of FeOx Islands Grown on Pt(111) for Use in Catalysis

Abstract:

Supported ultra-thin metal oxide films grown on metallic surfaces garner substantial interest in their ability to solve hosts of engineering problems, ranging from micro-electronics, protective coatings, magnetic devices and catalysis. Much of research being performed on these materials involves understanding the long-range structures, magnetic and optical properties of these films. For their use in catalysis, however, it is generally accepted that the reactions of interest occur at the interface created by the edge of the film and the substrate. A clear example of this is that FeO_x films supported on Pt(111) have shown improved performance for the CO oxidation reaction. Normally, a Pt(111) surface is rather inactive for the oxidation of CO, but the addition of a monolayer FeO_x film on the Pt enhances the catalytic properties by several orders of magnitude. Several experimental and computational studies infers that this reaction occurs through a bifunctional mechanism with the reaction is occurring at the interface between FeO and Pt(111). But the nature of this interface is not well understood and as with all catalysts, it is essential that one gains an understanding of the nature of the active site to elucidate how the reaction could proceed. This work first uses Density Functional Theory to examine a series of possible edge and corner structures that could be present for FeO_x islands supported on Pt(111). This information can be then translated into the creation of 2D Wulff constructions, allowing for the determination of the oxidation state and ratio of the different edges or a range of conditions. In addition, the thermodynamically predicted size of the FeO_x islands is also studied, where it is observed that the calculated island size drops as the island becomes more oxidized. This methodology would allow for an estimate of the conditions that would facilitate the edges that are the most reactive towards certain reactions.

503210 Molecular Simulation of Crystallization of Chain Molecules from the Melt

Abstract:

The properties and performance of commercial polymers depend strongly upon both the molecular scale architecture of the chains and the manner in which they are processed to form engineering materials. Polyolefins in particular owe much of their utility to the spectrum of semicrystalline morphologies that can be realized. The development of semicrystalline morphology is controlled through processes of crystal nucleation and growth as the polymer is cooled from the melt, in either the presence or absence of process flows. Due to the small spatiotemporal scales involved, however, it remains a challenge to examine these processes directly and determine their underlying physical mechanisms.

In this talk, we describe the use of molecular dynamics and Monte Carlo simulations to characterize nucleation and growth of a new crystal phase from the polymer melt. Homogeneous nucleation is analyzed in terms of classical nucleation theory using a method based on mean first passage times (MFPT) of crystalline clusters of varying size. This MFPT analysis is then applied in two dimensions within layers near a crystal surface, to examine the so-called secondary (or surface) nucleation theory for chain molecules; within this analysis, crystal growth can be decomposed into sequential and simultaneous processes of surface cluster nucleation and spreading of the crystal cluster across the growth front. Finally, we introduce the effects of flow fields typical of polymer processes, and examine the phenomenon of flow-enhanced nucleation (FEN). In accord with experiments, substantial acceleration of the nucleation rate is observed as the system approaches a Weissenberg number (Wi) of unity. The acceleration is consistent with reduced entropy of the flow-oriented melt, but one also observes significant amplification of the rate pre-factor above Wi=1, indicative of enhanced transport effects. Application of these approaches to the study of crystallization of polymers under other conditions of contemporary interest appears promising, and may serve to shed new light on fundamental questions within the field of polymer crystallization in engineering applications.

495425 Anomalous Crystallization Behavior of Ring Polymers

Abstract:

We have modeled the crystallization of cyclic polyethylenes using coarse-grained Langevin dynamics simulations and compared with the behavior of linear polyethylenes. We have studied various features of polymer crystallization including early stage nucleation kinetics, lamellar thickness, growth kinetics, and melting. Contrary to expectations from equilibrium thermodynamics, ring polymers melt at lower temperatures than the corresponding linear polymers. The rings crystallize faster and exhibit several metastable lamellar thicknesses. The topological effect arising from the absence of chain ends is found to be significant even at large molecular weights. Comparison of our simulation results with available experimental data help to resolve several long-standing puzzles in the determination of equilibrium melting points and crystallization kinetics.

497899 Thin Film Crystallization of Cyclic Polymers and Their Linear Analogues

Abstract:

While the crystallization of linear and branched polymers has received much attention in the literature due to the importance of polymers like linear poly(ethylene glycol) to the biological sciences and branched polyethylenes to the commodity polymers market, the effect of ring architecture on polymer crystallization is less well understood, in part due to the challenges associated with synthesizing well-defined cyclic polymers. However, recent advances have enabled the synthesis of cyclic polymers with exact linear analogues using an azide-alkyne "click" reaction for ring closure and prompted additional research into the properties of cyclic polymers. With respect to semi-crystalline polymers, literature on bulk materials indicates that cyclic polymers have faster crystallization kinetics and greater thermal stability than linear polymers. In this work, we focus on the differences between linear and cyclic polymers under thin film confinement, which leads to deviations from bulk behavior. Specifically, we compare the thin film stability, thermal transitions (melting temperature and equilibrium crystallization temperature), crystallization kinetics, and morphologies of cyclic and linear thin films. As a model system, we have chosen poly(ε-caprolactone) (PCL), a semi-crystalline, aliphatic polyester that has applications in biomedical engineering including anti-adhesion biomaterial films, drug delivery media, and others.

484060 Synthesis, Characterization, and Structural Evolution of Designer Block Polyelectrolyte Complexes

Abstract:

Polyelectrolyte complexes,Â assemblies of oppositely-charged polymers in aqueous solutions, can address broad societal needs as key ingredients in designing multifunctional nanomedicine carriers and gels. However, because of the interconnected roles of chemically-driven interactions and electrostatic associations, the assembly mechanism and kinetics of formation remain poorly understood and variable in terms of reliability. In this work, we employed RAFT polymerization strategies to investigate structure-property relationships in tunable, ionic block polymers. This facilitates opportunities to judiciously vary the ratio of water-soluble neutral to charged block lengths, tailor specific chemical attributes towards stronger or weaker complexation, and functionalize labile Ï‰-chain ends with nucleophilic markers (such as fluorescent dyes) for elucidating self-assembly dynamics. Small libraries of charged systems were prepared in a parallel synthesizer from poly(ethylene oxide) (PEO) methyl ether as RAFT macromolecular chain transfer agents. As an initial model polyelectrolyte platform, we synthesized PEO-*b*-poly(vinyl benzyl trimethylammonium chloride) (PEO-PVBTMA) and PEO-*b*-poly(sodium 4-styrenesulfonate) (PEO-PSS) as the positively- and negatively-charged polymers, respectively, modulating structural parameters, molecular architecture, and ionic character between systems. Subsequently, dynamic light scattering, microscopy, and small-angle X-ray scattering were employed to examine ordering into various morphologies as a function of polymer concentration and salt. The rational pairing of well-defined polyelectrolytes ultimately enables pathways of complexation-driven assemblies to be understood, thereby guiding future prediction capabilities and advancing the development of more sophisticated biomaterial properties and applications.

498081 Solid and Liquid Core Polyelectrolyte Complex Micelles

Abstract:

Polyelectrolyte complexes form by mixing oppositely charged polymers in solution. The resultant complex phase separates from solution into either irregularly shaped solids (or rather glasses), called precipitates, or micron sized liquid droplets that can coalesce into a distinct phase, called a coacervate. Using oppositely charged polypeptides, one can tune between solid and liquid complexes by manipulating the chirality of the polyelectrolyte. Homochiral complexes form precipitates with hydrogen bonded b-strand structure. In contrast, if one or more polypeptide is composed of both L and D monomers, coacervates are formed with no secondary structure. This inability to form secondary structure is attributed to steric hindrance of the racemic polypeptide impeding hydrogen bond formation. Therefore, since both types of complexes are formed using weak polyelectrolytes, the ability of the homochiral molecules to hydrogen bond causes the difference in phase. Using oppositely charged block-copolyelectrolytes that contain neutral blocks covalently linked to charged blocks allows the phase separation to be stabilized on the nanoscale, creating self-assembled micellar structures in dilute solutions. These polyelectrolyte complexes micelles can be used as drug and gene delivery vehicles for charged therapeutics like nucleic acids and proteins. Using diblock copolypeptides of varying chirality the resulting micellar structure can have both solid and liquid polyelectrolyte cores. This presentation will focus on characterization of polypeptide based polyelectrolyte complex micelles using scattering techniques (light, x-rays, neutrons), circular dichroism, and electron microscopy to reveal structural differences of solid and liquid micellar cores. In addition, the exchange kinetics of solid and liquid core micelles will be investigated by mixing solid and liquid core micelles that contain FRET pairs. Insight will be provided as to how differences in solid and liquid cores influence the design of drug delivery vehicles. If time permits, investigations into related micelles with dynamic coronas will be discussed.

489488 Uniqure Crystallization Behavior of Isotactic Polypropylene in the Presence of L-Isoleucine and Its Inhibition and Promotion Mechanism of Nucleation

Abstract:

A novel and efficient anti-nucleating agent L-isoleucine (L-I) was found and 0.08 wt % L-I could significantly decrease the the peak crystallization temperature (T_{cp}) of isotactic polypropylene (iPP) up to 8 ^{o}C at the cooling rate of 20 ^{o}C/min. To our knowledge, so highly efficient anti-nucleating agent of iPP was reported for the first time. Furthermore, L-isoleucine exhibited two opposite nucleating ability, aiti-nucleation and nucleation ability, i.e., hindering the crystallization of iPP at low content and promoting the crystallization of iPP at high content. The unique crystallization behavior of iPP in the presence of L-I were investigated by Differential Scanning Calorimetry (DSC), Polarized Optical Microscopy (POM) and rheological mesurement. It was found that low content of L-I was completely dissolve in iPP melt, while high contnt of L-I did not by the observation of POM. Therefore, a mechanism which L-I inhibits and promotes nucleation of iPP was proposed. L-I moleculars dissoving in the iPP melt hindered the homogeneous nucleation of iPP moleculars for solvent effect, while with increase of L-I content, L-I could not be completely dissolved in molten iPP, the residual crystals of L-isoleucine provided heterogeneous nucleation for iPP and further promoted the crystallization. Experimental evidence from rheology and POM supported the mechanism.

499194 Scalable Nanocomposites Synthesis Via Electrospray-Mediated Electroemulsification and Flash Nanoprecipitation

Abstract:

The objective of this work is to develop a robust nanocomposite synthesis platform, which incorporates electrospray techniques to overcome the challenges associated with current, small-scale nanocomposite synthesis methods. Conventional methods (i.e. sonication, film rehydration, dialysis) for producing nanocomposites are limited to mostly batch scales, confined to a low product volume and a long processing time, diminishing commercial application of these technologies. Recently, coaxial electrospray-enabled AEROsolization and subsequent self-assembly through Interfacial inStability (Aero-IS) and Flash NanoPrecipitation (FNP) were introduced to address these challenges. These approach are both used to fabricate nanoparticles comprised of amphiphilic block copolymers (i.e. PS-PEO) and hydrophobic encapsulants (i.e. nanoparticles, drug molecules). Aero-IS utilizes the coaxial electrospray technique, which generates oil in water emulsion droplets in aerosol form. The collected emulsion droplets undergo interfacial instability[1], which promotes the self-assembly of amphiphiles to encapsulate the hydrophobic cargoes. Flash NanoPrecipitation[2] utilizes either a confined jet mixer or a multi-inlet vortex mixer to promote nucleation and growth of amphiphiles by achieving rapid mixing of a water-miscible organic phase and an aqueous phase. Despite the significantly increased throughput achieved by both methods, challenges still remain. The Aero-IS process requires the addition of surfactants that are present at concentrations of more than 90X those of the product, presenting a purification challenge. Similarly, in FNP, solvent must be removed from the final product. In addition, in FNP high flow rates are required to ensure good mixing, which may limit the ability to process intermediate (pilot scale) quantities of material (i.e. ml min-1 scale in FNP). Hence, the goal of this research was to develop an improved nanocomposite synthesis platform, Liquid-Liquid Electrospray (LLE), that can operate over a range of flow rates with reduced purification requirements.

Unlike traditional electrospray techniques, in the LLE configuration, the electrospray needle is placed directly inside an aqueous medium, along with a grounding electrode. The needle is insulated with tightly fitted PTFE tubing, which prevents short-circuiting upon the application of a high voltage to the needle. The organic phase with amphiphlic block copolymers and the hydrophobic encapsulants is processed through the needle, and, depending on the miscibility of the organic phase in the aqueous phase, the applied high voltage enables either electroemulsification, or micromixing of the organic phase. For water-immiscible organic solvents (i.e. chloroform), instantaneous emulsification is followed by interfacial instability self-assembly of micelles. For water-miscible organic solvents (i.e. tetrahydrofuran), the high degree of mixing achieved using LLE promotes nucleation and growth of the hydrophobic components introduced into the aqueous medium. Nanocomposites synthesized via LLE-mediated electroemulsification and flash nanoprecipitation include micelles encapsulating quantum dots (QDs), and/or superparamagnetic iron oxide nanoparticles (SPIONs). Overall, this new nanocomposite synthesis platform provides the advantages of scale-up and high controllability without the requirement for surfactants, improving methods to develop nanocomposites at scale.

1. Zhu, J., & Hayward, R. C. (2008). Spontaneous generation of amphiphilic block copolymer micelles with

multiple morphologies through interfacial instabilities. Journal of the American Chemical Society, 130(23), 7496-7502.

2. Akbulut, M., Ginart, P., Gindy, M. E., Theriault, C., Chin, K. H., Soboyejo, W., & Prud'homme, R. K. (2009). Generic method of preparing multifunctional fluorescent nanoparticles using flash nanoprecipitation. Advanced Functional Materials, 19(5), 718-725.

487602 Engineering Vapor-Deposited Polymers for Energy Conversion and Storage

Abstract:

Globally, we are pacing towards a cleaner energy economy as we recognize the limits on fossil fuel supplies and the environmental pollution they produce. To mitigate these issues, new technologies harvest energy from cleaner albeit intermittent natural resources like the sun, wind, and water. These energy resources in turn require new technologies for storing energy that can power anything from electronics to vehicles on demand.

With new technologies for energy capture and storage come significant challenges related to the intrinsic materials and architectural design of these energy devices. Polymers, with their vast chemical and physical tunability, offer a unique platform for enabling more efficient alternative energy solutions in solar cells, fuel cells, batteries, and capacitors.

Many of these electrochemical systems rely on high surface area architectures containing nanoscale pores or features to enhance conversion efficiency and energy density. However, with these nano-architectures, it is often difficult to find effective ways to interface them with polymeric materials. Conventional methods, like spin casting or dip coating, that use pre-made polymer liquid solutions frequently suffer from surface tension and wettability issues that lead to non-conformal coatings and poor integration.

To overcome these challenges, my laboratory has been developing new methodologies for engineering electronic polymers for energy conversion and storage that improves materials interfacing and device integration. Central to these efforts is the use of liquid-free chemical vapor deposition (CVD) pathways that effectively integrate polymers in nanostructured electrochemical energy systems and improve on device efficiency and operability.

In this context, my talk will discuss our efforts in engineering polymer electrolytes for next-generation solar cells and conducting polymers for supercapacitors. Critical to this work are two unique CVD approaches, initiated CVD (iCVD) and oxidative CVD (oCVD), that enable (1) the direct chemical synthesis of true polymers from their monomer building blocks â€“ this is typically not achievable with other CVD methods like plasma polymerization â€“ and (2) the tandem formation of the polymers as conformal thin films with well-controlled thickness and morphology â€“ this is generally difficult with liquid phase polymer processing.

Fundamentally, a solid understanding of the competing processes of reactant diffusion and polymerization kinetics is key for enabling conformal growth of polymers on nanoscale pore surfaces and features. Our hypothesis is that better polymer interfacing and integration improves interfacial dynamics in these nanostructured energy systems. My talk will demonstrate how this ultimately translates to enhanced device performance and brings us closer to more viable and robust sustainable energy technologies.

492670 Charge Transfer Mechanisms in Organic Radical Polymer Batteries

Abstract:

Organic radical polymer batteries offer an intriguing path towards plastic energy storage. Organic radical polymers often consist of an aliphatic backbone and pendant stable radical groups. These electroactive polymers store charge by a reversible redox reaction, usually with near-100% Coulombic efficiency and extremely fast charge transfer kinetics. However, their performance is complicating by the generally insulating nature of the polymer backbone. The general nature of this reaction is not completely understood, as it is fundamentally different from more traditional battery electrode materials that undergo oxidation. Here, we present electrochemical quartz crystal microbalance with dissipation (EQCMD) monitoring as a means to quantify ion transport, dynamic swelling, and mechanical properties of an organic radical polymer during electrochemical interrogation. We focus upon EQCMD characterization of the redox-active nitroxide radical polymer, poly(2,2,6,6-tetramethylpiperidinyloxy methacrylate) (PTMA). Upon oxidation, PTMA becomes positively charged, which requires the transport of a complementary anion into the polymer for electroneutrality. By EQCMD, we quantify anion transport and resultant swelling upon oxidation, as well as decoupling of contributions attributed to the ion and the solvent. We explore the effect of different lithium electrolyte salts in which each salt gives different charge storage and mass transport behavior. This is attributed to varied polymer-dopant and dopant-solvent interactions. The work was supported by the grant DE-SC0014006 funded by the U.S. Department of Energy, Office of Science.

487907 Free Volume Enhanced Anion Exchange Membranes from Triptycene Poly(Arylene Ether Sulfone) Copolymers

Abstract:

As an attractive energy conversion technology, fuel cells generate electric power from fuels with high energy density and slight chemical emissions. Anion exchange membrane fuel cells (AEMFCs) have been developed as a desirable alternative to Proton exchange membrane fuel cells (PEMFCs) because materials cost-effectiveness and comparable performance. As a pivotal module of AEMFCs, anion-exchange membranes (AEMs) serve as both the fuel barrier and hydroxide carrier. It is required to make membranes with properties of superb hydroxide conductivity, good dimensional and mechanical stability, and prominent alkaline resistance.

Here we present that we have achieved high performance AEMs from the functionalized triptycene copolymers. Quaternized triptycene copolymer were casted into a large scale film. Hydroxide conductivity of OH^{-1} activated films were measured to be over 0.1 S/cm for working temperature ranges (60-80 °C). Advantages of using triptycene containing polymers are many folds: 1) triptycene containing moieties have high stability to heat and chemicals for long term usability, 2) triptycene containing polymers have nanoscale phase separation which would facilitate nanoscale binding and transport and 3) degree of functionalization would be readily controlled by reaction conditions. Extensive studies from X-ray scattering, TGA, DSC corroborated the above findings.

SURFACTANT-POLYMER SYSTEM OPTIMIZATION IN HETEROGENEOUS MODEL WITH MOBILITY CONTROL

Nai Cao, Pingchuan Dong, China University of Petroleum Beijing.102249, China
Brian McPherson, Civil& Environmental Engineering, University of Utah, Salt Lake City, UT.84102 United States
XiaoxiaoLiu,Puguang Gas Production Plant; SINOPEC. Dazhou Sichuan Province. 635000 China.

Introduction

According to data collected from case studies in China, thought flooding has been utilized to improve oil recovery, high water content still remains. After flooding, oil still can make nearly half remaining in the reservoir, meaning there is still a long way to run to make it a effective oil production. This kind of mentioned remained oil has a scattered dispersion which results in a more intensified heterogeneity of the operated reservoir. To solve problem like this, surfactant flooding may be an useful tool, but it is not quite enough.

Mobility control can help a lot to make it work: When injected without mobility control, surfactant-polymer may block low permeability layers and thereby increase the remaining oil amount. At the same time, cross flow happens in high permeability layers, in that case, the injection can hardly do its job. To improve the utility rate of surfactant-polymer and oil recovery rate, also, to initiate the production of lower permeability layers, mobility control with surfactant-polymer can be a potential tool to settle all these down.

Overview

In today's China, remaining oil after flooding is studied in different oil fields, based on geology characteristics, in-time monitoring and numerical simulation methods, an overall situation about the remaining oil after flooding is clear: 27 percent lies on the low permeability layer, where the polymer can not reach, 19.5 percent is in the dead flows, 16 percent is kept in tiny holes because of capillary pressure, 13.5 percent leaves on the rocks in reservoirs as a film cover on them. In the other words, remaining oil exists in the sections blocked from displacement because the heterogeneity. Since 1950s, reservoir simulation started to be one of the musts for reservoir engineers. And from early 1980s, surfactant-polymer has become an effective tool for oil recovery.

Theory

Traditional opinion suggested that the polymer improve oil recovery by increasing the viscosity of injection fluid and enlarging the swept volume. But in-door experiments shows that besides enlarging the swept region, it also improve the efficiency of displacement. And Recovery efficiency rate E_R can be described as follow:

$$E_R = \frac{N_P}{N} = E_V \times E_D \tag{1}$$

N_P -output, N - initial oil reserve, E_V -swept coefficient, E_D -flooding efficiency.

Because of the absorption of polymer, water's mobility decrease but it has little effect on oil phase. So under the same oil saturation, oil phase has a higher permeability than the water phase, thereby making the remaining oil start to flow. At the same time, with the viscous force of polymer, viscoelasticity on the oil-water interface can drag oil drops and make them transform (into drops and filaments), so that they can pass through smaller holes. Viscous fingering phenomenon can be overcome in this process, because of the high viscosity.

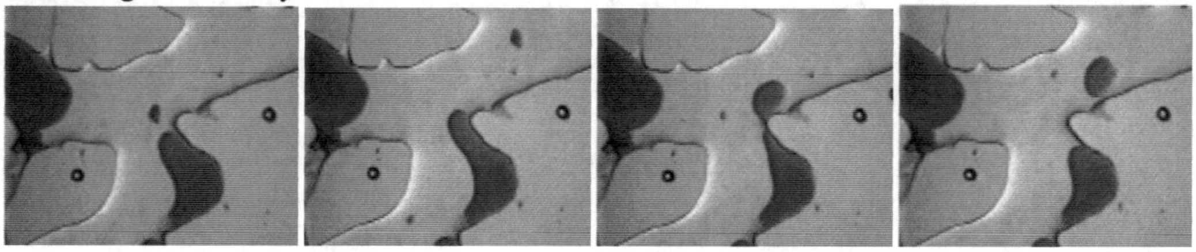

Fig.1 remaining oil dragged into drops

Oil can be captured in holes is largely because of capillary pressure and viscous force, decreasing in interfacial tension will amplify the interaction between oil and water captured in holes, and decrease the absorption, thereby increasing the water's driving effect. As expressed in capillary pressure expression:

$$P_c = \frac{2\sigma \cos \vartheta}{r} \quad (2)$$

Oil mobility has a negative correlation with capillary pressure, while capillary pressure has a positive correlation with oil-water interfacial force.

In the flooding process, oil drops can accumulate into bigger ones and at last to be connected. And in the microscope surfactant polymer flooding, emulsion phenomenon can be observed, which means the surfactant can emulsify big oil drops- which can hardly move in the holes. Water-in-oil emulsion and oil-in-water emulsion dispersed in different sections. The emulsion has changed the status of remaining oil, changed the inter-relation of capillary pressure, driving force and absorption force, which makes the dead remaining oil start to flow again.

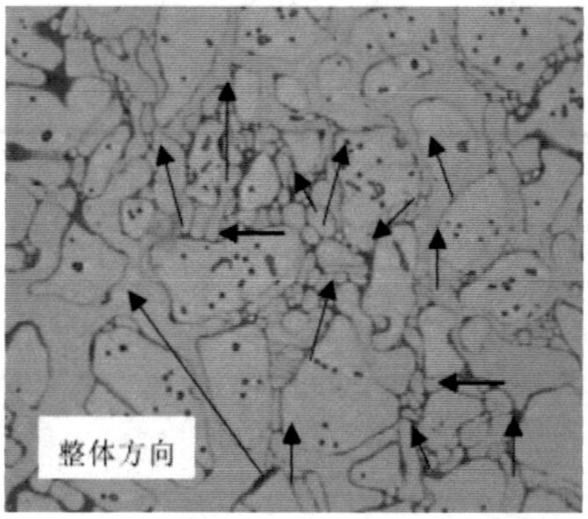

Fig.2 Water-in-oil emulsion

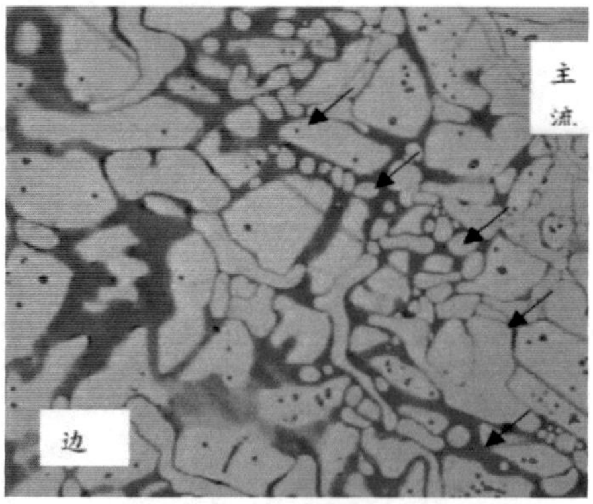

Oil-in-water emulsion

Mobility Control

The heterogeneity is intensified after flooding, so if directly inject surfactant polymer, it may enter into high permeability layers leaving the lower permeability layer incapable to be recovered. To solve this, we introduced mobility control: To optimize surfactant-polymer system, two parallel cores models are adopted with different permeability contrast. High concentration polymer is injected in the first place to prevent subsequent solution being injected to higher permeability layers. Thereby, the swept volume is enlarged and the remaining oil in the lower permeability layer is initiated. With physical experiment as a validation for the simulation work, a preferred scheme is determined.

Physical Experiment

Model: two parallel cores models with permeability contrast 1:10 and 1:5, the size for each layer is 4.5×4.5×30 cm, the fixed effective permeability for the low permeability layer is200×10-3μm2;
Oil: made up by raw oil and coal oil from Daqing Oil Field, viscosity 10mPa•s;
Water: Salty water with mineralization 3700mg/L
Experiment temperature: 45℃
Experimental Parameter Selection: Since the higher the molecular weight the higher the viscosity, and polymer solution with high viscosity can effectively block the higher permeability layer to increase the swept volume. So here, polymer solution with molecular weight is 2500×104 and 80mpa • s viscosity is selected. Based on in-door experiment, polymer with high viscosity may be not compatible with surfactant, so solution with1900×104 molecular weight, 40mpa • s viscosity is selected. And From Form 1, we can see when surfactant concentration is 0.2% and 0.3%,the interfacial tension can reach 10^{-3}mN/m, so combined with consideration on absorption in reservoir, 0.3% is selected.

surfactant concentration-%	0.05	0.1	0.2	0.3
interfacial tension -mN/m	0.010416	0.01005	0.006054	0.007297

Form. 1 relation between interfacial tension and surfactant concentration

Numerical Modeling

The theory numerical modeling based is also mobility control: two parallel cores model is established with 1: 5 and 1: 10 as permeability contrast respectively. In vertical direction, 7 layers are divided, taking 1.5cm as the effective thickness. As ineffective layer, the 4th layer divides these layers into two groups. Layer 1to 3 share the same effective permeability and the effective permeability from layer 5 to layer 7 is the same. Average porosity is 0.24, ground water viscosity is 0.59mpa • s, oil viscosity is 11mpa • s, meshing is 20×3×7 with one - well injection and one - well production. Under each permeability contrast, 0.1 to 0.3PV high concentration polymer and 0.2to 0.4PV Surfactant flooding scheme are used in different combination.

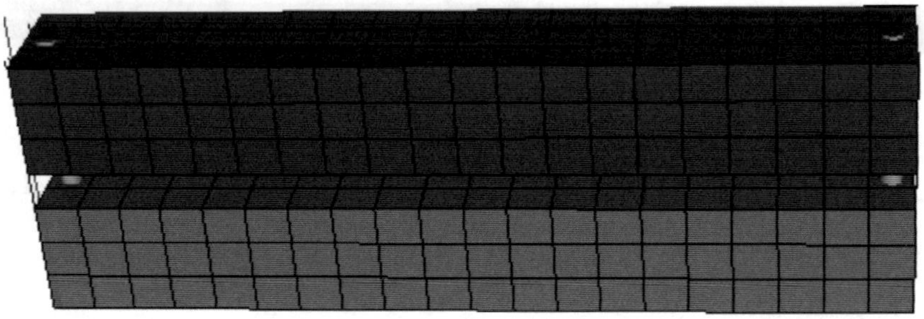

Fig. 3. Numerical Model Sketch

Results

Submerged degree after water flooding:
Fig.4 and 5 show the submerged degree after water flooding in different permeability contrast:

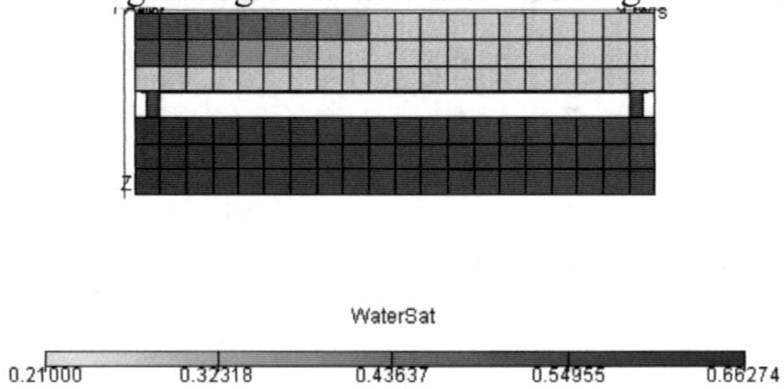

Fig.4 submerged degree with permeability contrast 1:10

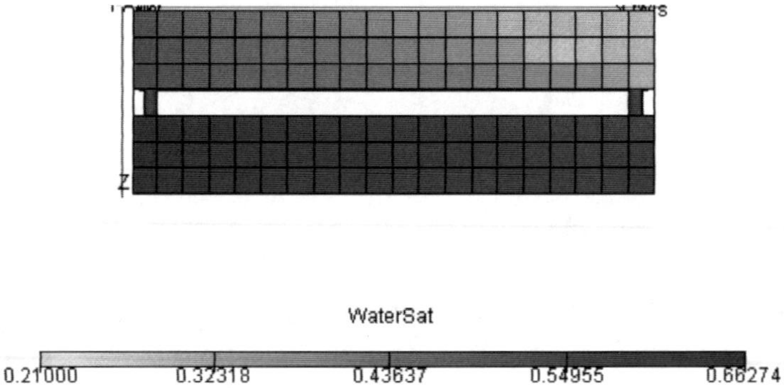

Fig.5 submerged degree with permeability contrast 1:5

It is clearly shown that, high permeability layer has a higher submerged degree and smaller contrast also has a higher submerged degree. This is because the heterogeneity is severe and the flooding coefficient decreasing with it, so it is with the swept coefficient, which means, injection water can't reach low permeability layer. So, surfactant polymer should be used to make the injection go through low permeability layer.

Effect of Surfactant-Polymer Flooding for 1:10 permeability contrast:

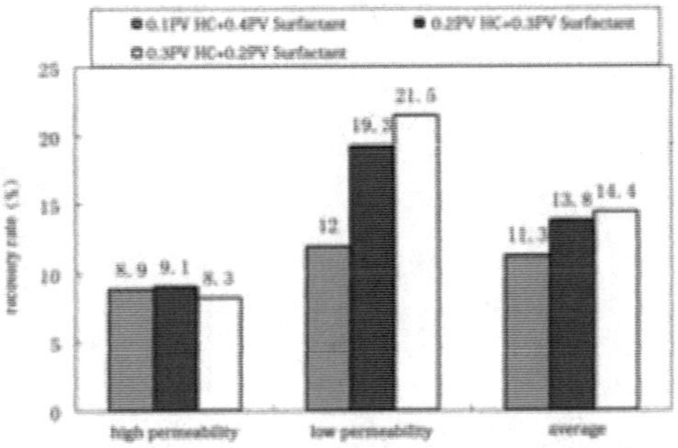

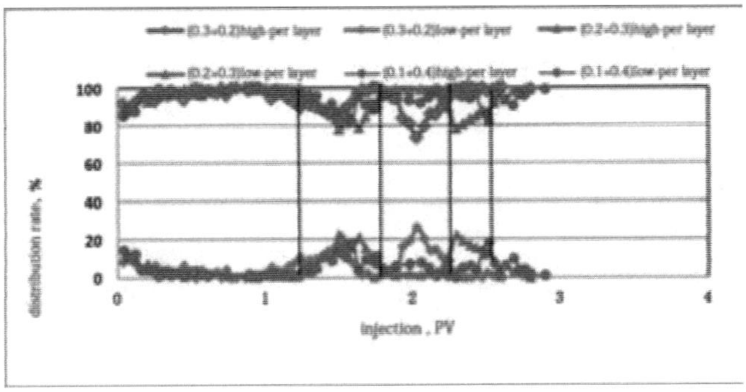

Fig.6 recovery rate

Fig. 7 distribution rate

When the slug size changes from 0.1PV to 0.3PV, the production rate is increasing. The optimized scheme is 0.3 PV high concentration polymer add 0.2PV surfactant flooding. But when it changes from 0.2 to 0.3, the increase is tiny, especially for the low permeability layer, which is only 0.1%. So may be the best value lies between 0.2 PV to 3PV.

From the distribution curve, it can be seen that in the water flooding section, low permeability layer stays almost intact and cross flows happens a lot. With surfactant flooding, the situation is eased.

Effect of Surfactant-Polymer Flooding for 1:5 permeability contrast:

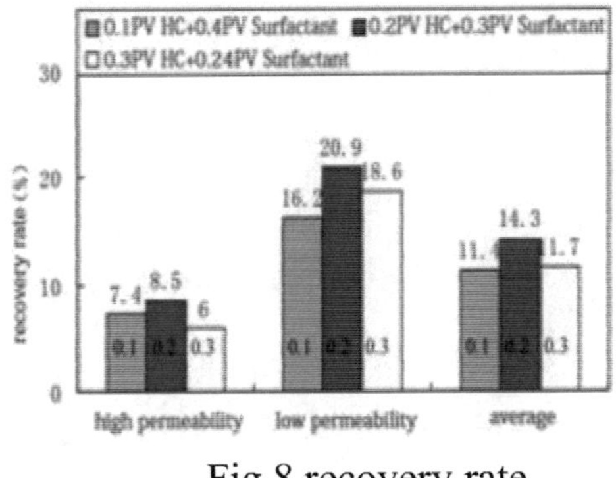

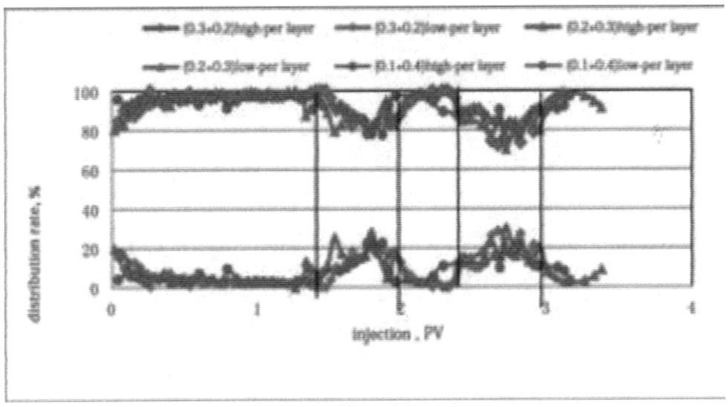

Fig.8 recovery rate

Fig. 9 distribution rate

From the recovery rate curve, it can be seen that under 0.2 PV, it has the highest recovery rate. And from the distribution, we can see the surfactant flooding can do the job to adjust the flooding rate in different layers. Since the difference in distribution is so tiny that it can be concluded that 0.2 PV has already reached a fine result in adjusting flooding in with layer different levels of permeability.

Conclusions

A tool need to be used in the right way to do its job most effectively. So mobility control can be taken as the right way to use surfactant flooding here. With scheme, both high and low permeability layers can be source of oil production. A way to optimize surfactant polymer flooding and to get low permeability layer initiated is provided here, it might be an solution for the low recovery rate in oil fields with high heterogeneity.

References

Duerksen J H, Hsueh L. Steam distillation of crude oil[R]. Soc Pet Eng J, 1983, 265-271.
Greaves M, Ren S R, Xia T X. New Air Injecting Technology for IOR Operations in Light and Heavy Oil Reservoirs[R]. SPE 57295-MS, 1999.

Sorbie K S. Polymer-Improved Oil Recovery [M]. Boca Ration, Florida: CRC Press Inc. 1991.

Mungan N. Enhanced Oil Recovery Using Water as a Driving Fluid-Part4:Fundamentals of Alkaline Flooding [J]. World Oil(June 1981):9-20.

Mungan N. Role of wet ability and inter-facial tension in water flooding[Z]. SPE 705, 1963.

Silpgarmlers N, Gluer B,E rtek in, Development and testing of two- phase relative permeability predictors using artificial neural networks[C],SPE54, 2002:52- 56.

Healy R.N., Reed R. L. Immiscible Micro emulsion Flooding. Society of Petroleum Engineers Journal. 1977, 17:129-139.

Raimondi Pet al. Alkaline Water Flooding Design and Implementation of a Field Pilot [J]. J. pet. Tech. (Oct. 1977) :59-68.

Ryles. RGE valuated temperature testing of mobility control. SPE, 12008.

Greaves M, Ren S R, Xia T X. New Air Injecting Technology for IOR Operations in Light and Heavy Oil Reservoirs[R]. SPE 57295-MS, 1999.

AllenE, Boger.DV. The influence of rheologieal properties on mobility control in polymer augmented water flooding. SPE, 18097.

Foster WR. A Low-Tension Water flooding Process. Journal of Petroleum Technology. 1973, 25:205-210.

489190 Breaking the Compensation Effect within the Vogel-Tammann-Fulcher Equation for Polymer-Based Electrolytes

Abstract:

Single-ion conducting polymer electrolytes have long suffered from poor conductivity at low to moderate temperatures, though their ability to eliminate deleterious concentration gradients in lithium ion batteries has continued to attract research interest. We have synthesized a new random copolymer of oligomeric poly(ethylene glycol) and ion containing sulfone monomers (PSf-co-PEG) that possesses a wide miscible and accessible composition window, thereby allowing careful elucidation of the interplay between the impacts of segmental motion and ion content on the final observed conductivity. Here we demonstrate those effects, and highlight the importance of optimized data fitting in the commonly used Vogel-Tammann-Fulcher (VTF) equation. Particularly, we show that the use of a fixed Vogel temperature 50 - 100°C below the glass transition may lead to misleading trends in the VTF prefactor and activation energy. Utilizing an optimized fitting method applied to the PSf-co-PEG system, we have found a strong positive correlation between the VTF equation prefactor and apparent activation energy for polymers in this electrolyte class. This relationship, termed the compensation effect, has been observed in many related activated processes governed by the Arrhenius equation, and we demonstrate the effect for several additional polymer electrolyte classes. Given that conductivity scales inversely and exponentially with apparent activation energy, maximum conductivity must be achieved in samples where the activation energy is small. For a system in which the compensation effect exists, low activation energies are only found at low values of the prefactor, clearly indicating the limiting nature of the compensation effect for any electrolyte. We find that blending of small molecules, in this case short chain poly(ethylene glycol), breaks the apparent compensation effect, lending new motivation for a common method to increase ionic conductivity and demonstrating a clear route to high Li^+ transference number, high conductivity electrolytes.

492575 Carbon Derived from Polymerized Ionic Liquids

Abstract:

The development of new carbon materials with exquisite control over surface chemistry, structure and form, surface area, pore size, and conductivity is of significant interest for numerous applications, including energy storage (*e.g.*, capacitors). Recently, a few studies have demonstrated the ability to produce carbon from ionic liquids (ILs) and polymerized ionic liquids (PILs), both of which have unique physiochemical properties and result in carbon with various surface chemistries depending on the chemistry of the IL or PIL. However, to date, few studies have explored the relationship between the possible diverse PIL chemistries and the resulting PIL-derived carbon properties. In this study, PILs with various backbone/cation pairings (backbones: ethyl methacrylate, styrene; covalently attached cations: butylimidazolium, trimethylammonium, butylpyrrolidinium) were successfully synthesized as carbon precursors. PIL-derived carbons with metallic luster were produced from this set of PILs *via* pyrolysis at 900 °C. Higher graphitic content was obtained by PIL-derived carbons compared to polyacrylonitrile (PAN)-derived carbon. Different structures were also observed between PAN-derived carbon (powder form) and PIL-derived carbons (sheet form). Furthermore, results show that cation type has a significant impact on surface area, carbon yield, graphitic content, and surface chemistry of the resulting carbon materials. The design of carbon materials based on a diverse set of PILs with various cation and polymer backbone chemistries may be an effective strategy to optimize carbon properties for its subsequent application in energy storage devices.

498884 Tailoring Surface Functionalization of Silica Nanoparticles in Nafion Nanocomposites for Improved Ion Selectively in Vanadium Redox Flow Batteries

Abstract:

Vanadium redox flow batteries (VRFBs) have emerged as a promising solution for grid-scale energy storage for intermittent energy generation such as wind and solar. However, the polymer electrolyte membrane (PEM), typically a perfluorinated ionomer (e.g., Nafion), utilized in VRFBs suffers from high vanadium crossover, which decreases the lifetime and efficiency of the battery. In response, the introduction of nanoparticles in Nafion has proven to be a successful modification method to reduce the vanadium crossover, though there are conflicting theories on how they act to reduce crossover as recent studies have shown that the nanoparticles reside in both the hydrophobic (Teflon-like) and hydrophilic (ionic) domains. In this study, silica (SiO_2) nanoparticle loading, size, and surface chemistry were systematically tailored to elucidate its influence on nanoparticle dispersion in the membrane and examine the role of nanoparticle dispersion in mitigating vanadium crossover. Specifically, several types of interactions between silica nanoparticle surface and Nafion structure were investigated, including electrostatic and covalent interactions within the ionic channels of Nafion, and electrostatic interactions with the hydrophobic Teflon backbone. The effect of these interactions, as it applies to nanoparticle dispersion, was obtained by electron imaging, and vanadium ion permeability through the membrane over time was measured by ultraviolet-visible (UV-Vis) spectroscopy. In addition, the bridging support between silica nanoparticle surface and end group was also investigated, ranging from a relatively flexible, short saturated chain to a significantly less flexible chain containing aromatic groups. Results from this study indicate that the surface functionalization of the silica nanoparticles plays an important role in controlling the permeation of vanadium ions across these nanocomposite membranes.

Thermally Cross-Linked Poly(acrylic acid) / Reduced-Graphene Oxide Aerogels as a Replacement for Metal-Foil Current Collectors in Lithium-Ion Batteries

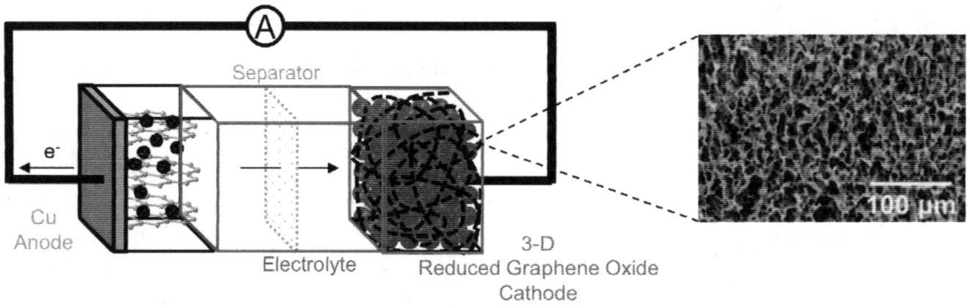

Han Xiao[1], *Joshua P. Pender*[2], *Mackenzie A. Meece-Rayle*[2], *J. Pedro de Souza*[2], *Kyle C. Klavetter*[2], *Heonjoo Ha*[1], *Jie Lin*[2], *Adam Heller*[1], *Christopher J. Ellison*[1] , *and C. Buddie Mullins*[2]

(1) Dept. of Chemical Engineering and Materials Science, the University of Minnesota, Minneapolis, MN
(2) Dept. of Chemical and Petroleum Engineering, the University of Texas at Austin, Austin, TX,

17AIChE Annual Meeting, Minneapolis, MN

UNIVERSITY OF MINNESOTA Driven to Discover℠

Motivation: Li-ion batteries lighter and higher capacity

Schematic of lithium-ion battery (LIB)

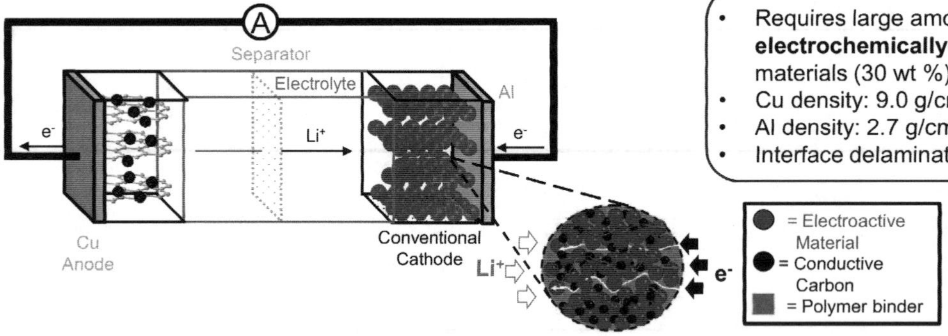

Issues

- Requires large amount of **electrochemically-inactive** materials (30 wt %)
- Cu density: 9.0 g/cm^3
- Al density: 2.7 g/cm^3
- Interface delamination

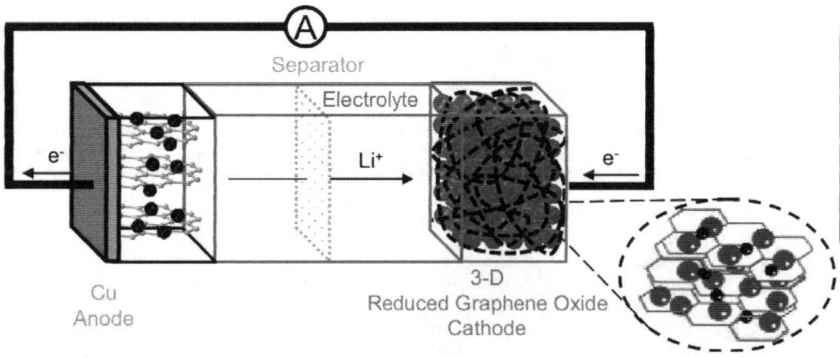

Motivations

- Aerogel density: 0.005 g/cm^3
- Porous framework facilitates large amount of electrolyte in contact with electrode for improved Li$^+$ transport
- Conductive scaffold alleviates dependency on conductive additive

1

Graphene Oxide Synthesis

Graphite
(Stacks of **Graphene**)

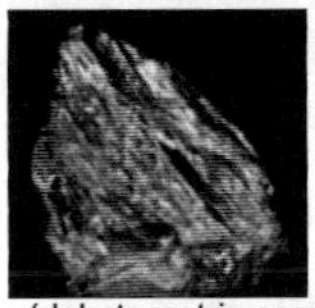

(dakotamatrix.com)

H_2SO_4,
$KMnO_4$
DI
water,
H_2O_2
HCl

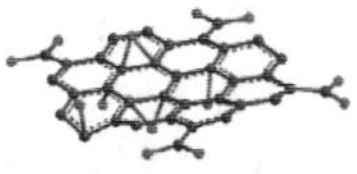

Graphene Oxide (GO)
Insulating
Soluble in H_2O
Easily processable

Reduction Conditions
e.g., heat, HI, hydrazine, etc.

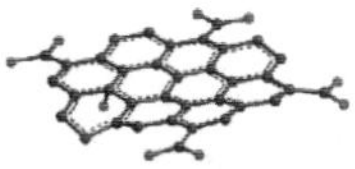

Reduced Graphene Oxide (rGO)
Conductive
Insoluble in H_2O

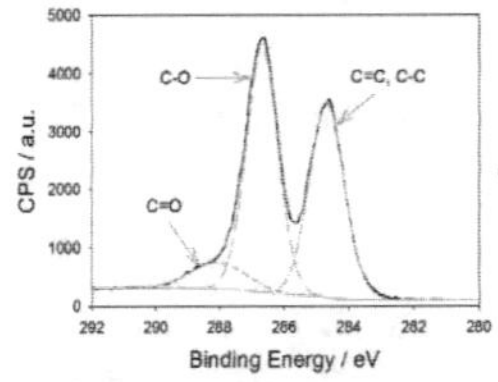

XPS C 1s GO aerogel

HI @ RT
24 h

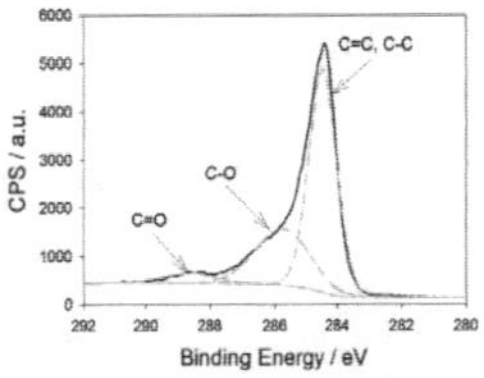

XPS C 1s rGO aerogel

Ha et al., *ACS Appl. Mater. Interfaces* 2015, 7, 6220–6229

2

Aerogel Synthesis

- Sponge-like, synthetic material
- Porous, composed mostly of air

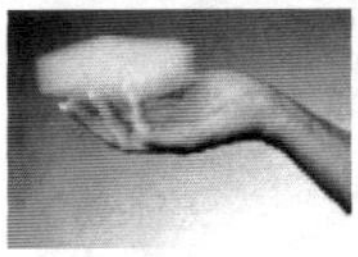

(Courtesy of Wikipedia)

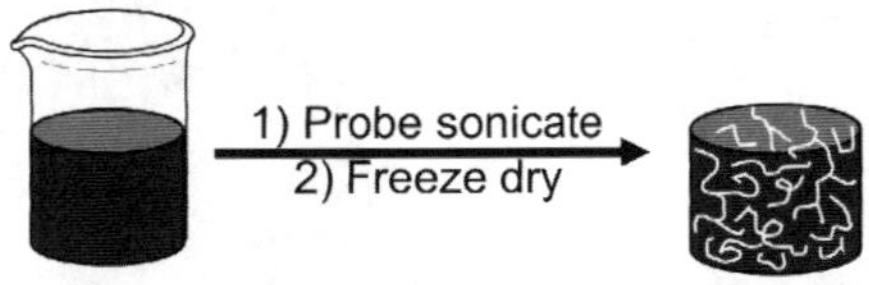

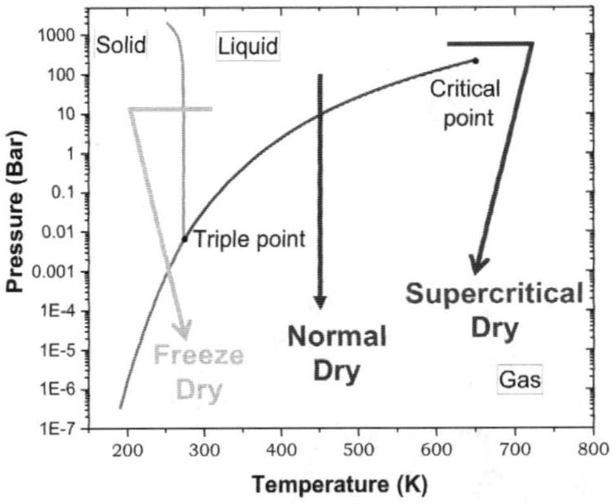

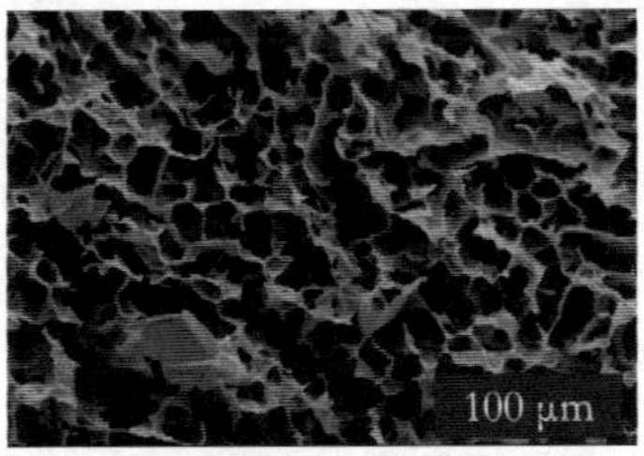

Ha et al., *ACS Appl. Mater. Interfaces* 2015, 7, 6220–6229

Aerogel density [mg/cm³]	Aerogel porosity [%]
3-6	99.8

3

Aerogel Synthesis

- **GO aerogel in its pure form is delicate and fragile → add polymeric crosslinker**

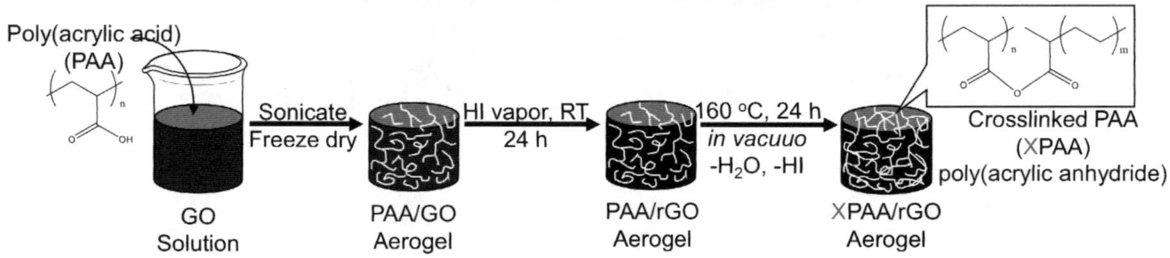

GO [mg]	PAA [mg]	Mw of PAA [kDa]	Aerogel density [mg/cm^3]	Aerogel porosity [%]
50	0	-	3.67	99.8
50	12.5	450	4.38	99.8
50	25	450	5.08	99.7
50	37.5	450	5.48	99.7
50	12.5	1,250	4.18	99.8
50	25	1,250	5.12	99.7
50	37.5	1,250	6.69	99.6

Average density **5 mg/cm^3** 200x lighter than water, **porosity >99.6%, surface area of 100 m^2/g**

Ha et al., *ACS Appl. Mater. Interfaces* 2015, 7, 6220–6229

4

Aerogel Morphology and Mechanical Properties

Morphology of XPAA/rGO Aerogel

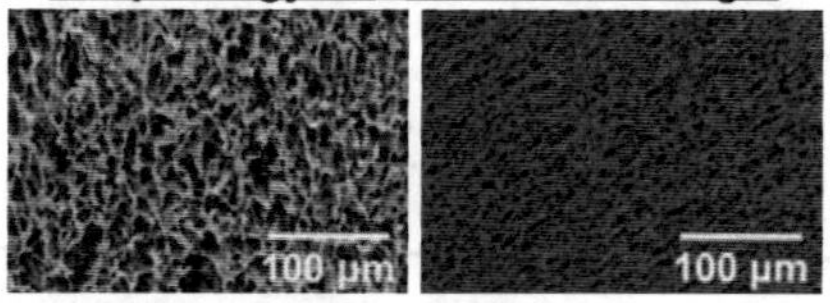

rGO/XPAA = 100:50 rGO/XPAA = 100:75

Before Apply After

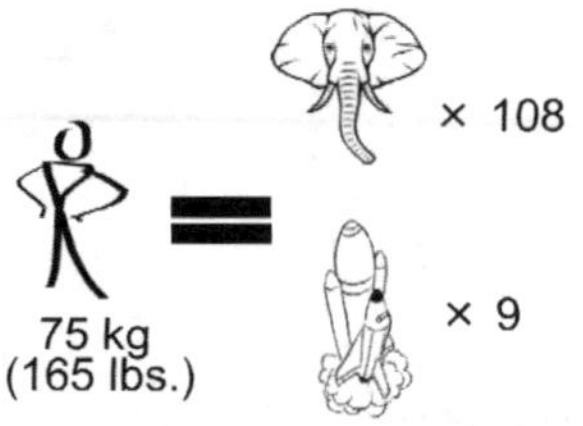

rGO/XPAA (1250 kDa)
2:1 w/w
Some Permanent Deformation

rGO/XPAA (450 kDa)
2:1 w/w
Full Recovery

5

Aerogel Electrodes syntheses

Proof-of-concept Study

"Active material"

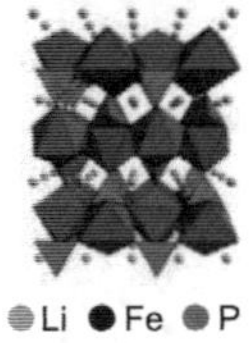

$LiFePO4$ **(LFP)**
- Commercial product
- D_{avg} ~ 1μm

"Inactive material"

Carbon black **(Super P)**
- Commercial product
- D_{avg} ~ 200 nm
- Conductive material

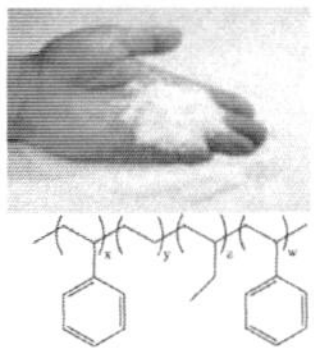

Styrene-ethylene-butylene-styrene (**Kraton**)
- Commercial product
- Thermoplastic elastomer
- Binder

3-D XPAA/rGO composite cathode

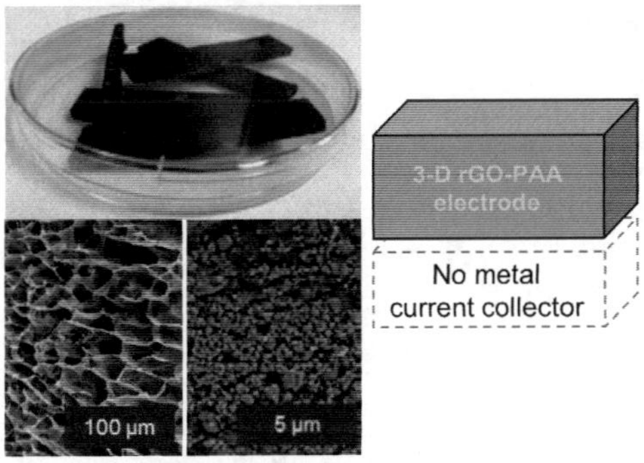

92.5% **LFP**
5% **Super P**
2.5% **Kraton**

Max stable loading: ~75 mg_{LFP}/cm^2 at ~420 μm thick

Conventional Li-ion cathode

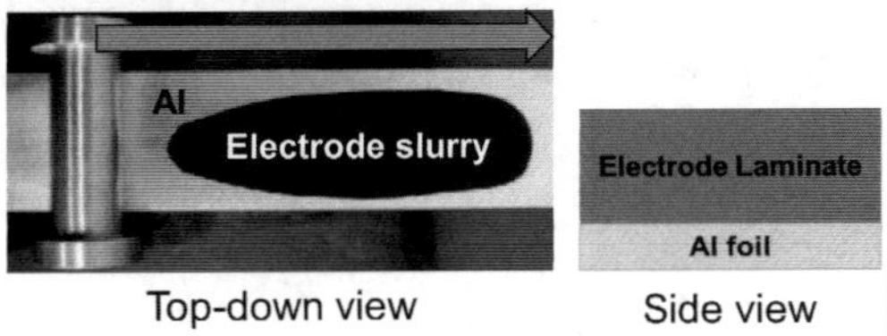

92.5% **LFP**
5% **Super P**
2.5% **Kraton**

Max stable loading: ~16 mg_{LFP}/cm^2 at ~100 μm thick

Courtesy of Wikipedia
Xiao et al., *ACS Appl. Mater. Interfaces* 2017, 9, 22641−22651

Discharge Capacity

- Assembled in CR 2302 coin cell
- Electrolyte: 1:1 EC:DEC
  1 M $LiPF_6$
- Anode: Li/Li^+

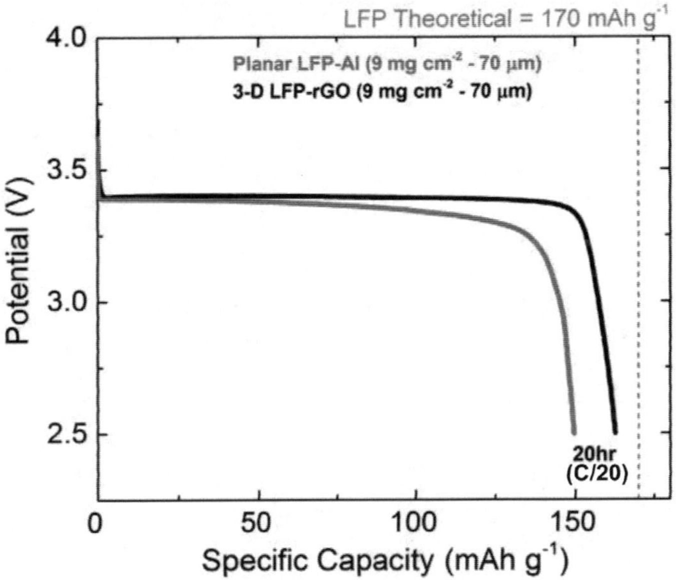

Xiao et al., *ACS Appl. Mater. Interfaces* 2017, 9, 22641−22651

Discharge Capacity

- Assembled in CR 2302 coin cell
- Electrolyte: 1:1 EC:DEC
  1 M $LiPF_6$
- Anode: Li/Li^+
- LFP-rGO: ~ 79 wt% LFP by mass
- LFP-Al: ~ 68 wt% LFP by mass

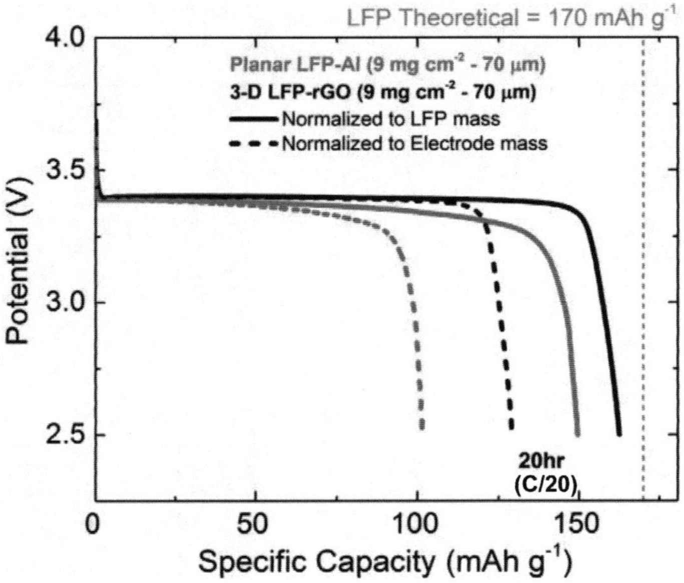

Xiao et al., *ACS Appl. Mater. Interfaces* 2017, 9, 22641−22651

8

Discharge Capacity

- Assembled in CR 2302 coin cell
- Electrolyte: 1:1 EC:DEC
  1 M $LiPF_6$
- Anode: Li/Li^+
- LFP-rGO: ~ 79 wt% LFP by mass
- LFP-Al: ~ 68 wt% LFP by mass

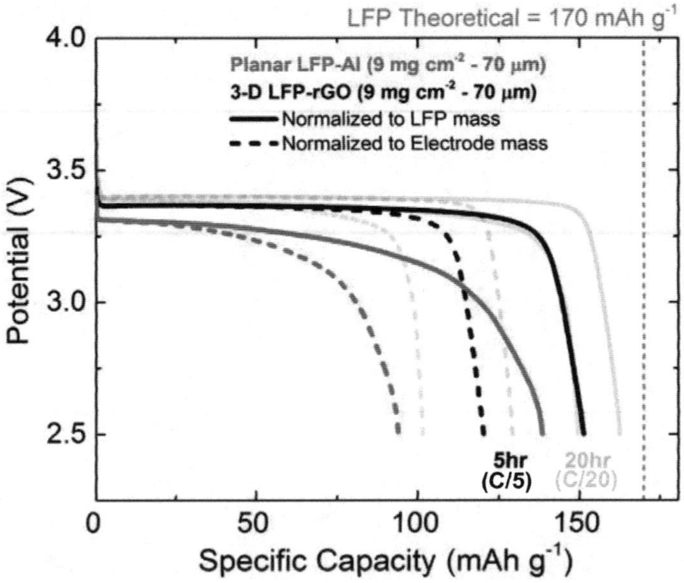

Xiao et al., *ACS Appl. Mater. Interfaces* 2017, 9, 22641−22651

9

Discharge Capacity

- Assembled in CR 2302 coin cell
- Electrolyte: 1:1 EC:DEC
  1 M $LiPF_6$
- Anode: Li/Li^+
- LFP-rGO: ~ 79 wt% LFP by mass
- LFP-Al: ~ 68 wt% LFP by mass

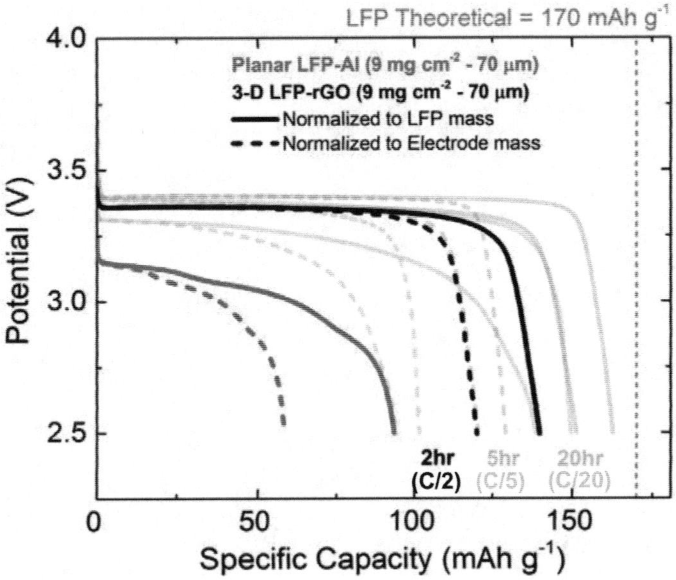

Xiao et al., *ACS Appl. Mater. Interfaces* 2017, 9, 22641–22651

10

Discharge Capacity

- Assembled in CR 2302 coin cell
- Electrolyte: 1:1 EC:DEC
  1 M $LiPF_6$
- Anode: Li/Li^+
- LFP-rGO: ~ 79 wt% LFP by mass
- LFP-Al: ~ 68 wt% LFP by mass

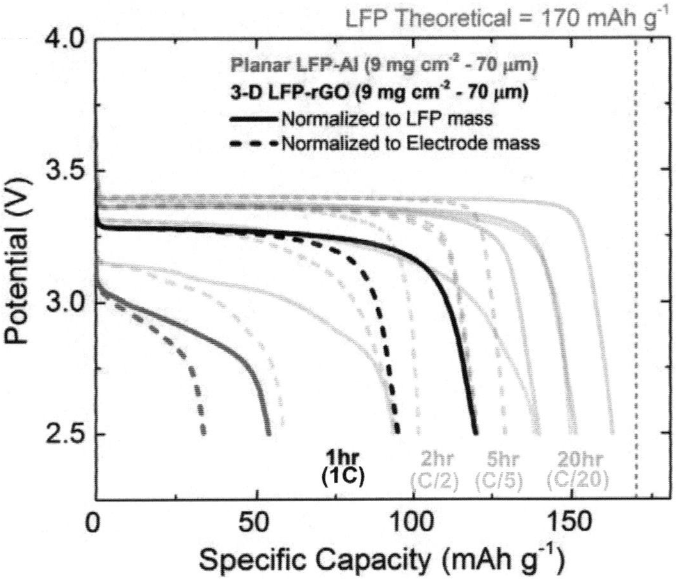

Xiao et al., *ACS Appl. Mater. Interfaces* 2017, 9, 22641–22651

11

Discharge Capacity

- Assembled in CR 2302 coin cell
- Electrolyte: 1:1 EC:DEC
  1 M $LiPF_6$
- Anode: Li/Li^+
- LFP-rGO: ~ 79 wt% LFP by mass
- LFP-Al: ~ 68 wt% LFP by mass

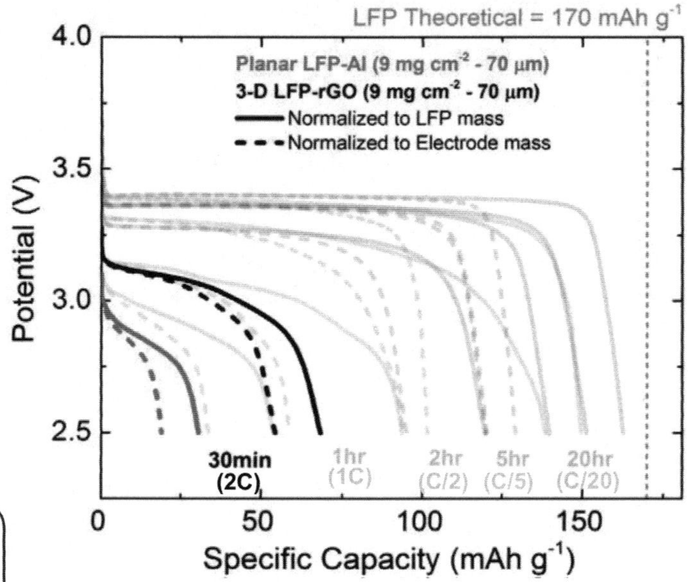

Major Conclusions

Aerogel electrodes have…
- Higher capacity per LFP/electrode mass
- Steadier discharge voltage

Xiao et al., *ACS Appl. Mater. Interfaces* 2017, 9, 22641−22651

12

Discharge Capacity

Gravimetric Discharge Plot

Long-Term Stability

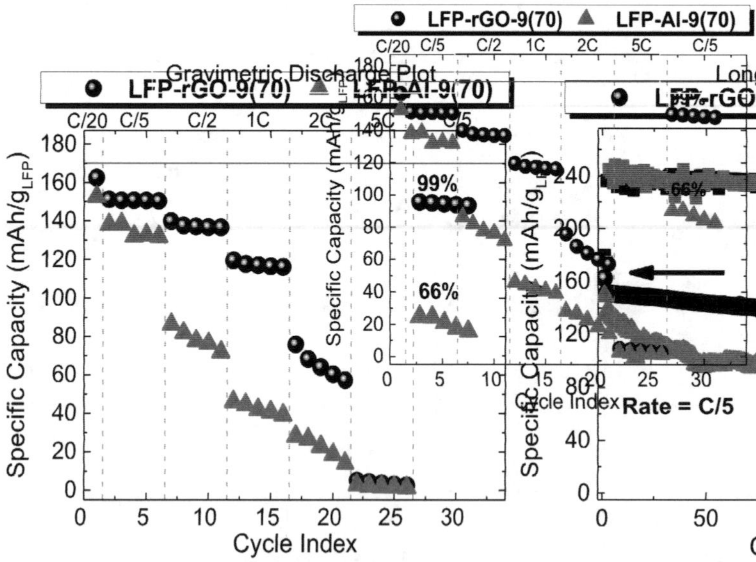

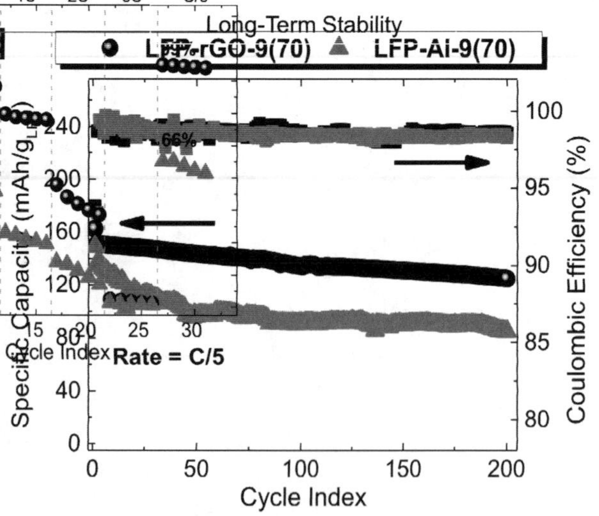

LFP-rGO has better capacity retention after cycling at a fast rate

The capacity of LFP-rGO decays at a slower rate

13

Correlation of Effective Surface Area and Performance

$\textit{Ratio of LFP } \underline{\textit{Effective Surface Area}}$
$= \mathbf{2.5x\ increase\ for\ LFP - rGO}$

Electrochemical Impedance Spectroscopy

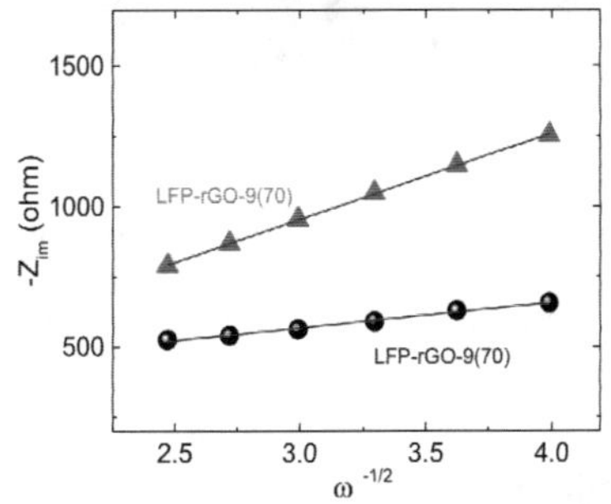

Electrodes with high effective surface areas:

- Possible better ionic contacts between the electrolyte and active materials
  - Allows electrolyte to be in closer proximity with the active materials
  - Large surface area for electrolyte penetration
  - Improved Li^+ diffusion and rate performance

Xiao et al., *ACS Appl. Mater. Interfaces* 2017, 9, 22641−22651
Ji et al., *Nano Lett. Interfaces* 2012, 12, 2446–2451

14

Summary

✓ Developed a method for fabrication and processing of robust 3-D LFP electrodes with a wide range of mass loading and thickness

✓ Improved specific energy of LFP cathode with rGO-PAA electrode

✓ Increased effective surface area of LFP by 2.5x upon replacing Al-foil with rGO-PAA

❑ Future work:

(1) Run the rate performances using a full cell (with rGO-PAA **anode**)

(2) Optimize the aerogel electrode design (loading/thickness/porosity)

15

Acknowledgements

Prof. Chris Ellison

Dr. Heonjoo Ha

The Ellison Group

Prof. Buddie Mullins

Prof. Adam Heller

Joshua Pender

Dr. Kyle Klavetter (Sandia)

Pedro de Souza (MIT)

THE Welch FOUNDATION

NSF

Kraton Giving Innovators Their Edge

CELGARD

UNIVERSITY OF MINNESOTA
Driven to Discover℠

UNIVERSITY OF MINNESOTA
Driven to Discover®

YouTube UniversityofMinn   UMNews   UofMN

PI:
Prof. Chris Ellison, cellison@umn.edu

Speaker:
Han Xiao, xiaox366@umn.edu

Group website: ellison.cems.umn.edu

UNIVERSITY OF MINNESOTA
Driven to Discover℠

Reduction Step and Thermal Annealing

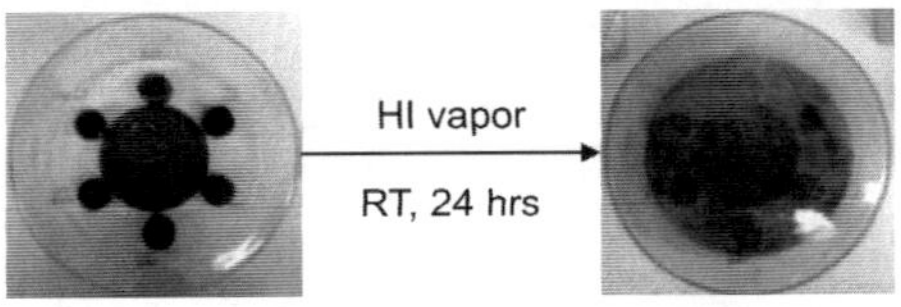

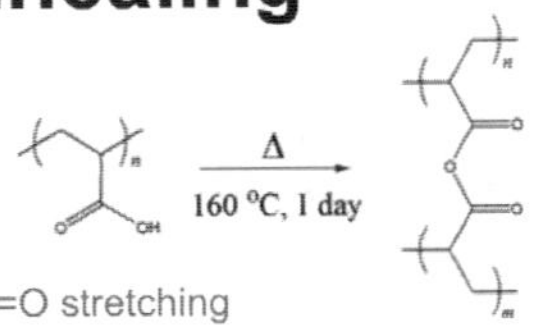

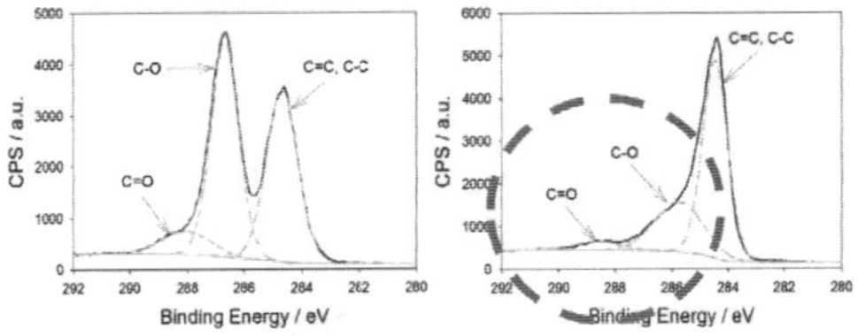

XPS C 1s GO aerogel   XPS C 1s rGO aerogel

- Effectively reduces to rGO
- **Residual groups enable hydrogen bonds and covalent bonds w/XPAA**

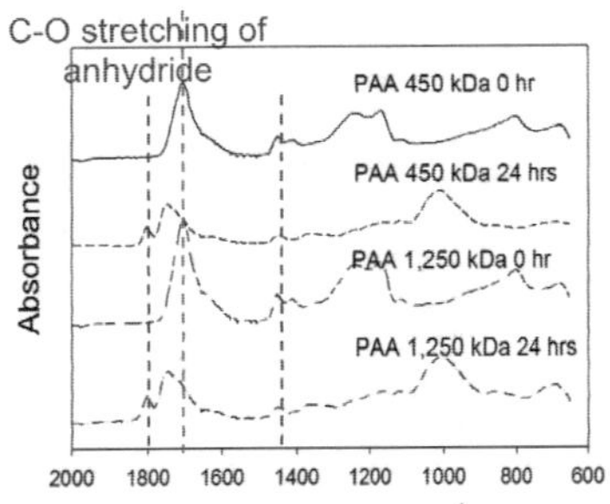

FT-IR Absorbance Spectra

UNIVERSITY OF MINNESOTA
Driven to Discover

Influence of Mass Loading/Thickness on Rate Performance

Electrodes with LFP loading spanning
~ 4 – 75 mg cm⁻²

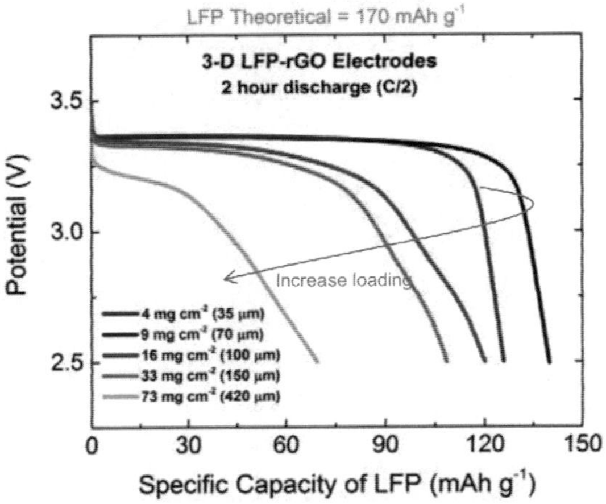

Speculation: best performance at lowest overall resistance

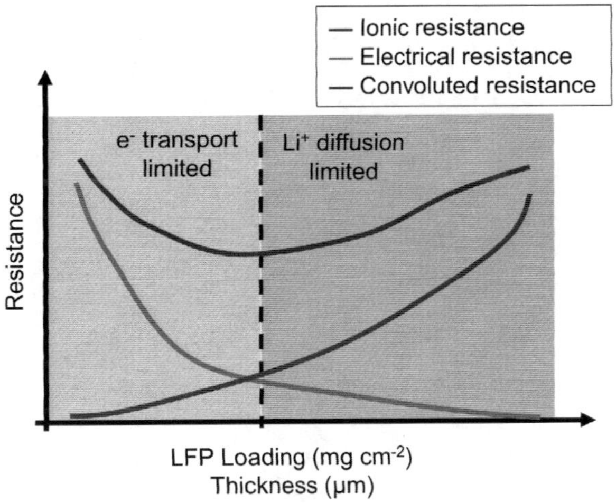

Xiao et al., *ACS Appl. Mater. Interfaces* **2017, 9, 22641−22651**
Cuevas et al., *J. Porous Mater.*, **2009**, 16, 675-681
Zheng et al., *Electrochim. Acta*, **2012**, *71*, 258-265

15

Cyclic Voltammetry (CV)

✓ ***Stability of pristine rGO-PAA in LIB voltage window (2.5 – 4.0 V vs. Li/Li⁺)***

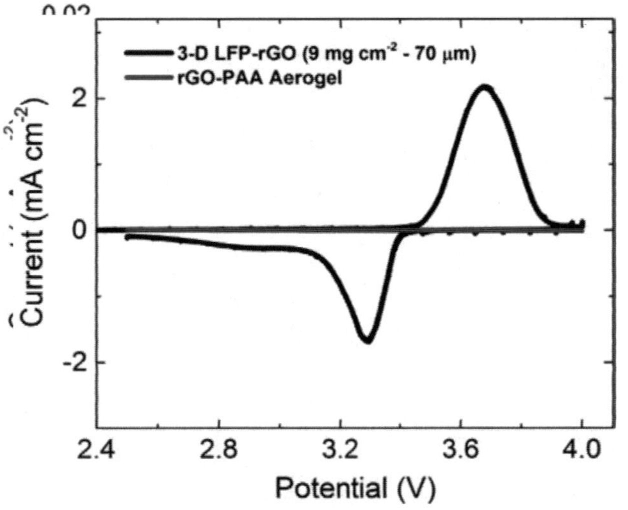

Pristine rGO-PAA Peak Current: ±14 μA/cm²

Electrolyte: 1M $LiPF_6$ in 1:1 Ethylene Carbonate (EC): Diethyl Carbonate (DEC)

Potential Window: 2.5 – 4.0 V (vs Li/Li⁺)

Scan rate: 0.1 mV/s

Xiao et al., *ACS Appl. Mater. Interfaces* **2017, 9, 22641−22651**

University of Minnesota
Driven to Discover

Discharge Capacity at 16 mg cm^{-2}

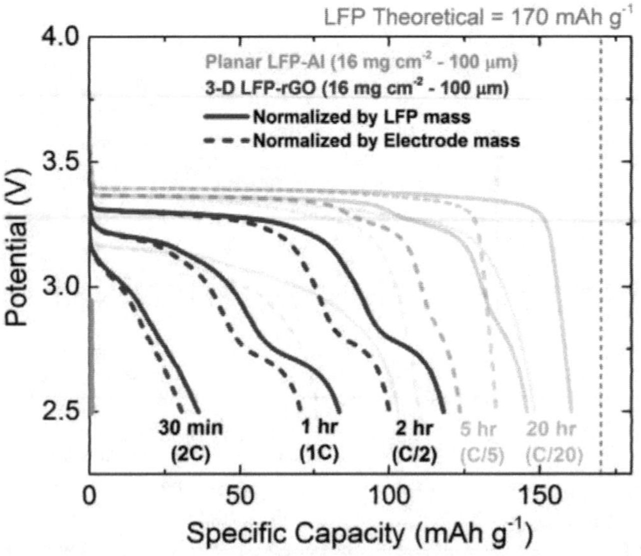

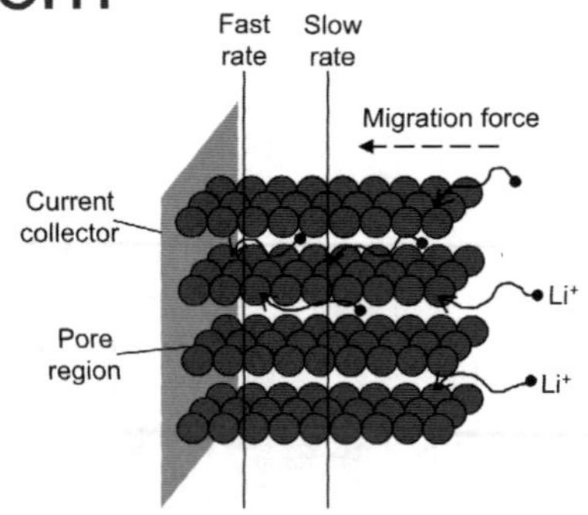

LFP-rGO: ~ 79% LFP by mass

LFP-Al: ~ 68% LFP by mass

LFP-rGO at 16 mg cm^{-2} exhibits cycle stability until 5C

LFP-Al at 16 mg cm^{-2} is overcome by transport limitations at C/2

Xiao et al., *ACS Appl. Mater. Interfaces* **2017, 9, 22641−22651**

University of Minnesota
Driven to Discover

12

Long term stability at C/5

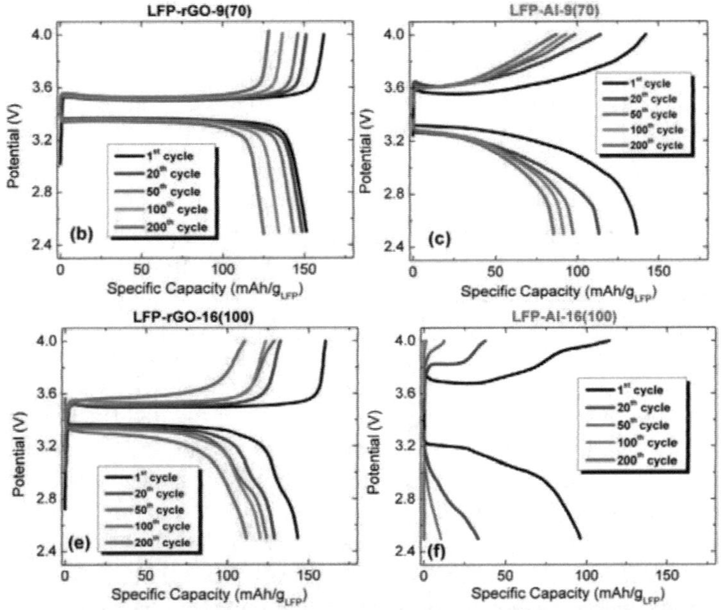

Xiao et al., *ACS Appl. Mater. Interfaces* **2017, 9, 22641–22651**

Alternative Electrode Scaffolds

Metal foams

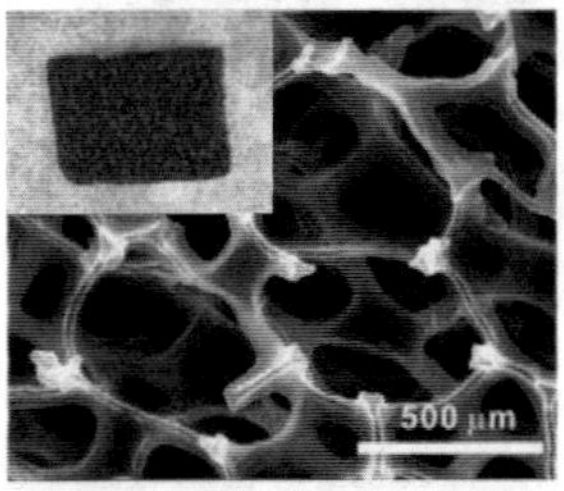

Li, J. et al. *Nano Lett.*, **2012**, *12*, 2446-2451.

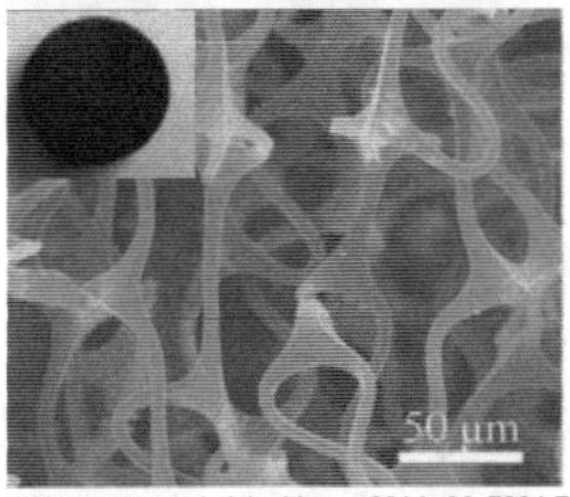

Huang, X. et al. *Adv. Mater.*, **2014**, *26*, 7264-7270

- x **High processing cost**
- x **Poor flexibility**
- x **High density**

- x **Complex syntheses**
- x **Mechanical Fragility**
- x **Metal-foil required**
  - x **≤ 5 mg cm^{-2}**

Porous Carbons

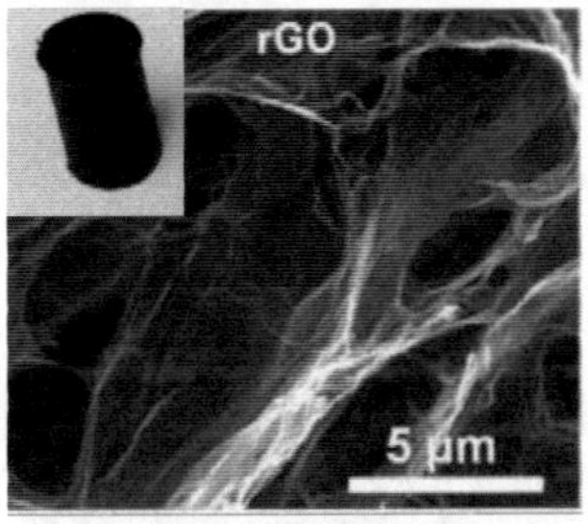

Zhou, G. et al. *Adv. Energy Mater.*, **2016**, *6*, 1-9.

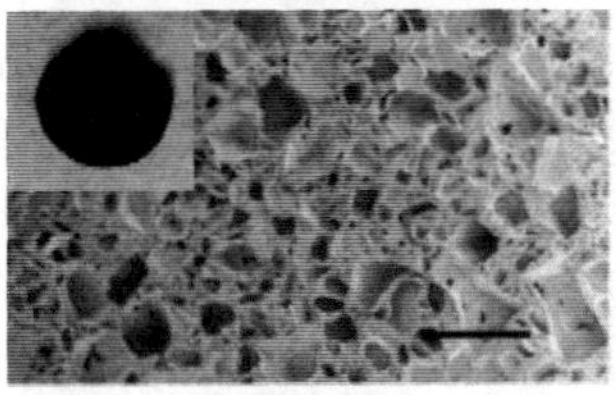

Liu, W. et al. *Adv. Mater.*, **2016**, *28*, 3578-3583

499826 Graphene-Oxide/Polybenzimidazole Nanocomposite Membrane for High Temperature Fuel Cell Application

Abstract:

Polymer electrolyte membrane fuel cells (PEMFCs) are ideally suited for future transportation applications because they are lightweight and low-cost. Current PEMFCs use Nafion membranes that require humidification equipment to maintain proton conductivity and cannot work at temperatures above about 80Â°C. Yet high temperature is crucial in achieving favorable reaction kinetics, high efficiencies, high power densities, and reduced sensitivity to Carbon Monoxide (CO) poisoning in PEMFCs. Acid-doped polybenzimidazole (PBI) membranes were found to meet the DOEâ€™s targets for high temperature PEMFCs operating under no humidification on both anode and cathode sides. The proton conductivity of acid-doped PBI is derived from the acid and thus eliminating the need for water humidification equipment. It is expected that graphene-oxide/PBI (GO/PBI) nanocomposite membranes will further improve the performance of the fuel cell. GO nanosheets can increase proton conductivity of the PBI by forming hydrogen bonds with acid and water. In addition, GO can decrease acid-induced swelling of PBI and increase mechanical strength of PBI membrane. In this study, we incorporated different amounts of GO nanosheets (whose edges are 4 - 10% oxidized) in PBI solution prior to membrane casting. The nanocomposite membrane was doped in PA (85 wt%) at room temperature. The properties of the nanocomposite membrane were characterized using XRD, TGA and SEM. In XRD, graphite diffraction peak (002) was found at 2-theta = 26.4Â°. H_3PO_4 doping level (number of H_3PO_4 molecules per repeat unit of PBI) was found to increase with increase in acid concentration for all samples and the doping level was maximum for 10% GO/PBI compared to 5% GO/PBI and neat PBI. The membrane electrode assembly (MEA) was prepared by hot-pressing carbon electrodes (20 wt% platinum, loading of 0.5 mg/cm^2) on both sides of the GO/PBI nanocomposite membrane. The MEA was evaluated using Scribner Associateâ€™s Fuel Cell Testing Module 850e. The polarization curves of the MEAs were obtained with non-humidified H_2/air at 70°C, 100°C and 120°C, respectively. At cell potential of 0.6 V, we observed highest current density at 120°C for both 5% GO/PBI and 10% GO/PBI (80 mA/cm^2 at 0.6V). At 120°C, 5% GO/PBI exhibited peak power density as high as 92.8 mWcm^{-2}. At 100°C, 10% GO/PBI exhibited peak power density of 121.6 mWcm^{-2}. All this was promising and addition experimental data of the GO/PBI study will be needed in order to optimize the membrane.

496133 Autoperforation of 2D Materials for Generating Two Terminal Memresistor Janus Particles with Nonvolatile Memory

Abstract:

Due to its inherent stochasticity, brittle fracture is seldom used as a nanofabrication method. However, the imposition or templating of a specific strain field can guide fracture along a pre-set design. Herein, we show that this autoperforation provides a means of spontaneous assembly for 2D surfaces. Chemical vapor deposited mono- and bi-layer graphene, molybdenum disulfide, or hexagonal boron nitride (hBN) autoperforate into circular envelopes when sandwiching a microprinted polymer spot, allowing lift-off and assembly into solution. The resulting composite microparticles have two independently addressable, external Janus faces that we show can function as an intraparticle array of parallel, two terminal electronic devices. As an example, we print a 0.9 wt% black phosphorous (BP) nanoflake in polystyrene (PS) latex ink into mono-layer graphene sandwich particles, resulting in micro-particles possessing non-volitale, 15 bit memory storage via a spatially addressable memresistor array throughout the particle interior. Such particles form the basis of free floating devices capable of collecting and storing digital information in their environment. The 2D envelopes demonstrate remarkable chemical and mechanical stability during 4 months of preservation in aqueous media. They also survive a highly acidic gastrointestinal environment, as well as aerosolization over 0.3 meters and re-collection. Autoperforation of 2D materials into such envelope structures open the door to precise compositing of particulate devices, extending nanoelectronics into previously inaccessible environments

489451 Synthesis and Properties of Polymer/Graphene Oxide (GO) Thermosets with Multifunctional GO As a Crosslinker

Abstract:

Flexible polymers have wide applicability in many areas including wearable devices, flexible displays, and devices for monitoring physiological signals. These high-end applications typically require the use of an elastomer, such as crosslinked poly(dimethylsiloxane) (PDMS) or thermoplastic elastomer, to provide flexibility and mechanical robustness in addition to comfort when interfaced with the human body. While a variety of commercialized choices are available in the market, such precursors used to synthesize elastomers are generally high in viscosity and modulus, limited to several grades and the recipe for preparing the final product can be complex. Therefore, developing new methods for forming elastomers that exhibit unique combinations of properties (*e.g.*, mechanical, electrical, gas barrier, and gas separation, *etc.*) from only a few components could be important for practical use.

Herein, we summarize two different mechanisms for synthesizing polymer/graphene oxide (GO) composite elastomers that use GO as a multifunctional crosslinker as well as mechanical property enhancer. Both methods utilize the epoxy group on the surface of GO, allowing the synthesis of highly crosslinked but highly extensible materials. The first method [1] uses primary amine functional groups on the ends of a telechelic PDMS precursor (a liquid) which can undergo post-processing reactions with epoxy groups on the surface of GO to form a solid robust elastomer during simple heating. Experiments indicate that the elastomer formed from primary amine-terminated telechelic PDMS and 1 wt % GO is crosslinked with more than 75 wt % gel, but flexible such that it can be stretched up to 300% of its original length. Gas barrier properties were also examined in detail in order to develop fundamental understanding of the connection between composite architecture and gas transport properties. By incorporating less than 4 vol. % of GO resulted in a 99.9% reduction in gas transport, thereby highlighting its potential use as a gas barrier membrane.

The second method [2] explores similar polymer/GO composite elastomers that were synthesized from secondary amide containing polymers; in this case, the PDMS precursors were terminated with pyrene ends connected to the main polymer chain through a secondary amide linkage. These pyrene ends form dynamic Ï€-Ï€ interactions with GO surfaces, which reduce the intermolecular spacing between secondary amide groups near epoxides on the surface of GO. In this case, pyrene ends act as catalysts during the crosslinking reaction, thereby molecularly templating the reaction. From these studies, we conclude that the simplicity and generality of the reactions coupled with wide-tunability in the final composite elastomer properties could enable new composite elastomers to satisfy high performance needs in many application areas.

[1] Gas Permeation and Selectivity of Poly(dimethylsiloxane)/Graphene Oxide Composite Elastomer Membranes, H. Ha, J. Park, S. Ando, C.B. Kim, K. Nagai, B.D. Freeman and C.J. Ellison, *Journal of Membrane Science*, 518, 131-140 (Jun. 2016). http://dx.doi.org/10.1016/j.memsci.2016.06.028

[2] Molecularly-Templated Reaction for Forming Poly(dimethyl siloxane)-Graphene Oxide Composite Elastomers, H. Ha, K. Ha, and C.J. Ellison, *under review*.

499184 Polymer Silica Composite Nanofibers Via Sol Gel Electrospinning

Abstract:

In recent years, polymer-silica composites with enhanced thermal and mechanical properties have found applications in a variety of fields including catalysis, adsorption, pervaporation, sensors, and enzyme encapsulation. Scope of application of these composites can be further broadened by transforming them to nanofibrous structures with high surface to volume ratios. In this work, we have employed a controlled solâ`'gel synthesis incorporated with electrospinning to produce polymer silica composite nanofibers. We used polyvinyl alcohol (PVA) and polyacrylonitrile (PAN) as the polymer phase and by mixing PAN or PVA with silica precursor, we produced electrospun composite fibers having varying properties depending on the polymer phase and nature of the sol gel processed silica. We systematically investigated the effects of i) silica precursor mixture composition, ii) silica-polymer ratio, and iii) aging time of the silica precursor mixture, on the fiber properties.

Stabilized carbonization of PAN silica composite fibers at 800 Â°C transformed the fibers to carbon silica composite nanofibers with smooth morphology and diameter ranging from 400 to 700 nm. FTIR analysis of the fibers confirmed the presence of silica in the as-spun as well as the carbonized material. Role of silica in modifying the graphitic character of the carbon-silica fibers is studied through Raman analysis. When soaked overnight in water, the PVA silica composite nanofibers remain intact, essentially maintaining their morphology, even though PVA is soluble in water. FTIR analysis of the fibers further helps us to propose that mixing silica precursor with PVA in solution initiates the participation of the silica precursor in cross-linking of PVA, so that its â`'OH group becomes unavailable for hydrogen bonding with water. The ability to form cross-linked nanofibers of PVA using thermally stable and relatively inert silica broadens the scope of these materials in various technologies.

492888 Graphene Oxide/Lipid Composite Material Towards a Multifunctional Drug Delivery Vehicle

Abstract:

A graphene oxide [GO]/lipid composite for an enhanced and multifunctional drug delivery vehicle has been devised. In this study, liposomes were prepared using the gentle hydration method and integrated with GO, to advance the loading of payloads via noncovalent bonding—so as not to deform the confirmation of the bound molecule. Fluorescent integration of the liposome and GO was analyzed and monitored under light microscopy and UV-vis spectroscopy. This work can also be extended to theranostic processes, such as detection and delivery schemes.

Localizing Graphene at the Interface of HDPE/PLA Polymer Blends

Sung Cik Mun[1], Min Jae Kim[1], Liangliang Gu[1], Monica Cobos[2], Christopher W. Macosko[1]

[1] *University of Minnesota, Minneapolis, MN, United States*
[2] *University of the Basque Country (UPV/EHU), San Sebastian, Spain*

Abstract

Conductive polymer nanocomposites have been developed for potential uses in electrostatic discharge and electromagnetic interference shielding. Although high filler loadings improve electrical conductivity, there are potential problems originating from elevated melt viscosity, poor appearance and contamination by fillers sloughed off from the surface as well as high material cost. Among various strategies, the localization of graphene at the interface of polymer blends has emerged as a promising way to reduce the percolation threshold.

We investigated two approaches to the interfacial localization of graphene nanosheets in co-continuous high density polyethylene (HDPE)/polylactic acid (PLA) blends prepared by melt compounding. First, localization of graphene at the interface of HDPE/PLA blends was achieved by controlling kinetic factors. Three different mixing sequences were carried out by (i) mixing HDPE, PLA and graphene at the same time, (ii) diluting pre-mixed HDPE/graphene nanocomposites with PLA, and (ii) diluting pre-mixed PLA/graphene nanocomposites with HDPE. In the case of (iii), a very low percolation threshold (~ 0.05 vol%) was achieved by migration of graphene from PLA to the interface. On the other hand, mixing sequences (i) and (ii) gave percolation thresholds around 0.5 vol%, as expected for preferential localization of graphene in the HDPE phase; the percolation threshold of HDPE/graphene nanocomposites is 1 vol% and the blend was 50/50 by volume.

In our second approach, polystyrene (PS) was added as a third component to HDPE/PLA binary blends. In HDPE/PLA/PS ternary blends (45/45/10 or 40/40/20 by volume), PS layer was located between HDPE and PLA phases and exists as a thin, continuous interfacial layer. When graphene meets the interface of HDPE/PS or PLA/PS, it was predicted that it prefers being in PS phase or at the interfaces, which implies graphene will be kept in the PS layer during melt compounding. To demonstrate our prediction, PS and graphene were pre-mixed then diluted with HDPE and PLA to fabricate HDPE/PLA/PS/graphene nanocomposites. Graphene was found to be localized in the PS layer only, and the percolation threshold was reduced to 0.1~0.2 vol% accordingly. Relying on these experimental results along with thermodynamic calculations, we suggest general ways to achieve high electrical conductivity from co-continuous polymer blends.

Introduction

Graphene, a two-dimensional layer consisting of sp^2-hybridized carbon lattice has attracted much attention in industry and academia owing to its superb mechanical, electrical, and thermal properties together with flexibility. The thickness of well-exfoliated graphene sheets ranges from 0.35 to a few nanometers, which leads to very high aspect ratio compared with other conventional carbon materials such as carbon black or carbon fiber. Among various attempts to make use of excellent properties of graphene, conductive polymer nanocomposites with graphene or its derivatives have been developed to meet requirements for plastic industry. Depending on its potential applications, there are several issues to be dealt with; excessive cost, elevated melt viscosity, sloughing of graphene from the surface, and optical opacity at high concentrations. For this reason, there has been much efforts to reduce concentration of graphene for percolation, which represents three-dimensional, space-spanning network of incorporated fillers through the whole matrix. Polymer/graphene nanocomposites were fabricated by either solution blending or melt compounding. It should be noted that melt compounding requires higher graphene loadings than solution blending, as summarized in Figure 1 [1-13]. Although it is much preferred for industrial applications accompanying mass production, melt compounding generally results in poor dispersion of inorganic fillers in polymer matrices. Several strategies have been developed to improve the interfacial interaction between graphene and matrix polymer by modifying the surface of graphene [14-16] or the chemical nature of polymer resin [6,17,18]. On the other hand, localizing graphene at the interface of polymer blends has been regarded as the most promising way to reduce percolation threshold as illustrated in Figure 2 [19]. In this study, we suggest two approaches to localizing graphene at the interface of co-continuous blends. HDPE/PLA co-continuous blends were used as a model in this study. First, we compare three different mixing sequences to control kinetics of graphene migration toward the blend interface. Second, we introduce PS, a third polymeric component, as the interfacial continuous layer between major components in which graphene was localized exclusively. It was demonstrated that both approaches lead to percolation thresholds as low as 0.05~0.2 vol%, drastically reduced from 1 vol% of HDPE/graphene nanocomposites.

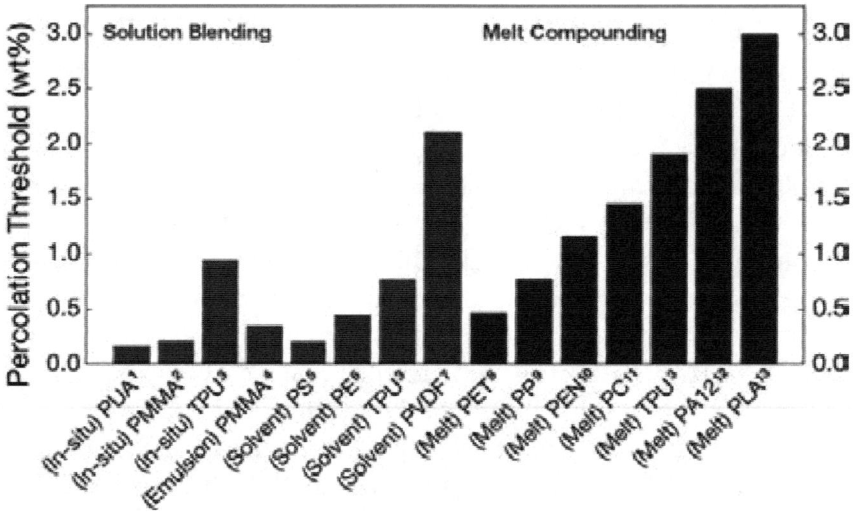

Figure 1. Percolation threshold of polymer/graphene nanocomposites prepared by solution blending and melt compounding.

Experimental

Materials

HDPE and PLA, PS were provided from J-REX (HD KF 251A), Nature Works (Ingeo^{TM} Biopolymer 2003D) and Dow Chemicals (STYRON 666D), respectively. Graphene nanoplatelets in the form of powder (N002-PDR) were purchased from Angstron Materials. PLA and PS were dried in a vacuum oven at 80 °C for overnight to remove any residual moisture.

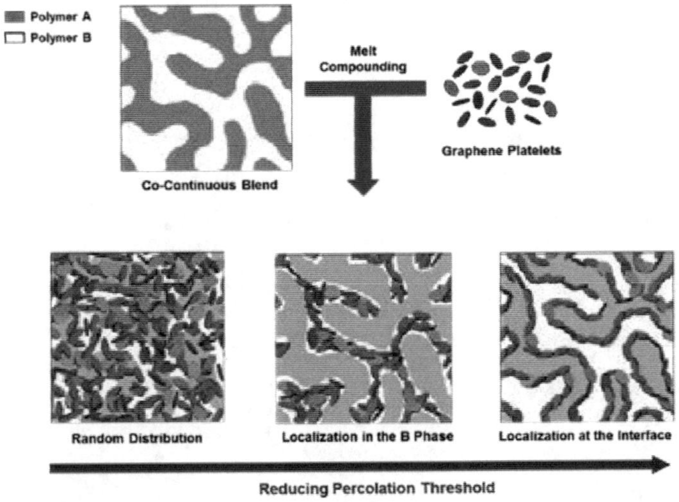

Figure 2. Schematic of graphene distribution in a co-continuous blend

Preparation of HDPE/PLA/Graphene Nanocomposites

There are three mixing sequences used in this study; (i) mixing HDPE, PLA, and graphene at the same time, (ii) diluting pre-mixed HDPE/graphene nanocomposites with PLA, and (iii) diluting pre-mixed PLA/graphene nanocomposites with HDPE. In the case of (i), dry-mixed resin and graphene were put into a microcompounder (MC5, Xplore Instruments) at 180 °C for 5 min at 50 rpm. Then, the mixture was mixed at 180 °C for extra 5 min at 200 rpm, followed by extruding the melt from the microcompounder. In the case of (ii) and (iii), the experimental procedure is the same, except pre-mixed PLA/graphene or HDPE/graphene nanocomposites was dry-mixed and diluted with pure HDPE or PLA, respectively. Melt compounding was carried out under N_2 gas protection. The pre-mixing of one polymer and graphene was carried out in the processing condition as same as the dilution process. Sample codes were assigned for each nanocomposite; (i) HDPE/PLA/G, (ii) HDPE/(PLA/G), and (iii) (HDPE/G)/PLA.

Preparation of HDPE/PLA/PS/Graphene Nanocomposites

The pre-mixed PS/graphene nanocomposites were prepared by solution blending. PS was dissolved in dimethylformamide (DMF) at 50 °C for 2 h. Graphene powder was dispersed in DMF at room temperature and sonicated with a tip ultrasonic homogenizer for 2 h. Then the graphene dispersion was added to the PS solution for 24 hr with magnetic stirring. The mixture was precipitated in methanol and the solid component was filtered, washed with ethanol for three times, and dried under vacuum at 80 °C. The pre-mixed PS/graphene nanocomposites were diluted with pure HDPE and PLA resins and processed in the microcompounder at 180 °C for 5 min at 50 rpm and for extra 5 min at 200 rpm. Melt compounding was carried out under N_2 gas protection.

Characterization

Prior to measuring electrical conductivity, nanocomposites were put in a metal mold between two metal plates, placed on a compression molder heated to 180 °C for 3 min, and then compressed for 10 min. Copper electrodes were attached on the surface of samples. The current-voltage (I-V) curves were collected within -20 V to 20 V with an interval of 1 V by two-probe method. The reciprocal of the average slope of I-V curve was used as the electrical resistance. It was converted to volume conductivity in the unit of S/cm.

The morphology of HDPE/graphene and HDPE/PLA/graphene nanocomposites were observed by transmission electron microscope (TEM). Samples were microtomed by a diamond knife at -140 °C, then subsequently the microsections were placed on 300 mesh Cu grids. For scanning electron microscopy (SEM) analysis, the surface of HDPE/PLA/PS/graphene nanocomposites was trimmed by a diamond knife at -140 °C then PS layer was removed by cyclohexane at 50 °C for overnight. The samples were sputter-coated with Iridium before SEM images were collected.

Table 1. Wetting coefficients of graphene in polymer pairs of HDPE/PLA, HDPE/PS, and PLA/PS

polymer pair	Wetting coefficient (ω_a)	
	Owens and Wendt	Wu
HDPE/PLA	-0.57	-0.57
HDPE/PS	-1.16	-0.87
PLA/PS	0.95	0.79

Table 2. Spreading coefficients in HDPE/PLA/PS ternary blends

polymer pair (i/j)	Spreading coefficient ($\lambda_{i/j}$)	
	Owens and Wendt	Wu
PLA/PS	-11.0	-13.1
PS/PLA	6.2	4.2
HDPE/PLA	-15.2	-13.1

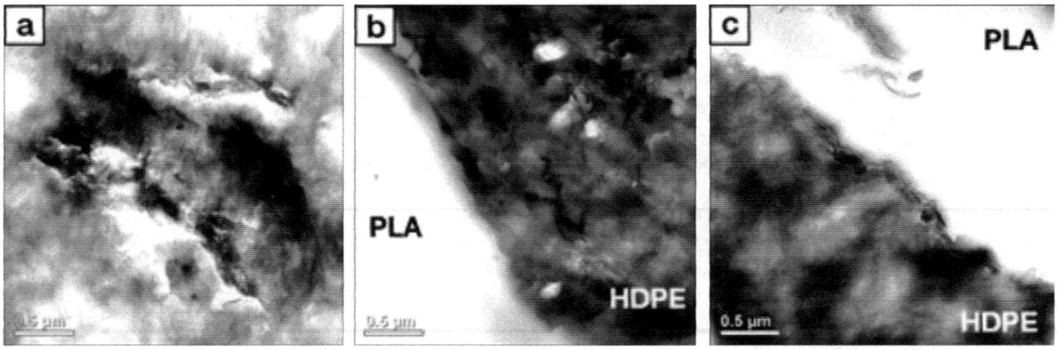

Figure 3. TEM images of HDPE/graphene and HDPE/PLA/graphene nanocomposites. (a) HDPE/G (0.2 vol%), (b) HDPE/PLA/G (1 vol%) and (c) HDPE/(PLA/G) (0.2 vol%)

Results and Discussion

Wetting Coefficients in HDPE/PLA/Graphene Nanocomposites

Calculated upon Young's equation, wetting coefficients (ω_a) predict the distribution of particles in polymer blends in thermodynamic equilibrium at certain temperature [20]. In HDPE/PLA/graphene nanocomposites, ω_a is between -1 and 1, which indicates graphene is expected to be localized at the interface of HDPE and PLA (Table 1). However, it was observed that graphene was localized in HDPE phase, not at the interface (Figure 3b). This inconsistency implies that the mixture did not reach the equilibrium. It cannot be attributed to the viscosity difference during melt compounding because HDPE showed one-order-magnitude higher viscosity than PLA. It should be noted that negative sign of ω_a means that graphene has lower interfacial tension with HDPE than with PLA. Thus, it was postulated that graphene prefers to be wetted by thermodynamically favorable HDPE phase when the system is not in the equilibrium state.

Spreading Coefficients and Wetting Coefficients in HDPE/PLA/PS/Graphene Nanocomposites

Spreading coefficients ($\lambda_{i/j}$) indicate the tendency of component i to encapsulate or spread around component j in a matrix of component k [21]. Table 2 summarizes spreading coefficients of polymer pairs (HDPE/PLA, HDPE/PS, and PLA/PS) at 180 °C. We estimated that the morphology of HDPE/PLA/PS ternary blends is co-continuous for all three components and the minor PS phase forms an interfacial layer between HDPE and PLA.

Wetting coefficients was calculated to estimate filler localization in HDPE/PLA/PS ternary blends. During melt compounding, graphene premixed with PS could meet two different interfaces; HDPE and PS or PLA and PS. From wetting coefficients listed in Table 1, it was predicted that graphene prefers PS or interfaces, which implies that graphene would remain in PS phase after the dilution process.

HDPE/PLA/Graphene Nanocomposites: Effects of Mixing Sequence

TEM images were taken to clearly observe graphene sheets in nanocomposites (Figure 3). Figure 3a shows HDPE/graphene binary nanocomposites. Graphene sheets were observed in HDPE (cloud-like background). In HDPE/PLA/G (Figure 3b), graphene sheets were located in HDPE phase, which is inconsistent with the prediction from wetting coefficients, presumably due to thermodynamic inequilibrium as discussed in a previous section. In HDPE/(PLA/G) (Figure 3c), graphene sheets were localized at the interface and some sheets were observed in PLA phase. During melt compounding, graphene sheets pre-dispersed in PLA phase migrated toward the interface and trapped because of thermodynamic preference. It is also worth noting that planar-shaped graphene is more likely stable at the interface of blends compared with one-dimensional fillers with high aspect ratio [22].

Figure 4 present the electrical conductivity of nanocomposites. The percolation threshold of HDPE/G is 1 vol%. In the case of HDPE/PLA/G and (HDPE/G)/PLA, the effective concentration is halved by localization of graphene in HDPE phase, thereby resulting in the percolation threshold at 0.5 vol%. The percolation threshold of HDPE/(PLA/G) is as low as 0.05 vol% because of the localization of graphene at the interface, although some graphene sheets remain in PLA phase. Provided that graphene is most stable at the interface in equilibrium state, we conclude that premixing graphene with less preferred phase and diluting the mixture with preferred phase is a promising way to achieve the interfacial localization.

HDPE/PLA/PS/Graphene Nanocomposites: Graphene Localization in the Interfacial PS Layer

Figure 5 shows the morphology of HDPE/PLA/PS/graphene nanocomposites. A portion of graphene sheets might be removed together with the interfacial PS layer, but some were observed in which the PS layer was etched out. They were not seen in either HDPE or PLA phase. Thus, we conclude that graphene sheets were encapsulated by PS layer in which graphene was premixed and did not migrate to HDPE or PLA phase. Figure 6 presents the electrical conductivity of HDPE/PLA/PS/graphene and premixed PS/graphene nanocomposites. The percolation threshold is around 0.1~0.2 vol%, which was reduced from 1 and 0.5 vol% for HDPE/graphene and HDPE/PLA/graphene nanocomposites (HDPE/PLA/G, mixed at the same time), respectively.

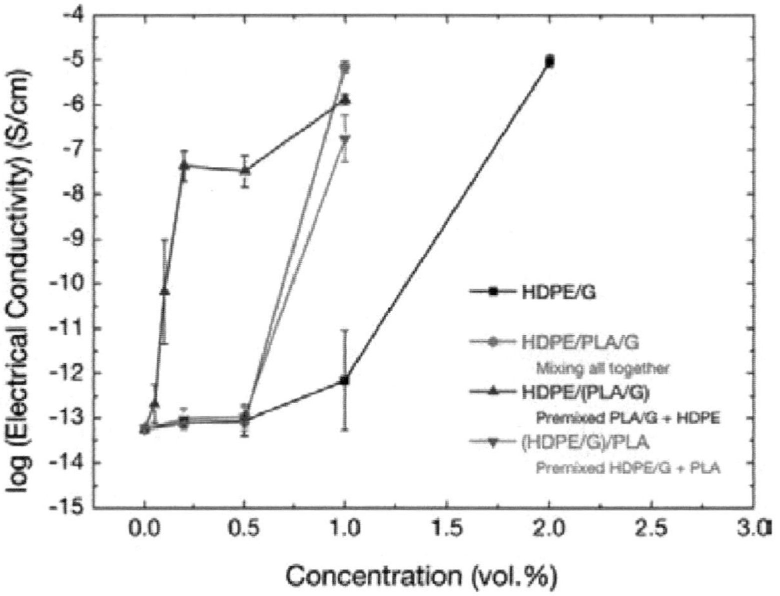

Figure 4. Electrical conductivity of HDPE/graphene and HDPE/PLA/graphene nanocomposites.

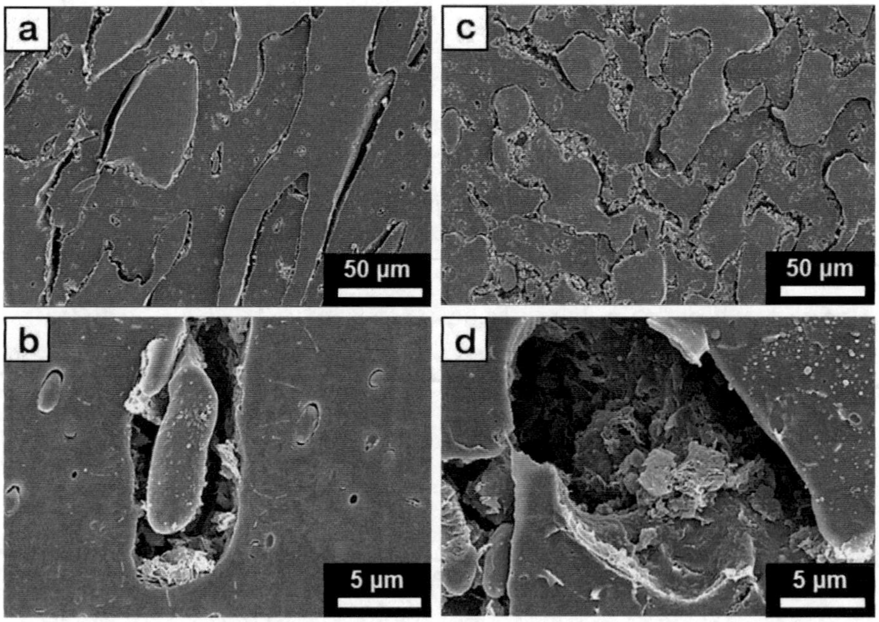

Figure 5. SEM images of HDPE/PLA/PS/graphene nanocomposites with volume ratios of (a-b) 40/40/9.9/0.1 and (c-d) 40/40/19.8/0.2

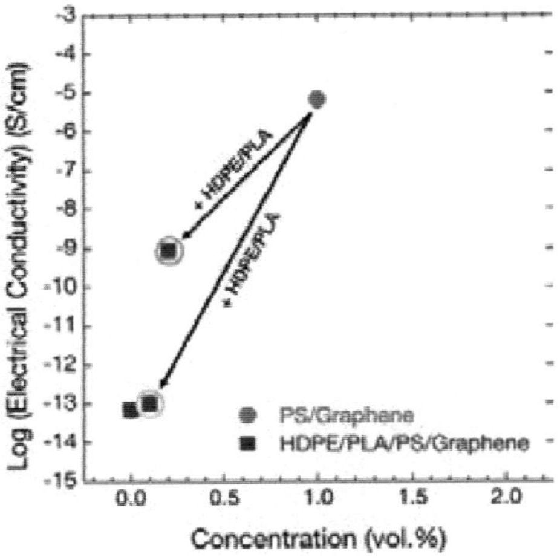

Figure 6. Electrical conductivity of PS/graphene and HDPE/PLA/PS/graphene nanocomposites. (volume ratio: yellow and green circles for 40/40/9.9/0.1 and 45/45/19.8/0.2, respectively)

Conclusion

We demonstrated that two strategies for interfacial localization of graphene are effective to achieve a very low percolation threshold in polymer/graphene nanocomposites fabricated by melt compounding. Thermodynamics and kinetic factors give insight into the way to the interfacial localization of fillers. In this study, we have limited ourselves to HDPE/PLA co-continuous blends with graphene, but these approaches could be better evaluated and utilized to improve various properties of co-continuous polymer systems in which percolating network formation by nano- and micro-sized particles is demanded at very low concentrations.

References

1. K. H. Liao, Q. Qian, C. W. Macosko, “Ultralow percolation graphene/polyurethane acrylate nanocomposites”, *Polymer* **2012**, *53*, 3756-3761.
2. J.-M. Thomassin, M. Trifkovic, W. Alkarmo, C. Detrembleur, C. Jérôme, C. Macosko, “Poly (methyl methacrylate)/graphene oxide nanocomposites by a precipitation polymerization process and their dielectric and rheological characterization”, *Macromolecules* **2014**, *47*, 2149–2155.
3. H. Kim, Y. Miura, C. W. Macosko, “Graphene/polyurethane nanocomposites for improved gas barrier and electrical conductivity”, *Chem. Mater.* **2010**, *22*, 3441–3450.
4. J. Wang, H. Hu, X. Wang, C. Wu, M. Zhang, X. Shang, “Preparation and mechanical and electrical properties of graphene nanosheets–poly (methyl methacrylate) nanocomposites via in situ suspension polymerization”, *J. Appl. Polym. Sci.* **2011**, *122*, 1866–1871.
5. S. Stankovich, D. A. Dikin, G. H. B. Dommett, K. M. Kohlhaas, E. J. Zimney1, E. A. Stach, R. D. Piner, S. T. Nguyen, R. S. Ruoff, “Graphene-based composite materials”, *Nature* **2006**, *442*, 282-286.

6. H. Kim, S. Kobayashi, M. A. AbdurRahim, M. J. Zhang, A. Khusainova, M. A. Hilmyer, A. A. Abdala, C. W. Macosko, "Graphene/polyethylene nanocomposites: effect of polyethylene functionalization and blending methods", *Polymer* **2011**, *52*, 1837-1846.
7. S. Ansari, E. P. Giannelis, "Functionalized graphene sheet—Poly (vinylidene fluoride) conductive nanocomposites", *J. Polym. Sci. Part B: Polym. Phys.* **2009**, *47*, 888-897.
8. H.-B. Zhang, W.-G. Zheng, Q. Yan, Y. Yang, J.-W. Wang, Z.-H. Lu, G.-Y. Ji, Z.-Z. Yu, "Electrically conductive polyethylene terephthalate/graphene nanocomposites prepared by melt compounding", *Polymer* **2010**, *51*, 1191-1196.
9. S. H. Ryu, A. M. Shanmugharaj, "Influence of long-chain alkylamine-modified graphene oxide on the crystallization, mechanical and electrical properties of isotactic polypropylene nanocomposites", *Chem. Eng. J.* **2014**, *244*, 552-560.
10. H. Kim, C. W. Macosko, "Morphology and properties of polyester/exfoliated graphite nanocomposites", *Macromolecules* **2008**, *41*, 3317–3327.
11. H. Kim, C. W. Macosko, "Processing-property relationships of polycarbonate/graphene composites", *Polymer* **2009**, *50*, 3797-3809.
12. D. Hofmann, M. Keinath, R. Thomann, R. Mülhaupt, "Thermoplastic carbon/polyamide 12 composites containing functionalized graphene, expanded graphite, and carbon nanofillers", *Macromol. Mater. Eng.* **2014**, *299*, 1329–1342.
13. Y. Gao, O. T. Picot, E. Bilotti, T. Peijs, "Influence of filler size on the properties of poly (lactic acid)(PLA)/graphene nanoplatelet (GNP) nanocomposites", *Eur. Polym. J.* **2017**, *86*, 117-131.
14. G. Tang, Z.-G. Jiang, X. Li, H.-B. Zhang, S. Hong, Z.-Z. Yu, "Electrically conductive rubbery epoxy/diamine-functionalized graphene nanocomposites with improved mechanical properties", *Compos. Part B: Eng.* **2014**, *67*, 564-570.
15. Y. T. Park, Y. Qian, C. Chan, T. Suh, M. G. Nejhad, C. W. Macosko, A. Stein, "Epoxy toughening with low graphene loading", *Adv. Funct. Mater.* **2015**, *25*, 575-585.
16. S. Park, S. He, J. Wang, A. Stein, C. W. Macosko, "Graphene-polyethylene nanocomposites: Effect of graphene functionalization", *Polymer* **2016**, *104*, 1-9.
17. M. Z. Iqbal, A. A. Abdala, V. Mittal, S. Seifert, A. M. Herring, M. W. Liberatore, "Processable conductive graphene/polyethylene nanocomposites: Effects of graphene dispersion and polyethylene blending with oxidized polyethylene on rheology and microstructure", *Polymer* **2016**, *98*, 143-155.
18. A. U. Chaudhry, V. Mittal, "High-density polyethylene nanocomposites using masterbatches of chlorinated polyethylene/graphene oxide", *Polym. Eng. Sci.* **2013**, *53*, 78-88.
19. L. Bai, S. He, J. W. Fruehwirth, A. Stein, C. W. Macosko, X. Cheng, "Localizing graphene at the interface of cocontinuous polymer blends: Morphology, rheology, and conductivity of cocontinuous conductive polymer composites", *J. Rheol* **2017**, *61*, 575-587.
20. M. Sumita, K. Sakata, S. Asai, K. Miyasaka, H. Nakagawa, "Dispersion of fillers and the electrical conductivity of polymer blends filled with carbon black", *Polym. Bull.* **1991**, *25*, 265-271.
21. J. Zhang, S. Ravati, N. Virgilio, B. D. Favis, "Ultralow percolation thresholds in ternary cocontinuous polymer blends", *Macromolecules* **2007**, *40*, 8817-8820.
22. A. Göldel, A. Marmur, G. R. Kasaliwal, P. Pötschke, G. Heinrich, "Shape-dependent localization of carbon nanotubes and carbon black in an immiscible polymer blend during melt mixing", *Macromolecules* **2011**, *44*, 6094-6102.

487556 Cocontinuous Ternary Polymer Nanocomposites with Interfacial Graphene Nanoplatelets

Abstract:

Cocontinuous ternary polymer nanocomposites with interfacial graphene nanoplatelets

Lian Bai[1], Radhika Sharma[1], Catherine Esnaashari[1,2], Christopher W. Macosko[â€ 1], Xiang Cheng*[1]

1. Department of Chemical Engineering and Materials Science, University of Minnesota, 421 Washington Ave. SE, Minneapolis, Minnesota 55455, United States
2. Department of Plastics, Iran Polymer and Petrochemical Institute, Tehran, 1497713115, Iran

â€ Email: macosko@umn.edu

* Email: xcheng@umn.edu

ABSTRACT: Interfacial localization of graphene at the interfaces in cocontinuous polymer blends is shown to be effective in stabilizing the cocontinuous morphology and increasing conductivity with a low electrical percolation threshold. In this study, we created ternary polymer nanocomposites composed of cocontinuous polylactic acid (PLA) and polystyrene (PS) blends intercalated with interfacial graphene nanoplatelets. The resulting conductive nanocomposites show dramatically improved conductivity at low filler loadings and an ultralow percolation threshold. By systemically exploring different melt-compounding sequences and times, we showed that the platelets can be transferred from premixed in the PLA phase to the interface during melt compounding and form a spanning 3D network during annealing. The coverage of graphene platelets on the interface effectively suppresses the coarsening of the cocontinuous structure during annealing and dramatically improves the electrical conductivity of the blends. We quantitatively correlated the transfer of graphene with the change of the morphological, electrical and rheological properties of the polymer nanocomposites.

THE HONG KONG UNIVERSITY OF
SCIENCE AND TECHNOLOGY

AIChE

3D $ReSe_2$ /Graphene Foam Architecture for Hydrogen Evolution Reaction and Lithium Ion Battery

THE HONG KONG UNIVERSITY OF SCIENCE AND TECHNOLOGY

Zhenjing Liu, Zhengtang Luo

THE HONG KONG UNIVERSITY OF
SCIENCE AND TECHNOLOGY

AIChE

Contents

- Introduction
- Material synthesis and characterization
- Catalytic activity towards hydrogen evolution reaction
- Anode performance of lithium ion battery
- Conclusion

THE HONG KONG UNIVERSITY OF SCIENCE AND TECHNOLOGY

Introduction

- **Two-dimensional layered transition metal dichalcogenide (TMDC)**

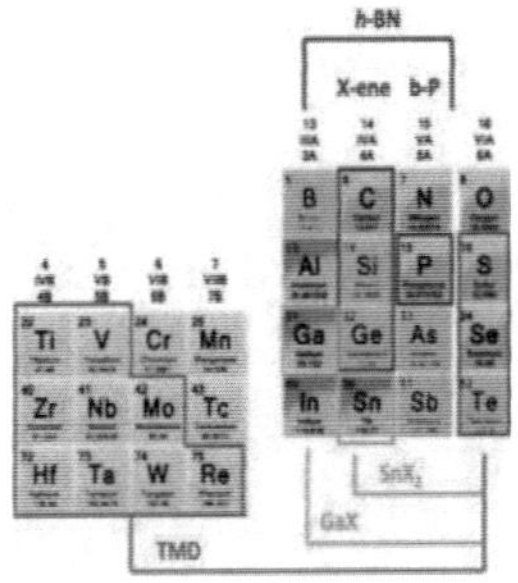

Xie L M. **Nanoscale**, 2015, 7(44)

- **Potential application: hydrogen evolution catalysis**
  - Hydrogen economy
  - Water splitting

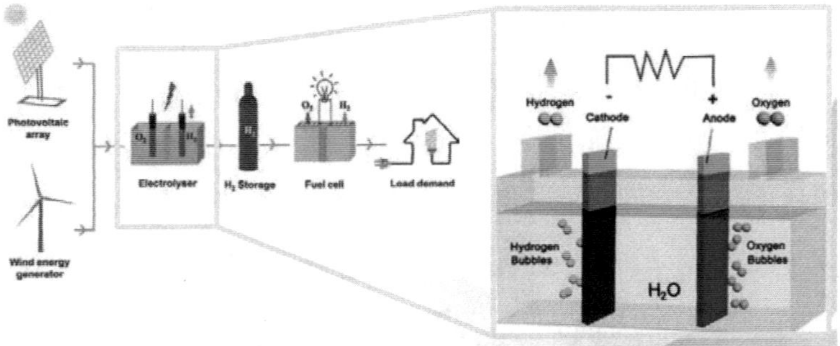

Wang J, Xu F, and Wang Y. **Adv Mater**, 2017, 29(14)

Zou X, Zhang Y. **Chemical Society Reviews**, 2015, 44(15)

THE HONG KONG UNIVERSITY OF SCIENCE AND TECHNOLOGY

Introduction

- **Potential application: hydrogen evolution catalysis**

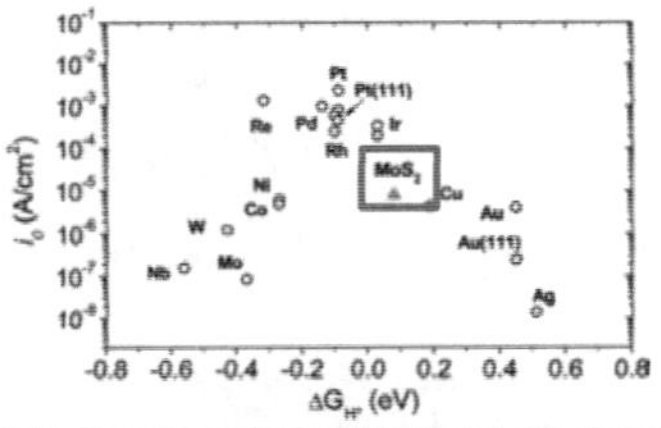

Jaramillo T F, Jørgensen K P and Chorkendorff I, **Science**, 2007, 317(5834)

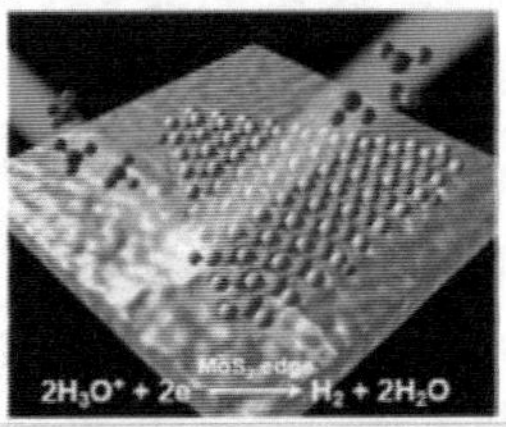

Ji Q, Zhang Y, Shi J, and Liu Z. **Adv Mater**, 2016, 28(29)

- **Anode material of lithium ion battery**

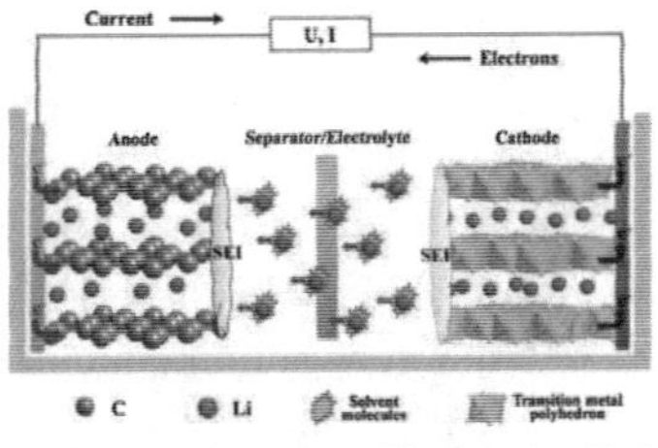

Xu B, *et al.* **Materials Science and Engineering: R: Reports**, 2012, 73(5)

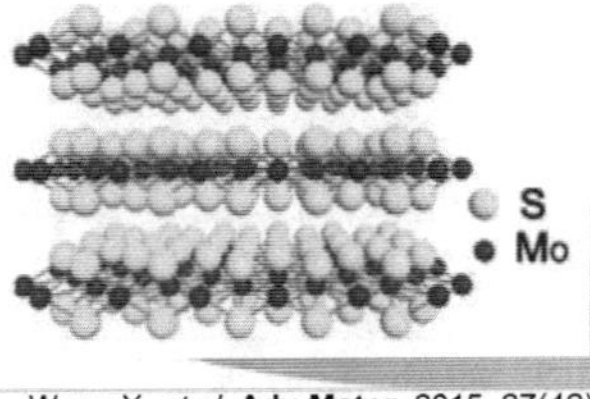

Wang X, *et al.* **Adv Mater**, 2015, 27(42)

THE HONG KONG UNIVERSITY OF SCIENCE AND TECHNOLOGY

Introduction

□ Objective

- Grow $ReSe_2$ on three dimensional graphene foam(GF) by chemical vapor deposition(CVD);
- Investigate the catalytic activity towards hydrogen evolution reaction(HER);
- Investigate the performance as the anode material of lithium ion battery(LIB).

➢ Active material: $ReSe_2$

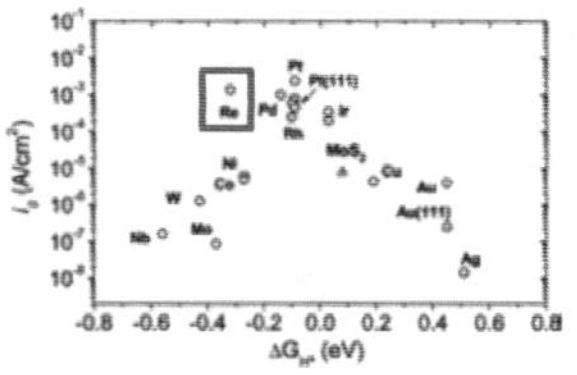

Jaramillo T F, Jørgensen K P and Chorkendorff I, **Science**, 2007, 317(5834)

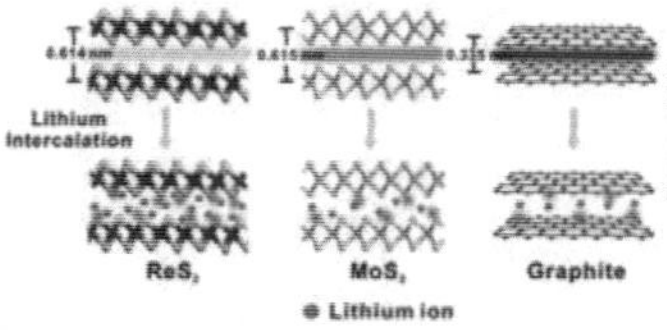

Zhang Q, Tan S and Fei L, **Adv. Mater**. 2016, 28(13)

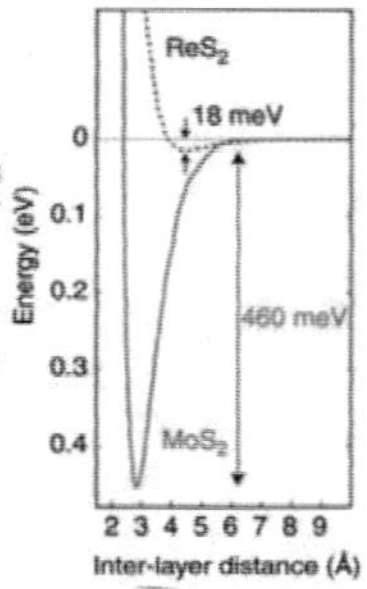

Tongay S, *et al.* **Nat. Commun**, 2014, 5(3252)

➢ Growth template: Graphene foam

- High conductivity
- High specific surface area
- Free-standing

THE HONG KONG UNIVERSITY OF SCIENCE AND TECHNOLOGY

Material Synthesis and Characterization

□ Step 1: GF

- Grow graphene on Ni foam through CVD process

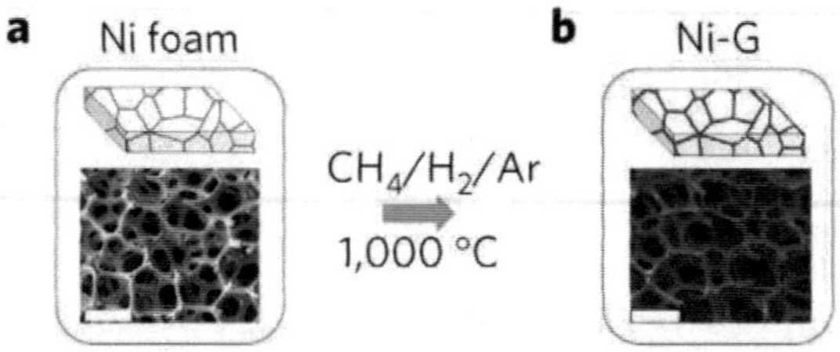

Chen Z, Ren W and Cheng H M. **Nat. Mater**, 2011, 10(6)

□ Step 2: $ReSe_2$ /GF

- Grow $ReSe_2$ on GF through CVD process

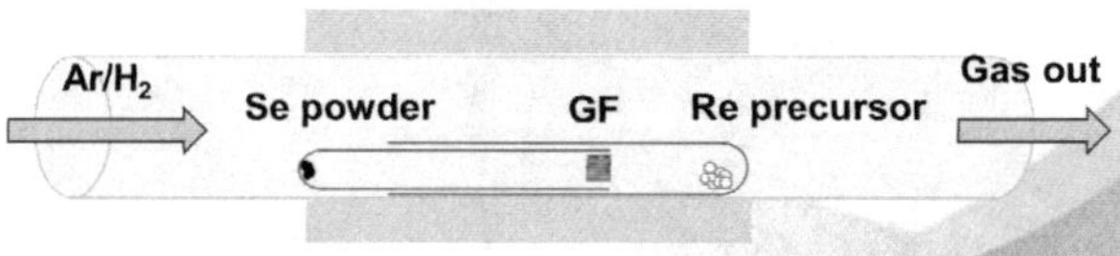

THE HONG KONG UNIVERSITY OF SCIENCE AND TECHNOLOGY

Material Synthesis and Characterization

- **SEM**
  - High nucleation density
  - Uniform distribution
  - Average size of $ReSe_2$ nanosheet: 191±65 nm
- **TEM**
  - Crystalline information of graphene and $ReSe_2$

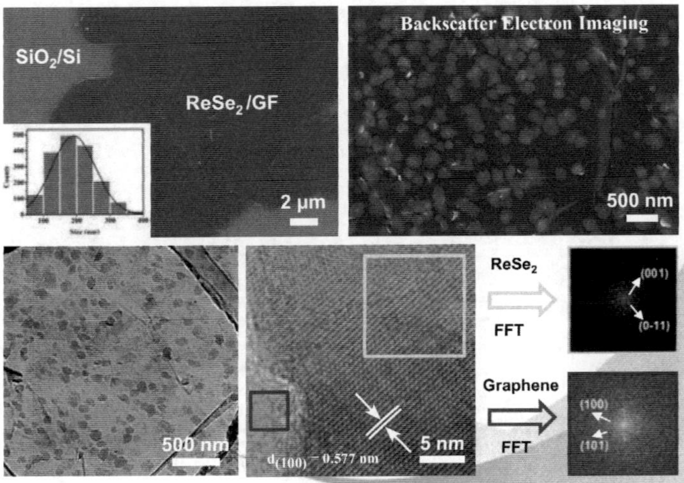

THE HONG KONG UNIVERSITY OF SCIENCE AND TECHNOLOGY

Material Synthesis and Characterization

- **Raman**
  - Typical peaks of graphene and $ReSe_2$ are observed

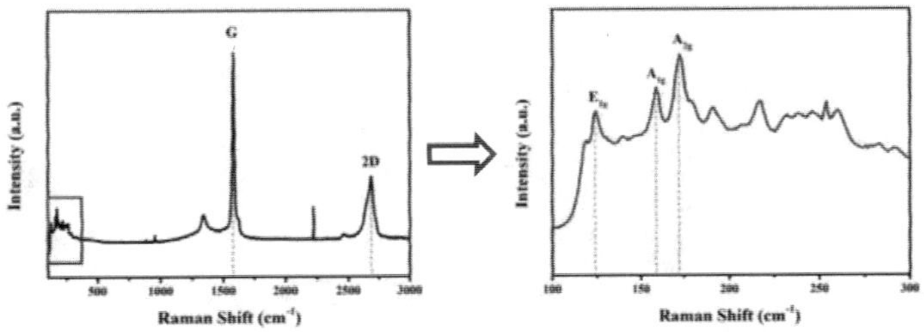

- **XPS**
  - Chemical state of Re and Se

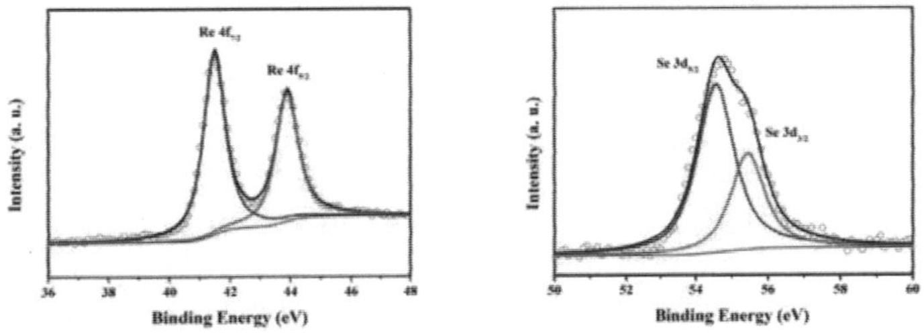

THE HONG KONG UNIVERSITY OF SCIENCE AND TECHNOLOGY

Catalytic Activity towards HER

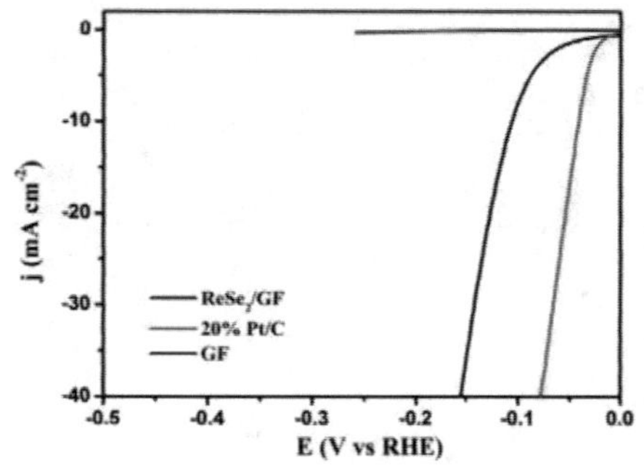

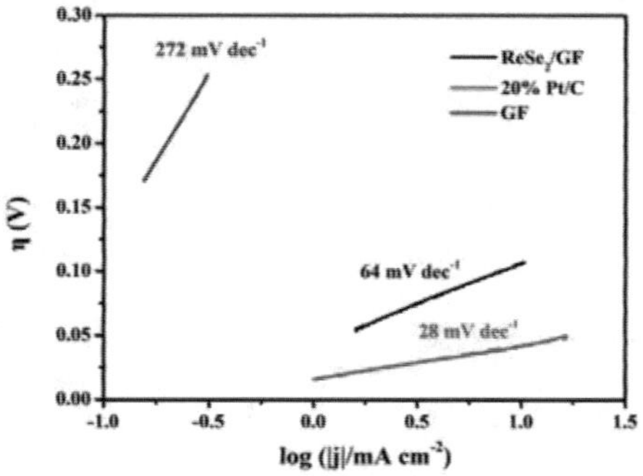

- ☐ A small overpotential of only **104**mV to achieve the current density of -10 mA cm^{-2};
- ☐ A small Tafel slope of **64** mV dec^{-1};
- ☐ A large exchange current density of 312.3 μA cm^{-2}.

THE HONG KONG UNIVERSITY OF SCIENCE AND TECHNOLOGY

Anode Performance of LIB

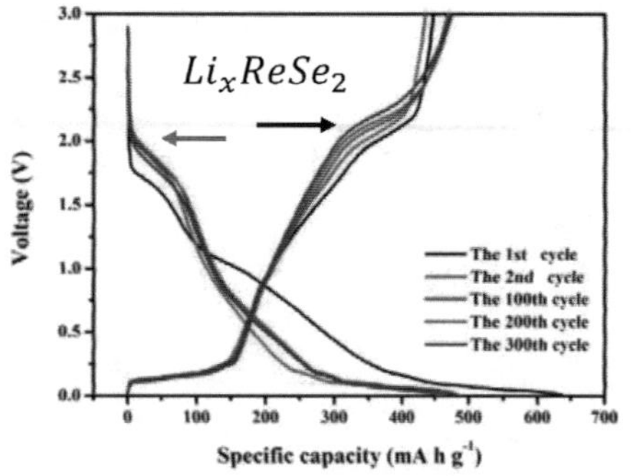

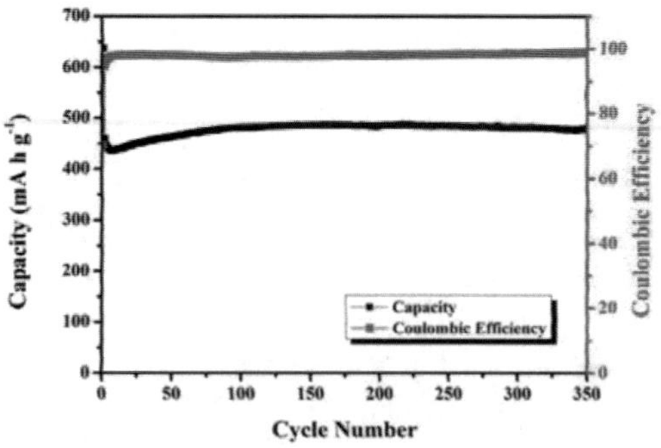

- ☐ The ultrahigh cycling stability comes from two main factors:
  - ➢ Large interlayer spacing;
  - ➢ Extremely weak van der Waals interaction.

THE HONG KONG UNIVERSITY OF SCIENCE AND TECHNOLOGY

Conclusion

- The 3D $ReSe_2$ /GF architecture is synthesized by CVD method successfully with an average size of 191nm, high density and uniform distribution.
- This 3D framework has shown excellent catalytic activity towards HER with a small overpotential of 104mV to achieve the current density of -10 mA cm^{-2} and extremely large exchange current density of 312.3 μA cm^{-2}.
- Additionally, this material exhibits ultrahigh stability after 300 cycling due to the large interlayer spacing and weak van der Waals interaction.

THE HONG KONG UNIVERSITY OF SCIENCE AND TECHNOLOGY

Thanks for your attention!

Contact us:
zliubp@connect.ust.hk
keztluo@connect.ust.hk

香港科技大學 THE HONG KONG UNIVERSITY OF SCIENCE AND TECHNOLOGY

AIChE The Global Home of Chemical Engineers

UGC 大學教育資助委員會 University Grants Committee

BAI XIAN 百賢亞洲研究院 ASIA INSTITUTE

Asian Future Leaders Scholarship Program

THE HONG KONG UNIVERSITY OF SCIENCE AND TECHNOLOGY

AIChE The Global Home of Chemical Engineers

Q&A

492756 Peg Hydrogels with Tunable Biodegradation Rate for Sustained Delivery of Platelet-Rich Plasma for Treatment of Osteoarthritis

Abstract:

Knee osteoarthritis (KOA), a disease caused by chronic inflammation and wearing of articular cartilage, affects over 50 million Americans. Sustained delivery of platelet-rich plasma (PRP) is a promising KOA therapy and has the potential to slow, reverse or halt its progression. PRP is a concentrated assortment of growth factors and cytokines derived from the patientâ€™s own platelets and is currently used in the clinic as a direct injection in the knee. However, such bolus injections have shown mixed results due to a rapid clearance of the growth factors. Here, we tested the therapeutic effects of a sustained PRP release from polyethylene glycol (PEG) hydrogels on primary articular chondrocytes.

Poly(ethylene glycol) (PEG) has been studied widely for the design of biodegradable hydrogel networks because of its exceptional biocompatibility and inertness. For temporal control of PRP release, we first designed and characterized PEG hydrogels with a wide range of properties. We made hydrogel using a combination of multiarm PEG-acrylate (PEGAc) of varying molecular weight and number of arms and three categories of dithiol crosslinkers, namely, ester-containing, non-ester-containing and crosslinkers with neighboring functional groups of different electronegativities. By changing these parameters, we obtained hydrogels with controlled degradation ranging from 10 h to 32 d and mesh size in the range of 9-14 nm. Uniquely, the hydrogel storage moduli could be controlled by the dithiol crosslinker chemical identity independent of the degradation time or mesh size. Further in vitro biocompatibility testing showed suitability for PRP delivery.

We then tested the PEG hydrogels for encapsulation and release of PRP. Specifically, to model the release of multicomponent PRP through PEG hydrogels, we examined bulk diffusion of total PRP proteins and model proteins having a hydrodynamic radius and molecular weight in a range corresponding to the growth factors found in PRP. Model proteins or PRP at 2-10% w/v were encapsulated during gelation of the hydrogels. Our results indicated that protein size and hydrogel degradation controlled diffusion of all proteins and secondary structure of proteins encapsulated during gelation remained unaffected post-release. Multiplex analysis showed release of therapeutically relevant proteins from PRP (endothelial derived growth factor (EGF) and platelet derived growth factor (PDGF)) at all time points until complete degradation of the hydrogels. Further investigation of the mechanism of protein release from high PRP concentration containing hydrogels using fluorescence correlation spectroscopy (FCS) indicated anomalous diffusion which may be due to protein crowding or protein-protein interactions. PRP released from hydrogels promoted the proliferation of human dermal fibroblast (HDF) cells, indicating retained bioactivity upon encapsulation and release. PRP releasates/ fractions collected at specific time intervals also showed increased proliferation in HDF cells comparable to that of a bolus PRP injection.

Lastly, we evaluated the therapeutic effects of sustained release of PRP from PEG hydrogels on primary articular chondrocytes. We examined the release profiles of different PRP fractions and their effects on chondrocyte proliferation, expression of anabolic and catabolic gene transcripts post-treatment with PRP fractions released at different time from the PEG hydrogel. We also examined the effect of the released PRP on glycosaminoglycan (GAG) production in cartilage explants. We observed a sustained release of several growth factors including EGF and PDGF for PRP-loaded PEG gels. In particular, PRP released from hydrogel promoted chondrocyte growth *in vitro* with different fractions showing differential ability to promote chondrocyte proliferation correlating to their growth factor content. PRP released from the PEG gels also led to significant reduction in gene expression for genes related to matrix degradation and upregulation of anti-inflammatory genes. Treatment of cartilage explants with different fractions of released PRP showed an increase in GAG production. Taken together, these results indicate that sustained delivery of PRP from PEG hydrogels had a significant beneficial effect on chondrocytes, in terms of cell proliferation and matrix production, as compared to a bolus PRP dose or no PRP.

494348 Control over the Temporal Profile and Sequence of Anticancer Therapeutics from Magnetically Responsive Hydrogels

Abstract:

ABSTRACT

CONTROL OVER THE TEMPORAL PROFILE AND SEQUENCE OF ANTICANCER THERAPEUTICS FROM MAGNETICALLY RESPONSIVE HYDROGELS

Tania T. Emi[1], Tanner Barnes[2], Emma Orton[2], Anne Reisch[1], Zachary Silveira[2], Miranda Mitchell[2], Celia Dunn[2], Anita Tolouei[1], Stephen M. Kennedy[1,2]

[1] Department of Chemical Engineering, [2] Department of Electrical, Computer and Biomedical Engineering, University of Rhode Island, Kingston, RI 02881

Problem: Cancer is the second leading cause of death in the US. It is estimated that, in 2017, 1.7 million American patients will be diagnosed with new cancer cases. Traditional chemotherapies involve systemic deliveries which can have difficulties in maintaining therapeutic drug concentration at tumor sites and can cause off-target side effects. Biomaterials-based chemotherapeutic strategies can circumvent these issues by providing localized and sustained delivery at tumor sites; however, the temporal drug delivery profiles provided by traditional biomaterial implants may not be optimal. Moreover, many current and emerging anticancer treatments involve the delivery of multiple therapeutics. Traditional biomaterials cannot coordinate temporally complex sequences of multiple therapeutics. We *hypothesized* that hydrogels designed to deform in response to magnetic stimuli could provide control over more optimized temporal profiles and sequences of anticancer drugs. Additionally, magnetically responsive hydrogels would provide real-time, remote control over drug delivery regimen. This flexibility would be critical in clinical settings where altering the course of therapy according to updated patient prognoses and patient history are critical. Here, we demonstrate the ability to magnetically control the temporal profile and sequence of anticancer deliveries from magnetically responsive hydrogels. Furthermore, we demonstrate that specific temporal delivery profiles and sequences generated by these magnetically responsive hydrogels are more effective at killing melanoma cells than deliveries associated with traditional hydrogels.

Methods: Magnetically responsive hydrogels were made by casting 7 wt% Fe_3O_4 in 1 wt% alginate with 2.5 mM Adipic acid dihydrazide crosslinker in the presence of a magnet to yield biphasic constructions. Gels of 8 mm diameter were cut using a biopsy punch, rinsed for 3 days in deionized water, frozen at -20^{0}C overnight and lyophilized to yield porous, magnetically deformable structures. Gels were soaked with 125 ug of chemotherapeutic drugs (mitoxantrone, dacarbazine, 5-fluorouracil, or irinotecan) for 24 hours, rinsed to get rid of the excess drug, and placed in 1 mL of PBS. Drug-loaded gels were placed in a magnetic piston system and stimulated for various durations at various stimulation frequencies. During this magnetic stimulation, samples were taken periodically by collecting and replacing PBS. The concentration of collected samples was quantified using BioTek Cytation3 microplate reader to measure optical absorbance at different wavelengths, depending on the chemotherapeutic. In cell studies, B16-F10 mouse melanoma cells were seeded at 500 cells per cm^2 and allowed to grow for 1 to 3 days before treatment. In treatments that involved exposing melanoma cells to various chemotherapeutic temporal profiles, each treatment schedule used the same integrated amount of drug (666 ng per mL per day of mitoxantrone). In treatments that involved exposing melanoma cells to multiple drugs, cells were exposed to different sequences of 5-fluorouracil (5FU) and irinotecan at 10 and 25 micrograms per mL, respectively. To assess viability, both endpoint LIVE/DEAD viability staining was used as well as real-time cell indexing using an xCELLigence system. For statistical analyses, ANOVA with Tukeyâ €™s post-hoc tests was used with $p < 0.05$ being the benchmark of significance.

Results: Our *in vitro* cell studies demonstrated that pulsatile temporal delivery profiles significantly enhance the toxicity of mitoxantrone exposure to melanoma cells when compared to constant delivery profiles produced by traditional hydrogels ($p < 0.01$). Real-time cell viability analyses revealed that melanoma cell populations were most affected during drug pulsing, suggesting that increasing the number of pulses may be more effective. Critically, these more effective pulsatile mitoxantrone delivery profiles were capable of being reproduced using our magnetically responsive hydrogels. These hydrogels were capable of producing a wide range of pulsatile deliveries where pulse height was controlled through the frequency of magnetic stimulation and pulse timing/duration through the timing/duration of magnetic stimulation. Additional *in vitro* cell studies demonstrated that when delivering sequences of irinotecan and 5FU (a common combination cancer treatment), that initial delivery of 5FU followed by irinotecan yielded more efficient melanoma death than irinotecan followed by 5FU ($p < 0.01$). Experiments revealed several strategies for magnetically controlling sequences of 5FU and irinotecan. Compartmentalization of our magnetically responsive hydrogels into (1) a cylindrical shaped inner compartment nested inside (2) a ring-shaped outer compartment enabled independent deformations of inner and outer compartments when stimulating using cylindrical- and ring-shaped magnets, respectively. Individual drugs could be released at specific times by using appropriately shaped magnetics to trigger the compartment containing the desired drug. Additionally, by analyzing the release of 5FU and irinotecan as a function of magnetic stimulation frequency, it was discovered that 5FU could be released in earnest at early time points without magnetic stimulation (i.e., it diffused out of our alginate gels) whereas irinotecan could be retained in the gel until magnetically triggered to release. These 5FU and irinotecan release characteristics enable the sequenced release of 5FU followed by irinotecan, as shown to be optimal in our *in vitro* studies.

Conclusions: We successfully developed a hydrogel system that can deliver chemotherapeutics in a pulsatile manner through the use of externally applied magnetic fields. The temporal profiles of these deliveries were remotely controlled by the timing, duration, and frequency of the applied magnetic fields. *In vitro* cell studies demonstrated that the pulsatile deliveries made possible by these magnetically responsive hydrogels were significantly more effective in killing melanoma cells than the relatively continuous delivery profiles produced by traditional hydrogels. Additional *in vitro* studies demonstrated that delivery of 5FU followed by irinotecan was more effective than the delivery of irinotecan followed by 5FU. Further experiments performed on our magnetically responsive gels revealed several promising strategies for coordinating sequences of 5FU and irinotecan deliveries. We believe these magnetically responsive hydrogels will provide useful tools for treating solid tumors by enabling localized deliveries of more advanced drug temporal profiles and sequences. Their ability to be regulated remotely also enables real-time control over treatments, which adds greatly to their potential clinical utility.

Contact Information: Name: Stephen M. Kennedy, Institution: University of Rhode Island, Department: Electrical, Computer, & Biomedical Engineering and Chemical Engineering, Phone #401-874-5295, Email: smkennedy@uri.edu.

495742 Electrospun Gelatin Nanofibers As Carrier for Controlled and Sustained Release of a Hydrophobic Drug

Abstract:

Controlled drug release systems have gained a lot of attention in pharmaceutical industry, due to its improved therapeutic efficacy by delivering drug molecules in a predetermined rate at the targeted site for a prolong time. Designing polymer based drug delivery system involves in-depth understanding of drug-device interactions and good control over the parameters which can modulate transportation of drug molecules as per the treatment. Recently, electrospun nanofibers have attracted interest as a controlled drug delivery device due to their unique structural features. Electrospinning has become the most common nanofiber production method in the past decade due to its capability of fabricating ultrafine fibers (nano to micron level) from a variety of materials like synthetic and natural polymers, polymer blends. This is the most cost-effective way to produce continuous fibers with very high surface-to-volume ratio, interconnected porosity with tuneable pore dimensions. The versatility in morphology in nanometer level accuracy of fibers can be achieved by optimizing electrospinning process parameters, the polymer solution properties. This kind of nanofibrous structure has tremendous applications in different biomedical fields.

The purpose of the present study is to fabricate large surface area to volume polymeric electrospun nanofibric device for drug delivery application. The work includes main three steps i.e. (a) fabrication of hydrophobic drug (piperine) encapsulated hydrophilic natural polymer (gelatin) based nanofiber film, (b) modulating the stability of hydrophilic film by crosslinking with saturated vapor of glutaraldehyde (GTA) with minimal exposer time and (c) finally modifying the vehicles with different crosslinking strategy and different layer by layer deposition of polymer to achieve sustained and prolonged release of hydrophobic drug piperine.

In this work, first biodegradable polymer gelatin is electrospun to produce continuous long nanofibers and then the possibility of using them as a suitable carrier of hydrophobic drug (piperine) is explored. Releasing hydrophobic drug is a challenge which significantly depends upon the nature of drug and polymer and their interactions. Thus, drug-polymer interactions are checked using Fourier Transform Infrared spectroscopy (FTIR) and it is found that piperine is intact in gelatin nanofiber mesh without reacting with polymer matrix. There is no interference of polymer matrix is found from FTIR and the characteristic peaks of piperine are present clearly. The morphology of the fibers is also examined using Scanning Electron Microscopy (SEM) imaging instrument after gold coating fibers for 10 sec. Long and continuous fibers with average diameter of 150 nm and characteristic peaks of piperine in gelatin nanofiber film are evidence which confirm the encapsulation of drug in polymer matrix.

As a next step, drug loaded gelatin nanofiber film is crosslinked using saturated vapor of GTA only few minutes. *In-vitro* degradation study of crosslinked samples (6 and 8 min) is performed to check the stability of the vehicle in different physiological pHs for more than 24 h. The fusion of nanofibers at the crossing points of fiber network is due to the water vapor in GTA solution which is evident from SEM images. Interestingly, it is found that the presence of hydrophobic drug can significantly reduce the fusion of gelatin fibers while crosslinking with GTA vapor. Crosslinked samples shows significant amount of improvement in terms of thermal stability after performing Thermogravimetric analysis. Thus we have successfully fabricated piperine encapsulated gelatin nanofiber based device which is chemically, thermally and mechanically stable.

After the fabrication of the device, *in-vitro* release study is done to analyze the effect of release medium and degree of crosslinking on the drug release from the nanofiber mat. While the amount of drug release decreases with increase in crosslinking time, the same increases with higher pHs. Release profile shows bi-layered curves i.e. sudden release of drug followed by sustained release which is far from zero-order kinetics. One very quick fit is to increase the crosslinking timing which is not desirable because of the toxicity of GTA and also it will reduce the overall total release. Thus new strategies are incorporated to control the rapid release of drug during the initial hours. Sandwiched structured nanofiber mesh with 6 min and 8 min crosslinking with GTA is the next attempt in order to increase the diffusional barrier between the drug molecules and release medium to tone down the initial rapid release. Piperine loaded gelatin nanofiber layer is basically put in-between two layers of only gelatin nanofiber layers. It is evident that, increase in barrier layer decreases the initial release and also decreases the drug loading. As a consequence different variation of barrier and core layer has been tried and this greatly improves the release profile. In the present vehicle, with higher drug loading the barrier is not able to release molecule in zero order manner. Thus, the concept of sequential crosslinking comes into the scenario to give a near zero order release profile. There is a hierarchy of crosslinking degree within the layers in case of single time crosslinking for 6 min. To increase the compactness of the fiber mesh uniformly within the layers, fiber mesh is crosslinking each time for 2 min after depositing barrier layer, core layer and finally at the end of opposite side barrier layer. This drastically improves the release kinetics of higher drug encapsulated device. These results suggest potential use of biodegradable electrospun nanofibers as an oral drug delivery system with promising controlled release features.

496134 Engineering of Degradable Biopolymer Films Loaded with Imiquimod for Controlled Released in a Mucosal Environment

Abstract:

Oral cancer is one of the most common neoplasia, with over 450 thousand new cases worldwide being diagnosed each year. In the United States only, it is estimated that a total of approximately $3.2 billion is spent yearly on treatment of head and neck cancers. One of the most common procedures for this type of cancer is the surgical resection of the lesioned tissue, usually when the disease has already evolved to oral squamous cell carcinoma (OSCC), which may affect mouth and throat lining cells. A major drawback of this aggressive approach is the postoperative functional and esthetic sequelae that can result from the procedures, such as the loss of the ability to swallow due to the impairment of essential tissues. Given these potential problems associated with surgical resection, in this study we aimed to develop a degradable mucoadhesive film, composed of the biopolymers chitosan (C) and alginate (A), loaded with the antineoplastic agent imiquimod (IM).

Both C and A are mucoadhesive polymers, being capable to interact with the glycoproteins of the buccal mucosa. Through buccal adhesion it is possible to tailor drug delivery to oral lesions *in situ*, both improving and prolonging the contact of the bioactive agent with the tumor, increasing drug bioavailability and uptake. The drug utilized in this system, IM, is a Toll-like receptor 7 (TLR7) agonist that up-regulates cytokines and generates an immune response. This compound is extensively used in the treatment of skin neoplasms, but reports on its effects upon the proliferation of OSCC are still scarce. Focusing on optimizing the controlled release of the drug, several formulations of the films were analyzed, both based on the isolated biopolymers or involving their combination in different proportions. The films were characterized with respect to morphology, infrared spectra, thickness, mucoadhesion, stability in physiological solutions, thermal stability, mechanical properties, drug incorporation efficiency, IM release kinetics and antitumor potential in OSCC.

The drug was efficiently loaded into thin films (incorporation efficiencies above 60% in films with average thickness below 0.14 mm). IM was distributed in the polysaccharide matrix as amorphous aggregates or as rods, depending on the conditions of film production. In general, IM incorporation reduced the tensile strength of the films, but did not alter their mucoadhesion properties. For most formulations developed, efficient release of the drug was noticed in less than 8 hours, as a result of film swelling and degradation after exposure to simulated saliva, suggesting promising results *in vivo*.

The authors are grateful for the financial support of this work by the Brazilian National Council for Scientific and Technological Development (CNPq), the Coordination for the Improvement of Higher Education Personnel (CAPES), and by the São Paulo Research Foundation (FAPESP), all from Brazil.

497203 Microparticles for the Delivery of Anti-Diabetic Drugs to the Adipose Tissue

Abstract:

Many drugs that show promise for metabolic diseases, such as obesity or diabetes, suffer from off-target effects and low bioavailability. Furthermore, efficacy may be decreased if the patient is unable or unwilling to follow the prescribed administration schedule. This presentation will focus on the development of a biodegradable microparticle delivery system for sustained, localized release of resveratrol to the adipose tissue. Resveratrol is a polyphenol with anti-diabetic properties that promote anti-inflammatory and anti-oxidant gene expression; however, the molecule is quickly metabolized in the blood and liver making it difficult to achieve therapeutic concentrations within adipose tissues. The microparticle delivery system was fabricated using a single emulsion/solvent evaporation technique to encapsulate resveratrol into poly(lactide-co-glycolide) (PLG), an FDA approved biodegradable polymer. We optimized drug encapsulation by varying the solvent, surfactant, homogenization speed, vessel geometry and resveratrol concentration in the organic phase. We then determined the effect of microparticle drug loading, size distribution, and morphology on the resveratrol release profile using UV-Vis spectroscopy, laser light scattering, and electron microscopy. To investigate the efficacy of drug delivery, we compared the effect of free resveratrol versus microparticles on gene expression, lipolysis rates and glucose uptake in 3T3-L1 adipocytes. Future work will address the translational potential of this system by investigating the effect of microparticle delivery to the inguinal and subscapular adipose tissues on glucose tolerance and adiposity in high fat diet fed mice. Collectively, we expect these studies to establish this microparticle delivery system as a means to modulate adipose tissue function as a potential treatment for diabetes.

500293 Template Assisted Micro-Patterned Electrospun Nanofibrous Mat As a Potential Carrier for Controlled Drug Release

Abstract:

Controlled drug release by manipulating the surface wettability and degradability of the drug carrier is essential for medical treatments.[1] Usually, the wettability of a surface can be altered by surface coatings and by changing the surface roughness. In this work we have fabricated micro-patterned cellulose acetate electrospun nanofibrous mats using template assisted electrospinning (Figure 1A). This approach produces mats with tunable wettability. Nylon meshes (Figure 1B) with different grid spacing (50 µm, 100 µm and 200 µm) were used as a template to fabricate micro-patterned nanofabric over a large area (10×10 cm^2) in a single step. This non-conductive nylon mesh in strong electrostatic field polarizes and generates a negative static charge on the nylon mesh by static induction and polarization.[2] Further, the nylon grid attracts fibers onto grid lines and forms micropatterned fiber mats as shown in Figure 1 (A). Initially, fibers are preferably deposited on nylon grid lines. After 10 min of deposition, the fibers repel incoming material resulting in a random deposition to yield micro-patterned nanofabric surfaces as shown in Figure 1C.

Water contact angle measurements were performed on these micro-patterned nanofibrous mats fabricated with different spacing to study the effect of micropatterning on wettability. As shown in Figure 1C, these micro-patterned surfaces form non-communicating air gaps and air trapped in this gaps minimizes the solid-water interface and opposes wetting by capillary action. This non-wetting pressure plays a key role in the wettability of non-commuting air trapped systems.[3] The Water Contact Angle (WCA) changes from 30° for the non-patterned surface to 138° for a micro-patterned surface with a 50 µm spacing. This change in WCA can be attributed to capillary pressure increases with decreasing spacing. WCA measurements are summarized in Figure 1D.

Further, the effect of surface wettability on drug release kinetics was investigated by loading Diclofenac sodium into cellulose acetate precursor solutions prior to electrospinning. Transdermal drug releases study shown a significant change in drug release kinetics with a change in surface wettability. Micro-patterned surfaces with a WCA of 138° showed zero order release kinetics up to 12 hours, whereas non-patterned hydrophilic (30°) samples only showed controlled drug release for 1 hr.

References:

1. Ajiro, H., Takahashi, Y., Akashi, M. and Fujiwara T. *Polymer,* **2014,** *55,* 3591-3598.

2. Zhao, S., Zhou, Q., Long, Y.Z., Sun, G.H. and Zhang, Y. *Nanoscale* **2013,** *5,* 4993-5000.

3. Dash, S., Alt, M. T. and Garimella, S. V. *Langmuir, **2012**, 28*, 9606-9615.

4. Kakunuri, M., Khandelwal, M., Sharma, C.S., Eichhorn, S.J. *Materials Letters*, ***2017***, 196, 339-342.

5. Kakunuri, M., Wanasekara, N.D., Sharma, C.S., Khandelwal, M., Eichhorn, S.J. *J. Appl. Poly. Sci.*, ***2017***, 134, 44709.

500460 PÎ•G-Based Nanocomposite Hydrogels: Thermoresponsive Sol-Gel Transitions and Decomposition Rates Regulated By the LÎ'/GA Ratio of PLGA-PÎ•G-PLGA

Abstract:

Our group developed a nanocomposite made of PLGA-PEG-PLGA (poly ($_{D,L}$-lactide-*co*-glycolide)-*b*-poly (ethylene glycol)-*b*-poly ($_{D,L}$-lactide-*co*-glycolide)) and laponite-clay nanoparticles (laponite/PLGA-PEG-PLGA nanocomposites), exhibiting thermoresponsive sol-gel transitions and biodegradability [1, 2]. The laponite/PLGA-PEG-PLGA nanocomposites could exhibit higher biocompatibility by their high PEG/PLGA ratio, and the higher drug-encapsulation efficiency by the low solute concentration. To establish an efficient laponite/PLGA-PEG-PLGA nanocomposite as a material for the drug-delivery system (DDS), the decomposition rate of the laponite/PLGA-PEG-PLGA nanocomposite should be properly controlled. In this study, the thermoresponsive sol-gel transition and the decomposition rate of the laponite/PLGA-PEG-PLGA nanocomposites were optimized by varying the LA/GA ratios (LA/GA=1.1, 4.0, 8.8) in the PLGA parts of the PLGA-PEG-PLGA with the constant molecular weights of 600-1000-600, respectively. The sol-gel transition temperatures were found to be in the temperature range between 25°C and 37°C regardless of the LA/GA ratios when the polymer concentration was 4 wt% and the laponite concentration was 0.75 wt%. In fact, the sol-gel transition temperatures were 34°C for LA/GA=1.0, 35°C for LA/GA=4.0, and 35°C for LA/GA=8.8. As for the decomposition of the gels at 37°C, the laponite/PLGA-PEG-PLGA nanocomposite gels with the LA/GA=4.0 and the LA/GA=8.8 both showed ~30% of the mass loss after 10 days, while the laponite/PLGA-PEG-PLGA nanocomposite gels of LA/GA=1.1 showed ~45% of the mass loss after 10 days. Additionally, the protein release from the gel was also examined as a preliminary test for the DDS application. The results revealed the highly versatile features of the aqueous laponite/PLGAPEG-PLGA nanocomposites with prospective thermoresponsive sol-gel transitions and with finely-controlled decomposition behavior especially for the DDS application.

[1] Naho Oyama, Hiromasa Minami, Daichi Kawano, Makoto Miyazaki, Tomoki Maeda, Kazunori Toma, Atsushi Hotta, Koji Nagahama, A nanocomposite approach to develop biodegradable thermogels exhibiting excellent cell-compatibility for injectable cell delivery, Biomaterials Science, Vol.2 (8), 1057-1062 (2014).

[2] Makoto Miyazaki, Tomoki Maeda, Kenji Hirashima, Naruki Kurokawa, Koji Nagahama, Atsushi Hotta, PEG-based nanocomposite hydrogel: Thermoresponsive sol-gel transition controlled by PLGA-PEG-PLGA molecular weight and solute concentration, Polymer, Vol.115, 246-254 (2017).

501007 Electrospun Collagen Scaffold for Peripheric Nerve Regeneration

Abstract:

Damage of peripheral nervous system directly interferes with the communication network between brain, spinal cord and the rest of the body, and around 20 million people in United States suffer this [1]. Donor site morbidity and limited graft material are limitations confronted by solutions such as allograft and autograft [2]. Otherwise nerve conduits have the capacity to develop a successful nerve recovery , and most important return a better quality of life to patients. The lost of extracellular matrix (ECM), local growth factors and cells is what limits the regeneration of nerves.The proposed research consists of the engineer of 3D collagen nanofiber scaffold through electrospinning, a versatile technique for nanofibers development . The scaffold, composed of the most abundant protein in our body collagen, mimics ECM and facilitates cell adhesion and elongation to promote nerve regeneration, providing biological and mechanical support[3][4]. The development of the collagen nanofiber structure is optimized via electrospinning parameters in order to modify nanofiber orientation which is directly related towards the direction of nerve regeneration. A random and aligned pattern has been developed and confirmed by Scanning Electron Microscope (SEM). In vitro assays and fluorescene microscope analysis with Schwann and PC-12 cells, both key cell lines in nerve tissue regeneration, resulted on successful cell adhesion to electrospun collagen nanofiber, random and aligned. Cellular tissue has been observed on collagen scaffold through SEM, after 24h cell culture and proper cell fixation method. Present experiments consist on the addition of nerve growth factor (NGF) to PC-12 cell line in order to promote differentiation. The differentiation of PC-12 cells will be analyzed towards the influence of collagen scaffold orientation on the direction of cell growth elongation. A future approach is a co-culture between Schwann and PC-12 ,on collagen scaffold,with the vision to prove that the naturally secreated NGF by Schwann cells promote the differentiation of PC-12 . Results from this co-culture will validate the use of collagen electrospun nanofiber scaffold impregnated with Schwann cells as a vehicle to deliver the proper natural components in a injured nerve and achieve nerve regeneration.

References:

1. National Institute of Neurological Disorders and Stroke: Peripheral Neuropathy Fact Sheet (Bethesda, Maryland, 2014).

2. W. Z. Ray and S. E. Mackinnon: Management of nerve gaps: Autografts, allografts, nerve transfers, and end-to-side neurorrhaphy. *Exp. Neurol.* **223**(1), 77 (2010).

3. S. Previtali, M. Malaguti, N. Riva, M. Scarlato, P. Dacci, G. Dina, D. Triolo, E. Porrello, I. Lorenzetti, R. Fazio, G. Comi, A. Bolino, and A. Quattrini: The extracellular matrix affects axonal regeneration in peripheral neuropathies. *Neurology* **71**(322–331) (2008).

4. G. a. Di Lullo, S. M. Sweeney, J. Körkkö, L. Ala-Kokko, and J. D. San Antonio, "Mapping the ligand-binding sites and disease-associated mutations on the most abundant protein in the human, type I collagen," *J. Biol. Chem.*, vol. 277, no. 6, pp. 4223–4231, (2002).

501165 3D Printing of Nerve Guidance Channels for Peripheral Nerve Repair

Abstract:

The peripheral nervous system (PNS) is a complicated and extensive network of nerves that are the means by which the brain and spinal cord control the rest of the body. The PNS is fragile and can be easily damaged by injuries or trauma. Surgical treatment is the only remedy currently available, with the gold standard for defects greater than 8 mm being autologous nerve grafts; however only around 40% of the 1.8 million US PNS patients each year regain normal function. In addition, nerve grafts have been particularly ineffective at repairing critical-size nerve defects (> 3 cm).[1] Scaffold-based strategies where a tubular nerve guidance channel (NGC) is used to bridge the nerve defect have been promoted as a potential alternative that could avoid the additional surgeries and associated donor site morbidity involved in the harvest of nerve grafts.[2] Clinicians have thus increased the use of NGCs combined with current surgical therapeutics. However, current NGCs lack patient-specific tunability and are only approved for small-gap (< 3cm) injuries by the U.S. Food and Drug Administration (FDA). Current research efforts are focused on creating more complex NGCs that can support regeneration of critical-size defects.

In this context, our research seeks to use additive manufacturing technologies to create bioactive and cellular NGCs on demand for the repair of critical-size nerve defects. Recently, 3D printing has been increasingly used in research and medical therapeutics for rational, computer-aided design of biomaterial-based scaffolds with complex architecture. Furthermore, printing with co-axial extruders can enable the direct printing of layered tubular structures for use as NGCs. The NGCs should contain an outer flexible shell that seeks to mimic the mechanical properties of the surrounding biological tissue and enable diffusion of nutrients to support encapsulated cells. The use of biodegradable block copolymers with both hydrophilic and relative hydrophobic functions can provide a flexible, partially-hydrated, biocompatible and bioresorbable NGC shell.

In this study, an A-B type diblock copolymer of poly(ethylene glycol) methyl ether (mPEG) combined with poly(l-lactide) acid (PLLA) and another A-B-A type triblock copolymer of PLLA-PEG-PLLA were synthesized using varied ratios of mPEG, PEG and PLLA. The resulting block copolymers of mPEG-PLLA and PLLA-PEG-PLLA containing different lengths of the PLLA hydrophobic chain were characterized with gel permeation chromatography (GPC), differential scanning calorimetry (DSC), and nuclear magnetic resonance (NMR) to determine molecular weight, polymer structure, and thermal behavior. In addition, equilibrium water content, degradation rates, and cell response were all evaluated and correlated to polymer structure. With the 3D printing of copolymers using coaxial printing needles, a scalable process for the production of cellular NGCs could be developed.

References

1. Burkhard Schlosshauer, L.D., et al., Neurosurgery, 59 (2006), 740-748.

2. Cinteza, D., et al., Maedica â€“ A Journal of Clinical Medicine, 10:1 (2015), 65-68.

489637 Combined Physical and Biochemical Cues Direct the Growth of Inner Ear Neurites

Abstract:

Cochlear implant (CI) devices help to restore hearing to individuals who suffer from sensorineural hearing loss. While CIs provide basic auditory perception, implant patients only experience limited voice and tonal perception. Due to spatial separation between the electrode array and the neural receptors, significant current spread and nonspecific stimulation of the spiral ganglion neurons (SGNs), or inner ear neurons, occurs. Thus, devices with advanced electrical design are unable to recapitulate the tonotopic stimulation of neurons because of spatial resolution. Strategies to decrease the distance between CI electrodes and primary neural receptors have focused on regenerating the SGN neurites. To be effective, the spatial order of the neurites must remain intact and thus cellular cues must orient and guide the pathfinding of SGN neurites. In this work, physical and biochemical cues are used to direct the regrowth of SGN neurites. Utilizing the inherent temporal and spatial control of photopolymerization, physical microgrooves are fabricated using a photomask in a single step process. The spatial control of the photon-induced reaction forms ridges under transparent band and depressions under reflective regions with channel amplitudes ranging from 250 nm to 10 μm. Biochemical patterns are generated by adsorbing laminin, a cell adhesion protein, to acrylate polymer surfaces followed by irradiation through a photomask with UV light to deactivate protein in exposed areas and generate parallel biochemical patterns. Laminin showed deactivation with increased exposure time as measured by immunofluorescence. Adding photoinitiator decreased the time required to deactivate the adsorbed laminin. To evaluate neural response to the patterns, SGNs were first cultured on physically patterned substrates and showed increased alignment with higher amplitude and lower periodicity polymers. SGNs were also cultured on biochemical patterns and showed excellent alignment to the parallel bands confirming the photodeactivation process was effective in creating active laminin patterns. To further evaluate the ability of biochemical and physical features to be used in overcoming conflicting cellular cues, competing patterns were tested. Physical microgroove channels were first fabricated followed by a perpendicular laminin protein pattern. The strength of the physical cue was independently varied by changing the amplitude or the periodicity. Lower periodicity, or more frequent, surfaces induced a greater degree of alignment along the channels. Higher amplitudes, or deeper channel depths, also increased the guidance of the SGNs. When SGN neurites were seeded onto substrates with perpendicular biochemical and physical patterns, alignment was strongly dependent on both amplitude and periodicity. At low amplitudes neurites tended to grow along the laminin stripes. As the amplitude was decreased, alignment was disrupted and neurite segments followed neither the biochemical or physical pattern. Similarly, at lower periodicity, SGN neurites were not strongly aligned to either pattern. As the periodicity was increased, alignment shifted to follow the biochemical laminin pattern. These results demonstrate the ability of photopolymerized micro-features to modulate alignment of inner ear neurites even in the presence of conflicting physical and biochemical cues laying the groundwork for next generation cochlear implants and neural prosthetic devices.

495713 Harnessing Multi-Functional Microbial Cells for Designing Sweat-Responsive Bio-Hybrid Wearables

Abstract:

We found a variety of microbial cells and their cellular components have hygroscopic behavior. By depositing these biomaterials on top of a humidity-inert film, a multi-layered bio-hybrid material is obtained, which can reversibly change shape and biofluorescence intensity in response to environmental humidity change. We characterized the performance of the film, which guide the design of a wearable running suit and a fluorescent shoe prototype. Those bio-flaps on wearables can dynamically modulate ventilation in synergy with the bodyâ€™s need for cooling.

497286 Development of Silk Protein Conjugates for Mucoadhesive Applications

Abstract:

Polymers that adhere to wet surfaces that are biocompatible, degradable, and devoid of animal-sourced materials are desired for applications as tissue adhesives. Silk from the domesticated silkworm *Bombyx mori* is a naturally occurring biopolymer with charged hydrophilic terminal regions that endcap a blocky core dominated by hydrophobic domains. Silk was selected for modification for use as an aqueous adhesive because of its high molecular weight, tunable mechanical and degradation properties, aqueous processability, and for its hydrophobicity that was envisioned to afford control over the mechanical properties and aqueous swelling of the modified biopolymers. This work sought to exploit the unique structure and properties of this protein by synthesizing gels that are stable in an aqueous environment and adhere to mucus-covered surfaces. To prepare adhesive hydrogel coatings that can be applied as a liquid and subsequently gel, silk was successfully conjugated with tyramine, a phenol-containing compound capable of being oxidized to a catechol functional group using mushroom tyrosinase (MT). As both aqueous and organic phase syntheses are possible with silk, several solvent systems were investigated to determine the conditions that led to successful coupling without protein aggregation. The silk polymers were characterized to determine composition, adhesive strength using lap shear measurements, cell compatibility and support of proliferation, and their ability to form beta sheets, which physically crosslink silk. Importantly, the conjugation of tyramine showed no effect on silkâ €™s ability to form beta sheets, which may be used to further tune the materialâ€™s mechanical properties and degradation profile. These materials afford a new route to synthesize silk-based gels, which we envision to have utility as biodegradable adhesives and sealants.

499466 Laser-Activated Sealants for Skin Tissue Repair

Abstract:

Sutures and staples are ubiquitous in surgery, but present their own complications and cause local trauma to tissue during application. Glues, adhesives and sealants have been investigated as means for approximating wounds, but many suffer from toxicity and poor strength. Photothermal tissue sealing and repair, facilitated by near-infrared laser-responsive nanoglues, has emerged as a potential alternative to existing tissue apposition methods. Here, we demonstrate the photothermal sealing of lacerted skin and other ruptured tissues using laser activation of novel nanoglues in which, gold nanorods (GNRs) are embedded throughout a polypeptide matrix. Critical factors including laser power density, exposure time, and laser type (continuous wave or pulsed wave) were investigated in order to determine optimal treatment parameters that led to the highest recovery in sealing strength. Laser-activated sealing of full thickness lacerations in skin of live mice resulted recovery of tissue mechanical properties and integrity that were comparable to that seen with commercial sutures. Local inflammation caused by photothermal sealing approach was minimal and was compared to that seen with commercial sutures. Furthermore, laser-activated sealing was also employed to reinforce suture lines in skin, as a combination approach that can provide significant benefit for rapid sealing and healing of skin lacerations. Our results demonstrate that polypeptide-GNR nanoglues may have applications in photothermal sealing and repair of ruptured tissues and may emerge as attractive alternatives to sutures.

497772 Layer-By-Layer Assembled Thin Film Biomaterials As Porous Biomolecular Delivery Systems

Abstract:

Biomaterials capable of delivering controlled quantities of bioactive agents, while maintaining mechanical integrity, are needed for a variety of cell contacting applications. We describe here two strategies toward porous, polyelectrolyte-based thin films capable of controlled biomolecular loading and release. In one approach, termed "nanoparticle templating," films are formed via the layer-by-layer assembly of charged polymers and nanoparticles (NP), then chemically cross-linked to increase mechanical rigidity and stability, and finally exposed to tetrahydrofuran to dissolve the NP and create an intra-film porous network. In another approach, termed "solution shock," films are subjected briefly to an extreme acidic environment following LbL assembly, causing expansion and ultimately a spinodal-like breakup of the film into a porous structure. We report here on film structure and mechanics, on biomolecular agent loading and release, and on the behavior of cells cultured on these films. While both strategies yield mechanics and biomolecular agent loading that are suitable for applications such as tissue engineering, the solution shock method offers advantages in simplicity and controllability.

496274 Molecular Simulations of the Influence of Interfaces on the Dynamics of Polymers

Abstract:

After multiple decades of work, it is widely accepted that interfaces in a polymer severely modify both the dynamics and the thermodynamics of polymers and other glass-forming materials. This is a problem not only in the design of thin films, where the material is confined in one dimension, but also in polymer nanocomposites, which can have a complicated confining geometry in multiple dimensions. Despite extensive work in the area, there are many outstanding issues that remain poorly understood. In this talk I will summarize our work where we attempt to understand the dynamics of polymers in a variety of confining geometries. Under simple planar confinement, we find dramatic changes in the dynamics associated with the glass transition, and we find a slight reduction in the entanglement density as the films become thinner than the characteristic size of the polymers. These results are compared to the dynamics of polymers within polymer nanocomposites where the nanoparticle loading approaches random close packing. In such a composite, interfaces are ubiquitous within the material and dominate the behavior relative to the bulk dynamics of the polymer. In these materials, we provide a molecular view of the spatial variations in the polymer dynamics relative to the bulk material. We expect the insights provided from our simulations to impact a variety of membrane and thin film technologies.

497145 Princess and the Pea Behavior in Polyelectrolyte Multilayers: Influence of the First Layer on Polyelectrolyte Multilayer Assembly and Properties

Abstract:

In many applications of polyelectrolyte multilayers (PEMs), a base polycation layer is adsorbed to promote adhesion of the PEM to the substrate. The effect of the first polyelectrolyte adsorbed in a PEM was investigated by assembling PEMs with first layer polycations of different chemistries and molecular weights. In this study, quartz crystal microbalance with dissipation monitoring (QCM-D) was used to monitor the PEM assembly process. First layer choice affects the total mass accumulation of the PEM as well as the stoichiometry of the PEM, although linear growth was observed in all cases. PEM thickness was also affected by first layer choice, although not consistent with changes in mass. Combined with the stoichiometry results, these findings indicate that the structure of a PEM is fundamentally different depending on first layer chemistry and molecular weight. PEM topography is also affected by first layer choice. Selection of appropriate first layer material is therefore an important consideration in the design of a PEM, and changing first layer material may be a facile way to tailor the structure and properties of PEMs. We propose that this represents an ideal model system for investigating individual, synergistic, and competing effects of topography, surface chemistry, and stiffness on cell adhesion.

494885 Instability of Liquid Crystal Thin Film on Topographically Patterned Substrates

Abstract:

Abstract:

Ultrathin liquid films tend to rupture and form isolated droplets by destabilizing Van der Waals interaction on a defect free smooth solid surface. This phenomenon is termed as dewetting [1]. Dewetting is a sequential process starting from rupturing and creation of holes at different locations on the thin films. Then holes grow and coalesce with the neighbouring holes to form cellular like interconnected ribbons. Finally these ribbons disintegrated into isolated random droplets [1].Dewetting can be use as a very effective tool to fabricate different nano scale or meso scale structures of different materials like homo polymer, block co-polymer, liquid crystal etc. For example, dewetted and aligned liquid crystal droplets can be use in opto-electronic applications. Here, we have reported the dewetting mechanism of liquid crystal (5CB) thin film on flat and topographically patterned (one dimension grating) polymer substrates. We observed that the dewetting mechanism of liquid crystal thin films on topographically patterned substrates is different than its homo-polymer counterpart. Dewetting of liquid crystal thin film on topographically patterned surface is independent of surface heterogeneity. But patterned directed dewetting was obtained for homo polymer on topographically patterned substrates [2]. Different intermediate liquid crystal structures were created on spin coating platform by changing different coating parameters. We have obtained a transition concentration of liquid crystal solution for discontinuous to continuous film on flat as well as 1-D topographically grating substrates. Here, we have also reported the spin dewetting process for liquid crystal on flat as well as topographically patterned substrates. Spin dewetting is a very unique process where an ultra thin film rupture and form isolated droplets during the spin coating process itself by coating with very dilute solution and high rpm[3]. Dewetting process of liquid crystal thin films were captured under optical microscope in both bright field and polarised modes. The morphology, macroscopic physical orientation and phase transition of dewetted and spin dewetted liquid crystal droplets were also observed under polarised microscope and atomic force microscope platform.

References:

[1].Reiter,G.,*Phys. Rev. Letts.*, 1992, 68, 75.

[2]. Roy, S., Mukherjee, R., *ACS Appl.Mater. Interfaces.*, 2012, 4(10), 5375.

[3]. Bhandaru, N., Das, A., Salunke, N., Mukherjee, R., *Nano Lett.*, 2014, 14(12), 7009.

487634 Electroless Deposition of Copper on Quaternized Chitosan Coatings for Antibacterial Application

Abstract:

Surface modification techniques are useful for the development of various functional materials, while preserving the bulk properties. In this work, we seek to develop antimicrobial surface coatings that can help to reduce the transmission of pathogens within a clinical environment. Our focus is on chitosan derivatives since chitosan is an inexpensive, nontoxic naturally derived polysaccharide which is known to possess broad antimicrobial activity. The quaternization of chitosan greatly increases its solubility in water as well as its antimicrobial efficacy. Quaternized chitosan (QCS) modified with acrylate groups were grafted on polymer films under UV irradiation to give highly stable, non-cytotoxic QCS coatings. When bacteria-contaminated droplets were placed in contact with these coatings, about 90% of the bacterial cells were killed after 2 h. We hypothesized that the antibacterial efficacy could be further increased by the introduction of copper on the QCS-grafted films. Electroless deposition of copper on the QCS coatings was carried out, with and without catalytic pre-treatment using acidic $PdCl_2$. The total copper content of the coatings with and without $PdCl_2$ pretreatment was similar at ~ 4 $\mu g/cm^2$, after 1 h of electroless deposition. Thus, the positively charged quaternary ammonium groups of QCS promoted deposition of copper on the surface without the necessity of catalytic pre-treatment. XPS analyses showed that majority of the copper on the surface was Cu(0) species. Copper-containing QCS coatings reduced bacterial count by 95% and 100%, when placed in contact with contaminated droplets for 1 h and 1.5 h respectively. Additionally, the copper-containing films were found to have no significant cytotoxic effects on fibroblast cells placed in contact with these films. Thus, these copper-containing films with high antibacterial efficacy can potentially be placed over high-touch areas in hospitals to prevent pathogen transmission.

485884 Patterning Buckles in Polymer/Metal Thin Films By Laser Irradiation

Abstract:

Patterning Buckles in Polymer/Metal Thin Films by Laser Irradiation

Kunal Mondal[#], Ying Liu[#], Michael D Dickey*, Jan Genzer*

Department of Chemical & Biomolecular Engineering, NC State University, Raleigh, NC 27695-7905

Abstract

Buckling of thin films is a promising way to fabricate micro-scale structures without the use of expensive lithographic processes or clean rooms. Laminated films on a rigid substrate will form buckled topography to accommodate the stresses arising from the mismatch in thermal expansion of the individual layers. Buckles form over the entire film if heated uniformly. Here, we induce buckles at specified locations of a metal/polymeric laminate (supported by a silicon wafer) by delivering heat locally to the surface of the film via a laser beam. Upon exposure with a laser beam, the thin bilayer film buckles locally. Rastering the laser beam across the surface can form shapes such as circles, lines, and squares. The characteristic buckling wavelength is 43 µm, as measured by atomic force microscopy and image analysis. Although the wavelength of the CO_2 laser is 10 µm, the resolution of features formed by conventional laser ablation is closer to 100 µm; thus, buckling offers a way of patterning structures with 'sub-ablation' dimensions. Buckles have prospective applications in optoelectronics, biology, biomedical engineering, and small-scale fabrication.

*Corresponding authors E-mail: jan_genzer@ncsu.edu, mddickey@ncsu.edu

#Contributed equally to this work

488062 High Performance Conducting Polymer Coatings for Corrosion Protection

Abstract:

Conducting polymer coatings were electrochemically synthesized on different metal surfaces, i.e., carbon steel, using different sulfonic acids as reaction media, including sulfuric acid (SA), (±) camphor sulfonic acid (CSA), sodium dodecyl sulfate (SDS) etc. The effect of acidic dopants on passivation of metal surface was investigated and compared with morphology and corrosion protection performance of the coating produced. The types of the dopants used were significantly affecting the protection efficiency of the coating against chloride ion attack on the metal surface. The corrosion performance depends on size and alignment of dopant in the polymer backbone. Some dopants have large aromatic rings that may stack together to form a lamellar sheet-like barrier to corrosive ions, thus making them appropriate dopants for the polymer coating in suppressing the corrosion at significant level. Further, adhesion performance was enhanced by adding long chain carboxylic acids directly in the reaction system.

499925 Enhanced Wetting Stability of Initiated Chemical Vapor Deposited (iCVD) Polydivinylbenzene Thin Films By Thermal Annealing

Abstract:

Modification methods for turning surface energy and wetting behavior have enabled many applications, including fouling-resistant materials, anti-icing coatings, wettability-based liquid separations, friction reduction, condensation control, etc. However, the durability of wettability is still a challenge for functional coatings. Here, we report a study on the wetting stability of polydivinylbenzene (PDVB) thin films deposited via initiated chemical vapor deposition (iCVD). We found that surface oxidation during air exposure leads to a decrease of receding water contact angles and an increase of contact angle hysteresis of iCVD PDVB over time. FTIR and XPS results indicate the formation of carbonyl and epoxy groups and the reduction of pendant vinyl groups during the oxidation process. We analyzed the oxidation process with a modified first-order kinetic model, and obtained the apparent rate constant k_{app} = 0.0238 ~ 0.0294 h^{-1}. In order to inhibit the oxidation in ambient air and stabilize the wettability of iCVD PDVB, we developed an *in situ* thermal annealing process to further crosslink PDVB thin films in the vacuum reactor after deposition. Our thermal annealing method effectively improves the crosslinking degree of iCVD PDVB, reduces the rate of oxidation in air, and consequently enhances the long-term wettability. The advancing and receding water contact angles of annealed iCVD PDVB remain above $100^{\hat{a}\tilde{}\tilde{}}$ and $70^{\hat{a}\tilde{}\tilde{}}$, respectively, even after air exposure for 2 months. With increased degree of crosslinking, the elastic modulus of iCVD PDVB films is also improved from 4.5Â±0.3 GPa to 6.4Â±0.4 GPa after thermal annealing.

500926 Nanoscale Characterization of Water Penetration through Plasma Polymerized Coatings and Water at the Coating/Substrate Interface

Abstract:

Plasma enhanced chemical vapor deposition (PECVD) is a versatile technique to deposit plasma polymerized thin films on various substrates without volatile organic solvents for a wide range of applications including protection of metal against corrosion. Penetration of water to the metal/coating interface is often the first step in the corrosion of coated surfaces. While past research has focused on macroscopic corrosion phenomena, understanding on the nanoscale of some fundamental aspects of the corrosion process at the metal interface is lacking. Gaining nanoscale information on the movement of water and structure of water at the interface is important to understand and design new protective coatings and also predict corrosion failure in a short time. Off specular X-ray measurements have been used to show that the plasma polymerized coatings are conformal to an underlying Al metal substrate with its native oxide, consistent with good coverage of the substrate and minimization of void space at the interface. Neutron reflectivity (NR) and infrared-visible sum frequency generation spectroscopy (SFG) have been used to characterize the water penetration through two plasma polymerized coatings on the nanoscale.

NR can determine the depth profile of a substance near an interface with a resolution of 1-2 nm. To simulate a practical corrosion process, an in situ experiment in which the sample is in the presence of water or water vapor has been used. By replacing H_2O with deuterium oxide, D_2O, the contrast has been enhanced. X-ray reflectometry (XR) measurements provide complementary information about interfacial roughnesses and changes in thickness upon exposure to water. Such measurements have revealed that the hydrophobic plasma polymerized hexamethyldisiloxane (pp-HDMSO) coating can prevent water penetration, while a hydrophilic plasma polymerized maleic anhydride (pp-MA) coating absorbs water and swells to approximately 1.5 times the dry coating thickness.

SFG measurements have directly probed the water between the coating and a sapphire substrate mimicking the native oxide on Al. These measurement have also shown that a pp-HMDSO coating prevents water penetration to the coating/sapphire interface over the time scales probed. They also reveal the hydrogen-bonded water network that forms at the interface when water penetrates through the pp-MA coating.

492894 Surface Topology and Modulus Effects on Adhesion in Novel Polyorganosilozane-Based Coatings

Abstract:

One of the ongoing effects of global climate change is the opening of previously inaccessible waterways in the Arctic. The availability of these areas to shipping and resource exploitation necessitates an increased coast guard and naval presence in these regions. Operating in conditions of extreme cold has leads to a number of issues, not the least of which is the build-up of ice on ship superstructures. As a result, the development of coatings that minimize ice adhesion is an on-going priority. We have studied the impact of coating properties on ice-adhesion for a series of novel polyorganosiloxane-based coatings. These coatings, based on the Hybridsil platform developed by Nanosonic, Inc (Pembroke, VA,) have exhibited exceptionally low ice adhesion and durability properties. The current work has focused on understanding the relationship between the surface morphology, chemistry, and modulus of the coatings on the ice adhesion behavior. Attenuated total reflectance Fourier-transform infrared spectroscopy (ATR-FTIR) was carried out to examine the chemical nature of the coating surface. In general, only small differences were observed in the IR spectra between samples. Atomic Force Microscopy (AFM) was used to image the coatings and to measure the modulus at different locations. Significant differences in film roughness and homogeneity are observed in the AFM imaging, both in the height images and the phase contrast images, depending on the specific formulation and method of coating application. These lead to significant differences in ice adhesion. In addition, force curves were obtained for a subset of samples and were analyzed to determine the film moduli. We observed differences in the film moduli both between different samples and at different locations for a single sample (i.e. between low and high points on the film.) These spatial variations also impact the strength of ice adhesion on the coatings. The coating chemistry is important in order to create a durable film with a low affinity for interactions with water ice; however, it is vitally important to understand and control the coating surface properties in order to achieve the desired low values for ice adhesion.

488428 Templated Synthesis of Polymer - Gold Nanocomposites with Pluronic Gels

Abstract:

Templated Synthesis of Polymer - Gold Nanocomposites with Pluronic Gels

Gold nanoparticles with uniform size and shape have many applications in a variety of fields such as biomedical, sensing and plasmonic materials. Frequently, ordered assemblies of nanoparticles are sought to fabricate practical devices utilizing the unique properties of these nanoparticles. For example, assemblies of gold nanoparticles with complex shapes can show emergent behavior such as metamaterials. Obtaining fully 3 dimensional assemblies of gold nanoparticles with macroscale ordering has been a significant challenge. One of the methods to create such ordered structures is the in situ formation of nanoparticles within polymer matrices, especially block copolymer matrices which act as templates for the formation of nanoparticles and their assembly. Nucleation and growth of the nanoparticles in the polymer matrices can be tuned by pH, choice of reducing agent, concentrations of reactants involved and the reaction temperature. Ordered nanoparticle assemblies also typically require high concentration of monodisperse nanoparticles in the assembly. In this regard, we present some results on the formation of high concentration of monodisperse gold nanoparticles in Pluronic gels. Pluronic block copolymers solutions in water are known to form liquid crystal lattices at high polymer concentration at room temperature, and the formed liquid crystals have been used to template the formation of mesoporous inorganic oxides. In addition, Pluronic is also known to reduce chloroauric acid to gold nanoparticles in solutions. Here we present the formation of gold nanocomposites at high concentration in Pluronic solutions and gels. Different conditions were tested for the formation of uniform gold nanoparticles without the loss of gelation behavior of the polymer at room temperature. Effect of pH and added reducing agent on the kinetics of the redox reaction, gelation, nanoparticle yield and morphology were studied. Basic pH was found to aid both the gelation, and fast redox reaction. Fast reduction resulted in the formation of uniform gold nanoparticles in the gel with diameter of 9.4 ± 1.9 nm. Under other tested conditions, formation of gold triangles and rods was observed. Addition of trisodium citrate to the gel was found to improve the results of the reaction considerably relative, but was found to be not as effective as synthesis at high pH with sodium hydroxide. Our results indicate a competition between Pluronic and added reducing agents for reducing gold ions which can be tuned by adjusting solution pH. Differences in observed results with pH and added salt were explained on the basis of Hofmeister effect and the effect of pH on the hydration of PEO blocks and the redox reaction. SAXS and cryoTEM will be used to characterize the assembly of the particles in the gels. Using these results we aim to achieve controlled and selective deposition of metal nanoparticles and metal films in domains of Pluronic block copolymer gels.

Figure 1: TEM images of gold nanoparticles synthesized in 22.7 wt% Pluronic gels in 54mM sodium hydroxide with 1.6mM $HAuCl_4$. The particles have a narrow size distribution (9.4 ± 1.9 nm).

496270 Predicting Surface Area in Green Synthesis of Sol-Gel Materials

Abstract:

Rigidity Theory (J.C. Maxwell, 1864) explains why some physical structures are rigid and others not, by balancing the degrees of freedom of movement of the component parts, with the constraints on their movement effected by connections between them. We applied the theory to porous amorphous materials recently prepared *without* template molecules or organic solvents and can predict which precursors will and will not produce high surface area materials useful for adsorption and catalysis.

In sol-gel routes to porous materials, the precursors are hydrolyzed and then condensed to form a 3D polymeric network. During drying, the network substantially collapses due to strong capillary forces, unless something is done to prop it up. Our hypothesis is that, in the absence of pore-templating guest molecules or complex solvent-removal processes, the network must reach a *critical level of rigidity* in order to maintain a porous structure. We used rigidity theory to calculate the level of condensation present at the critical level.

For such a network to become critically rigid before collapse, it must form quickly (kinetics) and be stable (thermodynamics) under the conditions of drying. We developed descriptors for the time it takes for the network to achieve rigidity and for the level of solvent remaining in the gel at the rigidity transition. We further hypothesize that the amount of solvent remaining when rigidity is reached determines the total porosity as the originally solvent-filled volume becomes empty.

We show that the theory successfully distinguishes between precursors that exhibit very high levels of porosity and surface area (> 1000 m^2/g), those that exhibit intermediate levels (~ 250 m^2/g), and those that substantially collapse (< 50 m^2/g). The theory is simple to implement for many precursors, and is being used to predict new precursors that might make useful materials.

501403 Leveraging Nanoparticle Template Assembly and Interfacial Phenomenon for Multi-Scale Control over Polyimide-Derived Carbon Molecular Sieve Films

Abstract:

Synthetic challenges common to carbon molecular sieve (CMS) and other inorganic membrane materials include the need to minimize defect-free membrane thickness while simultaneously tailoring membrane texture and function in order to break through permeability-selectivity trade-offs. In this talk, we will present a facile hard-templating approach [1], which we have recently extended to produce ultra-thin microporous polyimide-derived CMS films that are self-supported on three-dimensionally ordered mesoporous (3DOm) carbon layers. This approach exploits convectively deposited, size-tunable silica nanoparticle colloidal crystal films as a sacrificial template and poly (pyromellitic dianhydride-co-4,4â€™-oxydianiline) (PAA) as the carbon source used for replicating the template voids. Replication is carried out by tuning viscosity of the PAA molecular replica precursor solution to facilitate infiltration into the template voids by simple immersion, followed by controlled spin-off of excess precursor. Precursor spin-off rates and the multi-layer thickness of the sacrificial template itself provide handles for independently controlling the thickness of the CMS over-layer and the supporting 3DOm underlayer, respectively. Beyond scaffolding film mesostructure and asymmetry, we will demonstrate how the three-dimensionally distributed template-replica interface can be exploited as an additional handle for tuning microstructure of the carbon matrix [2]. Specifically, we will discuss complementary high-resolution TEM, Raman, XPS, physisorption, and permeation data demonstrating how interfacial chemistries can be employed to tailor carbon allotrope distribution (e.g., sp^2 carbon allotrope content) and thus microporosity and its distribution in CMS films. In doing so, we will aim to develop critical synthesis-structure-function relations that should help establish the versatility of the simple nanoparticle crystalline film platform for facile templating of hierarchically micro-mesostructured thin films and membranes.

References

[1] Z. Tian, M. A. Snyder, *Langmuir,* 30 (2014) 9828-9837.

[2] M. A. Snyder, *MRS Bulletin*, 41(9) (2016) 683-688.

494162 Toward Rational Design of Hierarchical Beta Zeolites Via Cost-Effective Approaches

Abstract:

Saudi Aramco has the worldâ€™s largest crude oil production at more than 10 million barrels of oil per day. During the traditional downstream refinement process, zeolites are often used as catalysts due to their excellent hydrothermal stability and acidity in order to crack the hydrocarbons from crude oil into lower boiling point small molecules. Conventionally used zeolites are limited by their small micropore sizes and are unable to accommodate the larger substrates of crude oil, resulting in a vacuum residue that is often times flared off or burned for electricity. To address these challenges, hierarchical zeolites were conceived with the purpose of introducing additional porosity, usually in the mesopore range, by either top-down or bottom-up approaches. The top-down approach uses microporous zeolites that are post-synthetically leached to create hierarchical structures, while bottom-up approaches often involve more complicated templating routes, such as hard- or soft-templating strategies. As compared to the previous bottom-up templating methods, the top-down approach by post-synthetic base leaching is a highly reproducible, convenient, and cost-effective approach for the creation of hierarchical zeolites, which is favorable for commercial zeolite production. In this talk, the strategies of tailoring of the hierarchical beta structures of zeolites with different Al contents will be highlighted by using top-down post synthetic base leaching. The feasibility of achieving rational design of hierarchical zeolites will be discussed focusing on the mesopore creation and catalytic activities.

502316 Nanoscale Control of Homoepitaxial Growth on a Two-Dimensional Zeolite

Abstract:

Zeolites, crystalline microporous solids are of interest for a variety of applications including catalysis, adsorption, ion-exchange, separation membranes etc. Given a pore size, improving the performance of zeolites depends to a great extent on the ability to tune their crystal morphology, size and shape. Advances in the understanding of zeolite synthesis and crystallization have facilitated the development of core-shell catalysts[1], hierarchical materials[2], exfoliated two-dimensional nanosheets[3] and thin films[4]. These developments reiterate the importance of controlling zeolite growth and crystallization at a nanoscale to tailor their microstructure. However, studies focusing on development of methods to control zeolite crystal growth at a dimension approaching single-unit-cell are still in their infancy.

In this work, we report on solution-based growth conditions that lead to slow and controllable epitaxial growth of two-dimensional zeolite nanosheets[5]. Growth rates on the order of few nanometers per day were achieved. Anisotropic growth in the absence of misoriented domains was facilitated by use of an organic additive which suppresses the non-classical crystal growth pathway predominant in most zeolites to favor a classical pathway involving growth by addition of molecular silica species. Controlling and monitoring growth at a nanoscale also revealed novel crystal growth phenomenon associated with the lateral size and local surface curvature of 2D zeolites.

The growth method described herein is enabling us to tune the microstructure of zeolite thin films and hierarchical catalysts at a scale approaching single-unit-cell dimensions which offers great potential to improve their performance in a plethora of commercial applications. Recently, using this growth method we were able to obtain zeolite MFI thin films on gold coated Si wafer that demonstrated exceptionally low dielectric constant and great mechanical strength. The method is transferrable and can be used to grow zeolite crystals irrespective of their size and the support material used. This paves way for development of high performing separation membranes.

References

[1] D. Van Vu, M. Miyamoto, N. Nishiyama, S. Ichikawa, Y. Egashira, K. Ueyama, *Microporous Mesoporous Mater.* **2008**, *115*, 106–112.

[2] X. Zhang, D. Liu, D. Xu, S. Asahina, K. A. Cychosz, K. V. Agrawal, Y. Al Wahedi, A. Bhan, S. Al Hashimi, O. Terasaki, et al., *Science* **2012**, *336*, 1684–1687.

[3] K. Varoon, X. Zhang, B. Elyassi, D. D. Brewer, M. Gettel, S. Kumar, J. A. Lee, S. Maheshwari, A. Mittal, C.-Y. Sung, et al., *Science* **2011**, *334*, 72–75.

[4] N. Rangnekar, M. Shete, K. V. Agrawal, B. Topuz, P. Kumar, Q. Guo, I. Ismail, A. Alyoubi, S. Basahel, K. Narasimharao, et al., *Angew. Chemie Int. Ed.* **2015**, *54*, 6571–6575.

[5] M. Shete, M. Kumar, D. Kim, N. Rangnekar, D. Xu, B. Topuz, K. V. Agrawal, E. Karapetrova, B. Stottrup, S. Al-Thabaiti, et al., *Angew. Chemie Int. Ed.* **2017**, *56*, 535–539.

492619 Thermodynamics of Charging in Weak Polyelectrolytes

Abstract:

Applications in water desalination, drug delivery, and the design of microstructured functional materials involve extensive use of charged polymers. In particular, there has been significant interest in "smart" materials, comprised of weak polyelectrolytes, which are responsive to environmental pH and salt content, and take advantage of these conditions to preferentially separate solutes or release therapeutic payload. When dendritic polymers, polymer brushes, or condensed phases are present, the local structure and charging state exert enormous influence over the system's thermodynamics. Motivated by recent experiments demonstrating unexpected charging behaviors in weak polyelectrolyte brushes and functionalized membranes, we utilize Molecular Dynamics simulations coupled with Grand Canonical Reaction Ensemble Sampling methods to investigate the subtle interplay between salt concentration, pH, and associative interactions that control polymer microstructures. In particular, we are able to elucidate significant pKa shifting effects induced by polymeric topology and confinement. Importantly, we will discuss the consequences of these findings in the formation of associative phases, material uptake in polyelectrolytes and the effective operating ranges of smart materials.

498847 Thermodynamics and Kinetics of Ordered, Stronly Segregated Diblock Copolymers

Abstract:

The formation of ordered states in diblock copolymers is based on the kinetic capability of the system to achieve a thermodynamically equilibrium structure. Strongly segregated, low molar mass (unentangled) diblock copolymers (< 10 kg/mol) eliminate the kinetic limitations of conventional polymer systems and are therefore an ideal platform for understanding equilibrium structures. Although the dynamics in these systems are relatively facile, phase transitions to and from complex particle-forming morphologies, such as a dodecagonal quasicrystal and the Frank-Kasper sigma phase, have been shown to take months. There is motivation, then, to elucidate the underlying dynamics that govern these transitions.

A series of model systems containing compositionally asymmetric hydrocarbon and polylactide blocks were synthesized to probe this strongly segregated regime. Methods such as X-ray photon correlation spectroscopy, small angle X-ray scattering, and rheology were utilized to provide a full physical characterization of the dynamics and phase behavior of these systems. Conformational asymmetry is shown to promote the formation of the Frank-Kasper sigma phase, which often forms through a metastable dodecagonal quasicrystal. Additionally, the Frank-Kasper phase is shown to emerge in blends of two neat diblocks with disparate compositions and degrees of polymerization. These results experimentally validate theoretical predictions, and point towards new controls of complex phase behavior in particle-forming diblock copolymers.

496611 Effect of pH on the Interaction between Poly(vinyl alcohol) and Ice

Abstract:

The development of improved cryopreservative materials is necessary to enable complete recovery of living cells and tissue after frozen storage. Remarkably, poly(vinyl alcohol) (PVA) is very effective at halting the Ostwald ripening of ice, a process believed to impart mechanical damage to cells, but the mechanism through which it does so remains unclear. Herein we quantitatively evaluate ice growth kinetics in the presence of PVA at different pH conditions and in the presence of a range of neutral salts. We demonstrate that pH, but not salt identity, plays a crucial role in PVA's ability to halt ice grain coarsening. However, we show that favorable pH conditions are not sufficient to impart ice-halting properties to similar polyols. These findings highlight the crucial role of hydrogen bonding in PVA/ice interactions. Additionally, these results provide a convenient on/off switch for PVA/ice interactions which is a valuable tool for elucidating PVA's mechanism of action.

488139 Self Consistent Field Theory Study of Multivalent Cation Effect on Semiflexible End-Grafted Random Polyelectrolytes

Abstract:

Biopolymers such as DNA, aptamers and proteins have long been a matter of interest for diverse applications. Modeling these biopolymers requires indepth understanding of random (heterogeneous) polyelectrolytes due to their resemblance in structure and property. Randomness of the sequence creates more variation in the local environment surrounding a random polyelectrolyte chain than a regular homogeneous polyelectrolyte. Presence of multivalent cations largely effects the structure and chemistry of these polyelectrolytes due to strong electrostatic interactions and salting out effect [2] and is critically important in biosensing and drug delivery applications. This study aims to elucidate the underlying physics of multivalent cation effect on surface-grafted random polyelectrolytes from a statistical thermodynamic viewpoint.

This theoretical study implements a Self Consistent Field Theory (SCFT) approach to build the molecular model for the random polyelectrolytes grafted to a surface and surrounded by multivalent cations. Structural rigidity of the polyelectrolyte chain is taken into account by using the Wormlike Chain Model of semi-flexible polymers. The molecular model properly accounts for the structural, thermodynamic and electrostatic properties of all the species present in the system. The highly non-additive coupling between various interactions creates a varying local environment near each polyelectrolyte chains that ultimately results in collapse of the chain. Formation of salt bridge between different chains with the aid of multivalent cations is considered in this theory by using a chemical reaction formalism. Poor solvent quality due to presence of thermodynamically controlled salting out effect is accounted with a Flory-Huggins Chi parameter . Variation of the cation effect is analyzed at different pH, salt concentration, polymer grafting density and polymer chain length. This model gives an exhaustive framework to demonstrate multivalent cation induced structural change in random polyelectrolytes tethered to nano surfaces for drug delivery and biosensing applications and also in colloid chemistry.

References

1. Borukhov, I., Andelman, D., and Orland, H. (1998). Random polyelectrolytes and polyampholytes in solution, *Eur. Phys. J. B 5*, 869â€“880.
2. Djabourov, M., Nishinari, K., and Ross-Murphy, S. B. (2013).Physical Gels from Biological and Synthetic Polymers, *Cambridge University Press*.

489873 The Sequence-Dependence of the Persistence Length of DNA

Abstract:

While the sequence-dependence of the DNA bending rigidity for short molecules has been studied for many decades, how this short-scale behavior is connected to the persistence length of the chain at long length scales remains unclear. Most studies involving long DNA presume that the persistence length is independent of sequence, based on the assumption that the sequence-dependence is averaged out at larger length scales. We test this conventional wisdom by measuring the extension of human DNA in nanochannels. By labeling the DNA with a sequence-specific probe, we are able to map these molecules back to the reference genome and thus determine the DNA sequence between the probes. The resulting data set consists of more than 10 million such probes, separated by distances ranging from 2.5 kilobase pairs (kbp) to 500 kbp, and containing a wide range of GC content. By converting the fractional extension of these chains to a persistence length by Odijkâ€™s theory, we found that persistence length was positively correlated with GC content, with a variation from 55 nm to 70 nm for sequences containing 31.25% GC and 56.25% GC, respectively. These results, obtained using human DNA and 41 nm nanochannels, were corroborated by additional competitive binding experiments using T4 bacteriophage DNA (35% GC) and lambda-phage DNA (49% GC) in 100 nm nanochannels. We will discuss the importance of these results in the context of sequence-dependent theories for the DNA bending rigidity.

Anomalous Hydrodynamic Radius of Polyethylene Glycol Molecules in Mixed Solvents Containing a Hydrotope

Xiong Zheng[†,††], *Mikhail A. Anisimov*[†,†††,*], *Jan V. Sengers*[†], *and Maogang He*[††]

[†]Institute for Physical Science and Technology and Department of Chemical and Biomolecular Engineering, University of Maryland, College Park, MD 20742, USA

[††]Key Laboratory of Thermo-Fluid Science and Engineering, Ministry of Education, Xi'an Jiaotong University, Xi'an, Shaanxi Province, 710049, P.R. China

[†††]Oil and Gas Research Institute of the Russian Academy of Sciences, Moscow, 117333, Russia

Abstract: We present results from an experimental dynamic light-scattering study of the mesoscopic behavior of polyethylene oxide (PEO) in a mixed solvent (tert-butanol+water). In the mixed solvent, the dimension of PEO exhibits an anomalous dependence on the concentration of tert-butanol, similar to anomalous mesoscopic behavior of other polymers in alcohol+water mixtures reported in the literature. We have also investigated possible aggregation of PEO in the mixed solutions. It is found that aggregation of PEO is absent in all solutions of PEO with molecular weights less than 20,000 g mol^{-1}, but supramolecular clusters of PEO do appear in both the pure and the mixed solvent for molecular weights larger than 100,000 g mol^{-1}. Intermolecular hydrogen bonding is assumed to be responsible for this cluster formation.

INTRODUCTION

Polyethylene oxide (PEO), whose chemical formula is $H–(O–CH_2–CH_2)_n–OH$, is a very important water-soluble polymer. It is commercially used in many areas, like cosmetics, agricultural applications, and pharmaceutical materials[1]. Because of the specific interactions, originating from its hydrophobic ethylene units and hydrophilic oxygen atoms, similar to those found in biomolecular interactions, the polymer is also relevant to bioengineering[2]. The purpose of the present work is to investigate how the behavior of this polymer in aqueous solutions is modified upon the addition of a hydrotrope. A hydrotrope is a substance consisting of amphiphilic molecules that are too small to self-assemble in aqueous solutions, but that can form short-lived molecular structures H-bonded with water molecules[3-6]. For this study, we selected a typical hydrotrope, tertiary butanol, denoted by TBA[5].

Even though PEO is a relatively simple polymer, the results of studying aggregation in PEO solutions have been controversial in the literature. Some investigations have indicated the existence of aggregation of PEO molecules in water[2, 7-14]. Polverari and van de Ven *et al.*[9] found that PEO clusters and free coils coexist in thermodynamic equilibrium because of the hydrophobic interactions of PEO molecules. Hammouda *et al.*[2,10,11] found clusters in PEO aqueous solutions due to hydrogen bonding and hydrophobicity; they systematically studied the influence of these two effects on the formation of aggregations. However, other investigators[15-20] have concluded, that there is no aggregation in aqueous PEO solutions; they have suggested that aggregation behavior may occasionally appear as a result of the presence of impurities or crystallites[15,17,18], or associated with air bubbles[20].

Another important issue is the concentration dependence of the dimension of polymer chains in solutions. Accurate data on the concentration dependence of the dimension of polymer chains are not only helpful in understanding the fundamental nature of polymer solutions, but are also useful for obtaining information about physical properties of polymer solutions, like viscoelastic properties[21]. In spite of the importance of PEO solutions, studies of the concentration dependence of the dimension of PEO, in particular in mixed solvents are still somewhat limited[20,22-25].

EXPERIMENTAL SECTION

Materials. Six polyethylene oxide (PEO) samples with molecular weights of 200, 600, 2,000, 20,000, 100,000 and 200,000 g mol^{-1} used in this study were all purchased from Sigma-Aldrich. Analytical grade tert-butanol (TBA) with a purity of 99.5 % was purchased from Alfa-Aesar. Deionized water was freshly made from a Millipore Direct-Q 3 Ultrapure Water System with a resistivity of 18.2 MΩ cm at 25 °C.

Solution preparation. The PEO aqueous solutions were prepared gravimetrically by using a weight balance. After PEO was dissolved in deionized water, the solution was gently shaken, but no stirring or ultrasonic agitation was applied. The solutions with PEO molecular weight 200, 600, 2,000, 20,000 g mol^{-1} were filtered through Millipore 20 nm filters. Those with 100,000 and 200,000 g mol^{-1} molecular weight were filtered through Millipore 100 nm filters. The TBA+H_2O solvents were filtered through 20 nm Millipore filters at a temperature below 10 °C in order to remove any trace amount of hydrophobic impurities in TBA, which could cause aggregation in aqueous solutions of TBA[4,27].

Light-Scattering Technique and Procedure. A dynamic light-scattering (DLS) technique was used to study the PEO solutions. The setup was the same one used in our laboratory by Jacob *et al.*[28] and by Subramanian *et al.*[4,27,29]. The temperature of the samples was controlled with an accuracy of ±0.1 °C. The time-dependent autocorrelation function was determined by using a commercial photon-correlator[30,31].

In dilute solutions of PEO with lower molecular weight in a pure solvent the intensity autocorrelation function $g_2(t)$, arising from the Brownian diffusion of PEO molecules, exhibits an exponential decay with time t:

$$g_2(t)-1=\left[A\exp(\Gamma t)\right]^2 = A^2\exp\left(2\Gamma t\right) , \tag{1}$$

where A is the amplitude and Γ the decay rate of the relaxation mode associated with the diffusion of the polymer chains[32]. For a diffusive relaxation process the decay rate is related to the diffusion coefficient D as

$$\Gamma = Dq^2 , \tag{2}$$

where q is a wave number related to the refractive index of the medium and the angle of the scattered light by the Bragg-Williams relation[33]. The refractive index of the solvents in this study was measured at room temperature with an Abbe refractometer. In view of the limited temperature range of our experiments, we have neglected the influence of temperature on the refractive index.

For monodisperse, non-interacting, spherical Brownian particles, the hydrodynamic radius of the particles is related to the diffusion coefficient and, hence, to the decay rate, by the Stokes-Einstein relation

$$R = \frac{k_B T}{6\pi\eta D} = \frac{k_B T}{6\pi\eta} \frac{q^2}{\Gamma} , \tag{3}$$

where k_B is Boltzmann's constant, T the temperature and η the shear viscosity of the solvent. Values for the shear viscosity of water[34,35], of TBA[36] and of TBA+water[36] were taken from the literature. In practice, Eq. (3) is also used for concentrated polymer solutions to determine from the experimental decay rate Γ a characteristic length scale R referred to as *dimension* of the polymer[37-39].

In PEO with larger molecular weight and in PEO solutions in a mixed solvent two diffusive relaxation processes are observed, so that

$$g_2(t) - 1 = \left[A_1 \exp\left(\Gamma_1 t\right) + A_2 \exp\left(\Gamma_2 t\right) \right]^2 , \tag{4}$$

where A_1 and A_2 are the amplitudes and Γ_1 and Γ_2 the decay rates of the two relaxation processes. Thus, the light-scattering experiments then yield two characteristic length scales

$$R_{1,2} = \frac{k_B T}{6\pi\eta} \frac{q^2}{\Gamma_{1,2}} . \tag{5}$$

RESULTS AND DISCUSSIONS

Anomalous dimension of PEO in mixed solvents of TBA+H_2O. The DLS measurements were performed in solutions of 1 % mass fraction PEO with molecular weight 20,000 g mol^{-1} as a function of the TBA concentration c_{TBA} at 25 °C. The DLS correlation function reveals the presence of two relaxation modes: a relatively fast mode associated with the TBA concentration fluctuations yielding values for the correlation length ξ of these fluctuations and a slower mode associated with the relaxation of the PEO molecules yielding values for the dimension R of PEO. The experimental results obtained for both the dimension R of the PEO and for the correlation length ξ of the TBA concentration fluctuations are shown in Figure 8 as a function of the TBA concentration. Note that the mixture of PEO and TBA at 25 °C is in the solid phase. The DLS technique could only be used for the solutions in the liquid phase, *i.e.*, in solutions with $c_{TBA} < 0.9$. Our data for the correlation length associated with the TBA concentration fluctuations are in basic agreement with the values earlier determined by Euliss and Sorensen[32]; these authors have attributed the somewhat unusual behavior of the correlation length in aqueous solutions of TBA to the possible presence of a virtual critical demixing temperature above the boiling temperature of the mixture and possibly again below the temperature of crystallization[32].

Most interestingly, our measurements show an anomalous behavior of the dimension of the PEO chains. The PEO chains first precipitously collapse with increasing TBA concentrations, then re-swell, pass through a maximum size, and ultimately decrease again upon further increase of the TBA concentration.

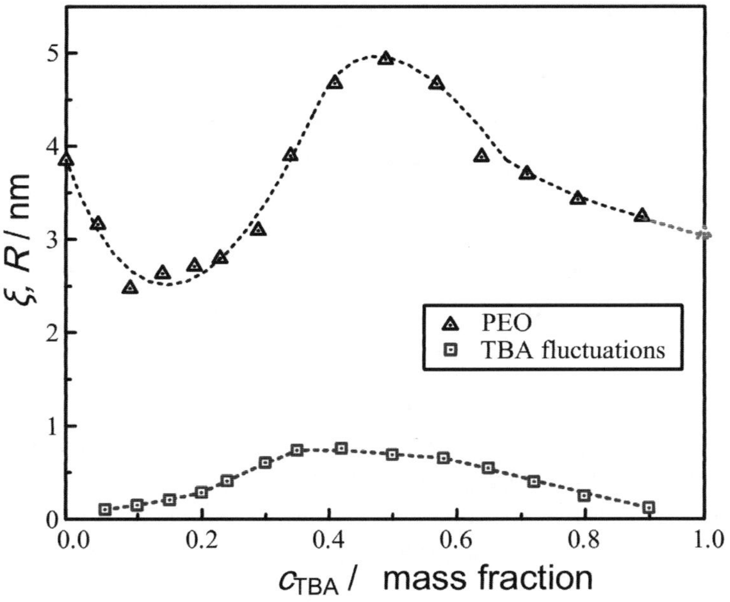

Figure 1. Dimension *R* of PEO 20,000 and correlation length ξ of the TBA concentration fluctuations in PEO+TBA+H_2O as a function of the TBA concentration at 25 °C. The dotted curves represent empirical representations of the data.

A theory for the polymer chain dimensions in a mixed solvent has been formulated by Schultz and Flory[33] and further developed by Noel *et al.*[34]. This theory yields an equation for the expansion factor as a function of the interaction parameters of the components in the ternary mixture and, hence, accounts for the non-ideality of the mixture interactions. The interaction parameter of PEO and H_2O is available in the literature[35], while the interaction parameter of TBA and H_2O can be readily calculated from experimental data[36] reported for the excess Gibb energy for TBA+H_2O. However, with any realistic estimate for the interaction parameter of PEO and TBA, this theory yields an expansion factor that is positive for all TBA concentration and does not allow for any shrinking of the dimension of PEO from its value in pure water.

Collapse of polymer dimensions has been encountered in solvents with a critical point of mixing due to preferential adsorption induced by critical fluctuations[37-41]. Although the fluctuations of TBA in water are reminiscent of pre-critical fluctuations[32], they are too small to account for the rapid collapse of the PEO at low TBA concentrations.

A rapid collapse of polymer coils followed by a re-swelling, like the behavior found by us for PEO in TBA+H_2O, has been observed in polymers in alcohol+water mixtures by many investigators, such as in methanol+water[42-54], in ethanol+water[42-44,47,55], in 1-propanol+water[42,47], in 2-propanol+water[43,44], and also in TBA+water[42,47]. This initial rapid collapse, followed by re-swelling, is commonly referred to as a co-nonsolvency phenomenon. This co-nonsolvency phenomenon has been widely debated in the literature[51-54,56-62]. A generally accepted explanation is that the initial collapse of the polymer is a generic phenomenon driven by preferential binding of the polymer with one of the two solvent components[26]. We conclude that we have encountered the same phenomenon when PEO is dissolved in TBA+H_2O. The detailed microscopic mechanism for the preferential adsorption of one of the solvent components is still somewhat controversial[51-53].

Aggregation of PEO with large molecular weight in TBA+H_2O. While aggregation was absent for PEO with molecular weights $M_w \leq 20{,}000\ \mathrm{g\ mol^{-1}}$ in all aqueous solutions, we did find aggregation of PEO 100,000 not only in H_2O, but also in the mixed solvent TBA+H_2O. Figure 2 shows our experimental results for the dimension *R* of both PEO and of the PEO clusters in TBA+H_2O as a function of the PEO concentration. Two features are to be noticed. First, the magnitude of the dimension of the PEO clusters in pure TBA is smaller than of the PEO clusters in pure H_2O, providing additional evidence that hydrogen bonding is responsible for the cluster formation; it is easier to form hydrogen bonding in H_2O than in TBA. Second, there appears to exist a close correlation between the dimension of the PEO chains themselves and of the PEO clusters. This may explain why the overlap concentrations for PEO 100,000 and PEO 200,000 are smaller than the values of the overlap concentration to be expected in the absence of aggregation.

Aggregation of PEO with large molecular weight in TBA+H_2O. While aggregation was absent for PEO with molecular weights $M_w \leq 20{,}000\ \mathrm{g\ mol^{-1}}$ in all aqueous solutions, we did find aggregation of PEO 100,000 not only in H_2O, but also in the mixed solvent TBA+H_2O. Figure 2 shows our experimental results for the dimension *R* of both PEO and of the PEO clusters in TBA+H_2O as a function of the PEO concentration. Two features are to be noticed. First, the magnitude of the dimension of the PEO clusters in pure TBA is smaller than of the PEO clusters in pure H_2O, providing additional evidence that hydrogen bonding is responsible for the cluster formation; it is easier to form hydrogen bonding in H_2O than in TBA. Second, there appears to exist a close correlation between the dimension of the PEO chains themselves and of the PEO clusters.

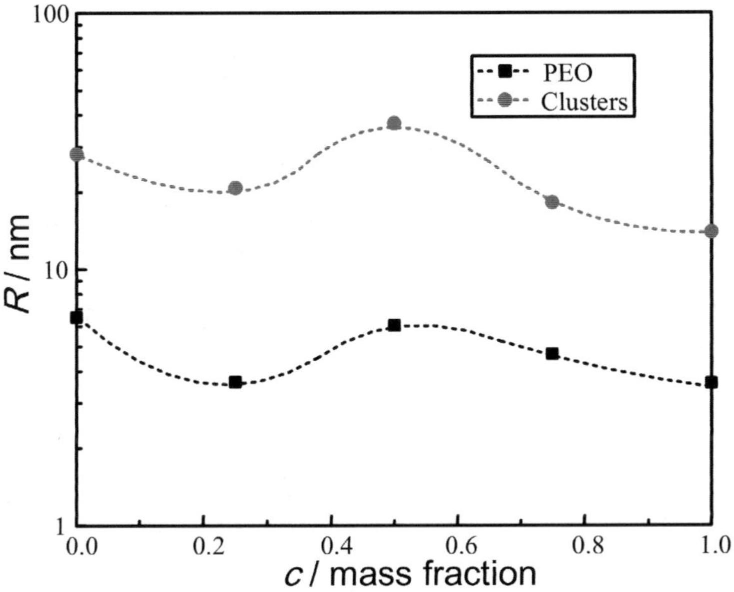

Figure 2. Dimension of clusters and chains of PEO 100,000 in TBA+H_2O at 55 °C as a function of the PEO concentration.

CONCLUSIONS

We have investigated the mesoscopic behavior of PEO both in the pure solvent H_2O and in the mixed solvent TBA+H_2O over a large range of concentrations.

In the mixed solvent TBA+H_2O we found an anomalous dependence of the dimension *R* of PEO as a function of the TBA concentrations. It turns out that this anomalous behavior is similar to that observed for other polymers in alcohol+water mixtures[42-62]. Specifically, the initial collapse of the polymer dimension upon adding TBA to appear to be caused by preferential adsorption in hydrogen-bonding solvent mixtures[26].

We have found that the possibility of aggregation of PEO in aqueous solutions strongly depends on the magnitude of the polymer molecular weight. For PEO with molecular weights $M_w \leq 20{,}000\ \mathrm{g\ mol^{-1}}$, no aggregation was observed either in the pure solvent H_2O, or in the mixed solvent TBA+H_2O, even at very high polymer concentrations. However, PEO clusters do form both in H_2O and in TBA+H_2O for PEO molecular weights $M_w \geq 100{,}000\ \mathrm{g\ mol^{-1}}$. The temperature dependence of the size of the polymer clusters in water, and the reduction of this size upon the addition of TBA, suggest that intermolecular hydrogen bonding is responsible for the cluster formation.

REFERENCES

(1) Bailey, F. J.; Koleske, J. V. *Poly (ethylene oxide)*, Academic Press: New York, 1976.
(2) Hammouda, B.; Ho, D. L.; Kline, S. R. SANS from poly (ethylene oxide)/water systems. *Macromolecules* **2002**, *35*, 8578-8585.
(3) Kunz, W.; Holmberg, K.; Zemb, T. Hydrotropes. *Curr. Opin. Colloid Interface Sci.* **2016**, *22*, 99-107.
(4) Subramanian, D.; Boughter, C. T.; Klauda, J. B.; Hammouda, B.; Anisimov, M. A. Mesoscale inhomogeneities in aqueous solutions of small amphiphilic molecules. *Faraday Discuss.* **2013**, *167*, 217-238.
(5) Robertson, A. E.; Phan, D. H.; Macaluso, J. E.; Kuryakov, V. N.; Jouravleva, E. V.; Bertrand, C. E.; Yudin, I. K.; Anisimov, M. A. Mesoscale solubilization and critical phenomena in binary and quasi-binary solutions of hydrotropes. *Fluid Phase Equilibria* **2016**, *407*, 243-254.
(6) Novikov, A. A.; Semenov, A. P.; Monje-Galvan, V.; Kuryakov, V. N.; Klauda, J. B.; Anisimov, M. A. Dual Action of Hydrotropes at the Water/Oil Interface. *J. Phys. Chem. C* **2017**, 1*21*, 16423-16431.
(7) Kalashnikov, V. N.; Kudin, A. M. Size and Volume Concentration of Aggregates in Drag-reducing Polymer Solution. *Nature* **1973**, *242*, 92-94.
(8) Polik, W. F.; Burchard, W. Static light scattering from aqueous poly (ethylene oxide) solutions in the temperature range 20-90 °C. *Macromolecules* **1983**, *16*, 978-982.
(9) Polverari, M.; van de Ven, T. G. Dilute aqueous poly (ethylene oxide) solutions: clusters and single molecules in thermodynamic equilibrium. *J. Phys. Chem.* **1996**, *100*, 13687-13695.
(10) Ho, D. L.; Hammouda, B.; Kline, S. R. Clustering of poly (ethylene oxide) in water revisited. *J. Polym. Sci. Part B: Polym. Phys.* **2003**, *41*, 135-138.
(11) Hammouda, B.; Ho, D. L.; Kline, S. R. Insight into clustering in poly (ethylene oxide) solutions. *Macromolecules* **2004,** *37,* 6932-6937.
(12) Khan, M. S. Aggregate formation in poly (ethylene oxide) solutions. *J. Appl. Polym. Sci.* **2006**, *102*, 2578-2583.
(13) Hammouda, B. The mystery of clustering in macromolecular media. *Polymer* **2009**, *50*, 5293-5297.
(14) Schwahn, D.; Pipich, V. Aqueous Solutions of Poly (ethylene oxide): Crossover from Ordinary to Tricritical Behavior. *Macromolecules* **2016**, *49*, 8228-8240.
(15) Devanand, K.; Selser, J. C. Polyethylene Oxide Does Not Neccessarily Aggregate in Water. *Nature* **1990**, *343*, 739-741.
(16) Faraone, A.; Magazu, S.; Maisano, G.; Migliardo, P.; Tettamanti, E.; Villari, V. The puzzle of poly (ethylene oxide) aggregation in water: Experimental findings. *J. Chem. Phys.* **1999**, *110*, 1801-1806.
(17) Kinugasa, S.; Nakahara, H.; Fudagawa, N.; Koga, Y. Aggregative Behavior of Poly (ethylene oxide) in Water and Methanol. *Macromolecules* **1994**, *27*, 6889-6892.
(18) Porsch, B.; Sundeltif, L. O. Apparent aggregation behavior of poly (ethylene oxide) in water as a result of the presence of an impurity. *Macromolecules* **1995**, *28*, 7165-7170.
(19) Venohr, H.; Fraaije, V; Strunk, H.; Borchard, W. Static and dynamic light scattering from aqueous poly (ethylene oxide) solutions. *Eur. Polym. J.* **1998**, *34*, 723-732.
(20) Wang, J. Q. The Origin of the Slow Mode in Dilute Aqueous Solutions of PEO. *Macromolecules* **2015**, *48*, 1614-1620.
(21) Kalwarczyk, T.; Ziebacz, N.; Bielejewska, A.; Zaboklicka, E.; Koynov, K.;

Szymanski, J.; Wilk, A.; Patkowski, A.; Gapinski, J.; Butt, H. J.; Hołyst, R. Comparative analysis of viscosity of complex liquids and cytoplasm of mammalian cells at the nanoscale. *Nano letters*, **2011**, *11*, 2157-2163.
(22) Devanand, K.; Selser, J. C. Asymptotic behavior and long-range interactions in aqueous solutions of poly (ethylene oxide). *Macromolecules* **1991**, *24*, 5943-5947.
(23) Rangelov, S.; Brown, W. Microgel formation in high molecular weight poly (ethyleneoxide). *Polymer* **2000**, *41*, 4825-4830.
(24) Xu, M.; Xu, M.; Chen, Q.; Zhang, S. M. Investigation on extremely dilute solution of PEO by PFG-NMR. *Colloid Polym. Sci.* **2010**, *288*, 85-89.
(25) Linegar, K. L.; Adeniran, A. E.; Kostko, A. F.; Anisimov, M. A. Hydrodynamic radius of polyethylene glycol in solution obtained by dynamic light scattering. *Colloid J.* **2010**, *72*, 279-281.
(26) Mukherji, D.; Marques, C.M.; Kremer, K. Polymer collapse in miscible good solvents is a generic phenomenon driven by preferential adsorption. *Nature Comm.* **2014**, *5*, 4882. 6pp.
(27) Subramanian, D.; Anisimov, M. A. Resolving the mystery of aqueous solutions of tertiary butyl alcohol. *J. Phys. Chem. B* **2011**, *115*, 9179-9183.
(28) Jacob, J.; Anisimov, M. A.; Sengers, J. V.; Dechabo, V.; Yudin, I. K; Gammon, R. W. Light scattering and crossover critical phenomena in polymer solutions. *Appl. Opt.* **2001**, *40*, 4160-4169.
(29) Subramanian, D.; Ivanov, D.; Yudin, I. K.; Anisimov, M. A.; Sengers, J. V. Mesoscale inhomogeneities in aqueous solutions of 3-methylpyridine and tertiary butyl alcohol. *J. Chem. Eng. Data* **2011**, *56*, 1238-1248.
(30) Yudin, I. K.; Nikolaenko, G. L.; Kosov, V. I., Agayan, V. A; Anisimov, M. A.; Sengers, J. V. A compact photon-correlation spectrometer for research and education *Int. J. Thermophys.* **1997**, *18*,1237-1248.
(31) http://www.photocor.com
(32) Euliss, G. W.; Sorensen, C. M. Dynamic light scattering studies of concentration fluctuations in aqueous t-butyl alcohol solutions. *J. Chem. Phys.* **1984**, *80*, 4767-4773.
(33) Shultz, A. R.; Flory, P. J. Polymer chain dimensions in mixed - solvent media. *J. Polym. Sci.* **1955**, *15*, 231-242.
(34) Noel, R.; Patterson, D., Somcynsky, T. Polymer chain dimensions in mixed solvents. J. *Polym. Sci:* **1960**, *42*, 561-570.
(35) Venoir, H.; Fraaije, V.; Strunk, H.; Borchard, W. Static and dynamic light scattering from aqueous poly (ethylene oxide) solutions. *Eur. Polym. J.* **1998**, *34*, 723-732.
(36) Koga, Y.; Siu, W.W.; Wong, T.Y.H. Excess partial molar free energies and entropies in aqueous tert-butyl alcohol solutions at 25. degree. C. *J. Phys. Chem.* **1990**, *94*, 7700-7706.
(37) Brochard, F.; De Gennes, P. G. Collapse of one polymer coil in a mixture of solvents. *Ferroelectrics* **1980**, *30*, 33-47.
(38) He, L.; Cheng, G.; Melnichenko, Y. B. Partial Collapse and Reswelling of a Polymer in the Critical Demixing Region of Good Solvents. *Phys. Rev. Lett.* **2012**, *109*, 067801.
(39) Grabowski, C. A.; Mukhopadhyay, A. Contraction and reswelling of a polymer chain near the critical point of a binary liquid mixture. *Phys. Rev. Lett.* **2007**, *98*, 207801.
(40) Nierlich, M.; Cotton, J. P.; Farnoux, B. Observation of the collapse of a polymer chain in poor solvent by small angle neutron scattering. *J. Chem. Phys.* **1978**, *69*, 1379-1383.

(41) To, K.; Choi, H. J. Polymer conformation near the critical point of a binary mixture. *Phys. Rev. Lett.* **1998**, *80*, 536-539.
(42) Mukae, K.; Sakurai, M.; Sawamura, S.; Makino, K.; Kim, S.W.; Ueda, I.; Shirama, K. Swelling of poly (N-isopropylacrylamide) gels in water-alcohol (C1-C4) mixed solvents. *J. Phys. Chem.* **1993**, *97*, 737-741.
(43) Zhu, P.W.; Napper, D.H. Coil-to-globule type transitions and swelling of poly (N-isopropylacrylamide) and poly (acrylamide) at latex interfaces in alcohol–water mixtures. *J. Colloid Interface Sci.* **1996**, *177*, 343-352.
(44) Crowther, H.M.; Vincent, B. Swelling behavior of poly-N-isopropylacrylamide microgel particles in alcoholic solutions. *Colloid Polymer Sci.* **1998**, *276*, 46-51.
(45) Zhang, G.; Wu, C. The water/methanol complexation induced reentrant coil-to-globule-to-coil transition of individual homopolymer chains in extremely dilute solution. *J. Am. Chem. Soc.* **2001**, *123*, 1376-1380.
(46) Zhang, G.; Wu, C. Reentrant coil-to-globule-to-coil transition of a single linear homopolymer chain in a water/methanol mixture. *Phys. Rev. Lett.* **2001**, *86*, 822-825.
(47) Hiroki, A., Maekawa, Y., Yoshida, M.; Kubota, K.; Katakai, R. Volume phase transitions of poly (acryloyl-l-proline methyl ester) gels in response to water–alcohol composition. *Polymer* **2001**, *42*, 1863-1867.
(48) Scherzinger, C.; Lindner, P.; Reert, M.; Richtering, W. Cononsolvency of Poly(N,N-diethylacrylamide) (PDEAAM) and Poly(N-isopropylacrylamide) (PNIPAM) Based Microgels in Water/Methanol Mixtures: Copolymer vs Core−Shell Microgel. *Macromolecules* **2010**, *43*, 6829-6833.
(49) Walter, J.; Sehr, J.; Vrabec, J.; Hasse, H. Molecular dynamics and experimental study of conformation change of poly (N-isopropylacrylamide) hydrogels in mixtures of water and methanol. *J. Phys. Chem.* **2012**, *116*, 5251-5259.
(50) Kojima, H.; Tanaka, F.; Scherzinger, C.; Richering, W. J. Temperature dependent phase behavior of PNIPAM microgels in mixed water/methanol solvents. *Polym. Sci. B Polym. Phys.* **2013**, *51*, 1100-1111.
(51) Mukherji, D.; Wagner, M.; Watson, M.D.; Winzen, S.; de Oliveira, T.E.; Marques, C.M.; Kremer, K. Relating side chain organization of PNIPAm with its conformation in aqueous methanol. *Soft Matter* **2016**, *12*, 7995-8003.
(52) van der Vegt, N.F.A.; Rodriguez-Ropero, F. Comment on "Relating side chain organization of PNIPAm with its conformation in aqueous methanol" by D. Mukherji, M. Wagner, M. D. Watson, S. Winzen, T. E. de Oliveira, C. M. Marques and K. Kremer, Soft Matter, 2016, 12, 7995. *Soft Matter* **2017**, *13*, 2289-2291.
(53) Mukherji, D.; Wagner, M.; Watson, M.D.; Winzen, S.; de Oliveira, T.E.; Marques, C.M.; Kremer, K. Reply to the 'Comment on "Relating side chain organization of PNIPAm with its conformation in aqueous methanol"' by N. van der Vegt and F. Rodriguez-Ropero, Soft Matter, 2017, 13, DOI: 10.1039/C6SM02139E. *Soft Matter* **2017**, *13*, 2292-2294.
(54) Dalgledir, C.; Rodriguez-Ropero, F.; van der Vegt, N.F.A. Computational Calorimetry of PNIPAM Cononsolvency in Water/Methanol Mixtures. *J. Phys. Chem. B* **2017**, *121*, 7741-7748.
(55) Hore, M. J.; Hammouda, B.; Li, Y.; Cheng, H. Co-nonsolvency of poly (n-isopropylacrylamide) in deuterated water/ethanol mixtures. *Macromolecules* **2013**, *46*, 7894-7901.
(56) Schild, H.G. Poly (N-isopropylacrylamide): experiment, theory and application. *Progr. Polym. Sci.* **1992**, *17*, 163-249.
(57) Hao, J.; Cheng, H.; Butler, P.; Zhang, L.; Han, C. C. Origin of cononsolvency, based on the structure of tetrahydrofuran-water mixture. *J. Chem. Phys.* **2010**, *132*,

154902.
(58) Tanaka, F.; Koga, T.; Kojima, H.; Xue, N.; Winnik, F. M. Preferential adsorption and co-nonsolvency of thermoresponsive polymers in mixed solvents of water/methanol. *Macromolecules* **2011**, *44*, 2978-2989.
(59) Tanaka, F.; Koga, T.; Winnnik, F. M. Temperature-responsive polymers in mixed solvents: competitive hydrogen bonds cause cononsolvency. *Phys. Rev. Lett.* **2008**, *101*, 028302.
(60) Zhang, Q.; Hoogenboom, R. Polymers with upper critical solution temperature behavior in alcohol/water solvent mixtures. *Progr. Polym. Sci.* **2015**, *48*, 122-142.
(61) Budkov, Y. A.; Kolesnikov, A. L.; Georgi, N.; Kiselev, M. G. A statistical theory of cosolvent-induced coil-globule transitions in dilute polymer solution. *J. Chem. Phys.* **2014**, *141*, 014902.
(62) Heyda, J.; Muzdalo, A.; Dzubiella, J. Rationalizing polymer swelling and collapse under attractive cosolvent conditions. *Macromolecules* **2013**, *46*, 1231-1238.

483770 Energy Saturation: An Alternative Mechanism for the Glass Transition

Abstract:

Treating monomer-monomer energetics of a tangent sphere chain model as a square well potential in a quasi-chemical approximation yields configurational energy saturation prior to reaching zero temperature. At saturation, a balance exists between repulsive and attractive forces that stabilize the liquid density, which thereafter becomes effectively temperature independent. These low temperature, force stabilized states are identified as ideal glass states. Energy saturation at the glass temperature (Tg) appears to obtain in 23 of 25 polymers analyzed. Among 14 polymers for which fragility indices have been measured, a good correlation between fragility and saturation degree was found; fragile polymers tend to be less saturated. The repulsion-attraction force balance raises the question as to whether it defines a unique repulsive state. To address this question, electron density was used to measure the strength of repulsive forces; global electron densities were calculated at Tg and averaged 0.61 mol/cc for 15 polymers that contain oxygen and 0.53 mol/cc for 7 hydrocarbon polymers. In comparing 2 polymers of similar structure at the same temperature above the Tg of either, one expects the polymer with the higher electron density to have the higher Tg.

Molecular Dynamics Simulation of Polymerized Stockmayer Fluids: Effects of Chain Length and Connectivity on Saturated Dipoles Near Ions

Issei Nakamura and Lijun Liu

Department of Physics
Michigan Technological University

Michigan Technological University

Background: Dielectric value

Key concept: Reorientation of solvent dipoles near the ions (so-called dielectric saturation), *Debye, P. Polar Molecules; Dover: New York (1929)*

$$E = \frac{q_1 q_2}{4\pi\varepsilon_0\varepsilon_r r^2} \quad \varepsilon_r \to \varepsilon_r(r)$$

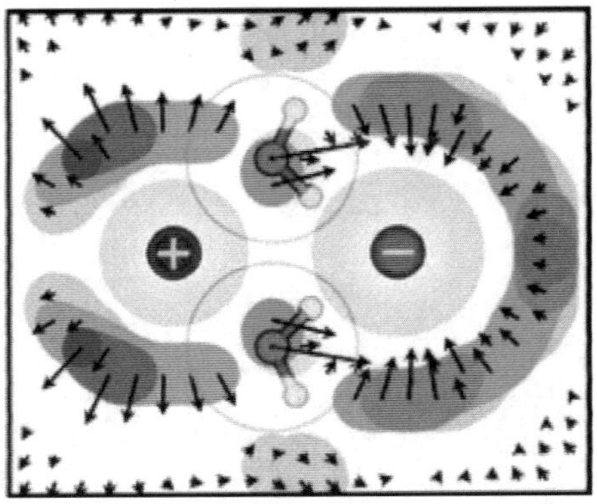

C. Fennell, et al., J. Phys. Chem. B, 113, 6782, 2009

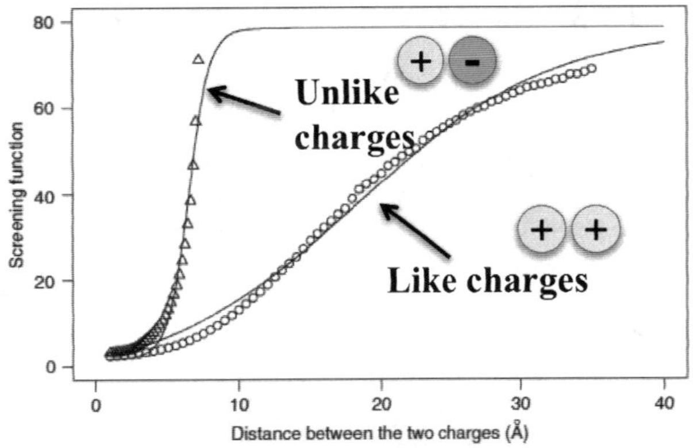

H. Gong and K. F. Freed, Phys. Rev. Lett., 102, 057603, 2009

multi-body interaction among cation, anion, and solvent molecule

Michigan Technological University

Background: Dielectric value

Experiment vs. MC simulation

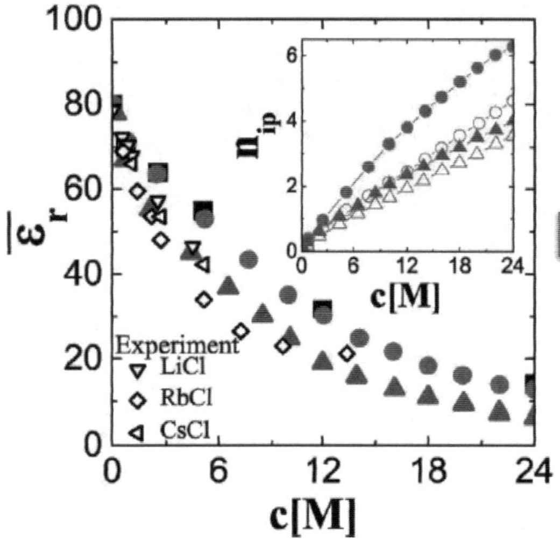

But then, the Born solvation energy is a good approximation?

$$\Delta G = \frac{q^2}{8\pi\varepsilon_0 a}\left(\frac{1}{\varepsilon_r} - 1\right)$$

X. Duan, X. and I. Nakamura, Soft Matter, 11, 3566 (2015)
[*Exp. data from Y. Wei and S. Sridhar, J. Chem. Phys., 96, 4569 (1992)*]

Saturation effect from microscopic to macroscopic scale

Michigan Technological University

Background

Isseu1: ***The dielectric constant may not be a good parameter for the dielectric response in ion-containing liquids.***

Issue 2: ***Does the dielectric saturation correlate with chain architecture? If so, is the solvation energy of the ions altered?***

[cf. J. E. Mark and P. J. Flory (*JACS, 1966, 88, 3792, 1966*), W. H. Stockmayer (*Pure Appl. Chem. 1967, 15, 539*), etc. without the ions]

Motivation: ***We need to develop coarse-grained simulation methods to capture the dielectric saturation of polymers.***

Michigan Technological University

Coarse-grained molecular dynamics simulation

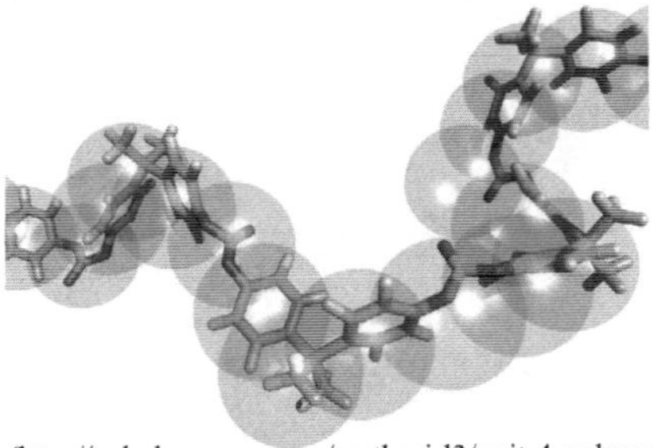

(http://nebulouszone.com/earthsci-l3/unit-4-polymers/)

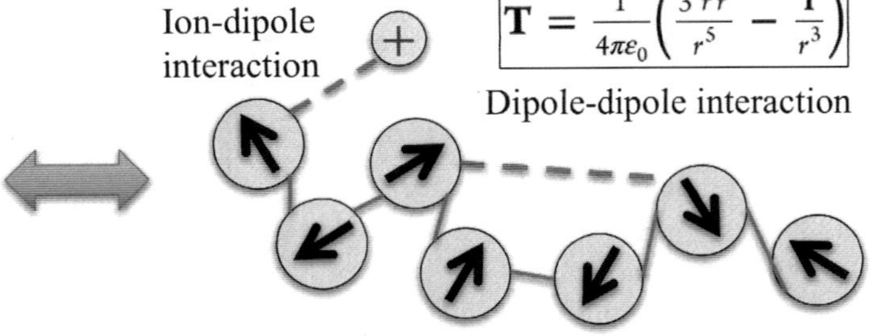

$$\mathbf{T} = \frac{1}{4\pi\varepsilon_0}\left(\frac{3\,\vec{r}\vec{r}}{r^5} - \frac{\mathbf{I}}{r^3}\right)$$

Polymerized Stockmayer fluids with the permanent dipole moment $\vec{\mu}$.

Translational motion $$m\ddot{\vec{r}}_i = \vec{\mu}_i\cdot\sum_{j\neq i}\nabla\mathbf{T}\cdot\vec{\mu}_j + \vec{\mu}_i\cdot\nabla\vec{E} + \sum_{j\neq i}\vec{F}_{ij}^{WCA} + \vec{F}_i^{FENE}$$

Rotational motion for dipoles $$\frac{I}{\mu^2}\vec{\mu}_i = \sum_{j\neq i}\mathbf{T}\cdot\vec{\mu}_j + \vec{E} + 2\lambda_i\vec{\mu}_i$$

S. H. Lee, J. C. Rasaiah, and J. B. Hubbard, J. Chem. Phys., 96, 2383 (1987)
L. Liu and I. Nakamura, J. Phys. Chem. B, 121, 3142 (2017)

Michigan Technological University

Results: Non-polymeric particles

Peak value of g(r) vs. Dipole moment

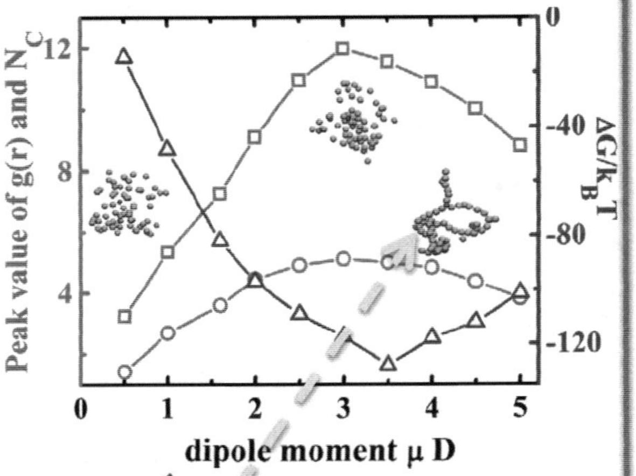

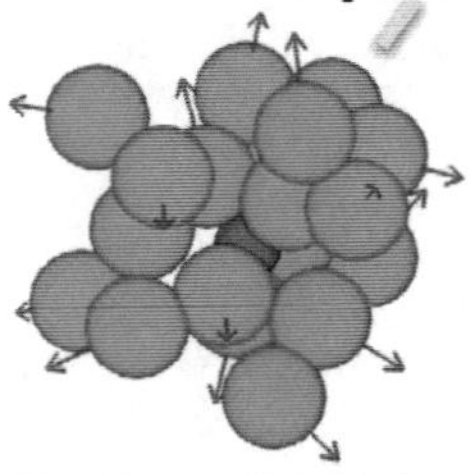

Spiky configuration

Mixture of higher- and lower-dielectric particles

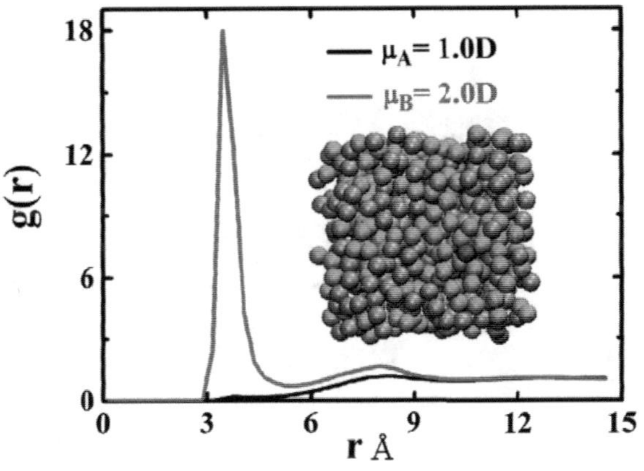

Local enrichment of higher-dielectric component near the ions

• *Z.-G. Wang, J. Phys. Chem. B, 112, 16205 (2008)*

L. Liu and I. Nakamura, J. Phys. Chem. B, 121, 3142 (2017)

Michigan Technological University

Results: Polymers

Chain length vs. Solvation energy of an ion

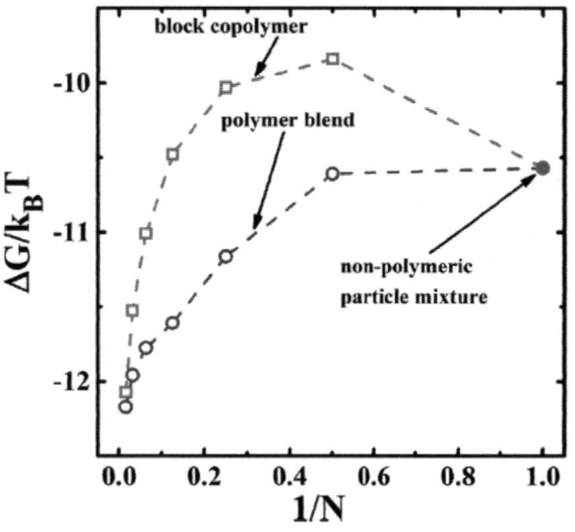

Coarse-grained mean-field theory

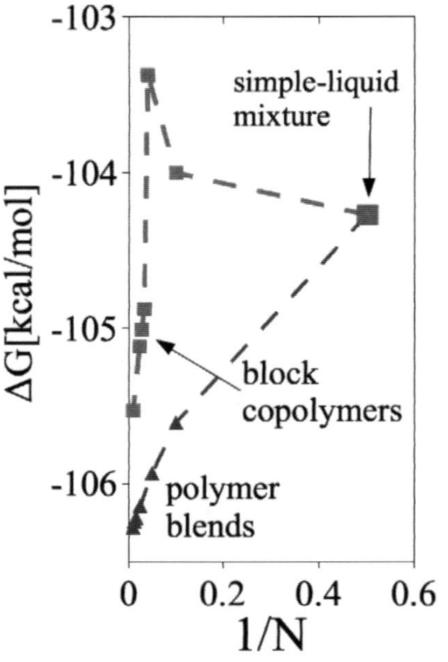

Significance of chain architecture

I. Nakamura, J. Phys. Chem. B, 118, 5787 (2014)

L. Liu and I. Nakamura, J. Phys. Chem. B, 121, 3142 (2017)

Michigan Technological University

Results: Polymers

Single-component polymer vs. polymer mixtures

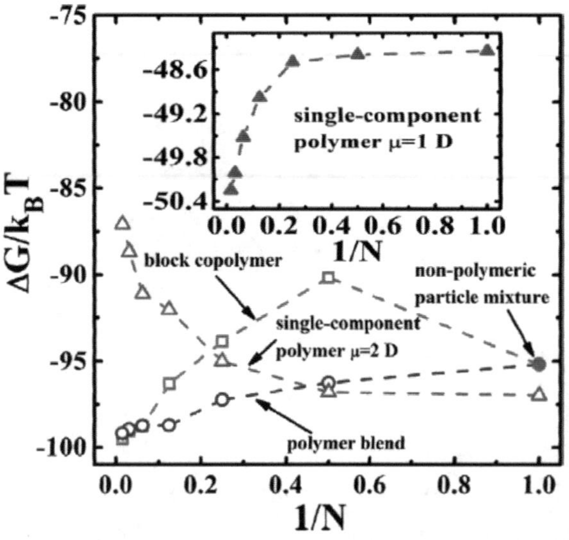

Significance of chain architecture

Strong ion-dipole and dipole-dipole correlations

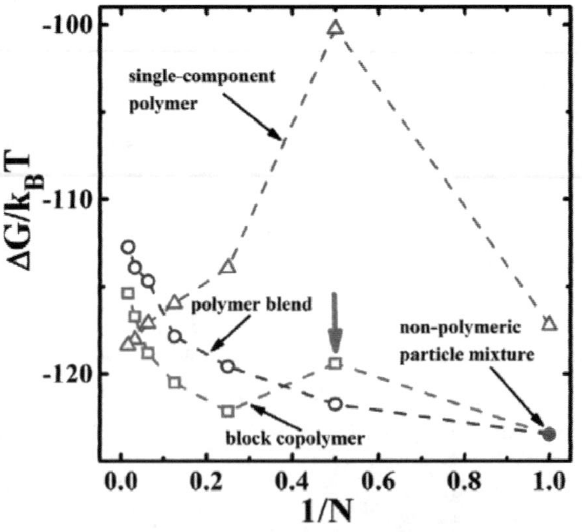

Beyond the mean-field theory

L. Liu and I. Nakamura, J. Phys. Chem. B, 121, 3142 (2017)

Michigan Technological University

Summary and Conclusion

- MD simulation for polymerized dipolar molecules (Stockmayer-like fluids)
- Our new MD simulation **vs.** the recent coarse-grained mean-field theories

  → Local enrichment of the higher-dielectric component near the ions ☑ ***consistent***

  → Effects of chain architectures ☑ ***consistent***

  → Chain-like association of monomeric units near the ions ☑ *new*

  → ***New ΔG trend***

Conclusion

- **Chain architecture vs. Dielectric saturation**
- **Multibody correlation through polymer chains**

  → ***Image charges in polymers?***

  → ***Polymerized ionic liquids?***

[Acknowledgement: NSFC(21504093 and 21474112), start-up funds of Michigan Tech, and MTU High-Performance Computing (Superior)]

Michigan Technological University

Background: Dipole response of polymers

- **Dielectric properties of polar chain properties**
  W. H. Stockmayer (*Pure Appl. Chem. 1967, 15, 539*)

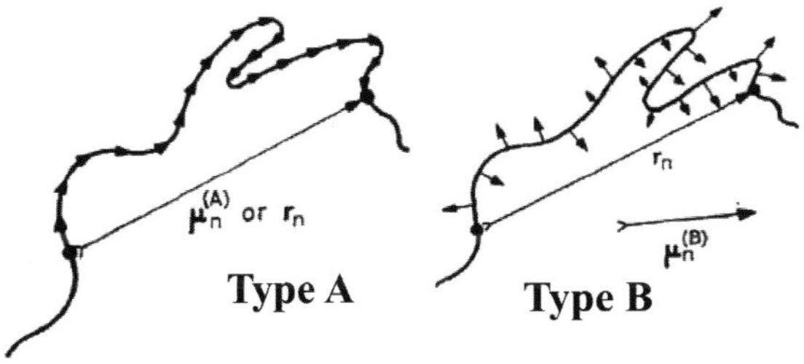

Example	Type A	Type B	Type C
Poly(vinyl halides) Poly-p-halostyrenes Poly(ethylene oxide)		X	
Poly(vinyl acetate) Poly(methyl methacrylate) Poly(m-halostyrenes)		X	X
Polyalanine Poly--propiolactone	X	X	
Cellulose acetate Poly-y-benzylglutamate	X	X	X

- **Dependence of polymerization on the mean-square dipole moment of polymers**
  J. E. Mark and P. J. Flory (*JACS, 1966, 88, 3792, 1966*)

With ions?

494731 Probing Polymer Blend Phase Diagrams Via Oligomer Molecular Simulations

Abstract:

The ability to predict polymer blend phase diagram from their molecular structures alone is crucial in developing new multi-component polymeric materials. In our previous work, a simulation framework has been established to compute the Flory-Huggins chi parameter of polyolefin pairs from the Gibbs ensemble Monte Carlo simulations. In this work, we extend the simulations to blends with larger chi values and explore influence of molecular weight and dispersity. The phase diagrams obtained from the molecular simulations are compared to experimental data. Analysis of the trajectories provides microscopic-level information on spatial distributions of end and middle segments and on the molecular-weight dependent partitioning in disperse samples.

501786 Silica-Peg Gel Encapsulation Platform for Isolation of Dormant Cancer Cells

Abstract:

A major barrier to improving treatment of breast cancer is the ability of a subset of tumor cells to leave the tumor early in disease progression and enter a dormant state at other sites in the body. Since most chemotherapies target rapidly growing cells, these dormant cells can evade treatment and remain undetectable within the body for months or even years before reawakening and causing disease recurrence. Our understanding of the basic mechanisms regulating dormancy and reawakening has been severely limited by the lack of *in vitro* model systems that can maintain cells in a dormant state. To address this need, we have developed a method for encapsulation of cancer cells in highly porous yet stiff silica-PEG (SPEG) gels that physically confine cells, inhibiting their ability to proliferate, while still enabling nutrient transport. Furthermore, the SPEG gels are mechanically reversible (shear-thinning), enabling release of the encapsulated cells without the need for chemical or thermal treatment of the gel. Alamar blue assays demonstrated that following encapsulation, metabolic activity of MCF-7 breast cancer cells, which are known to exhibit dormant phenotypes, decreased to 30% of the initial value by Day 3 and remained constant up to Day 7. In contrast, MDA-MB-468 cells, which are highly aggressive and are not known to transition into dormancy *in vivo*, exhibited a sharp decrease in metabolic activity, with no activity observed after Day 2. Comparison of post-encapsulation viability of MCF-7, MDA-MB-468 and MDA-MB-231 cells, which are also more proliferative than MCF-7, confirmed the metabolic activity results, with a population of live MCF-7 cells persisting until Day 7, while all MDA-MB-468 and MDA-MB-231 cells were dead by Day 3. Co-staining of Calcein AM with Ki67, a marker of actively cycling cells, indicated that the persisting population of live MCF-7 cells exhibited a significant reduction in Ki-67 expression, which is a hallmark of tumor dormancy. Importantly, the cells resumed proliferation after removal from the gels, indicating that the proliferation-arrested cells were quiescent and not senescent. Taken together, the preliminary data show that the SPEG gel encapsulation platform is a promising tool for identification and analysis of dormancy-capable sub-populations of cancer cell lines through physical inhibition of proliferation. Future efforts are focused on exploring the utility of this platform in understanding mechanisms of dormancy and reawakening as well as in development of clinical diagnostic tools.

491934 Biodegradable Nano-Film Coated Self-Floating Hollow Glass Microspheres for Rapid Cell Isolation and Recovery

Abstract:

The ability to accurately and efficiently isolate specific cell populations from heterogeneous mixtures has enabled researchers to advance the fields of experimental cell biology and translational medicine. Established cell isolation and purification techniques such as fluorescence-activated cell sorting, microfluidic devices and immunomagnetic micro/nano-particles are either expensive and non-portable, involving sophisticated microfabrication and excessive lab equipment or having negative impact on cell viability and function, which limit their practical usage in resource-limited settings. We present a simple yet effective method for cell isolation and recovery that requires neither specialized lab equipment nor any form of power source. Specifically, self-floating hollow glass microspheres were coated with an enzymatically degradable Nano-film, and conjugated with antibodies to allow both fast capture and release of subpopulations of cells from a cell mixture. Targeted cells were captured by the microspheres and allowed to float to the top of the hosting liquid thereby isolating targeted cells. An antifouling polymer brush layer was grafted onto the Nano-film to minimize non-specific cell adhesion. Using the EpCAM expressing cancer cell line, PC-3, in blood as a model system, we have demonstrated the isolation and recovery of cancer cells without compromising cell viability or proliferative potential. The whole process takes less than one hour. To support the rational extension of this platform technology we introduce extensive characterization of the film fabrication and application. Our approach is expected to overcome practical hurdles and provide viable targeted cells for downstream analyses in resource-limited settings.

493160 Design of Electrohydrodynamic Sprayed Polyethylene Glycol Hydrogel Microspheres for Cell Encapsulation

Abstract:

Electrohydrodynamic spraying (EHS) has recently gained popularity for microencapsulation of cells for applications in cell delivery and tissue engineering. Some of the polymers compatible with EHS are alginate, chitosan, and other similar natural polymers, which are subject to ionotropic or physical gelation. It is desirable to further extend the use of EHS technique beyond such polymers for wider biofabrication applications. Here, for the first-time we utilized the principles of EHS to fabricate cell-laden polyethylene glycol (PEG) hydrogel microspheres. The gelation of PEG hydrogel microspheres was achieved by forming covalent crosslinks between multiarm PEG acrylate and dithiol crosslinkers via Michael-type addition. We conducted a detailed investigation of the critical parameters of EHS, such as the applied voltage, inner needle diameter (i.d. needle), and flow rate, to obtain PEG microspheres with high cell viability and tightly-controlled diameters in the range of 70-300 Î¼m. The polydispersity of cell-laden PEG hydrogel microspheres as measured by % coefficient of variation (CV) was between 6-23% for all condition tested. We established that our method was compatible with different cell types and that all tested cell types could be encapsulated at high densities of 10^6 - 10^9 and â‰¥90% encapsulation efficiency. We observed cell aggregation within the hydrogel microspheres at applied voltage >5 kV. Since PEG is a synthetic polymer devoid of cell attachment sites, we could overcome this limitation by tethering Arg-Gly-Asp-Ser (RGDS) peptide to the PEG hydrogel microspheres; upon RGDS tethering, we observed uniform cell dispersion. The microencapsulated cells could be cultured in the PEG hydrogel microspheres of different sizes for up to one week without a significant loss in cell viability. In conclusion, the EHS technique developed here could be used to generate cell-laden PEG hydrogel microspheres of controlled sizes for potential applications in cell delivery and organoid cultures.

493418 Quantum Dot-Based Biomarkers of Neuroinflammation in the Developing Brain

Abstract:

Introduction: Neuroinflammation, mediated by activated microglia and astrocytes, is an important disease hallmark within the development of central nervous system (CNS) diseases, including hypoxic-ischemia encephalopathy (HIE) and autism spectrum disorders (ASD). Thus, comprehension of neuroinflammation is considered essential for both understanding CNS diseases and developing neurological disease therapeutics. However, current studies of biomarkers of neuroinflammation have been difficult due to (1) highly restricted barriers such as the blood brain barrier (BBB) and the brain parenchyma penetration barrier, and (2) the dynamic and heterogeneous neurological disease microenvironment, which is dependent on disease etiology and disease progression, and is variable from patient to patient. Relevant biomarkers and high-throughput platforms that can assay neurological disease severity representative of the *in vivo* environment can overcome these challenges, but are currently lacking. Our previous research showed that small fluorescent nanoparticles (*e.g.* 4 nm dendrimers) could target activated microglia in the brain upon systemic administration, and were able to only localize in the region of injury in newborn rabbits with cerebral palsy (CP). This suggested that nanoparticles can be targeted to the regions of the brain that contain diseased cells, as well as specific cell types within those regions, especially in the presence of inflammation. Nano-sized fluorescent semiconductor quantum dot (QD) particles come to light as potential neuroinflammation biomarkers and have several advantages over traditional biomarkers including small size (5-10 nm), tailorable excitation and emission spectra, tailorable surface functionalities, and high photoluminescence and photostability, which are ideal characteristics for *in vivo* imaging. In this study, we evaluated QDs as biomarkers of inflammation in the developing brain. Our overall hypotheses are that QD cell-specific uptake within the brain will be governed by (1) QD stability in the brain microenvironment, (2) QD surface functionality, and (3) the extent of neuroinflammation, evaluated by microglia and astrocyte activation.

Materials and Methods: We investigated QD stability, toxicity, and cellular uptake in both organotypic brain slices and *in vivo* in a neonatal rat model. QDs with different surface functionalities were incubated with saline, artificial cerebrospinal fluid (aCSF) and brain tissue homogenate to characterize their stability in relevant media. For QD toxicity, lactate dehydrogenase (LDH) assays, propidium iodide (PI) staining, and Fluoro-Jade C staining were utilized following QD incubation with 300-400 μm organotypic brain slices obtained from neonatal wide type (WT) rats. QD cellular uptake in brain slices was evaluated by using immunohistochemistry (IHC) and confocal microscopy to characterize the colocalization of QDs with cell populations in the brain, including microglia, astrocytes, and neurons. We also evaluated QD biodistribution and cellular localization following systemic administration in the presence of neuroinflammation in a newborn rat model with hallmarks of ASD (metabolic glutamate receptor 5 knockout rats (mGluR5 KO)), compared to age-matched WT controls. Biodistribution, confocal microscopy and fluorescence-activated cell sorting (FACS) were utilized for quantitative analysis of QD uptake in the brain.

Results and Discussion: We show that QDs functionalized with polyethylene glycol (PEG) chains are stable up to 24 h in all media at both room temperature and at physiological temperature. QDs without a PEG coating (*e.g.* QD-carboxyl) rapidly aggregate in saline and aCSF at 23ËšC and 37ËšC, independent of the composition and ion concentration of the biological fluid. This stability result correlates with QD toxicity and cellular uptake imaging, where QDs without PEG aggregate in tissue and do not localize in cells, exhibiting overall lower toxicity and lower cell internalization compared to ones with PEG. Between the two most stable QDs, QD-PEG-NH_2 (PEG-amine) is found to specifically localize into neurons, while QD-PEG-OMe (PEG-methoxy) remains in the extracellular space (ECS). This suggests that QD surface functionality can affect the stability of QDs, and therefore determine their cytotoxicity and cell co-localization. Our *in vivo* results demonstrate that the QD-PEG-NH_2 localizes in activated microglia in the mGluR5 KO rat brain, and does not penetrate the BBB or uptake in cells in age-matched, litter-matched WT controls. Furthermore, FACS sorted microglia cells contained QDs, demonstrating that these cells are a pro-inflammatory phenotype. Based on these results, there is an association between QD uptake in the brain and neonatal neuroinflammation severity, suggesting that QDs can be a biomarker of neuroinflammation presence and severity.

Conclusions: Biomarkers for characterizing the severity of neonatal neuroinflammation are in need for understanding developing brain injuries and furthering therapeutic development. QD-based biomarkers have unique imaging advantages over traditional organic dyes and have great potential as biomarkers in the brain. Our study systemically evaluates QD stability, toxicity and cellular uptake in the developing brain, which begins to fill the critical knowledge gap in developing ideal QD-based biomarkers for investigation of neuorinflammation etiology and progression. Exploration of QD penetration of the BBB, specific cell internalization, intracellular trafficking, and diffusion in the CNS would also bring about a detailed molecular-level understanding for the future *in vivo* application of QDs in the developing brain.

CrystEngComm

PAPER

View Article Online
View Journal

Cite this: DOI: 10.1039/c5ce01656h

The culprit of gout: triggering factors and formation of monosodium urate monohydrate†

Meng Hsiu Chih, Hung Lin Lee and Tu Lee*

Gout is a crystal-induced inflammatory arthritis brought about by the deposition of monosodium urate monohydrate (MSUM). Hyperuricemia is considered an essential factor for gout formation. However, the *in vivo* MSUM crystallization mechanism is still uncertain, and only some of the patients with hyperuricemia are able to develop gout. Attempts were made to answer those unsolved questions based on the experimental methods developed by Perrin and Swift including (1) morphological studies of MSUM under various Na^+ ion levels, (2) effect of pH and conversion from uric acid dihydrate (UAD) to MSUM, and (3) the synergistic effect of Na^+, K^+, and Ca^{2+} ions and hyaluronate chains on MSUM crystallization. Various MSUM morphologies such as "beachballs", "urchin-like aggregates" and "bow-like aggregates" could be prepared at different Na^+ ion concentrations at 37 °C. Consequently, the pathogenesis of gout might be related to the MSUM morphological transformation from a "beachball" structure to a needle. The conversion of UAD to MSUM was studied by adjusting the pH value. The determined UAD-to-MSUM pathway was thought to be followed by the MSUM deposition during phagocytosis when lactic acid was present. In addition, a new type of MSUM "fishtail" morphology was observed in the hyaluronate-, Na^+ ion-, and Ca^{2+} ion-containing solutions. The synergistic effect of hyaluronate and the cations on the inhibition of MSUM crystallization was further verified based on the crystal yields. The higher water solubility of the hyaluronate–Ca-urate complex compared to the one of the urate ion might explain why only a fraction of hyperuricemic patients developed gout. Although there was no reported evidence about the complex, it was probable for it to minimize the risk of gout formation in hyperuricemic patients. Finally, the puzzle of gout formation was pieced together and illustrated.

Received 17th August 2015,
Accepted 19th November 2015

DOI: 10.1039/c5ce01656h

www.rsc.org/crystengcomm

Introduction

Gout is a common inflammatory arthritis induced by the deposition of monosodium urate monohydrate (MSUM) and might cause serious joint deformity and disability. MSUM originates from uric acid (pK_a = 5.75), which is an end-product of purine metabolism.[1,2] Although MSUM has been identified for over 100 years[3] and its molecular structure and solid-state properties have also been well characterized,[4–11] the prevalence and incidence of gout have continued to rise in the US,[12] UK[13] and Taiwan ROC.[14] However, the *in vivo* crystallization mechanism of MSUM is still uncertain to date.

A MSUM crystal has a crystallographically triclinic structure.[15] Its crystal habit usually exhibits a needle shape, and rarely a spherical morphology and two medical doctors, Fiechtner and Simkin, dubbed the sphere "beachball".[7,8] However, the "bow-like shaped" MSUM needle-clusters often appear *in vivo* under the influence of biological substances.[9] Besides, the MSUM clusters were also reported to be produced in a buffer solution exceeding 100 mM of Na^+ ions.[16] The many interesting morphological appearances of MSUM crystals have prompted us to investigate the influence of Na^+ ions on MSUM crystal morphology. Interestingly, the "beachball-like" MSUM morphology has been seldom discussed and has been regarded as an amorphous form by others.[6] For the biological strategy of making the mineralized skeletons, a recent study suggested that stable crystals in living organisms are usually developed by the transformation of the first-formed solids which are usually disordered deposits.[17] This crystallization strategy is widely discovered in nature such as in chiton teeth, mollusk shells and sea urchin spicules and spines.[18] Therefore, we speculate that MSUM "beachballs" may serve as an essential transient precursor during MSUM morphological evolution. Although the MSUM beachballs have been prepared by quenching, the preparation method was completely deviated from the *in vivo* conditions.[6,7] Consequently, the first aim of our study is to

Department of Chemical and Materials Engineering, National Central University, 300 Jhong-Da Road, Jhong-Li District, Taoyuan City 32001, Taiwan, Republic of China. E-mail: tulee@cc.ncu.edu.tw; Fax: +886 3 425 2296; Tel: +886 3 422 7151 ext. 34204

† Electronic supplementary information (ESI) available. See DOI: 10.1039/c5ce01656h

reproduce MSUM beachballs and its other morphologies under various Na^+ ion levels and to shed some light on the pathogenesis of gout.

Accordingly, there were several factors which were considered to facilitate the MSUM crystallization: (1) hyperuricemia,[9] (2) temperature,[19–21] (3) ionic strength,[4,19,21,22] (4) pH,[21–24] and (5) promoter/inhibitor molecules.[22,25–27] Among these factors, the temperature effect was used to explain why MSUM preferred to crystallize out mainly in human joints and connective tissue. This is because the lower local temperatures in joints and connective tissue than in other positions led to the decrease in MSUM solubility. Some cations, biomolecules and proteins could also affect the MSUM precipitation and be regarded as promoters or inhibitors. For example, the presence of K^+ ions would stabilize urate ions in an aqueous solution to inhibit the crystallization of uric acid and MSUM.

The pH effect is related to the inflammatory response[28] and acidosis.[29] Noticeably, the pH–solubility relationships of uric acid and MSUM in Fig. S1† established by Wang and Königsberger[24] have shown that the solubility of MSUM increases as the pH value decreases.[22,24] However, this unusual observation is against the fact that the decline in pH would promote MSUM nucleation *in vitro*.[22]

Moreover, uric acid is quite sensitive to pH fluctuation as reported in the renal system.[30] There were three identified phases of uric acid, including uric acid anhydrate (UAA), uric acid monohydrate (UAM), and uric acid dihydrate (UAD). All of them have been observed in renal stones,[31] and the relationships among UAA, UAD and MSUM are shown in Scheme S1 of the ESI.† Therefore, the second aim of our study is to investigate the pH effect on MSUM formation and the conversion from uric acid to MSUM in a simulated body fluid (serum) system.

Hyperuricemia is an abnormally high level of uric acid in the blood when the concentration of urate in the serum is greater than 6.8 mg dL^{-1}. It is a prerequisite for MSUM deposition, but intriguingly, only a fraction of hyperuricemic subjects suffer from gout disease.[9] There seems to be physiological substances capable of inhibiting urate crystallization. Individual synovial fluid components such as hyaluronic acid had no significant effect on the growth rate constant of MSUM.[26] However, the interactions among hyaluronate chains and Na^+, K^+, and Ca^{2+} ions,[32] which are abundant in synovial fluid, may synergistically influence the *in vivo* MSUM crystallization. Therefore, the third aim is to explore the synergistic effect of hyaluronate chains and various ions on MSUM crystallization.

Experimental section

All chemicals were used as received without further purification and are listed in Table S1.†

Morphological studies of MSUM under various Na^+ ion levels

The method developed by Perrin and Swift[11] was modified and then used in our studies. Solutions A and B were first prepared in a 37 °C water bath. 15 mL of aqueous solution A was composed of (1) 13.3 mM uric acid and (2) 15 mM NaOH. 5 mL of aqueous solution B was composed of (1) 20 mM KCl, (2) 8 mM $CaCl_2·2H_2O$ and (3) 400, 2400 or 6000 mM NaCl, and the resulting solutions were called B1, B2 and B3, respectively. The pH values of those solutions were adjusted to 7.4 using 1 M HCl (aq) or 1 M NaOH (aq). Upon mixing of both solutions A and B, the supersaturated solutions of MSUM containing (1) 10 mM uric acid, (2) 5 mM KCl, (3) 2 mM $CaCl_2·2H_2O$ and (4) 100, 600 or 1500 mM NaCl were produced and were called C1, C2 and C3, respectively (Table 1). Mixing the pre-dissolved solutions of A and B together was necessary to avoid the occurrence of heterogeneous nucleation induced by undissolved NaCl, KCl and $CaCl_2·2H_2O$ solids. In a separate experiment, the concentration of NaCl was fixed to 1500 mM and MSUM crystals would be harvested after 2 months. All MSUM crystals were filtered, rinsed with reversible osmosis water, oven dried at 40 °C for 24 h, and characterized using polarized optical microscopy (POM), scanning electron microscopy (SEM), Fourier-transform infrared spectroscopy (FT-IR), powder X-ray diffraction (PXRD), and thermogravimetric analysis (TGA).

Based on the pH–solubility relationship shown in Fig. S1,† the saturation of MSUM was around 0.26 mM in 150 mM Na^+ aqueous solution under the *in vivo* normal condition of pH = 7.4 and 37 °C as a reference.

Effect of pH and conversion from UAD to MSUM

Presores and Swift had discussed the conversion and transformation of uric acid in urine-like systems[33] such as McIlvaine buffer and artificial urine solutions.[34] In this study, we chose simulated body fluid (SBF)[35] to mimic synovial fluid instead. SBF comprising 92.45 mM NaCl, 8.81 mM $NaHCO_3$, 19.3 mM Na_2CO_3, 3.02 mM KCl, 1.01 mM $K_2HPO_4·3H_2O$, 1.50 mM $MgCl_2$, 0.5 mM Na_2SO_4, and 50 mM HEPES was prepared accordingly.[35] The pH value of the SBF was adjusted to 7.4 with 1 M NaOH (aq).

The very same procedure in the section "Morphological studies of MSUM under various Na^+ ion levels" was followed except for the Na^+ ion concentration which was fixed to 140 mM for simulating the *in vivo* conditions.[11] Instead of using 1 M HCl (aq), an about 10 mM aqueous solution of lactic acid[36] was chosen as an acidosis factor[28] to lower the pH value to 5.0 to 5.8. Lowering the pH to an extreme value of 5.0 was to accelerate the UAD formation within a reasonable time frame. UAD solids were isolated after a few minutes, filtered, air dried for 10 min and stored at −4 °C to prevent the

Table 1 The compositions and conditions of the supersaturated MSUM solutions under the various Na^+ ion levels for MSUM morphological studies

Composition/condition	Solution C1	Solution C2	Solution C3
Uric acid	10 mM	10 mM	10 mM
NaCl	100 mM	600 mM	1500 mM
KCl	5 mM	5 mM	5 mM
$CaCl_2·2H_2O$	2.5 mM	2.5 mM	2.5 mM
pH	7.4	7.4	7.4
Temperature	37 °C	37 °C	37 °C

dehydration of UAD to UAA.[37] About 30 mg of UAD solids was dispersed in 3 mL of SBF at 37 °C.[35] Solids were harvested every 12 h in a total period of 36 h, filtered, rinsed with RO water, air dried and characterized immediately using POM and PXRD.

Synergistic effect of Na^+, K^+ and Ca^{2+} ions, and hyaluronate chains on MSUM crystallization

Since the concentration of hyaluronate in normal joints ranges from 1.45 to 2.94 mg mL^{-1},[38] 5 mL of aqueous solution D1 was composed of (1) 560 mM NaCl and (2) 40 mg of sodium hyaluronate. 5 mL of aqueous solution D2 was composed of (1) 560 mM NaCl, (2) 20 mM KCl, and (3) 40 mg of sodium hyaluronate. 5 mL of aqueous solution D3 was composed of (1) 560 mM NaCl, (2) 8 mM $CaCl_2 \cdot 2H_2O$, and (3) 40 mg of sodium hyaluronate. 5 mL of aqueous solution D4 was composed of (1) 560 mM NaCl, (2) 20 mM KCl, (3) 8 mM $CaCl_2 \cdot 2H_2O$, and (4) 40 mg of sodium hyaluronate. All aqueous solutions were prepared at 37 °C and their pH values were adjusted to 7.4 using 1 M HCl (aq) or NaOH (aq). The supersaturated solutions of MSUM, called E1, E2, E3 and E4, were prepared by mixing solution A with D1, D2, D3, and D4, respectively, and their compositions are provided in Table 2. MSUM crystals would be harvested in 5 days, filtered, rinsed with RO water, oven dried at 40 °C for 24 h and characterized using POM and PXRD.

The constant composition method,[39] developed by Nancollas *et al.*, could overcome the problems associated with the dynamically changing solution composition during precipitation for many crystallization and mineralization kinetics approaches. The concentrations of lattice ions were maintained constant by the simultaneous addition of reagent solutions and controlled by a glass electrode probe. However, in our experiments, only the pH values of all the abovementioned solutions were checked to be around 7.4 by a pH meter due to its simplicity and robustness.

Results and discussion

Morphological studies of MSUM under various Na^+ ion levels

POM images in Fig. 1 show that MSUM crystals grown under various Na^+ ion levels at 37 °C were crystalline because of the strong birefringence. 200–300 μm radiating bow-like aggregates, 20–50 μm urchin-like aggregates, and 20–60 μm "beachballs" were obtained under 100 mM, 600 mM, and 1500 mM Na^+ ion levels, respectively. It took two months for the "beachballs" produced from a 1500 mM solution of Na^+ ions to gradually evolve into the "urchin-like" and "bow-like aggregates" (Fig. 1d). The SEM images in Fig. 2 show that the "beachball" structure was composed of many 1–2 μm-sized platelets filled with primary nanometer-sized needles.

Table 2 The compositions and conditions of supersaturated MSUM solutions for the synergistic effect of Ca^{2+}, K^+ and Na^+ ions and hyaluronate chains on MSUM crystallization

Composition/condition	Solution E1	Solution E2	Solution E3	Solution E4
Uric acid	10 mM	10 mM	10 mM	10 mM
NaCl	140 mM	140 mM	140 mM	140 mM
KCl	—	5 mM	—	5 mM
$CaCl_2 \cdot 2H_2O$	—	—	2.5 mM	2.5 mM
Sodium hyaluronate	2 mg mL^{-1}	2 mg mL^{-1}	2 mg mL^{-1}	2 mg mL^{-1}
pH	7.4	7.4	7.4	7.4
Temperature	37 °C	37 °C	37 °C	37 °C

The FTIR spectra (Fig. S2†) indicate that neither UAD nor UAA was formed for all solids prepared at pH 7.4. All of the MSUM crystals were verified according to the characteristic IR peak assignments given in Table S2.† The relatively broad peaks with lower intensities in the PXRD pattern (Fig. S3†) further confirmed that the "beachball" structure was composed of nanometer-sized crystallites with poor crystallinity based on Scherrer's formula.[40,41] In TGA scans (Fig. S4†), the first weight losses for both the "bow-like aggregates" and "beachball" structure were about 7.2% approximating to the weight percent of hydrate within the crystal lattice of MSUM. However, the dehydration temperature for the "beachball" structure was around 100 °C, which was lower than the one for the "bow-like" structure at 150 °C. It might be attributed to the relatively large surface area of the MSUM beachball structure for accelerating the dehydration process. They began to degrade when heated to over 300 °C.

Our *in vitro* experiments suggested that the "beachball" structure, "urchin-like aggregates" and "bow-like aggregates" were related to the Na^+ ion levels (Fig. 1) and not just only to the biological[9] and/or macromolecular[42] components in synovial fluid as claimed by others.[9,42] All MSUM morphologies were mutually related to one another because the first-formed metastable "beachballs" could transform to "urchin-like aggregates" and then to "bow-like shaped" clusters or needle-shaped crystals in solution if given long enough time. According to the SEM images in Fig. 2, we proposed that the MSUM "beachball" structure might be a mesocrystal having a hierarchical structure[43] *via* an aggregation-mediated pathway by nano-clusters instead of merely amplification in crystal growth.[44] However, more studies for the exact pathway would still be needed in the future.

Since "beachballs" were only discovered in a hypertensive female patient who also required weekly peritoneal dialysis,[8] it was speculated that the local fluctuation of the Na^+ ion level or other conditions raising the MSUM supersaturation level in her synovial fluid might play a dominant role in "beachball" structure formation. The pathogenesis of gout may be related to the MSUM morphological transformation from "beachball" to needle.[8]

Effect of pH and conversion from UAD to MSUM

Within a few minutes of the addition of lactic acid, the pH was first reduced to 5.0 and then increased to 5.8 gradually as the UAD plates were being precipitated out. Lowering the pH value to 5.0 was to accelerate the UAD crystallization and UAD crystals would precipitate after the pH value was

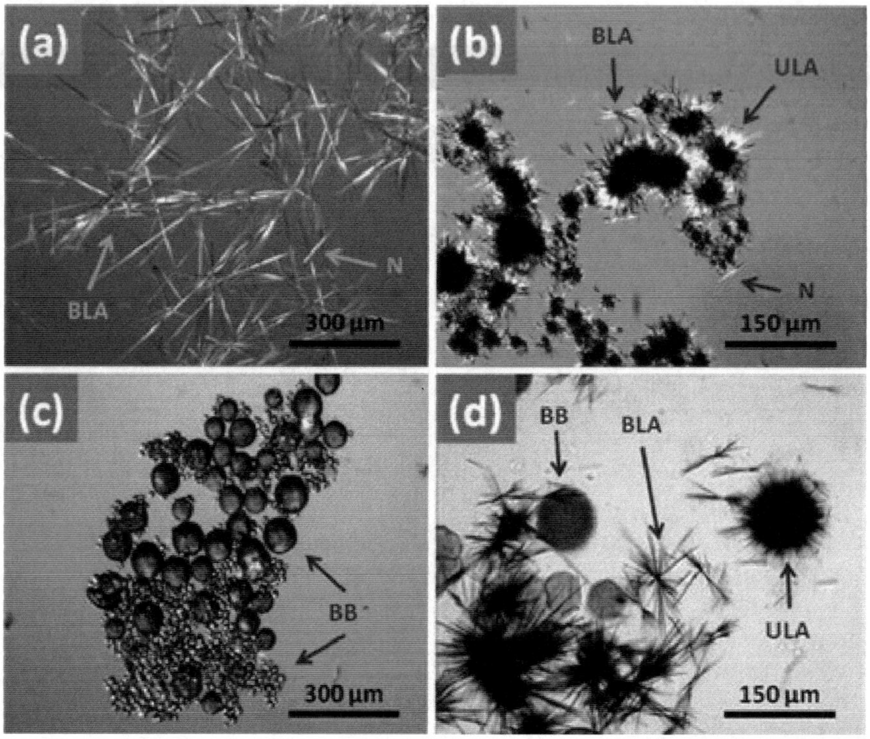

Fig. 1 POM images of MSUM grown: (a) in 100 mM Na^+ ions for 2 days, (b) in 600 mM Na^+ ions for 2 days, (c) in 1500 mM Na^+ ions for 2 days, and (d) in 1500 mM Na^+ ions after 2 months (N: needle; BLA: bow-like aggregate; ULA: urchin-like aggregate; BB: beachball).

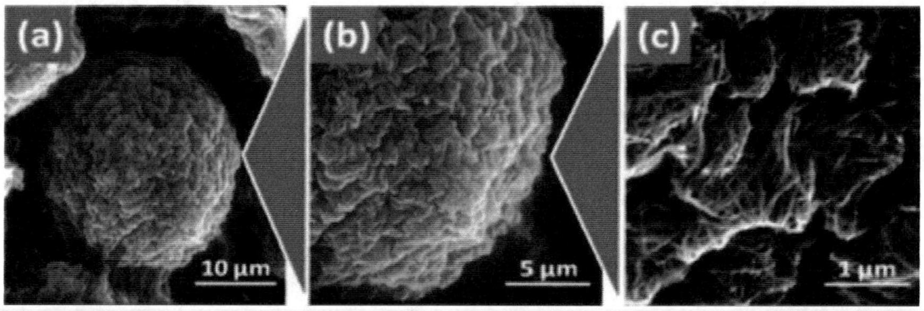

Fig. 2 SEM images of an MSUM beachball structure grown in 1400 mM Na^+ ions for 2 days at various magnifications: (a) a full view of the "beachball" structure, (b) oriented platelets, and (c) primary nanometer-sized needles.

lowered to 5.5. UAD plates were then isolated and converted in a freshly prepared SBF environment to induce the transformation from UAD to MSUM. The POM images and PXRD patterns for UAA, UAD and MSUM are also provided in Fig. S5† for comparison. Based on the crystal habits, the POM images in Fig. 3 (ref. 45) show that the integrity of the UAD plates was partially maintained during dissolution in the first 12 h. The UAD plates were bristled and covered with many tiny crystals 24 h later. UAD plates disappeared after 36 h and needle-shaped MSUM crystals were found 48 h later.

The PXRD patterns in Fig. 4 illustrate the conversion of UAD to MSUM thoroughly. Intense diffraction lines for UAD[33] are at 2θ = 10° and 20°, for UAA[33] the diffraction line is at 2θ = 13.5°, and for MSUM[46] the lines are at 2θ = 11.6°, 18.9° and 28.3°. The POM images and the PXRD patterns in Fig. 3(d) and 4(d) also reveal, respectively, that UAD took 36 h to convert to MSUM, and UAA was an intermediate phase coexisting with UAD at 24 h (Fig. 4(c)). Finally, MSUM needles would be grown completely after 48 h (Fig. 3(e) and 4(e)). TGA scans of UAD and UAA are shown in Fig. S6.†

Under the normal physiological Na^+ ion concentration of approximately 140 mM,[22] the dominant MSUM morphology was "bow-like aggregates" (Fig. 1) which were strikingly similar to the native crystals formed in cartilage.[9] As for the gout formation mechanism in humans, a fundamental question still remains: How does the MSUM crystal appear initially? Most likely, the triggering factor is the local pH reduction in connective tissues and joints[25] brought about by the

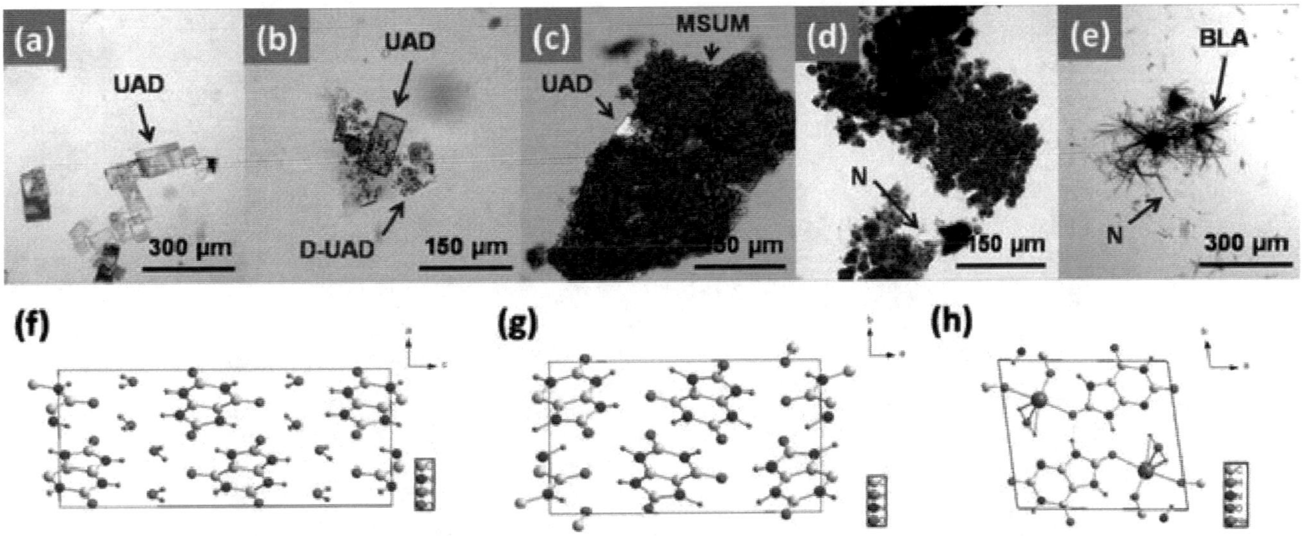

Fig. 3 POM images during the conversion of UAD to MSUM: (a) initial UAD plates, (b) UAD plates with partially dissolved fragments at 12 h (D-UAD: dissolved UAD fragments), (c) UAD plates covered with tiny MSUM crystals at 24 h, (d) aggregation of tiny MSUM needles at 36 h, (e) well-grown MSUM needles at 48 h (N: needle; BLA: bow-like aggregate), (f) view of the crystal structure of UAD showing the top view along the *b*-axis, constructed from fractional coordinates in refcode: ZZZPPI02,[45a] (g) view of the crystal structure of UAA showing the top view along the *c*-axis, constructed from fractional coordinates in refcode: URICAC,[45b] and (h) view of the crystal structure of MSUM showing the top view along the *c*-axis, constructed from fractional coordinates in refcode: NAURAT.[15]

inflammatory reaction[28] and acidosis.[29] The sudden decline of local pH did favor the rapid formation of UAD plates more than MSUM crystals according to their solubility behaviors.[24,37] However, as the pH was returned to about 7.4,[47] the conversion from UAD to MSUM would be activated and finished after 36 h (Fig. 4). Therefore, UAD crystals are likely to be the triggering factor for *in vivo* gout formation.

This UAD-to-MSUM conversion could offer a more reasonable time frame in human life for gout deposition by acidosis than the time required for direct MSUM crystallization. The growth rate of an MSUM crystal at its favorable pH[23] of 7.0 and a urate concentration of 0.6 mM (*i.e.* >6.8 mg dL^{-1}, hyperuricemia)[1] was calculated to be only 10^{-11} μm min^{-1}.[48] However, UAD crystals could precipitate immediately with the temporary decline of local pH value, and UAD would then convert to MSUM at a normal pH value (pH = 7.4) during a relatively short time. The UAD-to-MSUM pathway offered an explanation for the appearance of MSUM crystals in cartilage, synovial and tendon sheaths[25] because of the pH gradient between human plasma and tissue.[28] Furthermore, acute gouty arthritis would develop when MSUM needles were ingested by leucocytes. During phagocytosis, the production of lactic acid would further provoke MSUM deposition according to the UAD-to-MSUM pathway. This self-sustaining cycle provided a stimulus to a greater inflammatory response.[28]

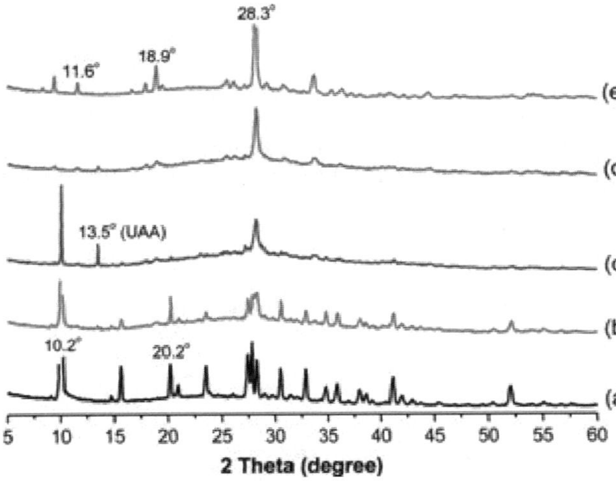

Fig. 4 PXRD patterns of solids harvested from SBF during the conversion of UAD to MSUM at (a) 0 h (UAD), (b) 12 h (UAD), (c) 24 h (UAD and UAA), (d) 36 h (MSUM), and (e) 48 h (MSUM).

Synergistic effect of Na^+, K^+ and Ca^{2+} ions, and hyaluronate chains on MSUM crystallization

Solids grown in aqueous solutions with sodium hyaluronate and various cations were identified to be MSUM using PXRD (Fig. S7†) and the crystal yield for each condition was weighted. Because the hyaluronate chain would interact with Na^+, K^+ and Ca^{2+} ions,[32] we broke down the study of those interactions among hyaluronate and cations into four parts. POM images in Fig. 5 showed that the number of MSUM crystals decreased as more types of cations were added into the hyaluronate-containing solutions. A new type of MSUM "fishtail" morphology was observed in the hyaluronate-, Na^+ ion-, and Ca^{2+} ion-containing solutions (Fig. 5). The synergistic effect of hyaluronate and the cations on the inhibition of MSUM crystallization was further verified based on the

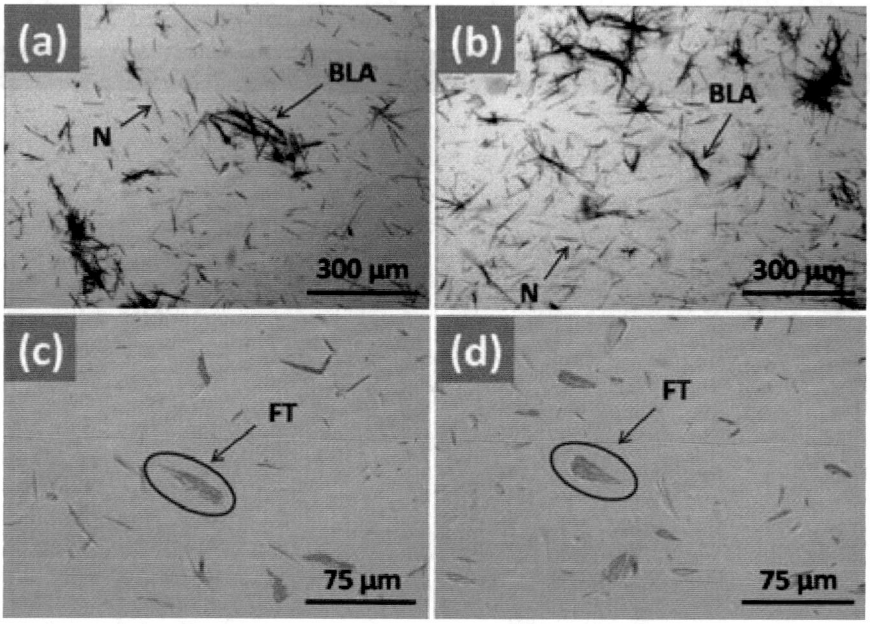

Fig. 5 POM images of MSUM crystals grown in solutions: (a) E1: 2 mg mL^{-1} sodium hyaluronate and 140 mM Na^+ ions, (b) E2: 2 mg mL^{-1} sodium hyaluronate, 140 mM Na^+ ions and 5 mM K^+ ions, (c) E3: 2 mg mL^{-1} sodium hyaluronate, 140 mM Na^+ ions and 2 mM Ca^{2+} ions, (d) E4: 2 mg mL^{-1} sodium hyaluronate, 140 mM Na^+ ions, 5 mM K^+ ions and 2 mM Ca^{2+} ions (N: needle; BLA: bow-like aggregate; FT: fishtail).

crystal yields (Table S3†). SBF with the coexistence of physiological concentrations of hyaluronate, Na^+, K^+ and Ca^{2+} ions gave the minimum MSUM crystal yield of 7.2 wt% which was at least six times less than the crystal yield of 47.3% from the SBF having only hyaluronate and Na^+ ions. The saturation values of MSUM in solutions E1 to E4 were determined and are summarized in Table S3.†

Moreover, Ca^{2+} ions were normally believed to reduce the solubility of MSUM, and the initial nucleus for gout was thought to be calcium urate.[22,49] However, our findings of the synergistic effect of hyaluronate chains and Ca^{2+} ions and even with Na^+ and K^+ ions have shown otherwise.[32] The dominant role of Ca^{2+} ions might have been a crystal inhibitor for MSUM deposition instead (Table S3†). Since hyaluronate chains contain hydrophilic hydroxyl (–OH) and carboxylate ($-COO^-$) pendent groups, hyaluronate chains are water soluble. With the presence of Ca^{2+} ions and other cations acting as salt bridges between the $-COOH^-$ groups of the hyaluronate chains and the N3 (pK_a = 5.75) sites[1,50] on the six-membered ring of the urate ion, the slightly soluble urate ion has now become a solubilized complex in water thanks to the high solubility power of hyaluronate. The higher water solubility of the hyaluronate–Ca-urate complex compared to the one of the urate ion might be one of the reasons for only a fraction of hyperuricemic patients developing gout. Although there was no reported evidence about the complex, it was probable for it to minimize the risk of gout formation in hyperuricemic patients. If the complex was disrupted, MSUM deposition would occur.

Furthermore, the interactions among the hyaluronate chains, Na^+ and Ca^{2+} ions produced fishtail MSUM clusters (Fig. 5). Interestingly, the fishtail clusters had never been reported in gouty patients. Fishtail clusters should have appeared if hyaluronate, Na^+ ions and Ca^{2+} ions were all abundant in gouty patients' synovial fluid. This implied that the decrease in the production, polymerization, molecular size and concentration of the hyaluronate chains during inflammation[38] had prevented the formation of the hyaluronate–Ca-urate complex, and thus, only the usual bow-like MSUM aggregates were grown.

Finally, our hypothesis of gout formation is illustrated in Fig. 6. Local pH decline and the disruption of the hyaluronate–Ca^{2+}-urate complex are the triggering factors for gout formation. Local pH decline could induce UAD crystal formation at first, and UAD would convert to MSUM after the return of the pH value. The dissolution of UAD may promote MSUM crystallization due to the increase of the local urate level. The rise of the concentration of Na^+ ions (or the increase of the urate ion level) could increase the degree of supersaturation, and this driving force may promote the "beachball" structure formation. In addition, MSUM crystals would induce acute gouty arthritis due to leucocytes, and this could decrease the local pH again. This self-sustaining vicious cycle provided a stimulus to an even greater inflammatory response.

Conclusions

Seven key lessons were learned in this study: (1) various morphologies of MSUM: "beachball", urchin-like and bow-like aggregates could be prepared under different Na^+ ion levels at 37 °C. (2) Morphological transformation from the "beachball" structure to the urchin-like aggregate to the bow-like aggregate can be achieved. (3) The "beachball" structure

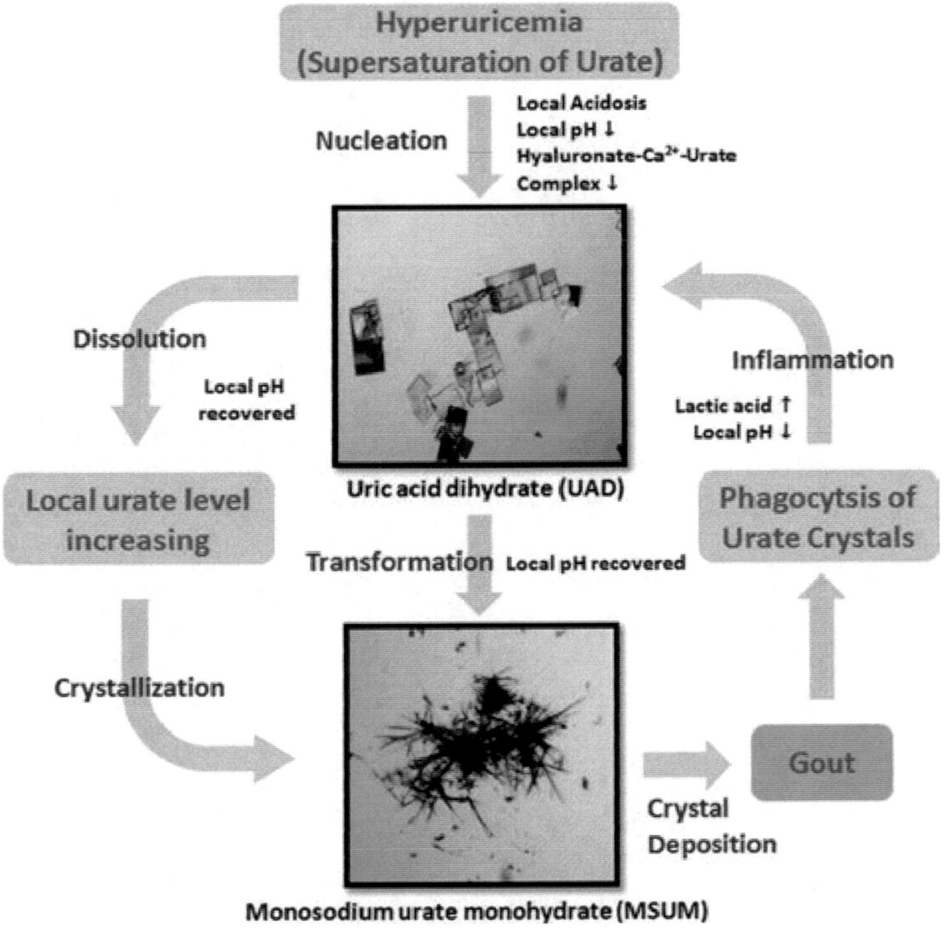

Fig. 6 Triggering factors, the proposed mechanism and the self-sustaining cycle for the crystallization of MSUM and gout.

is a metastable first-formed precursor for all MSUM morphologies. (4) A new type of MSUM fishtail morphology was observed in the hyaluronate, Na^+ and Ca^{2+} ion containing solutions. (5) The dilemma of MSUM deposition at thermodynamically unfavorable low pH conditions was plausibly solved by the kinetic pathway for the phase transformation of UAD to MSUM in the gout study. (6) The UAD-to-MSUM pathway can also be used to explain the effects brought about by the inflammatory response and acidosis, and why MSUM deposition occurs in cartilage and connective tissue. (7) The question of why only a fraction of hyperuricemic patients have gout might be answered by the solubilization of urate through complex formation with hyaluronate chains and Ca^{2+} ions.

Acknowledgements

This research was supported by grants from the Ministry of Science and Technology of Taiwan, ROC (MOST 102-2221-E-008-071-MY2 & 104-2221-E-008-070-MY3). We are greatly indebted to Ms. Ching-Tien Lin for the assistance with SEM, Ms. Shew-Jen Weng for PXRD, and all associated with the Precision Instrument Center at National Central University.

References

1 F. Oliviero, A. Scanu and L. Punzi, *Reumatismo*, 2011, **63**, 221–229.
2 A. Chhana and N. Dalbeth, *Clin. Rheum. Dis.*, 2014, **40**, 291–309.
3 W. A. Katz, *Arthritis Rheum.*, 1975, **18**, 751–756.
4 H.-K. Tak, W. R. Wilcox and S. M. Cooper, *J. Colloid Interface Sci.*, 1980, 77, 195–201.
5 N. S. Mandel, *Arthritis Rheum.*, 1980, 23, 772–776.
6 J. E. Seegmiller, R. R. Howell and S. E. Malawista, *JAMA, J. Am. Med. Assoc.*, 1962, **180**, 469–475.
7 J. J. Fiechtner and P. A. Simkin, *Adv. Exp. Med. Biol.*, 1980, **122**, 141–143.
8 J. J. Fiechtner and P. A. Simkin, *JAMA, J. Am. Med. Assoc.*, 1981, **245**, 1533–1536.
9 N. W. McGill, A. Hayes and P. A. Dieppe, *Scand. J. Rheumatol.*, 1992, **21**, 215–219.
10 H. Paul, A. J. Reginato and H. R. Schumacher, *Ann. Rheum. Dis.*, 1983, **42**, 75–81.
11 C. M. Perrin and J. A. Swift, *CrystEngComm*, 2012, **14**, 1709–1715.
12 Y. Zhu, B. J. Pandya and H. K. Choi, *Arthritis Rheumatol.*, 2011, **63**, 3136–3141.

13 C.-F. Kuo, M. J. Grainge, C. Mallen, W. Zhang and M. Doherty, *Ann. Rheum. Dis*, 2014, 74, 661–667.

14 C.-F. Kuo, K.-H. Yu, L.-C. See, I.-J. Chou, Y.-S. Ko, H.-C. Chang, M.-J. Chiou and S.-F. Luo, *Rheumatology*, 2013, **52**, 111–117.

15 N. S. Mandel and G. S. Mandel, *J. Am. Chem. Soc.*, 1976, **98**, 2319–2323.

16 H.-K. Tak, S. M. Cooper and W. R. Wilcox, *Arthritis Rheumatol.*, 1980, **23**, 574–580.

17 S. Weiner, I. Sagi and L. Addadi, *Science*, 2005, **309**, 1027–1028.

18 A.-W. Xu, Y. Ma and H. Cölfen, *J. Mater. Chem.*, 2007, **17**, 415–449.

19 D. J. Allen, G. Milosovich and A. M. Mattocks, *Arthritis Rheumatol.*, 1965, **8**, 1123–1133.

20 J. N. Loeb, *Arthritis Rheumatol.*, 1972, **15**, 189–192.

21 I. Kippen, J. R. Klinenberg, A. Weinberger and W. R. Wilcox, *Ann. Rheum. Dis.*, 1974, **33**, 313–317.

22 W. R. Wilcox and A. A. Khalaf, *Ann. Rheum. Dis.*, 1975, **34**, 332–339.

23 W. R. Wilcox, A. Khalaf, A. Weinberger, I. Kippen and J. R. Klinenberg, *Med. Biol. Eng.*, 1972, **10**, 522–531.

24 Z. Wang and E. Königsberger, *Thermochim. Acta*, 1998, **310**, 237–242.

25 W. A. Katz and M. Schubert, *J. Clin. Invest.*, 1970, **49**, 1783–1789.

26 H. M. Burt and Y. C. Dutt, *Ann. Rheum. Dis.*, 1986, **45**, 858–864.

27 H. M. Burt and Y. C. Dutt, *J. Cryst. Growth*, 1989, **94**, 15–22.

28 J. E. Seegmiller and R. Rodney Howell, *Arthritis Rheum.*, 1962, 5, 616–623.

29 M. A. Martillo, L. Nazzal and D. B. Crittenden, *Curr. Rheumatol. Rep.*, 2014, **16**, 400.

30 H. Iwata, S. Nishio, M. Yokoyama, A. Matsumoto and M. Takeuchi, *J. Urol.*, 1989, **142**, 1095–1098.

31 G. Schubert, G. Reck, H. Jancke, W. Kraus and C. Patzelt, *Urol. Res.*, 2005, 33, 231–238.

32 J. K. Sheehan, C. Arundel and C. F. Phelps, *Int. J. Biol. Macromol.*, 1983, 5, 222–228.

33 J. B. Presores and J. A. Swift, *CrystEngComm*, 2014, **16**, 7278–7284.

34 T. C. McIlvaine, *J. Biol. Chem.*, 1921, **49**, 183–186.

35 A. Oyane, H.-M. Kim, T. Furuya, T. Kokubo, T. Miyazaki and T. Nakamura, *J. Biomed. Mater. Res., Part A*, 2003, **65**, 188–195.

36 K. Kaneko and M. Maru, *Anal. Biochem.*, 2000, **281**, 9–14.

37 F. Grases, A. I. Villacampa, A. Costa-Bauzá and O. Söhnel, *Scanning Microsc.*, 1999, **13**, 223–234.

38 R. K. Strachan, P. Smith and D. L. Gardner, *Ann. Rheum. Dis.*, 1990, **49**, 949–952.

39 (*a*) M. B. Tomson and G. H. Nancollas, *Science*, 1978, **200**, 1059–1060; (*b*) P. Koutsoukos, Z. Amjad, M. B. Tomson and G. H. Nancollas, *J. Am. Chem. Soc.*, 1980, **102**, 1553–1557; (*c*) T. J. Halter, B. M. Borah, B. Xie and G. H. Nancollas, *Colloids Surf., A*, 2015, **482**, 300–305.

40 H. P. Klug and L. E. Alexander, *Crystallite size and lattice strains from line broadening*, Wiley, New York, 2nd edn, 1974, ch. 9, pp. 657–661.

41 F. Qian, J. Tao, S. Desikan, M. Hussain and R. L. Smith, *Pharm. Res.*, 2007, **24**, 1551–1560.

42 S. N. Kalkura, E. K. Girija, M. Kanakavel and P. Ramasamy, *J. Mater. Sci.: Mater. Med.*, 1995, **6**, 577–580.

43 A.-W. Xu, M. Antonietti, H. Cölfen and Y.-P. Fang, *Adv. Funct. Mater.*, 2006, **16**, 903–908.

44 (*a*) M. Niederberger and H. Cölfen, *Phys. Chem. Chem. Phys.*, 2006, **8**, 3271–3287; (*b*) T. Lee and C. W. Zhang, *Pharm. Res.*, 2008, **25**, 1563–1571.

45 (*a*) H. Ringertz, *Acta Crystallogr., Sect. B: Struct. Sci.*, 1966, **20**, 397–403; (*b*) S. Parkin and H. Hope, *Acta Crystallogr., Sect. B: Struct. Sci.*, 1998, **54**, 339–344.

46 B. B. Parekh, S. R. Vasant, K. P. Tank, A. Raut, A. D. B. Vaidya and M. J. Joshi, *Am. J. Infect. Dis.*, 2009, **5**, 225–230.

47 P. S. Treuhaft and D. J. McCarty, *Arthritis Rheum.*, 1971, **14**, 475–484.

48 P. D. Calvert, R. W. Fiddis and N. Vlachos, *Colloids Surf.*, 1985, **14**, 97–107.

49 A. A. Khalaf and W. R. Wilcox, *J. Cryst. Growth*, 1973, **20**, 227–232.

50 S. Satter, M. J. Carroll, A. A. Sargeant and J. A. Swift, *CrystEngComm*, 2008, **10**, 155–157.

499202 Fabrication of Polyethylene Glycol-Based Templated Macroporous Hydrogels for Cell Encapsulation

Abstract:

In recent years, macroporous hydrogels are sought after for a wide range of tissue engineering and drug delivery applications. There are various approaches to create porosity in gels such as lyophilization or foam formation, to name a few. However, most do not allow a precise control over pore size and interconnectivity and are not compatible with *in situ* cell encapsulation. Here, we developed novel templated macroporous hydrogels by encapsulating uniform degradable hydrogel microspheres produced via microfluidics into a hydrogel slab. The microspheres degrade completely leaving macropores behind. Uniquely, the degradable microspheres are biocompatible and when laden with cells or other molecules of interests, the cells and/or molecules were deposited in the macropores upon microsphere degradation. Microsphere degradation was dependent on the degradation medium, microspheres size, microsphere constraint from the hydrogel as well as cell encapsulation. The encapsulated cells were cultured in PEG hydrogel microspheres for up to 7 days with a cell viability higher than 90%.

495148 Multicomponent Transport Models for Non-Electroneutral Solid Electrolytes

Abstract:

The most popular multicomponent transport models used by chemical engineers are based in the theory of 'irreversible thermodynamics', which combines the theory of equilibrium with continuum dynamics to produce an entropy balance equation that guides the formulation of transport constitutive laws. Our group has put significant effort into the development of consistent transport theories to describe concentrated electrolytes of various types, including ionically conductive liquids, ionomer membranes, and even solid ion conductors like glasses or ceramics. In the past we have extended multicomponent transport theory to account for excluded-volume effects, which arise from the thermodynamics of material volume, and have developed transport constitutive laws applicable to domains containing space charges.

When modeling materials in which mass or charge transfer occurs in more than one spatial dimension, the addition of a local thermodynamic state equation for volume makes standard multicomponent models based on Onsager–Stefan–Maxwell theory underspecified. To achieve closure it is necessary to introduce a momentum balance alongside the other transport governing equations. This raises a number of fundamental physical questions about how to handle the potential interactions between mechanical and electrochemical transport.

This talk will address the formulation of multicomponent, multiphysics transport models applicable to ion-conductive solids in which simultaneous mass, heat, charge, and momentum transfer occurs. We will touch on a number of thermodynamic measurements that may be needed to characterize elastic solid electrolytes for accurate modeling, as well as describing how the laws governing diffusion can be modified to include the effects of stress and strain. A number of practical examples will be discussed to illustrate the signatures of electrochemical/mechanical coupling. We will show that the high-frequency impedance of ionically conductive materials is controlled in part by mechanical forces, and discuss the Gibbs–Donnan effect, in which charge separation leads to a pressure gradient that can drive flow.

486180 Diffusion and Sorption Phenomena of Organic Vapor Penetrants in Unmodified and Ethylenediamine Vapor Phase Cross-Linked Matrimid Thin Films

Abstract:

Sorption of organic penetrants in polymeric films and membranes can have a significant impact on diffusion and induced relaxative sorption of the penetrants in the polymer. Matrimid is a commercial polyimide with a glass transition temperature of 320 °C and is known for its high chemical and thermal stability. This work examines C1-C7 alcohol vapor sorption in unmodified and vapor phase ethylenediamine cross-linked Matrimid thin films ranging from 0.1 to 2.0 μm in thickness. A marked dependence of diffusion coefficient at infinite dilution on film thickness for unmodified films was observed spanning approximately two orders of magnitude for each respective penetrant over the 0.1 to 2.0 μm thickness range. The presence of alcohol vapor even at low activity causes a high degree of plasticization in unmodified Matrimid. Swelling allows continued sorption and consequently increases the apparent diffusivity of the alcohol penetrants over an order of magnitude at activities near unity. Ethylenediamine cross-linking is used to mitigate plasticization effects by preventing polymer chain motion. Films were cross-linked using ethylenediamine vapor and therefore were not exposed to any additional solvents prior to sorption and diffusion experiments in contrast to traditional polymer cross-linking methods. Ethylenediamine cross-linking reduced C6 and C7 alcohol sorption by 90% compared to unmodified Matrimid, while C1-C5 sorption was largely unchanged. Diffusion coefficients for C3 and higher alcohols decreased an order of magnitude in cross-linked films. Relaxation induced sorption behavior was significantly reduced for C4 and higher alcohols. This work demonstrates that ethylenediamine vapor can be used as a surface-only cross-linking technique for polyimide films and membranes and can have a significant reduction in plasticization effects for molecularly larger penetrants.

498437 Segmental Dynamics and Water Transport in Nafion-SiO_2 Hybrid Membranes

Abstract:

Perfluorinated ionomers, specifically Nafion, are the most widely utilized polymer electrolyte membranes for vanadium redox flow battery (VRFB) applications due to their high proton conductivity and good chemical stability. However, Nafion suffers from high vanadium ion crossover, an undesired event that reduces energy efficiency and lifetime of the battery. Incorporation of silica (SiO_2) nanoparticles into the Nafion membrane has proven to be effective in suppressing crossover of vanadium ions. However, the mechanism of this reduction in crossover is still unclear, making it impossible to rationally design cheaper and bettering performing membranes for VRFB applications. To understand the effect of SiO_2 nanoparticles on both segmental dynamics and transport properties of the hybrid membranes, several spectroscopic techniques were employed to characterize hydrated Nafion-SiO_2 hybrid films. Hybrid membranes were prepared by an *in situ* sol gel condensation process of SiO_2 in Nafion 117 membranes, as well as by solution casting Nafion-SiO_2 dispersions on Teflon substrates. Water transport was characterized by *in situ* time-resolved Fourier transform infrared-attenuated total reflectance (FTIR-ATR) spectroscopy, which provides molecular-level contrast between the diffusant and the polymer in real time during diffusion experiments, allowing for the simultaneous capture of both water transport and water-induced polymer relaxation (swelling). The diffusion of water through the hybrid membranes was found to be non-Fickian and the kinetic uptake data were regressed to a diffusion-relaxation model to obtain both water diffusivity and polymer relaxation dynamics. Further, the spectra were deconvoluted to resolve the morphological rearrangements within the membrane due to water sorption. In addition, the chain relaxation dynamics of the hybrid membranes, which is attributed to fluctuations on the length scale of inter-backbone spacing, were investigated using neutron spin-echo (NSE) spectroscopy. NSE data were collected at two temperatures, 298 K and 323 K, and these results were compared to unmodified Nafion with varying thermal treatments.

499718 Designing Core/Shell Metal Organic Framework/Polymer Films As Scalable Barrier Layers for Enhanced Protection on Photovoltaics

Abstract:

Preventing module deterioration due to environmental stressors including moisture, mechanical stress, temperature and solar irradiation is critical in wide-ranging packaging applications, especially for the coatings used to ensure the life span and reliability of Photovoltaics cells. A simple, scalable, and cheap hybrid barrier layer system has been built based on Metal Organic Framework (MOF)/Cyclic Olefin Copolymer (COC) films which simultaneously achieved high light transmittance and mechanical stability, all while maintaining limited permeability of the reactive species into the device.

We previously developed highly transparent hybrid films based on a water-scavenging MOF and a commercial hydrophobic polymer COC fabricated by doctor blading, a commonly used industry method. Incorporating the hygroscopic MOF desiccants (a hydrophilic MOF capped with oleic acid for retaining an organic solubility) into the COC leads to a 10-fold decrease in WVTR, while maintaining an outstanding transparency over most of the solar spectrum.[1]

In order to further optimize this successful system using MOF/COC films as barrier layers, we adapted a simple chemical functionalization method and achieved hygroscopic core/shell MOF nanoparticles providing a uniquely robust high-performance system at low cost. Besides using oleic acid for surface functionalization of MOF in the COC/MOF film, we adapted a layer-by-layer method to coat the MOF with ZIF-8 (hydrophobic MOF with smaller pore size). This ZIF acts as hydrophobic barrier, while also enhancing the compatibility with any hydrophobic polymers. This approach allows the MOF core to maintain its water-scavenging property and gain more stability by the shell protection. Controlling shell thickness and coverage area allows us to optimize design of the barrier layers and better understand the diffusion process. SEM, FTIR, TGA and XRD has been used for understanding the core/shell structures and compatibility between polymers and MOFs in the films. Small angle X-ray scattering (SAXS) has been used to investigate the nanocomposite morphology-moisture barrier relationship, which will further allow determination of the nanoparticle-nanoparticle spacing, nanoparticle size and copolymer lamella spacing and extent of multilayer order. These morphology studies will show how water is absorbed into the nanocomposite, and this understanding will allow design of more optimal copolymer-nanoparticle composites.

1. Youn Jue Bae, Eun Seon Cho, Fen Qiu, Daniel T. Sun, Teresa E. Williams, Jeffrey J. Urban, Wendy L. Queen â €œTransparent Metal-Organic Framework/Polymer Films as Water Vapor Barriersâ€ *ACS Appl. Mater. Interfaces* **2016**, 8, 10098-10103

494603 PEO-Based Semi-Interpenetrating Polymer Networks (S-IPNs) for CO_2-Selective Membranes

Abstract:

PEO-based membranes are attractive for CO_2-related gas separations due to their high selectivity towards CO_2. However, PEO-rich membranes are frequently challenged by weak mechanical properties and a high crystallization tendency of PEO that hinders gas transport. While crosslinked PEO membranes represent an effective strategy to address these issues, their limited mechanical and thermal stability are still not satisfactory for some practical applications. In this talk, we report a new design strategy of semi-interpenetrating polymer networks (S-IPNs) that are fabricated from linear polyimides entwined by crosslinked PEO network to address abovementioned issues. This type of network structure can synergistically combine the mechanical properties of glassy polyimides with the excellent CO_2 separation ability of crosslinked PEO. Specifically, these new S-IPNs are consisted of crosslinked networks of Jeffamine® polyetheramines penetrated by triptycene-containing linear (co)polyimides. In all, six unique S-IPNs were produced in various network-polyimide combinations to systematically investigate the dependence of membrane properties on the composition and membrane morphology. In particular, the PEO networks are varied in crosslink density, and the linear polyimides used include homopolyimide and PEO-*co*-polyimide copolymers of varied PEO content. Tested S-IPNs showed significantly improved mechanical properties relative to crosslinked pure PEO membranes. For example, the tensile strength of the low crosslink density S-IPN incorporating the homopolyimide improved by around 600%. Moreover, all S-IPNs featured greatly improved CO_2 permeability (up to 25-fold) as well as improved CO_2/H_2 and CO_2/N_2 selectivity compared to their counterpart homopolyimides and copolyimides, which place S-IPNsâ€™ separation performance approaching and surpassing the 2008 Robeson upper bound for CO_2/H_2 and closely approaching the CO_2/N_2 upper bound. Given the same overall PEO content, S-IPNs consisted of linear copolyimides with moderate PEO content and PEO network of low crosslink density showed the best separation performance, suggesting the formation of the most desired membrane morphology in regards to PEO domain connectivity for selective and fast CO_2 transport. In this talk, the synthesis, fabrication and characterization of these new S-IPN membranes will be presented. Discussions will be focused on relating variations in microscopic structures (linear polyimide composition, crosslink density, and phase compatibility) with macroscopic physical and transport properties.

497459 An Experimental and Triple-Mode Sorption Modeling of Sorption and Diffusion in Polymers

Abstract:

An experimental and triple-mode sorption modeling of sorption and diffusion in polymers

Hom N. Sharma, Stephen J. Harley, Yunwei Sun, and Elizabeth A. Glascoe

Lawrence Livermore National Laboratory

L288, 7000 East Ave., Livermore, CA 94550

925-423-5368

sharma@llnl.gov

Sorption and diffusion in polymeric materials is of great interest in wide range of applications (e.g. electronic components to microfluidics).(1-3) To many application, moisture sorption and diffusion can be catastrophic due to the change in chemical and mechanical properties of materials over time. The uptake and outgassing of moisture is also associated with aging and compatibility issues in a system, which can directly alter the lifetimes and viability of system assemblies and screening new materials for future designs. The process is dynamic and consists of different sorption modes and varies dramatically in different materials. Therefore, a detailed understating of the moisture uptake and diffusion is crucial.

In this study, we investigate the moisture sorption and diffusion phenomena using a combined experimental and modeling approach. Polymeric materials (for example: KaptonH and Kapton HN) are investigated over a wide range of temperatures and relative humidities (RH) to quantify the moisture transport mechanism as shown in the figure 1 below. Gravimetric type dynamic vapor sorption (DVS) experiments were employed to measure the moisture uptake and used for the modeling. A reactive transport model is used which includes a triple-mode sorption model(3) that includes absorption, adsorption, and pooling of species, molecular diffusion, and chemical reaction kinetics. Our sorption experiments and modeling results quantify the differences between the two example materials and demonstrate a substantially larger adsorption capacity and moisture affinity for Kapton HN compared with Kapton H. A distinct pooling mode can be seen above 70% RH in both materials, however the pooling mode is larger in Kapton HN. We also show a simultaneous simulation of the transport and chemical reactions of moisture through those example materials. Density functional theory calculations are performed to investigate the impacts on moisture uptake due to $CaHPO_4$ filler, which is present only in Kapton HN.

This work performed under the auspices of the U.S. Department of Energy by Lawrence Livermore National Laboratory under Contract DE-AC52-07NA27344.

Fig 1: Experiments and modeling results of dynamic moisture uptake by Kapton HN at a range of relative humidities at 40 .

References:

1. Harley SJ, Glascoe EA, Lewicki JP, Maxwell RS. Advances in Modeling Sorption and Diffusion of Moisture in Porous Reactive Materials. Chemphyschem. 2014;15(9):1809-20.

2. Harley SJ, Glascoe EA, Maxwell RS. Thermodynamic Study on Dynamic Water Vapor Sorption in Sylgard-184. J Phys Chem B. 2012;116(48):14183-90.

3. Sun YW, Harley SJ, Glascoe EA. Modeling and Uncertainty Quantification of Vapor Sorption and Diffusion in Heterogeneous Polymers. Chemphyschem. 2015;16(14):3072-83.

499946 *In situ* Monitoring of Emergent Transport in Polymer Membranes

Abstract:

The permeation of aqueous solutes through membranes is of critical importance in many applications, from traditional membrane separations to fuel cells. Transport of small-molecule solutes through polymeric membranes is typically characterized in single-component permeation experiments. Traditionally the techniques used to measure organic permeation through a membrane, e.g. gas chromatography, require aliquot sampling and are limited by their ability to resolve the analytes of interest and water. Here, we utilize in-situ ATR FTIR spectroscopy to examine emergent transport phenomena, where the transport of a small molecule is affected by the presence of another solute, in real time without aliquot sampling. First, we quantitatively determine the concentration of single and multicomponent mixtures of various solutes including methanol, formate and acetone. So validated, we apply this methodology to monitoring the permeation of single and multicomponent mixtures across commercially available membranes, such as Nafion 117, in situ using a diffusion cell. Membrane permeabilities are extracted from time-resolved half-cell concentration data using free volume models. We demonstrate the capability of this methodology to simultaneously measure permeation of multiple components, provided that each analyte has a distinct IR signature. Membrane permeabilities and selectivities obtained from the single component permeation experiments are compared to those calculated for solutes in multicomponent permeation experiments. In some instances, the membrane permeability to an organic solute substantially deviates, sometimes by orders of magnitude, when co-solutes are present. Similarly, membrane selectivities calculated from multicomponent diffusion experiments differ significantly from those calculated from single component permeabilities.

501096 The Role of Chlorine Substituent in Gas Transport Properties of Polychlorotrifluoroethene (PCTFE)

Abstract:

Polymeric membranes have been widely used for industrial gas separation due to their inherent advantages such as high energy-efficiency. There is a great interest in understanding the polymer structure/property relationship to design high performance membranes with high gas permeability and selectivity. Halogenated polymers have emerged as interesting membrane materials for gas separation. For example, perfluoropolymers govern the upper bounds for He/H_2, He/N_2, and He/CH_4 separation in the Robeson's plots, which is ascribed to the more favorable interactions for He than other gases with fluoropolymers, as shown in the Figures above. Herein, we demonstrate that chlorine groups exhibit unique separation properties for the removal of noble gas from mixtures, as exemplified in polychlorotrifluoroethylene (PCTFE, $-CF_2-CFCl-$). At 35 ^oC, PCTFE exhibits a He permeability of 26 Barrers and pure gas selectivity of He/H_2, He/CO_2 and He/CH_4 of 6.2, 53 and 1,100, respectively. These selectivity values are among the highest reported, and the performance is near or above the upper bounds for these separations. Interestingly, PCTFE also shows a H_2/CO_2 selectivity of 8.5 at 35 ^oC, an attractive performance for H_2 purification and CO_2 capture. The effect of Cl substituents on CO_2 solubility can be interpreted using the regular solution theory. Similar behavior for CO_2/CH_4 solubility selectivity is also observed in halogenated organic liquids. This presentation will also discuss potential applications of halogenated polymers for the separation of other noble gases (such as Argon) from their mixtures.

TEXAS TECH UNIVERSITY

Thermodynamic Modeling for Swelling of Poly(N-isopropyl acrylamide) Hydrogels in Solvent Mixtures

Sheik Tanveer[a], Fazle Hussain[b], Chau-Chyun Chen[a]

[a] Department of Chemical Engineering, TTU

[b] Department of Mechanical Engineering, TTU

11/02/2017

Talk outline

17AIChE Annual Meeting, Minneapolis, MN

- Overview of hydrogel swelling
- Experimental data from Hüther et al. [1]
- Prior thermodynamic modeling works
- Thermodynamic model
- Results
- Summary
- Future research plan

[1] Fluid Phase Equilib. 219 (2004) 233-243.

Overview of hydrogel swelling

Hydrogels: three dimensional elastic cross-linked hydrophilic polymers

Hydrogel swelling depends on

- Temperature
- Pressure
- Polymer structure
- Solvent compositions
- Salinity
- pH etc.

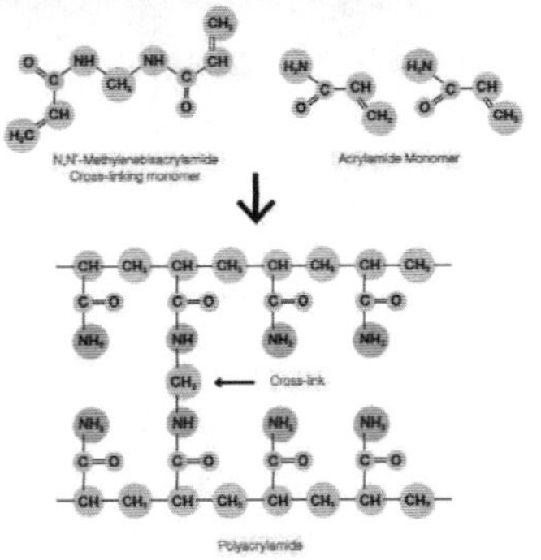

[2] Polymerization of acrylamide monomers and bis-acrylamide

A comprehensive thermodynamic model is required to correlate and predict the swelling behavior with minimal number of interaction parameters

[2] http://www.bio-rad.com/en-us/applications-technologies/introduction-polyacrylamide-gels

Experimental data from Hüther et al.[1]

Chemicals used in preparation:

- Monomer: N-isopropyl acrylamide
- Crosslink agent: N,N′-methylenbisacrylamide
- Solvent: water
- Mass of polymerizable substances: 8 mass% of feed
- Crosslink: 0.015 (mole/mole) of polymerizable material

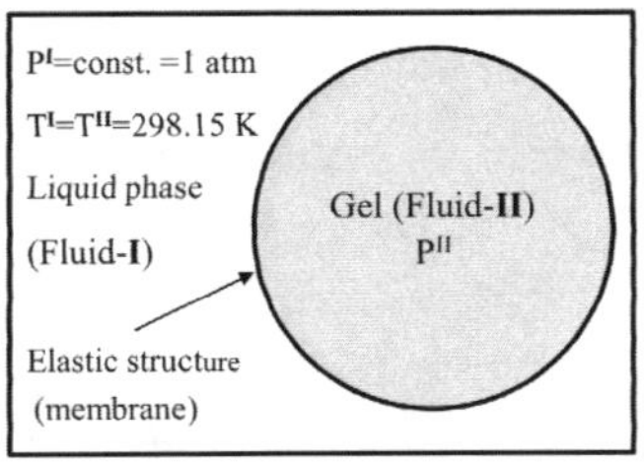

$$\textbf{degree of swelling} = \frac{\text{mass of swollen gel}}{\text{mass of dry gel}}$$

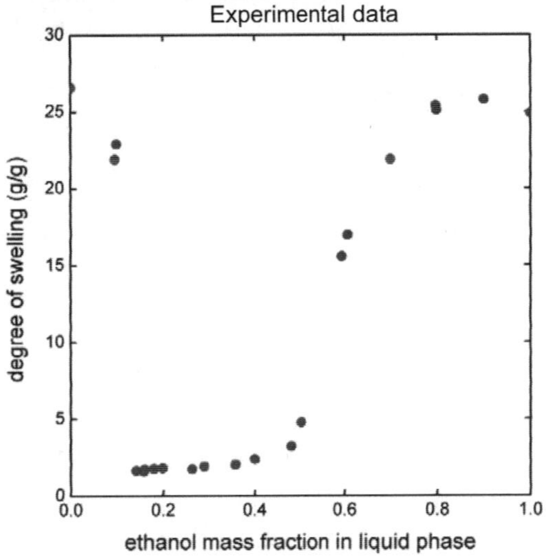

[1] Fluid Phase Equilib. 219 (2004) 233-243.

Experimental data from Hüther et al.[1]

Chemicals used in preparation:

- Monomer: N-isopropyl acrylamide
- Crosslink agent: N,N′-methylenbisacrylamide
- Solvent: water
- Mass of polymerizable substances: 8 mass% of feed
- Crosslink: 0.015 (mole/mole) of polymerizable material

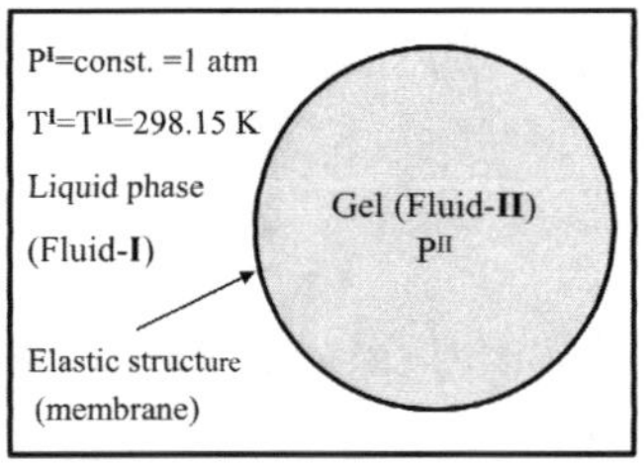

degree of swelling

$$= \frac{\text{mass of swollen gel}}{\text{mass of dry gel}}$$

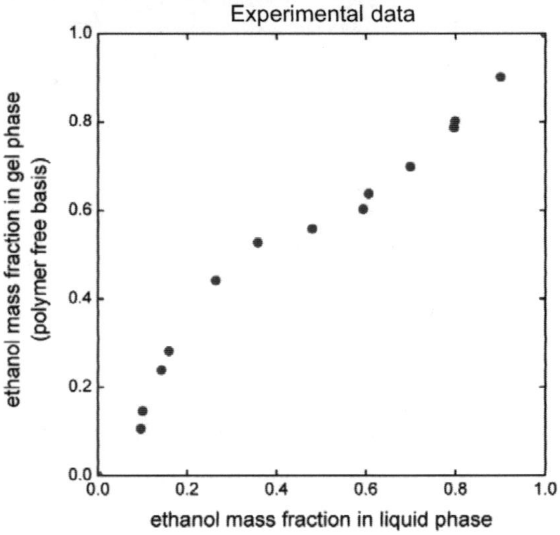

[1] Fluid Phase Equilib. 219 (2004) 233-243.

Prior thermodynamic modeling works

Model by Hüther et al.[1] (extended UNIQUAC)

- At **equilibrium**[3]

  $\left[G^{I}(T, P^{I}, n_j^{I}) + G^{II}(T, P^{I}, n_j^{II}) + A^{m}(T, V^{II})\right]$ is minimum

- At constant T, P^I, phase equilibrium criterion[3] :

  $$\mu_i^{I}(T, P^{I}, n_j^{I}) = \mu_i^{II}(T, P^{I}, n_j^{II}) + \upsilon_i^{II}\left(\frac{\partial A^{m}}{\partial V^{II}}\right)_T$$

- If the same reference chemical potential is chosen for component i in both phases, then

  $$\ln a_i^{I}(T, P^{I}, n_j^{I}) = \ln a_i^{II}(T, P^{I}, n_j^{II}) + \frac{\upsilon_i^{II}}{RT}\left(\frac{\partial A^{m}}{\partial V^{II}}\right)_T$$

[1] Fluid Phase Equilib. 219 (2004) 233-243.
[3] Fluid Phase Equilib. 115 (1996) 113-133.

Prior thermodynamic modeling works

17 AIChE
Annual Meeting, Minneapolis, MN

Model by Hüther et al.[1]
(extended UNIQUAC)

$$G^E = G^{E,UNIQUAC}\ [4] + G^{E,FV}\ [5]$$

$$G^{E,UNIQUAC} = G^{E,comb} + G^{E,res}$$

Model deficiencies:

- Too many adjustable parameters. Additional parameters required to account free volume contribution
- Free volume term has no physical significance at low pressure
- Prediction of solvent compositions in gel phase is not satisfactory

[1] Fluid Phase Equilib. 219 (2004) 233-243.
[4] AIChE J. 21 (1975) 116-128.
[5] J. Am. Chem. Soc. 86 (1964) 3507-3520.

Prior thermodynamic modeling works

17 AIChE
Annual Meeting, Minneapolis, MN

Model by Hüther et al.[1]
(extended UNIQUAC)

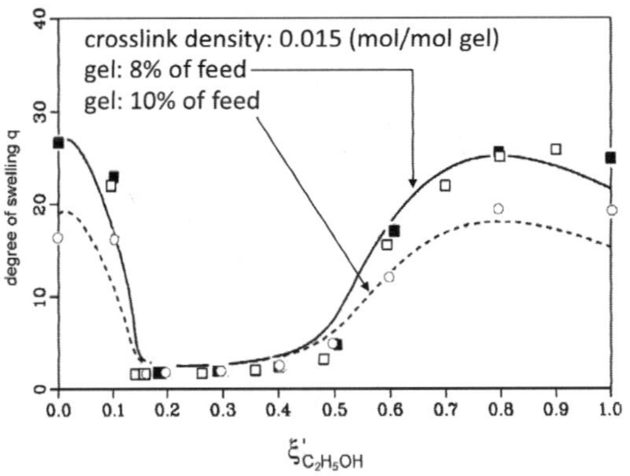

Model deficiencies:

- Too many adjustable parameters. Additional parameters required to account free volume contribution
- Free volume term has no physical significance at low pressure
- Prediction of solvent compositions in gel phase is not satisfactory

[1] Fluid Phase Equilib. 219 (2004) 233-243.
[4] AIChE J. 21 (1975) 116-128.
[5] J. Am. Chem. Soc. 86 (1964) 3507-3520.

Prior thermodynamic modeling works

Model by Hüther et al.[1]
(extended UNIQUAC)

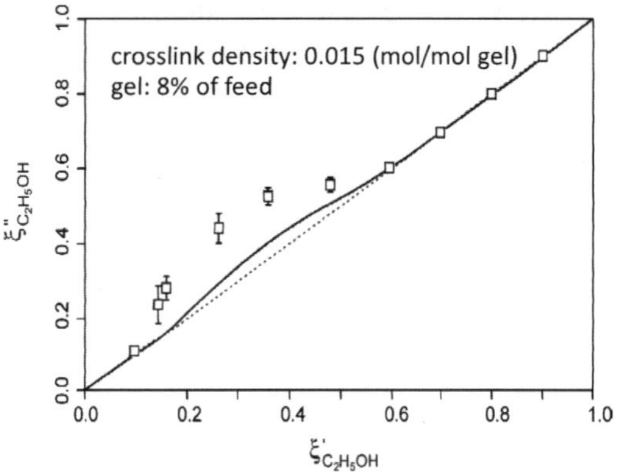

Model deficiencies:

- Too many adjustable parameters. Additional parameters required to account free volume contribution
- Free volume term has no physical significance at low pressure
- Prediction of solvent compositions in gel phase is not satisfactory

[1] Fluid Phase Equilib. 219 (2004) 233-243.
[4] AIChE J. 21 (1975) 116-128.
[5] J. Am. Chem. Soc. 86 (1964) 3507-3520.

Prior thermodynamic modeling works

Model by Arndt and Sadowski[6]
(extended PC-SAFT)

$$A^{res} = \begin{cases} A^{hard-chain} \\ A^{dispersion} \\ A^{association} \\ A^{elastic} \end{cases}$$

(first three terms: PC-SAFT[7] equation of state)

$$\frac{A^{\text{elastic}}}{Nk_BT} = x_{\text{polymer}}\frac{\Phi - 2}{\Phi}\left[\frac{3}{2}\times\frac{\left(\frac{V}{V_0}\right)^{2/3} - 1}{1 - \left(\frac{V}{V_{\max}}\right)^{2/3}} - \ln\left(\frac{V}{V_0}\right)\right]$$

Model deficiencies:

- Too complicated for practical engineering calculations
- Prediction of degree of swelling in ethanol enriched solution is not satisfactory

[6] Macromolecules 45 (2012) 6686-6696.
[7] Ind. Eng. Chem. Res. 40 (2001) 1244-1260.

Prior thermodynamic modeling works

17AIChE Annual Meeting, Minneapolis, MN

Model by Arndt and Sadowski[6] (extended PC-SAFT)

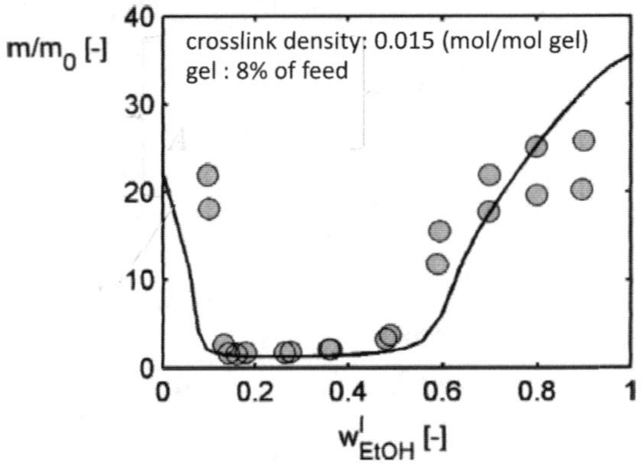

Model deficiencies:

- Too complicated for practical engineering calculations
- Prediction of degree of swelling in ethanol enriched solution is not satisfactory

[6] Macromolecules 45 (2012) 6686-6696.
[7] Ind. Eng. Chem. Res. 40 (2001) 1244-1260.

Thermodynamic model

17AIChE Annual Meeting, Minneapolis, MN

- At **equilibrium**[3]

$$\left[G^{I}(T,P^{I},n_j^{I}) + G^{II}(T,P^{I},n_j^{II}) + A^{m}(T,V^{II})\right] \text{ is minimum}$$

- At constant T, P^{I}, phase equilibrium criterion[3] :

$$\mu_i^{I}(T,P^{I},n_j^{I}) = \mu_i^{II}(T,P^{I},n_j^{II}) + \upsilon_i^{II}\left(\frac{\partial A^{m}}{\partial V^{II}}\right)_T$$

- If the same reference chemical potential is chosen for component i in both phases, then

$$\ln a_i^{I}(T,P^{I},n_j^{I}) = \ln a_i^{II}(T,P^{I},n_j^{II}) + \frac{\upsilon_i^{II}}{RT}\left(\frac{\partial A^{m}}{\partial V^{II}}\right)_T$$

Pressure difference in both phases:

$$P^{II} - P^{I} = \left(\frac{\partial A^{m}}{\partial V^{II}}\right)_T$$

[3] Fluid Phase Equilib. 115 (1996) 113-133.

Thermodynamic model

17AIChE Annual Meeting, Minneapolis, MN

Local composition concept
Individual segments and solvents molecules[8]

Flory-Huggins expression
Configurational entropy of mixing molecules of different sizes

$$G^E = G^{E,LC} + G^{E,FH}$$

$$\ln \gamma_i = \ln \gamma_i^{NRTL} + \ln \gamma_i^{FH}$$

$$G^{E,LC} = f(\tau_{ij})$$

Helmholtz energy of elastic network
Semi-empirical equation from affine[9] and phantom[10] network theory

$$\frac{A^m}{RT} = n_{poly} C \left[\left(\frac{V^{II}}{V_0} \right)^{2/3} - 1 \right] + f\left(\frac{V^{II}}{V_0} \right)$$

$$\left(\frac{\partial A^m}{\partial V^{II}} \right)_T = \frac{2}{3} \mathrm{RT} \frac{\mathrm{C}}{V^{II}} n_{poly} \left[\left(\frac{V^{II}}{V_0} \right)^{2/3} \right]$$

- V_0 = volume in the stress free state
- V^{II} = volume in the stressed state
- C = network parameter
- n_{poly} = number of polymer chains between crosslinks

[8] Fluid Phase Equilib. 83 (1993) 301-312.
[9] J. Phys. Chem. 18 (1950) 108-111.
[10] J. Polymer. Sci. 4 (1949) 153-182.

Model parameters and correlations

17AIChE Annual Meeting, Minneapolis, MN

Pure component/polymer parameters

Network parameter		1.51[1]
Effective molar volume, v^* (cm^3/mol)	Water	18.06
	Ethanol	18.06
	Polymer	1664

Binary Interaction parameters (τ_{ij})

i	j		
	Water	Ethanol	NIPAAm
Water	0	1.486[a]	3.805
Ethanol	0.031[a]	0	3.087
NIPAAm	-0.982	-1.218	0

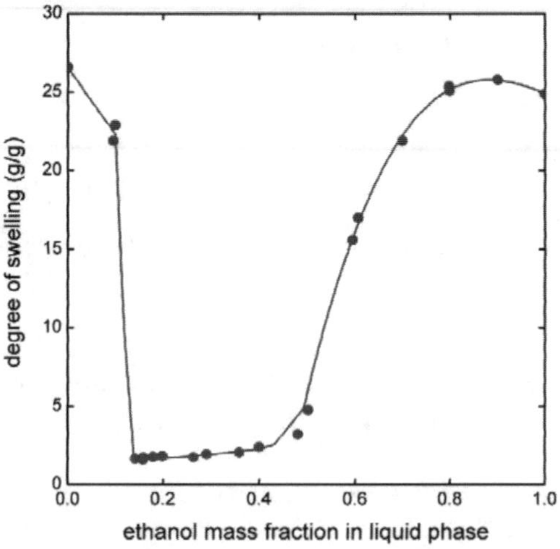

[1] Fluid Phase Equilib. 219 (2004) 233-243.
[a]Databank, Aspen physical property system, version 8.8

Model parameters and correlations

17AIChE Annual Meeting, Minneapolis, MN

Pure component/polymer parameters

Network parameter		1.51[1]
Effective molar volume, v^* (cm^3/mol)	Water	18.06
	Ethanol	18.06
	Polymer	1664

Binary Interaction parameters (τ_{ij})

i	*j*		
	Water	Ethanol	NIPAAm
Water	0	1.486[a]	3.805
Ethanol	0.031[a]	0	3.087
NIPAAm	-0.982	-1.218	0

[1] Fluid Phase Equilib. 219 (2004) 233-243.
[a]Databank, Aspen physical property system, version 8.8

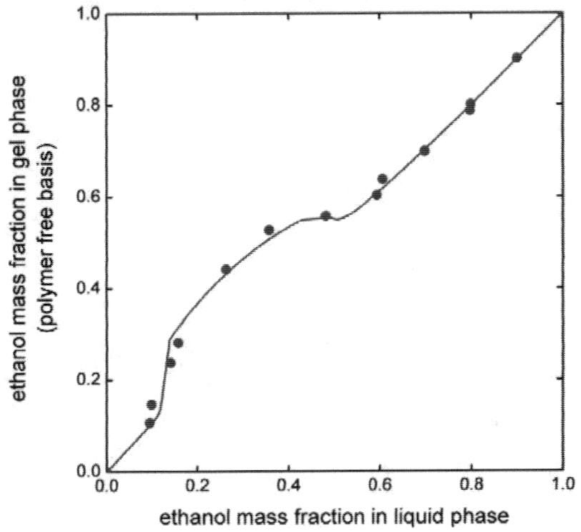

Model correlations on solvent activities

17AIChE Annual Meeting, Minneapolis, MN

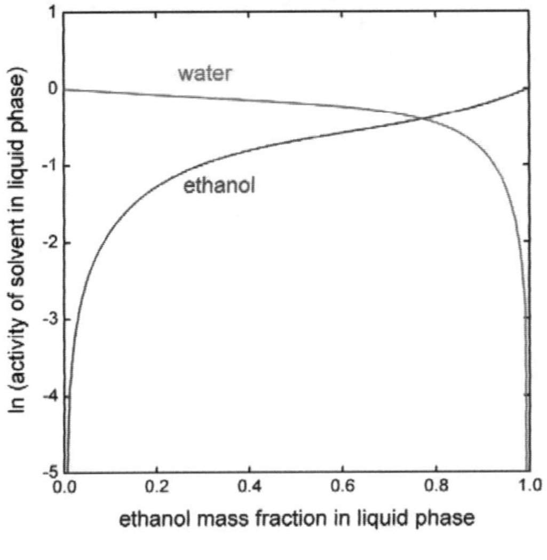

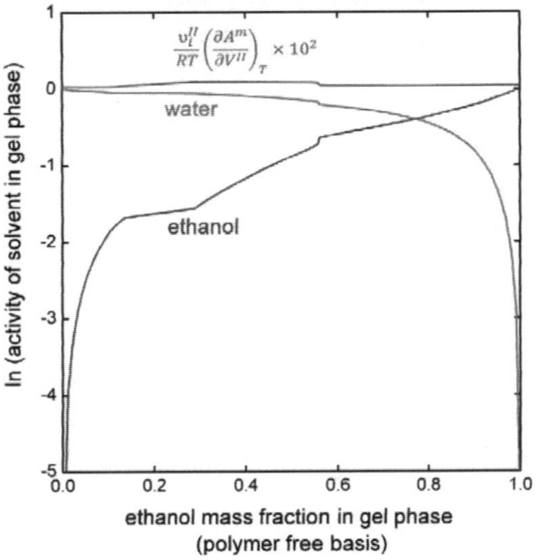

Model sensitivity

Sensitivity on effective molar volume (v^*) of PNIPAAM

Network parameter		1.51
Effective molar volume, v^* (cm^3/mol)	Water	18.06
	Ethanol	18.06
	Polymer	1664

i	*j* Water	Ethanol	NIPAAm
Water	0	1.486	3.805
Ethanol	0.031	0	3.087
NIPAAm	-0.982	-1.218	0

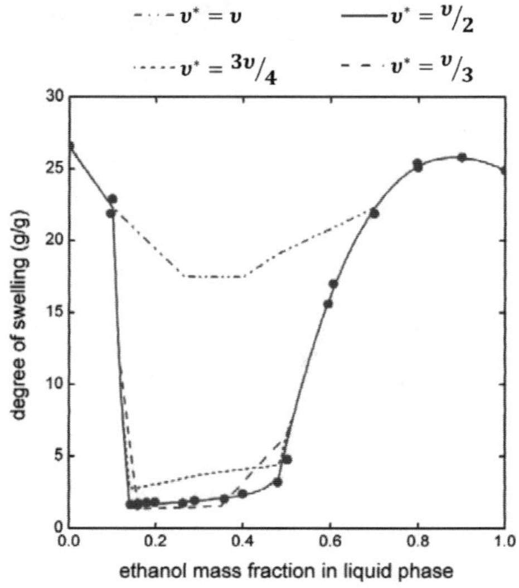

Model sensitivity

Sensitivity on effective molar volume (v^*) of PNIPAAM

Network parameter		1.51
Effective molar volume, v^* (cm^3/mol)	Water	18.06
	Ethanol	18.06
	Polymer	1664

i	*j* Water	Ethanol	NIPAAm
Water	0	1.486	3.805
Ethanol	0.031	0	3.087
NIPAAm	-0.982	-1.218	0

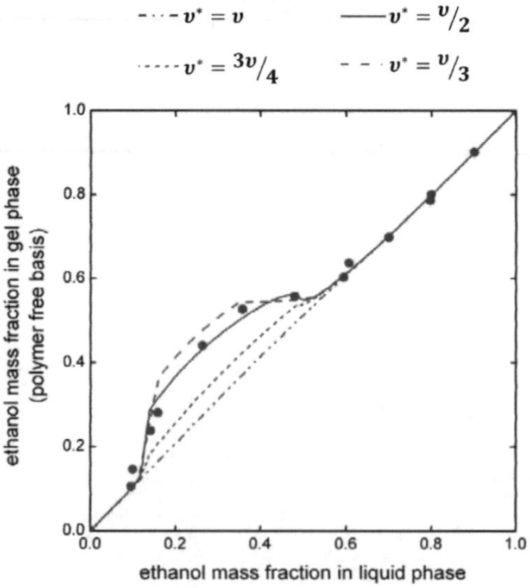

Model predictions

Crosslink density: 0.01 (mole/mole) of polymerizable material

Network parameter, C=1.35[1]

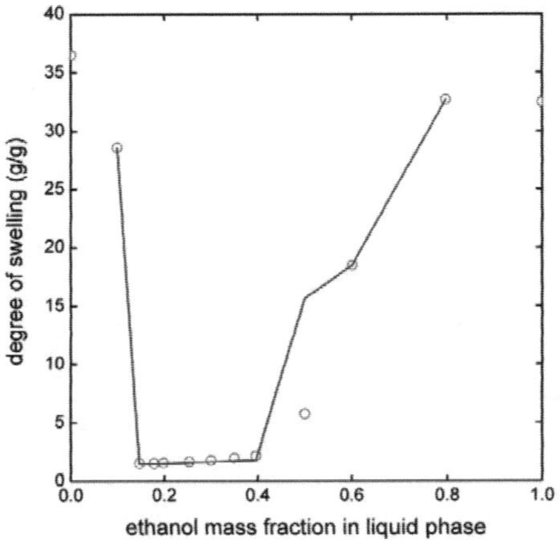

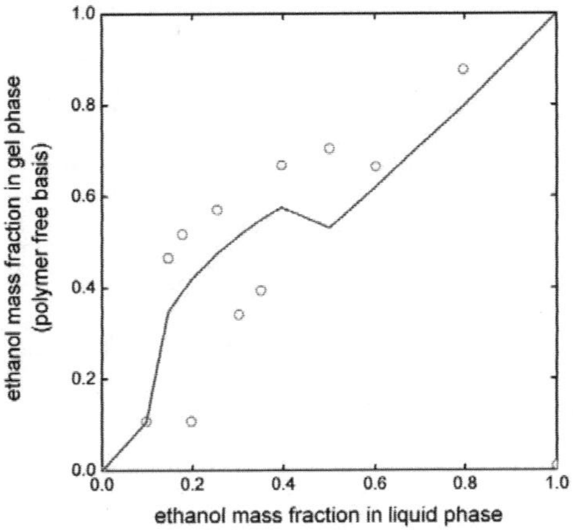

[1] Fluid Phase Equilib. 219 (2004) 233-243.

Summary

- Deswelling occurs when ethanol mass fraction is within the range (0.1 - 0.5) in coexisting liquid phase. Model clearly captures this deswelling phenomena.
- Swelling of N-isopropyl acrylamide hydrogel in aqueous ethanol solution is dominated by solvent-solvent interactions and solvent-segment interactions.
- Excess Helmholtz energy of the hydrogel network is found insignificant compared to the excess Gibbs free energy of the gel.
- Effective molar volume of the PNIPAAm polymer in hydrogel network is found half of the value use in linear PNIPAAm polymer.

Future research plan

17AIChE
Annual Meeting, Minneapolis, MN

Extend the thermodynamic model of hydrogel for

- Higher polymer concentrations
- Water-acetone mixed solvents
- Water-Alcohol (C_1-C_4) mixed solvents

Integrate the model with eNRTL activity coefficient model for

- Water-electrolyte solutions

Thank you for your attention

J.F MADDOX FOUNDATION Significantly improving the quality of life in southeastern New Mexico
INVESTING IN EDUCATION, COMMUNITY DEVELOPMENT, AND SOCIAL PROGRAMS

501683 Nanostructure-Driven Fatigue Resistance and Dynamic Recovery in Thermoplastic Elastomer Hydrogel Networks

Abstract:

Creative efforts in network design have led to numerous impactful improvements in hydrogel mechanics. Our group has been focused on the synthesis of hydrogel networks capable of rapid and complete elastic recovery coupled with a high intrinsic resistance to mechanical fatigue under continuous cyclic loading. Much of our motivation is derived from a desire to synthesize biphasic materials capable of replicating the biomechanical demands of musculoskeletal soft tissues, which has continued to present a formidable materials science challenge. Reduction to practice has been plagued by the gap in mechanical performance that continues to differentiate biological and synthetic systems. The most prominent advances in hydrogel design have employed the inclusion of interpenetrating networks, hydrophobic interactions,[2, 3] ionic interactions,[4, 5, 6] deformable domain structures,[7] sliding junction points,[8, 9, 10] and multifunctional macromers of prescribed structure[2, 11, 12] to produce intriguing network designs pushing the boundaries of mechanical achievement. These designs are responsible for stunning demonstrations of modulus, [4, 6] stretchability, [3, 5, 6, 8, 9] ultimate strength,[1, 2, 11] and overall toughness [3, 5, 6, 8, 9] that were inconceivable less than two decades ago. However, most of these systems are plagued by combinations of limited moduli at small strain, unacceptably high levels of energy dissipation and fatigue, or slow and incomplete elastic recovery that are incompatible with the cyclic biomechanical loading profiles of most musculoskeletal soft tissues. Intrinsic susceptibility to permanent covalent bond rupture and significant plastic deformation preclude a network from returning elastically to its original configuration, while recovery dynamics in the minutes to hours range[5, 6] preclude it from doing so at physiologically relevant (often sub-second) time scales. Unfortunately, none of these advanced systems would yet be acceptable, for example, as substitutes for the fibrocartilage comprising menisci of the knee, the annulus fibrosis of the intervertebral disc, or connective tissue of the cardiovascular system, which rely on high levels of elasticity to rapidly absorb and transfer (not dissipate) strain energy into efficient body movement or pulsatile blood flow.[13, 14, 15]

Here, we report a new paradigm in hydrogel design based on prefabrication of an efficient nanoscale network architecture using the melt-state self-assembly of amphiphilic block copolymers. Rigorous characterization and mechanical testing reveal that swelling of these preformed networks produces hydrogels with physiologically relevant moduli and water compositions, negligible hysteresis, sub-second elastic recovery rates, and unprecedented resistance to fatigue over hundreds of thousands of compression cycles. Furthermore, by relying only on simple thermoplastic processing to form these nanostructured networks, the synthetic complexities common to most solution-based hydrogel fabrication strategies are completely avoided. We use melt blends of sphere morphology forming AB and ABA diblock copolymers to form a network of tethered micelle units. During melt-state self-assembly, A blocks form spherical aggregates while the B blocks form the corona or matrix in which the spherical A domains sit. The triblock copolymer acts as a bridge, mechanically linking adjacent spherical domains. Hydrogel formation relies on two important phenomena that dictate the choice of A and B blocks. One, the A block must be able to become a vitrified solid upon cooling. We use polystyrene (PS) with a glass transition temperature near 80 ËšC. Two, the B block must be selectively soluble in the swelling media, while the A block remains impervious to it. We use poly(ethylene oxide) (PEO) as the water soluble B block.

In terms of network structure this method of hydrogel fabrication produces key features we believe are critical for eliminating fatigue, producing a high modulus, and ensuring rapid recovery. First we counter the effects of strand length inhomogeneity by keeping the density of fixed junction points ($\ddot{I}_{\text{x-link}}$, glassy spherical domains) low, and the strand molecular weight adjoining them high ($M_{\text{x-link}}$,180kDa) and narrowly distributed (PDI < 1.06). This would normally favor soft, weak materials, but we use extremely high junction point functionalities ($f \sim 230$, block copolymer chains per spherical aggregate) in the 200 â€“ 300 range. Thus even moderate triblock copolymer compositions produce a high numbers of strands emanating from each sphere. Finally, because assembly takes place in the melt, the density of topologically trapped entanglements among these bridging (and looping) triblock copolymers is intrinsically quite high. In the model PS-PEO systems used to develop our preliminary systems, we estimate the number of entanglements per fixed junction point numbers in the thousands. These entanglements act much like the sliding junction points introduced by Ito in his slide ring-gels, and provide the network with both a significant modulus and an ability to rapidly redistribute stress, the latter being the keystone to the unusual fatigue resistance exhibited by these systems.

References

1. Gong JP, Katsuyama Y, Kurokawa T, Osada Y. Double-network hydrogels with extremely high mechanical strength. *Adv Mater* 2003, **15**(14)**:** 1155-+.

2. Kamata H, Akagi Y, Kayasuga-Kariya Y, Chung UI, Sakai T. "Nonswellable" hydrogel without mechanical hysteresis. *Science* 2014, **343**(6173)**:** 873-875.

3. Li WB, An HY, Tan Y, Lu CG, Liu C, Li PC*, et al.* Hydrophobically associated hydrogels based on acrylamide and anionic surface active monomer with high mechanical strength. *Soft Matter* 2012, **8**(18)**:** 5078-5086.

4. Henderson KJ, Zhou TC, Otim KJ, Shull KR. Ionically Cross-Linked Triblock Copolymer Hydrogels with High Strength. *Macromolecules* 2010, **43**(14)**:** 6193-6201.

5. Sun JY, Zhao X, Illeperuma WR, Chaudhuri O, Oh KH, Mooney DJ*, et al.* Highly stretchable and tough hydrogels. *Nature* 2012, **489**(7414)**:** 133-136.

6. Sun TL, Kurokawa T, Kuroda S, Bin Ihsan A, Akasaki T, Sato K*, et al.* Physical hydrogels composed of polyampholytes demonstrate high toughness and viscoelasticity. *Nat Mater* 2013, **12**(10)**:** 932-937.

7. Brown AEX, Litvinov RI, Discher DE, Purohit PK, Weisel JW. Multiscale Mechanics of Fibrin Polymer: Gel Stretching with Protein Unfolding and Loss of Water. *Science* 2009, **325**(5941)**:** 741-744.

8. Bin Imran A, Esaki K, Gotoh H, Seki T, Ito K, Sakai Y*, et al.* Extremely stretchable thermosensitive hydrogels by introducing slide-ring polyrotaxane cross-linkers and ionic groups into the polymer network. *Nature communications* 2014, **5:** 5124.

9. Ito K. Novel cross-linking concept of polymer network: Synthesis, structure, and properties of slide-ring gels with freely movable junctions. *Polym J* 2007, **39**(6)**:** 489-499.

10. Okumura Y, Ito K. The polyrotaxane gel: A topological gel by figure-of-eight cross-links. *Adv Mater* 2001, **13**(7)**:** 485-+.

11. Sakai T, Akagi Y, Matsunaga T, Kurakazu M, Chung U, Shibayama M. Highly Elastic and Deformable Hydrogel Formed from Tetra-arm Polymers. *Macromol Rapid Comm* 2010, **31**(22)**:** 1954-1959.

12. Sakai T, Matsunaga T, Yamamoto Y, Ito C, Yoshida R, Suzuki S*, et al.* Design and fabrication of a high-strength

hydrogel with ideally homogeneous network structure from tetrahedron-like macromonomers. *Macromolecules* 2008, **41**(14): 5379-5384.

13. Andrews S, Shrive N, Ronsky J. The shocking truth about meniscus. *Journal of Biomechanics* 2011, **44**(16): 2737-2740.

14. Cortes DH, Elliott DM. The Intervertebral Disc: Overview of Disc Mechanics. In: Shapiro IM, Risbud MV (eds). *The Intervertebral Disc: Molecular and Structural Studies of the Disc in Health and Disease*. Springer Vienna: Vienna, 2014, pp 17-31.

15. Wagenseil JE, Mecham RP. Vascular Extracellular Matrix and Arterial Mechanics. *Physiological reviews* 2009, **89**(3): 957-989.

497511 Interfacial Crystallization of Polyolefins: An Improved Outlook for Polymer Blends

Abstract:

Although isotactic polypropylene (*i*PP) and polyethylene (PE) are the two highest production volume polymers in the world, interfacial adhesion between these chemically similar, yet immiscible, polymers is poorly understood. Past work has highlighted the consequences of catalyst selection (i.e. multi-site Ziegler-Natta vs. site-specific metallocene) on interfacial adhesion in laminated films. Metallocene catalyzed PE/*i*PP pairs often possess interfacial weld strength considerably greater than the very poor adhesion obtained with Ziegler-Natta catalyzed PE/*i*PP pairs. This result has been attributed to the dispersity in molecular weight and branch content obtained with Ziegler-Natta catalysts leading to amorphous build-up at the PE/*i*PP interface which thwarts interfacial crystallization. Here, we utilize multiple manufacturing techniques to evaluate the role of processing parameters, catalyst selection, and chain architecture on PE/*i*PP adhesion. First, we demonstrate a dual-bore capillary rheometer adaptation to investigate interfacial properties in melt-coextruded multilayer structures using small quantities of material (~10 g), a technique that mimics traditional laboratory-scale (> 2 kg) multilayer coextrusion. We demonstrate a 3-fold increase in interfacial adhesion between metallocene high-density polyethylene (mHDPE) and metallocene *i*PP (m*i*PP) when the interface is rapidly quenched compared with slow cooling. By altering the PE chain architecture and substituting a metallocene linear low density polyethylene (mLLDPE), adhesion between mLLDPE/m*i*PP is improved more than 12x, highlighting the role of chain architecture in PE/*i*PP adhesion. Finally, PE/*i*PP blends were fabricated to investigate the role of interfacial adhesion on mechanical properties. In the case of weakly adhering pairs, blend mechanical properties were worse than those of the individual components, while blends fabricated from strongly adhering pairs possessed superior mechanical properties than the constituent polyolefins. Understanding how molecular parameters control polymer-polymer interfacial adhesion has enabled us to produce blends of commercial polymers with desirable end-use properties that inform product design.

492341 Ultra-Stable Amorphous Teflon: Extreme Fictive Temperature Reduction As a Means to Probe Sub-T_g Dynamics

Abstract:

We have succeeded in making ultra-stable films through physical vapor deposition of amorphous Teflon® on to substrates held at temperatures from 30.0 °C (303.2 K) to 116.7 °C (389.9 K) corresponding to approximately 0.75 to 0.97 times the glass temperature T_g of the vapor deposited materials. We find a maximum fictive temperature T_f reduction (compared to the T_g) of 57 °C and this opens a window for experimental investigation of deep glassy state behavior that is even greater than the 43.6 °C reduction seen in 20 million year old amber [1]. In that work samples could be tested using macroscopic rheometric methods to establish that the upper bounds to the relaxation times below the glass transition temperature deviate strongly from the notional Vogel-Fulcher type of super-Arrhenius behavior expected for glass-forming materials. These results remain to be reproduced. The present success in creating a low fictive temperature glass of amorphous Teflon® provides an opportunity to expand the window of investigation to even further below the glass temperature. However, unlike the amber samples which were large enough to use normal rheological measurements, the vapor deposition process produces samples of mass in the nano- to micro-gram range. Therefore nano-mechanical measurement methods are required to interrogate the viscoelastic response of the ultra-dense amorphous Teflon®. In the present investigation we are using the TTU nano-bubble inflation method [2,3] to determine the viscoelastic response of the ultra-dense material. By making measurements from the reduced T_f to above the T_g we are able to explore the deviations from super-Arrhenius behavior seen in the ancient amber, but now with a synthetic material that is readily reproduced, albeit in small quantities. Our current state of knowledge is that we have produced the ultra-stable amorphous Teflon® and carried out the first viscoelastic measurements of the material response in the temperature regime of interest. The full set of results and comparisons with the amber results will be presented and discussed.

[1] J. Zhao, S.L. Simon, and G.B. McKenna, "Using 20-million-year-old amber to test the super-Arrhenius behaviour of glass-forming systems," Nat. Comm., **4,** 1783, (2013). DOI: 10.1038/ncomms2809

[2] P.A. O'Connell and G.B. McKenna, "Rheological Measurements of the Thermoviscoelastic Response of Ultrathin Polymer Films," Science, **307**, 1760-1763 (2005).

[3] P.A. O'Connell and G.B. McKenna, "Novel Nanobubble Inflation Method for Determining the Viscoelastic Properties of Ultrathin Polymer Films," Rev. Sci. Inst., **78**, 013901-1 – 013901-12 (2007).

497811 Influence of Strucure and Dynamics in Matrix-Free Polymer-Grafted Nanocomposite Memebranes

Abstract:

Recent works have exhibited enhanced molecular transport for membranes comprised solely of polymer grafted nanoparticles compared to that of bulk polymer or un-grafted, “traditional” mixed composites. Additionally, these enhancements appear non-monotonic as a function of grafted chain length. We postulate chemically grafting polymers to the surface of nanoparticles restricts the ability of the chains to sufficiently pack the spaces between the particle cores. In this regime, there exist locations which are essentially “starved” of polymer at the edges of the brush. To support these claims and investigate the origin of the non-monotonic behavior in diffusion we use both linear and non-linear rheology to probe the chain dynamics. In addition, small angle X-ray and contrast variation small angle neutron scattering (SAXS, SANS) are employed to investigate how structure may influence the unexpected performance of these membranes. The elucidation of the mechanism behind this phenomenon may lead to simple rational design of high performance membrane technology for gas purification.

487205 Role of Interfacial Adhesion in Rate-Dependent Deformation and Failure of Model Ring Opening Metathesis Polymer (ROMP) Filled Composites

Abstract:

Glassy polymers formed by ring opening metathesis polymerization (ROMP) such as poly(dicyclopentadiene) (PDCPD) have unique potential as structural and protective materials due to their combination of high glass transition temperature, remarkable toughness, and outstanding low-temperature performance. However, these polymers tend to be highly nonpolar and thus do not adhere well to glass, limiting their application in fiber reinforced composites (FRPCs), laminates, and structural adhesives. Commercial glass fibers feature proprietary sizing packages optimized for conventional resins, which are challenging to remove without compromising fiber roughness and mechanical integrity. Therefore we have systematically investigated rate-dependent deformation and failure mechanisms of model composites comprising ROMP polymers containing well-defined rigid particulate fillers, focusing on the effects of filler-matrix interfacial chemistry and adhesion. Monodisperse micron-sized silica spheres are modified with a library of organosilanes and characterized using surface spectroscopic and wettability measurements. Mechanical effects of filler surface treatments presenting high, moderate, and inert reactivity toward the polymer network formed during curing are evaluated using a combination of high rate impact testing and quasistatic tensile and fracture testing in controlled environmental conditions, combined with in-situ and post-mortem imaging of failure processes. Preliminary results suggest well-bonded, reactively functionalized fillers suppress void growth in highly ductile resins with low crosslink density, facilitating deformation by shear. This results in a 10-fold increase in ultimate elongation relative to the unfilled polymer, and a 5-fold increase relative to an equivalent formulation containing poorly bonded fillers. These results are in contrast to our previous work on semi-brittle poly(methyl methacrylate), in which well-bonded fillers severely restrict plastic deformation and degrade toughness. The results suggest a rich design space for optimizing impact energy dissipation of protective systems based on interfacial properties and matrix ductility.

500950 Mechanically Ductile and Stiff, Triazole-Based Glassy Photopolymer Network

Abstract:

Photopolymerization is an indispensable technique broadly utilized in numerous applications including coatings, adhesives, additive manufacturing, and dental composites. Dominantly, conventional thermosetting photopolymers in the industrial applications are based on acrylate free radical polymerization using multi-functional monomeric systems. While these polymer networks provide rapid *in situ* photo-curing kinetics with desirable mechanical performance, the nature of chain-growth polymerization often limits the maximum conversion achieved in these polymers and generates heterogeneous network structures that are highly brittle; thereby, frequently causing mechanical failure.

Step-growth polymerizations utilizing “click” chemistries, which are classified as efficient and orthogonal reactions, possess numerous potential advantages. While the nature of “click” chemistries supports orthogonal and selective reactions without forming byproducts, step-growth polymers yield more homogeneous networks with narrow glass transition temperatures in comparison with chain-growth polymers. Moreover, variable backbone functionalities in step-growth polymers as well as high reactive functional group densities are readily achievable. This behavior consequently enables systematic modification of vitrification-induced polymerization kinetics, gel-point conversion, and overall mechanical performance and enables structure-property relationship determination.

Herein, we present a densely crosslinked glassy polymer network formed via photo-initiated CuAAC (copper-catalyzed azide-alkyne cycloaddition) polymerization in bulk that exhibits considerably improved mechanical stability, through both high tensile toughness and strength. For various monomer structures, a universal testing machine (MTS) was utilized to evaluate mechanical properties of CuAAC-based polymer films. The scope of the research is extended to not only the mechanical performance of the CuAAC polymers but also the physical aging behavior and shape memory-recovery properties of the polymer networks. Furthermore, the application in resin-based composites as high-performance materials is studied.

500560 Controlled Topology Toughning Epoxy Via Incorporation of Partially Reacted Substructures

Abstract:

A method of toughening amine-cured epoxy systems was proposed and the resulting epoxy systems were characterized. Accordingly, a Monoamine-functionalized Partially Reacted Substructures (mPRS) resin was formed by partially curing a mixture of tetraglycidyl of diaminodiphenylmethane (TGDDM) and polyether monoamine (PEMA). Four different degree of polymerization of mPRS (0%,70%,75% and 80%) were added to a mixture of Diglycidyl Ether of Bisphenol A (DGEBA) at different concentration (0% to 25%) and the mixtures were cured. Based on the results of the quasi-static tensile test, the mechanical properties are tunable by control the conversion and concentration of mPRS, Youngâ€™s modulus (1.69 to 2.54 Gpa), yielding strength (39 to 59 Mpa) and failure strain (15% to 44%). Besides, the toughness improvement was also observed on fracture toughness and quasi-static compression test. The improved toughness was found to be due to an enhanced plastic microvoid growth as a result of the presence of the mPRS in the crosslinked structure of the mPRS-modified epoxy systems.

500887 Thermoplastic Polydimethylsiloxane Realized By Hydrogen-Bond Networks through L-Phenylalanine Terminals

Abstract:

Thermoplastic polydimethylsiloxane (PDMS) containing L-phenylalanine at the end of the molecule was synthesized by the liquid-phase synthesis method with an Fmoc-protected amino acid, followed by the condensation polymerization. In order to investigate the mechanical and the thermoplastic properties of the newly synthesized PDMS with L-phenylalanine, the tensile testing and the dynamic mechanical analysis (DMA) were conducted. From the results of the tensile testing, the stress at break of the PDMS with L-phenylalanine nearly became 7 times higher than that of the PDMS without L-phenylalanine. The DMA results revealed that the PDMS could maintain its storage modulus for 3 cycles, while the melting temperature was also kept at ~120°C. Moreover, the mechanical property was analyzed through 2 remolding cycles, and it was found that the PDMS could almost maintain its mechanical property after the heat remolding. In detail, the strain at break and the stress at break of the sample after the second cycle maintained above 90% of the values observed for the first remolded sample. As for PDMS without L-phenylalanine, the Youngâ€™s modulus and the stress at break of the sample after the second cycle maintained above 90% of those of the original (non-remolded) sample. As for PDMS with L-phenylalanine, the stress at break of the sample after the second cycle also maintained over 90% of that of the original (non-remolded) sample.

501514 In *vitro* and *silico* Characterization of Grafted Hydrophobic Brush Membranes

Abstract:

Tethered polymers have been implemented in several engineering/biomedical applications to stabilize colloids, to reduce friction between surfaces, to design novel fluidic devices, to fractionate proteins, etc. The Belfortâ€™s lab is currently working on a new class of selective brush membranes, where a selective layer comprising â€œhydrophobic brushesâ€ is grafted onto a non-selective permeable membrane.

The membranes were synthesized by graft-induced polymerization of hydrophobic vinyl monomers onto poly(ether sulfone) nanofiltration support membranes. The novel grafting method is based on the combination of atmospheric pressure plasma treatment and Activators Regenerated by Electron Transfer (ARGET) with Atom Transfer Radical Polymerization (ATRP). Several hydrophobic vinyl monomers were tested to investigate the effect of chain length, the presence of linear versus branched chains, etc. Membranes were characterized by ATR-FTIR (chemistry) and AFM (morphology, roughness, hydrophobicity). To gain a better understanding of the selective layer properties, the hydrophobic brush layer was grafted onto silica surfaces, following the same ARGET-ATRP method. Surfaces were characterized by a multitude of additional techniques: XPS (atomic composition and heterogeneity), Ellipsometry, AFM and DLS (thickness), QCM-D (polymerization reaction and layer viscoelasticity), ToF-SIMS (brush layer density). Complementary atomistic simulations have been used to quantify how polymers at different grafting densities used in the brush interact with the solvent (aqueous/organic mixture).

The novelty here is the use of hydrophobic brush-like structures as a selective skin or dense layer attached to a non-selective support membrane. However, if this new class of hydrophobic brush membranes is to reach its full potential, it is critical that we determine how it selects for one species (e.g. isobutanol) over another (e.g. water). Thus, we need to know how brush microstructural properties, chemistry and solution conditions influence the properties and eventually the filtration performance of the selective brush layer. This knowledge will allow structure-by-design for the first time.

502574 Direct Synthesis and Morphology Control of Metal Organic Framework Nanosheets

Abstract:

Metal-organic frameworks (MOFs) are promising candidates for gas separation membranes due to their regularly arranged pores, tunable pore sizes and functionalities. In particular, MOFs nanosheets, which possess few nanometer thickness and large lateral dimensions, are highly desirable for membranes as mass transport is increased along the thin dimension and the orientation of pores can be readily controlled during the fabrication. However, it remains challenging to synthesize free-standing MOF nanosheets with uniform size and shape. MOFs nanosheets are usually fabricated through delamination of layered MOFs. This top-down method has a low yield meanwhile the resultant nanosheets often suffer from fragmentation and reaggregation. Herein, we demonstrate a direct synthesis approach for zinc benzimidazole-acetate (Zn(Bim)OAc) MOF nanosheets. Surfactant and judicious selection of dispersing solvents are found critical to prevent reaggregation of Zn(Bim)OAc during the synthesis. The obtained nanosheets have a lateral dimension of 500 nm when synthesized at room temperature. The morphology of thin Zn(Bim)OAc can be tuned from nanosheets to nanobelts by adjusting the synthesis temperature. To measure the thickness of nanosheet, a method was developed using high angle annular dark field scanning transmission electron microscopy (HAADF-STEM). A thickness of 7 nm is determined for Zn(Bim)OAc nanosheets which would be promising building blocks of gas separation membranes.

497458 Mixed-Linker MOF Synthesis and Remediation of Acid-Gas Degradation Using Linker-Exchange Techniques

Abstract:

Solvent Assisted Linker Exchange (SALE) is known as a useful tool to modify existing MOFs and thereby obtain otherwise difficult-to-synthesize structures, or improve the stability and functional properties of MOFs. By partially exchanging the native linker in a MOF, this technique can also be used to synthesize mixed-linker MOFs. However, the structural properties of such materials, including the spatial distributions of linkers in these mixed-linker materials and how they differ from mixed-linker MOFs made via *de novo* routes, are not well known.

This work consists of two main parts. First, we compare and contrast a series of mixed-linker ZIF-8-90 materials made via both routes. ^1H CRAMPS NMR spectroscopy, water and nitrogen adsorption, and native fluorescence confocal microscopy are used to elucidate the differences in the outcome of these syntheses routes. A generalizable mechanism for SALE in ZIFs is proposed by studying the impact of crystal size, temperature, and SALE duration. We observe that under moderate conditions, SALE in ZIFs is diffusion-limited, and results in a core-shell type structure. Harsher conditions lead to a deviation from this behavior including linker etching resulting in morphological changes to particles. Under appropriate reaction conditions, SALE techniques appear to be capable of producing controlled core-shell ZIF structures of good morphological quality that complement the well-mixed structures obtained by *de novo* routes.

Second, linker exchange was explored as a means of recovering crystallinity and porosity of ZIFs damaged by acid-gases. We developed a technique called Solvent Assisted Crystal Redemption (SACRED) wherein the damaged crystals of a ZIF are treated with a solution of its native linker. The structural damage to ZIFs induced by acid-gases were found to be partially or completely reversed by this treatment. The recovered material is shown to be chemically identical to the pristine material using FTIR spectroscopy and X-ray diffraction. Adsorption studies show that the functional performance of ZIF-8 is retained after the recovery. The use of deuterium-substituted linkers and ^2H NMR spectroscopy provides mechanistic insight into the recovery process. This method has the potential to reduce costs and improve lifetime of ZIF-based adsorbents or membranes.

References

[1] Highly Tunable Molecular Sieving and Adsorption Properties of Mixed-Linker Zeolitic Imidazolate Frameworks; K. Eum, K.C. Jayachandrababu, F. Rashidi, K. Zhang, J. Leisen, S. Graham, R.P. Lively, R.R. Chance, D.S. Sholl, C.W. Jones, S. Nair; *J. Am. Chem. Soc.;* 2015; 137(12); 4191-4197

[2] Structure Elucidation of Mixed-Linker Zeolitic Imidazolate Frameworks by Solid-State ^1H CRAMPS NMR Spectroscopy and Computational Modeling; K.C. Jayachandrababu, R.J. Verploegh, J. Leisen, R.C. Nieuwendaal, D.S. Sholl, S. Nair; *J. Am. Chem. Soc.;* 2016; 138(23); 7325-7336

[3] Structural and Mechanistic Differences in Mixed-Linker Zeolitic Imidazolate Framework Synthesis by Solvent Assisted Linker Exchange (SALE) and *de novo* Routes; K.C. Jayachandrababu, D.S. Sholl, S. Nair; *J. Am. Chem. Soc.;* 2017; *Just Accepted Manuscript*

490184 Rapid Microwave-Assisted Synthesis of Hybrid Zeolitic-Imidazolate Frameworks with Mixed Metals and Mixed Linkers

Abstract:

Separating a crude component into a more valuable purer product is a major part of many industrial chemical plants. Typically these processes utilize distillation, which account to 10 – 15 % of world's energy consumption.[1] One alternative solution that can reduce energy consumption up to 90 % relative to distillation is through membrane separation using porous solid material such as zeolites and metal organic frameworks (MOFs).[2] Zeolitic-imidazolate frameworks (ZIFs),[3] a subclass of MOFs, in particular have attracted many attentions due to their chemical/thermal stabilities, their ultra-microporosities, and high surface area when compared to other MOFs material.[3] ZIFs possess zeolite-like topologies in which the tetrahedral Si or Al and the bridging O in zeolite structure are replaced by transition metals (such as Zn,[3] Co,[4] and Cd[5]) and imidazolate-derived ligands. A common drawback for membrane gas separation is the limited availability of aperture pore size and functionalities of the porous materials. Studies have shown that through mixing metals[6,7] and ligands,[8,9] one can continuously tune the MOFs/ZIFs (termed hybrid ZIFs) properties to match with the characteristic of specific gas mixture. However, the common synthesis of hybrid ZIFs has generally been through slow diffusion techniques,[4,10] conventional hydrothermal,[7,11] and solvothermal methods,[9,12,13]requiring several hours up to days of synthesis time. In addition, the yields of those synthesis protocols are often low, an economical issue for hybrid ZIFs to be used in a large industrial scale.

Herein we plan to present a new microwave-assisted synthetic strategy to rapidly prepare hybrid zeolitic-imidazolate frameworks (ZIFs): ZIFs with mixed metal centers and/or mixed linkers.[14]The microwave-based method significantly shortens synthesis time, produces higher yield, substantially reduces the amount of ligand, and eliminates the use of deprotonating agents. X-ray diffraction pattern reveals a mixed metal CoZn-ZIF-8 (i.e., ZIF-8 with both Co and Zn centers) maintains the sodalite (SOD) zeolitic topology from ZIF-8 parent. Elemental mapping using Energy-dispersive X-ray spectroscopy (EDS) and electronic/geometric information obtained from X-ray absorption spectroscopy (XAS) confirm the uniform distribution of tetrahedral Co and Zn metal centers within the same framework of the mixed-metal ZIF. The metal to nitrogen (M-N) stretching frequencies on IR band were observed to be systematically blue-shifted as the Co/Zn ratio in the mixed metal ZIF increases. Furthermore, for the first time, a hybrid ZIF with both mixed metal centers (Co and Zn) and mixed linkers (2-methylimidazolate and benzimidazolate) was prepared through one-step microwave synthesis. Finally, a mixed metal CoZn-ZIF-8 with the Co/Zn ratio of ~ 1 was grown as membranes on porous α-Al_2O_3 supports, showing higher propylene/ propane separation factor (~120) when compared to pure Zn-ZIF-8 membranes (~63) prepared with similar method.

References:

1. *Materials for Separation Technologies: Energy and Emission Reduction Opportunities*; Technical Report: Oak Ridge, TN **2005**.

2. Sholl, D. S.; Lively, R. P. Seven Chemical Separations to Change the World. *Nature* **2016**, *532*, 435-437.

3. Park, K. S.; Ni, Z.; Cote, A. P.; Choi, J. Y.; Huang, R.; Uribe-Romo, F. J.; Chae, H. K.; O'Keeffe, M.; Yaghi, O. M. Exceptional chemical and thermal stability of zeolitic imidazolate frameworks. *Proc. Natl. Acad. Sci. U. S. A.* **2006**, *103*, 10186-10191.

4. Banerjee, R.; Phan, A.; Wang, B.; Knobler, C.; Furukawa, H.; O'Keeffe, M.; Yaghi, O. M. High-Throughput Synthesis of Zeolitic Imidazolate Frameworks and Application to CO2 Capture. *Science* **2008**, *319*, 939-943.

5. Karagiaridi, O.; Bury, W.; Sarjeant, A. A.; Stern, C. L.; Farha, O. K.; Hupp, J. T. Synthesis and characterization of isostructural cadmium zeolitic imidazolate frameworks via solvent-assisted linker exchange. *Chem. Sci.* **2012**, *3*, 3256-3260.

6. Wang, C.; Yang, F.; Sheng, L.; Yu, J.; Yao, k.; Zhang, L.; Pan, Y. Zinc-Substituted ZIF-67 Nanocrystals and Polycrystalline Membranes for Propylene/Propane Separation. *Chem. Comm.* **2016**, *52*, 12578-12581.

7. Kaur, G.; Rai, R. K.; Tyagi, D.; Yao, X.; Li, P.-Z.; Yang, X.-C.; Zhao, Y.; Xu, Q.; Singh, S. K. Room-temperature synthesis of bimetallic Co-Zn based zeolitic imidazolate frameworks in water for enhanced CO2 and H2 uptakes. *J. Mater. Chem. A* **2016**, *4*, 14932-14938.

8. Eum, K.; Jayachandrababu, K. C.; Rashidi, F.; Zhang, K.; Leisen, J.; Graham, S.; Lively, R. P.; Chance, R. R.; Sholl, D. S.; Jones, C. W.; Nair, S. Highly Tunable Molecular Sieving and Adsorption Properties of Mixed-Linker Zeolitic Imidazolate Frameworks. *J. Am. Chem. Soc.* **2015**, *137*, 4191-4197.

9. Thompson, J. A.; Blad, C. R.; Brunelli, N. A.; Lydon, M. E.; Lively, R. P.; Jones, C. W.; Nair, S. Hybrid Zeolitic Imidazolate Frameworks: Controlling Framework Porosity and Functionality by Mixed-Linker Synthesis. *Chem. Mater.* **2012**, *24*, 1930-1936.

10. Phan, A.; Doonan, C. J.; Uribe-Romo, F. J.; Knobler, C. B.; O'Keeffe, M.; Yaghi, O. M. Synthesis, Structure, and Carbon Dioxide Capture Properties of Zeolitic Imidazolate Frameworks. *Acc. Chem. Res.* **2010**, *43*, 58-67.

11. Roshan, R. K.; Babu, R.; Jeong, H.-M.; Hwang, G.-Y.; Jeong, G.-S.; Kim, M.-I.; Kim, D.; Park, D.-W. A solid solution zeolitic imidazolate framework as a room temperature efficient catalyst for the chemical fixation of CO2. *Green Chem.* **2016**, *18*, 6349-6356.

12. Wu, T.; Bu, X.; Zhang, J.; Feng, P. New Zeolitic Imidazolate Frameworks: From Unprecedented Assembly of Cubic Clusters to Ordered Cooperative Organization of Complementary Ligands. *Chem. Mater.* **2008**, *20*, 7377-7382.

13. Rashidi, F.; Blad, C. R.; Jones, C. W.; Nair, S. Synthesis, characterization, and tunable adsorption and diffusion properties of hybrid ZIF-7-90 frameworks. *AIChE J.* **2016**, *62*, 525-537.

14. Hillman, F.; Zimmerman, J. M.; Paek, S.-M.; Hamid, M. R. A.; Lim, W. T.; Jeong, H.-K. Rapid microwave-assisted synthesis of hybrid zeolitic-imidazolate frameworks with mixed metals and mixed linkers. *J. Mater. Chem. A* **2017**, *5*, 6090-6099.

488991 Micro-, Meso-, and Macro-Scale Defects in Porous Organic Cages

Abstract:

Porous organic cages (POCs) are single molecules with permanent porosity built from organic linkers. They have attracted significant interest in adsorption and separation studies because of their modularity, solubility, and processability. In this study, we analyzed a variety of defect behaviors inside imine-based POCs, including defects that arise from packing frustration, grain boundaries, and molecular defects. POC molecules can pack in the solid state to form crystalline interconnected porous systems. The accessible surface area of these crystals arises from contributions of both the internal and external pores. By controlling the removal rate of the solvents, one type of POC, CC3-R, showed total amorphization and a 2.5-fold increase in surface area, which is attributed to external pore contributions from random packing. The amorphization of cage crystals was also achievable via vertex functionalization—i.e., by introducing mixed diamine linkers into the cages, a variety of amorphous scrambled porous organic cages (ASPOCs) were synthesized, which can be used as molecular porous substrates and membrane fillers. Conventional CC3-R/CC3-S crystals were examined with confocal fluorescence microscopy and the mesoscale grain boundaries were revealed in these as-synthesized crystals. We utilized inter-cage chiral recognition with stronger packing affinity to eliminate the grain boundaries, and achieved increased acid stability of the resulting mixed-chirality cage crystals. Lastly, we explored (both experimentally and computationally) the likelihood of molecular point defects in pristine POCs. To study the effects of the point defects on the material's properties, we applied non-solvent induced crystallization to introduce a non-trivial amount of defects for magnified detection of the resulting properties. We also designed and synthesized a "missing-linker" cage to introduce extra functionality.

498194 Computational Studies of the Enhanced Acidity of Defect MOF 808: The Effect of Activation Process

Abstract:

The hexanuclear zirconium metal-organic frameworks are nowadays considered as promising materials exhibiting different structures, excellent stability, and interesting textural properties useful for adsorption, gas separation and catalysis. Among these materials, the tricarboxylic acid based material named as MOF-808 a has emerged as a promising MOF for catalysis applications like esterification and dehydration. Moreover,Â the pore size of MOF-808Â can even be increased by the creation of missing linker defects, which would enhance catalytic performance.Â The partial substitution of organic ligands by modulator agents like monocarboxylic acids during synthesis allows the engineering of the MOF-808 structure towards better accessibility/availability of zirconium centers. Activation treatment implies the use of higher temperatures than the one used during synthesis, allowing the nodeâ€™s dehydroxylation and the modulator removal, which involves proton mobility into the zirconium node. In this regard, a comprehensive understanding of the activation process is needed to describe the resulting defective structure, which would allow the correlation between the missing linker defects and the MOF catalytic activity.

In this work, we performed state-of-the-art density functional theory (DFT) and molecular dynamics (MD) calculations for studying the activation processes of the MOF-808. By starting from a defect-free structure, we added different modulators (formic acid, acetic acid) and evaluated the effect of its addition with adsorption calculations of probe molecules (carbon monoxide, pyridine). For the acid sites created, before and after adsorption of probe molecules, we performed vibrational frequency calculations and charge distribution calculations to elucidate their acidic nature and quantify their acidic strength. It is demonstrated that the creation of defect centers might help to increase the acidity of the MOF-808. We also found out that the removal of labile coordinated species did not lead to open metal sites as expected; instead, it was the redistribution of charges which induced the acidity. This theoretical rationalization of the activation process can serve as a basis for engineering of defects in MOF-808 materials.

495672 Crystallization Process Development of Metal-Organic Frameworks By Linking Secondary Building Units, Lattice Nucleation and Luminescence: The Insight into Reproducibility

Abstract:

The interdependence of crystallinity, secondary building unit (SBU) formation, linker vacancy, crystal habit, photoluminescence, and surface area for UiO-66 and In-MIL-68, respectively, produced by different processing modes has been fully investigated and linked together to comprehend the relationships among SBU synthesis pathways, structures, and functional properties. The solid-state properties of UiO-66 were found to be process-sensitive, but those of In-MIL-68 except for crystal habits were mode independent, most likely due to the versatile $Zr_6(\mu_3\text{-}O)_4(\mu_3\text{-}OH)_4(CO_2)_{12}$ SBU formation steps with more competing events for UiO-66 and the relatively straightforward $\{In(\mu\text{-}O_2CR)_2(\mu\text{-}OH)\}_\infty$ chain SBU formation steps for In-MIL-68. The reproducibility of the desired properties of metal–organic frameworks (MOFs) relies on the crystallization process designed for specific SBU formation and nucleation trajectory.

490333 Fast Mechano-Chemical Synthesis N-Doped Uio-66 with Dopamine to Enhance Chlorobenzene Competitive Adsorption Under Humid Air

Abstract:

In this work, we initiated a novel mechanochemical strategy to prepare N-doped UiO-66(Zr) using dopamine (UiO-66(Zr)@DA) with competition coordination with Zr^{4+}, and then investigated their adsorption towards chlorobenzene/acetaldehyde under humid air conditions. To further confirm the speculation, the prepared composites UiO-66(Zr)@DA were characterized by N_2 adsorption, X-ray diffraction, scanning electron microscopy and Fourier transform infrared analysis. Meanwhile, temperature programmed desorption (TPD) experiments were conducted to estimate the desorption activation energy of chlorobenzene/acetaldehyde on the UiO-66(Zr)@DA.

In comparison with the UiO-66(Zr) previously reported, the composites particles were more irregular texture, suggesting a lower crystallinity of the M-UiO-66(Zr) [01]. The particle size of it in this work was about 100 nm, which was similar to the crystals reported in the literature [02]. After being composited with DA, pronounced changes can be found in particle size and morphology of the UiO-66(Zr) composites. No any micro-sized aggregation of spheres appeared in all the M-UiO-66(Zr) composites, indicting no occurrence of dopamine self-polymerization during process of composite formation. Their crystal particle size firstly increased and then decreased with increasing DA content (< 7.0 wt%). More interestingly, the composites particle depicts a more regular cubic-like shape in compared to parent UiO-66(Zr). Among which, the M-UiO-66(Zr)@$DA_{3.0}$ possessed the largest crystal particle size (250-300 Î¼m) with high crystallinity (Fig. 1c). Different from other composite MOFs [03, 04], it contributed to UiO-66 crystallization by adding a fraction of dopamine into the UiO-66 framework. This may arise from the deprotonation of the amino group of DA and thus promoted the coordination between the oxygen groups of terephthalic acid and Zr^{4+} mental center in the UiO-66(Zr).

Obviously, the M-UiO-66(Zr)@$DA_{3.0}$ possessed the highest adsorption uptake towards acetaldehyde covering the whole measured pressure range. The uptake value of acetaldehyde on the N-doped UiO-66 was increased by 33 and 140% compared to that on the UiO-66 and UiO-66-NH_2, respectively. The significant increase of adsorption capacity can be mainly attributed to the increase of electron density of Zr-O cluster caused from N coordination, which efficiently enhanced the interaction towards benzene with attract electron group.

498985 High Yield Synthesis of ZIF-8 Nanoparticles Using Stoichiometric Reactants in a Jet-Mixing Reactor

Abstract:

Nanomaterials are increasingly important for many fields such as Zeolitic Imidazolate Frameworks for use in membrane and adsorption applications. A significant challenge for greater utilization of ZIFs is creating a continuous synthesis method to produce these materials with uniform, nanometer-scale particle size, and high surface areas while achieving high yields. Current continuous methods tend to produce low yields, particles larger than 100 nm, and/or low quality materials. We demonstrate a novel method to produce ZIF-8 nanoparticles with uniform diameters in the range of 60-80 nm involving a jet-mixing reactor. While many ZIF nanoparticle syntheses utilize excess ligand, the jet-mixing reactor can be operated at ambient conditions using preferred solvents and stoichiometric quantities of ligand and metal (i.e., ligand:metal of 2:1) to synthesize ZIF-8 with yields as high as 89%. The improved yield can be attributed to the homogeneous conditions created in the reaction volume as a result of the turbulent mixing generated by the high velocity flow through the jets. The jet-mixing reactor is studied to determine the effect of different synthesis parameters including ligand to metal ratio, base concentration, solvent, and mixing intensity. The particle properties including the size, morphology, and yield are affected most strongly by the base concentration. Further, we demonstrate that the jet-mixing method can be scaled to increase productivity by designing, constructing, and testing a larger reactor.

497856 Polymer Semiflexibility Induces Non-Universal Phase Transitions in Block Copolymers

Abstract:

The order-disorder phase transition and the associated phase diagrams of semiflexible diblock copolymers in the presence of concentration fluctuations are investigated using the wormlike chain model. A free-energy functional up to quartic-order in the concentration fluctuations is developed with chain-rigidity-dependent expansion coefficients, which are evaluated using our exact results for the wormlike chain model, and one-loop renormalization is used to account for the fluctuation effects. We find that degree of polymerization N and the monomer aspect ratio *\alpha* directly control the order-disorder transition and the resulting microstructures at different chemical compositions f_A. We find that when monomers are infinitely thin (*i.e.* large aspect ratio), finite degree of polymerization N lowers the critical phase transition *\chi N*. However, fluctuation effects become important when chains have finite radius, and a decrease in the degree of polymerization N elevates the phase transition *\chi N*. Phase diagrams of diblock copolymers at a full range of N and *\alpha* are calculated based on our fluctuation theory. We find that both finite N and *\alpha* enhance the stability of the lamellar phase above the order-disorder transition.

489982 A Strongly Coarse-Grained, Charge-Fluctuating Model for Polyelectrolytes

Abstract:

Coarse-graining techniques are a powerful approach for the modeling of uncharged polymeric materials over many length scales. However, for charged polymers (polyelectrolytes), the ability to coarse-grain over large length scales has been limited by the ability of the coarse-grained model to faithfully describe the underlying charge distribution and fluctuations of the fine-grained model. Here, we develop a theoretical and computational formalism for coarse-graining polyelectrolytes over non-trivial length scales. Our method begins on a discretized grid to properly account for the vital entropic contributions associated with a polyelectrolyte's counterions, and uses carefully chosen approximations to develop an off-lattice version. Effective particle interactions and charge fluctuations in our model are parameterized by simulations of the underlying fine-grained model. We demonstrate the successes and failures of our approach and outline important strategies going forward in refining such a model, as well as potential applications in polyelectrolyte rheology.

495739 Development of New Coarse-Grained Water Models Using Particle Swarm Optimization

Abstract:

Research on understanding the structure and dynamics of water molecules in various processes is a subject of considerable interest in the fields ranging from biology, chemistry, to chemical engineering. Coarse-grained (CG) Molecular Dynamics (MD) simulations, conducted with an accurate model, is a powerful method to study the structure and dynamics of a large system up to micro- to millisecond timescales. In the present study, we have developed polarizable and non-polarizable water models using 2:1 mapping scheme (two water molecules are represented by a single interaction center). The interaction parameters between the CG water beads were developed using a systematic two step approach: 1. Global search was conducted using particle swarm optimization (PSO) method and 2. A local minimum was identified using modified gradient descent algorithm. The new CG models could reproduce the experimentally reported thermodynamic, physical, and chemical properties of water with accuracy. These new models are further utilized to study the phase transitions in stimuli-responsive polymers.

497794 Unified Polymer Erosion Model

Abstract:

Hydrolytically degradable polymers have become standard in many applications and are being enthusiastically investigated for many more. When modeling the erosion of these polymers, they are typically first categorized as either bulk-eroding or surface-eroding, then modeled with assumptions tied to that assumed behavior. In fact, these are simply the extremes in a continuum of erosion behavior dictated by the polymer's hydrolysis rate, moisture permeability, and thickness, along with modifying factors such as autocatalysis. Few polymers actually erode with perfect bulk- or surface- behavior, particularly among newer copolymers, blends, and composites. Moreover, some applications require not only bulk property information, but detailed information regarding the polymer's structure—particularly when used as a barrier for controlled release applications.

This presentation introduces a stochastic, Monte-Carlo style model which calculates the erosion profile of a degradable polymer monolith based on its moisture diffusion coefficient, monolith thickness, and hydrolytic degradation rate constant, generating erosion profiles spanning the continuum from strongly bulk-eroding to strongly surface-eroding without any *a priori* assumptions about the erosion behavior. The location of the polymer on this surface-to-bulk erosion continuum is quantified by the Bulk Erosion Factor, which compares the polymer's mean erosion time to the pure surface- and bulk-eroding ideals. The model generates time-evolving surface roughness parameters and the monolith's average percolation time. Finally, it is applied to controlled release applications in calculating the release profile for an underlying drug layer.

498473 Development of a Fused-Sphere SAFT-γ Mie Force Field for Polymers and Application to Poly(vinyl butyral) Adsorption to Silica

Abstract:

Introduction

SAFT-γ Mie, a group-contribution equation of state (EoS) rooted in Statistical Associating Fluid Theory[1], provides an efficient framework for developing accurate, transferable coarse-grained polymer force fields for molecular dynamics (MD) simulation. Using this EoS, a Mie potential governing intermolecular interactions can be parameterized to reproduce experimental vapor-liquid equilibria (VLE) data of small-molecule analogues over a range of state points. SAFT-γ Mie force fields have been successfully developed for a variety of small molecules modelled as tangentially bound spheres[2,3]. Here, we address two key issues in extending the approach to polymers: 1) the treatment of chain rigidity neglected by the first-order thermodynamic perturbation theory used to derive SAFT-γ Mie, and 2) the disparity between the structure of linear chains of tangent spheres and the structure of the real polymers.

Modelling

We use a fused-sphere version of SAFT-γ Mie as the basis for nonbonded interactions, and derive effective bond-stretching and angle-bending potentials from all-atom oligomer MD simulations using Boltzmann inversion. The introduction of an additional free energy parameter characterizing the degree of overlap between Mie spheres leads to a degeneracy when fitting to monomer vapor pressure and saturated liquid density, which we resolve by matching polymer density from coarse-grained MD simulation with that from all-atom simulation. The result of our hybrid top-down/bottom-up coarse-graining approach is a chain of monomers rigorously parameterized to experimental VLE data and with structural detail consistent with all-atom simulations.

We apply our coarse-graining methodology to poly(ethylene) and poly(vinyl alcohol), and test its limits on a structurally complex copolymer system, poly[(vinyl butyral)-co-(vinyl alcohol)] (PVB), at an interface with amorphous silica. PVB is of interest industrially as the principal component of the interlayer in laminated safety glass, and to the best of our knowledge this is the first molecular model for it. Interaction forces between the monomers and silica slab are mapped from the Optimized Potentials for Liquid Simulations (OPLS) all-atom force field[4,5] to an external potential acting on the coarse-grained beads. The tacticity of the underlying polymer chain structure is preserved by assigning stereo-specific angle potentials to triads of coarse-grained beads, and co-monomer sequence-dependent behavior can be studied, features not achievable with SAFT-γ Mie alone. Coarse-grained bulk polymer melt simulations successfully reproduced experimental density, glass transition temperature, and local structural distributions mapped from the all-atom level. For the PVB-silica interfacial system, we investigate the effects of overall chemical composition and co-monomer sequence distribution on adhesion strength and interfacial structure.

Conclusions

We present a new approach for developing coarse-grained force fields for polymer melts using the fused-sphere SAFT-γ Mie EoS. Nonbonded interactions parameterized to monomer experimental VLE data and bonded potentials derived from all-atom chains result in a model that is capable of accurately reproducing both thermodynamic and structural properties of polymer melts. We demonstrate the utility of our approach for studying complex copolymer systems in which co-monomer sequence and stereochemistry play important roles.

References

1. Dufal, S. *et al. J. Chem. Eng. Data* **59,** 3272–3288 (2014).

2. Lafitte, T. *et al. Mol. Phys.* **110,** 1189–1203 (2012).

3. Avendaño, C. *et al. J. Phys. Chem. B* **117,** 2717–2733 (2013).

4. Jorgensen, W. L., Maxwell, D. S. & Tirado-Rives, J. *J. Am. Chem. Soc.* **118,** 11225–11236 (1996).

5. Black, J. E., Iacovella, C. R., Cummings, P. T. & McCabe, C. *Langmuir* **31,** 3086–3093 (2015).

Effects of Coarse-Graining on Simulations of Mechanical Properties of Polymers

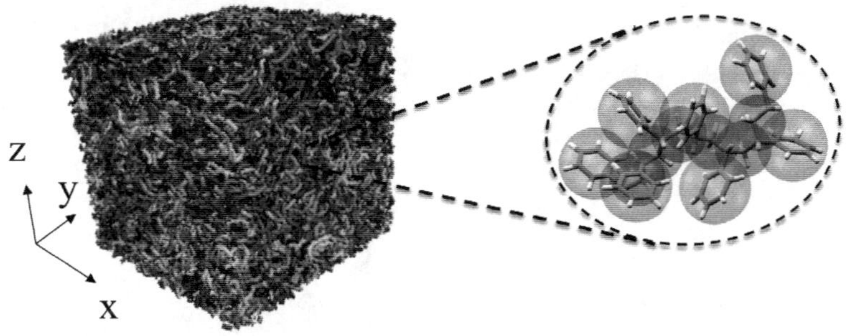

Ting Ge
University of North Carolina at Chapel Hill
Mark O. Robbins
Johns Hopkins University

2017 AIChE Annual Meeting
Oct. 29 – Nov. 3, 2017, Minneapolis, MN

1

General Features of Polymer Stress-Strain Curves

- Elastic response at small strains (few %)
- Peak at yield stress τ^y
  overcome barriers to local segmental rearrangements
- Strain softening as material "rejuvenated" by shear
- Hardening as chains pulled taut between entanglements
- Fracture

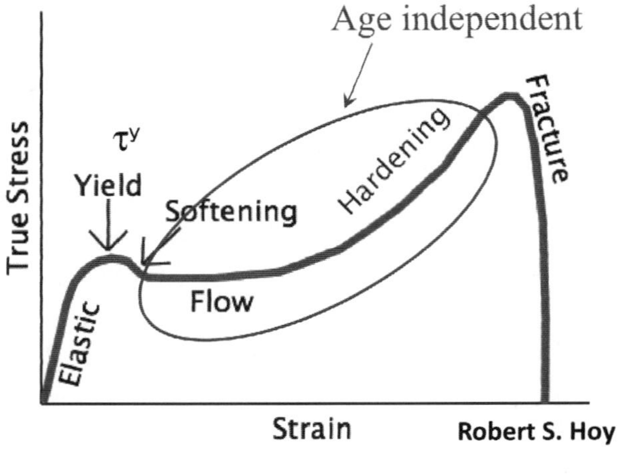

How do we develop coarse-grained models for non-equilibrium flow and hardening?

2

Coarse-Grained Models for Chemically Specific Polymers

Existing approaches match equilibrium properties

- Configurations – unique mapping between pair correlation functions
- Forces – average forces between coarse-grained configurations over fine-grained degrees of freedom

Compare response of AA, UA, and CG models of polystyrene

All Atom (AA) United Atom (UA) Coarse-Grained (CG)

Polystyrene (PS)

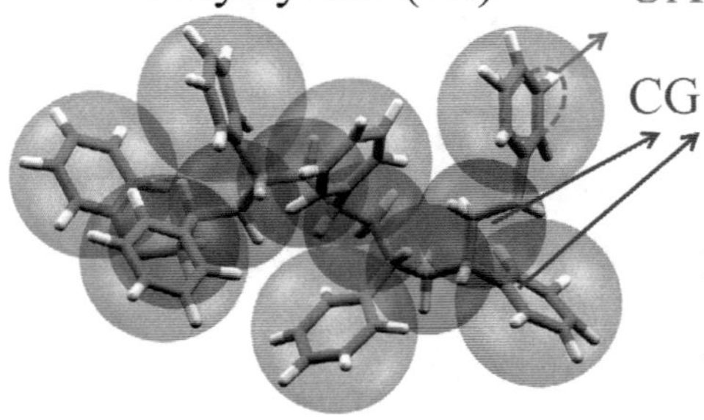

UA replaces CH_n by single sphere

CG replaces clusters of atoms by beads consistent with AA pair correlation functions in equilibrium

Fritz, Harmandaris, Kremer, and van der Vegt *Macromolecules* **42**, 7579 (2009)
Agrawal, Aryal, Perahia, Ge, and Grest *Macromolecules* **47,** 3210 (2014)

3

Simulation Method

- Generate CG PS samples (M_w=50,000g/mol, M_e=17,000g/mol) using EMC builder Enhanced Monte Carlo (Pieter J. in 't Veld, BASF)
- Fine grain to UA or AA
- Equilibrate PS melts at T=503K using LAMMPS MD code Large-scale Atomic/Molecular Massively Parallel Simulator (S. J. Plimpton, Sandia NL)
- Quench to $T<T_g\approx 373K$ for mechanical tests

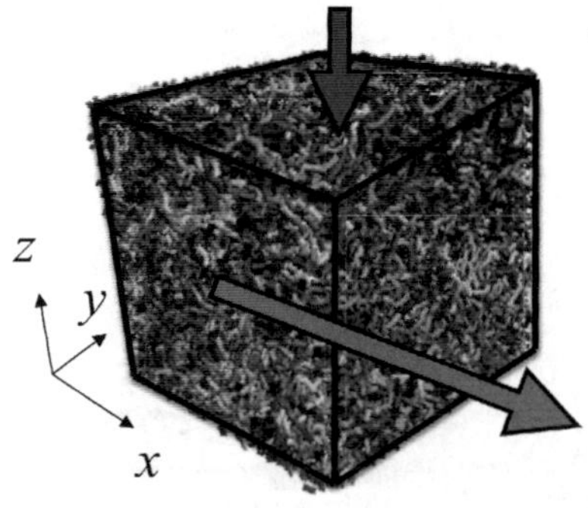

Uniaxial compression along z at constant strain rate $\dot{\varepsilon}_z$

Simple shear along x at constant rate $\dot{\gamma}$

4

Stress After Coarse-Graining

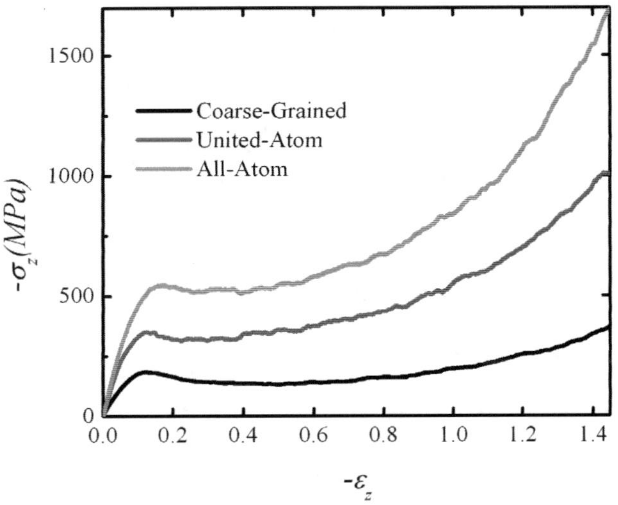

Uniaxial compression along z
T=1K

Coarse-graining lowers stress
T=1K removes thermal activation

5

Rescaled Stress

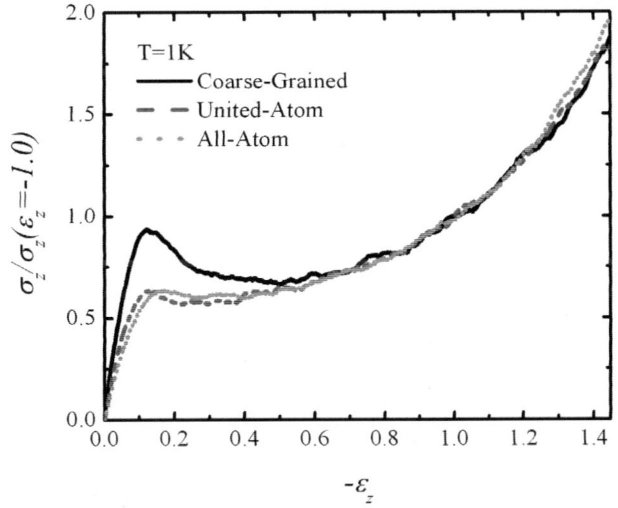

Uniaxial compression along z
T=1K

Coarse-graining rescales energy barriers
Initial yield stress age dependent
Flow stress and strain hardening for different degrees of CG collapse with a single factor until large strain

6

Rescaled Energetic Stress

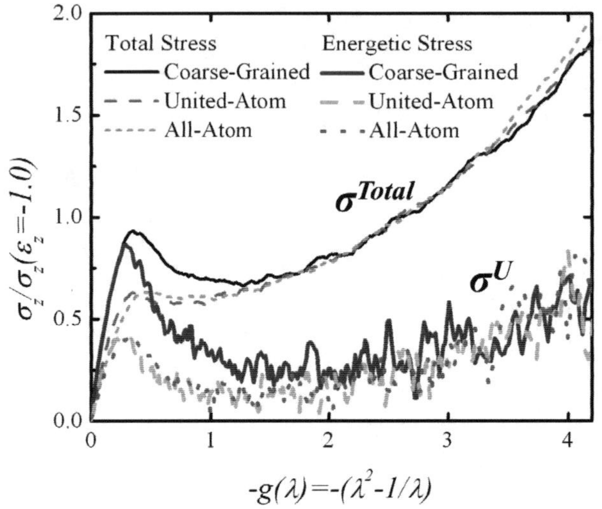

Uniaxial compression along *z*
T=1K

- Total stress $\boldsymbol{\sigma^{Total}}$ = dW/dε = energetic stress $\boldsymbol{\sigma^{U}}$ + dissipative stress $\boldsymbol{\sigma^{d}}$
  $\boldsymbol{\sigma^{U}}$ = dU/dε where U is internal energy density
- **$\boldsymbol{\sigma^{U}}$ scales with same factor as $\boldsymbol{\sigma^{Total}}$ – rescaled energy landscape**
  except: initial yield and late stage where bonds stretch

7

Rescaled Dissipative Stress

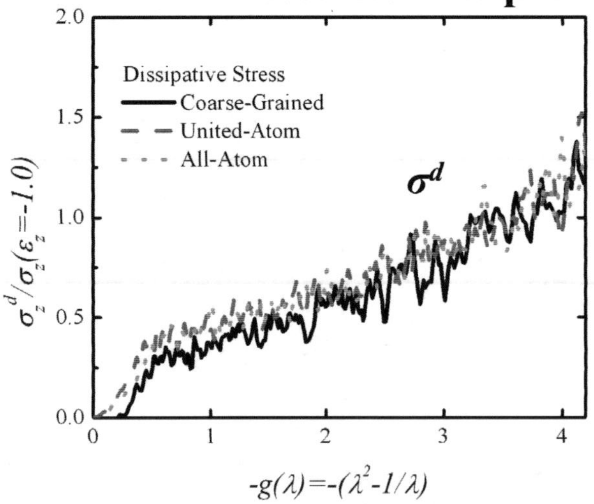

Uniaxial compression along *z*
T=1K

Dissipative stress $\boldsymbol{\sigma^{d} = \sigma^{Total} - \sigma^{U}}$
$\boldsymbol{\sigma^{d}}$ ~ rate of local LJ bond breaking × energy barrier
$\boldsymbol{\sigma^{d}}$ scales with same factor as $\boldsymbol{\sigma^{Total}}$
– the same rate of LJ breaking (plasticity) in all systems just rescale energy barrier

8

Simple Shear

Shear Stress

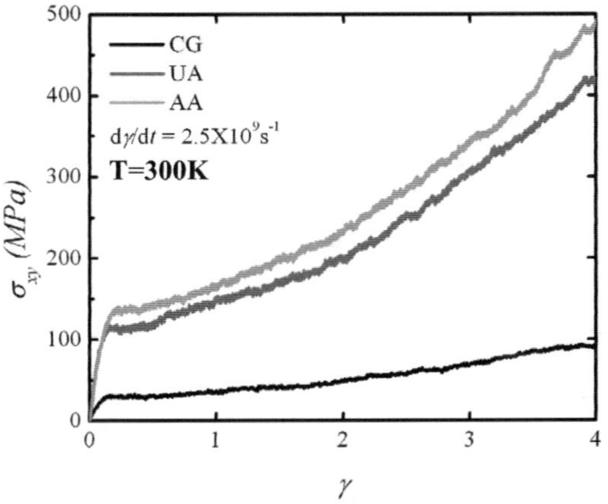

Rescaled Shear Stress

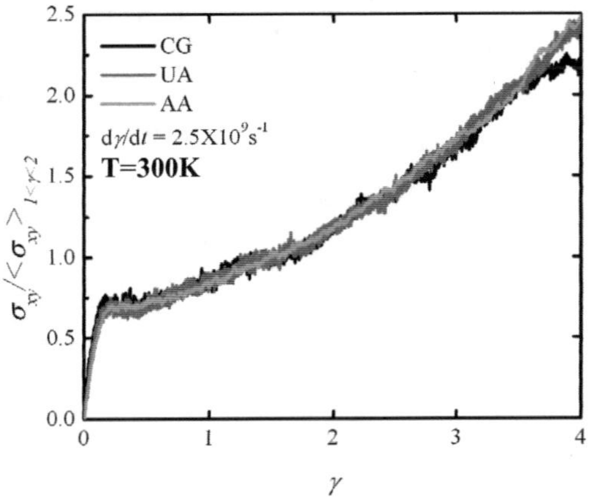

- Coarse-graining lowers shear stress
- Coarse-graining rescales energy barriers
  - Shear stress curves for different degrees of CG collapse with a single factor until large strain

9

Accelerating Simulations by Fine Graining after Strain

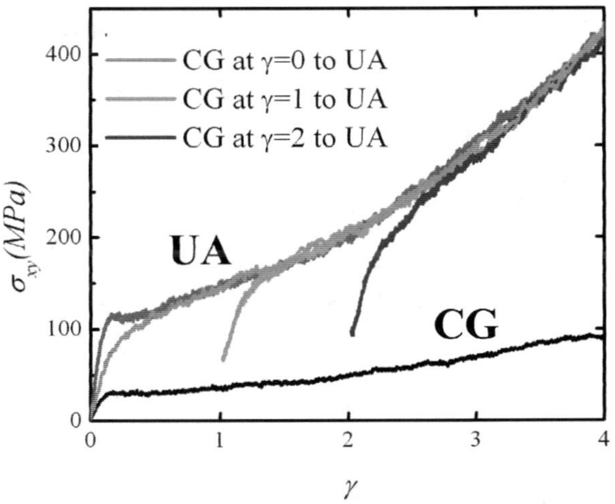

Simple shear
along x
T=300K

Shear to $\gamma = 1$ or 2 using CG model
Fine grain to UA description and start to shear
Agree with UA result eventually
– ***CG model captures large scale changes during deformation***

10

Accelerating Simulations by Fine Graining after Strain

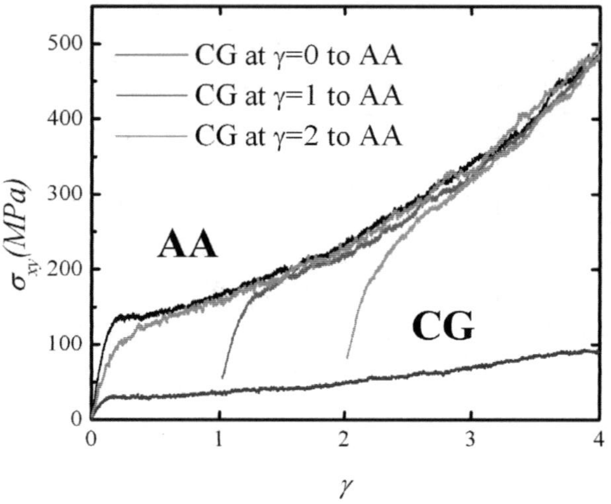

Simple shear along x T=300K

Can also fine grain UA or CG states to AA description

11

Accelerating Simulations by Fine Graining after Strain

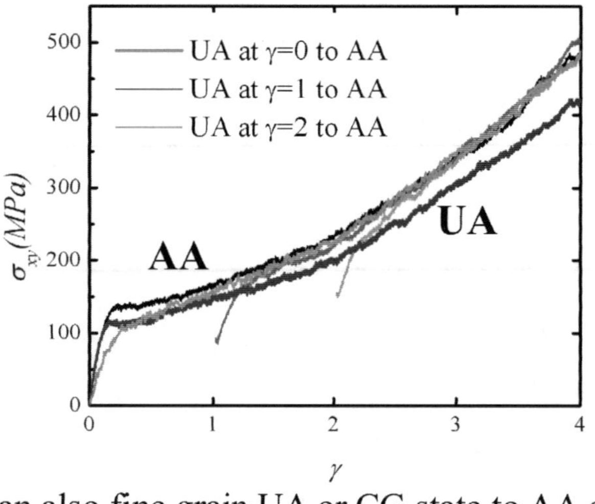

Simple shear along x T=300K

Can also fine grain UA or CG state to AA description
AA result is recovered eventually
– ***All models have same large scale deformations***
Fine scale correlations need strain interval to develop
– ***hysteretic system – no unique stress at given strain***

12

Conclusions

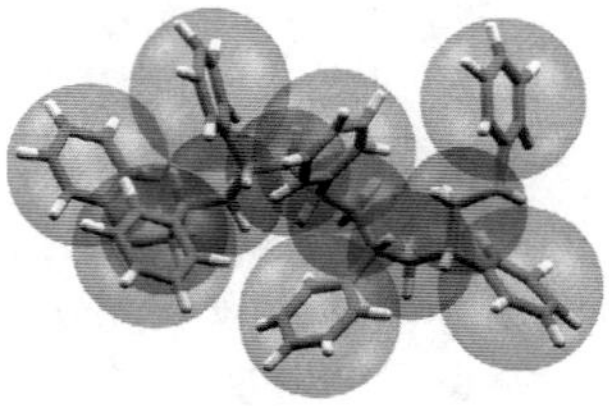

Coarse-graining approaches for equilibrium

- lower shear stress in glassy state
  - – rescale energy landscape, affect energetic stress
  - – remove degrees of freedom that can absorb heat during mechanical deformation, affect dissipative stress

Stresses for different models related by simple scaling factors

- separate energetic and dissipative terms (local plastic events)

CG simulations capture large scale changes in configurations

Coarse-graining provides acceleration method

– generate strained CG states, then do fine-grained simulations

13

501258 New Computational Methods for Rapid Simulation of Hydrodynamic Interactions in Polymer Solutions

Abstract:

Coarse-grained simulations of soft materials and complex fluids such as polymer solutions require rapid calculation of the hydrodynamic forces exerted on suspended, microscopic constituents (particles). Because of the small length scales associated with polymers, low Reynolds number flows dictate these forces. Inertia of the fluid and of the constituents is negligible on time scales that are relevant for rheological measurements. Many coarse-grained simulation methods for modeling soft materials construct an explicit model of the solvent, determining both the position and momentum of the solvent and suspended particles as a function of time. These include Lattice-Boltzmann, Dissipative Particle Dynamics, Stochastic Rotation Dynamics, Multi-particle Collision Dynamics, and others. Always, degrees of freedom associated with the solvent are neglected, and a stochastic force satisfying the fluctuation dissipation theorem must be exerted on all components of the dispersion.

The explicit solvent models were developed in part because of difficulties performing rapid calculations with implicit solvent models, such as Brownian Dynamics, in which the hydrodynamic interactions between particles are modeled deterministically and depend only on the relative configuration of particles. The chief advantage of explicit solvent models comes from sampling of the stochastic forces, which is no more expensive computationally than generating a string of random numbers. For implicit solvent models, the stochastic forces must be drawn from a normal distribution whose covariance is a complicated function of the particle configuration. For a system of interacting N particles, drawing a single sample requires O(N^3) operations, if numerically exact techniques from linear algebra are employed. So-called â€œfastâ€ methods can approximate the sampling with roughly O(N^mÂ log N) computational complexity, where m is a coefficient greater than one which depends on the configuration of the particles. Typical values of m range from 1.25 to 1.5. This puts a serious limit on application of implicit solvent methods to large-scale simulation.

In the presented work, we will demonstrate an optimal and spectrally accurate, fast method for calculation of thermal forces in implicit solvent simulations of soft materials such as Brownian Dynamics. The computational complexity of this approach is O(N (log N)^(d/(d+3))), where d is the fractal dimension of the polymer microstructure being modeled. Remarkably, this new approach adapts to the structure of the material under study by properly balancing the computational effort spent evaluating the hydrodynamics on to different spatial scales. The separation between local and global scales is set by computational necessity and enables optimization of the algorithms performance. We term this algorithm optimal because in the periodic geometries needed to model viscometric flows of polymer solutions, log-linear computational scaling is the best available. Up to 10^6 discrete elements can be simulated with this methodology on a single consumer grade graphics processing unit. Applications of this approach to macromolecular solutions and systematic coarse-graining of suspended polymers are demonstrated.

501320 Coarse-Grained Molecular Dynamics Simulations of PNIPAM Grafted Graphene Systems in an Aqueous Environment

Abstract:

Individually, graphene nanostructures and the thermo-sensitive polymer Poly(N-isopropylacrylamide) (PNIPAM) have shown the potential to aid in drug delivery applications. A viable drug delivery vehicle will likely take the form of a macromolecule consisting of a graphene architecture functionalized with PNIPAM chains. Designing a delivery vehicle will require a detailed understanding of how such a structure behaves in an aqueous environment. All-atom Molecular Dynamics (MD) simulations have been conducted on systems comprised of a PNIPAM chain 30 monomer units in length and a graphene sheet measuring up to 40 Å by 40 Å in the presence of water. These initial simulations have already required moderate computational time at the traditional all-atom MD time scale. Coarse-graining of graphene, PNIPAM, and water is required if a macromolecule-water system is to be studied on a time scale representative of a biological system. In this work, we have employed Particle Swarm Optimization (PSO) and traditional coarse-graining techniques to develop a model that was utilized to examine graphene and PNIPAM architectures in an aqueous environment.

493731 Gel Assemblies of Colloidal Nanocrystals

Abstract:

Colloidal nanocrystals have distinctive optical properties that are broadly tunable by varying size, shape, and composition. When nanocrystals are assembled into ordered superlattices or disordered gel networks, the resulting materials exhibit properties inherited from their nanocrystal building blocks that also depend on their structural organization at the mesoscale. I will describe our efforts to make reconfigurable, dynamic assemblies of nanocrystals by making controlled, reversible bonds between them. Low volume fraction gels are assembled from indium tin oxide nanocrystals, which exhibit tunable plasmonic absorption of infrared light. Gelation is induced by combining the nanocrystals with low molecular weight polyethylene oxide (PEO). At moderate PEO concentrations, the polymer can act as a cross-linking agent, connecting adjacent nanocrystals in the gel. At high PEO concentrations, depletion forces can induce physical bonds between nanocrystals. Gels can be caused to flow again by changing the relative concentration of nanocrystals and PEO, suggesting reversible bonding between nanocrystals. Additional strategies for reversible bonding between assembling nanocrystals are under development to introduce chemical specificity and chemo-responsivity. These materials offer an outlook toward structurally and optically responsive mesoscale assemblies.

493212 Examining the Optical Effects of Chiral Carboxylic Acids Bound to the Surface of CdSe Nanoparticles

Abstract:

Chiral quantum dots (QDs) have recently emerged as an attractive platform for chiral fluorescent sensors, as well as tunable chiral emitters for nanophotonic applications. However, a fundamental understanding of the chiroptical response of these materials is still in its infancy. For example, a number of mechanisms have been proposed to explain the observed chirality in these nanoparticles, likely originating from the surface capping ligands. One such mechanism has suggested that electronic coupling between the chiral surface ligands and achiral QDs may play an important role. In order to probe this mechanism, it is necessary to compare the chiroptical response of QDs bound by chiral ligands with a range of different electronic energies. However, the vast majority of ligands utilized for the synthesis of chiral QDs have been derived from the amino acid, cysteine, where binding likely occurs through the thiolate sulfur atom. In contrast, there has been much less work carried out on chiral ligands where primary binding occurs through other atoms, which would yield different electronic energies and should thus provide noticeable differences in the chiroptical properties of the resulting QDs.

To investigate this, we have synthesized and characterized a new family of chiral CdSe QDs based on monocarboxylic and dicarboxylic acids, including lactic acid, malic acid and tartaric acid, where binding to the CdSe surface occurs through the carboxylate oxygen atom. We have examined the chiroptical response of these materials through circular dichroism (CD) experiments and compared the CD signals to those associated with CdSe QDs bound by cysteine ligands. The results suggest that the intensity and broadness of the CD peaks may be tuned by slight changes to the structure of the chiral carboxylic acid chosen. In addition, the commercial availability of a wide number of these chiral carboxylic acids has allowed for the first time a detailed and systematic study in which the number of stereocenters on the bound chiral ligand can be varied and the resulting CdSe QDs examined by CD spectroscopy. Our initial results suggest an increase in the intensity of CD signals for CdSe QDs bound by chiral ligands containing two stereocenters versus one stereocenter. We have also carried out a detailed study of the binding of these ligands to the surface of the CdSe QDs, suggesting the presence of an extended hydrogen-bonded network of ligands surrounding and stabilizing the nanoparticles in polar solutions. This model is supported by ^{1}H NMR and FTIR spectroscopies, dynamic light scattering and transmission electron microscopy experiments.

Our results establish a new family of chiral CdSe QDs based on chiral carboxylic acids, as well as a novel method of stabilizing these QDs in polar solutions. These results will not only provide an increased understanding of ligand binding to CdSe surfaces by small monocarboxylic and dicarboxylic acids but will also be critical for the development of both chiral fluorescent sensors and chiral emitters for nanophotonic applications such as security and anti-counterfeiting technologies.

497388 Radial Elemental Distribution Analysis of Spherical Core/Shell Nanocrystals with STEM/EDX

Abstract:

Radial Elemental Distribution Analysis of Spherical Core/Shell Nanocrystals with STEM/EDX

Jacob T. Held[1], Katharine I. Hunter[2], Uwe R. Kortshagen[2], K. Andre Mkhoyan[1]

[1.] *Department of Chemical Engineering & Materials Science, University of Minnesota, Minneapolis, MN 55455.*

[2.] *Department of Mechanical Engineering, University of Minnesota, Minneapolis, MN 55455.*

Semiconductor quantum dots exhibit many useful size-dependent optoelectronic properties.[1] Shells can be used to protect such nanocrystals from oxidation and surface trap states, and by managing strain and band alignment between the core and shell, they can be used to control optoelectronic and catalytic properties.[2,3] However, many of these properties are sensitive to the interface between the core and shell.[4] Despite its importance, characterizing the radial composition of core/shell nanocrystals remains a significant challenge.[5] In this study, we demonstrate a method for quantifying the radial distributions of elements in spherical core/shell nanocrystals by fitting simulated distributions to radially-averaged scanning transmission electron microscopy/energy dispersive X-ray (STEM/EDX) spectrum images of the nanocrystals.

Plasma-grown spherical Ge/Si core/shell nanocrystals with well-controlled core and shell dimensions[6] were used as an ideal test case for demonstration of this analysis. These crystals were directly deposited onto holey/thin double carbon grids and transferred under Ar into an FEI-Titan G2 60-300 equipped with a Super-X EDX system. The microscope was operated at at 60 kV and 125 pA beam current with a 25 mrad convergence angle. STEM/EDX spectrum images were acquired with frame-by-frame drift correction with a dwell time of 3 μs/pixel and a pixel resolution of 0.03 nm/pixel. The resulting spectrum images (Figure 1) were radially-averaged around the centroid of a fit ellipse (aspect ratio <1.05) to produce 1D datasets for further analysis.

An error function-based model of the radial composition of spherical core/shell nanocrystals was developed, accounting for interface broadening, surface roughness, and residual core material in the shell. The modelled spherical radial distributions were numerically projected down the cylindrical Z axis and convoluted with a Gaussian probe function to account for STEM electron beam spread, producing a fit to the experimental EDX data. The simulation parameters (core and shell radii, interface broadening, surface roughness, and shell composition) were then optimized by minimizing the sum-squared error of the fit, defining the uncertainty as the maximum change in each parameter necessary to increase the fitting error by 5%. Figure 2 shows the results of this analysis for the crystal in Figure 1.

Because only a single parameter is used to define the alloying and roughness at each interface, it is impossible to decouple these features through the analysis demonstrated here. However, the interface broadening values, expressed as standard deviations, are upper bounds on the total broadening of each interface (including non-sphericity, roughness, and alloying). For example, the analysis in Figure 2 quantitatively shows that the alloyed region between the Ge core and Si shell shown here is at most only 1-2 unit cells ($s_{Ge/Si}$ ˜ 0.5 nm). This technique can be readily applied to other spherical core/shell systems with a wide variety of chemistries, and could be expanded for use in other well-defined geometries.

1. Mangolini, L. Synthesis, properties, and applications of silicon nanocrystals. *J. Vac. Sci. Technol. B Microelectron. Nanom. Struct.* **31,** 20801 (2013).

2. Reiss, P., Protière, M. & Li, L. Core/shell semiconductor nanocrystals. *Small* **5,** 154–168 (2009).

3. Strasser, P. *et al.* Lattice-strain control of the activity in dealloyed core–shell fuel cell catalysts. *Nat. Chem.* **2,** 454–460 (2010).

4. Bae, W. K. *et al.* Controlled alloying of the core-shell interface in CdSe/CdS quantum dots for suppression of auger recombination. *ACS Nano* **7,** 3411–3419 (2013).

5. Tschirner, N. *et al.* Interfacial alloying in CdSe/CdS heteronanocrystals: A Raman spectroscopy analysis. *Chem. Mater.* **24,** 311–318 (2012).

6. Hunter, K. I., Held, J. T., Mkhoyan, K. A. & Kortshagen, U. R. Nonthermal Plasma Synthesis of Core/Shell Quantum Dots: Strained Ge/Si Nanocrystals. *ACS Appl. Mater. Interfaces* **9,** 8263–8270 (2017).

Figure 1. Spectrum image data of a single Ge/Si core/shell nanocrystal. (a) High-angle annular dark field (HAADF) image of the crystal obtained after spectrum image acquisition. (b) Composite Ge (green) and Si (red) EDX map. (c,d) EDX maps of the Ge core and Si shell.

Figure 2. Analysis of the radial distribution of elements in a Ge/Si core/shell nanocrystal. (a) Radially-averaged EDX data from the crystal shown in Figure 1 and the corresponding fit profiles. (b) Spherical radial concentrations of Ge and Si from the best-fit model. (c) Optimized values for each fitting parameter.

498041 Directional Carrier Transfer in Strongly Coupled Binary Nanocrystal Superlattice Films Formed By Assembly and *in situ* Ligand Exchange at a Liquid-Air Interface

Abstract:

Two species of monodisperse nanocrystals (NCs) can self-assemble into a variety of complex 2D and 3D periodic structures, or binary NC superlattice (BNSL) films, based on the relative number and size of the NCs. BNSL films offer great promise for both fundamental scientific studies and optoelectronic applications; however, the utility of as-assembled structures has been limited by the insulating ligands that originate from the synthesis of NCs. Here we report the application of an *in situ* ligand exchange strategy at a liquid-air interface to replace the long synthesis ligands with short ligands while preserving the long-range order of BNSL films. This approach is demonstrated for BNSL structures consisting of PbSe NCs of different size combinations and ligands of interest for photovoltaic devices, infrared detectors, and light-emitting diodes. To confirm enhanced coupling introduced by ligand exchange, we show ultrafast (~1 ps) directional carrier transfer across the Type-I heterojunction formed by NCs of different sizes within ligand-exchanged BNSL films. This approach shows the potential promise of functional BNSL films, where the local and long-range energy landscape and electronic coupling can be adjusted by tuning NC composition, size, and inter-particle spacing.

Reference: Y. Wu, S. Li, N. Gogotsi, T. Zhao, C.R. Kagan, C.B. Murray and J.B. Baxter, "Directional Carrier Transfer in Strongly Coupled Binary Nanocrystal Superlattice Films Formed by Assembly and in Situ Ligand Exchange at a Liquid-Air Interface," *J. Phys. Chem. C,* **121**, 4146 (2017).

496529 The Effect of CdS Shell Thickness on the Complex Index of Refraction of CdSe/CdS Core/Shell Nanocrystal Films

Abstract:

The Effect of CdS Shell Thickness on the Complex Index of Refraction of CdSe/CdS Core/Shell Nanocrystal Films

Dana B. Dement, Mayank Puri, Vivian E. Ferry
University of Minnesota, Minneapolis MN USA

Understanding how structural factors influence the complex refractive index of quantum dot (QD) solids is crucial to tailoring the light-matter interactions of QD-containing photonic and optoelectronic devices. QD films are challenging to accurately model as they are a mixture of the core/shell materials and the surrounding organic ligands. Furthermore, QD packing density varies based on particle size, ligand length, and the deposition process. Currently, effective medium approximations are often utilized to approximate the refractive index of QD solids. However, these models are inadequate to describe QD solids as they generally do not account for quantum confinement effects and are often optimized for dilute solutions rather than neat films. Here, we have undertaken a systematic study of the effect of shell thickness and ligand type on the refractive index of neat CdSe/CdS core/shell QD films by using variable-angle spectroscopic ellipsometry to derive the complex refractive index directly from fabricated films. This information allows us to more accurately predict and understand the behavior of QD based devices.

The use of CdSe/CdS heterostructures provides many advantages over bare CdSe QDs, including increased quantum yield and stability, as well tunable Stokes shifts. We have successfully synthesized zinc-blende CdSe QDs on a multigram scale through a reproducible, non-hot-injection synthesis, yielding monodisperse particles with an average diameter of 3.5 nm and standard deviation of 10%. The large scale of this synthesis allows us to maintain the same CdSe core particles while systematically varying the thickness of the CdS shell, leading to a family of CdSe/CdS QDs of varying diameters dependent only on the monolayers of CdS shell grown. We have realized between 3 and 12 monolayers of CdS growth, producing a range of CdSe/CdS particle sizes from 3.5 nm to 11 nm and emission peaks ranging from 585 nm to 645 nm.

Spin-coating a concentrated solution of the CdSe/CdS QDs onto an Al_2O_3 surface forms thin films with thicknesses ranging from 20 nm â€“ 60 nm, depending on the coating conditions. Variable angle spectroscopic ellipsometry measurements on films containing CdSe QD cores and CdSe/CdS QDs with 3-9 monolayers of CdS shell have been carried out. For the case of CdSe surrounded by 5 monolayers of CdS, a refractive index, n, and extinction coefficient, k, of 1.925 and 0.022, respectively, were found at 600 nm with appropriate dispersion throughout the full spectrum, characterized from 400 to 1000 nm. The ellipsometry models include an empirical Sellmeier fit at energies below the band gap, and multiple Gaussian oscillators to model the QD absorption features throughout the visible. To model the strong UV absorption, the tail of a Tauc-Lorentz oscillator is used, which accounts for the bandgap of the QDs. These

models show evidence of quantum confinement, having features where the first excitonic peak of the CdSe/CdS QDs would be expected based on absorption data taken by UV-Vis spectroscopy. We also have the ability to correlate ellipsometry measurements to 2D photoluminescent emission and lifetime maps of the QD films by using a piezo-controlled stage with nanometer control of sample position. This gives further insight into packing effects and other structural difference within the QD solid.

A systematic trend between the CdS shell thickness, ligand length, and the refractive index of the film will be presented to inform the development of computational models incorporating CdSe/CdS QDs, and to enable the design of tailored optoelectronic devices that use nanophotonic elements.

487497 Engineering the Light-Matter Interactions of Ultrasmall CdSe Quantum Dots Via Modification of Surface Species for Light Harvesting Applications

Abstract:

The highly tunable nature of the electronic band structure of quantum dots (QDs) is promising for many reasons, among them engineered light harvesting materials for solar energy capture. For a given composition, alteration of the size of the QD alone will change the light absorbed and emitted by the QD. Upon entering the ultrasmall regime, sub 2 nm radii, CdSe QDs have been shown to emit white light, i.e. all wavelengths of visible light, upon exposure to UV light. Two main theories have arisen to explain this phenomenon: 1) a variety of surface traps yielding various energies of emitted wavelengths, or 2) a constantly fluctuating band gap due to a high degree of disorder on the surface. This second theory is known as fluxionality. In this research, the applicability of the two theories is compared based upon alteration of the ultrasmall CdSe band gap via modification of the surface. At this size, approximately 90% of the atoms in the QD are on the surface, thereby allowing for significant control over the QD behavior via design of different surface conditions. The conditions that we have implemented include placing the QD in mediums with a range of dielectric constants, addition of capping ligands, and addition of shell materials. Techniques used to probe the nature of the light-matter interactions of these systems include time correlated single photon counting (TCSPC), ultraviolet-visible (UV-vis) and photoluminescence (PL) spectroscopy, transmission electron microscopy (TEM) and x-ray photoelectron spectroscopy (XPS) for crystal structure information, and linear sweep voltammetry for device performance information of these QDs in preliminary solar cells. Continuing work will include simulations of the QDs in the various environments to garner better understanding of the alterations of the electronic band gap based on the conditions imposed.

496782 Computational Prediction and Evolutionary Design of Polymer Glass-Formation Behavior

Abstract:

In most polymers, the glass transition is one of the most important phenomena determining performance properties including mechanical response, processability, and transport behavior. For this reason, predicting and achieving rational control of the glass transition is a longstanding goal of polymer science. However, the vast range of timescales associated with glass formation, coupled with a lack of an agreed-upon theoretical description of the problem, have posed major challenges to achieving this goal. Here I describe a new approach to this problem, combining efficient molecular dynamics simulations, physics-based heuristics, machine learning, and evolutionary algorithms to predict and design the polymer glass transition.

The authors acknowledge the W. M. Keck Foundation for generous financial support of this research.

494230 Domain Spacing and Phase Behavior of Salt-Doped Block Copolymers from Fluids Density Functional Theory

Abstract:

Salt-doped block copolymers have potential as mechanically robust, nonflammable electrolytes in batteries, among other applications. Typically, one monomer type has a low dielectric constant but provides mechanical strength to the material, while the other phase has a higher dielectric constant and can transport lithium ions. The salt strongly prefers to exist in the higher dielectric constant phase, and adding salt is known to significantly alter the order-to-disorder transition temperature, interfacial width, and, in some cases, the preferred microphase morphology. All of these impact both ion conduction and mechanical properties, and better understanding of these trends would increase the ability for rational design within this class of materials. We model these materials at the coarse-grained level to more clearly delineate how basic molecular characteristics such as ion contact strength and ion solvation (or dielectric constant differences) drive their phase behavior and morphology trends.

In particular, we aim to capture monomer and ion local packing and correlations as well as overall polymer chain behavior in a coherent way by using fluids density functional theory (fDFT) with correlation functions calculated from liquid state theory. Within liquid state theory, we consider a restricted primitive model (a mixture of ions and solvent of identical size) that represents a reference fluid for the ion-containing microphase of the overall system. The ions have Coulomb interactions with each other as well as shorter ranged solvation interactions that are meant to account for the amount of preferential solvation of ions in the higher dielectric microphase (the strength of these interactions can be varied). With the Ornstein-Zernike equation and the hypernetted-chain closure, we find direct correlation functions that include features of local ion and solvated monomer packing that would not be captured at the mean field level. The fDFT uses these direct correlation functions and also accounts for bonding and interactions with the other type of polymer to predict the overall equilibrium structure. We capture the initial sharpening of the interface as a function of salt concentration as well as the slight broadening of the interface at high salt concentrations seen in some experiments. These trends depend significantly on the choice of solvation strength and ion contact energy.

493182 Atomistic Simulations of Lamellae-FormingÂ PS-b-P2VP

Abstract:

Development of advanced functional polymeric materials is making use of multicomponent systems that phase separate into multiple domains, where interfaces play a critical role on the functionality of the devices created by such materials. A precise understanding of the relationship between molecular details and interfacial properties is a fundamental step towards rational material design. Computational studies of phase-separated polymer systems usually consist of field-theory or other coarse-grained approaches, where atomistic details are averaged out and lost. However, as modern devices and applications rely on phenomena where microscopic details can tip the balance toward faster, greener, and cheaper devices (photovoltaics, solid-state batteries, etc.), atomic-level resolution becomes important. With this in mind, we employ a multi-scale strategy to study the interfaces in a block copolymer lamellae morphology.

A backmapping approach was employed to generate an atomistic lamellae configuration from a coarse-grained lamellae morphology. The morphology was generated using the theoretically-informed coarse-grained (TICG) model and atomistic detail corresponding to poly(styrene-*block*-2-vinylpyridine) (PS-*b*-P2VP) was mapped onto it. Molecular dynamics (MD) simulations were performed on the atomistic PS-*b*-P2VP lamellae system. We focus on the fluctuations of the interface as well as the differences in mobility between the interface and bulk regions of the morphology. Dynamics for each block are compared to the dynamics of each respective homopolymer and to the disordered PS-*b*-P2VP system. Finally, we briefly discuss ion transport on a functionalized PS-*b*-P2VP lamellae system.

489794 Rapid Conformational Fluctuations in a Model of Methylcellulose

Abstract:

Methylcellulose is a biopolymer derived from cellulose that is widely used in the food, construction, cosmetic and pharmaceutical industries. An important feature of methylcellulose is that it can form a hydrogel upon heating. The gel morphology has recently been studied by Lott et al. [*Macromolecules* **2013**, *46*, 9760], who demonstrated that methylcellulose forms a fibrillar structure with a uniform diameter of around 14 nm. The detailed gelation mechanism, however, remains unknown. We have investigated the formation of fibrils using the coarse-grained simulation model proposed by Huang et al. [*Macromolecules* **2016**, *49*, 1490], wherein an isolated methylcellulose chain collapses into a ring-like structure with an outer diameter consistent with the experimental result. Although this ring-like structure appears to be a precursor to a high-aspect ratio fibril, the model also leads to many "misfolded" shapes, such as bundles and hairpins. Using Langevin dynamics simulation with the LAMMPS package, we found that "flipping events", characterized by a rapid conformational fluctuation, play an important role in rearranging the shape of the collapsed states, which are essential to fibril formation. We further found that the frequency of flipping events depends strongly on the dihedral potential. We will discuss the implication of these rapid conformational fluctuations on fibril formation.

496479 Multi-Phase Coarse-Grained Models of Rod-like Polymers from Iterative Boltzmann Inversion

Abstract:

Iterative Boltzmann Inversion (IBI) is a widely used computational method for developing coarse-grained models. However, the main criticism of this technique is the poor transferability across concentrations, temperatures, and phases. The poor transferability is due to overfitting of the non-bonded potential to the radial distribution function. Previous work has shown that a transferable potential for short alkanes can be derived by using multiple points in phase space as inputs into IBI. In this work, we extend this idea to include different phases such as the amorphous and crystalline phases for polymer-like materials. We apply this method to the case of poly-3-alkyl thiophene-type molecules, where coarse-grained models are derived in the amorphous state but are used to describe the crystalline state, which is the relevant phase at the conditions of interest.

497822 Coarse-Grained Model of Exciton Dynamics on Long-Chain Conjugated Polymer System

Abstract:

Conjugated polymers are important components in optoelectronic devices such as light-emitting diodes, field-effect transistors, and photovoltaic cells. A comprehensive understanding of exciton dynamics in conjugated polymers is challenging, given the effects of electron-electron interactions , electron-nuclear coupling, and disorder on a wide range length scalesâ€"from molecular level up to the device scaleâ€"has on electronic and optical properties of polymers.

Here we present a new phenomenological model for simulating the dynamics of excitons in long-chain organic conjugated molecules. In our model, the polymer is described as a time-dependent array of ring-ring torsion angles. Torisonal angles define the conjugation environment that determines the position, size, and energy of the excitons. Exciton dynamics arise in direct response to the evolution of this torsional landscape along its excited state potential energy surface, which includes exciton-induced forces (such as those that lead to self-trapping). The framework for generating an accurate description of these heterogeneous excited state forces was developed based on the analysis of mixed QM/MM simulations known as QCFF/PI with a semi-empirical Pariser-Parr-Pople (PPP) Hamiltonian to describe the pi-electron system to accurately reproduce the local structure and dynamics of an atomistic long-chain polymer (e.g. polythiophene)

We show that this model can reproduce transient pump-probe experiments; we remark on the importance on the excited state force field when describing these systems. Then we go on to present molecular-level physical insights into exciton dynamics in these polymer materials, which have been previously speculative, to help better engineer organic solar cell devices.

495550 Molecular Simulation of Thermoplastic Polyurethanes Under Large Mechanical Deformation

Abstract:

Analysis of the effect of topological features at the inter-domain surfaces within a heterogeneous polymer on its mechanical behavior is important to understand and design these materials for various applications, yet such detailed analysis remains elusive so far. In this study, using atomistic Monte Carlo and molecular dynamics simulations, we reveal how the interphase topology affects the mechanism of mechanical deformation for a set of thermoplastic polyurethanes (TPUs). TPUs are versatile polymers that have rigid and flexible segments that segregate on the nanometer length scale into hard and soft domains that are coupled mechanically through prolific interfacial zones between regions of differing compliances. Using a Monte Carlo method originally developed for semicrystalline polymers, we construct a detailed, thermodynamically rigorous, atomistic model of the interfacial zone between hard and soft domains within a TPU, with realistic distributions of bridges, loops and tails. We then deform these structures using nonequilibrium molecular dynamics simulations, identify the evolution of important topological features such as bridge during mechanical loading, and correlate the molecular behavior to system observables like yielding, plastic flow and toughing. We have found several distinct mechanisms of mechanical deformation, depending on the categories of interphase topology and loading directions.

499443 Harnessing Virtual High-Throughput Screening and Machine Learning for the Discovery of Novel High Refractive Index Polymers

Abstract:

Organic materials with refractive index (RI) values higher than 1.7 have attracted significant interest in recent years due to the tremendous potential for their application in optical and optoelectronic devices. According to the Lorentz-Lorenz equation, the incorporation of substituents with a high molar refractivity and low molar volume can increase the RI values of these materials. Polymers with a compact structure (such as aromatic polyimides and poly(aryleneethynylene)s) are based on low molar volume building blocks, and are thus generally promising candidates for high-RI materials. However, their molar refractivity has to be maximized as well in order to obtain a viable HRI material. The molar refractivity depends on the polarizability of the elements or moieties comprising the compound of interest. As carbon has only a moderate atomic polarizability, typical carbon-based polymers have RI values of around 1.5. Highly polarizable elements (such as sulfur, phosphorus, bromine, iodine, etc.) can significantly improve the RI values of the corresponding polymers. Hydrogen and strongly electronegative elements (such as oxygen and fluorine) have very low polarizability and exhibit the inverse effect.

We present a computational study of novel high-RI polymer candidates proposed according to the guidelines discussed above. Various paths to introducing highly polarizable elements into carbon-based frameworks are being explored. We compute static polarizabilities at the density functional theory (DFT) level and benchmark the predictive performance of different model chemistries and computational protocols. We obtain RI values based on the Lorentz-Lorenz equation.

To accelerate the discovery process of next-generation high-RI materials, we employ a high-throughput screening, materials informatics, and data-driven rational design approach that we have been developing in the group. It is a powerful tool and has shown to be highly promising for rapidly identifying polymer candidates with exceptional RI values.

496883 Development of Reaction Ensemble Monte Carlo (REMC) Algorithms to Study the Kinetics of Polymerization

Abstract:

Various fascinating materials are synthesized from the molecular precursor of silica by altering synthesis protocol or controlling polymerization parameters. The knowledge of molecular events is essential to control the properties of end products. In this work, a model for silica polymerization is developed using reaction ensemble Monte Carlo (REMC) which enable us to probe events at different length and time scale. Algorithms for molecular events are designed to simulate translation, rotation and reaction events, in accordance with realistic time so that kinetics of the system is captured, appropriately. The developed algorithm was applied to four coordinated silica polymerization, where excellent agreement between the simulation and experimental data was observed.

The algorithms when used for three functional silica precursor, show excellent agreement with experimental observation, thereby confirming that the algorithms are robust. The algorithms were then extended to study the effect of functionality by studying the combinations of two, three and four functional silica precursor to span entire phase space. The effects of functionality on the final silica structure were analyzed in detail.

488219 Controlled Liquid-Liquid Phase Separation of Recombinant Oleosin

Abstract:

Protein engineering enables the creation of materials with designer functionality and responsiveness. In this work, we designed a protein that phase separates into liquid droplets upon oxidation of a cysteine residue. The phase transition temperature can be tuned via the location of the cysteine residue in the protein backbone. Our work is based on the naturally occurring plant protein, oleosin. The protein is amphiphilic, with distinct hydrophilic and hydrophobic domains: an N-terminal hydrophilic segment, a highly hydrophobic central core, and a C-terminal hydrophilic segment. From wild type oleosin, a truncated version was engineered that reduced the hydrophobic core from 87 amino acids to 30 and added 5 glycines. Circular dichroism indicates that this mutant has less secondary structure than the wild type and approximately 65% of the protein is disordered. A single amino acid mutation was made to introduce a cysteine residue in the hydrophilic arm. Oxidation of this residue results in a dimer. Placing the cysteine closer to the N-terminus raises the transition temperature. The phase behavior of the protein can be eliminated by reducing the protein. Because the transition temperature varies with the location of the added cysteine, this protein provides a novel way to control protein liquid droplet formation.

500753 Evaluation of the Procedure to Obtain Chitosan Based Gels with Potential Use As Bone Adhesive on Clinical Settings

Abstract:

Comminuted fractures are defined as fractures in which an anatomical region is broken into multiple pieces. Medical treatments today have problems dealing with this kind of fractures because of the complications associated with these injuries. Among the techniques used nowadays to treat such wounds are the closed reduction, external fixation of pins, internal fixation with wires and internal reduction and fixation with the use of plates, screws, and wires[1]. Some of these treatments end up in a wrong fixation of fragments in the fracture [2]. Furthermore, in most cases, there are fragments so small that they cannot be fixated by any of the aforementioned techniques and are therefore discarded[3]. This only worsens the treatment, since the fact of having missing pieces may require a shortening in the bone to make an appropriate fixation with traditional methods and avoid an improper fixation[4].

Acknowledging this issue, the use of synthetic osseous adhesives can be considered as an alternative. Such adhesives usually present a great adhesion power, but they fail on being biocompatible [5–8]. Due to the lack of biodegradation and biocompatibility properties, the use of synthetic adhesives is restricted to big fragments of bone. On the other side, some biological based adhesives have been under research, but none of them have been able to reach the desired strength to hold through the stress a bone is constantly under [5, 9–11]. Therefore, this work proposes the use of chitosan as the main component for a bio adhesive that allows fixation of small bone fragments. Additionally, this adhesive is designed to allow cellular growth and biodegradation in the human body, which allows faster recovery without harming the patient.

Previous works have achieved to formulate this bio-adhesive, using three other components apart of chitosan. Two of the components are biocompatible agents that allow the adhesive to be more similar to the bone extracellular matrix, giving it a better chance of cellular growth and differentiation [12–14]. The third component is used as a cross linker to link the polymer chains and form a gel[15]. Due to the chemical interaction of the components with the polymer, the preparation method design is in stages. This allows the different components to interact with chitosan, and to restrict the crosslinking process giving properties to the hydrogels intended to work as bone adhesives (including wanted gelation time, malleability, wettability and mechanical strength.) [14, 15]. Accordingly, the method developed has two mixing stages. Each of these stages uses a different set of stirrers and impellers. The first stage is mixed with a DISPERMAT® LC30 and a Lightweight stainless steel dispersion impeller of 25mm. The second stage is done with a Mechanical stirrer EUROSTAR power control-visc 6000 (IKA) and a three-blade boat propeller of 35mm[14, 15].

While performing preclinical trials, we identified the need to modify the preparation process of the osseous bio adhesive. This was done since the equipment used to produce the gels on laboratory settings could not be sterilized and transported to a clinical setting. Therefore, in this study, a throughout evaluation of chemical and mechanical properties of adhesives prepared with different agitation methods is conducted. In it, different small and easy to sterilize home appliances' motors with fixed velocities and impellers were tested. Additionally, the most viable motor was tested with the original impellers (dispersion and three blade boat) to analyze what the critical factors in the production process are. The analysis of these gels has demonstrated that rheological properties do not reflect the difference in mechanical strength as believed before this work. Nevertheless, the change of the equipment used during the preparation of the adhesive changed the mechanical properties of it, decreasing the adhesive strength up to 62%. Additionally, it has been proven that one of the agitation stages has a bigger impact on the adhesive properties of the product. The mixing velocity resulted to have a more significant impact on the properties of the bioadhesive compared to the impeller used. Based on the given results, we propose a novel mixing method to produce gels on clinical settings and decrease the difference in the mechanical properties comparing the adhesive prepared on clinical and laboratory settings.

The results of the chemical and mechanical tests on the gels produced using both the laboratory and the clinical set of equipment will be presented in detail. Furthermore, we present the novel formulation procedure for the osseous chitosan-based bio-adhesive.

References

1. Holz U (2015) The elements of fracture fixation. Indian J Orthop 49:682. doi: 10.4103/0019-5413.168763

2. Shah JP, Patel SG, Singh B, Shah JP (2012) Jatin Shah's head and neck surgery and oncology. Elsevier/Mosby

3. Schatzker J, Tile M (1987) The rationale of operative fracture care. Springer-Verlag

4. Herscovici D, Scaduto JM (2014) Assessing Leg Length After Fixation of Comminuted Femur Fractures. Clin Orthop Relat Res 472:2745–2750. doi: 10.1007/s11999-013-3292-0

5. Porter JR, Ruckh TT, Popat KC (2009) Bone tissue engineering: A review in bone biomimetics and drug delivery strategies. Biotechnol Prog 25:NA-NA. doi: 10.1002/btpr.246

6. Arora M, Chan EK, Gupta S, Diwan AD (2013) Polymethylmethacrylate bone cements and additives: A review of the literature. World J Orthop 4:67–74. doi: 10.5312/wjo.v4.i2.67

7. Toriumi DM, Raslan WF, Friedman M, Tardy ME (1990) Histotoxicity of Cyanoacrylate Tissue Adhesives: A Comparative Study. Arch Otolaryngol - Head Neck Surg 116:546–550. doi: 10.1001/archotol.1990.01870050046004

8. Fedak PWM, Kolb E, Borsato G, Frohlich DEC, Kasatkin A, Narine K, Akkarapaka N, King KM (2010) Kryptonite Bone Cement Prevents Pathologic Sternal Displacement. Ann Thorac Surg 90:979–985. doi: 10.1016/j.athoracsur.2010.05.009

9. Farrar DF (2012) Bone adhesives for trauma surgery: A review of challenges and developments. Int J Adhes Adhes 33:89–97. doi: 10.1016/j.ijadhadh.2011.11.009

10. Donkerwolcke M, Burny F, Muster D (1998) Tissues and bone adhesives—historical aspects. Biomaterials 19:1461–1466. doi: 10.1016/S0142-9612(98)00059-3

11. Heiss C, Kraus R, Schluckebier D, Stiller A-C, Wenisch S, Schnettler R (2006) Bone Adhesives in Trauma and Orthopedic Surgery. Eur J Trauma 32:141–148. doi: 10.1007/s00068-006-6040-2

12. Swetha M, Sahithi K, Moorthi A, Srinivasan N, Ramasamy K, Selvamurugan N (2010) Biocomposites containing natural polymers and hydroxyapatite for bone tissue engineering. Int J Biol Macromol 47:1–4. doi: 10.1016/j.ijbiomac.2010.03.015

13. Fujihara K, Kotaki M, Ramakrishna S (2005) Guided bone regeneration membrane made of polycaprolactone/calcium carbonate composite nano-fibers. Biomaterials 26:4139–4147. doi: 10.1016/j.biomaterials.2004.09.014

14. Pinzon LM, Cedano FJ, Castro CI, Briceño JC, Casas JP, Tabima DM, Salcedo F (2017) Formulation and Characterization of Chitosan-Based Biocomposites with Potential Use for Bone Adhesion. Int J Polym Mater Polym Biomater 00914037.2016.1263948. doi: 10.1080/00914037.2016.1263948

15. Cedano Serrano FJ, Pinzón LM, Narváez DM, Castro Paéz CI, Moreno-Serrano CL, Tabima DM, Salcedo F, Briceno JC, Casas-Rodriguez JP (2017) Evaluation of a water-resistant and biocompatible adhesive with potential use in bone fractures. J Adhes Sci Technol 31:1480–1495. doi: 10.1080/01694243.2016.1263055

486710 Deep Space Drug Shielding

Abstract:

Space medications help prevent astronauts from being ill and provide acute treatment in medical emergency situation for them to fulfill the primary goal of a successful long duration space mission: maintaining astronauts' health in a unique, isolated, and extreme space environment. However, recently, National Aeronautics and Space Administration (NASA) have reported a shorter shelf life of space medications caused by chronic ionizing radiation and space medications' long-term efficacy in space became questionable. Hence, to help extend the space medications' stability and protect pharmaceuticals from the indirect radiation damage, Trolox, free radical scavenging antioxidant, were bio-conjugated with PLL (Poly-l-lysine) on the surface of drug encapsulated PLGA (Poly-lactic-co-glycolic acid) nanoparticles (NPs) using carbodiimide crosslinking chemistry. In this study, Melatonin, a sleep aid medication used in ISS, was chosen as a model drug. Radiation induced hydroxyl radicals, highly reactive free radicals, will be oxidized by Trolox before it could react with the pharmaceuticals in PLGA-PLL-Trolox composites' core. The mean size of PLGA-PLL-Trolox composites were 491.3 $\pm$ 33.09 nm with zeta potential value of 26.4 $\pm$ 10.18 mV. A successful conjugation of PLL-Trolox on the surface of melatonin encapsulated PLGA NPs were determined by FT-IR and H^1-NMR analysis. Hydroxyl radical scavenging capacity assay (HOSC) was performed to test Trolox's free radical quenching ability. These results suggest a new molecular approach to shield pharmaceuticals from indirect radiation damage using antioxidant conjugated biocompatible polymer.

501173 Synthesis of Designer Lipids Using "Click" Chemistries

Abstract:

Liposomes have great potential in the fields of drug delivery, imaging and in simplified cell membrane studies. However, to achieve their full potential in these applications, it frequently becomes necessary to incorporate synthetic constructs and new functionalities into the membrane architecture. We have created a simplified method for incorporation of functional groups of interest into the phospholipid structure through the covalent attachment of a second aliphatic tail onto a lysolipid backbone. This addition is now possible through the Copper-catalyzed Azide-Alkyne Cycloaddition (CuAAC), thiol-acrylate and/or thiol-Michael reactions using lysolipids functionalized with an alkyne, thiol, or acrylate group, respectively. This wide range of chemistries enables more simplified synthesis of lipids containing variable chemical moieties and enables the covalent attachment of these functionalized tails using a range of catalysts. In the end, this approach will ease the testing of liposome systems containing new behaviors and characteristics.

501181 Loading and Mobility of RNA in Porous Silica Nanoparticles for Delivery to Insects

Abstract:

RNA interference (RNAi) technology, which enables the suppression of targeted genes, is of wide interest to insect pest management through the silencing the genes that relate to the insect fitness or mortality. Identifying efficient methods to deliver the dsRNA to insects is the key challenge to achieve effective RNAi. Porous silica nanoparticles (pSNPs), which have tunable pore morphology, large surface area, biocompatibility and ease of surface functionalization, are potential carriers for biomolecules. The goal of this work is to demonstrate the loading, release, and protection of RNA in porous silica nanoparticles. pSNPs with pore sizes appropriate for biomolecule loading have been successfully synthesized to encapsulated the dsRNA within the pores. The silica particles were further functionalized with aminosilanes, providing amine groups with positive charge on the surface necessary to load negatively charged nucleic acids. To investigate the RNA interaction with the particles, direct visualization of the loading of RNA on micron-sized silica particles with similar pore sizes to the pSNPs was performed under confocal laser scanning microscopy. The interaction of dsRNA with the pSNP and the accessibility of the pores of the pSNP can be described from the diffusivity (mobility) of the dsRNA in RNA-loaded pSNPs. Using Fluorescence Recovery after Photobleaching (FRAP), the effect of both pore size (nonporous, 3.6 nm, 7.4, and 11.8 nm diameter) and length of dsRNA (84 bp and 282 bp of fragments within the inhibitor of apoptosis (IAP) genes of *Spodoptera frugiperda*) on the diffusivity at the surface and in the core of dsRNA-loaded pSNPs has been quantified. The importance of pores to a mobile dsRNA network is demonstrated by the lack of measurable mobility for both lengths of RNA on nonporous materials. In addition, when the dsRNA could not penetrate the pores, dsRNA mobility is also not measurable at the surface of the particle. Thus, the pores seem to serve as a â€œsinkâ€ in providing a mobile network of dsRNA on the surface of the particle. The mobility of dsRNA in the nanopores of the pSNPs is expected to have a functional effect on delivery of dsRNA to insects and the delivery of the IAP genes to *Spodoptera frugiperda* using pSNPs is interpreted from loading and mobility as a function of pore size. This work successfully demonstrates the loading of RNA on functionalized pSNPs and identified factors that affects RNA loading and releasing, which provides basis for the delivery of RNA-loaded silica particles *in vivo*.

505175 Engineering and Functionalizing Protein-Based Materials

Abstract:

The development of materials with diverse mechanical and functional properties enables a broad range of applications. For materials composed of protein, well-established molecular biology techniques can theoretically be used to alter the mechanical and functional properties of materials. However, repetition of amino acid sequences in the constituent proteins can complicate engineering the mechanical properties. Furthermore, most protein-based materials are assembled under denaturing conditions, thus precluding incorporation of full-length proteins in an active state. Our lab discovered that the *Drosophila melanogaster* transcription factor Ultrabithorax (Ubx) self-assembles into materials. These materials are stabilized by specific dityrosine bonds that spontaneously form during assembly. The location of the participating tyrosines in unique sequence motifs enables the reliable tuning of the materials by mutagenesis to be either weaker or stronger. Because Ubx materials rapidly form in mild, aqueous buffers, a surprisingly wide variety of proteins can be incorporated via gene fusion without harming either materials assembly or the function of the appended protein. Indeed, when Ubx is fused to angiogenic growth factors, the resulting fibers control cell signaling and cell behavior *in vitro*, and can instigate and guide blood vessel formation *in vivo*. Finally, Ubx proteins in materials retain the ability to bind specific DNA sequences with high affinity. Ubx materials provide an inexpensive and facile platform for customizing materials for a variety of applications.

485053 Synthesis and Applications of Bio-Inspired Oligotea Peptidomimetics

Abstract:

Precise control over the composition, sequence and architecture of synthetic polymers is critical for mediating a high level of control over their chemical and physical attributes. Naturally occurring biomacromolecules including peptides, nucleic acids and oligosaccharides, have evolved over millennia to exhibit this level of control. However, synthetic analogs with comparable sequence and structural control remain scarce. Motivated by these opportunities and the need for sequence-control and structural diversity in polymer research, we present a versatile methodology for the synthesis of a new class of sequence-defined abiotic oligomers. Our approach to sequence-defined oligomer assembly involves the sequential and iterative addition of *N*-allylacrylamide and dithiol monomers to build sequence-defined oligothioetheramides (oligoTEAs) on a soluble fluorous support. The combination of fast solution-phase reaction kinetics with a robust fluorous solid-phase extraction (FSPE) technology allows for the efficient assembly of molecularly encoded synthetic oligoTEAs with broad structural diversity and tunability. Strategic placement of complementary reactive monomers during the oligoTEA assembly process facilitates the assembly of a wide variety of sequence-defined architectures. These novel sequences and structures will be presented and used to *i*) investigate the effect of primary sequence and structure on the potency and toxicity of antimicrobial oligoTEAs, and *ii*) create cleavable heterotrifunctional oligoTEA cross-linkers for quantifying endosomal-specific bond cleavage kinetics and for building a new class of antibody drug conjugates.

493216 A Polyester Based Photoluminescent Hydrogel As Extracellular Matrix Mimics

Abstract:

Photoluminescent hydrogels have emerged as 3D materials with potential applications in many fields such as imaging, tissue engineering, and drug delivery. Current photoluminescent hydrogels are fabricated by either conjugating or doping a fluorescent dye, fluorescent protein, lanthanide chelate or quantum dot into polymeric hydrogel matrix. Their scope of applications has been severely limited by drawbacks such as poor photostability or substantial toxicity. There is an urgent need for a synthetic hydrogel with degradable, injectable, biocompatible and photoluminescent properties, but a successful development has not yet been reported. Herein, we report a successful development of biodegradable and photoluminescent hydrogel with synthetic polymers. A series of polyester oligomers with tunable intrinsically fluorescent properties have been successfully synthesized from biocompatible monomers including citric acid, poly(ethylene glycol) (PEG)-diol and various amino acids via a facile polycondensation reaction. The synthesized oligomers were further modified by introducing multiple thiol functional groups using a high efficient enzyme assisted transesterification reaction. A library of hydrogels with diverse physiochemical properties can be formed by combining thiol group functionalized polymers with different molecular weight multi-armed PEG acrylates or PEG maleimides through a convenient Michael addition reaction. Preliminary data show that the chemical, optical, mechanical and biological properties of the intrinsically photoluminescent hydrogels can be tuned by changing a series of parameters such as gel composition, ionic strength, polymer molecular weight, gel forming temperature or pH. The hydrogel can be further modified with an extracellular matrix (ECM) moiety to mimic ECM microenvironment for cell growth and tissue formation. The hydrogel shows good biocompatibility when cultured with NIH 3T3 cells. We then conducted *in vivo* florescence bioimaging of the hydrogels that were injected under the skin of nude mice to demonstrate their potential use in injectable biomaterials and noninvasive implant tracking.

501827 Alginate Nanoparticle Platform for Controlled Release of Biotherapeutics

Abstract:

The design of novel drug delivery vehicles has been focused on the synthesis and development of materials that can load and release biomolecules for therapeutic or prophylactic use with a controlled behavior. These materials need to be biocompatible, and depending on their application stimulate the immune system appropriately. Alginates are anionic polymers that have been used for drug delivery and tissue regeneration since they are bioresorbable and non-immunogenic. These materials can be ionically or chemically crosslinked, creating different physicochemical characteristics. This ability is critical to tailor the properties of alginate-based drug delivery vehicles. By combining ionic and chemical crosslinking methods on alginate nanoparticles the release profiles of biotherapeutics can be modified to develop platforms with sustained delivery mechanisms.

The studies described herein were developed to synthesize a drug delivery platform based on alginate nanoparticles using dual crosslinking mechanisms. Sodium alginate was functionalized with methacrylic anhydride via the esterification of its hydroxyl groups. Successful functionalization of alginate was confirmed via Fourier transform infrared spectroscopy and NMR spectroscopy. Alginate nanoparticles were synthesized using water-in-oil microemulsions. Briefly, a functionalized-alginate solution (up to 15% (v/v)) was emulsified in toluene with the biomolecule, 1% (w/w) Span80® as a surfactant, and 1% (w/w) Irgacure 2959 as a photoinitiator. Nanoparticle solutions were then crosslinked for various times using a UV point-source at 140 mW/cm^2. Resulting nanoparticles were purified and then suspended in a calcium chloride solution (5M) over different periods of time. Particle characterization was performed using light and electron microscopy, dynamic light scattering and zeta potential measurements confirming the morphology and charge of the functionalized alginate nanoparticles. Most importantly, the ability of these nanoformulations to release proteins with a sustained release profile were evaluated. The release profile of three different proteins with either similar molecular weights or isoelectric points were analyzed under solutions with decreasing levels of ionic groups. The degree of chemical crosslinking combined with the molar concentration of the ionic solution affected the release profiles of the protein molecules. These results show that combination of different crosslinking methodologies can confer natural polymer systems, such as alginate nanoparticles, the ability to be delivery vehicles for biotherapeutics with controlled release behaviors.

501294 Complex Bone Regeneration Via Controlled Release of Simple Signaling Molecules

Abstract:

A growing number of patients are suffering from disabilities related to bone defects such as fracture nonunion. Tissue engineering offers a promising approach to enhance host bone regeneration with most of these methods have focused on developing osteoinductive biomaterials. However, because of complexity of bone tissue and its vast neurovascular network, the vast majority of these strategies have not been successfully translated to the clinic. The goal of this research is to develop novel biomaterials which are capable of inducing neurogenesis and angiogenesis along with osteogenesis in critical size bone defects. Upstream, non-proteinaceous inductive chemicals termed simple signaling molecules were utilized in this research because they are less expensive and easier to process and manipulate compared to more commonly utilized downstream growth factors. Specifically, we have investigated osteoinductive calcium and phosphate ions, angioinductive hydrogen peroxide, and neuroinductive carbon monoxide and hydrogen sulfide. In the current research, we have first focused on the concentration-dependent effects of simple signaling molecules on *in vitro* differentiation of stem cells. We have then leveraged the therapeutic windows determined from this work to design and synthesize unique polymers capable of achieving desirable controlled release of these inductive molecules to achieve optimal cell differentiation outcomes.

500789 Biomimetic Growth of a Pathologic Biomineral in Hydrogels

Abstract:

Biominerals found in living organisms have diverse biological functions and distinct properties arising from the particular structure, orientation, and morphology of the constituent minerals. A systematic investigation of the morphological evolution of biominerals grown in organic matrices, and the interaction of such matrices with the inorganic crystals are essential for understanding biomineralization processes and for developing bioinspired materials. We employ here hydrogels as model organic matrices and a â€˜double-diffusionâ€™ method to mimic the formation of calcium oxalate, the major constituent of kidney stones. Gel-mediated crystallization provides a great way to vary the local supersaturation through controlling the diffusion rate of the reacting ions. We study the nucleation, growth, type of polymorph, orientation and aggregation of the calcium oxalate particles as affected by the gel density, reservoir concentrations, and molar stock solution ratio. We also explore the inhibitory effects of anionic macromolecules on calcium oxalate formation. These findings help our understanding of the mechanism of polymer incorporation in biomineral aggregates, and provide insights into the formation of organic-inorganic composite materials. Furthermore, they guide the design and synthesis of therapeutic agents (modifiers with molecular recognition for crystal faces) to control biomineralization-associated diseases, such as kidney stone.

487325 The Influence of Intrinsic Framework Flexibility on Adsorption in Nanoporous Materials

Abstract:

For applications of metal–organic frameworks (MOFs) such as gas storage and separation, flexibility is often seen as a parameter that can tune material performance. In this work we aim to determine the optimal flexibility for the shape selective separation of similarly sized molecules (e.g., Xe/Kr mixtures). To obtain systematic insight into how the flexibility impacts this type of separation, we develop a simple analytical model that predicts a material's Henry regime adsorption and selectivity as a function of flexibility. We elucidate the complex dependence of selectivity on a framework's intrinsic flexibility whereby performance is either improved or reduced with increasing flexibility, depending on the material's pore size characteristics. However, the selectivity of a material with the pore size and chemistry that already maximizes selectivity in the rigid approximation is continuously diminished with increasing flexibility, demonstrating that the globally optimal separation exists within an entirely rigid pore. Molecular simulations show that our simple model predicts performance trends that are observed when screening the adsorption behavior of flexible MOFs. These flexible simulations provide better agreement with experimental adsorption data in a high-performance material that is not captured when modeling this framework as rigid, an approximation typically made in high-throughput screening studies. We conclude that, for shape selective adsorption applications, the globally optimal material will have the optimal pore size/chemistry and minimal intrinsic flexibility even though other nonoptimal materials' selectivity can actually be improved by flexibility. Equally important, we find that flexible simulations can be critical for correctly modeling adsorption in these types of systems.

497750 Towards a Generalized Understanding of Acid Gas Interactions with ZIF Materials

Abstract:

Zeolitic Imidazolate Frameworks (ZIFs) are a set of nanoporous metal-organic frameworks (MOFs) with tunable porosity and functionality. Among MOFs, they also show relatively good stability with respect to temperature and humidity. These characteristics lead to their potential applications as membranes or adsorbents in separation processes, in which they would be exposed to a variety of molecular species including acid gases. However, there is very little systematic knowledge of the effects of acid gas exposure on the structure and separation properties of ZIFs, in particular the mechanistic aspects of ZIF degradation by acid gases as well as the effects of differences in ZIF crystal topology and linker composition on their stability. [1-2] The objective of this work is to develop a generalized understanding of ZIF stability in acid gases. The stability of ten ZIF materials (of differing structural topologies such as SOD, RHO and ANA) under different environments such as humid air, liquid water, dry, humid and aqueous SO_2, dry and humid CO_2 are investigated in detail. Combined characterization by several techniques (PXRD, N_2 physisorption, EDX, XPS, and FTIR) allowed us to track the structural and compositional properties before and after exposure. All ZIFs studied were found to be stable under dry acid gas environments in this study while a few ZIFs underwent transitions into non-porous phases under humid air or liquid water exposure. Most ZIFs were found to be unstable under long-term humid SO_2 or CO_2exposure. We present quantitative kinetic data and parameters for the degradation of ZIFs, and discuss in detail the effects of topology and linker composition. This work provides a comprehensive guide to the acid gas stability of ZIF materials, and the main features could be extended to other classes of MOFs and to understand stability under exposure to other kinds of reactive species.

References

1. Bhattacharyya, S.; Pang, S. H.; Dutzer, M. R.; Lively, R. P.; Walton, K. S.; Sholl, D. S.; Nair, S., Interactions of SO_2-Containing Acid Gases with ZIF-8: Structural Changes and Mechanistic Investigations. *J.Phys.Chem.C* **2016**, *120*, 27221-27229.

2. Pang, S. H.; Han, C.; Sholl, D. S.; Jones, C. W.; Lively, R. P., Facet-Specific Stability of ZIF-8 in the Presence of Acid Gases Dissolved in Aqueous Solutions. *Chem. Mater.* **2016**, *28*, 6960-6967.

489910 Heat-Treatment of Defective Uio-66 from Modulated Synthesis:Adsorption and Stability Studies

Abstract:

Metal-organic frameworks (MOFs) are well-known for their diversity in terms of pore size, geometry and functionality. Recently, defect engineering in MOFs has drawn intense research interest for both fundamental and practical reasons. However, in the area of defective MOFs, studies evaluating adsorption performance of large molecules and elucidating the influence of defect on material stability are relatively sparse. In this work, two versions of defective UiO-66 structures containing supermicropores or micro-mesopores have been prepared through a joint of modulated synthesis and post-synthetic heat treatment at 200 °C and 320 °C under vacuum, respectively. The effect of defects inside UiO-66 on guest interaction is evaluated using several molecule physisorption with different sizes, including N_2, SO_2, benzene, and cyclohexane. Gas uptake experiments as well as GCMC isotherm calculations show that the defective UiO-66 (200 °C) with the incorporation of trifluoroacetate groups displays a significant increase in large size molecules, specifically cyclohexane. In addition to adsorption properties, we also systematically studied defect concentration and its correlation with deviation in stability from pristine UiO-66 in water and acidic environments. We propose a potential degradation pathway for both pristine and defective UiO-66 upon water attack for stability analysis. Interestingly, both experimental and computational results demonstrate that the defective UiO-66 (200 °C) with the incorporation of trifluoroacetate groups displays excellent stability upon exposure to water and acidic environments.

493404 Design of Stratified Hybrid Metal Organic Frameworks for Chemical Detection and Destruction

Abstract:

There is a pressing need for materials capable of rapidly detecting and destroying chemical warfare agents. Stratified metal-organic frameworks (MOFs) containing plasmonic nanoparticles may be able to meet this need. Stratified MOFs consist of layered materials having different functional groups in each layer, the purpose of which is to provide a gradient of functional groups to direct transport of target analytes to the center of the MOF. We seek to develop a detailed understanding of the fundamental properties of sorption, transport, photodetection, and photocatalytic degradation of target chemical species in stratified hybrid MOF-nanoparticle systems. The first step to achieve this is to design, synthesize, and evaluate MOFs to identify promising functional groups to produce differential binding of analytes. We use quantum mechanical methods to identify functional groups for binding target species, Monte Carlo methods to perform fluid adsorption isotherms in candidate MOFs, and ab initio molecular dynamics to sample configurations and determine optimal binding sites in MOFs. In addition, we synthesize robust Zr-based MOFs with the identified functional groups, measure isotherms of target molecules to compare with calculations, and use temperature programmed desorption to measure loading and binding energies of target molecules in MOFs.

500444 Understanding Structure, Metal Distribution, and Water Adsorption in Mixed-Metal MOF-74

Abstract:

We present a joint computational and experimental study of Mg–Ni-MOF-74 and Mg–Cd-MOF-74 to gain insight into the mixing of metals and understand how metal mixing affects the structure of the undercoordinated open-metal sites that exist in these materials. Our calculations predict that metal mixing is energetically and entropically preferred in these materials. Recent experimental work has demonstrated that Mg–Ni-MOF-74 shows much greater surface area retention in the presence of water than Mg-MOF-74. To probe this effect, we study H_2O adsorption in Mg–Ni-MOF-74, finding that the adsorption energetics and electronic structure do not change significantly at the metal sites when compared to Mg-MOF-74 and Ni-MOF-74, respectively. We conclude that the increased stability of Mg–Ni-MOF-74 is a result of a M–O bond length distortion in mixed-metal MOF-74, consistent with recent work on the stability of MOF-74 under water exposure.

490781 Extraction of Rare Earth Elements from Geothermal Brine Solution Using Magneric Core Shell Microspheres

Abstract:

Rare-earth elements (REEs) have significant commercial uses in the millitary and high-tech sector. However, difficulties associated with REE extraction in the conventional mining process has resulted in the search for REEs from alternative resources in a cost effective, environmentally-friendly manner. Among various sources, geothermal brine, used for generating geothermal energy was found to possess high concentration (~ppm level) of REEs in +3 ionic forms along with other dissolved metal ions. A system that can selectively trap the REEs using the existing setup of a geothermal power plant will lead to an additional revenue stream that will accelerate the development and deployment of geothermal plants in the United States and the rest of the world. Here, i present a magnetic core-shell approach that can effectively extract REEs in their ionic form from aqueous solution with up to 99.99% removal efficiency. The shell, composed of thermally and chemically stable metal-organic frameworks (MOFs) surrounds a Fe_3O_4 magnetic core. Majority of the Fe^{2+} species of the magnetite (Fe_3O_4) microsphere is oxidized to Fe^{3+} species (maghemite) during *in-situ* coating of the MOF materials as evident by magnetometry, Mössbauer spectroscopy, and X-ray absorption spectroscopy. The core-shell particles can be completely removed from the mixture under an applied magnetic field, offering a practical, near-term efficient REEs-removal process.

495939 Cycloaddition Reaction of Epoxides with CO_2 in a Copper Metal-Organic Framework: A Density-Functional Theory Study

Abstract:

A large number of metal-organic frameworks (MOFs) have been reported to be efficient catalysts in Lewis acid catalyzed reactions such as the cycloaddition reaction of epoxides with CO_2 to carbonates under mild condition. Herein, we report a density functional theory (DFT) study to investigate the cycloaddition reaction in a Cu-MOF. Two competitive pathways are considered in the channel of the Cu-MOF associated with a co-catalyst tetrabutylammonium bromide (TBAB). Both pathways are found to be energetically more favorable compared with other systems. The intermediates and transition states have low free energies with high stability, as attributed to the long-range electrostatic interactions with the Cu-MOF. In both two pathways, the rate-determining step is revealed to be the ring opening of epoxide through the necleophilic attack of Br^- of TBAB. Furthermore, the concerted reaction by inserting CO_2 into the ring opening step has a lower energy barrier and is a preferential pathway. This theoretical study provides microscopic insight into the cycloaddition reaction mechanism, and could facilitate the rational screening and development of new catalysts.

499559 Nanostructured Materials for Separations Based on Reactive Block Polymers

Abstract:

Block polymers are remarkable hybrid materials that can self-assemble on nanoscopic length scales. By controlling the composition, architecture, and molar mass, synthetic chemists can finely tune the morphologies adopted by these materials. Of the typical block polymer morphologies, bicontinuous phases such as the gyroid structure have been targeted due to their special utility in various applications that require interpenetrating domains structured on the nanoscale. While block polymers can be designed to self-assemble into such bicontinuous nanostructures, the window of thermodynamic stability is often quite narrow, and thus such structures can be difficult to experimentally access. In this presentation I will discuss the design, synthesis, self-assembly and applications of block polymer modified materials that adopt bicontinuous nanostructures through a polymerization induced microphase separation process. The enabling concept is the in situ formation of block polymers in a thermosetting system using controlled polymerizations such that their chemical synthesis and self-assembly occur in a single process. One of the important elements necessary for the adoption of bicontinuous structures is that the block polymer is crosslinked during synthesis to form a thermosetting material that results in chemical fixation of the final morphology. The central technique we utilize is controlled radical polymerization from a macromolecular chain transfer agent in the presence of a mixture of monomers that can undergo crosslinking reactions. I will discuss how we developed this approach, mechanistic consideration associated with the formation of bicontinuous structures, and applications of the resulting nanostructured materials in, for example, polymer electrolyte membranes for use in batteries and fuel cells.

501187 Block Copolymer Derived Nanostructured Surfaces: Templating Confined Surface Reactions

Abstract:

Polymer-inorganic hybrid materials that yield highly crystalline and hierarchically structured composites remain a challenging to realize, yet continue to entice for their applications from new synthetic biomaterials to improved catalysis. Block copolymer assembly, with the ability to produce periodically ordered structures over long ranges, has been used to structure direct a variety of inorganic materials, but relies primarily on spatial confinement. In contrast, proteins that assemble biological hierarchical composites such as nacre and bone utilize self-assembly pathways in which the chemical moiety localization in the protein matrix (determined by the amino acid sequence and chain folding) directs the crystal phase and orientation of inorganic materials. Here we introduce a system based on a reactive block copolymer that can be tailored with a variety of chemical moieties to introduce independent chemical and spatial control on the 10s of nanometer length scale. The versatility of this system is demonstrated with localized thin film functionalization by amino acids and ultrasmall silica nanoparticles, and this platform could produce new approaches to catalyst and sensing technologies. Implications for inorganic growth control over multiple length scales will be discussed.

498609 Nanostructure Formation on Collapse of Polyelectrolyte Brushes

Abstract:

Polyelectrolyte brushes have been used to modify surfaces for applications in lubrication, colloidal stabilization and tailoring interactions with biological substances. A majority of these applications rely on the extended state of the polyelectrolyte brush, where the polymer chains extend from the surface due to strong osmotic pressure from counterions trapped within the brush volume. However, polyelectrolytes are sensitive to many stimuli in their local environment including pH, ionic strength, solvent quality and counterion valency, which can disrupt the extended morphology of the polyelectrolyte brush. Using a combination of theory, simulations and experiments, we demonstrate that the polyelectrolyte brush collapses into nanoscale â€œpinned micellesâ€ in the presence of multivalent ions.Â Previous studies of polyelectrolyte brush interactions with multivalent ions demonstrated a sharp collapse in the brush height and attractions between opposing brushes when multivalent ions are present, which is hypothesized to be due to the bridging of polyelectrolyte chains by the multiple charges of multivalent ions. However, due to high levels of counterion condensation for multivalent ions, which can effectively neutralize the polyelectrolyte, solvophobic effects from the polyelectrolyte backbone must also be considered. The interplay between the multivalent ion bridging and solvophobic interactions leads to formation of nanoscale structures on brush-coated surfaces in the presence of multivalent ions. This fundamental understanding leads to capabilities for rational design of new, stimuli-responsive materials that take advantage of this alternate polyelectrolyte brush morphology, nanoscale pinned micelles.

497128 iCVD Deposition and Integration of Poly-(1H,1H,2H,2H-Perfluorodecylacrylate) (PPFDA) Under High Loading of TiO_2 Nanoparticles

Abstract:

A low surface energy fluoropolymer, poly-(1H,1H,2H,2H-perfluorodecylacrylate) (PPFDA), has received much attention due to its hydrophobicity and oleophobicity. Its potential applications in latent heat transfer, self-cleaning, barrier protection, and biomedicine demand suitable integration of the polymer with nanoscale features, particularly as devices shrink in size [1-3]. To cater to such demands and to probe the polymer properties particularly under high loading of nanostructures as required by many of these applications, initiated chemical vapor deposition (iCVD) is utilized to enable conformal growth and tight integration of the polymer with the nanostructures. As a solvent-free approach, iCVD overcomes the issues of solution viscosity, liquid surface tension, and solvent residues with conventional solution casting, sol-gel, or blending methods. These issues have limited the loading of nanofillers to less than 15 wt% or 10 vol% in conventional liquid processing, while iCVD has been shown to enable a much higher loading, e.g. for TiO_2 nanoparticles, a loading of 82 wt% and 54 vol% can be achieved [4]. As a result of such high nanoparticle loading, polymers can display unique properties and behavior compared to bulk because of the significant interactions between the polymer and nanoparticles at their interfaces. For example, the iCVD deposition of polyglycidol (PGL) within the mesoporous TiO_2 nanoparticle network resulted in a significant increase in the glass transition temperature of PGL compared to bulk PGL without any nanofiller, which is attributed to the hydrogen bonding interactions between PGL and surface hydroxyl groups on the TiO_2 nanoparticle surfaces [5]. Here, the iCVD deposition of PPFDA in a mesoporous network of TiO_2 nanoparticles will be demonstrated. The effect of the high loading of TiO_2 nanoparticles on PPFDA properties will be discussed based on results from dynamic scanning calorimetry (DSC), thermogravimetric analysis (TGA), X-ray diffraction (XRD), and contact angle measurements. In particular, polymer crystallinity is found to decrease appreciably. Such a decrease in crystallinity is attributed to the tortuous mesoporous TiO_2 network as well as the significant interfacial interactions of the polymer, which disrupt chain alignment. The ability to alter polymer crystallinity can ultimately influence its hydrophobicity and therefore offers a potential way to control the anti-aging, drag reduction, and heat transfer properties for its various applications [5].

1. Ã–lÃ§eroÄŸlu, E.; Hsieh, C.-Y.; Rahman, M. M.; Lau, K. K. S.; McCarthy, M., Full-Field Dynamic Characterization of Superhydrophobic Condensation on Biotemplated Nanostructured Surfaces. *Langmuir* **2014,** *30* (25), 7556-7566.

2. Feng, L.; Li, S.; Li, Y.; Li, H.; Zhang, L.; Zhai, J.; Song, Y.; Liu, B.; Jiang, L.; Zhu, D., Super-Hydrophobic Surfaces: From Natural to Artificial. *Adv. Mater.* **2002,** *14* (24), 1857-1860.

3. Tian, X.; Verho, T.; Ras, R. H. A., Moving Superhydrophobic Surfaces toward Real-World Applications. *Science* **2016,** *352* (6282), 142.

4. Hsieh, C.-Y.; Lau, K. K. S., Growth of Polyglycidol in Porous TiO_2 Nanoparticle Networks via Initiated Chemical Vapor Deposition: Probing Polymer Confinement Under High Nanoparticle Loading. *Adv. Mater. Interfaces* **2015,** *2* (17), 1500341.

5. Coclite, A. M.; Shi, Y. J.; Gleason, K. K., Controlling the Degree of Crystallinity and Preferred Crystallographic Orientation in Poly-Perfluorodecylacrylate Thin Films by Initiated Chemical Vapor Deposition. *Adv. Funct. Mater.* **2012,** *22* (10), 2167-2176.

492785 Development of Resin-Containing Polymer Particles for Thermoset Powder Coatings

Abstract:

Due to their superior durability, mechanical properties, and chemical resistance, thermoset coatings are broadly used across various industries, such as, transportation, food processing, healthcare, marine, and power generation. Unfortunately, the thermoset resins and hardeners that are used in commercial coating applications are solvent-based and thus, they emit volatile organic compounds (VOCs). Due to the significant environmental and human health risks that VOCs pose, the emphasis on reducing VOCs is increasing. Cold sprayed coatings exhibit numerous improvements over the current industrial coatings, including their faster curing behavior, zero-VOCs, unlimited shelf life, low cost, and expanded operational conditions that include application at any temperature. The objective of this project is to synthesize reactive particles that have a low glass transition temperature (T_g) and that when cold sprayed, form high-performance industrial coatings with a high T_g. In this study, we utilized suspension polymerization to synthesize two different reactive thermoplastic particles, one with thermosetting resins and the other with hardeners. The thermal and mechanical properties of the particles were tuned for powder coating processes by introducing a core-shell structure and by varying the monomer composition. Additionally, the cold spray processing parameters were modified to optimize coating properties. The expected outcome of this study is a novel coating that is easy to apply, has high performance, significant environmental benefits, and a commercial feasibility.

493974 Roll-to-Roll Micromolding of UV Curable Thiol-Ene Based Coatings

Abstract:

In micromolding, a microstructured coating is created by pressing a stamp into a coating, UV curing with the stamp in place, and then removing the stamp. The process can be implemented in roll-to-roll systems and nanometer to micrometer features can be created at low costs; however, UV-micromolding processes are limited by the availability of UV-curable coating materials which possess high cure speeds and easily tunable properties that allow for flexibility in processing conditions and product performances. In this study, a thiol-ene based UV curable coating system with tunable moduli and surface energies was prepared for roll-to-roll micromolding applications. The thiol-ene chemistry possesses a variety of characteristics desirable for the pattern replication, including fast curing, oxygen insensitivity, and low volume shrinkage. Elastic moduli of cured coatings were adjusted over two orders of magnitude by changing the acrylate content. A fluorinated polymer was designed and synthesized for the surface modification of thiol-ene coatings, which effectively reduced the coating surface energy at a loading of 1 wt%. Micron-scale surface patterns, including square dots and microchannel arrays, were successfully replicated on a roll-to-roll imprinting pilot line using the thiol-ene based coatings. The roll-to-roll process reached a web speed of 16 inches per minute without noticeable defect formation.

491504 Elastic Networks for Shape-Memory Contact Printing

Abstract:

Current methods for microscale pattern transfer fail to combine high feature resolution, ability to pattern large areas, high throughput and cost effectiveness. Over the past few years, several researchers have demonstrated how films of shape-memory polymers can exhibit topographical memory, leading to switchable adhesion or optical properties. We are designing and developing new stimuli-responsive materials to ultimately enable high fidelity pattern transfer by thermal modulation of the elastomeric contact printing process. Glass forming, semicrystalline, and hydrogen bonding shape-memory networks are prepared and characterized in the bulk and have been molded into films with simple surface features such as spheres, posts, and cones. The influence of temperature and temperature history on the contact mechanics of molded shape-memory thermosets is studied using custom-built materials testing / contact printing apparatus. The apparatus is capable of controlling temperature as well as the compressive strain between two soft surfaces while measuring stress and contact area (optically). Contact mechanics experiments are modeled and interpreted using finite element analysis to offer deeper insight to the nature of contact and detachment. Micro-indentation and atomic force microscopy studies have also yielded new relationships and valuable insight to enabling shape-memory contact printing.

501238 Dynamics of Electric Double Layer Formation and Dissipation in Polyethylene Oxide:$LiClO_4$ on Graphene Transistors

Abstract:

A convenient tool for exploring the electronic properties of two-dimensional (2D) materials is the formation of an electric double layer (EDL). The EDL is formed at the interface between the 2D material and an electrolyte where cations or anions induce image charge in the 2D material, doping it n- or p-type respectively. This approach is powerful because it can induce ultrahigh sheet carrier densities up to 10^{14} cm^{-2}, surpassing traditional gate oxides by more than one order of magnitude. To extend the use of EDL doping beyond basic materials exploration to practical devices, what is needed is a better understanding of the dynamics of EDL formation/dissipation. In this work, epitaxial graphene Hall bar devices are gated with a solid polymer electrolyte, polyethylene oxide (PEO):$LiClO_4$. Hall measurements are used to quantify sheet carrier density and mobility in the electrolytic-gate bias range of ± 2 V. Ion dynamics of EDL formation and dissipation are measured as a function of temperature. At an electrolyte thickness of 1 Î¼m, the room temperature EDL formation time (~1 â€“ 100 s) is longer than the dissipation time (~10 ms). The EDL dissipation is modeled by a stretched exponential decay, and the temperature-dependent dissipation times are described by the Vogel-Fulcher-Tammann equation, reflecting the coupling between polymer and ion mobility. The measured temperature-dependent relaxation times qualitatively agree with COMSOL multiphysics simulations of time-dependent ion transport in the presence of an applied field. Additional results showing EDL response to pulsed bias measurements will also be presented.

This work was supported in part by the Center for Low Energy Systems Technology (LEAST), one of six centers of STARnet, a Semiconductor Research Corporation program sponsored by MARCO and DARPA.

497561 Redox-Active Organometallic Polymers for Small Molecule Separations

Abstract:

Electroactive polymers are a class of stimuli-responsive materials with great potential as energy storage materials, drug release platforms, and electrocatalysts. Organometallic polymers are particularly attractive for selective binding processes due to the ability to molecularly tune their Faradaic processes based on ligand and metal design.[1] Here, we present a series of redox-active metallocene polymers for the selective separation of small molecules for environmental and chemical process applications.

For the anode, investigations with poly(vinyl)ferrocene (**PVF**) have previously shown significant ion-selective binding properties of these redox-electrodes towards various organic functional groups, especially carboxylates. During oxidation, poly(vinyl)ferrocenium specifically binds to carboxylates with >300-fold separation factors and >250 mg/g adsorption capacities.[2] We show that through organic modification of the ligand, these anion-metallocene interactions can be enhanced even further to achieve fine chemical separations between very similar, substituted benzoic acids, through modulation of the redox-properties. In parallel, at the cathode, cobalt-based polymers [based on Cobaltocenium (**PMAECoCp$_2$**) and (Î·^{5}-cyclopentadienyl)-cobalt(Î·^{4}-tetraphenyl-cyclobutadiene) (**PCpCoCb**)Â complexes] are utilized to asymmetrically couple with the anode ferrocene to enhance electrochemical performance.[3] In particular, during reduction PCoCpCb can become negatively charged and selectively remove a range of cations, including heavy metals and aromatic organics. Finally, we show recent work on how these tandem asymmetric systems can be used for efficient bulk deionization.

Thus, whether used individually or in tandem, properly designed redox-active polymers have shown to be efficient materials for achieving selective ion-binding in both aqueous and organic media, at moderate to low voltage windows - pointing to their unique capability as a bridge between energy and environmental applications.

References

(1) Su, X.; Hatton, T. A. *Advances in Colloid and Interface Science* **2016**. 10.1016/j.cis.2016.09.001.

(2) Su, X.; Kulik, H. J.; Jamison, T. F.; Hatton, T. A. *Advanced Functional Materials* **2016**, *26*, 3394.

(3) Su, X.; Tan, K. J.; Elbert, J.; Ruttiger, C.; Gallei, M.; Jamison, T. F.; Hatton, T. A. *Energy & Environmental Science* **2017**, *ASAP article*.

501148 Disorder, Nonequilibrium Transport, and the Origin of Deep Traps in Quantum Dot Solids

Abstract:

Using a combination of ultrafast spectroscopy, time-resolved optical microscopy, and kinetic Monte Carlo simulation, I will illustrate the effects of structural and energetic disorder on charge and exciton transport in colloidal quantum dot solids. In particular, I will show how analysis of early-time nonequilibrium transport phenomena can yield especially useful insight into fundamental charge and exciton transport processes. Additionally, I will present experimental evidence pointing to an unexpected origin of deep electronic traps in PbS nanocrystal solids.

497315 On the Molecular Origin of Intra-Gap Emission from $CuInSe_{2-X}S_X$ Quantum Dots

Abstract:

$CuInSe_{2-x}S_x$ (CISeS) quantum dots (QDs) have recently gained attention as heavy-metal free alternatives to Cd- and Pb-based QDs for applications in next generation solar and solid-state lighting technologies. CISeS QDs exhibit several unusual photophysical properties that are distinct from those of structurally similar II-VI QDs, and have broad implications for their use in photonic devices. The large Stokes shifts ($\Delta s \geq 250$ meV), for example, inhibits reabsorption losses in luminescent solar concentrators by reducing the spectral overlap between the emission and absorption spectra. On the other hand, the broad linewidths limit their use in high-color-definition displays. A possible explanation for these spectral features is the involvement of intra-gap states containing a localized hole capable of coupling with a conduction band electron for a radiative transition. However, the origin of these intra-gap states, and their role in both emissive and non-emissive decay channels still remain poorly understood. Here, we address these questions with quantum chemical calculations supported by spectral electrochemistry (SEC) and ultra-fast spectroscopy experiments. Density-functional theory (DFT) calculations predict the formation of intra-gap states, which arise from either copper anti-site swapping (Cu_{In}^{2-} + In_{Cu}^{2+}), or from excess positive charge accumulated on a lattice copper in order to charge-balance a copper vacancy (V_{Cu}^{-} + Cu_{Cu}^{+}, which is a Cu^{2+} atom on a Cu^{+} site). The density of states (DOS) for both defect-free QDs and QDs with copper defects are then calculated and used to predict the "theoretical Stokes shift", which is determined by the energy offset between the DOS localized at the Cu-based defect and the valence band. The theoretical Stokes shifts for the combined set of copper defects vary from ~0.3 eV to ~0.7 eV based on their proximity with their corresponding charge-balancing paired defect (In_{Cu}^{2+} in the case of Cu_{In}^{2-} and V_{Cu}^{-} in the case Cu_{Cu}^{+}). These results are in excellent agreement with experimental studies, which show that individual CIS QDs have narrow emission linewidths with Stokes shifts that vary from QD to QD. Finally, we compare DFT calculations to SEC and ultra-fast spectroscopy measurements to determine the relationship between the copper defect's oxidation state (Cu^{+} in the case of Cu_{In}^{2-} and Cu^{2+} in the case of Cu_{Cu}^{+}) on carrier trapping and nonradiative decay. The DOS for the V_{Cu}^{-} and Cu_{Cu}^{+} clearly show magnetization and intra-gap states below the Cu_{Cu}^{+} state, which can trap photoexcited holes prior to emission. This suggests that Cu^{2+} emission occurs via a pre-existing hole due to its unfilled *d* shell, which leads to paramagnetism and requires capture of a valence band hole at a separate intra-gap defect state prior to emission. Consequently, quantum yields for CISeS QD ensembles with a high relative concentration of Cu_{Cu}^{+} defects are not significantly affected by hole trapping. This differs from QD ensembles with a high relative concentration of Cu_{In}^{2-} defects (filled *d* shell) where a valence band hole must be captured at the Cu-defect site before emission. Further considering that the emission transition arises from a conduction band electron to a Cu defect regardless of its origin or initial oxidation state, nonradiative excited-state deactivation for

either Cu_{In}^{2-} or Cu_{Cu}^{+} defects primarily occurs through electron traps. Each of these predictions are confirmed by SEC and ultra-fast spectroscopy experiments comparing stoichiometric QDs (predominantly Cu_{In}^{2-} defects) and Cu-deficient QDs (predominantly Cu_{Cu}^{+}). Lastly, we show that electron traps form as a result of inadequate passivation of the QD surface and can be removed by overcoating with a ZnS shell to achieve emission quantum yields nearing 80%.

501421 Accelerating Emission Dynamics in Perovskites Plasmonic Nanolasers

Abstract:

Halide perovskites are an emerging class of semiconductors with their long exciton lifetimes that drives perovskites' outstanding photovoltaic performance. However, such slow exciton recombination brings a fundamental limitation for nanophotonic and optoelectronic devices, eg. lasers, in applications such as ultrafast optical information processing and communications. Here we demonstrate an ultrafast perovskite plasmonic nanolaser with high emission enhancement at room temperature by a scalable solution process. The accelerated emission dynamics has been enabled by the strong plasmonic confinement of laser modes via perovskite emission hybridized with surface plasmons. It not only downscales the device but also facilitates its emission efficiency dramatically exhibiting a clear single mode lasing behavior at room temperature. Such accelerated emission dynamics also significantly facilitates laser gain compensation through Einstein equation. The capability to combine plasmonic confinement with ultrafast amplification through perovskites not only promises fundamental physics for accelerating light–matter interactions but also opens new avenues for applications from ultrafast integrated optical communications to implantable computing microchips.

491447 Synthesis and Characterization of Plasmon-Resonant Hollow Gold Nanoshells

Abstract:

Spatio-temporal controlled delivery of siRNA, proteins, and small molecules is the main goal of new drug delivery systems. Activation by highly penetrating, physiologically friendly, near infrared (NIR) light can provide sub-millisecond initiation of delivery with sub-micron spatial control. We have developed new synthesis routes to produce monodisperse plasmon-resonant, hollow gold nanoshells (HGN) with strong optical adsorption from 700 â€“ 950 nm. This covers the near-infrared window over which physiological material has minimal adsorption. We have made HGN as small as 10 nm, with < 2 nm thick gold shells as well as larger HGN with cubic and spheroidal shapes. The adsorption maxima can be tuned by controlling the ratio of shell thickness to shell diameter to make HGN of a given size with a range of NIR adsorption maxima.

Adsorption of picosecond NIR light pulses induces large, transient temperature increases in the HGN, which in turn generate transient vapor nanobubbles, similar to the cavitation bubbles generated by ultrasound. The nanobubbles can be used to rupture endosomes, liposomes and even entire cells. We show that the threshold laser fluence needed to initiate nanobubbles increases linearly with HGN radius at a given wavelength for HGN with similar adsorption maxima. Higher fluence is required for NIR wavelengths off resonance, which allows us to differentiate HGN with different adsorption maxima using a combination of NIR wavelength and laser fluence to initiate nanobubble formation. When continuous NIR irradiation is used we find that the temperature difference between the nanoparticle and the surroundings is minimal and there is little difference in performance between HGN of different sizes and shapes and differentiating between various HGN based on wavelength or fluence is difficult.

490504 Solution-Based Synthesis of Cesium Bismuth Halide Perovskite Nanocrystals for Optoelectronic Applications

Abstract:

Lead halide perovskite materials, such as methylammonium lead iodide, are interesting for a variety of optoelectronic applications. However, halide perovskites suffer from instability upon exposure to atmospheric conditions, thus much research in the area seeks to improve their stability. Bismuth halide perovskites are a potential alternative that are relatively unexplored, but have promising initial performance and are predicted to have similar properties to lead-based perovskites.

This work reports the synthesis of cesium bismuth halide perovskite nanocrystals using solution-based methods. The Bi-perovskite nanocrystals show improved phase-stability compared to Pb-based perovskite nanocrystals. Transmission electron microscopy and X-ray diffraction were used to characterize nanoparticle morphology and crystal structure. Absorbance and photoluminescence measurements of these nanocrystals reveal tunable optical properties. These Bi-based perovskite nanocrystals have potential for high performance, stable, and non-toxic solution-processable optoelectronics.

493289 Integrated Logic Gate Devices Fabricated Using Non-Toxic $CuInSe_2$ Quantum Dots

Abstract:

Copper indium diselenide ($CuInSe_2$) quantum dots (QDs) are an environmentally friendly alternative to heavy-metal based QDs for solution-processed devices. The key to using chemically synthesized QDs in electronic circuits is understanding charge-carrier transport in QD films, which is known to be influenced by factors such as surface chemistry, inter-dot coupling and doping levels. Here we study charge carrier transport in solids of $CuInSe_2$ QDs exposed to various types of surface treatments and explore their applications in solution-processed electronic devices such as field effect transistors (FETs) and logic gates. We show that using appropriate surface modifications, the QD-film transport polarity can be tuned from *p*-, to ambipolar, and *n*-type. Our initial findings suggest that as prepared $CuInSe_2$ QD films have a high density of surface defects, which, however, can be dramatically reduced by in-filling via atomic layer deposition (ALD) with a-Al_2O_3. The ALD-treatment also significantly increases carrier mobilities, reduces the free carrier density and improves filmâ€™s stability. Using ALD-in-filled QD solids, we demonstrate high performance *n*-channel FETs with good electron mobilities (up toÂ ~1 cm^2/Vs) and the on/off ratio of ~10^4.Â Furthermore, we fabricate logic gate devices such as complementary metal-oxide-semiconductor inverter and negative AND (NAND) gate digital circuits by combining unipolar *p*- and *n*-type $CuInSe_2$ QD-films. These device demonstrations point towards a considerable potential of these environmentally friendly QD materials for applications in solution-processible electronics.

491581 Synthesis and Surface Functionalization of Group IV Quantum Dots

Abstract:

Using photons instead of electrons to transport information on chips would enable wider telecommunication bandwidth and generate less waste heat compared to conventional electronic circuits. However, so-called integrated photonic circuits have been challenging to implement because Si, the cornerstone of modern electronics, has an indirect bandgap and therefore does not emit or absorb light efficiently. As a plausible alternative, quantum-sized Group IV nanocrystals, or quantum dots (QDs), can be tuned to emit and absorb light in relevant telecommunication bands by controlling size and/or composition. The synthesis of Group IV QDs with controlled size and composition has been challenging for many reasons, including a need for high temperature to crystallize Group IV elements and poor understanding over factors that govern precursor decomposition and crystallite nucleation/growth.

Here, we report the synthesis of Group IV quantum dots (QDs) that have near-infrared absorption and emission that can be tuned by changing the size or composition. We demonstrate colloidal and solid-state routes to making Group IV NCs with controlled size and composition. We also explore new surface passivation strategies that facilitate processing Group IV OQs into semiconducting solids that are useful for NIR optoelectronic applications.

487028 Understanding Rate Dependent Mechanical Properties of Supramolecular Hydrogels through Real Time SAXS Measurements during Stretching

Abstract:

Energy dissipation mechanisms are critical to the development of tough hydrogels with the double network being the most well established route. One challenge with the double network design is the toughness is derived from the breaking of covalent bonds in the sacrificial 2nd network. One alternative to the rupture of covalent bonds is the use of non-covalent interactions, such as ionic bonds, hydrogen bonds or hydrophobic aggregates, to act as the sacrificial network. In this case, the bonds can be reversible, so the energy dissipation should also be reversible. However, the exact mechanisms associated with the breaking and reforming of these non-covalent bonds are not clear. In this talk, I will discuss the use of a model hydrogel material that is based on a random amphiphilic copolymer. The hydrophobe is perfluorinated and this leads to aggregation of the fluorinated moieties into well-defined nanoscale aggregates. The higher electron density of the fluorinated domains provides x-ray contrast to enable the characterization of the nanostructure. As the crosslinks are dynamic and only based on hydrophobic association, the evolution of the nanostructure as the hydrogel is uniaxially stretched is strongly dependent on the stretching rate. Changing the stretching rate by 3 orders of magnitude leads to significant differences in the nanoscale anisotropy developed in the sample during deformation. This difference in the change in the nanostructure can be directly related to the mechanical response of the hydrogel (stress-strain curve). If time permits, a brief description of the structural recovery on cessation of deformation will be provided to provide insight into the reversibility of these tough hydrogels and the relevant time scales.

Microstructure and mechano-electrical property of ultra-stretchable iono-elastomer via small angle scattering and rheology

Ru Chen, Carlos R. López-Barrón, Norman J. Wagner
Department of Chemical and Biomolecular Engineering, University of Delaware, Newark, Delaware, USA

1. Introduction

Ionic liquids (ILs) and block copolymers (BCPs) are two representative classes of "designer compounds", which are known for tunability and controllability of their physical and chemical properties via careful selection of their components.[1] Physically or chemically crosslinking IL with BCPs can lead to the formation of solid electrolytes, also termed "ion gels". One way to produce ion gels with enhanced mechanical properties is to combine the self-assembly provided by BCPs with subsequent chemical crosslinking steps.[2] A clear benefit of this approach is that by careful design of the microstructure one could finely tune the conductivity required for a given application, and the final product is a solid electrolyte that can sustain mechanical loads without permanent deformation, *i.e.,* without flowing. In this work we use an established synthetic route[2] to prepare Pluronic F127-diacrylate (FDA) from commercial Pluronic F127. The self-assembly and rheological response of FDA in the protic ionic liquid ethylammonium nitrate (EAN) are studied and compared to those corresponding to the non-functionalized F127/EAN system.[3–5] As recently reported, F127 forms spherical micelles in EAN that assemble into a face-centered cubic (FCC) supramolecular crystal upon increasing concentration or temperature.[6] The transition from disordered to crystalline (ordered) structure manifests as a sol−gel rheological transition. Furthermore, the effects of photo-crosslinking on the microstructure, mechanical response and conductivity of F127/EAN solutions and gels are thoroughly studied. Our results provide a platform for the design of materials with enhanced viscoelasticity or with elastomeric properties to be used in electrolyte applications. The main advantage of the synthetic strategies proposed here is the use of inexpensive, commercially available copolymers along with the ease in all the synthetic steps. This work has been recently published and patented, and the readers are referred to elsewhere for detailed information.[3–5]

2. Experimental section

Sample synthesis and preparation. FDA was synthesized following the procedure reported by Cellesi *et al.*[2] FDA/dEAN solutions with compositions ranging from 5 to 24 wt % were prepared by weighing corresponding quantities of BCP and dEAN in capped vials, followed by strong mechanical mixing and centrifugation until obtaining clear, transparent solutions.

Rheology. Rheological characterization was performed using a stress-controlled rheometer (DHR-3, TA Instruments) using parallel plates of diameter = 40 mm. Dynamic temperature ramps were performed using angular frequency $\omega = 1$ rad/s and strain amplitude $\gamma_0 = 5\%$. Dynamic frequency sweeps were carried out at different temperatures ranging from 15 °C to 40 °C with $\gamma_0 = 5\%$. Dynamic time sweeps were performed during photo-curing in the ARES-G2 rheometer with 20mm parallel plate geometry with $\omega = 10$ rad/s and $\gamma_0 = 1\%$.

Static small angle neutron scattering (SANS). SANS measurements covering a q-range from 0.01 nm^{-1} to 4 nm^{-1} were performed at the National Institute of Standards and Technology (NIST) Center for Neutron Research (NCNR) in Gaithersburg, MD, on the NGB 30 m SANS instrument.

Sentmanat extensional rheometer (SER) in SANS. The tensile properties and in-situ microstructure evolution during uniaxial extension of the ionoelastomers are measured using the SER (Fig. 1) mounted in the NG 7 SANS instrument at NCNR NIST.

3. Results and discussion

Sol-gel transition. Dynamic temperature ramps were used to measure sol-gel transition, where the dynamic moduli is measured as a function of temperature for FDA/dEAN solutions with different composition. A sol-gel rheological transition upon heating is characterized by the sharp increase in five orders of magnitude in the elastic modulus. This behavior is virtually identical to that observed in the F127/dEAN system, as recently reported.[7] Note that the gel point is not affected by the presence of the acryloyl groups in the copolymer. This is expected, as these groups represent less than 1 wt. % of the copolymer. Hence, the overall hydrogen bonding strength between the oxygens in the PEO blocks and the amine group in dEAN is virtually unchanged after the acrylation of F127.

Structure and tensile properties. The self-assembly microstructure of F127DA/dEAN is studied via SANS. Analogous to the non-acrylated Pluronics (F127),[7] F127DA/dEAN solutions form spherical micelles. We found that this morphology remains unchanged upon crosslinking the solutions in presence of a UV-curing initiator. Crosslinking of concentrated F127DA/dEAN solutions produces soft elastomers (ionoelastomers) with remarkable large elongation at break and high ionic conductivity. Uniaxial deformation via SER induces a transition from isotropic FCC lattice grains to a hexagonal close packed (HCP) crystal structure in the F127DA/dEAN. This transition is typically observed under strong shear flows, however it is unprecedented under extensional flow. The FCC-HCP transition is reversible after unloading (Fig. 1).

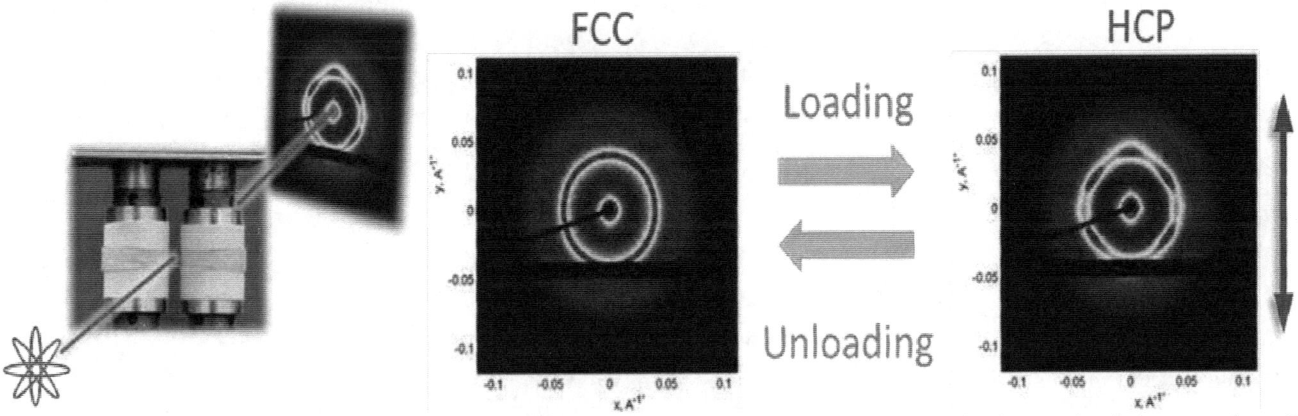

Fig.1 Left panel: shcematic representation of the in-situ SANS experiment during uniaxial extension. Rifht panel: typical 2D SANS profiles during loading and unloading cycles.

4. Conclusion

We report the fabrication of mechanically robust and highly conductive ion gels based on the protic IL, EAN, and an end-functionalized amphiphilic triblock copolymers (Pluronic F127-diacrylate) via UV-initiated crosslinking. This approach takes advantage of the self-assembly of the triblock copolymer in the IL, to design the desired micellar structure, and the subsequent chemical bonding to enhance the mechanical strength. We found that the acrylation of F127 has no effect neither on the formation of micelles, nor on the sol-gel transition, nor in the lattice structure of the concentrated gels. Moreover, the thermal response of the structure for the solutions with compositions below the gel transition are unaffected by crosslinking. These ionoelastomers display outstanding tensile responses and high ion conductivity.

5. Acknowledgement

We acknowledge the support of the National Institute of Standards and Technology (NIST), U.S. Department of Commerce, in providing the neutron research facilities used in this work. The statements, findings, conclusions and recommendations are those of the author(s) and do not necessarily reflect the view of NIST or the U.S. Department of Commerce. N.J.W and R.C. acknowledge support of cooperative agreements 70NANB12H239 and 70NANB15H260 from NIST, U.S. Department of Commerce. R.C. also acknowledges support from the National Science Foundation Graduate Research Fellowship Program under Grant No. 1247394.

References:

(1) Xie, R.; López-Barrón, C. R.; Wagner, N. J. Self-Assembly of Block Copolymers in Ionic Liquids. In *Ionic Liquids: Current State and Future Directions, ACS Symposium Series 1250*; Shiflett, M. B., Scurto, A. M., Eds.; Oxford University Press, 2017; pp 83–142.

(2) Cellesi, F.; Tirelli, N.; Hubbell, J. A. Materials for Cell Encapsulation via a New Tandem Approach Combining Reverse Thermal Gelation and Covalent Crosslinking. *Macromol. Chem. Phys.* **2002**, *203* (10–11), 1466–1472.

(3) López-Barrón, C. R.; Chen, R.; Wagner, N. J.; Beltramo, P. J. Self-Assembly of Pluronic F127 Diacrylate in Ethylammonium Nitrate: Structure, Rheology, and Ionic Conductivity before and after Photo-Cross-Linking. *Macromolecules* **2016**, *49* (14), 5179–5189.

(4) López-Barrón, C. R.; Chen, R.; Wagner, N. J. Cross-Linked Ionoelastomers with Outstanding Tensile Responses and High Ion Conductivity. U.S. Patent Serial No. 62/393,133, 2016.

(5) López-Barrón, C. R.; Chen, R.; Wagner, N. J. Stretchable Iono-Elastomers with Mechano-Electrical Response, Devices Incorporating Iono-Elastomers, and Methods of Making Thereof. PCT/US17/26621, 2017.

(6) López-Barrón, C. R.; Li, D.; Wagner, N. J.; Caplan, J. L. Triblock Copolymer Self-Assembly in Ionic Liquids: Effect of PEO Block Length on the Self-Assembly of PEO-PPO-PEO in Ethylammonium Nitrate. *Macromolecules* **2014**, *47* (21), 7484–7495.

(7) López-Barrón, C. R.; Wagner, N. J.; Porcar, L. Layering, Melting, and Recrystallization of a Close-Packed Micellar Crystal under Steady and Large-Amplitude Oscillatory Shear Flows. *J. Rheol. (N. Y. N. Y).* **2015**, *59* (3), 793–820.

491437 Toughening Isotactic Polypropylene with Block Copolymer Micelles

Abstract:

A series of compositionally symmetric poly((ethylene-*alt*-propylene)-*b*-(ethylene-*ran*-ethyl ethylene)) (PEP-*b*-PEEE) block copolymers (BCPs) with molecular weights ranging from 50 to 240 kg/mol were synthesized by anionic polymerization of isoprene and butadiene followed by heterogeneous catalytic hydrogenation. These materials were melt blended with isotactic polypropylene (*i*PP) at loadings from 1.25-20 wt.% and the resulting morphologies were investigated using transmission electron microscopy (TEM) and scanning electron microscopy (SEM). Discrete BCP micelles with an average diameter of approximately 100 nm were uniformly dispersed in *i*PP matrix of 5 wt.% blends. Tensile testing was used to study the blend mechanical properties. The nano-sized BCP micelles led to a 25-fold increase in the strain at break (Îµ$_b$ â‰ˆ 20% for pure iPP), compared to Îµ$_b$ â‰ˆ 100% in the blend containing 5 wt.% PEP homopolymer (a model analogue of ethylene propylene (EPR) rubber). A BCP loading of just 2.5 wt.% produced an optimal combination of high toughness with little or no discernable loss in elastic modulus or tensile strength. The blend mechanical properties also have been evaluated with notched Izod impact testing. Blends containing 5 wt.% and 10 wt.% BCP displayed impact strength that was 5 and 12 times that of the pure *i*PP (23.6 J/m), again significantly outperforming the PEP based counterpart. The toughening mechanisms have been investigated using electron microscopy, which revealed cavitation induced shear yielding and multiple crazing. A published theory that accounts for this type of toughening has been employed to model yielding the rubber modified solids when subjected to different stress states leading to predictive criteria for cavitation and shear yielding.

501590 The Importance of Crystalline Structure on the Tensile Properties of Uhmwpe

Abstract:

A.J. Pennings experimentally established a set of gel spinning conditions to produce stiff and strong ultra-high molecular weight polyethylene fibers in a series of papers that concluded slow spinning speeds and low draw ratios produce crystal morphology better suited for post drawing. Recent research has suggested that the initial structure of gel-spun fibers is irrelevant and with proper post-drawing parameters, fibers comparable to state of the art systems can be produced. In this work, we resolve the literature contradictions by utilizing a novel gel-spinning apparatus with precise control of extensional flow, and timescale properties of the spinning process. This allows for detailed analysis of the as-spun fiber in terms of conditions experienced during initial processing, specifically strain rate, and draw ratio. The crystalline structure of various as-spun fibers will be classified by small angle x-ray scattering, and correlated to the processing conditions. Post drawing and final mechanical property measurement is performed on a modified extensional rheometer. This information will provide an understanding of the effects of gel-spinning on fiber morphology and mechanical properties, and to answer if certain gel-spun fibers produce superior post drawn fibers.

496531 Controlling Polymerization Induced Phase Separation (PIPS) Using the Nonlinear Optical Properties of Light

Abstract:

The organization of polymer blends and polymer solutions is critical for their processing into functional materials, which are attractive for applications in lithium-ion and proton transport; thermo-electrics; solid-state lighting; and advanced composites. Scalably and precisely controlling their morphology remains a challenge, yet is crucial for the advancement of their structure-property relationships. To this end, photopolymerization-induced phase separation (PIPS) is a promising route to morphology control, and uniform irradiation and holographic polymerization are currently the two major approaches. We report our recent work on controlling PIPS of both photoreactive polymer blends and photopolymer-solvent mixtures by eliciting nonlinear optical interactions between the mixtures and the irradiation source. Photopolymerization induces changes to the refractive index of the mixtures that result in strong modulations to light propagation, which creates spatially varying photopolymerization rates. Thereby, a variety of optical patterns spontaneously form, and PIPS is induced in the regions of high intensity. As a result, the mixtures evolve into a morphology that mimics the profile of the transmitted light. We harness this mutual, dynamic interaction between polymer and light to direct morphology evolution in photopolymers, binary polymer blends, and photopolymer-solvent mixtures. Precise control over morphology is achieved over large areas and depths. The new types of morphologies achieved are discussed in terms of their potential applications as solid polymer electrolytes, separation membranes, and anti-wetting surfaces. Coupling nonlinear optical dynamics to the nonlinear chemical dynamics of phase separation represents a fundamentally new processing approach to control phase evolution in polymer mixtures.

495562 Engineering Polymer Structure and Dispersity to Access Designer Materials with Exquisite Properties

Abstract:

We recently reported a versatile and scalable approach to access a library of monodisperse oligomers from a wide range of monomer families.[1] Such monodisperse materials have allowed mixtures with tailored dispersity to be prepared and their properties examined with unprecedented accuracy. In one study, a class of conjugated oligomer was separated into a library of discrete species, each with distinct optoelectronic properties. Upon mixing these discrete species, a series of artificially polydisperse samples having similar average number of molecular weight were found to give remarkably different fluorescence (Figure 1). In another study, we examined the effect of block dispersity on the self-assembly behavior of short block copolymer. In contrast to the disordered state of the polydisperse sample, the monodisperse sample formed an ordered assembly with a feature size of < 10 nm. Such findings reinforced the importance of dispersity for designing next-generation polymeric materials.

1. Lawrence, J.; Lee, S.-H.; Abdilla, A.; Nothling, M. D.; Ren, J. M.; Knight, A. S.; Fleischmann, C.; Li, Y.; Abrams, A. S.; Schmidt, B. V. K. J.; *et al.* A Versatile and Scalable Strategy to Discrete Oligomers. *J. Am. Chem. Soc.* **2016**, *138*, 6306â€“6310.

499526 Thermal Bridges for Phonon Transport through Short Polymer Chains in a Hydrogen Bonded Polymer Composite

Abstract:

Most work to develop thermally conductive polymer composite are based on traditional fillers (metallic/ceramic/carbonic) and matric system which requires high loading of fillers and in turn suffers with various cost and fabrication issues. In this work, thermal highways for heat conduction were introduced by engineering intermolecular interactions within the polymer chain, which in turn drives thermal conduction in polymers films without using any traditional fillers. Thermal conduction pathways were developed in a blend film of long chain Poly(vinyl alcohol) and short chain Poly(ethylene glycol) which lead to thermal conductivity enhancement of around 1.6 times the neat polymer. The critical factor responsible for thermal conduction in these films was found to be homogeneous distribution of â€œthermal bridgesâ€ formed by hydrogen bonding between polymer chains. These thermal bridges leads to the formation of continuous thermal network for the efficient phonon transport in blend films. This work presents a fascinating yet promising non-conventional method to develop thermally conductive polymer based material without using traditional fillers for thermal management applications.

496327 The Effect of Pendant Alkyl Chain Length in Epoxy/Amine Thermosets Materiel Properties and Water Barrier Properties

Abstract:

Aliphatic amido amine crosslinkers have been commercially used in epoxy coatings to enhance water barrier properties for corrosion prevention. However, there is only an empirical understanding of how these aliphatic polyamine crosslinkers affect properties of thermoset polymers, which limits our ability to design newer and more environmentally friendly coating systems. In this work, a series of model aliphatic amido amine curing agents are synthesized and used to create model epoxy/amine networks found in commercial coating formulations. These model systems were interrogated to understand the effects of pendant alkyl chain length material properties including storage modulus, glass transition temperature, solubility, diffusivity, and permeability of water. This work shows that the pendant alkyl chains form an â€œideal mixtureâ€ with the polymer backbone, where material properties such as storage modulus and solubility of water are observed to be linearly dependent on the pendant alkyl chain volume fraction. Figure 1 shows our proposed hypothesis, where addition of pendant alkyl chains reduces the solubility of water by excluding some of the available free volume. We propose that the alkyl chains simultaneously enhances the diffusivity of water by the dispersal of higher water diffusivity segments (alkyl chains).

Novel Chemistries for the Replacement of Methylenedianiline in Polymer Composites

Jayson D. Cosgrove, Kevin M. Schmalbach, Owen M. Stecca, Alexander W. Bassett, ***Joseph F. Stanzione, III, Ph. D.***

Rowan University Department of Chemical Engineering

John J. La Scala, Ph. D.

Joshua M. Sadler, Ph.D.

United States Army Research Laboratory

2017 AIChE Annual Meeting

Minneapolis, MN - October 29th – November 3rd

Structure and Properties in Polymers

MDA Needs a Re cement

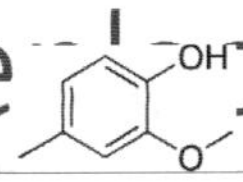

- 4,4'-methylenedianiline (MDA) used in polyimide resin (PMR-15) for aircraft engines and rocket casings
- Problem: MDA is a known carcinogen and liver toxin
  - Objective: develop a novel alternative to MDA
- Requirements for new monomer: eliminate toxicity, maintain thermal and mechanical properties, simplify composite pro[redacted]ng, and reduce lifecycle costs

$(H_2N)_n$-S-Ring-(Connector-Ring')$_p$-S-$(NH_2)_m$ (with substituents R', R'', R''', R'''')

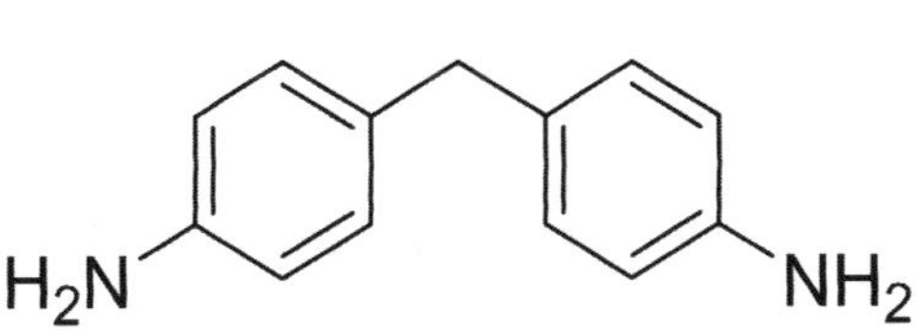

High Performance Alternative PMR Composites

Approach to Finding Alternatives

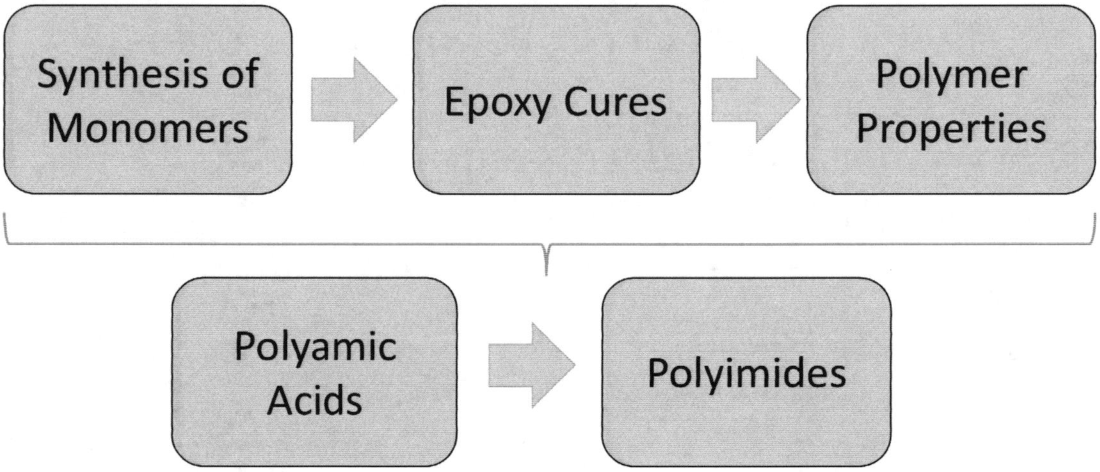

Synthesized MDA Alternatives

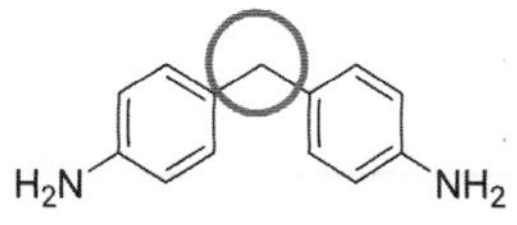

4,4′-methylenedianiline (MDA), MP = 89 °C

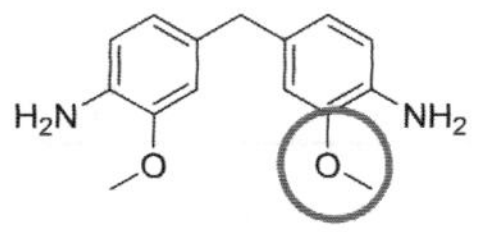

4,4′-methylenebis(2-methoxyaniline), MP = 102 °C

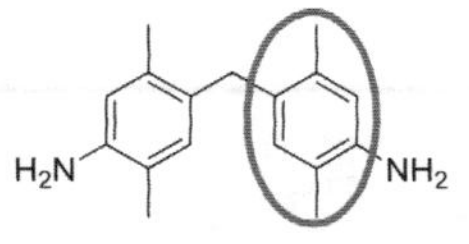

4,4′-methylenebis(2,5-dimethylaniline), MP = 143 °C

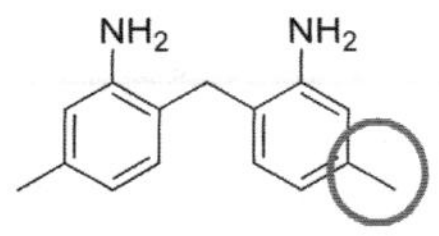

6,6′-methylenebis(3-methylaniline), MP = 124 °C

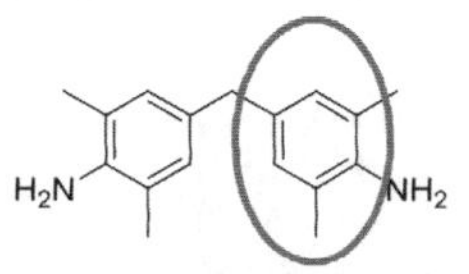

4,4′-methylenebis(2,6-dimethylaniline), MP = 122 °C

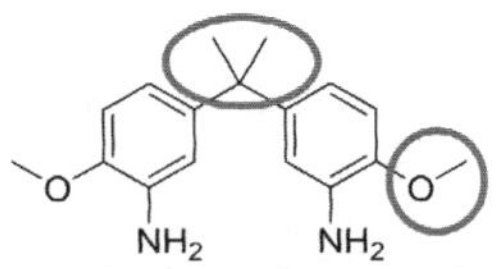

5,5′-(propane-2,2-diyl)bis(2-methoxyaniline), MP = 104 °C

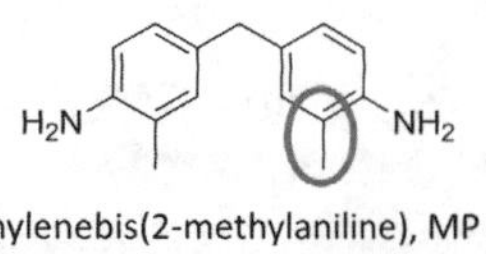

4,4′-methylenebis(2-methylaniline), MP = 160 °C

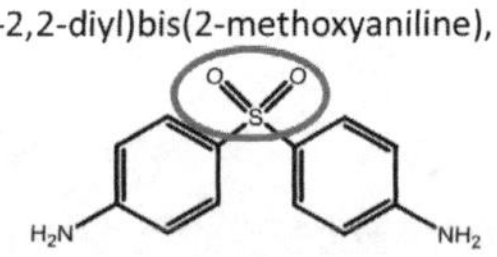

4,4'-sulfonyldianiline (Dapsone), MP= 175 °C

Synthesis of Substituted Dianilines

- Purified using flash chromatography (ethyl acetate and hexanes gradient)
- ~ 40 % yield

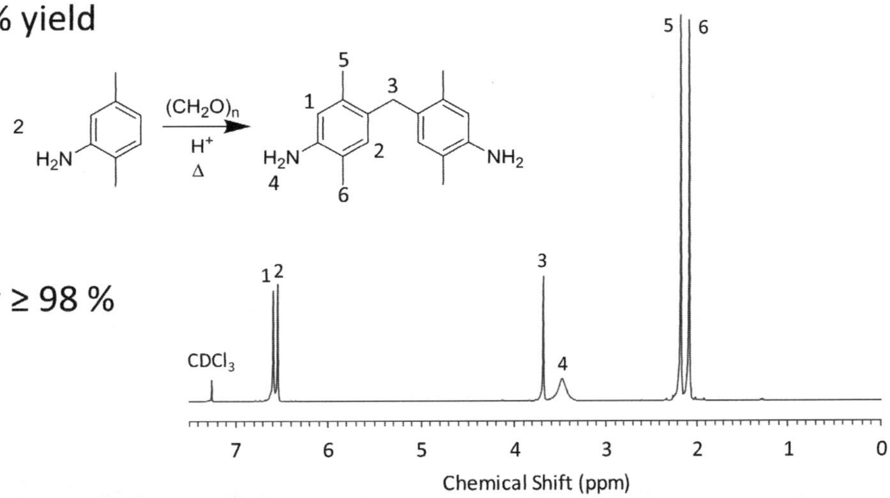

- Purity ≥ 98 %

Epoxy-Amine Curing

- Blended dianiline in equal weight ratio of Epikure W
  - Mixed with stoichiometric equivalents of Epon 828
- Cure cycle: 140 °C for 5 hr, 230 °C for 3 hr
- Post-cured samples cut and sanded into thin (1.5 mm) and thick (2.5 mm) dimensions[1]

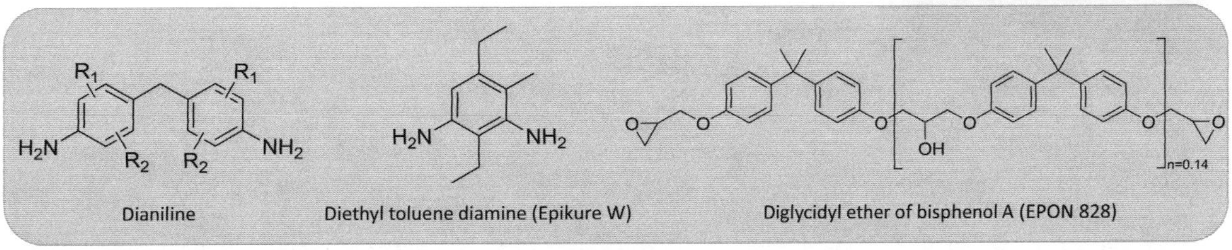

[1] McAninch, I. M.; Palmese, G. R.; Lenhart, J. L.; La Scala, J. J. Polym. Eng. Sci. 2015, 55, 2761.

High Extents of Cure Achieved

- Spectra acquired both before cure and after post-cure
- All polymers cured to ≥ 99 %
  - Epoxide ring **4530** and **6060** cm^{-1}
  - 1° amine **4925** cm^{-1}
  - 1° & 2° **amines** 6535 cm^{-1}
  - Hydroxyl **6600-7000** cm^{-1}

$$\text{Extent of Cure} = \frac{\left(\frac{Abs_{4530}}{Abs_{ref}}\right)_{precure} - \left(\frac{Abs_{4530}}{Abs_{ref}}\right)_{postcure}}{\left(\frac{Abs_{4530}}{Abs_{ref}}\right)_{precure}}$$

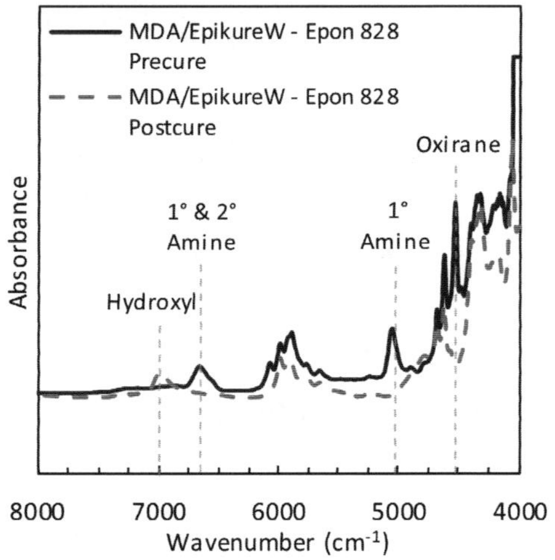

Reference peaks around 6000 cm^{-1}

[2] E. D. Hernandez, A. W. Bassett, J. M. Sadler, J. J. La Scala, J. F. Stanzione, *ACS Sustainable Chemistry & Engineering* **2016**.

Methyl Groups Increase T_g

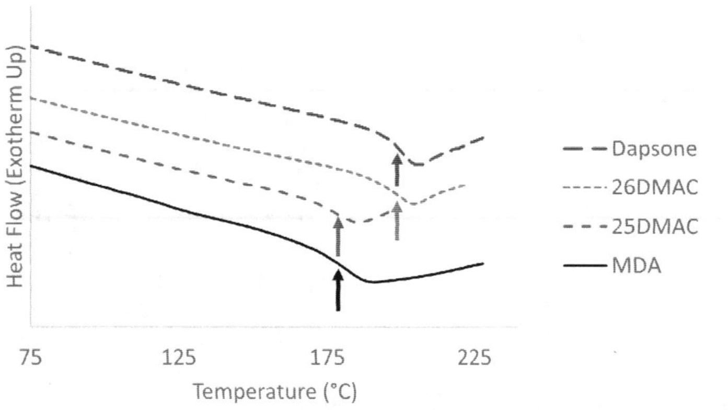

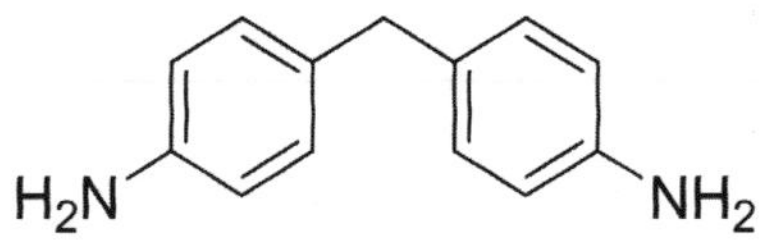

T_g= 178 °C

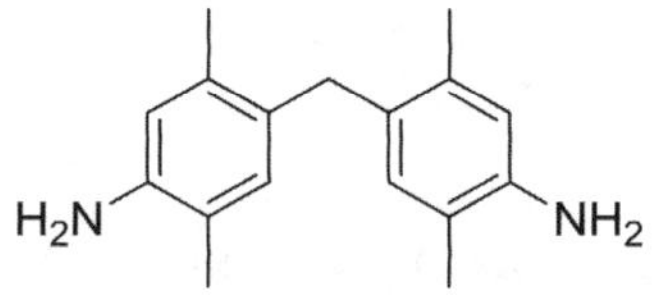

T_g= 179 °C

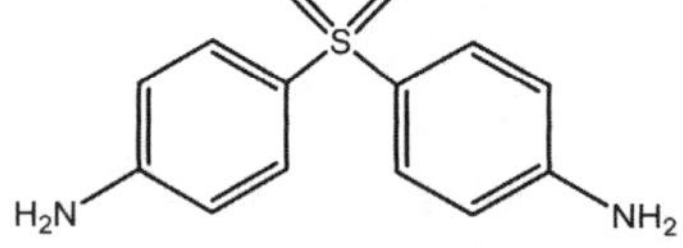

T_g= 198 °C

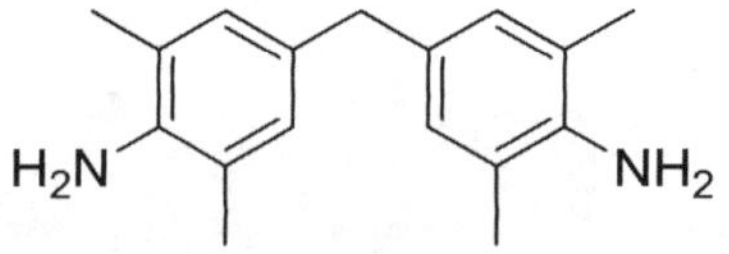

T_g= 199 °C

Methyl Groups Increase Glassy Modulus

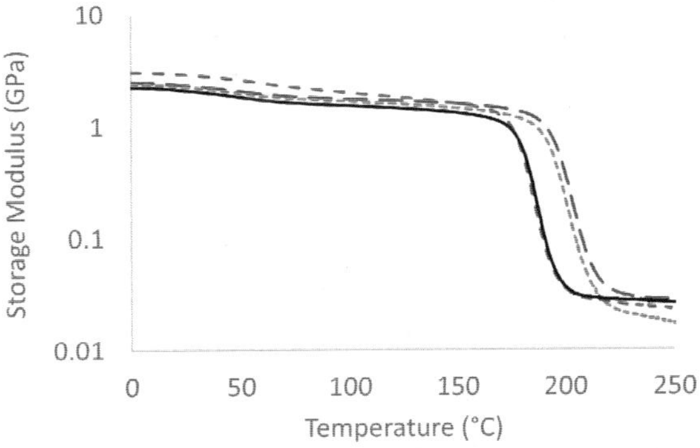

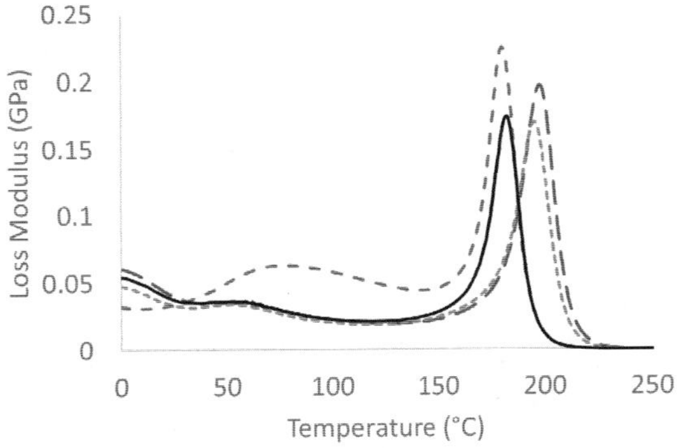

(Not shown) MDA: E'= 2.1 GPa, T_g = 182 °C

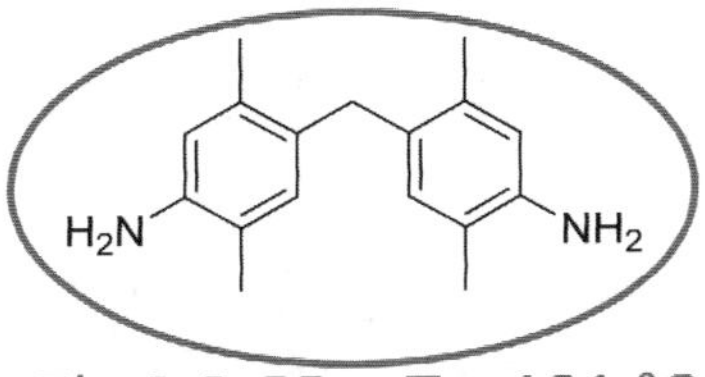

E'= 3.0 GPa, T_g= 181 °C

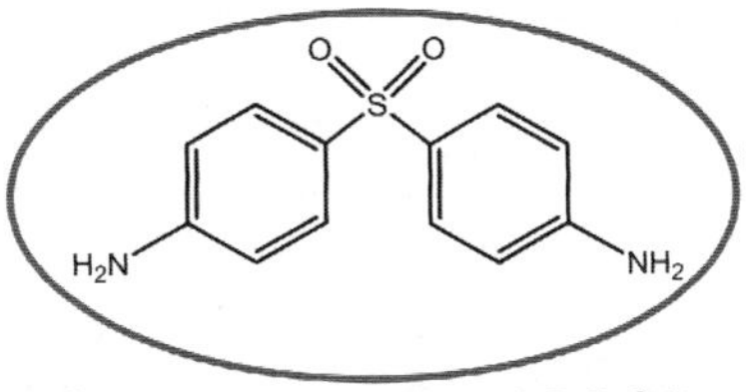

E'= 2.4 GPa, T_g= 198 °C

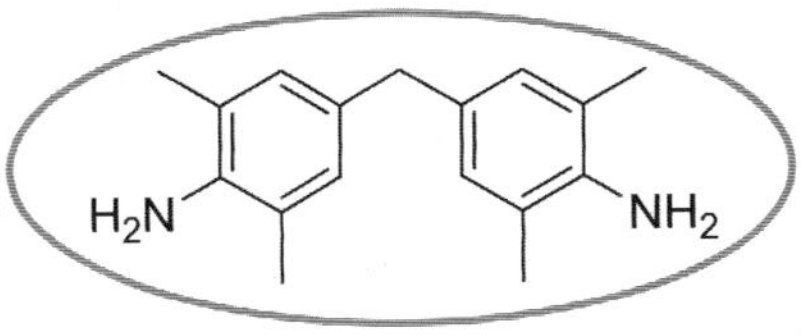

E'= 2.3 GPa, T_g= 196 °C

Substituents Do Not Affect Thermal Properties

- All initial degradation temperatures ≈ 365 °C
  - Exception, Dapsone = 375 °C
- All $T_{50\%}$ ≈ 395 °C
- All char content ≈ 10 %
- T_{max} ≈ 390 °C

TGA Trace for 2,5-DMAC
Atmosphere = N_2

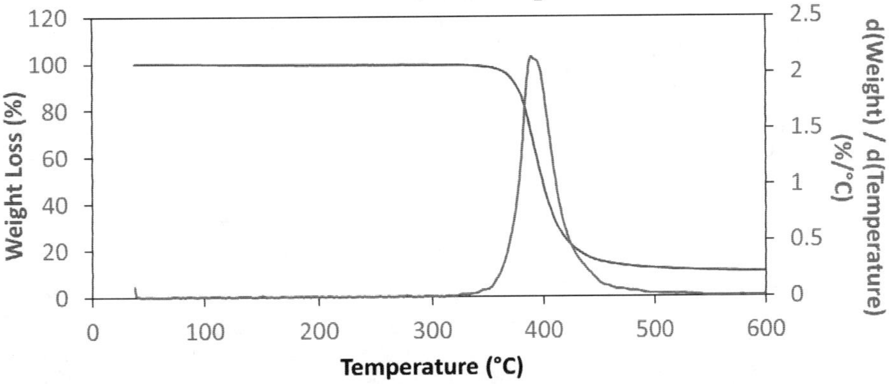

From Dianilines to Polyamic Acids

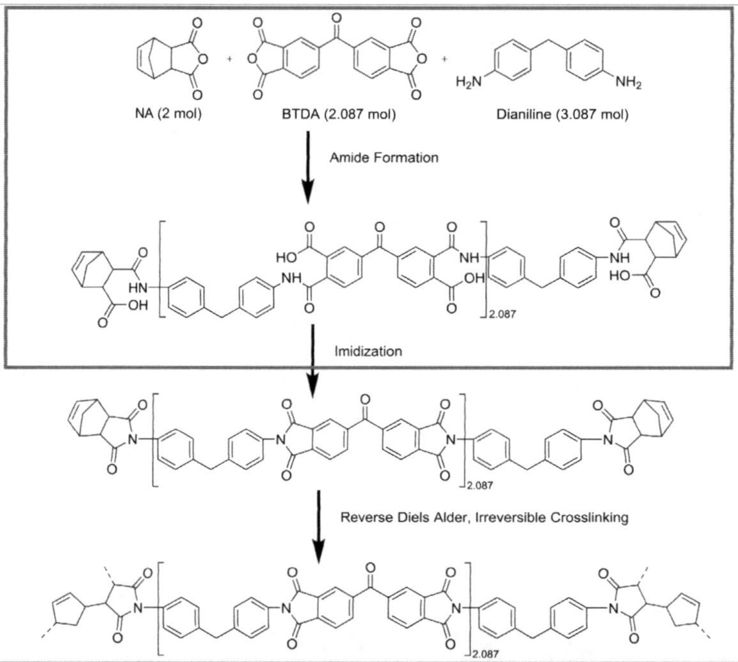

Synthesized Polyamic Acids Are of Appropriate Molecular Weight

- Nadic anhydride (NA) and 3,3', 4,4'-Benzophenonetetracarboxylicdianhydride (BTDA) blended at 90 °C in methanol for approximately 2 hours
  - Reaction cooled before adding dianiline

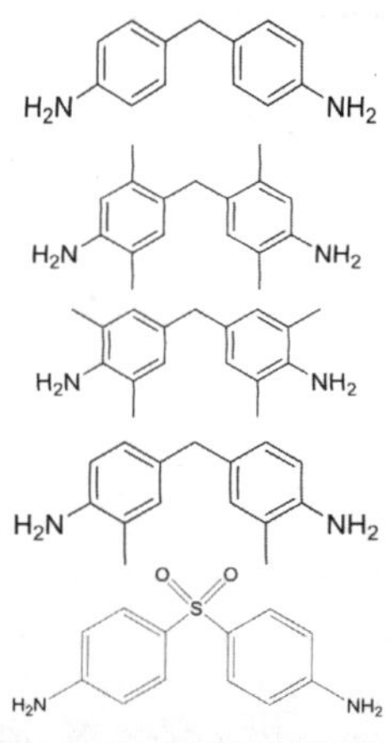

Dianiline Used	Highest MW (Da)
MDA	970
2,5-DMAC	980
2,6-DMAC	560
2-MAC	970
Dapsone	650

Conclusions

- Synthesized a variety of substituted dianilines
- Epoxies cured with substituted dianilines
  - Epoxies proxy for polyimides
- Achieved properties similar to MDA
  - Several dianilines displayed similar T_gs to MDA
  - All dianilines displayed higher glassy modulus
- Those with similar T_gs will undergo further investigation

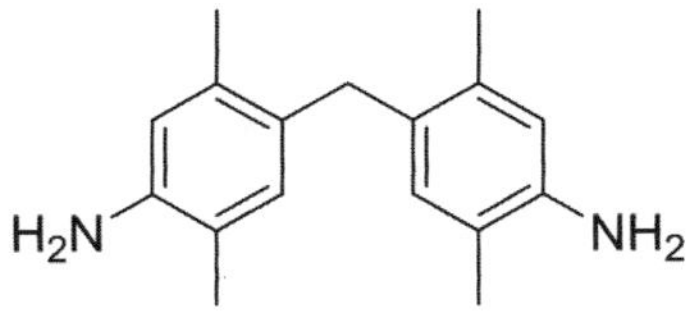

4,4'-methylenebis(2,5-dimethylaniline)

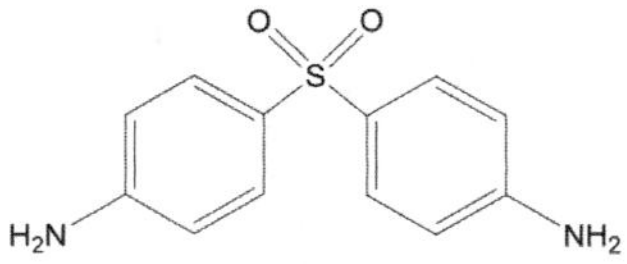

4,4'- sulfonyldianiline (Dapsone)

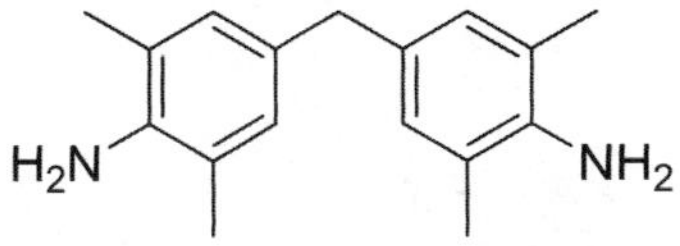

4,4'-methylenebis(2,6-dimethylaniline)

Completed Steps

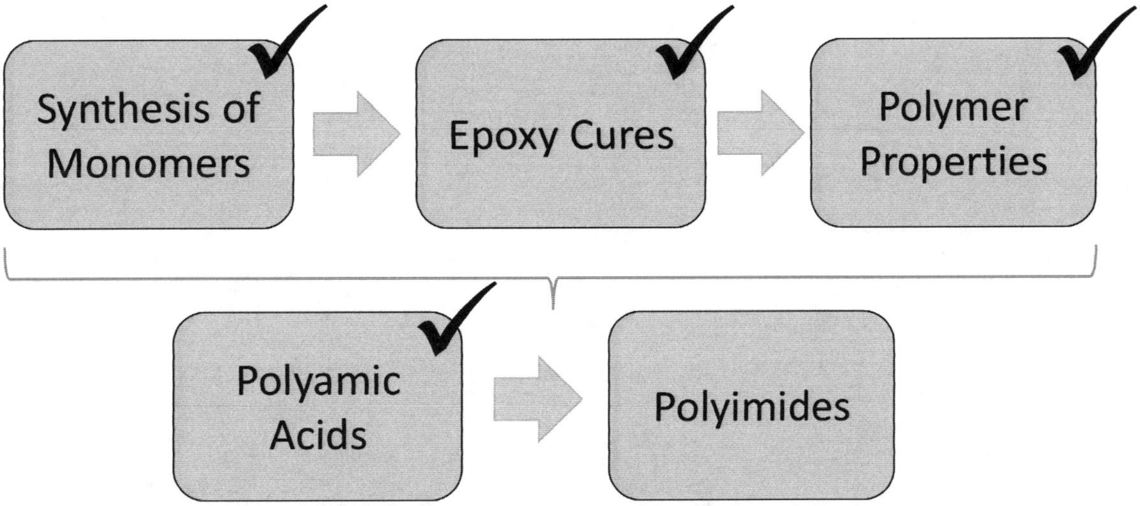

Next Steps

- Convert polyamic acids to polyimides
- Mechanical and thermal properties of polyimides
- Monomer toxicity studies

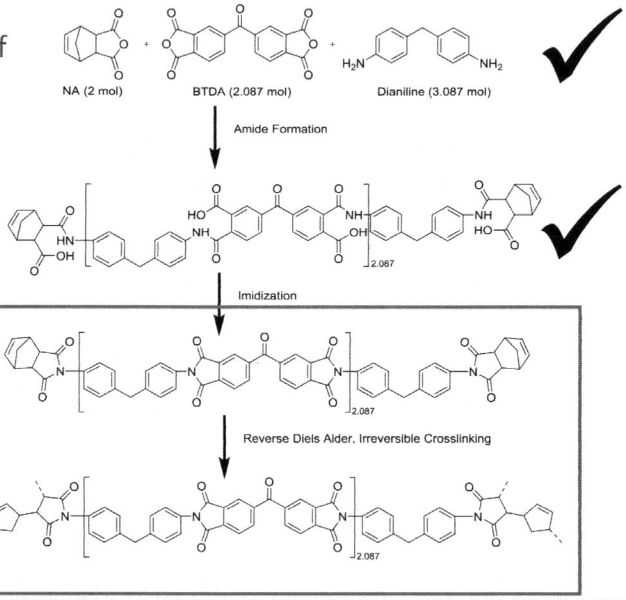

Acknowledgements

- Sustainable Materials Research Group
- Drs. Joshua M. Sadler, Craig M. Paquette, US Army Research Laboratory
- Dr. William S. Eck, US Army Public Health Command
- Rowan University Department of Chemical Engineering
- Rowan University Department of Chemistry
- Financial Support Provided Through SERDP WP-2402 and ARL Cooperative Agreement W911NF-14-2-0086

488883 Improvement of Thermal Conductivity for POSS Functionalized BN Fillers/Polyphenylene Sulfide Composites

Abstract:

Polyphenylene sulfide (PPS) presents excellent comprehensive properties and has been widely applied in the fields of electron, automobile, aerospace and machinery, etc. However, relatively low thermally conductive coefficient (*Î»*) of pristine PPS matrix has limited its wider application. In our present work, the combining method of aminopropyllsobutyl polyhedral oligomeric silsesquioxane (NH_2-POSS) and silane coupling reagent of Î³-aminopropyl triethoxy silane (KH-560) was performed to functionalize the surface of boron nitride (BN) fillers, aiming to fabricate the corresponding *f*-BN/PPS composites with enhanced thermal conductivities and optimal dielectric properties. Results revealed that the NH_2-POSS molecule had been grafted on the surface of the BN fillers. And the surface functionalization was benifit for improving the thermal conductivities and decreasing dielectric constant of the PPS composites. The obtained *f*-BN/PPS composite with 60 wt% *f*-BN fillers is an excellent polymeric composite with high *Î»*, ideal dielectric constant (*Îµ*) and outstanding thermal stability, *Îµ* of 3.99, *Î»* of 1.122 W/mK (increased by 400% compared to that of pristine PPS matrix), and $T_{Heat\text{-}resistance\ index}$ (T_{HRI}) of 275°C, which holds potential for packaging in continuous integration and miniaturization of microelectronic devices.

References:

1. J. Gu, C. Liang, X. Zhao, B. Gan, H. Qiu, Y. Guo, X. Yang, Q. Zhang, D. Wang, *Compos. Sci. Technol.*, 2017, **139**, 83-89.

2. J. Gu, X. Yang, Z. Lv, N. Li, C. Liang, Q. Zhang, *Int J Heat Mass Tran*, 2016, **92**: 15-22.

3. J. Gu, Y. Guo, X. Yang, C. Liang, W. Geng, L. Tang, N. Li, Q. Zhang, *Compos. Part A*, 2017, **95**, 267-273.

4. J. Gu, Z. Lv, Y. Wu, Y. Guo, L. Tian, H. Qiu, W. Li, Q. Zhang, *Compos. Part A*, 2017, **94**, 209-216.

5. J. Gu, X. Meng, Y. Tang, L. Yang, Q. Zhuang, J. Kong, *Compos. Part A*, 2017, **92**, 27-32.

506140 Structural Composites with Multiple Functionalities

Abstract:

Polymer-based composites are susceptible to many different types of mechanical damage that may reduce their structural reliability, thermal conductivity, optical transparency, and surface qualities, thereby potentially decreasing the overall lifetime of these products.[1-5] We have previously reported on the fabrication of epoxy-based transparent composite coatings consisting of tailored halloysite nanotubes (HNTs) orientations.[6, 7] Mechanical reinforcement (i.e., stiffness and hardness) and wear-resistance of the resultant composites were correlated to the specific alignment of the HNTs. Composites displaying a high transparency of 90% even at a high halloysite concentration of 20 wt% can be achieved by matching the refractive index between the HNTs and polymer matrix ($n \sim 1.5$ for both) [6, 7]. These transparent coatings with well-controlled thickness are very attractive for fabricating multi-functional composites. However, it is challenging to find polymer additives that can be incorporated into epoxy resins with appropriate epoxy miscibility, and that result in high transparency, and unaffected mechanical robustness of the final composite.[8] The currently used self-healing additives in industry are not compatible with epoxy at macromolecular level, thus blocking the light transmittance and leading to low transparence in coatings. Examples of the self-healing additives include polycaprolactone (PCL),[9, 10] polyethylene-co-methacrylic acid (EMAA),[11, 12] and poly(bisphenol A-co-epichlorohydrin) (PBE).[12-14]

In this research, we studied the self-healing mechanisms that recovers mechanical robustness and light transparency in epoxy-based composite coatings. By demonstrating that the inclusion of cellulose derivatives can be utilized to produce self-healing composites, self-cleaning properties were also studied. Cellulose derivatives in our material system is an ideal candidate to compound with epoxy because of its solubility in a wide range of solvents compatible with our epoxy precursor formulations, excellent miscibility with the epoxy resin, high mechanical rigidness, and manageable self-healing transition temperatures and viscosities.[15] Importantly, the refractive index, ~ 1.5, closely matches each component of the epoxy/HNT composites. We showed a useful strategy for producing multifunctional material systems with enhanced mechanical durability, transparent optical properties, reversible self-healing behavior, as well as self-cleaning ability, which is practically applicable in optical systems.

1. Trask R, Williams H, and Bond I. Bioinspiration & Biomimetics 2007;2(1):1-9.

2. Tuteja A, Choi W, McKinley GH, Cohen RE, and Rubner MF. MRS bulletin 2008;33(08):752-758.

3. Youngblood JP and Sottos NR. Mrs Bulletin 2008;33(08):732-741.

4. Yuan Y, Yin T, Rong M, and Zhang M. Polym. Lett 2008;2(4):238-250.

5. Kessler M. Proceedings of the Institution of Mechanical Engineers, Part G: Journal of Aerospace Engineering 2007;221(4):479-495.

6. Song K, Polak R, Chen D, Rubner MF, Cohen RE, and Askar KA. ACS Applied Materials & Interfaces 2016;8(31):20396â€“20406.

7. Song K, Chen D, Polak R, Rubner MF, Cohen RE, and Askar KA. ACS Applied Materials & Interfaces 2016;8(51):35552-35564.

8. Murphy EB and Wudl F. Progress in Polymer Science 2010;35(1):223-251.

9. Rodriguez ED, Luo X, and Mather PT. ACS applied materials & interfaces 2011;3(2):152-161.

10. Luo X and Mather PT. ACS Macro Letters 2013;2(2):152-156.

11. Kalista SJ and Ward TC. Journal of the Royal Society, Interface 2007;4(13):405-411.

12. Pingkarawat K, Wang C, Varley R, and Mouritz A. Composites Part A: Applied Science and Manufacturing 2012;43(8):1301-1307.

13. Hayes S, Jones F, Marshiya K, and Zhang W. Composites, Part A: Applied Science and Manufacturing 2007;38(4):1116-1120.

14. Meure S, Wu DY, and Furman S. Acta Materialia 2009;57(14):4312-4320.

15. Gooch JW. Cellulose Acetate Butyrate. Encyclopedic Dictionary of Polymers, vol. 1. New York: Springer, 2011. pp. 128-128.

494954 Fabrication of High-Dispersed Strawberry-like P(St-*co*-MAA)/SiO_2 Composite Microspheres and Their Applications in Mortar

Abstract:

Silica nanoparticles (SiO_2 NPs) can be used in cementitious materials to improve their workability and mechanical properties. But avoiding the agglomeration of SiO_2 NPs within the cementitious matrix are still the challenges to play the role of them effectively. In this work, the poly(styrene-co-methyl acrylic acid) (P(St-*co*-MAA)) latex particles with hydrophilic carboxylic chains grafted from the surface were first fabricated via soap-free emulsion polymerization, and then high-dispersed strawberry-like composite microspheres were prepared with the P(St-*co*-MAA) latex partially covered by SiO_2 NPs through lay-by-layer (LBL) assembly process. The effects of such composite microspheres on the properties of cementitious materials were investigated afterwards.

The surface carboxylic content of P(St-*co*-MAA) latex particles was measured by conductometric titration. The morphologies of the latex particles and P(St-*co*-MAA) / SiO_2 composite microspheres were characterized by transmission electron microscopy (TEM) and scanning electron microscopy (SEM). Thermos gravimetric analyzer (TGA) and dynamic light scattering (DLS) were employed to characterize the composition and the dispersibility of the composite microspheres, respectively. After adding the composite microspheres into cementitious matrix, the hydration process of the fresh mixture was monitored by cement hydration heat analyzer. Its workability and mechanical properties were tested via conventional methods. The porosity of the hardened specimens was characterized by mercury injection method.

The results show that the surface carboxylic content of P(St-*co*-MAA) latex was about 1.5×10^{-4} *mol*/g. For the P(St-*co*-MAA) / SiO_2 composite microspheres, SiO_2 NPs with the mass ratio about 5.0 *wt*% were anchored on the latex surface, leaving above 50 % area of the surface uncovered. Such strawberry-like composite microspheres could be easily dispersed in aqueous phase without further treatment where the poly-dispersity index (PDI) was about 0.3. The composite microspheres had no obvious effects on the fluidity of the mortar and the water absorption of the hardened specimen. But it could promote the condensation of the mortar and improve their early compressive strength and flexural strength. Compared to the blank control, the 3d compressive strength of the specimens in which the mass of composite microspheres was 4 *wt* % of the cement was 5 MPa higher, and the flexural strength of those was 1 MPa higher. The modified mechanism of the composite microspheres for mortar was proposed.

501772 Hydrogen Bonded Thermal Highways Driving Heat Conduction in Polymer and Polymer Blends Films

Abstract:

The fundamental factors driving thermal conduction in neat polymers is significantly less explored. Though most work is available in thermally conductive filler-polymer composite, clear understanding of the heat conduction mechanism in polymers without adding inorganic/metallic filler is a subject of great investigation especially for the promotion and development of advanced thermal management technologies. This study shows an interesting tale of intermolecular interaction in governing thermal conduction in polymers. Here by engineering intermolecular interaction in polymer chains a significant enhancement in thermal conductivity was achieved and similarly decrement was observed by breaking interactions. Thermal conductivity of free standing PVA films with different molecular weights and its blends with PEG and PVP was investigated. Effect of the absorption of vapor molecules of water, ethanol, methanol and ethylene diamine on polymer filmâ€™s thermal conductivity was studied. Penetration of water molecule in films was found to drastically alter inter-molecular interaction contributing to remarkable 250 % increment in thermal conductivity. The thermal conductivity of PVA films and its blends as a function of moisture content, molecular weight and drying temperature was studied. Films with lower molecular weight had higher thermal conductivity due to less hindered polymer chains and ease of interaction to form continuous thermal highways for phonon transport. This study unveils some of the important yet fascinating drivers of thermal conductivity in polymers & polymer blends which can pave the way for development and fabrication of polymer based materials for thermal management applications.

501122 Low-Temperature Synthesis of Mn-Based Mixed Metal Oxides with Novel Fluffy Structures As Efficient Catalysts for Fto Reaction

Abstract:

Lower olefins, particularly ethylene (C_2H_4), propylene (C_3H_6), and butylenes (C_4H_8), are important raw materials for chemical production, which are mainly produced by catalytic cracking of crude oil. Since China is rich in coal but lack of oil, it is meaningful to develop technologies for production of light olefins directly from coal-based syngas via the Fischer-Tropsch reaction (FTO), while the key factor is the design and synthesis of catalysts with high catalytic activity and good olefine selectivity.

Herein, low-temperature crystal splitting has been developed to synthesize Mn-based mixed metal oxides with large surface area, impressively active surface and excellent catalytic activity. The catalytic activities of as-synthesized Mn-based catalysts for FTO reaction have been tested and the relationship between surface structures of the catalysts and their catalytic activities have been investigated by mean of Operando spectroscopy. The mechanism of the catalytic reaction has been revealed through the determination of the catalytic active center, thus providing guidelines for further synthesis of the catalysts.

486725 Structural Control of Polybenzoxazine/Epoxy Composites with Dual Crosslinking Network for Corrosion Protection

Abstract:

With the advent of corrosion protection technology, composite coatings have aroused much more attention among the area due to the apparent enhancement on their performance after combing with different components. Polybenzoxazine/epoxy composites were developed recently as a novel coating material for corrosion protection. The reaction between the phenolic –OH groups of the ring-opened benzoxazine and epoxide groups in epoxy can effectively increase the cross-link density of the composite matrix and improve coatings' corrosion resistance by strengthening their barrier ability against corrosive medium, which has been confirmed by our previous work[1]. The polybenzoxazine/Epoxy composite coating with dual crosslinking network exhibit a good corrosion protection improvement, a reduction of one decade on corrosion rate was observed by the incorporation of Epoxy into polybenzoxazine matrix, as well as a good hydrophobic ability with a water contact angle of 100° even after the corrosion test, which could be a fascinating characteristic for the anticorrosion application. The dual crosslinking network structure is believed to be a vital part of the coatings' performance.

Therefore, mono-, bi-, and trifunctional benzoxazine monomers are designed and synthesized in this work. Composite coatings based on these benzoxazines are prepared by combining with epoxy in order to fabricate coatings with controllable cross-link density. Thus, the relationship between coatings' crosslinking structure and performance for corrosion protection could be investigated thoroughly. The reaction between two components was characterized by Fourier transform infrared spectroscopy. The effects of composites' structure on the corrosion resistance of coated aluminum were investigated by open circuit potential and electrochemical impedance spectroscopy, respectively. Molecule stimulation was performed to calculate the cross-link density of different samples. The results of this work could provide guidance for the design of high-performance polybenzoxazine-based coatings used for corrosion protection.

Changlu Zhou, E-mail: changluzhou@ecust.edu.cn

*Correspondence Zhong Xin, Professor, Tel.: +86 21 64240862; Fax: +86 21 64251772, E-mail: xzh@ecust.edu.cn

1. Zhou C, Lin J, Lu X, et al. Enhanced corrosion resistance of polybenzoxazine coatings by epoxy incorporation.RSC Advances. 2016, 6(34): 28428-28434

488528 Multifunctional Epoxy Conductive Nanocomposites

Abstract:

The traditional epoxy conductive adhesive has many disadvantages in some aspects, such as poor dispersibility in the epoxy matrix, weak interfacial interaction, high conductivity thresholds and thin adhesive strength. Therefore, multifunctional and high-performance conductive adhesive is required urgently. In this work, the multifunctional epoxy conductive adhesive filled with a kind of filler composed of nano-silver and carbon. The polydopamine was introduced on the surface of carbon structure to improve the dispersibility of carbon materials within epoxy matrix and the interfacial interaction between nanofillers and epoxy matrix. The nano-silver was used to further increase the conductivity of epoxy adhesives. The improved adhesive property, conductivity and mechanical property of epoxy conductive adhesive was obtained due to the isolated structure which forms a bridge to transfer the charge carrier efficiently. The adhesive strength of this adhesive was enhanced compared with that of pure epoxy. The conductivity thresholds of this adhesive were decreased in comparison with the inorganic metal filler.

REFERENCES

[1] H Cui, J Jiu, et. Ultra-fast photonic curing of electrically conductive adhesives fabricated from vinyl ester resin and silver micro-flakes for printed electronics. RSC Advances, 2014, 4, 15914

[2]Â N Pua, Y Peng, et. Application of nitrogen-doped graphene nanosheets in electrically conductive adhesives. Carbon, 67(2014)449-456.

[3]Â H Gu, C Ma. An overview of multifunctional epoxy nanocomposites. J. Mater. Chem. C, 2016,4, 5890.

[4] C Yang, C Wong, et. Printed electrically conductive composites: conductive filler designs and surface engineering. J. Mater. Chem. C, 2013, 1,4052.

[5]Â T Jia, L Chena, et. In-situ reduction of Ag nanoparticles on oxygenated mesoporous carbon fabric: Exceptional catalyst for nitroaromatics reduction. Applied Catalysis B: Environmental 182 (2016) 306â€"315.

492741 Nano-Structured Ceramic ALD Coatings to Stabilize SiC Against Reaction in High Temperature Steam

Abstract:

Silicon Carbide (SiC) is an ideal material for many high-temperature applications due to its resistance to thermal shock and high thermal conductivity. However, SiC degrades in water-rich environments limiting its applications in extreme oxidative environments such as combustion engines, heat exchangers, and high temperature reactor materials. Current deposition methods for environmental barrier coating (EBCs) such as plasma spraying and aerosol spraying generate micron-scale films with inherent grain boundaries and cracks. Atomic layer deposition (ALD) generates nano-scale films that are dense, crack-free, and chemically bonded to the surface without the use of excess precursor. We have found that the application of mullite nanostructured films grown with ALD significantly improves the oxidation resistance of SiC in extreme environments with a film much thinner than standard applications. In addition to mullite, we have targeted a variety of high temperature coating materials chosen based on thermal properties, structural characteristics, and stability in oxidative environments. We have grown ALD layers of our desired materials on high surface area SiC particles using a fluidized bed particle ALD reactor. Using thermogravimetric analysis, it has been shown that these coatings improve the oxidation resistance of SiC by up to 64% at 1000°C. We have applied density functional theory (DFT) to model the performance of these mullite coatings in an oxidative environment. Computational modeling of oxygen diffusion through mullite and other candidate materials allows identification of promising coatings to further extend the lifetime of SiC. We expect to use this understanding to develop more advanced stabilizing coatings for SiC and the myriad of other applications that require high-temperature oxidation resistant materials.

508553 Multi-Scale Metrology for Visualization and Characterization of Interphase Failure

Abstract:

With recent focus on modeling efforts at the micro scale it is important to augment and validate such efforts with experimental efforts at the same scale. This paper will provide a general overview of activities at NIST focused on characterizing the micro scale mechanical properties at the fiber matrix interphase including:

i) Characterization of structure and mechanical properties of interphase
ii) Visualization of interphase failure

The objective of this work is to characterize the structure property relationship of glass fiber composites using recently improved metrology tools at macro and nano scale. The approach to characterize the interface is via interfacial shear strength measurements using single fiber fragmentation and multiple fiber fragmentation test; interphase modulus measurements using AFM; and failure mode visualization using multi-functional molecular probes and advanced fluorescence lifetime imaging microscopy. Due to the complexity of glass fiber sizing systems, model systems were used to investigate interfacial properties and fracture failure modes. It is our belief that creating a mimic to characterize and visualize interface at micro scale is important for validating algorithms for modeling and simulation. Additionally, it aids in fundamental understanding of adhesion and toughening around the interface that is critical for better performance and service life.

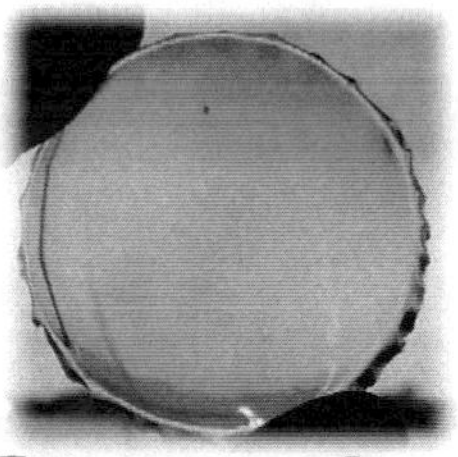

Strategic Assemblies of Wood-Derived Building Blocks for the Sustainable Redesign of BPA-Based Polymers

Joseph F. Stanzione, III, Ph.D.

Department of Chemical Engineering

Sustainable Materials Research Laboratory

2017 AIChE Annual Meeting
Minneapolis, MN October 29 – November 3 2017
"Bio-Based Polymers"

 1

About Rowan University (Glassboro, NJ)

 2

About Rowan University (Glassboro, NJ)

- Comprehensive public research university
- Three campuses:
    - ❖ Glassboro
    - ❖ Camden
    - ❖ Stratford
- ~ 17,500 students
- Schools/Colleges:
    - ❖ Business
    - ❖ Biomedical Science & Health Professions
    - ❖ Communication & Creative Arts
    - ❖ Cooper Medical School
    - ❖ Education
    - ❖ Engineering
    - ❖ Global Learning & Partnerships
    - ❖ Humanities & Social Sciences
    - ❖ Performing Arts
    - ❖ School of Osteopathic Medicine
    - ❖ Science & Mathematics

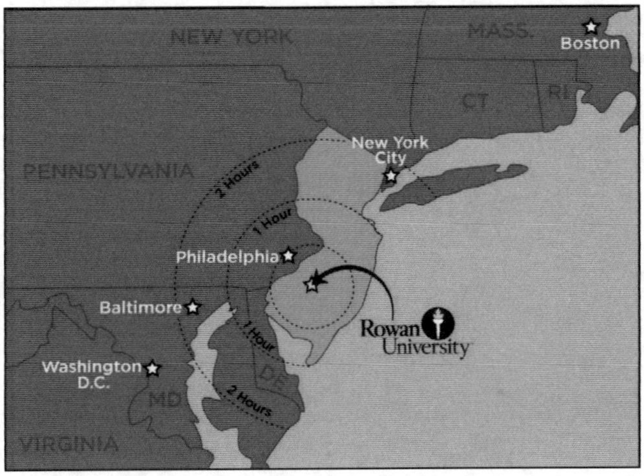

RowanUniversity
HENRY M. ROWAN
COLLEGE OF ENGINEERING

Plastics: The Facts

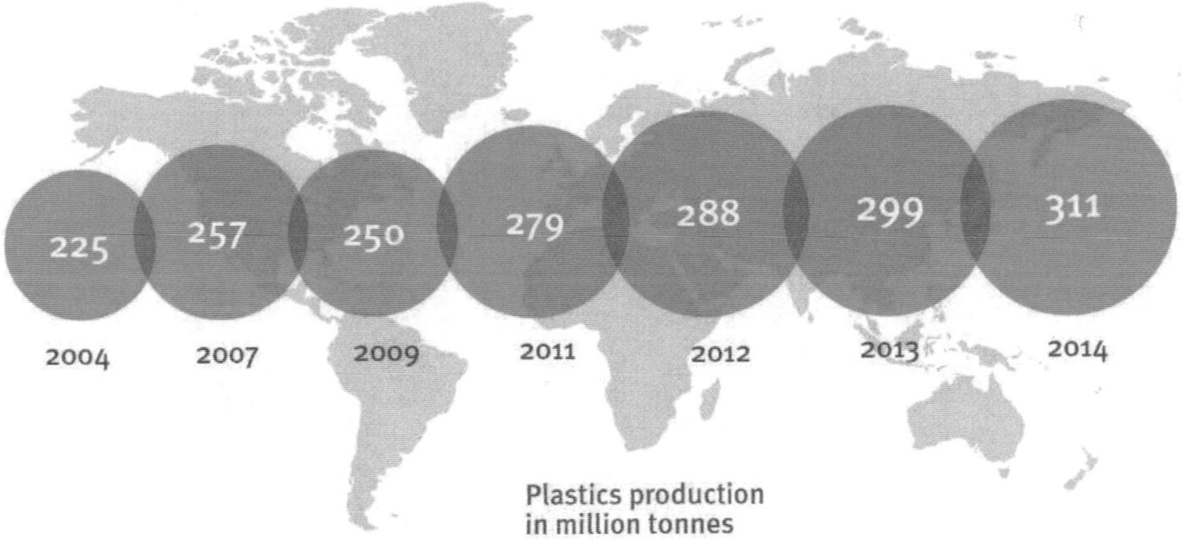

- **Plastic production is growing globally[1]**
- **Currently around 330 Mtonnes**

[1] PlasticsEurope, *Plastics - the Facts* (2015)

RowanUniversity
HENRY M. ROWAN
COLLEGE OF ENGINEERING

Plastics: The Future

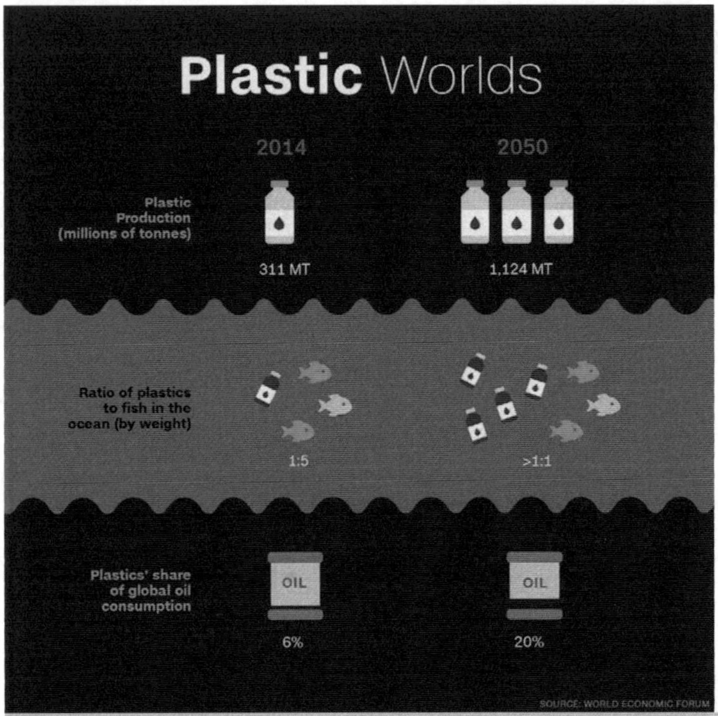

By 2050:

- Expected to triple
- More plastic than fish in the ocean
- 20% of global annual oil consumption

 5

RowanUniversity
HENRY M. ROWAN
COLLEGE OF ENGINEERING

Plastics: The Facts

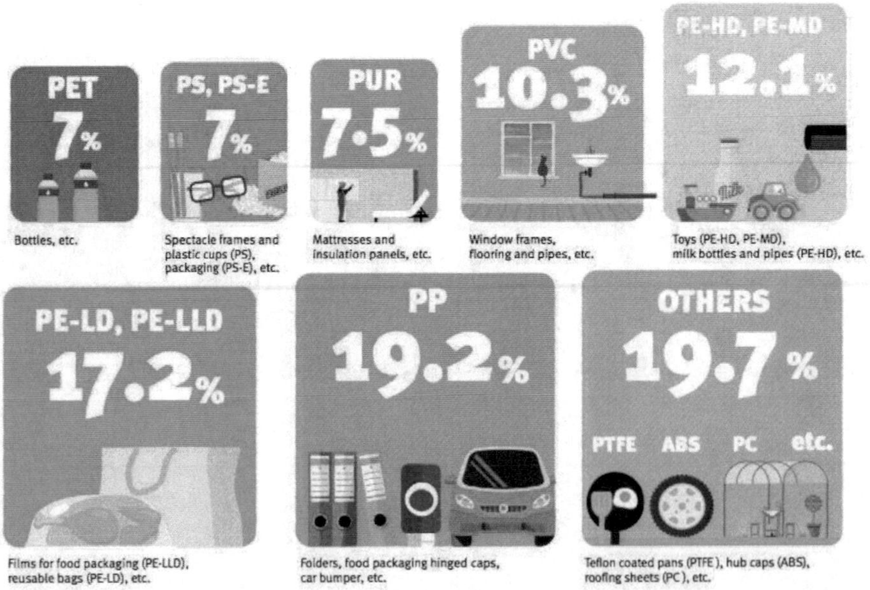

Less than 1.5% of plastics is renewably sourced[1]

[1] European Bioplastics, *Bioplastics facts and figures* (2017).

 6

RowanUniversity
HENRY M. ROWAN
COLLEGE OF ENGINEERING

Paradigm Change in Polymer Manufacturing Occurring

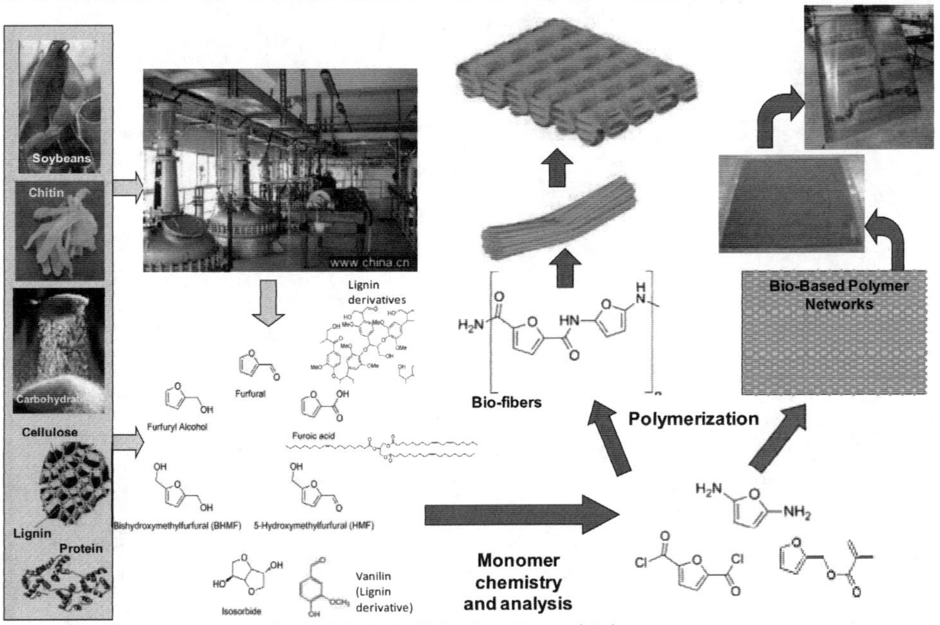

© John J. La Scala, Ph.D., Army Research Laboratory

 7

RowanUniversity
HENRY M. ROWAN
COLLEGE OF ENGINEERING

SMRL's Biomass Utilization to Date

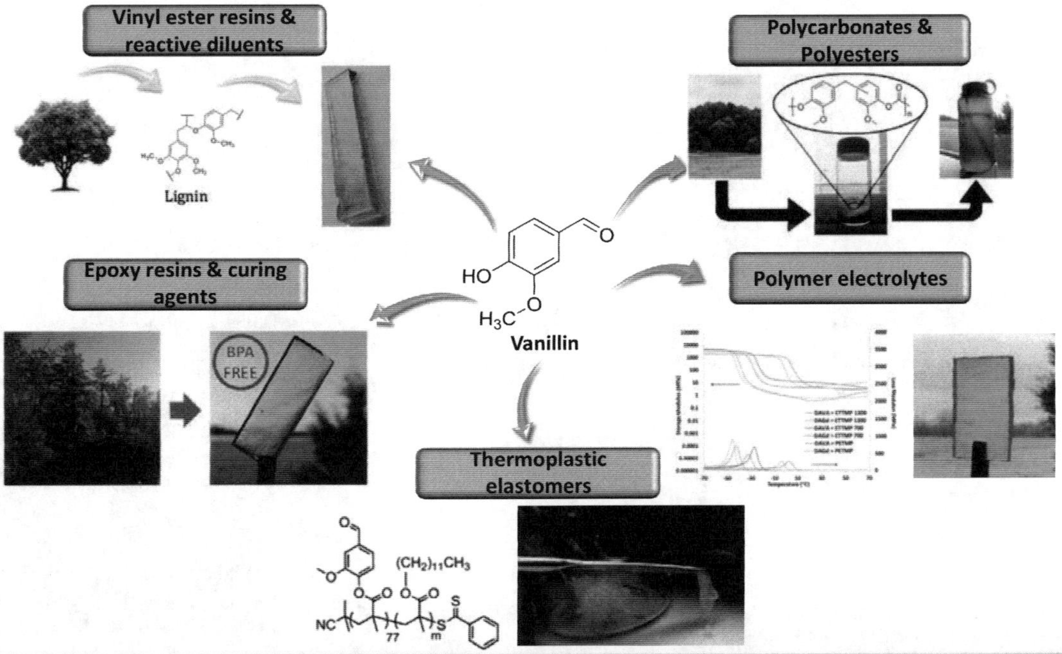

 8

RowanUniversity
HENRY M. ROWAN
COLLEGE OF ENGINEERING

Vanillin-based resin created[1]

Reaction 1

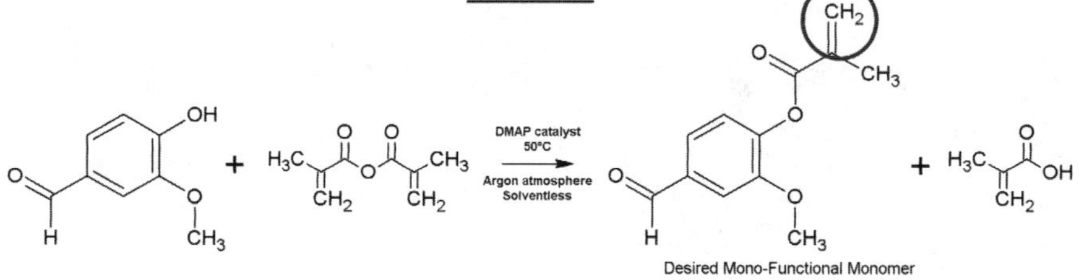

Reaction 2

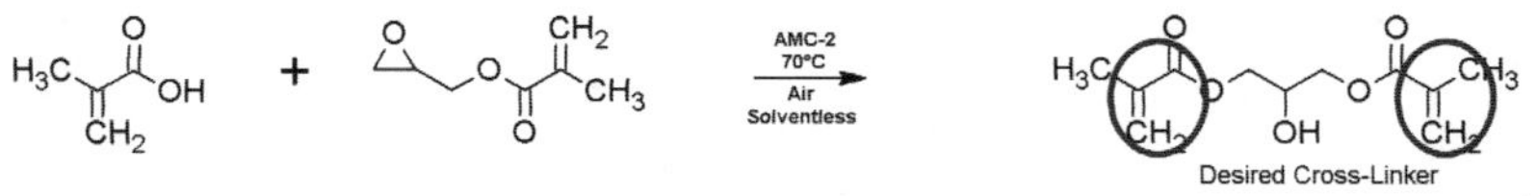

- Green Attributes:
  - All reactants can be bio-based
  - No by-products
  - Solventless reactions
  - Catalyzed reactions
  - Low reaction temperatures
  - **Atom economy/efficiency = 100 %**

Resin viscosity (at 25°C) = 87.25 ± 0.5 cP

Glass transition temperature = 160°C

Glassy storage modulus = 3.6 ± 0.3 GPa

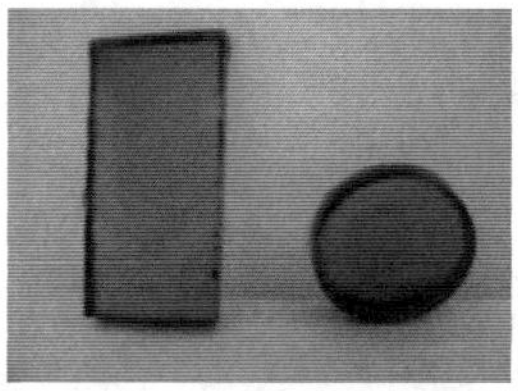

Bio-based Content = 34.8 %

1. Stanzione III, J.F.; et al. *Green Chemistry* 14 (2012) 2346-2352.

 9

RowanUniversity
HENRY M. ROWAN
COLLEGE OF ENGINEERING

Diglycidyl ether of bisphenol-A (DGEBA) alternatives synthesized[1-2]

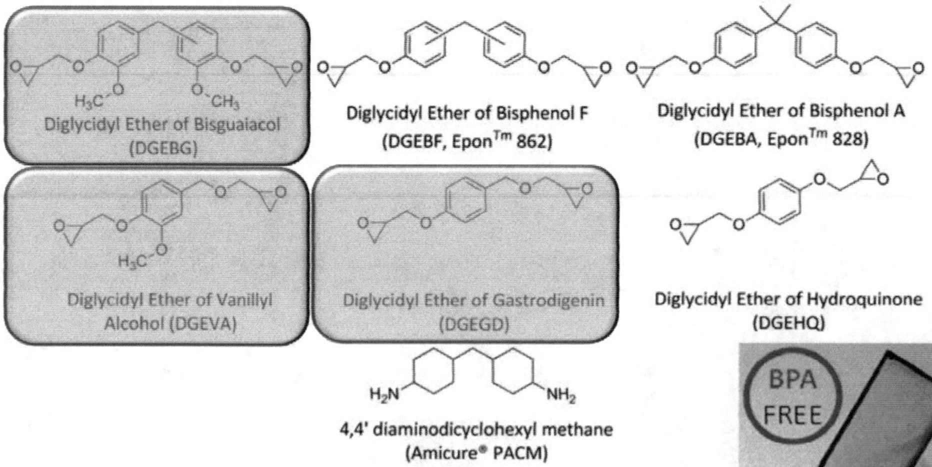

1. Hernandez, E.D.; et al. *ACS Sustainable Chemistry & Engineering* **2016** 4 (8) 4328-4339.
2. U.S. Patent Application Serial # 15/032.583.

 10

RowanUniversity
HENRY M. ROWAN
COLLEGE OF ENGINEERING

Thermogravimetric properties comparable to commercial epoxy resins[1-2]

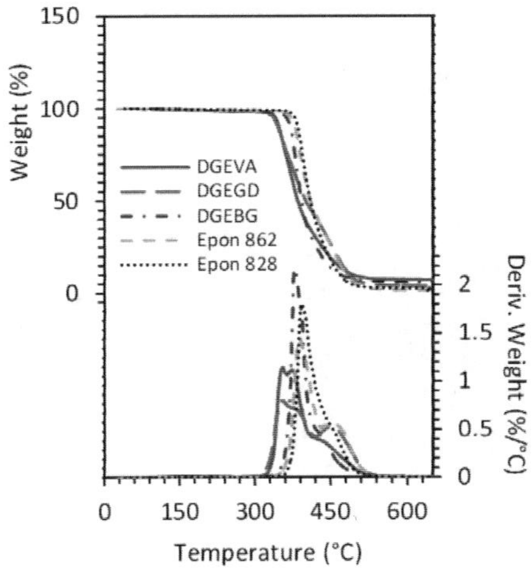

Epoxy Resins	IDT^a (°C)	$T_{50\%}$ (°C)	T_{Max} (°C)	Char Content (%)
DGEVA	341	384	352	6.9
DGEGD	337	400	354	3.6
DGEBG	363	392	376	5.6
Epon 862	371	410	388	1.2
Epon 828	379	410	392	2.3

[a] *IDT* measured as the temperature at 5 % weight loss.

Resins cured with stoichiometric equivalents of PACM

1. Hernandez, E.D.; et al. *ACS Sustainable Chemistry & Engineering* **2016** 4 (8) 4328-4339.
2. U.S. Patent Application Serial # 15/032.583.

RowanUniversity
HENRY M. ROWAN
COLLEGE OF ENGINEERING

Thermomechanical properties comparable to commercial epoxy resins[1-2]

Epoxy Resins	T_g (°C)	E' (GPa)
DGEVA	100	3.02
DGEGD	123	2.29
DGEBG	104	3.35
Epon 862	131	2.57
Epon 828	149	2.37

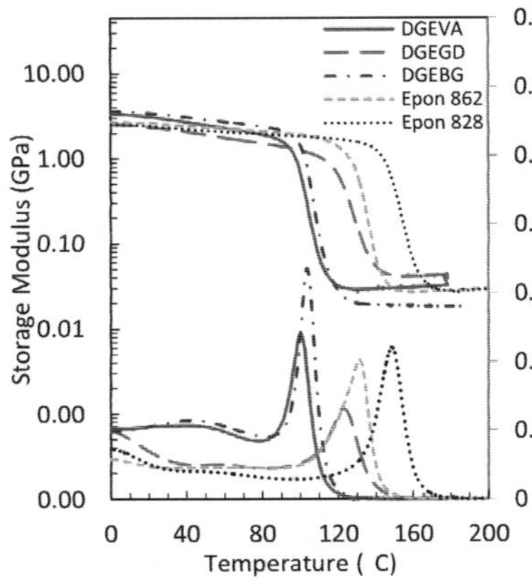

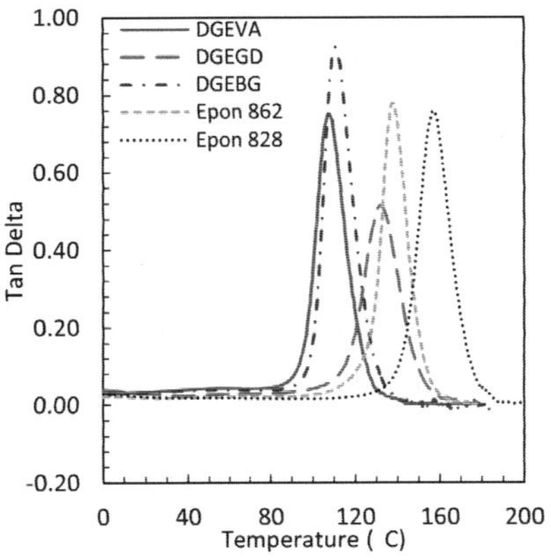

Resins cured with stoichiometric equivalents of PACM
E' measured at 25 °C

1. Hernandez, E.D.; et al. *ACS Sustainable Chemistry & Engineering* **2016** 4 (8) 4328-4339.
2. U.S. Patent Application Serial # 15/032.583.

RowanUniversity
HENRY M. ROWAN
COLLEGE OF ENGINEERING

High bio-based content cured epoxy-amine resins produced[1-4]

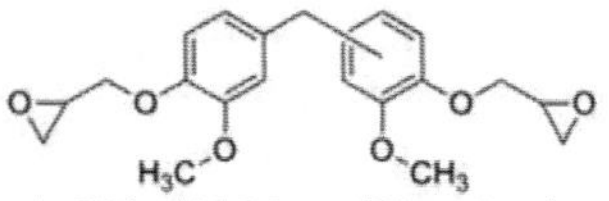

Diglycidyl Ether of Bisguaiacol (DGEBG)

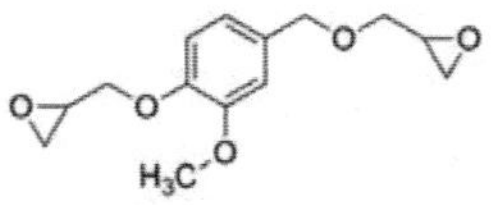

Diglycidyl Ether of Vanillyl Alcohol (DGEVA)

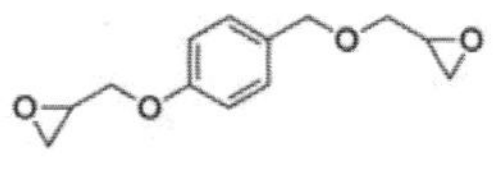

Diglycidyl Ether of Gastrodigenin (DGEGD)

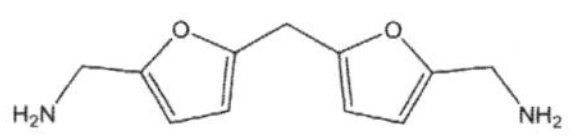

5,5′-methylenedifurfurylamine (DFDA)

> 96.5 % bio-derived carbon content

1. Mauck, J.R; et al. *Macromolecular Chemistry & Physics* **2017** DOI:10.1002/macp.201700013 .
2. U.S. Patent Application Serial # 15/032.583.
3. Hu, F.S.; et al. *Macromolecules* **2014** 47 (10) 3332-3342.
4. Hu, F. S.; et al. *Macromolecular Chemistry & Physics* **2015** 216 (13) 1441-1446.

Drexel UNIVERSITY

13

RowanUniversity
HENRY M. ROWAN
COLLEGE OF ENGINEERING

Thermomechanical properties comparable to commercial epoxy resins[1-4]

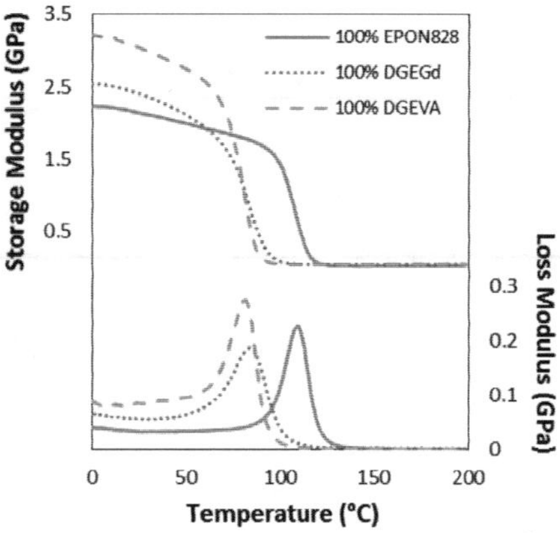

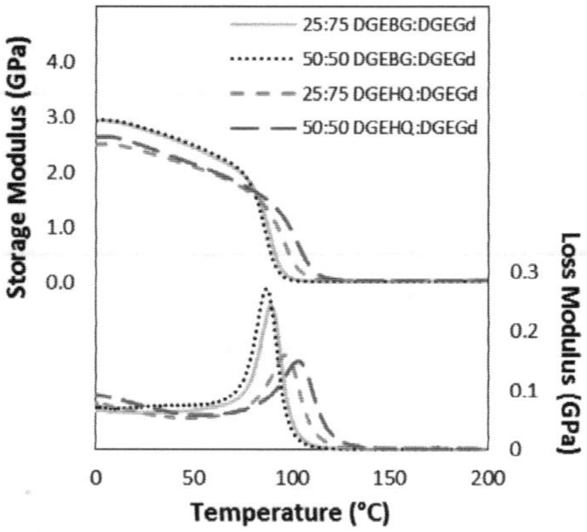

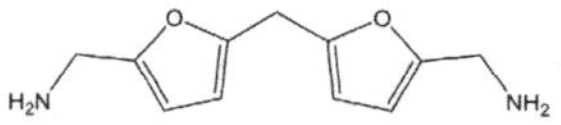

5,5′-methylenedifurfurylamine (DFDA)

1. Mauck, J.R; et al. *Macromolecular Chemistry & Physics* **2017** DOI:10.1002/macp.201700013 .
2. U.S. Patent Application Serial # 15/032.583.
3. Hu, F.S.; et al. *Macromolecules* **2014** 47 (10) 3332-3342.
4. Hu, F. S.; et al. *Macromolecular Chemistry & Physics* **2015** 216 (13) 1441-1446.

Drexel UNIVERSITY

14

Bisguaiacol (BG)-based Polycarbonate Synthesized[1]

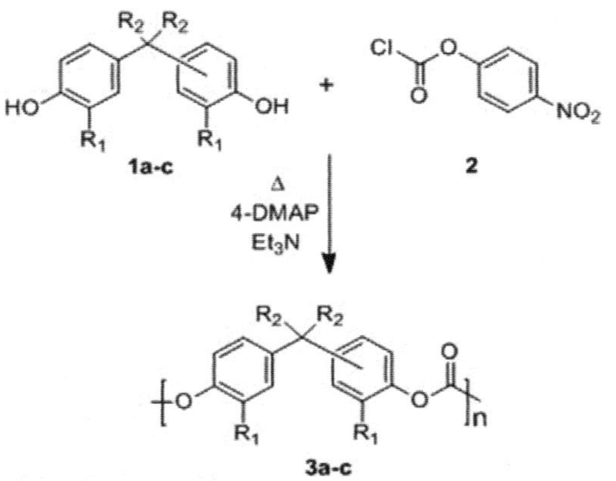

1a,3a: R_1=H, $R_2=CH_3$
1b,3b: R_1=H, R_2=H
1c,3c: $R_1=OCH_3$, R_2=H

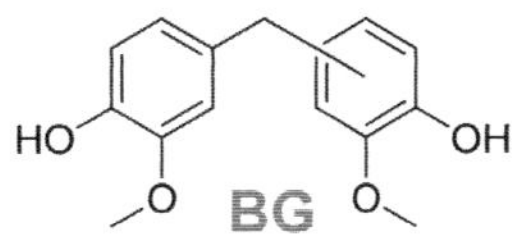

BG

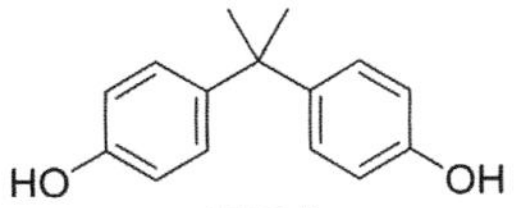

BPA

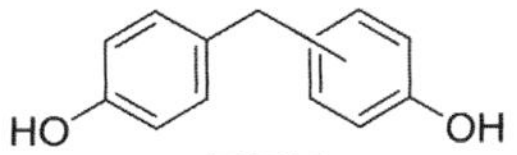

BPF

[1] J. Mauck *et al.*, *J Biobased Materials & Bioenergy* (2017) In Press.

BG-based Polycarbonate Possesses Comparable Thermal Properties[1]

Sample	*IDT* (°C)	$T_{50\%}$ (°C)	T_{max} (°C)	Char Content (%)	T_g (°C)
BG-PC	375	455	450	24.2	100
BPA-PC	440	543	545	23.2	150
BPF-PC	369	546	501	40.5	119

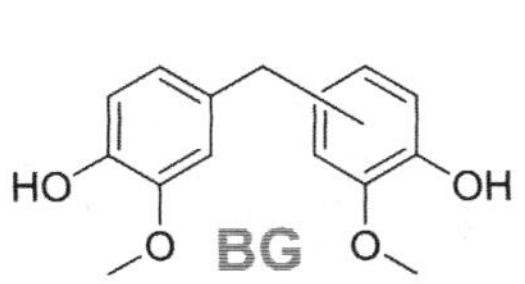

BG

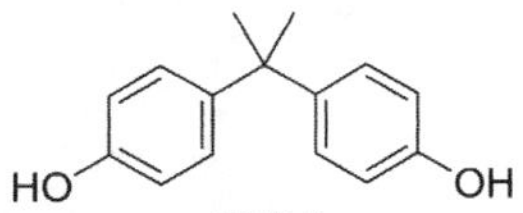

BPA

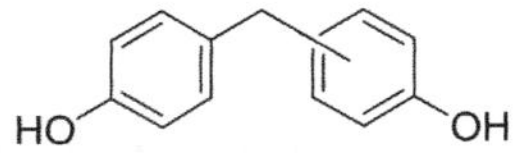

BPF

[1] J. Mauck *et al.*, *J Biobased Materials & Bioenergy* (2017) In Press.

RowanUniversity
HENRY M. ROWAN
COLLEGE OF ENGINEERING

BG-based Polyesters Synthesized[1]

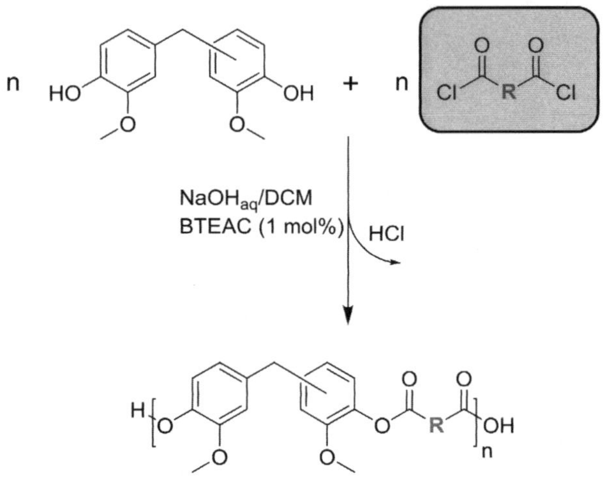

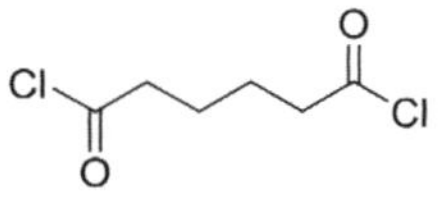

Adipoyl Chloride

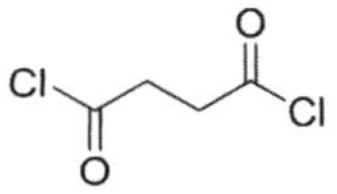

Succinyl Chloride

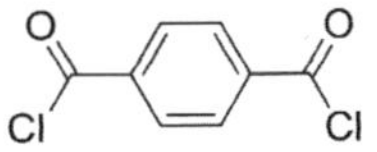

Terephthaloyl Chloride

[1] S. Curia *et al.*, *In preparation* (2017)

RowanUniversity
HENRY M. ROWAN
COLLEGE OF ENGINEERING

BG-based Polyesters Synthesized[1]

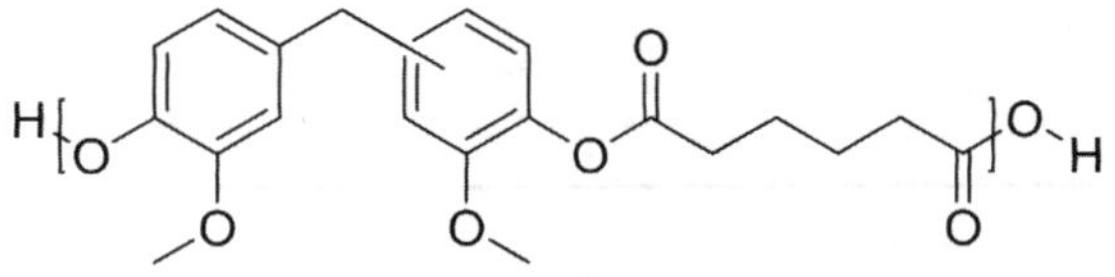

Poly(BG adip)
M_n=6.6 kg mol^{-1} (Đ=1.82)

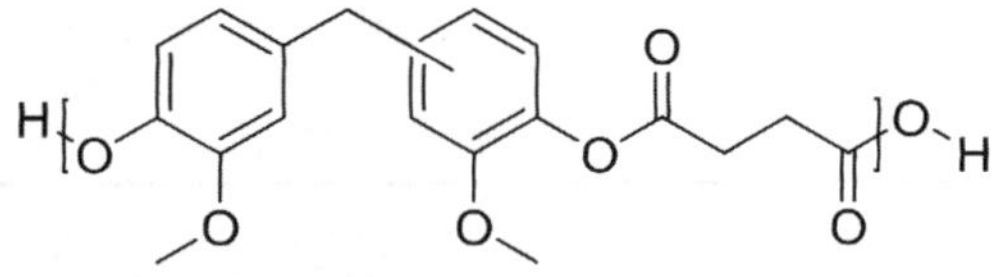

Poly(BG succ)
M_n=7.2 kg mol^{-1} (Đ=2.16)

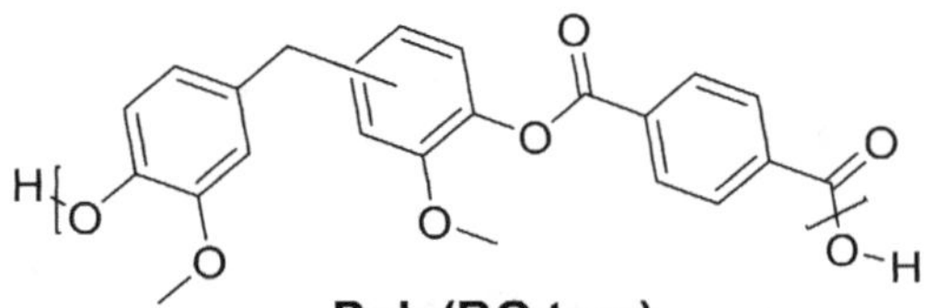

Poly(BG tere)
M_n=26.9 kg mol^{-1} (Đ=2.29)

[1] S. Curia *et al.*, *In preparation* (2017)

RowanUniversity
HENRY M. ROWAN
COLLEGE OF ENGINEERING

BG-based Polyesters Synthesized[1]

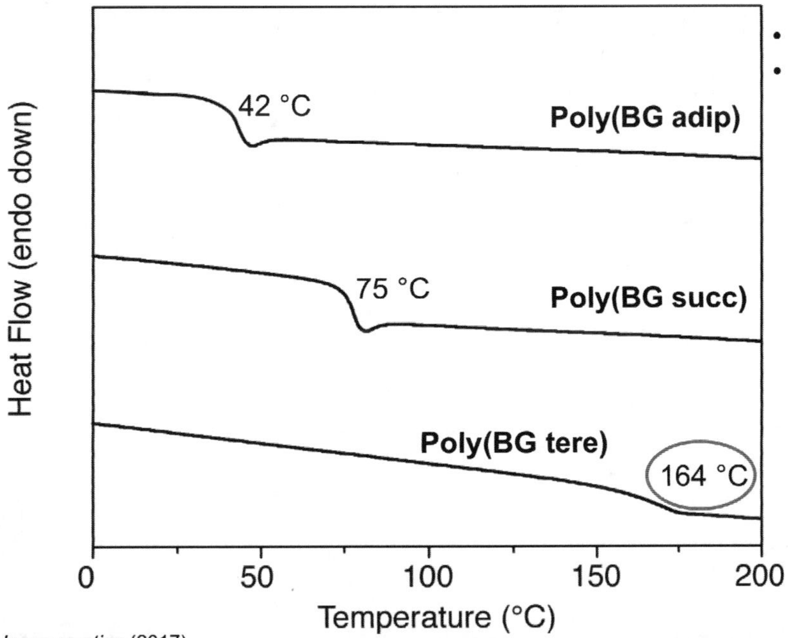

- 10°C/min
- N_2 50 mL/min

[1] S. Curia *et al.*, *In preparation* (2017)

 19

RowanUniversity
HENRY M. ROWAN
COLLEGE OF ENGINEERING

Thermal stability a strong function of comonomer type (in N_2)[1]

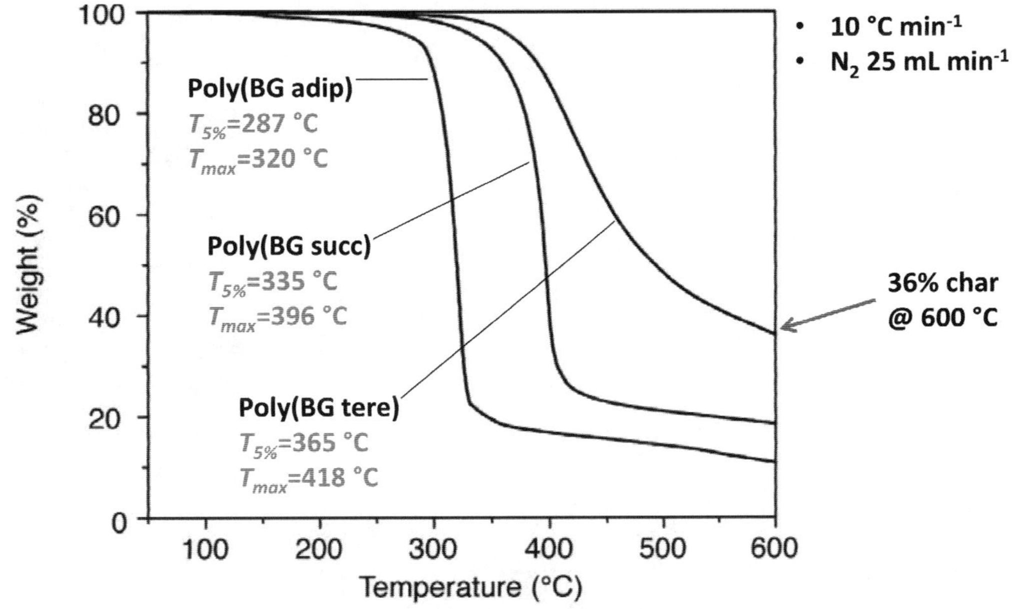

- 10 °C min^{-1}
- N_2 25 mL min^{-1}

[1] S. Curia *et al.*, *In preparation* (2017)

 20

RowanUniversity
HENRY M. ROWAN
COLLEGE OF ENGINEERING

Thermal stability a strong function of comonomer type (in Air)[1]

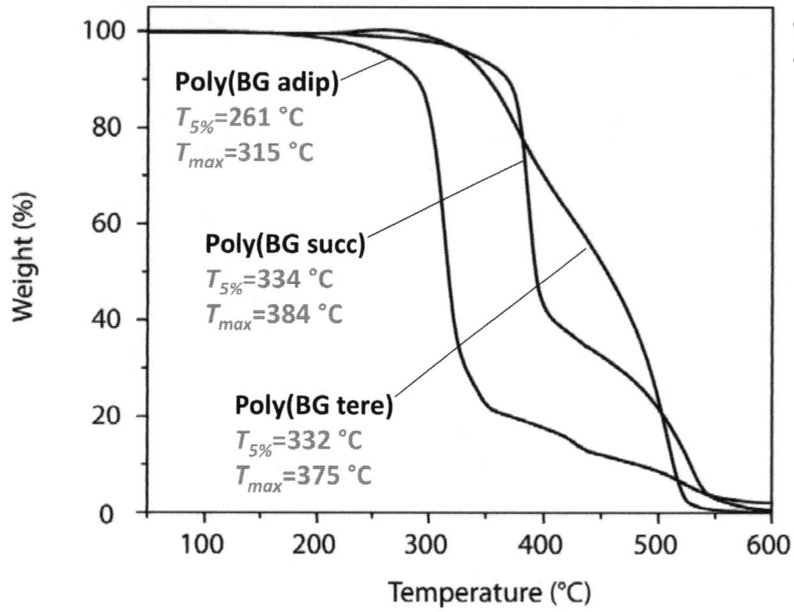

- 10 °C min^{-1}
- Air 25 mL min^{-1}

T_{max} still above 300 °C

[1] S. Curia *et al.*, *In preparation* (2017)

RowanUniversity
HENRY M. ROWAN
COLLEGE OF ENGINEERING

Fully Bio-based Polyaryate Synthesized[1]

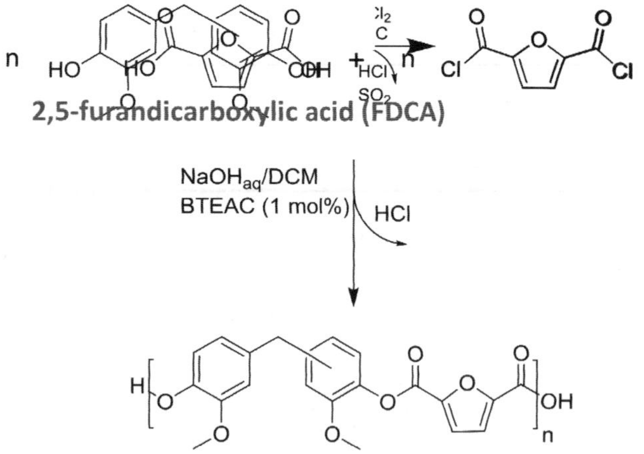

Poly(BG FDCA)
M_n=6.3 kg mol^{-1} (Đ=1.57)

[1] S. Curia *et al.*, *In preparation* (2017)

RowanUniversity
HENRY M. ROWAN
COLLEGE OF ENGINEERING

Promising Thermal Stability Demonstrated[1]

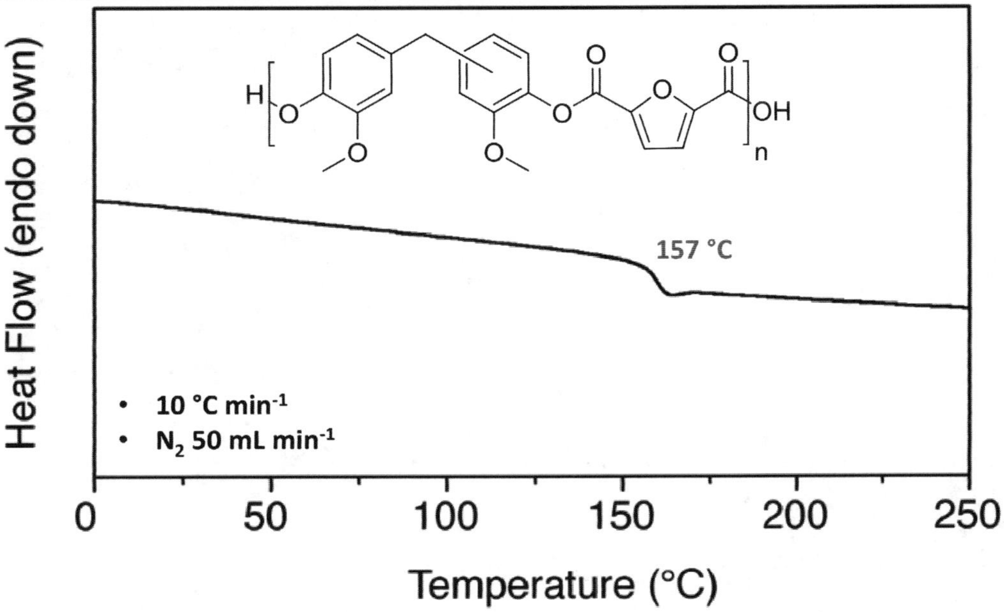

[1] S. Curia *et al.*, *In preparation* (2017)

RowanUniversity
HENRY M. ROWAN
COLLEGE OF ENGINEERING

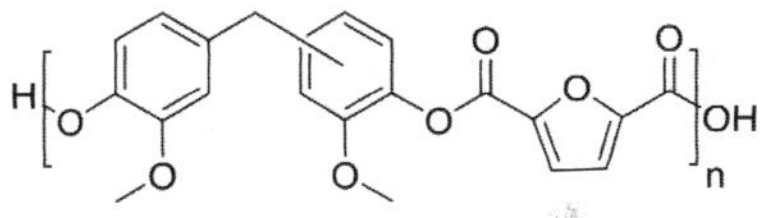

Promising Thermal Stability Demonstrated[1]

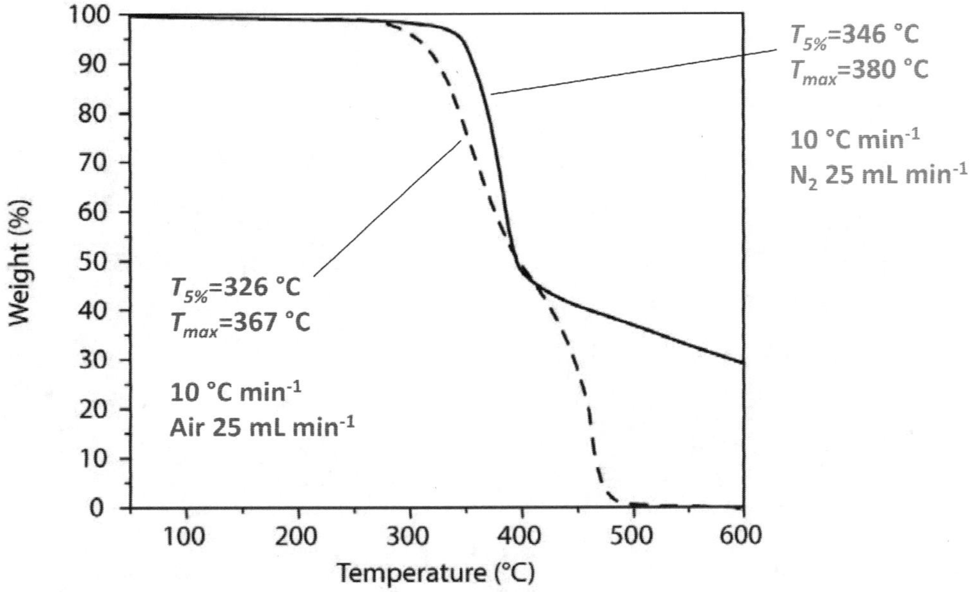

[1] S. Curia *et al.*, *In preparation* (2017)

Rowan University
HENRY M. ROWAN
COLLEGE OF ENGINEERING

Conclusions & Acknowledgements

- Sustainable Materials Research Laboratory
  - Eric D. Hernandez
  - Joseph R. Mauck
  - Silvio Curia, Ph.D.
- U.S. ARL for Financial Support via Cooperative Agreements
  - W911NF-06-2-001
  - W911NF-14-2-0086
  - W911NF-16-2-0225

ARL
U.S. ARMY RDECOM
TECHNOLOGY DRIVEN. WARFIGHTER FOCUSED.

DREXEL UNIVERSITY
Chemical and Biological Engineering
College of Engineering

Rowan University
HENRY M. ROWAN
COLLEGE OF ENGINEERING

Rowan University
HENRY M. ROWAN
COLLEGE OF ENGINEERING

Questions?

Crater Lake National Park

Rocky Mountain National Park

500886 Preparation and Characterization of Cardanol Based Vinyl Ester Resins As Cross-Linker Units

Abstract:

Abstract

Cardanol which is the main component of the thermally threated cashew nutshell liquid is a versatile monomer which combines a phenyl ring with a C15 alkyd chain with different degrees of unsaturation. Because of its unique structure and reactive functionalities, cardanol based building blocks has a growing potential to supply aromatic building blocks for many applications such as corrosion prevention coatings. In this study, cardanol glycidyl ether (LITE 2513HP, CardoliteÂ®) was chemically modified from the unsaturation sites of the side chain by means of epoxidation in the presence of hydrogen peroxide and formic acid. The structure of the modified cardanol resin (SCECGE) was confirmed via proton nuclear magnetic resonance, Fourier Transform infrared spectroscopy, and epoxy equivalent weight titrations. This modified resin was cured with petroleum based epoxy hardener 4-4â€™-methylene biscyclohexanamine (PACM). Also for comparison, another cardanol based di-phenol di-epoxy resin NC514 (Cardolite®) and petroleum based epoxy resin diglycidyl ether of bisphenol-A (DGEBA) were thermally cured with listed epoxy hardeners at stoichiometric ratios. The extent of curing reaction and curing mechanism of different epoxy/amine systems were explored via mid-IR techniques and differential scanning calorimetry (DSC) to determine that the secondary epoxides are not as reactive as the primary glycidyl epoxides. The glass transition, modulus, and strength of the cured SCECGE resins were lower than that of the NC514 and DGEBA because of the lower extent of reaction of the secondary epoxides. In addition, the water barrier properties of this cardanol based SCECGE resin were also determined via contact angle and, water and vapor sorption studies to explore the coating performance of the material.

499647 Ionic Liquids: Green Solvents for Dry Native Cellulose and Chitosan

Abstract:

Cellulose and chitosan (the deacetylated form of chitin) are the most abundant renewable polymeric natural resources on our planet. Native cellulose and chitosan can dissolve in certain ionic liquids. Having nearly zero vapour pressure and being almost completely recyclable make these liquids strong alternatives for the traditional solvents with quite large environmental impacts and derivative effects on the polymers. It is crucial to gain better understanding of the physical state of cellulose and chitosan in ionic liquids. We report solution rheology of six native cellulose samples of different molecular weight in two different ionic liquids. Based on the concentration dependences of viscosity and longest relaxation time, 1-ethyl-3-methylimidazolium acetate (EMImAc) and 1-butyl-3-methyl imidazolium chloride (BMImCl]) are Î¸-solvents for cellulose. The strange failure of the Cox-Merz rule (shear viscosity larger than linear complex vsicosity in the shear-thinning region) for cellulose solutions in ionic liquids suggests that cellulose is not a simple flexible polymer in solution. This was hypothesized to be related to intra-chain hydrogen bonding at shear rates that align and stretch the longest chains. Trace amounts of water compete for hydrogen bonds and can impart a yield stress to the solution rheology, which grows in magnitude as more water is incorporated. However, simply heating the solution to 80°C drives off the water and reverts the solutions to viscoselastic liquids with no yield stress. Rheology of five different chitosan samples (with deacetylation ratio ~75% and of different molecular weight) was also studied in EMImAc. Unlike those of dry cellulose, chitosan solutions contained non-crystalline aggregates, identified using linear viscoelastic tests. This was related to the chemical crosslinks present in the samples due to reaction between amine and hydroxyethyl groups on the chitosan backbone while being extracted from crustacean shells at elevated temperatures in strong base. The chitosan solutions were much less sensitive than the cellulose ones to the presence of water in the system. Upon adding water (<3 wt.%) to the system. the solutions of dry cellulose transition from having terminal behaviour to weak gels, whereas the chitosan solutions already show non-terminal characteristic and do not tend to change at small water content (<20 wt.%).

490787 Environmentally Friendly Flame Retardants Based on Adhesive Catecholamine

Abstract:

There has been much effort expended over the last few decades to develop a wide variety of flame retardant (FR) additives for the purposes of preventing fire spread and reducing the risk of fire-related injuries from everyday-use materials such as polymers and textiles. Some of the most frequently used FR chemicals are halogenated and organophosphorous materials due to their cost-effectiveness and extensive industrial availability. However, there have been serious concerns raised regarding their impact to human health and the environment. This has spawned efforts in industry and academia to development replacement FR materials. In this presentation, we will describe a novel surface coating FR system based on a catecholamine derivative called polydopamine (pDA) that was applied to polyurethane (PU) foam in a single step [1, 2]. PU foam is a particularly attractive test material because it is highly flammable without FRs and it is one of the most widely-used polymers in our daily life. The dopamine precursor to pDA is a naturally occurring chemical found in our body and other living organisms, and this material has an ability to undergo self-polymerization into pDA under mild basic pH conditions. The conformal pDA coating was applied to PU during a simple aqueous dip coating procedure. The catechol functional group in pDA inherently gives this material a nearly universal surface adhesion capability and imparts radical scavenging capability, one of many important features of additives that serve as FRs. pDA coated PU foam exhibited a remarkably reduced flammability compared to pure PU foam and also self-extinguished in some cases. Data regarding the mechanism of flame retardancy of pDA, including morphological observation using scanning electron microscopy (SEM) and quantitative thermal analysis using micro combustion calorimetry, cone calorimetry and thermogravimetric analysis, will also be discussed. We believe this study introduces new opportunities for utilizing naturally derived materials for future green FR applications.

References:

[1] Bioinspired Catecholic Flame Retardant Nanocoating for Flexible Polyurethane Foams, J.H. Cho, V. Vasager, K. Shanmuganathan, S. Nazarenko, and C.J. Ellison, *Chemistry of Materials,* 27, 6784â€“6790, 2015.

http://pubs.acs.org/doi/10.1021/acs.chemmater.5b03013

[2] Development of Kinetic Parameters for Polyurethane Thermal Degradation Modeling Featuring a Bioinspired Catecholic Flame Retardant, B.C. Roberts, A.R. Jones, O.A. Ezekoye, C.J. Ellison, and M.E. Webber, *Combustion and Flame*, 177, 184-192, 2017.

http://www.sciencedirect.com/science/article/pii/S0010218016303777

490792 Soybean Oil Based Thermoset Materials with High Biorenewable Content

Abstract:

In recent years, there has been growing interest in substituting petroleum-based materials with bio-renewable feedstock due to concerns of environmental pollution and the depletion of fossil fuel resources. Thus, the conversion of various bio-sourced monomer precursors into useful polymeric materials has emerged as an attractive strategy. Vegetable oils are a prime example onto which many useful chemical functionalities have been introduced to make them suitable alternatives to petroleum-based monomers. While a large number of vegetable oil derivatives have been blended/incorporated with petroleum-based materials to prepare thermosetting polymers, existing examples normally incorporate low total biorenewable content into the final product. With the goal of generating thermosets with high biorenewable content, recent work [1] from our group will be described where two different soybean oil derivatives with different functional groups (e.g., one as a multifunctional thiol and the other as a multifunctional ene) were photocured to prepare cross-linked thermosets. For example, thiol functional groups were introduced to produce mercaptanized soybean oil by reaction of commercially available epoxidized soybean oil with cysteine functionalities. Then, as-prepared mercaptanized soybean oil was mixed with the commercially available acrylated counterpart, diluent, and photoinitiator, followed by UV irradiation under ambient conditions to produce crosslinked films and fibers. Not only do the resulting materials possess high biorenewable content (e.g., as high as 90 % biosourced carbon) but the fact that they are thermosets makes them potentially suitable for high performance applications.

[1] Soybean Oil Based Fibers Made Without Solvent or Heat, D.W. Janes, K. Shanmuganathan, D.Y. Chou and C.J. Ellison, *ACS Macro Letters*, 1, 1138-1142 (Sept. 2012).

http://pubs.acs.org/doi/abs/10.1021/mz300325g

492684 Self Assembly of Coil-Hyperbranched Poly(styrene-block-acrylated epoxidized soybean oil) Block Copolymers

Abstract:

Biorenewable thermoplastic elastomers have shown its economic potential to substitute their petroleum counterparts. In this study we preset a series of coil-hyperbranched poly(styrene-b-acrylated epoxidized soybean oil) (PS-PAESO) diblock copolymers through reversible addition-fragmentation chain transfer (RAFT) polymerization, which demonstrate its competitiveness toward poly(styrene-b-butadiene-b-styrene) (SBS) in applications such as pressure sensitive adhesives and bitumen modification. Even with significant dispersity in both molecular weight and molecular architecture, triglyceride-based diblock copolymers undergo microphase separation. Through tuning the degree of acrylic functionality in AESO monomer, the influence of degree of branching of PAESO block can be shown on the phase diagram by constituting an additional parameter which has never been discussed before. By thoroughly understanding the self assembly behavior of PS-PAESO, the ability to adjusting the performance in applications will be enhanced. We have constructed a library of PS-PAESO polymers sampling the parameter space defined by the molecular weight, composition, and functionality of AESO. The self assembly behavior is demonstrated through dynamic shear rheology, transmission electron microscopy and the corresponding small angle X-ray scattering.

482356 Physicochemical Properties of Polylactide/Delta-Valerolactone/Organosolv Lignin Atactic Terpolymers

Abstract:

Compostable graft terpolymers of stereotactic lactide, delta-valerolactone, and switchgrass organosolv lignin were synthesized via ring-opening polymerization to improve on polylactide homopolymer properties for commercial applications. Organosolv lignin from swtichgrass has been shown to have properties that improve some of the deficiencies of polylactide, including polylactideâ€™s limited use in food, beverage and medical applications due to its ease to degrade hydrolytically. Delta-valerolactone has been shown to add flexibility to copolymers. It was found that controlling the ratios of the stereoisomers of lactide had significant effects the tensile properties of the resultant terpolymers. Terpolymers containing various ratios of L(-) lactide to D(+) lactide showed greatly increased elasticity and lower Youngâ€™s modulus compared to terpolymers containing only L(-) lactide.

جامعة الملك فهد للبترول والمعادن
King Fahd University of Petroleum & Minerals

Oil Field Chemicals from Macromolecular Renewable Resources: A Case Study of Date Pit Nanoparticles for Drilling Fluid Formulation

Jimoh K. Adewole, Musa O. Najimu

College of Petroleum Engineering & Geosciences
King Fahd University of Petroleum & Minerals,
Dhahran 31261, Saudi Arabia

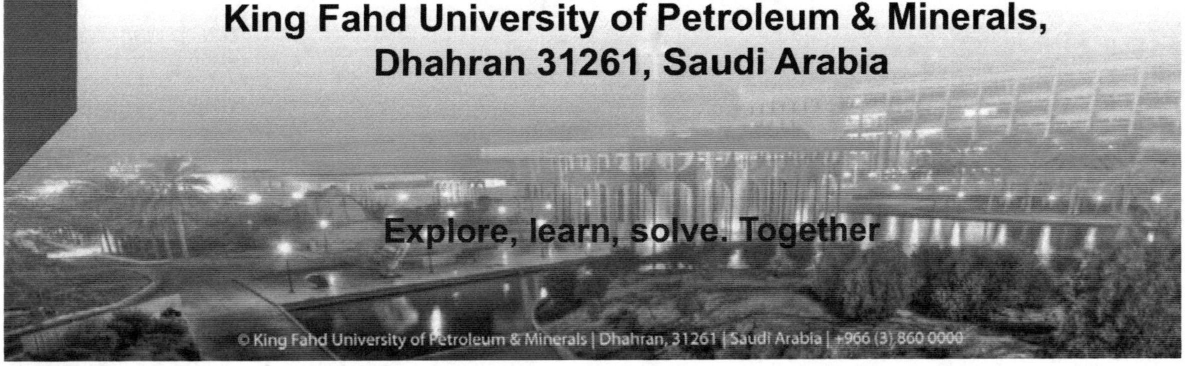

Presentation Outlines

- Innovative Drilling Fluid Technologies
- Drilling Fluid and its Composition
- Date Pit as a Multicomponent Polymer
- Nanoparticle Based Drilling Fluids/Additives
- Preliminary Results and Discussions
- Conclusion & Recommendations

KFUPM

Smart Drilling Fluid Technologies

Time

Costs

Environmental

Improved Drilling Operations:
- ✓ **Physically small**
- ✓ **Chemical/Thermally stable**
- ✓ **Biologically degradable**
- ✓ **Environmentally benign**

Drilling Fluid and its Composition

- Drilling fluids are mixture natural and synthetic chemical compounds
- Used as circulating fluid in drilling operations
- Successful completion of wells depends to a large extent on the properties of the drilling fluid used during the operation
- Types
  - ✓ Water-base mud
  - ✓ Oil-base mud
  - ✓ Gas

Drilling Fluid and its Composition

Functions of Drilling Fluids

- ✓ Transportation of cuttings
- ✓ Cooling and Cleaning of bit
- ✓ Friction Reducer

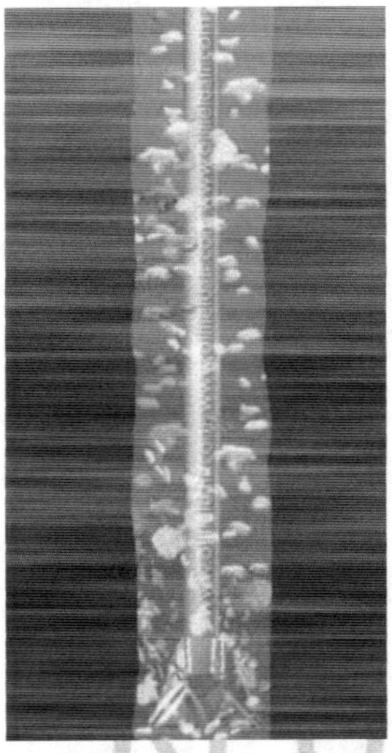

Drilling Fluid and its Composition

Functions of Drilling Fluids

- ✓ Maintaining borehole stability
- ✓ Prevent inflow of fluids from permeable rocks that have been penetrated
- ✓ Formation of thin filter cake to seal pores/other openings

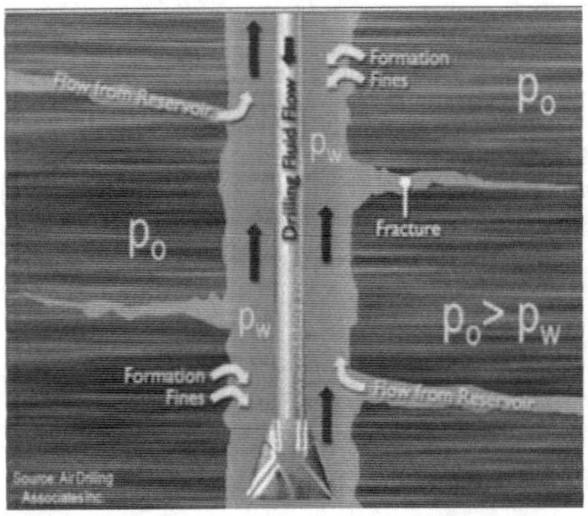

Drilling Fluid and its Composition

Performance Requirements of Drilling Fluids

- Non-damaging/offensive
- No expensive completion methods
- No interfere with productivity of the fluid bearing formation
- No corrosion/excessive wear of drilling equipment

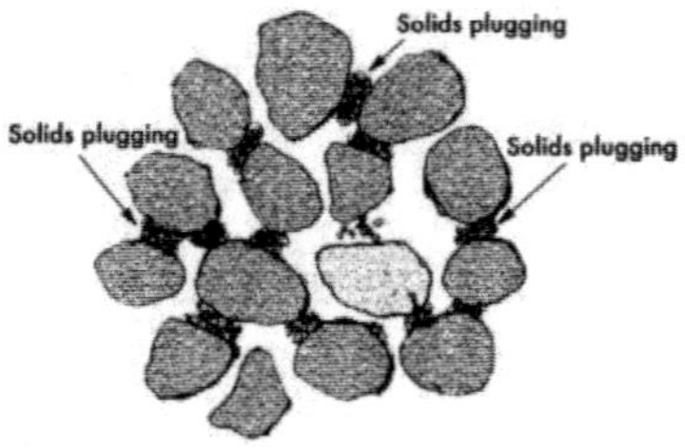

Formation damage caused by solids plugging (petrowiki.org)

Drilling Fluid and its Composition

Composition of drilling mud

Drilling Fluid Additives	Function	Possible Source
Water	Base fluid	-
Bentonite	Base solid	Clay
Barite	Weighting agent	barium sulfate
Calcium carbonate	Weighting agent	Inorganic
Xanthan Gum	Thermal stability/Viscosity control	Modified bio materials
Starch/cellulose and its derivatives	Viscosifiers/clay stabilizer	Natural polymers
Lignin and its derivatives	Viscosity and gel strength control	Modified bio materials
Quebracho	dispersant	heartwood
Polyol	shale inhibitor	Bio-materials

Date Pit as Multicomponent Polymer

Date Fruit Production in KSA

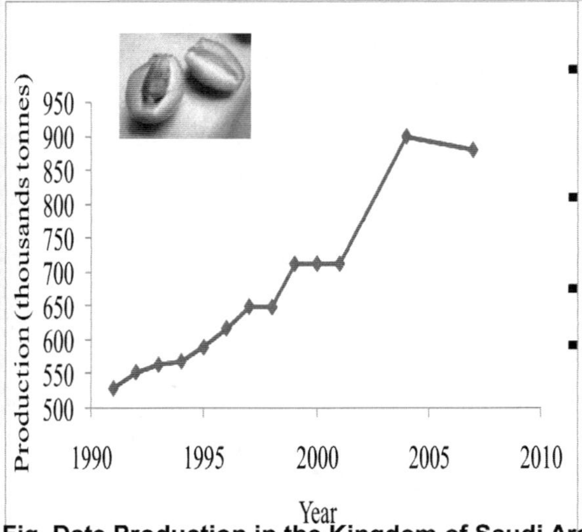

- Saudi Arabia ranked among the top ten date producing countries in the world
- 4th largest producer in 2001 (13.3% of global production)
- 3rd largest producer in 2007
- Improved Agriculture Tech can increase production

Fig. Date Production in the Kingdom of Saudi Arabia between 1990 and 2007 (Adewole et al 2013)

Date Pit as Multicomponent Polymer

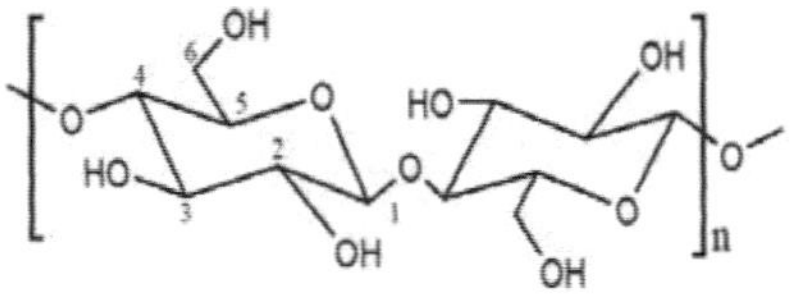

Fig. Molecular Structure of Cellulose

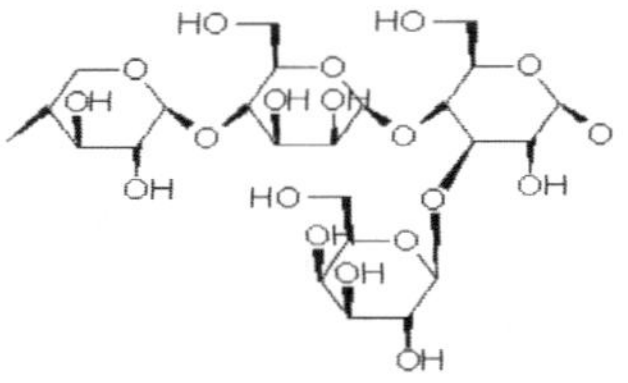

Fig. Molecular Structure of Hemicellulose

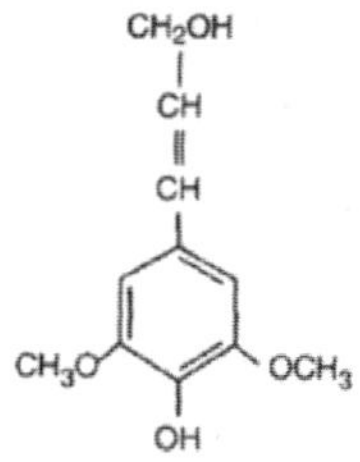

Fig. Molecular Structure of Lignin

Table. Chemical Composition of Date Pit Polysaccharides Fraction (Adewole & Sultan (2014)

Component	Molecular Formula	Composition (Wt%)
Hemicellulose	(C5H8O4)n and (C6H10O5)n	23.3 - 55.5
Lignin	$C_9H_{10}O_2$, $C_{10}H_{12}O_3$, $C_{11}H_{14}O_4$	14.7 - 23.2
Cellulose	$C_6H_{10}O_5$	20.18 – 56.7
Ash	-	1.1 – 5.3

Date Pit as Multicomponent Polymer

Cellulose in Drilling Fluid

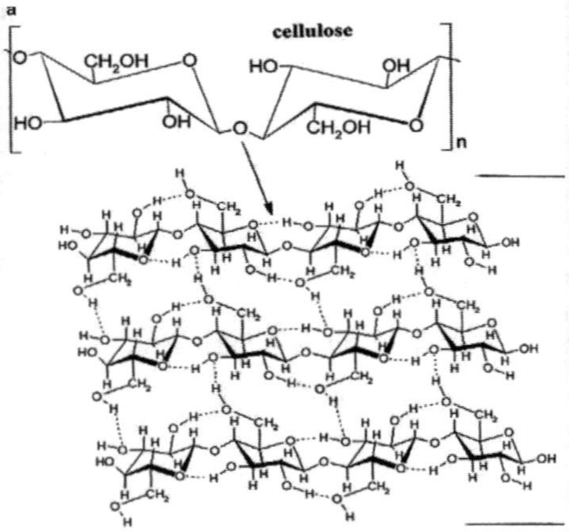

Fig. Structure of Cellulose

Kamel, 2007

Cell-MPs & NPs as:

Natural morphology of fiber
Cellulose whiskers:
15nm in diameter & ~1um long
Microfibrils (vegetable):
2-10nm diameter & several um in length
Cellulose Crystallites:
Length 300 – 2000 nm
Diameter 5 – 20 nm
Aspect ratio 20 – 60
Tensile strength 10 000MPa
E-modulus 150 GPa
Used for Viscosity enhancement and Filtration Control

KFUPM

Date Pit as Multicomponent Polymer

Lignin in Drilling Fluid

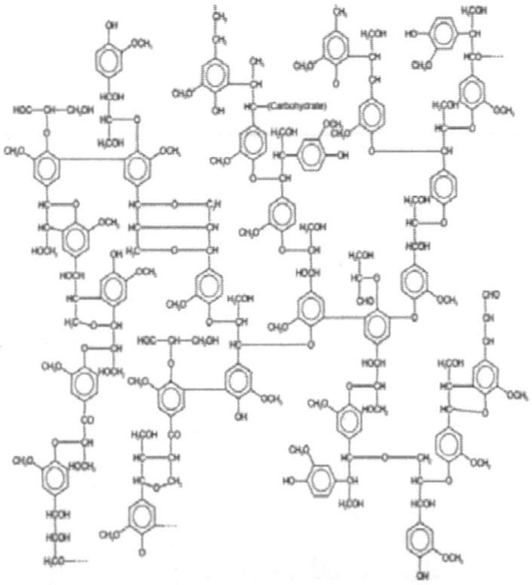

Fig. Structure of Lignin

Lignin

High contents of aromatics moieties
Could be spliced into monoaromatic monomers
Serves as stiffening materials for cellulose and hemicellulose
Its derivative:
Lignosulfonate
Polyanionic lignin
Uses:
Thinner
Deflocculants

KFUPM

© King Fahd University of Petroleum & Minerals | Dhahran, 31261 | Saudi Arabia | +966 (3) 860 0000

Date Pit as Multicomponent Polymer

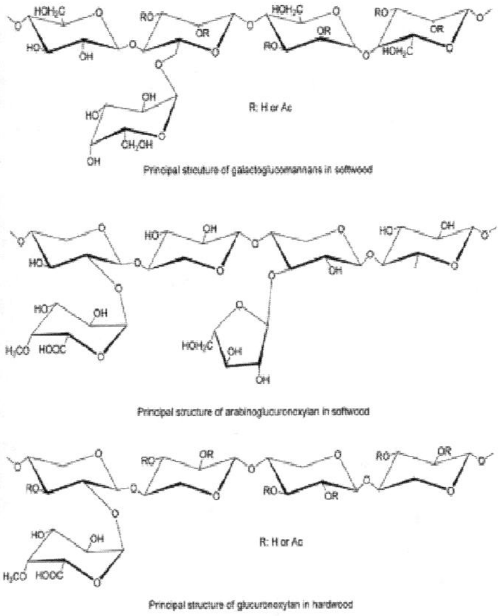

Hemicellulose in Drilling Fluid

Hemicellulose

Different composition & Structures
Combination of : pentose, arabinose, galactose, hexoses
Used:
Polyelectrolyte
rheology modifiers

Fig. Structure of Hemicellulose

Date Pit as Multicomponent Polymer

Processing

Functionalization

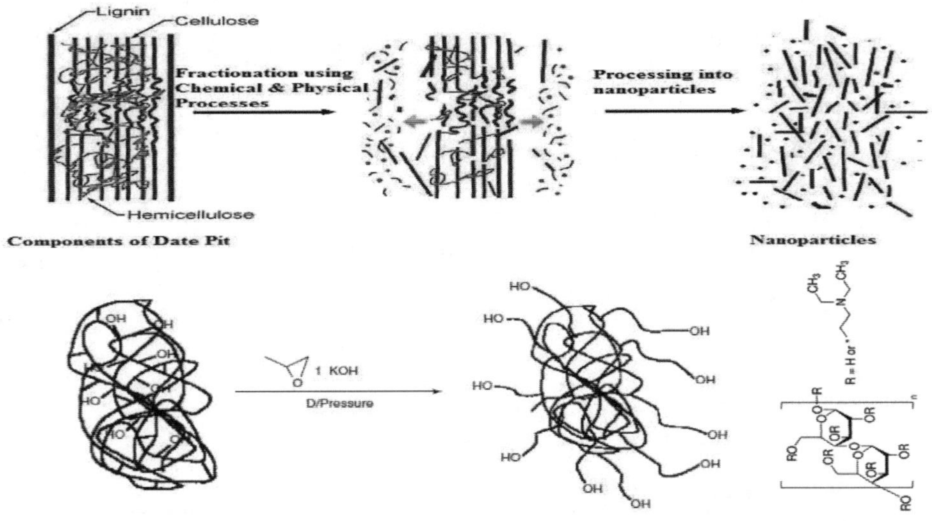

Lignocelluose processing into Nanoparticles (Gandini & Belgacem, 2008; Lejeune & Deprez, 2010)

Polymeric Nanoparticles for Drilling

Why Nanoparticles?

- Competitive performance properties
  - ✓ Generally nano-based drilling fluids will form compact and less permeable mud cake that will reduce fluid invasion/spurt losses
  - ✓ Penetrant particles during spurt loss will be small enough to prevent throat blockage
  - ✓ It will also contribute to the plastering capability
  - ✓ Thin mud cake

KFUPM

Polymeric Nanoparticles for Drilling

Why Bio-based Multicomponent Polymeric Nanoparticles?

- Naturally occurring composite
- Renewable nature
- Availability and Sustainability of sources
- Nonfood agricultural based
- Low cost
- Low density
- High specific strength and modulus
- Relatively easily functionalized
- Non-toxic and Biodegradable

KFUPM

© King Fahd University of Petroleum & Minerals | Dhahran, 31261 | Saudi Arabia | +966 (3) 860 0000

Nano-materials for Drilling Fluids: Literature Review

- Studies on Nano-based drilling fluids using commercial nanoparticles (Amanullah et al, 2011 & Zakaria et al, 2012)
  - ✓ Significant improvement in performance using Low concentration
  - ✓ Reduce the tendency of the formation of internal mud cake (facilitate effective cleaning of borehole before completing the well)
  - ✓ Deposition of thin mud cake may lead to decrease in differential pipe sticking in highly permeable formations
  - ✓ Formation of low solid content of unweighted nana-based drilling fluids will enhance the rate of penetration in hard rock formations

KFUPM

Nano-materials for Drilling Fluids: Literature Review

- Cellulose Nanoparticles investigated as modifiers for Rheology and Fluid loss in bentonite water-based fluids (Li et al, 2015)
- Cellulose nanocrystals (CNC) was prepared (by hydrolysis) from microfibrillated cellulose (MFC)
  - ✓ CNC has better surface interaction with bentonite layers
  - ✓ Superior rheological properties
  - ✓ Higher thermal stability
  - ✓ Less fluid loss volume
  - ✓ Thinner filter cake

KFUPM

Project Objectives

- Main objective is to investigate the potential of locally available multicomponent (non-petrochemical) raw materials for applications as oil field chemicals.
  - ✓ Drilling Fluid Additives

Date Pit/Seed

- Minimize the cost of production
  - ✓ Particle size (conserve the composition)
  - ✓ Weight percent date pit
  - ✓ Date pit oil content

Drilling Fluids

- Maximize project efficiency
  - ✓ Filtration properties
  - ✓ Borehole stability/Density
  - ✓ Rheology/ Flow properties
  - ✓ Thermal properties

Date Pit Processing Method

Fig. Date pit overall processing sequence

Preliminary Results and Discussions

Drilling Fluid Density

- $P_{static\ mud\ column}\, f(density, depth)$
- Density: the weight per unit volume
- Units: lb/gal or ppg, $^{psi}/_{ft}$ or $^{kg/cm^2}/_{m}$
- Balanced drilling: $p_w - p_o \geq 200psi$
- Prevent inflow of formation fluids
- Lay down a thin, low permeability filter cake on the walls
- Maintain borehole stability
- Buoyant effect of mud $\propto$ density

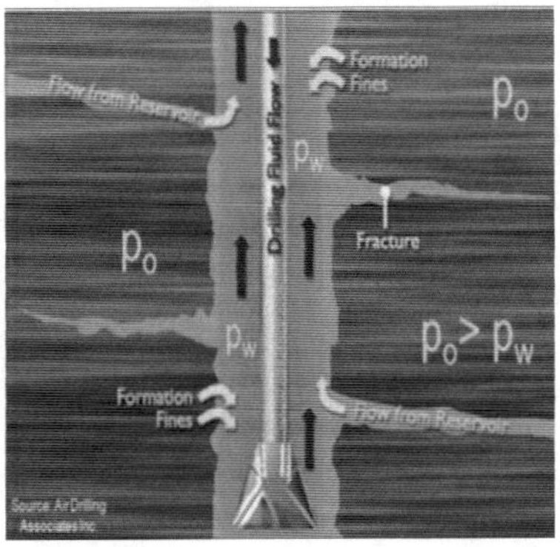

Preliminary Results and Discussions

Effect Date Pit Concentration on Density

Densities of Common Mud Components (lb/gal)	
Oil	6.66
Water	8.33
Salt	18.3
Clay	20.8
Barite	34.2

Change observed at loading greater than 10%

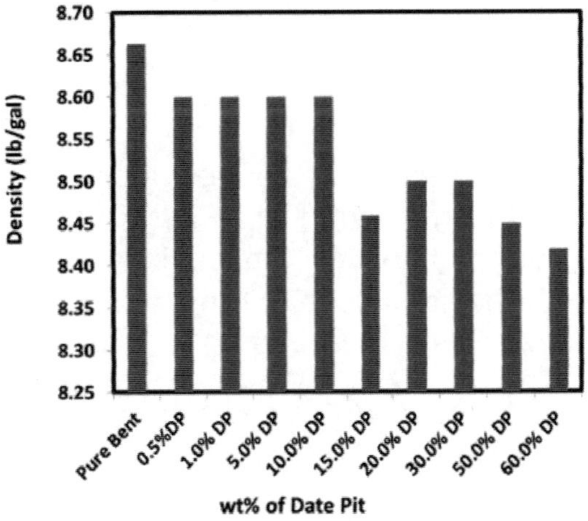

Fig. Change in density with date pit loading

Preliminary Results and Discussions

Rheology & Flow Properties

- **Flow properties is responsible for removal of drill cuttings**
- Unsatisfactory flow properties can lead to:
  - ✓ Filling bottom of hole with drill cuttings
  - ✓ Reduced penetration rate
  - ✓ Hole enlargement
  - ✓ Stuck pipe
  - ✓ Loss of circulation
  - ✓ Blow out

A Blowout Scene

Preliminary Results and Discussions

Rheology & Flow Properties

- The shear stress/ shear rate relationship of drilling fluids depends on its compositions
- Behaviour close to Bingham plastics
- Yield point and Plastic viscosity

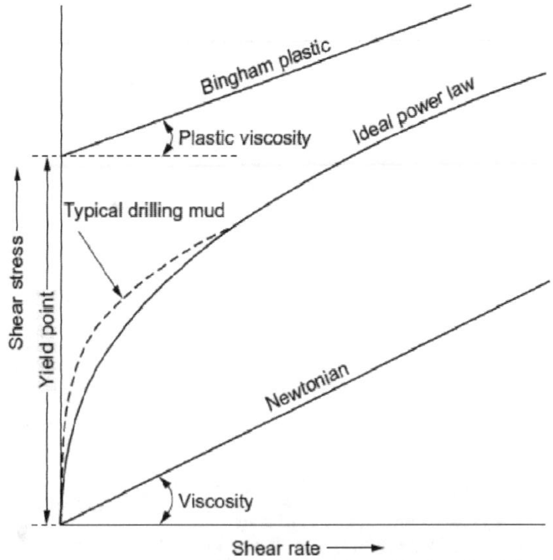

Fig. Consistency curves for common flow model

Preliminary Results and Discussions

Rheology & Flow Properties

- Decrease in viscosity due to increase in shear rate (shear thinning)
- high shear rate in the drill pipe reduces pumping pressures
- high viscosity in the annulus increases cuttings carrying capacity

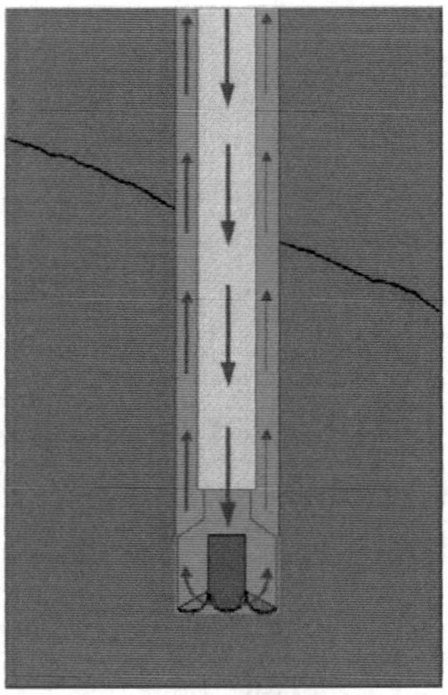

Fig. Typical flow during drilling operation

Preliminary Results and Discussions

Effect Date Pit Loading & Particle Size

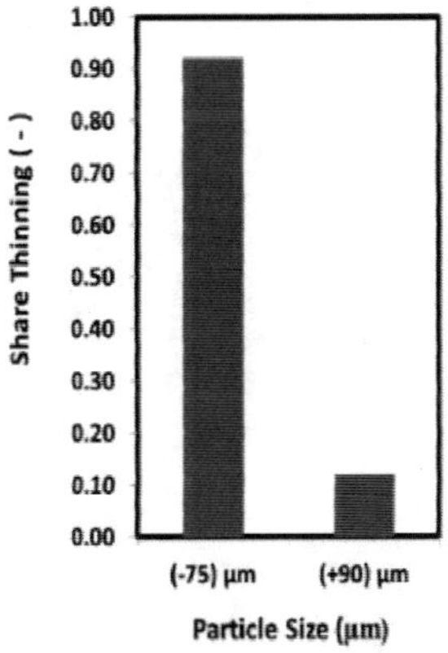

Fig. Effect of Particle Size

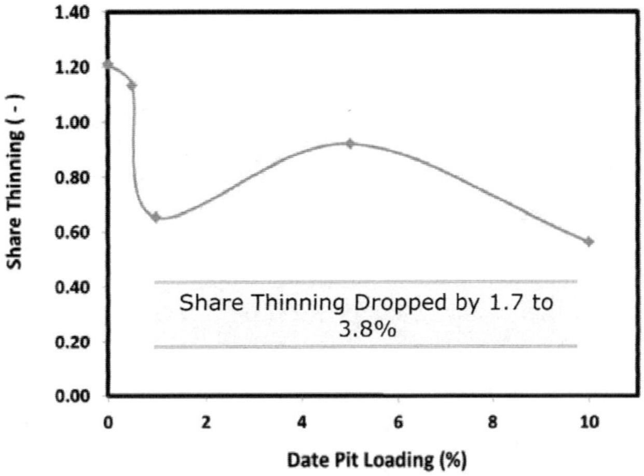

Fig. Effect of wt% Date Pit Loading

Preliminary Results and Discussions

Effect Date Pit Particle Size Distribution

Date pit has the potential to be used as Rheological modifier if used as smaller particles (such as nanoparticles)

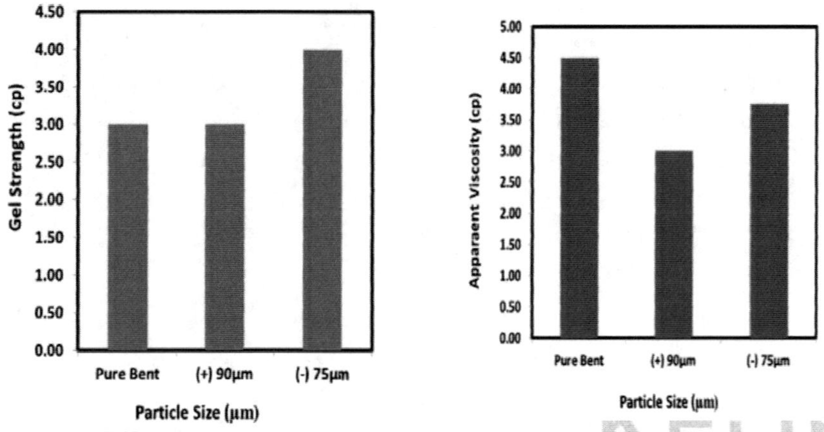

Fig. Effect of particle size on Gel Strength and Viscosity

Preliminary Results and Discussions

Rheology & Flow Properties: Effect Date Pit Loading

At 60wt% date pit loading:
- Gel strength dropped by 50%
- Apparent viscosity dropped by 27% drop

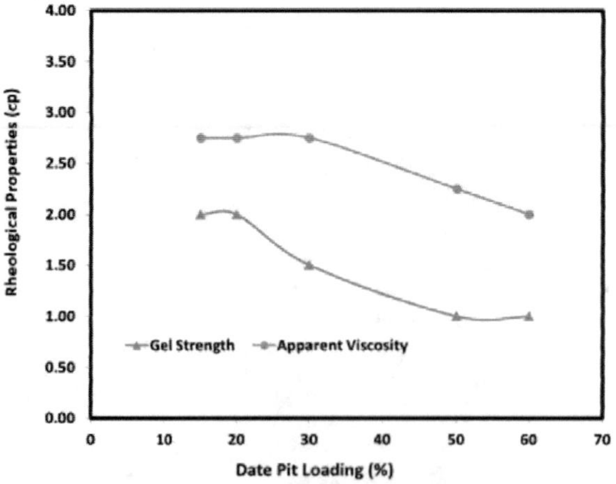

Fig. Effect of date pit loading

Adewole and Najimu, (2017) Journal of Energy Resources Technolog

Preliminary Results and Discussions

Filtration Properties

- Another major requirement for successful hole completion is the ability of the drilling mud to seal permeable formations exposed by the bit
- Done by using low permeability filter cake
- Permeability of filter cake depends on:
  - ✓ Particle size distribution in the mud
  - ✓ Electrochemical conditions (Caenn et al 2011)
- Filtration properties depends on the nature of formation to be drilled
- In permeable formations; control are required to prevent thick filter cake:
  - ✓ borehole gauge reduction
  - ✓ differential sticking
- In unconsolidated sands; to prevent slumping into the hole by rapid formation of filter cake

Preliminary Results and Discussions

Filtration Properties: Effect Date pit oil content

For fluid loss prevention, de-oiling process can be eliminated to improve date pit performance and reduce cost of production

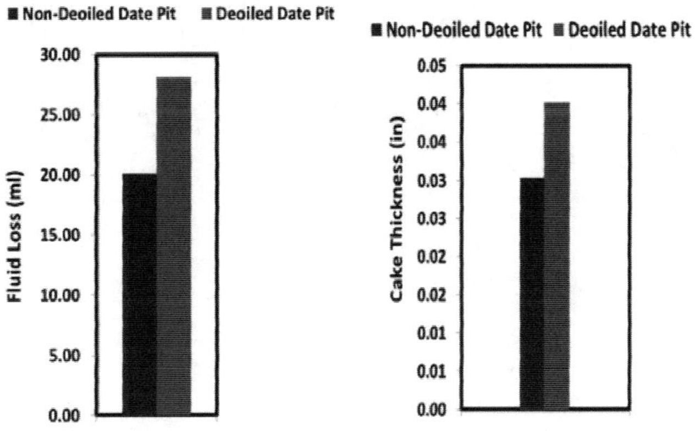

Fig. 15 Effect of Oil Content

Adewole and Najimu, (2017) Journal of Energy Resources Technology

Preliminary Results and Discussions

Filtration Properties: Effect Date Pit Oil Content

The effect of oil content increased with time

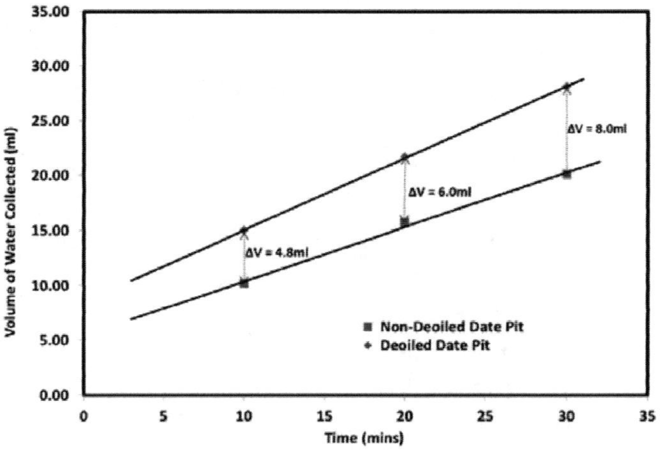

Fig. 16 Effect of Oil Content with time

Adewole and Najimu, (2017) Journal of Energy Resources Technology

Preliminary Results and Discussions

Filtration Properties: Effect Date Pit Particle Size Distribution

Table. Effect of size on Filtration properties (Adewole and Najimu, 2017)

Sample	Fluid Loss after 30 mins (ml)	Residue Thickness (in)
Pure Bentonite	38	0.034
(-) 75µm	24.8	0.035
(+) 180µm	28.1	0.040
(+) 710µm	30.2	0.044

Adewole and Najimu, (2017) Journal of Energy Resources Technology

Preliminary Results and Discussions

Effect of Date Pit Loading (Low Concentration)

- Low wt% date pit loading
- Continued drop in Fluid Loss and residue thickness
- Agreement with published results (Amanullah et al 2016), J. Petrol Sc & Eng

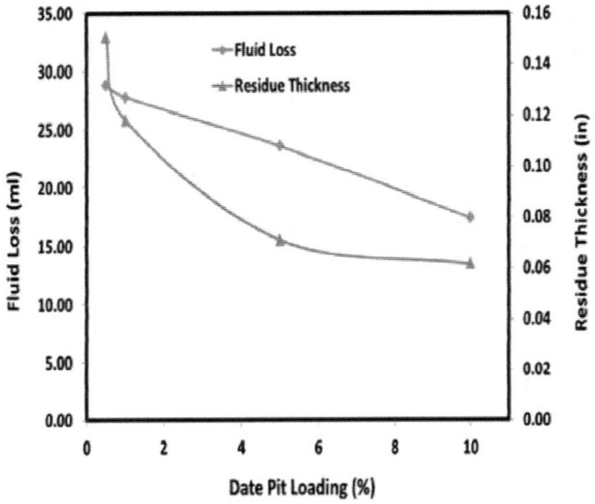

Fig. Effect of date pit concentration on filtration properties

Preliminary Results and Discussions

Effect Date Pit Loading (High Concentration)

- High wt% date pit loading may lead to formation of soft and thick mud cake which increase the potential of differential sticking.
- Reduce the ease of recovery of stuck pipe

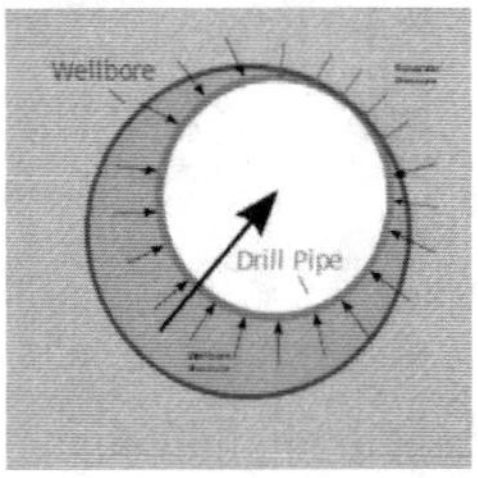

Differential sticking (wikipedia.org)

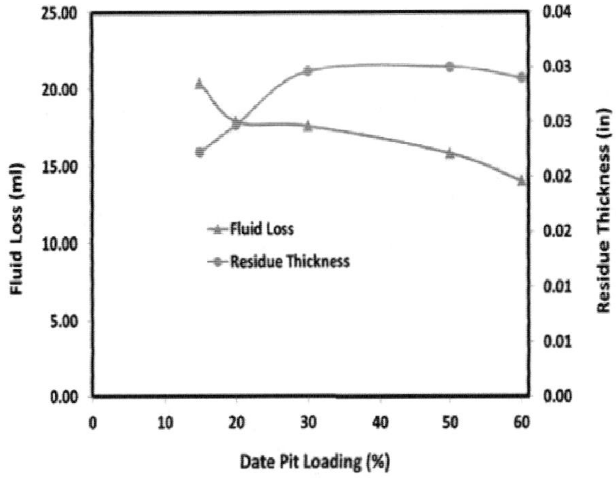

Fig. Effect of date pit concentration on filtration properties

KFUPM

© King Fahd University of Petroleum & Minerals | Dhahran, 31261 | Saudi Arabia | +966 (3) 860 0000

Preliminary Results and Discussions

Thermal Properties: Effect Date Pit Loading

Temperature and rate of degradation are used to specify thermal stability

- Critical Temperature
- ✓ Starch 107°C
- ✓ Cellulosic polymers 135°C
- ✓ Polyacrylates 204°C
- ✓ Certain acrylic polymers 260°C

Onset Degradation Temperature

Date Pit	**250**
Bentonite	**455**

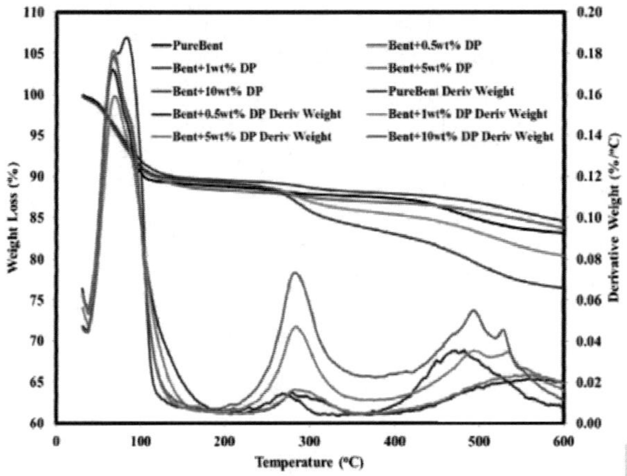

Fig. Thermogravimetric curves for date pit composite

Adewole and Najimu, (2017) Journal of Energy Resources Technology

Preliminary Results and Discussions

Thermal Properties: Effect Date Pit Oil Content

- Weight loss difference due to oil content
- Smoke Points (oxidative stability) of vegetable oils range between 107 - 271°C

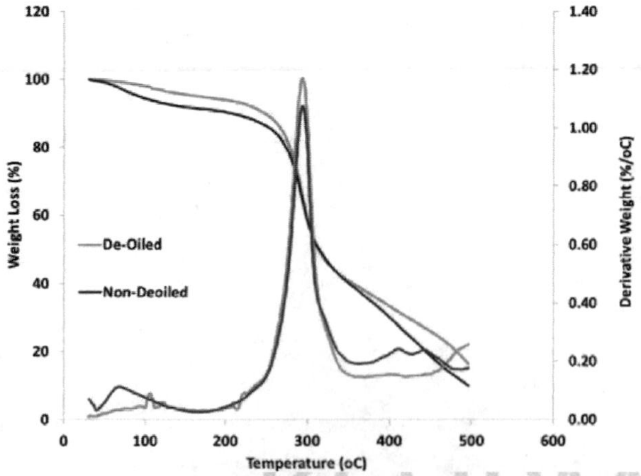

Fig. Thermogravimetric curves for de-oiled and non de-oiled date pit

Preliminary Results and Discussions

Smart Behaviour of Natural Composite

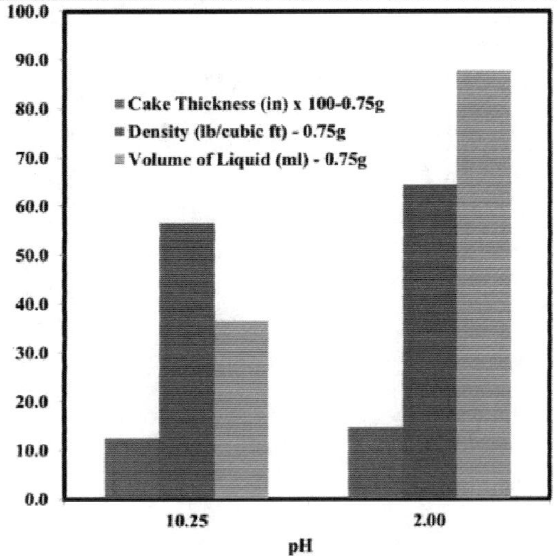

Fig. pH Responsiveness of date pit based drilling fluid

Conclusion and Recommendations

- De-oiling processes could be eliminated to reduce cost
  - ✓ reduction in energy requirement for processing
  - ✓ reduced time of heating
- Multifunctional at lower particle size
  - ✓ fluid loss prevention
  - ✓ rheological property modifier
- Better at low concentration (overbalance drilling)
  - ✓ date pit loading of 1-10wt%
  - ✓ higher wt% above 15% will affect density
  - ✓ increase mud thickness
- Demonstrated potential for use in high temperature environment

Research Collaboration Needs

- Compositional Analysis of naturally occurring multicomponent materials (Lignocellulose)
- Functionalization of these materials to enhance its relevant properties

Acknowledgement

- **College of Petroleum Engineering and Geosciences, King Fahd University Petroleum Minerals, Saudi Arabia**
- **Dr Ahmad A. Adewunmi, CIPR/CPG, KFUPM**
- **Dr Shahzad M. Kamal, CIPR/CPG, KFUPM**
- **Dr Hakeem Abbas, CENT, KFUPM**
- **Mr Musa O. Najimu, ChE, KFUPM**
- **Mr Mobeen Murtaza, CPG, KFUPM**
- **Mr Idrees Bakare, CENT, KFUPM**

This presentation should be cited as:

Adewole and Najimu (2017)

Referenced as:

Jimoh K. Adewole and Musa O. Najimu (2017), Oil Field Chemicals from Macromolecular Renewable Resources: A Case Study of Date Pit Nanoparticles for Drilling Fluid Formulation, AIChE Annual Meeting in Minneapolis, MN, USA, Oct 29 - Nov 3, 2017

KFUPM

485856 Designing Synthetic Extracellular Matrices for the Creation of Controlled Culture Systems in the Study of Disease

Abstract:

Biomaterials increasingly are used as extracellular matrix (ECM) mimics for controlled cell culture applications. In particular, synthetic hydrogel-based ECMs have been designed that afford precise control of the biophysical and biochemical properties of the matrix for probing cell response and hypothesis testing. Here, we present an approach for the creation of well-defined, hydrogel-based synthetic ECMs formed by light-mediated thiolâ€“ene click chemistry. Matrices with homogeneous nanostructure were formed by photoinitiated step growth polymerization of multi-arm poly(ethylene glycol) functionalized with thiols and cell-degradable peptides functionalized with alkenes (allyloxycarbonyl (alloc)-protected lysine). Hierarchical nanostructures reminiscent of the native ECM were incorporated within these materials using assembling peptides. Self-assembling collagen mimetic peptides based on variants of Proline-Hydroxyproline-Glycine repeats [$(POG)_n$] were decorated with alloc reactive groups. These multifunctional CMPs assemble to form fibrils (~10-250 nm wide, TEM and AFM) that are subsequently â€˜lockedâ€™ into place by photopolymerization. We are utilizing both the homogeneous and hierarchically-structured matrices for controlled cell culture to study the role of microenvironment cues in cell activation during disease. Specifically, activation of breast cancer cells and fibroblasts in response to microenvironment cues is being examined with these synthetic ECMs toward a better understanding of metastatic disease and fibrosis, respectively.

499558 The Influence of Matrix Stiffness on the Behavior of Brain Metastatic Breast Cancer Cells

Abstract:

Our research focuses on understanding the behavior of metastatic breast cancer cells using physiologically relevant engineered environments. Metastatic cancer cells interact with the microenvironment at a distant organ site during the process of colonization. This organ specific host microenvironment provides mechanical, chemical and biological cues that dictate whether the cancer cell will survive and outgrow into metastasis. Recapitulating the physical and chemical aspects of organ specific host microenvironment using biomimetic materials and investigating metastatic cancer cell-material interactions could further understanding of the mechanisms that govern metastatic colonization. As a model system, we are studying breast cancer brain metastasis, which is associated with poor prognosis and a median survival time of only 4-9 months. A major hurdle in developing therapeutic strategies to tackle breast cancer brain metastasis is our limited understanding of mechanisms involved in metastatic progression of breast cancer to the brain. This is, in part, due to lack of biomimetic *in vitro* models to study the interactions between metastatic breast cancer cells and brain microenvironment. To address this need, we employed hyaluronic acid hydrogels to recapitulate the native brain microenvironment and examined how hydrogel stiffness impacted the behavior of brain metastatic breast cancer cells. Hyaluronic acid was chosen as it is a glycosaminoglycan predominantly found in the brain and is known to interact with the CD44 receptors on metastatic breast cancer cells.

Hyaluronic acid hydrogels of varying stiffness (0.2 kPa - 4.6 kPa), including the range relevant to the brain environment (0.2 kPa - 1kPa) were prepared by functionalizing hyaluronic acid with methacrylate groups followed by crosslinking with dithiothreitol. Hydrogels were also engineered to promote cell adhesion by conjugation of the RGD peptide. MDA-MB-231 Br brain metastatic human breast cancer cells were used in our studies and were seeded on top of hydrogels. We first studied the dependence of metastatic breast cancer cell adhesion on the hydrogel stiffness and surface concentration of RGD and observed that the breast cancer cell adhesion increased with increasing hydrogel stiffness and RGD concentration. We also observed that the cell morphology was significantly affected by the hydrogel stiffness and cell spreading area increased with increasing hydrogel stiffness and RGD concentration. In addition, we noted that cell proliferation increased with increasing hydrogel stiffness. Our observations, thus far, indicate the activation of mechanotransduction pathways in brain metastatic breast cancer cells. We are currently investigating the role of focal adhesion kinase (FAK) - phosphatidylinositol 3-kinase (PI3K) signaling in stiffness dependent behavior of brain metastatic breast cancer cells. Ultimately, such systems could provide mechanistic insight into the pathways involved in breast cancer brain metastasis enabling development of new therapeutic targets for metastatic disease.

493309 The Combined Effect of Matrix Microenvironment and Hypoxia on the Activity of Glioblastoma Stemcells

Abstract:

Introduction Glioblastoma (GBM) is the most common and lethal form of brain cancer. GBM is known to diffusely infiltrate throughout the brain tissue, which contributes to poor patient prognosis. The complex and heterogeneous cell population, structural and nutrient gradients along the tumor margin, complicates the study of GBM. Despite the standard-of-care treatment, which includes surgical debulking tumor mass as much as safely for the patient normal brain function followed by chemo- and/or radiotherapy, the short 14.6 months median survival rate, low long term (5+ year < 5%) survival rate and the fast and high recurrence rate (~6.9 months after standard treatment) suggesting the little-to-no effect of current therapy. Currently, there are limited tools approaches to rigorously assess the relationship between tissue microenvironment and the growth, invasion, and therapeutic outcomes. Several subsets of cells within the heterogenous GBM tumor niche has been considered to contribute to the poor prognosis. GBM stem cells (GSCs) also known as GBM-initiating cells are thought to be therapeutic resistant and hence are important for curing and avoiding recurrence of GBM. Tumor-associated-macrophages/microglias are immune cells recruited into tumor sites and the polarization (proinflammatory M1 and anti- inflammatory M2) is recently thought to be important to tumor growth and related to GBM invasive phenotype. This study focuses on developing an engineered hydrogel platform to mimic the brain tumor environment to investigate how matrix composition, metabolic environment and drug treatments alter the cellular heterogeneity of patient-derived GBM specimens.

Materials and Methods Hydrogels are generated from methacrylamide-functionalized gelatin (Gel-MA) and methacrylated hyaluronic acid (HAMA). Patient-derived xenograft (PDX) GBM cells (GBM6, GBM10, GBM12, GBM15, GBM26, GBM39) were obtained from our collaborator (Dr. Jann Sarkaria, Mayo Clinic). These specimens maintain patient-specific morphological and molecular characteristics and span a range of invasive and therapy resitant phenotypes. Hydrogels were maintained for 4 wt% and made for with or without 15 w/w% matrix-immobilized HA and photo-initiator (PI) was adjusted (-HAMA: 0.1%, +HAMA:0.02%) for maintaining elastic modulus. Pre-polymerized solutions were then mixed with cell suspension homogenously and polymerized using UV lamp (7.1 mW/mm^2, 30 sec). Hydrogels were then put into culture for either normoxia (20% O_2) or hypoxia (1% O_2) and culture media were made with different treatments: +Dimethyl sulfoxide (0.1 vol%, control), +Erlotinib (Erl, receptor tyrosine kinase inhibitor, 10 Î¼M) or Temozolomide (TMZ, standard oral chemotherapy drug, 30 Î¼M). Therapeutic outcomes were determined by overall metabolic reactivity MTT assay (ThermoFisher). The heterogeneity of the GBM cell populations were determined using flow cytometer before each experiment and after degradation of hydrogels at timepoints throughout culture. Hydrogels were obtained after culture and put into collagenase type IV (100 unit/ 500uL) and hyaluronidase (100 unit/ 500uL) containing PBS for 30 minutes for fully degraded hydrogel. Four cell populations were stained after degradation: GSCs (CD133), TAMs (CD68), astrocytes (GFAP) and endothelial cells (ECs, CD31).

Results and Discussion Hyrogel matrix microenvironment significantly influences the kinetics of glioblastoma stem cells within the PDX specimens. Interestingly, the subfraction of GSCs within the invasive GBM6 PDX popualtion showed a significant increase under hypoxia (1% O_2) culture for up to 7 days compare to groups cultured in normoxia (20% O_2) condition. Specimens cultured in the presence of matrix-bound HA showed a significant higher fraction of GSC$^+$ cells, suggesting that GBM GSC% is influenced by both matrix composition as well as the metabolic environment. To study the complex heterogenous cell population between different PDX cells, a screening experiment of GSC%(CD133$^+$) and TAM%(CD68$^+$) was performed on 4 PDX lines: GBM6, GBM10, GBM12 and GBM39, each having different invasive potential (lo:0; 7:hi), Erlotinib response (0:lo, 100:hi) and molecular subtype (M: mesenchymal, C:classical). These speimens showed a wide range of initial and final cell compositions, suggesting profiling the heterogeneity of PDX specimens in biomimietic hydrogels may yield new, therapeutically actionable data. Ongoing work is looking at expansion of cell markers into four interesting cell groups (GSCs, TAMs, astrocytes and ECs) as well as in the presence of two therapeutic molecules (erlotinib, Temozolomide) to cell how the heterogeneous cell population changes as a response to microenvironmental cues and therapies.

Conclusion This study provides a rigorous way to study cell population, matrix influence and therapeutic outcome in a defined and controlled biomaterial in vitro platform. Data showed a huge heterogeneity existing in GBM and future work will examine more.

498973 Glioblastoma/Astrocyte Co-Culture on Polyelectrolyte MultilayerÂ Films: A Template for Studying the Role of Astrocytes in Glioblastoma Progression

Abstract:

Over 50% of all brain tumors are Glioblastoma Multiforme tumors (GBMs), with a median survival time of only one year post-diagnosis. These tumors are extremely aggressive, and have a poor prognosis, contributing to the distress surrounding a brain tumor diagnosis. GBMs arise from astrocytes and are characterized by their invasion into surrounding tissues and recruitment of healthy tissues into tumor tissue. Astrocytes are the most abundant glial cell in the brain, primarily functioning to maintain ion homeostasis and modulate energy production through both direct and indirect communication. While recent studies indicate that glial cells in close proximity to the tumor microenvironment undergo alterations in their responses, the underlying mechanisms driving these changes remain largely unknown. Elucidating the particular mechanisms of cell-cell communication between GBMs and astrocytes is key to the development of therapeutic strategies to combat this deadly disease. Current research on cell-cell communication utilizes three methods: conditioned media, transwell chamber co-culture, and random co-culture. However, none of these methods capture the full scope of GBM-astrocyte communication. As such, an accurate and representative *in situ* model is needed to study this communication. Using layer-by-layer assembly, we have engineered patterned co-cultures to precisely mimic both the physical and chemical signaling found in the brain. This novel platform provides new opportunities for studying the various signaling modalities of GBMs and Astrocytes.

Biofabrication

PAPER • OPEN ACCESS

Three-dimensional tissues using human pluripotent stem cell spheroids as biofabrication building blocks

To cite this article: Haishuang Lin *et al* 2017 *Biofabrication* **9** 025007

Biofabrication

PAPER

RECEIVED
12 October 2016

REVISED
27 February 2017

ACCEPTED FOR PUBLICATION
13 March 2017

PUBLISHED
20 April 2017

Three-dimensional tissues using human pluripotent stem cell spheroids as biofabrication building blocks

Haishuang Lin[1,5], **Qiang Li**[1,2,5] **and Yuguo Lei**[1,2,3,4]

[1] Department of Chemical and Biomolecular Engineering, University of Nebraska, Lincoln, Nebraska, United States of America
[2] Biomedical Engineering Program, University of Nebraska, Lincoln, Nebraska, United States of America
[3] Mary and Dick Holland Regenerative Medicine Program, University of Nebraska Medical Center, Omaha, Nebraska, United States of America
[4] Fred & Pamela Buffett Cancer Center, University of Nebraska Medical Center, Omaha, Nebraska, United States of America
[5] These authors contribute equally.

E-mail: ylei14@unl.edu

Keywords: human pluripotent stem cells, spheroids, tissue biofabrication, organ bioprinting

Abstract

A recently emerged approach for tissue engineering is to biofabricate tissues using cellular spheroids as building blocks. Human pluripotent stem cells (hPSCs), including human embryonic stem cells (hESCs) and induced pluripotent stem cells (iPSCs), can be cultured to generate large numbers of cells and can presumably be differentiated into all the cell types of the human body *in vitro*, thus are an ideal cell source for biofabrication. We previously developed a hydrogel-based cell culture system that can economically produce large numbers of hPSC spheroids. With hPSCs and this culture system, there are two potential methods to biofabricate a desired tissue. In Method 1, hPSC spheroids are first utilized to biofabricate an hPSC tissue that is subsequently differentiated into the desired tissue. In Method 2, hPSC spheroids are first converted into tissue spheroids in the hydrogel-based culture system and the tissue spheroids are then utilized to biofabricate the desired tissue. In this paper, we systematically measured the fusion rates of hPSC spheroids without and with differentiation toward cortical and midbrain dopaminergic neurons and found spheroids' fusion rates dropped sharply as differentiation progressed. We found Method 1 was appropriate for biofabricating neural tissues.

1. Introduction

Tissue engineering combines engineering principles and life science to repair, regenerate or restore the formats and functions of injured or dysfunctional tissues or organs [1–4]. In scaffold-based tissue engineering, a porous material is fabricated using biocompatible and/or biodegradable polymers [5, 6]. Living cells are then seeded onto the scaffold and the construct is cultured in a bioreactor *in vitro* to achieve the desired cell density, mechanical properties and/or functions before implantation. This classic approach has achieved many successes, but also has limitations [1, 5–11]. For instance, it is very challenging to generate vasculature; to achieve the organ-level cell density; and to precisely place specific cell types at the desired position in the construct. The mechanical properties of the scaffold and the products resulting from the scaffold degradation sometimes change the cell viability, proliferation, phenotype and physiology. In addition, due to the slow scaffold degradation and tissue morphogenesis, this approach is labor, time and reagent consuming.

A recently emerged approach for tissue engineering is to biofabricate tissues using cellular spheroids as building blocks [5, 6, 12]. Large numbers of human cells are first obtained and processed into multicellular spheroids, with a diameter of hundreds of micrometers, through methods including re-aggregation, microfluidics, hanging drop, micro-molds, spinner culture and rotating wall vessels [5, 6, 12, 13]. With the re-aggregation method, cells are centrifuged in a tube to form a pellet that is then cut into small fragments that are subsequently cultured to form spheroids. With the hanging drop method, small volumes of cell suspensions are placed on a lid that is subsequently inverted to form hanging drops. Cells in each hanging drop aggregate to form one spheroid. With the microfluidic and micro-mold methods, cells are placed in micro-wells to form one spheroid per well. With the spinner culture and rotating wall vessel methods, large numbers of cells are cultured in suspension in the spinner flasks or rotating wall vessels to simultaneously generate many spheroids. Next, the cellular spheroids are dispensed into a desired geometry, using a bio-printer or a mold. Upon contact, neighboring spheroids fuse to form a cohesive tissue due to the cell-cell and cell-matrix interactions. Lastly, the fused tissue is cultured in a bioreactor to mature before implantation.

This scaffold-free approach has been investigated for biofabricating many types of human tissue [13]. However, there are some challenges to be resolved [13]. The first is to obtain large numbers of human cells for the biofabrication. It is known that isolating large numbers of primary cells from a patient is challenging and only very limited human cell types can be cultured *in vitro*. In addition, when cultured, primary human cells quickly change phenotype and lose proliferation capability. Assuming we have sufficient cells, the second challenge is to quickly and reproducibly bioprocess them into uniform spheroids. The aforementioned re-aggregation, microfluidics, hanging drop, and micro-molds lack the capability to produce spheroids in large scale [13]. And with spinner culture and rotating wall vessel methods, the size and characteristics of the produced spheroids are not uniform [14, 15]. In addition, the numbers of spheroids that can be produced per volume of culture is low, resulting in high production costs [14, 15]. The third challenge is that, after being placed in position, spheroids fuse slowly, which will negatively impact the subsequent tissue maturation [13].

We explored whether these challenges could be resolved using human pluripotent stem cells (hPSCs) as the starting cell source. hPSCs, including human embryonic stem cells (hESCs) [16] and induced pluripotent stem cells (iPSCs) [17], can be cultured *in vitro* over a long term to generate large numbers of cells, and they can presumably be differentiated into all the cell types of the human body [18]. For instance, hPSCs have been successfully differentiated into cortical neurons [19, 20], gamma-aminobutyric acid interneurons [21–23], midbrain dopaminergic neurons [24, 25], endothelial cells [26–28], mesenchymal stem cells [29, 30], cardiomyocytes [31–33], hepatocytes [34–36], beta cells [37, 38] and other human cell types in the past decade. Additionally, patient-specific iPSCs and their derivatives (i.e. cells differentiated from iPSCs) induce minimal or no immune rejection in the patient. These characteristics make hPSCs an ideal cell source for tissue biofabrication.

Conventionally, hPSCs are cultured in two-dimensional (2D) dishes or flasks. However, these 2D culture systems are considered unsuitable for economic large-scale cellular biomanufacturing due to their low yield, scalability, and reproducibility [39, 40]. In addition, cells produced in 2D cell culture systems need further bioprocessing to form spheroids. To overcome this challenge, we previously developed a simple, efficient, scalable, and current Good Manufacturing Practice (cGMP) compliant three-dimensional (3D) culture system for expanding and differentiating hPSCs to generate hPSC spheroids and tissue spheroids (e.g. neuron spheroids) [14, 15, 40]. The system utilizes a thermoreversible PNIPAAm-PEG hydrogel as scaffolds for culturing hPSCs. Single hPSCs are suspended in liquid PNIPAAm-PEG polymer solution at ~4 °C. Upon heating to 37 °C, the polymer solution gels to form an elastic matrix, resulting in single hPSCs encapsulated in the hydrogel matrix that grow into uniform hPSC spheroids within the hydrogel. With further culturing in a differentiation medium, these hPSC spheroids can be differentiated into tissue spheroids. This culture system enables us to economically produce large numbers of hPSC and tissue spheroids for tissue biofabrication.

With hPSCs and this culture system, there are two potential methods to biofabricate a desired tissue. In Method 1, hPSC spheroids are utilized to biofabricate an hPSC tissue that is subsequently cultured in a medium capable of differentiating the hPSC tissue into the desired tissue. The key for Method 1 to succeed includes: (1) hPSC spheroids can quickly fuse to form an hPSC tissue, and (2) hPSCs in the hPSC tissue can be differentiated into the desired tissue cells. In Method 2, hPSC spheroids are first converted into tissue spheroids in the PNIPAAm-PEG hydrogel culture system and the tissue spheroids are utilized to biofabricate the desired tissue. Since we have shown hPSC spheroids could be efficiently differentiated into tissue spheroids in the PNIPAAm-PEG hydrogel [15, 41, 42], the key for Method 2 to succeed is that tissue spheroids can quickly fuse to form a cohesive tissue. In this paper, we will use the biofabrication of 3D neural tissues as examples to explore these questions and to find out which method is more appropriate for tissue biofabrication using hPSCs as cell source.

2. Materials and methods

2.1. Culturing hPSCs in 2D

iPSCs (iPSCs reprogrammed from human mesenchymal stem cells) were obtained from George Q. Daley laboratory (Children's Hospital Boston, Boston) [43]. H9 hESCs were purchased from WiCell Research Institute. hPSCs (iPSCs and H9s) were maintained in a 6-well plate coated with Matrigel (BD Biosciences) in Essential 8^{TM} medium (E8, Invitrogen) [18]. Cells were passaged every 4 days with 0.5 mM EDTA (Invitrogen) and the medium was changed daily. Cells were routinely checked for the expression of pluripotency markers, Oct4 and Nanog, their capability to form teratomas in immunodeficient mice, their karyotypes and bacterial contaminations.

2.2. Biomanufacturing hPSC spheroids in PNIPAAm-PEG hydrogels

To transfer the culture from 2D to 3D PNIPAAm-PEG hydrogels, hPSCs maintained in Matrigel-coated 6-well plate were treated with Accutase (Life Technologies) at 37 °C for 5 min and dissociated into single cells [14, 15]. Dissociated cells were mixed with 10% PNIPAAm-PEG (Cosmo Bio, USA) solution dissolved in E8 medium on ice and cast on tissue culture plate, then incubated at 37 °C for 10 min to form hydrogels before adding warm E8 medium containing 10 μM ROCK inhibitor (Y-27632, Selleckchem). Medium was changed daily and cells were routinely passaged every 5 days. To passage cells, medium was removed, and 2 ml ice-cold PBS was added to dissolve the hydrogel for 5 min. Cell spheroids were collected by spinning at 100 g for 3 min. Cells were incubated in Accutase (Invitrogen) at 37 °C for 10 min and dissociated into single cells, and dissociated into single cells for re-encapsulation, as mentioned above. The NucleoCounter NC-200 (Chemometec) was used to count cell number.

2.3. hPSC spheroids differentiation and fusion experiment

hPSCs (~3000 cells) in 200 μl E8 medium was placed in one V-shape low attachment well of a 96-well plate for 48 h to form one hPSC spheroid. For cortical neuron differentiation, the hPSC spheroid was cultured in a cortical neuron induction medium containing 50% DMEM/F12 + 50% Neurobasal medium, 1% N2 (Life Technologies), 2% B27 (Life Technologies), SB431542 (10 μM, Selleckchem) and LDN193189 (100 nM, Selleckchem) for 11 days to differentiate hPSCs into cortical progenitor cells. Medium was changed daily. The progenitor cells were matured in a neural differentiation medium containing Neurobasal® Media (Life Technologies), BDNF (20 ng ml^{-1}, PeproTech), GDNF (10 ng ml^{-1}, PeproTech), L-ascorbic acid (200 μM, Sigma), Dibutyryl-cAMP (0.5 mM, Santa Cruz Biotechnology) and 2.5 μM DAPT (Selleckchem) for another 20 days. For ventral midbrain dopaminergic neuron differentiation, the hPSC spheroid was cultured in a ventral midbrain neuron induction medium containing 50% DMEM/F12 + 50% Neurobasal medium, 1% N2, 2% B27, SB431542 (10 μM), LDN193189 (100 nM), CHIR99021 (0.7 μM, Selleckchem), purmorphamine (2 μM, Selleckchem) and SHH (200 ng ml^{-1}, R&D Systems) for 11 days to differentiate hPSCs into DA progenitor cells. Medium was changed daily. The progenitor cells were matured in a neural differentiation medium containing Neurobasal® Media (Life Technologies), BDNF (20 ng ml^{-1}), GDNF (10 ng ml^{-1}), L-ascorbic acid (200 μM), Dibutyryl-cAMP (0.5 mM) and 2.5 μM DAPT for another 20 days. For the spheroid fusion experiments, on varied days along the differentiation, two spheroids were placed in one V-shape well of the 96-well plate. The fusion process was then imaged with microscopy.

2.4. Neural tissue biofabrication

Spheroids containing about 5–10 million hPSCs were harvested from the PNIPAAm-PEG hydrogel on day 5, suspended in E8 medium and placed in a 6.5 mm Polyester Membrane Transwell Insert (8 μm pore, Corning). Cells were then cultured in the cell culture incubator. For some samples, 5 h later, the medium was removed and a thin layer of ECM (5 μl) (Matrigel, BD Biosciences) was overlaid on top of the fused tissue. Medium was added back to culture the cells. To biofabricate the cortical neural tissue, the fused hPSC tissue was cultured in the cortical neuron induction medium for 11 days and matured in the neural differentiation medium for another 20 days. To biofabricate the ventral midbrain dopaminergic neural tissue, the fused hPSC tissue was cultured in the ventral midbrain neuron induction medium for 11 days and the neural differentiation medium for another 20 days.

2.5. Staining and imaging

To assess the pluripotency of hPSCs cultured in hydrogels, hPSC spheroids were harvested and dissociated into single cells with Accutase and stained in suspension. Cells were fixed with 4% paraformaldehyde (PFA) at room temperature for 15 min, permeabilized with 0.25% Triton X-100 for 10 min, and blocked with 5% goat serum for 1 h before incubating with primary antibodies at 4 °C overnight. After extensive washing, secondary antibodies and 10 μM 4′,6-Diamidino-2-Phenylindole (DAPI) in 2% BSA were added and incubated at room temperature for 4 h. Cells were washed 3 times with PBS before imaging with a fluorescence microscope. The percentage of Oct4+ or Nanog+ nuclei was quantified with Image J. Around 10 000 cells were analyzed.

To stain, the 3D tissues were fixed with 4% PFA at room temperature for 2 h, then incubated with PBS + 0.25% Triton X-100 + 5% goat serum + primary antibodies at 4 °C for 3 days. After extensive washing (3 × 8 h), secondary antibodies and 10 μM DAPI in 2% BSA was added and incubated at 4 °C for 1 day. Cells were washed with PBS for 3 times (8 h each) before imaging with confocal microscope. Antibodies used: Oct4 (1:200, Santa Cruz Biotech); Nanog (1:200, Santa Cruz Biotech); PAX6 (1:300, Covance); Otx2, Tuj1 (1:5000, Sigma); Tbr1, FOXA2 (1:200, Santa Cruz Biotech); Lmx1a (1:500, Millipore) and TH (1:500, Pel-Freez). The images were analyzed with Image J for quantification. Around 5000 cells were analyzed.

3. Results

We first confirmed that large numbers of uniform hPSC spheroids could be economically produced in

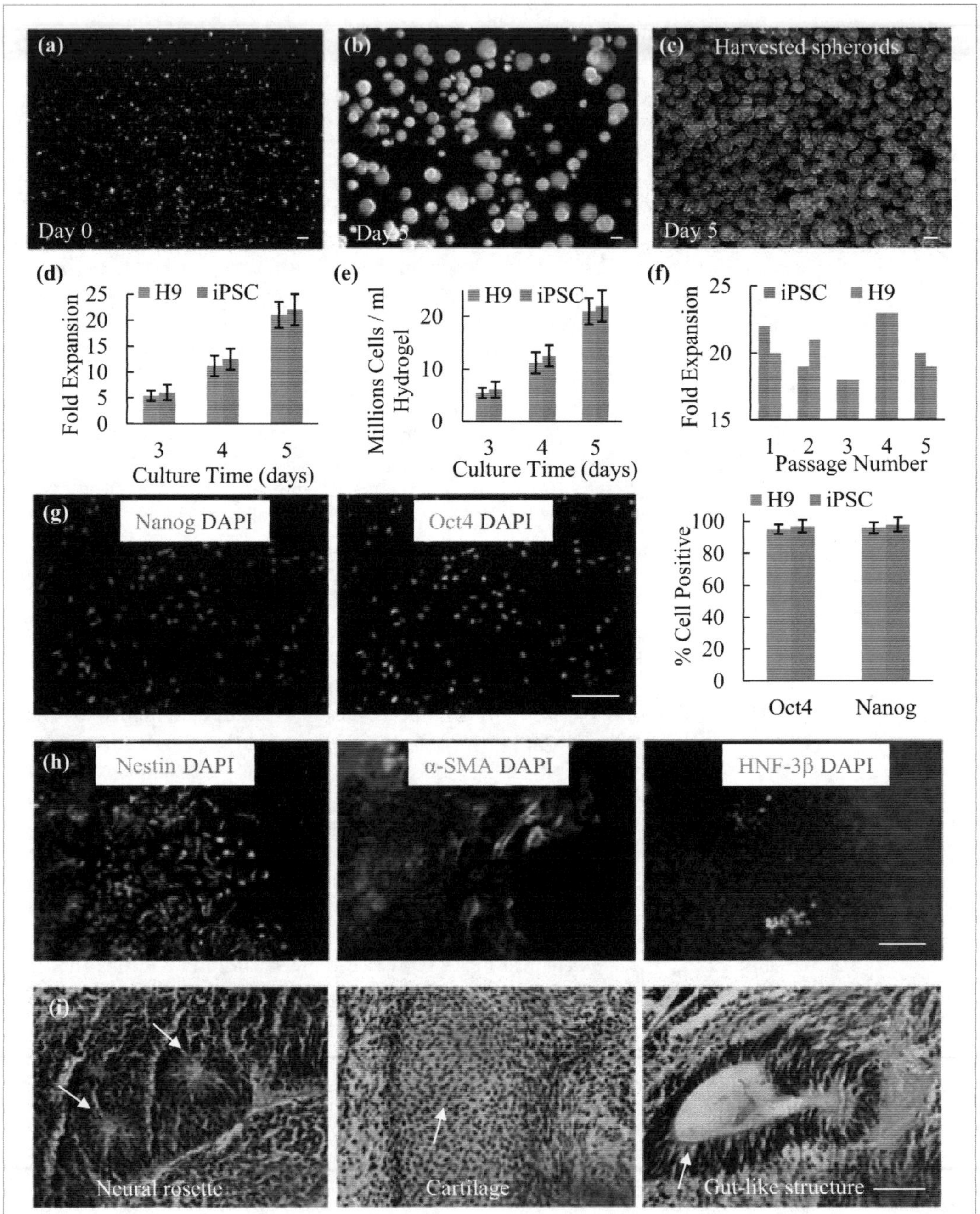

Figure 1. Biomanufacturing hPSC (H9 and iPSC) spheroids in thermoreversible PNIPAAm-PEG hydrogels. Single hPSCs were encapsulated in the PNIPAAm-PEG hydrogel and cultured in the Essential 8 (E8) medium. Single hPSCs (a) grew into uniform spheroids (b). On day 5, spheroids were harvested (c). Around 5-, 10- and 20-fold expansion was achieved, yielding ~5, 10 and 20 million cells per ml of hydrogel on day 3, 4 and 5, respectively ((d), and (e)). Consistent 20-fold expansion per 5 days was seen from passage 1 to 5 during the long-term hPSC culture within the hydrogel (f). After long-term culturing, above 95% cells were positive for the pluripotency markers, Oct4 and Nanog (g). (h) These hPSCs could be differentiated into the Nestin+ ectodermal, α-SMA+ mesodermal and HNF-3β+ endodermal cells in the *in vitro* embryoid body assay. (i) When transplanted to mice, they formed teratomas containing all the three germ layer tissues, such as the ectodermal neural rosette, the mesodermal cartilage and endodermal gut-like structure (arrows). Scale bars: 100 μm.

the thermoreversible PNIPAAm-PEG hydrogel [15]. We used both H9 hESCs and iPSCs to test the generality of the culture system. Single hPSCs grew into uniform spheroids on day 5 (figures 1(a)–(c)). hPSCs expanded around 5-, 10- and 20-fold on day 3, 4, and 5 of the culture, yielding around 5, 10, and 20×10^6 cells per ml of hydrogel, respectively (figures 1(d) and (e)). hPSC spheroids could be harvested by liquefying the hydrogel through cooling their temperature, i.e. adding ice-cold PBS for 5 min hPSCs could be cultured for multiple passages within the hydrogel with around 20-fold expansion every 5 days (figure 1(f)). After long-term culture, >95% of the cells remained pluripotency as shown by the expression of pluripotency markers, Nanog and Oct4 (figure 1(g)). Their pluripotency was further confirmed with the embryoid body assay *in vitro* and teratoma assay *in vivo*. hPSCs could be differentiated into Nestin+ ectodermal cells, α-SMA+ mesodermal cells and HNF-3β+ endodermal cells in the embryoid body assay (figure 1(h)). When transplanted to NOD-SCID mice, hPSCs formed teratomas containing all the three germ layer tissues, such as the ectodermal neural rosette, the mesodermal cartilage tissues and the endodermal gut-like structures (figure 1(i)). In summary, large numbers of hPSC spheroids could be readily made in the thermoreversible PNIPAAm-PEG hydrogels.

Based on the literature [44], we then developed protocols for differentiating hPSC spheroids into cortical neural tissue spheroids and ventral midbrain dopaminergic neural tissue spheroids. For cortical neural tissue differentiation, a cortical neuron induction medium containing small molecules LDN193189 and SB431542, which inhibit the dual SMAD signaling [44], was applied for 11 days to induce hPSCs into cortical progenitors. The spheroids were further cultured in a neural differentiating medium for an additional 20 days to mature the progenitor cells into neurons (figure 2(a)). On day 7, immunostaining showed ~90% cells in the spheroid were positive for the neural stem cell makers, PAX6 and Nestin (figures 2(b) and (d)). On day 30, ~80% cells in the spheroid were Tuj1 + neurons (figures 2(c) and (d)). For dopaminergic neural tissue differentiation, a neural induction medium containing small molecules LDN193189, SB431542, purmorphamine and CHIR99021, as well as protein SHH was applied to the hPSC spheroid for 11 days [24]. LDN193189 and SB431542 can induce hPSCs toward neural lineage, while SHH and purmorphamine can pattern cells to the ventral fate and CHIR99021 can pattern cells to the midbrain fate. Together, these molecules pattern hPSCs to the ventral midbrain dopaminergic neuron progenitor cells [24]. The spheroid was further cultured in the aforementioned neural differentiating medium for an additional 20 days to mature the progenitor cells into neurons (figure 2(e)). On day 11, ~90% cells in the spheroid were positive for the ventral midbrain progenitor makers, FOXA2 and Lmx1a (figures 2(f) and (h)). On day 30, ~80% cells in the spheroid were Tuj1 + and ~50% tyrosine hydroxylase positive (TH+) ventral midbrain dopaminergic neurons (figures 2(g) and (i)). No significant cell death was observed during the differentiation.

Quick and efficient spheroid fusion is crucial for the subsequent tissue maturation and thus is required for rapid tissue biofabrication, especially through bioprinting. We then quantitatively studied the fusion rate of undifferentiated and differentiated hPSC spheroids (figures 3 and 4). One hPSC spheroid was placed in a V-shape low attachment well of a 96-well plate and differentiated to a cortical neural tissue spheroid using aforementioned protocol (figure 2(a)). At different days along the differentiation, two spheroids were placed in one V-shape well to allow them to contact quickly. Their fusion process was then recorded with microscopy (figure 3(a)). We used the $(\pi r^2)/(\pi R^2)$, wherein r and R are the radius of the interface of the two spheroids and radius of the spheroid, respectively, to quantitatively measure the percentage of the spheroid fusion completed (figure 3(b)). And $t_{1/2}$ is referred to the time when 50% of the fusion is completed or $(\pi r^2)/(\pi R^2) = 0.5$ (figure 3(c)). We found that undifferentiated hPSC spheroids completed the fusion in 5 h. As the differentiation progressed, the fusion rates of spheroids decreased. The fusion became extremely slow after 14 days of differentiation. The $t_{1/2}$ of undifferentiated hPSC spheroids, spheroids differentiated for 4, 7, 14 and 28 days is 2.4, 2.6, 8.1, 66.6 and 90.6 h, respectively (figure 3(c)). Similar trends were found for ventral midbrain dopaminergic neural tissue spheroids (figure 4). The $t_{1/2}$ of undifferentiated hPSC spheroids, spheroids differentiated for 3, 6, 13 and 27 days is around 4.0, 6.0, 9.0, 11.0 and 16.0 h, respectively (figure 4(c)). Based on these results, we concluded that using neural tissue spheroids to biofabricate neural tissues would be limited by the slow fusion rate, and the Method 1 described in the introduction would be more appropriate for biofabricating tissues from hPSCs.

We then studied whether the fused hPSC tissue could be differentiated and matured into neural tissues. When spheroids were placed in a Transwell insert, they sedimented to the bottom and tightly packed within 10 min due to gravity (figures 5(a) and (b)) without centrifugation. Phase images showed packed individual spheroids at 10 min after placing them in the Transwell insert (figure 5(b), left). After 3 h, they fused to form a continuous tissue with rough surface (figure 5(b), middle). After 5 h, the surface became very smooth, indicating the completion of spheroids' fusion (figure 5(b), right). Live/dead cell staining showed very few dead cells within the fused hPSC tissue (figure 5(c)).

We then treated this hPSC tissue with the cortical neuron induction medium for 11 days and the neural differentiation medium for additional 20 days as described in figure 2(a). On day 5, small islands with diameters around 100 μm were frequently observed in the developing tissue (figure 6(a)). On day 11, a thin tissue that could not be lifted from the underlying Transwell insert was formed (figure 6(b)). Immunostaining showed ~75% of the cells in this thin tissue were positive for the cortical progenitor makers, PAX6 and Otx2 (figures 6(c) and (d)). On day 30, more than 80% cells were Tuj1+ neurons and 50% were positive for the cortical neuron marker, Tbr1 (figures 6(f) and (g)). However, we were unable to make a thick, 3D

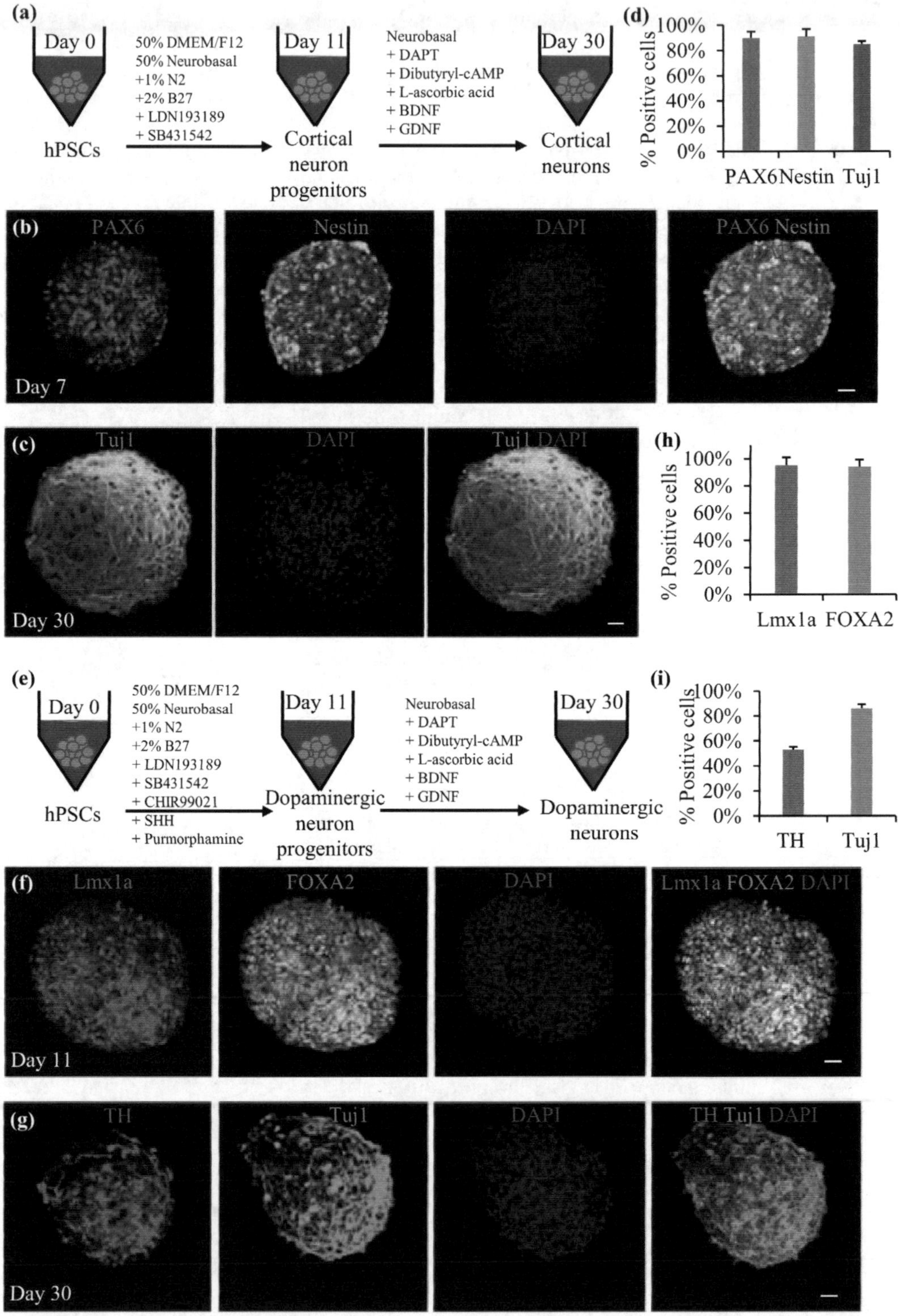

Figure 2. Differentiating hPSC spheroids into neural tissue spheroids. (a) The protocol for making cortical neural tissue spheroids. (b) ~90% cells in the day 7 spheroids were positive for neural stem cell makers PAX6 and Nestin (d). (c) ~80% cells in the day 30 spheroids were Tuj1+ neurons (d). (e) The protocol for making ventral midbrain dopaminergic neural tissue spheroids. (f) ~90% cells in the day 11 spheroids were positive for ventral midbrain progenitor makers Lmx1a and FOXA2 (h). (g) ~80% cells in the day 30 spheroids were Tuj1+ and ~50% TH + dopaminergic neurons (i). Scale bars: 25 μm.

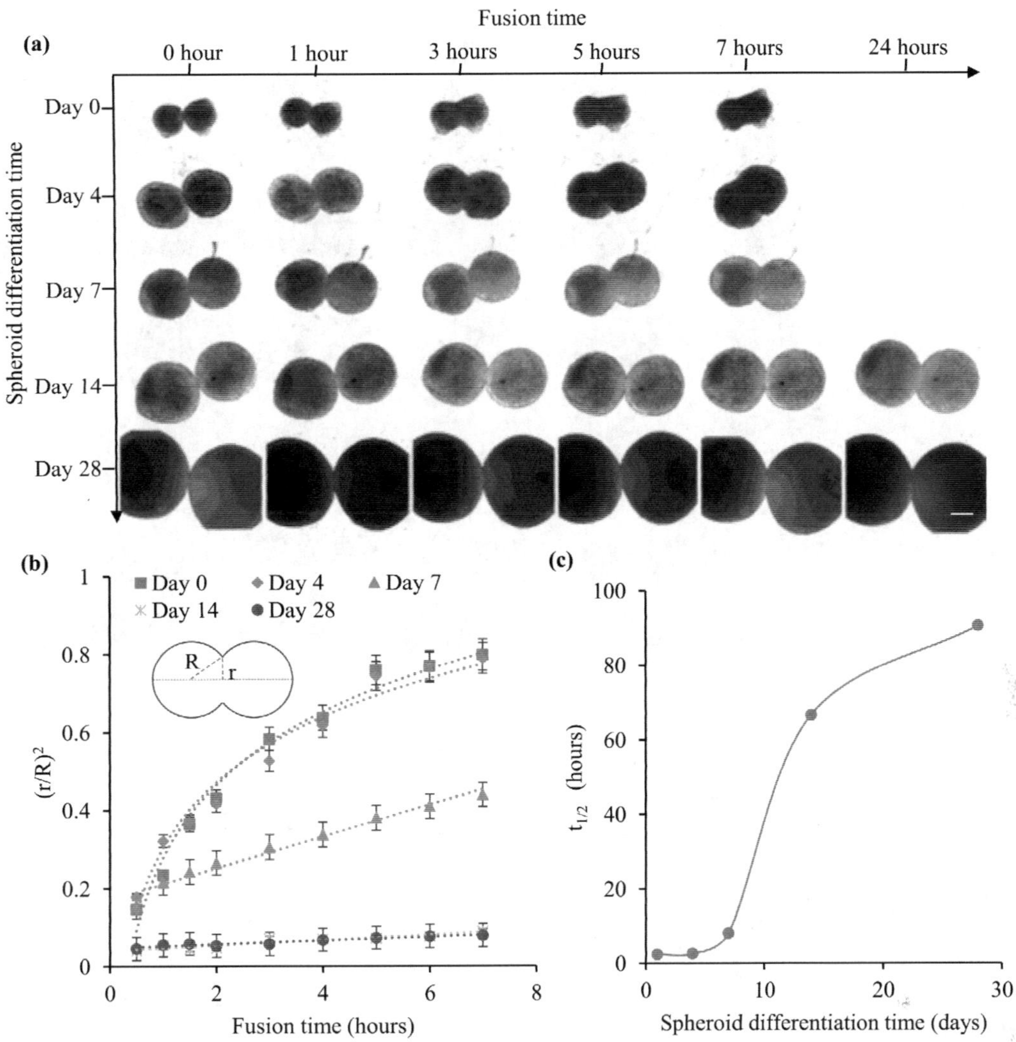

Figure 3. Spheroid fusion experiment. hPSC spheroids were differentiated toward cortical neural tissue spheroids, as described in figure 2(a). On day 0, 1, 4, 7, 14 and 28 along the differentiation, two spheroids were placed in contact and the fusion process was imaged at 0, 1, 3, 5, 7 and 24 h later. (a) Phase images of their fusion. (b) $(r/R)^2$ is used to measure their fusion rates. R and r is referred to the radius of a spheroid and the radius of the interface of the two fused spheroids, respectively. (c) $t_{1/2}$, the time to complete 50% fusion or $(r/R)^2 = ½$, is measured. Scale bars: 200 μm.

cortical tissue (figure 6(e)); instead, neurons clustered into aggregates (figure 6(f)).

We also treated the hPSC tissue with the ventral midbrain dopaminergic neuron induction medium for 11 days and the neural differentiation medium for additional 20 days as described in figure 2(e). Similar results were obtained. On day 5, cell islands were frequently seen in the developing tissue (figure 7(a)). On day 11, a thin tissue that could not be lifted from the underlying Transwell insert was formed (figure 7(b)). Immunostaining showed ~85% of the cells were positive for the ventral midbrain progenitor makers, FOXA2 and Lmx1a (figures 7(c) and (d)). On day 30, more than 90% cells were Tuj1+ neurons and 50% were positive for the ventral midbrain dopaminergic neuron marker, TH (figures 7(f) and (g)). Again, we were unable to make a 3D tissue (figure 7(e)); instead, neurons clustered into aggregates (figures 7(e) and (f)). In summary, although hPSCs could be differentiated into neurons, the fused hPSC tissues could not be converted into 3D neural tissues.

Literature studies showed that ECM were necessary for mammalian cells to assemble into 3D tissues [45]. We hypothesized that the loss of 3D structure during the hPSC tissue differentiation was due to the shortage of ECM. To test the hypothesis, a thin layer (~5 μm) of Matrigel was overlaid on the hPSC tissue prior to the neural induction. The hPSC tissue was then differentiated as aforementioned. Low concentration of Matrigel proteins (e.g. the commercial Matrigel product was diluted 100 fold) was supplemented to the culture medium during the 30 days. On day 5, no cell islands were observed (figure 8(a)), and on day 11, a soft, transparent and 3D tissue with

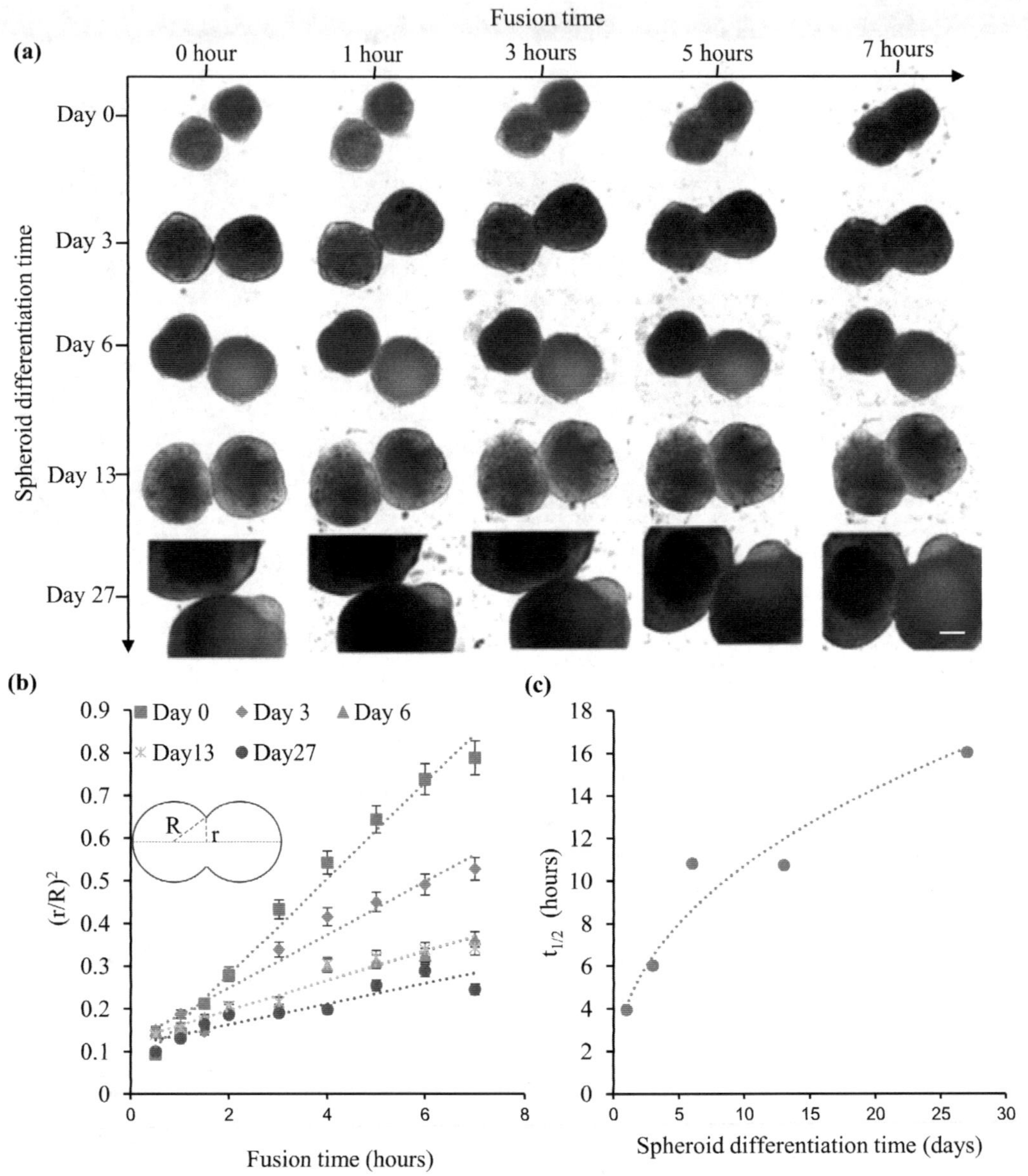

Figure 4. Spheroid fusion experiment. hPSC spheroids were differentiated toward ventral midbrain dopaminergic neural tissue spheroids, as described in figure 2(e). On day 0, 3, 6, 13 and 27 along the differentiation, two spheroids were placed in contact and the fusion process was imaged at 0, 1, 3, 5, 7 and 24 h later. (a) Phase images of their fusion. (b) $(r/R)^2$ is used to measure their fusion rates. R and r is referred to the radius of a spheroid and the radius of the interface of the two spheroids, respectively. (c) $t_{1/2}$, the time to complete 50% fusion or $(r/R)^2 = ½$, is measured. Scale bars: 200 μm.

thickness around 1 mm was formed (figures 8(b) and (c)). Immunostaining showed ~75% of the cells within the tissue were positive for the cortical progenitor cell markers, PAX6 and Otx2 (figures 8(d) and (e)). No Nanog+ or Oct4+ hPSCs were detected in the tissue (figure 8(f)). Further maturation in the neural differentiation medium, a 3D (~1 mm thickness), elastic and semi-transparent cortical tissue was produced (figures 8(g) and (h)). ~85% of the cells within the day 30 tissue were Tuj1+ neurons and ~50% were positive for the cortical neuron marker, Tbr1 (figures 8(i) and (j)).

Similar results were obtained when the hPSC tissue was overlaid with Matrigel and differentiated toward the dopaminergic neural tissue. On day 5, no cell islands were observed (figure 9(a)). On day 11, a soft, transparent and 3D tissue with thicknesses around 1 mm was formed (figures 9(b) and (c)). Immunostaining showed ~85% of the cells within the tissue were positive for the ventral midbrain progenitor makers, FOXA2 and Lmx1a (figures 9(d) and (e)). No Nanog+ or Oct4+ hPSCs were detected in the tissue (figure 9(f)). Further maturation in the neural differentiation medium, an elastic, semi-transparent

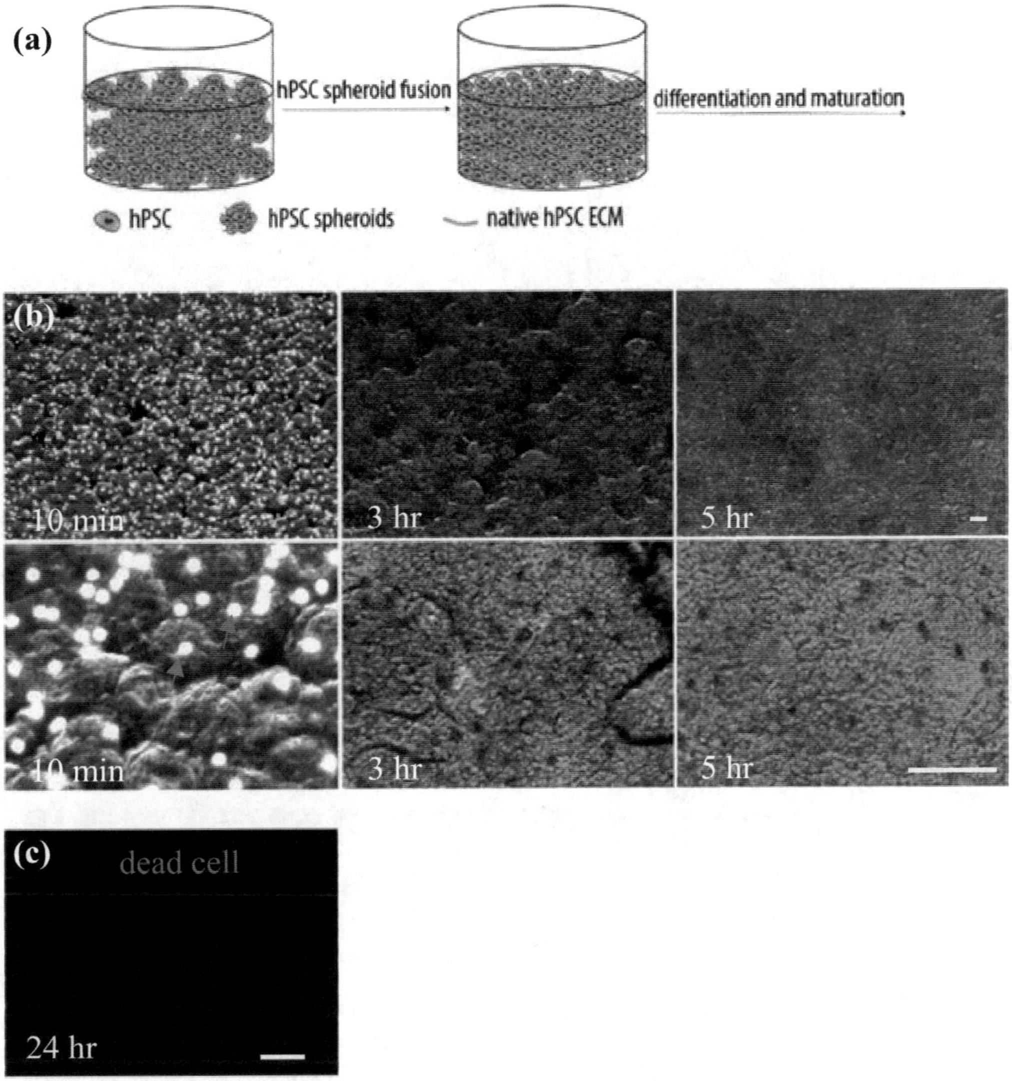

Figure 5. Biofabricating hPSC tissues. hPSC spheroids were harvested and placed in a Transwell insert to fuse into a cohesive tissue (a). Spheroids sedimented to the bottom of the insert within 10 min (b, left). The white dots (red arrows) were the pores of the insert. 3 h later, a continuous 3D tissue was formed (b, middle). 5 h later, the surface of the tissue became very smooth, indicating the completion of the fusion process (b, right). Top views of the samples were shown (b). Live/dead cell staining at 24 h showed very few dead cells within the tissue (c). Scale bars: 100 μm.

and 3D (~1 mm thickness) ventral midbrain tissue was produced (figures 9(g) and (h)). More than 90% cells were Tuj1+ neurons and 52% were positive for the ventral midbrain dopaminergic neuron marker, TH (figures 9(i) and (j)). In summary, hPSC tissues were successfully converted into 3D neural tissues in the presence of exogenous ECM.

4. Discussion

We biofabricate neural tissues including the cortical neural tissue and midbrain dopaminergic tissue to explore the tissue biofabrication methodology using hPSCs as starting materials in this study. Compared with neurons cultured in 2D, 3D neural tissues more closely resemble the *in vivo* brain tissues and thus are superior for investigating neurobiology and diseases, for drug discovery, and for treating neural diseases [46]. For instance, in the brains of Alzheimer's disease patients, large amounts of insoluble amyloid beta (Aβ) plaques are formed within the extracellular matrix. iPSCs derived from AD patients can be made and differentiated into brain neurons to model the disease *in vitro* [47]. However, the production of Aβ plaques in 2D cell culture is difficult due to the limited extracellular matrix in 2D cultures. A recent study showed that, in 3D neural tissues made from neural stem cells overexpressing three AD-causing genes, large amounts of Aβ plaques were made within their ECM [48]. Thus, 3D neural tissues made through the method developed in this study can be very valuable

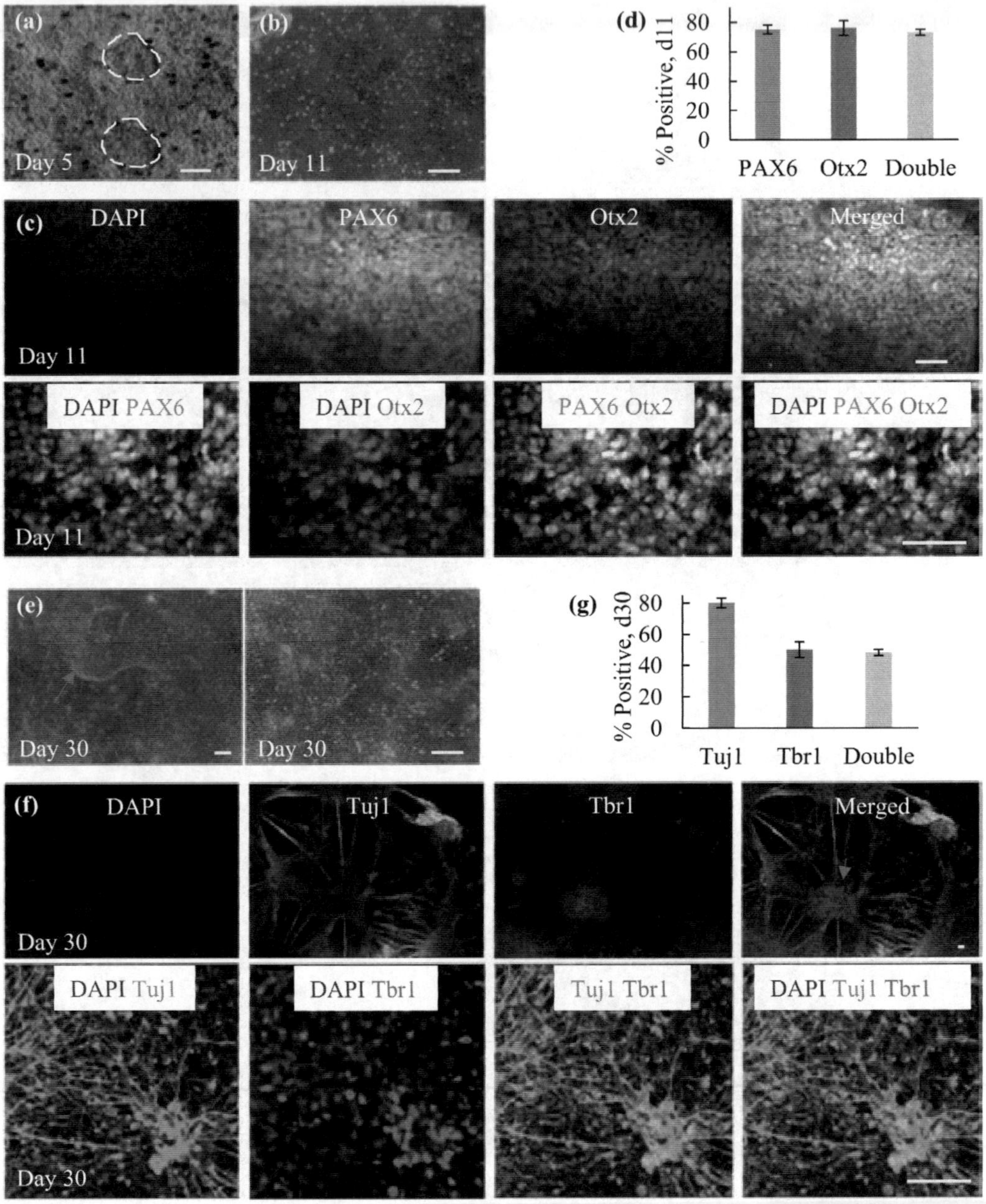

Figure 6. Biofabricating cortical neural tissues. The hPSC tissue was treated with neural induction and neural differentiation medium as described in figure 2(a). Cell islands (outlined with white dash lines) appeared on day 5 (a) during the differentiation. A thin cell sheet was formed on day 11 (b). The white dots seen in (b) were the pores of the insert. 75% of the cells were positive for the cortical neuron progenitor markers, PAX6 and Otx2 ((c) and (d)). On day 30, cells were differentiated into cortical neurons. 80% of the cells were Tuj1 + neurons and 50% were Tbr1+ cortical neurons ((f) and (g)). No 3D neural tissues were formed. Cells aggregated into clusters (red arrows) ((e) and (f)). ((a), (b) and (e)) Top views of the tissue. Scale bars: 50 μm.

for modeling brain diseases, particularly those with pathology in ECMs. In a second example, 3D neural tissues could be used to study glioblastoma, the most aggressive brain tumor. Glioblastoma cannot be completely removed through surgery because the tumor cells can infiltrate several centimeters into the surrounding brain tissues. The biofabricated 3D neural tissues can be used to study how tumor cells invade the brain tissue. In addition to studying the biology, these 3D neural tissues can be applied for discovering drugs that can prevent forming or dissolve Aβ plaques or stop the tumor cell invasion. Biofabricated 3D neural tissues can also be very useful for treating neural diseases and injuries. Transplanting of neural stem cells or neurons is being widely investigated for treating neural diseases and injuries. A current challenge with cell transplantation is the low survival of the transplanted cells *in vivo*. In general, about 5% to 10% of the cells can survive, limiting the efficacy of cell therapies [49]. Our laboratory research

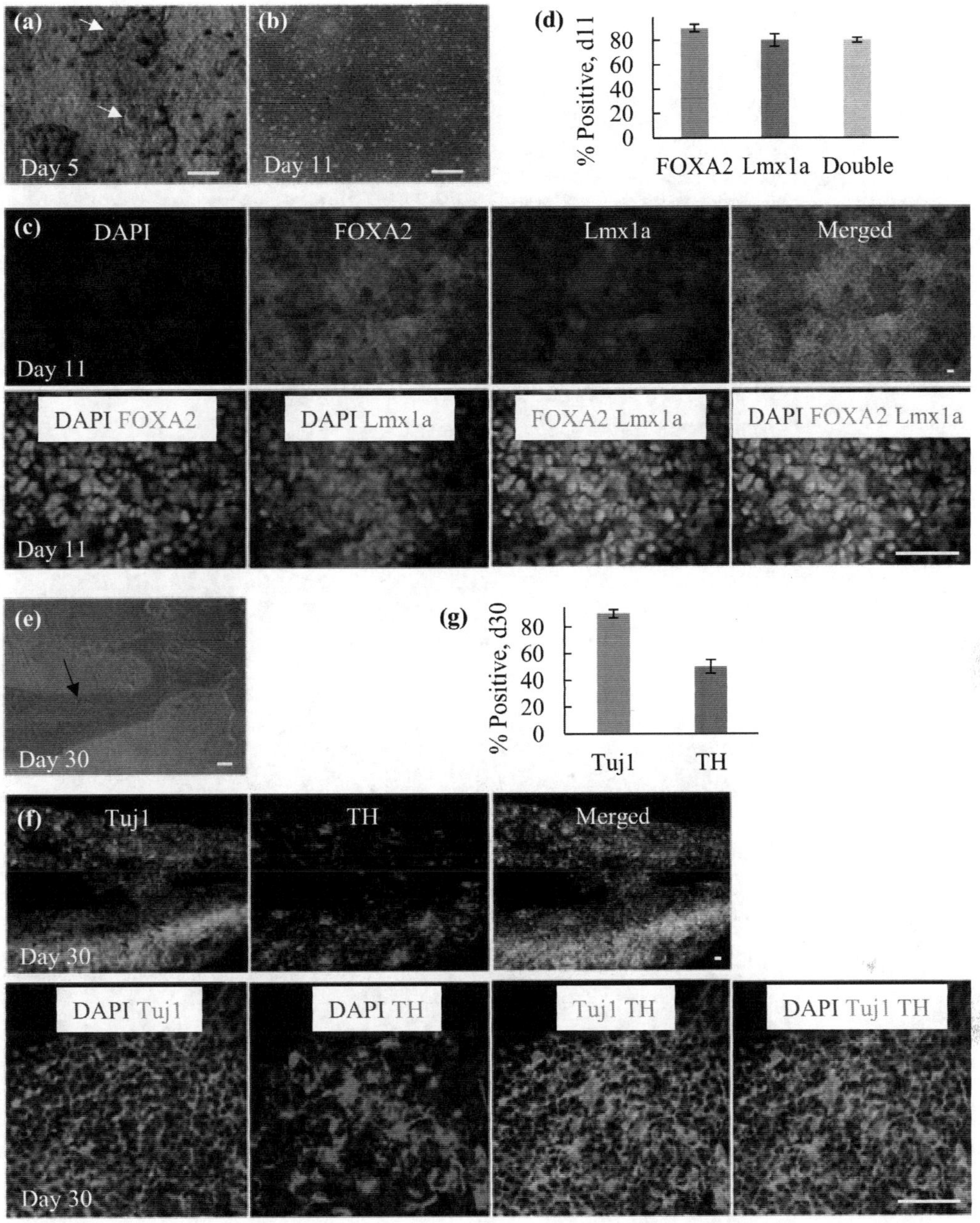

Figure 7. Biofabricating ventral midbrain dopaminergic neural tissues. The hPSC tissue was treated with neural induction and neural differentiation medium as described in figure 2(e). Cell islands (white arrows) appeared on day 5 (a) during the differentiation. A thin cell sheet was formed on day 11 (b). The white dots seen in (b) were the pores of the insert. Above 85% of the cells were positive for the ventral midbrain dopaminergic neuron progenitor markers, FOXA2 and Lmx1a ((c) and (d)). On day 30, cells were differentiated into dopaminergic neurons. 90% of the cells were Tuj1+ neurons and 50% were TH+ dopaminergic neurons ((f) and (g)). No 3D tissues were formed. Cells aggregated into clusters (black arrows) ((e) and (f)). ((a), (b) and (e)) Top views of the tissue. Scale bars: 50 μm.

as well as other research has shown that transplanting 3D neural tissues, rather than 2D cell cultures, could significantly improve cell survival and function *in vivo* [50]. Thus, biofabricated 3D tissues may be better than cells as cellular products for regenerative medicine.

As aforementioned, large numbers of hPSC spheroids are required for tissue biofabrication. To scale up the hPSC production, three-dimensional (3D) suspension culture systems, such as spinner flasks and stirred-tanks have been studied [40, 51]. Three components, cells; a medium; and stirring or shaking are within these systems. Ideally, the seeded single hPSCs initially associate to form small clusters that subsequently expand to generate uniform spheroids. In reality, these spheroids frequently aggregate during the culture to form large agglomerates, resulting in two negative consequences [52]. First, the produced spheroids varied significantly in their size and

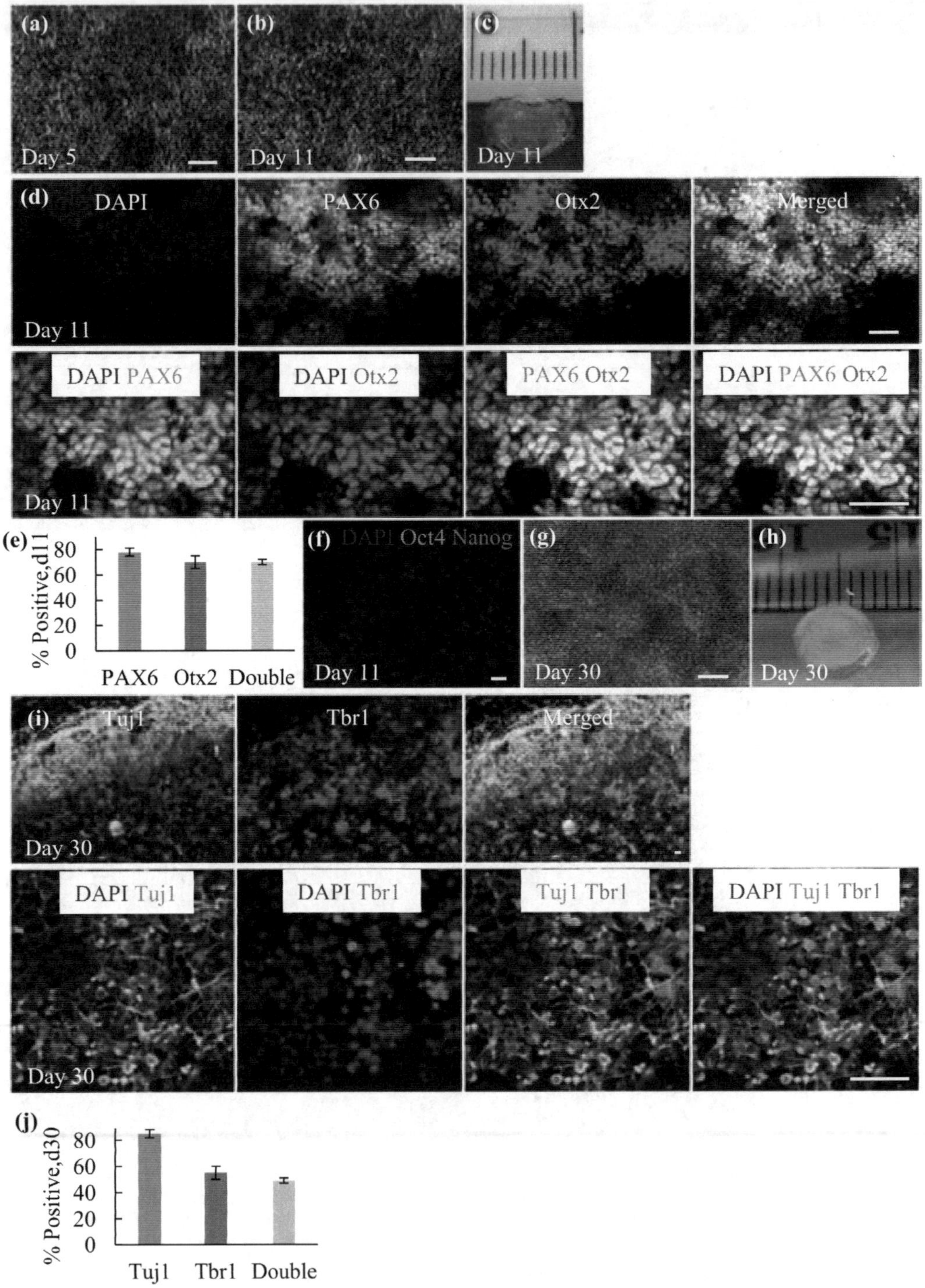

Figure 8. Biofabricating cortical neural tissues in the presence of exogenous ECM. The hPSC tissue was overlaid with a thin layer of Matrigel and treated with neural induction and neural differentiation medium as described in figure 2(a). Cell islands were not seen on day 5 (a) during the differentiation. A 3D, soft and transparent tissue was formed on day 11 ((b) and (c)). 75% of the cells were positive for the cortical neuron progenitor markers, PAX6 and Otx2 ((d) and (e)). No Nanog+ or Oct4+ hPSCs were found in the tissue (f). A 3D, elastic and semi-transparent cortical tissue was formed on day 30 ((g) and (h)). 85% of the cells were Tuj1+ neurons and 55% were Tbr1+ cortical neurons ((i) and (j)). ((a), (b) and (g)) Top views of the tissue. Scale bars: 50 μm.

characteristics. Second, it is well known that the transport of nutrients, oxygen and growth factors, and the metabolic waste from cells located at the center of agglomerates with diameters >500 μm become insufficient, leading to slow cell proliferation, apoptosis, and uncontrolled differentiation [53]. While stirring or shaking reduces spheroid agglomeration, they also generate hydrodynamic stress that negatively

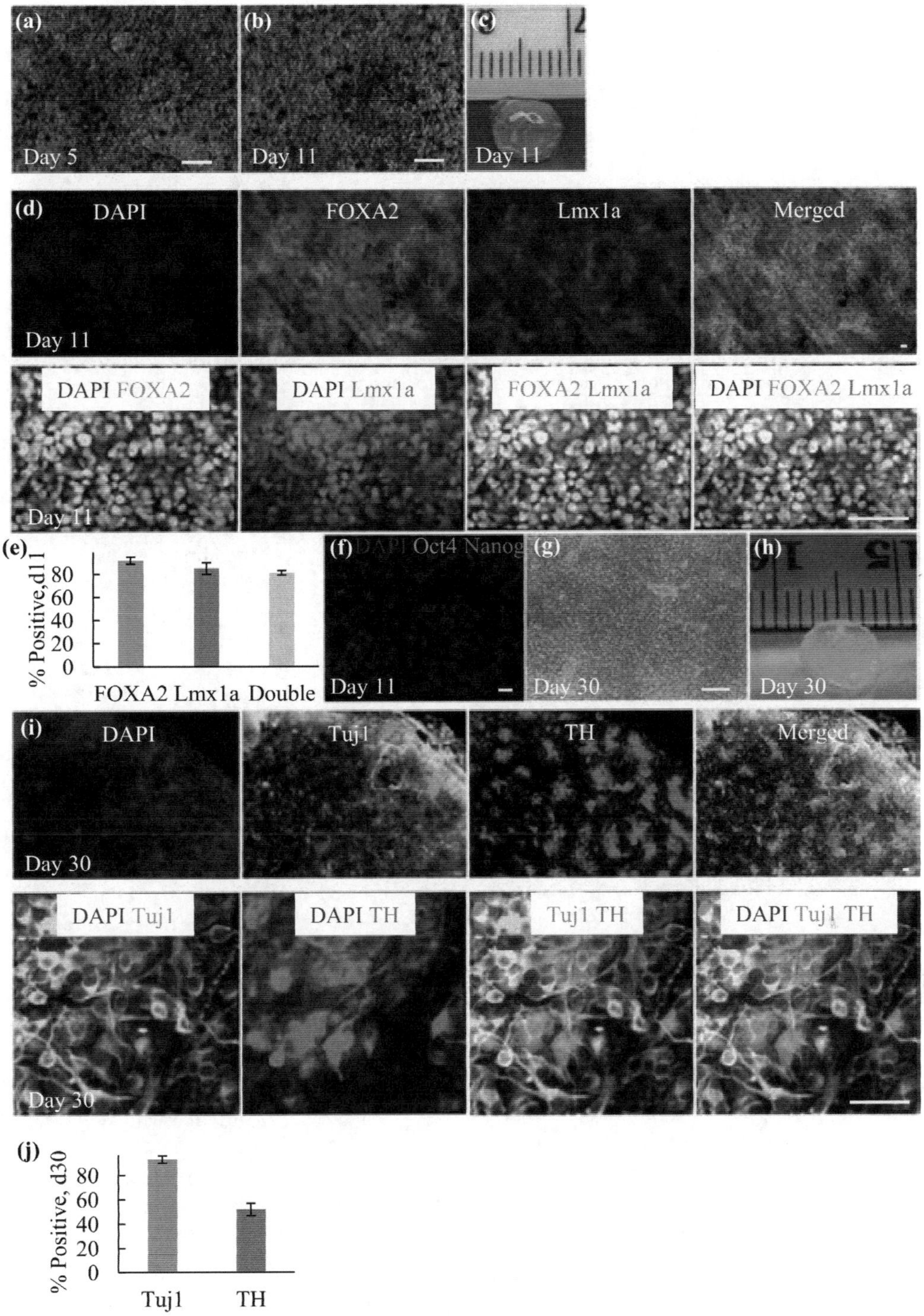

Figure 9. Biofabricating ventral midbrain dopaminergic tissues in the presence of exogenous ECM. The hPSC tissue was overlaid with a thin layer of Matrigel and treated with neural induction and neural differentiation medium as described in figure 2(e). Cell islands were not seen on day 5 (a) during the differentiation. A 3D, soft and transparent tissue was formed on day 11 ((b) and (c)). Above 85% of the cells were positive for the ventral midbrain dopaminergic neuron progenitor markers, FOXA2 and Lmx1a ((d) and (e)). No Nanog+ or Oct4+ hPSCs were found in the tissue (f). A 3D, elastic and semi-transparent ventral midbrain tissue was formed on day 30 ((g) and (h)). 93% of the cells were Tuj1+ neurons and 52% were TH+ dopaminergic neurons ((i) and (j)). ((a), (b) and (g)) Top views of the tissue. Scale bars: 50 μm.

affects cell growth [54, 55]. High cell density in the culture also promotes spheroid agglomeration [14]. Considering all these factors, in current suspension culture studies, hPSCs are generally seeded at low density (e.g., ~3 $\times$ 10^5 cells/mL) and stirred at 70 to 120 rotations-per-minute (rpm). Under even these optimized conditions, slow cell growth, significant cell death, phenotype change, genomic mutations, and low volumetric yield are common [40]. For instance, we and others showed iPSCs typically expanded 4-fold per 4 days to yield around 2.0×10^6 cells/mL in suspension cultures [14, 15, 40]. It is good to note that cells occupy <0.2% of the medium or bioreactor volume with this yield. This low yield results in extremely high biomanufacturing cost. This challenge can be answered with the PNIPAAm-PEG hydrogel culture system (figure 1). The hydrogel scaffold not only provides space for cell growth, but also acts as physical barriers to prevent the neighboring spheroids from agglomeration and isolate the hydrodynamic stress. These lead to high cell viability, proliferation and yield, as well as uniform spheroid size. For instance, 20-fold expansion per 5 days to yield 2.0×10^7 cells has been achieved.

An essential requirement for rapid tissue biofabrication is that spheroids can quickly fuse to form a cohesive tissue. Culturing hPSCs in 3D thermoreversible PNIPAAm-PEG hydrogels allowed the production of spheroids which contained both cells and their native ECM (figure 1) [56]. Spheroids placed in contact with one another quickly fused into 3D tissues. The driven force for tissue fusion is the minimization of the tissue surface free energy [57, 58]. Through tissue fusion, cells can maximize their interactions with neighboring cells and ECM. hPSC spheroids harvested from hydrogel sedimented, packed and fused within 3 h (figures 5(a) and (b)). No cell death was observed during the fusion (figure 5(c)). Notably, when hPSCs from 2D cultures or dissociated spheroids (e.g. single cells) were placed in contact, a minimal of 24 h were required for cells to fuse into a cohesive tissue and large percentage of cells went apoptosis (data not shown), indicting the importance of the native ECM for the spheroid fusion and highlighting the advantage of using the PNIPAAm-PEG hydrogels for culturing hPSCs.

An important finding of this study is that the fusion rate decreased significantly once hPSC spheroids were differentiated (figures 3 and 4). This is particularly obvious for the cortical neural tissue spheroids (figure 3). This phenomenon leads us to conclude that a more appropriate method to biofabricate neural tissues is to fuse hPSC spheroids first, followed by differentiating and maturing the hPSC tissue. Importantly, our study also showed that the hPSCs in the hPSC tissue could be differentiated into neurons with protocols similar to these used for differentiating hPSCs into neurons in 2D culture (figure 2). It should be noted that not all the hPSCs in the biofabricated tissues were differentiated into the desired neurons in this study. These biofabricated tissues are appropriate for modeling diseases and drug discovery. For *in vivo* application, the differentiation protocols should be optimized to quantitatively convert hPSCs into the desired neurons in the future.

The finding that extra ECM were needed to form 3D neural tissues during the differentiation of hPSC tissues can be explained by the inability of neurons to produce enough ECM (figures 6 to 9). *In vivo*, the ECM in the brain are produced by glia cells [59]. In the biofabricated tissues, there are no or little glia cells to produce ECM. Matrigel was supplemented as exogenous ECM to facilitate the 3D tissue formation in this research. Matrigel, a solubilized basement membrane preparation extracted from the mouse sarcoma, however, is animal-derived, not defined, and not qualified for biomanufacturing cells for clinical applications [60]. Future research to replace Matrigel is necessary in order to biomanufacture 3D tissues for clinical application. Notably, adding exogenous ECMs did not significantly change the differentiation process and efficiency. The cellular compositions in the day 11 and day 30 tissues produced without or with exogenous ECM were very similar (figure 6 versus 7 and figure 8 versus 9).

Biofabricating tissues through first fusing hPSC spheroids, followed by differentiation and maturation is simple, however, this also has limitations. First, the biofabricated tissues lack vasculature. A solution to this challenge is to use bio-printer to print a sacrificing hydrogel network in the hPSC tissue [61]. After the hPSC tissue becomes the mature neural tissue, the sacrificing hydrogel network is removed and endothelial and smooth muscle cells are seeded to the channels to form the vasculature. Both the PNIPAAm-PEG and alginate hydrogels can be used for this purpose. They can be removed by cooling the temperature or adding EDTA for a few minutes. The second challenge is to biofabricate neural tissues with multiple types of neurons. This challenge can be solved by using the progenitor tissue spheroids as building blocks to biofabricate the cohesive tissue. For instance, the day 11 cortical progenitor and the day 11 ventral midbrain progenitor tissue spheroids can be bio-printed into a cohesive tissue that can be further matured into a neural tissue with two types of neurons. Our study showed the progenitor spheroids could fuse, through their fusion rates were slower than hPSC spheroids (figures 3 and 4).

5. Conclusion

In summary, this study explored some fundamentals and methods for biofabricating tissues using hPSCs as the cell source. We found: (1) the PNIPAAm-PEG hydrogel culture system was excellent for biomanufacturing hPSC spheroids; (2) the spheroid fusion experiment was a quick and efficient method to evaluate the fusion capability of tissue spheroids; (3) the fusion rates of differentiated spheroids were significantly lower than hPSC spheroids; (4) hPSCs could be differentiated into tissue cells in the 3D tissues; and (5) the method by first fusing hPSC spheroids, followed with differentiation and maturation was appropriate for tissue biofabrication.

Acknowledgments

This work was partially funded by Nebraska DHHS Stem Cell Research Project (to Y.L.). Confocal microscope imaging was done in the Morrison Microscopy Core Research Facility at University of Nebraska, Lincoln. Drs. You Zhou and Christian Elowsky assisted the confocal imaging.

Author Contributions

YL, HL and QL conceived the idea and designed the study. HL and QL performed experiments and analyzed data. YL wrote the manuscript.

Competing financial interests

The authors declare no competing financial interests.

References

[1] Wobma H and Vunjak-Novakovic G 2016 Tissue engineering and regenerative medicine 2015: a year in review *Tissue Eng. Part B Rev.* 22 101–13

[2] Langer R and Vacanti J P 1993 Tissue engineering *Science* 260 920–6

[3] Page C O F, Griffith L G and Naughton G 2002 Tissue engineering-current challenges and expanding opportunities *Science* 295 1009–14

[4] Mironov V, Boland T, Trusk T, Forgacs G and Markwald R R 2003 Organ printing: computer-aided jet-based 3D tissue engineering *Trends Biotechnol.* 21 157–61

[5] Mironov V, Visconti R P, Kasyanov V, Forgacs G, Drake C J and Markwald R R 2009 Organ printing: tissue spheroids as building blocks *Biomaterials* 30 2164–74

[6] Mironov V, Kasyanov V and Markwald R R 2011 Organ printing: from bioprinter to organ biofabrication line *Curr. Opin. Biotechnol.* 22 667–73

[7] Novosel E C, Kleinhans C and Kluger P J 2011 Vascularization is the key challenge in tissue engineering *Adv. Drug. Deliv. Rev.* 63 300–11

[8] Mason C 2006 The time has come to engineer tissues and not just tissue engineer *Regen. Med.* 1 303–6

[9] Khademhosseini A, Langer R, Borenstein J and Vacanti J P 2006 Microscale technologies for tissue engineering and biology *Proc. Natl Acad. Sci. USA* 103 2480–7

[10] Lokmic Z and Mitchell G M 2008 Engineering the microcirculation *Tissue Eng. Part B Rev.* 14 87–103

[11] Langer R 2007 Editorial: tissue engineering: perspectives, challenges, and future directions *Tissue Eng.* 13 1–2

[12] Laschke M W and Menger M D 2016 Life is 3D: boosting spheroid function for tissue engineering *Trends Biotechnol.* 35 133–44

[13] Olsen T R and Alexis F 2014 Bioprocessing of tissues using cellular spheroids *Bioprocess. Biotech.* 4 1000e112

[14] Lei Y, Jeong D, Xiao J and Schaffer D V 2014 Developing defined and scalable 3D culture systems for culturing human pluripotent stem cells at high densities *Cell. Mol. Bioeng.* 7 172–83

[15] Lei Y and Schaffer D V 2013 A fully defined and scalable 3D culture system for human pluripotent stem cell expansion and differentiation *Proc. Natl Acad. Sci. USA* 110 E5039–48

[16] Thomson *et al* 1998 Embryonic stem cell lines derived from human blastocysts *Science* 282 1145–7

[17] Takahashi K, Tanabe K, Ohnuki M, Narita M, Ichisaka T, Tomoda K and Yamanaka S 2007 Induction of pluripotent stem cells from adult human fibroblasts by defined factors *Cell* 131 861–72

[18] Chen G *et al* 2011 Chemically defined conditions for human iPSC derivation and culture *Nat. Methods* 8 424–9

[19] Shi Y, Kirwan P, Smith J, Robinson H P C and Livesey F J 2012 Human cerebral cortex development from pluripotent stem cells to functional excitatory synapses *Nat. Publ. Gr.* 15 477–86

[20] Shi Y, Kirwan P and Livesey F J 2012 Directed differentiation of human pluripotent stem cells to cerebral cortex neurons and neural networks *Nat. Protoc.* 7 1836–46

[21] Liu Y, Liu H, Sauvey C, Yao L, Zarnowska E D and Zhang S-C 2013 Directed differentiation of forebrain GABA interneurons from human pluripotent stem cells *Nat. Protoc.* 8 1670–9

[22] Maroof A M *et al* 2013 Directed differentiation and functional maturation of cortical interneurons from human embryonic stem cells *Cell Stem Cell* 12 559–72

[23] Hunt R F, Girskis K M, Rubenstein J L, Alvarez-Buylla A and Baraban S C 2013 GABA progenitors grafted into the adult epileptic brain control seizures and abnormal behavior *Nat. Neurosci.* 16 692–7

[24] Kriks S *et al* 2011 Dopamine neurons derived from human ES cells efficiently engraft in animal models of Parkinson's disease *Nature* 480 547–51

[25] Kirkeby A, Grealish S, Wolf D A, Nelander J, Wood J, Lundblad M, Lindvall O and Parmar M 2012 Generation of regionally specified neural progenitors and functional neurons from human embryonic stem cells under defined conditions *Cell Rep.* 1 703–14

[26] Orlova V V, van den Hil F E, Petrus-Reurer S, Drabsch Y, Ten Dijke P and Mummery C L 2014 Generation, expansion and functional analysis of endothelial cells and pericytes derived from human pluripotent stem cells *Nat. Protoc.* 9 1514–31

[27] Prasain N *et al* 2014 Differentiation of human pluripotent stem cells to cells similar to cord-blood endothelial colony-forming cells *Nat. Biotechnol.* 32 1151–7

[28] Lian X, Bao X, Al-Ahmad A, Liu J, Wu Y, Dong W, Dunn K K, Shusta E V and Palecek S P 2014 Efficient differentiation of human pluripotent stem cells to endothelial progenitors via small-molecule activation of WNT signaling *Stem Cell Reports* 3 804–16

[29] Kimbrel E A *et al* 2014 Mesenchymal stem cell population derived from human pluripotent stem cells displays potent immunomodulatory and therapeutic properties *Stem. Cells Dev.* 23 1611–24

[30] Nishio M and Saeki K 2014 Differentiation of human pluripotent stem cells into highly functional classical brown adipocytes *Methods Enzymol.* 537 177–97

[31] Burridge P W *et al* 2014 Chemically defined generation of human cardiomyocytes *Nat. Methods* 11 855–60

[32] Lian X, Hsiao C, Wilson G, Zhu K, Hazeltine L B, Azarin S M, Raval K K, Zhang J, Kamp T J and Palecek S P 2012 Robust cardiomyocyte differentiation from human pluripotent stem cells via temporal modulation of canonical Wnt signaling *Proc. Natl Acad. Sci. USA* 109 E1848–57

[33] Lian X, Zhang J, Azarin S M, Zhu K, Hazeltine L B, Bao X, Hsiao C, Kamp T J and Palecek S P 2013 Directed cardiomyocyte differentiation from human pluripotent stem cells by modulating Wnt/β-catenin signaling under fully defined conditions *Nat. Protoc.* 8 162–75

[34] Si-Tayeb K, Noto F K, Nagaoka M, Li J, Battle M A, Duris C, North P E, Dalton S and Duncan S A 2010 Highly efficient generation of human hepatocyte-like cells from induced pluripotent stem cells *Hepatology* 51 297–305

[35] Hannan N R F, Segeritz C-P, Touboul T and Vallier L 2013 Production of hepatocyte-like cells from human pluripotent stem cells *Nat. Protoc.* 8 430–7

[36] Szkolnicka D, Farnworth S L, Lucendo-Villarin B and Hay D C 2014 Deriving functional hepatocytes from pluripotent stem cells *Curr. Protoc. Stem Cell Biol.* 30 1–1G

[37] Rezania A *et al* 2014 Reversal of diabetes with insulin-producing cells derived *in vitro* from human pluripotent stem cells *Nat. Biotechnol.* 32 1121–34

[38] Pagliuca F W, Millman J R, Gu M, Segel M, Dervort A V, Ryu J H, Peterson Q P, Greiner D and Melton D A 2014

Resource generation of functional human pancreatic β cells *in vitro* *Cell* **159** 428–39
[39] Polak J M and Mantalaris S 2008 Stem cells bioprocessing: an important milestone to move regenerative medicine research into the clinical arena *Pediatr Res* **63** 461–6
[40] Serra M, Brito C, Correia C and Alves P M 2012 Process engineering of human pluripotent stem cells for clinical application *Trends Biotechnol.* **30** 350–8
[41] Lin H, Li Q and Lei Y 2017 An integrated miniature bioprocessing for personalized human induced pluripotent stem cell expansion and differentiation into neural stem cells *Sci. Rep.* **7** 40191
[42] Li Q, Lin H, Wang O, Qiu X, Kidambi S and Lei Y 2016 Scalable production of glioblastoma tumor-initiating cells in 3 dimension thermoreversible hydrogels *Sci. Rep.* **6** 31915
[43] Park I H, Zhao R, West J A, Yabuuchi A, Huo H, Ince T A, Lerou P H, Lensch M W and Daley G Q 2008 Reprogramming of human somatic cells to pluripotency with defined factors *Nature* **451** 141–6
[44] Chambers S M, Fasano C A, Papapetrou E P, Tomishima M, Sadelain M and Studer L 2009 Highly efficient neural conversion of human ES and iPS cells by dual inhibition of SMAD signaling *Nat. Biotechnol.* **27** 275–80
[45] Sasai Y, Eiraku M, Suga H and Glossary B 2012 *In vitro* organogenesis in three dimensions: self-organising stem cells *Development* **4121** 4111–21
[46] Ravi M, Paramesh V, Kaviya S R, Anuradha E and Solomon F D 2014 3D cell culture systems—advantages and applications *J. Cell. Physiol.* **230** 16–26
[47] Kondo T *et al* 2013 Modeling Alzheimer's disease with iPSCs reveals stress phenotypes associated with intracellular Aβ and differential drug responsiveness *Cell Stem Cell* **12** 487–96
[48] Choi S H *et al* 2014 A three-dimensional human neural cell culture model of Alzheimer's disease *Nature* **515** 274–8
[49] Brundin P, Karlsson J, Emgård M, Schierle G S, Hansson O and Petersén A C R 2000 Improving the survival of grafted dopaminergic neurons: a review over current approaches *Cell Transpl.* **9** 179–95
[50] Clarkson E D, Zawada W M, Adams F S, Bell K P and Freed C R 1998 Strands of embryonic mesencephalic tissue show greater dopamine neuron survival and better behavioral improvement than cell suspensions after transplantation in parkinsonian rats *Brain Res.* **806** 60–8
[51] Gouzheng W, Zhang W, Rich M and Gossain V 2009 Growing CHO cells in a CelliGen ® BLU benchtop, stirred-tank bioreactor using single-use vessels *Biotechnol. Prog.* **1** 1–4 (http://www.biotechniques.com/multimedia/archive/00074/New_Brunswick-FP-Cel_74610a.pdf)
[52] Hajdu Z, Mironov V, Mehesz A N, Norris R A, Markwald R R and Visconti R P 2010 Tissue spheroid fusion-based *in vitro* screening assays for analysis of tissue maturation *J. Tissue Eng. Regen. Med.* **4** 659–64
[53] Grimes D R, Kannan P, McIntyre A, Kavanagh A, Siddiky A, Wigfield S, Harris A and Partridge M 2016 The role of oxygen in avascular tumor growth *PLoS One* **11** e0153692
[54] Hu W, Berdugo C and Chalmers J J 2011 The potential of hydrodynamic damage to animal cells of industrial relevance: current understanding *Cytotechnology* **63** 445–60
[55] Brindley D, Moorthy K, Lee J-H, Mason C, Kim H-W and Wall I 2011 Bioprocess forces and their impact on cell behavior: implications for bone regeneration therapy *J. Tissue Eng.* **2011** 620247
[56] Ngangan A V and McDevitt T C 2009 Acellularization of embryoid bodies via physical disruption methods *Biomaterials* **30** 1143–9
[57] Ray H J and Niswander L 2012 Mechanisms of tissue fusion during development *Development* **139** 1701–11
[58] Fleming P A, Argraves W S, Gentile C, Neagu A, Forgacs G and Drake C J 2010 Fusion of uniluminal vascular spheroids: a model for assembly of blood vessels *Dev. Dyn.* **239** 398–406
[59] Wiese S, Karus M and Faissner A 2012 Astrocytes as a source for extracellular matrix molecules and cytokines *Front. Pharmacol.* **3** 1–13
[60] Kleinman H K and Martin G R 2005 Matrigel: basement membrane matrix with biological activity *Semin. Cancer Biol.* **15** 378–86
[61] Kang H, Lee S J, Ko I K, Kengla C, Yoo J J and Atala A 2016 A 3D bioprinting system to produce human-scale tissue constructs with structural integrity *Nat. Biotechnol.* **34** 312–9

Dynamic Culture of Trabecular Meshwork Cells in 3D Biomimetic Scaffolds

Matthew Osmond, Mina B. Pantcheva, and Melissa D. Krebs

AICHE 2017

SCHOOL OF MEDICINE
Department of Ophthalmology
UNIVERSITY OF COLORADO ANSCHUTZ MEDICAL CAMPUS

COLORADO SCHOOL OF MINES
EARTH • ENERGY • ENVIRONMENT

1

Motivation for Glaucoma Research

- Glaucoma is a complicated disease characterized by progressive damage to the optic nerve
- It is the leading cause of irreversible blindness worldwide
- As of 2010, there were 60.5 million people with glaucoma globally
- Prevalence of glaucoma is expected to increase over the next several decades

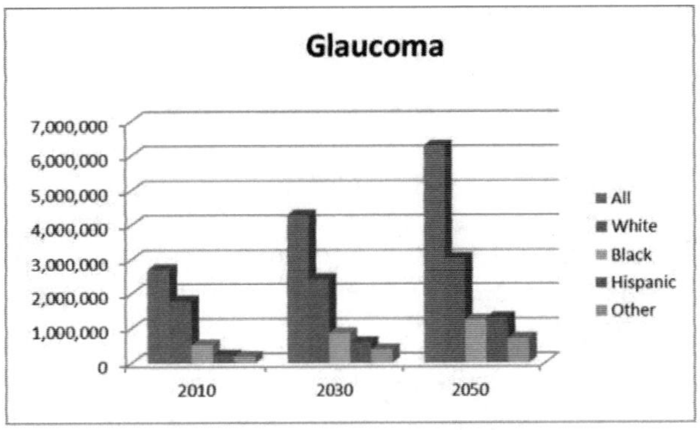

Glaucoma Prevalence in the US

Quigley, H. A., and A. T. Broman. "The Number of People with Glaucoma Worldwide in 2010 and 2020." *The British Journal of Opthamology* 90.3 (2006): 262-67. Web.

"Glaucoma, Open-angle." *Statistics and Data [NEI]*. National Institutes of Health, Web. 21 Jan. 2014.

2

Importance of Trabecular Meshwork in Glaucoma

- The Intraocular Pressure (IOP) is the only modifiable risk factor of glaucoma and is determined by the balance between the rate of aqueous humor production and its outflow
- Trabecular Meshwork (TM) is believed to account for the bulk of aqueous humor drainage
- Any structural change in the TM would likely influence the resistance to aqueous humor flow

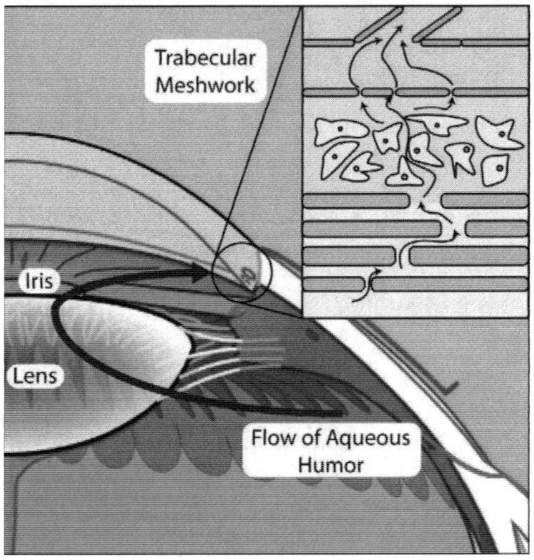

3

Trabecular Meshwork is a Complex System

- Majority of *in vitro* studies are oversimplified
- Cell culture on a flat, plastic culture dish does not mimic the environment found *in vivo* and results in decreased expression of TM - specific protein (myocilin)
- TM cells exist in a complex environment of extracellular matrix

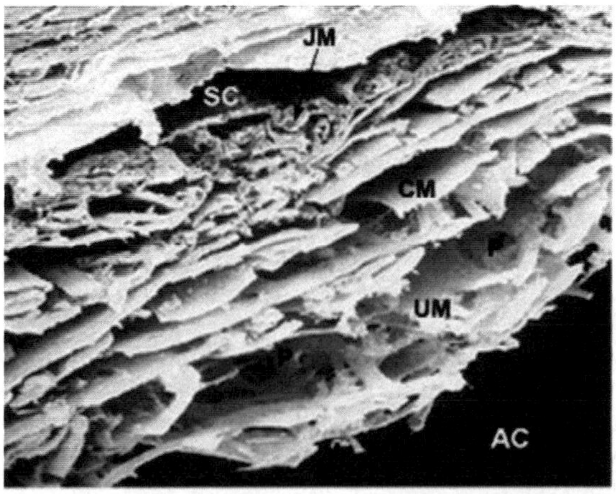

http://www.glaucoma-eye-info.com/trabecular-meshwork.html

4

Research Objectives

- Develop a 3-dimensional matrix from native TM extracellular components
- Use temperature controlled directional freezing to induce controlled pore structure
- Seed human TM (hTM) cells onto scaffolds and grow them under static and dynamic conditions
- Quantify the growth, proliferation, and gene expression of hTM cells

hTM Cells Under Static Culture Conditions

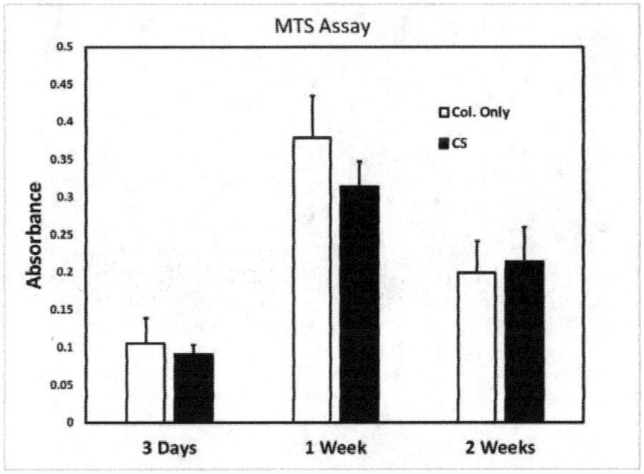

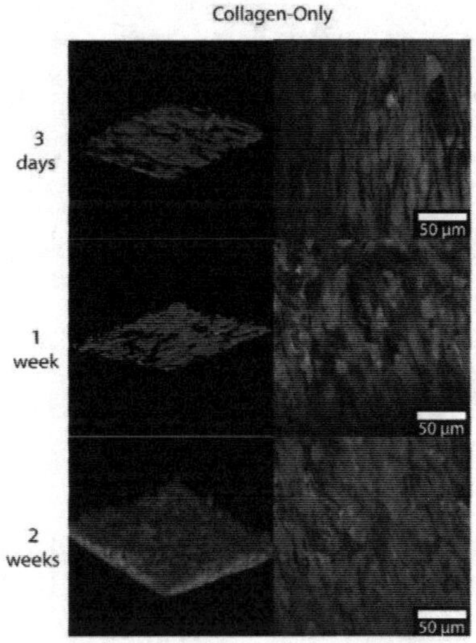

Osmond, Bernier, Pantcheva, Krebs, *Biotechnology and Bioengineering*, 114(4), 915-23, 2017

Myocilin Expression of hTM Cells Grown in Static Culture

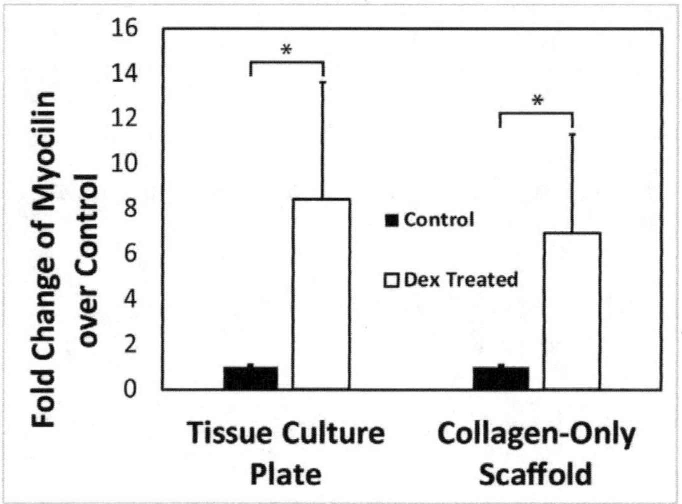

Osmond, Bernier, Pantcheva, Krebs, *Biotechnology and Bioengineering*, 114(4), 915-23, 2017

7

Directional Freezing for Controlled Pore Size and Alignment

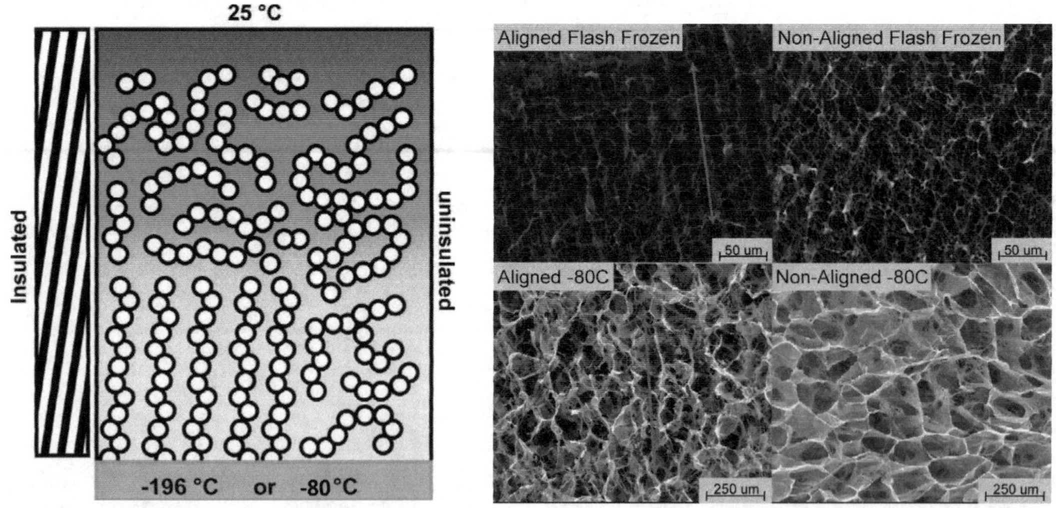

Harding, Osmond, Krebs. Tissue Engineering Part A. August 2017, ahead of print

8

hTM Cell Proliferation on Varied Pore Architecture

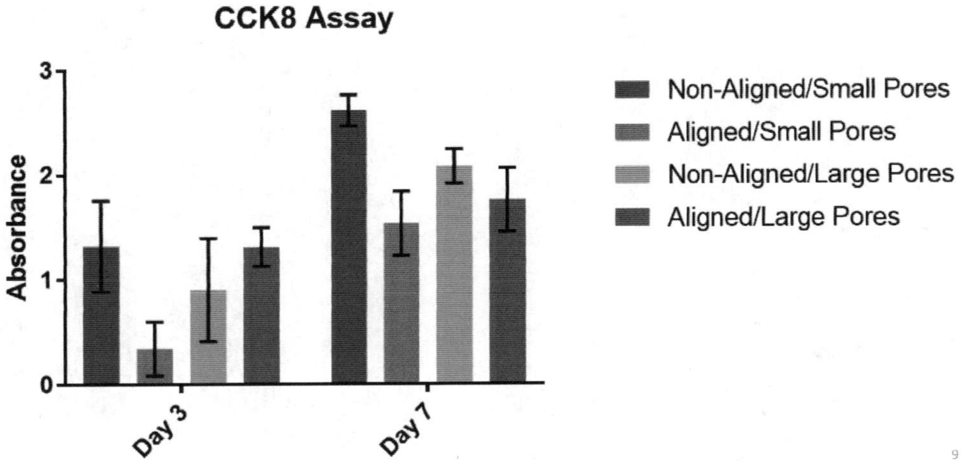

9

Effect of Pore Architecture on the Proliferation and Cell Morphology of hTM Cells in Static Culture

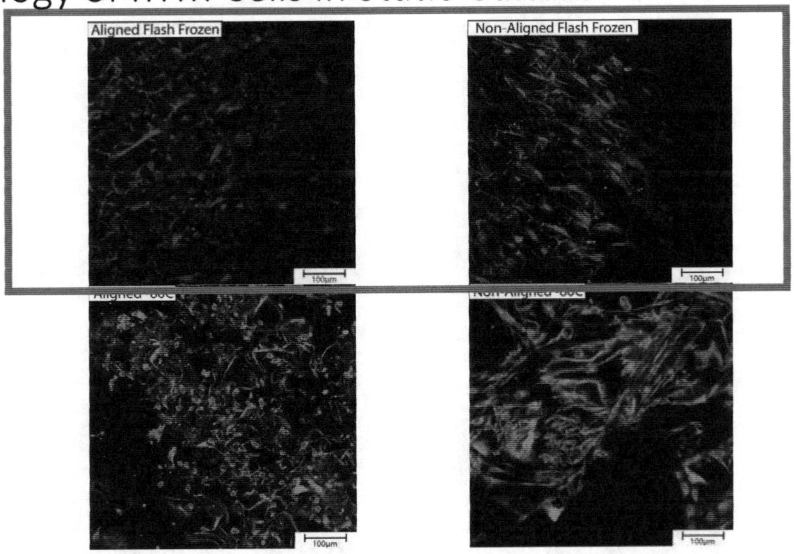

10

Effect of Pore Architecture on the Proliferation and Cell Morphology of hTM Cells in Static Culture

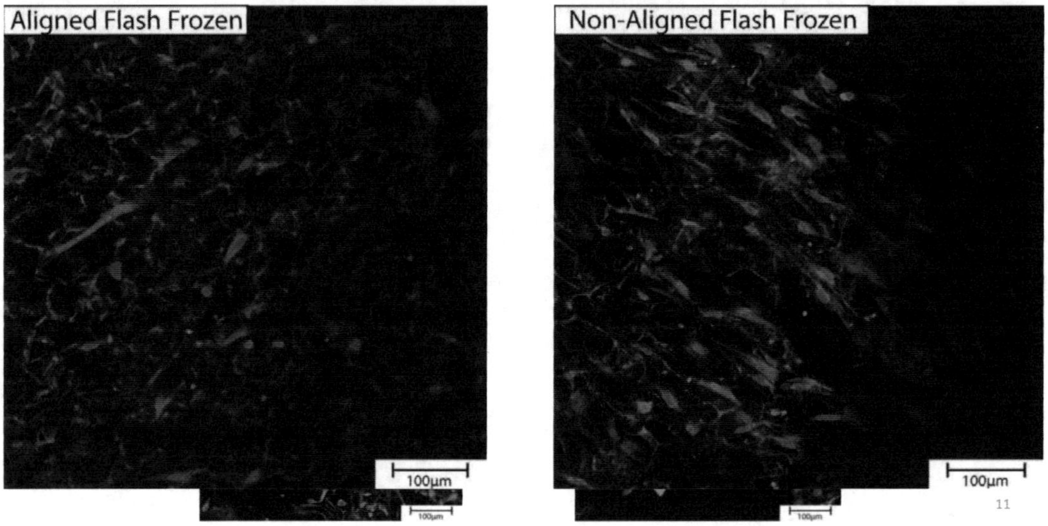

11

Effect of Pore Architecture on the Proliferation and Cell Morphology of hTM Cells in Static Culture

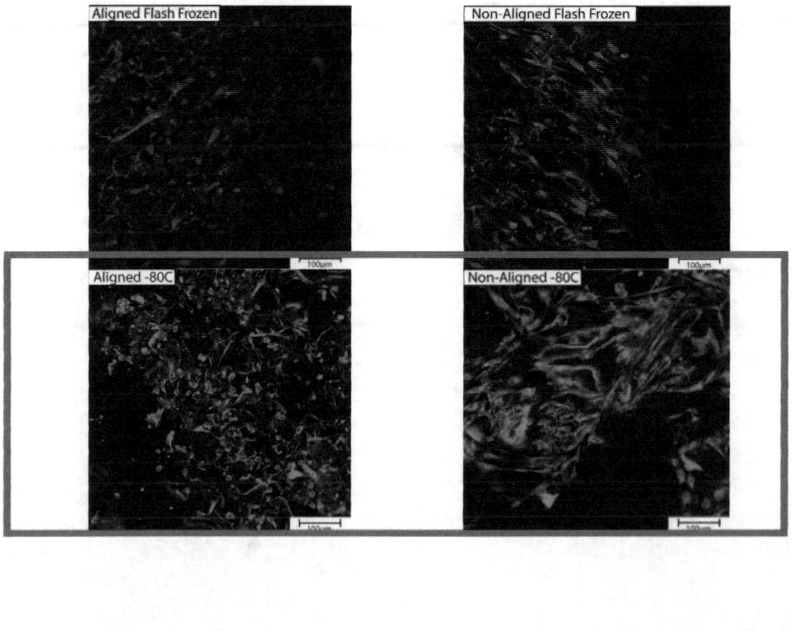

12

Effect of Pore Architecture on the Proliferation and Cell Morphology of hTM Cells in Static Culture

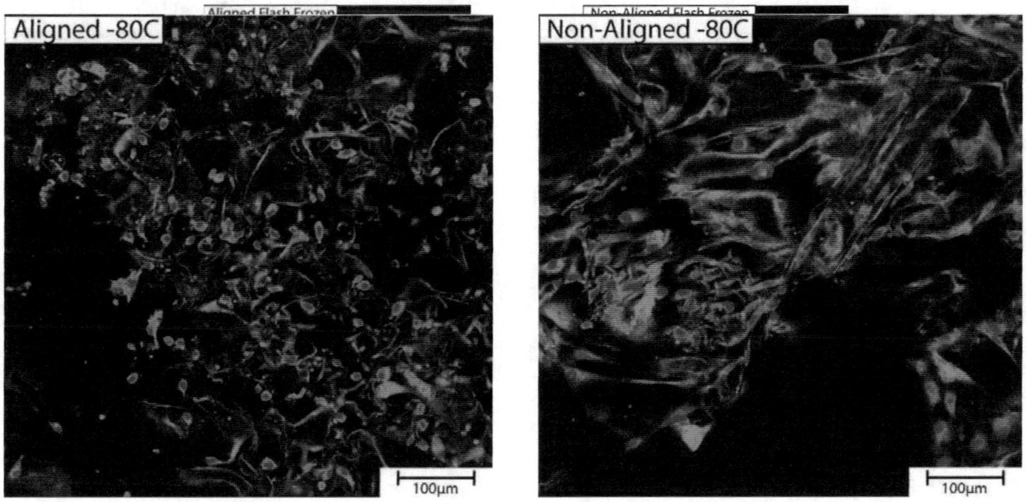

13

Culture of hTM Cells Under Perfusion

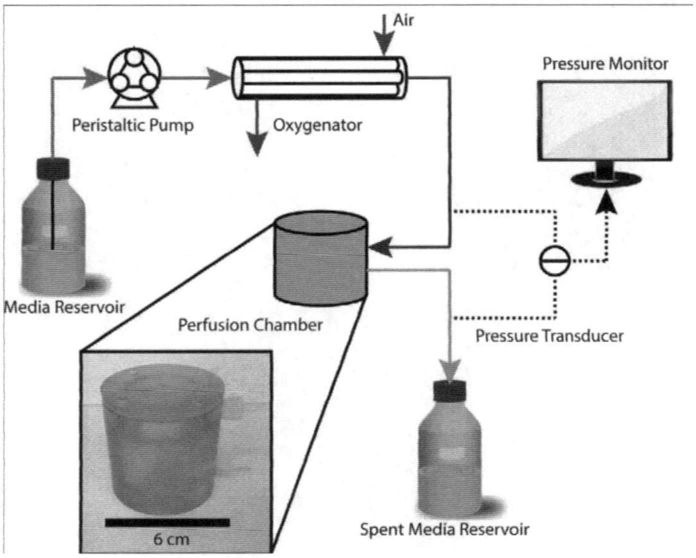

14

Importance of Choice of Flow Rate

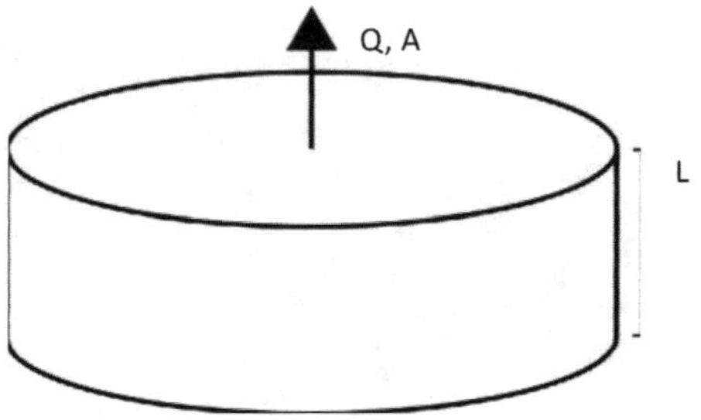

- Peclet Number: Ratio of convective to diffusive transport
- Pe > 10 is convectively controlled
- Both oxygen and glucose are important for cell survival
- Glucose is the limiting factor in terms of diffusion
- Q = 31.6 μL/min for convective transport
- Examined 40 μL/min for convective and 10 μL/min for conductive transport

$$Pe = \frac{QL}{AD}$$

D_{oxygen} is 2.3 x10^{-5} cm^2/s
$D_{glucose}$ is 6.7x10^{-6} cm^2/s

15

Effect of 10 μL/min Perfusion on the Proliferation and Cell Morphology of hTM Cells

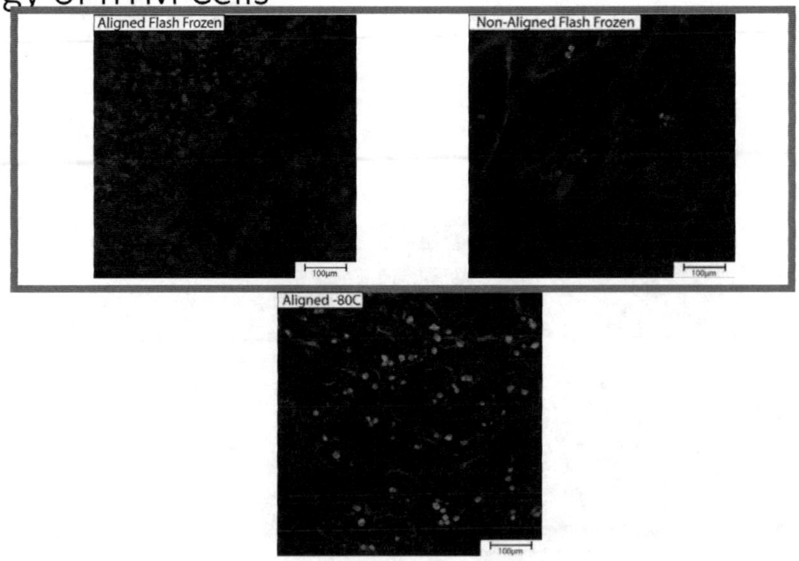

16

Effect of 10 μL/min Perfusion on the Proliferation and Cell Morphology of hTM Cells

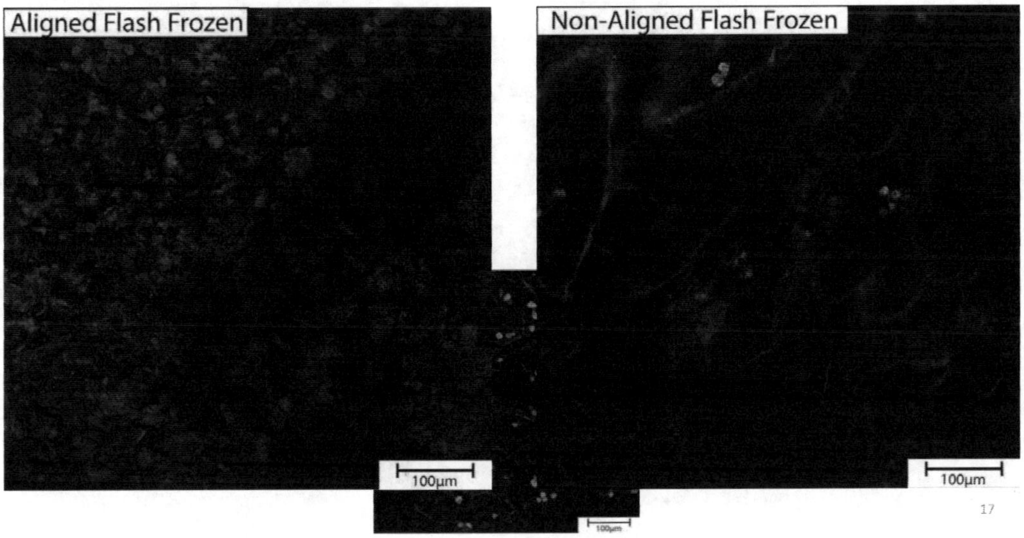

17

Effect of 10 μL/min Perfusion on the Proliferation and Cell Morphology of hTM Cells

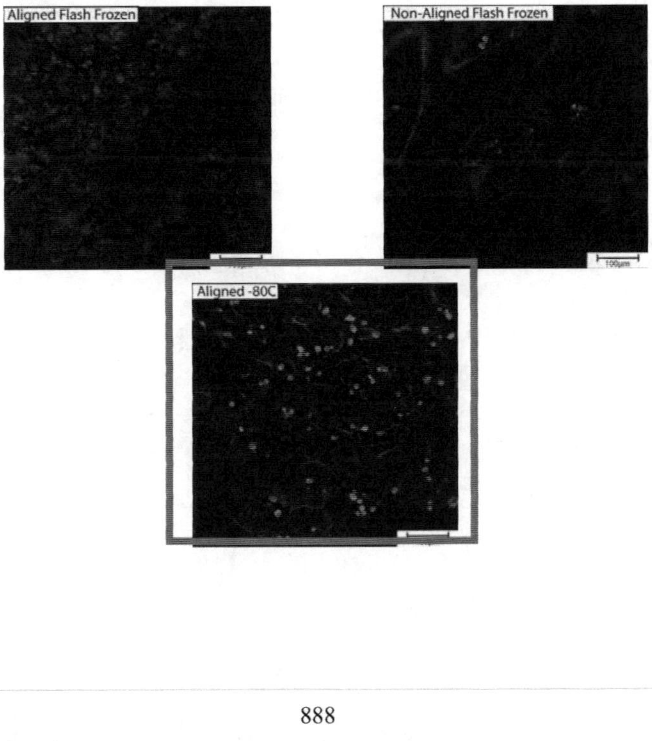

18

Effect of 10 μL/min Perfusion on the Proliferation and Cell Morphology of hTM Cells

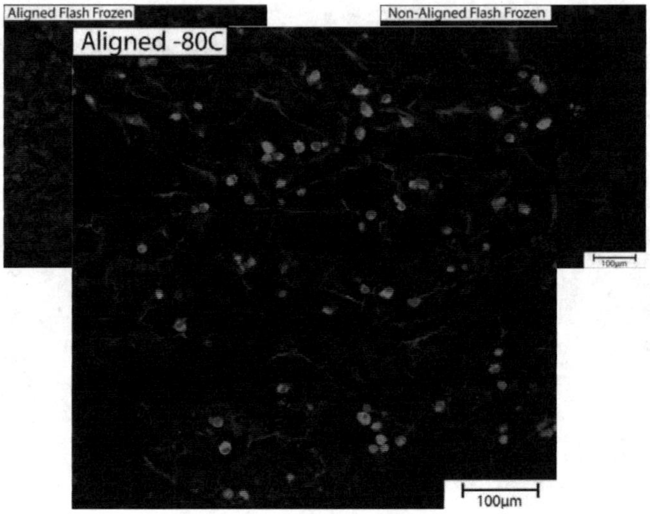

19

Conclusions

- Scaffold architecture affects the proliferation and morphology of human TM cells
  - Small pores support initial cell growth
  - Large pores support prolonged growth
  - Aligned pores restrict cell growth to the surface of the scaffold
  - Non-aligned pores allow penetration into the interior of the scaffold
- Perfusion increases cell growth, if cells attach.
  - Small pores support growth, but quickly become confluent
  - Non-aligned pores allow cells to be swept away
  - Large, Aligned pores allow attachment and more room for proliferation

20

Acknowledgements

BOETTCHER FOUNDATION

NIH National Institute of Arthritis and Musculoskeletal and Skin Diseases

NSF

NIH National Eye Institute

AO Foundation

Children's Hospital Colorado Research Institute

DELTA DENTAL

Colorado Office of Economic Development and International Trade

COLORADO SCHOOL OF MINES
EARTH • ENERGY • ENVIRONMENT

486707 Cellular Hitchhiking on Microparticles to Alleviate Skin Injury

Abstract:

Regenerative medicine holds great potential for the treatment of tissue damage. However, there are currently very few clinical applications of cell-based therapies; numerous studies have encountered complications with keeping transplant cells alive. To overcome this, the present study focuses on engineering polymer microparticles. These microparticles provide mechanical support for adherent cells in suspension, a means of culturing enormous amounts of cells in small volumes, and receptor-ligand specific signal-mediated cell growth. In our study, a simple flow-focusing device was developed to synthesize poly(lactic-co-glycolic acid) (PLGA) microparticles with the solvent diffusion method. Particle diameters of approximately 200 μm were achieved. Additionally, particle surface chemistry was modified to promote cell adhesion, and human fibroblasts cells were cultured with these particles in suspension. We call this *cellular hitchhiking*. The specific goal of this research is treatment of skin injury, but the technique has versatile tissue regeneration applications.

485866 Sugar-Coating the Answers to Virus Binding: Glycocalyx-Mimetic Interfaces

Abstract:

The glycocalyx is a membranous structure composed of glycoproteins and carbohydrate residues. This sugary forest that sheaths our cells is responsible for protecting us from pathogen invasions and forms a crucial component of our cellsâ€™ immune machinery. Virus-glycocalyx interactions are poorly understood and a detailed study of these interactions is hindered by the tendency of the native glycocalyx to collapse duringÂ in vitroÂ studies. We have developed a bio-inspired model system for the glycocalyx to identify and understand physicochemical parameters that shape its interactions with viral pathogens. Like the native glycocalyx, our tunable surface engages in both specific and non-specific modes of viral binding. We have developed a library of carbohydrate-based polymer brushes, including mannose, glucose and galactose, which recapitulate the chemical complexity of the glycocalyx. Though our carbohydrate-functionalized surfaces are inert to non-specific protein adsorption, they recognized lectin. Spatially controlled immobilization of brushes helped us achieve precisely patterned presentation of lectins and viruses. In addition to specific affinity based interactions, we have also engineered non-specific electrostatically mediated interactions into our surface to elucidate the role of surface charge on viral binding. Our model surface is composed of positively charged affinity sites and polymerization initiation sites for the growth of polymer brushes bearing carbohydrate residues. By varying the ratio of affinity sites to initiation sites, we were able to observe changes in adsorption profiles of model proteins, influenza viruses and adenoviruses. Quartz crystal microbalance, scanning electron microscopy and fluorescence microscopy were employed to visualize these interactions and develop design rules for virus-resistant coatings

488359 Glucose-Derived Cationic Block Poly(beta-peptides) Reverse Intrinsic Antibiotic Resistance in Gram-Negative Pathogens

Abstract:

The rapid rise of multi-drug resistant (MDR) bacteria together with the decrease of effective antibiotics has been a mounting problem especially for pathogenic Gram-negative bacteria, since no new family of drug for the latter was found in the last past 50 years. Gram-negative bacteria possess intrinsic resistance to many antibiotics, particularly hydrophobic ones, because of the presence of the impermeable outer membrane which acts synergistically with their multi-drug efflux pumps to lead to high levels of multi-drug resistance in many pathogenic clinically relevant bacteria. The translation of the well-studied class of natural antimicrobial Î±-peptides to clinic use has been hampered by their relative toxicity and also their proteolytic instability. Herein, we have developed a novel glycosylated Î^{2}-peptide series that is more proteolytically stable and acts as effective potentiating agents for several conventional antibiotics which Gram-negative bacteria are intrinsically resistant to. Our glycosylated cationic block Î^{2}-peptides were synthesized through a one-step polymerization reaction due to the vastly differing reactivity of the monomers following by one-pot global deprotection. Our optimized glyco-Î^{2}-peptide (named as PAS8-*b*-PDM12) acts synergistically with several antibiotics not commonly used for treating Gram-negative bacteria to reverse the intrinsic resistance of wild type *Pseudomonas aeruginosa* as well as various clinically relevant multi-drug resistant *P. aeruginosa* against these antibiotics. This is particularly significant since several recent reports show that *P. aeruginosa* has a highly impermeable outer membrane that eludes synergy effect of other combinations. We also found that the potentiating effect is broad spectrum towards various other Gram-negative bacteria including multi-drug resistant *Escherichia coli*, *Klebsiella pneumoniae* and *Acinetobacter baumannii* with the MIC reduction factor ranging from 8 to 128 times decrease with 16μg/mL polymer addition. Also, *in vivo* synergistic effect was proven in a murine wound model. After 4h infection and 24 h treatment, the synergistic combination with rifampicin reached 99.5% eradiation against PAO1.

494553 Mechanism Study of Selective Killing of Cationic Peptidopolysaccharide Nanoparticles with *in vitro* and *in vivo* Efficacy Against Multi-Drug Resistant Bacteria

Abstract:

Antimicrobial peptides (AMPs) are active against many multi-drug resistant (MDR) bacteria but only a limited number of these compounds are in clinical use due to their unselective toxicity. Herein, we show that short peptidopolysaccharide, specifically chitosan-*graft*-oligolysine (hereafter denoted as CSM5-K5) with molecular weight of about 1450 Daltons, suppresses the growth of methicillin-resistant *Staphylococcus aureus* (MRSA) bacteria by 3.5 orders of magnitude in a murine wound model, an efficacy comparable to that of the last resort MRSA antibiotic vancomycin; it is also non-inflammatory with little/no activation of neutrophils. *In vitro* studies show that the polymer is also bactericidal against oxacillin-resistant *Staphylococcus aureus* (ORSA) and pathogenic strains of Gram-negative *Escherichia coli* and *Pseudomonas aeruginosa* and has good fibroblast biocompatibility and hemocompatibility. Light scattering and TEM show that CSM5-K5 aggregate into nanoparticles. Computer simulation shows that hydrogen bonding between chitosan backbones led to nanoparticle formation. However, the other constituent of the copolymer - the oligolysine side chains - interacts with the head groups (PO_4^-) of bacterial membrane through electrostatic interaction. The aggregation of multiple polymers into a single nanoparticle, rather than individual molecules, accentuates the bacterial membrane damage by synergistic clustering of anionic lipids from the zwitterionic lipids. The small nanoparticle size with small cationic charge (zeta potential of + 15 mV) and without hydrophobic residues seems evasive to fibroblasts. This work validates a critical link between the design of a new class of selective antimicrobial nanoparticles formed from short cationic peptidopolysaccharides and its efficacy towards treating MDR infections.

499720 Tether Supported Biomembranes with Phase Composition and Orientation Control for Biomaterial Ligand Displays and Membrane Protein Assays

Abstract:

Integral membrane proteins are highly sensitive to the composition and phases within biomembrane microenvironments of a cell and have a wide variety of functionalities, ranging from cleavage to transport to cell-to-cell signaling. The proteins often contain large extramembrane domains leading to difficulties in reconstitution in isolated systems. These proteins require a confluent bilayer, proper protein and substrate orientation, and appropriate chemical environment, along with adequate spacing to relieve outside stresses on the membrane. Despite extensive studies across a variety of integral membrane proteins, a high throughput systematic assay has yet to be developed for such proteins. In these studies, a supported lipid bilayer containing tethering and cushioning moieties is developed to isolate the biologically relevant transmembrane orientations of the membrane bound enzymatic system. This work shows two biomembrane platforms used to build new multifunctional systems, planar biomembranes and also spherical proteolipobeads that facilitate high throughput analysis. We have characterized the structures, phase localization and compositions of enzymes in proteolipobead and planar systems by employing correlative optical and surface microscopy from the micro- to the nanoscale. The primary imaging modalities are superresolution microscopy and atomic force microscopy performed in tandem. This work is directed at the at the intramembrane protease gamma-secretase, an enigmatic enzyme with major public health relevance. It is proposed that a platform containing biologically active proteins in appropriate configurations and lipid compositions with proper support structures can be developed for the functional analysis of gamma-secretase and other integral membrane proteins.

500235 Understanding the Interaction of the Polysaccharide, Chitosan, with a Novel Cuticular Protein, CPR 27, Using Quartz Crystal Microbalance with Dissipation

Abstract:

The exoskeleton of the insects is made up of cuticle, which contains the polysaccharide, chitin, and several cuticular proteins [CPs]. Chitin is the second-most abundant biopolymer, and its N-deacetylated derivative, chitosan, is a potential source of polysaccharide. Chitosan is biocompatible, biodegradable, and non-toxic; naturally, it finds a number of applications that includes the development of biomimetic materials. It has been used in cosmetics, in making artificial skin, in formulating wound-healing products and as drug-delivery systems for controlled release. In this report, we attempt to understand the interaction of chitosan with a novel cuticular protein, named CPR 27, using Quartz Crystal Microbalance with Dissipation [QCM-D]. Abundant quantities of CPR 27 are found in the elytra of the beetle, *Tribolium castenium*. It has been hypothesized that a conserved sequence of amino acids in CPR 27, the Rebers and Riddiford [RR] sequence, helps CPR 27 to bind with chitin. Therefore, we speculate that chitosan binds similarly to CPR 27. This hypothesis was tested using QCM-D, where a change in frequency of the resonating quartz crystal is noted with the adsorption of the material of interest. Chitosan was first adsorbed onto a silica surface, and CPR 27 was subsequently adsorbed to chitosan. Increase in mass of CPR 27 was observed with an increase in concentration. However, the opposite trend was observed in the case of rising temperature, where less amount of CPR 27 adsorbed with an increase in temperature. A possible explanation is the influence of temperature on the RR sequence with the gradual unfolding of the protein. However, further studies along with kinetic evaluation of the binding are necessary to have a better understanding of the interaction. The ultimate goal here is to come up with a novel chitosan-CP composite that will be suitable for biomedical applications.

500777 Probing Phase Transitions in Dynamic Biopolymer Complexation

Abstract:

In nature, biopolymers partition into dynamic compartments to facilitate and regulate their interactions. These dynamic compartments are referred to as membraneless organelles (e.g., nucleolus, P bodies, stress and germ granules) and consist of biopolymer rich interiors that rapidly assemble and disassemble to form liquid droplets, hydrogels or fibril structures. The physical interactions that affect the formation, dissolution, and regulation of these assemblages are poorly understood, yet vital in determining their function in normal and disease states. Therefore, there is a strong need to investigate the underlying mechanisms that drive dynamic biopolymer complexation and develop an understanding of their biochemical function. Recent studies have identified several significant interactions that drive phase-separation, including specific multivalent protein interactions, RNA-protein interactions and non-specific interactions between intrinsically disordered regions (IDRs). IDRs are disordered protein domains with low amino acid sequence diversity that can self-associate into oligomeric structures via non-specific weak interactions. These sequences are repetitive and rich in charged, polar and aromatic side-chains and highlight the importance of the interplay between electrostatic, dipole-dipole and short-range aromatic interactions.

Membraneless organelles formed in vivo and in vitro are strikingly similar to polyelectrolyte complexes formed using much simpler synthetic polyelectrolytes and charged homopolymer polypeptides (e.g., (poly)-lysine and (poly)-glutamic acid). In light of these observations, we designed a series of synthetic polypeptide sequences that mimic IDRs to elucidate the forces that govern membraneless organelle formation. Using solid-phase synthesis we are able to control the polypeptide sequence precisely to create alternating, block and gradient polypeptide copolymers to systematically investigate the influence of charged, polar and aromatic groups on phase separation using a variety of techniques. Our work holds the potential to develop a fundamental understanding of complex biopolymer complexation, impacting design of novel functional biomimetic materials for various applications.

500928 Photo-Induced Pinocytosis in Synthetic Liposomes

Abstract:

An essential step in the development of artificial cells is making these cells capable of taking up resources for sustained transcription and translation. Previous work on incorporation of appropriate machinery into the membrane to make this possible has proven to be complex and complicated. Here, we present the introduction of photocleavable units into the phospholipid structure through which photo-induced pinocytosis behavior is achieved. Upon exposure to 365nm or 405nm light, cleavage products build up, leading non-spherical vesicles to engulf external fluid during their transition to a more spherical shape. This process would allow future artificial cellular systems to uptake resources that range from dNTPs and energy to whole proteins from the external environment.

501015 Protein Protection and Purification through Virus-like Particle Encapsidation: A Rapid 2-Step Cell-Free Protein Synthesis Approach

Abstract:

Nature has evolved to bear a micro-compartmentalization strategy which provides protection and delivery of cargo molecules as well as regulation of chemical reactions. Biotechnologists have re-engineered these capabilities in synthetic systems and a small number of viral capsids have been used to encompass enzymes to build nanoreactors. These nanoreactors enable the substrates and products to diffuse through pores in the viral capsid, while protecting and stabilizing the encapsidated enzyme. Here we demonstrate a rapid 2-step cell-free protein synthesis approach to encapsidate and purify a desired enzyme in a few hours. Enhanced stability and retained activity of the encapsidated model enzyme CalB is demonstrated when subjected to biologically harsh environments.

498576 Measurement and Modeling of Carrier Dynamics in Photovoltaic Cztse

Abstract:

Understanding the relationship of photoexcited carrier lifetimes, mobilities, and recombination mechanisms to structure and processing of photovoltaic (PV) absorber materials is critical to the design of efficient solar cells. Carrier dynamics in PV absorbers have conventionally been characterized by time-resolved photoluminescence (TRPL). However, not all materials are strongly emissive or otherwise suitable for TRPL. Alternative non-contact probes could enable measurement of ultrafast carrier dynamics for a wider range of materials. Here we demonstrate the use of time-resolved terahertz spectroscopy (TRTS) and surface-sensitive transient reflectance (TR) spectroscopy coupled with numerical modeling of the transport equations to elucidate photoexcited carrier dynamics in photovoltaic absorbers. We show that critical parameters including surface recombination velocity, mobility, and Shockley-Read-Hall lifetime can be obtained. We used a high-quality $Cu_2ZnSnSe_4$ single crystal with device efficiency of 6.2% as a model case. The single-crystal sample eliminates complications arising from grain boundaries and high densities of point defects and secondary phases associated with thin film growth. Surface recombination velocity could be reduced by at least two orders of magnitude, to $3x10^3$ cm/s as determined by TR, with appropriate chemical and mechanical polishing. Mobility values of 80 cm^2/Vs from TRTS and modeling were validated with Hall effect measurements. Shockley-Read-Hall lifetimes beyond 10 nanoseconds indicate high material quality and promise for high photovoltages. The mobility-lifetime product would result in a diffusion length of ~700 nm, which is similar to the thickness necessary for light absorption and high photocurrents.

503051 Impact of the Active Layer Morphology on Bimolecular Recombination Dynamics in Organic Solar Cells

Abstract:

We present a reaction-diffusion model to describe the impact of the morphology of the active layer and charge-transfer (CT) lifetime on the bimolecular recombination kinetics in organic solar cells. The morphologies we consider, range from a bilayer with flat interface to a bilayer with rough interface and bulk heterojunctions with coarse and fine intercalated domains of donor and acceptor molecules. Using the reaction-diffusion model in kinetic Monte Carlo simulations, we find that only the density of states affects the order of bimolecular recombination kinetics. The results show that the morphology of the active layer only influences the charge carrier lifetime and the average delay time between the exciton dissociation and the onset of bimolecular recombination. The results also show that the same effect on the charge recombination dynamics obtained by a change in the donor or acceptor domain size, can be achieved by a change in the degree of disorder. Overall, our results reveal that the intrinsic lifetime of the CT states in conjunction with active layer morphology and nature of disorder can explain experimentally observed deviation of recombination dynamics from Langevin model.

501426 Atomistic Origin of the Concentration Dependence of Si Dopant Mobility in III-V Semiconductor Alloys

Abstract:

Compound semiconductors made of elements from groups III and V in the periodic table have desirable properties such as high electron mobility and small band gaps. Such properties position them as excellent candidates for a wide range of electronic and photonic applications from high-performance transistors to lasers and optical detectors. Silicon is a typical dopant in III-V materials, but displays highly unusual diffusion behavior whether it is grown-in, implanted or deposited in those materials to make *n*-type devices. Interestingly, and somewhat counter-intuitively, silicon diffusivity in In(Ga)As alloys improves considerably at higher Si concentrations. Previous studies assume that this is related to a vacancy-mediated mechanism, but this is open to debate. Using InAs/GaAs alloys as a model system, we investigate preferred diffusional pathways of silicon dopants and intrinsic defects (vacancies and self-interstitials) using *ab initio* calculations. We show that Si-defect complexes, especially split interstitials, play an important role in the experimentally observed concentration-dependent diffusion of Si, as well as compensation effects at high dopant concentration. Vacancy-mediated mechanisms are less important as they are often kinetically inaccessible. We can also explain the differing roles of In versus Ga to affect dopant activation and modify trends of Si diffusion. These results significantly improve our understanding of diffusion and dopant activation in III-V semiconductors and could aid in the design of next-generation III-V based devices.

500830 Characterizing Defects in Photovoltaic Semiconductors with Optical Spectroscopy

Abstract:

Understanding the defect properties of photovoltaic absorbers is critically important for high performance device fabrication. Such properties play a crucial role in the charge carrier dynamics which determine device performance. In this work we demonstrate how time-resolved photoluminescence (TRPL) is a useful technique to investigate charge carrier dynamics and defect properties in photovoltaic absorbers and solar cells. Useful material parameters, such as minority carrier lifetime and mobility, are often reported from TRPL measurements. Additionally, photoluminescence (PL) decay times are often used as a metric for absorber quality during process optimization. However, accurately interpreting the PL decay time can be quite complicated as the charge carrier dynamics are influenced by various recombination, transport, and charge trapping effects. Such metrics are particularly relevant for chalcopyrite materials, which often employ complicated band structures for high performance, as well as for kesterite materials, where short PL decay times are commonly reported for this highly defective material. Therefore, careful consideration is needed when analyzing TRPL data of such non-ideal absorbers and devices to correctly interpret the limiting absorber defect properties.

In this work, high-resolution voltage-, temperature-, and intensity-dependent TRPL data for chalcopyrite, kesterite, and perovskite devices and absorbers are analyzed to determine the origin of the PL decay time. Additionally, a supercontinuum excitation source is used to investigate the excitation energy-dependence of the PL decay time. Such measurements allow us to distinguish the various PL decay rate limiting mechanisms for these materials. We find chalcopyrite absorbers – including $CuInSe_2$ (CISe) and $Cu(In,Ga)Se_2$ (CIGSe) – from a variety of processing conditions and compositional profiles follow the expected behavior for Shockley–Read–Hall (SRH) recombination limited decay times. The impact of carrier diffusion and surface recombination are demonstrated. In contrast, TRPL analysis of kesterite devices – including $Cu_2ZnSnSe_4$ (CZTSe) and $Cu_2ZnSn(S,Se)_4$ (CZTSSe) from various processing conditions – suggests that PL decay times are not directly related to the recombination lifetime for CZTSSe. A model for minority carrier trapping in shallow donor states is proposed to explain the measured TRPL behavior for kesterites. These results illustrate the importance of measurement conditions and careful analysis when interpreting TRPL data, and further elucidate the defect related device limitations in chalcogenide solar cells.

501057 CIGS Nanocrystal Solar Cells on Plastics and Paper

Abstract:

Solar power has made enormous inroads in the past decade as a renewable energy source, primarily due to decreased Si solar panel cost. Additional opportunities are available to utilize solar power by integrating with buildings and transportation, powering low-power portable devices and wearables, and meet emergency and non-grid energy generation needs. However, bulky and rigid Si solar panels are ill fit to meet these needs. Low-cost, lightweight, flexible photovoltaics (PVs) are needed to extend the reach of solar power.

This presentation summarizes our work using Copper Indium Selenide (CIS) nanocrystal inks to build solar panels on a variety of low cost, heat intolerant substrates. CIS nanocrystal PVs have been fabricated using a variety of print methods on plastics such as polyethylene terephthalate (PET) and cellulose nanofiber paper, with champion devices reaching 2.8% and 2.25% respectively. Devices on paper demonstrated no performance loss even when flexed up to 100 times at a 5 mm radius. Additionally, devices with over 2% efficiency were demonstrated with all printed window layers.

496467 Solution Processed Thin Film Photovoltaics Using Amine-Thiol Chemistry

Abstract:

Future development of the thin film electronics industry relies on cost effective solution processing techniques. Solution processing enables us to print electronic materials on flexible substrates using roll-to-roll techniques, which is a challenge with high temperature vacuum processes. Along with ease of scalability, solution processing also reduces the cost of production leading to cheap, flexible, wearable electronic devices like photovoltaics, batteries, sensors and thermo-electrics. Solution processing involves either making nanoparticles of the desired material and then casting films with a nanoparticle ink or using a molecular precursor approach where the precursor solution is first cast as a film and then reacted to make the desired material. The molecular precursor approach is easier to scale up, reduces the steps in making films and hence the cost.

Many researchers working in this field are developing different solvent systems for dissolution of precursors and had most success with hydrazine based solution system.[1] Scaling up a process with such a highly toxic chemical is both costly and challenging. Hence, we have developed a mixture of commonly used amines and thiols to dissolve a host of materials including various elemental metals (Cu, Zn, In, Ga), elemental chalcogens (S, Se, Te), metal chalcogenides (Cu_2S, SnS, Cu_2Se, SnSe, In_2Se_3), metal oxides (ZnO, Cu_2O) and metal salts (CuCl, $CuCl_2$, $InCl_3$, GaI_3), which were otherwise insoluble in either amines or thiols alone.[2–6] This system allows us to adjust the cation to anion ratio in the solution during precursor preparation by choosing the appropriate ratio of amine or diamine with mono or dithiol. Taking advantage of this mixture's dissolution ability, we synthesized and cast thin films of electronic materials specifically for applications in photovoltaic devices. Our typical fabrication of a solar cell starts with the dissolution of appropriate metal salts or metal chalcogenides in a selected amine-thiol mixture and then casting a thin film (around 1 micron thick) from this precursor solution on to a Mo coated glass substrate with an optimized coating and drying procedure. This film is then selenized in a tube furnace and the device is finished with subsequent deposition of CdS, ZnO, and ITO layers followed by Ni/Al metal grids. For the CZTSSe material system, we used a mixture of hexylamine and propanethiol to dissolve metal chlorides ($ZnCl_2$, $CuCl_2$, $SnCl_2$) and chalcogens (S, Se) while for the CIGSe material system ethanedithiol is instead used along with hexylamine to dissolve a metal chalcogenide (Cu_2Se) and metal salts (Ga-acetylacetonate, In-acetate). Films obtained by using these inks yielded a power conversion efficiency of 7.86% for CZTSSe and 12.2% for CIGSe.[4,7] SEM cross section imaging of both devices showed a fine grain layer in the final architecture, which is thinner, compared to that seen in other solution processing routes for these materials. Using similar amine thiol chemistry, we also fabricated the first solution processed CdTe solar cell, with an initial power conversion efficiency of 0.5%.[8]

We have found that the use of metal salts in precursor solutions introduces unnecessary species in the film, which act as impurities affecting the performance of the cell. Here we will present our solution to this issue by using different combinations of amine and thiol to dissolve pure metals instead of metal salts for making inks. A challenge that emerges with the use of dissolved metals in as synthesized amine-thiol inks is the appearance of thick carbon residue in the film, deteriorating the cell performance. To address this challenge, we have developed alternative routes to utilize these inks while maintaining little carbon residue in the final film. In addition to the traditional photovoltaic material systems, we will also present use of amine thiol chemistry to synthesize lead chalcogenide nano/microstructures at room temperature, which can be used in photovoltaic and thermoelectric devices. Indeed our results are very promising and exciting.

References

(1) Zhang, T.; Yang, Y.; Liu, D.; Tse, S.; Cao, W.; Feng, Z.; Chen, S.; Qian, L. High Efficiency Solution-Processed Thin-Film $Cu(In,Ga)(Se,S)_2$ Solar Cells. *Energy Environ. Sci.* **2016**, *9* (12), 3674–3681.

(2) Webber, D. H.; Buckley, J. J.; Antunez, P. D.; Brutchey, R. L. Facile Dissolution of Selenium and Tellurium in a Thiol–amine Solvent Mixture under Ambient Conditions. *Chemical Science* **2014**, *5* (6), 2498.

(3) Zhang, R.; Cho, S.; Lim, D. G.; Hu, X.; Stach, E. A.; Handwerker, C. A.; Agrawal, R. Metal–metal Chalcogenide Molecular Precursors to Binary, Ternary, and Quaternary Metal Chalcogenide Thin Films for Electronic Devices. *Chem. Commun.* **2016**, *52* (28), 5007–5010.

(4) Zhang, R.; Szczepaniak, S. M.; Carter, N. J.; Handwerker, C. A.; Agrawal, R. A Versatile Solution Route to Efficient $Cu_2ZnSn(S,Se)_4$ Thin-Film Solar Cells. *Chemistry of Materials* **2015**, *27* (6), 2114–2120.

(5) Webber, D. H.; Brutchey, R. L. Alkahest for V2VI3 Chalcogenides: Dissolution of Nine Bulk Semiconductors in a Diamine-Dithiol Solvent Mixture. *Journal of the American Chemical Society* **2013**, *135* (42), 15722–15725.

(6) Agrawal, R.; Zhang, R.; Walker, B. C.; Handwerker, C. Homogeneous Precursor Formation Method and Device Thereof. US20160333200 A1, 2016.

(7) Zhao, X.; Lu, M.; Koeper, M. J.; Agrawal, R. Solution-Processed Sulfur Depleted $Cu(In,Ga)Se_2$ Solar Cells Synthesized from a Monoamine–dithiol Solvent Mixture. *J. Mater. Chem. A* **2016**, *4* (19), 7390–7397.

(8) Miskin, C. K.; Dubois-Camacho, A.; Reese, M. O.; Agrawal, R. A Direct Solution Deposition Approach to CdTe Thin Films. *J. Mater. Chem. C* **2016**, *4* (39), 9167–9171.

500331 Photonic Mirrors for Enhanced Optical Transport in Luminescent Solar Concentrators

Abstract:

Luminescent solar concentrators (LSCs) increase the performance of solar cells by concentrating both direct and diffuse sunlight onto small photovoltaic devices. A typical LSC consists of a large plastic sheet embedded with luminophores such as dye molecules or luminescent nanocrystals. Incident sunlight is absorbed by the luminophore and emitted at a longer wavelength. Emitted light propagates to the edges of the plastic sheet via total internal reflection, and a small photovoltaic cell is mounted at the concentrator edge to collect the guided light. The solar cell design can then be optimized for a spectrally narrow and focused light source. Moreover, these colorful LSCs form components of architectural materials and may serve as window replacements.

Nanocrystal luminophores have recently emerged as candidate luminophores for LSCs, offering improved tunability, the ability to engineer the Stokes shift, narrow emission bandwidths, and high quantum yields. Nevertheless, challenges remain for using these luminophores in LSCs: the nanocrystals aggregate or suffer from reduced quantum yields when incorporated into polymer matrices, and non-Cd based materials with broader emission bands are needed for architectural applications.

In this work we have created LSCs based on the combination of large Stokes shift nanocrystals with various photonic light-guiding designs. We study how the combination of different types of photonic mirrors, integrated onto the front, back, or throughout the LSC structure, influences the overall concentration factor. By using different types of photonic mirrors, we show how these designs can overcome luminophore non-idealities such as reduced quantum yields or reabsorption losses. We use a combination of ray-tracing Monte Carlo simulations and full-field electromagnetic calculations to design the mirrors in these LSCs and to separately track losses to the escape cone and reabsorption.

To integrate photonic mirrors throughout the LSC structure we design a photonic LSC consisting of a periodic stack of polymer and titania layers with semiconductor nanocrystals embedded in one set of the layers. This creates a photonic bandgap, or a spectral region where light is forbidden. The photonic bandgap can be tuned via the thickness of the layers to match the emission of the luminophore. Using this design, we predict a decrease in escape cone loss from 41% to 5% for luminophores embedded in the center polymer layer that emit at 1.11 times the photonic band gap of the LSC. A decrease in escape cone loss from 20% to 6% is also predicted for luminophores embedded in the center titania layer that emit at 1.15 times the photonic band gap of the LSC. We find that it is preferential to embed nanocrystals with significant reabsorption losses in the high index layer, whereas nanocrystals with low reabsorption losses should be embedded in the low index layer. This is due to the high concentration of emitted radiation that propagates through the polymer layer when the luminophore is embedded in the polymer or the titania layers. Although these systems were designed using the optical properties of CdSe/CdS nanocrystals, the results are extendable to other types of nanocrystals with broader emission spectra.

These theoretical predictions are also corroborated by experimental measurements of optical transport inside LSCs.

CdSe/CdS nanocrystals with varying shell thickness are incorporated into different polymers, including poly (lauryl methacrylate). Optical spectroscopy measurements of the radiative lifetime indicate that the materials are incorporated without significant changes in the optical quality. Transport measurements indicate good agreement with the predictions from simulation.

This work provides a pathway to high performance LSCs through enhanced optical transport. By controlling the emission angle we work to experimentally design LSCs with large collection areas that efficiently guide light to the edges, thus increasing the optical efficiency and the concentration factor of the LSC.

495781 Dopant Mediated Assembly of Nanorods into Atomically Coupled 2D Sheets in Solution

Abstract:

Self and directed assembly of nanorods into ordered superstructures is an area of intense research interest as in such architectures, the discrete size defined properties of each nanostructure can be collectively harnessed at a scale amenable to real application. The formation of vertically aligned nanorods into hexagonal superlattices is a recent development leading to structured arrays suited for high-density applications. Spontaneous supercrystal organization in nanorods requires low polydispersity in both diameter and length, which can complicate the assembly compared to that of spherical nanocrystals. These nanorods can be pre-assembled as 2D sheets during synthesis by incorporating certain amount of foreign ion or dopant in the nanorod structure. Further, dispersed solution containing 2D sheets can be deposited in any substrate by either spin coating or drop casting to form thin film of micrometre-sized superlattice. In this work, we focus on understanding how dopant incorporation on the nanorod modifies their surface chemistry to from vertically aligned nanorod superstructure (2D sheets) in solution. We demonstrate by using synchrotron small-angle x-ray scattering (SAXS) and dynamic light scattering, that optimum concentration of dopant is required to tune the nanorod-nanorod interaction (repulsive and attractive) in solution to nucleate the self-assembly process. These 2D sheets of nanorods were extensively studied by SAXS, XPS, HRSTEM, HRSEM and SAED.

499034 Tailored Polymers for Structural Application in Large-Scale Extrusion Systems

Abstract:

The explosive growth of inexpensive polymer-based 3D printers across the economy has been driven in part by the promise of true freeform production coupled to digital design. However - as anyone with experience in 3D printing will attest - the hype surrounding the set of technologies known as additive manufacturing often precludes the significant shortcomings and trade-offs inherent in currently available products. In order to realize the full potential of 3d printed products, significant issues in polymer material systems must be addressed.

Material defects are often introduced in 3d printers due to process variability, complex thermal histories due to variable part geometries, sub-optimal toolpaths, process artifacts from machines and software, interfacial chemical reactions between layers and reinforcements, operator errors, and a variety of other variables. Additionally, the set of polymers available for 3D printing is further constrained by practical considerations such as rheology, shelf-life, environmental stability, commercial availability, and cost. These challenges are primarily addressed through either process modification or material system development.

This talk will focus primarily on polymer system development for very large-scale (100 pounds per hour deposition up to about 1000 cubic feet build volume) but will also highlight examples where process and material development are highly dependent on one another. Topics covered will include desired material properties for printing, the role of composite reinforcement materials, composite reinforcement interface and surface chemistry, advanced magnetic multifunctional composite systems, a survey of material systems under development, and highlights of results from recent R&D. Results and highlights will focus on work performed via public-private partnership by the Manufacturing Demonstration Facility (MDF) at Oak Ridge National Laboratory with support from the Advanced Manufacturing Office at the U.S. Department of Energy.

497744 Modeling Flow Phenomena in Fused Filament Fabrication Geometry

Abstract:

The additive manufacturing (AM) field has recently been growing at an exponential rate and is currently transitioning from prototype visualization to functional prototypes and end product development. One of the greatest hurdles to advancing AM is the lack of available materials that can be printed into usable components. Extrusion based additive manufacturing, also known as fused filament fabrication (FFF) and the trademarked name fused deposition modeling (FDM), is the most common method of AM and has the largest catalog of useable materials. However, far fewer options are available when compared to traditional manufacturing methods, which limits the mechanical and chemical properties of parts created by this process. We propose a screening process to efficiently expand the catalog of available materials and reduce the need for extended and expensive testing. The screening process considers the most common failure modes for material extrusion: filament buckling and annular backflow, and integrates a continuum based transport approach to describe the physics of material extrusion through common FFF geometries. Our approach successfully describes the backflow phenomenon, which has not been previously investigated or reported but has been observed as we are developing new materials for the FFF process, including ionomers. We demonstrate the sensitivity of backflow to a materialâ€™s viscosity, the feed rate, and system geometry. Using this approach, we are able to rapidly predict the ability of a polymer to successfully print in an FFF system based on a single experimental measurement and the system geometry. We present the sensitivity of process conditions to the potential for failure to guide process condition selection, including temperature and feed rate.

496776 Influence of Processing on Additively Manufactured Mechanically Adaptive Cellulose Nanocrystal Polymer Composites

Abstract:

Cellulose nanocrystal (CNC) polymer composites have the potential to provide enhanced strength and impart stimuli responsive, mechanically adaptive behavior to a number of commercially available polymeric materials. We are currently investigating the potential to transition these unique characteristics from solvent cast polymer films to functional parts generated using extrusion based additive manufacturing (AM). The relationship between structure, process and properties of functional CNC polymer composites has not been explored when utilizing material extrusion based AM technologies. We are establishing relationships between key material properties such as mechanical strength, thermal transitions and morphology to aid in describing this complex relationship. Developing this understanding of the processing dependent material behavior will aid in developing novel CNC based materials for AM technologies.

In this work, we study a CNC / thermoplastic polyurethane (TPU) nanocomposite. Dynamic mechanical analysis (DMA) experiments were conducted with a submersion chamber attachment, which allows for analysis of mechanical properties in both dry and water saturated environments. Preliminary DMA data indicate processing has a significant impact on dry mechanical properties of the cast films (385 MPa), compression molded films (92 MPa) and extruded fibers (63 MPa). The wet mechanical properties of the cast film, compression molded film and extruded fiber are all very similar (25 MPa ± 5 MPa) indicating that thermal history and orientation has a significant impact on the dry mechanical properties. Geometry also has a substantial impact on rate of diffusion and corresponding rate of mechanical switching, with thin film samples adapting in 350 seconds compared to 15 hours for a 6 mm diameter fiber. Small angle X-ray scattering is used to qualitatively analyze the degree of orientation induced by processing the cast film and fiber samples; despite machine direction orientation in an extruded fiber, we observe reduced dry state storage modulus compared to the unoriented solvent cast film. These results suggest the potential impact of thermal history on surface functionality of the CNCs. We deconvolute the separate, competing contributions of CNC orientation and thermal decomposition; the former is expected to substantially increase modulus, while the latter is expected to reduce coupling between CNCâ€™s.

484122 3D Printed Shape Memory Objects Based on Olefin Ionomer of Zinc-Neutralized Poly(ethylene-co-methacrylic acid)

Abstract:

3D printing enables the net shape manufacturing of objects with minimal material waste and low tooling costs, but the functionality in these objects is generally limited by available materials, especially for extrusion based printing such as Fused Deposition Modeling (FDM). Here we demonstrate shape memory behavior in a complex object, a helix, fabricated by 3D printing with FDM using a commercially available olefin ionomer, Surlyn 9520, which is zinc neutralized poly(ethylene-co-methacrylic acid). The spring-like helix can be mostly straightened to a temporary shape and then recover in 3 dimensions to the initial shape on heating to 70 °C. The performance of 3D printed samples were compared to compression molded analogs using standard shape memory cycles using dynamic mechanical analysis (DMA) with a constant 1 MPa stress for generating the temporary shape. The initial fixity was similar between the printed and compression molded sample, but the recovery was much lower for the 3D printed sample due to the higher strain (lower modulus in the 3D printed part) and higher residual stress. Interestingly, the fixity and recovery on subsequent cycles to the initially recovered permanent shape is greater for the 3D printed than compression molded part. Moreover, the programmed strain can be systematically modulated by inclusion of porosity in the printed part without adversely impacting the fixity or recovery. FDM can be used to directly form thermoplastic shape memory polymers into complex shapes that can be recovered with the appropriate trigger.

494163 Three-Dimensional Printing By Multiphase Silicone/Water Capillary Inks

Abstract:

Three-dimensional (3D) printing of polymers is accomplished easily with thermoplastics as the extruded hot melt solidifies rapidly during the printing process. Printing with liquid polymer precursors is more challenging due to their longer curing times. One curable liquid polymer of specific interest is polydimethylsiloxane (PDMS). We demonstrate a new efficient technique for 3D-printing with PDMS by using a capillary suspension ink containing PDMS in the form of both pre-cured microbeads and uncured liquid precursor, dispersed in water as continuous medium. Owing to the capillary binding of the microbeads, their suspensions bound by small fractions of liquid precursor behaved like pastes, which are flowable at high shear stress and possess high storage moduli and yield stresses that are needed for direct ink writing. The resulting three-phase capillary ink could be 3D printed and cured both in air and under water. The liquid PDMS bridges were thermally crosslinked after printing, resulting in structures that were remarkably elastic and flexible. Their porosity and mechanical properties, such as tensile modulus, could be controlled by the fraction of liquid precursor in the original multiphasic dispersion. Self-standing elastic structures could also be made by directly printing and curing the ink under aqueous solutions. As this ink is made of porous biocompatible silicone, it can be used in 3D printed biomedical products, or in applications such as direct printing of bio-scaffolds on live tissue. The patent-pending method (prov. filed 2016) may also open new opportunities in soft robotics and stimuli-responsive materials.

493399 3D Printing of Polymer-Bonded Magnets Using a Combination of Extrusion Direct Write and Stereolithography Methods

Abstract:

Polymer-bonded magnets have recently gained popularity as a novel magnet manufacturing technique due to its low-cost, design freedom, and magnetizing patterns. The maximum energy product of polymer-bonded magnets is generally lower than that of sintered magnets. However, the fabrication of polymer-bonded magnets does not require high sintering temperature and thus can be more cost effective.

This presentation reports the use of a combination of direct write (DW) and stereolithography 3D printing methods with appropriate process controls to fabricate polymer-bonded magnets at room temperature. Additionally, the presentation will describe and discuss the ink formulation for controlled printing of conformal and 3D magnets. The performance of 3D-printed magnets will be presented and compared to those produced from other manufacturing methods.

490532 PDMS-Based Ink Development for 3D Printing Applications

Abstract:

Additive manufacturing (AM) is an emerging research field due to its ability to manufacture materials with controlled geometry and properties using computer-based 3D printing technology.[1-2] Currently, the main challenge with this technology deals with limited materials, or ink, available for processing, specifically in the field of polymers. Only a single commercially available silicone resin (SE1700) possesses an appropriate viscosity and flow suitable for 3D printing capabilities, severely limiting advancements for improved performance of silicone-based materials for aerospace, engineering, and biomedical applications. Advancements in resin development rely on both new printable formulations and integration of novel fillers in order to alter the materialâ€™s properties. Typical fillers include fumed silica particles; however, other fillers (carbon nanotubes, graphene, surface-modified silica) can be incorporated to target specific applications relying on thermal, mechanical, and/or electronic properties.[3]

In this study, we first report novel 3D printable PDMS-based inks that possess appropriate rheological properties and wide temperature range applicability. Furthermore, compared to commercial silicone kits (e.g. SE1700), these inks also exhibit a very clear composition of ink formulation to broaden our application spectrum. Elastic architectures, which were patterned in a face-centered tetragonal (FCT) structure, were obtained by extrusion from a micro-nozzle using pressurized ink from a syringe, followed by a post-heat treatment to cure the part. The PDMS-based inks were prepared by mixing PDMS with three types of silica fillers having different surface characteristics to achieve three main attributes: i) sufficient elasticity (storage modulus) in the liquid state to withstand typical sagging and flowing issues during 3D printing; ii) excellent processing conditions and engineering properties for 3D printing applications, such as an appropriate pot life (working life or printable time), thermal properties over a wide temperature range (glass transition/crystallization), and a clear ink formulation of the 3D printed products; iii) full understanding of structure-property-performance relationship between silica fillers in the PDMS matrix and their 3D printed products, offering tunable properties to further extend their field of applications. [4]

References

1. S. L. N. Ford, Additive Manufacturing Technology: Potential Implications for U.S. Manufacturing Competitiveness, *J. Int. Commer. Econ.* (2014)
2. *Additive Manufacturing: Opportunities and Constraints: A Summary of a Roundtable Forum Held on 23 May 2013 Hosted by the Royal Academy of Engineering*, Royal Academy of Engineering, (2013)
3. E. B. Duoss, T. H. Weisgraber, K. Hearon, C. Zhu, W. Small, T. R. Metz, J. J. Vericella, H. D. Barth, J. D. Kuntz, R. S. Maxwell, C. M. Spadaccini, T. S. Wilson, *Adv. Funct. Mater.* **24**, 4905 (2014).
4. K.-S. Lee and A. Labouriau *US Provisional Patent Application.* (2017).

500765 3D Printed Active Microfluidic Elements for Portable Bioanalysis Assays

Abstract:

There is an ongoing need for simple, inexpensive, and robust systems that replicate functions of conventional bioanalysis instruments in a format with sufficient portability to deploy in resource-limited settings. A key challenge in the development of these systems, however, involves the ability to dispense precisely metered quantities of liquid reagents and drive their transport through microfluidic networks. Active mechanical pumps offer considerable flexibility compared with passive methods (e.g., capillary wicking in paper-based systems), but sacrifice operational complexity electrical power consumption. Here we describe a novel approach to overcome these limitations by leveraging shape memory polymers to construct a library of fluidic control elements that can be assembled in various arrangements to execute sequential fluidic operations associated with field-based diagnostic assays. These elements (pumps, valves) are easily produced using conventional 3D printing technology, and display a shape memory response within a temperature range that allows them to be operated by exposure to heated water, eliminating the need for battery power and permitting operation by personnel with minimal training.

501190 3D Printed Polymer-Based Bio-Inspired Neural Systems

Abstract:

Polymers are critical materials for the design of biomedical devices and physiological models of biological systems. Additive manufacturing now provides the ability to process biocompatible polymers into complex geometries for a variety of biological applications. However, the design of functional and higher order scaffolds and platforms for regeneration and modeling of the nervous system requires the ability to interweave multiple polymers into bio-inspired architectures. Here, we address the manufacturing of multi-functional anatomical scaffolds for peripheral nerve regeneration and *in vitro* platforms for modeling higher order pathophysiological responses in the nervous system via a novel multi-material microextrusion-based 3D printing process. Printing of polymers with disparate material properties is discussed with a focus on critical polymer properties, such as polymer rheological properties, melting temperature, and gelation temperature. The integration of multiple polymers into bio-inspired architectures, including elastomers, thermoplastics, and hydrogels, will be discussed. The manufacturing process is also compatible with the 3D printing of neuronal and non-neuronal cells at high viability. 3D structured-light scanning is used to construct anatomical models of peripheral nerve that inform the programming of path information for scaffold printing. Ultimately, polymer-based additive manufacturing appears poised as a unique processing approach for the design and manufacturing of novel multi-functional microphysiological neural systems.

AUTHOR INDEX

Acar, Handan 236, 897
Adamovicz, Jeffrey 530
Adams, Justin 460
Adedoyin, Adedokun 351
Adewole, Jimoh K. 838
Adil, Maroof M. 80
Afeworki, Mobae 650
Agarwal, Sushant 366
Agrawal, Kumar Varoon 653
Agrawal, Pradeep K. 156
Agrawal, Rakesh 571, 905
Ahrens, Caroline 395, 677
Ailianou, Artemis 316
Aksan, Alptekin 676
Al Katheeri, Abdul Majed 183
Alabi, Christopher A. 355, 403, 528, 764
Alamgir, Faisal M. 156
Albert, Julie 576
Alexandrova, Anastassia 786
Alfonso, Dominic 4
Alhassan, Saeed 101, 106
Aliakbarighavimi, Soheila 767
Allen, Brittany 12, 767
Allison, Evan 301, 305
Almas, Qandeel 154
Almodovar, Jorge 322, 633
Alpert, Carol Lynn 278
Alsewailem, Fares 205
Altalhi, Abdulmajeed 228
Al-Thabaiti, Shaeel 653
Alvarez, Nicolas J. 797
An, Qi 87
An, Yaxin 525, 729
Anderson, Daniel G. 78, 408
Andreev, Marat 402
Andresen, Corinne 361
Anilionyte, Oksana 636
Anisimov, Mikhail A. 659
Annamalai, Prakasam 350, 630
Anseth, Kristi S. 307, 318
Anthamatten, Mitchell 782
Aoki, Reyn 679
Aqad, Emad 164
Arbelo, Noelia Almodovar 420
Archer, Lynden A. 342
Ardila-Suarez, Carolina 723
Armstrong, Matthew 18
Arora, Akash 356
Arturo, Steven G. 52
Arvidson, Sara A. 608
Asano, Yusuke 194
Asatekin, Ayse 370
Askar, Shadid 425
Aslam, Umar 61
Assary, Rajeev 95
Asteasuain, Mariano 198
Atsumi, Hiroshi 636
Azarin, Samira M. 538, 676
Bae, Tae-Hyun 69
Bafana, Adarsh 271
Bai, Baojun 216
Bai, Lian 285, 286, 618
Bailey, Travis S. 708
Bain, Erich 313, 714
Balasanthiran, Choumini 59
Balbuena, Perla B. 723
Ball, Rebecca 526
Balwani, Apoorv 691
Bao, Lei 196
Bao, Wei 787
Bao, Zhenan 39, 304, 358
Bara, Jason E. 346
Baranauskas, Vince 648
Baranek, Austin 715, 761
Barberio, Antonio E. 240
Barboun, Patrick 160
Bardhan, Rizia 42
Barnes, Tanner 627
Barnharst, Tanner 556
Barteau, Katherine P. 777
Bartel, Christopher J. 3
Bartholomew, Reid 635
Barua, Sutapa 760, 891
Basahel, Sulaiman N. 653
Baskaran, Durairaj 518
Bassett, Alexander W. 542, 802
Bates, Frank S. 341, 356, 711, 750, 796
Batra, Surinder 8
Bavarian, Mona 193
Baxter, Jason B. 58, 303, 744, 900
Bazan, Guillermo C. 893
Beckingham, Bryan S. 695
Beckman, David W. 436
Beers, Kathryn L. 76
Behazin, Ehsan 555
Beitelshees, Marie 463
Bejagam, Karteek K. 525, 729, 740
Belcher, Angela M. 636
Belfort, Georges 718
Bellaire, Bryan H. 437, 462
Beltran, Saul Perez 723
Beltran-Villegas, Daniel J. 368
Benamara, Mourad 417
Bender, Elizabeth C. 348
Beniah, Goliath 367
Benicewicz, Brian C. 713
Berg, Gayla 381
Berlinger, Maya 231
Berry, Carter 423, 729, 740
Berry, Joseph J. 561

AUTHOR INDEX

Berry, Keith 83
Berry, Vikas 416
Berthod, Mikael 183
Beyer, Frederick L. 313
Bhan, Aditya 155
Bharti, Bhuvnesh 912
Bhattacharjee, Ujjal 566
Bhattacharyya, Souryadeep 720, 770
Bhethanabotla, Venkat R. 277
Biagioli, Madeleine 344, 359
Bilchak, Connor 713
Binkhodor, Yazeed 205
Biria, Saeid 798
Bischof, John C. 538
Bishop, Brittany 679
Biswas, Pratim 560
Black III, Lauren D. 348
Blocher, Whitney C. 459
Blumeyer, Jack 378
Bodinger, Carter 83
Boissiere, Michel 639
Bollini, Praveen 155
Bonacina, Luigi 417
Bondos, Sarah 763
Boone, Evan 507
Borges, Fernando T P 202
Borguet, Eric 772
Bortner, Michael J. 909, 910
Bose, Arijit 279
Bose, Suman 78, 408
Botvinick, Elliot 317
Boudouris, Bryan W. 420
Bourdon, Buchanan 783
Bowen, Alec S. 511
Bowman, Christopher N. 50, 201, 325, 381, 391, 715, 761, 898
Boyd, Peter 769
Boyne, Robert W. 571
Bredas, Jean-Luc 901
Brenza, Timothy 255, 438
Brettmann, Blair Kathryn 778
Brinson, Catherine 818
Broadbelt, Linda J. 52
Broer, Dick J. 325
Brown, Jonathan R. 748
Brown, Joseph 355, 403
Brown, Lauren 507
Brown, Tobin E. 307
Brunelli, Nicholas 153, 726
Bryant-Friedrich, Amanda C. 401
Buenning, Eileen 713
Buenrostro, Denise 535
Bundy, Bradley C. 899
Burbach, Brandon 538
Burke, Kelly A. 637
Burkey, Aaron A. 656
Busch, Markus 200
Buss, Hilda G. 590
Byrley, Peter 60
Cabezas, Camila Saez 741
Cain, Laurance 571
Cairns, Johnnie 277
Calabro, David C. 650
Camargo, Lucas Garcia 629
Camci-Unal, Gulden 347
Cao, Han 658
Cao, Lixia 176
Cao, Nai 584
Cao, Zhiqiang 531
Capeling, Meghan 237
Carpenter, Ryan 323
Casas, Juan Pablo 757
Case, Natasha 626
Castaldi, Marco J. 746
Castilla, David 322
Castro, Camila 757
Castro, Carlos E. 62
Chai, Song-Hai 156
Chaikittisilp, Watcharop 5, 71
Chakraborty, Aishik 896
Champion, Julie A. 79, 464
Chan, Christina 9
Chan, Mary 894
Chang, Ji Woong 59
Chang, Liang 93
Chang, Li-Wei 400
Chang, Michael Yu Zarng 237
Chang, Yun Hsuan 724
Chan-Park, Mary B. 893
Chapman, Karena W. 773
Charles, Bayliss 189
Chatham, Camden 909
Chauvel, Paul 174
Chawla, Aseem 103
Chen, Chau-Chyun 272, 697
Chen, Gang 570
Chen, Gong 38
Chen, Jee-Wei Emily 393, 861
Chen, Jerry J. Y. 533
Chen, Qile 675
Chen, Ru 793
Chen, Sinn-Wen 280
Chen, Tao 177
Chen, Tzu-Hsuan 680
Chen, Tzu-Ling 233
Chen, Xi 49, 387
Chen, Yahua 893
Chen, Yang-Yuan 280
Chen, Zhengtao 779
Chen, Zhifeng 59

AUTHOR INDEX

Cheng, Chi-Hui 257
Cheng, Chin-Yi..636
Cheng, Chong 237, 326
Cheng, Elise 361
Cheng, Guanzhi..812
Cheng, Kenneth..392, 892
Cheng, Wei-I 237
Cheng, Xiang 285, 286, 618
Cheng, Yongqiang..157
Chesniak, Olivia 9
Chiarot, Paul R. 270
Chih, Meng-Hsiu 680
Chinzei, Nobuaki..626
Chiu, Yun..257
Cho, Joon Hee 834
Cho, Junghyun 270
Chodankar, Nilesh R. 269
Choi, Heechul 268
Choi, Jae-Soon..158
Choi, Joshua 565
Choi, Sunho 67
Chorpening, Benajmin T. 98
Chou, Daniel Y. 835
Choudhury, Pabitra 281
Choudhury, Snehashis 342
Christau, Stephanie 435
Christians, Jeffrey A. 561
Chu, Liang-Yin..315
Chu, Ping-Hsun 162
Chu, Weiwei 519, 749
Chuang, Hui-Min 658
Chung, Wook-Jin 226, 289
Churchill, Stuart W. 170
Chwatko, Malgorzata 48
Cima, Michael 411
Clancy, Paulette 902
Claure, Micaela Taborga..156
Cleveland, Iver J. 563
Cobos, Monica..610
Cochran, Eric W. 383, 836
Cockcroft, Jeremy..180
Coffel, Joel 730
Coffey, Aidan..265, 817
Cogswell, Christopher 67
Colby, Ralph H. 340, 833
Coleman, Maria 354
Collier, Graham 507
Colon, Yamil J. 511, 519, 749
Commisso, Alex..237, 326
Connell, Ryan 906
Cook, Jonathan 645
Coppens, Marc-Olivier 180, 382
Coropceanu, Veaceslav 901
Correa, Santiago 240
Corti, David S. 420
Cosgrove, Jayson D. 802
Cottrill, Anton 291
Cove, Matthew..643
Crane, Matthew 379
Crum, Jerry 252
Cui, Honggang 396, 532
Cummings, Chad 526
Curia, Silvio 542
Curtis, Chad D. 239
Dadelahi, Alexis 530
Daemen, Luke L. 157
Dahoumane, Si A. 271
Dai, Gance..89
Dai, Sheng 156
Dai, Xiaofeng 374
Dang, Binh 679
D'Angelo, Anthony 96
Dardona, Sameh 913
Darling, Ross..462
Darugar, Qusai 209
Dasgupta, Dwaipayan 377
Dastidar, Subham.. 58, 303, 562
Davidson, Bruce A. 463
Davis, E. James 379
Davis, Eric M. 592, 691
De La Torre, Roger 530
De Laporte, Laura..321
De Pablo, Juan J. 43, 402, 511, 519, 728, 749
De Rosa, Claudio 195
De Silva, Udaka K. 401
De Souza, Fernanda C. Bombaldi 241, 253
De Souza, Pedro..593
De Souza, Renata F. Bombaldi..241, 253
De Villers, Bertrand Tremolet..561
Deegan, Michael 18
Deforest, Cole A. 406
Dehankar, Abhilasha 62, 278
Demello, Andrew J. 564
Dement, Dana 745
Deml, Ann M. 3
Deneff, Jacob I. 475
Deng, Shikai..416
Deng, Yulin 151
Dennis, Grayson P. 346
Desai, Tejal 166
Deshmukh, Sanket A. 423, 525, 729, 740
Deshmukh, Swapnil Dattatray 905
Deshpande, Nitish..153
Desimone, Joseph M. 74
Devore, David 254
Dhar, Prajnaparamita..896
Dhara, Palash..642
Dhinojwala, Ali 647
Diannan, Lu 15
Diao, Ying 509

AUTHOR INDEX

Dichtel, William 367
Dickey, Michael D. 644
Diederich, Vincent 312
Diederichsen, Kyle M. 590
Dillon, Andrew D. 58
Dinari, Mohammad 229
Ding, Fuchuan 187
Ding, Zhenya 395, 677
Dinic, Jelena 344, 359, 508
Dittmer, Neal T. 896
Dixit, Marm 504
Dobyns, Breanna 695
Dodam, John 530
Dolog, Rostyslav 209
Dominguez, Kevin 746
Dong, Hongfeng 273
Dong, Juyao 606
Dong, Pingchuan 584
Dong, Ziye 395, 677
Donta, Maxsam 319
Dorfman, Kevin D. 356, 658, 750
Doufas, Antonios 339
Doughty, Benjamin 157
Douglas, Jack F. 402
Dreaden, Erik 240
Drennan, Corinne 158
Du, Lin 300, 375, 415, 418
Du, Yuyang 781
Duan, Yuhua 98
Dubensky, Thomas 462
Dumont, Joseph H. 914
Dunklin, Jeremy 83
Dunn, Celia 627
Durbin, Marlow M. 346
Durning, Christopher J. 713
Dziubla, Thomas 14, 314, 476
Eastman, Stephen 66
Echeverri, Juanita 757
Edgar, James H. 64
Edwards, Brian J. 85, 284, 298, 345
Eichhorn, Stephen 631
Eickman, Erin 862
Ekenseair, Adam 351
Elabd, Yossef A. 233, 591
Elbert, Johannes 474
Ellebracht, Nathan 68
Ellison, Christopher J. 194, 227, 288, 388, 593, 607, 834, 835
Elmore, Bill B. 204
El-Sayed, Mostafa 191
Emi, Tania 627
Endo, Fuyuaki 195, 522
Engstrom, James R. 431
Eniola-Adefeso, Omolola 534
Epps, Thomas H. 324
Escobar, Francisco 318
Escobedo, Fernando 521, 751
Esnaashari, Catherine 286, 618
Estroff, Lara A. 777
Ewart, Sean 51
Facchetti, Antonio 505
Fafarman, Aaron T. 58, 303, 562
Fairbanks, Benjamin D. 50, 201
Fallon, Jacob 909, 910
Fan, Wei 104
Faraone, Antonio 691
Favero, Claudio Vilas Boas 173
Fein, Katherine 526
Feng, Moon Tay Yue 893
Feng, Yafei 276, 287
Fernandez, Sergio 652
Ferrier Jr., Robert C. 48
Ferry, Vivian E. 412, 742, 745, 906
Finn, M. G. 458
Fiore, Andrew 739
Flake, Tanner 252
Fleury, Blaise 744
Foglio, Mary Ann 242
Foo, Guo Shiou 154
Forcherio, Gregory T. 83, 417
Ford, Hunter 97
Forster, Colleen L. 527
Fortman, David 367
Foster, Earl J. 910
Foster, Mark 647
Francis, David 536
Francis, Lorraine F. 44, 781
Francis, Matthew 81
Frazier, Deonante 204
Freeman, Benny D. 607
Freeman, Charles 204
Freire, Stephan 186
Frisbie, C. Daniel 44
Fromen, Catherine A 534
Frye, John G. 158
Fuhr, Addis 786
Fullerton, Susan 783
Funkenbusch, Maylin 231
Furst, Ariel 81
Gaber, Dina 101
Gaber, Safa 101
Galaska, Alexandra 85, 184, 284, 298
Gallazzi, Fabio 12, 530
Gamelin, Daniel 379
Gan, Yunn-Hwen 893
Gand, Adeline 639
Gao, Hanyu 52
Gao, Jian 716
Gao, Zhaoli 413
Garcia, Caterina Bartomeu 433

AUTHOR INDEX

Gartner III, Thomas 235, 424
Ge, Ting 732
Gebbie, Matthew A. 57
Gehlen, David B. 321
Gehrke, Stevin H. 896
Geiger, Brett C 434
Geiger, Mackenzie 327
Geldmeier, Jeffrey 191
Geng, Jiaming 216
Geng, Xiaohua 56
Genzer, Jan 644, 731
Ghorbanpour, Arian 568
Ghosh, Deepanjan 252, 638
Gilchrist, Lane 895
Gillespie, Adam 837
Gilman, Jeffrey W. 330, 818
Gilmer, Eric L. 909
Ginzburg, Valeriy 164
Giri, Gaurav 568
Glascoe, Elizabeth 694
Gleason, Karen K. 646
Glotzer, Sharon C. 169
Go, David 160
Gogotsi, Natalie 744
Gong, Tao 761
Goodenough, Isabella 772
Goodrich, J. Taylor 50
Goodrich, Johnathan 201
Goto, Eisuke 799
Goto, Renata Nishida 629
Gounaris, Chrysanthos E. 4
Gower, Michael 350, 539, 630
Goyal, Ritu 254
Grady, Michael 53
Grasman, Jonathan M. 348
Grayson, Scott 576
Greeley, Jeffrey 573
Green, Brian J. 320
Green, Matthew D. 419
Green, Micah 192
Greenberg, Everett Peter 893
Greene, Ashlee 540
Greenfield, Michael L. 206
Greenwald, Andrew 767
Griego, Charles 281
Griffith, Linda 407
Grim, Joseph 307
Grodzinsky, Alan 434
Grover, Martha A. 41, 162, 512
Grunlan, Jaime C. 397
Gu, Hongbo 86, 88, 184, 295, 477, 816
Gu, Junwei 86, 810
Gu, Kevin L. 304, 358
Gu, Liangliang 610
Gu, Xiaodan 304, 358
Gu, Zhuxiao 297
Guan, Yi-Xin 215
Gudapati, Vamshi 818
Guironnet, Damien 47
Guo, Chengchen 252, 638
Guo, Jiang 85, 184, 284, 298
Guo, Ruilan 693
Guo, Xiaocui 17
Guo, Yongqiang 810
Guo, Zhanhu 85, 184, 284, 298
Gupta, Devanshi 91
Gupta, Rakesh K. 366
Guymon, C. Allan 320, 361, 369, 635
Ha, Heonjoo 227, 288, 388, 593, 607
Ha, Kiryong 292, 388, 607
Hachmann, Johannes 754
Hadi, Atefe 563, 789
Hafez, Islam 293
Hages, Charles J. 903
Haley, Adrian 380
Hall, Lisa M. 234, 404, 748
Hallinan Jr., Daniel T. 570
Hamdeh, Umar H. 566
Hamid, Mohamad 260, 721
Hammer, Daniel A. 756
Hammond, Karl D. 378
Hammond, Paula 10, 11, 231, 240, 394, 407, 434
Han, Duanduan 915
Han, Rebecca 770
Han, Yi-Fan 814
Handwerker, Carol 905
Hanselman, Christopher L. 4
Hansen, Marlan 361, 635
Haque, Farihah 576
Haranczyk, Maciej 769
Harley, Brendan A. 393, 861
Harley, Stephen 694
Harper, Kaid 159
Harris, Michael A. 527
Harris, Stephanie 837
Hartman, Dave 818
Hartwell, Brittany 541
Harvey, Steve 561
Harvey, Taylor B. 904
Hasa, Erion 369
Hasegawa, Urara 533
Hashemnejad, Seyed Meysam 308, 385
Hatton, T. Alan 474, 784
Hatzell, Kelsey 504
Hatzell, Marta 504
Hawker, Craig J. 799
He, Hongkun 231, 407
He, Mao-Gang 659
He, Siyao 87, 296
He, Yanpu 11, 231

AUTHOR INDEX

Heath, William 367
Hedrick, James L. 426
Heichel, Danielle L. 637
Held, Jacob 743
Heller, Adam 593
Helton, Tyler 18
Hendley, Michael 350, 630
Hendrikx, Matthew 325
Henry, Christopher 797
Herceg, Eldad 174
Herman, Chase 891
Hernandez, Nacu 836
Hershman, Rebecca 459
Herwig, Cara 910
Herzog-Arbeitman, Abraham 577
Hickey, John 537
Hicks, Jason C. 160
Higginson, Cody 458
Hildahl, Kate 679
Hill, Andrew 463
Hill, James C. 170
Hillman, Febrian 260, 721
Hillmyer, Marc A. 194, 232, 422, 776
Hilt, J. Zach 14, 314, 476
Hindie, Mathilde 639
Hohmann, Austin D. 836
Holder, Aaron M. 3
Holewinski, Adam 68
Hollingsworth, Jennifer 282
Holloway, Michael 254
Holmberg, Vincent C. 99, 379, 679
Holmes, Gale 818
Hong, Celestine 231
Hosein, Ian 798
Hoskins, Amanda 265, 817
Hossain, Md. Tashfin Zayed 605
Hotta, Atsushi 195, 207, 208, 386, 522, 632, 717
Hou, Qingqing 275
Houlihan, William 895
Howard, Micah J. 796
Howard, Tyler 83
Howe, Joshua D. 773
Hu, Bo 556, 557
Hu, Dong-Dong 196
Hu, Peng 725
Hu, Yun Hang 93
Hu, Zhiqi 510
Huang, Emily 186
Huang, Jun 72
Huang, Mengfei 780
Huang, Sophia Chao-Wei 257
Huang, Wenhan 186
Huang, Zi-Yang 280
Huda, Md Masrul 385
Huelsenbeck, Luke 568
Hule, Rohan 339
Hung, Jui-Hsiang 360, 747
Hunt, Heather K. 378
Hunter, Katharine I. 743
Hussain, Fazle 697
Hustad, Phil 164
Hwang, Kiseob 292
Hyun, Woo Jin 44
Ibanez, Sergio 264
Ida, Junichi 290, 488
Ilias, Shamsuddin 605
Imaninezhad, Mozhdeh 389, 410, 688
Imbrogno, Jennifer 48
Inturi, Siva Nagi Reddy 473
Irvine, Darrell J. 11, 231, 240
Isapour, Golnaz 188
Isely, Christopher 630
Ishii, Hiroshi 636
Ishikawa, Ai 488
Islam, Syed Z. 572
Ismail, Issam 101
Iyer, Kiran S. 575
Jackson, Enrique M. 346
Jackson, Nicholas 43, 511, 728
Jackson, Richard 382
Jacoby, Jeremy M. 563
Jadrich, Ryan B. 741
Jaekel, Esther 321
Jaffer, Shaffiq 711
Jahan, Merina 657
Jain, Era 626, 678
Jallorina, Jerel 513
Jamison, Timothy 784
Janes, Dustin 835
Jang, Yeongseon 79
Jansto, Allison 592
Jawahery, Sudi 769
Jayachandrababu, Krishna Chandran 720
Jayan, B. Reeja 569
Jayaraman, Arthi 235, 368, 424
Jeffryes, Clayton S. 271, 609
Jeon, Ju-Won 191
Jeong, Hae-Kwon 260, 261, 721
Jerke, Amber C. 438
Jessop, Julie L. P. 369
Jiang, Dawei 184
Jiang, Jianwen 775
Jiang, Mingzhe 189
Jiang, Peng 297, 306, 421
Jiang, Shaoyi 428
Jiao, Yang 771, 773
Jimenez, Jairo A 757
Jimenez, Leidy N. 344, 359, 508
Jin, Kailong 49, 387
Jin, Xin 266

AUTHOR INDEX

Jing, Benxin 399
Jinrui, Liu 378
Johnson, A. T. Charlie 413
Johnson, Blake 916
Johnson, J. Karl 772
Johnson, Joshua 62
Jones, Amanda 834
Jones, Charles H. 463
Jones, Christopher W. 65, 68, 156, 722
Jones, Susanne 158
Joo, Yong Lak 518
Jordan, Alex M. 711
Jordan, Carolyn T. 14
Joshi, Anup 354
Joshi, Rutuja 153
Ju, Xiao-Jie 315
Kagan, Cherie R. 744
Kakunuri, Manohar 631
Kalaga, Eswar Arunkumar 255
Kalluru, Sri Harsha 383
Kang, En-Tang 185, 643
Kang, Wooram 100
Kanie, Kiyoshi 102
Kanost, Michael R. 896
Kaplan, David L. 348
Kar, Suvrajyoti 206
Karapetrova, Evguenia 653
Karnik, Rohit 636
Kasiewicz, Lisa 390, 457
Katabathini, Narasimharao 653
Kavadiya, Shalinee 560
Keane, Danny 41
Keating, John J. 718
Kelley, William 534
Kelly, Giovanni 576
Kelly, Sean 462
Kennedy, Stephen 627
Kenttamaa, Hilkka 571
Keogh, Damien 893
Keoh, Sye Hoe 71
Khabashesku, Valery N. 209
Khademhosseini, Ali 167
Khaleel, Maryam 101, 106
Khalimonchuk, Oleh 862
Khan, M. Arif 572, 762
Khan, Saad A. 327, 608
Khandelwal, Mudrika 631
Khomami, Bamin 85, 284, 298, 345
Kidambi, Srivatsan 862
Kilbey, Michael 507
Kilbourne, Jacquelyn 638
Kim, Do-Heyoung 269
Kim, Donghun 653
Kim, Dongsub 799
Kim, Hanim 834
Kim, Hanna 292
Kim, Hannah 760
Kim, Jihan 6
Kim, Jongsik 160
Kim, Kang-Min 50, 201
Kim, Kyungtae 356, 711
Kim, Min Jae 610
Kim, Sanggon 60
Kim, Soyoung 268
Kim, Sung-Soo 194, 835
Kim, Yoonseob 583
Kimura, Takaya 182
Kimura, Yoshihiko 182
Kinaci, Emre 832
Kitagawa, Midori 207, 386, 632
Kitchens, Christopher L. 189
Klavetter, Kyle 593
Klier, John 780
Klimov, Victor I. 786, 790
Kline, Gregory 693
Kloxin, April M. 859
Kner, Peter 278
Knight III, Paul 463
Knorr Jr., Daniel B. 714
Knutson, Barbara L. 572, 762
Koenig Jr., Gary M. 91
Koh, Yung P. 712
Kohler, Jens 321
Kokini, Jozef 309
Kokkoli, Efrosini 235, 527
Kolar, Grant 678, 688
Kolb, Ben 910
Kollengode, Ramanathan 643
Koman, Volodymyr 606
Konda, Shailesh 364
Konetski, Danielle 391, 761, 898
Konstantinov, Ivan 52
Kopyeva, Irina 392, 892
Korgel, Brian A. 904
Kornfield, Julia A. 316
Kortshagen, Uwe R. 743
Koshimura, Tomoyuki 524
Kossuth, Mary Beth 316
Kou, Yangming 285
Kovalenko, Maksym V. 564
Kozawa, Daichi 606
Kramer, Jake 12, 767
Kratzer, Domenic 392, 892
Krebs, Melissa 880
Kretzschmar, Ilona 746
Kuang, Huihui 235, 527
Kubal, Joseph 573
Kubo, Masaki 524
Kuhn, J. N. 277
Kuljanishvili, Irma 389, 410

AUTHOR INDEX

Kumar, Ashish 377
Kumar, Manjesh 653
Kumar, Prashant 105, 106, 719
Kumar, Ramya 392, 892
Kumar, Sanat K. 713
Kumar, Satish 341
Kumar, Shishir V 271
Kumar, Sushant 390, 457
Kunai, Yuichiro 291
Kundu, Santanu 204, 308, 385
Kunitake, Yusuke 182
Kurokawa, Naruki 522
Kuznetsov, Anatoliy 106
Kwon, Ohmin 6
La Scala, John J. 542, 801, 802, 819
Labbe, Matthew 343
Labelle, James L. 236
Labouriau, Andrea 914
Lacko, Christopher 409
Laha, Anindita 628
Lahann, Joerg 392, 435, 892
Lam, Tiffany 538
Lampe, Kyle 349
Langenberg, Marcel 728
Langer, Robert 78, 408, 411
Lany, Stephan 3
Lapitsky, Yakov 401
Larson, Ronald G. 173, 357
Lattuada, Marco 188, 312
Lau, Kenneth 581, 779
Law, Matt 790
Lawrence, Jimmy 799
Lawton, Carl 256
Le, Thuy T. 181
Lee, Elizabeth M. Y. 302, 752
Lee, Hak Rae 676
Lee, Hansol 514
Lee, Ho Nyung 157
Lee, Hung-Lin 680, 724
Lee, Jungwoo 323
Lee, Kangtaek 258, 259
Lee, Kil Ho 278, 580
Lee, Kwan-Soo 914
Lee, Suh-Jane 158
Lee, Tu 680, 724
Leeper, Caitlin 12
Legge, Kevin 461
Lei, Yuguo 863
Leigh, Braden 361, 635
Leite, Michelle Franz Montan Braga 629
Lekse, Jonathan W. 4
Lenhart, Joseph L. 313, 714
Leo, Sin-Yen 306, 421
Leon, Lorraine 578
Leopoldino, Andreia Machado 629
Levi, Adam 310
Lewis III, Ronald M. 356, 655
Lhost, Olivier 711
Li, Dongyang 373
Li, Fei 582
Li, Hongbo 786
Li, Huajian 812
Li, Hua-Min 783
Li, Jiahe 10, 11, 231, 394
Li, Jian 273
Li, Lei 176
Li, Lingqiao 49, 387
Li, Lu 402
Li, Min 185
Li, Mingqi 164
Li, Qiang 863
Li, Quanchang 650
Li, Rui 103, 186
Li, Siming 303, 744, 900
Li, Wei 395, 677
Li, Xiaolan 750
Li, Xiaolei 12
Li, Xiyi 275, 478, 489
Li, Xu 775
Li, Xue 520
Li, Yanxiang 176
Li, Yi 463
Li, Yingzhong 11
Li, Yueming 895
Li, Yun 177
Li, Zhipeng 263
Li, Zhong 478, 489
Liang, Chaobo 86
Liao, Neng 275
Liddle, J. Alexander 330
Lignos, Ioannis 564
Lim, Hui Wen 893
Lim, Hyuntaek 395, 677
Lim, Jaehoon 790
Lim, Kwang-Hee 292
Lim, Woo Taik 260, 261, 721
Limjuco, Lawrence A. 226
Lin, Fang-Yi 836
Lin, Haiqing 364, 696
Lin, Haishuang 863
Lin, Jie 593
Lin, Jui-Che 257
Lin, Kehua 399
Lin, Yi-Ching 257
Lin, Yu-Chuan 783
Lind, Fatin 405
Lindquist, Beth A. 741
Ling, Sanliang 769
Linic, Suljo 61
Lipomi, Darren 73

AUTHOR INDEX

Little, Steven R. 540
Liu, Albert Tianxiang 291, 299, 606
Liu, Bin 64
Liu, Danqing 325
Liu, Dongxia 175
Liu, Jia 39
Liu, Jian 274, 774
Liu, Jing 812
Liu, Jingjing 187
Liu, Kunwei 87, 296
Liu, Lijun 670
Liu, Luman 406
Liu, Pengwei 299
Liu, Pingwei 606
Liu, Song 64
Liu, Tao 196
Liu, Xiaoxiao 584
Liu, Xiaoze 787
Liu, Xiufeng 177, 273
Liu, Yalin 343
Liu, Yang 771
Liu, Yidong 276, 287
Liu, Ying 644
Liu, Zheng 38
Liu, Zhenjing 619
Liu, Zhuang 315
Lively, Ryan P. 65, 722
Livingston, Dana A. 174
Lloyd, Michael L. 900
Lodge, Timothy P. 232, 422, 675
Long, Brian 507
Long, Timothy E. 909
Long, Tyler R. 313
Loo, Y. L. Lynn 75
Lopez, Francisco M. 39
Lopez, Gustavo 746
Lopez-Barron, Carlos R. 793
Lou, Emil 676
Lou, Jianzhong 605
Love, Dillon 50, 201
Lowe, Christopher J. 529
Lu, Diannan 38, 263
Lu, Hao 783
Lu, Yanfei 15
Lu, Yang 279, 531, 609
Lueckheide, Michael 398
Luo, Tianyi 772
Luo, Zhen 55
Luo, Zhengtang 190, 199, 619
Luther, Joseph M. 561
Luther, Yeboah Martin 294
Lutkenhaus, Jodie L. 506, 582
Lutterman, Daniel 157
Luu, Bryan C. 436
Lynd, Nathaniel A. 46, 48, 656
Lytle, Tyler 400
Lyu, Xuejian 641
Ma, Anson 913
Ma, Chao 295
Ma, Cunliang 276, 287
Maceiczyk, Richard 564
Macosko, Chris 296
Macosko, Christopher W. 87, 163, 285, 286, 341, 610, 618, 711
Macpherson, Quinn 727
Madinya, Jason 400
Maeda, Tomoki 207, 208, 386, 522, 632, 717
Maher, Michael 194
Mahmoud, Mahmoud A. 191
Maiti, Debtanu 277
Majumdar, Saptarshi 628
Makarov, Nikolay 786
Malani, Ateeque 755
Maldovan, Martin 283
Malhotra, Abhinav 283
Mallam, Gopichand 768
Mallapragada, Surya 460, 649
Mallapragada, Surya K. 8, 319
Mantovani, Diego 241
Mantripragada, Shobha 605
Mao, Hai-Quan 537
Mao, Shifan 727
Marar, Abhijeet 278
Marciel, Amanda B. 328, 897
Maroudas, Dimitrios 300, 375, 377, 415, 418
Marozas, Ian 307
Marr, James M. 330
Marshall, Blake 908
Martin, Stephen M. 648
Martinez, Carol Rivera 633
Matranga, Christopher 4
Matsumoto, Atsushi 488
Matsuyama, Tatsushi 290, 488
Matthew, Howard W. T. 251
Mauck, Joseph 542
Mazoni, Julia Natalia Oliveira 250
McBride, Matthew K. 325, 381
McBride, Michael 41, 162, 512
McCandless, Brian E. 900
McCloskey, Bryan D. 590
McCormick, Alon 781
McDaniel, Hunter 786
McEnnis, Kathleen 435
McGrail, B. Peter 274, 774
McKenna, Gregory B. 712
McKenna, Mike 239
McKittrick, Michael 161
McLoughlin, Sean 435
McPherson, Brian 584
Meece-Rayle, Mackenzie 593

AUTHOR INDEX

Meek, Kelly M. 591
Meenakshisundaram, Venkatesh 360, 747
Mehra, Nitin 800, 813
Mehraeen, Shafigh 55, 901
Mehta, Prateek 160
Melo, Mariane B. 240
Melosh, Nicholas A 57
Mendez, Janet 633
Meng, Bo 814
Mengel, Shawn 58
Menon, Vinod 746
Meredith, J. Carson 513
Merkel, Alyssa 535
Merrill, Laura 97
Meyer, Pimphan Aye 158
Miller, Daniel J. 695
Miller, Darren 909
Miller, David C. 4
Miller, Jeffrey T. 571
Millican, Samantha L. 3
Milliron, Delia J. 741, 907
Millward, Dan 164
Min, Yong 276, 287
Mineart, Kenneth 426
Miodownik, Mark 382
Miraz, Md. Alamin 609
Mirmajlesi, Keana 529
Miskin, Caleb 571, 905
Misra, Manju 555
Misra, Mayank 521
Mitchell, Doug 264
Mitchell, Miranda 627
Mitra, Debirupa 185, 643
Mittenthal, Max 346
Miyazaki, Makoto 386
Mkhoyan, K. Andre 105, 743
Mofrad, Amir M. 378
Mohammadi, Erfan 509
Mohanty, Amar K. 152, 555
Mohraz, Ali 317
Mok, Jorge 510
Molinaro, Alessandra 225
Moller, Martin 321
Momani, Brian 343
Mondal, Kunal 644
Monroe, Charles W. 689
Montoya, Joseph H. 1
Mony, Sujyot 833
Morad, Viktoriia 564
Moraes, Angela Maria 241, 242, 253, 629
Morehead, Sam 904
Morelock, Cody R. 773
Morton, Logan 12
Moschetta, Eric G. 159
Mou, Tong 92
Mrozek, Randy A. 313
Mu, Bin 567
Mu, Liwen 800
Mukherjee, Rabibrata 642
Muller, Marcus 728
Mullins, C. Buddie 593
Mullis, Adam 437
Mun, Sung Cik 610
Muniz, Andre R. 415
Muramatsu, Atsushi 102, 182
Muraoka, Koki 5, 71
Murata, Hironobu 526
Murphy, Kendall 539, 630
Murray, Christopher B. 744
Murria, Priya 571
Musgrave, Charles B. 3, 50, 201, 265, 817
Muthukumar, Murugappan 575
Muto, Fumiya 102
Nadeau, Emily 762
Nagahama, Koji 386
Nagpure, Suraj 572
Nair, Sankar 720, 770
Najimu, Musa O. 838
Nakamura, Issei 670
Nakamura, Nathan 569
Nakase, Ikuhiko 533
Nakaya, Masafumi 102
Nance, Elizabeth 239, 679
Nanda, Jagjit 570
Narasimhan, Balaji 437, 460, 461, 462
Narkhede, Akshay 860
Nautiyal, Amit 645
Nayak, Srikanth 649
Nayerhoda, Roozbeh 463
Naylor, Carl H. 413
Nazarenko, Sergei 834
Nazari, Behzad 340, 833
Nealey, Paul F. 517
Nejati, Siamak 193
Nelson, Rainie D. 563, 566, 789
Neo, Darren Chi Jin 790
Neoh, Koon Gee 185, 643
Ng, Fay 45
Nguon, Helen 186
Nguyen, Julie 530
Nguyen, Quoc T. 171
Nieh, Mu-Ping 187
Nisola, Grace M. 226, 289
Nowak, Christian 521, 751
Nowottny, Jonas 200
Nuckolls, Colin 40, 45
Nuxoll, Eric 730
Oberhauser, James Paul 316
Ochi, Masanori 488
O'Flanagan, Stephen 538

AUTHOR INDEX

Ogilvie-Battersby, James 225
Oh, Hyeonji 833
Oh, Jinyoung 39
O'Harra, Kathryn 346
Ohodnicki, Paul R. 98
Okamoto, Yoshi 696
O'Keefe, Sean 669
Okolie, Chukwuemeka 154
Okoshi, Ryusuke 522
Okubo, Tatsuya 5, 71
Olarte, Mariefel V. 158
Oleske, Katharine W. 777
Olsen, Bradley D. 384
O'Neill, Sean C. 311
Onstott, Ellie K. 896
Opembe, Naftali 264
Oquendo, Luis E. 194
Orbey, Nese 225, 256
Orton, Emma 627
Oschmann, Bernd 799
Osmond, Matthew 353, 880
Ostraat, Michele L. 652
Osuji, Chinedum O. 165
Ou, Jifei 636
Ousley, Evan 18
Ozgur, Derya Oncel 100
Pachidis, Pavlos 412
Padmanaban, Munirathna 518
Paek, Seung-Min 260, 721
Palmese, Giuseppe 797, 801
Panchal, Kushal 373
Pandres, Elena P. 99, 379
Pang, Simon H. 65
Pantcheva, Mina 880
Panthani, Matthew G. 414, 563, 566, 789, 791
Panzer, Matthew J. 96, 405, 523
Papait, Andrea 348
Papavassiliou, Dimitrios V. 171
Parekh, Dishit 912
Park, Chihoon 914
Park, Jaesung 607
Park, Jong Keun 164
Park, Sanghoon 6
Park, Seungdoo 264
Park, Yongkuk 323
Parkin, William M. 413
Parohinog, Khino J. 226, 289
Parulkar, Aamena 153, 726
Patel, Pooja 254
Patra, Tarak Kumar 747
Pauthe, Emmanuel 639
Pauzauskie, Peter 379
Pawar, Prasad P 271
Pearce, Timothy R. 527
Peixoto, Caio 378
Pekkanen, Allison M. 909
Pelaez, Francisco 538
Peles, Yoav 172
Pender, Joshua 593
Peng, Xiaoshan 579
Pengo, Thomas 527
Pepa, Kristi 746
Peppas, Nicholas A. 436, 766
Perera, Ayomi S. 180, 382
Perez, German 352
Perez, Maritza Perez 329
Peroutka-Bigus, Nathan 437, 462
Perreault, Luke R. 348
Perry, Sarah L. 343, 400, 459
Persson, Kristin 1
Persson, Nils 41, 162, 512
Peterson, Amy M. 641
Peterson, Brian K. 650
Pethe, Kevin 893
Petrecca, Katherine 636
Petrich, Jacob W. 566
Pfeifer, Blaine 463
Pflug, Kristina M. 200
Pfromm, Peter 690
Phan, Ngoc 528
Phan, Vu 609
Pi, Yunhong 478, 489
Pietryga, Jeffrey M. 790
Pinezich, Meghan 349
Pinge, Shubham 518
Pinhero, Patrick J. 301, 305, 380
Pintos, Esteban 198
Pires, Ana Luiza Resende 242
Pirzada, Tahira 327, 608
Plawsky, Joel L. 172
Podlaha, Elizabeth 56
Polanco, Ashli 256
Ponder, James 191
Potter, Matthew 68
Prabhu, Vivek M. 426
Prakash, Nikhil 54, 84, 230
Preciado, Julian 676
Priestley, Rodney D. 429
Protesescu, Loredana 564
Pu, Jingyang 216
Pu, Yuji 371
Puri, Mayank 742, 745, 906
Purwar, Rahul 390, 457
Qayyum, Anisa 678
Qi, Zhaoxiang 91
Qian, Yuqiang 87, 296
Qin, Huan 523
Qiu, Bin 184
Qiu, Fen 692
Qu, Ge 509

AUTHOR INDEX

Rachagani, Satyanarayana....8
Ragauskas, Arthur J....150
Ragelle, Heloise....411
Rahimi, Khosrow....321
Rahman, Ashiqur....271, 609
Rai, Muhammad....626
Rai, Neeraj....385
Rajendran, Aravindan....557
Ramachandran, Karthik....316
Ramirez, Julia Vela....766
Ramirez-Caballero, Gustavo....723
Ranadive, Pinaki....726
Rangnekar, Neel....105, 653
Rankin, Stephen E....572, 762
Rao, Shreyas....860
Rappe, Andrew M....53
Ratay, Michelle....540
Rebollar, Luis....405
Reed, Ellen H....756
Rege, Kaushal....252, 638
Reichmanis, Elsa....41, 162, 512
Reid, Daniel....43, 511
Reifenberger, Jeffrey G....658
Reis, Cristiano....557
Reisch, Anne....627
Ren, Jing Ming....799
Ren, Liwei....263
Reveil, Mardochee....902
Reynolds, John....191
Rezac, Mary....690
Rezaei, Fateme....66
Rezende, Gabriela Souza....629
Rhoades, Alicyn M....340
Riazi, Hossein....53
Ricarte, Ralm....232, 422
Richard, Melissandre....772
Richter, Steve....159
Riggleman, Robert A....640
Rimer, Jeffrey....103, 181, 653
Rinaldi, Carlos....409
Rioux, Robert M....59
Robbins, Mark....732
Roberts, Jesse....352
Robinson, Joshua....783
Rodrigues, Mariana V....154
Rodriguez, Christina G....48
Roh, Jeongkyun....790
Roh, Sangchul....912
Rohner, Nathan A....536
Rokke, David....905
Roper, D. Keith....83, 417
Rosa, Paulo De Tarso Vieira E....250
Rose, Jonas C....321
Rosi, Nathaniel L....772
Ross, Kathleen....461
Roth, Charles....254
Rother, Gernot....68
Rouf, Tahrima B....90
Roy, Manosi....760
Ruffley, Jonathan....772
Rumptz, John R....3
Russell, Alan....526
Russell, Lauren....349
Rutkevicius, Marius....327
Rutledge, Gregory C....574, 753
Ryan, Bradley J....566
Sacco Jr., Albert....107
Saccone, Max....606
Sadeghi, Morteza....229
Sadler, Joshua....801, 802
Saed, Mohammad....192
Saifuddin, Indira....833
Sajib, Md Symon Jahan....331
Sakaguchi, Donald S....319
Sakaguchi, Moe....102
Sakai, Risako....290
Sakwa-Novak, Miles....68
Salcedo, Felipe....757
Salehi, Ali....357
Salehi, Mahsa....229
Sampath, Janani....234, 404
Sanchez, Isaac C....669
Sandell, Linda....626
Santiso, Erik E....731
Santos, Daniela Barrios....766
Santos, Stephany Di Carla....629
Santosa, Daniel....158
Saquing, Carl D....608
Saravi, Sina Hassanjani....272
Sarkar, Mohammad Shahadat Hussain....609
Sarkaria, Jann N....393, 861
Sarmiento, Paula A....757
Satam, Chinmay C....513
Satish, Nune....774
Sautet, Phillippe....786
Schaake, Richard P....340
Schaefer, Jennifer....97
Schaffer, David V....80
Schieber, Jay D....338
Schiffman, Jessica D....780
Schloemer, Tracy H....561
Schmalbach, Kevin M....802
Schmalzer, Andrew M....914
Schmidt, Christine....409
Schneck, Jonathan....537
Schneider, William F....160
Schudel, Alex....458, 536
Schultz, Kelly M....318
Schulz, Philip....561
Schulze, Morgan W....356

AUTHOR INDEX

Seabaugh, Alan....783
Seabaugh, Matthew....264
Seaton, Nicholas....293
Seeger, Madeleine....837
Sefiddashti, Mohammad Hadi Nafar....345
Segura, Tatiana....168, 238
Sell, Scott A....626, 678
Sellinger, Alan....561
Semenchenko, Liya....261
Senapati, Sujata....460
Seney, Robert....189
Sengers, Jan V....659
Seo, Jiho....340
Servoss, Shannon L....352
Setaro, Angelo....718
Sevimli, Sema....535
Sfeir, Charles....540
Shadish, Jared A....406
Shae, Daniel....535
Shah, Rishabh....476
Shah, S. Sakhawat....608
Shahvari, Amin Vossoughi....251
Shamsabadi, Ahmad Arabi....53, 229
Shan, Bohan....567
Shanmuganathan, Kadhiravan....288, 834, 835
Shao, Qi....538
Sharma, Chandra....628, 631
Sharma, Hom....694
Sharma, Megha....63, 651
Sharma, Radhika....618
Sharma, Vivek....344, 359, 508
Shell, M. Scott....13
Shen, Alan....913
Shen, Chunyin....89
Shen, Haiyang....94
Shen, Jian....266
Shen, Jiaxin....276, 287
Shen, Yufeng....181
Shere, Inderdip....755
Sheridan, Richard....818
Shete, Meera....653
Sheth, Saahil....626
Shi, Di....433
Shi, Kun....315
Shim, Jong Hyun....270
Shimizu, Yoji....538
Shimojima, Atsushi....208, 717
Shin, Jeongeun....788
Shkrob, Ilya....95
Sholl, David....720
Sholl, David S....770, 771, 773
Shreiber, David I....529
Si, Zhangyong....893
Siepmann, J. Ilja....675
Sievers, Carsten....154
Silcox, Benjamin....95
Silveira, Zachary....627
Simmons, David S....360, 747
Simon, Sindee L....427, 712
Sing, Charles E....400
Singaravel, Gnana Pragasam....183
Singh, Ajay....282, 907
Singh, Ishita....409
Singh, Samrendra....525, 729
Slater, Ben....769
Smirniotis, Panagiotis....473
Smit, Berend....769
Smith, Andrew....186
Smith, Daniel....891
Smith, Ethan D....648
Smith, Josiah....12, 530
Smith, Mitch....9
Snyder, Mark A....63, 651
Sofman, Marianna....407
Soltani, Iman....341
Song, Han Byul....715
Song, Kenan....811
Song, Ping....86
Song, Xueyan....366
Song, Young-Geon....258, 259
Song, Zhuonan....692
Sorci, Mirco....718
Soroush, Masoud....53, 229
Sousa, Ilza Maria De Oliveira....242
Spakowitz, Andrew J....727
Spencer, David S....436
Srinivasan, Samanvaya....171
Srivastava, Samanvaya....310, 328, 372, 402, 577
Stanford, John P....690
Stanke, Kimberly M....862
Stansbury, Jeffrey W....369
Stanzione III, Joseph F....542, 802, 819
Stecca, Owen M....802
Stefanov, Zdravko....174
Steigerwald, Michael....45
Stein, Andreas....87, 296
Stein, Gila....365
Steiner, Helene....636
Stephane, M....183
Sterling, Julie A....535
Stevanovic, Vladan....3
Stoppel, Whitney L....348
Storti, Giuseppe....312
Stottrup, Benjamin....653
Stoyanov, Simeon....912
Stoykovich, Mark....50, 201
Strano, Michael....291, 299, 606
Stretz, Holly A....228
Stromsdorfer, Jessica....767
Su, Hao....396, 532

AUTHOR INDEX

Su, Xiao 474, 784
Subramanion, Jo Thy Lachumy 893
Suleiman, David 329
Sultana, Khondker 605
Sun, Haotian 237, 326
Sun, Jingze 261
Sun, Luyi 186, 187
Sun, Rui 591
Sun, Yan 17
Sun, Yangyang 521
Sun, Yunwei 694
Sutjianto, James 103
Swager, Timothy 583
Swan, James 739
Swartz, Scott 264
Sweeney, Charles 192
Swihart, Mark T. 364
Tafen, De Nyago 4
Taheri, Mohammad Mehdi 58
Tajvidi, Mehdi 293
Takatani-Nakase, Tomoka 533
Tan, Anthony 425
Tan, Kai-Jher 474
Tan, Shuai 157
Tang, Lin 810
Tang, Shuo 314
Tanimoto, Keishi 386
Tanveer, Sheik 697
Tata, Ram Rao 767
Tazawa, Shunsuke 208, 717
Tekin, Rumeysa 107
Teymour, Fouad 202
Thacker, Zachary 301, 305
Thakkar, Harshul 66
Thallapally, Praveen K. 774
Thate, Karine 278
Thelakkaden, Mitzi 252, 638
Thomas, Susan N. 458, 536
Thompson, Jessica 320
Thompson, Joshua A. 70
Thompson, Levi T. 95
Thompson, Matthew S. 366
Thompson, Nicole 679
Thorson, Todd 317
Thurman, Derek W. 339
Tian, Qian 427
Tian, Zheng 63
Tibbitt, Mark W. 411
Tiffany, Douglas 557
Ting, Jeffrey 372, 577
Tinkham, Jonathan 561
Tirrell, Matthew V. 77, 236, 310, 328, 372, 398, 402, 577, 897
Tisdale, William A. 302, 752, 785
Tokatlian, Talar 240
Tolouei, Anita E. 627
Tom, Karjala 51
Topuz, Berna 653
Torkelson, John M. 49, 367, 387, 425
Trefonas III, Peter 164
Triana, Juan Carlos Briceno 757
Trogadas, Panagiotis 180
Troung, Kristy 635
Truong, Quoc 225, 256
Truskett, Thomas M. 741
Tsapatsis, Michael 105, 106, 653, 719
Tschirner, Ulrike 837
Tsianou, Marina 768
Tsoras, Alexandra 464
Tsukada, Takao 524
Tsukruk, Vladimir V. 191
Tu, Raymond 311
Tucker, Budd A. 320
Tumas, William 3
Tung, Siu On 95
Turasan, Hazal 309
Tze, William T. Y. 293
Ugaz, Victor M. 915
Ulery, Bret 12, 530, 767
Uline, Mark J. 657
Um, Kiju 258, 259
Underhill, Patrick T. 718
Unold, Thomas 903
Urban, Jeffrey 692
Urie, Russell 252, 638
Utomo, Nyalaliska 833
Uz, Metin 8, 319
Vaidya, Amol 818
Vaithilingam, Balasubramanian 183
Vaknin, David 649
Valdehuesa, Kris Nino G. 226
Van Der Vlies, Andre 533
Van Lehn, Reid 16
Van Tassel, Paul R. 639
Vargas, J. German 757
Varma, Arvind 100
Vattipalli, Vivek 104
Velev, Orlin D. 912
Venkataraman, Shrinivas 426
Ventura, Darryl 209
Verburg, Alex 789
Verduzco, Rafael 365, 510
Vergara, John 801
Vicente, Fernando 537
Vieregg, Jeffrey 77, 398
Vikhe, Yogesh 894
Vlassopoulos, Dimitris 713
Vocelle, Daniel 9
Voggu, Vikas Reddy 904
Vogt, Bryan D. 792, 911

AUTHOR INDEX

Volkov, Dmytro 256
Waechter, Andreas 52
Wagner, Norman J. 793
Waldschmidt, Thomas 461
Walker, Christopher 731
Walsh, Dylan 47
Walton, Krista S. 475, 771, 773
Walton, S. Patrick 9
Wan, Haiqing 89
Wang, Alex 407
Wang, Bin 92
Wang, Boya 570
Wang, Chen 381
Wang, Daniel I. C. 636
Wang, Feihu 396, 532
Wang, Guanyun 636
Wang, Huamin 158
Wang, Jiechen 39
Wang, Jingyu 556
Wang, Junyan 89
Wang, Luda 636
Wang, Minghui 646
Wang, Muzhou 330
Wang, Nathan G.-J. 39
Wang, Ping 516
Wang, Qing 237
Wang, Rui 384
Wang, Shaoyang 582
Wang, Sheryl 231, 434
Wang, Sihong 39
Wang, Song 291, 606
Wang, Wade 10, 11, 394
Wang, Wei 315
Wang, Wen 636
Wang, Wenjie 649
Wang, Yujie 791
Wang, Yuzhu 396, 532
Wang, Zhaofeng 186
Wannemuehler, Michael J. 460, 462
Warzywoda, Juliusz 107
Wawrousek, Chris 636
Webb, Bruce 762
Webster, Megan 746
Webster, Thomas J. 433
Wegener, Evan C. 571
Wei, Huige 184
Wei, Min 94
Wei, Suying 85, 184, 298
Wei, Tao 331
Wei, Tong 425
Weidman, Jennifer 693
Weimer, Alan W. 3, 265, 817
Wells, Kacie M. 187
Westin, Cecilia Buzatto 242
Weston, Simon C. 650
Whitaker, Mariah 153
White, Tommi 12
Whitehead, Kathryn A. 7, 390, 457, 526
Whitham, Patrick 379
Whitmer, Jonathan K. 654
Wiesner, Ulrich 777
Wijayapala, Rangana 204, 308
Wilhite, Benjamin 397
Willard, Adam P. 302, 752
Williams, Christopher B. 909
Williamson, Grant A. 99
Willson, C. Grant 194
Wilson, Christina 862
Wilson, John 535
Winey, Karen I. 430
Winter, H. Henning 343
Winter, Jessica O. 62, 278, 580
Witman, Matthew 769
Wittenberger, Steven J. 159
Wong, Bryan M. 2, 376
Wong, Hoi Lun 190
Wong, Minhao 606
Woodcock, Cory 172
Woodcock, Jeremiah 818
Worrell, Brady 50, 325, 381, 761
Worthington, Kristan S. 320
Wu, Connie 10, 394
Wu, Gang 94
Wu, Hao 372, 577
Wu, Janet 164
Wu, Jianzhong 38
Wu, Qing-Xi 215
Wu, Wei 634
Wu, Yadong 89
Wu, Yaoting 744
Wu, Yun 237
Wu, Yuning 98
Wujcik, Evan K. 279, 609
Wyslouzil, Barbara E. 580
Xi, Li 373
Xi, Weixian 50
Xia, Yan 187
Xiao, Han 288, 593
Xiao, Jing 275, 478, 489
Xie, Rui 315
Xie, Shuyi 675
Xie, Xiaofeng 520
Xie, Yongjian 812
Xie, Zelong 67
Xin, Zhong 815
Xu, Dandan 106, 653
Xu, Honghong 177
Xu, Jun 796
Xu, Ke 783
Xu, Wenqian 719

AUTHOR INDEX

Xu, Xiaojiang 816
Xu, Xiaoyang 765
Xue, Feng 719
Yadav, Santosh Kumar 801
Yamamoto, Hideo 488
Yan, Hongping 304, 358
Yan, Ruoxue 60
Yan, Yushan 432
Yang, Chuanfang 176
Yang, Guang 570
Yang, Guozhen 780
Yang, Han-Seung 293
Yang, Seung Ook 899
Yang, Shendu 301, 305, 380
Yang, Sui 787
Yang, Szu-Ming 227
Yang, Xutong 810
Yang, Yeokyung 292
Yang, Yi Yan 426
Yao, Lining 636
Yao, Shan-Jing 215
Yarger, Jeff 252, 638
Yau, Mai Kwan 458
Yavari, Milad 696
Yin, Deqiang 364
Yokoi, Toshiyuki 102, 182
Yoon, Heedong 712
Yoshizawa, Sayuri 540
Young, Walter W. 194
Yu, Dan 395, 677
Yu, Haiyue 180
Yu, Xiangfei 172
Yu, Xiaoyuan 186
Yuan, Jun-Jie 215
Yun, Hyeong Jin 786, 790
Yung, Matthew M. 154
Zacharias, Zeb 461
Zacher, Alan H 158
Zasadzinski, Joesph A. 788
Zawaski, Callie 909
Zembrzuski, Michael 791
Zeng, Songshan 186, 187
Zhang, Baoquan 177, 273
Zhang, Boyuan 40
Zhang, Dawei 391, 761, 898
Zhang, Dianyun 186
Zhang, Guojian 463
Zhang, Han 105
Zhang, Hongyuan 88
Zhang, Huaiying 82
Zhang, Jing Jing 95
Zhang, Jinju 176
Zhang, Junyan 175
Zhang, Ke 652
Zhang, Lin 17
Zhang, Lu 95
Zhang, Mengying 679
Zhang, Na 216
Zhang, Qinnan 693
Zhang, Rui 12
Zhang, Ruihong 905
Zhang, Teng 636
Zhang, Xiang 787
Zhang, Xinyu 645
Zhang, Yanmei 556
Zhao, Haoyu 427
Zhao, Junjie 646
Zhao, Ling 196
Zhao, Mengqiang 413
Zhao, Shicheng 579
Zhao, Tianshuo 744
Zhao, Xin 905
Zhao, Xuanhe 636
Zhao, Yang 520
Zhao, Zhenxia 725
Zhao, Zhongxing 725
Zheng, Hou 371, 894
Zheng, Xinguo 812
Zheng, Xiong 659
Zhong, Yu 40, 45
Zhou, Changlu 815
Zhou, Jing 191
Zhou, Kang 636
Zhou, Shanshan 762
Zhou, Sunshine X. 194
Zhou, Yang 647
Zhu, Guanghui 722
Zhu, Huiming 273
Zhu, J. Y. 149, 515
Zhu, Jiahua 184, 800, 813
Zhu, Lingxiang 364
Zhu, Shuze 753
Zhu, Xiaoyang 40, 45
Zhu, Y. Elaine 399
Zhu, Yangzhi 60
Zhu, Yaqun 270
Zhuang, Minghao 199
Zimmerman, John 260, 721
Zoto, Christopher 225
Zustiak, Silviya Petrova 389, 410, 626, 678, 688

AIChE
120 Wall Street, FL 23
New York, NY 10005-4020

ISBN 978-1-5108-5797-1

9 781510 857971

CURRAN ASSOCIATES INC.
proceedings.com